NORTH AMERICAN TUNNELING
2018 PROCEEDINGS

NORTH AMERICAN TUNNELING
2018 PROCEEDINGS

Edited by Alan Howard, Brett Campbell,
Derek Penrice, Matthew Preedy, and Jim Rush

PUBLISHED BY THE
SOCIETY FOR MINING, METALLURGY & EXPLORATION

Society for Mining, Metallurgy & Exploration Inc. (SME)
12999 E. Adam Aircraft Circle
Englewood, Colorado, USA 80112
(303) 948-4200 / (800) 763-3132
www.smenet.org

The Society for Mining, Metallurgy & Exploration (SME) is a professional society whose more than 15,000 members represent professionals serving the minerals industry in more than 100 countries. SME members include engineers, geologists, metallurgists, educators, students, and researchers. SME advances the worldwide mining and underground construction community through information exchange and professional development.

ISBN 978-0-87335-466-0
eBook 978-0-87335-467-7

On the Cover: Herrenknecht TBM named "Lady Bird," after Lady Bird Johnson, being lowered into the future DC Water's Clean Rivers Project screening shaft at Blue Plains. The award-winning Blue Plains Tunnel was completed on time and under budget by design builder Traylor Brothers/Skanska/Jay Dee JV with design support by CH2M/Halcrow.

Contents

Preface xiii

TRACK 1: TECHNOLOGY

Session 1: Project Controls 1

5D BIM Applied to Cost Estimating, Scheduling, and Project Control in Underground Projects 3
   G. Venturini, Federico Maltese, and G. Teetes

Analysis of Concrete-Lined Tunnels Crossing Active Faults 11
   Stephanie Lange, H. Benjamin Mason, Michael H. Scott, and Scott A. Ashford

Atmospheric Cutting Tools Replacement in EPB TBM "Bertha": Innovative Approach
for Face Pressure Tunneling 19
   Juan Luis Magro, Ruben Piqueras, and Juan Garnero

Logistics and Performance of a Large-Diameter Crossover TBM for the Akron Ohio
Canal Interceptor Tunnel 28
   Pablo Salazar and Connor Maxon

Job-Site Security Risk Management Beyond Gates, Guards, and Guns 36
   Eric Jacobs

Session 2: Innovation 47

A Preliminary Investigation for Characterization and Modeling of Structurally
Controlled Underground Limestone Mines by Integrating Laser Scanning
with Discrete Element Modeling 49
   Juan J. Monsalve, Jon Baggett, Richard Bishop, and Nino Ripepi

Study of the Correlation Between RMR and TBM Downtimes 58
   Omid Frough and Jamal Rostami

Cross Passage Construction: Mechanized Solutions and Fields of Application 66
   Gerhard Lang

Communications, Tracking, Vehicle Telemetry, and Proximity Solutions 72
   Guido Perez Manfredini

Underground Communication Innovations 83
   Jason Jarrett, Eric Hansen, and Steve Harrison

SCMAGLEV Project—Fast and Innovative Mode of Transportation in the Northeast Corridor 92
   *Vojtech Gall, Sandeep Pyakurel, Nikolaos Syrtariotis, Cosema Crawford,
   Liviu Sfintescu, and David J. Henley*

Session 3: TBM Technology 1 97

Downtime Data Collection, Analysis, and Utilization on Blacklick Creek Sanitary
Interceptor Sewer Tunneling Project in Columbus, OH 99
   Amanda Kerr and Max Ross

EPB TBM Performance Prediction on the University Link U230 Project 108
   *Mike Mooney, Hongjie Yu, Soroush Mokhtari, Xiaoli Zhang, Xu Zhou,
   Ehsan Alavi, Lisa Smiley, and William Hodder*

Interpretation of EPB TBM Graphical Data 117
   Keivan Rafie

TBM Procurement—Owner's Dilemma 127
    Dan Ifrim and Derek Zoldy

TBM Tool Wear Analysis for Cutterhead Configuration and Resource Planning
in Glacial Geology. 135
    Ulf Gwildis, Jose Aguilar, and Kamyar Mosavat

The EPB Chamber Air Bubble: What Causes It? 144
    Yuanli Wu, Ali Nazem, and Mike Mooney

Session 4: TBM Technology 2 155

Digitization in Mechanized Tunnelling Technology 157
    Karin Bäppler

Incorporating Geological and Geotechnical Spatial Variability into TBM
and Ground Settlement Risk Assessment 163
    Jacob Grasmick, Michael Mooney, and Whitney Trainor-Guitton

Interpretation of Tunneling-Induced Ground Movement 171
    Mazen E. Adib, Matthew Crow, Allen Marr, and Shemek Oginski

Pre-Conditioning of Hard Rocks as Means of Increasing the Performance
of Disc Cutters for Tunneling and Shaft Construction 185
    Philipp Hartlieb and Jamal Rostami

The Karlsruhe Monster—Highly Flexible Steel Formwork, Operated Under
Hyperbaric Conditions: Design and Application of a High-Tech Formwork
in a Tapered Tunnel Section 190
    Rainer Antretter

Instrumentation Design Framework for Large-Diameter Tunneling Projects. 198
    Anil Dean, Jon Pearson, and Marlene Wong

Session 5: Fresh Approach 207

Hands-on Training and Technology Transfer in Tunneling. 209
    Matias J. Lazcano, Juan P. Merello, and Nicolás P. Zegpi

High-Precision Pilot Hole Drilling Utilizing Real-Time Optical Guidance
and Verification Techniques. 218
    Aaron C. Reel and Stefano Alziati

Innovative Concrete for Aggressive Ground in Qatar. 227
    Francois G. Bernardeau and Jacek B. Stypulkowski

In-Depth Inspection of a Century-Old San Francisco Water Tunnel. 236
    R. Fippin, J. Sketchley, T. Redhorse, and D. Tsztoo

Steel Shaft Liner Plate Segments. 246
    John Mulvoy

Teck BRE Raw Coal Conveyor Tunnel Rehabilitation 250
    Irwan Halim, David Forrester, Nicolas K. Oettle, and Brian Wong

TRACK 2: DESIGN

Session 1: Resiliency 261

In-Situ Fire Test for the Design of Passive Protection for the Lafontaine Tunnel 263
    Jean Habimana, Reza Showbary, and Jos Bienefelt

Preliminary Assessment—Rehabilitation and Expansion of the Central City Tunnel System
in Minneapolis, MN. 271
    Greg Sanders, Michael Gilbert, Mahmood Khwaja, and Bill Lueck

Contents

Recent Passive Fire Protection Strategies for US Highway Tunnels 279
Wern-Ping Nick Chen

The Use of Numerical Analysis in Rehabilitation of Baltimore's Howard Street Tunnel. 291
Saeid Rashidi and Verya Nasri

Evaluation of a Segmental Tunnel Lining Response to a Strike Slip Fault Rupture. 298
Angel Del Amo, Robert Goodfellow, Moustafa Awad, Bingzhi Yang, and Kurt Braun

Session 2: Sequential Excavation Methods 309

Design and Construction Challenges in Urban Settings for the NATM Tunnels' Line 2
of the Riyadh Metro 311
Sandeep Pyakurel, Kurt Zeidler, Andreas Gerstgrasser, and Iain Thomson

Downtown Bellevue Tunnel—Analysis and Design of SEM Optimization 321
Christoffer Brodbaek, Derek Penrice, Jake Coibion, and Chad Frederick

Elasto-Plastic Design of Fiber-Reinforced Concrete Tunnel Linings 332
Axel G. Nitschke

Ground-Liner Interaction During Seattle Northgate Link Cross-Passage Construction 341
Tamir Epel, Mike Mooney, Kurt Braun, Mike DiPonio, and Nate Long

Design and Construction of the Downtown Bellevue Tunnel 350
Mike Wongkaew, Mike Murray, Jake Coibion, Chad Frederick, and Mun-Wei Leong

The Influence of Material Models for the Effective Design of the Primary Lining 363
Martin Bakoš, Juraj Ortuta, and Peter Paločko

Shallow Cover SEM Tunneling for the Purple Line Project 368
David Watson, Philip Lloyd, Rafael Villarreal, and Richard Taylor

Session 3: Rock Tunneling 379

Design and Construction of South Hartford Conveyance and Storage Tunnel 381
Verya Nasri, James Sullivan, Vincent Prestia, and Andrew Perham

Approach to Characterizing BIM Rocks for Tunnel Design Using As-Built Tunnel Data 390
Shawna Von Stockhausen, David J. Young, and Christopher Slack

High In Situ Stress and Its Effects in Tunnel Design: An Update Based on Recent Project
Experience from WestConnex M4 East and New M5 Tunnels, Sydney, Australia. 401
D. Tepavac, P. Mok, Y. Sun, D. Oliveira, H. Asche, and S. Simmonds

Design and Construction Considerations for Atwater Water Intake and Conveyance Tunnel
in Montreal 411
Jean Habimana, Benoit Rioux, and Stephane Dusablon

Final Design of the River Des Peres Tunnel in St. Louis 419
Patricia Pride, Mike Robison, Jon Bergenthal, and Wojciech Klecan

Successful Implementation of "SEM—Single Shell Lining Concept" for the Emergency
Tunnel at the Brenner Base Tunnel Project—Theory and Praxis of an Innovative Shotcrete Model 425
Thomas Marcher

Session 4: Challenging Design Issues 435

Design and Monitoring of an Effective Deep Excavation Cut-off 437
Eric Sekulski, Jon Leech, Michael Chendorain, John Yao, and Mark Ramsey

A New Approach to Hydraulics in Baffle Drop Shafts to Address Dry and Wet Weather Flow
in Combined Sewer Tunnels 448
Michael J. Seluga, Frederick (Rick) Vincent, Samuel Glovick, and Brad Murray

Challenges of Design and Construction of Slurry Walls in a Congested Site 462
Ravi Jain, Mina Shinouda, and Angelo Colasante

Support of Excavation Structural Challenges and Steel Design for the
Crenshaw/LAX Transit Project 470
    Bradley Hoffman, Bingzhi Yang, and Paul Leduc

Tunnel Ventilation Systems Harmonization for Optimization of Tunnel Construction 476
    Sean Cassady and David Parker

Vibration Study on California High-Speed Rail Tunnels Based on Earthquake Data
and Numerical Analysis. 482
    Xiaomin You and Douglas Anderson

Tunnel Cross Passage Seismic Analysis Considering 3D Wave Propagation. 490
    Yue Shi, Peter Chou, and Danny Lin

Session 5: Precast Concrete Tunnel Linings 501

Design of the San Francisco Public Utilities Commission's Channel Tunnel. 503
    Michael Deutscher, Samer Sadek, and Manfred Wong

Designing a Bored Tunnel for the Unthinkable—A Marine Vessel Collision Assessment. 516
    David Watson, Chris Pound, Christoph Eberle, and Botond Beno

Designing Reinforcement for Precast Concrete Tunnel Lining Segments 524
    Jimmy Susetyo, Michael Dutton, Tomas Gregor, and Michel Mongeau

Evolution and Challenges of Segmental Liner Design and Construction for SR 99 Tunnel. 532
    Yang Jiang, Gordon Clark, and Jerry Wu

Guide for Optimized Design of Tunnel Segmental Ring Geometry 541
    Mehdi Bakhshi and Verya Nasri

Steel Fiber Reinforced Concrete (SFRC) for TBM Tunnel Segmental Lining—
Case Histories in the Middle East 550
    Guido Castrogiovanni, Gianpaolo Busacchi, and Gianni Mariani

TRACK 3: PLANNING

Session 1: Risk Management Challenges and Solutions 557

CEVP-RIAAT Process—Application of an Integrated Cost and Schedule Analysis 559
    Philip Sander, Martin Entacher, and John Reilly

Freezing of Glacial Soils for Cross Passage Construction in North America and Europe—
A Comparison 570
    Ulf Gwildis, Helmut Hass, Mike Schultz, and Mahmood Khwaja

Unique Design and Construction Challenges for Near-Surface Large-Span
Transit Station Caverns in Rock 579
    Charles A. Stone and Eric C. Wang

Bergen Point WWTP Outfall Tunnel Shafts—Risk-Mitigated Design for Excavation Support. 587
    Mahmood Khwaja and Michael S. Schultz

BART Silicon Valley (BSV) Phase II, Tunneling Methodology—Comparative Analysis
Independent Risk Assessment 596
    Saqib A. Saki, James J. Brady, Robert J.F. Goodfellow, Angel Del Amo,
    Alfred Moergeli, and Krishna Davey

Session 2: Project Delivery—Water and Wastewater 607

DigIndy Tunnel System: Pleasant Run Deep Tunnel Optimization 609
    Olivia Hawbaker, Maceo Lewis IV, and Leo Gentile

Planning, Design, and Construction of CSO Pumping Station Structures 616
    Geoffrey A. Hughes, Rafael C. Castro, Carlton M. Ray,
    Moussa Wone, Ronald E. Bizzarri, and John F. Cassidy

Contents

Emerging Demand for Subsea Tunnels in Chile 625
Victor Figueroa and Nicolás Zegpi

Implementing Major CSO Solutions via Deep Rock Tunneling: Louisville
Ohio River Tunnel (ORT) 633
Jonathan Steflik, Adam Westermann, Donnie Ginn, Todd Wanless,
Jacob Mathis, and Greg Powell

Narragansett Bay Commission Phase III CSO Abatement Program—Pawtucket Tunnel 643
Todd Moline and Christopher Feeney

Ship Canal Water Quality Project: Meeting Federal Requirements 652
Shannon M. Goff, Gregg W. Davidson, and Dylan Menes

The Coxwell Bypass Tunnel, Cleaning up Toronto's Waterfront 660
Daniel Cressman, Olive Cantina, David Day, Samantha Fraser,
Robert Mayberry, and Caroline Kaars Sijpesteijn

Session 3: Geotechnical, Environmental, and Sustainability 669

Assessing Resilience Impacts from Integrated Above- and Below-Ground Urban Infrastructure 671
Priscilla P. Nelson

Design, Construction, and Risk Management Strategies for Shallow Tunnels in Urban Settings 678
Kumar Bhattarai

Small-Diameter TBM Tunneling: Risk Management Approach to Face Geological Uncertainties 688
Giuseppe M. Gaspari and Andrea Lavagno

Rehabilitation of Tunnels: An Owner's Perspective 698
David Tsztoo, Anthony Yu, and Teena Redhorse

Sewer Tunnel Beneath Meramec River to Fulfill Regional St. Louis Treatment Plan
and Environmental Vision 705
Everett L. Litton, Mark J. Stephani, and Jerry L. Jung

Sustainable Infrastructure Tunneling: Construction Materials Considerations
from the Early Project Stage 714
Fabio Pellegrini, Brendan Daly, Andreea Enescu, and Nicolas Swetchine

Project Plans for the California WaterFix Tunnels 718
John Bednarski, Jay Arabshahi, Sergio Valles, and Shanmugam Pirabarooban

Session 4: Project Delivery—Transit and Highways 727

The Future of Transit Tunneling in Washington, D.C. 729
Brian H. Zelenko, Harald Cordes, and William H. Hansmire

Subsurface Investigations, Design, and Construction Considerations for the
Montreal Transportation Agency Côte-Vertu Underground Storage and Maintenance Garage 739
Giovanni Osellame and Jean Habimana

Windsor–Detroit Tunnel: Application of State-of-the-Practice Maintenance and Safety Standards. 746
Eric C. Wang, Ruben Manuelyan, and Paul Mourad

Consideration of Single Bore as a Construction Option for VTA's BART
Silicon Valley Phase II Extension Project 754
S. Zlatanic and K. Davey

The Scarborough Subway Extension (SSE)—Large Single-Bore Transit Tunneling in Toronto 767
Matthew Geary, Tomas Gregor, and Edward Poon

The Next Big Tunnel Under Downtown Seattle 775
Gordon Clark and Joseph Gildner

Crossing the Chesapeake Bay 21st Century Style: The Parallel Thimble Shoal Bored Tunnel 781
Enrique Fernandez, Alejandro Sanz, Juan Luis Magro, and Enrique Alcanda

Session 5: Planning for Success: Risk Management and Contracting Strategies 793

The Risks Associated with TBM Procurement and the Next Steps Towards Industry Change 795
Lok Home and Gary Brierley

Alternative Delivery Drives Alternative Risk Allocation Methods. 802
John Reilly, Randall Essex, and David J. Hatem

Construction Cost Estimating Using Risk-Based Approach 812
Michael S. Schultz and Greg Sanders

Procurement of and Contracting for Underground Construction Projects in North America 819
Gary Brierley and David Corkum

Current Trends in Procurement Delivery of Major Tunnel Projects 825
Steven R. Kramer

Fort Wayne Utilities Three Rivers Protection and Overflow Reduction Tunnel—
Project Bidding Successes and Lessons Learned 833
T.J. Short, Mark Gensic, Leo Gentile, and David Day

Interlake Tunnel—A Future Design-Build Project 839
Ronald D. Drake

TRACK 4: CASE HISTORIES

Session 1: Sewer/Water 1 849

A Case Study of Risk Mitigation Measures on the West End Trunk Line Microtunnel 851
Alex Prieto, Rory Ball, Jason Marie, and Gerald DeBalko

High-Capacity Hoisting at Rondout West Branch Tunnel Project 858
Donald Brennan and Derek Brennan

Boring Hard, Abrasive Gneiss with a Main Beam TBM at the Atlanta Water Supply Program. 869
Tom Fuerst and Don Del Nero

World's Largest Tunnel Gates and Reservoir Connection Go Online as Part
of Chicago's Tunnel and Reservoir Plan (TARP) 875
Miguel Sanchez, Faruk Oksuz, Dave Schiemann, and Patrick Jensen

Establishing Access into Chicago's Main Stream Tunnel Under Live CSO Flow 883
Lukasz Dubaj

Session 2: Transportation 1 891

After the Tunnel Drive, Finishing the Highway to Replace SR 99 Below Seattle 893
Gregory Hauser, Susan Everett, and Joseph Clare

Cross Passage Freezing from the Ground Surface 899
Aaron K. McCain, Brenton Cook, Larry Applegate, and Nate Long

Lessons Learned in Dry Ground Excavation Using an EPBM 908
David C. Girard and Ran Chen

Phase 1—Second Avenue Subway Project, New York: Light at the End of the Tunnel:
Delivering a Mega Project on Schedule While Maintaining Budget 915
Michael Trabold, Anil Parikh, and Richard Giffen

North Hollywood Station West Entrance—A Successful Connection of Metro Red Line
Subway and Orange Bus Line 922
Tung Vu, Milind Joshi, Alex Gonzalez, and Richard Silos

Risk Management in Early Subaqueous Transit Tunneling. 928
Vincent Tirolo Jr.

Shallow SEM Tunneling Under Major Roadways and Through Active Landslides
in Edmonton Glacial Tills 934
John Kuyt, May ElKhattab, Eden Almog, and Ian Cisyk

Contents

Session 3: Sewer/Water 2 949

Indianapolis Deep Rock Tunnel Connector and Pump Station Start-up 951
John Morgan and Alexander Varas

South Coast Water District's Wastewater Tunnel Rehabilitation Project, Complexity
Coupled with Environmental Sensitivity 956
Shimi Tzobery, Kevin Kilby, Trimbak Vohra, Richard McDonald, and Rick Shintaku

Norris Cut Force Main Replacement Tunnel, Miami, Florida 966
Robin Dill, Roger Williams, Lin Li, and Eloy Ramos

Operations and Maintenance of Waller Creek Flood Control Tunnel 979
John Beachy and Ramesh Swaminathan

Prefabricated Tunnel Pipe Liners—A Modern-Day Approach to Efficient Installation 988
Jesse Schneider and Brian Kelley

Pre-Excavation Grouting at the Hemphill Site–Atlanta WSP Tunnel, Atlanta, Georgia. 995
Adam L. Bedell, Konner Horton, Don Del Nero, and Brian Jones

Hard Rock Tunnel Design Improvements over 42 Kilometers Spanning Multiple Projects. 1003
Alston M. Noronha and Mark H. Bradford

Session 4: Transportation 2 1011

Risk Mitigation Through the Use of In-Tunnel Ground Freezing for Seattle
Light Rail Tunnel Cross Passage Construction 1013
Rick Capka and Nate Long

TBM Passing Under Existing Subway Tunnels in Los Angeles, California 1020
Jason Choi, Matthew Crow, Gary Baker, Ron Drake, Patrick Jolly, and Christophe Bragard

What Comes Down Must Go Up—Dewatering-Induced Ground Movement on the SR-99
Bored Tunnel Project, Seattle, WA, USA 1029
Joseph Clare

Widening of a Road Tunnel Without Interruption in Service 1037
Marco Invernizzi and Pooyan Asadollahi

The Crenshaw Corridor—Cross Passages: Design vs. Construction 1042
Luis Piek, Harnaik Mann, John Kuyt, and Patrick Finn

Humboldt Bay Power Plant Decommissioning Shaft—A Case Study in Cutter Soil Mix (CSM)
Wall Construction 1049
Zephaniah Varley

SEM Tunneling of City Trunk Line Across Tujunga Wash. 1058
Wolfgang Roth, S. Nesarajah, Bei Su, Philip Lau, and Ruwanka Purasinghe

Session 5: Sewer/Water 3 1067

Preparation for Tunneling, Blacklick Creek Sanitary Interceptor Sewer Project
in Columbus, Ohio. 1069
Ehsan Alavi, Ed Whitman, Valerie Wollet, and Matthew Anderson

The Columbus OARS CSO Tunnel from Design Through Construction—
An Experience in Solution-Filled Karst 1077
Paul Smith, Matthew Anderson, Raisa Pesina, Jeff Coffey, and Michael Hall

Urban Hard Rock Tunneling and Blasting in Baltimore City 1089
Todd Brown and Jordan Bradshaw

Various Deep Shaft Construction Techniques Used in Atlanta Water Supply Program 1097
Tao Jiang, Wayne Warburton, Yong Wu, and Brian Jones

South Hartford Conveyance and Storage Tunnel Project—Successful Use
of Large-Diameter Secant Piles for Shaft Support 1105
Andrew Perham, Jim Sullivan, Mike Brune, David Belknap, and Clay Haynes

An Innovative Approach with Granite Block—Mud Mountain Dam 9-Foot Tunnel
Rearmoring Project 1110
    Madison Brunk, James Carroll, Joel James, Terry Gilliland, and Ellen Engberg

Contractor and Engineer Collaborate on Shaft Design, Sewer Tunnel Stabilization Project 1118
    David Jurich, Zsolt Horvath, Evan Friedman, Brett Mainer, and Joseph McDivitt

Index 1127

Preface

The North American Tunneling (NAT) Conference chair, conference, and program committees as well as the Underground Construction Association (UCA) Executive Committee and UCA of SME staff are pleased to welcome the sponsors, exhibitors, authors, attendees, and guests to Washington, D.C., for NAT 2018.

The Conference Committee selected "A Capital Idea" for this year's conference theme because it expresses the current state and direction of our industry and conveys its important role in our society and economic future. North America, particularly the United States, is experiencing an infrastructure investment deficit due to deferred maintenance of aging infrastructure and because of a growing population needing new and improved infrastructure. At the same time, there are critical demands for fixing old environmental problems, sustainably constructing new infrastructure to improve quality of life, and avoiding impacts to third parties during construction of necessary major public infrastructure projects.

More capital needs to be invested to keep up with the demands, so we assembled this year in the nation's capital, in the shadow of The Capitol, where our elected officials debate where and how much capital will be invested. The papers collected in this book share our experiences, ideas, innovation, and technology for work where infrastructure capital is invested in tunnels and all the necessary related underground structures. This year we celebrate subsurface infrastructure lessons and successes yet again with conference tracks focused on technology, planning, design, and case histories comprising 126 papers divided into 20 sessions.

The success of the conference is codified in these proceedings, which are made possible by the commitment and enthusiasm for our craft by the authors published in these pages. They have dedicated their valuable professional time outside their already-busy professional lives to share their innovations, new projects, and design and construction "lessons learned" to the benefit of our industry and our peers who strive to constantly deliver better value when delivering underground projects. The authors' employers—owners, contractors, engineers, manufacturers, suppliers, other vendors, and academia—have all supported these writers to share their knowledge, and most have backed the conference in many other ways and deserve recognition.

The Program Committee wishes to thank not only the authors and their employers, but also the full Conference Committee, program session chairs and co-chairs, the UCA Executive Committee, and UCA of SME for all the support and hard work required to promote and deliver this important biennial conference for our industry, our colleagues and associates, and our various industry customers and beneficiaries.

Leon "Lonnie" Jacobs, Conference Chair
Jon Hurt, Conference Vice Chair
Alan Howard, Program Chair
Brett Campbell, Track Chair/Technology
Matthew Preedy, Track Chair/Planning
Derek Penrice, Track Chair/Design
Jim Rush, Track Chair/Case Histories

NAT 2018 Program and Track Chairs

Alan Howard, Program Chair
Brett Campbell
Derek Penrice
Matthew Preedy
Jim Rush

NAT 2018 Session Chairs

Michael Brethel
Michael Burnson
Andrew Bursey
Rick Capka
Bernard Catalano
Emily Chavez
Peter Chou
Claudio Cimiotti
John Criss
Bill Dean
Ben DiFiore
Dawn Dobson
Bruce Downing
Brent Duncan
Elmar Feigl
Patrick Finn
Lizan Gilbert
Brian Harris
Paul Headland
Joel Kantola

Nick Karlin
Mun Wei Leong
Steven Lotti
Dave Mast
Aaron McClelland
John McCluskey
Taehyun Moon
Hamed Nejad
Nancy Nuttbrock
James Parkes
Seth Pollak
Julian Prada
Carmen Scalise
Zuzana Skovajsova
Andre Solecki
Michael Torsiello
Rick Vincent
Desiree Willis
Clint Wilson
Mohamamed Younis

UCA of SME 2018
Executive Committee

Michael Roach, Chair
Robert Goodfellow, Vice Chair
Artie Silber, Past Chair
Jack Brockway
Ted Dowey
Leon Jacobs
Colin Lawrence
Mike Mooney
Erika Moonin
Pam Moran
Tony O'Donnell
Matthew Preedy
Richard Redmond
Michael Rispin
Robert Robinson
Paul Schmall
Michael Smithson
Mike Vitale
Leonard Worden

NAT 2018
Program Committee

Leon Jacobs, Chair
Jon Hurt, Chair Elect
Michael Smithson, Past Chair
Alan Howard, Program Chair
Colin Lawrence, Liasion UCA of SME
Brett Campbell
Matthew Crow
Jonathan Klug
John Morgan
Paul Madsen
Derek Penrice
Matthew Preedy
Carlton Ray
Jim Rush

TRACK 1: TECHNOLOGY

Session 1: Project Controls

Mohamamed Younis and Bernard Catalano, Chairs

5D BIM Applied to Cost Estimating, Scheduling, and Project Control in Underground Projects

G. Venturini and Federico Maltese
SWS Engineering

G. Teetes
Schnabel Engineering

ABSTRACT: Building Information Modeling (BIM) is a progressive delivery method that integrates multiple project sources into a 3D model toolbox. The construction schedule becomes the 4th dimension (4D) with the cost component becoming the 5th dimension (5D). 5D BIM allows the project lifecycle, from concept to construction, to be harnessed and shared in a common environment. The toolbox generates multi parameter models that permit to manage several strategic project data. BIM precision is related to the level of development (LOD), which increases with each single project life phase (from preliminary to detail and construction design). This paper provides useful case studies of BIM applied to cost estimating, scheduling and project control in underground projects.

BIM AND TUNNELING INDUSTRY

Introduction

Today, hundreds of technical and academic papers published about BIM can be easily found on Internet. In order to introduce BIM, "Analytical review and evaluation of civil information modeling" (Cheng et al., 2016), provides in our opinion the clear and comprehensive overview to the argument. In this paper, the authors state that "Since civil infrastructure projects are often large projects involving huge capital investment and intricate stakeholder relationships, it is especially important to integrate all the information and data analysis for better design, better construction and better operation for these complex structures."

BIM is a combination of software technologies and processes that generates a "methodology to manage the essential building design and project data in digital format throughout building life-cycle" (Penttilä, 2016). This object-based modeling not only changes how a facility is created from traditional CAD solutions, but also remarkably alters the key delivery processes involved in constructing a facility. Therefore, BIM is not only a technology change, but also a process change. (Cheng et al., 2016).

BIM was first introduced to building projects to facilitate complex systems and has achieved wide penetration in the building market, allowing to reduced costs and to improve productivity, exactness, efficiency and information transfer (McGraw-Hill, 2012).

Meanwhile the topographic, geotechnical and environmental condition have little impact on the construction of the building, except for the foundation components, civil infrastructure projects are subject to every nuance of the terrain. Therefore, building projects are also called "vertical projects," while civil infrastructure projects are usually called "horizontal projects." In fact, for civil infrastructure facilities like roads and bridges, all the specific entities are placed horizontally relative to the axis or the reference line.

Civil information modeling (CIM) is a term commonly used in the architecture, engineering and construction (AEC) industry to refer to the application of BIM for civil infrastructure facilities, such as bridges and tunnels. Project information modeling (PIM) is another term similar to CIM recently introduced by many researchers and practitioners. Other terms such as "Horizontal BIM" and "Heavy BIM" are also used to represent BIM for civil infrastructure (McGraw-Hill, 2012). InfraBIM is another synonym that easily describes the specific application of BIM to infrastructures (SWS, 2017).

State of the Art

BIM is often described as an approach or method rather than simply the latest technology or tool. Indeed, BIM is a Multi-User Platform where the design and cost data is stored, centralized and automated (Keane, 2012). As BIM has achieved wide adoption in the building industry, increasing effort

Table 1. BIM uses in the different life-cycle facility delivery phases

| | Facility Delivery Phases | | | |
| | Phase 1 | Phase 2 | Phase 3 | Phase 4 |
BIM Uses	Conceptual Design	Detailed Design and Documentation	Construction	Operation and Maintenance
Visualization	X	X	X	X
Lifecycle information management	X	X	X	X
Design review	X	X	X	
Computational fluid dynamics	X	X		
Structural analysis		X		
Sunlight analysis		X		
Traffic flow simulation	X	X		
Environmental simulation and analysis		X	X	
Clash detection		X	X	
Schedule modeling (4D)		X	X	
Cost estimation (5D)		X	X	
Quantity takeoff		X	X	X
Constructability analysis			X	
Crane operation simulation			X	
Virtual facility Inspection				X

Source: Cheng et al., 2016, modified

Table 2. BIM case histories subdivided per type, from industry and academic papers

| | Industry Cases | | | | | Academic Papers | | |
	Europe	North and South America	Asia	Oceania and Africa	Total	Total	Grand Total	
Bridges	3	8	10		21	27	48	21%
Roads	9	17	7	2	35	8	43	18%
Power generation	2	7	20	3	32	3	35	15%
Water and wastewater	3	14	9	2	28	1	29	12%
Railways	5	4	8	1	18	4	22	9%
Tunnels		2			2	12	14	6%
Utility infrastructure		2	2	2	6	3	9	4%
Oil and gas	1	1	4		6	2	8	3%
Recreational facilities		3	4		7		7	3%
Airports	1	1	3		5	1	6	3%
Mine	1	2	1	2	6		6	3%
Dams, canals and levees		4			4		4	2%
Ports and harbors	1				1		1	0%
Total	26	65	68	12	171	62	233	100%

Source: Cheng et al., 2016, modified

has been put into "Horizontal BIM." For example, Cho et al. (2012) proposed a holistic BIM library system which contains the geometry, properties and product information based on parametric modeling for efficient quantity takeoffs of tunnels constructed using the New Austrian Tunneling method (NATM).

Today in the AEC industry, companies world-wide have utilized BIM technology for various civil infrastructure projects. Most of the software developed BIM solutions to help leverage existing Geographic Information System (GIS) and survey data to design and simulate construction of several types of infrastructures. Clash detection and 4D schedule simulation have also become a standard in many projects.

Cheng et al., (2016) considered 233 case histories, finding that horizontal BIM has been mainly applied for Bridges (21%), Roads (18%), Power

Table 3. BIM case histories subdivided by type, from industry and academic papers

BIM Characteristics	LOD (Level of Development)					
	100	200	300	350	400	500
Approximate information	YES	YES				
Existence of the component	X	X	X	X	X	X
Generic placeholder		X	X	X	X	X
Geometric representation		X	X	X	X	X
Shape		X	X	X	X	X
Size		X	X	X	X	X
Quantity			X	X	X	X
Location			X	X	X	X
Orientation			X	X	X	X
Non-graphic information			X	X	X	X
Interfaces with other objects				X	X	X
Detailing					X	X
Fabrication, assembly					X	X
Installation						X
Field verified						X

Source: BIM Forum, 2016

generation facilities (15%) and Water and wastewater facilities (12%). Tunnel represent 6% of the case histories of this research. Nevertheless, considering the large numbers of papers and presentations that have been recently published in the proceedings of the WTC (2015, 2016, 2017), RETC (2017), TAC (2016), Cutting Edge Conference (2016, 2017), this percentage is likely to increase in the near future.

Several Countries are implementing BIM. In the UK, the government succeeded in meeting its target of having all public sector projects use BIM by April 2016. Spain has mandated BIM to be used on all public sector projects by 2018 and on infrastructure projects by 2019. Australia has set a target for BIM to be used on all government infrastructure projects with a value in excess of AUD50 million. In Singapore, since July 2015, plans for all projects over 5,000 sq. have to be submitted in BIM format.

Almost all large and medium engineering companies are now structured to develop in-house BIM models and projects. Most of the contractors experienced in vertical structures (social buildings, terminals, etc.) have also in-house BIM departments. The International Tunneling Association (ITA) has recently activated a Working Group #22 on Information Modelling in Tunneling. Finally, BIM seems to be designated to become the standard procedure in the future, mainly for tunneling in urban areas or large underground projects like deep and large tunnels.

BIM AND COST ANALYSES

Level of Development vs. Cost Analysis

Cost estimating is defined as the process of predicting the probable costs of a project, of a given and documented scope, to be completed at a defined location and point of time in the future (Amos, 2010). Therefore, the accuracy and usefulness of the estimate, as a basis for business decisions and budgeting, is directly related to the definition of the project scope (Keane, 2012).

A reliable cost analysis depends from (a) the precision of the Project's quantity takeoff, (b) the applied methods of construction and (c) several physical parameters (e.g., distances, access, interferences). For this reason, the reliability of the estimate is directly correlated to the accuracy of the informations that are delivered with the project. The exactness of the whole information represents the Level of Development (LOD) of the BIM model.

The LOD Specification is a reference that enables practitioners in the AEC Industry to specify and articulate with a high degree of clarity the content and reliability of Building Information Models (BIMs) at various stages in the design and construction process (BIM Forum, 2016). The LOD ranges varies from 100 to 500. An increase of LOD means that more graphic and non-graphic information have been added to the model.

Table 4 illustrates the minimal characteristics required for a model to proceed to a correct cost estimate during the different phase of the life-cycle of the project. During early stages of the design phase, some basic information (LOD 200) can allow the project to proceed to a first project budget estimate, while the Final Design usually generates enough information to proceed with a quite accurate estimate (LOD 300).

During the bid phase, the requested precision of the model depends on the project procurement

Table 4. BIM characteristics versus cost analysis and cost control activities

	Cost Analysis				Cost and Quality Control				
Activity Phase Action	Design Estimate		BID Estimate		Construction				
					Costs		Control		
BIM Characteristics	Budget	Final	Standard	Detail	by Work Package	by WBS	QC/QA	Subs	O&M
Approximate information	X	X	X	X					
Existence of the component	X	X	X	X					
Generic placeholder	X	X	X	X					
Geometric representation	X	X	X	X	(X)				X
Shape	X	X	X	X	(X)		X		X
Size		X	X	X	(X)		X	X	X
Quantity	X	X	X	X	X	X	X	X	X
Location		X	X	X	X	X	X	X	X
Orientation				X	(X)		X	X	X
Non-graphic information			X	X	X	X	X	X	X
Interfaces with other objects				X	X	X	X	X	X
Detailing				X	X	X	X	X	X
Fabrication, assembly							X	X	X
Installation				X	X	X	X	X	X
Field verified					X	X	X	X	X
LOD	200	300/350	300	350	400/500		500		500

Source: this paper; BIM characteristics from Table 3.

and delivery procedure. In case of design-bid-build processes, the detail corresponds to the final design except that usually the contractors adds a number of addition elements (non-graphic information) regarding construction's methods, equipment, site installations, subcontractors, etc.

In case of design-build projects, the contractor is requested to develop, through his Engineer, his own project solution. In this case BIM becomes particularly efficient, allowing the decision makers to analyze at the same time different solutions, looking at the results in terms of construction schedule and costs. For this procedure, an LOD 350 or higher is recommended.

Once the project has been awarded, BIM becomes a very useful tool for cost control either whether the project costs are managed through work packages or WBS. In fact, all information developed in the previous stages can be implemented and detailed with site reports, as well as with all fabrication, assembly and installation details provided by the specialized subcontractors or material suppliers.

The major benefit for estimators and the design team that adopts BIM is the acceleration in performe quantity take-offs and to ease the cost analysis realized by the inherent capabilities of BIM. The cost consultant no longer has to perform the take-offs of building structure and finishes manually. The BIM model generates the quantities of each building component at the point they are added to the model. All changes that are made in the model are reflected in the construction cost immediately. This allows estimators the freedom to carry out other analyses on parameters that affect project costs, and to explore the impact of various systems and design alternatives. Owners especially appreciate the ability to see design ideas illustrated and estimated almost in real time. They receive an exhaustive, concise summary of the building data including a record of the materials, fixtures, and specifications incorporated into the building, reducing operation and maintenance expenses in the long term (Keane, 2012).

These information, all implemented in the same model, can also be used to comply with the Quality Assurance plan. Finally, some information can be used to implement the interface with the different subcontractors. At this stage, the LOD of the model reaches his maximum expected level (500). The model is also a powerful tool to commission the project, to manage and close all the non-conformities and to manage both technical and contractual documentation. Once the project is completed the BIM model can be transferred to the Owner/client, becoming the core of the continued operation and maintenance project life phase.

Another aspect of BIM to be considered is the Project Maturity Level, which is a metric of the ability of the construction supply chain to operate and exchange information (BIMForum, 2017). Maturity levels range from 0 (basic CAD) to 3 (Figure 1). Most of the BIM project are at present

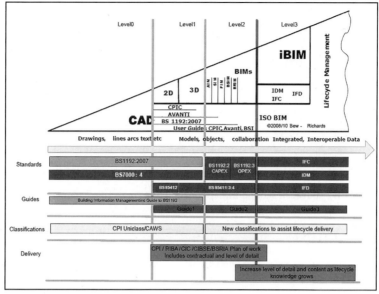

Source: http://bimtalk.co.uk/bim_glossary:level_of_maturity

Figure 1. Project maturity levels and related British standards

developed at level 2 and they can reach level 3 at project commissioning.

BIM EXPERIENCE ON BRENNER BASE TUNNEL, ITALIAN SECTION

Project Overview

The Brenner Base Tunnel (BBT), core of the new European High-Speed Railway Corridor 1, runs for 64 km between Tulfes/Innsbruck (Austria) and Fortezza (Italy), making it the longest underground railway stretch in the word. The BBT will be entirely developed below the Alpine Metamorphic Belt. A new rescue tunnel is being built running parallel to the bypass. The two-tube tunnel system between Innsbruck and Fortezza is 55 km long.

The project develops through several metamorphic units composed of granites, micaschists, gneiss, metamorphic limestones and metapelites. Several strongly tectonized shear zones will be crosscut under up to 2,000 meters of cover. Moreover, significant water inflows have been predicted.

One of the largest sections of the projects, with a contract value of about $1.35B, has been recently tendered and awarded to a Joint Venture leaded by Astaldi with Ghella, Oberholser, Cogeis and Pac. SWS is the construction engineer of the contractor.

In the next seven years the Contractor will excavate about 77 km of exploratory and main tunnels plus ancillary works (bypass, caverns, cross passages, connecting shafts). Forty five km of tunnels will be realized with the aim of 3 Tunnel Boring

Machines (TBM), meanwhile the remnant part will be excavated applying conventional methodologies.

In 2017 SWS also supported Strabag-Salini Impregilo JV to prepare the bid documents and to do the cost analysis of the main section that has to be executed from the Austrian side. Also in this case a 5D BIM model was developed. The two different experiences on very similar scopes of work have been compared and will be discussed in the following paragraphs.

BIM Applications for Estimation Purposes

Quantity Takeoff

In the first stage of the bid phase, all the project was digitalized at an LOD 350 and correlated with GIS data and non-graphic information. Most of the input data came from PDF, DOC and EXL type sources and they were transferred in a comprehensive data base linked to the 3D model generated from the drawings. Although the model was continuously enhanced during all the BID stage, after less than one month from the beginning the estimators were able to receive almost all detailed quantity takeoffs. All expected drill blast profiles were modeled and applied to the project chainages, following the geotechnical baseline report (GBR) inputs received from the Client (Figure 3). At the same time, the precast concrete lining was modeled for the TBM sections of the project, distinguishing the different type of lining and accessories requested by the project and depending from different expected ground conditions.

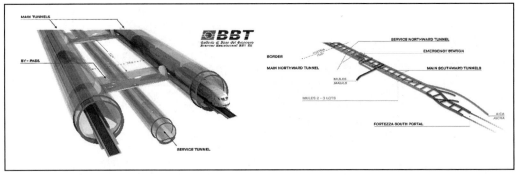

Source: BBT SE (2017)

Figure 2. Layout of BBT southern section under construction (in orange color)

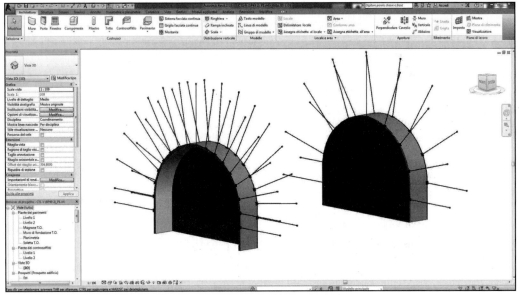

Source: SWS Engineering

Figure 3. Example of typical drill blast section, including all details and elements

Material Management

The procurement process has been developed through a Design-Bid-Build scheme. Nevertheless, several item determinations have been left to Contractors: among them the excavated material disposal and management, which represented one of the most challenging subjects that have been analyzed and determined by the estimators. In fact, By-law, the excavated materials have to be re-used as much as possible for the scopes of the project, e.g., aggregates, filling, temporary and final works, etc. in the case of the BBT all the material will be transferred to temporary stockpiles through the exploratory and service tunnel with the aim of conveyor belts and then selected, reused or sent by train to final disposal areas (Figure 4).

The combination between the location of the excavated material, his temporary stockpile and later re-use as aggregate for concrete or pre-cast segments represents a complex 4D equation. Moreover, the material management cost can dramatically vary depending on changes of the circumstance's mentioned above. Also for this reason the Contractor J/V decided to develop a 5D BIM Model to manage the bid phase. The model has been later enhanced to face with the construction needs and is now implemented on the daily base.

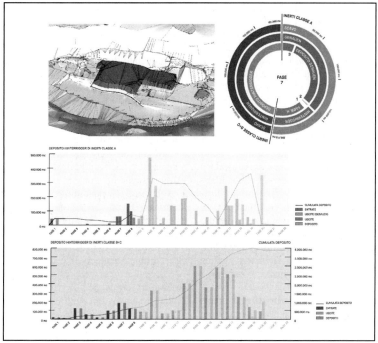

Source: SWS Engineering

Figure 4. Output of the material management related to a specific stage of the project development (Phase 7). The upper left 3D model represent the stockpile of the unusable disposal material (blue) and the stockpile of materials for aggregates. These same information are represented in form of circular graphic and histograms for each construction phase

Site Layouts, Organization and Management

BBT is a very complex project composed of two main railway tubes, one exploratory/service tunnels, bypass, shafts and connections from the main tubes to the exploratory/service tunnel, TBM's assembling and launching caverns, ventilation and utilities underground spaces and so on. The external facilities (crushing and batching plants, disposal areas, warehouses, segment yard etc., are also confined in small and predefined spaces, subject to several environmental and operational restrictions (Figure 5).

BIM models allowed estimators to become familiar with the different layout in very short term, giving them several useful tools to develop the direct and indirect cost analyses and/or dealing with material suppliers and subcontractors in order to receive highly developed and optimized technical and financial offers.

Lessons Learned

During the preparation of the first 5D BIM model for the Italian side (2016) SWS decided to develop all the project items at the same LOD or in any

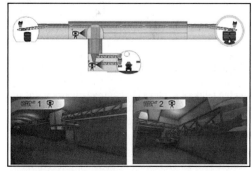

Source: SWS Engineering

Figure 5. Example of underground site layout (shaft for material management purposes)

case with an LOD higher than 350. This approach allowed to deliver a homogeneous and highly developed model but required a lot of investments and man-hours. During the bidding phase only part of the model was later used to generate direct and indirect cost analysis meanwhile other parts were used only for infographic matters. For examples, all the cross passages and connection between the two tubes and the exploratory tunnels were developed in detail, although the costs analyses were done only on cross passage elements and later adopted to the others.

In the northern section which was developed later the LODs of the different items were discussed in a kickoff meeting led by the proposal manager and attended, from the Contractor' side, by the technical manager, the estimate manager and the procurement manager as well as by SWS' project manager and BIM manager. Thus, the project was developed at LOD 350 for main tubes, exploratory tunnel, main caverns, adits, meanwhile cross passages, secondary enlargements and bypass as well as other secondary structures were developed at LOD 200 without limiting the Estimators in their detail cost analysis. In term of planning, the project was connected with the Level 3 items of the construction schedule. In terms of Maturity Levels SWS developed the project with an intermediate level 2-1, although some themes were developed with a maturity level 2, others with level 1 to 1.5.

Comparing the two BBT 5D BIM projects (southern and northern section), considering that the scope of work for the two tendered sections was very similar, SWS experienced a substantial improvement in the project efficiency: meanwhile the first bid (Southern section) required a considerable manpower investment to be fully implemented, the second one reached the same goals with less than 50% of the effort. This optimization was due to the following reasons: (A) The project developed at different LOD. Cross passage and other unique secondary structures developed at LOD 200, (B) use of different levels of maturity (C) availability of templates, models and automation and (D) increased experience of the BIM manager and the working team. This approach allowed to optimize the Return on Investments (ROI) during the Bid phase and deliver the 5D BIM by steps to support contractor's estimate team during the tendering. Once the contract awarded, the 5D BIM is reviewed and becomes the reference project, which will be further developed following the daily inputs and changes.

CONCLUSIONS

BIM applied to infrastructures, also called Horizontal BIM or CIM (Cheng et al, 2016) is becoming a common practice in several markets. The tunneling industry is rapidly filling the gap with the other more advance sectors like bridges, roads and so on (for reference see WTC 2016, WTC 2017, TAC 2016, RETC 2017 proceedings).

BIM approach, and as consequence 5DBIM, require an organic and multidisciplinary approach were several different departments (technical, estimate, procurement, methods depts.) interact and feed the model with graphic (3D) and not-graphic data. For this reason, this methodology requires a strong endorsement of the Contractor's upper management.

The investment to move from a traditional approach to the multidisciplinary BIM methodology is mainly due to (A) personnel training, (B) software licenses (B) learning curve due to lack of experience. A new working team usually take six to nine months and three to four projects to reach a reliable working BIM experience. In any case the core of the team has to be an experienced BIM manager. BIM is not for architects only. geotechnical and civil engineers can easily access and deal with BIM environments. Also, BIM model output represent and excellent tool to generate physical model for structural engineers to perform 2D and 3D FEM.

5D BIM is highly recommended in case of alternative project delivery like design-build and P3 Projects. BIM's best added value is appreciated in complex projects like urban tunneling (also due to frequent interferences and clash detections) and complex projects like underground hydropower plants, railway and highway twin tube tunnel projects and repository underground structures.

REFERENCES

Amos, S.J. (2010)—*Skills & Knowledge of Cost Engineering*. 5th Edition Revised,: AACE International Transactions, Morgantown, pg. A.4.

BIMForum (2015)—*Level of Development Specification for Building Information Models*. Version 2016. www.bimforum.org/lod.

Cheng, J., CP Lu Q., Deng Y. (2016)—*Analytical review and evaluation of civil information modeling*. Automation in Construction 67, 31–47.

Cho, D., Cho N.-S., Cho, H., Kang, K.-I. (2012)—*Parametric modelling based approach for efficient quantity takeoff of NATM-tunnels*. Terotechnology 11 (2) (2012), 70–75.

Edward, K. (2012)—*Building Information Modeling: The Cost Estimator's Role*. International Transactions, BIM837, 1–14.

McGraw-Hill (2012), in: E. Fitch (Ed.)—*The Business Value of BIM for Infrastructure: Addressing America's Infrastructure Challenges with Collaboration and Technology Smart Market Report*, McGraw-Hill Construction.

Penttilä, H. (2006)—*Describing the changes in architectural information technology to understand design complexity and free-form architectural expression*. ITcon 11, 395–408.

INTERNET HYPERLINKS

http://bimtalk.co.uk/bim_glossary:level_of_maturity (2017)

Analysis of Concrete-Lined Tunnels Crossing Active Faults

Stephanie Lange, H. Benjamin Mason, Michael H. Scott, and Scott A. Ashford
Oregon State University

ABSTRACT: It is known that earthquakes with magnitudes greater than M6.0 can cause significant damage to tunnels. In particular, large strains due to fault offsets lead to severe damage in the tunnel lining, (e.g., concrete spalling), which can lead to potential closure and disruptions to the transportation network.

Examining the response of the soil-structure interaction between tunnels and active fault zones, where damaging earthquake ruptures with significant offsets are possible, is critical to ensure resilient design and safe operation. Preliminary results of a calibrated 2D model of a circular reinforced concrete lined tunnel crossing an active fault show earthquake magnitude, fault geology, and structural properties of the tunnel are influential parameters. These results imply that novel tunnel design strategies are necessary.

INTRODUCTION

Urbanization of large cities with dense populations results in an increased use of the underground for storage, shopping, and transportation. Fast growing urban areas depend on transportation of individuals, which is one of the top priorities for city authorities. The construction of tunnels for individual traffic allows cities to grow and support their citizens' lives. Many large urban areas on the west coast of the United States are built in seismically active regions (e.g., Los Angeles, San Francisco Bay area, Seattle), and as a result, it is important to consider the effects of active faulting on tunnel design. In general, seismic impacts on tunnels can be categorized into two groups: (1) ground shaking; and (2) ground failure. The term ground failure usually refers to liquefaction, slope instability, and fault displacement, among others (Hashash et al., 2001). The primary focus of this study is on the latter group, ground failure due to fault displacements.

During fault rupture, large strains in the tunnel lining can cause spalling and potential closure. Accordingly, building tunnels through active fault zones that create damaging earthquakes is risky and is not considered during the planning phase of a tunnel project. In fact, conventional wisdom is to not construct tunnels through fault zones, if at all possible, and to find alternative routes. If alternative routes are not available, especially in urban areas or in cases of vital infrastructure, prior research suggests to designing the underground structure to withstand small predicted fault displacements and to account for repairs of damaged structural elements after an earthquake event (Hung et al., 2009; Hashash et al., 2001).

Predicting the response of a tunnel crossing a seismically active fault zone is complex. Along with long term maintenance, serviceability of a tunnel after an earthquake event is among the highest priorities for stakeholders, clients, and the public. The selection of the tunnel lining structure, material, and thickness all contribute to the response of the tunnel structure in an earthquake event with a considerable amount of fault offset. Many parts of a tunnel structure are affected by seismic excitation, such as the portals, water and ventilation systems, electronic systems, emergency exits, above ground buildings, and neighboring underground structures. However, this research will concentrate on the analysis of concrete-lined circular tunnels crossing active faults under large fault offset; specifically, the response of structural and geotechnical elements will be analyzed. The contribution to knowledge and understanding of tunnel engineering specifically under severe seismic and geological conditions such as active fault offsets will be accomplished by means of numerical analyses and parametric studies using the finite element framework OpenSees (McKenna et al., 2010) with appropriate constitutive models, boundary conditions, and initial conditions. Insights from the parametric studies may provide researchers and practitioners with important information on considerations for the design of tunnel projects.

BACKGROUND

A thorough literature review on the subject of concrete-lined tunnels crossing active faults reveals that the topic is a relatively young research area with great future potential. Although tunnel engineers have been working on the topic for decades, much of the knowledge is still proprietary or confined to site-specific cases. In summary, the reviewed literature consisted of reconnaissance reports of damaged tunnels, numerical case studies of specific future tunnels, tunnels under construction, or in retrospect, and design approaches to accommodate active fault offsets. We will utilize the findings of the reconnaissance reports to inform our numerical modeling efforts. Notably, in the context of this paper, failure of an underground structure is considered to be a structural failure, cave-in, or otherwise unsafe for human use, requiring closure and repair.

Kontogianni and Stiros (2003) compiled a list of tunnels affected by earthquakes with fault offsets over the previous 100 years. Their report shows underground structures such as railroad tunnels (mostly built during the nineteenth century) that cross active faults were damaged during earthquake rupture. They also discussed that road tunnels built after 1930 were damaged during earthquake rupture. In particular, the 1989 Loma Prieta earthquake (moment magnitude (M) 6.9), the 1995 Kobe, Japan earthquake (M6.8), and the 1999 Düzce (M7.1), Turkey earthquake damaged several underground structures crossing active faults.

Ulusay et al. (2002) reported damage that occurred to several road and rail tunnels during the 1999 Kocaeli (M7.4) and Düzce earthquakes in Turkey. Little damage occurred at the portals of the short TEM (Trans European Motorway) tunnels on the northern side of the Izmit Gulf due to the 1999 Kocaeli earthquake. On the opposite side, the Bolu tunnels suffered active fault shear failure from the Düzce earthquake while under construction, as well as ovaling due to the increase of stresses from additional dynamic loads during the earthquake event.

Prentice and Ponti (1997) report that the Wrights Tunnel beneath the Summit Ridge in the southern Santa Cruz Mountains, California was severely damaged with cave-ins and an offset of 1.5 m from the San Andreas Fault due to the 1906 San Francisco earthquake. Similar to the Bolu tunnels in Turkey, the damage on the Wrights tunnel was very localized.

Desai et al. (1991) discussed design considerations for the Los Angeles water tunnels crossing active faults. The authors give special attention to flexible joints between single tunnel elements to accommodate differential movements and to keep leakage at a minimum. Construction through fault zones is a difficult task as seen with the Izmir Metro Tunnel in Turkey. Kun and Onargan (2013) discussed the excavation and construction of the metro tunnel through a fault zone, which raised problems for above-ground structures due to poor rock characteristics. The construction team used additional bolts, face nails, and jet grouting to strengthen the existing tunnel support system for safer above-ground structures. The construction through the geology of a fault

zone is a critical point in the process of building a tunnel, but not further discussed in this writing.

BASIC FEATURES OF NUMERICAL MODEL

Numerical modeling of tunnel response is an important design tool, because the soil-structure interaction between the tunnel lining and the surrounding soil is complicated. Many researchers and practitioners use simplified beam-spring models, finite element (FE) methods, or finite difference (FD) methods to perform numerical modeling for tunnels (e.g., Keshvarian et al. 2004; Shahidi and Vafaeian 2005; Caulfield et al. 2005; Gregor et al. 2007; Anastasopoulos et al. 2008; Lanzano et al. 2008; Corigliano et al. 2011; Wang et al. 2012; Luo and Yang 2013; Varnusfaderani et al., 2015).

Modeling Structure

Based on the practical ranges of previously defined parameters, a baseline model was specified in this study with the foregoing characteristics. Figure 1 shows a schematic of the baseline model. Note that in Figure 1, the springs are shown only on one side of the tunnel to allow a cleaner image. During numerical modeling, the springs are modeled on both sides of the tunnel beam. The numerical model represents a simplified plan view of the tunnel interacting with the movement of a strike-slip fault.

Baseline model characteristics:

- M7 earthquake with a fault offset of 2.0 m;
- Strike-slip fault with a dip angle of 90°, which is characteristic for a vertical fault plane and an important characteristic of the San Andreas and Hayward Faults in the San Francisco Bay area;
- Rock properties for the surrounding ground of medium strengths ($\sigma = 50$ MPa), and rock properties of weak strengths to simulate the fault gauge ($\sigma = 0.25$ MPa);
- Fault offset occurs on a single rupture plane;
- 90° crossing angle of the tunnel structure (i.e., the tunnel crosses perpendicular to the fault plane;
- Circular tunnel cross-section;
- Ratio of tunnel lining thickness, t_w, to inner tunnel diameter, D, of 1/20 , with $t_w = 0.30$ m and $D = 6.0$ m;
- Normal concrete strength of C25/30 with characteristic cylinder strength f'_{ck} of 25 MPa at 28 days (corresponding to f'_c of 4000 psi);
- Normal steel strength of 275 MPa (corresponding to steel grade 40).
- Tunnel length, $L = 200$ m, with fault offset in middle area.

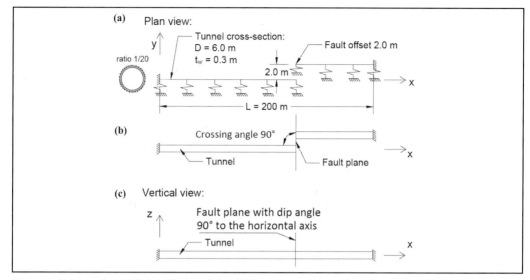

Figure 1. Schematics of 2D baseline model: (a) Plan view on spring-beam model with cross-section, (b) plan view ot tunnel crossing strike-slip fault, (c) vertical view of tunnel crossing strike-slip fault

The open-source, finite-element software framework OpenSees was chosen to simulate the tunnel response under imposed ground displacements. OpenSees is able to model simple 1D elements, like beams and springs, as well as 2D plane strain elements with a variety of constitutive models. Further, OpenSees is able to apply multiple point displacements simultaneously to simulate offsets at different locations of a tunnel crossing a fault zone. The scripting capabilities of OpenSees will be utilized to conduct parametric studies with varying elements including fault characteristics such as fault offset and fault zone width.

To inform the modeling decisions, information about practical ranges of specific tunnel parameters were given by half a dozen tunnel engineering design and construction companies in the United States. The primary consensus of the parameters was:

- Lining thickness-to-inner diameter ratio ranging from 1/15 to 1/30, with 1/20 being the most found ratio;
- Concrete cylinder strengths f'_{ck} between 25 MPa and 40 MPa (4000 psi to 6000 psi);
- Steel strengths between 275 MPa and 500 MPa (grade 40 and 75);
- Initial support could be considered as structurally bearing, but is generally not considered due to quality issues with shotcrete during construction.

Together with the industry input, the structural model resulted in the following numerical elements:

- Force-based beam elements for the tunnel structure;
- Zero length elements for the springs;
- Circular fiber section for the tunnel cross section with inner and outer rebar.

For the purpose of obtaining preliminary results, the concrete and steel materials are defined as elastic. The spring materials are defined as compression only with no tension.

The numerical model was validated with known closed-form analytical solutions of simple beams, such as maximum moment and deflection of beam with single point load, reaction forces and maximum moment of beam with prescribed vertical displacement of one end support, or maximum moment and deflection of beam on elastic ground after Hetényi (1946). To further calibrate the numerical model, the radial and circumferential discretization of the concrete lining in the tunnel cross section was varied to obtain adequate results without the need for excessive computations. The same procedure was performed with the number of springs, or distance of two beam nodes along the tunnel length. The distance of two beam nodes was set to 10 m after calibration, which coincides with a typical length of a cast-in-place tunnel lining segment.

Modeling Fault

The ground offset is based on a strike-slip fault. Therefore, the numerical model is a horizontal view of the tunnel cutting through the strike-slip fault. Kasahara (1981) found that the displacement fields

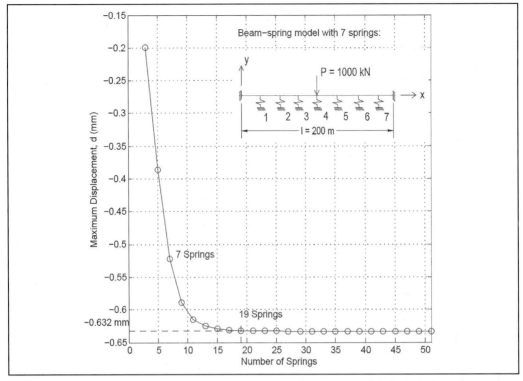

Figure 2. Calibration plot of beam-spring models with varying number of springs under a single point load, with maximum displacement as calibration parameter. Chosen: 19 springs over a tunnel length of 200 m results in 10 m element length between two springs.

associated with a strike-slip fault can be viewed as a shear dislocation. Surface displacement measurements, such as from the San Andreas Fault after the 1906 San Francisco earthquake (M7.8), show that the strain accumulation away from the fault goes to zero at around 20 km. With general tunnel lengths in the order of hundreds of meters to some kilometers, it can be assumed that one side of the tunnel stays and the complete other side of the tunnel will be dislocated according to the fault offset. This detail can be seen in the figure showing the schematics of the 2D baseline model. The fault offset is introduced into the numerical model as nodal displacements of one side of the tunnel. Figure 3 shows the preceding modeling simplification.

This simplification of offset in one direction instead of half the offset in each direction as observed in strike-slip faulting, puts the same relative displacement on the structure, but may influence the reaction of the surrounding ground, which will be examined in the parametric studies.

Herein, we assume that the minimum moment magnitude required to cause damage to tunnels is M6. The literature on correlations of various rupture

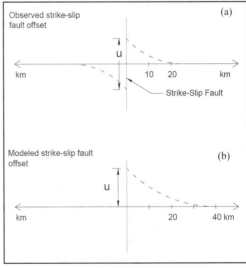

Figure 3. (a) Schematic strike-slip offset of the San Andreas Fault after 1906 earthquake with u = 5 m offset, (b) simplification of strike-slip fault for computation. Offset, u, is exaggerated.

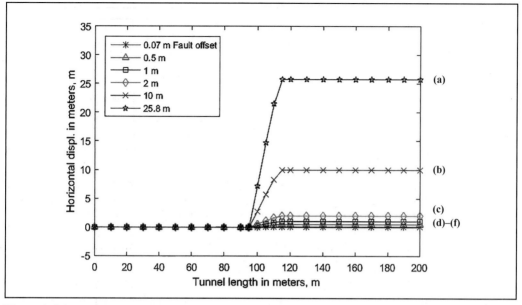

Figure 4. Schematic tunnel beam displacement with 20 m fault zone width and different fault offset values: (a) 25.80 m for earthquake M8.0 (b) 10 m for M7.4 (c) 2.0 m for M7.0 (d) 1.0 m for M6.8 (e) 0.50 m for M6.4 (f) 0.07 m for M6.0 (medium strong rock properties along the tunnel alignment outside the fault zone)

parameters was analyzed to describe earthquake magnitudes with respect to specific fault offsets relatively close to the earth's surface where tunnels are typically located. The literature review resulted in a conservative adoption of the well-known correlations of Wells and Coppersmith (1994) with additional earthquake data from more recent events from Wesnousky (2008). The following formulations were used to correlate the parameters, maximum and average fault offset and earthquake magnitude:

$$M = 6.712 + 0.912 * \log(MD)$$

$$M = 6.996 + 0.948 * \log(AD)$$

where MD is the maximum surface displacement in meters (m), and AD is the average surface displacement in meters (m). Notably, to simplify the modeling of static fault rupture, the propagation of the rupture path from the bedrock through the overburden soil to the surface is not included in this study.

Modeling Rock Material

We consulted the tunneling professionals to determine the type of ground that tunnels through seismically active areas normally traverse and found that tunnels built in earthquake zones in the Pacific Coast area are mainly in softer ground, such as sandy/gravelly soils, up to medium strong rock materials.

The preceding knowledge formed the basis of the range stiffness of the rock surrounding the tunnel, which is modeled with springs.

The springs are defined as elastic no-tension materials with static Young's moduli, E, for different rock types. Since the model is a horizontal cut, it is assumed that there is no gap between the tunnel lining and the surrounding ground, as it might be above the crown of a tunnel. The following parameters used in the analyses are based on ISRM (1987), Johnson and DeGraff (1988), and Marinos and Hoek (2001):

- Fault gauge with uniaxial compressive strength of 0.25 MPa and $E = 100$ MPa
- Clayey sandstone or shale with uniaxial compressive strength of 5 MPa and $E = 1000$ MPa
- Sandstone with uniaxial compressive strength of 50 MPa and $E = 15000$ MPa

The uniaxial compressive strengths are given as an indicator for fault gauge material, low strength, and medium hard rocks. The foregoing values indicate three different levels of possible ranges of ground conditions along the Pacific Coast of the United States and are not exact measurements.

PRELIMINARY RESULTS

Preliminary results show that shearing of the tunnel structure at different fault zone lengths and with

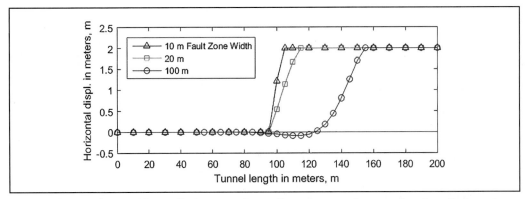

Figure 5. Schematic tunnel beam displacement for medium strong rock properties along the tunnel outside the fault zone; with 2.0 m fault offset for varying fault zone widths of 10 m, 20 m, and 100 m

different amounts of fault offsets is the primary failure characteristic.

Preliminary results show a direct influence between the amount of fault offset and the quantity of strain in the tunnel lining which is anticipated. Further, the preliminary results show an influence of fault zone length beyond the fault zone in terms of strains in longitudinal direction of the tunnel alignment. With longer fault zones, this result diminishes; the strains in the tunnel elements beyond the fault zone do not show a long influence for a longer fault zone, which means that the strains and therefore damages to the tunnel lining are very localized. This effect still has to be verified by comparison with short fault zones.

These preliminary results are in an early stage of research and need further evaluation. Research is ongoing and further results are forthcoming.

CONCLUSIONS

This research focuses on the structural lining and soil-structure-interaction of a circular tunnel and aims to expand the knowledge on the subject. First steps to accomplish the main goal of understanding the behavior of concrete-lined circular tunnels crossing active faults were taken by varying fault offset and with it the earthquake magnitude impacting the tunnel lining.

Behavioral insides from the parametric studies may provide researchers and practitioners with important information on considerations for the design of tunnel projects. Ulusay et al. (2002) described active fault damages in the existing Bolu tunnels and found that the damages were very localized and not dangerous to the long-term stability of tunnel structures. This could coincide with the preliminary results of a larger fault zone with more localized strain accumulation in the tunnel lining.

A construction solution to accommodate fault offset with the knowledge of a larger fault zone and localized damages was analyzed by Russo et al. (2002). These authors analyzed the design of the rehabilitated and finished Bolu tunnels, which uses the articulated design solution with seismic joints in the concrete lining to accommodate a fault offset of 50 cm over a length of 100 m. The case study of a previously damaged tunnel, such as the Bolu tunnels with an adaption of the final lining is an example of executed earthquake engineering.

Beyond the scope of this research study, which is still ongoing, are considered parameters for improvements in future work.

- Varying the dip angle of fault from 0°, 30°, up to 90°;
- Including different fault types: dip-slip normal, reverse, thrust faults;
- Simulating static and dynamic (creep) fault displacement for numerical modeling;
- Combine static and dynamic fault displacement and dynamic tunnel shaking due to earthquake waves;
- Consider the propagation of rupture path from the bedrock through the overburden soil to the surface interacting with the tunnel.

Investigation of this last point in connection with a future immersed tunnel had begun by Anastasopoulos et al. (2008), but further research is necessary to understand the underlying behavior of underground structures in soft ground.

ACKNOWLEDGMENTS

The support of six different companies is gratefully acknowledged. The authors would like to express their thanks to Michael McRae for supporting this research. The support given to this project by Oregon State University professors David Trejo, Edmund Dever, Burkan Isgor, John Nabelek, Eric Kirby, and Andrew Meigs is greatly appreciated.

This material is based upon work supported by the National Science Foundation Graduate Research Fellowship Program under Grant No. 1314109-DGE.

REFERENCES

Anastasopoulos, I., Gerolymos, N., Drosos, V., Georgarakos, T., Kourkoulis, R. and Gazetas, G., 2008. Behaviour of deep immersed tunnel under combined normal fault rupture deformation and subsequent seismic shaking. *Bulletin of Earthquake Engineering*, 6(2): 213–239.

Barton, N., 1978. Suggested methods for the quantitative description of discontinuities in rock masses. *ISRM, International Journal of Rock Mechanics and Mining Sciences & Geomechanics Abstracts*, 15(6): 319–368.

Caulfield, R., Kieffer, D.S., Tsztoo, D.F. and Cain, B., 2005. Seismic design measures for the retrofit of the claremont tunnel. In *Jacobs Associates. Rapid Excavation and Tunneling Conference (RETC) Proceedings. San Francisco: Jacobs Associates*, p. 1–11.

Corigliano, M., Scandella, L., Lai, C.G. and Paolucci, R., 2011. Seismic analysis of deep tunnels in near fault conditions: a case study in Southern Italy. *Bulletin of Earthquake Engineering*, 9(4): 975–995.

Desai, D.B., Merritt, J.L. and Chang, B., 1991, November. 'Shake and slip to survive'- tunnel design: Proc 1989 Rapid Excavation and Tunnelling Conference, Los Angeles, 11–14 June 1989: 13–30. Publ Littleton: SME, 1989. In *International Journal of Rock Mechanics and Mining Sciences & Geomechanics Abstracts* (Vol. 28, No. 6: A392). Pergamon.

Gregor, T., Garrod, B. and Young, D., 2007, May. Analyses of underground structures crossing an active fault in Coronado, California. In *Proceedings of the World Tunnel Congress*, p. 445–450.

Hashash, Y.M., Hook, J.J., Schmidt, B., John, I. and Yao, C., 2001. Seismic design and analysis of underground structures. *Tunnelling and Underground Space Technology*, 16(4): 247–293.

Hetényi, M., 1971. *Beams on elastic foundation: theory with applications in the fields of civil and mechanical engineering*. University of Michigan.

Hung, C.J., Monsees, J., Munfah, N. and Wisniewski, J., 2009. Technical manual for design and construction of road tunnels–civil elements. *US Department of Transportation, Federal Highway Administration, National Highway Institute, New York.*

Johnson, R.B. and DeGraff, J.V., 1988. *Principles of engineering geology*. Wiley.

Kasahara, K., 1981. *Earthquake mechanics*. Cambridge university press.

Keshvarian, K., Chenaghlou, M.R., Emami Tabrizi, M. and Vahdani, S., 2004, July. Seismic isolation of tunnel lining-a case study of the Gavoshan tunnel in the Morvarid Fault. In *Tunnelling and Underground Space Technology. Underground Space for Sustainable Urban Development. Proceedings of the 30th ITA-AITES World Tunnel Congress, Singapore, 22–27 May 2004* (Vol. 19, No. 4–5), H19.

Kontogianni, V.A. and Stiros, S.C., 2003. Earthquakes and seismic faulting: effects on tunnels. *Turkish Journal of Earth Sciences*, 12(1): 153–156.

Kun, M. and Onargan, T., 2013. Influence of the fault zone in shallow tunneling: A case study of Izmir Metro Tunnel. *Tunnelling and Underground Space Technology*, 33: 34–45.

Lanzano, G., Bilotta, E. and Russo, G., 2008. Tunnels under seismic loading: a review of damage case histories and protection methods. *Fabbrocino & Santucci de Magistris eds*, p. 65–74.

Luo, X. and Yang, Z., Finite element modeling of a tunnel affected by dislocation of faults. In *Proceedings of 5th Asia Pacific Congress on Computational Mechanics & 4th International Symposium on Computational Mechanics (APCOM & ISCM)*, Singapore.

Marinos, P. and Hoek, E., 2001. Estimating the geotechnical properties of heterogeneous rock masses such as flysch. *Bulletin of engineering geology and the environment*, 60(2): 85–92.

McKenna, F., Scott, M.H. and Fenves, G.L., 2009. Nonlinear finite-element analysis software architecture using object composition. *Journal of Computing in Civil Engineering*, 24(1): 95–107.

Prentice, C.S. and Ponti, D.J., 1997. Coseismic deformation of the Wrights tunnel during the 1906 San Francisco earthquake: a key to understanding 1906 fault slip and 1989 surface ruptures in the southern Santa Cruz Mountains, California. *Journal of Geophysical Research: Solid Earth*, 102(B1): 635–648.

Russo, M., Germani, G. and Amberg, W., 2002, October. Design and construction of large tunnel through active faults: a recent application. In *Proceedings International Conference of Tunneling and Underground Space Use, Istanbul.*

Shahidi, A.R. and Vafaeian, M., 2005. Analysis of longitudinal profile of the tunnels in the active faulted zone and designing the flexible lining (for Koohrang-III tunnel). *Tunnelling and Underground Space Technology, 20*(3): 213–221.

Ulusay, R., Aydan, Ö. and Hamada, M., 2002. The behaviour of structures built on active fault zones: examples from the recent earthquakes of Turkey. *Structural Engineering/Earthquake Engineering, 19*(2): 149s-167s.

Varnusfaderani, M.G., Golshani, A. and Nemati, R., 2015. Behavior of circular tunnels crossing active faults. *Acta Geodynamica et Geomaterialia, 12*(4): 363–377.

Wang, Z.Z., Zhang, Z. and Gao, B., 2012. Seismic behavior of the tunnel across active fault. In *The 15th World Conference on Earthquake Engineering, Lisbon, Portugal, Sept*, p. 24–28.

Wells, D.L. and Coppersmith, K.J., 1994. New empirical relationships among magnitude, rupture length, rupture width, rupture area, and surface displacement. *Bulletin of the Seismological Society of America, 84*(4): 974–1002.

Wesnousky, S.G., 2008. Displacement and geometrical characteristics of earthquake surface ruptures: Issues and implications for seismic-hazard analysis and the process of earthquake rupture. *Bulletin of the Seismological Society of America, 98*(4): 1609–1632.

Atmospheric Cutting Tools Replacement in EPB TBM "Bertha": Innovative Approach for Face Pressure Tunneling

Juan Luis Magro
DRAGADOS

Ruben Piqueras and Juan Garnero
DRAGADOS USA

INTRODUCTION

The need for tunnels everywhere and in every kind of ground and environment is growing rapidly, especially for tunnels excavated in soils in congested urban areas by means of face pressure Tunnel Boring Machines (TBMs), which demand access to the Cutterhead and cutting tools for inspection and replacement, always problematic and sometimes very difficult or not even possible.

Cutterhead and cutting tools inspections are needed to replace worn tools, but also to learn how tools wear evolves and as such, adjust the frequency of those inspections. Besides, the traditional approach of anticipating areas with stable ground conditions to perform Cutterhead inspections or even built ground improved areas or "safe havens" has become highly complicated these days, for reason such as acquisition of right of ways and complex permitting processes everywhere, or just because the area to be treated is too deep or not accessible at all from the surface or from any closer location.

Concurrently, the quality of the cutting tools and its capability to self-monitoring have dramatically improved over the last years but the need to inspect and replace them, as well as any other component of the Cutterhead, remains the same.

It has been customary to perform Cutterhead maintenance to face pressure TBMs by means of shutting down TBM and entering the excavation chamber at atmospheric or hyperbaric conditions, which is called an "intervention," spaced certain length according to the ground conditions expected and encountered. However, those interventions are not always doable and there is usually no plan B for them and even if they are, they could be time consuming and to some extent, risky to the crews and to the surrounding and above ground structures and utilities.

The concept of being able to access some of the cutting tools from within the Cutterhead hollow arms was born in response to this situation, to provide a supplementary way to inspect and replace tools at any time and any location, without having to empty

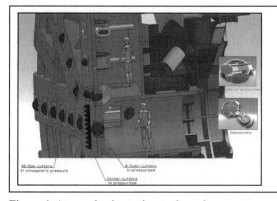

Figure 1. Atmospheric cutting tools replacement concept in "Bertha" by Hitachi Zosen

the excavation chamber and to expose TBM crews and divers to hyperbaric conditions.

This paper addresses the background of this system and the use that Seattle Tunnel Partners, JV (STP) made of it to dig the SR99 Bored Tunnel in Seattle, WA, and highlights some of its benefits and difficulties that will contribute to the continuous and steady growth and development of our tunneling industry.

PREVIOUS CASES

Atmospheric Cutter Changing Devices (ACCDs) were first used in soil TBMs back in the mid-nineties by Herrenknecht and right after by Asian manufacturers, as shown next in Figure 2.

ACCDs holding precutting bits and cutter discs had been tested for performance up to 20 bar for

TBM Diameter (m)	Project	Location	TBM Type	Manufacturer	Year	Tunnel length (m)
14.20	4th Elbe River Highway Tunnel	Hamburg, Germany	Slurry	Herrenknecht	1996	2,560
2.95	Egawa No.4 Stormwater trunk line	Kawasaki City, Kanagawa, Japan	EPB	JIMT	1998	532
9.90	Subway Line 11 Sumida river section	Tokyo, Japan	Slurry	JIMT	2000	1,258
3.48	Myoseiji river No.2 sewage trunk line	Tokyo, Japan	Slurry	JIMT	2001	2,416
13.05	Metropolitan Expressway, Shinjuku Line (road tunnel)	Matsumizaka-Yoyogi, Japan	Slurry	Kawasaki	2001	2,646
5.84	High speed railway, Tozai line, Ishida-Daigo Section	Kyoto, Japan	EPB	JIMT	2002	1,143
5.84	High speed railway, Tozai line, Ishida Kita section	Kyoto, Japan	EPB	JIMT	2002	1,143
8.99	Katsushimna sewage tunnel	Katsushima, Japan	Slurry	Kawasaki	2002	383
7.16	Horikawa central sewage trunk line (phase 1-2)	Kyoto, Japan	Slurry	JIMT	2003	2,700
5.85	Eba area sewage tunnel	Hiroshima, Japan	EPB	JIMT	2004	2,700
3.48	Kobe mega capacity water supply tunnel (Ouji section)	Kobe city, Hyogo, Japan	EPB	JIMT	2005	3,454
4.53	Okayama Nishi utility tunnel section 2, phase 4	Okayama, Japan	EPB	JIMT	2005	830
15.43	Shanghai Changjiang Under River Tunnel Project	China	Slurry	Herrenknecht	2006	7,472
15.43	Shanghai Changjiang Under River Tunnel Project	China	Slurry	Herrenknecht	2006	7,476
14.93	Nanjing Yangtze River Tunnel	China	Slurry	Herrenknecht	2008	3,026
14.93	Nanjing Yangtze River Tunnel	China	Slurry	Herrenknecht	2008	2,996
10.30	Katsushimna sewage tunnel	Katsushima, Japan	Slurry	Kawasaki	2009	980
12.55	Metropolitan Expressway, Shinagawa highway project	Shinagawa, Japan	EPB	Kawasaki	2009	8,030
4.19	Hori-river left bank storm drainage system Sewage	Nagoya City, Japan	EPB	Hitachi Zosen	2010	4,087
12.47	Hanshin Expressway, Yamatogawa highway project	Yamatogawa, Japan	EPB	Kawasaki	2010	4,012
15.43	Hangzhiu Qiantiang River Tunnel	China	Slurry	Herrenknecht	2010	6,200
13.61	Nanjing Metro	China	Slurry	Herrenknecht	2011	3,600
15.47	West Changjiang Road River Tunnel	China	Slurry	Herrenknecht	2011	3,072
14.93	Nanjing Weisanlu river access Road tunnel S line	Nanjing, China	Slurry	CCCCTH/JIMT	2012	4,158
15.93	Nanjing Weisanlu river access Road tunnel N line	Nanjing, China	Slurry	CCCCTH/JIMT	2012	3,557
13.66	Instanbul Strait Road Crossing	Turkey	Slurry	Herrenknecht	2013	3,350
14.93	Shouxshiu Lake Tunnel	China	Slurry	Herrenknecht	2013	1,276
15.43	Hangzhiu Qiantiang River Tunnel	China	Slurry	Herrenknecht	2013	5,210
17.48	SR99 Bored Tunnel	Seattle, WA, USA	EPB	Hitachi Zosen	2013	2,826
15.43	Sanghai West Changjiang Yangtze River Road Tunnel	China	Slurry	Herrenknecht	2014	5,264
15.76	Wuhan Metro road / Metro river crossing	China	Slurry	Herrenknecht	2015	2,590
15.76	Wuhan Metro road / Metro river crossing	China	Slurry	Herrenknecht	2015	2,590
12.51	Wuhan Metro	China	Slurry	Herrenknecht	2016	3,172
13.61	Foshan Donguan Intercity	China	Slurry	Herrenknecht	2016	4,900
15.53	Bei Heng Motorway	China	Slurry	Herrenknecht	2016	6,400
11.67	Hangzhou Guangjiang Road Tunnel	China	Slurry	Herrenknecht	2017	1,800
11.71	Hangzhou Guangjiang Road Tunnel	China	Slurry	Herrenknecht	2017	1,800
12.01	Sutong pipe gallery	China	Slurry	Herrenknecht	2017	5,486
12.61	Beijing Zhanjilakou Railway	China	Slurry	Herrenknecht	2017	2,000
12.61	Beijing Zhanjilakou Railway	China	Slurry	Herrenknecht	2017	1,750
15.00	Shantou Sea Crossing	China	Slurry	Herrenknecht	2017	3,047
15.43	Nanjing Maizizhou Tunnel	China	Slurry	Herrenknecht	2018	2,326

Figure 2. Some of the bored tunnel projects whose TBMs mounted atmospheric cutting tools

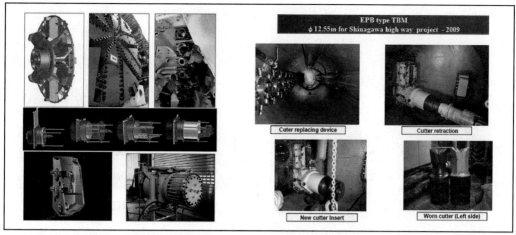

Figure 3. ACCD by Herrenknecht (Slurry shield) and Kawasaki (EPB)

Figure 4. TBMs featuring ACCDs by JIMT

the time being and mainly take advantage of large diameter soil TBMs where the size of the Cutterhead allows to hold this type of configuration and where interventions are perceived to be especially risky, e.g., in difficult geology and in urban environment, as discussed above. This system enables workers to change cutters at any time, entering into the Cutterhead from its rear part and always in atmospheric pressure, resulting in safer operation for both, the crews and adjacent buildings and structures since pressurized chamber or face support conditions are not affected by this access activity (no need to change from slurry or earth pressure face support to compressed air face support with all the associated difficulties, such as refilling chamber in EPB mode).

The system has been used in smaller diameter machines, down to 3 meters in diameter or less and there is some advantage when ACCDs are implemented in Slurry TBMs versus EPB machines, since ACCDs may require wider Cutterhead regardless of the diameter for workers to fit inside and that interferes to some extent with the flowing of the soil into the mixing chamber through Cutterhead openings, slightly more relevant on EPB machines. That being said, proper flow of muck is still achievable in EPB machines while having the ACCDs in the Cutterhead if using the right soil conditioning program and TBM design.

That being said, the interior of the Cutterhead shall be considered a confined space work environment and the applicable safety rules must followed all the time, regardless of which country or safety code any given TBM is working under.

In addition, entering the excavation chamber to perform direct observation of Cutterhead and cutting tools is still needed to perform a proper control and maintenance of the Cutterhead and the ACCDs are by no means deemed to replaced it but to mitigate its risks by reducing the amount of those interventions and its frequency. Both experience and common sense indicate that reducing the number of interventions equates to less risk for Projects, as long as TBM Cutterheads are properly maintained.

EXPLANATION OF THE SYSTEM

The SR99 Bored Tunnel Project has been one of the most publicized and real-time broadcasted projects in tunneling history, from inception just because of featuring the largest TBM ever built by that time and even more soon after when the world-record TBM was shut down for over two years.

Figure 5. Cutterhead of "Bertha"

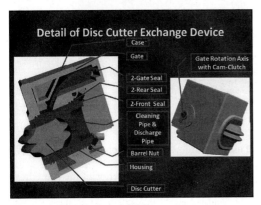

Figure 6. Detail of an ACCD

"Bertha," the 57.5 ft. world-record diameter EPB that Hitachi Zosen built for Seattle Tunnel Partners (a Joint Venture made of Dragados and Tutor Perini) successfully completed the excavation of the 9,272 ft. length and 56 ft. OD / 52 ft. ID tunnel named SR 99 under downtown Seattle (WA) in April 4, 2017 and it featured a key component to its final success to bore this challenging tunnel, which is the atmospheric cutting tools.

Forty-seven Atmospheric Cutter Changing Devices for inner or "face" cutters, holding each of them two precutting bits (RPB) or a double 19" cutter disc and two more ACCD for the gage or perimeter area, mounting as well one RPB or a single 19" cutter, were mounted in its Cutterhead.

The steps to follow to replace one of this ACCD, starting with retraction for inspection, were as follows:

Figure 7. RPB cage and long jack #1 retracting

- Rotating Cutterhead so the RPBs (or Discs) to be inspected were located at 9 pm (direction of mining).
- Pressurization of the gate chamber with bentonite or grease, to verify water tightness.
- Bring the mounting cage to the spoke and bolt it to the ACCD frame.
- Assemble long jack #1 and push the RPB box to remove securing wedges.
- Then retract the RPB box while injecting grease or bentonite through injection ports to ensure that the gate cavity remains full and dirt does not fill it up.
- When jack #1 has retracted enough the RPBs, the gate would be closed and secured.
- After this, verify that the door seals have not been damaged and can hold the ground that is pushing from the outside by releasing the pressure in the cavity through the top port of the RPB box.
- As the RPB box is settled, replace jack #1 by two shorter jacks #2 and then fully retract the RPB box.

Once inside the Cutterhead arm, the RPBs were inspected using the adequate gauges to measure wear and the set thresholds will indicate whether or not to change them.

The steps to exchange the RPB are as follows:

- Remove the two short jacks #2 and take the cage holding the RPB box with the worn RPBs out of the Cutterhead.
- Bring the cage with the new RPBs inside the Cutterhead and bolt it to the housing of the RPB.
- Push the box with the two short jacks #2 until there is room for longer jack #1 and then, continue pushing with jack #1 until RPBs (Cutter Disc, in this case) reach the gate.

Figure 9. New Cutter Disc in the cage, ready to be installed, pushed by the two shorter jacks #2

Figure 8. Cage with the retracted RPBs taken out of the spoke

- Pressurize the cavity again with grease or bentonite beyond pressure outside (in the excavation plenum) to ensure the gate opens.
- When the pressure is built, stop grease or bentonite injection and open the gate.
- Continue pushing the box until it is in place.
- Mount fixation wedges and remove jack #1 and cage.

STP APPROACH

Bertha's Cutterhead featured 736 cutting tools, including 32 scrapers or bucket teeth, 32 trim bits (adjacent to bucket teeth), 388 cutting bits, 178 welded precutting bits and 106 replaceable precutting bits or RPBs, whose cutting profile was according to Figure 11.

In terms of ground penetration, the first line of attack was the fix precutting bits (for which the Cutterhead was retractable to inspect and replace) and then come the atmospheric replaceable precutting bits, as shown above. And then the cutting bits and trim bits, protected by the precutting bits.

In the early stages of the project, STP had planned for 19 Cutterhead inspections, five of them supposed to be in clayey and impermeable soils and the remaining fourteen in permeable, granular and sandy soil conditions, as per Figure 13, although three interventions of this fifteen were planned in Safe Havens in the first portion of the alignment (learning curve).

Figure 10. RPBs (or Cutter Disc) box pushed out to its final position

Effective working time in the Cutterhead was tentatively planned to be 1,218 hours or almost 51 days of TBM shutdown, to replace 136 Double cutter discs, 146 trim bits or bucket teeth and 314 Cutting Bits.

Figure 11. Detail of the gate from the outside (face) halfway open

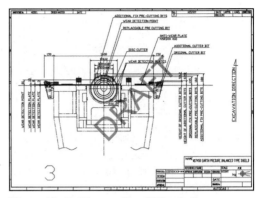

Figure 12. Cutting profile of "Bertha"

Nonetheless, "Bertha" featured the best hyperbaric interventions means available by its time, including three double chamber 6-persons manlocks and a couple of tool-locks, all of them designed to work with a breathable air pressure of up to 7 bar, as well as a 6-persons hyperbaric shuttle and a medlock facility/hyperbaric habitat up in the surface, to be able to treat potential injured divers still under pressure, as well as and to perform saturation diving, if needed at any given time.

However, actual ground conditions did not guarantee that all those interventions could be performed in a reasonable time and safe manner and some of them could be anticipated to be risky, to say the least. Besides, presence of boulders at any location along the alignment and not being able to perform safe havens after the first 1,500 ft. of the 9,272 ft. drive because of lacking access from the surface (downtown Seattle), among other reasons, led us to change our plan.

As a result, Hitachi Zosen as TBM supplier was tasked with designing and manufacturing a system to allow inspection and replacement of a number of cutting tools from within the Cutterhead hollow arms, strategically distributed to cover as much of the Cutterhead as possible, taking advantage of its large size. Thus, Bertha featured those 106 atmospheric cutting tools, out of the 736 total, which could be

replaced at any time without performing interventions. In addition, those tools could be precutting bits or rippers (RPBs) or Cutter Discs.

In any event, it was soon recognized that to be able to successfully complete the tunnel excavation, it had to be implemented a RPBs replacement program including replacing RPBs more often than theoretically needed to protect and extent the life of the other cutting tools that had to be inspected and replaced at hyperbaric conditions.

Once again, atmospheric cutting tools were not meant to replace hyperbaric interventions but to supplement them, since direct observation of the rest of the cutting tools, injection ports and Cutterhead structure and protections was still necessary and recommended.

Hitachi Zosen designed a system customized for this TBM, based on previous experiences that was fully tested in Factory and finally, added to the final configuration of the TBM.

ACCD PERFORMANCE

Again, frequency of atmospheric cutting tools inspection and replacement was increased, in order to protect and extent the life of the rest of the cutting tools in the Cutterhead, which had to be inspected and replaced during a hyperbaric intervention. At the end of the day, we had the opportunity to access 106 out of 736 cutting tools in the Cutterhead from within the Cutterhead arms at any given time, without having to empty the excavation chamber to perform hyperbaric interventions, and being those longer than cutting bits and bucket teeth.

Bertha's ability to retract its Cutterhead was inhibited at some point in time before it resumed mining in January 2016, thus replacing the first line of protection, the fix precutting bits, became extremely difficult, if not impossible and at this point, the atmospheric cutting tools (Bertha's second

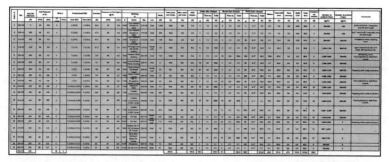

Figure 13. Earliest draft for Interventions Plan envisioned for SR99 BT Project

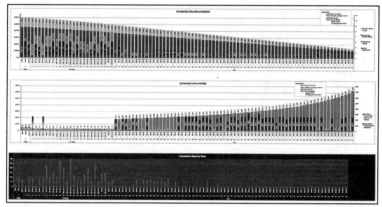

Figure 14. Estimated vs. actual RPBs replaced in atmospheric conditions and cumulative wear per track

line of defense in the original design) became even more critical to succeed.

Consequently, a dedicated crew was set to perform ACCDs inspection and replacement, specifically trained for it and TBM was shut down on weekends, at a minimum, for that crew to perform.

As a result of both efforts, bounce diving and utilization of ACCDs, a total of 6 hyperbaric interventions were carried out from inception (out of the 19 originally planned) with a total of 315 dives. 432 Cutting tools were replaced during those interventions, among other works, and effective working time was 1,135 hours.

In addition, 113 RPBs and Cutter Discs were replaced since "Bertha" resumed mining in January 2016, in 33 stoppages to perform ACCDs inspections, which took 97 days of TBM shut down.

It is clear that the amount of down time because of cutting tools inspection and replacement planned versus actual cannot be compared because the original projection relied on a different approach but given some of the actual conditions encountered, such as high abrasivity and permeability of the

ground in some locations of the alignment, it would be safe to say that having the atmospheric cutting tools and using them the way STP did not only was a major contributing factor to the final completion of the tunnel but prevented severe damage in the Cutterhead from happening caused by dealing with boulders along the tunnel alignment.

Moreover, the ACCDs were of great help at the end of the alignment, when presence of boulders in the TBM path damaged the Cutterhead in the gage area. ACCDs control plan set to make sure Cutterhead was properly controlled until the end of the drive let us notice the damage, control it and eventually, mitigate it by replacing the RPBs in the believed damaged area, confirmed once TBM broke through at the disassembly shaft (see Figure 15).

FOAM PORTS

Soil conditioning was performed in Bertha's excavation chamber primarily by means of foam and a given concentration of polymer, like in any other EPB machine. However, in this TBM it was particularly important that the foam was injected right in the

Figure 15. Worn area in the Cutterhead perimeter

face by means all of the 22 ports in the Cutterhead, as opposed to in the excavation plenum and that all the ports could be used all the time.

A world-record TBM this size and the heterogeneous soil conditions along the alignment required large capacity of soil conditioning but also flexibility to deal with changing ground. In addition, Bertha's Cutterhead design featured these ACCDs, which could have made Cutterhead arms slightly wider than they would have been otherwise, so proper soil mixing was an issue to be paid even more attention than in any other EPB machine and lacking injection ports in the Cutterhead because of they getting clogged (as unfortunately, they usually do) was something "Bertha" could not afford.

We soon realized that having the option to replace atmospheric precutting bits any time also gave us the opportunity to add foam ports in the precutting bits' boxes that were going to be mounted in the Cutterhead, replacing that way any clogged injection port that we could not recover by any other mean, at any given time and without hyperbaric interventions.

As a result, "Bertha" could actually use on the 22 foam ports in the Cutterhead to condition the ground for the most part of the alignment, which was key for the good progress of the TBM.

CONCLUSIONS

Atmospheric Cutting Tools or ACCDs are not only a risk mitigation tool for challenging underground projects but a feature that improves changes to be successful on boring tunnels in challenging mixed faced and abrasive soils, with limited or no access from the surface and as such, they will definitely play a key role for the growth and development of our industry.

It is true that they come with a price tag and with side effects, such as interference with soil conditioning, key for the success of any EPB or slurry machine. But that hurdle can be jumped, as proven in the SR99 Bored Tunnel by "Bertha" and its crew.

Most importantly, Atmospheric Cutting Tools are not meant to replace the necessary hyperbaric interventions but to supplement them, especially in those not that uncommon cases when the conditions to perform hyperbaric interventions cannot be successfully achieved, for whichever the reason is (low coverage/shallow tunneling, under water, permeable granular soils).

For the time being, atmospheric tools may be less cost or time effective than hyperbaric interventions, but they still could save a lot of time, money and headaches by preventing major Cutterhead damage from happening if properly used, such as an alternative for hyperbaric interventions when they can't be performed. An ACCD system shall permit the replacement of enough cutting tools to continue mining until finding a better location for the TBM to stop, where conditions for hyperbaric interventions are more likely to be achieved.

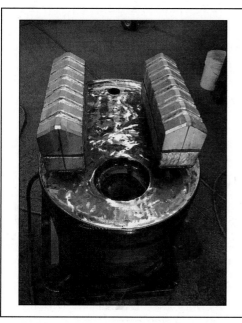

Figure 16. Foam ports in an ACCD (precutting bits' box)

It is clear that working from within the Cutterhead, which is a confined space is not easy and that it works better for large diameter machines and that further development of these systems is still needed. However, with proper planning it is not only doable but also safer for our crews than being exposed to potential unstable ground conditions during interventions and therefore, we believe that replacing cutting tools from inside the Cutterhead at atmospheric conditions should be the way to go in the near future.

At the end of the day, atmospheric cutting tools are and will be helpful to all of us to make tunnel projects safer and to mitigate costly ground improvements in very congested urban areas and their associated public disturbance, but they come with a price

and a commitment that we all should share and hence, every party should commit to invest on R+D to continue developing technologies like this one, for the benefit of our industry.

ACKNOWLEDGMENTS

The authors would like to express grateful Acknowledgment to everyone involved in the construction of the SR99 Bored Tunnel, one of the most challenging tunnel projects ever built, especially to Seattle Tunnel Partners, Dragados and Tutor Perini and Hitachi Zosen; field managers and the crews that made this great success happen, as well as to every other TBM manufacturer, particularly to Herrenknecht, JIMT and Kawasaki.

Logistics and Performance of a Large-Diameter Crossover TBM for the Akron Ohio Canal Interceptor Tunnel

Pablo Salazar
Robbins

Connor Maxon
Kenny-Obayashi JV

ABSTRACT: The Ohio Canal Interceptor Tunnel (OCIT) below the city of Akron is utilizing the first large diameter Dual Mode, "Crossover" type TBM in the United States. The 30.4 ft diameter machine is excavating in variable conditions including soft ground and shale rock. Due to the unique conditions, the TBM has been designed with features including a flexible cutterhead design and abrasion-resistant plating on the cutterhead and screw conveyor. As part of a predictive maintenance plan, measurements for the screw conveyor's exposed features will be taken along the drive to report on the wear rate of these components in shale. This paper will concentrate on the logistics and process of the TBM launch, and component wear and performance at the jobsite in variable ground conditions.

INTRODUCTION

The OCIT is the key component of the city's long-term control plan aimed at reducing Combined Sewer Overflows (CSOs) into the Little Cuyahoga River and surrounding streams. The tunnel will be combined with drop shafts, diversion structures, consolidation sewers, and related appurtenances that will decrease the flow of untreated CSO and direct wastewater and storm water through Akron's water reclamation facility. Consent decree timing stipulates that the OCIT must be complete and operational by December 31, 2018 and the rest of the system and its structures must be online by 2028.

The city looked at rainfall records and identified 1994 as a typical year. The 6,200 ft long OCIT will thus be able to handle 450 gallons of CSO annually in a 27 ft finished inside diameter tunnel. If rainfall is above typical, a further 17 million gallons of overflow can be diverted through an Enhanced High Rate Treatment (EHRT) process that will allow the water to be directly released into the Little Cuyahoga. The EHRT is a separate project that will be completed in the future. The tunnel will pull from nine main regulators referred to as racks (overflow spots) along a narrow corridor to achieve the water storage, while the city's other overflow spots are being controlled through storage basins, sewer separations, maximizing conveyance and green infrastructure (see Figure 1).

GEOLOGY

The Storage Tunnel will pass through three zones or reaches with distinctly different ground conditions. Starting at the construction portal, the generalized ground conditions in these reaches will consist of soft ground (Reach No. 1), mixed face conditions with soft ground overlying bedrock (Reach No. 2), and bedrock with two sections of low rock cover (Reach No. 3).

Reach No. 1—Soft Ground

Reach No. 1 extends from the construction portal to Sta. 13+50. Ground surface after the placement of fill between the construction portal and about Sta. 12+50 will range from approximately El. 845 to El. 885, which corresponds to a thickness of cover over the crown of the tunnel ranging from about 25 to 65 feet.

Reach No. 2—Mixed Face

Reach No. 2 extends from Sta. 13+50 to Sta. 19+50. Existing ground surface ranges from approximately El. 868 to El. 893, which corresponds to a thickness of cover over the crown of the tunnel ranging from about 45 to 70 feet. Ground conditions in Reach No. 2 are expected to range from a full face of soft ground to a full face of bedrock in the vicinity of Sta. 19+50, with the contact between soil and bedrock rising from north to south. Although geotechnical studies predict a relatively uniformly sloping bedrock surface in this reach, it is possible that the

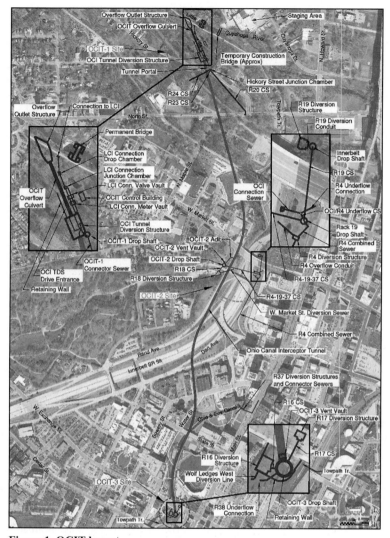

Figure 1. OCIT layout

bedrock surface has a stepped or staircase configuration as a result of the horizontal and vertical jointing in the rock mass.

Within Reach 2, four obstructions will be encountered within the soft ground section consisting of naturally deposited boulders with UCS no greater than 45 ksi. The bedrock in Reach No. 2 will consist primarily of shale with a slightly higher degree of weathering and more closely spaced joints than the overall rock mass due to the proximity of the top of bedrock. A zone of completely to highly weathered bedrock will be encountered at the top of the bedrock surface. For baseline purposes, the thickness of completely to highly weathered rock at the top of bedrock will be no greater than 10 feet in

Reach 2. Very widely spaced horizontal clay seams were observed in the rock core, parallel to bedding, and ranged from about 1 to 3 inches thick along the tunnel alignment in Reach 2.

Overall bedrock quality in Reach No.2, in terms of RQD is predominantly fair (50 < RQD < 75) to very good (RQD > 90), with some zones of poor (25 < RQD <50), to very poor quality rock (RQD < 25). These zones of poor to very poor quality rock are associated with closely spaced, decomposed bedding joints.

Reach No. 3—Bedrock

Reach No. 3 extends from Sta. 19+50 to the OCIT-3 Drop Shaft. Existing ground surface in Reach No.

Table 1. TBM operating modes in varying ground conditions

Akron OCIT TUNNEL REACHES AND OPERATING MODE							Robbins
START STATION	STOP STATION	FOOTAGE	GROUND CONDITIONS	OPERATION CONDITION	FOOTAGE	TOTAL FOOTAGE	
11 +40	14 +50	310	Full face soft ground, shale below invert	EPB, Alpha 20		310	
14 + 50	19 + 50	500	Mixed ground, Partial face shale	EPB, Alpha 25			
19 + 50	27 + 50	800	Full face shale, low cover	EPB, Alpha 25			
27 + 50	51 + 00	2350	Full Face shale, good cover	Open Mode, 40 MPa shale			
51 +00	56 + 50	550	Full face shale, low cover	EPB, Alpha 25		1850	
56 + 50	73 + 50	1700	Full Face shale, good cover	Open Mode, 40 MPa shale	1300	3650	
				Open Mode, 70 MPa shale	400	400	
TOTAL		6210	feet			6210	feet

3 ranges from approximately El. 893 to El. 992, which corresponds to a total thickness of cover over the crown of the tunnel ranging from about 70 to 165 feet. The thickness of bedrock cover over the crown will range from a few feet adjacent to Reach No. 2 to about 90 feet, with two zones of generally lower bedrock cover of between about 10 to 30 feet. The limits of the zones of low bedrock cover are from Sta. 19+50 to Sta. 27+50 and from Sta. 51+00 to Sta. 56+50 as shown on Figure 2. In addition, the Storage Tunnel will pass below the existing St. Vincent-St. Mary's landfill between approximate Sta. 26+00 to Sta. 33+00.

The Storage Tunnel in Reach 3 will be entirely in shale and siltstone, with minor amounts of sandstone. A zone of completely to highly weathered bedrock will be encountered at the top of the bedrock surface. For baseline purposes, the thickness of completely to highly weathered rock at the top of bedrock will be no greater than 15 feet in Reach 3. Very widely spaced horizontal clay seams were observed in the rock core, parallel to bedding, and ranged from about 1 to 3 inches thick along the tunnel alignment in Reach 3.

Overall bedrock quality in Reach No.3, in terms of RQD will be predominantly good (75< RQD <90) to very good (RQD >90), with some zones of poor (25< RQD <50) to very poor quality rock (RQD < 25) above the tunnel crown and near the top of bedrock. These zones of poor to very poor quality rock are associated with closely spaced, decomposed bedding joints.

TBM Operating Modes in Various Sections of the Tunnel

The geotechnical plans and profile from the contract documents, as well as the GBR, were studied to define the approximate sections of tunnel and operating modes, as displayed in Table 1.

TBM CUSTOM SETUP

The OCIT Project was awarded to the Kenny/ Obayashi JV with a Notice to Proceed date of November 4, 2015. The contractor chose the 30.4 ft Robbins Crossover (XRE) TBM due to varying geology that begins in full-face soft ground, giving way to partial face shale and mixed conditions, and finally full-face shale rock. The machine is capable of operating in both hard rock and soft ground (EPB) configurations. The TBM has been customized in a number of ways, from cutting tools to abrasion-resistant wear plating (see Figure 2).

Crossover Design for Abrasive Ground

One of the main problems in mixed ground for TBM tunneling is related to abnormal flat and multi-flat cutter wear, causing the cutter to wear flat on part of its surface. This is because the soft ground material in the mixed face cannot provide sufficient rolling

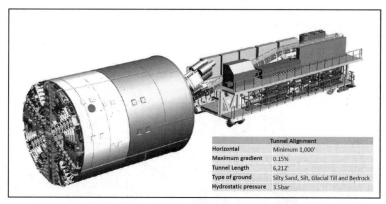

Tunnel Alignment	
Horizontal	Minimum 1,000'
Maximum gradient	0.15%
Tunnel Length	6,212'
Type of ground	Silty Sand, Silt, Glacial Till and Bedrock
Hydrostatic pressure	3.5bar

Figure 2. 3D Model of Akron OCIT TBM

Figure 3. View of the OCIT TBM cutterhead

force for cutters to overcome the pre-torque of cutter bearings. The XRE TBM cutterhead is equipped with 56 housings that can be dressed with either knife bits/rippers or 17-inch disc cutters. Due to geologic variability Kenny/Obayashi and Robbins decided that disc cutters would be beneficial from the outset, launching the machine with a full dressing of discs.

Since the disc cutters are operating in soft ground, the pre-torque of the cutter bearings has been reduced by 25% to require less rolling force for the cutter to rotate evenly. This could result in a shorter cutter life once in rock; however, the shale is not expected to be harder than 70 MPa so cutter ring wear is estimated to be very limited. Furthermore in order to avoid hyperbaric intervention in the first 1,610 ft (490 m), sacrificial rippers have been welded to intervene in case of ring wear greater than 0.6 in (15 mm). In the event a cutter gets blocked these rippers should be able to cut the face until the machine

reaches a section where it can operate in open mode. In consideration of the 65% drive in rock, the cutterhead has also been dressed with Hardox 450 faceplates and peripheral grill bars to reduce the risk of abrasive wear (see Figure 3).

The screw conveyor is another customized component: it is a shaft-type design, 64.5 ft long, 47 inches in diameter, with a tapered front nose to 30 inches. In consideration for the OCIT geology and the necessity to muck out shale bedrock, the single shaft-type screw conveyor required a much higher speed than would normally be provided. The hydraulic power unit has an output of 5×110 kW, which brings the max theoretical speed of the conveyor to 16 rpm at a limited torque of 232 kNm (see Figure 4).

In rock mode and mixed ground conditions, and to a lesser degree in EPB mode as well, the auger and the casings are in contact with abrasive material, creating wear. The screw conveyor features

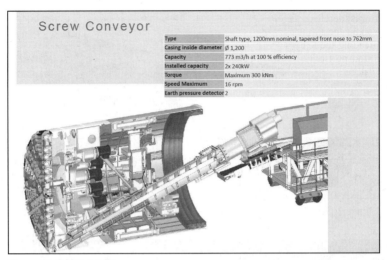

Screw Conveyor	
Type	Shaft type, 1200mm nominal, tapered front nose to 762mm
Casing inside diameter	Ø 1,200
Capacity	773 m3/h at 100 % efficiency
Installed capacity	2x 240kW
Torque	Maximum 300 kNm
Speed Maximum	16 rpm
Earth pressure detector	2

Figure 4. OCIT TBM screw conveyor

the following characteristics to limit wear due to abrasion:

- The leading face of the front auger flight and the outside diameter is covered with welded-in wear plates "inserts" made of Chromeweld 600™ and hardfacing in a crosshatch pattern
- The auger shaft is covered in hardfacing in a crosshatch pattern
- The inside diameter of the casing is lined with welded 0.4 in thick Chromeweld 600™ for the first ⅓ of the casing
- The inside diameter of the remaining ⅔ of the casing is covered with hardfacing
- Four ports have been incorporated into the design for lubricant injection of foam, bentonite, etc.

Chromeweld 600™ is a premium grade of chromium carbide wear plate, produced with a mild steel base plate and hardfaced/overlayed chromium carbide wire (see Figure 5).

TBM LOGISTICS, ASSEMBLY, EXCAVATION, AND UPCOMING CHALLENGES

On May 30th, 2017 the commissioning of the TBM was performed in Solon, Ohio with representatives of the city of Akron, the contractor and the TBM manufacturer. During the commissioning in the factory all the main systems were tested. Once the TBM test was finished the manufacturer disassembled the machine into pieces most convenient for shipping to the jobsite.

A shipping plan was developed between the contractor and the TBM manufacturer to use the shop assembly location as an intermediate place to

Figure 5. Screw conveyor designed with welded-in wear plates and hardfacing

store the TBM components. Components would then be shipped in the right assembly sequence to reduce double handling at the jobsite. Due to the fact the cutterhead was designed to avoid any field welds, it was shipped to be assembled and mounted directly to the cutterhead support (main drive); this setup proved to be a great benefit for the shipment and assembly sequence (see Figure 6).

One of the most critical items shipped to Akron was the cutterhead support assembly: a piece weighing approximately 142,000 kg. The main challenge for the transportation of the cutterhead support assembly was to find the right solution to cross a temporary bridge built inside the jobsite that was designed for a 23,500 lbs/ft² load per axle, or about 136,078 kg total (see Figure 7).

Another limiting factor was the local regulation limiting the maximum weight per axle to under

23,000 lbs/ft². It was therefore decided to use a special truck that spread the load across two different lanes, further spreading the load per axle. The truck selected for the transport of the cutterhead support was a 15-axle dual lane truck plus additional tractor in the back (see Figure 8–9).

Figure 6. TBM assembly site and setup

The 15-axle truck was used for the transportation between the Robbins facility in Solon, Ohio to the jobsite in Akron. Once the cutterhead support arrived at the jobsite it was transferred again before crossing the bridge to a 19-axle truck that spread the load longitudinally and respected the maximum load that the bridge was designed for (see Figure 10).

TBM Launch

The TBM launched from its portal site October 20, 2017 with a depth to invert of approximately 40 ft through a jet grout plug installed to provide a controlled launch environment. The TBM is boring at a uniform slope of 0.15 percent through ground conditions that consist of soft ground, mixed face soft ground over bedrock, and bedrock.

The OCIT is being lined with a steel fiber reinforced precast concrete segmental liner installed concurrent with the mining advance. The low cover environment of the launch itself was the project's first challenge: There is only 20 ft of cover from the top of the TBM, in soft soils, so the challenge has been to ensure there is no impact to the ground above the TBM (see Figure 11).

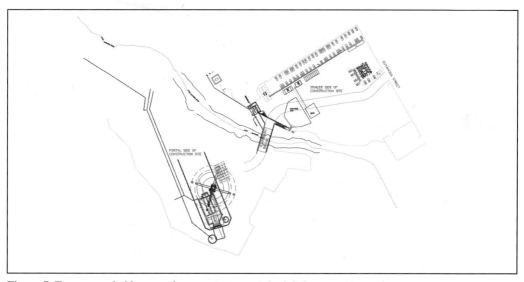

Figure 7. Temporary bridge crossing a waterway at the jobsite

Figure 8. Cutterhead support on the 15-axle dual lane truck

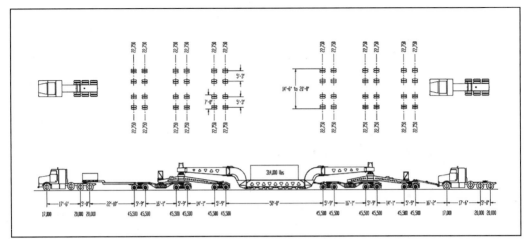

Figure 9. Schematic of transport using the 15-axle dual lane truck

Figure 10. Transfer of the cutterhead support to a 19-axle truck

Figure 11. Launch of the TBM under low cover in October

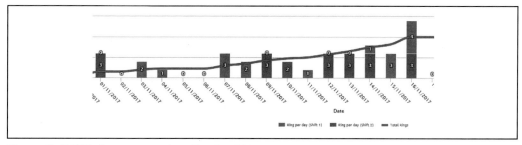

Figure 12. OCIT advance rates since October 2017

By mid-November 2017 the TBM had bored over 180 ft of tunnel, with the best performance on November 1st, 2017 when five rings were placed in two shifts. The performance of the TBM has been controlled and without any major issue since the start of tunnel boring (see Figure 12).

As previously mentioned the second challenge will come later, in the second half of the mixed face zone, when crews will encounter rock in more than 50% of the face. This challenge will be overcome with several strategies that have been evaluated to allow for multiple mix combinations of additives, bentonite and/or foam solutions to stabilize any face condition encountered. Additionally, the design of the cutterhead can be changed to maximize efficiency in all anticipated geology using soft ground tooling and lower bearing torque cutters. It is expected that the TBM will reach this point by the end of November or beginning of December 2017.

The TBM excavation will end at a retrieval shaft with approximately 180 feet from the surface to the invert of the tunnel. Tunneling is expected to take between seven and ten months. Once complete the tunnel will be connected to the network of associated structures that make up the OCIT and to the existing sewers. There are a total of four shafts, six diversion structures, four hand-mined tunnels, and a microtunnel involved in this work. Associated construction is being performed simultaneously in order to meet the project goals for the City of Akron.

CONCLUSIONS

While tunneling has just begun at the Akron OCIT, much can be learned thus far about design, transport, assembly and launch of a large diameter, mixed ground machine in an urban setting. From the design point of view Robbins has identified three different types of geologies that affect machine design, and make the use of a Crossover TBM advantageous in the mixed ground conditions. In addition, having the jobsite relatively close to the manufacturer's assembly yard has been beneficial from a logistical standpoint, since all the big components did not require double handling; this reduced the cost at the jobsite and had minimal effect on the traffic of the City of Akron. Lastly, thoroughly planning the staging and delivery logistics of the Tunnel Boring Machine allowed crews to speed up the assembly and consequently shorten the period of time required for launching the TBM.

Job-Site Security Risk Management Beyond Gates, Guards, and Guns

Eric Jacobs
MSAG

ABSTRACT: Economics and safety are driving technology into tunnel project monitoring and tunnel control systems at an accelerating pace. GPS, SCADA systems, Industrial Control Systems (ICS), security systems, traffic signaling systems, operational control and monitoring systems connected via wires and wirelessly are critical to successful and safe tunnel projects and operations. While this rapid technology infusion facilitates significant efficiencies, provides enormous amounts of data and enhances safety, it also adds new challenges. Today, there are new safety and economic risks on the job site created by cyber security vulnerabilities associated with the proliferation and integration of technology. The hazards and risks created must be addressed and understood. This paper presents the scope of the challenges faced, identifies examples of cyber security vulnerabilities on the job site during construction as well as once operational and in service. Concluding discussions will address and discuss remediation/mitigation strategies and best practices.

CHALLENGES FACED

Building and operating tunnels has been a high-stakes endeavor in terms of costs, complexity, safety and impact on contractors, owners and the public for a long time. There are many variables that need to be closely monitored and managed to construct a tunnel on-time, on-budget and safely. Once built, there are also many elements that need to be monitored and managed to operate a tunnel safely and efficiently.

Fundamentally, the risks to the safe and efficient construction and operation of a tunnel are a function of the degree of harm and the likelihood of the harm occurring. From a cyber security perspective, risk assessments used to inform decision makers and support risk responses are typically based on identification of relevant threats; internal and external vulnerabilities; the impact that may occur given the potential for threats to exploit vulnerabilities, and the likelihood that harm will occur. This paper focuses on creating awareness of vulnerabilities associated with technology used in tunnel construction and operation and some of the challenges associated with remediation of these vulnerabilities.

As we all know, technology has been evolving at an accelerating pace and infiltrating every aspect of our lives in new and often exciting ways. In the world of tunneling, sensors and systems that provide position and time have become accurate, rugged and cost-effective enough to come out of the research labs and into the equipment and systems used every day. There are many opportunities to employ connected computing devices and sensors to cost-effectively capture, integrate and analyze key data with minimal latency to manage risk better, improve safety and avoid surprises.

A representative example where these opportunities are leveraged includes the metro project Cityringen in Copenhagen (Denmark) (Chmelina, et al, 2016)[*]. This urban tunneling project will be a completely new, fully automatic, driverless metro ring line 15.5 km long and situated under downtown Copenhagen, the 'bridge quarters' area, and Frederiksberg. The line is to have 32 km twin tubes bored by TBMs, 17 cut and cover station structures, and several emergency and ventilation shafts, cross-over caverns and stub tunnels for future extensions. To manage risks the project monitors hydro and geologic conditions; existing infrastructure such as roads, water, sewer, utilities and buildings as well as project progress data such as tunnel alignment, ring location, TBM position and operating parameters and cavern excavation rounds. A real-time monitoring system on the TBM collects data that includes face pressures, grouting pressures and other operating parameters and sends this data to the contractor as well as the owner. Other parameters measured across the job include:

- Structural monitoring of stations, shafts, mined tunnels/caverns and bored tunnels including convergence measures
- Geotechnical monitoring including surface leveling, automatic 3D surveying and in-ground instrumentation from piezometers,

---

[*]Klaus Chmelina, Klaus Rabensteiner *Monitoring and Data Management Challenges at the Metro Project Cityringen Copenhagen* World Tunneling Conference 2016, 911.

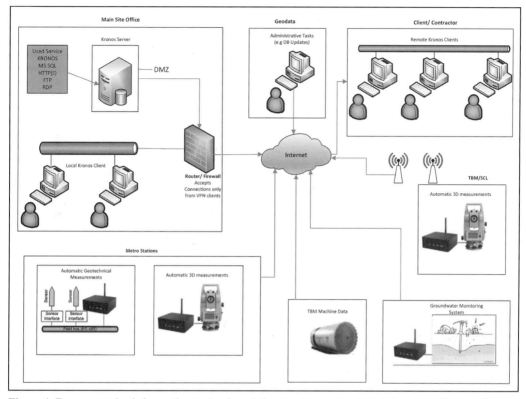

Figure 1. Representative information technology infrastructure and network topology of a tunnel project monitoring network

multi-point borehole extensometers and other sensors
- Building monitoring
- Groundwater monitoring including dewatering, discharge and recharge water yields and volumes, levels, conductivity soundings and chemistry data
- Environmental monitoring
- Tunnel alignment including ring location, TBM position and cavern excavation rounds
- Construction progress by excavated rings and face chainage
- Ground treatment such as geometry, drilling parameters, advance rates and injection data

For safety, a multi-parameter system measured acceleration, acoustic emissions, electric potential, temperature, humidity and air pressure to monitor the response and stability of the tunnel during and after blasting operations. The following diagram depicts the information technology infrastructure and network topology of the monitoring network employed.

Correspondingly, the safe and reliable operation of modern tunnels incorporate a number of critical control systems distributed throughout the tunnel typically interconnected with one or more of the owner's tunnel control centers or transportation management centers and other stakeholders (local government, law enforcement, emergency response organizations, equipment/system vendors, etc.). Typical control systems critical to tunnel safety and operations include:

- Fire protection and life safety systems
- Fire, smoke and ventilation systems
- Monitoring for hazardous materials
- Kinetic energy management systems
- Intelligent Lighting Control System (ILCS)
- Closed-Circuit Television systems (CCTV)
- Tunnel Communications Systems (TCS)—data, voice, video
- Radio communications systems—Highway Advisory Radio (HAR), AM/FM commercial station overrides, etc.
- Standpipe and water supply system monitoring
- Drainage system monitoring
- Traffic sensors—Loop Detectors, etc.

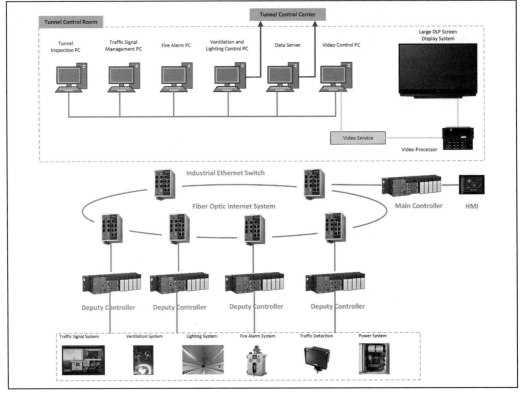

Figure 2. Representative information technology infrastructure and network topology of a tunnel control center

- Traffic incident detection systems
- Environmental monitoring system—wind speed, carbon monoxide (CO) and volatile hydrocarbon (HC) information
- Physical security
- Over-Height vehicle sensors
- Traffic control systems
- Variable Message Signs (VMS)—Changeable Message Signs (CMS), changeable Lane Use Signs (LUS), traffic signals, etc.
- Programmable Logic Controllers (PLCs)

Further complicating these highly intertwined system of systems, we have the Intelligent Transportation Systems (ITS) on the visible horizon. ITS now introduces complex vehicle-to-vehicle (V2V), vehicle-to-infrastructure (V2I) and even vehicle-to-anything (V2X) communications.

It is easy to see the evolution of these systems. In the past, these systems were physically and electronically isolated systems running proprietary protocols using purpose-built hardware and software. Now these systems employ widely available low-cost wired and wireless devices along with industry standard computers, operating systems and network protocols. While these capabilities offer almost unthinkable advances in the safety and efficiency of tunnels, they also carry along some significant challenges when looking to deliver protection from cyber threats.

Prior to exploring examples of cyber vulnerabilities associated with these systems, it is prudent to present four additional, often overlooked, challenges faced when trying to secure these systems against cyber-attacks:

- *Differences in safety risk management versus cyber risk management*—tunnel project monitoring and tunnel control systems lie within the intersection of cyber security and safety management. Fundamental differences between these two disciplines include that safety typically deals with random and unintentional events whose probability of occurrence can be reasonably quantified based on established techniques whereas cyber security deals with intentional, targeted attacks whose probability of occurrence is difficult to accurately

Table 1. Summary of IT system and tunnel project monitoring/tunnel control system differences

Category	Information Technology	Tunnel Project Monitoring/Tunnel Control Systems
Performance Requirements	• Non-real-time • Response must be consistent • High throughput is demanded • High delay and jitter may be acceptable • Less critical emergency interaction • Tightly restricted access control can be implemented to the degree necessary for security	• Real-time • Response is time-critical • Modest throughput is acceptable • High delay and/or jitter is not acceptable • Response to human and other emergency interaction is critical • Access should be strictly controlled, but should not hamper or interfere with human-machine interaction
Availability Requirements	• Responses such as rebooting are acceptable • Availability deficiencies can often be tolerated depending on the system's operational requirements	• Responses such as rebooting may not be acceptable because of process availability requirements • Availability requirements may necessitate redundant systems • Outages must be planned and scheduled days to weeks in advance • High availability requires exhaustive pre-deployment testing
Risk Management Requirements	• Manage data • Data confidentiality and integrity is paramount • Fault tolerance is less important, momentary downtime is not a major risk • Major risk impact is delay of business operations	• Control physical world • Human safety is paramount, followed by protection of the process or safe tunnel traffic flow • Fault tolerance is essential. even momentary downtime may not be acceptable • Major risk impacts are loss of life, harm to humans, regulatory non- compliance, environmental impacts, loss of equipment or production, social and psychological impacts on citizens
System Operations	• Systems are designed for use with typical operating systems • Upgrades are straightforward with the availability of automated deployment tools	• Differing and likely proprietary operating systems often without security capabilities built in • Software changes must be carefully made usually by software vendors. because of the specialized control algorithms and perhaps modified hardware and software involved
Resource Constraints	• Systems are specified with enough resources to support the addition of third- party applications such as security solutions	• Systems are designed to support the intended monitoring or control function and may not have enough memory and computing resources to support the addition of security capabilities
Communications	• Standard communications protocols • Primarily wired networks with some localized wireless capabilities • Typical IT networking practices	• Many proprietary and standard communication protocols • Several types of communications media used including dedicated wire and wireless (radio and satellite) • Networks are complex and sometimes require the expertise of control engineers or engineers with specialized knowledge
Change Management	• Software changes are applied in a timely fashion in the presence of good security policy and procedures. The procedures are often automated.	• Software changes must be thoroughly tested and deployed incrementally throughout a system to ensure that the integrity of the control and monitoring system is maintained. • Outages often must be planned and scheduled days or weeks in advance. • May use operating systems that are no longer supported
Managed Support	Allow for diversified support styles	Service support is typically via individual vendors for each sub-system with little if any awareness or concern for other sub-systems
Component Lifetime	Lifetime on the order of 3 to 5 years.	Lifetime on the order of 1 to 15 years.
Component Location	Components are usually local and easy to access	Components can be isolate, remote and require extensive physical effort to gain access to them.

quantify. Further, the level of interconnectivity of modern tunnel monitoring or control systems, often overwhelms traditional hazard-analysis techniques such as failure modes and effects analysis (FMEA) and fault tree analysis (FTA). Lastly, where both safety engineers and cyber security professionals are working together to assess risk, frequent miscommunications arise as a result of a lack of familiarity with each other's discipline and terminology.

- *Differences between cyber risk management approaches for IT information systems versus tunnel project monitoring/tunnel control systems*—tunnel project management/tunnel control systems have many different characteristics, risks and priorities from IT information systems and different consequences in the event of compromise. The nature of the systems' devices and network connectivity makes it difficult, if not impossible to employ even basic protection approaches used in typical IT information systems such as patching and anti-virus software. Security must be implemented such that system integrity is maintained during daily operations as well as during cyber attack. Table 1, adapted from NIST Special Publication 800-82 Revision 2, *Guide to Industrial Control Systems (ICS) Security* offers a summary of typical differences between IT systems and tunnel project monitoring/tunnel control systems.

The impact these differences can have on operations can be seen in two representative examples where well-intentioned organizations and experienced IT system cyber professionals impacted the availability and integrity of an ICS/SCADA system causing economic impact and compromising safety. In the first incident, a vulnerability scan performed on a SCADA network caused a three meter robotic arm to come out of standby mode and swing 180 degrees. In separate incident, a well-intentioned penetration test locked up an oil and gas company's SCADA system resulting in the inability to send gas through its pipelines to customers for four hours. (Sandia National Laboratories, 2005)[*]

- *Oversight of the design, implementation and operations of tunnel project monitoring/ tunnel control systems*—IT departments within contractors and owners typically employ cyber security savvy professionals. A paradox exists, however, that these

professionals often have little to no experience with tunnel project monitoring/tunnel control system's ICS/SCADA systems and their design, configuration, maintenance or protection. Tunnel project monitoring/tunnel control systems are typically designed, installed and configured by contractors with little, if any cyber security awareness and maintained by staff with little, if any cyber security awareness. Further, the common use of joint ventures, alliance partners, and outsourced services in the industrial sector has led to a more complex situation with respect to the number of organizations and groups contributing to security of the industrial automation and control system. These practices must be taken into account when developing security for these systems.

- *The need for a security mind*set—the fourth often overlooked, challenge faced trying to secure these systems from cyber-attacks is the need for a security mindset. A security mindset involves thinking about how something can be made to fail, thinking like an attacker or criminal. Thinking like this is not natural for most people, especially engineers. Engineers are typically focused on how things can be made to work and withstand unintentional and extreme events. Only recently are efforts beginning to emerge to engineer-in cyber resiliency into systems as a functional requirement.

EXAMPLES OF VULNERABILITIES

Identification of vulnerabilities associated with tunnel project monitoring/tunnel control system infrastructure can be best obtained from open sources including Industrial Control Systems Cyber Emergency Response Team (ICS-CERT) advisories (https://ics-cert.us-cert.gov/advisories), NIST's National Vulnerability Database (NVD) (nvd.nist.gov) and SCADA Strangelove (scadastrangelove.blogspot.com). Publicly known cyber security vulnerabilities are stored in the NVD and are assigned a Common Vulnerabilities and Exposures (CVE) identifier. CVE Identifiers are used to discuss or share information about a unique software or firmware vulnerability; provide a baseline for tool evaluation, and; enable data exchange for cybersecurity automation.

The number of vulnerabilities in ICS and SCADA components keeps growing. With increased attention to ICS/SCADA security over the last few years, more and more information about vulnerabilities in these systems is becoming public. That said, due to the nature and service life of these components of tunnel project monitoring/tunnel control systems, it's likely that vulnerabilities are present in

---

[*]Sandia Report SAND2005-2846P, Penetration Testing of Industrial Control Systems; Sandia National Laboratories; Duggan.

these products for years before they are revealed. In 2015, 189 vulnerabilities in ICS components were published. Of these 189, 49% of them are classified as critical severity and 42% as medium severity. (Andreeva et al.)[*]

Vulnerabilities are exploitable. Exploits are readily available for 26 of the vulnerabilities published in 2015 and for many vulnerabilities (such as hard-coded credentials) an exploit code is not needed at all to obtain unauthorized access to the vulnerable system. It is also important to note that a majority of ICS security assessments show that default credentials in ICS components are often not changed and could be used to gain remote control over the system.

Based on 2015 data from the NVD and ICS-CERT, the most widespread vulnerability types for ICS components often found in tunnel project monitoring/tunnel control systems were buffer overflows (9% of all detected vulnerabilities), use of hard-coded credentials (7%) and cross-site scripting (7%). The top ten most widespread types of vulnerabilities are presented in Figure 3[†] (ibid).

While detailed descriptions and explanations of these vulnerability types are outside the scope of this paper, a brief overview of these top ten vulnerabilities adapted from Kapersky Lab's *Industrial Control Systems Vulnerabilities Statistics* are presented below.

Buffer Overflow—Buffer overflow is a programming error, in which software, while writing data to a buffer, overruns the buffer's boundary and overwrites adjacent memory locations. Writing outside the bounds of a block of allocated memory can corrupt data, crash the program, or cause the execution of malicious code. In total, 17 buffer overflow vulnerabilities were found in ICS components in 2015, eight of them have a high risk level. These security flaws were discovered in different components, including SCADA systems, HMI, controllers, DCS and others. Four of these vulnerabilities have the highest CVSS score—10, corresponding to the maximum impact (high-privileged access), which could have been carried out by a remote unauthenticated attacker.

Hard-Coded Credentials—Hard-Coded Credentials, such as a password or cryptographic key, typically create a significant hole that allows an attacker to bypass the authentication that has been configured by the software administrator. This vulnerability was discovered in 14 different ICS components (HMI, PLC, network devices and others),

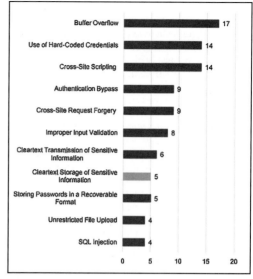

Figure 3. Top 10 vulnerabilities of ICS components in 2015

and in most cases it has a high risk level. Almost all of the identified vulnerabilities of this type could be exploited by a remote attacker.

Cross-Site Scripting—Cross-site scripting enables attackers to inject client-side scripts into web pages viewed by users, which could be used to steal user authentication data (cookies), perform social engineering attacks, or spread malware. Vulnerabilities of this type are present in 14 ICS components (most of them are SCADA systems).

Authentication Bypass—An attacker exploiting authentication bypass vulnerabilities may be able to capture or modify privileged information, inject code, or bypass access control. Depending on the vulnerable system, such flaws can have a different nature, for example the use of a vulnerable servlet (NVD CVE-2015-6480[‡]), a web server allowing an attacker to directly access the information by ignoring the location header (NVD CVE-2015-7910[§]), or incorrect file system architecture (NVD CVE-2015-1599[¶]). Authentication Bypass vulnerabilities were found in eight different types of ICS components, including HMI, a network device, RTU and others.

Cross-Site Request Forgery—The cross-site request forgery vulnerability exists when a web server is designed to receive a request from a client

[*]Industrial Control Systems Vulnerabilities Statistics; Andreeva, Gordeychik, Gritsai, Kochetova, Potseluevskaya, Siderov and Timorin; Kaspersky Lab.
[†]Industrial Control Systems Vulnerabilities Statistics; Andreeva, Gordeychik, Gritsai, Kochetova, Potseluevskaya, Siderov and Timorin; Kaspersky Lab.

[‡]NIST National Vulnerability Database, https://nvd.nist .gov/vuln/detail/CVE-2015-6480.
[§]NIST National Vulnerability Database, https://nvd.nist .gov/vuln/detail/CVE-2015-7910.
[¶]NIST National Vulnerability Database, https://nvd.nist .gov/vuln/detail/CVE-2015-1599.

without any mechanism for verifying that it was sent intentionally. Then, it might be possible for an attacker to trick a client into making an unintentional request to the web server, which will be treated as an authentic request. This can be done via a URL, image load, XMLHttpRequest, etc. and can result in the exposure of data or unintended code execution. Four of nine vulnerabilities discovered are present in SCADA systems.

Improper Input Validation—Products containing the improper input validation vulnerability do not validate, or incorrectly validate, inputs that can affect the control flow or data flow of a program. Most of these flaws are related to arbitrary code execution. Eight vulnerabilities are present in HMI, SCADA system, RTOS and OPC server components.

Clear Text Transmission of Sensitive Information—Clear Text Transmission of Sensitive Information vulnerabilities allow an unauthorized actor to sniff sensitive or security-critical data in a communication channel because the software transmits data in clear text. Communications, including login credentials are exchanged in easily readable clear text when tunnel project monitoring/tunnel control systems employ devices that do not support cryptographic protocols that provide communications security over a computer network such as SSL (Secure Sockets Layer) or these protocols are not enabled.

Clear Text Storage of Sensitive Information—When information is stored in clear text, attackers could potentially read it. Even if information is encoded in a way that is not readable by humans, techniques could determine which encoding is being used, and then decode the information. This type of medium level vulnerability can be found in tunnel project monitoring/tunnel control systems' ICS components, Human Machine Interfaces (HMI), SCADA systems, Web servers, and pumps.

Storage of Passwords in a Recoverable Format—The storage of passwords in a recoverable format makes them subject to password reuse attacks by malicious users. Essentially, recoverable encrypted passwords provide no significant benefit over plain text passwords, since they are subject not only to reuse by malicious attackers, but also by malicious insiders. If a system administrator can recover a password directly, or use a brute force search on the available information, the administrator can use the password on other accounts. HMIs of tunnel project monitoring/tunnel control systems are the most affected by this vulnerability.

Unrestricted File Upload—Unrestricted file upload vulnerabilities in software allow an attacker to upload or transfer files of dangerous types that can be automatically processed within the product's environment. These vulnerabilities have been

discovered in ICS and SCADA components. For example, (NVD CVE-2015-7912)[*], documents a high-level vulnerability where, through a servlet, it is possible to upload arbitrary Java code to a particular SCADA product and allow application properties to be imported through uploaded files that could allow arbitrary code and command execution.

SQL Injection—SQL injection is the insertion of an SQL query via data input from the client to the application. Successful SQL injection exploits can read sensitive data from the database, modify database data (Insert/Update/Delete), execute administration operations on the database (such as shutdown the DBMS), recover the content of a given file present on the DBMS file system and in some cases issue commands to the operating system. SQL injection vulnerabilities are common and their presence has been documented in ICS and SCADA components.

Finally, no discussion on cyber vulnerabilities would be complete without discussing email phishing type vulnerabilities. Attackers use email phishing to obtain sensitive information such as account credentials; knowledge about how systems or processes work, and; to install malware (software written with the intent of doing harm to data, devices or to people). The malware can exfiltrate data and provide a platform for further lateral movement within the targeted system. The latest Verizon Data Breach Report (Verizon, 2017)[†] presents that these social attacks were utilized in 43% of all breaches in this year's dataset and that 66% of malware linked to data breaches or other incidents was installed via malicious email attachments.

Up to this point, the discussion has been around known vulnerabilities. The 2017 Verizon Data Breach Report (Verizon, 2017)[‡] once again supports the validity of this focus as, of the 1,935 breaches analyzed, 88 percent were accomplished using a familiar list of nine attack vectors. These breaches could probably have been prevented by a few simple measures. That said, tunnel project monitoring/tunnel control systems' stakeholders need to also be aware of zero-day vulnerabilities. Especially in the case of Tunnel operations, the consequences of a compromise can be large enough to justify a significant, focused, organized effort in attacking this critical infrastructure. Zero-day vulnerabilities are vulnerabilities unknown to those who would be interested in remediating or mitigating the vulnerability (including the vendor of the target software). The fewer the days since Day Zero, the higher the chance no fix or mitigation has been

[*]NIST National Vulnerability Database, https://nvd.nist .gov/vuln/detail/CVE-2015-7912.
[†]Verizon, 2017 Data Breach Investigations Report, 10th Edition.
[‡]Verizon, 2017 Data Breach Investigations Report, 10th Edition.

developed. Even after a fix is developed, the fewer the days since day zero, the higher the probability that an attack against the afflicted device or software will be successful, because not every user of that device or software will have applied the fix. For zero-day exploits, the probability that a user has patched their bugs is of course zero, so the exploit should always succeed. Based on research conducted by the Rand Corporation (Ablon, 2017)[*], once an exploitable vulnerability has been found, the time to develop a fully functioning exploit is relatively fast, with a median time of 22 days. Additionally, a recent Trend Micro TrendLabs Research Paper[†] presents that "the average time between disclosing a bug to a SCADA vendor to releasing a patch reaches up to 150 days." Therefore, tunnel project managers and operators are faced with the fact that it's most likely that an exploit will be circulated significantly before a patch is available from the vendor. The hard to patch nature of many of these fielded assets further amplify this challenge.

Other approaches to gaining an understanding of vulnerabilities associated with tunnel project monitoring/tunnel control systems components include exploring Shodan and obtaining a better understanding of how adversaries operate. Shodan (https://www.shodan.io/) is a search engine for internet connected devices. As opposed to a search engine like Google that scans webpages, Shodan scans the internet for connected devices and collects data from banners, which are metadata about a software that's running on a device. Shodan collects and indexes data from connected devices that include ICS and SCADA components, IP-enabled cameras, refrigerators and traffic signals. One way of obtaining a better understanding of how adversaries operate is to explore The MITRE Corporation's.

Common Attach Pattern Enumeration and Classification (CAPEC)™ community resource for identifying and understanding attacks located at https://capec.mitre.org. CAPEC™ is a comprehensive dictionary and classification taxonomy of known attacks that can be used to increase awareness of vulnerabilities within tunnel project monitoring/ tunnel control systems and enhance defenses.

REMEDIATION/MITIGATION STRATEGIES AND BEST PRACTICES

The foundation for remediating or mitigating tunnel project monitoring/tunnel control systems' cyber vulnerabilities is a comprehensive and accurate identification of system components including devices, operating systems, applications and communications and how they are interconnected. Once these components are identified, known vulnerabilities associated with each component can be retrieved from open sources mentioned earlier, including ICS-CERT advisories (ics-cert.us-cert .gov), NIST's NVD (nvd.nist.gov) and SCADA Strangelove (scadastrangelove.blogspot.com) and the vendors themselves.

From that point, threat vectors that identify where an attacker may come from need to be explored. Threat vectors can include direct access, wireless, virtual private network (VPN) connections, other network connections and computers with dual NIC cards. In considering threat vectors, think like adversary and assume connectivity.

In considering priorities for protecting tunnel project monitoring/tunnel control systems infrastructure, it is helpful to consider two important concepts. The first is that knowledge of the process is essential for high-consequence disruption. The second is that an attacker can generally have one of four effects on a compromised system component to achieve their overall objective data exfiltration, denial of service, establish a platform for lateral movement and to generate misleading information to operators or other systems/components. Application of the first concept in tunneling means to put a high priority on protecting control room and HMI components. Application of the second concept provides for assigning a lower priority for concern about a denial of service attack to a sensor collecting data. Typically safety hazards associated with this scenario have already been addressed in the design of the system. Also, the exfiltration of data from sensors (with the exception of configuration information that can be used to further enumerate the target system) is typically another low priority concern whereas causing a sensor to generate valid, but misleading, information or as a platform (trusted within the target system) to access other systems/components is typically a significant concern. Once there is a good grasp of the infrastructure's components, connectivity, known vulnerabilities and threat vectors contractors and operators can begin to see their exposure and make informed choices to methodically improve their security posture.

The fact that most cyber security breaches are based on known vulnerabilities is clearly evident in this paper and is readily supported through additional research. Based on this knowledge some reasonably simple, low-cost activities can be started that would have a significant effect on the security posture of tunnel project monitoring/tunnel control systems.

- Increase and reinforce general cyber security awareness among all stakeholders and users (remember that email phishing is a significant attack vector).

---

[*]RAND Corporation, Zero Days, Thousands of Nights— The Life and Times of Zero-Day Vulnerabilities and Their Exploits.
[†]Trend Micro, TrendLabs Research Paper, The State of SCADA HMI Vulnerabilities, 2017.

- Within IT organizations, increase awareness of the unique and different characteristics of tunnel project monitoring/tunnel control systems with respect to traditional IT systems.
- Be relentless implementing and enforcing strong password policies. See new standards for password security in NIST SP 800-63 Digital Identity Guidelines (https://pages.nist.gov/800-63-3) and use two-factor authentication wherever possible.
- Be sure you know accurately your assets: How they are connected (logical and physical topologies) and how they are configured with respect to access control, protocols enabled and connectivity permissions—white-list versus black-list.
- Remediate known vulnerabilities to the extent possible, mitigate the remaining vulnerabilities including email.
- Assess cyber security impacts when any change is made to the system and periodically review and document security posture.
- Become familiar with and constantly monitor best practices guidelines and resources from organizations such as NIST (https://www.nist.gov/cyberframework) and DHS ICS-CERT (ics-cert.us-cert.gov)
- Be prepared for "when" an incident occurs.

From these foundational efforts, organizations should also establish a closed-loop systems approach to cyber security that includes policy, risk management, assurance (evaluating continued effectiveness, identifying new risks/vulnerabilities) and security culture (awareness). Organizations typically find NIST's cyber security framework core a valuable foundation. NIST's framework core consists of five concurrent and continuous functions—Identify, Protect, Detect, Respond and Recover. NIST's cyber security framework website, https://www.nist.gov/cybersecurity-framework, offers a vast amount of useful information and tools to assist organizations. Overlaying the guidance included in NIST 800-82, Guide to Industrial Control Systems (ICS) Security extends the foundation for organizations focused on tunnel project monitoring/tunnel control systems.

Suggested additional specific activities under this approach include:

- Getting a free assessment from ICS-CERT[*]
- Downloading the free Cyber Security Evaluation Tool[†] (CSET®) from ICS-CERT.

---

*ics-cert.us-cert.gov/sites/default/files/FactSheets/ICS-CERT_FactSheet_Private_Sector_Assessments_S508C.pdf.
†ics-cert.us-cert.gov/sites/default/files/FactSheets/ICS-CERT_FactSheet_CSET_S508C.pdf.

CSET is a systematic and repeatable approach to assessing security posture that includes both high-level and detailed questions related to all industrial control and IT systems.

- Minimizing the systems' attack surface.
- Establishing a process for Continuous Diagnostics and Mitigation (CDM) (www.dhs.gov/cdm)
- Periodically, exploring emerging alternative approaches for analyzing and securing systems and used by other critical infrastructure domains, DoD and the intelligence community. Examples of these include employing data diodes to assure one-way information transfer and restricting network access and connections between allowed elements through the use of Software Defined Perimeters (SDP).
- Consider collaborating with other tunnel operators to establish a formal or informal Information Sharing and Analysis Center (ISAC). ISACS in other sectors are trusted entities that collect, analyze and disseminate actionable threat information to their members and provide members with tools to mitigate risks and enhance resiliency.

CONCLUSION

Tunnel project monitoring and tunnel control systems are increasing safety and efficiency at a dramatic pace. These systems are rapidly becoming more complex and integrated with other systems and stakeholders. To economically and rapidly develop components of these systems, vendors are employing commercial-off-the-shelf (COTS) components and resources that include operating systems, applications, network protocols and devices. These components often carry with them cyber security vulnerabilities as awareness of the security and safety consequences of compromise is only recently being recognized. Further there are accelerating efforts from individuals and organizations with malicious intent to exploit vulnerabilities in these systems to achieve their objectives. Defensively, there is a gap in the availability of cyber security professionals with experience protecting traditional IT assets. This gap is exacerbated by the unique and distinctive nature of the ICS/SCADA components which make up a significant part of tunnel project monitoring and tunnel control systems. Further amplifying the challenge is that most of the tunnel project monitoring and tunnel control systems are installed and maintained by operators and owners with staff distinct from these organizations' IT support staff.

While a formidable challenge, risk to tunnel project managers and operators associated with hazards created by known and unknown cyber security vulnerabilities can be considerably and cost-effectively reduced. Significant risk reduction can be realized from a few simple, low-cost, quick to implement actions as presented in this paper. Furthermore, there are a growing number of resources available to tunnel stakeholders to provide guidance and resources to help further reduce risk.

All that said however, tunnels are part of the transportation systems critical infrastructure sector, one of 16 critical infrastructure sectors whose assets, systems, and networks, whether physical or virtual, are considered so vital to the United States that their incapacitation or destruction would have a debilitating effect on security, national economic security, national public health or safety. It is likely they are specifically targeted by well-educated, well-funded, resources with malicious intent. Defenders need to be right all the time, attackers only once. Therefore, in spite of valuable risk reduction efforts, incidents are inevitable. An incident response plan that facilitates rapid recovery is essential.

REFERENCES

Andreeva, O., Gordeychik, S., Gritsai, G., Kochetova, O., Potseluevskaya, E., Sidorov, S., & Timorin, A. (2016). *Industrial Control Systems Vulneraabilities Statistics.* Kaspersky Security Intelligence Service.

Chmelina, K., & Rabensteiner, K. (2016). Monitoring and Data Management Challenges at the Metro Project Cityringen Copenhagen. *World Tunneling Conference,* (p. 911).

Department of Homeland Security, National Cybersecurity and Communications Integration Center. (2017). *Cyber Security Evaluation Tool.* Retrieved from Industrial Control Systems Cyber Emergency Response Team (ICS-CERT): ics-cert.us-cert.gov/sites/default/files/FactSheets/ICS-CERT_FactSheet_CSET_S508C.pdf.

Department of Homeland Security Industrial Control Systems Cyber Emergency Response Team (ICS-CERT). (2017). *ICS-CERT Advisories.* Retrieved from The Industrial Control Systems Cyber Emergency Response Team (ICS-CERT): https://ics-cert.us-cert.gov/advisories.

Department of Homeland Security, National Cybersecurity and Communications Integration Center. (2017). *ICS PRIVATE SECTOR CRITICAL INFRASTRUCTURE ASSESSMENTS.* Retrieved from The Industrial Control Systems Cyber Emergency Response Team (ICS-CERT): https://ics-cert.us-cert.gov/sites/default/files/FactSheets/ICS-CERT_FactSheet_Private_Sector_Assessments_S508C.pdf.

NIST. (2015). *CVE-2015-1590 Detail.* Retrieved from National Vulnerability Database: https://nvd.nist.gov/vuln/detail/CVE-2015-1590.

NIST. (2015). *CVE-2015-6480 Detail.* Retrieved from NIST National Vulnerability Database: https://nvd.nist.gov/vuln/detail/CVE-2015-6480.

NIST. (2015). *CVE-2015-7910 Detail.* Retrieved from Nantional Vulnerability Database: https://nvd.nist.gov/vuln/detail/CVE-2015-7910.

NIST. (2015). *CVE-2015-7912 Detail.* Retrieved from National Vulnerability Database: https://nvd.nist.gov/vuln/detail/CVE-2015-7912.

NIST. (2015). *NIST Special Publication 800-82 Revision 2, Guide to Industrial Control Systems (ICS) Security .* NIST.

NIST. (2017). *Cybersecurity Framework.* Retrieved from Cybersecurity Framework: https://www.nist.gov/cybersecurity-framework.

NIST. (2017, June). *NIST SP 800-63 Digital Identity Guidelines.* Retrieved from NIST SP 800-63 Digital Identity Guidelines: https://pages.nist.gov/800-63-3/.

NIST. (2017). *NIST's National Vulnerability Database (NVD).* Retrieved from NIST's National Vulnerability Database (NVD): https://nvd.nist.gov/.

RAND Corporation. (2017). *Zero Days, Thousands of Nights—The Life and Times of Zero-Day Vulnerabilities and Their Exploits.* Santa Monica, CA: RAND Corporation.

Sandia National Laboratories. (2005). *Sandia Report SAND2005-2846P, Penetration Testing of Industrial Control Systems.*

SCADA Strangelove. (2017). *SCADA Strangelove.* Retrieved from SCADA Strangelove: http://scadastrangelove.blogspot.com.

The Mitre Corporation. (2017). Retrieved from Common Attack Pattern Enumeration and Classification (CAPEC): https://capec.mitre.org/.

Trend Micro. (2017). *Hacker Machine Interface The State of SCADA HMI Vulnerabilities.* Trend Micro, Incorporated.

Verizon. (2017). *2017 Data Breach Investigations Report, 10th Edition.* Verizon.

TRACK 1: TECHNOLOGY

Session 2: Innovation

Bruce Downing and Paul Headland, Chairs

A Preliminary Investigation for Characterization and Modeling of Structurally Controlled Underground Limestone Mines by Integrating Laser Scanning with Discrete Element Modeling

Juan J. Monsalve, Jon Baggett, Richard Bishop, and Nino Ripepi
Virginia Polytechnic Institute and State University

ABSTRACT: Stability of large opening underground excavations in jointed rock masses primarily depends on the distribution and properties of the geological discontinuities. Conventional methods for structural mapping may not be optimal for rock mass characterization. Laser scanning is a technology that rapidly sends out laser pulses in order to measure the position of certain objects by generating a massive point cloud with millimeter precision. This paper reviews the application of laser scanning along with discrete element method (DEM) modelling, to produce a more realistic response of the rock mass behavior during excavation compared to analytical approaches. Additionally, a methodology to evaluate the stability in a structurally controlled underground limestone mine with these two technologies is proposed.

INTRODUCTION

Over the past ten years, 40% of underground mining fatalities were caused by ground control issues related to roof, rib collapses, and pillar bursts (MSHA, 2016). Over the same time period, the underground stone mining industry had the highest fatality rate in four of those ten years (MSHA, 2016), more than any other sector. Underground limestone mines in the eastern U.S. have become more common over the past decade and typically there is less underground experience compared to underground mines in the western part of the country, resulting in a need for more engineering, education and training. In addition, the large openings and great spans between pillars can create instability conditions, leading to a large collapse of the roof or pillar if the mining system is not designed properly or if pillars deteriorate over time.

In 2011, the National Institute for Occupational Safety and Health (NIOSH) developed empirical guidelines for designing underground stone mines. This was limited to room and pillar mines in flat-lying bedded formations located in the eastern and Midwestern United States (NIOSH, 2011). However, these guidelines are not necessarily applicable to mines that do not share the same conditions as the mines in which the study was performed.

A case study mine (CSM) in which NIOSH guidelines does not apply will be evaluated. The CSM is a dipping (approximately 30°), underground room and pillar limestone mine. This rock has UCS of 159.2 MPa ± 21.25 MPa, a tensile strength of 6.3 MPa ± 1.99 MPa and a Young's modulus of 64.11 GPa ± 2.37 MPa. The deepest point in the mine is approximately 700 m below ground surface. This mine leaves 24 m by 24 m pillars and the dimensions of the drifts and crosscuts are generally 12.8 m wide and 7.6 m high. The mine's main failure mechanism is structurally controlled failure, evidenced by the jointing pattern and spacing observed in the tunnels, the amount of fallen blocks observed on the floor, and other geological structures such as faults and contacts that may generate a rock fall in the absence of ground support. In this case, further analysis may be performed in order to ensure the stability of the excavation and safety during operation. Even though general guidelines may not apply for this case, analysis methodologies may be proposed in order to analyze mines that present similar modes of failure.

In recent years, the development of digital mapping techniques through Terrestrial Laser Scanning (TLS) and photogrammetry have allowed the development of three-dimensional digital terrain models as tools for characterizing, modeling, and designing underground structures, not only in tunneling projects but also in mining environments (Cacciari & Futai, 2017; Rogers, Bewick, Brzovic, & Gaudreau, 2017; Fekete & Diederichs, 2013; Slaker, Westman, Fahrman, & Luxbacher, 2013; Grenon, Landry, Hadjigeorgiou, & Lajoie, 2017). In addition, it has been proven that this technology is useful, faster, and cheaper than conventional mapping methods, and also that with the right analyses and data interpretation the results may offer a better representation of the rock mass structure than conventional methods.

Not only have great advances been made in digital mapping techniques during the past decade,

Figure 1. Geotechnical conditions observed in the CSM

Table 1. Intact rock property test results from CSM

Lithology	Density (ton/m³)		UCS (MPa)		Brazilian Tensile Strength (MPa)		Young's Modulus (GPa)	
	Mean	SD	Mean	SD	Mean	SD	Mean	SD
Hanging wall	2.69	0.01	163.74	37.84	11.96	3.14	61.02	6.79
Ore body	2.69	0.01	159.20	21.25	6.30	1.99	64.11	2.37
Footwall	2.72	0.01	217.29	36.12	13.72	2.62	61.43	3.15

but the advances in computing power allow numerical models to more realistically represent the behavior of rock masses (Lorig & Varona, 2013). Due to these current advances, the use of distinct element codes such as 3DEC, is proposed over conventional analysis methods to evaluate the stability of the mine workings in order to better understand the rock mass behavior and ultimately improve safety performance in the operation.

The following paper reviews both technologies: Laser Scanning for rock mass characterization and discrete element model software for analyzing the instability of a structurally controlled limestone mine. Additionally, a methodology that integrates these technologies is proposed for analyzing the stability in the case study mine.

STRUCTURALLY CONTROLLED INSTABILITY

Structurally controlled instability has been widely described by many authors (Goodman, 1989; Brady & Brown, 1985; Hudson & Harrison, 2000; Hoek, 2000). This mode of failure has been defined as the sliding or natural falling of rock blocks either from the roof or the walls due to gravity action (Brady & Brown, 1985). This type of instability occurs in rock masses that present two or more structural features that intersect, generating wedges; once the tunnel is excavated, these wedges generate blocks that tend to displace or rotate towards the opening. Additionally, researchers such as (Martin, Kaiser, & Christiansson, 2003) have proposed risk assessment

charts in order to identify whether the conditions of the rock mass allow for either stress controlled instability or structurally controlled instability based on the stress condition, strength of the intact rock (σ_{ci}) and the Geological Strength Index (GSI); they state that when the rock mass has a GSI grater than 40 and the σ_{ci} is greater than 2 times the vertical stress (σ_v), or when the GSI is greater than 40 and the principal stress (σ_1) is less than 0.15 σ_{ci}, the rock mass is under structurally controlled instability.

Figure 1 presents the geotechnical condition observed in the CSM, where the structurally controlled instability was defined as main failure mechanism. This was evidenced by the jointing pattern and joint spacing observed in the tunnels, the amount of fallen blocks observed on the floor, and other geological structures such as faults and contacts that may generate a rock fall in the absence of the required support. This failure mechanism was enhanced by the multiple karst formations present in the mine, which during excavation have generated rock blocks up to 4 m³ (144 ft³) that pose high risk for workers, equipment, and the overall mining plan. Table 1 presents intact rock properties obtained form previous studies in CSM. During recent visits, GSIs greater than 75 have been identified throughout the mine. Considering the minimum strength conditions (137.95 MPa) and the greatest stress condition (20 MPa) the stress/strength relationship (σ_1/σ_{ci}) is lower than 0.15, therefore, according to (Martin, Kaiser, & Christiansson, 2003) this mine is also defined as a structurally controlled mine.

ANALYSIS METHODS FOR STRUCTURALLY CONTROLLED INSTABILITY

As described by (Lorig & Varona, 2013), two main analysis methods exist for the failure mechanism present in the CSM (structurally controlled instability); the limit equilibrium theory, also known as key block theory, described by (Goodman & Shi, 1985; Brady & Brown, 1985; Hudson & Harrison, 2000); and the two and three dimensional discontinuous modeling methods, initially developed by (Cundall, 1971).

LIMIT EQUILIBRIUM METHOD

The limit equilibrium method is an analytical analysis method that consists of determining the geometry of the wedge formed by the intersection of 3 joint planes and the excavation using the block theory described by (Goodman & Shi, 1985). The forces acting on the wedge are then determined and its respective vectors calculated; The sliding direction is determined and so are the normal forces acting on each wedge plan; finally resisting forces due to joint shear strength and tensile strength are calculated, allowing for the calculation of a safety factor. These calculations may be performed by hand or using specialized software such as Unwedge (Rocscience, 2017).

Table 2 presents some capabilities and limitations of using a limit equilibrium software for stability analysis in jointed rock masses. Although, it is clear that this analysis method is not able to represent the real structural setting of the rock mass, it allows a general interpretation of the sizes, volumes and kinematics of the possible blocks to be performed.

Table 2. Advantages and disadvantages of using limit equilibrium method for analyzing structurally controlled instability (Rocscience, 2017)

Limit Equilibrium Method (UNWEDGE)	
Capabilities	Limitations
• User friendly • Reproduce multiple models in short time. • Ability to consider support such as rock bolts and shotcrete. • Allows to perform statistical analysis based on the mechanical properties of the discontinuities.	• Can only use three discontinuity sets at a time for the analysis. • Discontinuity surfaces are assumed to be persistent and to extend through the volume of interest. • Complex geometries may not be analyzed. • Discontinuities can occur at any location in the rock mass. • Only considered one joint of the actual joint set.

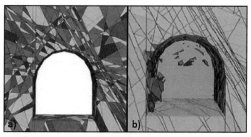

Figure 2. (a) Construction of a deterministic model in a tunnel (b) Deterministic model showing the block displacement in a tunnel, modified after (Fekete S., 2010)

TWO- AND THREE-DIMENSIONAL DISCONTINUOUS METHODS

A discrete or distinct element method (DEM) according to (Cundall & Hart, 1985) refers to a computer program that allows finite displacements and rotations of discrete bodies, including complete detachment; and recognizes new contacts automatically as the calculation progresses.

3DEC is a numerical code that simulates the response of discontinuous media (such as a rock mass) subjected to either static or dynamic loading. In order to perform the analysis, it applies the discrete element model (DEM) proposed by (Cundall, 1971), which uses a time-stepping explicit algorithm to solve the equations of motion of the blocks. This software divides the rock mass into a mesh where each block is free to break or maintain its connection with the blocks that surround it. Additionally, it recognizes the new contacts between the displaced blocks as the calculation takes place. The input to this software consist on a series of parameters including physical and mechanical properties of both intact rock and discontinuities (ITASCA, 2016). Figure 2 presents an example of a 3DEC model in a road tunnel in Norway evidencing the effects of input parameters such as persistence on displaced blocks after the excavation, modified after (Fekete & Diedrichs, 2013).

This numerical tool permits working with different constitutive models for both intact rock and discontinuities, generating models based on laboratory and field test data. Staged models can be performed using 3DEC due to its workflow methodology that allows running models stage by stage reproducing mining or excavation sequences. Complex geometries can be represented, thanks to its compatibility with CAD based software and other meshing plug-ins such as Griddle to facilitate modeling. However, one of the most important characteristics of 3DEC is that it has its own programming language allowing the user to program their own functions, in order to create automated models.

3DEC is able to reproduce more reliable models that provide a better representation of the rock mass. This is possible due to its capability of elaborating discrete fracture networks (DFN) which simulate the structural setting in the rock mass. A DFN is a three-dimensional geometric representation of a geological structure based on statistical information of its characteristics measured on the field (Pierce, 2017). The parameters required to reproduce a DFN are joint orientation, density of the fractures (number of fractures per unit volume) and joint size generally quantified by the joint trace length (Grenon, Landry, Hadjigeorgiou, & Lajoie, 2017). All of these parameters can be collected by manual rock mass characterization; however, in order to generate reliable DFN's, great amounts of statistical data may be required, which may be difficult to obtain through manual mapping alone.

Some of the drawbacks of the 3DEC software are that when the model is very complex, it may take several iterations to process and significant time (days to weeks), compared to other modeling software it is expensive, and the reliability on the results strongly depends on the quality of the input data from the rock mass characterization.

Table 3 presents a summary of the capabilities and limitations described for discontinuous analysis methods such as 3DEC (ITASCA, 2016) for stability analysis in jointed rock masses. The application of this analysis tool will help to obtain a better understanding of the failure mechanism present in the CSM under present geotechnical conditions and propose alternative engineering designs, support measures and excavation methods to reduce structurally controlled instability and enhance safety conditions in the operation.

ROCK MASS CHARACTERIZATION

In order to have sufficient information to perform any kind of analysis in a structurally controlled rock mass, a detailed survey of the geological structure must be done. Figure 3 presents a schematic representation modified after (Hudson & Harrison, 2000) of the properties to be measured in discontinuities present in a rock mass during a detailed rock mass characterization, according to (ISRM, 1978). These properties include spacing and frequency, orientation, persistence, roughness, aperture, seepage, filling, and wall strength. These properties are used in order to evaluate both geometrical and strength parameters of each discontinuity set along the rock mass.

According to (Voyay, Roncella, Forlani, & Ferrero, 2006), in many cases a traditional survey performed manually may be dangerous or difficult, provide a small sample size, and take a considerable

Table 3. Advantages and disadvantages of using discontinuous analysis method for analyzing structurally controlled instability (ITASCA, 2016)

Discontinuous Analysis Methods (UDEC & 3DEC)	
Advantages (Capabilities)	**Disadvantages (Limitations)**
• Works with different constitutive models, both for discontinuities and intact rocks. • Ability to input statistical parameters of discontinuities properties. • Can represent complex geometries and sequential excavations. • Ability to consider multiple discontinuity sets by generating Discrete Fracture Networks (DFN).	• Complex models are time consuming. • Representation of the behavior of the rock mass depends on the reliability on the geotechnical data.

amount of time. Alternatively, digital mapping techniques such as laser scanning and photogrammetry have been proposed by different authors. Although these technologies were initially proposed for slope stability analysis, today their applications have been extended to underground environments. In this study, laser scanning was selected over photogrammetry due to its time savings while acquiring and processing the data. Photogrammetry requires illumination in order to obtain the images, which may present a logistic complication during field work. It is worth remarking that photogrammetry can also offer an alternative to laser scanning and present great results as shown by (Grenon, Landry, Hadjigeorgiou, & Lajoie, 2017) (Rogers, Bewick, Brzovic, & Gaudreau, 2017) who have used photogrammetry to generate DFN to analyze wedge stability in mining drifts.

Laser Scanning

Laser scanning is a technology that rapidly sends out laser pulses in order to calculate the position of certain objects by generating a massive point cloud with millimeter precision (Kemeny, Turner, & Norton, 2006). Laser scans provide a three dimensional detailed image of the rock mass that allows one to quickly map, with more precision and less bias, the structural features present in the rock mass. Additionally, it has also been proven that laser scanning obtains qualitatively sound data in a short time without affecting the mining plan in an active tunnel environment; it provides detailed quality control

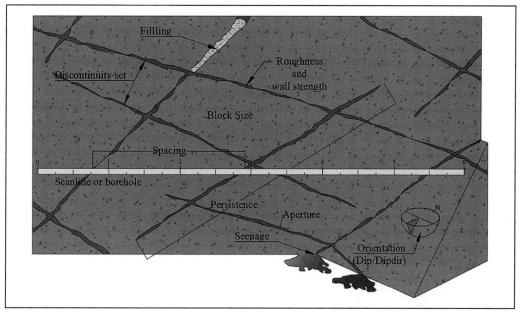

Figure 3. Schematic of the primary geometrical properties of discontinuities in rock (modified after Hudson & Harrison, 2000)

information on the precision excavation and installed support (Fekete & Diedrichs, 2013). Other authors such as (Slaker, Westman, Fahrman, & Luxbacher, 2013) have used laser scanning for deformation measurements in pillars in underground mines.

Even though laser scanning has proved to be a useful tool for analyzing instability in structurally controlled rock masses, its greatest challenges are related to structural data analysis. As mentioned before the main properties of interest for each joint set are joint orientation, density of the fractures and joint size. Many authors have been centering their research in identifying the best way to analyze the structural data in order to generate more reliable numerical models. (Fekete & Diedrichs, 2013) proposed a workflow for integrating structural data obtained from TLS into DEM analysis. They discuss some of the advantages and challenges of TLS data in underground rock mass evaluation considering some of the bias that can be present during the extraction of discontinuity properties. Additionally, they proposed two approaches (deterministic and stochastic) for modeling the excavation using TLS data, allowing the designer to have a widened view of the rock mass performance. (Cacciari, Morikawa, & Futai, 2015) generated a 3D block model using 3DEC based on a DFN obtained from discontinuity mapping analysis from laser scans. The results from the stability analysis were not presented. (Cacciari & Futai, 2015) analyzed different methods for estimating the mean trace length of the discontinuities from

laser scans; they defined that the trace length estimation method proposed by (Wu, Kulatilake, & Tang, 2011) presented better results than other methodologies proposed by other authors. (Cacciari & Futai, 2017) proposed a practical approach to create single and continuous DFNs considering the variations of the fracture density along the tunnel. Additionally, they performed numerical models based on the methodology obtaining similar behaviors in the model than those presented in the field.

PROPOSED METHODOLOGY

Figure 4 presents the workflow proposed in order to integrate laser scanning with numerical modelling in the CSM. This methodology is divided into seven phases: Preliminary evaluation and site selection, site scanning and rock mass characterization, structural data processing, sample collection, rock mechanics laboratory testing, numerical modeling and design considerations and monitoring. The stages are explained in detail below.

Preliminary Evaluation and Site Selection

Several visits were made to the mine. Initial visits were carried out in order to get familiar with the mining method and geotechnical conditions of the mine. The main areas of concern for this study are working areas and travel ways where miners are at risk of roof falls. The areas in which the research will be performed will not only be those areas presenting

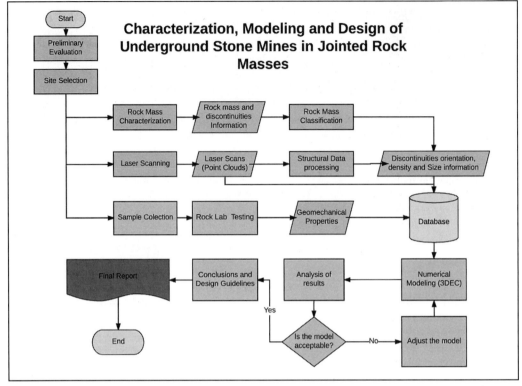

Figure 4. Proposed workflow for the integration of laser scanning and 3DEC modelling in the CSM

Table 4. Laser scan conditions used by different authors

Research	Equipment	Measurement Speed (points/s)	Distance Accuracy (mm)	Spacing Between Scan Stations (m)
(Fekete & Diedrichs, 2013)	Leica Geosystems HDS6000	500,000	0.6	1 Diameter
(Cacciari & Futai, 2015)	Faro Focus³ᴰ	976,000	2	10
(Slaker, Westman, Fahrman, & Luxbacher, 2013)	Faro Focus³ᴰ	976,000	2	10

greater geotechnical risks for the miners, but also those that present great interest for mine planning. These areas will be selected with mine management.

Site Scanning and Rock Mass Characterization

Once the locations for study are defined, the laser scans will be performed. In order to define what are the most appropriate operational conditions, a series of laser scans will be taken. Table 4 presents different operational conditions used by different authors, which will be used as a reference to set the scanning conditions. A Faro Focus³ᴰ laser scanner will be used for taking the scans, considering safety precautions defined by the manufacturer.

Additionally, conventional rock mass characterization will be performed in order to have control information that allows a comparison of results obtained from the scans and have a better understanding on the rock mass structure. This information also will be used in order to classify the rock mass with the conventional rock mass classifications such as Q, MRMR, GSI and RMi.

Structural Data Processing

In order to process the scans and extract discontinuities' properties the software I-Site Studio provided by MAPTEK will be used. I-Site is a point cloud processing and modelling software that extracts

Figure 5. Extraction of joint sets from a laser scans using the software I-SITE

geotechnical information from laser scans by using a set of geotechnical tools (MAPTEK, 2017). This software will be used in order to identify the orientation, density and size of the discontinuities. Figure 5 present a preliminary structural analysis with a scan obtained from CSM.

Sample Collection

Sample collection will take place in the areas previously selected. Due to the fact that the ore body lies between the two different types of limestone on both Hanging Wall and Footwall, samples from these formations will also be considered. The samples will be rock blocks which will be transported to the rock mechanics preparation lab, where NQ diameter cores will be extracted. The goal will be to obtain at least 3 samples for Triaxial test, 3 samples for direct shear, 3 samples for tensile strength, 3 samples for uniaxial compressive strength and 4 samples for point load, for each type of rock at each location. The results from this sampling stage will be the maps showing the different locations where the samples where extracted.

Rock Mechanics Laboratory Testing

Among the properties to be considered for the blocks include mass density, intrinsic deformability properties (bulk & shear modulus or Young modulus or Poisson's ratio), intrinsic strength properties (Constitutive models and strength parameters such as compressive strength and tensile strength) and post-failure properties. On the other hand, joint properties including physical properties for joint friction angle, cohesion, dilation angle, and tensile strength, as well as joint normal and shear stiffness. The joint cohesion and friction angle correspond to the parameters in the Coulomb strength criterion (ITASCA, 2016). Generally, these parameters are obtained from rock laboratory tests. The laboratory tests proposed are presented in Table 5. This table also shows the

intact rock properties and the discontinuity properties obtained from each test. These properties are the parameters that will feed numerical models.

The expected results from the laboratory plan will be descriptive statistics comparing the results for all the geomechanical parameters (UCS, tensile strength, E, & Poisson's) comparing the properties between different types of rock. Additionally, the spatial variability of this properties will be analyzed. Summary maps will be elaborated in order to present the variability of the geomechanical properties of the mine.

Numerical Modeling and Design Considerations

A modelling plan has been developed in order to define a route that allows to generate models that provide a better understanding on the usage of 3DEC, but also to obtain models that represent the actual conditions of the CSM. The models include a continuous single pillar, a single pillar with a single oblique discontinuity, a single tunnel considering a discrete fracture network, a single pillar considering a discrete fracture network, the mining sequence of the CSM without considering any discontinuities and the mining sequence of the same mine considering the geotechnical conditions represented using a DFN. The numerical models predict and anticipate the potential wedges before mining into a new area. The aim of these models will be to set guidelines and methodologies that can apply to any underground project and help them to obtain particular answers to each specific problem.

Monitoring

After numerical modeling results are obtained, a site evaluation will be performed in order to verify these results. Once the simulations coincide with the actual behavior observed in the rock mass, design considerations and engineering measures to control the instability mode will be proposed.

Table 5. Rock lab testing standards defined by the ASTM description and properties obtained

ASTM Standard	Description	Properties obtained
D2664-04	Standard test methods for Triaxial compressive strength of undrained rock core specimens without pore pressure measurements	• Hoek & Brown failure criteria • Young's modulus at different confinement pressures • Basic friction angle for artificial joints
D7012-14	Standard test methods for compressive strength and elastic moduli of intact rock core specimens under varying states of stress and temperatures	• Uniaxial compressive strength • Young modulus • Poisson's ratio
D5607-08	Standard test method for performing laboratory direct shear strength tests of rock specimens under constant normal force	• MohrCoulomb failure criteria for discontinuities (phi, c) • Shear stiffness
D5731-08	Standard test method for determination of the point load strength index of rock and application to rock strength classifications	• Uniaxial compressive strength (indirect)
D3967-16	Standard test method for splitting tensile strength of intact rock core specimens	• Tensile strength of the rock (indirect)
D2845-08	Standard test method for laboratory determination of pulse velocities and ultrasonic elastic constants of rock	• Young's modulus (indirect) • Density • Poisson's ratio
D57873-14	Standard test method for determination of rock hardness by rebound hammer method	• Uniaxial compressive strength (indirect)

CONCLUSIONS

Different authors have proven that TLS technologies are a powerful suite of tools for rock mass characterization. Not only are they time efficient, but they also provide detailed information, which processed adequately considering actual limitations in both processing and modeling software, allows to generate models that represent accurately jointed rock mass behavior. DEM software is a powerful tool to interpret and analyze structurally controlled instability in underground excavations. This software with detailed structural information may provide not only deterministic but also stochastic analysis allowing the engineer to have a better understanding of rock mass behavior and provide better engineering solutions. The integration of both technologies (TLS and DEM) may be potentially used in order to prevent rock falls in underground excavations to enhance worker's safety. The CSM is an ideal environment to apply both technologies, and to develop risk analysis methodologies based on these technological advances, ultimately reducing accidents related to rock falls in the underground stone mine industry. The proposed methodology will be applied in the CSM and results will be presented in further work.

ACKNOWLEDGMENTS

This work is funded by the National Institute for Occupational Safety and Health's Mining Program under Contract No. 200-2016-91300. The author's would like to thank Lhoist mining engineers and management for their support and guidance during this project. MAPTEK is acknowledged for providing a license of the software I-Site. Views expressed here are those of the authors and do not necessarily represent those of any funding source.

REFERENCES

Brady, B., & Brown, E. (1985). *Rock Mechanics for Underground Mining.* UK: Chapman & Hall.

Cacciari, P., & Futai, M. (2015). Mapping and Characterization of Rock Discontinuities in a tunnel Using 3D Terrestrial Laser Scanning. *Bulletin of Engineering Geology and the Environment.*

Cacciari, P., & Futai, M. (2017). Modeling a Shallow Rock Tunnel Using Terrestrial Laser Scanning and Discrete Fracture Networks. *Rock Mechanics and Rock Engineering*, 1217–1242.

Cacciari, P., Morikawa, D., & Futai, M. (2015). Modelling a Railway Rock Tunnel Using Terrestrial Laser Scanning and The Distinct Element Method. *Integrating Innovations of Rock Mechanics*, 101–108.

Cundall, P. (1971). A computer model for simulating progressive large scale movements in blocky rock systems. *Proc. Symp. Rock Fracture (ISRM)*, II-8.

Cundall, P., & Hart, R. (1985). "Development of Generalized 2-D and 3-D Distinct Element Programs for Modeling Jointed Rock," Itasca Consulting Group. *Misc. Paper SL-85-1.* US Army Corps of Engineers.

Fekete, S. (2010). *Geotechnical Applications of LiDAR for Geomechanical Characterization in Drill and Blast Tunnels and Representative 3-Dimensional Discontinuum Modelling.* Kingston: Queen's University.

Fekete, S., & Diedrichs, M. (2013). Integration of Three-dimensional Laser Scanning with Discontinuum Modelling for Stability Analysis of Tunnels in Blocky Rockmasses. *International Journal of Rock Mechanics & Mining Sciences*, 11–23.

Goodman, R. (1989). *Introduction to Rock Mechanics.* New York: Wiley.

Goodman, R., & Shi, G. (1985). *Block Theory and Its Application to Rock Engineering.* London: Prentice-Hall.

Grenon, M., Landry, A., Hadjigeorgiou, & Lajoie, P. (2017). Discrete Fracture Network Based Drift Stability at the Éléonore Mine. *Mining Technology*, 22–33.

Hoek, E. (2000). *Rock Engineering.* North Vancouver: A.A. Balkema Publishers.

Hudson, J., & Harrison, J. (2000). *Engineering Rock Mechanics.* Oxford: Imperial College of Science, Technology and Medicine, University of London, UK.

ISRM. (1978). Suggested Methods for the Quantitative Description of Discontinuities in Rock Masses. *International Journal on Rock Mechanics, Mining sciences & Geomechanics. Vol 15*, 319–368.

ITASCA. (2016). *3 Dimensional Discrete Element Code User's Guide.* Minneapolis: Itasca Consulting Group.

Kemeny, J., Turner, K., & Norton, B. (2006). LIDAR for Rock Mass Characterization: Hardware, Software, Accuaracy and Best-Practices. *Laser and Photogrammetric Methods for Rock Face Characterization*, 49–61.

Lorig, L., & Varona, P. (2013). Guidelines for numerical modelling of rock support for mines. *Ground Suport 2013*, 81–105.

MAPTEK. (2017, October 05). *MAPTEK I-Site Studio: Geotechnical Module.* Retrieved from http://www.maptek.com/pdf/i-site/Maptek_I-Site_Studio_Geotechnical_Module.pdf.

Martin, C., Kaiser, P., & Christiansson, R. (2003). Stress, instability and design of underground excavations. *Rock Mechanics and Mining Sciences*, 1–21.

MSHA. (2016). *Accident Injuries Data Set.* Mine Safety and Health Administration.

MSHA. (2016). *Metal/Nonmetal Daily Fatality Report.* Mine Safety and Health Administration.

NIOSH. (2011). *Pillar and Roof Span Design Guidelines for Underground Stone Mines.* Pittsburg: National Institute for Occupational Safety and Health.

Pierce, M. (2017). An Introduction to Random Disk Discrete Fracture Network (DFN) for Civil and Mining Engineering Applications. *ARMA e-Newslatter 20*, 3–8.

Rocscience. (2017, October 3). *Unwedge Online Help.* Retrieved from Program Assumptions: https://www.rocscience.com/help/unwedge/webhelp4/Unwedge.htm.

Rocscience. (2017, September 13). *Unwedge Theory.* Retrieved from rocscience Website: https://www.rocscience.com/help/unwedge/webhelp4/Unwedge.htm.

Rogers, S., Bewick, R., Brzovic, A., & Gaudreau, D. (2017). Integrating Photogrametry and Discrete Fracture Network Modelling for Improved Conditional Simulation of Underground Wedge Stability. *8th International Conference in Deep and High Stress Mining*, 599–610.

Slaker, B., Westman, E., Fahrman, B., & Luxbacher, M. (2013). Determination of Volumetric Changes from Laser Scanning at an Underground Limestone Mine. *Mining Engineering, Vol. 65 Issue 11*, 50–54.

Voyay, I., Roncella, R., Forlani, G., & Ferrero, A. (2006). Advanced techniques for geo structural surveys in modelling fractured rock masses: application to two Alpine sites. *Laser and Photogrammetric Methods for Rock Face Characterization*, 102–108.

Wu, Q., Kulatilake, P., & Tang, H. (2011). Comparison of rock discontinuity mean trace length and density estimation methods using discontinuity data from an outcrop in Wenchuan area, China. *Comput Geotech*, 258–268.

Study of the Correlation Between RMR and TBM Downtimes

Omid Frough and Jamal Rostami
Colorado School of Mines

ABSTRACT: In Mechanized tunneling, machine performance is very sensitive to ground conditions and can be substantially impacted by variations in rock type, strength, jointing, and rock mass classes. Adverse ground conditions have a great impact on TBM downtimes and can considerably reduce the machine utilization. In this paper, the influence of rock mass properties on machine downtimes were examined. Data from three long mechanized tunnels recently completed by D.B. TBM was used to evaluate the relation between engineering geology information and daily machine utilization and downtime. Various regression equations were developed to offer empirical models for estimation of machine utilization and RMR. Results showed that the relationship between RMR and rock mass related downtimes can reach correlation coefficient (R^2) of 0.62.

INTRODUCTION

The use of mechanized tunneling has substantially improved the tunneling practices in past three decades. Parallel to the expansion in use of TBMs there has been many attempts to develop reliable and accurate performance prediction models to estimate rate of penetration (ROP), machine utilization rate (U), advance rate (AR), and disc cutter life. These models have been in use since early use of TBMs in 1960s and have continuously been updated and modified as the capabilities of the machines have improved and they are used in ever more challenging ground conditions. Example of such models can be found in the technical literature (Tarkoy, 1975; Roxborough and Phillips, 1975; Graham, 1976; Ozdemir, 1977; Blindheim, 1979; Farmer and Glossop, 1980; Snowdon et al., 1982; Sanio 1985, Sato et al. 1991; Nelson, 1993; Rostami and Ozdemir, 1993; Palmstrom, 1995; Rostami 1997; Bruland, 1999; Barton, 2000; Yagiz, 2008). Among many different parameters affecting TBM performance, influence of rock mass characteristics on TBM performance, mainly from the point of view of penetration rate, has been investigated by many researchers such as Sapigni et al., 2002, Bieniawski et al., 2007, Hassanpour et al. 2010, and 2011, Khademi et al., 2010, Farrokh et al. 2012 and Zare Naghadehi and Ramezanzadeh 2016.

Some efforts have been made on the application of different soft computing techniques for estimating TBM performance. Benardos and Kaliampakos (2004) attempted to correlate TBM performance with geological and geotechnical site conditions by using artificial neural network. Grima et al (2000) and Kim (2004) developed TBM performance predictor models based on fuzzy logic method. Also numerical analyses were performed to explore the effect orientation and spacing of the discontinuities on rock fragmentation TBM boring process (Gong et al., 2005, 2006). Some attempts also have been made to correlate rock mass classification to TBM advance rate, utilization factor, and TBM downtimes. TBM rock mass related downtimes (GRRD) were studied to derive empirical formulas relating GRRD to ground condition parameters using statistical and RES approaches (Frough and Torabi 2013 and Frough et al., 2014).

Most of the available prediction models have only focused on penetration rate estimates. However a small subset of models such as CSM, NTNU and Q_{TBM} offer some relationships and charts for estimating utilization factor. It is well known to all parties involved with TBM operation that rock mass condition can have a major impact on machine downtime and utilization factor. It is very difficult to estimate TBM utilization accurately under variety of rock mass conditions due to mechanical and geological uncertainties. In fact, an imprecise estimate of TBM utilization would cause significant errors in estimates and can lead to project delays and increased cost. The current study focuses on the impact of ground conditions on downtimes and machine utilization. Some equations are introduced in this paper which has been developed based on statistical analysis of data from close monitoring of the machine performance, operational parameters, and ground conditions. These equations define the relationship between GRRD and RMR.

TBM UTILIZATION

For prediction of the machine performance, both penetration rate and utilization factor must be estimated (AR=ROP×U). ROP is defined as the distance excavated divided by the machine clock time which measures duration of a continuous excavation cycle. ROP is often expressed in m/hr (ft/hr). AR is the actual distance mined and supported in a specific period of time, often expressed in terms of m/day. Utilization factor is the ratio of boring time to total operation time, which determines the productive portion of operational time. TBM downtimes are the period when machine is not excavating the face and it include the times for support installation, regripping, grouting, maintenance, machine break down, cutter change, mucking delays, stoppages caused by geological adverse conditions, and other components such as shift changes and breaks. Human factor such as operator experience has a great effect on TBM Utilization factor.

During the excavation process, various factors including machine parameters, geological conditions, and site arrangements determine TBM utilization level (Kim, 2006). Rock mass related parameters have major impact on machine downtime, especially stoppages for ground support installation or stabilizing the face, and therefore they affect TBM utilization. They often constitute the main causes for lower utilization and loss of boring time and vary significantly from site to site.

Previous studies show that adverse geological conditions have a great impact on TBM downtimes. Instability of the ground and collapses at the face and walls can cause extended downtime due to the necessity for ground improvement and cleaning of rock falls in the tunnel or at the face. While the main controlling factor of cutter wear is the presence of quartz and other abrasive minerals, alternation soft and hard bedded of rocks can inflict severe damages to the cutter and increase the frequency of cutter change and related downtime. Presence of clay and sticky minerals is another reason of downtimes due to the issues of cutter head clogging and mucking problems. Similarly groundwater can negatively affect utilization and cause substantial delays in tunnel boring operations. Existence of poisonous gases can also cause delays and downtimes by decreasing the crew performance, damaging electronic devices and requiring extra ventilation. The activities that have major impact on TBM utilization and ROP are summarized in Table 1.

Although activities in these groups occur mostly in series with boring cycle, some of them can easily be performed at the same time or in parallel. The art of planning/performing these activities in parallel is part of the experience of the contractor and their crew and site management. In good operations the amount of parallel activities are higher and thus it prevents the TBM downtime and improves the utilization rate and daily advance.

CASE STUDIES

In order to evaluate the effect of geology and rock mass conditions (GRRD) on TBM downtimes, 3 long water conveyance tunnels, Karaj-Tehran water conveyance tunnel (lot-1 length: 16 km and lot-2 11 km of 14 km) and Ghomrood water conveyance tunnel (lots 3 and 4 length: 18 km), were studied. The collected data consist of the boring times, different downtimes and also their causes on a daily basis as well as engineering geology characteristic of each rock mass unit. This information was extracted from daily shift reports and tunnel geological back mapping and available initial geotechnical information at the design stage.

The Karaj-Tehran water conveyance tunnel was 30 km long, located in northwest of Tehran. The tunnel was divided into two sections with 16 and 14 km in length, respectively. The first Lot was excavated and lined with a Double Shield TBM and then Lot 2 was excavated and lined by the same TBM after overhaul in a valley along the alignment. In this study, data for only 11 km of lot 2 was analyzed. The 36-kilometer Ghomrood water conveyance tunnel is located in central part of Iran. Ghomrood tunnel (lots 3 and 4) was excavated and lined using a Double Shield TBM. The basic machine specifications of the TBMs are summarized in Table 2. The TBMs utilized in these tunnels are shown in Figure 1. Table 3 contains the recorded performance parameters for these tunnels.

The lithology of karaj tunnel lot 1 and 2 area consists of a sequence of Karaj formations. In lot 1 the litology is composed of variety of pyroclastic rocks, often interbedded with sedimentary rocks. The characteristic rock type is a green vitric to crystal lithic tuff, tuff breccias, sandy and silty tuffs with shale, siltstone and sandstone (SCE 2004). The lithology of lot 2 contain of variety of pyroclastic rocks, often interbedded with sedimentary rocks. The characteristic rock type in this section is gray tuff, siltstone, sandstone, monzodiorite and monzogabro (SCE 2009).

The lithology Ghomrood lot 3&4 area consists of a sequence of Jurassic-cretaceous formations containing massive limestone and dolomite, slate, schist and metamorphic shale and sandstone units (SCE, 2003). More details about these projects, geology, and machine performance can be found in (Frough and Torabi 2013).

Table 1. Activities having major effect on TBM utilization (Frough and Rostami 2014)

No.	Activity	Cause	Concurrent	Influence	Frequency
1	Re-gripping	Starting boring cycle	S*	U	After each boring cycle
2	Support installation	Tunnel instability	P*/S	U	After each boring cycle
3	Pea gravel and contact grouting	If segmental lining is installed	P/S	U	After ring installation
4	Cutter head cleaning	Presence of clay	S	U/ROP	Random
5	Rock fall/Face Collapse, Rock burst	Unstable blocks	S	U	Random
6	TBM release	Jamming or entrapment in Squeezing ground	S	U	When excessive convergence occurs
7	Ground improvement	Poor rock	S	U	Random, when weak zones are encountered
8	Dewatering	High ground water flow	S/P	U	Random, where water bearing zones are encountered
9	Probe drilling	Supplementary investigations	S/P	U	Contractual requirements or preventive measure
10	degassing	Existing of Poisonous gases	S/P	U	If/when such gas are encountered/detected
11	Extra cutter change	Highly abrasive rocks	S	U/ROP	A function of rock abrasion
12	Cutter inspections and routine cutter change	Normal wear of cutters	S	U/ROP	Every shift or if necessary
13	Routine maintenance	Normal operation of the TBM	S	U	Every shift (Electrical, Mechanical, Hydraulically: TBM and Back up)
14	Water pipe installation	Utility extension	S/P	U	Adding of pipes (every 50m)
15	Power cable	Utility extension	S	U	Extension of flexible cable (every 150 to 250m)
16	Ventilation duct	Utility extension	S	U	Extension of duct (every 100 to 250m)
17	Rail extension	Utility extension	P/S	U	Extension of rails (every 6 to 12m)
18	Train delay	Unexpected breakdowns or insufficient train sets, slow speed, by passes	S	U	Depends on excavation rate, distances, muck haulage capacity
19	Unloading problems	Clay or water at the heading	S	U	Less frequent, depends on rock mass condition
20	Conveyor belt maintenance	Normal wear	S	U	Less frequent, depends on rock mass condition
21	Changing survey stations	Range of surveying equipment, turns in tunnel	S	U	Depends on tunnel aliment (straight: every 250m, curves: every 50m)
22	Delays for supply or spare parts	Unavailable spare parts	S	U	Unexpected (sometimes, depend on site management)
23	Lunch time or break		S	U	Every day shift (depend on management)
24	Crew delay	Shift change	S	U	Every shift (depend on management)
25	Unexpected breakdowns	Mechanical/Electrical problems	P/S	U	Random, quality of manufacturing and effectiveness of routine maintenance

*S: Series to Boring, *P: Parallel with Boring

Using mentioned information a database of geological and machine performance parameters have been developed. The database includes records of all rock mass units from field mapping. Machine performance parameters and duration of downtimes were collected from shift reports of 1016 days of operation for 15684 m of Karaj Lot 1 tunnel, 682 days of operation for 10,639 m of Karaj 2 tunnel, and 911 days of operation (except long time stoppage) for 16,992 m of Ghomrood tunnel. Overview of information on tunneling operation such as work time, average ROP, AR and TBM utilization factor of different rock mass units are summarized in Table 4.

Table 2. Specifications of utilized TBMs for these 3 tunnels

Specification	Karaj Tunnels	Ghomrood Tunnel
Machine diameter (m)	4.66	4.53
Disc diameter (mm)	432	432
Disc/Cutter spacing (mm)	70	75
Number of cutters	31	36
Cutter head power (KW)	1250	1120
Cutter head speed (RPM)	11	12
Cutter head torque (kN-m)	2500	1600
Max cutter head thrust (kN)	17,000	18,000

Different adverse rock mass conditions caused different downtimes in these tunnels. Initial section of Karaj lot 1 was in raveling ground due to excessive jointing and shallow depth of tunnel. In Karaj lot 2, water flow was the main problem during the construction. In Ghomrood tunnel ground raveling and squeezing was the main reason of rock mass related downtimes (Figure 2).

STATISTICAL ANALYSIS

To evaluate the impacts of various adverse ground conditions on machine downtime and utilization, statistical analysis of TBM field data was used for this study. Using commercial statistical software SPSS various regression equations were obtained from the analysis of downtime against the input parameters. These equations were subsequently examined to develop the most reliable and accurate empirical equations to estimate GRRD as a function of RMR, which was selected to represent rock mass and ground condition. Some of the relationships developed based on the database of TBM field performance in the three tunnels are listed in Table 5. The results show that the quadratic equation between rock mass related downtime and RMR with maximum coefficient of determination equal to 0.62 (Equation 1) could be a reasonable way to estimate

Figure 1. TBMs utilized in these tunnels: (left) Karaj tunnel, (right) Ghomrood tunnel

Table 3. Comparing performance parameters of case studies

	Ghomrood	Karaj 1	Karaj 2
Max. penetration rate	12.595 m/h	7.3 m/h	6.63 m/h
Max. daily utilization factor	44.792%	53.8%	52.01%
Max. daily advance rate	52.738 m/day	38.9 m/day	39.03 m/day
Average penetration rate	4.93 m/h	3.32 m/h	2.77 m/h
Average advance rate	18.66 m/day	15.4 m/day	15.6 m/day

Table 4. TBM performance parameters for various ground conditions (Frough and Torabi, 2013)

Rock Units	RMR	Length (m)	Operation (days)	Ave. ROP (m/h)	Ave. AR (m/h)	U (%)	GRRD (h/m)
Gta-1-1	35	581.27	84	2.892	0.661	22.87	0.324
Gta-1-2	50	381.12	89	3.218	0.690	21.45	0.155
Gta-2	49	1333.58	114	3.860	0.824	21.36	0.201
Gta-3	64	3901.18	89	3.874	0.835	21.55	0.328
Gta-4	75	648.65	257	3.515	0.570	16.20	0.382
Sts-1	57	681.78	101	4.582	0.738	16.12	0.240
Sts-2	72	2255.3	112	3.879	0.656	16.90	0.254
Mdg	63	2502.28	34	4.009	0.622	15.51	0.163
Tsh	46	1790.11	51	3.563	0.557	15.63	0.232
Cz	21	507.46	85	3.235	0.285	8.81	1.881
MG	67	3843.93	300	3.409	12.813	15.66	0.272
SC	40	3901.18	254	3.353	15.359	19.08	0.294
SC2	45	648.65	38	2.477	17.070	28.71	0.275
MO	67	837.12	64	2.177	13.080	25.03	0.321
CZ2	30	381.12	18	3.745	21.173	23.56	0.182
GT	50	1027.44	56	2.765	18.347	27.64	0.225
K1	54	297.67	31	2.710	9.602	14.763	0.307
K2	41	595.34	38	3.311	15.667	19.715	0.195
K3	21	155.84	14	3.229	11.131	14.365	0.603
J1	38	10626.16	529	5.081	20.087	16.471	0.316
J2	41	4334.40	189	5.475	22.933	17.451	0.157
J3	24	983.27	110	4.338	8.939	8.585	1.271

Table 5. Result of different regression equations between GRRD (h/m) and RMR

Method	Rsq	Rsq Adj.	F	Sigf
LIN	0.27	0.24	7.45	0.013
LOG	0.39	0.37	13.29	0.002
INV	0.52	0.49	21.76	0.0001
QUA	0.62	0.58	15.28	0.0001
POW	0.34	0.31	10.50	0.004
EXP	0.21	0.18	5.45	0.030

Table 6. Result of different regression equations between U (%) and RMR

Method	Rsq	Rsq Adj.	F	Sigf
LIN	0.05	0.007	1.15	0.296
LOG	0.11	0.065	2.47	0.131
INV	0.18	0.134	4.25	0.052
QUA	0.31	0.238	4.29	0.029
POW	0.18	0.138	4.37	0.050
EXP	0.10	0.057	2.27	0.148

GRRD. Figure 3 shows the relationship between these two parameters.

$$GRRD\ (h/m) = 0.000911\ RMR^2$$
$$- 0.0996\ RMR + 2.84\ R^2$$
$$= 0.62 \quad (1)$$

Also the same data set was used in regression analysis to find relationship between U and RMR, as has been presented in Table 6. The most reliable relation to estimate GRRD as a function of RMR was also a quadratic equation but with maximum $R^2=0.31$ (Equation 2) that is illustrated in Figure 4. The correlation between measured GRRD and U and those predicted by equations 1 and 2 are shown in Figures 5 and 6.

$$U\ (\%) = -0.01\ RMR^2 + 1.03\ RMR$$
$$-5.24\ R^2 = 0.31 \quad (2)$$

As it is seen in graphs, the relation between RMR and GRRD has higher coefficient of determination than the relation between RMR and U. Histograms of Boring and downtimes in these three tunnels has been showed in Figure 7. This figure shows that the most of downtimes were due to other reasons such as maintenance, transportation delay, hydraulic/electrical/mechanical problems, and shift change. This explains the reason for low coefficient of determination between RMR and U. Also, the results show that for accurate estimation of utilization factor both machine and ground related downtimes must be used in the equations.

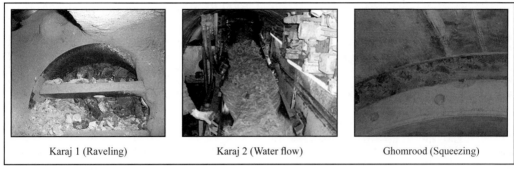

| Karaj 1 (Raveling) | Karaj 2 (Water flow) | Ghomrood (Squeezing) |

Figure 2. Examples of adverse ground condition in tunnels for this study

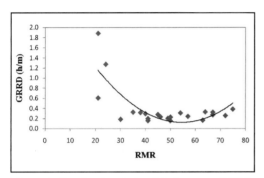

Figure 3. Quadratic relationship between GRRD (h/m) and RMR ($R^2 = 0.62$)

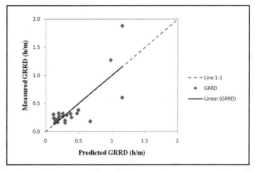

Figure 5. Comparison between the measured and the predicted GRRD (R = 0.78)

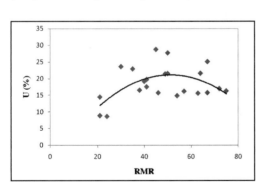

Figure 4. Quadratic relationship between U and RMR ($R^2 = 0.31$)

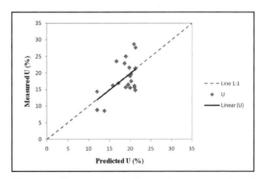

Figure 6. Comparison between the measured and the predicted U (R = 0.56)

One should keep in mind that only limited rock mass characteristics and ground condition parameters are considered in RMR, hence to develop a strong relationship between RMR and GRRD perhaps additional parameters can be considered and added to the analysis. Some adverse ground conditions such as Squeezing ground, abrasive rocks and presence of clay in joints, high water inflow, and detection of toxic gases are situations that could cause substantial downtimes and result in low utilization and are not considered in current study. To improve the capability for accurate prediction of GRRD, all adverse rock mass conditions must be included in the analysis.

CONCLUSIONS

Data obtained from three recent projects using double shield TBMs for mining 45-km of tunnels in various ground conditions was used to estimate the GRRD as a function of rock mass parameters,

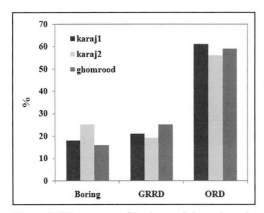

Figure 7. Histograms of boring and downtimes in studied tunnels

represented by RMR. A database containing detailed engineering geology information, mining rates, and daily boring/downtimes in different rock masses was used for this study. According to the data, the highest utilization and lowest downtimes was achieved in fair rocks with RMR of 40 to 60.This confirms the observation by other researchers such as Sapigni et al. (2002) and Frough et al. (2014). The highest rock mass related downtimes and lowest utilization factor have occurred in crushed rocks and faulted zones, with lowest RMR 20 to 30 present in the available database. It must be considered that skilled TBM Crew can reduce downtime in adverse geological condition based on their experience.

A set of empirical equations are proposed to estimate GRRD from RMR. The maximum correlation coefficient of $R^2=0.62$ was obtained for a quadratic equation. The analysis of data to relate RMR and U was not very successful with maximum $R^2=0.31$. This shows the influence of other parameters (such re-gripping, grouting, maintenance, machine break down, transportation, stoppages and human factor) that is not related to rock mass condition which could dominate the downtime and thus machine utilization factor. To improve the reliability and accuracy of proposed formulas perhaps a modified rock mass rating system in which, in addition to the existing items, squeezing, mixed face condition, abrasively, effect of sticky materials, ingress of hazardous gases, etc. are included.

ACKNOWLEDGMENT

The authors thank Seyed Rahman Torabi for his contribution to this paper.

REFERENCES

Barton N, 2000, TBM Tunneling in Jointed and Faulted Rock. 173p, Balkema, Brookfield.

Benardos AG, Kaliampakos DC, 2004, Modeling TBM Performance with Artificial Neural Networks. Tunneling and Underground Space Technology, 19, 597–605.

Bieniawski ZT, Celada B, Galera JM, 2007, TBM Excavability: Prediction and Machine and Rock Interaction, Rapid Excavation and Tunneling Conferences, p15–18, Toronto.

Bruland A, 1999, Hard Rock Tunnel Boring; Advance Rate and Cutter Wear. Volume 3 of 10, PhD Thesis, Trondheim Norwegian University of Science and Technology, Trondheim, Norway.

Blindheim OT,1979, Boreability Predictions for Tunneling, PhD thesis, Department of Geological Engineering, the Norwegian Institute of Technology, 406p.

Farmer IW, Glossop NH, 1980, Mechanics of disc cutter penetration. Tunnels and Tunnelling, 12(6), 22–25.

Farrokh E., Rostami J., Laughton Ch., 2012, "Study of Various Models for Estimation of Penetration Rate of Hard Rock TBMs," Tunneling and Underground Space Technology, doi:10.1016/j.tust.2012.02.012.

Frough O., Rostami J., 2014, Analysis of TBM Performance in Two Long Mechanized Tunnels, Case History of Karaj Water Conveyance Tunnel Project Lots 1.and 2 (Iran), Proceedings of the World Tunnel Congress—Tunnels for a better Life. Foz do Iguaçu, Brazil.

Frough O., Torabi S. R., 2013, "An Application of Rock Engineering Systems for Estimating TBM Downtimes," Engineering Geology, pp 157 122–123.

Frough O., Torabi S. R., Yagiz S., 2014, Application of RMR for Estimating Rock-Mass–Related TBM Utilization and Performance Parameters: A Case Study, Rock Mech Rock Eng, DOI 10.1007/s00603-014-0619-4.

Gong, Q. M., Zhao, J., Jiao, Y. Y., 2004, "Numerical modeling of the effects of joint orientation on rock fragmentation by TBM cutters," Tunnelling and Underground Space Technology.

Gong, Q. M., Jiao, Y. Y., Zhao, J, 2005, "Numerical modelling of the effects of joint spacing on rock fragmentation by TBM cutters," Tunnelling and Underground Space Technology.

Grima, M. A., Bruines, P.A., Verhoef, P.N.W., 2000, Modelling Tunnel Boring Machine Performance by Neuro-Fuzzy Methods, Tunnell. Undergr. Space Technol. 15 (3), 259–269.

Graham. P. C, 1976, Rock Exploration for Machine Manufactures, Symposium on Exploration for Rock Engineering. Johannesburg, South Africa, pp173–180.

Khademi Hamidi J, Shahriar K, Rezai B, Rostami J, 2010, Performance Prediction of Hard Rock TBM Using Rock Mass Rating (RMR) System. Tunnelling and Underground Space Technology, 25, 333–345.

Hassanpour J, Rostami J, Khamechian M, Bruland A, Tavakoli HR, 2010, TBM performance analysis in pyroclastic rocks: Case History of Karaj Water Conveyance Tunnel. Rock Mech and Rock Eng, 43, 427–445.

Hassanpour J., Rostami , J., Zhao J., 2011, A new hard rock TBM performance prediction model for project planning, Tunneling and Underground Space Technology, doi:10.1016/j .tust.2011.04.004.

Kim T, 2004, Development of a Fuzzy Logic Based Utilization Predictor Model for Hard Rock Tunnel Boring Machines, PhD Thesis, Colorado School of Mines, Colorado USA, 254.

Kim T, Marcelo GS, 2006, Fuzzy Modeling Approaches for the Prediction of Machine Utilization in Hard Rock Tunnel Boring Machines. 41st Industrial Application Society Annual Meeting, Industry Applications Conference, Record of the 2006 IEEE, pp. 947–954.

Nelson P., 1993, TBM Performance Analysis with Reference to Rock Properties, Comprehensive Rock Engineering, Hudson J (Ed), Volume 4 (Chapter, 10), 261–291.

Ozdemir L., 1977, Development of Theoretical Equations for Predicting Tunnel Probability, Ph.D. Thesis, T-1969, Colorado School of Mines, Golden Colorado.

Palmstrom A., 1995, RMi Parameters Applied in Prediction of Tunnel Boring Penetration. Chapter 7, RMI- A Rock Mass Characterization System for Rock Engineering Purposes. PhD thesis, Oslo, Norway, 400p.

Rostami J., Ozdemir L., 1993, A New Model for Performance Prediction of Hard Rock TBM, Proceedings of Rapid Excavation and Tunneling Conference, Boston, MA, pp.793–809.

Rostami, J., 1997. Development of a Force Estimation Model for Rock Fragmentation with Disc Cutters through Theoretical Modeling and Physical Measurement of Crushed Zone Pressure. Ph.D. Thesis, Colorado School of Mines, Golden, Colorado, USA.

Roxborough FF, Phillips HR., 1975, Rock Excavation by Disc Cutter, Int. J Rock Mech. Min. Sci., 12, 361–366.

SCE, 2003, Engineering Geology Report of Ghomrood Water Conveyance Tunnel (Lot 3&4), SCE.

SCE, 2004, Engineering Geology Report of Karaj-Tehran Water Conveyance Tunnel (Lot 1), SCE.

SCE, 2009, Engineering Geology Report of Karaj-Tehran Water Conveyance Tunnel (Lot 2), SCE.

Sato K, Gong F, Itakura K., 1991, Prediction of disc cutter performance using a circular rock cutting ring. In: Proceedings 1st International Mine Mechanization and Automation Symposium, June, Colorado School of Mines, Golden, Colorado, USA.

Sanio H.P.,1985, Prediction of the Performance of Disc Cutters in Anisotropic Rock, International Journal of Rock Mechanics and Mining Sciences & Geomechanics Abstracts, 22(3); 153–161.

Sapigni, M., Berti, M., Bethaz, E., Busillo, A., Cardone, G., 2002, TBM Performance EstimationUsing Rock Mass Classification, International Journal of Rock Mechanic and MiningSciences, Vol.39, pp 771–778.

Snowdon AR, Ryley DM., Temporal J.1982, Study of Disc Cutting in Selected British Rock, Int. J. of Rock Mech and Mining Sci., 19, 107–121.

Tarkoy, P.J., 1975, Rock Hardness Index Properties and Geotechnical Parameters for Predicting Tunnel Boring Machine Performance, PhD Thesis, University of Illinois at Urbana-Champaign, 326p, Illinois, USA.

Yagiz S., 2008, Utilizing Rock Mass Properties for Predicting TBM Performance in Hard Rock Condition, Tunneling and Underground Space Technology, 23; 326–339.

Zare Naghadehi M., Ramezanzadeh, 2016, Models for estimation of TBM performance in granitic and mica gneiss hard rocks in a hydropower tunnel, Bull Eng Geol Environ, DOI 10.1007/ s10064-016-0950-y.

Cross Passage Construction: Mechanized Solutions and Fields of Application

Gerhard Lang
Herrenknecht AG

ABSTRACT: With the growing world population and the rising transport volume, intelligent underground tunnel solutions become a necessity. At the same time, safety is becoming an increasingly important aspect in planning of these infrastructures. One of the measures in discussion is the construction of cross passages, mostly between parallel tunnels. In general, these tunnels are very short but their construction can be highly demanding, especially in poor ground conditions, and can have a big impact on the overall program. Various concepts have been worked out in relation to tunnel size, geology and hydrology. Cross passages in water-bearing ground with high water pressure are a special challenge. In the past, this could often only be realized with the help of ground freezing methods or with extensive grouting and consolidation measures. Mechanized approaches have been completed successfully in the past driven by the new increased demand, TBMs for cross passages are now discussed more often as a real alternative to save time and cost.

INTRODUCTION

Urbanization and the growing world population increases the demand for smart tunnel systems, especially in large cities. To handle the rising traffic volume in and between cities, the demand of tunnel systems is increasing. With rising demand and higher safety standards, more cross passages will be required for rescue uses in road and rail tunnels in the future.

Several design approaches for rescue tunnels are available. For a twin tunnel arrangement, a typical approach is the planning of cross passages to link parallel tunnels, whether in construction or already in operation. Connecting a shaft with a tunnel is another concept for emergency exits, often used for train and metro line or traffic tunnels. Cross passages can also play a role, where sewage collectors have to be connected, to extend existing systems or to connect new interceptors to the existing ones.

The major benefit of a completely mechanized approach to cross passage construction is the synchronisation of excavation and lining operation. In the past, sometimes tunnelling machines (e.g., road header) have been used for the excavation of cross passages, but the lining has been done with conventional shotcrete or cast in place methods.

CROSS PASSAGE CONCEPTS

The general approach to connect two underground structures by a cross passage is not new. The technical approach is very similar to standard pipe jacking, where shafts serve as launch and reception structures for the tunnelling equipment. Figure 1 gives an overview about different cross passage concepts, each of which are discussed in detail below.

Shaft to Tunnel

At the moment, the most common type of cross passages is to link a shaft and a tunnel. When emergency exits have to be installed to existing traffic tunnels, this can be done by conventional or a mechanized construction method. The shaft, at the same time, can serve as a final structure for safety or ventilation. Mechanized pipe jacking is considered the preferred technology, also assuring construction safety and installing the final liner. In this case, a tunnel boring machine starts excavation in the shaft and then

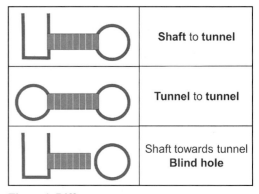

Figure 1. Different cross passage concepts

	Shaft to tunnel
	Tunnel to tunnel
	Shaft towards tunnel **Blind hole**

Figure 2. View into the Emisor Oriente main sewage collector tunnel with AVN 1800 breakthrough in the left half of the photo

breaks through in the tunnel, where it is dismantled and transported back to surface.

In Spain, nine shafts have been built by a vertical shaft sinking machine (VSM) to add emergency exits and ventilation shafts to the express train route from Montcada to Trinitat. The shafts are up to 57 m deep and were linked to the tunnel using conventional excavation methods.

For one of the world's largest sewage projects in Mexico City, Emisor Oriente, an AVN 1800 slurry pipe jacking machine was used to construct two link tunnels to the 8.7 m in diameter main sewer tunnel. With depth up to 42 m and groundwater pressure beyond 4 bar, this method offered a very safe solution.

Tunnel to Tunnel

Cross passage concepts are more commonly used as connection between two tunnels for rescue purposes. In general, twin traffic tunnels have to be connected to satisfy necessary safety standards.

The concept and design of a mechanized approach depends on whether one or both tunnels are still under construction, already finished and fully accessible, or in operation. A later section will present two reference projects and a mechanized solution.

Shaft Towards Tunnel—Blind Hole

In some cases, the tunnel has to remain unaffected by the construction of the cross passage as much as possible. A mechanized solution with a retractable machine allows the construction of the cross passage without interruption of the traffic in the tunnel and without impeding logistics when the tunnel is still under construction.

The tunnelling machine excavates the cross passage up to the predetermined end position close to the tunnel. Depending on the project requirements and the TBM selected, it can be either retracted or dismantled and transported back to the launch shaft. The final breakthrough into the main tunnel is being done conventionally within the shelter of a preinstalled grout block. A later section will present two reference projects and a mechanized solution.

TUNNELLING MACHINE CONCEPTS

In general, pipe jacking is the preferred technology for mechanized cross passage construction. Accordingly, different machine solutions can be considered to suit the geology and hydrogeology.

Auger Boring

Auger Boring Machines (ABM) use augers for transport of the cuttings. They are an efficient solution in displaceable, dry, soft soils. Augers transfer the torque from the drive to the cutterhead. At the same time, they transport the excavated material from the tunnel face to the launch shaft. In this way, the pilot drill is driven into the ground pipe by pipe. Afterwards, the back-reamer (hole-opener) widens the cross passage to the final diameter with a separate drive unit and finally the product pipes are being pulled in.

Hard Rock MTBM

The Herrenknecht Hard Rock MTBM concept is a simple and economical solution for tunnelling above the groundwater table. The excavated rock is transported by a conveyor belt, which reaches into the excavation chamber, to a muck skips in the rear part of the MTBM, which then is pulled outside the tunnel by e.g., hydraulic winches.

This machine concept combines the benefits of an open-face-tunnel boring machine and a remotely controlled MTBM system. The machine operator controls and supervises the excavation and muck transport from the control container on the surface via video cameras installed inside the MTBM.

Open-face Digger Machine

Open-face tunnelling machines can be applied in a wide variety of dry geological conditions. According to the geology, they are equipped with a roadheader or an excavator boom. The excavated soil is transported by a belt conveyer and muck skips. Open-face shields are operated underground, sometimes even under compressed air.

This machine type can excavate circular and non-circular tunnel cross sections. For the excavation of rectangular cross passages, e.g., for pedestrian underpasses, Herrenknecht designed an open-face machine, called MH Box machine.

Earth Pressure Balanced Machine

For soft, cohesive soils above or below the groundwater table, Earth Pressure Balance (EPB) machines

Table 1. Cross passage design criteria for TBM selection

Design Criteria for Cross Passage	
Geometry	• Shape: circular–rectangular
	• Dimension, cross section
	• Length
	• Position to tunnel axis

Construction Process	Cross passage construction
	• During tunnel excavation
	• During tunnel operation
	• As blind hole (machine refraction)
Geology	• Ground stability
Hydrogeology	• Groundwater level

are the preferred option. They use the excavated soil within the excavation chamber as the medium to support the tunnel face. This makes it possible to balance the pressure conditions at the tunnel face and to avoid uncontrolled inflow of soil into the machine. This not only avoids increased settlements on the surface but also allows for rapid tunneling progress.

The excavated soil is transported by a screw conveyor and muck skip or a belt conveyor throughout the tunnel. For cross passages, the machine length is typically minimized, thus the length of the auger and the maximum allowable water pressure may be limited.

AVN Slurry Machine

AVN Slurry Machines are closed, full-face excavation machines with a hydraulic slurry circuit for face support and soil disposal. They require a separation plant on the surface. The conical crusher inside the excavation chamber crushes the particles, which are small enough to fit through the cutting wheel openings, to a size which can be transported inside the discharge lines. Those lines are connected to the excavation chamber and transport the slurry/soil mix to the separation plant on the surface. Cobbles and boulders, which are too big to fit through the cutting wheel opening are broken up by the cutting tools on the cutting wheel.

AVN machines can be operated under high ground water pressures. For cross passages, this is the most flexible technology as the machine can be design very compact and can handle a wide geological range with high ground-water pressures. For working under the groundwater table, a launch seal and pipe brake is needed.

Horizontal Raise Boring

The horizontal raise boring process works similar to auger boring and can be used in stable ground or

rock conditions. A pilot hole is drilled first with conventional drilling equipment. Then a reaming head is attached to the drill string and pulled back to enlarge the pilot hole to the final diameter.

The reaming head is equipped with cutting tools similar to a TBM cutting wheel which allows efficient removal of the cuttings.

CRITERIA FOR CROSS PASSAGE TBM SELECTION

As described above, there are several construction options for a cross passage. The most efficient mechanized approach depends upon the criteria in Table 1.

Geometry

For a mechanized solution a circular shape is preferred. Non-circular shapes are generally limited to an open-face technology.

The main parameters to consider are the geometrical dimensions especially diameter and length.

For rescue purposes the position of the cross passage to the tunnel axis is determined by the level of the roadway. A large vertical offset needs to be avoided.

In water-bearing ground a circular shape tunnel helps to seal of the cross passage tunnel in the transition zone to the main tunnel.

Construction Process

A cross passage construction affects the operation in the main tunnel. Depending on the construction process, the impact can be minimized. One extreme would be the construction of a cross passage into a road or rail tunnel which is already in operation. The other extreme is the construction of the cross passage when the reception tunnel does not even exist.

Under such conditions, a dead-end installation (see Figures 1 and 4) would allow the operation of the main tunnel to continue during the main construction period. Access of the main tunnel would only be affected for the final breakthrough. In large tunnels, the cross passage construction can take place simultaneously to the construction of the significantly larger tunnel, excavated by a TBM. In correlation to the diameter of the cross passage tunnel, the diameter of the main tunnel needs to be significantly larger (min. 10 meters in diameter).

Geology and Hydrogeology

The geotechnical conditions are the major criteria to determine the most suitable cross passage technology. Chapter 3 describes the different technologies applicable to various geotechnical conditions. In unstable or wet conditions pipe jacking will likely be the most advantageous installation method. Only

Table 2. Machine selection according to geology/hydrogeology

Geology	Hydrogeology	Machine Concept
Stable	No GW	Horizontal Raise Boring
		Open-face Shield (with Roadheader)
		Hard rock TBM
Instable/mixed soil	No GW	Auger Boring (with Backreaming)
		Open-face Shield (Roadheader/Excavator)
		Hard rock TBM
		EPB
Instable/mixed soil	Up to 3 bar	Open-face Shield (under compressed air)
		EPB
		AVN Slurry
All soil	>3 bar	AVN Slurry

in dry and stable conditions conventional shotcrete or in-situ cast linings can be considered as feasible alternatives.

General Considerations

The cross passage will influence the main tunnel structure in a variety of ways. Whether it is excavated conventionally or using a mechanized approach, the structure of the tunnel needs to be adapted (Negro, Cecilio, Bilfinger, 2014). For a mechanized solution, the forces during construction (jacking forces) need to be handled by the tunnel structure.

In addition the loads from the groundwater, the force on the seal (launch seal) and the TBM (pipe brake) need to be taken into account. In the start and reception area the segments of the main tunnel have to be prepared for the breakthrough of the cross passage TBM. For example, the steel rebar reinforcement is usually replaced with steel fibre or carbon fibre, which can be cut by the standard cutting tools of a TBM. Furthermore, the segments surrounding the cross passage entry or exit points inside the main tunnel need to be strengthened too.

The restricted space in the launch and retrieval areas of the cross passage generally limits the flexibility in handling large and heavy components. The general objective is to build the cross passage as large as necessary and as small as feasible. The internal diameter of a cross passage for rescue purposes is generally in the range of 2.5 to 4 meters, depending on the main tunnel dimension and the safety requirements.

In addition to the above-mentioned criteria, further project specific requirements need to be considered.

SELECTION OF MECHANIZED SOLUTION

As summarized in Table 2, there is a wide range of available boring methods and concepts for cross passages for varying geology and design parameters.

Figure 3. Installation of the open-face shield in the 4th Elbe tunnel

A cross passage under the groundwater table can be a challenge. In the past, cross passages in water-bearing soils could often only be realized with the help of ground freezing methods or with extensive grouting and soil improvement measures.

However, a mechanized alternative can offer considerably benefits regarding construction performance and safety.

REFERENCE PROJECTS

Rescue Tunnel, Hamburg, Germany

In the late 1990s, the fourth tunnel under the River Elbe was built in Hamburg using the world's largest Tunnel Boring Machine (OD 14.2 m, ID 12.35 m) of its time. A main part of the comprehensive safety measures was the construction of two rescue tunnels to connect the new Elbe tunnel with the ventilation structure of the existing Elbe tunnel.

Neither the shipping traffic on the Elbe River nor the operation of the existing traffic tunnels would be affected by the tunnel and cross passage construction. Therefore, it was decided to use the pipe-jacked cross passage construction method with an open-face

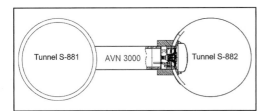

Figure 4. AVN 3000 used for the cross passage between the two tunnels under construction

machine to excavate the 70-m and 32-m long sections from the new 4th Elbe tunnel towards the ventilation structure. In order to handle groundwater pressure of up to 3.5 bar, the machine was operated under compressed air with a pressure wall between the operator cabin and the excavation chamber in the shield.

The tunnelling machine was launched from the new tunnel, which had already been prepared for the launching procedure by the integration of steel segments in the launch section of the tunnel.

As other work in the tunnel had to be continued during the cross passage construction, special solutions were necessary to install the jacking frame. Main reason for the selection of the custom designed cross passage tunnelling equipment was the small footprint.

Rescue Tunnel Twin-Tube, Hong Kong

The world's largest TBM, a Herrenknecht Mixshield (Ø 17.6 m) and two Mixshields (Ø 13.95 m) are constructing a gigantic twin-tube road tunnel in Hong Kong.

With a length of around 5 kilometers each, the Tuen Mun—Chek Lap Kok Link tunnels will connect the international airport and the Hong Kong-Zhuhai-Macau Bridge (currently under construction) with the mainland to the north. The tunnels will cross under a branch of the Pearl River Delta at depths of up to 50 meters.

From the northern launch shaft, the 17.6-m Mixshield first drove a 650 meter long section. The TBM was then scaled down to a diameter of 14 meters for the rest of the tunnel to the airport.

Due to the necessity to construct 44 cross passages under extremely difficult conditions with a very tight schedule, Herrenknecht and Dragages-Bouygues Joint Venture jointly developed a concept for the two AVN3000 custom-designed cross passage TBMs. In addition, special launch and reception structures were developed. These allowed for pipejacking directly from one of the large tunnel tubes while all support services for the large diameter TBMs could be maintained simultaneously.

The 44 cross passages are up to 14 meters in length. Two complete sets of equipment are being

Figure 5. AVN 3000 recovery in the tunnel, after breakthrough into reception bell

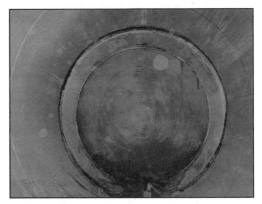

Figure 6. Blind hole cross passage built by retractable MTBM

in operation simultaneously to meet the tight time schedule. As logistics in both large diameter tunnels had to be maintained, the available space for the cross passage equipment is very restricted. Therefore a very compact design of AVN machine and the launch as well as reception structure was required.

On the launch side, a jacking trailer with a mobile gantry crane is used to allow quick repositioning of the equipment for the construction of the next cross passage. Due to the surrounding groundwater pressure of 5.5 bar a safe launch and breakthrough procedure was designed. While a special launch seal with emergency seal is installed on the launch side, safe breakthrough into the "target tunnel" is assured by using a closed 'pot construction' at the exit side.

Rescue Tunnel, Jenbach, Austria

Traffic volumes crossing the Alps are growing enormously. Therefore, a high-capacity railway connection is being built between Munich and Verona with the 55-kilometer-long Brenner Base Tunnel as the centerpiece. The tunnels of lots H3–4

Münster-Wiesing and H8 Jenbach are part of the access route's to the Brenner Base Tunnel.

A retractable AVND4000AB was excavating a total of 17 escape tunnel varying from 18 to 130 meters in length. First, the machine with an outer diameter of 4.8 meters was lowered into the shafts that had been sunk beforehand. Afterwards the machine tunnelled diagonally towards the tunnel axis to an area that had been grouted beforehand and was later expanded to a rescue space. The machine was equipped with a special mechanism to collapse the cutting wheel so that the TBM could be retracted through the tunnel back into the shaft.

CONCLUSION

In many cases, the construction of cross passages is one of the critical parts in the overall tunnel construction program. For mechanized cross passage construction, a detailed design is necessary for project preparation in order to achieve a fast and successful operation.

Similar to TBMs, there is a variety of different types of cross passage TBMs suited for certain soil conditions. The most suited type needs to be selected by evaluating the soil parameters as well as the project specific requirements.

In general, a mechanized approach for installing cross passages has significant benefits over conventional processes. A closed face mechanized approach increases the safety during construction and allows a high installation rate.

REFERENCE

Negro, A., Cecilio, M., Bilfinger, W. 2014. *Cross passages between twin tunnels. Preliminary design schemes. linings.* Tunnels for a better life, WTC 2014, p. 49.

Communications, Tracking, Vehicle Telemetry, and Proximity Solutions

Guido Perez Manfredini
MST Global

ABSTRACT: As with TBM and other tunneling machines, *communication technologies* have evolved significantly over the last few years. With today's digital technologies, safety is enhanced by means of personnel tracking and location, electronic access control, vehicle proximity detection, video transmission, gas monitoring, and most importantly the *people* connection. Any person underground can now be connected to the world, get internet, and send an immediate emergency alarm wherever they are underground via mobile phones

Digital technology increases productivity in tunnel projects. Thanks to *Wi-Fi* platforms, applications like ventilation on demand, vehicle tracking, TBM data transfer (excavation parameters), cutterhead interventions, geotechnical information and so on, can assist with project management, reduce construction costs, and achieve improved quality and production rates.

In tunnels with a diameter of less than 5 meters, where TBMs tend to be longer, the Leaky-Wi-Fi technology lays as the best technical solution to overcome the challenge of propagating the Wi-Fi signal through the TBM.

In the present day, there is no reason why someone should be in a tunnel without full access to modern means of communication. Everyone has the right to be connected, monitored, and protected, especially when working in an underground environment.

INTRODUCTION

Improving safety and achieving better productivity results has been occurring over the last few decades. From the mid-1990s through to the present-day, operations have persisted with trying to get the basic analogue systems, such as leaky feeder radio and copper data cabling, to be 'stretched' beyond the technology capabilities in an attempt to achieve the impossible: tracking people and assets, video surveillance and the like. Success was partial at best and in hindsight, persistence with this style of communication technology is likely throwing good money after bad.

The development of IT infrastructure into a rugged, fit for purpose operational technology (OT) infrastructure system for use in the underground environment, has enabled all the features and functionality required for success to be available at the tunnel face underground. Through the use of firewalls and VLAN segregation, protected corporate IT data can be available underground across the single piece of infrastructure delivering OT data applications such as production reporting and management, fleet management, and automation.

WI-FI TECHNOLOGY

The most useful digital technology for underground projects are Wi-Fi based systems (Figure 1). These systems provide a unified communications and data networks that allows the following:

- Two-way voice and texting via Voice over Internet Protocol (VoIP) portable handsets (Phones).
- Provides real-time location of people and equipment in the tunnel.
- Data communication (for TBMs, roadheaders and other equipment).
- Interface to existing radio systems to allow the systems to "talk to each other."
- Remote access to systems.
- Data acquisition.
- Video surveillance system in the tunnel.
- If a site has a private branch exchange (PBX) for an internal phone system and links to the outside phone network, then the interface to the PBX can be made to allow the underground VoIP phones the ability to call any other internal or external phone.
  - For example, this approach would enable a technician underground the ability to call and send images of a TBM issue to an OEM engineer in a remote location (e.g., Germany).

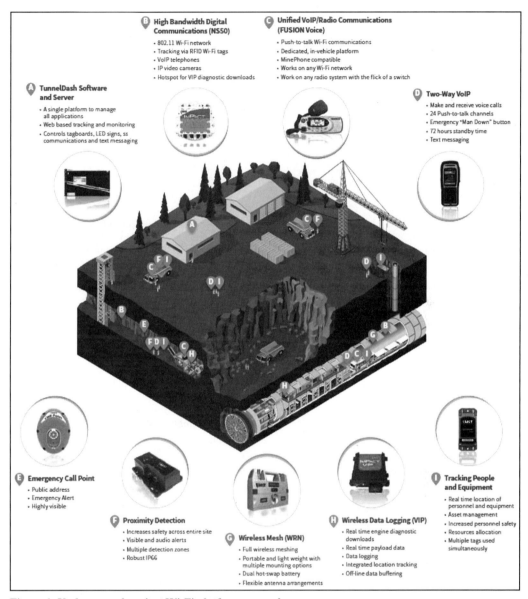

Figure 1. Underground project Wi-Fi platform example

Normally when installing a Wi-Fi backbone, the first stage of the network deployment would be at each required location on the surface sites, such as the offices and workshops. The network would then extend into the shafts/portal and tunnel environment ultimately reaching the tunnel and tunnel face. The network would simply be extended as the tunnel progresses.

Figure 2 provides an overview of how all the Wi-Fi solutions would integrate underground. These systems then connect to the surface for operational and safety purposes.

The Wi-Fi network can converge all data onto a single network. There is no need to run separate optic fiber or copper data lines to get data from critical equipment, such as TBMs. There are systems with up to 1 Gb bandwidth, more than enough capacity to handle all the data requirements of a typical tunneling operation.

A typical high bandwidth network topology is shown in Figure 3. This shows the logical connections

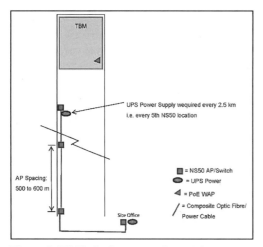

Figure 2. Wi-Fi platform tunnel extension

between the various IT and OT applications which can be utilized in an underground project. The core infrastructure components that comprise the communications platform are:

- Wireless network switches
- Wireless access points
- Composite cable
- Antennas
- Communications appliance.

A more detailed overview of some of the key components is provided in the following sections.

Wi-Fi Platform, Key Components

Wireless Network Switch

The 'workhorse' of the wireless infrastructure platform, is the wireless network switch (Figure 4). The switch design may vary. In general terms, it consists of four managed, fiber optic Ethernet switches, two

802.11 b/g (IEEE) wireless access points, and four Power over Ethernet (PoE) ports providing scalable wired and wireless network access.

It is a multiservice device to support automation systems and/or facilitate functionality such as VoIP, IP video streaming, remote PLC programming, mobile data acquisition, real-time vehicle diagnostics, and asset/personnel tracking.

There are switches with a rugged enclosure designed for the underground environment and hence do not require a separate environmental enclosure. They are installed directly onto the backs or side walls of the underground drives.

Composite Cable

Traditional enterprise networks have a star topology which requires power at every network node, which is not a cost-effective solution in the underground environment. The challenge of limited power availability can be overcome by using composite fiber cable, which carries both power and the single mode optical fiber cores.

This seemingly simple concept is a key to simplifying installation and maintenance of the network to semi-skilled labor. By incorporating both optic fibers and copper power cores so that power is transferred along the network by the single cable, the number of power insertion points required is greatly reduced.

Communications Appliance

The ability to remotely manage and monitor the network is critical to any network deployment. A server based hardware and software platform that constantly monitors the network and provides tools for managing the individual network components is required. There are many software applications that display tracking, voice communications, machine health data, system alerts etc.

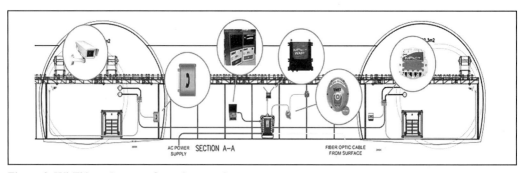

Figure 3. Wi-Fi based system for twin tunnel

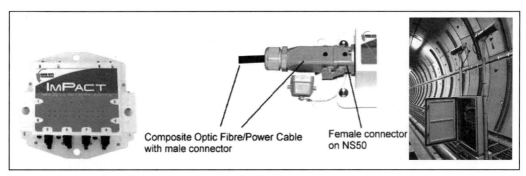

Figure 4. Wireless network switch

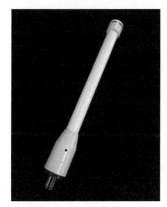

Figure 5. Antenna

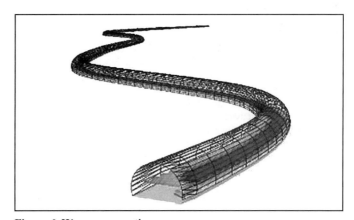

Figure 6. Wave propagation

Antennas

There are several tunnel specific antenna designs. The most appropriate design depends on the tunnel alignment. Antennas can be easily installed or relocated as required. Depending upon the final antenna propagation testing, you can also deploy other models of antennas to ensure saturation coverage for the tunnel. Figure 5 is a new tunnel antenna and on Figure 6 is an example of network prediction model inside a long and curved tunnel using propagation software.

Wi-Fi Technology Applications

Voice Communications

Voice over Internet Protocol (VoIP) phones (Figure 7) can connect directly to the tunnel and surface network using wired Ethernet or Wireless coverage. Every wireless network switch and wireless access point (WAP) can be used to carry voice calls and messaging. The network solution would be installed as a voice ready system and could be integrated to the site IP PBX phone system if it has SIP trunk capability. This would create a seamless network for voice communications when operating on site. Restrictions could be enabled based on each user and call groups established as required. This would provide the following call types:

- Call direct to any user
- Push to talk to any user
- Programmed channels
- Emergency broadcast channels
- Calls to outside line.

You can easily find VoIP phones that also have a push to talk (PTT) button and are able to communicate to a number of customisable PTT channels as well as a broadcast channel.

Tracking People and Equipment

Tracking Systems report the location and track the movement of people and equipment as they travel through the tunnel in real-time.

The tagging and tracking systems use strategically placed Wi-Fi Access Points as readers for RFID Tags. These are normally carried by workers or attached to vehicles & other equipment, so their

Figure 7. VoIP phone

real time position is monitored underground and on the surface (Figure 8).

Many of the RFID Tags have a "MAN DOWN" button. If the person holding the tag presses the button, an emergency alarm will be sent to the control room, the location will be known and the spot on the tracking software will be turned RED.

There are many systems that analyze and interpret location data to deliver measurement reporting and comparison against theoretical values, to allow the real-time management of various critical production processes. In tunneling, if any person or mobile equipment fall outside pre-set parameters for time and location, an alert can be raised to investigate the issue and manage corrective action, before any delay to the process becomes significant. The most common rules to trigger alarms are:

- Breach Site Rule: This alarm triggers if a tag enters or exits a specified restricted zone.
- Distribution Site Rule: This alarm triggers when too many are in a defined zone (such as the TBM).
- Out of Range Site Rule: This alarm triggers when a person has been out of zone and not come back.
- Motion Stop Site Rule: This alarm triggers when a person has been stationary for a long duration.

The functionality could be extended beyond simple location awareness, so in addition to the safety benefits, an investment in a Wi-Fi Tracking System can deliver tangible cost benefits.

Why should everyone be permanently located, connected and tracked while underground?

The first reason is because safety comes first. With today's technological capabilities there is no excuse for not tracking personnel when underground.

Although it's been decades since the tunneling industry started to take safety as the most important

aspect of a tunneling project, no major changes have been developed to monitor personnel in a tunnel.

Tunneling projects normally have security access control points where people are checked into and out of the tunnel. The challenge with this scenario is that beyond being checked in, managers are unable to locate people in real-time. Not to mention how challenging it is to control who is inside or outside of the tunnel if it has multiple access points.

Secondly, productivity is increased and costs are reduced. Tracking people underground is not only essential to safety but it also helps managers, superintendents and supervisors lead their teams more effectively and efficiently.

The amount of time used to locate someone to address an issue that requires immediate action can be reduced. On larger projects, unless a person is around the TMB, the portal or the shaft, or main area, you can easily spend a considerable amount of time trying to find or get in touch with them. All that wasted time equates to a lot of wasted money. If this scenario happens multiple times a week, the costs start to become significant.

Why should you track locomotives or tunnel service vehicles?

The first reason to track your locomotives or tunnel service vehicles is safety. In case of an emergency, relying on radio communications to locate where trains are in the tunnel is not effective enough. Should a rescue team need to get in or out, wouldn't it be amazing to know that the tunnel track is clear?

Secondly, production will improve. Traffic in the tunnel can be analyzed, reducing wait time and number of trips. What if you could monitor your

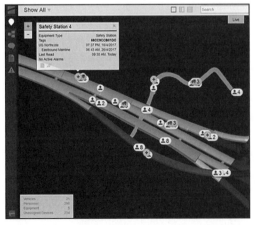

Figure 8. Plan view of personnel and equipment in tunnel

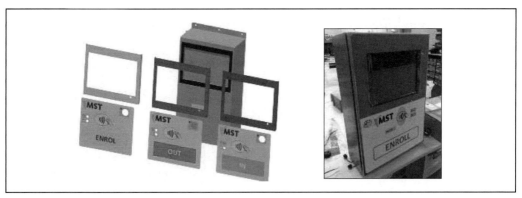

Figure 9. Kiosk

driver's behavior? Being able to analyze where the locomotive has stopped, for how long, who has been driving it, the time it took to supply the TBM, and the time it takes to get in or out of the tunnel.

Discussing when to install a Californian Switch?

By tracking vehicles underground, more accurate traffic data is available to optimize production which in turn, saves money. Unfortunately, TBM's don't work 100% of the time and one of the reasons is waiting for supplies. Bad traffic management causes most of delays, but in some cases, waiting times are caused by the distance from the shaft/portal to the TBM. If this is the case, then it's time to install a Californian switch. Let's suppose that for the total duration of your project you could avoid 30% of the supply waiting. How much time and money would that save?

Electronic Access Control

There is a wide variety of electronic access control systems, though, most of them use a rugged computer kiosk. These Kiosks (Figure 9) send the read information to the server and control room, either wirelessly or wired into the network.

Vehicle Intelligence Platform

There are data log devices that act as a Wi-Fi Bridge which can be installed on tunnel vehicles, drills and other machinery. The devices stream in real-time, critical parameters that gauge the performance and help maintain each machine. Data is streamed from the machine via the network back to the surface.

The status of a vehicle can be transmitted in real time from the tunnel to the control room or to the management team reading: engine temperature, ground speed, exhaust gases and the like. This basic data can be analysed, and reports can be created to quickly identify bottlenecks or inefficiencies in the

tunnel. The data log devices are able to accurately record production cycle times and trigger alarms if any cycle goes outside of pre-determined values.

Emergency Call Points and Sirens

To facilitate a dedicated fixed emergency alarm system, IP Help Points (Figure 10) are the best solution. These help points feature a single emergency call button. This button can call the control room or even broadcast the message to all connected devices. It provides direct communication between the control room and the IP Help Point.

Help points may contain two relay outputs. These relays can be controlled by the master control. Emergency sirens can also be installed to the call point, as shown in Figure 10. These sirens have configurable emergency tones and volume levels. From the control room, groups of emergency alarms can be trigged as deemed necessary and appropriate to the situation. For example, the group could be set up to trigger alarms for the complete evacuation of the tunnel in case of an emergency.

Video Surveillance and More

Any device with an ethernet output or Wi-Fi connectivity can be connected to the Wi-Fi platform. Including video cameras, gas detectors, traffic lights, geotechnical sensors and the like.

As an example, when the TBM is boring underneath a very sensitive foundation where instrumentation to measure ground movements has been installed, the measurements could be linked with people in the tunnel, triggering an alarm in case of any geological shift.

Proximity Detection

Proximity Detection Systems provide an additional level of control in managing the collision risks

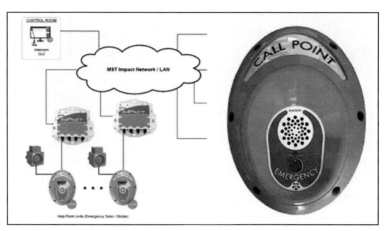

Figure 10. Emergency call points

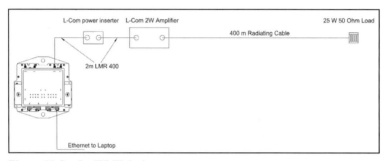

Figure 11. Leaky-Wi-Fi design

between people and vehicles, and between vehicles and vehicles.

With a proximity detection system, equipment operators feel more secure when their vision is partially obscured. The proximity detection system can be set up so the system controller displays people, vehicles or other objects detected on the radar screen.

Proximity detection systems provide detailed reporting capabilities enabling jobsites to not only improve safety via enabling peer to peer alerting, but also by providing management with data and reports to monitor driver behavior.

Leaky-Wi-Fi

For small diameter tunnels, where TBMs tend to be longer and the available section is much smaller due to the amount of steel, the Wi-Fi signal struggles to be propagated through the TBM.

For these circumstances, Leaky-Wi-Fi solutions (Figure 11) solve the challenge. It is still a Wi-Fi based system (Figure 12 & 13), with a radiating cable connected to the last wireless access point, instead of using an antenna to propagate the Wi-Fi signal through the TBM.

Temporary Wireless Connection

There are tunnel projects where full coverage along the tunnel is not needed, and people mainly work on the TBM and on the surface. Under these circumstances, if a temporary job needs to be done somewhere in the middle of the tunnel, wireless repeater nodes (WRN) are a great solution.

WRN's simplify the connectivity in the most dynamic areas underground, providing wireless connection to the working faces. A core requirement to leverage IoT is to have a reliable Wi-Fi connection. There are many WRN that meet this requirement and extend the Wi-Fi platform from the fixed infrastructure to the active working faces. These repeaters also include an 802.11n (IEEE) interface to allow easy access to the many Wi-Fi devices.

For cutterhead atmospheric intervention, even the leaky-Wi-Fi can't guarantee the signal in the cutting wheel compartment. Wireless repeaters nodes will extend the network to the cutting wheel chamber, allowing what the miners are seeing and what they are breathing to be monitored.

Figure 12. Leaky-Wi-Fi TBM wave propagation

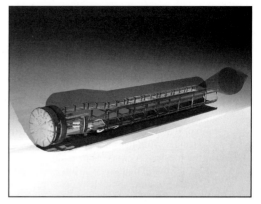

Figure 13. Leaky-Wi-Fi TBM wave propagation

BENEFITS OF DIGITAL TECHNOLOGY UNDERGROUND

There are many benefits associated with the use of a Wi-Fi system. Benefits depend on project characteristics and requirements. Projects implementing a Wi-Fi system will get the following benefits:

1. Save time.
   Time is the biggest waste on construction projects and the primary cause of project delays that result in cost overruns. With a Wi-Fi platform, construction professionals no longer have to waste time:
   – trudging back and forth from the jobsite to the tunnel
   – reaching out to someone on site to have a private conversation with him.
   – waiting for Method Statements' revisions to get distributed.
   Indeed, after reviewing numerous studies conducted over a 30-year period documenting levels of wasted time in construction activity, one meta-analysis concluded that an average of *49.6 percent of time in construction is devoted to wasteful activity. "Horman, MJ. "Quantifying Levels of Wasted Time in Construction with Meta-Analysis." Jan 2005. Journal of Construction Engineering and Management. 13:1"*
   – Construction productivity software can reduce time wasted allowing management to focus on what they do best—running a tunnel project.
   – Daily/routine meetings, tool box talks and safety meetings can be presented online to everyone located at the working place.
   – Being ready to start working after the talk and avoiding the breaks during the commute to their job location (TBM, Cross Passages, Shafts, Surface, office...) increases productivity.
   – Access to all project information (Risk Assessments, Method Statements, Reports, Tool box talks, flash cards...) from your smart phone or tablet
   – Engineers/inspectors/surveyors... can send reports and pictures from any point in the tunnel.

2. Analysis.
   – TBM stoppage times due to maintenance or any failure. Shift Mechanics and electricians could be doing a video-call with expert technicians from wherever the failure is to fix it and be assisted.
   – To monitor drivers' behavior and to analyze the traffic in the tunnel.
   – Waiting times analysis.
   – To know loco drivers' position and what they are doing, remotely.

3. Automate Administrative Tasks.
   There is software that can improve overall project efficiency by streamlining time-consuming tasks like job costing, crew allocation and timesheet processing. For example, tracking allows supervisors to keep track of employee-time data, significantly improving accuracy and reducing processing costs. Supervisors can also use mobile devices to send employee time-data wirelessly to the field office. Ultimately, by using tracking systems, workers are freed from time-consuming data entry or even paper reporting. Your firm's organization and overall productivity will increase when your staff spends less time on data entry or tracking down missing paperwork.

4. Improve Communication and Collaboration. Many projects experience cost overruns that can largely be attributed to unnecessary downtime or delays. Time waste and delays are often a by-product of inefficiencies that can be virtually eliminated with real-time communication and collaboration enabled by the digital technology. This technology can go where your team does. Notifications can prompt workers to take action or alert them of changes, eliminating idle or downtime on the jobsite. In short, your team is just a tablet or smartphone away from being able to access and communicate project information that will reduce time waste and increase efficiency.

5. Increase Visibility and Reduce Risk. Beyond eliminating manual work by automating routine administrative tasks, a Wi-Fi based system helps construction professionals improve workflow as well as create workflow analysis based on accurate real-time data. For example, you can enter updated data in accounting and payroll systems and view the overall performance of the company as well as the progress of each project. In short, digital technology allows managers to have more visibility into their projects so they can maintain control and reduce risk.

6. Eliminate Rework. A Wi-Fi platform improves communication and collaboration. One of the most common causes of rework and quality claims is poor communication and error from lack of document control.

   65% of construction professionals say the biggest problem they face is the cost of doing rework[1] and the median cost of rework due to poor document control costs the U.S. industry \$4.2 billion a year[2]. *1) TSheets. "10 Reasons to Use Apps in Construction." Infographic. 2) "Hwuang, Bon-Gang et al. "Measuring the Impact of Rework on construction Cost Performance." March 2009. Journal of Construction Engineering and Management"*

   The Wi-Fi technology keeps the tunnel and the office employees connected and everyone updated with the latest project information and workflow. The Wi-Fi platform will help you dramatically reduce rework and quality claims because every member of the team always has the most current set of documents on hand.

7. Improve Accountability and Mitigate Disputes.

   Traditional "paper" project documentation causes a lot of confusion on the jobsite that results in errors and leads to disputes, including indecipherable notes, inconsistent records, or missing or damaged reports. The Wi-Fi platform can help ensure better accuracy and accountability through time stamps, electronic signatures, tracking, and photographic or video recordings. In the case of incidents, disputes or litigation, the Impact system can help you easily recall information.

8. Use of Apps on smart phones and tablets, uploading/downloading data in real time.
   a. Construction Apps—you could take advantage of all the construction related apps to improve your productivity. In some cases, mobile construction apps can accelerate project schedules allowing construction firms to deliver ahead of schedule.
   b. Emergency Apps—i.e., Having everyone access to the American Red Cross app, could save lives in case of an emergency. Being able to deal with injuries swiftly and accurately on the jobsite may mean the difference between completing your project late or completing it early.

9. Monitoring.
   a. VIDEO Full HD, Video IP system for real time viewing and recording.
   b. Environmental monitoring: toxic gases, temperature, humidity, air flow and pressure integrated.
   c. GEO Geotechnical instrumentation monitoring: clinometers, pressure and load cells, hydrological instruments…
   d. ALARM Emergency signals for tunnel evacuation.

There are more benefits associated with the Wi-Fi based system like ventilation and lighting on demand, which save projects thousands of dollars each year; not to mention the intangible value added to SAFETY. Everyone will be better protected with this system, and that will be reflected in more confident workers, which eventually will also improve productivity.

FINANCIAL IMPACT

Wi-Fi Technology Benefits into Real Dollars

Assuming the following project characteristics:

- tunnel length of 25,750'
- *Working 20 hours a day, 6 days a week*
- *A team of 40 people*
- *3 Working locations on site*
- 20 VoIP phones

Estimate values would be:

1. Save time:
   21 hours a week saved across a team of 40 people
2. Analysis:
   5 h/week saved due to performance betterments
3. Automate Administrative Tasks:
   2 h/week saved thanks to a more efficient system
4. Improve Communication:
   6 h/week saved because of a reliable communication system
5. Increase Visibility and Reduce Risk:
   2.5 h/week by having a better control of info and works on site
6. Eliminate Rework:
   3 h/week due to less rework
7. Improve Accountability and Mitigate Disputes:
   4 h/week saved by avoiding investigation times
8. Use of Apps on smart phones and tablets:
   11 h/week saved by doing reports in the tunnel
9. Monitoring:
   5 h/week due to advance incidents detection

Total time saved per week	59.5 hours
Assuming an average labor cost of	$63/hour
Total savings per week	$3,748.5

CONCLUSION

Thanks to the Wi-Fi Technology, a project with the above assumed characteristics will save US $194,922 each year.

The project would increase its productivity and reduce its cost, but at the same time, the project would be protecting its workers integrity in a more secure way. Safety is first when it comes to companies who want to be lead the tunneling industry.

The estimate values have been based on data from the following resources:

- MST Global case studies:
  - Data Communications:
    - Safeguarding Reliable Fleet Management
    - Surface to Underground Data Integration
  - Proximity Detection:
    - Increasing Safety and Productivity
  - Asset & People Tracking:
    - Better storage utilization boosts profitability
    - Managing People Cost in Construction
    - Asset & People Tracking—Diesel Machinery
    - Asset & People Tracking—Longwall Move

- Global Construction Survey 2016. KPMG.
- Sriram Changali, Azam Mohammad, and Mark van Nieuwland. "The Construction Productivity Imperative." July 2015. McKinsey Global Institute.
- Filipe Barbosa, Jonathan Woetzel, Jan Mischke, et al. McKinsey Global Institute. "Reinventing Construction through a Productivity Revolution." February 2017.

RECOMMENDATIONS

The following general comments are offered regarding key considerations when investing in a Wi-Fi based communication system:

- Clear Goals & Objectives.
  Ensure clear goals and objectives are outlined early and scope changes are change managed via the stakeholders.
- Scale of the operation.
  For a large scale operation, the question that needs to be asked are:
  - What level of Wi-Fi coverage are you after and what are you willing to spend to achieve it?
  - Are you after the safety benefits of the system due to remote areas?
  - Do you need real time data for accurate grade or draw control? Or is it a straightforward operation, where there is limited remote locations and reliable production, and you are just confirming production results?
- Complexity.
  Consider the complexity of the operation. Location of the switches so as to service multiple areas.
  - What is the layout of the underground handling system?
  - Will the system need to expand further in the future?
  - Will the system be integrated into an existing network or phone system?
- Engagement.
  Ensure that the key stakeholders and end users are engaged throughout the implementation process. Engage the leadership team and workforce from the initial install to ensure reliability of system going forwards, the accuracy of data and results, and the longevity of the system.

- Leadership.
  Ensure that you have leadership of the project. This is important to ensure that the original outcomes of the system are achieved and that there is no scope drift, ensure that the workforce/end users are communicated and are kept engaged, and that the objectives are completed within the required time frames.
- Training.
  Ensure that you train as many people as possible in how to use the system so that it is used by all parties—tunnel control, supervisors, engineers, maintenance personnel.
- Opportunities.
  Be open to opportunities, possibilities and functionalities of the system that you did not originally consider.

The future opportunities and applications for Wi-Fi based communication Systems are only limited by our thinking!

Think safety, produce effectively.

REFERENCES

Asset & People Tracking—Diesel Machinery. www.MSTGlobal.com. http://mstglobal.com/wp-content/uploads/2014/12/Asset-People-Tracking_UGC_-Crinum-Mine.pdf.

Asset & People Tracking—Longwall Move. www.MSTGlobal.com. http://mstglobal.com/wp-content/uploads/2014/11/Asset-People-Tracking_-UGC_Asset-Util_-Longwall.pdf.

Better storage utilization boosts profitability. www.MSTGlobal.com. http://mstglobal.com/wp-content/uploads/2014/12/Asset-People-Tracking_All-Mining_Asset-Utilisation_Containers.pdf.

Filipe Barbosa, Jonathan Woetzel, Jan Mischke, et al. McKinsey Global Institute. "Reinventing Construction through a Productivity Revolution." February 2017.

Global Construction Survey 2016. KPMG. https://assets.kpmg.com/content/dam/kpmg/xx/pdf/2016/09/global-construction-survey-2016.pdf.

"Horman, MJ. "Quantifying Levels of Wasted Time in Construction with Meta-Analysis." Jan 2005. *Journal of Construction Engineering and Management. 13:1.*

Increasing Safety and Productivity. www.MSTGlobal.com. http://mstglobal.com/wp-content/uploads/2014/12/MST-Personnel-Proximity-System-Solution-Brochure-Ports.pdf.

Managing People Cost in Construction. www.MSTGlobal.com. http://mstglobal.com/wp-content/uploads/2014/11/Asset-People-Tracking_RC_Time-Attendance1.pdf.

Safeguarding Reliable Fleet Management. www.MSTGlobal.com. http:/mstglobal.com/wp-content/uploads/2014/11/Data-Communications_Surface_Pilbara.pdf.

Sriram Changali, Azam Mohammad, and Mark van Nieuwland. "The Construction Productivity Imperative." July 2015. McKinsey Global Institute.

Surface to Underground Data Integration. www.MSTGlobal.com. http://mstglobal.com/wp-content/uploads/2014/11/Data-Communications_UGHR_Newmont-Gold.pdf.

TSheets. "10 Reasons to Use Apps in Construction." Infographic. 2) "Hwuang, Bon-Gang et al. "Measuring the Impact of Rework on construction Cost Performance." March 2009. *Journal of Construction Engineering and Management.*

Underground Communication Innovations

Jason Jarrett, Eric Hansen, and Steve Harrison
Innovative Wireless Technologies

ABSTRACT: Tunnels are rugged environments with large moving equipment, explosives and hazards including water, dust and gas. To date, there are few underground communication and tracking systems available for tunnel projects that support OSHA (Occupational Safety and Health Administration) compliance. After the 2006 Sago Mine disaster, research and development was performed to improve communication systems for the underground environment. Today, over 250 US mines utilize these new technologies to improve productivity, reduce costs and enhance safety. This paper explores these new technologies and provides guidance for planners through the process of selecting a tunnel communication solution. It reviews the key requirements and constraints for consideration and provides an overview of various communication and networking technologies and discusses their strengths and weaknesses. [1]

TUNNEL NETWORK CONSIDERATIONS

This section reviews tunnel environment characteristics which are key considerations when selecting a tunnel communications network.

Tunnel Environment

Irrespective of the type of tunnel project (new construction, repair or remediation), the tunnel geometry, geology and work flow processes are key considerations. Tunnel dimensions and geology affect radio propagation and installation. A common practice is to use concrete for tunnel lining which is conducive to mounting equipment like radio nodes, antennas and power lines. In addition, depending on the composition of the concrete, the lining provides a circular waveguide path for Radio Frequency (RF) signals of certain wavelengths with the ideal wavelength being dependent on the tunnel dimensions [2, 3]. Alternatively, tunnels with porous geology are conducive to water and gas ingress, driving requirements for equipment with high Ingress Protection (IP) ratings and possibly Intrinsically Safe (IS) requirements.

Installation Considerations

New and existing tunnel projects include permanent and semi-permanent installations as well as quick, modular installations. If the project timeline is long enough and power is available, the best systems use line power and back-up batteries for uninterrupted communication. Rugged equipment is needed in addition to proper mounting to prevent obstruction or damage for a long life. Figure 1 shows two

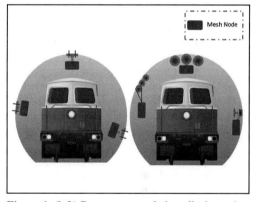

Figure 1. (left) Permanent node installation using brackets, (right) temporary node installation using existing infrastructure

examples of equipment mounting schemes in tunnels using existing infrastructure.

Using existing line power and built-in battery backup for normal operation minimizes maintenance. Short-term projects have similar mounting and monitoring requirements but require smaller size and less weight for mobility. If power availability is a problem or the project requires voice, data and tracking for a few weeks or months, the best solution is a fully battery powered solution. Ideally the system would offer the ability to hang equipment in various locations of the entry using minimal manpower and no heavy machinery as shown in Figure 1. Smaller and lighter equipment allows mounting with basic hooks or rope hangers or possibly placing on existing structures or the ground as shown in Figure 2.

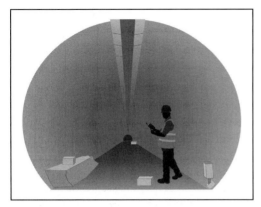

Figure 2. Temporary node installation for short term projects using the ground, outcrops or existing holes

If a project requires inspection in a branch or adit shown in Figure 3, the ability to set up temporary infrastructure extensions is useful.

Obstructions and Tunnel Impairments

A Tunnel Boring Machine (TBM) is commonly used for tunnel construction. Proactive communication solutions account for the TBM structure and provide effective data transmission along the TBM and to the surface [4]. In addition, other equipment in tunnels create time-varying obstructions in the middle of the network. Effective system solutions must account for environmental interference and fading effects

created by mobile and stationary equipment regardless of the tunnel size [1].

TUNNEL NETWORK APPLICATIONS

Personnel Communications, Tracking, and Compliance

Per OSHA, "the employer must maintain a check-in/check-out procedure to ensure that above ground personnel maintain an accurate accounting of the number of persons underground" [5]. Legacy tracking systems are manual with a check-In/check-out brass board or sign in list. This solution is inexpensive but prone to intentional or unintentional human error and may require a dedicated person. Some newer systems provide a situational awareness display (digital blueprint) of the tunnel and track people and equipment automatically using RFID-based technology, thus eliminating the manual check in/check out process. Tracking equipment includes battery operated tracking devices worn or carried by personnel and mounted on equipment. Solutions that are more sophisticated provide multiple tiers of devices including simple personnel and asset trackers as well as higher-tier integrated devices, supporting voice and text communications as well as tracking in a single portable radio. The ability to address different user needs through tiered devices and to provide specific services in specific locations is important since it enables flexibility when trading off equipment cost against operational needs.

OSHA requires tunnel projects to support a minimum of two forms of voice communication

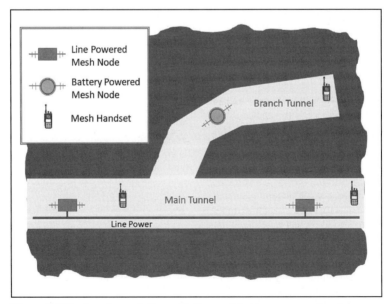

Figure 3. Mixed system solution with line and battery-powered nodes

and have an on-site personnel tracking system. In addition, other federal, state and local regulations may increase the regulatory requirements for underground projects. For example, MSHA requires two-way communication and tracking (C&T) of all mine personnel, 24 hour battery backup of all C&T systems, 14 days of data for personnel tracking and gas monitoring along with specific ventilation plans for each mine. The Department of Energy (DOE) requires 50' tracking accuracy in some of their underground facilities that store hazardous waste.

Determining the application requirements (voice, text, tracking, sensors) up front is key, since the system design varies based on these applications. In addition, for personnel and equipment tracking, one must consider the required tracking resolution; this varies from zonal indication to real-time GPS-like positioning. Invariably, underground tracking systems rely on anchor nodes. The most advanced systems provide nodes which operate as both communications nodes and anchor points as well as anchor point only nodes allowing the tracking resolution to be refined independent of wireless network coverage requirements. In a zonal system, the anchor points are readers that detect when a tracker is nearby and relays that information back to a server. The server will then state that the location of that tracker is "around" the reader. These systems are typically cheaper with tracking resolution in the range of 2000'. Alternately, real-time tracking systems compute location between anchor points with accuracy within a few hundred feet or less. For systems which include sensors, data sampling rates impact network requirements.

Asset Visibility and Equipment Monitoring

As tunneling project precision and schedules drive the profit picture for new construction, the need to effectively and efficiently deploy, maintain and utilize capital equipment (Overall Equipment Effectiveness (OEE) is an important consideration. In most tunneling programs, the Tunnel Boring Machine (TBM) is the most critical piece of equipment with data monitoring requirements for voltage, current, frequency, active and reactive power to voice and tracking coverage of equipment and personnel the entire length of the TBM. The TBM is typically the farthest distance from the entrance/exit point, supports a majority of underground miners and requires low-bandwidth (voice and tracking data) to high-bandwidth (machine statistics and video surveillance) support. Other tunnel equipment including shuttles, man trips and trains also have tracking and data requirements. The need for high-bandwidth in industrial applications has increased exponentially in recent years since equipment has become increasingly sensitive to even minute changes in the power supply voltage, current and frequency [6]. The TBM communications solution should support a wide range of bandwidth requirements while ensuring safety and minimizing maintenance. Greater tracking accuracy that leads to streaming data and large volume data is useful in key areas like the TBM, switches and the shaft for safety and operational efficiency [7]. Remote monitoring is useful to ensure proper operation throughout the life of the project [8].

Remote Monitoring and Control—Point Control, Sensors, IoT (Internet of Things) Devices, etc.

For voice, data and sensor communication, current systems range from manual tracking with pen and paper to fully automated data collection and analysis with alerts. Underground sensors monitor water, ventilation, gas, dust, frequency and power among other things [8]. Sensor data sampling varies from a few times a day to many times a second and is critical for efficient project planning [4]. Sensors are available in wired and wireless options where considerations are needed for cable, antennas, batteries and portability. Sensors and alarms sometimes require high data rates for streaming, spatial distribution, and near-real-time analysis for anomaly detection and emergency control [8]. Cameras are one type of sensor that allows remote monitoring at the expense of high power and bandwidth requirements. Similar to handset radios, it is convenient for sensors to combine multiple functions to save on labor, maintenance and cost.

Data Analysis and Response

Once the data arrives at a dispatch station, tablet or handset, the ability to analyze and determine inefficiencies, failure points and safety concerns is powerful. The top priority for data analysis is personnel and equipment safety but a properly designed solution addresses operational metrics as well. Basic monitoring includes power levels, air flow, gas levels, up/down time, water levels, etc. The faster that anomalies are detected, the quicker the situation is resolved [4, 6]. In addition, higher resolution is required to detect slight fluctuations [4, 6]. The ability to monitor any system from underground concurrently with a remote location for support is extremely important. The amount of data being passed between users, sensors, machines and the various computers dictates the type of network capabilities required for Data Mining and Big Data [4, 6].

Safety

Tunnel projects pose many hazards which requires a disaster resilient system that monitors personnel location and hazardous levels of air, gas and water [3, 8].

Personnel location is critical in case of a health emergency while it is also highly useful for operational and technical issues. Precise location in the event of an emergency could save someone's life since supervisors are able to determine personnel and equipment location to accurately determine time of arrival for rescue crews. Additionally, the ability to detect trends such as gas or dust build-up is useful so technical experts are alerted for appropriate adjustments [4]. Most underground environments are noisy and prevent standard level verbal communication. Public announcement devices along with audible and visual alarms at key locations ensure safety information is conveyed in the presence of high ambient noise. Wireless versions of these devices make installation and recovery simpler. Every underground project also needs to meet the law as dictated by MSHA or OSHA. These laws include communication, tracking and hazard monitoring regulations. From a safety perspective, the best system is one that meets the regulations which also providing operational efficiency gains.

PROJECT/SYSTEM REQUIREMENTS

This section builds on the challenges presented in the overview and discusses project requirements for data collection, transmission, analysis and safety. A designed solution accounts for all aspects of the project instead of providing a simple system to meet a subset of the requirements.

Service Type—"What Network Service, Why, and Where?"

Important considerations for choosing a networking or communication system are what type of service, for what purpose and where the service is needed. Systems deployed in other underground environments focus on regulatory requirements, service types, how the network supports workflows and cost implications. Network service types are divided into several categories including Internet Protocol (IP)-based data services (i.e., laptops, smart phones), voice services (Push-To-Talk (PTT) radios, Voice Over Internet Protocol (VOIP) phones, page phones), Supervisory Control And Data Acquisition (SCADA) data services (sensors, text, control) and personnel tracking/fleet management services. Underground projects monitor hazardous gasses, dust and water along with location and communication status of all underground personnel [9]. An overview of different technologies to provide these services is highlighted herein.

The page phone is a simple yet reliable technology that requires the phones along with twisted pair cable. This system meets the singular goal of voice communication but has limited accessibility (phone locations only), less re-use (throw-away cable) and is less reliable (i.e., single cable failure disrupts tunnel communications). Wireless handset radios using a variety of technologies give operators more freedom and mobility over a page phone system. Handheld radios support voice and texting capabilities while accommodating accessories like throat and speaker microphones.

Types of Underground Networking—Wired or Wireless?

The infrastructure network in the tunnel can consist of all wired equipment (infrastructure & phones), a hybrid system—wired infrastructure with wireless portables or all wireless infrastructure and portables. Some tradeoffs to consider when selecting your network include:

- Service(s) required
- Power availability
- Installation time
- Project duration
- Mobility
- Propagation environment, i.e., tunnel size
- Reliability & safety
- Data rate
- Total cost of ownership—initial price (cost per foot), maintenance, re-usability

Less expensive "all wired" systems utilize phones running over twisted pair wires or Ethernet. These are easy to use but restrict communications to a hardwired phone limiting mobility and impacting productivity. For low amounts of data, simpler systems like Mine Phones or Leaky Feeder are sufficient.

For wired data transmission, RS-485, Ethernet, twisted pair and fiber are popular choices and support page phones, sensors, ventilation and other point control devices. As with many other cabled solutions, twisted pair may have a single point of failure where all devices on one side of the break are down until repaired. Fiber solutions range from cheap to expensive based on number of modes and durability. This is the best wired option for high-bandwidth requirements at the cost of being difficult to maintain and repair. Fiber repair requires a clean environment and great precision which is unfortunate since it also has a single point of failure.

Hybrid systems are prevalent including walkie-talkies which can communicate on leaky feeder (wired) networks. Leaky feeder systems are wired systems using RF "leaky" coax connecting amplifiers spaced approximately 1000 feet apart throughout a tunnel. A headend unit is installed above ground which transmits RF signals down the coax cable. The benefit of these is that they are easy to use, but as shown in Figure 4, they suffer from poor reliability,

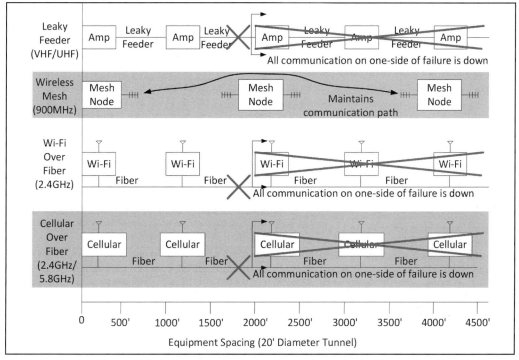

Figure 4. Technology comparison within a tunnel, including network response to failure

limited range and high installation and maintenance cost. Another consideration is the high recurring scrap component and cable cost. Leaky feeder systems are a popular way to provide voice but require laborious effort for installation and maintenance while generating high recurring cable costs [1, 9]. Leaky feeder systems are subject to noise with performance degradation each time the cable is damaged and repaired.

Hybrid systems sometimes utilize traditional Wi-Fi access points which rely on wired Ethernet or fiber infrastructure. Studies have shown that Wi-Fi propagation in tunnels and underground environments may have limited range which drives up solution cost [10]. Another emerging trend in the hybrid space to watch is Long-Term Evolution (LTE) or 4G pico-cells, with wired infrastructure and wireless smartphone/tablet clients. However, regulatory, spectrum and service model uncertainties still exist for private LTE solutions.

For high-bandwidth wireless solutions, it is a misnomer to consider Wi-Fi and LTE completely wireless as they only communicate between nodes using the cabled backbone—it is a hub and spoke architecture which relies on a wired backbone. Thus, it is categorized as a hybrid system. Features include the ability to use common cell phones and other Wi-Fi/LTE supported devices. Node and antenna

configuration are typically required for these systems and they have a single point of failure unless redundant paths are provided. A true wireless mesh system allows handset to be wireless without any wired backbone whereas Wi-Fi or Radio over Fiber (RoF) solutions still require fiber or wired Ethernet. [1]

"All wireless" systems have increased flexibility, mobility and may have better re-use factors to consider. However, as they are wireless by nature, careful consideration must be given to understanding impacts of tunnel aperture and equipment blockage, while care must be taken to ensure reliable connectivity. The data transmission distance dictates the size and type of network along with metrics like redundancy, power and data volume. One wireless system gaining popularity is wireless mesh networks. Mesh radios are like tiny micro-routers in a fire brigade which pass or "hop" network information along the tunnel. When properly designed, they are easy to install, maintain and deploy for underground applications.

Redundancy is typically inherent to mesh systems by means of scalability, flexibility along with ease of setup and maintenance whereas wired systems need parallel runs of cable to avoid a single point failures [3]. Mesh networks typically have a wireless backbone which is convenient for installation, recovery, redundancy and safety [3] but over

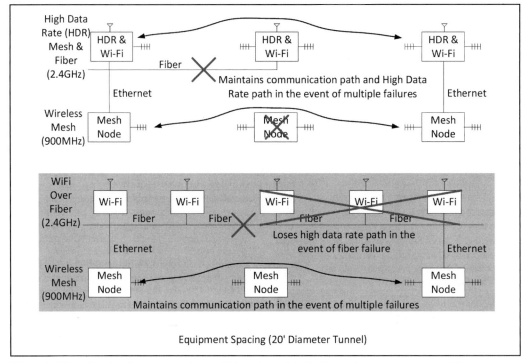

Figure 5. Hybrid technology comparison within a tunnel, including network response to failure

large distances may not support the high data rate or bandwidth of fiber. Newer technology systems extend communication between handsets by means of mesh and fiber [9]. Mesh networks use radios installed at periodic locations whereas Wi-Fi and LTE systems use fiber and Ethernet for the communication backbone. Current voice and data monitoring solutions range from these cabled systems to wireless mesh, Wi-Fi or cellular systems [9, 11].

Wireless mesh is typically more mobile which is convenient for installation and recovery with options for line and/or battery power. Some mesh networks have limited data throughput while others provide voice and high-bandwidth options. Research as far back as the 70s shows that high frequency propagation in tunnels is highly effective since the tunnel works like a waveguide and works well even with tunnel variations [2, 12]. High data rate mesh networks are much easier to install and maintain but don't have the maximum bandwidth afforded to fiber. Most mesh nodes run on battery and/or power where the computing power and bandwidth dictate battery life. Permanently installed systems prefer to use line power when available with battery backup. Mesh systems are digital which provide the benefits of combining voice clarity with texting, tracking and other data solutions. With large amounts of data, fiber or high-bandwidth wireless like mesh or

Wi-Fi are required. Research is ongoing to show how mixing wired (fiber) and wireless (mesh) systems, overcomes shortcomings for both and an example is given in Figure 5 [9].

SYSTEM SOLUTIONS

This section describes how to design a modular system to meet a project's requirements while minimizing the disadvantages of non-recoverable equipment, unnecessary labor or a minimally compliant system. Most systems on the market have strengths and weaknesses and understanding your particular case helps navigate the choices for an underground project. For example, a simple RFID tracking solution is an inexpensive way to track assets and personnel underground but is not capable of passing voice or interfacing with high-bandwidth equipment. A properly designed solution incorporates all the potential uses the tunnel system may be required to support (i.e., voice, data, tracking, etc.) while being fiscally responsible and safe.

Technical

All solutions on the market advertise voice communications, some have tracking and some also support data applications. The optimum solution takes into consideration the project environment, uses and total

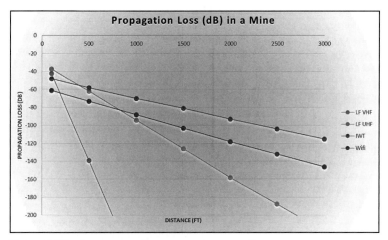

Figure 6. Wireless propagation comparison over frequency

cost of ownership. For example, if a particular tunneling project requires high-bandwidth data transmission but has areas that are subject to heavy traffic (adding breakage opportunities for the fiber) and is dirty and/or dangerous to cable, fiber alone may not result in the optimum network solution. Similarly, if you are running a short-term maintenance or inspection project a wired solution may not be cost effective to install for such a short duration. A hybrid system may be the best solution for a particular tunnel program. For example, running fiber in the TBM chassis may make sense but if you are worried about install cost/maintenance/breakage of fiber, a high data rate wireless mesh makes more sense. This solution is especially useful if machinery only operates in a particular area for a short period of time but needs to pass high-bandwidth data. This combined solution capitalizes on the bandwidth of both solutions while minimizing repair and labor costs [9]. As shown in Figure 6, wireless propagation provides an excellent solution in tunnels when properly designed to maximize node and antenna location along with antenna polarization and tunnel conditions [7].

The ideal solution is a hybrid that uses wireless nodes spaced at great distances to eliminate cable and labor costs while using small repeaters for adits and leaky feeder in areas of obstruction or interference. From a safety standpoint, a Wireless Mesh Network (WMN) is one of the best choices and is widely used for establishing communication networks during disaster recovery due to its scalability, flexibility and ease of setup and maintenance [3]. This proposed hybrid system utilizes redundancy and performance while keeping the cost and maintenance low. A system that is able to wield the power of a "leaky feeder antenna" along with omni-directional and directional antennas like Yagis afford the best flexibility. Care is needed when using leaky feeder since the radiated

signal is too weak for proper operation in tunnels where the distance between the cable and handsets is too great.

Project timeline and power availability vary from one project to the next so a system that uses line power and/or batteries is ideal. A tunnel might run power to installed nodes but for locations where work is planned for only a few weeks, batteries may be used. This hybrid "power system" is arguably the best way to minimize labor and unnecessary maintenance [7] while meeting any technical requirements. The ability to provide consistent, redundant communications throughout the tunnel at all times prevents unsafe outages [3]. If any type of data transmission is limited at the face due to the need to run cable, personnel and asset safety is compromised [3, 4]. With the hybrid solution, any tunnel project can maintain several layers of communication throughout the tunnel wirelessly while running cable as needed for operational or safety considerations.

Economics

Deployment of an underground network system is a critical criterion for the tunneling industry, particularly when teams are only using the system for the duration of the project. Thus, re-usability is an important consideration in the total cost of ownership picture. If 30–40% of the total cost is in cabling or installation, how much of that cost is recoverable and what can be re-used? If re-use is limited it should be included as a recurring cost for every new project. Integrity of cables, connectors and amplifiers becomes a consideration of maintenance cost of wired systems.

Throughput, propagation impairments and redundancy are important considerations for wired networking solutions. Be wary of networking providers which do not execute site surveys to understand your particular environmental challenges. After considering your tunnel environment and your service needs, the next step is to compare pricing on various options. We highly recommend an analysis based on total cost of ownership (TCO) over the networking equipment life cycle. IWT has worked with customers to develop TCO models of various networking solutions to optimize their budget and performance needs. Often the best solution is not the lowest initial equipment cost but results in higher operational efficiency, component recovery, lower install cost, reduced maintenance and has a better net present value than the "initial" lowest cost solution. Often the recurring cost analysis is left out which yields a long-term, lucrative return on investment (ROI).

Initial Expenses

Using the life cycle approach a hybrid technology solution might be optimum where a tunneling project purchases equipment modularly instead of using equipment from one system in the "One Size Fits All" model. This mix and match methodology allows the project team to use multiple technologies for use case optimization. If a project requires high-bandwidth, install high data rate wireless nodes to interface between a fiber port and the face of the operation then backfill the mesh network with fiber to limit downtime and unnecessary cable damage. For voice communication, use wireless antenna based propagation in open tunnel then leaky feeder in the presence of interference, blockage (e.g., rail system) or in small aperture spaces. Additionally, if an area is hazardous to cable or requires communication for a short period of time, use wireless propagation.

Many systems on the market provide voice, tracking and generic data transmission from sensors or machinery. Our experience shows that a hybrid model provides a lower sustained cost and better TCO. To get a full TCO picture, planners must consider the cost of scrap cables left behind, maintenance, installation, including the costs of running cables as well as the overall quality of the equipment as possible hidden costs that are left out of initial pricing. We have seen these costs approach 25–35% of the initial system price on an annual basis. Some questions to consider include:

- Are you using equipment designed for these environments—are your handsets designed to military or heavy industry standards (i.e., MIL-810)?
- Do I have the equipment to install and repair if it breaks on site (e.g., fiber splicer)?
- Will these products support multiple projects or do I have a heavy replacement cost?

Recurring Expenses

Cheap equipment minimizes re-use, yielding a significant recurring cost disadvantage. This analysis assumes that infrastructure for most systems is durable enough to use for multiple projects and the recurring cost is minimal. The ability to re-use cable after completion of a project is another factor. Cable that requires replacement, rework and/or preparation between jobs directly affect recurring costs [9]. Twisted pair, leaky feeder and Modbus cable all have relatively low initial purchase prices but continual repair and replacement cost accumulates quickly.

Table 1 shows how recurring costs from various cable types accumulate, minimizing or completely eliminating any advantages of a low purchase price. The types of cable that are typically non-recoverable range from twisted-pair and Modbus to Leaky Feeder and Fiber.

Labor Expenses

System equipment cost is typically the driving factor when purchasing a system but the previous section shows how recurring costs have a significant impact on ROI. Some inexpensive wired systems (Mine Phones) have minimal initial labor but often only meet one project requirement requiring substantial additional investment to support additional requirements. While many types of cable are easily installed with the proper equipment, the installation of a modular system with minimal cable is significantly faster. Installation labor is a considerable expense but many systems like Wi-Fi (which are also cabled) also require configuration performed by a trained, technical expert. If the Wi-Fi nodes or the IP addresses aren't configured correctly, part of the system or the entire system won't work. While maintenance labor to service some cable types like twisted-pair or AC/DC power cable is relatively low,

Table 1. Cost comparison of typical tunneling cable

Technology	Price/Foot (20' Diameter Tunnel)	Total Cost of 5 Miles Non-Recoverable Cable
Fiber Cable (Cheap)	$0.20	$5,280.00
Fiber Cable (Rugged)	$1.50	$39,600.00
Leaky Feeder Cable (50 Ohm)	$2.50	$66,000.00
Leaky Feeder Cable (75 Ohm)	$1.50	$39,600.00
Modbus	$2.00	$52,800.00
Tracking System Cable	$1.00	$26,400.00

other cable types such as leaky feeder cable and fiber are quite high. Repairing these types of cable often costs more than replacement due to the expertise and equipment required. When using fiber, a hybrid architecture limits fiber damage and the associated labor costs while meeting bandwidth requirements. In addition, the labor costs for technical experts to identify problems are quite high. A system that incorporates analytical analysis with the necessary alerts significantly minimizes these types of labor costs. Time spent identifying and repairing damage affects the bottom line and diverts attention from normal operational duties.

TECHNOLOGY RECOMMENDATION AND SUMMARY

Designed Solution

As shown in this article, the best solution for most projects is designed to meet the requirements while minimizing inherent system disadvantages. By purchasing a standard system to meet one or two specific needs like voice or tracking, the inherent disadvantages lead to other technical deficiencies and unnecessary initial or recurring costs. Most tunnel operations want to determine the best system and this is accomplished by finding a company that is not an equipment provider but a solution provider.

Build As You Go

In addition to designing a solution for a particular tunnel or project, it is very important that ongoing tunnel growth is considered. If the tunnel splits, has interference, or other anomalies, system flexibility is a must. Simply providing a system to meet the initial requirements will fail to meet ongoing challenges and changes throughout the project life. The correct solution accounts for the initial requirements while providing flexibility, scalability and system modularity to meet ongoing challenges.

Budgetary Friendly

Typically, the major disadvantage of a custom solution with modular capabilities exceeds the allocated project budget. We postulate that a properly designed solution combines the strengths of typical systems and allows the modular mindset. The properly designed solution is designed around Total Cost of Ownership, maximizing workforce safety and regulation compliance.

REFERENCES

[1] Tianluan Shuo, Ke Zhao and Hao Wu (2016). Wireless communication for heavy haul railway tunnels based on distributed antenna systems.

[2] Jiro Chiba, Tatsuo Inaba, Yoshitomo Kuwamoto, Osamu Banno and Risaburo Sato (1978). Radio Communication in Tunnels. IEEE Transactions on Microwave Theory and Techniques, Vol. MTT-26, No. 6.

[3] Yao H. Ho and William W. Y. Hsu (2014). Disaster Resilient Communication for Tunnels and Bridges.

[4] IBM Corporation (2012). Managing big data for smart grids and smart meters. IBM Software White Paper.

[5] OSHA, "Underground Construction (Tunneling). Internet: https://www.osha.gov/Publications/osha3115.html," Dec. 2017.

[6] Joseph Seymour. The Seven Types of Power Problems. White Paper 18, Revision 1. Schneider Electric white paper library. http://www.apc.com/salestools/VAVR-5WKLPK/VAVR-5WKLPK_R1_EN.pdf.

[7] Chen Peng, Liu Da-Tong and Yan Dong-Hui (2014). Research of Polarization and I-FEC on Wireless Communication in Mine Tunnel. 3rd Asia-Pacific Conference on Antennas and Propagation.

[8] Nanpeng Yu, Sunil Shah, Raymond Johnson, Rober Sherick, Migguo Hong and Kenneth Loparo (2015). Big Data Analytics in Power Distribution Systems.

[9] Pan Tao and Liu Xiaoyang (2011). Hybrid Wireless Communication System Using ZigBee and Wi-Fi Technology in the Coalmine Tunnels. Third International Conference on Measuring Technology and Mechatronics Automation.

[10] Andrej Hrovat, Gorazd Kandus and Tomaz Javornik (2011) Impact of Tunnel Geometry and its Dimensions on Path Loss at UHF Frequency Band. Recent Researches in Circuits, Systems, Communications and Computers.

[11] Andrej Hrovat, Ke Guan and Tomaz Javornik (2017). Traffic Impact on Radio Wave Propagation at Millimeter-Wave Band in Tunnels for 5G Communications. 11th European Conference on Antennas and Propagation.

[12] Donald G. Dudley, Samir F. Mahmoud, Martine Lienard and Pierre Degauque (2007). On Wireless Communications in Tunnels.

SCMAGLEV Project—Fast and Innovative Mode of Transportation in the Northeast Corridor

Vojtech Gall, Sandeep Pyakurel, and Nikolaos Syrtariotis
Gall Zeidler Consultants

Cosema Crawford and Liviu Sfintescu
Louis Berger

David J. Henley
Northeast Maglev

ABSTRACT: The Northeast Corridor Superconducting Maglev Project (SCMAGLEV) entails construction of a high-speed train system between Washington, D.C. and New York City, with the first leg between Washington and Baltimore, MD. The system operates using an electromagnetic levitation system developed and deployed in Japan that achieves an operating speed of 500km/h (311 mph).

SCMAGLEV is a technically challenging but innovative project that will shorten travel times between Washington D.C. and Baltimore to 15 minutes, and eventually cover Washington to New York City in an hour. The project will enhance mobility along the Northeast Corridor (NEC) and spur development and economic growth in the region.

This paper provides an overview of the SCMAGLEV project and discusses construction of the tunnel and elevated sections.

INTRODUCTION

The SCMAGLEV Project is proposing a high speed train system between Washington, D.C. and the City of Baltimore, approximately 60 km (37 mi) in length. This is the first leg of an envisioned route from Washington, D.C. to New York City. The SCMAGLEV system operates using a combination of electromagnetic levitation (support), propulsion and lateral guidance, rather than flanged wheels, axles and bearings as in conventional high-speed railways. The train system will cross several transportation corridors including interstate highways (I-95, I-195, MD295 Baltimore Washington Parkway, I-595, I-695, I-895), several state, city and local routes, and railroad lines, as well as the BWI Airport. All crossings will be grade separated. The project owner is the Northeast Maglev / Baltimore Washington Rapid Rail (TNEM / BWWR), with Louis Berger as the prime consultant and Gall Zeidler Consultants as the tunneling sub-consultant.

The project is in the preliminary engineering phase. An independent environmental review process was initiated in the fall of 2016 in accordance with the National Environmental Policy Act (NEPA), with a Record of Decision (ROD) anticipated in mid-2019. Construction is envisioned to commence in 2020 with an estimated total cost of over $10 billion.

TECHNOLOGY

The SCMAGLEV Project will provide new infrastructure, stations and facilities for the new high speed train system. The project will build on the safety practices and culture of system developer Central Japan Railway Company (JRC), which has operated the Tokaido Shinkansen bullet train between Tokyo and Osaka without a single fatality since 1964. JRC applied a similar safety approach to the development of the SCMAGLEV system. SCMAGLEV technology was fully evaluated by the Japanese government, and the system is currently operating on an approximately 27 mile long segment that is being extended to connect Tokyo and Nagoya. Safety systems for the Baltimore-Washington Maglev project will be developed through a collaborative process with the FRA Office of Safety and local emergency response forces.

The primary elements of the project include superconducting magnetic levitation rolling stock and systems using a proprietary technology developed by JRC and two guideways, one in each

Figure 1. Conceptual rendering of the SCMAGLEV

direction, borne by tunnel and viaduct structures. The system deploys technologies that are new to the U.S. or of previously limited application, including most notably an electromagnetic propulsion system. This technology is capable of accelerating trains to a top cruising speed of 500 km/h (311mph) in two minutes and allows for a driverless train operation. The total trip time from Washington D.C. to New York will be one hour, while a Washington to Baltimore trip will take 15 minutes with a stop at BWI Airport.

ALIGNMENT ALTERNATIVES

The project is located in Washington, DC and Maryland, traversing a distance of approximately 60 km (37 mi) with three underground stations in Washington D.C., at BWI Airport and in Baltimore.

The SCMAGLEV system runs on an independent grade-separated right-of-way. The ultra high speeds require relatively straight geometry with limited horizontal and vertical curvature. To accommodate the range of topographical and surface features, existing dense urban areas, utility mains, and existing structures, the proposed construction is expected to consist of below-ground (tunnel) for at over half of the route, and elevated structures (viaduct) for the remainder. The train system incorporates two main guideways, three stations, one rolling stock depot, electrical substations, tunnel ventilation plants and emergency egress facilities.

The environmental review process initially identified several possible alignment alternatives which generally follow existing transportation corridors such as Baltimore Washington Parkway (MD 295), Amtrak Northeast Corridor, Washington, Baltimore and Annapolis Trail or a combination thereof, as shown in Figure 4. Alignment alternatives have since been further screened to two, which traverse the eastern and western sides of the BW

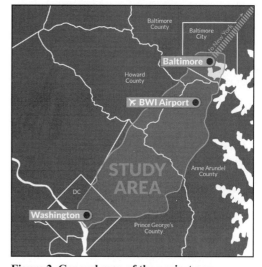

Figure 2. General map of the project area

Parkway. A preferred alternative will be identified in the Draft Environmental Impact Statement in late 2018.

Overall, approximately 75% of the alignment is anticipated to be in tunnel and the remaining 25% is on elevated viaduct.

Underground station locations in Washington and Baltimore are being assessed. The BWI Airport station will be located under the terminal building with direct access to the terminal and parking facilities. Stations will have a platform length of approximately 300 m (980 ft.), and total cavern length of up to 900 m (2,950 ft.) including switch chambers and other ancillary facilities. Stations will accommodate four guideways and two platforms, with a total width of up to 45 m (150 ft.).

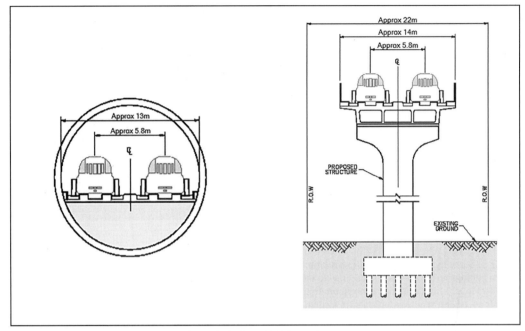

Figure 3. SCMAGLEV in tunnel (left) and viaduct (right) sections

GROUND CONDITIONS

The proposed alignments are located within the Coastal Plain Physiographic Province, consisting of relatively soft strata. These strata lie on top of crystalline bedrock and thicken to the southeast on the order of approximately 150 m (500 ft.) per 8 km (5 mi). These sedimentary deposits include clays, sands, and gravels with younger sediments composed of sand, silts and muds. In general, the soil profile along the alignment consists of fill, residual soils, loose granular soils and clays with interlayering and lenses along the alignment.

Groundwater conditions are expected to vary widely across the alignments, from dry conditions to groundwater levels ranging from relative shallow depths of less than 3 m (10 ft.), to depths in excess of 12 m (40 ft.). Fluctuations in groundwater levels across the alignment will occur seasonally due to variations in rainfall, evaporation, construction activity, surface runoff and proximity to adjacent streams and the Chesapeake Bay shoreline. Localized perched groundwater and isolated water-saturated sediment lenses can also be expected. Connectivity of the aquifers to rivers and creeks has been identified in various locations.

Construction is expected to occur primarily in soft ground for most of the alignment. The bedrock is deeper along the central part of the alignments and is expected to become shallower at the two ends in

Washington D.C. and Baltimore, although locally higher sections of the bedrock cannot be excluded.

A preliminary geotechnical exploration program is underway, comprising 24 borings and geotechnical testing. The program intends to investigate the soil formations and delineate the bedrock level, obtain ground properties to assess ground behavior along the alignment corridor and identify areas for further geotechnical investigations.

ELEVATED VIADUCTS

Elevated viaducts are proposed for each alignment in portions where development is less dense. A single viaduct structure approximately 14 m (46 ft.) wide will carry two guideways. The structure will be built with precast concrete superstructure elements supported on concrete piers with pile foundations. The typical span of the viaduct structure will be 30 to 35 m (100 to 115 ft.). Longer spans will be used at locations where the SCMAGLEV crosses waterway features or existing infrastructure.

TUNNELING

The proposed alignments include up to 48 km (30 mi) of deep tunnel sections. Considering the length of the tunnel sections and the required uniform geometry, it is anticipated that deep mechanized tunneling will be implemented for the majority of the alignment that will need to address the following challenges:

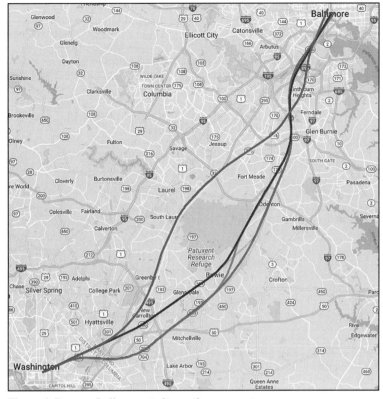

Figure 4. Proposed alignment alternatives

- Mostly tunneling in soft ground
- High groundwater level
- Tunneling across urban areas and under major roadways

Considering the soil types and groundwater conditions expected along the deep tunnel sections, which require an active face support, the use of a closed face Tunnel Boring Machine (TBM) will be required. Based on the available preliminary information on the geological and hydrogeological conditions and the critical impact of groundwater to the tunneling activities, implementation of an Earth Pressure Balance Machine (EPBM) is considered, at this stage, most appropriate for the anticipated subsurface conditions. Alternatively, a Mix Shield TBM could be considered, as the alignment could encounter sections of mixed geology with hard rock potentially shallower at the two ends of the alignment. The information acquired from the additional ground investigation program will be used to evaluate and select the TBM type and refine specifications.

TBM tunnels in soft ground are generally supported by pre-cast segments, which are erected at the tail end of the TBM producing a continuous lining over the tunnel length with a circular, uniform geometry. Segmental linings will be equipped with gaskets in the joints between the segments to inhibit groundwater inflow into the tunnel.

To minimize the construction footprint of the project, reduce surface disturbance and construction impact, and taking into consideration the spatial requirements for the train operation, a single bore TBM tunnel with an outside diameter of approximately 15 m (50 ft.) is being considered as optimal compared to twin bore tunnel configuration. Although tunneling with a large bore TBM is a challenge in itself, the technology and capabilities of present day TBMs allows for unimpeded tunneling and enhanced risk management. Additionally, TBM tunneling will be performed under at least one tunnel diameter of ground cover to minimize surface impact.

Subdivision of the TBM tunnel alignment into sections with a length of 5 to 6 km (3 to 4 mi) is currently considered to enable concurrent boring along various sections and provide flexibility for contracting and packaging of the project. Each TBM requires construction of a launch site, which is typically a cut-and-cover structure. In areas where space restrictions do not allow for construction of launch boxes, launch shafts of adequate size will be considered as an alternative. Ventilation shaft sites are planned to be used as launch shafts where possible to minimize cost and streamline construction. As the launch sites will be also used for stockpiling of the spoils, implementation of multiple launch sites along the alignment will allow efficient storage and transport of the spoils to the areas designated for disposition.

Short sections of cut-and-cover tunneling will be used for the stations, for the transitions between the viaduct and TBM tunnel sections and for TBM launch locations. Implementation of cut-and cover tunneling will involve installation of support of excavation, such as slurry walls, bored pile walls, soldier pile and lagging or shotcrete. Depending on the limits of disturbance, generally tie-back support or internal strutting is expected for deeper excavations. A waterproofing system will be installed to prevent groundwater inflow into the tunnel in the final permanent stage. The dense urban environment in Washington D.C. and Baltimore, coupled with the relatively deep alignment, will necessitate measures for construction of the stations with minimal surface impact and disruption to the city activities. Similarly, construction of the station under the BWI Airport will require significant coordination and phasing to avoid disrupting airport operations.

FIRE AND LIFE SAFETY

Design, construction and operations for the SCMAGLEV will be planned with a safety focus: safety of the traveling public, the construction and operations workforce, and the adjoining communities. These elements will be addressed in the planning and design of the infrastructure, core systems, facilities, and operating and maintenance practices for the SCMAGLEV system.

Fire and life safety will be given full consideration in all aspects of the system design, including linear infrastructure (viaducts and tunnels), passenger stations and operations and maintenance facilities. The fire and life safety considerations include elements and layout of egress and access paths in the tunnel system; definition of design fires for vehicles, cables, etc.

An emergency egress path for passengers to a point of safety will be provided in the underground sections. The proposed tunnel cross section allows the use of the space below the guideways as an emergency evacuation chamber. Additionally, ventilation plants and shafts are envisioned along the underground section of the alignment.

Due to the unique characteristics of the SCMAGLEV system, safety systems and practices researched and developed by JRC specifically for the SCMAGLEV system will be proposed to FRA Office of Safety for incorporation into the SCMAGLEV project in the U.S. to ensure that the highest standards for safety are deployed.

OUTLOOK

The Baltimore-Washington SCMAGLEV project will dramatically transform the region by reducing travel times between the two cities by a factor of two to four times. The proposed extension to New York will add stations in Wilmington and Philadelphia, and connect BWI Airport to the Philadelphia and Newark Airports, enhancing mobility and spurring growth and economic development. Construction is anticipated to start in 2020, with the start of operations later in the decade.

TRACK 1: TECHNOLOGY

Session 3: TBM Technology 1

John Criss and Taehyun Moon, Chairs

Downtime Data Collection, Analysis, and Utilization on Blacklick Creek Sanitary Interceptor Sewer Tunneling Project in Columbus, OH

Amanda Kerr and Max Ross
Michels Corporation

ABSTRACT: Budgetary and schedule related success in tunneling projects are often directly related to maximizing Tunnel Boring Machine (TBM) availability. This can be accomplished by increasing the amount of time spent on tunneling production, while utilizing resources to avoid known causes of downtime. It is very likely to encounter unforeseen conditions in tunneling, therefore preventable factors should be controlled as much as possible to maximize performance. The examination of opportunities to reduce downtime creates a potential for significant cost savings applicable to other tunneling operations. This paper will discuss the downtime analysis performed on the 3.7 meters (12 feet) diameter Herrenknecht Earth Pressure Balance (EPB) TBM used for the excavation of the Blacklick Creek Sanitary Interceptor Sewer Project in Columbus, OH.

INTRODUCTION

The Blacklick Creek Sanitary Interceptor Sewer (BCSIS) extends the existing 1.7 meters (66 inch) sewer with a new 3 meter (120 inch) gravity sewer from Blacklick Ridge Boulevard to Morse Road. This extension is required to support the development and services to the City of New Albany and Jefferson Water and Sewer District. Furthermore, this extension will enable the future connection between Rocky Fork Diversion to redirect sewer flow from the Big Walnut sewershed into the Blacklick sewershed.

The tunnel alignment initiates at the existing Manhole #12 (MH-12) outlet, just south of the Blacklick Ridge Boulevard and east of Reynoldsburg-New Albany Road. The tunnel extends in a northerly direction following the approximate alignment of Reynoldsburg-New Albany Road towards Morse Road (see Figure 1). The total length of the alignment is 7016 meters (23,020 linear feet), with 6894 meters (22,620 linear feet) to be constructed using an Earth Pressure Balance (EPB) Tunnel Boring Machine (TBM) and the rest by hand excavation. This project adds a new junction chamber at MH-12 south of Blacklick Ridge Boulevard. The tunnel will end upstream at the intersection of Reynoldsburg-New Albany Road and Morse Road, at a drop structure to accept future flow from the New Albany, Rocky Fork, and City of Columbus service areas. Along the sewer alignment, there are two intermediate shafts with permanent drop structures for future tie-in from the Jefferson Township service area, as well as for tunnel maintenance purposes. In addition to their formal purpose, these shafts serve as access points for TBM maintenance during the tunnel construction. Other associated work includes construction of connection pipes, manholes and ancillary facilities necessary for the operation and maintenance of the sewer. Blacklick Constructors LLC (BCL) was the lowest of four bidders on November 18, 2015 with a bid price of $108.9 million. The engineer's estimate for this project was $113.7 million.

GEOLOGY

A variety of geological conditions were encountered within the initial project phase. The hand tunneling and shaft excavation encountered fill and alluvial soils. The tunneling excavation in soil zones generally consisted of glacial till and outwash deposits, which are significantly weathered and consolidated. The transition between the soils and bedrock generally consisted of highly weathered to decomposed rock. Glacially-deposited erratic boulders were also anticipated along the transition zone. There were two major bedrock formations encountered during tunneling and shaft excavation, the Cuyahoga and Sunbury Shale. The Cuyahoga Formation has been subdivided into the Upper and Lower Cuyahoga with the Upper Cuyahoga containing fine sandstone and siltstone. This sand and siltstone has been shown to be absent from the shale of the Lower Cuyahoga. The contract documents classified the ground into: Fill, Alluvial, Glaciolacustrine, Glaciofluvial, Glacial Till, Transition material, Upper Cuyahoga, Lower Cuyahoga, and Sunbury Shale. These classifications were grouped into three main groups: soil, rock, and mixed ground. The tunnel alignment had an

Figure 1. BCSIS tunnel alignment

expected composition of 39% soil, 46% rock, and 15% mixed ground.

PROJECT LOGISTICS

In order to address the varied geology of this project, a 3.7 meter (12 feet) diameter EPB TBM manufactured by Herrencknecht was selected. The hydraulically driven TBM utilized a universal six piece precast segment ring as final tunnel lining. The cutterhead was designed to fit the requirements of both soil, rock, and the risks of boulder clusters. In conjunction, the 113 meters (370 ft) trailing gear followed on wheeled gantries and contained major components such as a two component grout system, a soil conditioning system, assorted grease pumps, an electrical transformer, and a utility installation deck.

In addition to the TBM and its support gear, BCL installed and commissioned all of the project ancillary specialty equipment and features at the Shaft 1 site. Relevant supplemental items included a breathable compressed air plant and a medical lock for hyperbaric interventions, water treatment systems, a substation and electric switchgear (to support the tunnel boring machine), Sagami Servo automated grout plants, a water cooling system, a muck pit and motion-activated wheel wash.

BCL utilized muck cars for material handling in the tunnel and a lattice boom crane for removal from the main shaft. Three California switches were planned for specific locations along the alignment to optimize tunneling operations. The phases of production were defined by the installation of each additional switch.

Project Systems

As with any large tunneling project, systems, locations, and equipment often interact in a complex manner. Systems are not necessarily restricted to a single location and can operate in multiple directions. Figure 2 shows the interaction between the physical components/equipment and their relationship to location and project system.

System interactions are vital to tunneling operations. The TBM advance system encompassed all pumps, motors, thrust cylinders, and articulation cylinders directly involved in the mining portion of the production cycle. In addition, the TBM advance system included equipment for primary muck haulage, from the working face to muck cars, as well as pumps involved with lubricating. Fully integrated into each function of the TBM, the Programmable Logic Controller (PLC) monitored and controlled a wide variety of sensors, valves and pumps. The TBM operator directly utilized this system to control mining operations. The grout system incorporated the surface grout batching plant to make the grout and steel conveyance pipe connecting to a secondary storage tank on the TBM. The two components of the grout were then mixed and injected by the use of progressive cavity pumps into the shield gap. The segment handling system included all equipment required to move the segments off the segment car to the segment feeder and to install segments by utilizing the erector. The tunnel support system included all necessary equipment and material to continually service and supply the production equipment and to haul muck away from the working face to the surface muck pit. The power supply system used a substation to provide constant 13.2 kilovolt power to the TBM.

Due to the limited clearances of the tunnel, the guidance system made use of a gyroscope and hydrostatic level to provide a location and azimuth for the operator. While not a constitutive component of the production cycle, the data collection and communication system provided valuable benefits for production optimization. The data collection system collected information from the guidance system and the TBM PLC and conveyed it via fiber optics to surface servers for future evaluation. The communication system facilitated contact between personnel working in each production zone with mine phones, radios via leaky feeder system, as well as a single direct phone line to the TBM operator cab. The personnel "system" referred to the consideration of the human needs as an arranged structure rather than as a composite group of unique individuals. This allowed for a more objective approach to resource management when viewed as a component, as personnel must be transported and injuries must be avoided.

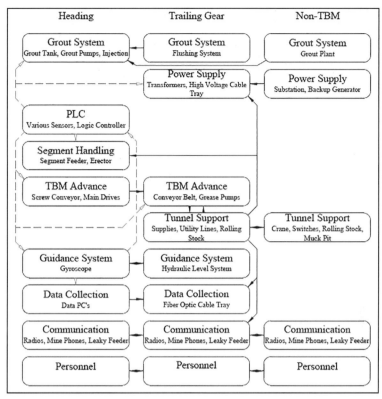

Figure 2. Project system interaction map

While efficient production required a close coordination of all systems present, some systems affected the TBM performance more directly than others. As shown in Figure 2, the tunnel support and PLC systems were more critical to production than the other systems. For example, the PLC controlled and monitored most of the advance related operation, and the locomotives carried all muck and materials in and out of the tunnel. Any issue with either of these two systems completely disrupted the production.

METHODOLOGY AND DATA CLASSIFICATION

Data Acquisition

All data points were collected by a heading engineer on the TBM, described to the best of his or her knowledge. This lowered the precision of the data by a few minutes per incident, but allowed for a deeper level of understanding compared to data strictly collected by Data Acquisition Software. The dates considered in this paper were selected to avoid the effects of startup operations and periods of non-mining days. Startup operations included initial mining completed

before the installation of the first switch, without the standardized support system utilized in the three main production phases. This study focused on the data from the first 1524 meters (5000 feet) of mining from shaft 1 to shaft 2 (82 working days). This period will be referred to as Phase One. The data was collected from a shift narrative which detailed the elapsed time associated with each event from the start of the shift until the engineer exited the tunnel at the end of the shift. For Phase One, the downtime data was comprised of a unique written description of each event based on the engineer's interpretation. The data collected for this analysis was retroactively classified to comprehensively describe the events experienced throughout Phase One. To increase the standardization of the data collection process this system was then utilized to create distinct menu options for future data collection in Phases Two and Three as a supplement to the unique description.

Preventable Downtimes

The analysis for this project aimed to minimize repeated occurrences and classified incidents through the use of the following definitions:

Table 1. Typical downtime events by project system

System Name	Description
Grout System	Blockages, pump failures, low supply, cleaning processes
Power Supply	Add on high voltage cable, power outages
Programmable Logic Controller (PLC)	Emergency stops, gas sensor alarm, liquid low level warning
Segment Handling	Segment erector malfunction, segment feeder malfunction, segment damage prior to installation, build area blockages
TBM Advance	Cutterhead inspections, grease pump malfunction, conveyor belt damages and maintenance, hydraulic system maintenance, screw conveyor malfunction
Guidance System	Software error, extended gyroscope start up, hydraulic level sensor malfunction
Tunnel Support	Crane malfunction, rolling stock derail, weather, extended utility install, rolling stock arrival lag, high muck volume
Data Collection	Add on fiber-optic cable, data communication error
Communication	Add on mine phones and cable, add on leaky feeder cable
Personnel	Extended safety meetings, extended travel time, unplanned cleaning

- A *Preventable* event was a period of downtime recorded for an activity that could have been mitigated without the halt of production.
- A *Non-preventable* event was a period of downtime recorded for an activity outside the regular project controls.

The preventability of the downtime was the primary method to distinguish which downtimes were controllable. By focusing on these controllable events, opportunities to minimize procedural and routine downtimes were identified.

TBM Availability

Through the incorporation of the preventability of an event, TBM availability was defined as the total hours worked without the time spent on events classified as non-preventable. This allowed for a more direct analysis of the effect of preventable downtime events on the TBM availability ratio, shown in the equation below.

$$A = \frac{E_{Production}}{E_{Production} + D_{Preventable}}$$

A = TBM Availability
$E_{Production}$ = Recorered Production, in hours
$D_{Preventable}$ = Recorded Preventable Events, in hours

Event Classification by System

Downtime events were primarily classified by the project system (illustrated in Figure 2) that was most directly affected (see Table 1).

In addition to system classifications of downtime, additional groupings were utilized to further isolate events in a consistent and precise manner. Tables 2 and 3 show other categories used in conjunction with the project system classifications. The physical location of a downtime event led to an increased awareness of the critical points within the project system as it affected the production cycle. The physical regions identified as the main zones in which downtime events initiated are denoted in Table 2.

Defining an issue as intrinsically mechanical or electrical does not provide enough detail to identify and amend a problem, Table 3. However, describing whether the issue stemmed from a mechanical or electrical component of a project system allowed for a more comprehensive representation of the issues.

This system of multilayered classification allowed for the consistent and precise isolation of issues with the intent of making the root cause of the events more apparent. For example, a recorded production stoppage caused by the routine change of a hydraulic filter of the TBM main drive would be classified as a preventable mechanical issue pertaining to the TBM advance system that took place in the heading zone.

RESULTS AND DISCUSSION

Several trends were identified in the analysis of the collected data. These trends were used to support the implementation of TBM availability optimization strategy. The most significant trends are discussed in the following figures. As shown in Figure 3, the average amount of reported downtime per week declined as the project advanced through Phase One, despite a few outliers. Overall, it is thought that this decrease in downtime was due to improved skill levels of

Table 2. Event location description

Location	Description
Heading	The downtime event initiated in the heading of the TBM (TBM Shield Body)
Gantries	The downtime event initiated in the gantries of the TBM
Non-TBM	The downtime event initiated in a location outside of the TBM (the tunnel, shaft, or surface)

Table 3. System component category

Category	Description
Electrical	Root cause of downtime event was an electrical component
Mechanical	Root cause of downtime event was a mechanical component

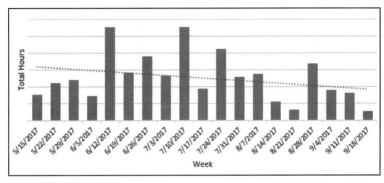

Figure 3. Recorded downtime per week

production tasks, as well as heightened awareness of systematic issues.

Preventable and Non-Preventable Events

As stated earlier, identifying the controllable downtime events allowed for better interpretation of data. As shown in Figure 4, there was a high amount of variance per day in the recorded amount of downtime. Routine preventable downtime events maintained a steadier occurrence, while non-preventable event duration varied dramatically from the verge of non-existence to significantly higher impact durations due to their unforeseen nature. The Non-preventable events, at 57% of total recorded downtime, clearly had an impact on production. However, discussion and implementation of measures to mitigate the impact of these events defined as non-preventable occurred on a less frequent basis than that of preventable occurrences. Mitigation measures of non-preventable issues required significant advanced planning and changes to project controls and setup. As an example, changing the shift start times would require coordination with the employees, unions, etc. which cannot be done in a short period of time. As such, preventable events were a better representation of the direct effects of regular downtime analysis. Preventable events accounted for

43% of the total recorded downtime in Phase One, and were the main focus of the study.

Figure 5 serves to demonstrate the number of times weeks with similar total preventable downtime were experienced, however the specific hours have been withheld to preserve proprietary knowledge. As shown, the recorded preventable downtime per week proved to be nearly normally distributed. This demonstrated that the distribution of the recorded downtime durations were well-varied within an upper and lower boundary, rather than clustering around an approached limit. This distribution indicates that improvements are still able to shift the average weekly downtime towards the lower limit. As the actual production was still well within our estimated rate, this distribution proved an invaluable management tool because it visually proved that performance could still have been improved during periods of high production.

Allocation of Preventable Downtime

Within these recorded preventable events, the allocation of downtime was most clearly viewed by the primary system classification. As shown in Figure 6, Phase One preventable downtime was dominated by occurrences within the Support system. This could have been skewed due to the uneven importance of the systems. For example, it was not possible to advance without first receiving and constructing the previous ring, but while not advisable, it was possible to mine without the data collection system. Neglecting the relative importance of system, the top three classifications of downtime made up over 90% of reported preventable downtime. BCL conducted weekly management meetings in which prevalent downtime events that occurred throughout the previous week were discussed to allow for a focused reaction to recorded events.

Tunnel Support encompassed all aspects of the supply chain system, and based on this data analysis was responsible for 81% of reported preventable downtime. The performance of this system was very vulnerable to human error. For example: Loci operator proficiency, the learning curve associated with the addition of a California switch, and the increased understanding required to successfully anticipate

and communicate the support needs of the TBM all had a dramatic impact on this category. As a result of these findings, it was demonstrated that further analysis was required on this specific project system. As such, additional data acquisition and studies have been performed to fine tune the project controls within this system.

While EPB TBM mining contains many automated components, the personnel remains a necessary aspect of the process. However, human error affected production of project systems outside of their direct operation, and at 6% was the second largest contributor to preventable downtime. The consideration of human needs and shortcomings as a system rather than as a composite group of individuals allowed for a more objective approach to resource management. The personnel category showed the effects of pre-shift meetings, discussions, and travel time to the TBM that extended

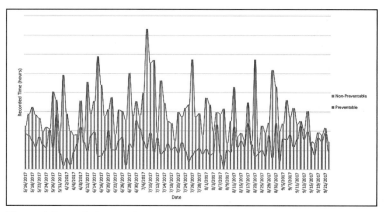

Figure 4. Recorded downtime per day

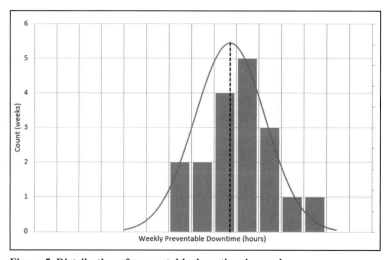

Figure 5. Distribution of preventable downtime by week

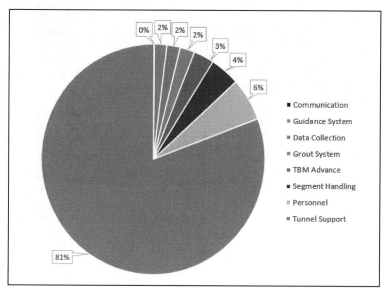

Figure 6. Allocation of preventable downtime by primary system classification

beyond the designated time allotment. In addition, daily safety discussions and pre-shift stretching were deemed essential, and a set duration for this activity was included as part of the production process. The downtime in this category was mitigated by mindful efforts to keep non-essential meetings brief, and the start and end of shift process and travel to and from the heading as efficient as possible.

Segment Handling referenced the process from the unloading of segments within the gantries to their subsequent installation. The productivity of this system was impeded by a range of factors, most notably the performance of the erection controls and cleaning the ring erection area. Through specific communication with operators, it was discovered that the segment lift and erector system had been experiencing frequent short delays. While only a few minutes per instance, this specific issue contributed to the majority of the 4% of the total downtime caused by the segment handling system. As a result, the root cause was investigated and a system update was coordinated with the TBM manufacturer. The update worked to avoid unnecessary lockouts, and gave more control to the operator. The identification of minor issues and the implementation of solutions contributed to an increase in production.

Location

The physical location of the downtime events provided an understanding of where improvement efforts should be focused. The analysis showed that 51% of the preventable downtime occurred in gantry

zones while 49% of non-preventable events occurred in the heading. The high level of cyclical activity and close proximity of equipment in the trailing gear led to an increased risk of preventable downtime events. However, downtime events in the heading zone stemmed from production essential components of each system and were primarily non-preventable, Figure 7.

Through the use of location-based analysis, efforts were made to strategically re-order and organize cyclic processes that took place outside of the heading zone. The low concentration of preventable events in the heading demonstrated that TBM optimization efforts had been moderately successful. While non-preventable events in the heading were not able to be eliminated by nature, a comprehensive maintenance program focused on the components of each system most at risk was implemented to decrease frequency and severity. As all zones contributed similarly to the total downtime experienced in this phase, the analysis demonstrated that each zone carried a similar proportion of the overall risk to production.

Electrical vs. Mechanical Trends

Demonstrated in Figure 8, downtime events electrical or mechanical in nature experienced clear trends. The peak of the electrically initiated events correlated to the previously described system updates to the segment handling system and the subsequent decline from a higher awareness of electrical issues. Better reporting procedures were implemented at the height of electrically caused downtime and the success in mitigating downtime reinforced the importance of heightened communication. The project electricians consistently conveyed information between shifts, as well as with electrical supervision staff. Prompt and thorough communication played an important role in the reduction of downtime.

TBM Availability

The overall goal in downtime analysis was to optimize tunnel operation and production. As seen in Figure 9, the overall availability of the TBM was 82% on average within the first 1524 meters (5000 feet) of mining. Although there were opportunities to optimize the operation, the overall experienced downtime was well within the acceptable range throughout Phase One.

Many factors contributed to the TBM availability experienced each week and even small fluctuations to this ratio affected the rate of production. Even in the initial weeks of regular production availability remained close to 80% and trended upward in general. The availability was dependent on ground

type, as the support system emptying muck cars could not perform as fast as the TBM was able to mine in soft ground and the relative location from trailing gear to the California switch. The distance from the end of the gantries to the switch approached a limit in September causing a lag in the arrival of the rolling stock.

In general production rates showed positive growth but experienced non-uniform behavior depending on the availability and ground type. This can be clearly seen in the week of June 12; the availability remained over 80% but production was lower due to extended mining times and a need for an additional muck car. In rock the availability percentage and production rate have a much larger gap than in soil. Higher production levels in soil was more directly correlated to higher mining speeds, with the limiting factor on production being the support system. In rock, mining speeds were slower, in addition to the complications caused by high quantities of water or strengths of rock higher than anticipated.

Production rate was also associated with the continuity of mining activity. The non-mining days

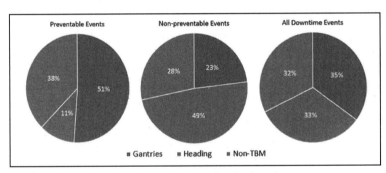

Figure 7. Allocation of preventable downtime by location

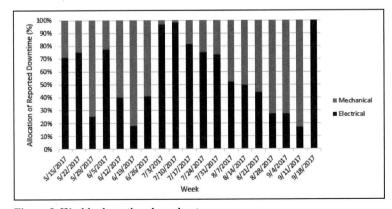

Figure 8. Weekly downtime by subcategory

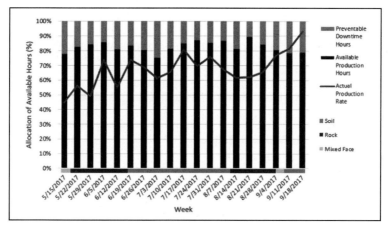

Figure 9. Weekly TBM availability

were not included in this analysis of tunneling procedures but the shifts leading up to and following planned shut downs experienced localized lower levels of production, as can be seen in the weeks of August 14th and August 21st. The implementation of improved communication, system updates, ground change to soft ground, and the heightened realization and expectation of production capabilities culminated to rapidly increase production. As anticipated, hitting production stride on a tunneling project took time; it required almost the entirety of Phase One to reach this period of acceleration in production. Consequently, it was imperative to identify what was hindering production capabilities and promptly mitigate downtime.

CONCLUSION

This paper summarizes the methodology that was used for downtime analysis on the Blacklick Creek Sanitary Interceptor Sewer Project in Columbus,

OH. In this paper, the downtime has been divided into two main categories, preventable and non-preventable events. The preventable downtime category was then sub-divided into TBM system, location, and electrical/mechanical categories. This analysis identified primary issues, such as the drawbacks of the tunnel support system, and directed efforts to amend controllable factors. Mindfulness of the impact of short downtimes led to increased availability and production. Improved communication of issues both routine and unexpected provided a basis for mitigating preventable downtime. This downtime analysis was implemented at the start of tunneling, and while discussion of downtime was a positive influence on availability and production, the specific feedback from this analysis has been given to the field management personnel to further optimize the operation. Based on the positive feedback of this study a more systematic supplement to the description of downtime events has been implemented for Phases Two and Three and will be discussed in future studies.

EPB TBM Performance Prediction on the University Link U230 Project

Mike Mooney, Hongjie Yu, Soroush Mokhtari, Xiaoli Zhang, and Xu Zhou
Colorado School of Mines

Ehsan Alavi, Lisa Smiley, and William Hodder
Jay Dee Contractors, Inc.

ABSTRACT: There has been significant effort over more than 20 years to predict advance or penetration rates of hard rock TBMs, but very little effort (and no publications) on advance rates of soft ground pressurized face EPB or slurry TBMs. This is due to the complexity that the pressurized excavation chamber and screw conveyor or slurry transport material discharge creates. This paper tackles EPB TBM performance prediction by considering the process of soft ground excavation and by learning from real project data, specifically using the Seattle University Link U230 tunnel project data.

INTRODUCTION

Pressure balance TBM tunneling, particularly earth pressure balance (EPB) TBM tunneling, has experienced significant growth worldwide. EPB TBMs are routinely used in urban soft ground environments for water and wastewater, transit, traffic and energy tunnels. The now-closely monitored and controlled pressure balance exerted by EPB TBMs allows for near zero deformation tunneling in congested infrastructure-laden urban environments.

The effective operation of EPB TBMs is quite complex. The fact that the in-situ soil and rock, often conditioned with foams, polymers and bentonite, is itself used to counterbalance the earth and water pressures, and control ground deformation, presents a considerable challenge. Maintaining the appropriate pressure balance for deformation control, guiding the TBM along the design tunnel alignment using thrust and articulation jack groups, scraping and imbibing the in-situ and often abrasive ground, and processing the muck through a de-pressurizing screw conveyor all combine to make the performance prediction of EPB TBMs quite challenging and complex.

Accordingly, there is no well-established formula for EPB TBM performance prediction or framework for achieving optimal advance rates (rate of penetration) of EPB TBMs. There has been extensive research to predict advance rates (aka penetration rates in the rock tunneling/mining community) in open mode hard rock TBMs, but no such models exist for EPB TBMs.

This paper addresses performance prediction (namely advance rate) of EPB TBMs through the lens of the metro-size University Link U230 tunnel project carried out in Seattle, Washington (Frank et al. 2013, Gharahbagh et al., 2013). The physical processes of EPB TBM tunneling that influence advance speed are introduced and discussed, as is the interplay between the physical processes. The TBM data is then presented for both northbound and southbound drives. We then employ artificial intelligence to learn about advance rate from the TBM data.

PROJECT BACKGROUND

The University Link (U 230) tunnel project, constructed by the Jay Dee–Coluccio–Michels joint venture, included 1.2 km (3800 ft) of twin tunnel construction with an excavated diameter of 6.44 m. The tunnels are located within areas of glacial and non-glacial sediments in Seattle, Washington (Northlink Tunnel Partners 2009). The geological profile along the tunnel is shown in Figure 1. The ground cover above the tunnel crown ranges from 4 m (13 ft) to about 40m (130 ft) with varying ground water pressure at the tunnel level (GWT varied from 0 to 40 ft). Exploration borings show at least three types of deposits from both glacial and non-glacial cycles, including clays (blue soil group), silts (turquoise soil group), and sand and tills (yellow/yellow 2 soil group) in various combinations, densities and consistencies. The soil properties of the U230 project are summarized in Table 1.

Tunneling began in mixed face sand and gravel/ sand and clay for 75 rings (each ring 1.5 m (5 ft) long), transitioned briefly into full face sand and gravel (30 rings) and hard clay (40 rings), and then

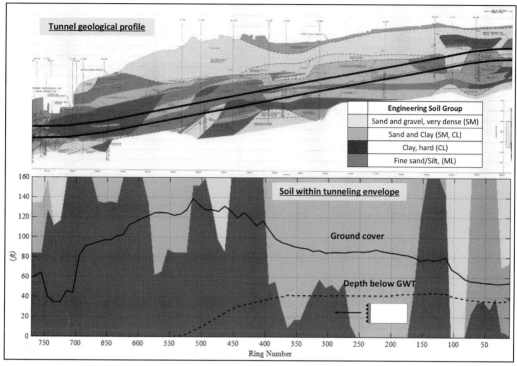

Figure 1. Tunnel geological profile (top, from Irish, 2009) and geology within the tunneling profile (bottom). Ground cover and groundwater table elevation above crown shown. Tunneling conducted right to left.

into mostly low plasticity silt with some clay for 200 rings. The second half of the tunnel excavation was mostly through the low plasticity hard clay soil with some areas of mixed ground with clay and silt or sand.

EARTH PRESSURE TBM BACKGROUND

A 6.64 m (21.1 ft) diameter Hitachi Zosen (Hitz) EPB TBM was used to excavate the 1.1 km (3800 ft) twin northbound and southbound tunnels in 2011 and 2012, respectively. Both tunnels were excavated beginning from the Capital Hill station. The Hitz TBM included forward and rear bodies with articulation jacks connecting the bodies. The TBM is shown in Figure 2. Sixteen thrust jacks in four groups (see Figure 2) pushed off the concrete segments to advance the TBM. The forward end of the thrust jacks was pin-connected to the forward shield. Each thrust jack could exert a 2500 kN force, providing a total thrust capacity of 40,000 kN. The cutterhead was outfitted with knife edge bits (pre-cutters) and scrapers only (no disc cutters). The opening ratio was 45%.

The TBM include six chamber pressure sensors mounted to the bulkhead at three elevations (see Figure 2). External shield pressure sensors were installed in the 11 and 5 o'clock positions on the forward and rear bodies to measure the external total pressure. Grout pressure was measured in three injection lines in the tail shield body as shown. Eight variable frequency electric drive (VFD) motors worked in parallel (one lead, remainder follow) to rotate the cutterhead. The equipped cutterhead torque capacity was 3107 kN-m at a max cutterhead rotation speed of 2.2 RPM. A 800 mm diameter two-piece ribbon screw auger was used to excavate material from the excavation chamber onto a belt conveyor (Figure 3). The length of screw conveyor sections 1 and 2 are 11.9 m and 16.9 m in length, respectively, and are inclined at 18° and 3°, respectively, from horizontal. They are each driven by 80 kN-m capacity torque motors capable of a peak rotation speed of 18.3 rpm.

TBM Operation

Understanding TBM operation is important to data processing and parameter selection for performance analysis. During ring build, the stationary TBM keeps a positive chamber pressure and thrust force. This maintains the needed face pressure to counterbalance the lateral earth and water pressures. The cutterhead

Table 1. U230 Average soil properties (Geotechnical Baseline Report, 2009)

Soil Group/USCS Classification	Sat. Unit Weight (kN/m³)	Effective Friction Angle, $\phi'(°)$	At-Rest Lateral Earth Pressure Coefficient, K_o	Cohesion, c' (kPa)
Purple/Hard clay (CL)	20.1	33	0.8	13
Turquoise/Fine sand/silt (ML)	19.6	37	0.8	0
Tan/Sand/Gravel (SM)	19.9	39	1.2	23.9
Yellow/Sand/clay (SM, CL)	19.9	39	0.37	0

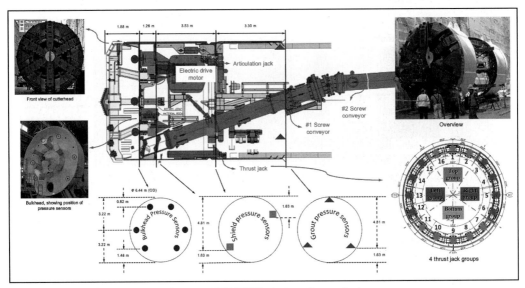

Figure 2. Hitachi-Zosen 6.44 m diameter EPB TBM used on the project

is not rotating during ring build, and therefore, the cutterhead torque is zero. The screw conveyor gates are closed and the screw conveyor auger rotation speed and torque are zero. When beginning the next excavation cycle, the cutterhead rotation speed increases until reaching the desired value, here on the order of 1.6–2.2 rpm. Cutterhead rotation speed is held constant through feedback control; the cutterhead torque varies to maintain the set rotation speed. The operator may change the desired rotation speed, and the servo-control system will adjust the torque accordingly. There are numerous other reasons why the torque will increase or decrease (see Godinez et al., 2015). When the desired cutterhead rotation speed is reached, the operator increases the thrust forces and the TBM begins to advance. Here, the operator manually inputs the hydraulic thrust pressures to each jack group; the operator does not set a speed. TBM advance rate (speed) is a system output, a result of many aspects discussed in this paper.

The operator initiates screw conveyor auger rotation and screw gate opening as the TBM begins to move forward. Similar to the cutterhead, the operator sets a desired screw conveyor auger rotation speed

and the servo-control motors deliver the required torque to achieve that speed. Screw conveyor rotation speed is adjusted by the operator based on a desire to achieve and maintain a target chamber pressure. The operator will increase screw conveyor rotation speed to decrease chamber pressure and decrease conveyor rotation speed to increase chamber pressure. To this end, the operator aims to maintain constant material flow through the excavation chamber. If the TBM advance rate increases then the operator must increase the screw conveyor rotation speed to maintain constant excavation chamber material flow.

Soil conditioning via foam and/or liquid distribution, to the cutterhead, chamber and/or screw conveyor are critical aspects of EPB TBM operation. The goal of conditioning is to transform the in-situ ground into a compressible, flowable medium that can be processed readily by the TBM, and that can be used as a pressure balance medium in the chamber. The desired recipe for soil conditioning depends on the primary ground types. In the U230 project, foam was injected through five cutterhead ports and water was injected into the chamber.

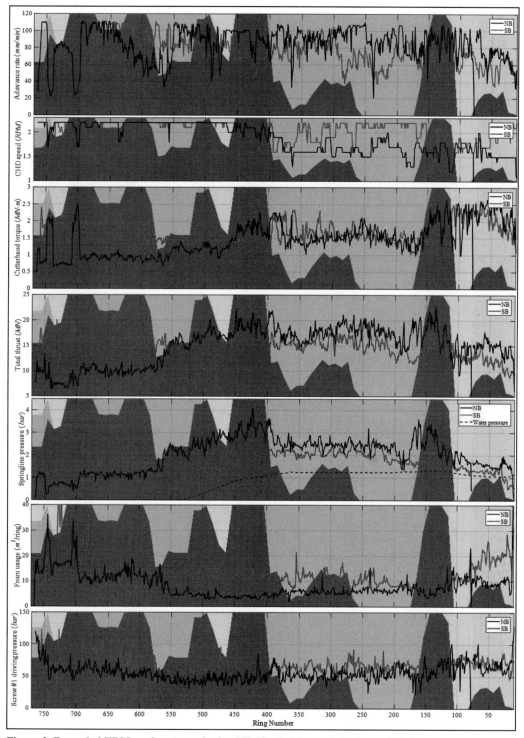

Figure 3. Recorded TBM performance during NB (first) drive and SB (second) drive

TBM Performance

The overall operational performance of the EPB TBM is shown in Figure 3 along the entire 1158 m (3800 ft) alignment for the northbound (NB) and southbound (SB) tunnels, expressed using a number of key performance parameters, including advance rate, cutterhead torque, total thrust, cutterhead rotation speed, etc. The NB tunnel was excavated first. The estimated geology within the tunneling envelope is also shown.

The average ring advance rate of the TBM varied considerably along the alignment, from a low of 20–30 mm/min to 110 mm/min over a number of stretches. The lowest advance rates observed from rings 650 until the end were due to purposeful slow down while tunneling under an interstate highway. The highest advance rate was achieved in a number of ground types and in mixed face ground conditions. The advance rate was generally higher for the first (NB) tunnel with some exceptions (e.g., ring 550–650).

Chamber pressures varied somewhat along the alignment, generally decrease with ring advance as the ground water table or ground cover decreased. Chamber pressures were decreased during SB tunnel excavation; accordingly, total thrust also decreased. Foam conditioning increased considerably during SB tunnel excavation. Cutterhead torque was the highest in the tan and yellow sand and gravel soils, then medium level in the fine sand/silt, and lowest in the hard clay. Cutterhead torque was generally similar from NB to SB tunnels, despite significant differences is cutterhead rotation speed, soil conditioning and thrust force.

The overall higher NB advance rates shown in Figure 3 can be misleading when considering production rates. The SB drive was actually completed in less time and therefore with higher production rates. NB tunneling was performed near fully capacity of the TBM but with higher down time. During SB tunneling, the decision was made to use more foam, decrease total thrust (and chamber pressure) based on the ambient pressures data measured by the TBM EPB cells, and increase the cutterhead rotation speed (lower penetration per revolution) to protect the cutters, and this resulted in less downtime.

PREDICTING PERFORMANCE USING MACHINE LEARNING

There are many interrelated parameters involved in EPB TBM tunneling. Their interaction and influence on performance is complex and likely nonlinear, making it difficult to discern relationships from visual inspection. One way to interrogate such relationships is through the use of artificial intelligence

techniques, particularly so-called machine learning techniques. A support vector regression (SVR) machine learning algorithm was used here to model TBM advance rate. The TBM parameters used to train the SVR were investigated using an attribute selection technique called RReliefF. And, the SVR was scrutinized to provide further insight into the causal relationships between the selected features and TBM advance rate.

Feature Selection Using RReliefF

TBM PLCs record hundreds of parameters, many of which are not related to key performance indicators such as advance rate. While machine learning methods can consider all parameters, inclusion of all parameters is computationally prohibitive. In addition, redundant or irrelevant parameters can adversely affect the prediction performance. Therefore, attribute evaluation and optimal feature-subset selection is essential for building an efficient and accurate model. In this study, we employed a feature evaluation technique called RRliefF, first introduced by Kira and Rendell (1992) for classification problems and expanded by Kononenko et al. (1997) to regression problems. The basis for the method is beyond the scope of this paper; the reader is referred to the aforementioned references for further information.

Using all of the TBM data from the NB and SB drives, RRliefF identified 10 most influential, nonredundant parameters, termed attributes in machine learning parlance. These included total thrust, cutterhead torque, foam flow rate, foam solution flow rate, chamber water flow rate, screw 1 and 2 rotation speeds, screw 1 and 2 torque, and front body rolling. We also included cover depth and depth below groundwater but these parameters did not weigh heavily on advance rate.

Advance Rate Modeling Using Support Vector Regression (SVR)

SVR was employed to train a model of advance rate using the TBM parameters identified by RRelieFf. For model training, the TBM dataset (nearly 400,000 observations of 10 parameters) was divided into training (80% of data) and testing (20% of data). To respect the distribution of data, fractional partitioning was applied wherein the range of advance rate (and associated parameters) was divided into 10 bins. 80% and 20% of each bin was used for training and testing, respectively. The SVR with RBF kernel was trained using sequential minimal optimization.

The results of training and testing are presented in Figure 4. The yellow lines convey a 1:1 relationship. The training data set (80% of the data) follows

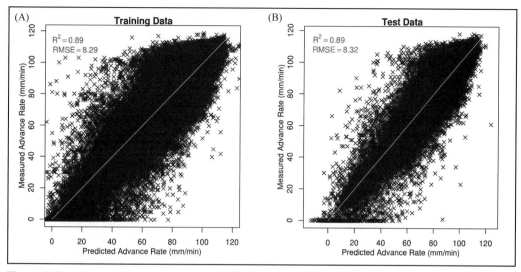

Figure 4. Resulting prediction capacity of the SVR model using the (a) training data and (b) test data. R^2 and RMSE (mm/min) statistics are shown for each.

the 1:1 relationship over the entire range of advance rate with some scatter. The deviation of the data from the line conveys the model misfit. The root mean square error in the training set was 8.29 mm/min. The R^2 was 0.89 indicating that a high percentage of the variability in advance rate was explained by the model.

The performance of the model against the test data was very good, with RMSE and R^2 similar to the training data. Recall that no test data was used to train the model. The lack of degradation of these statistics is important for prediction models. The level of fit exhibited by the model is reasonable considering that the geotechnical parameters are not explicitly introduced into the model. The ground is indirectly captured in the modeling through the TBM parameters.

Figure 5 presents a closer look at advance rate prediction by the SVR model. The test data shown spans 5–6 ring sections of advance in the NB and SB directions. As shown, the model captures the general magnitudes of advance rate and the changes in advance rate quite well. Note that the spikes shown are ramp up and ramp down at the beginning and end of ring advances; these are not of significant interest. There are clearly areas where the model performs better than other areas. For example, the model captures the magnitude and variation well during ring 302; however, the model over-estimates advance speed by 15–20% during ring 303 and 304. The model captures the fairly abrupt change observed

during ring 301 but overpredicts its magnitude by 10–15%.

Model Interpretation

Machine learning models such as SVR act as black boxes and offer little insight into how the parameters (features) influence the performance (here, advance rate). Here we use a combined partial dependence plot (PDP) and Individual Conditional Expectation (ICE) approach developed by Friedman (2001) and then Goldstein et al. (2015) to quantify the influence that features have on advance rate. The results of our analysis are presented in Figure 6 and the methodology is described as follows. Given all of the data used to fit the model, a sample of such data is shown as the dots in Figure 6. These capture a range of the parameter of interest (e.g., total thrust in Figure 6a) and a range of the advance rate. The magnitudes of the other nine parameters vary across their ranges for all these data. From each point in Figure 6a, a curve of advance rate is created over the range of total thrust values in the data set, assuming fixed values for the other nine parameters. The family of curves then represents an estimate of how advance rate changes with total thrust. At any thrust value on the horizontal scale, the observed range vertically represents the combined influence of the other nine parameters. The weighted average curve is also shown in yellow.

These results of model interpretation reveal that increases in thrust force over the range 5–18 MN

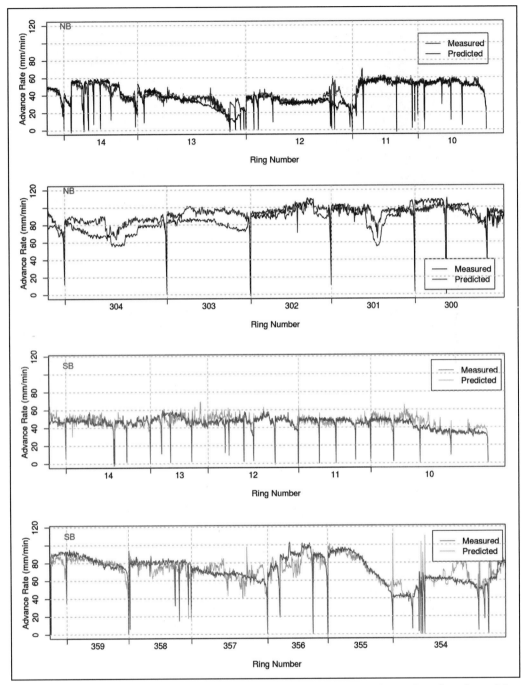

Figure 5. Example ring sections of measured and predicted advance rates along selected sections of the NB and SB alignments

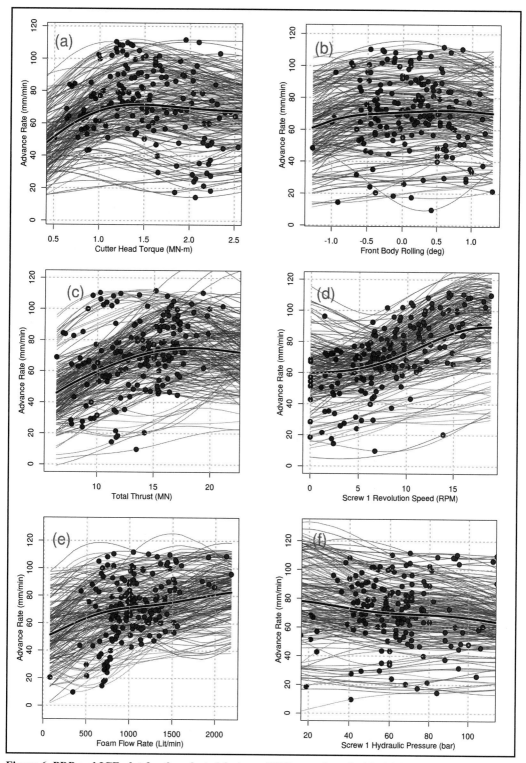

Figure 6. PDP and ICE plot for the selected features. PDPs are plotted with thicker lines.

correspond to increases in advance speed. Increases in thrust force when above 18 MN tend to not influence advance speed. Increases in cutterhead torque up to 1.3 MN-m correspond to an increase in advance speed, but further increases beyond 1.3 MN-m tend not to influence advance speed. As would be expected, increasing screw rotation speed correlates with advance speed, particularly over the range of 5–15 rpm. This is to be expected because the operator typically must increase/decrease screw rotation speed to extract more/less chamber muck that in turn would vary according to TBM advance rate. Increased foam flow corresponds mildly to an increase in advance speed. Surprisingly, the analysis shows that screw conveyor torque decreases with increasing advance rate. Though speculation, perhaps this is an indication of higher shear strength material that in turn would be more difficult to excavate. Finally, though rolling was identified as an important feature, the data shows that it has little correspondence with advance speed.

CONCLUSIONS

This study has shown that EPB TBM advance rate can be modeled using the TBM data itself and machine learning techniques. The analysis revealed that advance rate is most influenced by total thrust, cutterhead torque, foam flow rate, and screw conveyor rotation speed. The latter is expected given it is increased/decreased by the operator according with TBM advance rate. The analysis shows that these TBM parameters have an influence on advance rate only within certain ranges of thrust, torque, foam flow, etc. Geotechnical conditions, of course, matter significantly in tunneling. Here, they are indirectly considered in the TBM parameters. Ongoing research is explicitly incorporating geotechnical parameters in performance prediction.

REFERENCES

Frank, G., et al. (2013). Construction of the University Link Light Rail Tunnel U230 in Seattle, WA. *Rapid Excavation and Tunneling Conference 2013*. Washington, DC, Society for Mining, Metallurgy, and Exploration.

Friedman, J. H. (2001). "Greedy function approximation: a gradient boosting machine." *Annals of statistics*: 1189–1232.

Gharahbagh, E. A., et al. (2013). "Keeping the Chamber Full: Managing the Air Bubble in EPB Tunneling."

Godinez, R., Yu, H., and Mooney, M.A. (2015). Earth Pressure Balance Machine Cutterhead Torque Modeling: Learning from Machine Data. *Rapid excavation and tunneling conference*. New Orleans.

Goldstein, A., et al. (2015). "Peeking inside the black box: Visualizing statistical learning with plots of individual conditional expectation." *Journal of Computational and Graphical Statistics* 24(1): 44–65.

Irish, R.J. (2009). Prebid Engineering Gerologic Evaluation of Subsurface Conditions for the University Link Light Rail Tunnels, Capitol Hill Station to Pine Street Stub Tunnel, And Capitol Hill Station Excavation and Support, Contract U230, Seattle, Washington. Denver, CO, R J Irish Consulting Engineering Geologist Inc.

Kira, K. and L. A. Rendell (1992). *The feature selection problem: Traditional methods and a new algorithm*. Aaai.

Kononenko, I., et al. (1997). "Overcoming the myopia of inductive learning algorithms with RELIEFF." *Applied Intelligence* 7(1): 39–55.

Northlink Tunnel Partners (2009). *Geotechnical Baseline Report—Univerity Link 230*. Seattle, WA, Northlink Tunnel Partners. 6: 66.

Interpretation of EPB TBM Graphical Data

Keivan Rafie
Stantec

ABSTRACT: Tunnel construction using a tunnel boring machine (TBM) involves a highly complex operation. Such processes generate large amounts of data that can be used for monitoring, reporting and analysis. Major TBM manufacturers have developed software systems to support tunnel contractors and their site teams in both data management and analysis. These programs are mostly web-based and have many advantages.

Data acquisition cannot prevent breakdowns from occurring but can facilitate forensic investigations to quickly determine the root cause of a breakdown and provide basis for implementing corrective actions. This paper analyzes these data acquisition tools and presents case studies, primarily involving earth pressure balance (EPB) TBMs, to illustrate how the formation of critical interpretations can be made from user-defined charts and diagrams to diagnose issues and optimize TBM operational parameters.

INTRODUCTION

The storage and visualization of measured values acquired by sensors and recorders is a crucial element of TBM tunneling. All of the work being performed by the machine is documented in terms of the recorded data to allow the complete or partial tracing of the tunnel construction in real-time or after completion.

This information could help engineers and operators examine a vast and complex set of data related to TBM operation that cannot be ascertained in the field by the TBM engineers or work crews, particularly when visualized in a graphic format. The examples of measured data and sample graphs presented in this paper are mainly taken from Earth Pressure Balanced (EPB) TBMs, but the logic behind the interpretation of these examples can also be applied to data from hard rock or slurry TBMs.

TBM DATA ACQUISITION AND VISUALIZATION SYSTEM

The purpose of a TBM data and acquisition system can be summarized as the "acquisition, processing, storage, display and evaluation of all data connected to the tunneling machine." A TBM data acquisition system continuously records and visualizes all measured data in a pre-determined cycle. Logging, however, occurs only at specific times. The average time period between logs can be individually selected for each measuring point but is set to 10 s for most parameters.

The operating phases of the tunneling machine are generally classified according to three periods: advance, ring building and standstill. These three phases form a unit called a ring. The data for each ring are usually stored in separate consecutively numbered files. An immediate correlation to the respective construction phase can be made based on the ring number, file date and file time. The measured data acquisition program automatically opens after each system restart and loads all required program components into its memory. It then acquires, stores, and visualizes the currently available measured data. (See Figure 1.)

INTERPRETATION OF TBM OPERATIONAL GRAPHS IN CASE STUDIES

Some of the most common graphs representing the general status of TBM operations are ram extension, rate of advance (ROA), thrust force, cutter head torque, EPB/slurry pressures, weight/volume of excavated material, and grout volume. Of course, illustrating too many parameters on one chart makes interpretation more difficult, so there must always be a compromise between amount of information given and the clarity of the graphics.

EPB Pressure Graphs

Case Study 1

In successful EPB operation, face pressures should be maintained at all times and monitored with the data acquisition system. Pressure of material in the chamber could be assessed by information available from EPB cells. TBM operator closely monitors excavated material and adjusts the type and amount of water, bentonite, polymers, and foam to ensure that the material is properly forming a plug to resist piezometric and ground pressures.

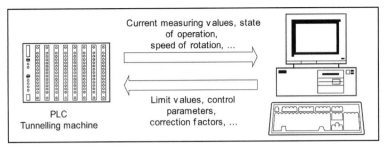

Figure 1. TBM PLC and data acquisition system

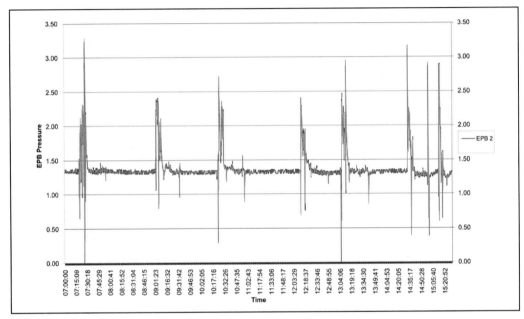

Figure 2A. Maintaining face pressures in proper TBM operation

Figure 2A and 2B graphs from pressure cell data are among the most used graphs in EPB tunneling and demonstrate the difference between correct (A) and incorrect (B) operation. Excavations similar to graph (A) result in safe and steady progress while performances similar to graph (B) are usually linked with significant loss of ground and surface settlement.

Case Study 2

The EPB pressures for the top, middle and bottom sensors used in this case study are presented in Figure 3, which shows that the bottom sensors record higher pressures due to the higher density and greater compaction of the excavated material. The top sensors record the least pressure and fluctuation because they have less direct contact with the soil and mud

in the chamber but middle and bottom sensors have more fluctuations as shown in Figure 3.

The proper estimation of material contact and density in an excavation chamber is important. It is common practice for the operator to perform and complete an excavation with full level of material in chamber when using EPB TBMs. However, TBMs must sometimes be operated in semi-open, in which only a portion of the face is balanced by excavated material. These operating conditions are generally determined by engineers based on the ground conditions and stoppage time. Compared to the semi-open mode, full material contact in a chamber requires more thrust and torque from the TBM and increases the equipment wear and the cost of replacing excavation tools on the cutter head. Working in a semi-open mode could alleviate these issues but is not advisable

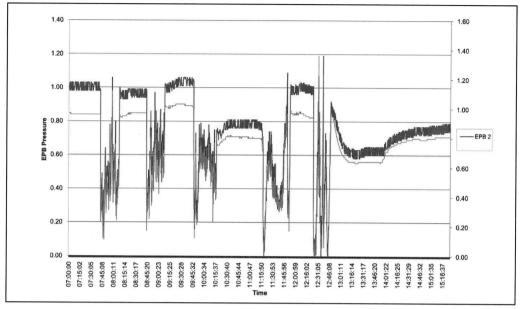

Figure 2B. Pressure drop during excavation

if there is a high risk of ground collapse and over-excavation. Other scenarios, such as preparation for cutter head maintenance or leaving the TBM unused for long periods of time, could also influence decisions regarding the level of material that should be present in a chamber during tunneling operations.

Case Study 3

The smooth rise (or drop) in EPB graph lines (Figure 4) indicates the passage of gaseous or liquid material into (or out of) the excavation chamber which occurs mostly during the TBM ring build phase. Soft rising curves may be the result of ground water filling the chamber or the injection of ground conditioning material (foam, water or compressed air). A smooth drop in cell pressure suggests the leakage of air or water through porous ground, a tail shield, purge line or screw conveyor.

Thrust, Cutterhead Torque, RPM, and Rate of Advance

Higher advance rates in TBMs are generally achieved in two ways.

A. A higher cutterhead rotation speed, which increases the distance that cutters or rippers travel and thus their work per unit time (mm per min). In this case, the cutterhead torque will increase. An increase in cutterhead torque can also result from other factors such as poor ground conditioning or high material density in the excavation chamber.

B. Higher forward forces in TBM cylinders to make cutters and rippers excavate more intensively, thereby increasing the cut depth per cutterhead rotation. In this case, both the TBM thrust force and its torque will increase. The TBM thrust can also be increased due to shield friction with the ground or the TBM's pulling force due to its weight (a factor discussed later in conjunction with contact force).

Case Study 4

Scenario (B) is illustrated in Figure 5. The cutterhead rotation speed is set at approximately 2 rpm, so the occasional increase in torque is due to higher thrust forces exerted by the propulsion cylinders at that moment and increases the rate of advance.

It should be noted that higher efficiency is usually achieved in soft ground with a lower RPM and higher thrust forces for deeper excavations, whereas cutters break into hard rock by rolling on it. Therefore, better advance rates occur with higher RPMs.

Cutter Head Contact and TBM Thrust Forces

The graphical representation of the relationship between TBM thrust force and contact force of cutter head is mainly used to identify any opposing forces to the TBM other than the excavation face.

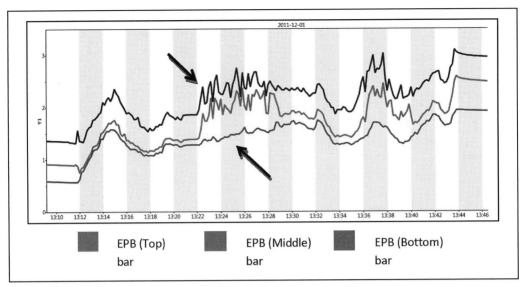

Figure 3. EPB pressures for the top, middle, and bottom sensors

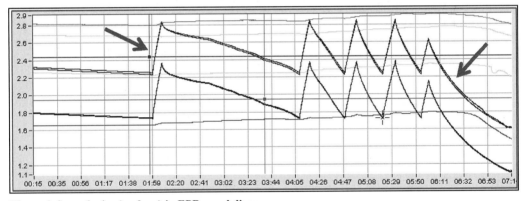

Figure 4. Smooth rise (or drop) in EPB graph lines

In general, the TBM thrust is used to maintain EPB pressure, push the material in the chamber, and pull the gantries and frictional forces of the shield.

The thrust left over from propulsion energy is consumed by the cutterhead in the form of the contact force required to cut through the ground. Because the parameters other than contact force are relatively constant during normal TBM operation, TBM contact and thrust forces are typically synchronized in their fluctuations. Therefore, any mismatch in the graphical patterns between these two forces suggests a status change in other parameters and usually indicates an obstacle during operation.

Case Study 5

The theoretical graphs shown in Figure 7 show a sudden drop in contact force despite a constant increase in the thrust force (Graph A). These data could indicate collapsed ground around the TBM shield or an entrapped gantry back in the tunnel. Variations between the contact and thrust force that are more gradual could result from a change in tunnel slope or the accumulation of heavy, dense material in the chamber (Graph B).

Identifying Overexcavation

Most EPB machines today are equipped with weight sensors and laser scanners to estimate the weight and volume of excavated ground. The theoretical weight and volume that a TBM data acquisition system is expected to show is usually calculated manually based on the TBM dimensions, ground properties and advancing distances. These figures are compared with the quantities shown on TBM graphs to check

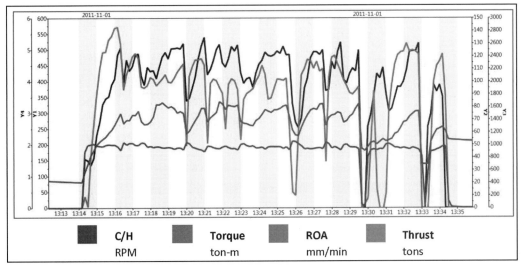

Figure 5. Higher thrust forces and increased rate of advance

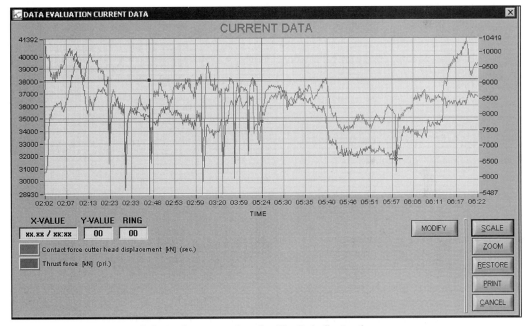

Figure 6. TBM contact and thrust forces synchronized in their fluctuations

for overexcavation. This information is also useful in the analysis of excessive volume loss and settlement.

Case Study 6

TBM advance with overexcavation can generally be recognized on TBM data graphs by a higher-than-normal grade in the excavation weight or volume line. For example, the following theoretical graph illustrates three sets of data from different advances. Line A, which is the typical advance at a constant rate of excavation, is usually the preferred scenario and ensures the consistency of other parameters, such as ground conditioning and screw conveyor speed during the shove.

Compared to Line A, Line B has a steeper increase at the beginning and end of excavation period and ends at a higher value. This condition

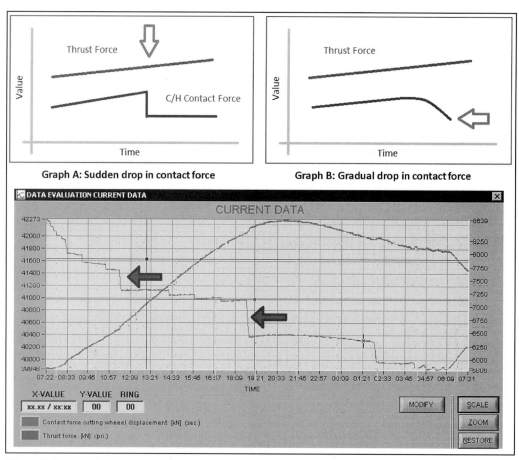

Figure 7. Sudden drops in contact force while TBM thrust force is increasing

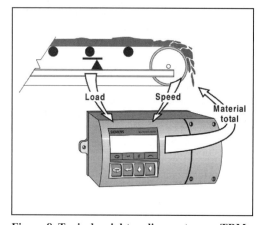

Figure 8. Typical weight scaling system on TBM conveyor

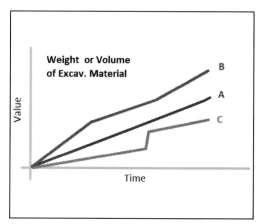

Figure 9. Three scenarios for excavation weight/ volumes

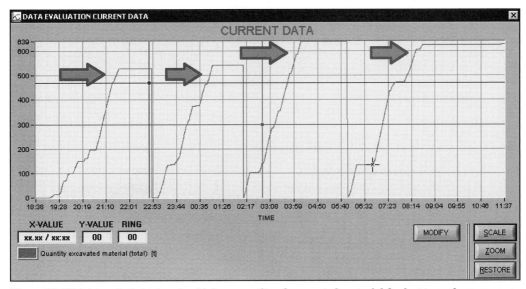

Figure 10. Weight scale data showing higher quantity of excavated material for last two advances

can be interpreted as general overexcavation. Line C represents a normal extraction scenario, except for a very sudden increase over short period of time that could indicate overexcavation with a sudden rush of material through the screw conveyor and out of the chamber. However, other aspects, such as those stated below, must be considered to achieve a realistic understanding of the data.

1. The higher slopes in the graph lines based on material weight and volume can be a result of higher advance rates. Thus, the final excavation values must be checked. The screw conveyor rotation speed in those time periods can provide insight as to whether the high extraction rates were intentional or due to ground conditions.

2. The theoretical material weight depends on the advance distance of the TBM and the density of the ground, but other factors, such as added water or ground conditioning agents, should also be considered. In regard to ground conditioning material, only the liquid portion will affect the material weight, so the foam expansion ratio (FER) should not be considered in calculations. The FER of the ground conditioning material added to the chamber has no effect on the weight calculations but should be considered with regard to the material bulking factor when scanned by a laser on a conveyor belt for volumetric data.

3. Comparing the weight and volume data/ graphs of different advances only makes

Figure 11. TBM operator checking TBM parameters to assess theoretical weight/volume

sense if the level of material in the chamber after each advance remains full or relatively constant. For example, if an operator has started an advance with a half-full chamber and decides to fill the chamber to its maximum level, less material will be extracted and shown in the data, even though the same amount of material has been excavated from the ground. On the other hand, when a TBM chamber must be emptied during an advance (e.g., the last advance before cutterhead maintenance), the TBM data will show a higher amount of extracted material than average. To eliminate this problem, engineers also look at the rolling average of values for several consecutive rings, which eliminates the effect of chamber space and gives a more realistic picture of the scenario to identify possible overexcavation.

4. Added water should be considered in theoretical calculations. Occasionally, depending on ground conditions, most of the injected water is absorbed by the ground, and sometimes added water only replaces the water in saturated soils.

5. The calibration of weight scales and laser scanners must be part of a contractor's regular maintenance program. Some weight sensors are very sensitive to misalignment and curves in the TBM conveyor belt, while laser scanners could have inaccurate readings

depending on their position and air/light interference, such as dust. Utilizing two belt scales and observing their averages can also aid in identifying errors and obtaining more realistic results.

Grouting System

Two-component grouting (A+B) systems through the tail shield have been one of the most problematic areas in TBM tunneling. Proper grouting is important to prevent ground movement and surface settlement due to volume loss at the tail void. Grouting also stabilizes segmental lining in the ground and improves a tunnel's watertightness.

Information available in TBM data acquisition systems can show early signs of system malfunctions and indicate which components require attention or which control settings need to be modified.

TBM data loggers typically record flow, pressure and volume parameters for each grout line. To check the quality of grouting behind segments and ensure that the correct dosage of accelerator (B) is mixed with part (A), TBM data for injected volumes should be checked against the theoretical volume of voids behind the segments. Gauge cutter wear should be considered in theoretical calculations, particularly for larger TBMs. Understanding the bore and cut diameters in hard and soft ground types can also lead to more accurate calculations.

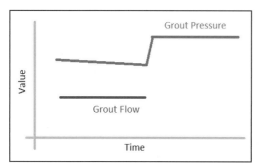

Figure 12. Grout pressure without flow, showing the blockage in lines

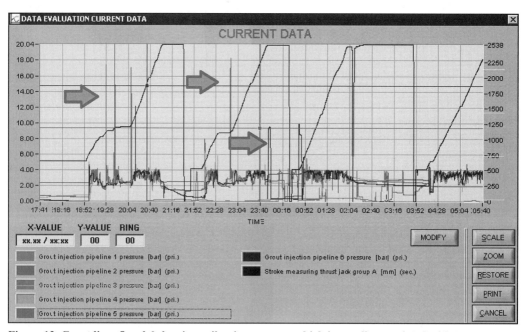

Figure 13. Grout lines 5 and 6 showing spikes in pressure, which is usually associated with temporary blockage

If the grouting volumes are lower than their theoretical values, other data must be checked to identify and solve the problem. The simultaneous spike in grout pressure and halt of grout flow in the Figure 12 is commonly an indication of blockage in the line. If grout volumes cannot be achieved when all lines are in operation, then the pre-sets and cutoff levels should be checked. Generally, grout pressures must overcome hydrostatic pressures by 1–2 bar behind segments.

Case Study 7

If grouting volumes exceed their theoretical values, assessments must be performed to identify any

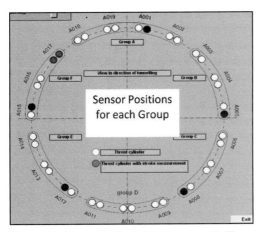

Figure 14. 19 rams collected in 6 groups (A-F); the location of a representative ram from each group is shown in black

channeling of grout to the surrounding environment or leakage through the tail shield. In some instances, high-pressure grout finds its way to the excavation chamber, mixes with the excavated material and exits through the screw.

Propulsion Cylinders and Ring Build

Information and graphs derived from TBM data on propulsion cylinders can be used to analyze several aspects of their operation, including ring build and steering. These data can also explain damage to segments that occurs after installation. TBM data acquisition systems generally display the pressures and extension of ram groups using the sensors on a representative ram from that group. Figure 14 shows information collected on 19 rams in 6 groups (A-F). The location of each group's representative ram is shown in black.

Case Study 8

Figure 15 shows the pressures applied to group of segments during ring erection. The lines representing each group show a sudden jump from zero, indicating that the rams have been extracted to hold each segment after its erection. The lines also indicate common slow pressure loss due to micro-movement of the TBM and ring compression in the tunnel. However, an excessive loss of pressure in any group could loosen the adjacent segment and cause vertical (step) and horizontal (gap) misalignments.

On the other hand, excessive pressures on cylinders can cause damage, particularly around the circumferential joints of the segments in front of the cylinders. Ring build reports from the guidance system (as shown in Figure 16) must be studied in

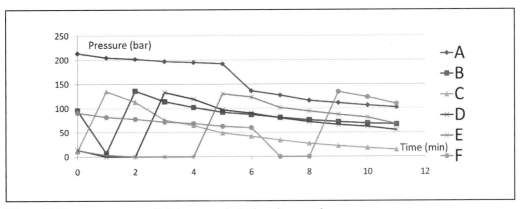

Figure 15. Ram pressures applied to segments during ring erection

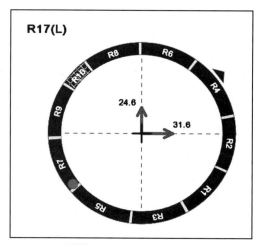

Figure 16. Sample ring build report from the guidance system

conjunction with ram pressure graphs to confirm the location of segments in relation to the propulsion cylinders and explain the damage incurred.

CONCLUSION

The graphical representation and measured values of TBM data can assist contractors by providing information that helps TBM crews increase the reliability and productivity of TBM operations. Such an advantage would ultimately lead to fewer breakdowns and lower tunneling expenses. Data acquisition and visualization alone does not benefit the contractor unless the data is accurately interpreted. The utilization and correct interpretation of data acquisition systems' outputs could greatly enhance the control of the excavation and operation of various tunnel boring machine systems. As TBMs grow in size and complexity, advances in data monitoring and presentation to optimize TBM parameters will likely continue as well. The correct interpretation of these data is essential for the effective utilization of these tools and to ensure efficient and productive tunneling operations.

The key to success in EPB tunneling is proper engineering combined with experienced operators. Data acquisition cannot prevent breakdowns from happening but allows the rapid identification of the root cause of a breakdown and the timely implementation of corrective actions.

TBM Procurement—Owner's Dilemma

Dan Ifrim and Derek Zoldy
Hatch Corporation

ABSTRACT: Tunnelling is a unique and challenging industry. This paper will discuss the importance in understanding technological innovations in the current TBM manufacturing market and provides suggestions for procurement options in design-bid-build and design-build project delivery approaches. The objective of the paper is to open a discussion within the industry with the goal of invoking forwarding thinking to improve Tunnel Boring Machine (TBM) specification language in contract documents.

INTRODUCTION

In the past decade, TBM configurations and operating challenges have been met with technological advancements of TBMs to the point where they have now entered new territories for use on a wider spectrum of infrastructure projects globally.

As TBM innovations are adopted by manufacturers the industry is seeing newer and wider capabilities of equipment to deliver more complex projects. The tunnel industry is seeing a renaissance, so to speak, in new opportunities, opening more serious discussions between Owners, TBM manufacturers, contractors and consultants. The bad news, is that consultants and Owners, are not progressively reviewing and updating existing procurement and contract documents to follow the advancement of TBM capabilities and to allow for new TBM innovations and new standards. This raises the question that needs to be debated now! When do we start accepting TBM innovations? The answer, should be NOW!

In order to capture consideration of new advancements in TBM manufacturing at the same time as looking at lessons learned over the past decade.

We understand that Owners are faced with compliance of their purchasing policies, responsible public spending while safeguarding public safety. These are serious challenges taken into consideration in the early stages of planning for the future growth, particularly in and surrounding Big Cities.

The Owner's predicament is that procurement of large underground infrastructure is complicated by the multiple source project funding and budgeting allocation. Assurance of cost certainty that meets long range planning of infrastructure development can create struggles and attract criticism for political and social support and ultimately community acceptance.

There are many technical and economic considerations and constraints when specifying the appropriate types of TBMs and excavation methodologies, along with supporting ancillary equipment, when encountering, for example "variable soil conditions."

Very often many tunnel alignments (horizontal and vertical) are considered, evaluated and options prioritized, based on a risk profile during the early design development stages. One of the major outputs of this evaluation process is an attempt to reduce unforeseen risks associated with geological and hydrogeological unknown conditions.

These considerations, evaluations and risk assessments are usually based on past project experiences (failures and successes) and in most cases, do not include consideration of TBM innovations and advancements that, if they were implemented in time, would have been a benefit to the project.

With all of these issues in mind, we have a few examples with suggestions for the tunnel industry to consider that may create an open discussion about how we should procure tunnel equipment, going forward.

TBM MANUFACTURING INNOVATIONS

Multimode / Cross Over TBMs

Multimode, or Cross Over TBMs have arrived in the tunnel market and have recently been considered for variable ground conditions tunnel alignments. The innovation of the Multimode TBM is that their multiple operation modes are pre-manufactured to allow for operator adjustments on the fly to address differing ground conditions while maintaining target excavation rates without compromise to safety.

The TBM multimode construction concept addresses all methodologies from Slurry to EPB to Open face modes.

Figure 1. Lake Meade Intake #3 Tunnel multimode TBM

- Combination of EPB and Open-face TBM
- Combination of Open-face and Slurry TBM
- Combination of EPB and Slurry TBM

Variable Density TBM

Variable Density TBMs is a totally unique tunnelling technology innovation that combines the advantages of both slurry and EPB modes in one machine. This TBM innovation addresses geological and hydrogeological changes along the alignment with operational flexibility.

Innovative Conversion Between Pipe Jacking and Segmental Lining Modes

There are many instances that a tail or starter tunnel is required to launch a TBM or MTBM from within a small diameter, deep shaft. Construction of starter or tail tunnels in soft ground conditions can be avoided by incorporating a recent TBM innovation that has the capability to start the tunnel excavation in pipe jacking mode, then switch to segmental lining mode once the TBM and trailing gear can be accommodated in the pipe-jacked section.

This TBM innovation, can be designed for specific project requirements, including the incorporation of a telescopic shield that would allow installation of jacked pipes for an initial 60m (approximately) of tunnel and then provide space to convert to launching of TBM, using segmental linings after the initial jacking mode.

The pipe jacking method of providing a starter tunnel also, provides another innovation, which utilizes a pipe thruster, similar to what is used with the Direct Pipe technology.

The Norris Cut Project, an example of technology and innovations at work, consisted of the replacement of Existing 54-inch Force Main from Central District Wastewater Treatment Plant (CDWWTP)

to Fisher Island, under Norris Cut; and was constructed by Nicholson Construction for the Miami Dade Water and Sewer Department (MDWASD) The 3m DIA TBM was supplied by the tunnelling subcontractor Bessac and included the following innovations:

- Variable Density TBM based on slurry muck disposal with integration of a Slurryfier box;
- Dual lining installation, pipe jacking from the launching pit for 170m and installation of segmental lining
- Use of a pipe thruster inside a launching shaft

Figure 2. Norris Cut Tunnel variable density TBM

Figure 3. Norris Cut TBM launching in pipe-jacking mode (courtesy of Bessac)

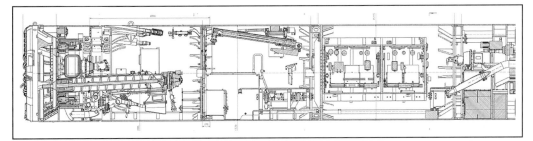

Figure 4. Norris Cut TBM general arrangement

- Integrated airlock
- Probe drilling capabilities (5 peripheral ports and 2 ports through the face)
- Cutter tools replacement in highly permeable ground, featuring a diver pit into the TBM and flooding the front of the TBM

Ground Prediction Systems

Exploratory drilling from within the tunnel at the tunnel face is frequently used to detect fissures or cavities ahead of the tunnel face.

The maximum predictive range of this methodology is about 50m. The method is usually associated with delays to excavation.

Tunnel Seismic Prediction (Amberg Technologies AG) is a new system that provides evaluation tools to measure elastic body waves excited by explosive charges. The system tools provide Real time data collection that measures body wave travel, as compression or shear waves, as they travel through the ground. Some of the outputs that are a benefit to procuring this type of system is that they provide information regarding reflection interfaces within different ground properties, including density and elasticity.

Cutterhead and Shield Design and Manufacturing Improvement Innovations

There are many factors to consider and verify during TBM component fabrication, including: strength and certification of steel materials, cutterhead modeling and Finite Element Analysis (FEA) stress analysis, geometrical accuracy, welding quality and application of wear protection.

Some innovative features that manufacturers can add to a TBM cutterhead, include: wear detection sensors for disc cutters and picks, disc cutter load monitoring and installation of permanent or removable cutterhead cameras.

Cutting Tools

As a suggested procurement option, specifications for cutterhead tools can make allowances for

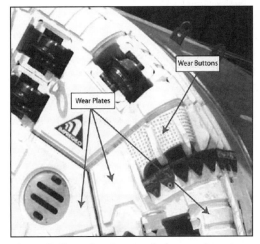

Figure 5. Cutterhead protective armor (courtesy of Robbins)

utilizing state of the art Disc Cutters and Drag Bits to be installed on the Cutterhead that has an improved lifespan.

Procurement documents shall include a submittal provision to show previous track records of tool manufacturers and verification documentation. Other features such as Disc Cutter Rotation Monitoring and Drag Bit wear detection systems can also be considered in procurement documents as these technological systems are now commercially available.

Cutterhead Maintenance Work in Free Air

Hyperbaric intervention shall always be plan B.

There are several innovations that are in the testing stages that will improve safety of workers in high earth pressure tunnels.

Among these innovations are:

- Automated (robotic) system for replacing disc cutters, currently being tested by NFM,
- Submerged wall gate for atmospheric crusher maintenance in Slurry TBMs, currently being

tested by Herrenknecht, with first implementation on Istanbul Strait Crossing TBM.

- Accessible cutterhead spokes for atmospheric cutter changes and soft ground tool changes, currently being tested by Herrenknecht, with first implementation on Istanbul Strait Crossing TBM
- Assisted Disc Cutter replacement by tool aids, already implemented by Robbins and Herrenknecht, should be considered at all times, as a safety measure, to reduce risks of injury.

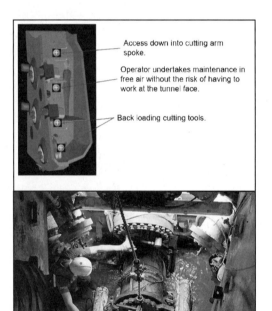

Figure 6. Disc cutter replacement in free air illustration (top) and Istanbul Strait Road Tunnel TBM intervention (bottom) (courtesy of Herrenknecht)

Note: These innovations would be worth considering for use on large TBMs.

DISCUSSION

Project Delivery vs. Procurement Method

The project delivery method (Design-Build, Construction Management at Risk and Design-Bid-Build) is usually a preference of the Owner in consultation with stakeholders responsible for the project delivery. It is understood that some delivery approaches are preferred by some Owners and may vary based on type of project, the risk approach to the project and scheduling and funding constraints.

The procurement method decision can impact the design and construction of the project and should take into consideration, risks, schedule and cost management factors, on a case by case basis. It is essential that the preferred procurement and project delivery methods, be compatible and consistent to ensure the optimum project performance and cost benefit.

TBM Procurement

The TBM procurement is an important aspect of the Contract and the approach needs to be consistent with the project delivery method, while ensuring important aspects of the TBM suitability and performance, safety and quality of essential systems requirements are clearly specified.

Owner's Procured TBM

The Owner procured TBM approach involves a full prescriptive specification developed by the Owner and the project designer. This approach, requires that the TBM, spare parts and tunnel liners be fully defined and procured in advance of a contractor being retained by the Owner. To successfully deliver this procurement process, the Owner would need to employ experts and consultants to manage all aspects of TBM operations to ensure that the TBM is not being abused and that regular inspections and maintenance is followed as specified by the TBM manufacturer.

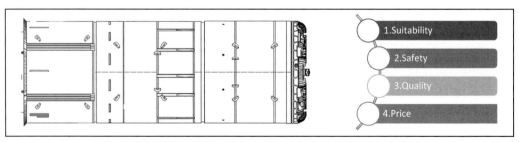

Figure 7. TBM sketch (left) Factors for TBM procurement (right)

Some of the advantages of this procurement method are:

- Minimized construction bid unknowns related to contractor TBM selection and cost estimating.
- Owner controls risk of TBM performance and warranties with the manufacturer
- Prescriptive Approach to specify major TBM characteristics specific to the project,
- The owner imposes specific means and methods for tunneling.
- Owner becomes substantially involved in the construction
- The Owner is engaged in TBM selection and evaluation of technologies to address Owner's risk management concerns
- Claim exposure related to equipment performance and maintenance is managed directly

Owners who have procured TBM's in the past, have developed a procurement strategy that can be used on multiple projects to save project delivery schedules and reduce unit cost of machines, and ensure TBM availability.

The method was applied on several projects including the St. Clair River Tunnel (1992), TTC Rapid Transit Expansion Program (1994), South East Collector Trunk Sewer (2009), Toronto York-Spadina Subway Extension (2009) and Eglinton Crosstown LRT (2010). This strategy was foreseen as revolutionary and considered a good risk mitigation tool at the time.

Mixed Procurement

This method requires customized procurement documents, usually prepared by an Owner and their design consultants.

There are several examples of mixed procurement strategies applied on past projects, which involved engaging TBM Manufacturers and Contractors.

- Early involvement of TBM manufacturers (San Diego Outfall Tunnel, 1996)
- Early involvement of Contractors (Burnaby Mountain Tunnel, 2016)

In this procurement process, a Risk Sharing approach is adopted, where the Owner shares the risk with the selected Contractor in a joint approach to specifying key TBM features are considered critical while giving the contractor TBM operation decisions to meet an agreed tunnel construction performance criteria/contract basis. The benefits of this procurement process include:

- Ensures an agreed TBM design, addressing the Owner's and Contractor's concerns mutually,
- Provides a basis for controlling TBM consumable and maintenance costs
- Provides an "agreed to" Decision Process parties TBM manufactured specifications can be incorporated into the contract to ensure the best fit for the purpose

Contractor Managed Procurement

- Owners with low risk tolerance are able to transfer the tunnel construction risk to Contractor, who is considered best capable to manage the project risks, efficiently
- Owners need to trust the technical capabilities of their design engineers and their selected contractor
- Performance based approach
- TBM characteristics are procured and sourced by the Contractor
- The Owner is responsible for the overall project funding and design compliance during construction as specified in contract documents can result in less preferred equipment used on the project (ie. refurbished TBMs), in some cases, TBMs that have been skinned up to match the desired diameter on the project.

Despite advances in TBM manufacturing and auxiliary systems innovations, the use of outdated procurement documents and specifications continue to expose Owner's and Contractors to contractual risks associated with known unknowns and unexpected ground conditions and TBM failures in response to selected methods. In these circumstances, all too often, contractual changes, claims and disputes regarding the existence of differing site conditions (DSCs) becomes litigative.

CONCLUSION

The success of the project is tied directly to the quality of the contract documents and specification language and the responsibilities of the contracting parties. Specifications related to the reliability and capability of TBMs should be project specific and include the most up to date information related to the equipment manufacturing environment and certifications (ie. ISO certifications).

While some of the TBM innovations may be outside of certain Owners, Contractors and Consultants comfort zone, it is suggested that familiarization with new and innovative technology be considered by attending industry seminars, workshops and conferences. The TBM manufacturers shall get engaged more in such educational workshops for the benefit of the industry.

For ensuring that the specified TBM will result in a robust quality TBM a number of recommendations for the TBM technical specification are suggested below.

TBM Manufacturer Verification

In the last decade, the TBM market has changed; some manufacturers got out of the business, some formed alliances and some new companies formed. The TBM manufacturer credibility and capability to design and fabricate the desired TBM needs to be verified for the project selected procurement. Essential information such as ISO certification, years in business, experience with the proposed type of TBM as well as quality track record of TBM produced needs to be verified.

Main Drive Design—Procurement Recommendation

The design of the TBM Main Drive, type of bearing, bearing life and lubrication are the most important considerations for reliability of any TBM. These considerations are not always included in procurement documents or specifications, however, TBM specifications should outline the need for this consideration to be addressed and included in the submittals part of the TBM specifications.

The typical requirement is to have the L10 bearing life specified to be a minimum of 10,000 hours. It is recommended that the bearing life calculation be completed as per the DIN ISO 281 calculation method. The calculation shall be carried out by the bearing Original Equipment Manufacturer (OEM) and should address expected loads and working conditions for the project.

The TBM bearing selection should be addressed in the TBM specifications, including the criteria of bearing diameter to tunnel diameter, so that no conflicts or constraints occur with the tunnelling methodology.

It is also recommended, that the selection of the gearboxes be provided by a proven manufacturer, and cross-referenced with the TBM manufacturer to ensure TBM reliability requirements. For example, tunnel projects that exhibit variable ground conditions (soft and hard), the gearboxes need to incorporate two-speed capability for the two modes of operation: low speed/high torque and high speed/low torque.

TBM Data Acquisition and Owner's interface

In the past decade, TBM data collection and data management systems have improved tremendously with the use of innovative software, capable of providing real time data access as well as customized dashboards to observe equipment performance, key performance indicators, customized reporting and shared viewports.

Options for data acquisition, can now include ground settlement monitoring, segmental lining manufacturing, installation records, measurement of tailskin clearances after ring building, TBM power consumption, integration of TBM inspection reporting, statistical reporting histories, tunnel inspectors' data dashboards and reports as well as camera surveillance in specific areas of interest. These technological options, should be updated in all contract documents and specifications.

New vs. Refurbished Consideration

At developing the TBM technical specification a minimum of prescriptiveness shall be considered regardless if a new or remanufactured/refurbished TBM is considered. The refurbished TBM shall carry the same warranty as new and shall include only new main/critical components such as main drive assembly (bearing, seals, motors and gearboxes) and other critical / job stoppers components.

The quality of the refurbishing process is essential for qualifying a refurbished TBM and as such the requirements should specify that the refurbishment must be carried under direct supervision of the original manufacturer and make the original manufacturer the bearer of the warranty.

TBM Technical Specifications

The minimum perceptiveness should include requirements that are and can be verified during the design phase and continue in fabrication. A factory verification testing, should be the time to verify all technical specification requirements and TBM overall construction and functionality. Further the technical specification shall address the requirements for manufacturer support for TBM site assembly and commissioning as well as warranty and service support during tunnelling.

The technical specification shall include requirements for:

Factory Test Verification

It is recommended that specifications include a TBM manufacturer audit process to ensure design solutions and manufacturing methods will achieve geometrical design assembly tolerances prescribed for the product. This audit process should also include raw material selection and certification.

Assembly of key systems, quality control documentation and certification of key components should be specified, including verification during fabrication and Factory testing.

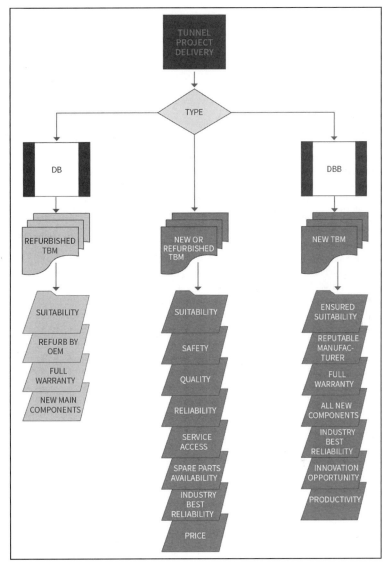

Figure 8. Tunnel project delivery and TBM procurement

Specifications should include a acceptance process for key components, such as the Main Drive Bearing, Motors, Gearboxes and Pumps.

Site Support Recommendation

Although these is all Contractor business it is important to ensure that the project will not encounter delays due to shortage of spare parts or lack of TBM manufacturer support. The provisions below will not only protect the Owner but the Contractor at the same time.

- Provision for timely technical support response. Technical support and technicians' availability to allow for a response of maximum 24 hours or as agreed with the Contractor and TBM manufacturer to minimize the TBM downtime.
- A minimum critical spare parts stock maintained by the tunnelling contractor on site or nearby location.
- Quick purchasing access to all non-critical spare parts assured by the TBM manufacturer with an adequate response time to minimize TBM downtime.

Reference Standards

Ensure that all applying standards, codes and regulations govern the specification are listed, where local standards or regulations are lacking industry standards shall be adopted.

Quality Control

Ensure that all training requirements, personnel qualifications, facilities certifications, products certifications, instrumentation equipment and calibration standards are listed clearly without ambiguities or use of terms "or similar."

Submittals

Submittals are the very important part of the specification and needs to ensure that all requirements for the qualification of the TBM, systems, equipment, materials, personnel, statements, schedules, work plans and methodologies have clear requirements for quality quantity and methods, time lines and deliverables to ensure that all contractors are bidding to the same standards. Submittals for safety plan and risk mitigation plans are equally important.

Products

All allowed products and materials shall be listed to ensure the level of quality required. All requirements for the TBM including type, ancillary systems, make and certifications need to be clearly specified.

For projects with long tunnel drives, difficult ground or extreme conditions the TBM quality requirements shall include terms like "industry best" "state of the art" for all critical components (i.e. Main Drive bearing, motors and gearboxes) to ensure that the respective components will last and perform throughout the tunnel project life. In some cases, listing the preferred manufacturers with no substitution can be used (i.e., SKF or Rothe Erde Main Drive bearings). Where substitutions will be acceptable it is advisable to add a requirement to submit reference projects where the respective substitutes were used to verify performance.

Execution

A prequalification of Contractors and proposed TBM is recommended to ensure that the design and work safety and quality are not compromised.

The diagram in Figure 8 a illustrates the differences and commonalities for TBM technical specification drivers and requirements in the two procurement approaches DB and DBB. The diagram is intended as a guide only.

REFERENCES

Jeffrey L. Beard (Author), Edward C. Wundram (Author), Michael C. Loulakis (Author), 2001, Design-Build: Planning Through Development, ISBN.

Dan Ifrim, Derek Zoldy, 2014, TBM Procurement Aspects, NAT 2014.

John Reilly, October 1996 , Owner Responsibilities in the Selection of TBMs, ITA.

http://www.ambergtechnologies.ch/en/.

TBM Tool Wear Analysis for Cutterhead Configuration and Resource Planning in Glacial Geology

Ulf Gwildis and Jose Aguilar
CDM Smith

Kamyar Mosavat
The Robbins Company

ABSTRACT: The abrasive nature of glacial geology generally results in Tunnel Boring Machine (TBM) cutting tool inspection and replacement needs that may require hyperbaric interventions and are a cost and risk factor. Correlation analysis of geotechnical conditions, TBM operational parameters, and tool wear measurements is a proven way to gain insight into the wear system behavior. This paper presents findings from various TBM drives in the Seattle and Vancouver, B.C. metropolitan areas on the performance of disc cutters and ripper-type tools in glacial and inter-glacial deposits. The authors provide recommendations for cutterhead configurations, tool management strategies, and the use of monitoring technology.

SUBJECT INTRODUCTION

TBM tool wear is a complex system behavior. The interacting system components include the subsoil characteristics (grain size distribution, coarse components, relative compaction, mineralogy, angularity of grains, etc.), tooling characteristics (tool type, tool materials, cutterhead design, etc.), the excavation process (TBM type using slurry method or earth pressure balance (EPB) method, the latter with variants of soil conditioning approaches), the way the TBM is operated (tool penetration rate, thrust, cutterhead revolution speed, etc.) and ambient conditions (hydrostatic head, temperature, salinity, etc.).

The term glacial geology encompasses a large variety of deposits ranging from fine-grained lacustrine and marine deposits to coarse grained meltwater (outwash) deposits to ice-contact and till deposits with a wide range of grain sizes including cobbles and boulders (Figures 1 and 2). The boundaries of these deposits are typically highly variable and discontinuous. Deposits overridden by glaciers are often very dense or hard. Generally glacial geology is known to be highly abrasive. Several geotechnical laboratory tests exist with the objective of quantifying soil abrasiveness (quartz content by x-ray diffraction, Miller number test for slurry abrasiveness, NTNU/SINTEF Soil Abrasion Test (SAT™), and others.); however, all these tests cover only a limited number of the soil characteristics that are considered causal factors of TBM tool wear. While these tests are valuable for providing contractual baseline descriptions of the subsoil conditions, the test results do not provide a sufficient basis for reliable tool wear prediction.

Another approach to gain insight into the system behavior of TBM tool wear is correlation analysis of the data provided by past TBM projects. In the metropolitan areas of Seattle and Vancouver, B.C. over the past decade numerous pressurized-face TBM drives have been completed in glacial geology (Table 1). This data base covers EPB and Slurry TBMs, cutterhead designs with a wide range of opening ratios, cutterheads equipped with disc cutters and ripper-type tools as primary cutting tools, and the full range of various glacial and inter-glacial deposits.

This paper analyses data sets of projects listed in Table 1 correlating tool wear data, subsoil data, cutterhead design, tool selection, and TBM operation. The correlation analysis aims at gaining insight into generally applicable wear mechanisms independent of which specific project has provided the data. The paper concludes with some helpful recommendations for future project planning, design, and execution.

CORRELATION ANALYSES OF GLACIAL GEOLOGY TBM PROJECTS

Plotting the accumulative wear of the various tool positions over the advance length of the TBM—the spread of the graphs reflecting the differences in travel path length as function of the tool position radius on the cutterhead—already shows changes in the graphs' gradients (Figure 3). The selected examples show that the changes in gradient are

Figure 1. Excavation exposing glacial deposits in the Seattle area

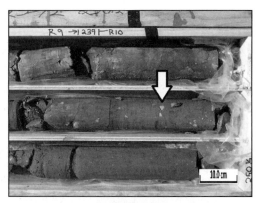

Figure 2. Core of glacially overconsolidated clay matrix with sharp-edged coarser components

Table 1. Seattle and Vancouver, B.C., area TBM tunnel construction contracts in glacial geology

Loc.	Project	Function	Year of Completion	TBM Type	Number of Drives	Length (m)	Diameter (m)
SEA	Beacon Hill Transit Tunnel	Light Rail	2008	EPBM	2	2 × 1,310	6.5
VAN	Canada Line	Light Rail	2008	EPBM	2	2×~2,500	6.1
SEA	Brightwater East Contract	Wastewater	2008	EPBM	1	4,231	5.9
SEA	Brightwater West Contract	Wastewater	2010	EPBM	1	6,424	4.7
SEA	Brightwater Central Contract	Wastewater	2011	STBM	2	6,651	5.4
SEA	Brightwater BT3C Contract	Wastewater	2011	EPBM	1	3,018	4.9
SEA	U-Link U220	Light Rail	2013	EPBM	2	2 × 3,475	6.6
SEA	U-Link U230	Light Rail	2013	EPBM	2	2 × 1,183	6.6
VAN	Port Mann	Water	2015	EPBM	1	~1,000	3.5
VAN	Evergreen Line	Light Rail	2015	EPBM	1	1,974	9.8
SEA	Alaskan Way Viaduct Replacement Tunnel	Highway	2017	EPBM	1	2,825	17.5
SEA	Northgate Link N125	Light Rail	2017	EPBM	6	5,617	6.5–6.6

Note: SEA = Seattle Metro Area; VAN = Vancouver, B.C. Metro Area; EPBM = Earth Pressure Balance TBM; STBM = Slurry TBM

synchronous and that the tool wear rates of linear sections, when normalizing them for tool travel path length, are quite similar. This indicates that the tool wear is a steady and continuous process although at changing rates over the TBM advance length leading through various soil types. In this scenario, the wear can be interpreted as being caused by variable soil abrasiveness characteristics, by changing TBM operational parameters, or by a combination of both. For normalizing the tool wear W (mm) for travel path length L (mm) a normalized wear parameter NWP is used, which is dimensionless and defined as NWP $= (10^8 \times W)/L$. Putting it in context with the specific energy consumption of the TBM (i.e., the product of cutterhead torque and rotation in relation to the excavated soil volume) as well as the average soil abrasiveness of the soil types excavated between two tool wear measurement locations provides a loose correlation (Figure 4). This correlation seems to confirm

the applicability of soil abrasiveness descriptors such as SAT™ values for quantifying tool wear (Gwildis et al., 2010). In this specific project example slurry TBMs with high-opening ratio cutterheads and disc cutters as primary cutting tools were utilized. During the two drives boulders were encountered. Observation of rock fragments in the spoils seemed to indicate that the disc cutters excavated the boulders by rock-chipping mechanism as long as they were held in place by the very dense or hard soil matrix of the glacially overconsolidated deposits. In one case the rock chipping reportedly continued till the time when the direction of cutterhead rotation was changed by the TBM operator and the boulder became dislodged, at which point an intervention for manual removal of the remaining boulder became necessary.

The results of Figures 3 and 4 could not be replicated at a TBM drive through glacially overconsolidated interglacial and then glacial deposits, where

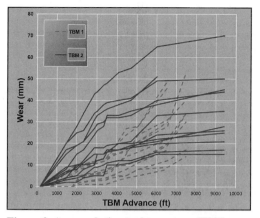

Figure 3. Accumulative tool wear over TBM advance length

Figure 5. Layout of replaced TBM tools per cutterhead positions

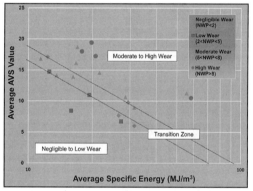

Figure 4. Normalized wear data (disc cutters) plotted over specific cutterhead energy consumption and soil abrasiveness descriptor (SAT™)

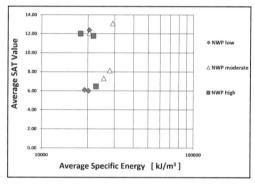

Figure 6. Normalized wear data (ripper-type tools) plotted over specific cutterhead energy consumption and soil abrasiveness descriptor (SAT™)

an EPB TBM with a cutterhead equipped with ripper type tools as primary cutting tools was utilized. Applying the same methodology as before does not show harmonious and continuous tool wear, which would have resulted in a symmetrical wear pattern over the face of the cutterhead (under consideration of tool changes) (Figure 5). Nor are there any correlative trends observable in the diagram of normalized wear over average specific cutterhead energy consumption and average soil abrasiveness descriptors (Figure 6) (Shinouda et al., 2011).

It becomes obvious that in this case a different wear mechanism applies and that soil abrasiveness as quantified by the soil abrasiveness descriptors mentioned earlier is not the driving factor. Comparing tool change intervals and average normalized wear rates with the results of tracking coarse component shards in the TBM muck on a per-ring basis indicates

a correlation while the comparison with specific energy consumption does not (Figures 7 and 8). This finding in conjunction with the non-symmetrical tool wear pattern over the area of the cutterhead points to impact damage sustained by the ripper-type tools when encountering coarse components (boulders, cobbles) as the dominating factor in this specific wear system behavior.

In the years following the publication of these early findings (Gwildis et al. 2010, Shinouda et al., 2011), several additional TBM drives have been completed in similar geotechnical conditions, for which relevant data sets were collected to allow further correlation analysis (Table 1). All these projects include tunnel drives beneath the groundwater table that were excavated in glacially overconsolidated, glacial and interglacial deposits. In all cases the geotechnical conditions were contractually described by a Geotechnical Baseline Report (GBR) grouping the

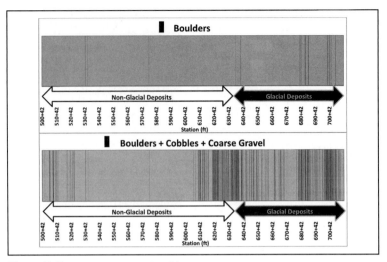

Figure 7. Coarse component tracking

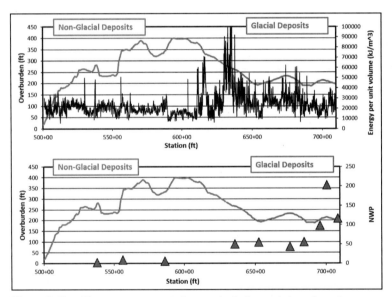

Figure 8. Specific energy consumption vs. tool change intervals and normalized tool wear

highly variable geologic units into soil groups with similar geotechnical characteristics and tunneling behavior and using terms such as Tunnel Soil Group or Engineering Soil Unit. Although defined based on geotechnical criteria, these soil groups nevertheless are a direct reflection of the geologic environment at the time of deposition. Lacustrine and marine sediments were deposited in low-energy environments resulting in mostly fine-grained soil materials. Meltwater (outwash) deposits of high-flow-velocity rivers generally consist of sands and gravels. In comparison, till and ice-contact deposits represent a wide range of grain sizes, often a silty and fine-sandy matrix with large amounts of coarser components that are distributed irregularly and include singular or locally accumulated (nested) cobbles and boulders. Due to the geologic processes related to glacial advances and retreats, boulders tend to accumulate at the boundaries of the till units to overlying or underlying deposits (till contacts). In the following the geotechnical conditions along the tunnel drives are categorized using a simplified terminology into

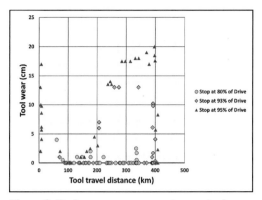

Figure 9. Tool wear measurements over tool travel distances

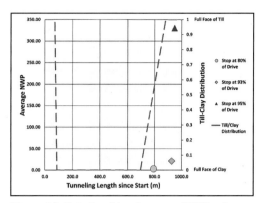

Figure 10. Relationship of average NWP values and geotechnical conditions

clay, sand/gravel, and till, with special mentioning of till contacts.

Case study project 1 includes an EPB TBM with ripper-type tools as primary cutting tools supplemented by disc cutters in gauge positions as well as scrapers. The tunnel alignment was excavated in clay and till. Significant differences in the tool wear rate were recorded during tool inspections at 80%, 93%, and 95% of the drive length. Plotting the wear measurements over the tool travel distance indicates low wear rates at the first inspection stop and significantly increased wear rates thereafter (Figure 9). Plotting the wear normalized for the tool travel distance as an average NWP value over the drive length provides a clearer picture of this wear behavior (Figure 10). Visualizing the advance lengths through the two soil units by the graph of a till-clay distribution factor with "1" indicting a full face of till and "0" indicating a full face of clay provides the geotechnical context and identifies the drive sections through the geologic contact. The relationship between tool wear rates and geologic contact sections does not point to the contact as the main contributor to the significant wear rate increase. This happens mostly in the full face of till near the end of the drive. Reported loss of soil conditioning functionality in this section would indicate lack of the associated mitigating effects (Hedayatzadeh et al., 2017) and therefore should be considered a potential factor.

Case study project 2 also includes an EPB TBM. Over the tunnel alignment section considered in this paper, the TBM cutterhead was almost exclusively equipped with ripper-type tools as primary cutting tools as well as scrapers. The tunnel section was excavated first mostly in clay followed by interfingering layers of sand and till. Plotting specific energy consumption and cutterhead torque over the TBM drive length shows zones with significantly decreased values in the transition between the clay and sand/till sections and within the latter

(Figure 11). The graphs show a fluctuating but overall constant specific energy consumption while the torque, fluctuating in sync, shows an upwards sloping trendline. The tool wear measurements normalized for tool travel length indicate generally low tool wear rates in clay and significantly increased tool wear rates in sand/till (Figure 12). The latter show a wide margin of variation, pointing in addition to continuous wear by soil abrasiveness to other wear mechanisms as causal factors. The presence of till contacts in the sections between inspection stops where the highest normalized wear values were determined points to boulder impact damage as a likely explanation. This explanation is also consistent with the observation that the highest normalized wear values were determined at distal tool positions (positions at large cutterhead radii away from the center) on the planar face area of the cutterhead, where the travel speed and impact forces are high in comparison with other tool positions.

Case study project 3 includes an EPB TBM whose cutterhead was dressed using varying tool configurations of disc cutters and ripper type tools as primary cutting tools in addition to scrapers. The tool types at individual tool positions were changed as the TBM advanced through mixed face conditions consisting of interfingering layers of till and sands and later a full face of sand. At the start of the drive exclusively ripper-type tools were used; however, the tool type was soon changed to disc cutters while the TBM advanced in face conditions with frequent till contacts and the resulting increased probability of boulder encounters. The tool type was changed back to ripper-type tools when the TBM was advancing predominantly through a full face of sands (interlayered with silts and gravels) (Figure 13). Plotting the average of all tool wear measurements and observations for each inspection stop over the drive length—in cases of asymmetric tool wear the maximum extent of material loss on a tool is considered—shows

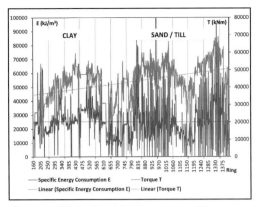

Figure 11. Cutterhead torque and specific energy over TBM drive length

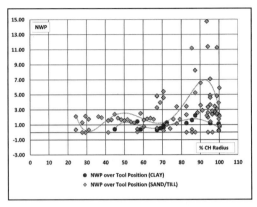

Figure 12. Normalized tool wear (NWP) over cutterhead radius

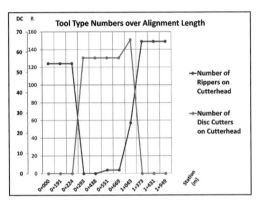

Figure 13. Dressing of the cutterhead with different tool types over the drive length

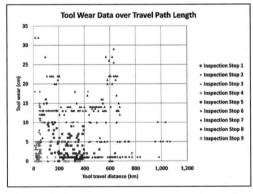

Figure 14. Tool wear measurements at inspection stops

generally higher wear in the mixed face sections of sands and tills as compared to the full-face sections in glacial sands. (Figure 14).

Taking a closer look at the disc cutters reveals in most cases of high wear measurement values that the wear is non-symmetrical due to flat-spotting of the cutter ring or tool damage significant enough that possible flat-spotting in the early stages of damage is not observable anymore. Plotting the disc cutter wear data over the cutterhead radius tool positions shows a trend of higher wear at the inner tool positions, a trend that becomes even more pronounced if the wear data are normalized for tool travel path length (Figure 15). Identifying those tool positions where tool damage or flat-spotting has occurred is mirroring the tool wear distribution pattern, as shown exemplarily for one of the inspection stops (Figure 16).

Flat-spotting of disc cutters is a phenomenon where the cutter loses its functionality of rotating which results in the ring and subsequently the whole tool to experience material loss only at the side that is in contact with the tunnel face. In the past this phenomenon has been associated mostly with fine-grained soils and high adhesive forces (stickiness potential). Observing this phenomenon in predominantly granular soils suggests other causal factors to explain the loss of tool functionality.

The observed phenomenon is interesting especially when compared to the symmetrical tool wear recorded for the slurry TBM operation described earlier (Figures 3 and 4), where flat-spotting and tool damage recordings were rare events despite significant sections of that drive involving fine-grained glacial soils. Further research may be directed at the interaction of EPB TBM operation, soil conditioning, and tool performance. Collecting tool performance data by continuous data logging would likely yield a valuable data basis for this research.

GENERAL TRENDS

The case study projects presented herein generally confirm that fine-grained glacial and interglacial deposits cause significantly less tool wear over the same tool travel path length than coarse grained ones irrespective of the specific energy required to excavate those soil types. This trend applies to slurry TBMs as well as EPB TBMs using either disc cutters or ripper-type tools as primary cutting tools. This general trend is consistent with the results of soil abrasiveness laboratory test procedures such as Miller Number and SAT™, at least qualitatively.

If the distribution of tool wear over the face of the cutterhead is symmetrical, there is an increased probability that some correlations between soil abrasiveness descriptors, TBM operational parameters, and tool wear measurements normalized for tool travel path length can be found (Figure 4). Examples for symmetrical tool wear are provided by Figure 3, where a slurry TBM equipped with disc cutters excavated a variety of glacial soil types including boulder conditions without experiencing significant tool damage, and by Figure 17, where an EPB TBM equipped with ripper-type tools had excavated a tunnel section through face conditions dominated by glaciofluvial sands and gravels.

Where EPB TBMs have been advanced through glacial soils that include till and till contacts, the latter often representing a depositional environment with an increased probability of the accumulation of boulders and cobbles, non-symmetrical tool wear distributions have been observed. Explaining irregular tool wear patterns of ripper-type tools with impact damage by boulders and subsequent accelerated soil abrasion seems likely in many cases (Figures 5, 7, 8, 12).

Where ripper-type tools are used, boulders encountered in the tunnel face are broken into smaller pieces by impact forces. Arguably the applicability of this mechanism is limited by the size of the boulder encountered in the face. When disc cutters are used, the boulder may be excavated by rock chipping similar to the application of disc cutters in hard rock. This requires that the boulder is not dislodged from its soil matrix during the excavation process. In glacially overconsolidated soils, boulders are generally assumed to be embedded in a soil matrix of sufficient strength to offer resistance against being easily dislodged. However, when a non-symmetrical wear and damage distribution of disc cutters on the cutterhead is observed, this suggests a different excavation mechanism than rock chipping. Furthermore, a high number of damaged tools at radius positions close to the center of the cutterhead also points to factors beyond the risk of shock-loading and bearing failure due to boulder encounters (Figure 18).

CONSIDERATIONS FOR CUTTERHEAD DESIGN AND TOOL MANAGEMENT PLANNING

Cutterhead design and tool selection by the tunneling contractor need to consider the requirements of the project specifications and the Geotechnical Baseline Report (GBR). The GBR contractually defines the subsoil conditions by providing geotechnical baselines. In addition to the distribution and engineering characteristics of the geotechnical and geologic units, baselines of specific relevance include soil abrasiveness descriptors, soil stickiness, boulder numbers, size, strength, and distribution, unit boundaries such as till contacts, matrix strength as it relates to the ease of dislodging boulders, as well as baselines for cobbles. The latter, although of lesser significance for cutterhead design, seem to be difficult to quantify. One approach in doing so is by using a Cobble Volume Ratio (CVR) for which empirical relationships to boulder distributions exist (Hunt 2017).

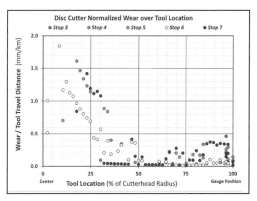

Figure 15. Normalized tool wear distribution of disc cutters by tool position

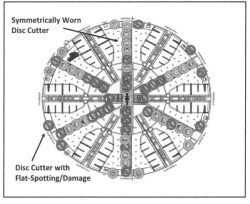

Figure 16. Example of distribution of disc cutter tool damage and flat-spotting on the cutterhead area

Figure 17. Cutterhead with symmetrical tool wear pattern (ripper-type tools)

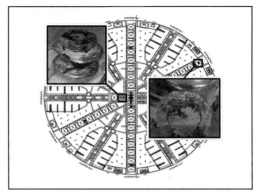

Figure 18. Cutterhead with non-symmetrical tool wear pattern (disc cutters)

For determining the opening ratio as a major parameter for cutterhead design, boulder baselines are a focus. Deciding between high-opening ratio vs. more closed-style cutterheads balances vulnerability regarding boulder impact vs. reduced muck flow and increased torque (Nishimura and Konda 2016). In glacially overconsolidated ground, scrapers are used for conveying the excavated material toward the openings but not as primary cutting tools. Primary cutting tools are either disc cutters or ripper-type tools or a combination of both, the tools being dressed at specific radii and spacing, with or without overlapping travel paths. Both of these primary tool types excavate the ground and deal with boulders differently. The mechanism for excavating boulders is impact crushing for ripper type tools and rock-chipping (as in a full face of rock) for disc cutters, at least theoretically. As presented in this paper, both tool types have shown effective excavation performance in glacial geology but seem to have limitations under certain operating conditions, which result in high wear rates and non-symmetric tool wear distribution over the cutterhead area.

Maintaining tool functionality as the tools wear down is crucial for an effective and ultimately successful TBM advance. This requires regular inspection and replacement stops due to the current lack of reliable tool wear prediction models. While large-diameter TBMs may have primary tools with individual tool locks and atmospheric access via the cutterhead spokes, for common excavation diameters of 5 to 8 m (15 to 25 ft) hyperbaric interventions or safe havens may be required for conducting tool inspections beneath the groundwater table and in unstable face conditions. The presented case study of high disc cutter damage rates in an EPB application in granular soils leads to the question of how to supplement the inspection schedule beyond common tracking procedures of face conditions and TBM operational parameters.

Figure 19. 3-D rendition of disc cutter housing with instrument box

Figure 20. Disc cutter housing with instrument box (rotation speed and temperature)

The authors of this paper suggest considering continuous monitoring of disc cutter functionality by technology that has been used in hard rock applications (Mosavat 2017) (Figures 19 to 21). RPM

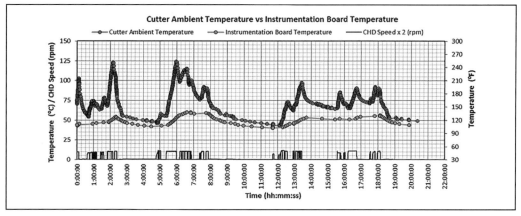

Figure 21. Example of disc cutter performance monitoring graph

recordings provide indications of flat-spotting while temperature readings may indicate the risk of tool blocking or particle cementation and baking, which inhibit effective excavation and material flow into the TBM excavation chamber. Response options could include changes in soil conditioning or adjustments of how the TBM is driven.

REFERENCES

Gwildis, U.G., Sass, I., Rostami, J., Gilbert, M.B., 2010. Soil Abrasion Effects on TBM Tunneling. *ITA-AITES World Tunnel Congress Proceedings*. Vancouver, B.C., June 2010.

Hedayatzadeh, M., Rostami, J., Peila, D., 2017. Impact of Conditioned Soil Parameters on Tool Wear in Soft Ground Tunneling. *RETC Proceedings*. San Diego, California, June 2017.

Hunt, S., 2017. Tunneling in Cobbles and Boulders. 10th Annual Breakthroughs in Tunneling Short Course, Chicago, Illinois, August 2017.

Mosavat, K., 2017. A Smart Disc Cutter Monitoring System Using Cutter Instrumentation Technology. *RETC Proceedings*. San Diego, California, June 2017.

Nishimura, G., Konda, S., 2016. Lessons Learned from EPB and Slurry Tunneling in Glacially Deposited Soils in Seattle, Washington, USA. *ITA-AITES World Tunnel Congress Proceedings*. San Francisco, California, April 2016.

Shinouda, M.M., Gwildis, U.G., Wang, P., Hodder, W., 2011. Cutterhead Maintenance for EPB Tunnel Boring Machines. *RETC Proceedings*. San Francisco, California, June 2011.

The EPB Chamber Air Bubble: What Causes It?

Yuanli Wu, Ali Nazem, and Mike Mooney
Colorado School of Mines

ABSTRACT: Contractors persistently deal with an air bubble in the chamber during EPB tunneling; however, the cause of this is unclear. This paper addresses this problem through a series of unique mixing tests coupled with the study of foam-soil interaction at the bubble and soil grain level. Testing reveals that foam is very stable when properly mixed with soil. Soil water content and foam injection ratio play important roles in influencing bubble migration from conditioned soil. The foam-soil mixing test results suggest that the shearing action of mixing tools creates a pathway for bubble migration. These results suggest that soil conditioning parameters, excavated soil properties, and pressure conditions in the EPB chamber should be all considered in design to eliminate the air bubble in the EPB TBM chamber.

INTRODUCTION

Proper soil conditioning is critical in effective EPB TBM tunneling. Soil conditioning is performed by injecting foam, polymer, water and bentonite at the tunnel face, mixing chamber (or excavation chamber) and screw conveyor (EFNARC, 2005; Alavi Gharahbagh et al., 2013). In EPB TBM tunneling, foam is one of the most commonly used soil conditioners to modify the in-situ soil properties. The desired properties of foam-conditioned soil include elasticity, high compressibility, low shearing strength and permeability, and flowability (or workability) (Budach and Thewes, 2015; Milligan, 2000; Mori et al., 2018; Mooney et al., 2016; Peila, 2014; Thewes et al., 2012; Vinai et al., 2008).

Contractors persistently notice and deal with the accumulation of air at the top of the mixing chamber during EPB excavation. This air bubble must often be purged via air release lines. There is limited literature discussing about the nature of air bubble in the mixing chamber. While the pressurized air bubble can counteract formation water pressure, without shearing resistance it cannot counterbalance the lateral effective stress. This can be an issue during stand still when EPB support pressure drops, leading to inadequate face support and possibility of local ground collapse and material flowing into the chamber (Alavi Gharahbagh et al., 2013; Mori et al., 2017).

This paper investigates the factors that cause the formation of air bubble in the chamber through a series of soil conditioning experiments. The stability of foam bubbles in conditioned soil was investigated by using a foam-soil capture device to obtain bubble size distribution with elapsed time. The volume change of foam-conditioned soil was measured via

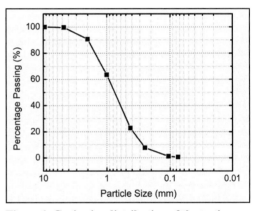

Figure 1. Grain size distribution of the testing sand

a transparent pressure cell. In addition, foam bubble migration from conditioned soil was assessed with different conditioning recipes. Moreover, a series of foam-soil mixing tests were conducted to study if the mixing/shearing process can lead to bubble migration.

STABILITY OF FOAM-CONDITIONED SOIL

The stability of foam-conditioned soil is an important characteristic for conditioned soil. It can be defined as the ability of foam-conditioned soil to maintain its structure and properties throughout residency time in the mixing chamber. The residency time can vary from 30–90 min depending on the diameter of the TBM, depth of the excavation chamber, advance rate, etc. In this study, we evaluate the stability of foam-conditioned soil through a micro-scale study

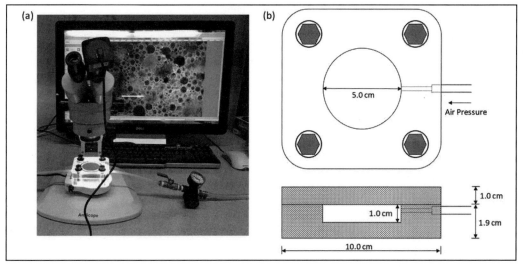

Figure 2. (a) Foam-soil capture device and optical microscope used to examine foam-soil microstructure and bubble size distribution under pressure; (b) schematic of the foam-soil capture device

on foam-soil microstructure and bubble size distribution as well as volume loss of conditioned soil at pressures with elapsed time. A common industrial surfactant was used to produce foam. The surfactant was diluted in water prior to foaming with a concentration of 5% by volume. A poorly graded medium sand (SP) was used for all conditioned soil tests. The grain size distribution curve for the sand is shown in Figure 1. The D10, D50 and D90 for the sand are 0.28, 0.76 and 2.0 mm, respectively. The minimum (emin) and maximum (emax) void ratios for the sand are 0.52 and 0.76, respectively, according to ASTM D4253 and D4254.

Micro-Scale Study on Foam-Conditioned Soil

To investigate the grain-bubble and bubble size distribution of foam-conditioned soil, a foam-soil capture device and an optical microscope with high resolution (1 μm/pixel) were used as shown in Figure 2. The foam-soil capture device is 5 cm in diameter and 1 cm thick (deep). In this test, foam and soil were first well mixed at atmospheric pressure, and then placed into the capture device. Air pressure was applied to the sample to a desired pressure for each test. The microscope was used to image foam-soil interaction with elapsed time.

Previous research has found that pressure has a significant influence on both foam and foam-conditioned soil properties (Williamson et al., 1999; Mooney et al., 2016; Mori et al., 2018; Wu et al., 2017). A foam-conditioned soil that shows ideal properties under atmospheric pressure ($p = 0$ bar) in terms of compressibility, shear strength and

abrasivity can behave poorly at higher pressure. The pressure dependent foam expansion ratio (FER_p) and foam injection ration (FIR_p) are defined in Equation 1–3 (Mooney et al., 2016; Mori et al., 2018). FER_p is the volumetric fraction of foam to the conditioning solution at pressure (p) as shown in Equation 1. FER_p is pressure dependent due to the highly compressible nature of air. According to the ideal gas law, the relationship between FER_p and FER_0 ($p = 0$ bar, atmospheric condition) is shown Equation 2. The volumetric fraction of foam that is mixed with a volume of soil to be excavated is depicted using FIR_p in Equation 3. In this study, both FER_p and FIR_p were kept constant at different pressures. Here, $FER_p = 10$ and $FIR_p = 50\%$ were used in the test. According to Equation 1 and 2, the soil conditioning parameters used in the tests are shown in Table 1. For comparison, foam samples with the same FER_p were also captured with the same pressures to obtain foam bubble size distribution.

$$FER_p = \frac{V_{foam}}{V_{solution}} = \frac{V_{air} + V_{solution}}{V_{solution}} \qquad (1)$$

$$FER_p = 1 + (FER_0 - 1)\frac{p_{atm}}{p + p_{atm}} \qquad (2)$$

$$FIR_p = FIR_0 \frac{FER_p}{FER_0} \qquad (3)$$

Figure 3 shows images of foam-conditioned sand at $t = 0$, 30 and 60 min under applied air pressure $p = 0$, 1 and 2 bar (gage pressure). As shown in Figure 3, foam bubbles appear in the pore space of soil after mixing and serve to expand the grain

Table 1. Soil conditioning parameters used for tests at different pressures

p (bar)	FER_p	FER_0	FIR_p (%)	FIR_0 (%)
0	10	10	50	50
1	10	19	50	95
2	10	28	50	140

structure of the soil. In addition, the number of bubbles decreases and bubble sizes increase with elapsed time, especially for bubbles connecting to each other. For bubbles that lie between soil particles, little change in bubble size was observed over time. The corresponding bubble size distribution curves with elapsed time are also shown in Figure 3. As shown,

bubbles become bigger and the bubble size distribution curve shifts to the right over time at atmospheric pressure ($p = 0$ bar).

The comparison of bubble size distribution of foam-only and foam-soil conditions at pressures is also shown in Figure 3. The results reveal that the increase in bubble size is more significant in the foam-only condition than the foam-soil condition. At $p = 1$ bar, there are negligible changes in bubble size distribution (bubbles of 80% smaller). But the sizes for bigger bubbles increase over time. Compared with the foam-only condition at $p = 1$ bar, there is much less change in bubble size distribution for foam bubbles in foam-soil mixtures. A similar trend

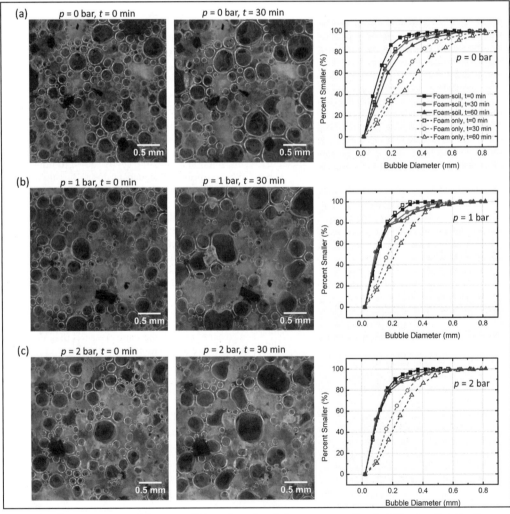

Figure 3. Microstructures of foam-conditioned sand and the corresponding bubble size distribution with elapsed time for bubbles in foam-soil and foam only conditions under pressure

of bubble size distribution was observed in the $p = 2$ bar condition.

Test results reveal that foam bubbles are more stable when foam is mixed with soil than foam-only in terms of less change in bubble size, indicating that soil particles help stabilize foam bubbles. The reason could be that soil particles provide a steric barrier to bubble coalescence and coarsening, and soil particles on the foam films between bubbles cause retardation of liquid drainage.

Volume Loss of Foam-Conditioned Sand

Another method to characterize the stability of foam-conditioned soil is to assess its volume loss over time. During EPB TBM tunneling, it is important for the conditioned soil in the mixing chamber to sustain its volume and engineering properties throughout the residency time. A transparent pressure cell is used to investigate the volume loss of foam-conditioned soil over time as shown in Figure 4. The liquid loss of conditioned soil cannot be recorded in this device. The pressure cell can be also used to study foam bubbles migration (or foam-soil separation) for conditioned soil. This test will be discussed later.

A suite of pressure cell tests was conducted to evaluate the stability of foam-conditioned soil in terms of volume loss over time. The testing sand was treated with water content of $w = 10\%$ and foam with $FIR_p = 30\%$ and $FER_p = 10$. Foam and the testing sand were first mixed at atmospheric pressure and then placed into the pressure cell. Compressed air was applied to the samples for testing under pressure. In order to simulate the movement of foam-conditioned soil in the EPB mixing chamber, the pressure cell was rotated with a constant rate to agitate the conditioned soil during each test, and the rotation direction is shown in Figure 4(a). The volume or height of the

conditioned soil was recorded every 5 min. The duration for each test was 60 min.

Figure 4(b) shows the pressure cell test results. The results show that the foam-soil volume loss is negligible (less than 2%) over 60 minutes. At higher chamber pressures, there was no volume loss of foam-conditioned sand over 60 min. Test results indicate that the foam-conditioned soil is stable in volume over time under pressure.

BUBBLE MIGRATION IN CONDITIONED SOIL

Pressure Cell Tests

To investigate the conditions that might cause foam bubble separation/migration from conditioned soil, a series of tests was performed using the pressure cell described in Figure 4. The testing sand was conditioned with different soil conditioning recipes to investigate bubble migration. Figure 5(a) shows the image of the testing sand with saturated water content of 19.5% without foam, consistent with an in-situ density of the testing sand = 2.09 g/cm³. When the saturated sand was placed into the pressure cell, there was about 1–2 mm of ponded water on the surface of the sand.

When the saturated sand was conditioned homogeneously with foam with $FIR_0 = 50\%$ at atmospheric condition, the separation of foam-water-sand was immediate as shown in Figure 5(b). One can observe that the height of the sand itself has increased 34% due to foam expansion. The amount of water ponding at the surface increases after mixing with foam. This suggests that foam is displacing the pore water. This is consistent with work by Bezuijen et al. (1999) who showed a portion of the pore water in saturated granular soils will be replaced by foam.

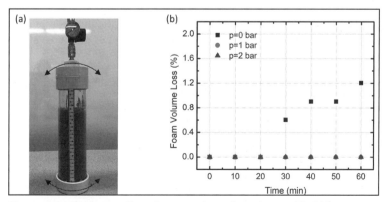

Figure 4. (a) Pressure cell used to examine volume loss and bubble migration of foam-conditioned soil (the arrows indicating the rotation direction for the test); (b) foam volume loss of foam-conditioned sand with elapsed time under pressure

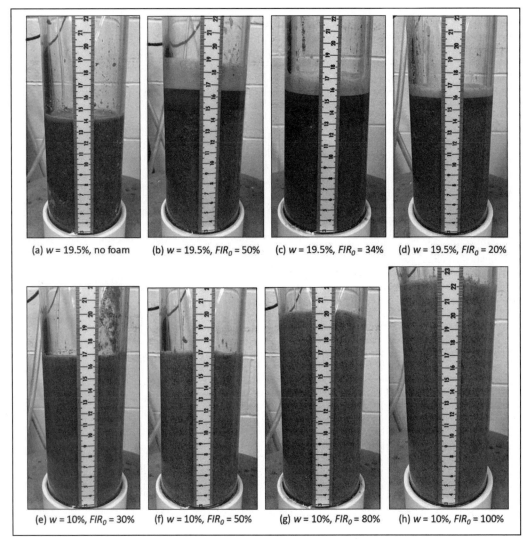

(a) w = 19.5%, no foam (b) w = 19.5%, FIR_0 = 50% (c) w = 19.5%, FIR_0 = 34% (d) w = 19.5%, FIR_0 = 20%

(e) w = 10%, FIR_0 = 30% (f) w = 10%, FIR_0 = 50% (g) w = 10%, FIR_0 = 80% (h) w = 10%, FIR_0 = 100%

Figure 5. Images of bubble migration from foam-conditioned soil with different conditioning parameters at $p = 0$ bar

The actual water content in the sand (not including the ponded water) is 10%. The rest of the foam (calculated as FIR_0 = 16%) floated on top of the water.

As FIR_0 was decreased to 34%, one can still observe the separation of foam-water-sand but with fewer amount of ponded water and foam as shown in Figure 5(c). There was FIR_0 = 22% of foam in the sand and the rest of the foam (FIR_0 = 12%) floated on top of the water. With further decreasing in FIR_0 to 20%, only a small amount of foam (FIR_0 = 6%) and ponded water were observed at the top of conditioned soil as shown Figure 5(d).

The sand was also treated with a lower water content (w = 10%) to study how water content influences bubble migration. There was no observed foam-soil separation in conditioned soil with less foam (FIR_0 = 30%) and water content (w = 10%) as shown in Figure 5(e). Further, there was no observed bubble migration when the testing sand was conditioned with high FIR (up to 100%) and low water content (w = 10%) as shown in Figure 5(f)–(h).

The influence of agitation on bubble migration in foam-conditioned soil was investigated in this study. The agitation was implemented by rotating the pressure cell with a constant rotation frequency of 30 per minute as the arrows indicate in Figure 4(a). The agitation test for the foam-conditioned sand lasted for 5 min to observe if there is any bubble migration.

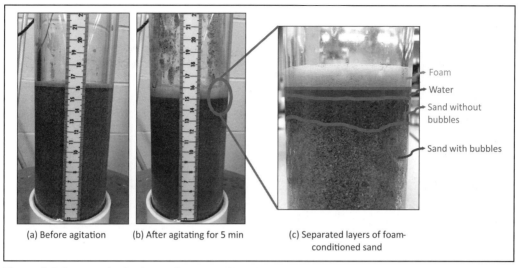

Foam

Water

Sand without bubbles

Sand with bubbles

| (a) Before agitation | (b) After agitating for 5 min | (c) Separated layers of foam-conditioned sand |

Figure 6. Influence of agitation on foam-conditioned sand ($p = 0$ bar)

| (a) Clean sand, $p = 0$ bar | (b) Silty sand, $p = 0$ bar | (c) Clean sand, $p = 1$ bar | (d) Silty sand, $p = 1$ bar |

Figure 7. Influence of soil types on foam bubbles migration at $p = 0$ and 1 bar

As shown in Figure 6(a), initially there was a very thin layer (3–4 mm in thickness) of foam accumulated on top the foam-conditioned sand ($w = 11.7\%$, $FIR_0 = 34\%$). After agitating the sample for 5 min, a 10 mm layer of foam accumulated and also a layer of water was observed between the foam and the soil (Figure 6(b)). The agitation process induces rearrangement of soil structure and thus causes bubble migration. Figure 6(c) shows a magnified image of the separated layers of foam-conditioned sand after agitation. As marked on the image, after agitating the foam-conditioned sand, foam bubbles in the top layer of the conditioned sand migrated upwards, and

therefore there was no observed bubbles in the top 1–2 cm of soil.

Figure 7 shows the influence of soil type on bubble migration. A silty sand was conditioned with foam using the same recipe ($w = 19.5\%$, $FIR_p = 50\%$) as the sand at both $p = 0$ and 1 bar. It was found that there was no observed bubble migration during the test at both pressures, in contrast to the significant foam and water accumulation on top of the clean sand sample. This is because there are more fines in the silty sand, so the pores are smaller than those of the clean poorly graded sand. Therefore, bubbles were trapped in the smaller pores of silty sand. In

Figure 8. Foam-soil mixing setup (dimensions in mm)

addition, more migrated foam was observed at p = 1 bar (gage) compared to p = 0 bar in the clean sand sample (see Figure 7(a) and (c)). This is because foam bubbles are smaller after the conditioned sand was compressed by 1 bar air pressure, though FIR_p was maintained the same. With similar pore size at both pressures (because effective stress is near zero), smaller bubbles will migrate more readily.

Foam-Soil Mixing Tests Under Pressure

Although sections presented above examined mechanics of bubble migration, the conditioned soil was not sheared; this section incorporates the role of shearing the foam-soil mixture on bubble migration. The soil type used here is the same as detailed in earlier sections and water content was kept constant at 19.5% for all tests. However, void ratio was not measured for these tests. A series of mixing tests was performed to investigate the influence of shear and total pressure on bubble migration in the chamber. A pressurized mixing chamber (PMC) was specifically designed to enable examination of foam-soil mixing and bubble migration by injecting foam via stationary ports into the pressurized environment while mixing tools shear the soil specimen. The tests were performed under total pressure and pore pressure was not recorded. A Plexiglas plate was placed to the bottom of the chamber to permit visualization of foam-soil mixing. The mixing tools are placed on a support plate and a central shaft, which transfers rotation to the mixing tools. The setup is then fastened on its side to a frame allowing for horizontal assessment of foam-soil mixing in the chamber. Figure 8 shows various parts and dimensions of the PMC.

This paper presents mixing tests under 0, 1, and 2 bar total pressure (gage). A high-resolution camera was placed in front of the Plexiglas plate to record the foam-soil mixing in a time-elapsed fashion.

Image analysis techniques were undertaken to quantify foam-soil mixing. The foam was dyed with a water-soluble green fluorescent powder to promote its trajectory once injected to the soil sample. The foam was then excited by ultraviolet lights positioned all around the PMC to illuminate the injected foam. A limited time of 6 minutes was set for each experiment and Matlab software was used to process the images taken at a certain time interval.

Although there are numerous indices for mixing quantification (Nalesso et al., 2015), a straightforward mixing index was used, defined as the ratio of the number of green pixels (occupied by foam bubbles) to the total number of pixels in the image (soil grains and foam bubbles). The FIR_p, FER_p, water content, and rotational speed of the mixing tools were kept constant at 50%, 10, 19.5% and 3 RPM, respectively, for all mixing tests. After foam was injected through the port, an image was taken every 20 seconds (1 revolution). Figure 9 shows series of images with the interval of 3 revolutions at various pressures.

As shown in Figure 9, the foam port is highlighted by a green disc and direction of rotation is counterclockwise. It is clear that the foam tends to flow towards the top of the chamber as the soil sample is mixed with foam. One possible reason is that the shearing action of the mixing tools restructures the soil grain network within the reach of the mixing tools, which opens up small fluid channels creating a pathway for bubbles to flow. This phenomenon is accompanied with the effect of gravity (Ottino and Khakhar, 2000), which tends to segregate the soil particles. The shear causes the finer particles to fall down into the voids created by bigger particles replacing portion of foam bubbles or water in the void.

Development of the mixing-induced bubble migration is depicted by a red dotted line on top of

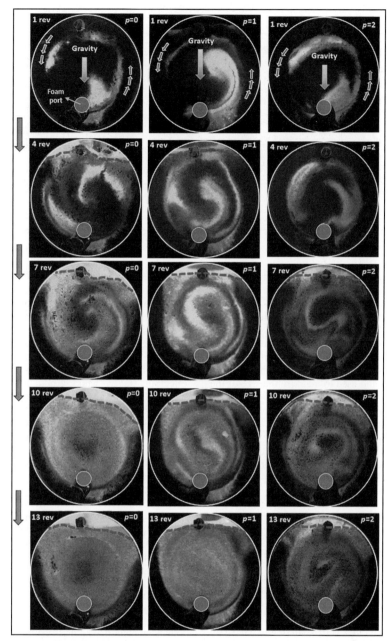

Figure 9. Mixing and bubble migration under various pressure through time

the chamber in Figure 9. Total pressure influences the onset and size of the air pocket accumulated on top of the chamber. The time of first appearance of the air bubble at the chamber top is indicated in Figure 10(a). Under 0, 1 and 2 bar total pressure, the air bubble emerges after 2, 4, and 7 revolutions, respectively. It reveals that it takes longer for the foam bubbles to migrate up the chamber as the total pressure rises. An increase in total pressure causes both effective stress and pore pressure to gain rise. The increased effective stress hinders bubble migration by interlocking the particles and by shrinking down the pore size trapping foam bubbles inside. This is a different scenario analogous to what was

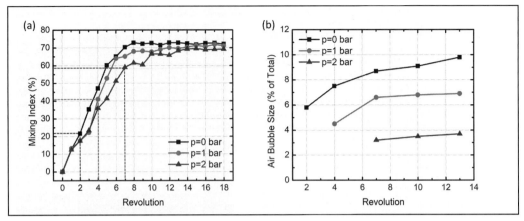

Figure 10. (a) Mixing efficiency and formation of the air bubble; (b) size development of the air bubble on top of the chamber due to mixing

observed in Figure 9, which concluded that induced pore air pressure resulted in more bubble migration. In this latter case, the pore air pressure merely compressed the foam bubbles without manipulating soil pore structures. Therefore, it became easier for the bubbles to migrate up the chamber at higher air pressure. The ratio of the area of the accumulated air on top of the chamber to the total area of the chamber was determined and used to quantify the size of the mixing-induced air bubble. The results are shown in Figure 10(b). In all tests, an air bubble always accumulated at the top of the chamber. The pressure suppresses the size of the air bubble due to lockage and reluctance of soil particles to segregate. Ultimate size of the air bubble under 0, 1 and 2 bar total pressure reaches 9.8%, 6.9%, and 3.7% respectively as shown in Figure 10(b).

CONCLUSIONS

Test results reveal that foam is more stable when it is mixed with soil than the foam only condition in terms of less change in bubble size and negligible volume loss, indicating that soil particles help stabilize foam bubbles. Therefore, the phenomenon of "air bubble" on top of the EPB mixing chamber is not due to the instability of foam in conditioned soil.

The pressure cell test results show that soil water content and *FIR* play important roles in influencing foam bubble migration (or foam-soil separation). Soil with high water content (i.e., saturate water content) and *FIR* could result in significant bubble migration. However, soil with relatively low water content (i.e., $w = 10\%$) but high *FIR* (up to 100%) shows no bubble migration. Soil pore water decreases the capacity of soil containing foam bubbles. In addition, test results show that agitation and soil types greatly influences bubble migration in conditioned soil. The agitation process could lead to the

rearrangement of soil structure and thus cause bubble migration. Soil with more fine particles shows much less bubble migration than coarser soil.

The foam-soil mixing test results indicate that the shearing action of the mixing tools restructures the soil grain network, which opens up small fluid channels creating a pathway for bubbles to migrate upward; this phenomenon is accompanied with the effect of gravity which tends to segregate foam bubbles from soil particles. Another finding from the mixing tests is that the increased total pressure hinders bubble migration by interlocking soil particles, taking longer for the bubbles to migrate upward. There are less accumulated foam bubbles at higher total pressure on top of the chamber.

In practice, to eliminate the "air bubble" phenomenon on top of the mixing chamber, soil conditioning parameters such as FIR_p and FER_p should be examined and selected based on the properties of excavated soil (i.e., particle size distribution, in-situ water content) and pressure conditions. Laboratory testing methods such as the pressure cell test and the mixing test can provide reliable results for investigating bubble migration in EPB soil conditioning.

ACKNOWLEDGMENTS

Financial support for this research was provided in part by BASF. We are very grateful to BASF for this support and for helping to make this research possible.

REFERENCES

Alavi Gharahbagh, E., Diponio, M.A., Raleigh, P., Hagan, B., 2013, Keeping the chamber full: managing the "Air Bubble" in EPB tunneling. Rapid Excavation Tunneling Confonference. pp. 1065–1073.

Budach, C., and Thewes, M., 2015, Application ranges of EPB shields in coarse ground based on laboratory research: Tunnelling and Underground Space Technology, v. 50, p. 296–304.

EFNARC, 2005, Specification and guidelines for the use of specialist products for mechanised tunnelling (TBM) in soft ground and hard rock.

Milligan, G., 2000, Lubrication and soil conditioning in tunneling, pipe jacking and microtunneling, a state of the art review: Geotechnical Consulting Group, G.W.E. Milligan.

Mooney, M., Wu, Y., Mori, L., Bearce, R., and Cha, M., 2016, Earth pressure balance TBM soil conditioning: it's about the pressure, World Tunnel Congress: San Francisco, CA.

Mori, L., Alavi, E., and Mooney, M., 2017, Apparent density evaluation methods to assess the effectiveness of soil conditioning: Tunnelling and Underground Space Technology, v. 67, no. Supplement C, p. 175–186.

Mori, L., Mooney, M., and Cha, M., 2018, Characterizing the influence of stress on foam conditioned sand for EPB tunneling: Tunnelling and Underground Space Technology, v. 71, no. Supplement C, p. 454–465.

Nalesso, S., Codemo, C., Franceschinis, E., Realdon, N., Artoni, R., Santomaso, A.C., 2015. Texture analysis as a tool to study the kinetics of wet agglomeration processes. Int. J. Pharm. 485, 61–69. doi:10.1016/j.ijpharm.2015.03.007.

Ottino, J.M., Khakhar, D. V., 2000. Mixing and Segregation of Granular Materials. Annu. Rev. Fluid Mech. 32, 55–91.

Peila, D., 2014, Soil conditioning for EPB shield tunnelling: KSCE Journal of Civil Engineering, v. 18, no. 3, p. 831–836.

Thewes, M., Budach, C., and Bezuijen, A., 2012, Foam conditioning in EPB tunneling, in Viggiani, ed., Geotechnical Aspects of Underground Construction in Soft Ground: London, Taylor & Francis Group.

Vinai, R., Oggeri, C., and Peila, D., 2008, Soil conditioning of sand for EPB applications: a laboratory research: Tunnelling and Underground Space Technology, v. 23, no. 3, p. 308–317.

Wu, Y., and Mooney, MA., 2017, An experimental examination of foam stability for EPB TBM tunneling. Tunnelling and Underground Space Technology, in review.

Mori, L., Alavi, E., and Mooney, M., 2017, Apparent density evaluation methods to assess the effectiveness of soil conditioning: Tunnelling and Underground Space Technology, v. 67, no. Supplement C, p. 175–186.

Mori, L., Mooney, M., and Cha, M., 2018, Characterizing the influence of stress on foam conditioned sand for EPB tunneling: Tunnelling and Underground Space Technology, v. 71, no. Supplement C, p. 454–465.

Nalesso, S., Codemo, C., Franceschinis, E., Realdon, N., Artoni, R., Santomaso, A.C., 2015. Texture analysis as a tool to study the kinetics of wet agglomeration processes. Int. J. Pharm. 485, 61–69. doi:10.1016/j.ijpharm.2015.03.007.

Ottino, J.M., Khakhar, D. V., 2000. Mixing and Segregation of Granular Materials. Annu. Rev. Fluid Mech. 32, 55–91.

Peila, D., 2014, Soil conditioning for EPB shield tunnelling: KSCE Journal of Civil Engineering, v. 18, no. 3, p. 831–836.

Thewes, M., Budach, C., and Bezuijen, A., 2012, Foam conditioning in EPB tunneling, in Viggiani, ed., Geotechnical Aspects of Underground Construction in Soft Ground: London, Taylor & Francis Group.

Vinai, R., Oggeri, C., and Peila, D., 2008, Soil conditioning of sand for EPB applications: a laboratory research: Tunnelling and Underground Space Technology, v. 23, no. 3, p. 308–317.

Wu, Y., and Mooney, MA., 2017, An experimental examination of foam stability for EPB TBM tunneling. Tunnelling and Underground Space Technology, in review.

Session 4: TBM Technology 2

Aaron McClelland and Dave Mast, Chairs

Digitization in Mechanized Tunnelling Technology

Karin Bäppler
Herrenknecht AG

ABSTRACT: A major step in the digital enhancement of products and services in mechanized tunnelling technology is to optimize the advance processes as far as possible. New and enhanced developments such as highly complex and powerful information systems, sensitive sensor systems and imaging techniques are particularly relevant for the users of TBM technology. They are key for the highest level of operational safety with measurable and verifiable progress.

The multitude of processes and data, in consideration of digital processes and tools, offer certain advantages including tactical and strategic options during tunnel advance. Machine performance and associated systems can be optimized at any time and precise real-time insights can be gained into the interaction of the machines with their geological or topographical environment.

This paper focuses on digitization in mechanized tunnelling technology and in particular on the interaction between machine, cutting tools and prevailing geology and thus the real-time networking of products and processes.

INTRODUCTION

The basis for the digitization of TBM technology is the availability of relevant information in real-time through the networking of all entities involved in the project, as well as the ability to derive the optimal value from the data at any time. By linking people, objects, processes, and systems, it is possible to achieve dynamic, real-time and self-organizing cross-company value-added networks which can be optimized according to different criteria such as cost, availability or resource consumption.

Digitization and networking in mechanized tunnelling technology is not a novelty. Key innovations developed and implemented in recent years have led to more and more value-added networks, with optimized processes related to information, economy and safety in tunnelling.

This paper emphasizes the features in digitization and networking in TBM technology such as data acquisition, processing, analysis and interpretation, that can achieve the values added in respect to safety, sensor technology, remote maintenance, and visualization. Recently developed high-tech systems enable the user and machine operator to have more precise information such as the real load and wear patterns of the cutting tools and cutting wheel structure in areas that are not accessible during the advancement process. These high-tech systems include:

- Wireless Disc Cutter Rotation Monitoring (DCRM) system for each disc cutter

(Figure 1): Real time monitoring of the rotational motion and the temperature of the disc cutter during the excavation process
- Wireless Disc Cutter Load Monitoring (DCLM) system: Monitoring of the rotary movement and temperature of the disc cutters, enabling optimization of the tool maintenance intervals. In addition, a camera system provides photos of the tunnel face to the monitors in the control cabin.
- Direct and indirect wear sensor systems
- Seismic and geological exploration in soft ground and hard rock
- Comprehensive process data management to monitor key construction and production processes at all times

Each tunnel project is a highly complex process in which all the individual steps must follow a precise master plan. Therefore, the continuous monitoring of the construction progress at every stage and providing real-time information for everyone involved are of enormous importance. The following sections of this paper focus on successful international tunnelling projects, which address the relevant and value-added key innovations and developments associated with the application of TBM technology towards more specific geological conditions and shallow or deeper alignments with high hydrostatic pressures for subsea crossings.

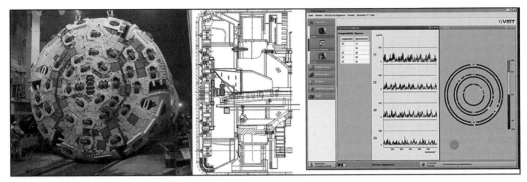

Figure 1. DCLM system; real time measurement of disc cutter load

DIGITIZATION WITH FOCUS ON DESIGN ENGINEERING AND PROCESS ENGINEERING ASPECTS

In addition to design engineering aspects, process-engineering questions are considered as the tunnelling projects of today are more complex in respect to prevailing geological conditions. Of particular relevance is the interaction of the geotechnical risks, choice of suitable machine type and design in order to safely and reliably excavate and line the tunnel. A detailed and thorough evaluation of all risks related to the tunnel works are of vital importance to the technical and economic success of TBM advance. Geotechnical engineering provides an important basis of decision making for the optimized choice of tunnelling technology and ensuring for example:

- Controlled support of the tunnel face
- Reliable cutting process of the ground with minimized clogging and wear of tools and cutting wheel structure
- Smooth material transport from the excavation chamber, handling of muck on jobsite, dumping or re-use of muck
- Required performance rates

Data Management System

In an urban environment, settlement controlled advance constitutes one of the most important decision criteria. A sensitive control of the support pressure to achieve face stability is essential in urban areas when crossing beneath buildings or other structures with low cover.

As an example, the 2.4-kilometer long twin-tube Waterview road tunnel project in Auckland was successfully built in an urban area with a settlement controlled boring process. The very large diameter tunnel drives were constructed using a 14.41m diameter EPB machine. The zones of influence above the drives encompassed a large number of buildings and utilities as the tunnel alignment passes beneath

residential properties, public reserve, and beneath the North Auckland Railway branch. The large diameter TBM operation was a focus of publicity and the demand for a settlement controlled boring process was of paramount importance. A data management system from VMT GmbH integrated settlement measurements and TBM data in a single data platform. This approach allowed instant cross-evaluation of the surface and building settlement and with particular focus on volume loss evaluation in near real-time, comparison with limiting and expected values, alarm function and correlation between TBM process data and settlement.

With this system, the information from all tunnelling related processes could be combined for monitoring, reporting, and analyses with the benefit of assisting management of the construction process to make decisions. Shift reports in the form of Gantt charts could be automatically generated by making use of machine bits indicating the current state. This could be either "advance," "ring build" or "downtime." Inside the control cabin on the TBM, the shift engineers used this automatic data to create comprehensive shift reports to enable the tunnel management at the job site offices to monitor progress. Moreover, using that data management system, the team was able to review and ensure efficiency of processes in quasi real-time.

Tunnels such as the Waterview Project in New Zealand, with demands associated with crossing beneath important sensitive structures, show that digitization in TBM technology and in real-time networking of products and processes help to facilitate the safe execution of such projects in sensitive environments.

The information and data management systems on TBMs provide constant quality of supervision and enable an efficient analysis of a huge volume of data including relevant information relating to settlement monitoring and critical conditions to ensure that all parties involved in the process are aware of the actual situation at all times.

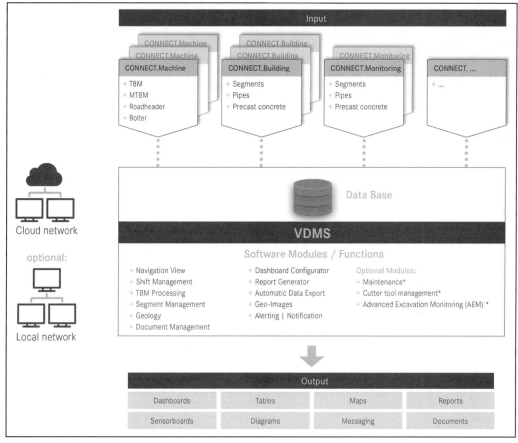

Figure 2. Comprehensive data processing with TBM tunnelling

Sensitive Sensor Systems

Mixed face conditions in mechanized tunnelling are not unusual. A variation of soil and rock constituents or encountering cobbles and boulders at the tunnel face can even vary from ring to ring so that generally the geological distribution along the tunnel alignment remains an uncertainty.

For the construction in 2003 of the Metro Line 9 in Barcelona (Spain), a 12.06-meter diameter EPB Shield excavated a 10-kilometer long section between the stations Gorg and Sagreda and between Zona Franca to Zona Universitaria. The subsurface conditions along the relevant section were characterized by variable geology comprised of sand in a clay matrix, clay, boulders, and a section with granite blocks. To deal with soft ground, hard rock and boulders, the cutting wheel was designed with disc cutters and soft ground tools. To support an efficient excavation and to avoid possible damage to the tools and the steel structure of the cutting wheel, Herrenknecht engineers developed an electronic tool monitoring

system for both, soft ground tools and buckets. Two buckets and four soft ground tools were equipped with a wear detection system that is capable of providing data on the tools online and thus alerting the TBM operators in real time.

In 2014 and 2015, the first road tunnel was built beneath the Bosporus Strait in Istanbul. A Mixshield with an excavated diameter of 13.71 meters excavated and lined a 3.34-kilometer subsea tunnel at a depth of more than 100 meters below sea level. The project encountered challenging tunneling conditions that were without precedent. The highly specialized and well-adapted TBM successfully dealt with the challenges of a large diameter bore in combination with challenging subsurface conditions and face pressures well above 10 bar.

The machine design incorporated newly developed systems for remote monitoring of the condition of the excavation tools and the cutting wheel structure. These systems reduced the number of hyperbaric chamber interventions and significantly increased their predictability. With such systems,

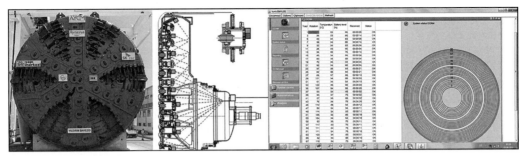

Figure 3. DCRM system; real time cutter temperature and detection of cutter blockages

as summarized in the following sections, time-consuming hyperbaric chamber interventions for tool inspection can be avoided or reduced. An added-value can be seen also with respect of the safety of the overall processes of maintenance and tool change as chamber interventions only take place when actually required and are not based on suspicion or guesswork.

These newly developed high-tech systems summarize the installed status control or "remote maintenance" monitoring system:

- Cutting Wheel—detection of the cutting wheel load and the eccentric load in real time using the operating data processing system
- Cutting Wheel Structure—regular ultrasound thickness measurement at defined measuring points
- DCRM system for each disc cutter—real time monitoring of the rotational motion and temperature of the dis cutter during the excavation process
- Monitoring Loops—include monitoring loops in certain sectors of the cutting wheel to detect any structural wear on the front and back of the cutting wheel
- Camera—permanently installed camera in the working chamber

Key aspects such as high hydrostatic pressure, large excavation diameter as well as complex geological conditions along the tunnel alignment beneath the Bosporus were the basis for the developed systems to gain precise real-time insights into the interaction of the machine with the geological environment. Such solutions, intended for the machine design, improved the job safety of personnel.

To ensure safety and security, sensitive gas detector sensor systems are installed on a TBM to detect possible special hazards such as the potential of methane gas. Between 2011 and 2013, one of the largest EPB machines excavated two tunnel sections for the Galleria Sparvo project in Italy. The machine had a diameter of 15.6 meters. A key hazard was

the potential presence of methane gas. The machine was specially designed to cope with this potential hazard with gas detectors that were coupled to switches that could shut down the power supply to the machine if gas concentrations were measured above the specified threshold levels. Portable measuring devices were used to measure the concentration of combustible gases. Further design aspects to control this potential hazard were focused on design features such as providing a continuous feed of large volumes of fresh air to dilute any gas that may be encountered. During TBM operation, the excavation chamber was always completely filled with excavated muck to prevent an inflow of material into the working chamber in case unstable tunnel face conditions were encountered. For this project, the main reason for an overall closed mode EPB operation was to prevent the possible risk of formation of an explosive mixture in the working chamber due to potential presence of gas in the rock mass. In the back-up area, technical measures were taken to prevent any concentration of methane. To eliminate the risk of explosion and to counteract the prevailing risk of any gas accumulating in the shield and back-up area where people were working, the machine design included a double-walled enclosure for the back-up conveyor belt. This was designed from the screw discharge gate to the transverse conveyor belt with permanent ventilation inside and outside of this system to adequately dilute any gas released from the excavated material that was discharged from the screw conveyor. The area from the transfer belt conveyor and loading chute to the tunnel belt was not covered, thus that from this point on tunnel was equipped with fully explosion proof equipment. The air quality and the effectiveness of the sealing system were continuously monitored. The constant level monitoring of the working chamber ensured that the chamber was always completely filled to avoid the danger of gas pocket formation. This allowed a controlled excavation process through the sections with a potential for gas presence.

Within the scope of digitization, it is of importance to have the relevant information, and in this

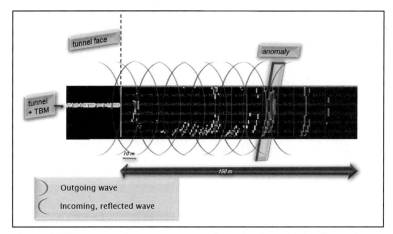

Figure 4. ISP system; advance accompanying seismic prediction in hard rock

case, real-time communication and monitoring ensured safe working conditions.

Imaging Techniques

Currently, projects are becoming increasingly complex with in respect of the prevailing geological conditions. Nevertheless, the applied technology needs to ensure safe and successful operation processes. Everybody involved in the tunnelling process is aware that the geology faced during tunnelling can differ significantly from the predicted subsurface conditions, which can result in complex and uneven procedural consequences in the project chain. With the completion of the excavation for the Coca-Codo Sinclair Project in Ecuador, two double shield hard rock TBMs (Ø9.04m) excavated the 24.8-kilometer long headrace tunnel. The project geology was characterized by medium to hard rock with UCS in the range of 30 to 130MPa, and a total of 27 fault zones were predicted along the tunnel.

Due to uncertainties in geology, one of the two double shield hard rock TBMs was equipped with a tunnel seismic prediction for hard rock conditions (Figure 4). The system called ISP (Integrated Seismic Prediction) System can be retrofitted to the TBM. It features a high investigation depth and can thus be used as an early warning system for the TBM drive. Possible geological anomalies can be identified and verified and investigated in more detail with conventional investigation methods such as probing drilling and, if necessary, treated by advance ground treatment ahead of the machine Systematic advance probing, which would be required in sections with uncertain ground conditions can be thus omitted, saving time and costs.

The ISP advance probing system can be used during tunnelling in hard rock and consists of three components: the source, the receivers with data

loggers and the computer unit with software for control, inspection and data acquisition. The system operates by using a pneumatically driven impact hammer to strike the tunnel wall to induce seismic waves into the rock mass. If the seismic waves are reflected by a seismic anomaly, they are received by three-component geophones, which are drilled into the tunnel wall. Then the data are logged and sent to the computer unit in order to process the data and to interpret the results. The ISP system works during the TBM advance and its high-level machine integration enables a partly automated process, which results in continuous availability of the system and a measuring activity that does not hinder TBM operation and advance.

Another imaging technique that can be used in soils with potential of natural or artificial obstacles or karstic features is the Seismic Softground Probing (SSP) system (Figure 5). This system is available for soft ground TBMs with liquid supported tunnel face and can be used to increase safety and to reduce costs that may result from unpredicted obstacles causing stoppages or damage to TBMs. This "added

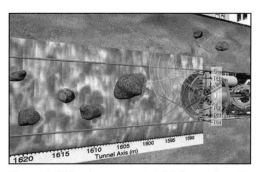

Figure 5. SSP system; structural information of the ground prior to its excavation

safety" system gives the machine operator a warning in advance and at sufficient distance to allow time to react. Changes in geology like layer boundaries, boulders, voids or contrasts in density may be recognized. The system was used for the first time in the Mixshield that excavated the 4th Tube Elbe Tunnel in Hamburg in 1997 and used since then in Mixshield operations in Europe, Russia, Malaysia and China.

The system functions by using a sound transmitter mounted on the cutting wheel that sends a frequency coded sound signal into the ground ahead of the shield and reflected signals are detected by receivers. From the measurement of values of acceleration, amplitude and travel time of the reflected sound waves, a 3-dimensional ground model can be created showing reflections and velocities in the ground up to 40 meters ahead of the tunnel. The measurement is carried out automatically during excavation. All data, digitized in the cutting wheel, is transferred via a slip ring to the processing system on the backup gantry.

Three computers, a controller and an amplifier, control data acquisition, processing and visualization. The final result is a 3-dimensional model of the ground ahead showing position and orientation of reflectors up to 40 meters away.

Tunnel boring machines, both for soft and hard rock, are designed and equipped with monitoring systems that are adapted to the specific project demands. The ability to complete the tunnelling process with sufficient monitoring information and methods of control are essential prerequisites for efficient TBM tunnelling. For this purpose, state of the art surveying and communication technology is used in the control cabins of the Herrenknecht tunnel boring machines. This allows the display and recording of all relevant data and the control of the machine, from microtunnelling machines to large-diameter TBMs.

CONCLUSION

Tunnelling projects are highly complex and follow a precise master plan. To monitor the different construction processes and interfaces, it is important that real-time information is available for everyone involved in the process at all times. Information and monitoring systems such as; tools for geological exploration; and detection systems for cutting tool and structural wear optimize all tunnelling processes with highly sensitive multi-sensor networks. They provide the basis for the contractors to monitor key construction and production processes at all times. Communication in the tunnel and on large jobsites is also playing an increasingly important role in the coordination of construction processes and providing for a safe working environment. Data management systems are integrated systems for process data management that link up all data concerning the machine technology, the navigation and the construction progress in real time, thereby increasing process expertise. This digitization in mechanized tunnelling is not a novelty but is increasingly gaining importance because the availability and networking of relevant information in real-time can enable criteria such as cost, availability and resource consumption to be optimized. This is not only of relevance to the tunnelling industry but the principles could also be applied to many other industrial processes.

Incorporating Geological and Geotechnical Spatial Variability into TBM and Ground Settlement Risk Assessment

Jacob Grasmick, Michael Mooney, and Whitney Trainor-Guitton
Colorado School of Mines

ABSTRACT: This paper describes a geostatistical approach for assessing the spatial variability and uncertainty in geology and geotechnical parameters from site investigations for urban tunneling in soft ground. Geostatistical analysis conveys important spatial trends that would otherwise be missed in classical statistical analysis (e.g., mean, standard deviation, range). To assess the levels of uncertainty in geology/geotechnical parameters, sequential indicator and Gaussian simulations were performed using site investigation data from a real tunneling project. The implications of measured uncertainty on tunnel boring machine performance risk are discussed. Furthermore, the paper demonstrates how results from sequential geostatistical analysis can be extended to assessing the risk for exceeding allowable ground deformation limits.

INTRODUCTION AND BACKGROUND

Geological/geotechnical spatial variability and uncertainty is a common risk in soft ground urban tunneling, largely due to sparse nature of borehole sampling. While current practice takes into consideration the uncertainty in geological/geotechnical parameters globally, spatial uncertainty is generally not considered. However, studies have shown that spatial variability can have a significant impact on ground deformation behavior in soft ground tunneling (Grasmick and Mooney 2017; Xiao, Huang, and Zhang 2017; Huber 2013). Spatial variability/uncertainty is also presumed to impact other tunnel performance like advance rate, cutter wear rate, etc. (Maria Stavropoulou, Xiroudakis, and Exadaktylos 2010).

In other applications (i.e., mining, petroleum, environmental science, hydrogeology, and hard/soft rock tunneling), spatial uncertainty is modelled using geostatistical methods (Chilès and Delfiner 2012; Webster and Oliver 2008; M. Stavropoulou, Exadaktylos, and Saratsis 2007; Pyrcz and Deutsch 2014). Geostatistical simulation is a well-accepted method for characterizing heterogeneous reservoirs or subsurface conditions, in part because it captures the heterogeneous character observed in many subsurface environments (Pyrcz and Deutsch 2014). This stochastic approach allows for the estimation of many equally probable realizations, which can be post-processed to quantify spatial uncertainty.

This paper presents an example geostatistical application where spatial uncertainty is estimated for geology from strata classification records and Standard Penetration Test (SPT, N1(60) values) from an urban soft ground tunneling project in Queens, NY. Variograms describing the spatial structure of the strata and SPT data are presented and interpretations discussed. A sequential indicator simulation approach was used to assess the levels of spatial uncertainty relating to geology using the strata classification records. A sequential Gaussian simulation approach was used to assess the levels of spatial uncertainty relating to geotechnical parameters using SPT N1(60) records. The implications of test locations, density and overall variance of the data have on measured uncertainty is discussed. The geostatistical results are extended to the risk assessment for ground deformation.

OVERVIEW OF GEOSTATISTICS

The variogram (or semi-variogram) is a model explaining the spatial variation (inverse of correlation) of a parameter (Figure 1). The sample variogram is derived by taking the average square difference of all pairs of data points at binned lag distances. The idea here is that as the distance between pairs increases, so does their variance (i.e., they become less correlated). The equation for calculating sample semivariance is:

$$\gamma(h) = \frac{1}{2N(h)} \sum_{N(h)} [z(u) - z(u+h)]^2 \qquad (1)$$

where: $\gamma(h)$ is the semivariance (one-half of the variance) at lag distance h, $N(h)$ is the number of data pairs separated by h, and $z(u) - z(u+h)$ is the parameter difference between two pairs separated by h, with u representing the location vector. For indicator techniques where categories or classes of values are

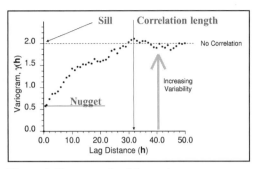

Figure 1. Components of the variogram

simulated, the data is coded as binary (1 if the category exists; 0 otherwise). The components of the variogram are:

- Nugget: The y-intercept, semi-variance at zero distance, and therefore, considered to be a combination of measurement error and micro-scale variability
- Correlation length (or Range): the lag distance beyond which the variance is constant and the data is no longer correlated.
- Sill: The maximum semi-variance of uncorrelated data, i.e., at distances greater than the correlation length.

Three dimensional variogram models enable modeling anisotropy in multiple directions. For example, due to the deposition process, the correlation length is often larger in the horizontal direction than in the vertical direction. The anisotropy ratio describes the ratio between horizontal to vertical correlation lengths and can vary from 1.0 to 100+. In addition, nested variogram models can account for different sill and nugget in various directions, or zonal anisotropy (Gringarten and Deutsch 2001).

While deterministic kriging methods can provide a conservative spatial interpolation of the data, sequential (or stochastic) geostatistical methods have further advantages. Geostatistical simulation methods such as sequential indicator and Gaussian can be utilized to estimate measures of uncertainty for a parameter(s) of interest (Webster and Oliver 2008). In addition, each realization in these sequential methods reproduce the variogram components while capturing the degree of heterogeneity that is otherwise 'smoothed out' in deterministic kriging methods. These realizations can be directly applied to engineering design problems for more accurately modeling performance (e.g., ground deformation). This is particularly the case for geotechnical parameters which are inherently variable even within a 'homogenous' soil unit.

PROJECT OVERVIEW

For this study, site investigation data from the East Side Access Queens bored tunnels project, constructed by the joint venture of Granite Construction Northeast, Inc., Traylor Bros. Inc., and Frontier-Kemper Constructors, Inc. (Robinson and Wehrli 2013) was used to demonstrate the applicability of geostatistical analysis. Along the tunnel alignment, the ground conditions consist of highly variable glacial till soils and outwash deposits that include various sandy soil with small lenses of clay, silt and gravel. In addition, large boulders (up to 2 ft in diameter) and cobbles were frequently encountered in the glacial till stratum. The majority of the tunnel drives were excavated in the glacial till soil. The geology is summarized in Table 1.

The project site layout with data colored according to borehole depth (plan view), strata classification, and SPT N1(60) records (both cross-sectional) is presented in Figure 2. Partial SPT N-values were extrapolated linearly to blows/30 cm (e.g., 50/15 cm = 100/30 cm).

Table 1. Subsurface strata summary

Strata	Description
F (Fill)	Very loose to very dense sands with silts/clays (10–30%), gravel and misc. debris. SPT N-values ranged from 3 to >60, but generally 25 to 30.
MGD (Mixed Glacial Deposits)	Loose to very dense, coarse to fine micaceous sands with silts/clays (2–15%). SPT N-values ranged from 5 to 63, but were generally 10 to 30.
GT (Glacial Till/ Outwash)	Heterogeneous mixture of medium dense to very dense sands with silts/clays (5–30%), gravel and boulders. SPT N-values ranged from 8 to >60, but generally were 30 to 60.
DR (Decomposed Rock)	Very stiff silts and clays (35–55%), with sands and gravel. SPT N-values generally >50.
BR (Bedrock)	Unweathered to moderately weathered, strong to very strong gneiss.
GC (Gardiners Clay)	Silts, clays and sands, fines contents generally ranged from 20 to 35%. SPT N-values generally exceed 40.
JG (Jameco Gravel)	Very loose to dense fine micaceous sands and silts/clays (9–62%) with gravels and boulders. SPT N-values generally >45.

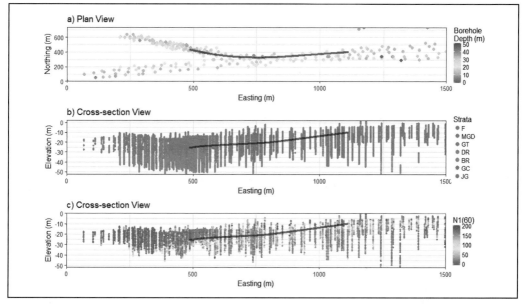

Figure 2. Site investigation configuration with designed tunnel alignment for one of the four tunnels: (a) Plan view; color corresponds to total depth of respective borehole. (b) Side view; color corresponds to strata. (c) Side view; color corresponds to reported N1(60). Elevation relative to NAVD88.

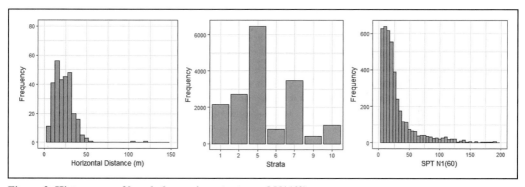

Figure 3. Histograms of borehole spacing, strata, and N1(60)

Figure 3 presents the histograms for the borehole spacing, strata classification and SPT N1(60) data. Additional statistics for the site investigation data are summarized in Tables 2 & 3. For this project, 301 boreholes were drilled for the site investigation. The typical horizontal (plan view) borehole spacing ranges between 1–50 m with an average spacing of 23 m. The borehole density within the zone of influence (ZOI: defined as within 100 m offset from the tunnel alignment) as 495 m²/borehole. The average vertical sample (strata and SPT) spacing was 1.2 m. Glacial till was the primary soil type encountered in this project. The majority of the tunnel drive encountered glacial till and mixed glacial deposits (see Figures 2a and 5a). The reported SPT N1(60) values follow a lognormal distribution and range from 0–540 with an average of 32 and standard deviation of 47.

VARIOGRAM ANALYSIS

Indicator and continuous variograms, calculated with Equation 1, for strata and for N1(60), respectively, are presented in Figure 4. For SPT N1(60), one model variogram representing the entire project site (all strata together) was used here. A more detailed geostatistical analysis would assess the spatial variation for each strata independently. By using one model for all strata, one assumes the variation in soil type is reflected in the variation in SPT N1(60).

Table 2. Summary of site investigation data

Length of Tunnel Drive (m)	# Boreholes	Average Borehole Spacing (m)	Average Sample Vertical Spacing (m)	Borehole Density in ZOI (m²/borehole)
650	301	23	1.2	495

Note: ZOI defined as within 100 m offset from tunnel alignment.

Table 3. Summary of N1(60) distribution

# SPT Records	Range	Average	Median	Standard Deviation
4364	0–540	32	19	47

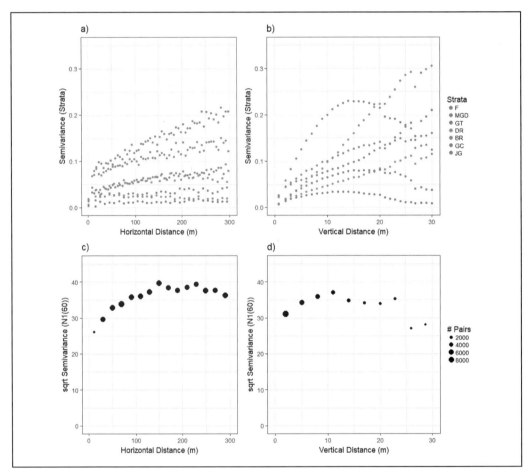

Figure 4. Sample variograms of strata (a, b) and SPT N1(60) (c, d)

Indicator variograms quantify the degree of spatial continuity for a particular category and distance. The semivariance represents the probability of transitioning from inside one strata to another. The sill of the indicator variograms relates to the overall proportions of each strata. Local predictions with kriging are interpreted as best estimates of the probability that the category prevails. From the indicator variograms presented here, it can be seen that more prominent and continuous strata include glacial till (GT), mixed glacial deposits (MGD), fill (F) and bedrock (BR). Furthermore, the cross variograms (not shown) convey the transition probability from one strata class to another. For tunneling, this provides insight into the most probable strata class to be encountered when tunneling in a particular stratum. For example, GT has a higher probability of transitioning to F or MGD than any other strata class, so

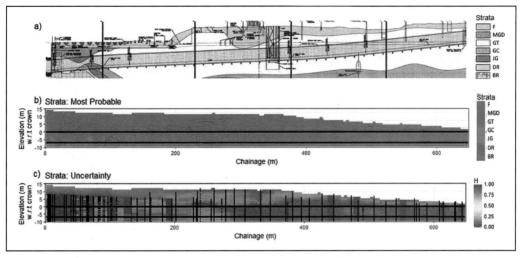

Figure 5. (a) Geologist's interpretation, (b) most probable, and (c) uncertainty (entropy). The black points in (c) correspond to borehole sample locations within 50 m of the alignment. Elevation in (b) and (c) are with respect to the crown of the tunnel alignment (solid black lines).

it should be expected that transitioning from GT-to-F and GT-to-MGD (or vice-versa) is more probable than transitioning from GT-to-JG.

The continuous variograms quantify the spatial continuity of the data, with the semivariance representing the variance (σ^2) between two data points at any separation distance. The sill relates the overall variance of the data. From the SPT N1(60) variogram models presented here, it can be inferred that the variance in the horizontal and vertical directions are similar (both have similar sill). Furthermore, the horizontal-to-vertical (H:V) anisotropy ratio is approximately 12:1, based on the ranges of the variograms for the two directions (~150 m and 12.5 m for horizontal and vertical directions, respectively). The horizontal range provides insight into the general rate or distance over which ground conditions (SPT N1(60)) will change by an order magnitude of 1 standard deviation. In the vertical direction, the range provides insight into how ground conditions may change across the face of the TBM at any given chainage during excavation. For example, given a ~7 m diameter tunnel, one can expect SPT N1(60) values to vary, on average, ±35 over the face, per Figure 4d.

SEQUENTIAL INDICATOR SIMULATION OF GEOLOGY

Sequential indicator simulation was performed using a random walk through all simulation nodes. At each node, a classification is selected based on the conditioning data (borehole data and previously simulated nodes) and the indicator variograms. The random walk is repeated until a desired number of realizations is obtained. Each realization is considered to be equi-probable and attempts to reproduce proportions of the categorical data (strata) as well as the variogram. A more detailed explanation of sequential simulations can be found in Pyrcz and Deutsch (2014).

In this study, 1000 realizations were simulated on a grid with a resolution of 10 m × 10 m × 1 m in the x, y, z directions, respectively. In general, the most probable geological model determined from the sequential simulation is in agreement with the geologist's interpretation (Figure 5). The uncertainty according to the 1000 sequential simulation realizations is presented in terms of the Shannon entropy index H (H=0 corresponds to zero uncertainty whereas H=1 corresponds to high uncertainty). An H value of 1.0 corresponds to the case where all categories are equi-probable. The uncertainty is notably low for this project (H = 0–0.25) due in part to the relatively high density of strata data from the borehole records, along with a clear stratified depositional environment in this project site.

SEQUENTIAL GAUSSIAN SIMULATION OF SPT N1(60)

Sequential Gaussian simulation follows the same approach as sequential indicator simulation. One advantage of performing sequential Gaussian simulations (as opposed to a simple kriging estimate) is the ability to capture the heterogeneity of SPT N1(60) or any parameter being simulated. 1,000 realizations were simulated on a 10 m × 10 m × 1 m grid. Three example realizations for each project are presented in Figure 6. It can be seen that for each single realization, the heterogeneity in SPT N1(60) is captured, as well as the variation in predicted values between realizations.

The average of all sequential Gaussian simulation realizations is presented in Figure 7. This result is equivalent to that of the simple kriging estimate. However, the mean of the realizations does not reveal the degree of heterogeneity that would be expected. The results could infer something about the macroscale (100+m) variability. For example, regions of higher N1(60) values suggest the soil's relative density and shear strength is higher, and therefore, it should be expected that more effort may be required to excavate the ground (e.g., slower advance speed, higher torque). Regions of lower N1(60) values suggest the soil generally exhibits a lower shear strength and stiffness, and therefore, could be more susceptible to ground deformation.

Uncertainty can be expressed in a number of different formats including absolute uncertainty, relative uncertainty, and misclassification. Absolute formats include a ± specified tolerance, ± a specified number of standard deviations, and a range based on specified percentiles (e.g., local p_{90}-p_{10}) (Pyrcz and Deutsch 2014). Any absolute format may be converted to a relative format by division with the mean predicted value (i.e., coefficient of variation CoV). A critical difference between relative and absolute formats is that the relative measures account to some extent for heteroscedasticity; the variability or uncertainty will scale with the predicted value.

Here, we quantify absolute uncertainty as ±1 standard deviation of N1(60) from the sequential Gaussian simulation realizations (Figure 8).

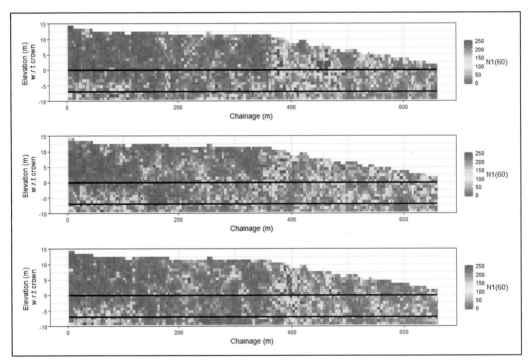

Figure 6. Three example realizations from the sequential Gaussian simulation. Elevation is with respect to the crown of the tunnel alignment (solid black lines).

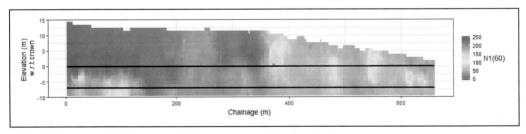

Figure 7. Average of 1,000 sequential Gaussian simulation realizations

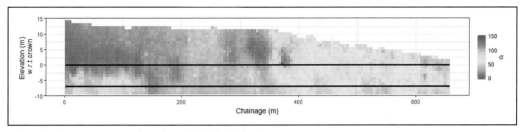

Figure 8. Absolute uncertainty (standard deviation)

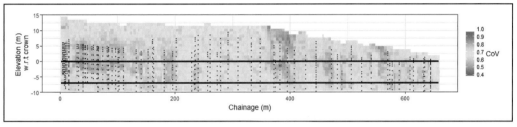

Figure 9. Relative uncertainty (CoV). Black points correspond to SPT record locations for boreholes within 50 m of the tunnel alignment.

Comparing the uncertainty with the sequential indicator simulation geology model (Figure 5), it can be seen that areas of higher uncertainty primarily correspond to glacial till, whereas regions of lower uncertainty correspond to mixed glacial deposits. This is explained by the lower standard deviations in N1(60) for mixed glacial deposits compared to glacial till. Standard deviations of N1(60) borehole records by strata F, MGD and GT are 40, 8 and 56, respectively.

The relative uncertainty in terms of CoV is presented in Figure 9. Overall, the CoV ranges from 0.4 to 1.0, with an average of 0.85. In general, it can be inferred that a higher density of samples yields lower uncertainty (e.g., chainage 0–200 vs. 350–600). However, this is not always the case as the data can vary significantly within one borehole or adjacent boreholes, resulting in a locally higher uncertainty in these areas.

GROUND SETTLEMENT RISK ASSESSMENT

This section presents an example of how the results from the geostatistical analysis can be incorporated into tunneling risk assessment. Individual realizations from the sequential Gaussian simulation could be incorporated into a Monte Carlo framework for modeling ground deformation. In a numerical modeling approach, each realization would serve as the input of geotechnical parameters for each element. Here, N1(60) was converted to geotechnical parameters shear modulus, friction and density using published correlations (Crespellani, Vannucchi, and Zeng 1991; Kulhawy and Mayne 1990; Cubrinovski

and Ishihara 2001). From the Monte Carlo analysis, a probability distribution of deformation can be derived, from which the probability of exceeding an allowable surface settlement can be obtained. See Grasmick and Mooney (2017) for a detailed description of this approach.

An example result of this application is presented in the Figure 10. Here, the cumulative density function (CDF) of maximum surface settlement from a 2D transverse cross sectional numerical model conveys the probability of exceedance. The deterministic result (vertical black line) represents the modeled settlement using the average geotechnical parameters based on select boreholes near the cross section analyzed. This conveys current practice in deformation prediction. The blue CDF represents the results of the Monte Carlo analysis using individual realizations from the sequential Gaussian simulation presented in this paper. This shows that the probability of exceeding the 8 mm deterministic estimate is 55%. The results convey the benefit of modeling the spatial heterogeneity, using results from the sequential analysis, for the risk assessment of ground deformation. Using the distribution of maximum settlement from the Monte Carlo analysis, the probability (or risk) of exceeding an allowable threshold can be quantified.

DISCUSSION

Geostatistical tools enable the characterization of spatial variability and the corresponding uncertainty, which can often be significant for soft ground

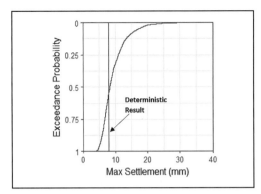

Figure 10. Cumulative distribution of probable max settlement from a 2D numerical model incorporating sequential Gaussian simulation realizations (blue) and the deterministic result using current practice (black)

tunneling. Variograms provide insight into the spatial structure of data and enable geostatistical modeling of this spatial variability and quantifying uncertainty. Here we use strata and SPT, but could extend to other parameters such as laboratory tests (depending on the sampling frequency for these tests) and geophysical data.

From the sequential indicator simulation, one can derive a most probable geological model as well as the corresponding spatial uncertainty. From the sequential Gaussian simulation, one can derive a measure of spatial uncertainty in geotechnical conditions, whether it be absolute or relative. In addition, sequential geostatistical simulation results can be extended to applications such as probabilistic ground deformation analysis presented in this paper, as well as uncertainty vs. sample spacing and site investigation design, tunnel boring machine performance prediction (advance rate, cutter wear, clogging), and perhaps others. Results can also be used to aid in the decision making for additional site investigation efforts if uncertainty exceeds an allowable tolerance, as determined on a project-specific basis.

REFERENCES

Chilès, J.-P., and P. Delfiner. 2012. *Geostatistics: Modeling Spatial Uncertainty: Second Edition. Geostatistics: Modeling Spatial Uncertainty: Second Edition.* https://doi.org/10.1002/9781118136188.

Crespellani, Y., G. Vannucchi, and X. Zeng. 1991. "Seismic Response Analysis." In *Seismic Hazard and Site Effects in the Florence Area.* Associazione Geotecnica Italiana.

Cubrinovski, M., and K. Ishihara. 2001. "Correlation between Penetration Resistance and Relative Density of Sandy Soils." *15th International Conference on Soil Mechanics and Geotechnical Engineering*, 393–96. http://ir.canterbury.ac.nz/handle/10092/2121.

Grasmick, J.G., and M.A. Mooney. 2017. "A Probabilistic Approach for Predicting Settlement Due to Tunneling in Spatially Varying Glacial Till." In *Geotechnical Special Publication.* https://doi.org/10.1061/9780784480717.028.

Gringarten, E, and Clayton V Deutsch. 2001. "Teacher's Aide Variogram Interpretation and Modeling." *Mathematical Geology* 33 (4):507–34. https://doi.org/10.1023/a:1011093014141.

Huber, Maximilian. 2013. *Soil Variability and Its Consequences in Geotechnical Engineering.*

Kulhawy, F H, and P W Mayne. 1990. "Manual on Estimating Soil Properties for Foundation Design." *Ostigov.* https://doi.org/EPRI-EL-6800.

Pyrcz, Michael J., and Clayton V. Deutsch. 2014. *Geostatistical Reservoir Modeling. Computers & Geosciences.*

Robinson, B, and J.M. Wehrli. 2013. "East Side Access-Queens Bored Tunnels Case Study." In *Proc. 21st Rapid Excavation and Tunneling Conference*, 1014–1141.

Stavropoulou, M., G. Exadaktylos, and G. Saratsis. 2007. "A Combined Three-Dimensional Geological-Geostatistical-Numerical Model of Underground Excavations in Rock." *Rock Mechanics and Rock Engineering* 40 (3):213–43. https://doi.org/10.1007/s00603-006-0125-4.

Stavropoulou, Maria, George Xiroudakis, and George Exadaktylos. 2010. "Spatial Estimation of Geotechnical Parameters for Numerical Tunneling Simulations and TBM Performance Models." *Acta Geotechnica* 5 (2):139–50. https://doi.org/10.1007/s11440-010-0118-z.

Webster, Richard, and Margaret A. Oliver. 2008. *Geostatistics for Environmental Scientists: Second Edition. Geostatistics for Environmental Scientists: Second Edition.* https://doi.org/10.1002/9780470517277.

Xiao, Li, Hongwei Huang, and Jie Zhang. 2017. "Effect of Soil Spatial Variability on Ground Settlement Induced by Shield Tunnelling," 330–39. https://doi.org/10.1061/9780784480717.031.

Interpretation of Tunneling-Induced Ground Movement

Mazen E. Adib
AZTEC Engineering Group

Matthew Crow
Los Angeles Metropolitan Transportation Authority

Allen Marr
Geocomp Corporation

Shemek Oginski
J.F. Shea Co., Inc.

ABSTRACT: An 8.5-mile-long light rail project in Los Angeles included twin 1-mile long tunnels, 21.5 feet in diameter. An Earth Pressure Balance (EPB) Tunnel Boring Machine (TBM) was used for excavating the tunnel in the Late-Pleistocene Alluvial Soils. An automated system was implemented for near-continuous measurement of tunneling induced ground movements from 53 No., Multi-Position Borehole Extensometer (MPBXs). Optical survey was performed to measure ground surface movements. This paper presents the subsurface conditions, the geotechnical instrumentation system, the measured ground movement along the first bored tunnel, and the interpretation of the movement as related to the performance of the EPB TBM and type of subsurface conditions along the alignment.

INTRODUCTION

The LA Metro Crenshaw/LAX Corridor project is part of a Los Angeles County Metropolitan Transportation Authority's (Metro) major capital expansion program. The design-build project was awarded to Walsh/Shea Corridor Constructions with HNTB as the Designer. Overall, the project has 8 stations, cut and cover box structures, and twin bored tunnels. The section of interest to this paper is Underground Guideway #4 on Figure 1, spanning the northern end of the project alignment with a total length of about 8,430 feet including three underground stations, which are from North to South: Expo, MLK, and Vernon. The ground movements presented in this paper are along the alignment shown on Figure 2.

The twin tunnels with a centerline separation of 39.13 feet were excavated by an EPB TBM in a Late-Pleistocene age deposit of the Los Angeles sedimentary basin, a deposit referred to as "Old Alluvium." The TBM had a 33-ft long shield, with "overcut" dimensions that increases along the shield from 0.6 inches just behind the cutter head, to 1 inch at the back of the shield. Tunnel excavation diameter was 21.49 feet. The precast concrete rings were 18.83-ft I.D. and 5 feet in length. Depth to the tunnel axis was 51.5 to 54.5 feet between Expo and MLK Stations, and 41.5 to feet between MLK and Vernon Stations.

Several challenges related to the optimization of the TBM control parameters were overcome during the initial stages of tunneling (Chan et al., 2017), from Expo Station halfway to MLK. An initial slow rate of advance was experienced, and was resolved by adjusting several TBM mining parameters related to the foam injection and the cutter head torque. In addition, the ground movements, especially those at depth, were larger than anticipated and exceeded the project performance limits for ground movements shown on Table 1. Design volume loss was 0.5%.

The exceedances correlated to periods of malfunction of TBM shield bentonite system (Chan et al., 2017). Although the 33-foot long shield prevents significant caving in, ground movement can occur in the overcut region above the shield, if a pressurized "envelop" is not maintained between the cutter-head and the tail (Cording, 2016). This TBM was equipped

Table 1. Performance limit for tunneling-induced ground movement

Location	Action Limit (inch)	Maximum Limit (inch)
Surface Movement	0.5	0.6
Bottom Anchor (5-feet above tunnel crown)	0.75	1.5

Source: Metro 2012c.

171

with a system to allow injection of bentonite around the shield skin. Because of the gravelly nature of the materials just above the shield, a bentonite "cake" must form in the granular material for the bentonite pressure to be effective (Diaz and Bezunjian, 2015). The ground movement was minimized by carefully

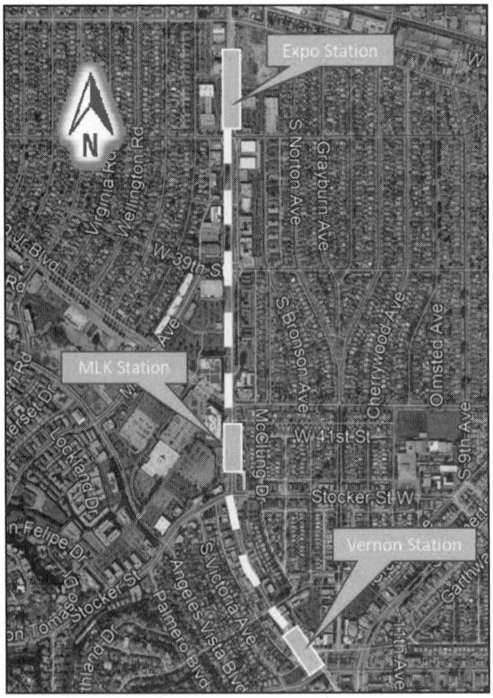

Source: Metro 2012a.

Figure 1. Location of underground guideway #4, LAX/Crenshaw Rail Project

controlling the bentonite injection volume and pressure. The injection was volumetrically controlled to match the advance rate, and had low and high-pressure limits to trigger pump operation. The injection pressure was set close to the calculated face support pressure at the tunnel crown (Chan et al., 2017). The effectiveness of this approach was confirmed by the ground movement monitoring system. Subsequently, ground movement at the surface was typically (¼") after each tunnel drive along the tunnel alignment.

GEOLOGICAL AND GEOTECHNICAL INFORMATION

The project is in the northern part of the Los Angeles Basin, which is directly underlain by unconsolidated Quaternary-age alluvial sediments (Metro, 2012a). These are subdivided into Holocene-age sediments (Young Alluvium) and late-Pleistocene Old Alluvium. The Old Alluvium is dissected by stream channel and floodplain deposits and includes the Lakewood Formation. The Young Alluvium is 30-ft thick, and consists predominantly of fines: Silt, lean Clay, and fat Clay, with frequent organic Clay strata. The upper 50 feet of the Old Alluvium (i.e., from 30 to 80 feet below ground surface) is coarse sediment including pebble-gravel and sand. Groundwater was reported at depths from 45 to 60 feet, below surface. Available laboratory test results, including moisture content, dry unit weight, and fraction of gravel, sand, and fines by weight, are shown on Figures 3 and 4. Figure 3 are for borings located along the section where relatively excessive ground movement occurred along the northern half of the alignment between EXPO and MLK; whereas Figure 4 are the results from borings located along the southern half.

Figure 2. Extent of alignment of interest and select MPBX locations

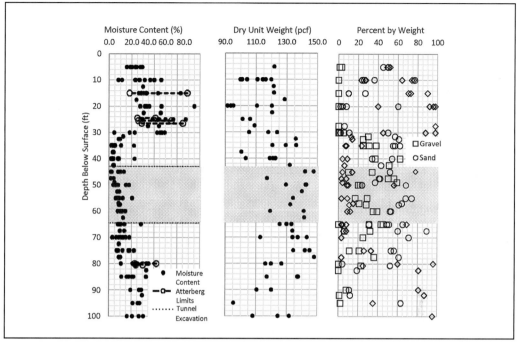

Source: Metro 2012b.

Figure 3. Geotechnical engineering properties of subsurface materials, MPBX 37 to 51

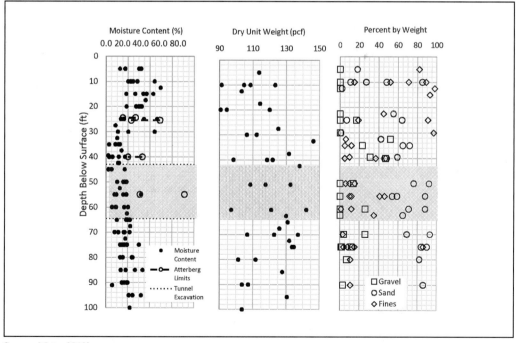

Source: Metro 2012b.

Figure 4. Geotechnical engineering properties of subsurface materials, MPBX 19 to 35

Grain-size distribution curves from the upper zone of the Old Alluvium, 35 to 50 feet below ground surface, are on Figure 5. The Shear modulus and Poisson's Ratio were obtained from two downhole geophysical logging tests, Figure 6. Downhole borehole locations are shown on Figure 2.

GEOTECHNICAL INSTRUMENTATION AND MONITORING

The geotechnical instrumentation system was procured, installed, and monitored by Geocomp Corporation under a contract with the design-build Joint Venture. The system was designed to monitor

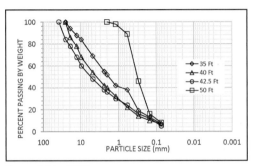

Source: Metro 2012b.

Figure 5. Grain size distribution of Old Alluvium just above tunnel excavation

construction induced movements and groundwater drawdown within the zone of influence of the station and tunnel excavations. Instrumentation consisted of both manually and automatically monitored instruments. Instrument data were logged into a cloud-based Geographic Information System (GIS) and made available to the project team. The cloud, GIS-based, settlement information was linked to the TBM data management system to correlate ground movement with TBM position.

Subsurface ground movement was monitored with Multi-Position-Borehole Extensometer (MPBX) with grouted anchors. Each of the 3 anchors houses a displacement transducer that measures the relative movement between the anchor and the head of the MPBX. The bottom anchor was located 5 feet above tunnel crown, the shallow anchor at 5 feet below ground surface, and the middle anchor halfway in between. The installed MPBX head is protected inside a well box (road box) and was "debonded" from the road box to allow the head to move independently of the road box (Figure 7). The movement at the surface was monitored using reference points (pk nails) and Feno-type anchors. The specification also required the monitoring of movement of a reference point at the MPBX head. Because the head is "buried" within a road box, it takes significant time and efforts to access the head, which also houses sensitive electronics. To allow less frequent monitoring of the MPBX head, a Position String

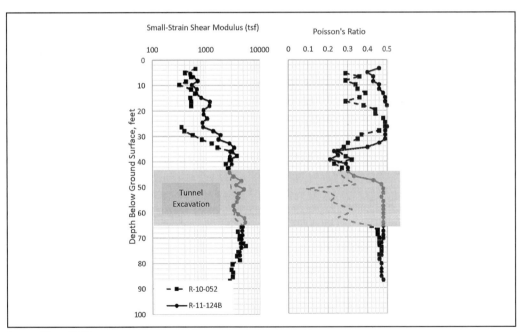

Source: Metro 2012b.

Figure 6. Shear modulus and Poisson's Ratio from downhole geophysical tests

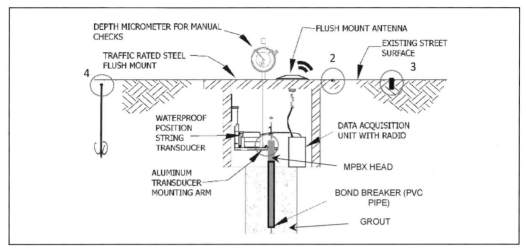

Figure 7. Schematic of data collection system of relative movement of MPBX head and location of reference survey points near ground surface: (1) Survey point for optical survey of MPBX head, (2) road box rim, (3) surface monitoring point, (4) Feno anchor

Transducer (PST) was installed to automatically monitor the relative movement of the head relative to the ground. Figure 8 shows arrangement of the PST system inside an MPBX road box. The use of the PST allowed for measurement of the surface movement rather than movement of the head itself, and thus there was far less traffic control effort and disruption to traffic than would have been otherwise required to monitor the head. The PSTs were installed to monitor the relative movement of the MPBX head when the Tunnel Boring Machine (TBM) cutter-head was within 200 feet of the MPBX. Optical survey was also performed during this period to measure the movement of four reference points (Figure 7).

The monitoring of the relative movement of the MPBX head proved very useful on several occasions where the head moved independently of the road box. This movement was added to (or subtracted from) the movement of the road box to obtain the movement of the MPBX head, which was then added to the relative movement of the anchors to calculate their total movement.

An instrumentation array, transverse to the direction of tunneling, was located at a nominal spacing of 200 feet along the tunnel alignment. Each array consisted of two MPBXs (one above each tunnel), and several Feno-type anchors as surface settlement reference points. Additional settlement reference points (pk nails in the asphalt) were spaced at 40 feet on centers along each tunnel alignment.

The MPBX data cables were "trenched" to a buried sidewalk box (communication module), where a secondary data logger, radio communication devices, and rechargeable batteries were installed. A mobile 30-gallon barrel was used as an enclosure to

safely store the master data logger, which was connected to a cellular phone modem.

The data loggers collected the ground movement from the MPBX anchors and the PSTs at 2- to 5-minute frequency, and the data were posted on Geocomp's GIS-based data management system to provide the tunneling engineers frequent ground movement data for assessing the performance of the TBM-ground system. The manual optical surveys were uploaded into the same system.

OBSERVED GROUND MOVEMENT

The ground movements reported in this paper were measured during mining of the SB tunnel between Expo and Vernon stations. The location of the reported MPBXs are shown on Figure 2. The frequent monitoring of the movement from the MPBX anchors allowed presentation of the relative anchor movement as a function of the TBM distance from the MPBX location.

Figure 9 shows the measured relative movements from the bottom anchor of MPBXs 51 to 37 (herein referred to as the northern set), where larger than anticipated ground movements occurred. Figure 10 shows the relative movements from the bottom anchor at MPBXs 35 to 19 (herein referred to as the southern set). The reduction in the relative movement at MPBX 35 to 19 is evident. The absence of relative movement means a uniform ground movement between the location of the bottom anchor and the ground surface.

Figure 11 shows a summary of the maximum absolute movements at the bottom anchor of each MPBX and the corresponding surface movements

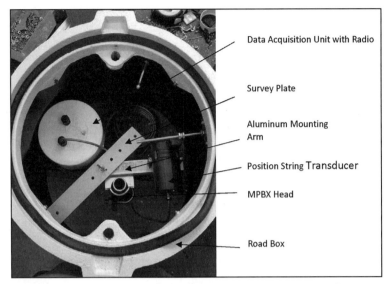

Figure 8. Arrangement of system to measure relative movement of the MPBX head inside the road box

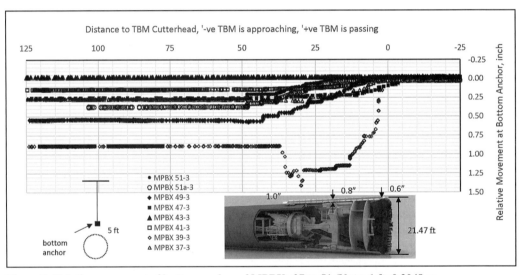

Figure 9. Relative movement of bottom anchor of MPBXs 37 to 51 (Note: 1 ft=0.3048 m, 1 inch=0.0254 m)

measured at the nearest surface settlement point (SMP)s and at the rim of the MPBX road box. The measurements are plotted as a function of the distance from the tunnel beginning at EXPO station. The following observations may be made from the information on Figures 9 to 11:

1. Most of the relative movement in the bottom MPBX anchors occurred immediately above the TBM shield after the cutter-head had passed.

2. The overall trend of the measured ground movement both at surface and at depth decreased following the implementation of the remedial measures discussed previously.

3. The relative movement of the bottom anchor at MPBX 39, exceeded the uniform maximum overcut thickness, supporting prior observations (Loganathan & Poulos, 1998,

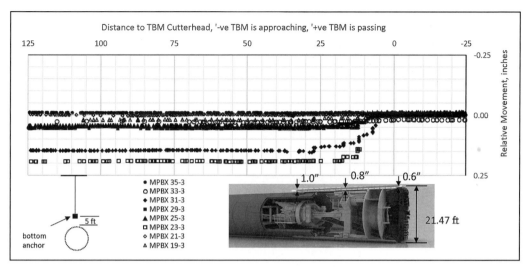

Figure 10. Relative movement of bottom anchor of MPBXs 19 to 35

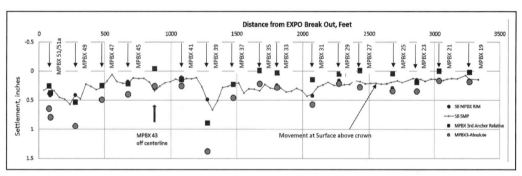

Figure 11. Measured ground movement between EXPO and MLK stations

and Pinto & Whittle, 2014) that the actual gap above the tunnel crown can exceed the thickness of the overcut above the shield.

4. The reduction in the ground movement at MPBX 39, just after the TBM shield tail had passed, confirms that tail grout pressure is capable of "lifting" the ground and reducing the movement. A reduction in settlement of about 0.5 inch was achieved by the tail grout.

Following the successful implementation of the improved bentonite injection and control system, the relative movement of the bottom anchor of the southern set became significantly less than before, even smaller than the surface movement near the MPBX location, Figure 12. This behavior contrasted with the observed behavior of the northern set, where the relative movement of the bottom anchor was about equal or more than the surface movement.

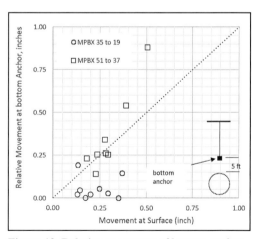

Figure 12. Relative movement of bottom anchor vs. surface movement

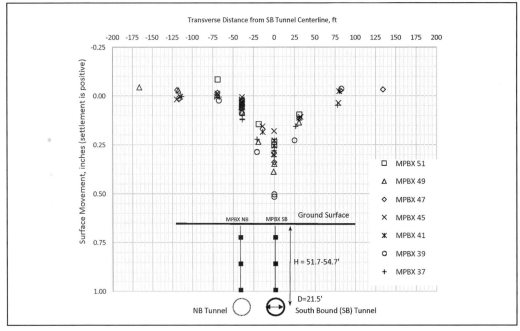

Figure 13. Transverse settlements at MPBX 37 to 51 arrays (Note: 1 ft=0.304 m, 1 inch=0.0254 m)

Further evidence of the change in ground movement behavior noticed by the bottom anchor of the MPBXs was provided by the transverse surface settlement troughs, Figures 13 to 16. Figures 13 and 15 shows the "nominal" measured movement values, which, as discussed herein, carry a potential variation of ±0.04 to ±0.06 inches. On Figures 14 and 16, the movement from each type of reference survey points was normalized with respect the ground movement, S_{max}, obtained from the SMP nearest to the location of the MPBX. There are important differences in the normalized settlement troughs of the two MPBX sets. The one obtained for the northern set is fitted with a Gaussian Curve (Peck, 1969) with an inflection point located at horizontal distance from the tunnel centerline of $X = 0.45H$, where H is the depth to the tunnel axis (H = 52.7 to 54.7 feet for the northern set). The one obtained from the southern set is best fitted with a curve having an inflection point at $X = 0.6H$ (H = 51.7 to 52.7 feet for the southern set), reflecting a wider settlement trough. The other important difference is the presence of heave in the northern MPBX set, away from the centerline.

To find out if there was a precedent of similar behavior, a review of published measured data from Seattle North Link and University Link projects (Salvati et al., 2016) was performed. The data presented by Salvati et al., 2016 were digitized to produce the plot on Figure 17, which shows many MPBXs with negligible relative movement, and/or significantly less relative movement than the total movement at the surface. This review showed that it is entirely possible in EPB TBM for the relative movement of the bottom anchor to be equal, more, or less than the total surface movement.

However, to confirm that the discrepancy in the behavior of the two sets of MPBXs is due to TBM-ground interaction behavior (and not an inherent issue with the MPBX system), 5 additional MPBXs, with double borros- point anchors, were installed 5 feet away from existing MPBXs 17, 13, 11, 9, and 7 to validate the measured movement from the grouted MPBX anchors. These additional MPBXs were installed between MLK and Vernon stations. Both types of MPBXs provided similar and equal ground movement response, and hence confirming that the observed change in the behavior of the MPBX sets is entirely due to TBM-ground interaction.

Attempts were therefore made to interpret the ground movement from the two MPBX sets, from a geotechnical perspective, to identify important ground deformation patterns that could explain the observed difference.

INTERPRETATION OF GROUND MOVEMENT

The analysis of tunneling induced ground movement is a complex TBM-operator-ground interaction problem that requires due consideration of the TBM control parameters, an understanding and proper

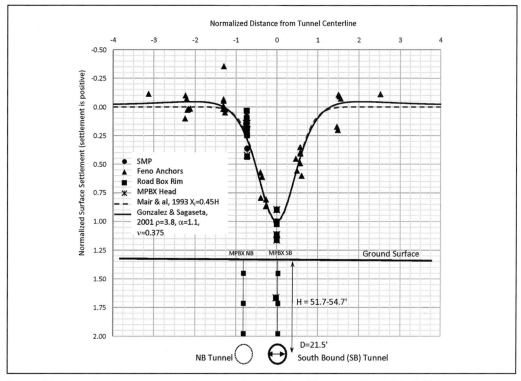

Figure 14. Normalized transverse settlement at MPBX 37 to 51 arrays (Note: 1 ft=0.304 m, 1 inch=0.0254 m)

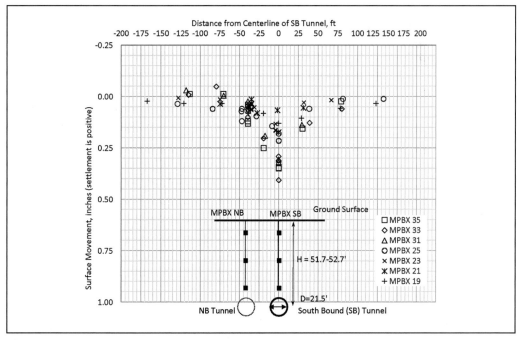

Figure 15. Transverse settlements at MPBX 37 to 51 arrays

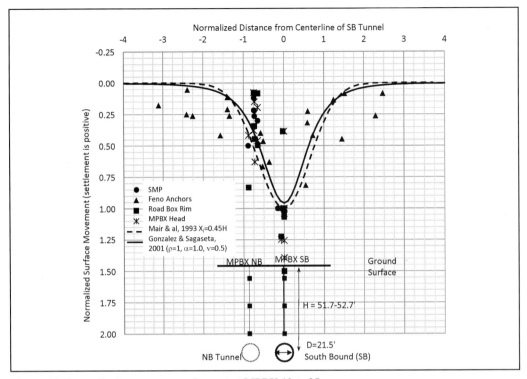

Figure 16. Normalized transverse settlement at MPBX 19 to 35 arrays

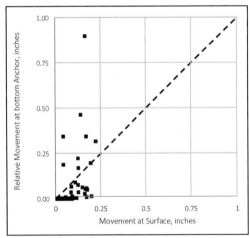

Source: Salvati et al. 2016.

Figure 17. Relative movement at bottom anchor vs. movement at surface

modeling of the TBM actions and the ground reactions, and realistic constitutive models for the subsurface materials, bentonite, and grout. Diaz and Bezunjian 2015 proposed a simulation technique that considers both the TBM operational parameters, and

the ground/bentonite/grout properties to model the complex interaction that occurs in EPB TBM excavation. This type of simulation is beyond the scope of this paper, and rather an attempt was made to rely on existing empirical correlations and analytical solutions to interpret the observed ground movements. Despite their shortcomings, these correlations and solutions are used in practice for estimating ground movement resulting from an assumed volume loss, as an expedient method to compare the output of the rigorous simulations, or to interpret observed ground movement.

There are several challenges, however, in interpreting the ground movement on this project, even with the simplified correlations and solutions. The interpretation must rely on relatively infrequent optical survey measurements (taken every 24 hours) and therefore there are important gaps in understanding the development of the total ground movement, both at surface and at depth, as the TBM excavation progresses. Also, while a specific optical survey measurement may be accurate to 0.001 inches, its day-to-day repeatability on this project was found to vary by up to 0.08 inches with most in the range of 0.04 to 0.06 inches (1 to 2 mm). This range was estimated from observation of the measurements trend of several reference points. This range maybe acceptable from a practical viewpoint, but renders the interpretation questionable when the total ground movement is of the same order. The interpretation by the analytical solutions is further complicated by the site stratigraphy, where up to 30 feet of Holocene-age deposits overlay the older, more competent, Pleistocene sediments. The analytical solutions assume a relatively uniform and homogeneous subsurface profile and therefore the interpretation is oversimplified at layered sites.

Empirical Correlations

Empirical correlations (Peck, 1969, O'Reilly & New, 1982, Mair et al, 1993) were used to fit the observed settlement troughs and calculate the volume loss ratio for clues on the observed ground movement. Previously, it was shown that the settlement trough from the northern MPBX set is fitted with a Gaussian Curve with a distance to point of inflection of 0.45H. A distance of 0.6H is appropriate for the southern MPBX set. Based on the fitted curves, it is possible to calculate the volume loss ratio (V_L), and then obtain the subsurface ground movement from Mair et al. 1993. The calculated deep movement for each of the MPBXs is plotted on Figure 18 against measurements. On average, this correlation underpredicted the movement of the bottom anchor of the northern MPBX set by 30%, and overpredicted the movement of the bottom anchor of the southern set by 75%. The values of the calculated volume loss ratio and the distance to point of inflection for each MPBX are summarized on Table 2. The volume loss ranged from 0.25% to 0.69% for the northern MPBX

set, and 0.23% to 0.67%, a remarkably similar range despite the difference in the observed subsurface ground movement at depth. Other than the change in the distance to the inflection point, this empirical correlation offered no insight for interpreting the ground movement, and therefore additional information was sought from the analytical solutions.

Analytical Solutions

Analytical solutions for interpretation of observed ground movement are discussed in Pinto & Whittle 2014, and Migliazza et al. 2009 and include those by

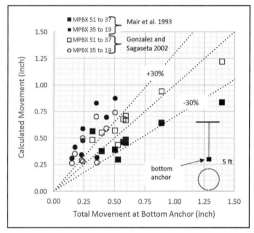

Figure 18. Comparison between total movement at bottom anchor and calculated movement

Table 2. Values of the parameters used in the empirical correlation and analytical solution

MPBX	Tunnel Axis Depth (ft)	S_{max} (inch)	Mair et al. Xi (ft)	V_L	Gonzalez-Sagaseta ν	ε	ρ	α
51	54.7	0.24	24.62	0.34%	0.375	0.14%	3.28	1.1
51a	54.7	0.28	24.62	0.39%	0.375	0.16%	3.28	1.1
49	53.7	0.39	24.17	0.55%	0.375	0.22%	3.28	1.1
47	53.7	0.29	24.17	0.41%	0.375	0.16%	3.28	1.1
45	53.7	0.18	24.17	0.25%	0.375	0.10%	3.28	1.1
43	53.7	0.34	24.17	0.48%	0.375	0.19%	3.28	1.1
41	52.7	0.23	23.72	0.31%	0.375	0.13%	3.28	1.1
39	52.7	0.50	23.72	0.69%	0.375	0.28%	3.28	1.1
37	52.7	0.28	23.72	0.39%	0.375	0.15%	3.28	1.1
35	52.7	0.35	31.62	0.64%	0.50	0.32%	1.00	1.0
33	51.7	0.29	31.02	0.53%	0.50	0.26%	1.00	1.0
31	51.7	0.37	31.02	0.66%	0.50	0.33%	1.00	1.0
29	51.7	0.25	31.02	0.45%	0.50	0.22%	1.00	1.0
27	51.7	0.20	31.02	0.36%	0.50	0.18%	1.00	1.0
25	51.7	0.14	31.02	0.26%	0.50	0.13%	1.00	1.0
23	51.7	0.13	31.02	0.24%	0.50	0.12%	1.00	1.0
21	51.7	0.17	31.02	0.31%	0.50	0.16%	1.00	1.0
19	51.7	0.13	31.02	0.23%	0.50	0.12%	1.00	1.0

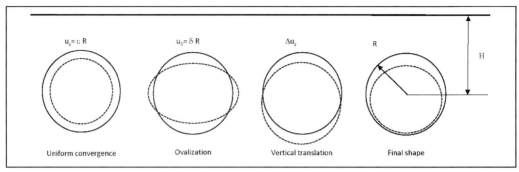

Source: Gonzalez and Sagaseta 2001.

Figure 19. Components of the deformation of the tunnel

Verrujit 1996, 1997, Verrujit & Brooker 1996, and Gonzalez & Sagaseta 2001. A common approach amongst these solutions is to assume that the deformation around the tunnel cavity is a combination of two hypothetical movements (Figure 19): convergence (characterized by a uniform squeeze, which causes the volume loss), and distortion (characterized by a vertical movement at the excavation roof without an associated volume loss). Vertical downward translation also occurs, but not explicitly considered in these solutions. Pinto et al. 2014 have shown that these solutions can provide a consistent framework for interpreting tunneling-induced ground movements, but offer little insight to explain the influence of soil properties on tunneling control parameters (Leronymaki et al. 2017).

The ratio of roof distortion (δ) to the uniform convergence (ε) is ρ. These two types of movements (i.e., convergence and distortion) are important for interpreting ground movement because each type produces different shapes of surface settlement troughs in uniform non-dilating elastic materials (Pinto & Whittle 2014). Convergence mode of deformation produces a relatively flat surface settlement profile, while distortion produces a narrow profile, with ground heave in the surface far field, away from the tunnel centerline (Pinto & Whittle 2014). If ε and ρ are known, the spatial ground deformation can be calculated from these methods and the only required soil property is Poisson's ratio (ν). However, Pinto & Whittle 2014 emphasized the need to include the effect of dilation for proper interpretation of ground movement when tunneling in dilating ground, as proposed by Gonzalez & Sagaseta 2001. Dilation is important to consider because it causes further narrowing of the settlement trough. The dilation is characterized by a factor (α), which is calculated from the dilation angle of the soil. An acceptable range for this parameter at this site is between 1.0 (non-dilating soil) and 1.42 (for an assumed dilation angle of 10 degrees). As noted by Pinto et al. 2014 there

are many combinations of ρ and α that can produce the same settlement trough, however, only distortion can produce a heave at the surface when the convergence and distortion are both downward at the tunnel crown.

These solutions have been used to interpret the ground deformation from several case histories (Gonzalez & Sagaseta 2001, Migliazza et al. 2009, Pinto et al. 2014, Leronymaki et al., 2017). Typically, three values of ground movement are required to carry out the interpretation: the maximum surface settlement, the distance to the inflection point of the settlement trough, and one deep vertical measurement or lateral movement from inclinometer. The vertical measurements are available at all the MPBX arrays and hence it is possible to estimate ρ, and ε from the field measurements at this site.

A systematic approach was suggested by Pinto et al. 2014 to calculate the "fitting" parameters, which are ε, ρ, and either α (for dilating soil) or ν, based on "least square solution" when subsurface movements are available. A different approach was followed herein due to the dearth of subsurface ground movement and outlined in Appendix I.

Because most of the data are vertical surface movement, the approach in Appendix I focuses on selecting fitting parameters that provide a reasonable match to the normalized surface settlement trough. Expert judgement is necessary to decide what reasonable is. The fitting parameters were then used to calculate the soil movement at the bottom anchor for comparison with the observed measurement. Because the tunnel excavation is in a potentially dilating material (sand and gravel), the analytical solution of Gonzalez & Sagaseta 2001 was used.

The normalized settlement troughs for the northern MPBX set were fitted by using the following parameters: $\rho=3.8$, $\alpha=1.1$, and Poisson's ratio $\nu=0.375$. These parameters were found by trial and error in accordance with the procedures outlined in Appendix I, while restricting Poisson's ratio in the range of 0.35 to 0.5, as obtained from the results of the downhole geophysical tests. These parameters provide the settlement trough shown on Figure 14, and overpredict the total movement of the bottom anchor, on average, by 15%, Figure 18. These parameters also produce heave in the far field of the settlement trough, consistent with the field observation.

The normalized southern MPBX set settlement troughs can be fitted with the following parameters: $\rho=1$, $\alpha=1$, and $\nu=0.5$. These parameters provide the settlement trough shown on Figure 16, and overpredict the movement of the bottom anchor by 55% on average. Table 2 summarizes the values of the parameters obtained for each MPBX array. It is possible to improve the prediction of the movement of the bottom anchor, but only if α is chosen lower than one (i.e., $\rho=0.3$, $\alpha=0.85$, and $\nu=0.5$). This low value of a is suggestive of "contraction" rather than dilation in the ground; it is outside the range stated by Gonzalez & Sagaseta 2001 and was hence eliminated as a credible value for (α). However, centrifuge tests of tunneling in sand by Zhou 2015 provided further insight into the observed ground movement, and indicated that contraction does occur if the movement near the tunnel crown is small enough. Contraction also produces uniform vertical movement profile above the tunnel crown all the way up to the ground surface, like the behavior of the southern MPBX sets.

CONCLUSION

This paper presented the tunneling instrumentation system for Crenshaw project and the tunneling-induced ground movement at 18 MPBX monitoring arrays. The calculated volume loss from the surface settlement trough of the first tunnel reach was between 0.23% and 0.69% with an average 0.43%. The maximum surface settlements were in the range of ⅛ inch to ½ inch. The deep movement, however, as measured by the bottom anchor of the MPBXs ranged from 0.15 inch to over 1.5 inches. Based on the observed ground movement pattern, it was necessary, for interpretation purpose, to aggregate these 18 MPBXs into two spatially distinct sets: a northern and a southern set.

An empirical correlation (Mair et al., 1993) was used to fit the observed spatial pattern of tunneling induced surface movement. The distance to the inflection point was 0.45H to 0.6H for the northern and southern sets, respectively. The empirical method offered no insight for the reason of the discrepancy in the behavior of the two MPBX sets, nor provided an explanation for the observed heave in the far field of the settlement troughs of the northern MPBX set.

An analytical solution (Gonzalez and Sagaseta 2001) was therefore used to interpret the observed ground movement. By fitting the observed settlement troughs, the solution indicated that the ground movement at the cavity may have been predominately distortion (Figure 19) at the northern MPBX set, which may also explain the observed heave in the surface far field. In contrast, far less distortion was obtained at the southern MPBX set, resulting in a relatively flat settlement trough. While the analytical solution overpredicted the movement of the bottom anchor by 15% and 55% for the northern and southern sets, respectively, it offered a possible explanation for the observed discrepancy in the ground movement.

Centrifuge tests by Zhou 2015 offered further insights into the contrasting behavior of the two MPBXs sets. At the northern set, the deformation above the ground was large enough to dilate the ground, with an associated attenuation of ground movement from the tunnel crown to the ground surface. At the southern set, the deformation above the crown was small: the ground contracted resulting in uniform vertical ground movement profile.

APPENDIX I

The procedures for calculating the fitting parameters ε, ρ, ν, and α from the measured data using Gonzalez & Sagaseta 2001 solution are as follows:

1. Normalize the settlements with respect to the maximum surface settlement, S_{max} (in this paper, the values from the settlement monitoring along the tunnel axis nearest to the MPBX was used), and distances (with respect to H), as done on Figure 14 and 16. This normalization is possible along this alignment for two reasons. First, the ratio H/R is constant, and secondly, the ground movement pattern from several MPBXs could be aggregated into two different sets (as reflected in the plot shown on Figure 12). In this manner, it was possible to obtain "global" fitting parameters for the remainder of the procedures.
2. Fit the settlement trough with Gaussian-type curve and obtain the volume loss, V_L.
3. Calculate the uniform convergence (ε) as follows (Pinto & Whittle 2014, and Migliazza et al., 2009):

$$\varepsilon = \frac{V_L}{4(1-\nu)}$$

4. Change the parameters ρ, α, and ν by trial and error while applying expert judgement, until a best fit is obtained for the settlement trough while targeting a match for the maximum normalized settlement (which has a value of 1).
5. From the fitting parameters ρ and α calculate the normalized vertical movement at the depth of the anchor of interest using the solution by Gonzalez and Sagaseta 2001.

6. Multiply the normalized movement obtained in (5) by the surface settlement at each MPBX to obtain the calculated total movement.

REFERENCES

Chan, R, Salai, J, and Shatz, B, (2017). "Station Excavation and TBM Tunnel on Los Angeles Crenshaw Project." *Rapid Excavation and Tunneling Conference*, San Diego.

Cording, E. J., (2016). "Monitoring, Observing & Controlling Ground Behavior at the Source." *Monitoring Control in Tunneling*, Society of Mining Engineers, San Francisco.

Dias, Tiago & Bezuijen, Adam. (2015). *TBM Pressure Models—Observations, Theory and Practice*. 10.3233/978-1- 61499-599-9-347.

González, C., and Sagaseta, C. (2001). "Patterns of soil deformations around tunnels. Application to the extension of Madrid Metro." *Comput. Geotech.*, 28(6–7), 445–468.

Leronymaki, E. S., Whittle, A. J., Sureda, D. S., (2017). "Interpretation of free-field ground movements caused by mechanized tunnel construction." *J. Geotech. Geoenviron. Eng.*: 10.1061/(ASCE)GT.1943-5606.0001632.

Loganathan, N., and Poulos, H. G., (1998). "Analytical Prediction for Tunneling-Induced Ground Movements in Clays." *J. Geotech. Geoenviron. Eng.*, 10.1061/ (ASCE)1090-0241(1998)124:9(846).

Metro (2012a), Crenshaw/LAX Transit Corridor, *Technical Reports Part 6.1, Geotechnical Baseline Report*, Hatch Mott MacDonald.

Metro (2012b), Crenshaw/LAX Transit Corridor, *Technical Reports Part 6.2, Geotechnical Data Report*, Hatch Mott MacDonald.

Metro (2012c), Crenshaw/LAX Transit Corridor, *Section 01 89 15.1—Performance Requirements: Property and Utility Monitoring and Protection*, Amendment 8: February 18, 2013.

Migliazza, M., Chiorboli, M., and Giani, G. P. (2009). "Comparison of analytical method, 3D finite element model with experimental subsidence measurements resulting from the extension of the Milan underground." *Comput. Geotech.*, 36(1–2), 113–124.

O'Reilly, M. P., and New, B. M. (1982). "Settlements above tunnels in United Kingdom—Their magnitude and prediction." *Proc., 3rd Int. Symp. Tunneling 82*, Institute of Mining and Metallurgy, London, 173–181.

Peck, R. B. (1969). "Deep excavations and tunnels in soft ground." *Proc., 7th Int. Conf. on Soil Mechanics and Foundation Engineering*, Sociedad Mexicana de Mecánica, Mexico, 225–290.

Pinto, F., and Whittle, A. J. (2014). "Ground movements due to shallow tunnels in soft ground. I: Analytical solutions." *J. Geotech. Geoenviron. Eng.*, 10.1061/(ASCE)GT.1943-5606.0000948, 04013040.

Pinto, F., Zymnis, D. M., and Whittle, A. J. (2014). "Ground movements due to shallow tunnels in soft ground. II: Analytical interpretation and prediction." *J. Geotech. Geoenviron. Eng.*, 10.1061/(ASCE)GT.1943-5606.0000947, 04013041.

Salvati, L., LaVassar, C., Sla, L. Theodore, J., Galissson, L. (2016). "Evaluation of Settlement Monitoring Over 6.75 Miles of Tunneling—Updating Prediction Methods and Designing Better Monitoring Programs." *World Tunnel Congress*, San Francisco, 1312–1321.

Verruijt, A. (1996). "Complex variable solutions of elastic tunneling problems." *Rep. COB 96-04*, Geotechnical Laboratory, Delft Univ. of Technology, Delft, Netherlands.

Verruijt, A.(1997)."A complex variable solution for a deforming tunnel in an elastic half-plane."*Int.J.Numer.Anal. Methods Geomech.*, 21(2), 77–89.

Verruijt, A., and Booker, J. R. (1996). "Surface settlements due to deformation of a tunnel in an elastic half-plane." *Geotechnique*, 46(4), 753–756.

Zho, B. (2015). "Tunneling-induced ground displacements in sand." PhD thesis, University of Nottingham.

Pre-Conditioning of Hard Rocks as Means of Increasing the Performance of Disc Cutters for Tunneling and Shaft Construction

Philipp Hartlieb
Montanuniversitaet Leoben

Jamal Rostami
Colorado School of Mines

ABSTRACT: Excavation of very hard intact rock by various mechanized systems is challenging and requires high cutting forces. This refers to tunneling through rock that is in excess of 200–300 MPa, which has also been considered difficult tunneling conditions by the ITA Workgroup 14 that is specialized in mechanical excavation. There are innovative concepts to improve the cuttability of hard and abrasive rocks by using pre-conditioning systems for introducing micro-fractures in the rock. This includes high-power microwave irradiation which leads to weakening of the rock by introducing micro fractures in the medium. The initial results of treating rock surface with a 24 kW microwave shows initiation of some micro-crack network and the preliminary results will be discussed in this paper. The findings will be linked to possible improvements of tunnel/shaft boring machines (TBM, SBM) performance, especially with respect to possibility of deploying a hybrid system to aid the center and gage cutters by pre-damaging the rock.

INTRODUCTION

Mechanical hard rock excavation machinery are widely used in underground mining and tunnelling for the development of major infrastructure projects. This includes a variety of machines used for horizontal, inclined, and vertical excavation into rock. Among these machines, tunnel boring machines (TBM) are the most frequently used machines that has been successfully used in variety of ground conditions for the past 60+ years. TBMs use disk cutters for rock fragmentation through applying the necessary forces to the rock to initiate fractures in the rock.

The efficiency of the processes is mainly governed by the mechanical properties of the rock mass, especially strength and abrasivity of the intact rock. While Roadheaders are used in rocks with UCS of up to roughly 150 MPa, especially for the development of underground mines, they cannot offer an efficient performance when used in hard rocks on extended stretches of a tunnel. Meanwhile, some applications require a flexible mechanized and automated system which is capable of excavating through very hard and mostly intact rocks without jointing and discontinuities there has been some movement to develop new machines for excavation of non circular openings using disc cutters. But the rule of thumb in excavation is: the harder the rock, the bigger the machine, the lower the flexibility. This is partially due to the higher cutting forces that is needed for penetrating into the rock by disc cutters. Thus there is a need to develop a hybrid system to weaken the rock for typical cutters, so that smaller and more flexible partial face machines can be used in hard rock.

TBMs are theoretically capable of cutting and penetrating rocks of any strength. However, in reality, there are some notable limits and in practice, there are still some disadvantages/restrictions in the strength of the rock that these machines can efficiently excavate. This is especially true for the center and gage cutters, where the discs experience high side forces and the trajectory of the disc cutter path is a tight radius circle. This causes quick wear and low life cycle for the cutting tools and as it can be seen in Figure 1, the conditions could be very severe and unforgiving, particularly at the center, where the center quad is installed.

The goal of some of the ongoing research activities around the world during the last decade has been to overcome the strength restrictions for TBMs and disc cutters. One of the approaches has been to artificially alter the rock properties and reduce rock strength, thus facilitate easier cutting of the rock by mechanical tools, in this case discs. This can be done by introducing fractures and micro-cracks in the rock which helps the subsequent cutting process with a cutting tool. The technologies under consideration are high-power laser (Parker et al., 2003; Batarseh et al., 2003; Graves and Bailo, 2005), microwaves (Hartlieb and Grafe, 2017; Toifl et al.,

Figure 1. Center cutters of a TBM after use

2017), high-pressure water jets (Ciccu and Grosso, 2010; Miller, 2016) and activated tools (Keller and Drebenstedt, 2017). Current R&D-activities focus on testing principles and methods by mainly qualitatively analysis of the results of tests performed on small scaled laboratory units. In some cases qualitative analysis of the influence of the alternative treatment method have been performed and clearly demonstrated the potential of the existing technologies for improving the cutting efficiency of the mechanical tools when a pre-treatment of the rock is applied (Hartlieb et al., 2017).

Hartlieb and Bock (2017) showed how well pre-conditioning would work by using the roadheader performance prediction models as an example. In a laboratory study Hartlieb and Grafe (2017) demonstrated that the cutting forces acting on conical picks can be reduced by a minimum of 10% when very hard granite has been pre-treated with microwave irradiation.

The majority of these studies in the past has been either qualitative assessment of the cracking and fracturing behaviour as reflected on rock strength, or was focused on conical cutting tools by measuring the cutting forces and specific energy of cutting various rock types with pick cutters with and without the pre-treatment with microwave irradiation. Thus, the aim of this study is to analyse how pre-damaging the rock would lead to decrease in cutting forces and potential improvement of the cutting performance of disc cutters, which could increase the production rate of TBMs and reduce the required machine thrust and torque.

PRE-TREATMENT OF THE ROCK BY MICROWAVE

Performance prediction models allow for analyzing and predicting the penetration rate of TBMs. They include various input parameters such as rock strength, as well as cutter geometry and machine specifications to estimate the penetration rate. The

most important and well-known models are those by Colorado School of Mines (Rostami et al., 1994; Rostami 2013, Rostami, 2016) and Norwegian model developed by NTNU in Trondheim (Bruland, 1999; Macias et al., 2014). This study is based on the CSM-model.

Pre-conditioning of the rock with microwave or any similar methods will influence the rock strength properties considered in these models. This mainly refers to introduction of micro-cracks that controls the rock strength (UCS), and large scale cracks/discontinuities that impacts machine penetration rate depending on joint spacing, aperture, and surface condition, which can be controlled by the application of alternative methods of rock pre-treatment. Figure 2 shows how a block of granite would looks after high-power microwave irradiation (24 KW @ 2450 MHz) of one spot (approximately 5 cm in diameter). This spot is the nucleus for a network of radial cracks reaching for tens of centimeters. Figure 3 presents a schematic representation of this crack network. It shows that the crack spacing is a function of the radius from the radial spot. The solid

Figure 2. Damage pattern after MW-irradiation of granite

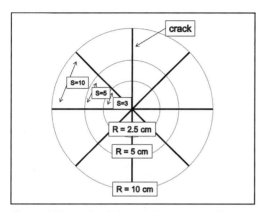

Figure 3. Resulting joint spacing (s=3, 5, 10 cm) depending on the radius from the central point (r=2.5, 5, 10 cm)

lines in the figure represent the cracks. A radius of 2.5, 5 and 10 cm translates to crack spacing of 3, 5 and 10 cm (1–4 inches), respectively.

RMR is a commonly used method for rock mass classification and as a way of expressing the strength of a certain rock mass. One of the main advantages of using rock mass classification, in particular RMR, is that different changes in the rock mass can easily be addressed and allows for estimation of the rock mass strength for use in various applications.

In our analysis a granite sample was used to represent strong rock to be cut by a TBM (UCS = 200 MPa, Tensile strength = 10 MPa and Abrasivity index (CAI) = 3.5; all other parameters are "excellent"—no cracks, no water, 100% RQD). However, when the rock is irradiated by microwaves the joint spacing is reduced to 5–10 cm, depending on the assumed influencing radius (see also Figure 3). This leads to a reduction of the associated RMR value from 20 down to 8 and a reduction of the total RMR from 97 to 85 (Table 1). Other parameters, especially crack width and orientation, cannot be controlled so easily and are still under investigation. For this study they are, therefore dealt with as a worst case scenario (very unfavorable conditions for cutting).

The RMR classification is not explicitly used in the CSM-model. Therefore, it is necessary to translate the described reduction of RMR to the parameters used in the model. This is done by UCS values for the rock mass using the available formulas to estimate UCS_{RM} from RMR values in CSM model. Different formulas are available for estimation of UCS_{RM} from UCS and RMR values. Table 2 shows the results of the commonly used formulas, which are listed below:

Hoek and Brown (1988):

$$UCS_{RM} = UCS_I \sqrt{\exp\left(\frac{RMR - 100}{9}\right)}$$

Table 1. Influence of changing the joint spacing on different rating parameters of the RMR

Rock Mass Rating Influencing Parameters	Rating	
	Original	Altered by Microwaves
UCS	12	12
RQD	20	20
Joint spacing	20	8
Joint conditions	30	30
Ground water	15	15
RMR	**97**	**85**

Table 2. Re-calculation of UCS as result of altered RMR for different models

UCS Original	RMR	Kalamaras and Bieniawski, 1993	Sheorey, 1997	Hoek and Brown, 1988
200	100	200	200	200
200	90	132	121	115
200	85	107	94	87
200	80	87	74	66
200	70	57	45	38
200	60	38	27	22
200	50	25	16	12

Kalamaras and Bieniawski (1993):

$$UCS_{RM} = UCS_I * \exp\left(\frac{RMR - 100}{24}\right)$$

Sheorey (1997):

$$UCS_{RM} = UCS_I * \exp\left(\frac{RMR - 100}{20}\right)$$

With UCS_{RM} = rock mass compressive strength (MPa) and UCS_I = UCS of intact rock.

If e.g., the UCS_I is 200 MPa, for the rock mass with RMR rating of 85 then the UCS_{RM} of 107, 94 or 87, can be estimated depending on which if the formulas was used. To be conservative, UCS value of 130 MPa was used for subsequent calculations to represent the rock that is pre-treated by microwave irradiation.

RESULTS AND DISCUSSION

Table 3 shows the parameters used for calculating the influence of pre-treatment of rock on the performance of a TBM in the described conditions. The calculations include an analysis of the performance in intact rock conditions, pre-treated/altered rock (with RMR = 85), and "mixed face" conditions where only the area around the center cutters is pre-conditioned.

For all three cases it is apparent that the center cutters are exposed to the highest cutting and rolling forces (Figure 4). For all other cutters, except for the gage cutters, the difference between the two different rock mass conditions is approximately 50 and 5 kN (approximately 20%), for normal and rolling forces, respectively.

The results of the presented analysis demonstrate how significant the introduction of artificial cracks in the rock could be and how it can contribute to improving the performance of a TBM. Especially the center cutters could be relieved and thus, longer cutter life could be expected, which can reduce the related cost and downtime for cutter changes, and hence higher Machine utilization and advance rate.

Pre-conditioning of the rock mass can be implemented using a high-power microwave system at critical locations (center/perhaps gage area) as previously described. Given the limited available space, total energy consumption, and the issue of overlapping system availability, perhaps it is not necessary to heat the entire tunnel face and spot-wise irradiation and cracking could be used to assist the rock cutting

in critical areas for the best exploitation of the energy consumed. This refers to selective heating and cracking at the areas where pre-conditioning and damage are necessary. Due to the ability of microwaves to be directed at a certain area and to penetrate the rock to certain depth (up to 30 cm in case of granite) the areas of pre-damage and heating can be well controlled. The crack density and spacing is governed by the distance of the irradiation spots to one another and duration of treatment.

In the experiments, irradiation times of up to 45 s has been used. This is much too long for continuous treatment of the entire tunnel face. Therefore, a procedure is suggested, where one to maximum five magnetrons (microwave sources) are placed around the center of the cutterhead to assist centre cutters. These magnetrons would be arranged in a way to guarantee suitable network of cracks that the cutters can then exploit. For granite a total effective depth of crack of 25–30 mm could be easily achieved with minimum efforts (Hartlieb et al., 2017). This is significantly lower than the theoretical penetration depth of up to 30 cm. Given the typical penetration depth for a single pass of 6 mm (Table 3) this would mean that after every 4th pass the procedure has to be repeated or a continuous treatment with lower power unit can be considered as an alternative. A rotational speed of 8 RPM this would allow for 30 s of irradiation (= 7.5 s per revolution * 4 = 30 s) as an average treatment time for the center area. The goal will therefore be to a) arrange the magnetrons in the best possible way to introduce the right amount of cracking in the right place and b) slightly reduce the necessary irradiation time by either increasing the microwave output power or improving the antenna system in order to generate higher specific power.

Table 3. Rock and machine parameters used for the calculations (cutting parameters from Rostami, 2008)

Parameters	Values
Spacing center (1–5)	90 mm
Spacing rest (6–42)	80 mm
Original UCS without MW	200 MPa
Pre-treated rock UCS after MW	130 MPa
Tensile strength	10 MPa
CAI	3.5
Penetration	6 mm
Disc diameter	17″

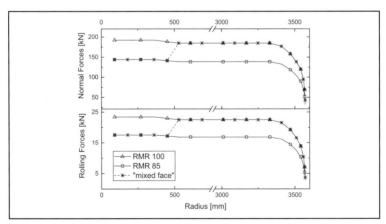

Figure 4. Forces acting on the discs depending on the distance from the center (radius) and several models of calculation. Face condition with intact rock without pre-treatment (RMR 100), face totally altered to RMR 85, face altered to RMR 85 only in central part (="mixed face").

A total of five magnetrons at 30 kW leads to additional electric power consumption of the machine of 150 kW. This is marginal compared to the total installed power of a TBM of several thousand kW and especially considering the mechanical and operational advantages that can be achieved.

CONCLUSIONS

A study on the theoretical performance increase for a TBM when the rock mass is pre-conditioned with high-power microwaves is presented. This preliminary study demonstrated that high-power microwaves can introduce an extended network of cracks in hard rocks such as granite that could be difficult to cut when higher strength rocks without joints and discontinuities are encountered at the face. This crack network is synonymous with a reduction of the intact rock strength, and lower mass rating strength. Calculations show that inducing micro and larger scale cracks can reduce RMR values and can have an impact on both rolling and normal forces of the cutting discs. The forces can be reduced by approximately 20% and depending on where the pre-conditioning is applied, it can lead to increased rate of penetration. Added advantage of applying the proposed system could be increased cutter life in strategic areas such as centre and gage areas.

REFERENCES

Batarseh, S., Gahan, B.C., Graves, R.M., and Parker, R.A., 2003, "Well Perforation Using High-Power Lasers," *Proc SPE Annu Tech Conf Exhib,* SPE Annual Technical Conference and Exhibition, Proceedings-Mile High Meeting of the Minds, Denver, CO, pp. 1–10.

Bruland, A., 1999, *Hard Rock Tunnel Boring.* Trondheim, Univ. Diss, 97 Bl.

Ciccu, R., and Grosso, B., 2010, "Improvement of the Excavation Performance of PCD Drag Tools by Water Jet Assistance," *Rock Mechanics and Rock Engineering,* Vol. 43, No. 4, pp. 465–474.

Graves, R.M., and Bailo, E.T., 2005, "Analysis of thermally altered rock properties using high-power laser technology," *Int. Pet. Technol. Conf. Proc.,* 2005 International Petroleum Technology Conference, pp. 1–8.

Hartlieb, P., and Bock, S., 2017, "Theoretical investigations on the influence of artificially altered rock mass properties on mechanical excavation," *Rock Mechanics and Rock Engineering.* in press, corrected proof.

Hartlieb, P., and Grafe, B., 2017, "Experimental Study on Microwave Assisted Hard Rock Cutting of Granite," *BHM Berg- und Hüttenmännische Monatshefte,* Vol. 162, No. 2, pp. 77–81.

Hartlieb, P., Grafe, B., Shepel, T., Malovyk, A., and Akbari, B., 2017, "Experimental study on artificially induced crack patterns and their consequences on mechanical excavation processes," *International Journal of Rock Mechanics and Mining Sciences,* Vol. 100, 100C, pp. 160–169.

Hoek, E., and Brown, E.T., 1988, "The Hoek-Brown failure criterion - a 1988 update," *Proc. 15th Canadian Rock Mech. Symp.,* J.C. Curran, ed., pp. 31–38.

Kalamaras, G.S., and Bieniawski, Z.T., 1993, "A rock mass strength concept for coal seams," *12th conference ground control in mining,* S.S. Peng, ed., pp. 274–283.

Keller, A., and Drebenstedt, C., 2017, "Hard Rock Cutting and Activated Cutting—Lessons Learned from the Laboratory," *BHM Berg- und Hüttenmännische Monatshefte,* Vol. 162, No. 2, pp. 67–71.

Macias, F.J., Jakobsen, P.D., Seo, Y., and Bruland, A., 2014, "Influence of rock mass fracturing on the net penetration rates of hard rock TBMs," *Tunnelling and Underground Space Technology,* Vol. 44, pp. 108–120.

Miller, H., 2016, "Pulsed waterjet technology —Recent advancements in excavation," *Proceedings of the 1st International symposium on mechanical hard rock excavation,* Bergmännischer Verband Österreichs, ed., 13–14 September.

Parker, R.A., Gahan, B.C., Graves, R.M., Batarseh, S., Xu, Z., and Reed, C.B., 2003, "Laser Drilling: Effects of Beam Application Methods on Improving Rock Removal," *Proc SPE Annu Tech Conf Exhib,* pp. 2579–2585.

Rostami, J., 2008, "Hard Rock TBM Cutterhead Modeling for Design and Performance Prediction," *Geomechanik und Tunnelbau,* Vol. 1, No. 1, pp. 18–28.

Rostami, J., 2016, "Performance prediction of hard rock Tunnel Boring Machines (TBMs) in difficult ground," *Tunnel Boring Machines in Difficult Grounds,* Vol. 57, pp. 173–182.

Rostami, J., Ozdemir, L., and Neil, D.M., 1994, "Performance prediction: a key issue in mechanical hard rock mining," *Mining Engineering,* Vol. 46, No. 11, pp. 1263–1267.

Sheorey, P.R., 1997, *Empirical Rock Failure Criterion,* Oxford IBH publishing co. and A.A. Balkema, Rotterdam, 176 pp.

Toifl, M., Hartlieb, P., Meisels, R., Antretter, T., and Kuchar, F., 2017, "Numerical study of the influence of irradiation parameters on the micro-wave-induced stresses in granite," *Minerals Engineering,* 103–104, pp. 78–92.

The Karlsruhe Monster—Highly Flexible Steel Formwork, Operated Under Hyperbaric Conditions: Design and Application of a High-Tech Formwork in a Tapered Tunnel Section

Rainer Antretter
BeMo Tunnelling GmbH

ABSTRACT: Connecting two stations of the new Light rail scheme in the heart of Karlsruhe, Germany the Kaiser-Friedrich-Tunnel had to be excavated and finally lined under compressed air conditions. An extreme widening of the tunnel on the last 40 meters made the inner lining works special with regard to both the selection and erection of the formwork as well as the construction of the lining. The nearly 200 ton formwork was erected and operated inside the pressurized tunnel, all works were performed in a 24/7 operation at up to 1.15 bar (17 psi) air pressure. The paper describes the challenges faced in connection with these unusual conditions.

PROJECT REQUIREMENTS

The last 50 m of the Kaiser Friedrich tunnel called for a significant enlargement in order to accommodate a rail switch, see Figure 1.

During the excavation phase of this tunnel portion in non-cohesive sand and fine gravel it was necessary to divide the most enlarged section horizontally in bench and crown as well as vertically in two sections by a shotcreted single side wall drift. The minimum excavation cross sections (block 21 in Figure 1) was 78 m² (840 sqft), the maximum (block A2 in Figure 1) was 170 m² (1,830 sqft) in size, an extension to more than double in size within just five inner lining concrete pours, each 8 m long. Two

vertical columns in the largest section A2 were to be poured in addition.

The inner lining design allowed the separation of invert and vault. Casting the invert was done concurrent with the erection of the reinforcement gantries, from the far tunnel end at the diaphragm wall back to the starting chamber with air locks, thus starting to pour with the largest tunnel section, see Figure 2.

Supply of material for erection of reinforcement gantries during pouring of the invert was established by means of a 10 ton monorail running in the crown of the tunnel. All parts required could be moved to the installation area by monorail after transporting the parts from atmospheric conditions via the 3.5 ×

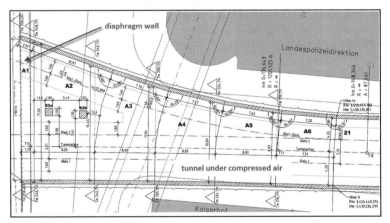

Figure 1. Plan view of the light railway link tunnel

4.5 m air lock into the pressurized starting chamber, see Figure 3.

Next step after having completed all invert pours was to finalize all reinforcement blocks in the vault and, simultaneously, to erect the huge tunnel formwork with its maximum size in the biggest block A2. Erection of the formwork took five weeks and could be finished within the estimated time frame.

Due to the limitation of working time under approx. 1,15 bar to only seven hours and max. 40 hours per week it was necessary to work in three shifts 24/6 with Sundays off.

The horizontal part of the ceiling formwork for the maximum size block A2 was created by use of special form parts exactly at the place required. Both halves of the formwork, left and right, were erected consecutively using the available space of block A3 and A4. Each of these was then moved into block A2 and linked with the horizontal ceiling formwork. Transport of parts was done on the finished invert using trailers and electrically driven loaders.

From block A2 to A6 the tunnel section shrank continuously and got smaller with each pour until the typical cross section of the running tunnel was reached (Block 21), see Figure 4.

After pouring block A6 all enlargement parts could get dismantled and the both formwork halves were linked together directly to form the typical cross section used from block 21 to block.

FORMWORK DESIGN

The significant costs for any day lost of compressed air tunnelling was the main driver to achieve a formwork design which enabled quick erection and demobilization as well as rapid conversion from block to block during the six enlargement sections. The owner's design foresaw different geometries in all of the enlargement blocks. This would have resulted in a significant extension of time and additional costs. BeMo's technical design department was able to convince the owner to accept an alternative inner

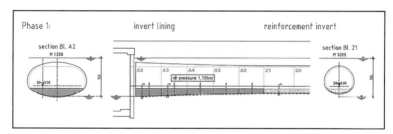

Figure 2. Situation after start of invert concrete installation

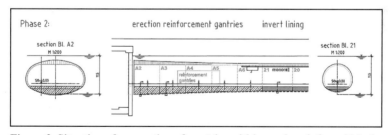

Figure 3. Situation after erection of gantries within ready reinforced blocks (hatched area)

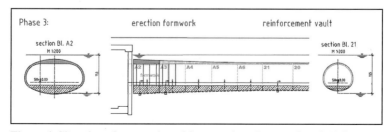

Figure 4. Situation after erection of formwork and start of vault lining

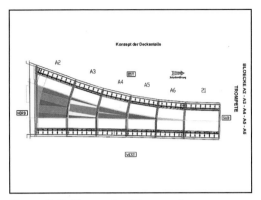

Figure 5. Ceiling parts of formwork

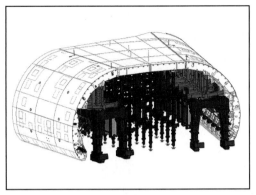

Figure 6. Formwork section

lining design which used the same geometry in all sections from the invert joint to the tunnel shoulder but applied a horizontal ceiling instead of a tapered vault.

BeMo's formwork supplier and partner, Kern Tunneltechnik SA of Switzerland presented a basic formwork solution with two formwork and carrier halves, each independently driven and self-supporting with special formed enlargement parts for multiple use in between the two formwork halves for pouring the enlarged sections.

Most of the ceiling parts were triangular and trapezoidal shaped in order to limit the total number of parts required. Their shape allowed to reuse them, see colored parts in different blocks shown in Figure 5.

The maximum size cross section of the formwork (block A2, shown in Figure 6.) shows the different elements of the construction. The two carriers left and right (blue) with the stiffening structure (red) and the circular shaped formwork skin structure (white) form the formwork-halves on either side. The strutting structure in between (blue) with the corresponding horizontal ceiling parts (white) form the enlargement structure.

Design Details

It was required by the owner to use a steel formwork with smooth surface for use in combination with watertight concrete, polypropylene microfibres for fire protection purposes, fully reinforced vault and a centralized radial rubber water stop between blocks. A separation membrane had to be used between shotcrete and lining concrete. Due to the tunnel design which included a 1,4 m (4.6 ft) thick crown part and heavy reinforcement it was not possible to shotcrete the inner lining as an alternative.

Structural Design Parameters for the Formwork

The formwork had to be designed to take loads and forces by the concrete pressure as follows:

- Typical cross section: pouring velocity limited to 3 m/hr (9.84 ft/hr), max. concrete pressure of 120 kN/m^2 (2,506 psf)
- Enlarged section: pouring velocity limited to 1,6 m/hr (5.25 ft/hr), max. concrete pressure of 75 kN/m^2 (1,566 psf)
- Concrete level difference: max. 0,50 m (1.64 ft)

Formwork Details

Uplift was to be expected due to the shape of the tunnel cross section, although it was not an issue since it could be counterbalanced by the weight of the formwork itself, still providing sufficient safety.

Longitudinal movement of the formwork due to the conical block shape formed by the side wings was another detail which had to be considered. Since the forces involved were quite significant (780 kN for block A2) it was necessary to anchor the formwork into the invert. As a solution *anchor bars* were installed from the supporting structure between carrier and formwork into the invert, see Figure 7.

The horizontal reaction force at each of the seven ribs was 225 kN and for quality reasons it was decided to limit horizontal deflection during pours to zero—to ensure this *shear pins* were installed between each rib toe and the invert, see Figure 8.

Based on the structural analysis the loads onto the bulk head were quite significant. In addition, the geometrical requirements called for a rigid solution with very high timber bulkheads to take the load from 1.100 mm (3.6 ft) concrete thickness in the vertical tunnel axis, see Figure 9.

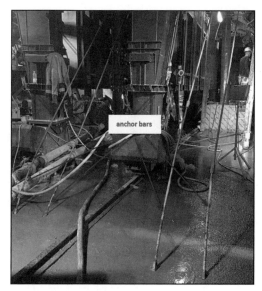

Figure 7. Anchor bars

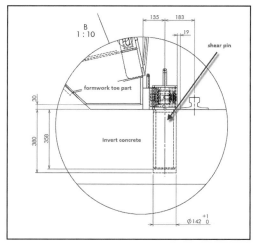

Figure 8. Shear pin

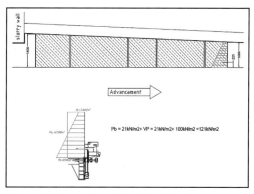

Figure 9. Stop end and bulkhead load

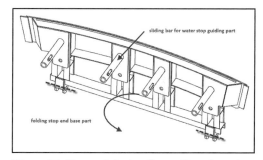

Figure 10. Stop end design (lower hinged part)

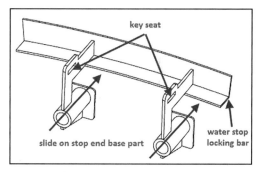

Figure 11. Stop end design (water stop guiding part)

As a standard the stop end design tried to eliminate heavy parts to handle manually. The BeMo standard stop end system with a folding lower part (Figure 10) and two attached parts which have to be wedged in position (Figure 11 and 12) was chosen.

While the lower folding part was heavy but hinged and therefore easy to operate, the upper attachment parts were relatively lightweighted and could be installed by one worker only.

The load distribution beams however had to provide a sufficient resistance resulting in long and heavy parts which had to be lifted into position, see blue colored beam in Figure 13. These heavy beams

were required only in the horizontal formwork ceiling part.

From tunnel shoulder to toe the typical concrete thickness of 500 mm (1.64 ft) determined the height of the bulkhead, which could—structurally—be designed without support extension beams, see Figure 14.

Deformation due to concrete pressure from theoretical geometry was calculated to be 18 mm (0.7 in) based on a load case "concrete filled to 4,5 m (14 ft) height." The geometry of the formwork was

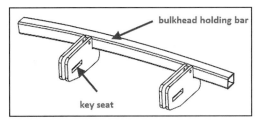

Figure 12. Stop end design (upper part/bulkhead form timber)

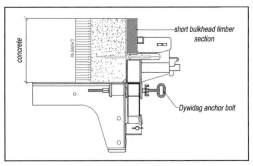

Figure 14. Bulkhead design (cut at typically concrete thickness)

therefore superelevated by 30 mm (~1.2 in) compared to the theoretical geometry. The calculated reaction forces at maximum loads by concrete pressure are shown in Figure 15.

The *horizontal ceiling formwork part* for pouring the enlargement blocks of the tunnel should be a quick and safe system structure which enables rapid changes from pour to pour. Kern Tunneltechnik SA has therefore chosen to use their TSK design for the bearing structure which consists of combineable system elements which allow to form almost any thinkable support structure nearly at will, see Figure 16 and 17.

Another advantage of this design was the possibility to preassemble the time-consuming ceiling parts exactly at the place required in the biggest block A2. Erection of the two formwork halves and the TSK ceiling parts could be done simultaneously, resulting in time savings.

The ceiling formwork was designed to be modular-shaped for multiple use and bolted together, see Figure 18. The ingeniously designed geometry

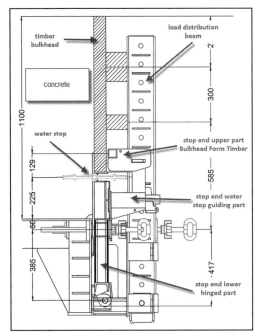

Figure 13. Bulkhead design (cut at tunnel axis)

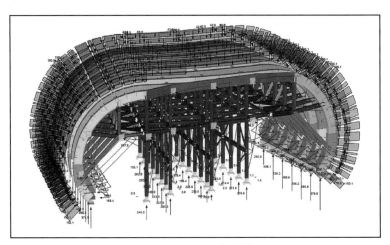

Figure 15. Loads due to concrete pressure (in kN)

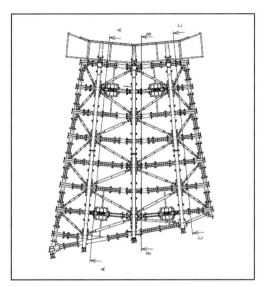

Figure 16. Plan view TSK supporting structure

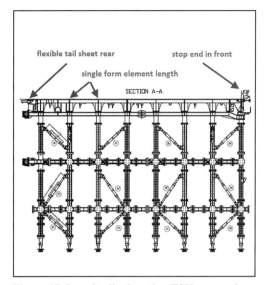

Figure 17. Longitudinal section TSK supporting structure

of each single part allowed quick dismantling and adjustment for the next pour.

Modification of Formwork—Handling

The sophisticated designed geometry of each single part allowed quick dismantling and adjustment for the next pour.

Overall Lining Works Time Schedule

Tight 22 weeks total time were scheduled for inner lining works consisting of the 21 pours (209 m = 228 yd) and 5 pours (40 m = 43 yd), divided into invert and lining concrete. Only 52 days were scheduled for pouring the enlargement blocks including formwork modification between each block. Focus was put on the enlarged blocks since construction time could be gained only there, see Figure 19 in red.

Planning of the single modification steps consisted of:

- preassembly of the TSK and horizontal ceiling system in advance
- stripping and dismantling of the formwork 12 to 18 hours after the pour
- unbolting of the two formwork halves and shifting those to the next block
- connecting the modified TSK and ceiling parts together with the two formwork halves
- placing the formwork for the next pour

These steps were identified as the most important ones for a highly efficient repositioning

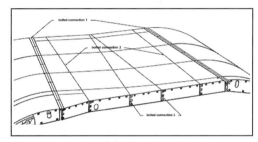

Figure 18. Modular horizontal ceiling formwork

procedure. An ambitious erection and modification time schedule was prepared resulting in an overall execution time of 52 days for lining of the entire enlargement section.

Erection of Formwork

It was decided to conduct a works commissioning with fully erected formwork prior to delivery. This can be considered an unusal step since off site formwork erections are no longer standard. In case of the TKF Tunnel however, with success dependent on every day gained, it was decided to accept the additional expense with the goal to gain certainty with regard to unexpected incidents on site, see Figure 20.

An advantage which went along with the shop assembly at the fabrication site was the possibility to dismantle the formwork not back to single pieces but into components which were able to pass the material locks doors (3,0 × 3,5 m (10 × 11.6 ft) and a weight lighter than 10 tons (limit of the monorail).

All mentioned provisions led finally to the desired result. With three erection teams working 7 hours per day in compressed air conditions for six days per week it was possible to erect the nearly 200 ton monster in its largest extent within 33 working days and 1.940 working hours. Pouring of block A2 could start in time.

Enlargement Lining Works—Time Schedule

Pouring of the first and largest block A2 (see Figure 21) was not only a technical challenge but put

also mental pressure on each of the members of the lining crew. It was a big relief for everybody when everything went well. Tolerances turned out to be well within the calculated limits and after stripping the formwork the surface was cavity free and of high quality.

The following challenge was to modify the formwork for the next smaller block A3. In this process it was realized that certain parts were not fabricated correctly and had to be rewelded on site. With more than a week of additional modification time the schedule was missed, see Figure 22.

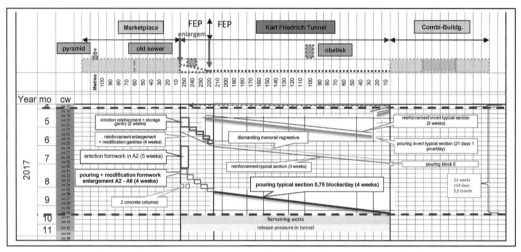

Figure 19. Time schedule overall lining works—red: lining enlargement incl. modification formwork

Figure 20. Formwork shop assembly in Slovenia

Figure 21. Formwork in largest block A2 during lining

Figure 22. Time schedule modification formwork and lining enlargement

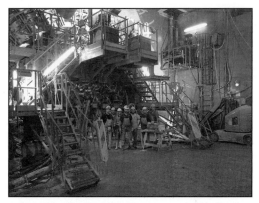

Figure 23. Arrival of formwork in last block of typical cross section

Of course, both teams, the Kern Tunneltechnik SA erection team and the BeMo lining team tried their utmost to mitigate the schedule impacts. It is in hindsight a great achievement that the crews did not only catch up with schedule but were even able to gain one day compared with the original schedule. Thus, the first block of the typical cross section (blocks 21 to 1) could start one day earlier and inner lining this section could be finished in time with one pour every day over 21 blocks, see Figure 23.

CONCLUSION

Tunnel lining works in hyperbaric conditions combined with extreme changes of the shape of the steel formwork and high overall time pressure on the project required thorough works preparation. Additionally, smooth work flow on site must be ensured, particularly under hyperbaric conditions. Nevertheless, the unexpected can happen and the weakest link within the chain turned out to be the quality control of a fabrication subcontractor which did not recognize a fabrication problem. This was outside of the reach of the works preparation team on site. However, it was possible to compensate the time loss and complete the job in time.

REFERENCE

Wechner, T. 2016. Complex inner-city tunnel excavation by means of the New Austrian Tunnel Method in combination with a hyperbaric atmosphere. RETC 2017. San Diego, CA.

Instrumentation Design Framework for Large-Diameter Tunneling Projects

Anil Dean, Jon Pearson, and Marlene Wong
Stantec

ABSTRACT: Large diameter tunneling projects are often located in developed and sensitive environments. As such, instrumentation layout, selection, monitoring and interpretation are essential to evaluate disturbance caused by tunnel construction. At the same time, there is limited modern guidance available in the industry relative to instrumentation design methodologies. This paper provides a framework of key principles and considerations relative to design of instrumentation programs for large diameter tunneling projects. Instrumentation discussed includes monitoring of ground and structure movements, groundwater levels, and construction induced vibrations.

INSTRUMENTATION SELECTION

Construction Impact Assessment Report (CIAR)

Selection of instrumentation is typically integrated into the design process. On large projects, or those that require thorough documentation of the instrumentation basis of design, development of a specific report may be advisable. This report can have several names, but is referred to as a Construction Impact Assessment Report (CIAR) for the purposes of this paper. The CIAR serves as a due diligence report that is designed to better quantify and qualify the identified risks as well as provide further refinement of the selected mitigation measures and justification for their specification in the contract documents. Regardless of whether a report is prepared, design considerations may include the following:

1. **Tunneling Means and Methods:** The CIAR will be used to help justify the decision to specify tunneling with TBMs that have pressurized face capability. Contractors generally prefer to tunnel with TBMs in open mode rather than pressurized face mode, all other things being equal. The CIAR can be used as the basis for a TBM specification to be included in the contract documents, which relates in part to open versus pressurized face requirements. Depending on results of the analyses, a CIAR may also be used to justify specification of pressurized face tunneling methods at specific locations along the tunnel alignments. As a result, a CIAR can be used to help the contractor to assess the potential for and the owner to evaluate requests to (a) provide TBMs that do not have pressurized face capability, or (b) operate TBMs in open mode.

2. **Shaft Construction Means and Methods:** The CIAR may be used to evaluate proposed temporary shoring systems at shaft locations.

3. **3rd Party Claims:** Settlement evaluations included in the CIAR may be used as the theoretical basis for expected behavior along the alignment resulting from impacts of tunnel and shaft construction. As such, the CIAR may be used to develop settlement and deformation monitoring requirements for the contractor for relevant structures and facilities along the alignment. Actual performance can then be compared to the theoretical performance. In doing so, third party claims may be evaluated more readily.

4. **Justifying Actions in Response to Monitored Conditions:** Information in the CIAR may be used to justify the selection of trigger and action levels in the Contract Documents presented in the Instrumentation & Monitoring specification for various types of instruments. This provides confirmation that the owner has exercised a significant amount of diligence in selection of the trigger and action levels provided to the contractor.

5. **Supporting Permit Applications as Appropriate:** The completion of the CIAR is anticipated to better support applications for certain permits, particularly those related to groundwater extraction during construction.

Figure 1. Automatic total station line of sight readings comprising a portion of an instrumentation design in an urban area (image courtesy of SIXENSE)

Specific Types of Instrumentation

There are a wide variety of instrumentation types in use on projects, and it is not possible to include all of them in this paper. Some of the most common instruments are discussed in this section. When specifying any instrument, it is important to estimate the zone of influence of the work (Dunnicliff, 1993). In a soft ground tunnel, the zone of influence (ZOI) is frequently estimated by conventional settlement estimates. Zones of influence for shafts in soft ground are generally related to the type of support installed and the depth of the shaft. Considerations in rock are different and generally less onerous unless there is soil overlying the rock. Overlying soil that is subject to dewatering induced settlement can be of particular concern in rock tunnels, as has been described in various case histories.

In any case, instrumentation outside of the ZOI would generally not be expected to show movement under normal circumstances. Instrumentation is therefore frequently specified or required within the zone of influence with survey closure, as needed, to stable points located outside of the ZOI. Specific types of instruments and their layout is predicated on an understanding of the location of the ZOI of the work, which would generally be the combined ZOI of the tunnel and any associated underground structures and shafts.

Where instrumentation is tied to utilities or structures, the contractor is frequently under contract to verify or modify designated lists of utilities and structures to be instrumented as necessary based upon their proposed means and methods of construction, and other factors. Information developed is ultimately used to inform the instrumentation design. Figure 1 (courtesy of SIXENSE) shows a sketch of

line of sight automatic total station (ATS) instrumentation design for a project in an urban area. The ATS instrumentation would be supplemented with other types of instrumentation as needed. The surface trace of a potential tunnel alignment is shown in green with other lines indicating ATS points that would be acquired by various stations in the area.

Building Monitoring Points (BMPs)

Monitoring points are available in a number of configurations, from mountable optical survey prisms (OSPs) to survey target decals that can be stuck onto windows or other parts of a structure and removed when they are no longer needed. Monitoring bolts can also be drilled into a structural façade.

Tiltmeters

Tiltmeters are used for measuring inclination of structures such as buildings, retaining walls, and piers. The tiltmeter is generally attached to the structure of interest with a bracket or otherwise bonded to the structure. They are available in both uniaxial versions, where tilt is measured in one direction, and biaxial, where tilt is measured in two directions. They can be manually read or can be read automatically or remotely. Most tiltmeters used for structural monitoring utilize a Microelectromechanical System (MEMS), which allows for more accurate measurements and can measure changes in inclination as low as one arc second.

Tiltmeters are typically placed on structures that are expected to fall within the zone of influence of the tunnel or nearby excavations. They can be paired with other building monitoring instrumentation such as building monitoring points to get data on both settlements and inclination.

Piezometers

For tunneling projects, piezometers are used to measure groundwater level changes in the vicinity of any tunneling or other excavation work. Monitoring of groundwater levels during construction is important as groundwater can have significant impacts on tunnel and excavation stability and loading, inflows into excavations, and dewatering induced settlements.

Ground Surface Monitoring Points (GSMPs)

Ground surface monitoring points are also available in a variety of configurations. One well-used system comprises a surveyor's nail or rebar driven flush or installed into new or existing concrete, masonry, or asphalt. These can be inscribed with labels as needed. GSMPs are typically installed over the centerline of each tunnel bore and around the perimeter of excavations for cut and cover structures, and at each cross-passage location or similar location where break-outs of the tunnel lining are required.

The use of GSMPs are sometimes criticized as they are not deep monitoring points. Settlement has to occur and propagate to the surface for detection to occur. This is of particular concern in a number of conditions, as asphalt, concrete and even desiccated ground can bridge underlying settlement, which may never be manifested at the GSMP itself until it is too late. However, GSMPs are one of the least expensive instruments to install and therefore are frequently used despite the fact that they can provide limited information.

Soil Deformation Monitoring Points (SDMPs) and Utility Monitoring Points (UMPs)

SDMPs and UMPs are often fabricated for each job, but frequently comprise one-inch (or similar) diameter steel rods driven or grouted in place below the pavement by around 5 feet or placed in sleeves resting on utilities of interest in the case of UMPs. This allows for deeper monitoring than the GSMPs described above, and therefore mitigates one of the main disadvantages of GSMPs.

SDMPs and UMPs are often useful when provided near the following facilities or features:

- Utilities identified as being at-risk of damage. An initial assessment of utilities identified as being at-risk of damage is ideally provided in the CIAR.
- Tunnel break-in and break-out locations. This is primarily due to the fact that ground movements tend to occur more frequently near break-in and break-out locations than at other locations along a soft ground tunnel, all else being equal. When ground improvement is used at a break-in and break-out locations, some risk is mitigated in this regard (Dean and Young, 2006).
- At intervals as appropriate along a soft ground tunnel to provide potential early warning for settlement.
- At soil and rock transitions to provide possible early warning for over-excavation in mixed face conditions.

In their simplest form, SDMPs are simpler to install and read relative to the multi point borehole extensometer, but are also less standardized and may not provide as much information, depending on configuration.

Multi Point Borehole Extensometers (MPBX)

MPBXs are standardized and can be purchased from instrumentation suppliers. One advantage is that they can be installed in multiple points, thereby allowing ground movement to be assessed at various depths through use of the same instrument.

MPBXs are often useful at the following locations:

1. Over the centerline of a tunnel bore at regular intervals. This can still be done cost effectively at relatively tight spacings where such data is important. This tends to be of particular concern for soft ground tunnels in urban areas (Dunnicliff 1993).
2. Over cross-passage excavations in soft ground, particularly where ground improvement is not used or is used in a limited way.
3. Near tunnel launch and reception areas. In these cases they may be used in lieu of, or in addition to, SDMPs.

As denoted by the name of the instrument, it is common for multiple points to be installed within the same sleeve. Frequently, a minimum of three anchors may be installed in each borehole. Depending on tunnel depth and instrumentation needs, it is possible, for example, to locate the deepest anchor 5 feet above the tunnel extrados, the shallowest at 5 feet below ground, and the other in the middle. MPBXs can also be automated and efficiently read and monitored when installed in this way. This can result in significant productivity and safety benefits. A variety of ranges are available, so it is important to specify the desired range of measurement for the instrument. For example: *MPBXs shall have a range of at least six inches for settlement measurements and two inches for heave measurements.*

Figure 2. Representative inclinometer installation (top) and probe at the surface of the inclinometer casing (bottom)

Inclinometers

Inclinometers are frequently used to evaluate ground movement on tunnel projects, particularly in the vicinity of shafts. On typical deep tunnel access shafts, two inclinometers are often specified to provide some redundancy and coverage around the shaft at various points since movements are not necessarily uniform when they do occur. For larger linear

excavations, such as station excavations for subway systems in urban areas, or rectangular tunnel launch shafts in urban areas, inclinometers can be installed at spacings of 100 to 250 feet along the excavation. Tighter spacings may be used where there are structures adjacent to the excavation and movement is of particular concern.

Inclinometers can be automated, but automation of this instrument is particularly expensive, depending on the spacing of cells desired along the depth of the inclinometer (Slope Indicator, 2004). On a deep shaft, it is common to take inclinometer readings using a probe at intervals of 2 feet. Therefore, the number of cells required for an automated system becomes large quite quickly. An inclinometer installation and probe is shown on Figure 2.

Seismographs

Seismographs used for construction vibration monitoring generally consist of a seismograph unit with a geophone. Some units allow for multiple geophones that can be installed at various locations or depths. Seismographs are installed with the geophones firmly mounted on the surface of a slab of concrete, directly affixed to a structure, or firmly set in undisturbed soil about 3 to 6 feet from the building façade. Seismographs provide continuous vibration monitoring, and can be installed with a modem to provide automated wireless transfer of vibration data to allow for cloud-based access of data. It is important to install the seismograph in accordance with the manufacturer's instructions and to set the trigger level (vibration level at which the seismograph will record a vibration event) at a level that is commensurate with the requirements of the project.

INSTRUMENTATION LAYOUT

Shaft Instrumentation Considerations

Several instrumentation considerations are coincident at shaft locations. Selected specific considerations are as follows:

- Support of Excavation (SOE) stability: This is frequently measured using inclinometers and/or arrays within the shaft that can be surveyed for convergence. For large shafts, several inclinometers may be used.
- Groundwater Conditions: Piezometers are frequently used near shafts to verify groundwater conditions prior to and during shaft sinking and construction.
- Settlement: At tunnel break-in and break-out locations, soil deformation monitoring points (SDMPs) are frequently required immediately outside of the shaft on the tunnel alignment. This is of particular concern

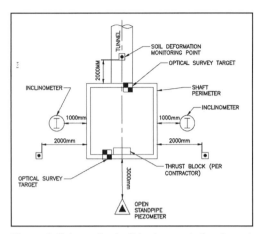

Figure 3. Schematic shaft instrumentation layout detail for surface based instrumentations for a rock tunnel (no SDMP)

in soft ground (soil) conditions, where over-excavation is known to occur at break-in and break-out locations.

Each project is unique and instrumentation needs will change depending on shaft location, ground conditions, parameters of interest, and so on. For example, an SDMP is not always deemed to be necessary, particularly when the tunnel heading is excavated in rock.

A typical detail used on selected projects is shown in Figure 3.

Note: Shaft convergence arrays located within the shaft are not shown on Figure 3, but are frequently specified when desired.

Instrumentation Considerations for Tunneling Induced Ground Movements

Ground movement monitoring above the tunnel alignment can consist of ground surface monitoring points, soil deformation monitoring points, multi point borehole extensometers, or a combination of these instruments. Typically, monitoring points should be placed every 100 to 250 feet along the alignment at a minimum in urban areas. Monitoring may be much less prevalent in undeveloped areas. In areas where settlement risk is high or the impacts of settlement would be significant, monitoring point density should be increased accordingly.

Where over excavation is a concern, such as a rock tunnel encountering a buried valley or a tunnel horizon expected to encounter an area of loose sand, it is also prudent to include MPBXs. MPBXs with an appropriate type of monitoring points near the tunnel horizon may capture ground settlement or ground loss deep in the ground, allowing the project team to take action before ground loss propagates to the surface.

Often, it is not practical or possible to install and manually monitor subsurface monitoring points such as SDMPs or MPBXs. An example of this is congested roads and highways where settlement would have significant impacts, but access to monitor and install points is restricted. In this case, reflectorless monitoring with an Automatic Total Station (ATS) can be used as shown on Figure 4. With reflectorless monitoring using an ATS, several points can be monitored above the tunnel alignment, with no need for entry onto the road surface to install or measure physical points. If installation of monitoring points is possible, but daily monitoring of points is not feasible, automated subsurface monitoring devices may be used. These can be read remotely and would provide data on subsurface movement that may be missed by surface monitoring points.

When monitoring points such as reflectorless or ground surface monitoring points are installed in arrays perpendicular to the tunnel alignment, the width of the array should be considered. At a minimum, the length of the array should cover the length of the estimated settlement trough. Sufficient array length is necessary to capture the actual lateral extents of tunneling induced settlements.

Tunneling may also induce settlement in certain ground, such as compressible clays, if dewatering occurs as part of the excavation process. Where the ground may be impacted by dewatering, it is

Figure 4. An ATS in use on a tunnel project for pavement surface monitoring of a major highway

recommended that piezometers be placed more frequently along the tunnel alignment in critical areas. These piezometers can then be used to monitor changes in the groundwater level and used as an early warning against dewatering induced settlements.

Building Instrumentation Considerations

As part of the design or CIAR process it is recommended that an inventory be taken of each building falling within the zone of influence along the tunnel alignment. This inventory could include:

- Information on the foundation type and depth. For example, a building with a mat foundation on soil may require more consideration than a building with piles extending to bedrock.
- Building age and type—When subject to ground settlement, a new steel frame building would be expected to perform better than an older masonry wall building.
- Historical or other sensitive structures—this includes buildings where small movements can have significant structural damage and even minor cosmetic damage is to be avoided.
- Structures where the risk of damage may appear low, but the consequences of damage would be high.

Based on this inventory, maximum allowable settlement and tilt can be estimated for each building and compared to estimated tunneling and shaft excavation-induced settlement at that building location. This then provides a framework for locating which buildings are at most risk and then choosing which buildings to monitor. This process usually requires discussion with the project owner, building owners, and other local authorities to obtain relevant information on each building of interest.

For some projects, it is often not possible to obtain detailed information on each building near the project site. The actual number of buildings to

monitor is dependent on several different project-specific details that cannot be covered fully in this paper.

Generalized recommendations for locating BMPs on conventional structures are as follows:

1. A minimum of one prism per story per address is recommended within the zone of influence.
2. BMPs are recommended on the outside of buildings within the ZOI at a convenient height above grade, on all corners and at each column or on walls, at a spacing not to exceed a distance deemed appropriate to measure tilt where BMPs are used in lieu of tiltmeters. BMPs can be set up for readings with an ATS, and in those cases, uniform layout is not necessary since the ATS will shoot the prisms automatically once calibrated.
3. Given the large linear extent of major tunnel projects, it is recommended that the contract allow for installation of an appropriate number of BMPs above what is specifically indicated on the plans and/or specifications to provide additional flexibility.
4. Historic, sensitive, or otherwise significant structures necessitate case-by-case consideration, and often present additional coordination needs through owners and regulatory agencies.
5. A sample layout of BMPs and other tunnel related instrumentation is shown on Figure 5.

MONITORING AND INTERPRETATION OF INSTRUMENTATION DURING CONSTRUCTION

The best instrumentation setups are irrelevant if they are not read and interpreted on a regular basis. This makes data management and interpretation a key issue, particularly as modern projects now include collection of significant amounts of electronic data. Automated systems can greatly facilitate this process, but it is rare that all instruments are automated

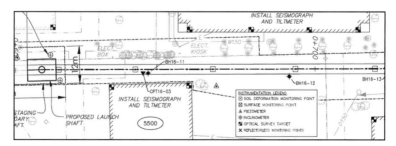

Figure 5. Example of tunnel alignment and building instrumentation layout

on a large project. Even when a high degree of automation exists, there is no substitute for human interpretation of the data. It is no longer possible for human review of each of the millions of data sets collected on a typical major tunnel project, select relevant data can and should still be checked, reviewed, and interpreted. The interval of time at which instrumentation is read and interpreted and the schedule at which it is shared among the project team is therefore important.

Setting instrumentation limits, referred to as trigger levels, is important. It is beneficial to have a multi-level system of triggers such that the project team has an opportunity to react, stabilize the ground, or even change excavation or support methodologies, if appropriate, prior to reaching a designated stop work level. There are various names in use for trigger levels. A three level system is not uncommon. The initial level may be referred to as the "threshold level" at which some action is required. The intermediate level may be called the "response level," which requires definitive action to attempt to avoid hitting the maximum allowable level of movement, and the maximum level allowed in the contract may be referred to as the "shutdown level." Of course, work cannot and should not be stopped in cases where adverse ground movements are still occurring. In these cases, 24/7 stabilization work is necessary and may be concurrent to the development of a mitigation plan.

Ground Movements

Selection of trigger levels for ground movements is somewhat dependent on the instrument and the criticality of each structure. Especially in urban areas, it is not practical or feasible to utilize different trigger levels for each structure, so generalizations usually need to be made with appropriate consideration of risk and likelihood of occurrence. Owners have different risk tolerances, so this also needs to be carefully considered. Vertical movements can be different, as appropriate, in the case of MPBXs with points installed at various depths below grade. There is usually some leeway between the shutdown value and the threshold and response values. For example, if the shutdown value on vertical movement of a structure is 1 inch, the threshold value may be 0.5 inch and the response value may be 0.75 inch. While 1 inch is a frequently used value, soft ground projects in urban areas have been constructed with shutdown values of 0.5 inch. Tiltmeters typically have levels expressed in degrees. Lateral movements, for example for a shoring system at a shaft, would be expected to have explicit response values specified that may be different from vertical movements.

Groundwater Levels

Groundwater levels need to be monitored to protect all parties and provide input on groundwater conditions during construction. The only exceptions may be where groundwater is clearly well below the elevation at which the work is constructed and would not rise into the work. Groundwater monitoring is often overlooked on projects for various reasons, but groundwater can obviously have a significant impact on construction of underground projects. Groundwater levels take on additional significance where groundwater resources may be impacted by pumping from wells for water supply, and also where dewatering may trigger consolidation settlement of soft ground or other ground movements.

Vibration

Humans are perceptive of very low levels of vibration, on the order of 0.03 in/sec, with vibration levels becoming disturbing to the human body at values around 0.1 in/sec, depending on the vibration frequency, source of vibration, and reference cited. This is especially notable when a new source of vibration, such as from construction, is experienced. Buildings, on the other hand, are less sensitive to such low levels of vibration, with the exception of buildings that house sensitive equipment, such as laboratories or hospitals. It should be noted that the appropriate levels of vibration thresholds for tunneling construction-related vibrations are typically above the level of vibration that are disturbing to humans, and therefore may generate complaints from people at vibration levels below the project threshold values.

There are three types of vibration sources associated with tunnel construction activities, continuous/intermittent, transient, and blasting. Equipment or activities that generate continuous vibration include track-mounted vehicles, excavation equipment, and vibratory pile drivers. Typical sources of transient vibration include impact pile drivers, jack hammers, and hoe rams.

Vibrations propagate through the ground causing vibration waves to propagate through the building structure. The type of building structure also plays a role in how vibration waves propagate. There is some reduction in the vibration levels due to the interaction between the ground and the building foundation, with the amount of reduction depending upon the mass and stiffness of the foundation. However, on the majority of tunnel projects there is no budget or scope to perform a detailed assessment of each building and its foundation along the tunnel alignment to make a building specific vibration assessment. This can and should be done for buildings and facilities that are deemed to be sensitive to vibration.

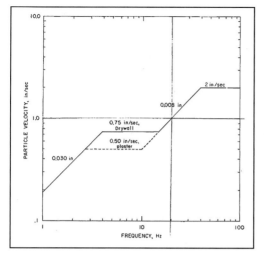

Figure 6. USBM RI 8507 alternative blasting level criteria (USBM, 1980)

The most commonly referenced vibration criteria in North America is contained in the U.S. Bureau of Mines Report of Investigations 8507 (USBM, 1980) and related publications, as presented in Figure 6. The criteria are based on vibration from a single blast experience at a house with plaster-on-lath wall construction of two stories or less in height. The USBM criteria indicate a vibration threshold of 2 in/sec for frequencies above 40 Hz, which is often inappropriately applied to construction vibration. Construction activities can generate continuous and semi-continuous vibrations, rather than a single blasting vibration event, so use of the USBM criteria is not necessarily applicable to tunnel construction, particularly in urban areas.

Applicable construction-based vibration criteria or standards need to be used for construction vibration monitoring. Some states or municipalities may have construction vibration guidelines, however, many of these standards tend to be based on the USBM criteria for blasting. The Federal Transit Administration (FTA) manual for vibration standards for road construction and traffic in the U.S. (FTA, 2006) defines three use categories for setting vibration limits—High-Sensitivity, Residential, and Institutional. For tunnel construction sites, there may be vibrations generated from both continuous sources and blasting. Another commonly used international standard, the Swiss Association of Standardization Vibration Damage Criteria, addresses both continuous (traffic and equipment) source and single-event (blasting) source vibrations and defines four types of building classification based on building structure and sensitivity (SN 640 312a-1992). Based on project experience, the Swiss Association Standard is a useful resource for setting vibration criteria for

tunnel construction projects as it takes into account the building structure, sensitivity, and vibration source, which all need to be considered when developing project-specific vibration criteria.

The selection of vibration threshold levels for a tunnel project should take into account the types of tunnel construction equipment to be used, the types of buildings, and their condition, that will be affected by the vibration, and any sensitive building occupancies, such as hospitals or art museums.

It is important to baseline the seismographs for a minimum of 7 days prior to the commencement of the tunnel construction activities. This allows for monitoring of the different types of ambient vibrations that may occur during a given week. Certain types of commercial vehicles, transit vehicles, and weather-related events may cause vibrations that exceed the project vibration thresholds. Commonly used vehicles, such as light rail, snow plows or garbage trucks, may trigger a vibration threshold exceedance reading, so it is useful to know the schedule of when these types of vehicles are in use near a vibration monitor. Thunder storms have also resulted in vibrations exceeding threshold readings.

CONCLUSIONS

Instrumentation and monitoring on tunnel and underground construction projects is increasingly important as tunnels are built in increasingly urban areas and in close proximity to sensitive buildings and infrastructure. This paper presents a brief rationale for selection of various types of instrumentation, starting with a review of instrumentation needs through a report, the CIAR, which can be included as part of the design process. Once instrumentation needs have been identified, guidelines for the installation, monitoring, and interpretation of some of the more commonly used instruments are presented. While there is no "one-size-fits-all" approach to instrumentation as each project is unique, it is hoped that some generalized guidelines may be helpful to inform the initial design of an instrumentation and monitoring program. Incorporation of a CIAR or similar report should allow for customization of instrumentation for any given project.

Instrumentation and monitoring designs do not exist in a vacuum. Even best instrumentation setups are irrelevant if they are not read and interpreted on a regular basis. It is the continuous review and interpretation of results that can drive the value proposition of an instrumentation program. When implemented in a comprehensive manner, an instrumentation and monitoring program can be useful for a variety of purposes including demonstration that the design intent is met, permit conditions have been complied with, and a basis for review of any claims related to building or infrastructure movement.

REFERENCES

California Department of Transportation, 2013 (Caltrans,2013). Transportation and Construction Vibration Guidance Manual. Caltrans Report No. CT-HWANP-RT-13-069.25.3.

Dean, A. and Young, D.J., 2006 (Dean and Young, 2006). A Framework for Preliminary Design of Tunnel Eyes. 2006 Tunnelling Association of Canada Proceedings.

DGSI - Slope Indicator. EL Beam Sensors and Tiltmeters Data Sheet. www.slopeindicator .com/pdf/el-tiltmeter-and-beam-sensor -datasheet.pdf.

Dunnicliff, John, 1993 (Dunnicliff 1993). Geotechnical Instrumentation Monitoring for Field Performance.

Federal Transit Administration, 2006 (FTA, 2006). Transit Noise and Vibration Impact Assessment. FTA-VA-90-1003-06.

Instantel. Digital image. http://www.instantel.com.

Siskind, D. E., Stagg, M. S., Kopp, J. W. and Dowding, C. H., 1980 (USBM, 1980). Structure Response and Damage Produced by Ground Vibration From Surface Mine Blasting. United States Bureau of Mines, Report of Investigations 8507 (RI 8507).

Slope Indicator, 2004 (Slope Indicator 2004). Guide to Geotechnical Instrumentation.

Swiss Association for Standardization, 1992 (SN 640 312a-1992). Schweizerische Normen-Vereinigung, Les ébranlements—Effet des ebranlements sur les constructions (Swiss Standard on Vibration Effects on Buildings), Zürich (in German) (SN 640 312a-1992).

TRACK 1: TECHNOLOGY

Session 5: Fresh Approach

Lizan Gilbert and Seth Pollak, Chairs

Hands-on Training and Technology Transfer in Tunneling

Matias J. Lazcano, Juan P. Merello, and Nicolás P. Zegpi
SKAVA Consulting S.A.

ABSTRACT: This paper describes three projects in which tunneling performance was improved significantly through Training and Technology Transfer. The projects included are the Andina Phase I project, which through modified equipment technology and construction methods, productivity more than doubled; the Cheves Hydropower project where training and mentoring the supervision team, improved the Owners' contractual control and enabled schedule recovery from a six month delay; and Ventilation Adits at El Teniente mine where through training of operators and workers on the Norwegian Tunneling Method productivity was increased by 80%. Experience has shown that leadership, having the correct methodology, planning and empathy, are key to succeeding with Training and Technology Transfer processes.

INTRODUCTION

Tunneling projects generally include the highest level of uncertainty in the construction industry. As shown in Figure 1 cost is extremely variable. Cost variability occurs mainly because of:

- Geological conditions
- Contractual disputes for different causes
- External events – including Force Majeure

Due to linear nature of tunneling projects, an hour lost in production is an hour added to the schedule, which therefore makes the ability to react ahead of unexpected events highly relevant. Unexpected geological conditions can only be effectively addressed by a prepared Contractor, otherwise the

risk of delays, schedule overruns, and related costs are very high for both the Owner and the Contractor.

A versatile team with technical capabilities, state of the art technology, and competent management for facing these situations therefore becomes critical to project success. The incorporation of hands-on training can provide significant benefit.

Case studies and experience using hands-on training are presented in this paper for different types of projects and a variety challenges among them.

CASE STUDY

Andina PDA Phase 1

Andina is a copper mine, owned by CODELCO, and located near to Los Andes, Chile. Andina is exploited partially as open pit mine and partially as

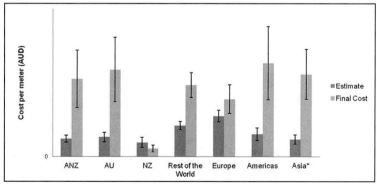

Figure 1. Estimate and final cost in tunneling projects (Efron, N., and Read M. 2012. Analyzing International Tunnel Costs. Worcester Polytechnic Institute)

underground mine. This condition gives Andina very unique operating conditions.

Problem Context

The PDA Phase 1 project was undertaken by VCP (Projects Corporate Vice Presidency of CODELCO) with the purpose of expanding the Andina Mine processing capacity of copper ore from 72,000 [tons/day] to 93,000 [tons/day]. This project included a series of underground openings for installation of processing equipment including crusher, mill, flotation plant, and other equipment.

Underground excavations have been divided in two groups with two different contracts, C1 and C2. Both contracts have been assigned to the same Contractor.

During the months of May and June 2008, the Client foresaw an important risk of delay in the underground excavations. SKAVA was hired to develop a schedule recovery plan.

The review of methods and progress rates confirmed the Client's concern of significant delays in the project. The consultant identified different sources of delay. The most relevant areas where improvements could be made were as follows:

1. Mining Production methods
2. Internal staffing organization of the Client
3. Definition of Contracts and Contract interfaces
4. Engineering and ground condition definitions
5. Contractor's productivity and work flow coordination

In parallel to this study, the quality of the work was under question. Shotcrete and excavation profiles were among the worst evaluated works. The Client initiated a quality audit on all aspects of the rock support system.

For the implementation of this new method, more efficient use of equipment was required. The Client acquired two Finish drilling rigs and also purchased two shotcrete robots in order to increase capacity and quality of the underground operations.

In order to achieve higher productivity both types of equipment needed well-trained operators. SKAVA contributed with providing highly trained crew for both, drilling, and shotcrete application.

Solution and Implementation

Once the main sources of delay were identified, the consultant began to look for solutions. From June 2008 to February 2009 SKAVA delivered around two hundred recommendations which can be grouped into the following categories:

Production methods. The methods planned for cavern and silo excavation were inefficient and were in complete contradiction with the capacities shown in schedules at that time. The planned methods involved horizontal drilling with tunneling jumbos for the majority of the construction sequences presented by the Contractor. In addition, the approach did not consider large working areas and were mostly focused on short working sequences/cycles.

Planning engineers from SKAVA changed these sequences to massive vertical excavations with long benches. The final bench length approximately doubled the original length from 5.5 to 10–14 meters. This modified approach requires expert drillers and appropriate equipment. Once blasted, this allows working over the muck and having a much larger working area.

Internal organization of the client. Just as a construction methodology is designed to produce an efficient work plan, the Client's organization of staff should be built up accordingly. In many respects the organization observed for the project did not comply with that principle.

The main deficiency of the organization was the excessive number of counterparts the Contractor had with equal levels of responsibility, which created confusion and bureaucracy. The approach impacted the capacity of the Contractor to focus on planning and production. Diminished focus on planning and production, increased dramatically improvisation, which directly leads to impaired quality and safety. The original organization structure is shown in Figure 2.

Instead of representing the Client, the EMPC company (Engineering, Procurement and Construction Management, BECHTEL) was in reality slowing down the flow of information and, therefore, VCP personnel started direct communication with the Contractor. However the EMPC company maintained its communication and information demand from the Contractor for several months. This approach obviously generated an overload of paperwork and confusion for the Contractor. The project real operating structure for the project before the changes proposed by the consultant is shown in Figure 3.

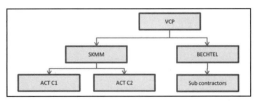

Figure 2. Real organization structure scheme

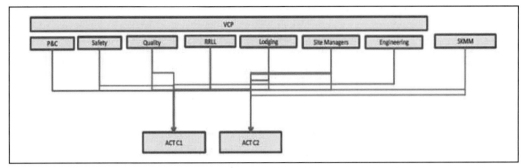

Figure 3. Real operation structure scheme

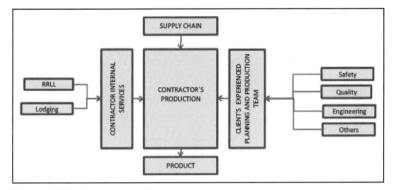

Figure 4. Ideal structure scheme

In an attempt to solve this problem SKAVA proposed a change in the organizational structure to focus the structure on production. As the crucial part of the organization is the production team. All the rest of the organization must be supporting it allowing this team to perform with quality, to go home safe, to have everything to work with, to deliver on schedule, to keep budget under control, etc. Figure 4 shows the ideal proposed structure.

Contracts and interdependencies. Contractual documents between Client and Contractor distributed accountability and merit depending on the outcome of the project. Operational costs were based in unit prices, but general costs and profit were fixed amounts. Additionally there were bonuses related to the accomplishment of certain specific milestones. These instruments also determined the capacity of the Contractor to solve its own problems, improve production capacity, and control its own economy and incentives.

When interdependencies are numerous and intricate, the distribution of accountability and merit which are directly connected with project economics become diffuse and difficult which was the situation on PDA Phase 1 project.

Interdependencies could be identified in:

- Logistics: housing, transport and food.
  - Long transportation times
  - Frequently late arrivals
  - Long lunch times
  - Effective work hours 50%–60%
- Construction supplies: concrete, explosives
  - Complicated supply structure
- Other contracts: paving, ventilation and instrumentation.
  - Those facilities and construction were cotemporaneous and not in the scope of the same Contractor, making the logistics difficult and the progress slower.
  - All these should have been part of the same Contractor's scope.
- There were also interfaces between existing mine installations and ongoing construction activities.
  - Blasting, the schedule and coordination could probably been handled in a better way the decreases the impacts between projects.

Engineering and geological definitions. Another source of delays in the project was lack of definitions from the designers. Drawings and changes continued to arrive during February 2009, even though the project should have been completed by the end of 2008. Some changes were so significant that the Contractor had to stop working in those areas for weeks or months. Although changes are normal in complex projects, the magnitude of changes should not impact the schedule as it did in this project.

The flow of information should be directly through the Client's site engineers and construction manager, as it should be in any project. Intermediaries present for this project created delays in the arrival of modified drawings and technical specifications. This was evident based upon the daily meetings where many engineering issues remained unresolved/open for weeks.

DAYG stands for "Design As You Go." This has been one of the key elements for high performance tunneling in Norway. Geologists are giving rock support definitions immediately as new rock surfaces are available for inspection. On several occasions in PDA Phase 1 these decisions took many weeks or even months. Good rock conditions helped; in poorer rock conditions the delay can have negative effects on the ground stability, support/lining quality, and personnel safety.

Contractor's internal coordination and productivity. Another major source for delays was the performance of the Contractor. The Contractor had an excess of crew and equipment on site. The lack of performance can generally be attributed to problems in coordination.

- Information Flow—the flow of information from Area Managers to Shift Leaders was insufficient. SKAVA operators and supervisors observed repeatedly that agreements made during meetings with Operation Managers were not implemented and managers and leaders were not informed.
- Shift Information Transfer—the information transfer between shifts was insufficient. It was not uncommon to see one shift discontinuing what the previous shift has completed, and several times this had a counter effect on production.
- Lack of Instruction—Lower in the organization chart, during any visit to site it was possible to find workers without instruction. This was most frequent during early hours in the shift or around lunch time. Improvisation and idling of people increased the probability of accidents and reduced the quality of the work.

SKAVA addressed these issues early in the project and options such as having a production manager, fully empowered, for C2 or taking out one or two levels in the line of command were proposed but none of these consultant recommendations were implemented.

Results

The main results of the assessment was that through modified equipment technology and construction methods, productivity more than doubled, even with a reduction of more than 30% in the staff.

One of the most important changes the consultant implemented was the change in construction methodology. This change dramatically increased the production capacity of the Contractor. It is important to note that from 31st May until 30th October the average production rate was 117 [m^3/day]. During this period the first massive bench in the Molienda Cavern was excavated with the previous methodology. If considering this bench separately productivity is similar to the 117 [m^3/day]. When excavation of the roof section in Flotation Cavern started the average production rate between 1st November and 13th February was 514 [m^3/day]. This is approximately 4.4 times the production rate of the previous period that includes the first bench of Molienda Cavern. As shown in Figure 5, the man-hours also reduced in this period.

To implement the change in methodology, expert drilling was needed. SKAVA was hired to provide highly skilled and experienced drilling operators. A group of six highly skilled people were provided in groups not bigger than four at a time. Some benefits of this activity are as follows:

- High drilling production rates. During the complete project an average of approximately 40 [m/h] was performed with peaks over 50 [m/h].
- Implementation of correct drilling patterns and sequences.

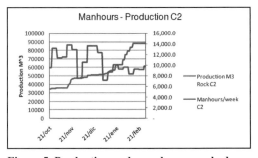

Figure 5. Production and man-hours worked

- Training of two local operators with good results.

An indirect, but no less important result was the higher production requirements placed upon the complete mining system cycle. This is not a minor effect, while SKAVA operators were drilling the rest of the personnel on site had to perform fast enough to keep up with the ongoing drilling and blasting operations.

Conclusions

In many aspects methodologies could have been improved even further. Suggestions were given, but for many different circumstantial reasons these were not implemented fully. SKAVA believes that if our contribution had happened in an earlier stage of the project, the number of implemented recommendations would have been greater. We also believe there could have been considerable improvements in final construction method/approach recommendations if implemented in an earlier stage.

For future projects it is considered important to ensure a proper excavation method from the beginning of the project as changes during construction require significant effort on the part of both Contractors and Client's organization. The client and his consultants also require the technical capacity to evaluate and improve the method presented in the proposals during the bidding process.

El Teniente Ventilation Adits

El Teniente is a copper mine located in the central region of Chile, about 50 [km] east to the city of Rancagua. El Teniente is one of the largest underground copper mines in the world with more than 3,000 [km] of underground galleries and more than 400,000 [tons] of copper production per year.

Problem Context

To maintain such a large amount of production, El Teniente needs to constantly expand and generate new production galleries, to operate these galleries there is an increasing need for additional ventilation systems as a crucial part of the expansion program. Currently the biggest expansion projects in El Teniente is the "Nuevo Nivel Mina" (NNM) project. To generate the level of ventilation needed, around 6 [km] of tunnels were needed, including the two main tunnel approximately 2.3 [km] long.

As the performance was not as expected, the Contractor decided to incorporate SKAVA to improve the performance results. During the first field visits SKAVA performed field measurements to understand where the problems were.

The results of the time measurement are shown in Figure 6 and Figure 7.

As can be seen in Figure 6 and Figure 7 the results show that in contrast with the Contractor's perception only 1% of the stand-by time was due to the Owner interruptions and the main problem came from improper internal coordination within the Contractor's team including inadequate planning of the personnel tasks related with the construction sequences, and inadequate or insufficient leadership.

Solution and Implementation

The solution proposed by SKAVA for this project came from three main recommendations:

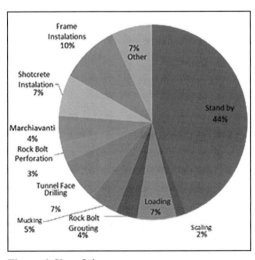

Figure 6. Use of time

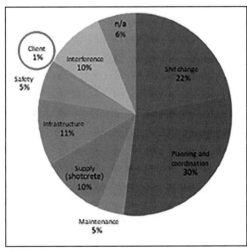

Figure 7. Source of interference

- Implementing latest generation equipment and procedures, and providing operators adequate training to operate properly this equipment.
- Improving the quality and training of the operator team encouraging multitasking of operators.
- Improving the relationship between the Client and the Contractor.

The main change in the equipment were:

- Implementation of automatized drilling jumbos, which gave the opportunity to implement new drilling technologies like Bever Control system and measurement while drilling.
- Using a Load Haul Dump with lateral discharge thereby improving the time needed to remove the muck from the tunnel face and directly reducing the construction cycle times.
- Scaling with mechanized equipment which provided improved safety for the personnel and a more efficient process.
- High capacity and autonomous sprayed concrete equipment was implemented improving production rates, efficiency, and worker safety.

In addition to the equipment used in the tunnel construction being enhanced, the procedures and rock support techniques were changed to get better results. The main idea was to implement a fast and simple support system to improve safety of the team working at the face and then install the final support to provide long term stability to the tunnel.

The main changes in the procedures were:

- Change the rigid steel frames for the Norwegian reinforced ribs of sprayed concrete
- Changes in the type and quality of the rock bolts
- Changes in the explosives from ANFO to emulsion
- Enhance the supply chain reliability

The second recommendation proposed by SKAVA to get better efficiency on the project was improving the quality and training to the operator team and encouraging multitasking of the operators. To make this possible, a number of changes in the hiring and training processes were made.

Firstly, specific professional profiles were established for each position. The most important characteristic was to learn fast and the possibility to adapt to changes as they occurred. Finding personnel who match with the established professional profile required a long interview and selection process. During this process only the adequate people with or without experience were selected.

To improve the training given to the operator, an initial phase was done in Norway and a second training was done in Chile but with the support of foreign operators.

In addition, the supervising personnel were retrained and educated in operational planning, leadership, conflict management, and negotiation.

The third issue requiring improvement was the relationship between the Client and the Contractor, which should improve the excavation time and costs. The goal was to create relationships based upon trust. To achieve this goal a fair risk distribution was established, a flexible support design, a change in the results measurements using equivalent times, and improving the use and clarity of the information obtained from the drilling registry.

Results

At the end of the improvement process the productivity was increased by 80%. The excavation rates started rounding 40 meters per month and finished with excavation rates rounding the 75 meters per month. The detailed advance rates can be seen in Figure 8.

In addition to the advance rates shown in the Figure 8 a detailed effect of each of the improvements had on the construction sequence are shown in Figure 9.

Conclusions

Changes are possible. Although the improvements were more difficult to implement than expected due to the regulatory framework.

The incorporation of cutting edge technology and procedures is possible, but it requires an important initial investment and a good training program and operator selection process to provide qualified operators, and external staff who can transfer knowledge and implementation of the technology.

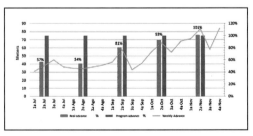

Figure 8. Monthly and weekly advance rates

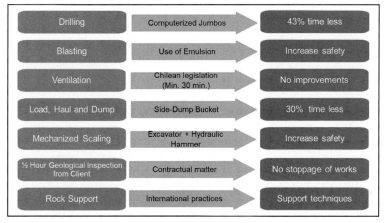

Figure 9. Detailed effect of the improvements

The main factors affecting the tunnel constructions process productivity were identified as the effective working time and the supply chain reliability, and still could have been improved substantially

To maintain these improvements the Contractor must generate capacity to replicate the methodologies and experience done in other projects.

There is not a one-size fits all program to transfer learning and information as the process depends of many factors including company culture, project type, available time, and latest technology.

Is very important that the Contractor have an open mind to evaluate changes and be prepared to try new technologies and construction methodologies, including incorporating experts that can transfer the information to them.

Underground projects have real opportunities to improve processes, production rates and lower costs. Important principles to realize these benefits are:

- Client and Contractor must approach projects with an open mind to try new methodologies
- Contractor must be disposed to invest
- External Agent must assess them both in the process
- Continuum training must take place during the project

Cheves Hydropower Project

Cheves is a hydropower plant wish produces 170 [MW] per year located in Peru, in the Huara river valley. For the power plant start-up, the construction included of a series of tunnels. In 2011 a tunnel construction project commenced using an international construction consortium with local contractor participation.

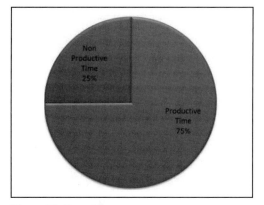

Figure 10. Use of time

Problem Context

The tunnel construction rates were not as estimated generating a delay projection of about 10 months. At that moment, the Client decided to hire SKAVA to develop a schedule recovery plan. To focus the plan in the most effective way, the first requirement was to identify the effect of the loss of productivity. This was performed by measuring the productive and non-productive times and identifying the task in which the time was spent or the cause why time was not being used properly. As can be seen in Figure 10 only 75% of the contract time was considered productive time.

One of the most important results was that, around the 90% of the non-productive time was related to 5 causes and the 75% of the productive time was used in three activities. The details of the non-productive time causes are presented in

Table 1. Average cycle times

	Average Cycle	Optimized Cycle	Difference
Cycle time (hr)	18.2	14.7	−3.5
Effective advance (m)	3.9	4.2	0.2
Production rate (m/week)	36.3	47.7	11.4

Figure 11 and the activity time distribution for the productive time is presented in Figure 12.

Once the time was evaluated the most important improvement opportunities were identified, and a solution developed with the aim to impact in the most non-productive time consuming activities.

Solution and Implementation

Solutions were implemented at both the operation and control level. As the most important time consumption activities were shotcreting, mucking and face drilling, the focus for operational changes with these activities resulting in a 20% decrease in the production cycle times. Table 1 shows the cycle times before and after optimization.

To improve the balance between productive time and non-productive time three control improvements were implemented:

- Information capture improvements
- Optimized information use
- Implementation of an automated reporting system

The implementation of these measurements resulted in an important decrease in the effort, and enhanced the quality of the project controls. This enabled the Client to have a more effective response to problems presented during construction.

Also, to regain project control a number of actions were taken:

- Continuum job supervision
- Non-conformance was emitted each time a sub-standard condition was detected
- Tracking of the non-conformances was implemented

With these improved processes the Client regained control in an effective way providing the opportunity to respond to unexpected situations during tunnel excavation.

Results

The improvements in time, monitoring, and control of the project was crucial to improve the excavation

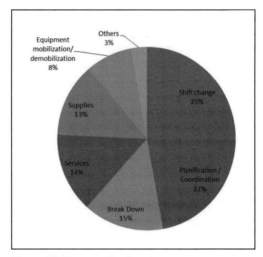

Figure 11. Non-productive time causes

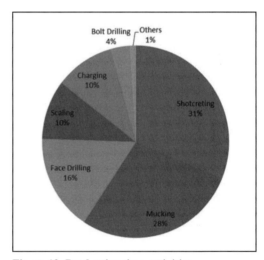

Figure 12. Productive time activities

rate and recover schedule during construction. The schedule delay was recovered within six months. The expected completion time evolution is presented in Figure 13.

Giving problem response capability and the project controls to the Client was a critical change that resulted in the maintenance of production rates, product quality, and worker safety. In addition, the mentoring on this project was also very effective.

Conclusions

Control and response capability are critical to maintain proper production rates and quality in the production. Proper training and mentoring to the

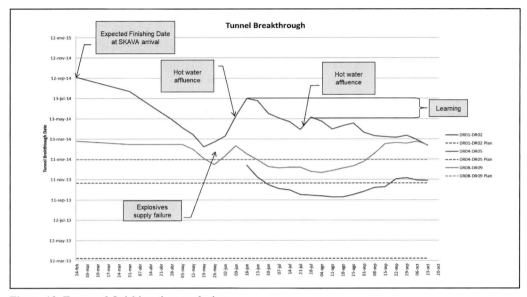

Figure 13. Expected finishing time evolution

supervision team is a very effective manner to regain the control of the project and improve the production rates.

Greater effect could be achieved if the measures were implemented in more levels and from the beginning of the project.

CONCLUSION

As shown in the present paper, including hands-on training in underground projects can help increasing productivity, through improving performance, reducing needed man-hours, or both.

The results of these projects show that leadership, having an adequate management, highly skilled and trained personnel and, finally, implementation of cutting edge technology are crucial to achieve a proper performance in tunneling projects. In many cases to achieve the expected results the

implementation of an external entity is needed to establish the source of problems without bias.

The incorporation of cutting edge technology and procedures is possible. However the major benefits come from training of personnel through the acquisition of new techniques, working methods, common understanding of the solutions within the crews and among the different crews.

As can be seen in the presented cases the source of the problems in the production can be found in different levels, organizational, technological, personnel, among others. For each one of these problems a measure can be taken to be solved and regain or even surpass the expected production.

The examples shown in this paper make a basis for Owners and Contractors to look for investing in getting hold of the above mentioned capabilities.

High-Precision Pilot Hole Drilling Utilizing Real-Time Optical Guidance and Verification Techniques

Aaron C. Reel
Coastal Drilling East (a division of Shaft Drillers International)

Stefano Alziati
North American Drillers (a division of Shaft Drillers International)

INTRODUCTION/BACKGROUND

Precision is the technological edge that engineers strive for as design tolerance become increasingly more stringent. The use of multiple pass drilling technologies using smaller diameter pilot holes as the initial pass to control and mitigate potential alignment deviation has long been recognized as a fundamental component of controlled drilling in deep foundations for civil, mining, and geotechnical construction. For those projects which require strict adherence to a specified alignment and have a narrow allowable deviation, the ability to accurately control the direction and inclination of the pilot hole is of paramount importance. It is the path of the pilot hole that ultimately dictates the final hole alignment.

What Is Directional Drilling and When Is It Needed?

When applied to a drilled hole, "alignment" typically refers to the direction and angle at which the hole has been, or will be, completed. In this context, the alignment of a drilled hole refers to either the "specified" alignment or the "as-drilled" alignment. Other terminology that is often used interchangeably with direction and angle when referring to alignment is azimuth and inclination. When the discussion of alignment is referenced to mechanically excavated shafts—either by blind bore or raise bore methods—the question of alignment is one of verticality. A hole that has been drilled truly vertical can be defined as hole that has neither direction (azimuth) nor angle (inclination). The measure of verticality can be defined as the deviation from a vertical line extending from the ground surface to depth of interest. In this manner, verticality is determined by measuring the position of the centerline of the as-drilled, or as-excavated, shaft and comparing this position to the specified vertical centerline.

Traditionally, a pilot hole is often completed first to control the alignment of a shaft. This initially drilled hole, as its name implies, will "pilot," or guide, the larger diameter shaft drilling equipment along its path and will therefore determine the ultimate alignment of the shaft itself. It is of critical importance therefore, that the pilot hole be drilled to the highest achievable level of verticality. When the conditions dictate, directional drilling methods are often used to provide a high level of verticality. Conditions that may require a directionally drilled pilot hole include strict alignment specifications and deep drilling or excavation depths in the presence of challenging subsurface conditions such as significantly broken or weathered rock, highly variable rock formations, or in highly congested sites. As mining depths continue to increase to access ever deeper reserves and civil construction projects face locations that are becoming more congested with new and existing features such as utilities and infrastructure, the demand for truly high-precision drilling has been driven by those benefits realized when stringent engineering specifications, design criteria and specified alignments are achieved. When designs, specifications, or project requirements dictate that a deep foundation or geotechnical element be developed adhering to a strict alignment criteria, directional drilling techniques are often used to provide a means of controlling down hole deviation and maintaining alignment.

Real-Time Optical Guidance and Verification Drilling

When designs, specifications, or project requirements dictate that a deep foundation or geotechnical element be developed adhering to a strict alignment criteria, directional drilling techniques are often used to provide a means of controlling down hole deviation and maintaining alignment. Traditional directional drilling methods have relied on drill fluid pulsation or other similar techniques to communicate the location of the drill tools in a measure-while-drilling (MWD) approach that utilize an error based measurement in order to detect deviations from the specified alignment. Once an error has been detected and a magnitude and direction established, corrective actions are implemented and the deviation is corrected by steering the development of additional borings back towards the specified alignment. These methods result in the development of hole path that sweeps in and out of alignment in a sinusoidal fashion over the entire length the drilled hole. Real-time Optical Guidance and Verification (RTOGV) directional drilling technology provides an opportunity for truly high-precision directionally drilled holes. Compared to the traditional, error-based, directional drilling technologies, RTOGV provides the significant advantage of optically measuring the location of the drilled hole in real-time and proactively maintaining alignment rather than just reacting to changes in deviation that must first become large enough to be measured. By actively controlling the drill hole alignment in real time, the RTOGV technology allows the operator to back up drilling bore and correct the deviation thus eliminating the error and subsequent sinusoidal hole path typical of the traditional technologies and provides superior pilot hole alignment. The Real-Time Optical Guidance and Verification (RTOGV) technology was developed for vertical and angled applications with relatively short drill lengths where a high degree of precision and accuracy was required. The technology was initially developed for a civil construction project to meet a maximum deviation tolerance of 1:150. That is, no more than 1 unit of deviation in any 150 units of hole length. The RTOGV technology is an optical system that is fundamentally based on Fermat's principle that light travels the shortest distance and in the least time between two points in a straight line. This principle is applied to directional drilling method by locating and measuring the position of the drilled hole by placing illuminated target just behind the drill bit. The illuminated target is made up of a series of high intensity Light Emitting Diodes (LEDs) arranged in a circular array. The light emitted by this target is made visible to the optical verification and measurement tools at the surface through the use of specialized dual tube drill rods making the position and location of the drill bit—and by extension, the drilled hole—known. The inner tube of the drill rod is pressurized to eliminate refraction and to maintain an unobstructed view from the surface to the illuminated target at the bottom of the drilled hole. As the drilling is advanced, the illuminated target remains fully visible to the optical verification tools indicating the position of the tools and allowing for a

precise location of the tools and the drilled hole. If the drilled hole begins to deviate from the designated alignment, the illuminated target projected onto the drill operator's screen will immediately indicate the deviation by showing a non-circular light pattern moving away from the center on the screen. By evaluating the target data, both the magnitude and relative direction from center the deviation has occurred is known. Corrective actions are facilitated by the directional drill operator's abilities to back the DTHH up the bore path and steer the "shovel" bit back through the alignment and make the necessary adjustments to bring the center of the guide hole back into tolerance. The RTOGV system closes the feedback loop on control by not only indicating that a deviation occurred, but provides the ability to backup and correct the deviation on in real time. After the RTOGV technology was repeatedly proven to significantly exceed the initial project deviation requirement, further applications of this method of directional drilling were investigated culminating in the successful application of this technology to the completion of pilot holes for blind bore shaft development having a tolerance of no more than 10-inches of total deviation in more than 550-feet of 1:660.

CURRENT DIRECTIONAL DRILLING TECHNOLOGIES

The market's demand for high precision drilling has fostered the drilling industry to develop several technical approaches for steering the pilot hole yielding a series of different directional drilling technologies. The directional drilling technologies most readily available and commonly used for pilot hole are:

- Rotary Vertical Drilling System (RVDS)
- Measure While Drilling (MWD)
- Directional Coring Drilling System (DCD)

Rotary Vertical Drilling System (RVDS)

Description of Technology

The Rotary Vertical Drilling System (RVDS) is an autonomous self-steering tool that is integrated into the Bottom Hole Assembly (BHA) of the drill string to steer the drilling bit back to the vertical position each time the onboard inclination technology detects a deviation off the vertical. The tool is typically fitted between the drilling bit and the first string of vertical stabilizer and consists of two modules.

The module closest to the drilling bit contains a set of non-rotating hydraulically activated steering ribs. These are responsible for pushing against the walls of the hole to create a lateral force pushing the drilling bit back towards the vertical position. The actuation of steering ribs is controlled by a processing unit which continuously receives inclination

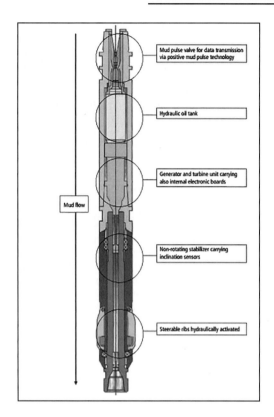

RVDS	Drilled distance by continent		Average deviation
7 3/4" (8 ½" – 9 ⅞" holes)	18,739 m	Europe	0.05 %
	–	Asia	–
	–	Australia	–
	–	Africa	–
	–	North America	–
	–	South America	–
Total	**18,739 m**	**Worldwide**	**0.05 %**
9 1/2" (12 ¼" – 13 ¾" holes)	7,948 m	Europe	0.13 %
	3,879 m	Asia	0.23 %
	1,738 m	Australia	0.08 %
	753 m	Africa	0.23 %
	1,010 m	North America	0.10 %
	–	South America	–
Total	**15,328 m**	**Worldwide**	**0.15 %**
10" (12 ¼" – 13 ¾ holes)	1,290 m	Europe	0.04 %
	1,104 m	Asia	0.04 %
	573 m	Australia	0.10 %
	–	Africa	–
	835 m	North America	0.08 %
	–	South America	–
Total	**3,802 m**	**Worldwide**	**0.06 %**
12 ⅞" (15" – 17 ½" holes)	3,056 m	Europe	0.05 %
	–	Asia	–
	3,315 m	Australia	0.07 %
	3,818 m	Africa	0.06 %
	2,062 m	North America	0.21 %
	1,399 m	South America	0.34 %
Total	**15,591 m**	**Worldwide**	**0.10 %**
Total all RVDS-Sizes	**53,460 m**	**Worldwide**	**0.09 %**

data from a set of 2-axis inclinometers located close to the drilling bit.

A second module is positioned directly above the first module. This module contains a turbine responsible for generating power for a hydraulic pressure units and the electronic instrumentation and control units in the first module. In addition, it also powers a pulse unit which transmits information back to the surface utilizing positive mud pulse technology. A receiver on surface decodes the pulse signals and displays the information on the computer. The turbine is driven by high pressure drilling mud flow through the drill string.

The RVDS functions independently and cannot be actively controlled by the operator or technician on surface. The steering capabilities can be adjusted by manipulating the bottom hole assembly and the manner in which the drill tools are stabilized or packed but the technology does not facilitate direct and active steering.

Applications

The RVDS technology is commonly found in use in the underground mining industry. This technology In conjunction with raise-boring and blind boring equipment it is used for drilling the vertical pilot holes used as the guide in the development of the larger production, ventilation and access shafts where vertical alignments are critical. The RVDS system is capable of drilling pilot holes with diameters ranging from 8.5" to 26". The depth of the pilot hole is not a limitation to the operation of the RVDS tool given the technology remains on board of the bottom tools, however depth may be a factor in the proper functioning of the positive mud pulse communication system. Other known applications include drilling of vertical boreholes for ground freezing applications, drilling of pilot holes for tunnel ventilation and access shafts as well as intake and surge shafts for hydro power plants.

Benefits

The RVDS technology has been around for close to two decades and has demonstrated significant benefits over other directional drilling technologies. The main benefits are:

1. The tool is designed to continuously measure inclination and when detectable variance from vertical, force drilling back towards the vertical direction. As a result the final product is a pilot hole with no or very minimal doglegs ("kinks"). This is extremely advantageous for the later reaming operations as the

drill string sees a less tortuous path and can rotate more freely reducing the mechanical losses and allows for higher reaming rates and better drilling performance.

2. The operation of the tool is relatively simple and requires little to no input from the operator on surface. In addition, it can be utilized in a range of different drilling rigs and configurations.

3. Most importantly, the technology has a proven track record working effectively. Typical deviation from the vertical at total depth is less than 0.1% of the depth.

Limitations

While the RVDS technology demonstrated the benefits of a more vertical and straighter pilot, it also exposed a series of limitations inherent to this approach.

1. The technology is only capable of measurements in the vertical plane and thus is limited to correcting drilling bit inclination. Each time the drilling bit deviates from the vertical inclination it will drill away from its centerline. The RVDS technology must first detect the error and then take corrective actions. By the time the steering ribs have brought the inclination back to vertical a new vertical centerline away from the original centerline is created. The RVDS is unable to redirect the drilling bit to return to the original centerline. This phenomena typically creates a spiral shaped pilot hole.

2. While the RVDS tool provides information back to the surface through mud pulses, the information received only contains data on the measured drilling bit inclination over time and general functioning of the tool's systems. The inclination data alone is insufficient to determine the pilot hole alignment form the top to the bottom. The result of this limited data is that the only way to determine the pilot hole alignment and if it is within the deviation limits is to the drilling operation and trip out the drilling tools and run an independent survey.

3. The RVDS technology is based on feedback from the inclination elements to detect an error from vertical, this inclination error must be of a magnitude large enough for the sensors to detect prior to any corrective actions taking place. The result of these inherent delays limits the technology to series of vertical and non-vertical elements strung together in series. Any pilot hole that is developed as the tool is deviating from vertical and for the period of correction back to vertical remains as an integral part of the final pilot hole.

4. The RVDS is limited to drilling vertical holes. The technology and systems will not operate on inclined holes.

5. While the average deviation from the vertical of the tool is better than 0.1%, the formation type and bedding angles can significantly alter that figure. This becomes an issue when the pilot hole needs to have a pre-determined exit location.

Measure-While-Drilling (MWD)

Description of Technology

Measure-While-Drilling (MWD) technology was developed in the 1960s as a means of overcoming the ineffectiveness of conventional logging tools at hole angles exceeding 60 degrees where the logging tools could no longer be pushed through the hole to retrieve information. This technology is typically utilized in the oil & gas industry which often requires deep and curved wells to tap oil and gas reserves in very specific areas of a geological formation.

The MWD uses a combination of sensors to determine the bore hole position. Three orthogonally mounted accelerometers measure inclination and three orthogonally mounted magnetometers measure magnetic direction from the north (azimuth) continuously and in real time. Using basic trigonometry, a three dimensional plot of the hole can be created. In addition, the MWD system also contains sensors capable of measuring rotational speed, downhole vibration, temperature, torque and drilling mud volumetric flow rate. All these sensors are incased in a down the hole unit that sits just above the drill bit and steering unit.

The data captured by the sensors is physically and digitally collected by a logic unit and transmitted to the surface by one of two ways: using a mud pulse system or an electromagnetic system. Decoding receiving systems on the surface decode the transmitted data and display all information to the driller in real time. Based on the information received, the driller can make adjustments to the parameters on the steering tools. The two most common steering tools that can be used in conjunction with MWD tools are the RSS (Rotary Steerable System) and the PDM (Positive Displacement Motor) also commonly referred to as steerable motor assemblies or mud motors.

Among the different PDM assemblies, the most commonly used deviation tool is the bent-housing mud motor. The bent housing mud motor utilizes a "tilted bit" (a bit whose face is misaligned in relation to the drill string axis) and bit side force to change the hole direction and inclination. When a

direction correction is required, the drill string rotation is ceased, the bent sub is oriented to the desired position and the mud motor powers the rotation of just the "tilted" drill bit. The rest of the drill string "slides" behind, following the curved path created by the drill bit. When steering is not required, the entire drill string rotates and drilling occurs typically in a straight path.

The RSS tools are an evolution of the PDM tools and are designed to drill directionally with continuous rotation from the surface, without the need to "slide." The directional correction is achieved using either a "push-the-bit" or "point-the-bit" tool. Push-the-bit tools use pads on the outside of the assembly which push on the walls of the hole causing the drill bit to push against the opposite side of the hole causing a change in direction. Point-the-bit tools trigger the orientation of the bit to change compared to drill string axis by bending the main shaft running through it. This requires a non-rotating housing in order to create the necessary deflection within the shaft.

Applications

MWD technology is widely used in the oil & gas industry for the development of inclined or horizontal oil and gas wells. However, there have been instances when the global positioning abilities of the MWD combined with the steering capabilities of PDM or RSS tools have been engaged in drilling directional pilot holes for the civil and mining industries.

Benefits

The greatest benefit of the MWD tools in drilling pilot holes is their ability to display the entire path of the hole and current location of the drill tools in real time. Based on that information it is possible to make fine and timely adjustments to the steering tools to keep the hole within the limits of the desired path and minimize the amount of doglegs and curves along its path. In addition, unlike the RTOGV, it does not

require straight line of sight to be able to locate the position of the drilling tools.

Limitations

The MWD and steering tools are very complex systems which required significant support, infrastructure and expertize to be able to use them effectively. These complexities contribute to their overall effectiveness in being used for drilling pilot holes.

1. While the magnetometers and accelerometers can measure inclinations in the order of magnitude of 0.1 degrees, this error margin steadily accumulates throughout the drilling of the hole. At full depth this accumulation can translate to a significant error in the overall deviation. While this error is acceptable in oil& gas wells, it typically is not for pilot holes.
2. The functionality of magnetometers inside the MWD tool package is heavily affected by the presence of steel or ferrous elements, which will lead to inaccurate information. In certain environments outside the oil & gas industry, where the presence of steel in the form of rebar or steel casing is normal, the use of MWD tools may not be possible.
3. The most typical MWD tools in the market work best when drilling in holes with angles in the range between 15 degrees and 80 degrees from the vertical. While they can be used to drill vertical holes, the accuracy and overall performance drops significantly. It is understood that over the past few years variations to the traditional MWD tools have been launched by directional tool providers specifically designed to drill vertical wells.

Directional Core Drilling (DCD)

Description of Technology

Directional core drilling technology, also commonly known as DCD, was developed specifically for core drilling applications however, its uses and application are not restricted to just core drilling applications. This technology has been used to develop directional guide holes which are then enlarged to pilot holes.

One of the more commonly known and used DCD tool in the market is the DeviDrill. The DeviDrill is a steerable wireline core barrel. Similar to RVDS tools, the Devidrill uses external expanding pads to steer the drilling bit. However, the expanding pads are located on only one side of the tool and are operated by a differential pressure. This ensures that the drilling tool is kept in a fixed orientation and steers the drill tools in a curved or straight alignment

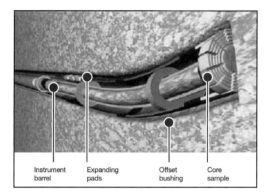

| Instrument barrel | Expanding pads | Offset bushing | Core sample |

as required. In addition to the steering capabilities, the inner assembly also carries an inner tube which is responsible for capturing and holding the core sample and contains an instrument barrel with the survey tool recording both inclination and orientation.

At approximately every three meter (10 feet) intervals, drilling is stopped and a wireline is used to extract the core sample from the inside of the tool and raise it to the surface. At the same time a wireline survey instrument is lowered into the hole. This survey's the latest section of the hole drilled and enables the operator to determine how the expanding pads need to be setup to drill the next section of the hole.

Applications

As expressed by the name, directional core drilling is typically used in core drilling operations for the mining, civil and oil and gas industry. Its greatest advantage is its ability to drill multiple holes from one single location on surface, many times being able to utilize some of the previously drilled hole before branching out in a different direction. DCD technologies steering and surveying capabilities can also be used for other applications. Underground mines have used the technology to drill guide holes with very specific entry and exit locations. These guide holes are then enlarged into pilot holes, which can then be reamed into shafts

Benefits

The use of DCD technology for developing pilot hole has a series of advantages over other technologies, namely:

1. The DeviDrill tool can be used to drill a guide to a specified end location. The continuous surveying and ability to steer allow the necessary adjustments to be made along the course of the hole to ensure the guide hole stay on track to reaching the target.

2. The DeviDrill tool can be used to drill holes to a wide variety of drill angles. Unlike the RVDS, which is restricted to a purely vertical hole, the DeviDrill can drill angled holes up to and angle of about 45 degrees.
3. The core samples extracted during the drilling operation gives a very precise indication of the geological formation that the successive pilot hole and reaming operations will traverse. In some cases it may indicate unfavorable drilling conditions, in which case the operators may decide to re-position the pilot hole.
4. DCD tools and instrumentation tend to be very compact and lightweight and can be easily handled and transported. In sites with limited access and limited infrastructure this can be a significant benefit.

Limitations

1. When configured to drill a straight hole, the DeviDrill tool will be allowed to naturally deviate from the intended straight path up to the point where it reaches the limits of the deviation tolerance. At this point a correction is made. The correction has the potential and often does introduce a "dogleg" to the hole that can be up to 20 degrees per 30 meters of pilot hole. Doglegs in pilot holes represent a risk to the successive reaming operation and

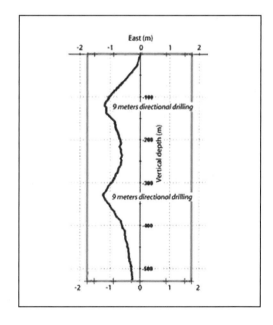

significantly increases the mechanical loading and reduces reaming rates.

2. The accuracy limits are typically determined by the geological formation but common limits are in the order of 1% deviation from the intended centerline at total depth. This is significantly higher than other technologies available.

3. Using DCD technology for a pilot hole necessarily means that at least two additional passes are required to get the hole to the final intended diameter. The DeviDrill tool is only configure to work with N-size core barrel (75 mm / 3"). This adds to the complexity of the overall operation and can lengthen the overall duration of the project.

4. As with all core drilling operations, penetration rates are typically slow. During steering mode, the DCD tool tends to further reduce penetration rates. Furthermore, core extraction using wireline frequent surveys contribute to increase the overall duration of the operation.

OPTICAL DIRECTIONAL DRILLING TECHNOLOGY

Introduction

The demand for higher precision drilling is being driven by the benefits realized when more stringent engineered specifications and design criteria are issued and specified drilling alignments need to be maintained over greater distances. Real-time optical technologies were developed to replace the more traditional error based measure-while-drilling (MWD) techniques due to the fact that MWD can only correct after a measureable deviation occurs and then only by steering a "new hole" back towards the specified alignment. This technique often results in the development of a drilled hole which "sweeps" into and out of the specified path. Optical technologies facilitate the instantaneous reading of an alignment deviation and allow the drill operator to pull back and correct the deviation by steering—in real time—to eliminate the error. The result of this proactive approach to directional drilling is a bored hole that remains within a very narrow deviation window.

Description of Technology

The Real-Time Optical Guidance and Verification (RTOGV) directional drilling technique a real-time optical verification and guidance system with the directional drilling capabilities of a "shovel" bit fitted on a pneumatically operated down-the-hole hammer (DTHH). The RTOGV system is based on Fermat's principles the light travels in the shortest distance from point to point which is a straight line. An illuminated target positioned just behind the DTHH and shovel bit provides the light source is made up of a series of high intensity Light Emitting Diodes (LEDs) arranged in a circular array. The light emitted by the target is made visible to the optical verification tools on surface through the use of specialized dual tube drill rods. The inner tube of the drill rod is pressurized to maintain an unobstructed view from the rotary drill head at the surface to the illuminated target at the bottom of the drilled hole. The optical verification tools are mounted on a survey base the surface which are positioned above the rotary drill head in such a manner as to obtain the direct "line of sight" from the surface to the illuminated target at the drilled depth. As the drilling is advanced, the illuminated target remains fully visible through the optical verification tools, indicating the position of the tools and allowing the drill operator to know the precise location of the tools and the drilled hole. If the hole begins to deviate from the designated alignment, the illuminated target projected onto the drill operator's screen will immediately indicate the deviation by showing a non-circular light pattern moving away from the center bull's eye on the screen. Reading the target data, the operator knows both the magnitude and relative direction from center the deviation has occurred and the necessary corrective actions to take. The DTHH and directional drilling "shovel" bit is designed to allow the bit to be steered in the direction desired. Corrective actions are facilitated by the directional drill operator's abilities to back the DTHH up the bore path and steer the "shovel" bit back through the alignment and make the necessary adjustments to bring the center of the guide hole back into tolerance. The RTOGV system closes the feedback loop on control by not only indicating that a deviation occurred, but provides the ability to backup and correct the deviation on in real time.

Benefits

Similar to RVDS, the primary benefit of the RTOGV technology for developing pilot holes is that this technology is able display the entire path of the hole and the current location of the drill tools in real time. The optical technology expands on this benefit by allowing the drill operator to go beyond just monitoring the path of the drilled hole as it advances and allows the operator to take an active role in controlling the path of the pilot hole. As stated previously, in traditional MWD technologies such as RVDS, the drill tools are merely reacting to the changes in deviation and only when the magnitude of the deviation is large enough to trigger a corrective action does the system attempt to bring the drill hole back into alignment. Conversely, RTOGV directional drilling is a proactive directional drilling technology that allows the drill operator to take a dynamic role in "steering" the path of the drill tools along the specified alignment.

Used together the ability to monitor the path and location of the pilot hole along the with the ability to steer the drill tooling and pre-empt significant, measurable deviation from the specified alignment from occurring, has resulted in deviation errors of less than 0.01%.

Limitations

Since the RTOGV system is, by definition, an optical system, it is limited by the ability of the of the verification tools mounted at the surface to be able to see the illuminated target at the bottom of drill hole. The current state of technology has limited the effective drilling depth for this system to approximately 200 m and pilot hole diameters of up to 250 mm. However, certain areas of improvement have been identified that may enable to the technology achieve similar levels of accuracy at greater depths.

Current Applications

Current applications for Real-Time Optical Guidance and Verification Directional drilling technologies include any project that requires that the strictest adherence to the tightest deviation tolerance where the specified alignment is a straight line in vertical or angles scenarios. This includes pilot or guide holes for mechanically excavated shaft development either by blind bore or raise bore methods, utility holes in congested environments where the entry and exit point is known, and precision drilling for subsurface instrumentation installation. This technology has also been successfully used on several heavy civil construction projects.

CASE STUDIES/PROJECT HIGHLIGHTS

Bluestone Dam Safety Assurance

Real-Time Optical Guidance and verification directional drilling technology was used to complete 278 pilot holes for high-capacity post-tensioned strand anchors for a dam safety project in West Virginia. The majority of the rock anchors on this project were 61-strands and up to 270 feet in length requiring up to a 15" diameter final hole that was drilled by following the directionally drilled pilot holes. This project required working above an active spillway from both elevated platforms and barges. Each pilot hole and final diameter hole to be drilled within a tolerance of 1:150 as measured against the specified centerline. By utilizing its real-time optical guidance and verification directional drilling technology, Coastal Drilling continually exceeded the specified tolerances.

Lake Mead

Real Time Optical Guidance and Verification (RTOGV) Directional Drilling Technology was utilized for drilling thirty four (34) pilot holes for the Southern Nevada Water Authority's Lake Lead Intake No. 3 Low Level Pumping Station (Underground Phase) in Boulder City, NV. The thirty four (34) 8" guide holes were initially directionally to a depth of 500' below surface. The guide hole was subsequently enlarged to a 17.5" diameter pilot hole. The Water Authority would ultimately be placing vertical turbine pumps into the well shafts and these pumps required extremely vertical shaft casings to ensure proper pump performance. As a result of the pump manufactures design criteria, deviation tolerances on the shaft and thus the guide and pilot hole were extremely tight. The specification for the pilot hole required the centerline of pilot hole to lie inside of a vertical cone with a 5" radius at the base from the apex at the entry point on surface.

All thirty four (34) pilot holes were successfully drilled to meet or exceed the deviation specifications despite the geological formation being significantly challenging including alternations between solid,

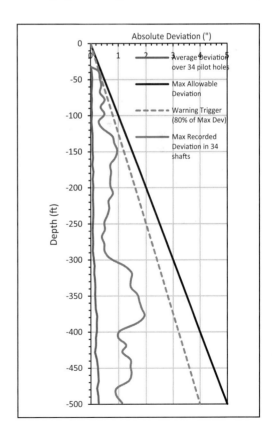

hard rock and altered, broken and fracture layers. Average deviation at a full depth of 500' was 0.28" and the maximum deviation at this elevation in the thirty four (34) shafts was 1.12". That translates to an average error of 0.005% and a maximum error of 0.018%.

CONCLUSIONS

The use of pilot holes to control verticality and mitigate alignment deviation from the vertical plane has long been recognized as a critical component of deep foundations for civil, mining, and geotechnical construction. For those projects that require a tight tolerance to the alignment specification, the ability to control the pilot hole alignment dictates the final hole verticality. The demand for high-precision drilling is driven by the benefits realized when stringent engineering specifications, design criteria and specified alignments are maintained over greater distances. Real-time Optical Guidance and Verification (RTOGV) directional drilling technology provides an significant advantage over conventional directional drilling technologies by optically measuring and proactively maintaining drill hole alignment rather than merely monitoring and just reacting to down hole changes in deviation. By actively controlling the drill hole alignment, the RTOGV technology provides a superior pilot hole alignment when compared to more traditional measure-while-drilling (MWD) technologies which require a measurable error in the vertical deviation prior to initiating corrective actions to bring the drill hole back into alignment. While the current state of technology has limited the effective drilling depth for this system to approximately 200 m and hole diameters of up to 250 mm, certain areas of improvement have been identified that may enable to the technology achieve accuracies of greater that 0.01% at greater depths in the future.

BIBLIOGRAPHY

https://en.wikipedia.org/wiki/Measurement _while_drilling.
https://en.wikipedia.org/wiki/Directional_drilling.
https://en.wikipedia.org/wiki/Mud_motor.
https://en.wikipedia.org/wiki/Rotary_steerable _system.
http://petrowiki.org/Directional_deviation_tools.
http://www.halliburton.com/en-US/ps/sperry/ drilling/directional-drilling/downhole-drilling -motors/default.page?node-id=hg98y5ku.

Innovative Concrete for Aggressive Ground in Qatar

Francois G. Bernardeau and Jacek B. Stypulkowski
CDM Smith

ABSTRACT: The Abu Hamour Tunnel is 9.5km long with 3.7m internal diameter (ID) storm water tunnel, located in a hot humid environment with high concentrations of chlorides and sulphates which are present in the ground and groundwater, as well as in the storm water and construction dewatering water that will flow through the tunnel. Most of the construction sites in Doha use dewatering as means of keeping the excavation dry. The original conceptual design achieved a 100-years design life and 50 years maintenance-free performance of the concrete for the tunnel by applying HDPE liner on the intrados and epoxy coating on the extrados. Abu Hamour (Musaimeer) Surface & Ground Water Drainage Tunnel—Phase I (ASHO) in Doha, Qatar proposed an alternative lining utilizing, amongst other things, steel fiber reinforcement and special concrete mix consisting of a triple blend concrete suitable for harsh ground and groundwater conditions. The compliance assessment focused on concrete durability, structural integrity of elements, and cost effectiveness considering the whole life cycle costs.

INTRODUCTION AND BACKGROUND

Durability is closely related to design life. The British Tunnelling Society Tunnel Lining Design Guide (2004) defines the following: "The durability of structures is their ability to perform satisfactorily during the specified design life. The 'durable' material used should be such as to maintain its integrity and, if applicable, to protect embedded materials." The design-build con-tractor on Abu Hamour was therefore contractually required to address issues that may arise during design and construction that have a clear durability implication in terms of the 100-year design life. The Abu Hamour Tunnel is located in a hot humid environment with high concentrations of chlorides and sulphates (Cl-30,000mg/l, SO_4^2-5,000mg/l) which are present in the ground and groundwater, as well as in the storm water and dewatering water that will flow through the tunnel. All the important deterioration mechanisms are governed by aggressive substances from the surrounding environment penetrating into the concrete. The substance either accumulates with time on the outer concrete layer until the concentration exceeds a critical value destroying the concrete (sulphates) or, alternatively, diffuses further inward towards the reinforcement (chlorides). Chloride induced corrosion and sulphate attack are considered to be the principal deterioration mechanisms that may influence the long term durability of concrete structures. Both chloride penetration and sulphate accumulation are accelerated by the cyclic wetting and drying of the structural concrete. The durability of concrete is also dependent upon the transport of gases, water and dissolved agents through this porous medium. Much of the mass transfer occurs through capillary pores and cracks, the continuity and occurrence of which are very important. This, in turn, is controlled by the water/cement ratio, cement type, presence of supplemental cementitous materials and adequacy of concrete curing. In terms of the environment, the most important parameters are: the availability of oxygen, moisture, temperature, surface levels of chlorides and sulphates, and pH levels. The deterioration of concrete structures is represented by the combined effect of the transport mechanisms, the accumulation of aggressive substance (e.g., sulphates) on the concrete, or the increase in chloride levels to threshold values at the reinforcement in case of traditionally carbon reinforced concrete. The design life is the time in service until a defined unacceptable state of deterioration is reached. Looking at concrete mixes resistance to chloride ingress and sulphate attack, it becomes apparent that a plain OPC concrete mix will not withstand the aggressive ground and groundwater conditions and needs to be enhanced. Triple blend concrete mixes are durable materials that typically require no major repairs throughout the 100 years design life. There are no known or published disadvantages of triple blend concrete mixes to-date. This implies that the function of the concrete structures will not be impaired over their service life provided that they are subject to due care and normal maintenance.

STRUCTURAL INTEGRITY

Structural integrity philosophy varied for tunnels and shafts. For tunnels, in accordance with CEB-FIP Model Code 2010, the structural design of Steel Fiber Reinforced Concrete (SFRC) elements was based on the post-cracking residual strength provided by the steel fiber reinforcement. Steel fibers improve the behavior at Serviceability Limit State (SLS) since they reduce the crack spacings and crack widths, thereby improving durability. The behavior in compression of SFRC is in general the same as that of plain concrete. The behavior in tension is determined using test methods. Nominal values of the material properties were determined by performing a 3-point bending test on a notched beam according to EN 14651. The applied force vs. deformation diagram was produced. The test results were used for deriving the stress-crack width relations. To classify the post-cracking strength, a linear elastic behavior was assumed by considering the characteristic flexural residual strength significant for serviceability (fR1K) and ultimate (fR3k) conditions. The grade of the SFRC used in this project is classified as C55 4c, i.e., concrete grade C45/55 with fR1K = 4.0MPa, which represents the strength interval, and the letter c corresponds to the residual strength ratio (0.9 < fR3K/fR1k < 1.1).

For shafts and adits permanent works, the structural integrity was verified using 3D finite element modeling. Ultimate Limit State (ULS) and SLS design checks, as well as global stability against uplift were all conducted in accordance with Eurocode. For SLS, the crack control requirement imposed by the contract documents (crack width <0.1mm) governed the design. The characteristic cylinder strength for the concrete was fck = 30MPa. The grade of the SFRC used for a portion of the shaft wall was classified as C37 3c, i.e., concrete grade C37/30 with fR1K = 3.0MPa, which represents the strength interval, and the letter c corresponds to the residual strength ratio (0.9 < fR3K/fR1k < 1.1).

CONCRETE MIX IMPROVEMENT

Concrete resistance to chloride ingress is highly dependent upon the type of cement used along with use of supplemental cementitious materials. This is due to differences in the pore structure and the chloride binding capacity of the concrete. Of the cement types commonly used structurally, plain Ordinary Portland Cement (OPC) tends to have the lowest resistance to chloride ingress. The addition of fly ash (FA), ground granulated blast-furnace slag (GGBS) and micro silica (MS) increases the resistance against chloride ingress. Optimal resistance may require triple blends of OPC, FA and MS or GGBS with a high slag content (>65%). The resistance to chloride ingress of cements containing these supplemental cementitious materials tends to continue to improve with time to a much greater extent than plain OPC. As a result of the ability of the concrete to develop an increased density and the chloride diffusion (migration) coefficient decreases with time. High amounts of sulphate in the groundwater or storm water can cause an expansive reaction in concrete surfaces. The speed of the sulphate attack depends on the amount of reactive aluminates in the cement matrix (chemical resistance), and the pore volume of the cement matrix (physical resistance). A reduction of the aluminate content of the concrete leads to a reduction of the potential reactant for sulphate, whereas a denser pore structure reduces the sulphate ingress into the concrete. To increase the chemical resistance, the calcium aluminate hydrates C3A content in the cement mix has to be limited. The influence of GGBS depends on how much of it is used in the mix. For GGBS content <60%, the sulphate resistance decreases with increasing slag content. For GGSS content >60%, a significant increase in sulphate resistance is observed due to the very dense pore structure. For GGBS content >65%, the concrete will display a very high resistance to sulphate attack due to the high physical resistance irrespective of the C3A content of its constituents. Implemented solution to achieve 100 years design life is presented in Table 1. Concrete mix requirements for the adopted solution are in Table 2.

REPLACING CARBON STEEL REINFORCEMENT WITH STEEL FIBER REINFORCEMENT

Two different design strategies can be followed to achieve durable structural concrete. One strategy is to avoid the degradation threatening the structure due to the type and aggressivity of the environment. An example of this strategy is the selection of non-reactive or inert materials such as: stainless steel reinforcement, epoxy-coated steel reinforcement, glass fiber reinforcement, non-reactive aggregates, sulphate-resistant cements or low alkali cements. Another strategy is to select an optimal material composition and structural detailing to resist, for a specified period of use, the degradation threatening the structure. This strategy allows for the deterioration to advance, albeit, only to a certain degree within the defined service life. This is also called a performance-based service life approach by using durability modelling. The mathematical modelling of the transport and deterioration mechanism is an integral part of this strategy. The design of reinforcement for the concrete structures for Abu Hamour Tunnel project implemented both strategies:

Table 1. Concrete mix

Structures	Materials	Other
Typical Tunnel Segments	Triple Blend Concrete:	Steel Fiber Reinforced Concrete (SFRC)
Special Tunnel Segments (Close to the Adits)	(OPC + GGBS 66–80% +MS) or (OPC+ FA 25–30% + MS) or	SFRC + Stainless Steel Bar Reinforcement
Shafts & Adits	(OPC min 50%+ FA + GGBS)	Carbon Steel Bar Reinforcement + PVC & Geotextile on Outer Surface

Table 2. Concrete mix requirements

Structural Element	Cement/Binder Type	Maximum Equiv. Water/ Cement Ratio	Minimum Cement/ Binder Content (kg/m^3)	Maximum Cement/ Binder Content (kg/m^3)	Maximum Chloride Migration Coefficient[5] (m^2/s)
Typical tunnel segment SFRC C55	OPC + GGBS[1]+MS[4] OPC + FA[2] + MS[4] OPC[3] + FA + GGBS	0.4	370	450	No requirement[6]
Special tunnel segments close to the adit SFRC+Stainless Steel C55	OPC + GGBS[1] + MS[4] OPC + FA[2] + MS[4] OPC[3] + FA + GGBS	0.4	370	450	No requirement[6]
Adits & Shafts C37 Carbon Steel Rebar	OPC + GGBS[1] + MS[4] OPC + FA[2] + MS[4] OPC[3] + FA + GGBS	0.4	370	450	2.2×10^{-12} @ 28 days 4.1×10^{-12} @ 28 days 3.1×10^{-12} @ 28 days
Special Shaft Section	OPC + GGBS[1] + MS[4]	0.4	370	450	No requirement

(1) GGBS = Ground granulated blast furnace slag, slag content of powder 66–80%
(2) FA = Fly ash, fly ash content of powder 25–30%
(3) OPC = Ordinary Portland Cement, cement content is minimum of 50%
(4) The addition of micro silica (MS) is limited to between 5–7% of the total binder
(5) Chloride migration test in accordance with NT Build 492 on laboratory samples (cast cubes/cylinders)
(6) Chloride migration testing is only for production monitoring purpose. No value larger than 10×10^{-12} m^2/s @ 28 maturity days. Testing performed on concrete without steel fibers.
(7) For all binder combinations, sulphate content (SO$_3$) of the concrete shall not exceed 3.6% of the binder content

- Included steel fiber reinforcement and stainless steel reinforcement bars to avoid deterioration was selected for bored tunnel typical and special segments.
- Conducted performance-based durability modelling for carbon steel rein-forced concrete to restrict the risk of corrosion to an acceptable limit for shafts and adits.

Several research investigations have shown favorable behavior of steel fiber reinforced concrete (SFRC) with respect to concrete durability (SALINI, et al, 2013). The durability of SFRC under chloride exposure is superior to that of traditional steel bar reinforced concrete of the following reasons:

1. The potential difference, which is a prerequisite for corrosion initiation and passivation, along a single fiber is minute, i.e., the formation of designated anodic and cathodic areas is very limited.
2. Due to the different casting conditions for steel fibers (floating in the concrete matrix) as opposed to traditional steel bar (fixed externally), the interface at the concrete-steel is entirely different. Consequently, the structure of the concrete-steel inter-face is entirely different for the two cases. The amount and magnitude of voids in the interface zone is largely reduced for the fibers compared to that for the traditional reinforcement. Such a dense interface further increases resistivity, evens out the passive layer and limits diffusion.
3. Fibers are short and discontinuous and rarely touch each other. Thus, there is no continuous conductive path and formation of macrocells is not possible. The missing continuous electric path also prevents any detrimental effects

from stray or induced currents on concrete durability, e.g., from electric systems in the vicinity of the tunnel, as opposed to traditional steel bar reinforced concrete.

4. For applications where SFRC is exposed to the atmosphere, fibers protruding from the concrete surface will depassivate and corrode. Hence, rust spots may occur at the surface, but corrosion will not penetrate deeper into the concrete. Such aesthetic considerations are irrelevant for drainage tunnels.

The bored tunnel special segments close to the adit consist of SFRC with additional stainless steel rebars. Stainless steel is available in grades that will not corrode in chloride-contaminated environments. To achieve compatibility of the SFRC used for the segments with the conventionally reinforced concrete used in the adit, the same percentage of GGBS, FA and/or MS has been used for the cement binders for both the segments and the adits. The presence of the fibers does not affect the chemical properties of the cement binder for the segments. The performance-based durability modelling for shafts and adits was based on the approach of fib Bulletin 34. For this approach, concrete quality represented by the concrete cover and the resistance to chloride migration was most important to ensure long-term durability.

CHLORIDE RESISTANCE

Chloride resistance verification for the tunnel is not required by the Eurocode for typical or special tunnel segments due to the use of SFRC. Based on the explanation presented herein, it can be concluded that minimal to none corrosion damage will occur in the cross-section of the SFRC segmental lining. The use of additional stainless steel reinforcement for the tunnel special rings close to the adit eliminated any durability related issues concerning chloride ingress and steel corrosion. The long-term performance of stainless steel in concrete structures as a replacement of carbon steel is well documented. For a sufficient corrosion resistance, austenitic chromium-nickel-alloy steel with minimum molybdenum content of

2 M-% or austenoferritic chromium-nickel-alloy steel with minimum chromium content of 22 M-% was specified. Corrosion of the carbon steel reinforcement is the greatest risk to the shafts and adits, in particular for the internal surface of the concrete exposed to saline water and oxygen. Performance-based durability modeling was conducted for the shafts and adits to restrict the risk of corrosion to an acceptable limit to fulfill the required service life. The selected approach, which is in accordance with fib Bulletin 34, has been used on a number of recent major projects in the Gulf area, e.g., Qatar-Bahrain Causeway, Oman airports (Seeb and Salalah International Airports), and the STEP tunnel project in Abu Dhabi. Of particular importance to the durability modeling are the concrete cover and the resistance to chloride migration, which represent the final quality of the concrete. The severe ground conditions, the high temperature and the high humidity create a harsh environment which adversely affects the durability modeling. To achieve a high chloride resistance concrete, the maximum chloride migration coefficients were determined for the types of triple blends of concrete consisting of OPC and GGBS and/or MS and/or FA. This was conducted for various proposed concrete covers to inhibit corrosion of the reinforcement depending on the exposure conditions. The required concrete covers for the various structural bar reinforced elements are presented in Table 3. These covers meet the durability requirements, using the concrete mixes provided in Table 2, and based on the exposure conditions detailed herein. Durability testing involved checking the compliance of the concrete mix with the target values for the chloride migration. Further compliance testing was performed during pre-testing using the batching plant. The pre-testing was done in two stages: (a) on separate cylinders/cubes cast and cured on the laboratory, (b) on cores drilled out from elements where the concrete is consolidated and cured under actual production conditions. Both specimens complied with the chloride migration coefficients determined from the durability modeling.

Table 3. Concrete cover requirements

Structural Element	Design	Structural Face	Cover to Carbon or Stainless Steel Reinforcement (mm)	
			Minimum	**Nominal**
Special tunnel segments close to the adit	Stainless Steel + SFRC C55	Internal	40	45
		External	40	45
Adits & Shafts	Carbon steel reinforced concrete C37	Internal	75	85
		External	60	70

SULPHATE ATTACK

Another type of deterioration is sulphate attack which can be resisted through proper selection of concrete composition and materials. To perform sulphate resistance verification for the tunnel The Centre for Building Materials (CBM) of the Technische Universität München, Germany, was commissioned to assess the binder compositions with regard to sulphate resistance using a flat prism (SVA) method. German DIBt Code SVA Testing Methods were implemented instead of the methods specified by QCS 2010-ASTM C1012. The test duration for the former method is 6 months while for the latter is 18 months. The aim of the test is to measure the expansion of flat prisms stored in sulphate solution. To assess the resistance of the specimens to sulphate attack, the length of the flat prisms is measured after storage periods of 0, 14, 28, 56, 91 and 180 days. Expert opinion was required to determine whether the test should stop after a duration of 180 days (for which a strain difference criterion of 0.5mm/m should be achieved), or continue for the full duration (for which a strain difference criterion of 0.8mm/m was achieved). Expert advice and assessment was also necessary as the German DIBt Code SVA Testing method is only applicable up to sulphate concentrations of 3,000mg/l since the sulphate concentration of the ground for the project is 5,000mg/l.

Sulphate resistance verification for the shafts and adits was based on the QRail document (Qatar Integrated Railways Project—Study of Groundwater Analysis Against Tunnel Concrete—Final Report," Version 0.2, May 30, 2012) where it was postulated that the higher water/binder ratio of the cast-in-situ concrete compared to the segmental lining will lead to lower physical resistivity against sulphate attack. This physical resistivity was achieved by production of a very dense matrix structure which reduced sulphate ingress into the concrete. For the segmental lining, significantly lower water/binder ratios were possible due to the industrial production conditions, allowing for better concrete compaction of low w/b concrete mixes, and smaller segments dimensions. As the cast-in-situ shaft had larger dimensions than the segments, a mix with a high amount of slag and/or fly ash was recommended in order to reduce the development of the heat of hydration. For AHSO, the same water/binder ratio was used for the segments as for the reinforced concrete permanents works for the shafts and adits. Admixtures were used to increase workability.

CONCRETE MIX CHALLENGES

The special mix required to satisfy durability requirements posed some issues regarding workability.

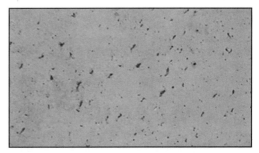

Figure 1. Defect Type C1—Pinholes and blowholes

Figure 2. Defect Type C2—Segregation/ Honeycombs

Due to the reduced workability of the concrete mix, defects were spotted in the shafts. In what follows, the defects are described together with the corresponding operative procedure for repair (SELI et al, 2015) of permanent concrete structures.

Defect Type C1—Pinholes and Blowholes

Pinholes and blowholes are relatively small voids of roughly spherical shape produced by air entrapment during the concrete placement and vibration. The affected surfaces were manually treated with the mortar. The product was applied in accordance with the manufacturer's recommendations. The application was manual, by trowel or wet spraying. Upon completion of the repair, in order to protect the fresh repair mortar from early dehydration, wind, rain and/ or direct sunlight, a plastic sheet was applied to the surface of the repaired concrete. See Figure 1.

Defect Type C2—Segregation/Honeycombs

Honeycombs appear as a rugged/coarse concrete surface where mortar is partially or completely absent. Honeycombing is usually caused by segregation of the coarse aggregate and mortar or a lack of adequate vibration. Repair was carried out as soon as possible after the formwork had been removed. This reduced differential shrinkage and improved bond between the original concrete and the repair material. The concrete surface to be repaired was marked on site with the lines to delineate the repair zone, including the defect areas plus an extra 50mm all around. The repair depth extended approx. 20–30mm inside sound concrete (i.e., that which is strong, rough and clean). The outline, as stated above, was cut with a masonry cutting disk or saw to ensure a square edge. The depth of the saw cuts was not to exceed the foreseen concrete cover over the rebars in order to avoid cutting the reinforcement. All the honeycombed concrete was removed by hammer and/or chisel. Repairs that did not require the use of a formwork were executed with a high strength non-shrink repair mortar. If the repair required the use of a formwork a non-shrink pourable micro concrete was used. Once the repair was completed, the final layer of repaired concrete was protected. See Figure 2.

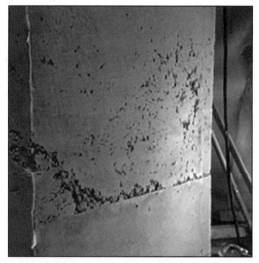

Figure 3. Defect Type C3—Cold joints

Defect Type C3—Cold Joints

Cold joints are discontinuities caused by delays in casting operations or bad workmanship. Such discontinuities generally are visible immediately after the formwork removal and can pass through the entire thickness of the defective structural member. As a first step for repairing this type of defects, the visible concrete surface surrounding the cold joint was removed by chipping 80–90mm on both sides of the joint line for the entire discontinuity length. The repair depth extended approximately 20mm inside sound concrete (i.e., strong, rough and clean). If the joint did not continue inside the member it was treated as a honeycomb. The repair was carried out after concrete hardening. Based on the location, orientation, width, length and depth of the cold joint, the necessary measures for treatment and repair were chosen. Width, length and depth of the joint was checked by means of specialized instruments such as crack-meters, micrometers etc. The repair process involved the use of thixotropic repair mortar and/or epoxy resin injection. At the end, the removed portion of concrete was repaired and final surface finished. See Figure 3.

Defect Type C4—Cracks

Two main different type of cracks affected the concrete structures:

- Surface shrinkage cracks: superficial cracks (generally extend less than 5mm in depth) due to inadequate concrete curing or windy conditions to which the concrete is generally

Figure 4. Defect Type C4—Cracks

exposed. They do not have structural or durability implications.
- Structural cracks: cracks extending inside the concrete members for a few centimeters (>50mm).

Surface cracks were repaired by cement bagging applied as soon as the crack development was noticed and prior to starting the concrete curing. Surface cracks developed on the extrados of the first stage shaft permanent base slab were rectified as they were covered later on by the second stage mass concrete invert. Structural cracks were treated as per Type C3 Defect—Cold Joint. See Figure 4.

Defect Type C5—Leaks

Upon completion of the permanent lining inside the shaft and subsequent cessation of dewatering

Figure 5. Defect Type C5—Leaks

Figure 6. Defect Type C6—Exposed rebar

Figure 7. Defect Type C7—Exposed spacers

activities, one of the following scenarios were observed:

1. The shaft was dry and no leakage or wet areas observed—no repair/injection was required;
2. Water was dripping/flowing from one of the injection tubes connected to the injection box which form part of the waterproofing system in the shaft final lining—repair/injection was required;
3. Wet patches/dripping were visible along the Construction Joint (CJ) between the base slab and the shaft wall Lift No. 1 or between

subsequent lifts—repair/injection was required;
4. water was dripping/flowing from the joint between the shaft permanent lining and the incoming micro-tunneling or GRP pipe—repair/Injection was required;

Waterproofing injection was required only in case the water leakage was exceeding the limits allowed by Class 3 of Table 16 in Clause 508.2 of the BTS/ICE "Specification for Tunneling" which is 0.1l/m^2. The injection was carried out using expansive acrylic resin. See Figure 5.

Defect Type C6—Exposed Rebar

This repair concerns the rebar exposed during the concrete chipping for repair of honeycombing, cracks or pour lines. These types of defects was repaired following the same methodology as proposed for honeycombs with a surface area $<1\text{m}^2$ and extending behind the steel reinforcement. See Figure 6.

Defect Type C7—Exposed Spacers

This repair concerns the plastic rebar spacers displaced during concrete placement and left exposed on the intrados of the concrete wall. This type of defects was repaired following the same methodology proposed for honeycombs presenting a surface area $<1\text{m}^2$ and extending behind the steel reinforcement. See Figure 7.

Defect Type C8—Shear Pockets / Cored Voids To Be Filled—1

The shear pockets moulds provided in the formwork were removed after 5 to 6 hours from completion of the concrete placement when the concrete started its initial setting. The exposed surfaces were roughened in order to improve bonding with the material subsequently used for repairing the pockets. The shear pockets or the cored voids were filled with a non-shrink micro-concrete compound. The surface of the area to be filled was ensured to be clean and free from oil, grease, dirt and paint. Priming of the substrate was required. Concrete surfaces were saturated with water prior to application of the repair mortar. See Figure 8.

Defect Type C9—Offsets/Steps in Concrete Lifts

Horizontal offsets/steps observed at the construction joint between Lift No.1 of the shaft wall and the base slab kicker were rectified only in the case that they were not covered later by the second stage mass concrete invert to be cast at shaft bottom and if they exceed 5mm in depth. In case repair was required, the side of the step protruding inside the shaft was

Figure 8. Defect Type C8—Shear pockets/cored voids to be filled—1

grinded down to form a smooth transition between the kicker and Lift No. 1. Cement bagging was applied to complete the finishing of the repaired area. Horizontal offsets/steps observed at the construction joint between consecutive shaft lifts were rectified only in case they exceed 5mm in depth. The side of the step protruding into the shaft was grinded down so as to form a smooth transition and eliminate the step. Cement bagging was applied to complete the finishing of the repaired area. Vertical steps noticed in the same lift at the location of the shaft formwork lap plates were rectified only in case they exceed 5mm in depth. The side of the step protruding into the shaft was again grinded to form a smooth transition and eliminate the step. Cement bagging was applied to complete the finishing of the repaired area. See Figure 9.

Defect Type C10—Discoloration

Discoloration of concrete was usually caused by rust or other debris which was present on the formwork panels due to lack of appropriate cleaning. This kind of defect is only aesthetic. Usually discoloration is surficial and only affects a very thin layer of concrete. The discolored concrete layer was removed by sandpaper or brush, until the surface was restored to its original color. See Figure 10.

PROJECT STATUS

Project handover was in December 2016 one month ahead of schedule.

CONCLUSIONS

The permanent works on several shafts have been completed. Since the Abu Hamour (Musaimeer) Surface & Ground Water Drainage Tunnel was the first application of advance tunneling technology

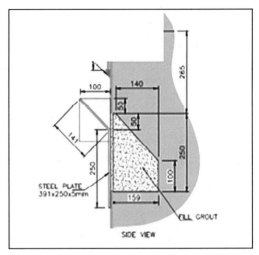

Figure 9. Cross section through shear pocket shown on Figure 8

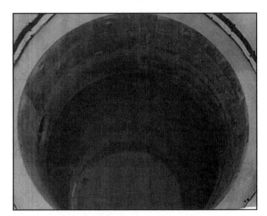

Figure 10. Defect Type C10—Discoloration

in Qatar our team wanted the engineering community to benefit from our experience in managing the design and construction.

ACKNOWLEDGMENT

The writers acknowledge Salini-Impregilo S.p.A who successfully excavated the shafts and tunnels.

REFERENCES

Jafari, M.R., Bernardeau, F.G., Stypulkowski, J.B., Siyam, A.F.M., "Abu Hamour Outfall Tunnel Project In Doha, Qatar," in RETC—Rapid Excavation and Tunneling Conference and Exhibit, June 7–10, 2015, New Orleans, LA, pp 401–411.

Pathak, A.K., Stypulkowski, J.B, Bernardeau, F.G., "Supervision of Engineering Geological Activities during Construction of Abu Hamour Surface and Ground Water Drainage Tunnel Phase-1 Doha, Qatar," in International Conference on Engineering Geology in New Millennium at IIT, New Delhi 27–29. Special Publication, J of EG pp 467–486, Oct 2015.

Siyam, A.F.M., Bernardeau, F.G., "Structural and Construction Challenges for Shafts on Abu Hamour Surface and Groundwater Drainage Tunnel" Arabian Tunneling Conference and Exhibition, Dubai, UAE, 23–25th November 2015.

Siyam, A.F.M., Stypulkowski, J.B., Bernardeau, F.G., "Improvement to Longevity of Tunnels in Aggressive Ground Conditions in the Middle East," Arabian Tunneling Conference and Exhibition Abu Dhabi, UAE, pp 163–185, 9–10th December 2014.

Stypulkowski, J.B., Pathak, A.K., Bernardeau, F.G.," Abu Hamour, TBM Launch Shaft, "A Rock Mass Classification Attempt for a Deep Shaft in Doha, Qatar." in EUROCK14, ISRM International Symposium, May 27–29, Vigo, Spain, CRC Press, Taylor and Francis Group, pp 117, 2014.

Stypulkowski, JB, Pathak, AK, Bernardeau, FG, "Engineering geology for weak rocks of Abu Hamour surface and ground water drainage tunnel Phase-1 Doha, Qatar," International Conference on Recent Advances in Rock Engineering, Specialized Conference of ISRM in Bengaluru, India, 16–18 November 2016, RARE2016, pp 85–90.

Stypulkowski, J. B., Siyam, A. A. F. M., Bernardeau, F. G. and Al Kuwari, N. G., "Abu Hamour Drainage Tunnel, First TBM Mined Tunnel in Doha, Qatar." The First Arabian Tunneling Conference and Exhibition, Dubai, pp. 300–314, 2013.

"Project References with Steel Fibre Reinforced Segmental Linings." Construction of Abu Hamour (Musaimeer) Surface & Groundwater Drainage Tunnel—Phase I, Report by SALINI-IMPREGILO S.P.A, 17 September 2013.

"Supplementary Concrete Specification for New Solution." Construction of Abu Hamour (Musaimeer) Surface & Groundwater Drainage Tunnel—Phase I, Report by SALINI-IMPREGILO S.P.A, 11 June 2013.

"Operative Procedure for Repair of Permanent Concrete Structures" Construction of Abu Hamour (Musaimeer) Surface & Groundwater Drainage Tunnel—Phase I, Report by SALINI-IMPREGILO S.P.A, 16 May 2015.

British Tunnelling Society/Institution of Civil Engineers, Tunnel Lining Design Guide, (London: Thomas Telford, Ltd., 2004).

Round Shaft Hydrogeological Analysis And Dewatering Assessment, SALINI-IMPREGILO S.P.A, 24 September 2013.

Hansel D. & Guirguis P., 2011. "Steel-fibre-reinforced Segmental Linings: State-of-the-art and Completed Projects."

QCS 2010, Qatar Construction Specification 2010.

fib (2006). "Model Code for Service Life Design." fib Bulletin 34, International Federation for Structural Concrete (fib), Lausanne, Switzerland, 1st edition.

fib (2012). "CEB-FIP Model Code 2010—Volume 1". fib CEB-FIP Bulletin 65, International Federation for Structural Concrete (fib), Lausanne, Switzerland.

fib (2012). "CEB-FIP Model Code 2010—Volume 2". fib CEB-FIP Bulletin 66, International Federation for Structural Concrete (fib), Lausanne, Switzerland.

"Quality Plan for Production of Pre-cast Tunnel Lining Segments ("New" Solution)." Construction of Abu Hamour (Musaimeer) Surface & Groundwater Drainage Tunnel—Phase I, Report by SALINI-IMPREGILO S.P.A, 23 September 2013.

"Qatar Integrated Railways Project—Study of Groundwater Analysis against Tunnel, Concrete—Final Report," Version 0.2, May 30, 2012.

"Plant Trial Test Result (Preliminary) for Concrete Mix Design Proposed for Permanent Works" Construction of Abu Hamour (Musaimeer) Surface & Groundwater Drainage Tunnel—Phase I, Report by SALINI-IMPREGILO S.P.A, 29 June 2014.

"Operative Procedure for Repair of Permanent Concrete Structures" Construction of Abu Hamour (Musaimeer) Surface & Groundwater Drainage Tunnel—Phase I, Report by SALINI-IMPREGILO S.P.A, 16 May 2015

In-Depth Inspection of a Century-Old San Francisco Water Tunnel

R. Fippin and J. Sketchley
McMillen Jacobs Associates

T. Redhorse and D. Tsztoo
San Francisco Public Utilities Commission

ABSTRACT: The 92-year-old, 18.9-mile-long (30.4 km) Mountain Tunnel is part of the San Francisco Public Utilities Commission Hetch Hetchy Water conveyance system. Only six years after commissioning, defects were observed in the concrete lining, which have grown over time to expose the rock behind. This paper presents the 2017 inspection of this water tunnel and data collected to perform an assessment of its integrity. The inspection's extensive data collection program contained a visual inspection, geologic mapping, concrete and rock coring, small-diameter drilled holes, GPR, LiDAR, and inflow measurements. The information obtained formed the basis for the condition assessment, which evaluated the tunnel's integrity and feasibility of its rehabilitation.

INTRODUCTION

San Francisco Public Utilities Commission (SFPUC) supplies drinking water to 2.6 million residents in the San Francisco and surrounding communities. Water is collected from the Hetch Hetchy watershed in Yosemite National Park and travels through a series of historic tunnels constructed in the early 1920s, known collectively as the Hetch Hetchy System, before crossing the San Joaquin Valley and feeding into the numerous downstream pipelines, tunnels and reservoirs. The Hetch Hetchy system was an engineering feat even by today's standard. The tunnels within the Hetch Hetchy system are non-redundant and serve as a singular route for 85% of the SFPUC water supply.

The Mountain Tunnel is one of the longest tunnel in the Hetch Hetchy system at 18.9 miles, with an internal diameter generally ranging between 10 feet and 15 feet (3.0 and 4.6 m). Water flows from east to west under gravity conditions between Kirkwood Powerhouse to Priest Reservoir (Figure 1). The eastern 7+ miles (11.3 km) of tunnel is predominantly unlined and the downstream western 11 miles (17.7 km) is continuously lined with unreinforced concrete. The Mountain Tunnel has been in service for 92 years and has been inspected numerous times over its life time with defects in the concrete lining noted as early as six years after commissioning. Inspections have reported an increase in defects within the tunnel concrete lining with some areas having exposed rock behind the lining.

In 2016, the San Francisco Public Utilities Commission (SFPUC) was undecided between constructing a new bypass tunnel to replace the lower 11 miles of tunnel that were experiencing lining deterioration or performing repairs over a series of outages. Part of the difficulty in selecting a path forward was that the condition of the tunnel was not consistently described by various consultants and experts. Given the importance of the Mountain Tunnel within the water delivery system, understanding the current condition of the tunnel was critical in formulating a decision and providing the SFPUC with sufficient data to evaluate the best solution for its asset to perform over the next 100 years.

To understand the current condition of the tunnel, a contract was awarded for an extensive tunnel inspection. This inspection focused on using quantitative methods to ensure that an informed engineering basis could be made to perform a condition assessment of the tunnel. These quantitative methods included defect quantities, core compressive strength results, concrete thickness measures, ground penetrating radar (GPR), inflow measurements, and other documented data. This paper presents the methods used in a multiday, multidisciplinary inspection of the tunnel to collect quantitative data and includes logistical implementation, lessons learned, and recommendations for other critical facility inspections.

2017 INSPECTION OVERVIEW

The 2017 inspection occurred as part of a 60-day tunnel outage starting in January 2017. The 2017 inspection took place over 18 days and was performed by a large multidisciplinary team. Table 1 summarizes the roles of the parties and their specific

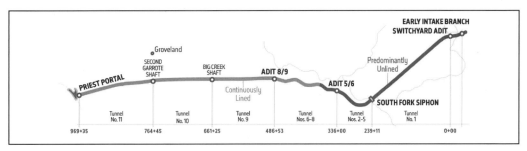

Figure 1. Mountain tunnel system

Table 1. Inspection team roles

Company (No. of Team Members)	Inspection Role	Inspection Activities
McMillen Jacobs Associates (15)	Prime consultant Inspection supervisor and inspection lead Participation and oversight of all inspection activities	Visual inspection (mapping, photographing defects, taking digital videos) South Fork Manifold inspection South Fork Siphon inspection Direct drill holes Key decision maker tour
Black & Veatch (3)	Collection of water samples and support of inspection activities	Water sampling
Gregg Korbin (1)	Assisting in inspection with expert insight	Key decision maker tour
GEI Consultants Inc. (1)	Collection of geologic structure data in unlined tunnel	Visual inspection (rock mapping in unlined section)
Structus (2)	Support of general inspection activities Logging of concrete core	Concrete coring
Robert Chew Geotechnical Consultants (2)	Logging of rock core	Rock coring
Agapito (5)	Subcontractor for concrete coring and rock coring	Concrete coring
Dibit (2)	LiDAR (with photogrammetry)	LiDAR with photogrammetry
Norcal Geophysical Consultants (3)	Ground Penetrating Radar (GPR)	GPR

inspection activity responsibility. A general contractor was responsible for providing overall tunnel safety and access during the inspection including ventilation, in-tunnel communication, transportation and opening bulkheads.

TUNNEL ACCESS AND LOCATION

The Mountain Tunnel is located in Tuolumne county, 140 miles (225 km) east of San Francisco, and is in a remote, mountainous location with restricted narrow road access to the tunnel entry points (Figure 2). Additionally, drivable access into the tunnel was restricted, with only one equipment access point at the downstream end of the tunnel at Priest Portal via a 10-foot-diameter (3 m) steel wye section. Four other personnel access points existed, but were

separated by several miles and access was via steep, narrow and mostly unpaved access roads. The access points are summarized in Figure 3.

Inspection Shifts

The Mountain Tunnel inspection occurred using 12-hour shifts. The first five days consisted only of day shifts for the inspection team, with night work allowed for contractor preparatory work. Following that inspection activities transitioned to both day and night shifts. The goal was for a maximum 12-hour shift with 10 working hours—an hour before and after were schedule for safety, equipment preparatory work, and the logistics of getting to each tunnel entry point. The plan included between one and four separate activities during each shift.

VISUAL INSPECTION

Concrete-Lined Tunnel Sections

The downstream 11 miles (17.7 km) of tunnel are continuously lined with unreinforced concrete. The visual inspection of the concrete-lined portions of the tunnel section included the following:

- Documenting station, type, size, and clock position of defects and grouping them into potential repair zones
- Photographing defects requiring repair and performing videography of entire lined section and siphon
- Documenting voids when and where observable
- Documenting infiltration by station, clock position, and estimated rates

- Visually assessing rock quality and whether the rock was self-supporting or ground support was used where observable within a defect
- Documenting areas where repair could be prioritized for future work
- Documenting size and shape of lining (e.g., width, height, concrete invert, footing)
- Documenting stationing at adits, bends, intermittently lined sections, or other elements of interest

The inspection team consisted of six to eight people and was divided into subteams each separated by approximately five hundred feet that included a lead inspector, stationing team, defect documenter, and photographer (Figure 4). Each subteam performed a QA/QC check of the previous team's inspection and identified defects individually if they had been

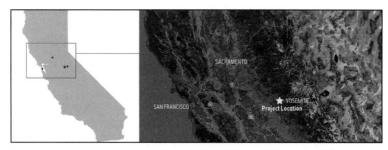

Figure 2. Vicinity map of Mountain Tunnel

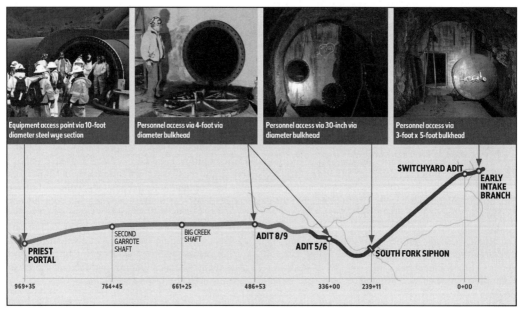

Figure 3. Tunnel access points

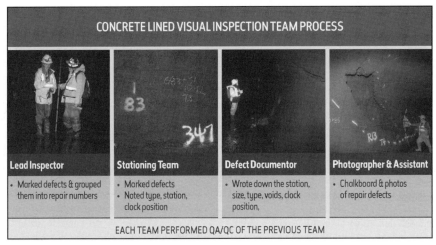

Figure 4. Process of documenting defects in concrete lined portion of tunnel

Figure 5. Geologic team performing rock mass data collection

missed. Given that the tunnel was only illuminated via headlamps, this process took considerable effort and a careful walking pace to locate all the defects.

Unlined Tunnel Sections

The upstream 7+ miles (11.3 km) of tunnel are predominantly unlined, and the geology team assessed the overall stability of the rock mass by performing in-tunnel geologic mapping. Data collected during the mapping included rock type and hardness; strike and dip of discontinuities; and persistence, surface roughness, type, surface shape, and spacing of discontinuities. Bedrock structure data were collected using a Brunton compass, and qualitative assessments of the relative strength of the rock were assessed with a rock hammer. However, one challenge was that the walls of the tunnel are blanketed with a continuous 0.25-inch-thick (6.4 mm) layer of black biofilm. This biofilm obstructs and artificially

smooths the walls of the tunnel, making it difficult to fully visually assess the existing discontinuities. As such, the floor of the tunnel was often the best place to identify lithologic and structural changes, as the floor typically was free of biofilm because the presence of continuously flowing water prevented its formation. Data were typically recorded at 500- to 1,000-foot (152–305 m) intervals and at pertinent structural changes, such as a distinct change in joint spacing or rock lithology. The geology team also noted the tunnel rock mass rating (RMR) at 1,000-foot (305 m) intervals, documented rock piles larger than 1 cubic foot (0.03 m^3) in the tunnel invert, and infiltration locations (Figure 5).

Lessons Learned from Visual Inspection of Lined and Unlined Tunnel

The visual inspection of the lined tunnel was the first scheduled activity and was scheduled to take four days with an inspection pace of 0.5 mile/hour (0.8 km/hr) and walking pace of 3 miles/hour (4.8 km/hr). The visual inspection included a substantial amount of walking due to the limited tunnel entry points. Given the quantity of defects and the desired data to be collected, the estimated inspection pace was too fast. The actual inspection pace was about 0.25 mile/hour (0.4 km/hr). Because of complications with surveying, ventilation, and a failure of the original communication equipment, the visual inspection of the lined tunnel took 6 days and was completed in a noncontinuous fashion.

Schmidt hammer testing was planned at 500-foot (152 m) intervals for possible correlation. However, because of the difficulty in removing the biofilm and obtaining measurements meeting the ASTM C805 criteria, this testing was abandoned.

The visual inspection of the unlined tunnel was scheduled to take four days and was accomplished in this time. It was completed on different days than planned because of the failure of the original communications system and logistics involved with implementing a backup plan. The walking pace while carrying equipment and walking on a rough, occasionally slippery invert was an average of 525 feet/hour (160 m/hr).

Key lessons learned from the visual inspection are summarized below.

1. Because of the limited access points and the intense walking requirements, it is important that inspectors working multiple days have a moderate to high level of fitness.
2. Establish and agree upon a feasible, quickly implementable backup communications system. Wireless nodes were initially planned for this inspection, but did not perform, which caused several days of delays or slow progress while mine phones were secured. This affected the schedule of the other activities and reduced planned rest days.
3. Perform a practice inspection with all relevant team members prior to entering the tunnel to ensure everyone has the appropriate equipment and understands their individual role.
4. The documented size of rock piles should be relative to the cross-sectional size of the tunnel. One cubic foot (0.03 m^3) was too small a criterion for a tunnel cross sectional area of 225 square feet (30 m^2).

CORING AND DRILLING OF LINING AND ROCK

Concrete Coring

Concrete coring was performed to evaluate the strength, thickness, and quality of the lining. Concrete coring operations commenced on January 11, 2017, and were completed on January 16, 2017, over the course of eight shifts. The equipment comprised a traditional core rig and small water tank affixed to the bucket of a skid steer. The core rig was powered from the auxiliary hydraulic hoses on the side of the skid steer, and the core rig was set up to be capable of articulating sufficiently to drill cores at various angles from springline and above (e.g., any angle between 3 and 9 o'clock positions). The core barrels used were 4 inches (100 mm) in diameter and up to 18 inches (455 mm) in length. An example of the coring operation setup is shown in Figure 6. The numbers of cores collected per shift ranged from 3 to 11. This was found to be a function of driller, engineer oversight, equipment performance, and travel

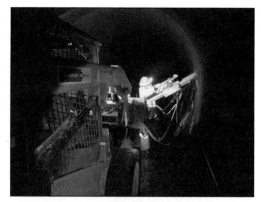

Figure 6. Concrete coring at Sta. 417+00

distance between cores scheduled for the day as the skid steer moved between core locations.

In total, 65 concrete cores were obtained from various locations within the Mountain Tunnel as represented in Figure 7, which exceeded the planned number of 50 cores. The coring for compressive strength conformed to all relevant ACI and ASTM standards (ASTM E105, ASTM C823, ASTM E122, ACI 214.4, and ACI 318-4). This means that the sampling locations are selected randomly within each reach of tunnel to provide for unbiased lower bound estimates, means, and standard deviations that could be used in a probabilistic assessment.

The core testing (Figure 7) established a statistical mean of 3,500 psi (24 MPa) for the concrete lining strength throughout the tunnel with a standard deviation of 755 psi (5.2 MPa). Petrographic testing also revealed that the risk of alkali-silica reaction was very low.

Direct Drill Holes

Direct drill holes were 1-inch-diameter (25.4 mm) holes drilled in the concrete lining to evaluate the lining thickness, to provide a view behind the lining for voids, and for calibration with the ground penetrating radar. Direct drill holes took place on January 12 through 16, 2017, during the day shifts and were performed using two battery-powered, hand-held, rotary hammer drills that drove a 1-inch-diameter concrete drill bit up to a depth of 16 inches (405 mm) into the tunnel lining.

The inspection team intended to collect up to 75 drilled hole measurements as a confirmation of lining thickness. However, the drill battery life and strength and the thickness of the concrete lining impeded progress. Therefore, the program shifted to primarily calibrating the ground penetrating radar (GPR) investigation with 28 holes being completed. A borescope with a 3-foot-long (0.9 m) probe was used to visually inspect and record each hole. At

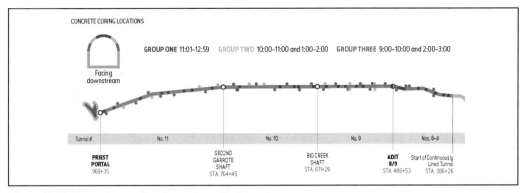

Figure 7. 65 Concrete core locations along the tunnel

some locations, the voids behind the lining could be successfully observed with a borescope.

Rock Coring

The team conducted in-tunnel rock coring on January 10, 13, and 14, 2017, over the course of four shifts. The intended rock drill (Ingetrol 75E) could not be used for drilling because of the unforeseen lack of availability of a supporting generator. Therefore, the original concrete coring drill (LA-100 air-powered drill) was converted for use as a rock drill utilizing an HQ gauge core barrel to extend the holes into the rock mass. It was acknowledged that the LA-100 would result in slower drilling, but the tight schedule for completing work did not allow for switching to the original equipment. Given the strength of rock, with a mean compressive strength of 18000 to 40000 psi (124 to 275 MPa), the progress was noticeably slowed.

The long equipment train needed for rock drilling (consisting of an underground all-terrain vehicle, a compressor, and a cart for the drill and power pack) took up to two hours to mobilize across the tunnel entry point. The angle into the tunnel required hitching and unhitching as the large equipment turned the corner. Once the equipment was in the tunnel, it still needed to continue the journey down the tunnel to the drill location. The setup of drill once at the drill location took approximately 1 hour. This left little time for actual drilling. The equipment remained in the tunnel when possible on adjacent shift, but often had to be demobilized for refueling outside the tunnel or because other activities needed to happen further upstream of the rock coring. While the travel time was accounted for in planning, the entry point proved more difficult than planned for the entry of larger equipment.

Overall, the rock core drilling suffered numerous setbacks with various equipment malfunctions (e.g., flat tires, hitch failures, difficult equipment train mobilization including cart toppling) combined

with a converted drill that was not powerful enough for the hard rock. This rock coring operation began to hamper other important data collection activities and it was determined that further efforts to obtain in-tunnel core would be halted due to the cost and schedule impacts.

A total of twelve 10-foot-long (3 m) holes and six 30-foot-long (9 m) holes were planned over six shifts. In the end, a total of three holes were drilled, all shorter than 10 feet in length. However, since the original purpose of the in-tunnel rock coring was to establish the blasting zone damage from the original construction and the near-tunnel rock quality, the obtained holes did reveal that blast zone damage was limited to less than 1 foot (0.9 m).

Lessons Learned on Coring and Drilling in the Tunnel

Overall, the concrete coring was a success, with 30% more cores obtained than originally planned. Although only 35% of the direct drill holes were obtained, they proved to be useful in correlating with the GPR data and providing video footage of the aggregate at the end of each hole and the voids behind the lining. With the in-tunnel rock coring drill not being correctly set up from the start, this was one activity that could be improved upon during future inspections. With the decision to abandon the rock coring effort, only 15 out of 23 planned shifts were performed for all drilling activities.

Key lessons learned from the coring and drilling operations are summarized below.

1. The concrete lining was substantially thicker than as-built design drawing of 6 inches (150 mm). An 18-inch (455 mm) core barrel did not always reach rock.
2. Coring required field adjustments because of equipment issues and modifications to coring locations to the presence of defects in the tunnel lining. A knowledgeable, decisive

engineer was useful for overseeing the drilling operations and documenting the cores.

3. Hammer drill equipment/strength was not adequate for penetrating the lining efficiently. Similarly, the rock encountered was too strong for the available drill.
4. The borescope was useful for seeing the end-of-hole aggregate conditions and occasionally behind the lining.
5. Since this was a water tunnel, refueling needed to take place outside the tunnel. This often required exiting the equipment from the tunnel at least every other shift. This required careful consideration in arranging the activity sequencing and entry order for each shift.
6. Productivity on the night shifts was substantially lower than on the day shifts and could be attributed to lack of quality engineering oversight.

NON-DESTRUCTIVE TESTING OF CONCRETE LINING

Ground Penetrating Radar

The primary purpose of the GPR was to obtain data on the thickness of the concrete lining with a secondary interest of evaluating typical voids behind the concrete lining. The subcontracted team performed GPR surveying on specific regions within the lined section of the tunnel and obtained 2-D shallow images of the reflection characteristics of the concrete tunnel lining, rock interface, and near-lining conditions.

The two GSSI SIR-4000 units used a 900 MHz frequency to scan into the concrete lining/rock during the night shifts on January 9 through 14, 2017, for a total of seven shifts. Two parallel longitudinal scans at about the 10 and 2 o'clock positions were obtained by personnel sitting on an all-terrain traveling at 1 mile per hour (1.6 km/hr), as shown in Figure 8. The sampling rate was 60 samples per second (about 40 samples per foot). A licensed geophysicist reviewed the data in real-time. Transverse circumferential scans on a clock face from the 3 to 9 o'clock positions were also obtained in each of the five scan regions. The GPR team planned to scan 17,500 lineal feet (5,334 m) of tunnel but instead captured 34,000 lineal feet (10,363 m) because of the efficiency of the setup.

The thickness data were captured every 0.25 inch (6.4 mm) for a total of 834,290 longitudinal data points and 663 circumferential data points. Based on an evaluation of these data, mean lining thickness was determined to be 14.2 inches (360 mm), with a standard deviation of 3.2 (80 mm) inches. A sample of the GPR processed data is shown in Figure 9.

Figure 8. GPR longitudinal setup

LiDAR with Photogrammetry

A subcontractor performed LiDAR scanning with photogrammetry within the Mountain Tunnel to generate a 3-D tunnel surface. The photographs are superimposed onto the 3-D tunnel surface. A viewing software allows for a visual walkthrough for future evaluation during the design of repairs or for review by persons who did not or could not physically enter the tunnel. LiDAR scanning took place during the night shift over the course of four shifts. The scanning unit comprised a two-level I-frame setup, with wheels located at the tips of the I, the scanning unit affixed to the front of the frame, and three digital single lens reflex (DSLR) cameras secured at varied angles along the center of the frame. The unit was hand-pushed through the tunnel, with continuous scanning and photography performed by the LiDAR unit and the cameras.

On the second shift, the LiDAR scanning unit was damaged in the tunnel because of the rough invert, and full, 3-D scanning within the tunnel was discontinued. The use of continuous photography was completed to the end of the tunnel lining. In total, approximately 4.2 miles (6.8 km) of lined tunnel were captured using continuous photography and 3-D scanning with the LiDAR unit. An additional approximately 6.9 miles (4.1 km) were captured using continuous photography only.

Lessons Learned on Nondestructive Testing in the Tunnel

The planned GPR setup included a scissor lift, but equipment was unavailable. The switch to a utility terrain vehicle (UTV) support was very successful

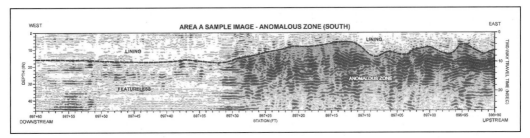

Figure 9. Sample longitudinal GPR scan (blue line represents lining/rock interface)

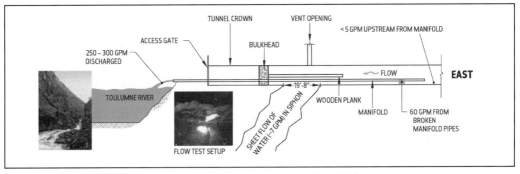

Figure 10. Water infiltration near South Fork East as measured in 2017

and likely allowed for faster progress as the vehicle could travel quickly between locations but was capable of being operated at controlled speed. The rig assembly for the LiDAR/photogrammetry was able to capture data quickly, but the rough invert caused problems for the equipment. The photogrammetry captured allowed for a postprocessed visual walkthrough of the tunnel that will aid in future repairs.

Key lessons learned from the GPR and LiDAR operations are summarized below.

1. Affixing the fiberglass poles to the UTV did not work as the bumps caused by the driving caused a pole to snap.
2. While the GPR team required a driver, both nondestructive teams were otherwise self-supporting requiring no engineering oversight beyond notification of shift priorities.
3. The roughness of the tunnel invert was underestimated placing stress on the LiDAR setup. Use of an all terrain vehicle (ATV) type setup may have been more successful. This was considered in planning but not implemented and not capable of being implemented without notice.

SOUTH FORK SIPHON INSPECTION

Water Inflow at the Manifold and Siphon

A portion of the tunnel, located immediately upstream of the South Fork East Adit, passes through a shear zone where water from the Tuolumne River infiltrates into the tunnel when it is running at low-to-zero flow conditions. To collect this water in the tunnel, a series of pipes and hoses connect into a manifold pipe mounted to the inside of the tunnel, which discharges the water back to the Tuolumne River.

As part of the inspection, inflows were measured (1) immediately upstream of the manifold; (2) in the immediate manifold vicinity from seepage into the tunnel using a sandbag dam; and (3) exiting the manifold pipe into the river. In general, the manifold system was in good shape except five connection hoses showed noteworthy damage; consequently, there was substantial leakage of river water into the tunnel. The river water entering the tunnel from the broken hoses was approximately 60 gpm (227 Lpm) and the discharge of the captured and redirected river water was estimated to be approximately 250–300 gpm (946–1,136 L/m), as depicted in Figure 10.

Figure 11. View from within the South Fork siphon

Siphon Inspection

A remotely operated vehicle (ROV) inspection of the South Fork siphon was originally scheduled, but subsequently canceled because of intense flooding in the area. Instead, the contractor partially dewatered the siphon to allow personnel entry to video its condition. Personnel entry was achieved with climbing ropes and ladders, and rafts were used to float through the siphon and take the video (Figure 11). The siphon water level was maintained at a pump rate of about 5 to 10 gpm (19–38 Lpm).

Lessons Learned on Siphon Inspection

The siphon access point is close to the Tuolumne River and subsequently is prone to flooding, and it was subjected to extreme weather in January 2017. Many delays were the result of dangerous conditions within the immediate vicinity of the access point. Although manned entry into the siphon was not planned, the rope access and raft with safety line were efficient and successful at confirming conditions of this section of tunnel. This provided critical information for the assessment of this section of tunnel.

WATER SAMPLING

Sixteen groundwater samples were taken to test infiltrating water for chemicals and turbidity. Groundwater samples were captured directly at the point of infiltration into the tunnel. Samples were photographed, and labeled by location with pH, temperature, and flowrate documented (Figure 12).

Lessons Learned on Water Sampling

Groundwater sampling points were to be selected based upon unusual conditions of infiltration such as odor, high temperature, high flow, accumulation of biofilm, or turbidity. Groundwater infiltration points were less common than expected, and because of a lack collectable infiltration, samples were generally taken wherever feasible.

OVERALL INSPECTION PLANNING AND IMPLEMENTATION LESSONS LEARNED

Overall, the Mountain Tunnel inspection in January 2017 was a success; however, even successful tasks can be used to learn, plan, and execute future endeavors more efficiently. The key lessons recommended for others attempting similar ventures fall under three main categories: planning, execution, and equipment, outlined below.

Planning

1. Ensure that the scheduled activities and associated float allow time for unanticipated breakdowns or delay in entry. Staff on-site should be capable of making quick decisions for reorganizing work priorities as situations arise.
2. Prior to site work, develop equipment schematics that indicate how the equipment will be mobilized into and out of the tunnel. This includes proper understanding of tow capacity and hitching capabilities.
3. Establish realistic safe backup communication practices in advance.
   - The established wireless node system failed to function properly. This caused extensive delays and last-minute procurement of 12 miles (19 km) of mine phones.

Figure 12. Water sampling process

4. With a large team, provide one person whose dedicated role is "Inspection Team Communication." This person should provide updates to the team at the start and end of both day and night shifts. Since anticipated work activities can change with little notice, effective communication to the supporting team is crucial.
5. Provide one person outside dedicated to performing errands and obtaining supplies.
6. To the extent possible, simulate inspection activities and roles and discuss roles as a team in a practice environment. On the Mountain Tunnel inspection, it was necessary for personnel to be capable of switching roles and important that more than one person understood other jobs.

Execution

1. Allow proper time for on-site contractor to set up in advance of the team. This includes site preparation, establishing in-tunnel communications, and establishing consistent ventilation.
2. Staging four activities per shift with one main tunnel access point requires daily and sometimes hourly planning to accomplish each individual goal.
3. Ensure each shift has a competent person who is authorized and capable of making decisions. These people should be aware of the overall project needs, have the knowledge that they will be expected to make decisions, and be comfortable coordinating with the contractor, owner, etc.
4. Strongly consider fatigue in decision making and scheduling decisions.
   – Personnel and team fatigue need to be addressed realistically. Schedule an unchangeable off-day or a six-day workweek. Staff extra personnel and have them available at site, as backup personnel offsite can be difficult to mobilize with immediacy, especially in remote locations.

Equipment

1. Require backup for key equipment. The small investment in a ready backup will outweigh the potential schedule delay and lost costs due to downtime.

INSPECTION CONCLUSIONS

The Mountain Tunnel January 2017 inspection required careful coordination among multiple parties as well as a large team to enact. While some activities such as rock coring and direct drilling required real-time adjustments of expectations, ultimately the intent of the inspection was achieved. The most successful elements included the concrete coring and GPR, which determined the mean compressive strength of the lining and the mean lining thickness, respectively. The comprehensive data that were collected during the inspection allowed a thorough condition assessment and structural analysis of the existing lining to be performed. This condition assessment was invaluable to the owner when evaluating the alternatives for repair or replacement from a cost and engineering perspective.

One important activity that was performed upon completion of the inspection activities was a tour comprising key decision makers from the SFPUC and the project technical advisory panel. The tour traveled the Mountain Tunnel to observe its current condition. This tour was vital in helping key decision makers visualize the condition of the tunnel and understand the collected data.

The money and time spent towards intensive study of the existing structure and surrounding rock paid dividends towards the final solution. It would be wise of owners to strongly consider such an approach to existing infrastructure before deciding future repairs or replacement.

REFERENCES

American Concrete Institute. ACI 214.4—Guide for Obtaining Cores and Interpreting Compressive Strength Results.

American Concrete Institute. ACI 318—Building Code Requirements for Structural Concrete.

ASTM International. ASTM C805—Standard Test Method for Rebound Number of Hardened Concrete.

ASTM International. ASTM C823—Standard Practice for Examination and Sampling of Hardened Concrete in Constructions.

ASTM International. ASTM E105—Standard Practice for Probability Sampling of Materials.

ASTM International. ASTM E122—Standard Practice for Calculating Sample Size to Estimate, With Specified Precision, the Average for a Characteristic of a Lot or Process.

Steel Shaft Liner Plate Segments

John Mulvoy
DSI Underground

TUNNEL SHAFT HISTORY

The history of tunnel shafts is as old as tunneling itself. Except where existing portal areas are conducive to tunnel entry or exit, shafts are necessary to enable any tunnel advance or exit.

Shafts are constructed in varying shapes depending upon restrictions at the surface or along the depth necessary to accommodate tunnel entry/exit. Shaft horizontal geometries range from simple square shapes, different rectangular shapes to simple circles morphing into ellipses. All shafts are vertical except for extreme necessity.

While primary structural shaft components are generally considered to be either driven sheet piling, soldier piles or rolled shaft ribs, the controlling component is typically the lagging selection between the primary structural members. Lagging is typically timber, sheet piling or liner plate material. Additionally, water infiltration is always a consideration for shaft design and functionality. Sometimes, water infiltration is beneficial to reduce shaft loads, but in some instances, water infiltration is prohibited due to operational control or pollutant discharge controls.

More recent history has seen the development and expansion of both drilled concrete secant piles and frozen ground applications to address water infiltration to advance shaft excavations. The advantages have proven to be in both the shaft diameter and depth of excavation available by these methods.

INNOVATIVE APPROACH

This paper addresses a hybrid design utilizing existing sheet pile products and reinforcement elements known as "spiders" in the tunneling process. As shown in details to follow, spiders replace another type of reinforcement referred to as "welded studs." Spiders offer additional contact points on sheet piling and additional structural contribution to the sheet piling design with additional concrete contact.

This paper represents ongoing work in developing a design for steel and concrete in large diameter, deep shafts requiring water tight control.

The simple premise is the use of existing slurry wall excavation with the introduction of the hybrid "spider sheeting" as the primary reinforcement and shaft barrier completed with a water resistant concrete pour to finish the assembly.

The application of existing design methodologies for shaft designs is maintained.

BACKGROUND

DSI Tunneling LLC has been in the underground support fabrication business for many decades dating back to the days of Commercial Shearing, Inc. DSI brings forth a comprehensive list of underground support systems for the unique ground conditions of your project. Through both field and design experience, DSI has identified an alternative design that potentially addresses problems with existing watertight shaft technology. Today shaft construction techniques can consist of utilizing steel four-flange liner plates with steel supports to support the surrounding ground during the excavation process including the use of gaskets to provide control of groundwater. Alternatively, more robust methods such as secant piles and slurry walls present an alternative to the traditional steel shaft supports and are suited to handle large diameter watertight construction shafts as well.

The need for larger tunnels and shafts has grown substantially adding to the complexity of tunnel and shaft design. Specifically, the shaft structures for these prospective projects have become much larger and developments in construction techniques have presented competing technologies to the traditional steel supports. Design of large diameter, watertight shafts requires an alternative design option for the underground contractors and designers to consider. Consider the following:

Spider Pile Shaft Components

Shown in Figure 1 is the complete spider pile steel segment shaft. The components of the spider pile shaft structure are as follows: straight-web sheet pile sections, a concrete wall, and DSI "spiders." A slurry wall is also used in the installation of the sheet piles/spider system, although it is not a permanent feature. Each of these components plays a critical role in the performance of the shaft system.

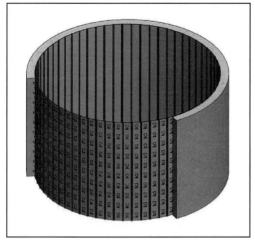

Figure 1. Completed spider pile steel segment shaft

Figure 2. Flat sheet pile profile

Sheet Piles

The straight-web sheet pile sections used in the design are conventional straight sheet piles with interlocking ends so that adjacent sheets may be joined together to form a pseudo-circular structure with a given inside diameter. The piles are typically driven into the ground with the aid of a hammer or vibratory system. A template is used to maintain the desired shape during the driving process. These types of sheets piles are commonplace in the construction industry and contractors have a very high familiarity with them and the process required to properly install them. See Figure 2 for a typical sheet pile.

DSI Spiders

The DSI spiders are 3 dimensional stiffeners created for use in lattice girders. The spiders are fabricated from either 0.39" or 0.47" diameter steel wire which

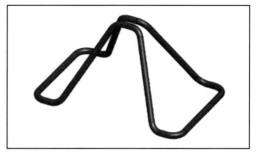

Figure 3. DSI standard spider

is bent into the desired position and then welded into a continuous loop. The spiders separate and connect either smoothback or billeted concrete reinforcing bars which are then completely encased in shotcrete, completing the reinforced concrete system. See Figure 3 for an illustration of a standard DSI spider.

Slurry Wall

Another vital component in the spider pile shaft system is the slurry wall. While this is not a component of the final system, the slurry wall allows for easier installation of the previously discussed sheet piles, followed by the placement of the concrete wall through the tremie concrete method. There are numerous ways to construct a slurry wall, any of which will suffice. The goal of the slurry wall is to provide a stable excavation for the installation of the sheet pile/spider system. The conventional methods of driving sheet piles are not suitable for sheet piles with attached spiders, hence the use of the slurry wall. The sheet piles may be installed in a conventional manner, either with a hammer or vibrator, but far less effort is required to drive the sheets to refusal as there is no soil resistance. The only resistance to driving the sheet is the friction at the interlocking joint.

Spider Pile Sub-Assembly

Once the DSI spiders are fabricated to specific shaft design demands and the flat sheet piles are acquired, the spiders are welded to the flat of the sheet pile with the long dimension of the spider parallel to the length of the sheet pile. Spiders are welded in 4 locations, where each leg contacts the sheet pile. Spiders are located along the length of the sheet pile at roughly 24" centers, although this dimension can vary according to project-specific demands. This welding configuration, combined with the space frame-like structure of the spider, allows the concrete and steel sheet pile to act as a composite unit, thereby increasing the strength and ductility of the system. See Figure 4.

Complete Spider Pile Concrete Shaft

Once delivered on-site and the slurry wall has been excavated, the spider pile wall panels are erected. The inside of the slurry wall should be roughly to the diameter of the steel segments, leaving the balance of the slurry wall to the outside of the steel segments. After the sheet pile/spider assemblies have been erected inside the slurry wall and around the entire perimeter of the shaft, conventional concrete may be placed by the tremie method. The thickness of the concrete is determined in design and shall be equal to the thickness of the excavated slurry wall, less the volume occupied by the steel segments. The concrete wall is unreinforced save for the bond to the sheet pile segments through the DSI spiders as shear connectors. The concrete must be placed from the bottom of the slurry wall up so that the slurry is displaced. This is a procedure familiar to all deep foundation and shaft contractors. After the slurry is displaced and the concrete reaches the top of excavation and cures, excavation of the inside of the shaft may commence. This steel segment concrete wall shaft is relatively watertight. Another benefit to the sheet piles on the inside of the shaft is that they lend themselves to the welding of hooks, ladders and other miscellaneous items often used in shaft construction.

Spider Pile Shaft Design

The design equations and approaches used for these structures have been proven time and time again to be safe, reliable methods to design the support of excavation (SOE) in soil. By applying these same equations and approaches to the steel segment shaft support system, DSI can offer a competitive, economical product that can safely support the soil so that below grade work may commence.

The anticipated loads are determined through conventional soil analysis methods with the coefficient of at-rest pressure playing a large role in the maximum anticipated load. The unit weight of the soil is also used, along with the height of soil to be retained. In cases where water pressure must be considered, the buoyant weight of the soil is used with the coefficient of at-rest pressure and then the hydrostatic head is added to the soil pressure. Lastly, surcharge pressures on the surface are considered in the design. Typically, DSI applies a standard surcharge pressure to the surface, but this can be superseded by larger loads, either from crane loading, soil stockpile, highway or rail loading, etc. The total pressure will change along the depth of the shaft, varying with coefficients of at-rest pressure, soil unit weight, water table and surcharge effects. The maximum total pressure is then found and compared to the allowable load on the structure.

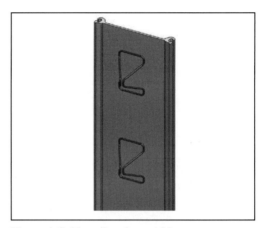

Figure 4. Spider pile subassembly

The determination of the allowable load is the lesser of 2 failure modes: arch thrust in the structure and the load required to buckle the structure out of plane. The lower of these 2 allowable loads will be the controlling load and will then be compared to the anticipated load. Each of these loads is affected by two things: the geometry of the structure and the materials of which it is composed. The spider pile shaft structure is made of hot-rolled steel and conventional Portland cement concrete. Geometric properties (moments of inertia, section moduli, etc) and material properties (compressive stress, unit weight, etc) are determined and used in the evaluation of the allowable load on the steel segment structure.

Arch Thrust

Arch thrust stems from the circular shape of the shaft and the radially applied loads. The thrust load, a force, is divided by the radius of the shaft and by the tributary height. When flexure must be considered, the section modulus is used to determine the available compression strength after the flexure demands are considered. The condition is not the case for the steel segment shaft design however, since the outside of the structure is continuously supported by the surrounding soil. The calculation for the allowable arch thrust load is a derivative of the interaction equation that maintains that the sum of the ratios of anticipated to allowable loads (or stresses) is less than 1.0. Figure 5 explains the effect that continuity of contact, or lack thereof, has on the available compressive load in the structure. The result is the allowable pressure on the structure due to arch thrust.

Per "Rock Tunneling with Steel Supports" by Proctor & White (1946) the allowable load due to arch thrust is as follows.

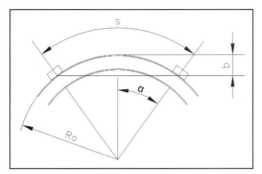

Figure 5. Blocking spacing

$$T_{arch} = (F_b \times A \times S_x)/(S_x + 0.86 \times b \times A)$$

where

F_b = Allowable bending stress in section (includes the factor of safety)

A = Area of section

S_x = Section modulus

b = Amount of rise between blocking points where

b = $(1 - \cos(\alpha)) \times R$ where

R = Outside radius of shaft support

α = Half of included angle between points of blocking where

α = $(s/2)/(2 \times \pi \times R) \times 360$ deg where

s = Distance between points of blocking measured along the outside arc of the shaft

As stated above, the wall is effectively continuously blocked around the perimeter. This leads to a value of 0 for the variable "s" which results in the structure developing no internal moments. Ongoing design development will include a simplified thick walled cylinder approach for comparative purposes.

The result is the allowable thrust load which is converted to an allowable thrust pressure by dividing the force by the outside shaft radius and by the tributary height. This tributary height may be taken as unity for the purposes of designing the spider pile shaft.

Buckling

The other failure mode evaluated is the buckling mode. The critical buckling load is calculated as follows and then the allowable pressure is found in the same manner as the arch thrust. See below for this determination.

Per "Theory of Inelastic Stability" by Timoshenko and Gere (1961), the allowable load due to buckling is as follows.

$$T_{buckling} = (3 \times E \times I_x)/(R^2 \times SF)$$

where

E = Young's modulus of section

I_x = Moment of inertia of section

R = Outside radius of shaft support

SF = Factor of safety applied to buckling

The result is the allowable buckling load which is converted to the allowable buckling pressure by dividing the force by the outside radius of the shaft and by the tributary height. Unity may be taken as the height for the purposes of this shaft design.

There are additional analyses available based on current research that will no doubt provide a more definitive process in determining the critical buckling load. One example is the US Corps of Engineers Manual on Tunnels and Shafts in Rock which provides alternative equations to determine the allowable buckling load for composite liners.

The lesser of the arch thrust pressure and the buckling pressure is compared to the maximum anticipated pressure. If the anticipated pressure exceeds the allowable pressure, then the structure must be re-evaluated to adequately meet the anticipated total pressure demands.

SUMMARY

Design for water-tight, shaft support structures using commonly implemented methods and materials is possible. The spider pile steel segments use simple flat sheet piles that interlock together, along with DSI standard spiders that are lowered into a slurry wall excavation. Once all the sheet piles are installed, then the concrete wall is placed through the tremie method. The result is a shaft that is water-tight and can safely support the anticipated soil, water and surcharge loads. Contractors are familiar with the products and DSI believes that the overall cost of the installed structure will be competitive with current methodology and ultimately safe for all involved.

REFERENCES

Proctor, R.V., White, T.L. 1946 *Rock Tunneling with Steel Supports.* Youngstown: Commercial Shearing Inc.

Timoshenko, S.P., Gere, J.M. 1961 *Theory of Elastic Stability.* 2nd ed. New York: McGraw-Hill Book Company Inc.

Teck BRE Raw Coal Conveyor Tunnel Rehabilitation

Irwan Halim, David Forrester, and Nicolas K. Oettle
AECOM

Brian Wong
Teck Resources

ABSTRACT: Continued operation of Elkview Mine involves open pit extraction of coal reserves under Baldy Ridge in Southeast British Columbia, Canada. An existing tunnel crossing under Baldy Ridge is the only means of raw coal conveyance to the process plant, so it must be preserved and remain operational to support future mining. This paper describes the rehabilitation of the tunnel that was deemed necessary to support close proximity mining activities. A thorough geotechnical investigation and test blasting programs were conducted to support the design of remedial measures for the tunnel. Static and dynamic numerical models were developed to evaluate the impacts on the tunnel from the proposed future mining and eventual backfill over the tunnel with waste rock. The rehabilitation was designed to be constructed without disruption to the continuously operating coal conveyor that runs through the existing tunnel.

PROJECT OVERVIEW

The Elkview Mine (EVO) property is owned and operated by Teck Coal Ltd (Teck). The site (see Figure 1) is located in the front ranges of the Rocky Mountains at elevations from 1,200 m to 1,950 m above sea level and is adjacent to the community of Sparwood in the southeast corner of British Columbia. The main area of the existing mine operation is about 13.5 km north-to-south and 4 km east-to-west. It is bounded by the Elk River to the west, Erickson and Harmer creeks to the east, and Michel Creek to the south; Highway 3 follows Michel Creek along the western and southern sides of EVO.

The Elkview mine produces approximately 7 million tonnes per annum of metallurgical coal for domestic and international markets. Coal is extracted from various seams via open cut pits, is passed through a rotary breaker to size the coal and remove large rocks, and then transported to the processing plant by belt conveyors, one of which goes through the Raw Coal Tunnel shown in Figure 1. Teck is executing a program of multiple activities required to facilitate the future recovery of coal from the Baldy Ridge mining areas, including rehabilitation of the Raw Coal Conveyor Tunnel. The current mine plan proposed mining activities for recovery of the coal seams above the Raw Coal Conveyor Tunnel. The existing tunnel ground support is deemed not adequate to withstand production blasting close to the tunnel and therefore requires rehabilitation of the ground support system to permit future mining activities. This paper describes the geological conditions along the tunnel, geotechnical investigations undertaken as a part of the rehabilitation effort, a test blast study performed, and the approach and criteria used for the tunnel rehabilitation design.

Existing Tunnel Features

The tunnel was originally built in 1969, is approximately 1,450 m long, and has a nominal width and height of 4.8 m to 5 m. The existing ground support along the tunnel alternates between tunnel sections of varying length consisting of either spot and pattern bolted sections with thin shotcrete and wire mesh cover, or steel arches supported sections with timber cribbing and lagging with some covered with thick shotcrete in the Primary Fault zone. The tunnel declines at 8.541 degrees (~15%) from horizontal and has a relatively smooth concrete floor along its whole length. A longitudinal profile of the tunnel is shown in Figure 2, and typical sections showing the two main methods of existing ground support are shown in Figure 3. Besides the conveyor structure, the tunnel houses some critical mine infrastructure including raw water, tailings, natural gas, compressed air and fire water pipelines, as well as a communications bundle, local electrical distribution and a leaky feeder radio system, all of which need to be preserved during the tunnel rehabilitation process.

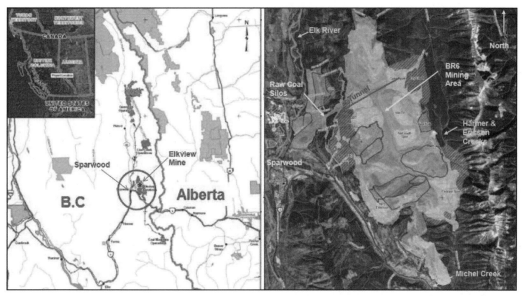

Figure 1. BRE project location

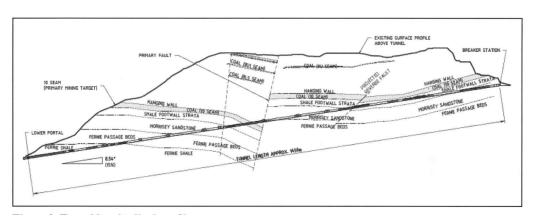

Figure 2. Tunnel longitudinal profile

SUBSURFACE CONDITIONS

Geologic Setting

The Elkview Mine is located within the Elk River Coalfields which extend northwest from the Canada-USA border. The region is underlain by Jurassic age Fernie Formation and the Jurassic-Cretaceous age Kootenay Group, which includes basal Morrissey Formation sandstone and the overlying, economically significant Mist Mountain Formation. The tunnel crosses Baldy Ridge in the east Kootenay mountain range. The rock lithology along the tunnel alignment is shown in Figure 2 and basically consists of (in sequence from the ground surface) (Poulton et al., 1994):

- Mist Mountain Formation of interbedded shale (mudstone), siltstone, sandstone, and coal. The hanging wall, 10 Seam, and footwall strata shown on the profile are part of this larger formation.
- Morrissey Formation sandstone.
- Fernie Passage Beds, the transitional zone between the Morrissey and Fernie Formations, consisting of thickly interbedded sandstone and shale.
- Upper Fernie Formation shale.

Structurally, the Elkview Mine is located within the Rocky Mountain foreland fold-and-thrust belt and is regionally situated on the eastern limb of the north trending, south plunging Sparwood syncline. Strata are cut by several north-trending, west dipping thrust faults as well as north-trending east and west-dipping normal faults. The Primary Fault zone and possible reverse fault have been observed and identified along the tunnel alignment as shown in Figure 2.

Geotechnical Investigations

Geotechnical investigations performed for the purpose of tunnel rehabilitation design consisted of:

- Two (2) surface geotechnical boreholes
- Ten (10) underground geotechnical boreholes
- In-Tunnel mapping of discontinuities for kinematic stability evaluation

In addition, geophysics logging and testing were performed in the boreholes. Plan locations of the boreholes along the tunnel are shown in Figure 4.

The surface drilling program was conducted using truck-mounted rotary diamond drill HQ triple tube coring with 3 m runs. Initially, the objective was to drill two boreholes from the surface down to

the elevation of the tunnel floor but offset to avoid intersecting it. Drilling of the first borehole was completed as planned to a depth of 110 m with piezometers installed to measure in-situ water pressures. The intent of the second borehole had been to intercept the Primary Fault zone but it was stopped at a depth of 123 m instead of the planned depth of 230 m. This was due to difficult ground conditions encountered, loss of drilling fluids and caving of the ground into it. Only limited geophysics testing was able to be performed (gamma only) due to the difficult ground conditions inside these surface boreholes, and no televiewer test could be completed as planned.

All ten (10) underground boreholes drilled from the tunnel were drilled vertically upward into the center of tunnel roof using an HQ diamond drill bit. The lengths of the holes ranged from 10.5 m to 78.5 m. A specially adapted compact and modular rig which was capable of drilling holes at various angles was required due to the tight space inside the tunnel. The typical arrangement of a drill site is shown in Figure 5. Six of the ten holes were closed with a cap

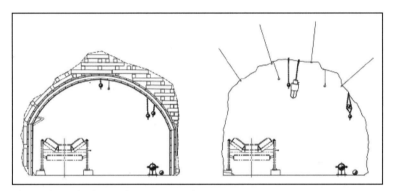

Figure 3. Typical tunnel support sections

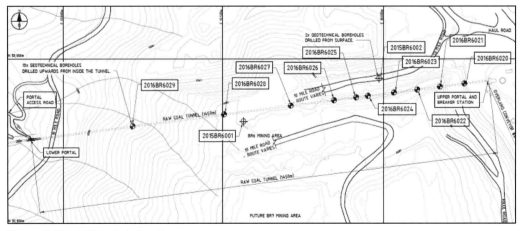

Figure 4. Plan of borehole locations

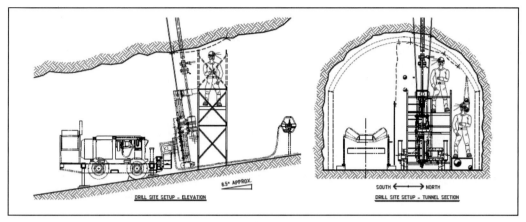

Figure 5. Typical arrangement of underground drill site

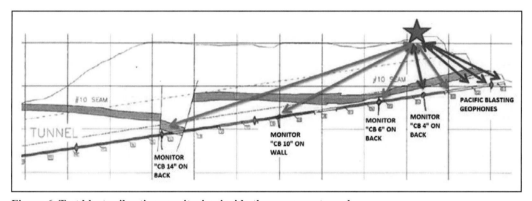

Figure 6. Test blasts vibration monitoring inside the conveyor tunnel

and valve system, allowing water pressures inside the hole to be measured and subsequently released if required. Four of the boreholes were fitted with MPBX instrumentation to allow measurement of relative displacement of the strata up to 20 m above the tunnel. Geophysical logging and testing were conducted on five holes to assess the lithology as well as jointing and bedding characteristics of the rock mass. Many of the boreholes drilled experienced significant loss of drilling fluids into the fractured ground.

Test Blast Study

To assess the future mine blasting impact on the tunnel, a test blast study was conducted for determination of the minimum safe distance from the tunnel for production blasting, and areas where limited blasting or no mining can take place around the tunnel. It comprised of nine signature holes including a waste hole, which were drilled, charged and fired similar to a typical mining sequence by the operation's blasting crews. The resultant blast vibrations were measured by nine seismographs—seven inside

the tunnel (four permanent and three portable) and two portable ones on the surface above the tunnel (see Figure 6). The test blasts were designed such that the tunnel did not experience a peak particle velocity (PPV) of more than ~50 mm/s, which is the currently established limiting criterion for the mine site. The test blast study created a refined and more robust regression formula for blast vibration prediction that would offer an additional level of protection for the tunnel during future production blasting. An important objective for the tunnel rehabilitation design was also to evaluate if the existing maximum vibration criterion can be increased to achieve more efficient production blasting during the future close-in mining. The test blast study results were used to support a recommendation to increase the maximum PPV limit from 50 mm/s to 100 mm/s.

Ground Characterization

Representative intact rock and rock mass properties used for the tunnel rehabilitation design are shown in Tables 1 and 2, respectively. They were estimated

Table 1. Intact rock and model design parameters

Rock Unit/Strata	Rock Mass Condition	Intact Rock (use HB/MC with GSI=100 in RocLab)					Rock Mass Index		Ubiquitous Joints				
		Density	E_i(Gpa)	v_i	σ_{ci} (MPa)	σ_{ti} (MPa)	GSI	D	$\varphi_j°$	c_j (MPa)	σ_{ti} (MPa)	v_j	G_j Gpa
Footwall	Poor	26	8	0.2	40	4	35	0	25	0.1	0.01	0.2	0.1
	Fair	26	12	0.2	60	6	45	0	32	0.3	0.03	0.2	0.3
Coal	Very Poor	12	2	0.2	9	0.9	15	0	Use HB / MC with RocLab				
Hanging Wall	Fair	26	12	0.2	60	6	45	0	Use HB / MC with RocLab				
Sandstone	Good	26	22	0.2	110	11	75	0	Use HB / MC with RocLab				
Fault Zone	Very Poor	22	5	0.2	25	2.5	15	8	Use HB / MC with RocLab				
Passage Beds	Fair	26	12	02	60	6	60	0	Use HB/MC with RocLab				
Fernie Shale	Poor	26	8	0.2	40	4	35	0	Use HB / MC with RocLab				

Table 2. Rock mass quality assessment from underground boreholes

Strata	RMR	Q	Description
Mist Mtn Hanging Wall	40–67	1.7–11.0	Average Fair
Mist Mtn Coal Seam	NA	NA	Not rated
Mist Mtn Footwall	25–59	2.1–5.6	Poor to Fair
Morrissey Sandstone	66–70	NA	Good
Fernie Passage Beds	59–62	2.0–12.7	Average Fair
Upper Fernie Shales	NA	NA	Not rated

and selected based on the field and laboratory testing of rock samples retrieved during the project geotechnical investigation, as well as from historic data. The rock mass conditions along the tunnel were rated and classified in accordance with the widely accepted rock mass classification indices, namely the RMR and Q, which were then correlated to estimate the Geological Strength Index (GSI) ratings typically used for numerical modelling purposes. During estimation of the rock ratings, the groundwater conditions were assumed to be dry since the groundwater impact was to be included in the numerical model itself. Furthermore, since the observed rock mass conditions in the underground boreholes would have already reflected any blasting damage during the original tunnel construction, a disturbance factor (D) of 0, for no to minimal disturbance, was assumed when deriving the rock mass parameters in the GSI system. Some of the geologic units listed in Table 2 could not be directly rated since either they were not intersected by the underground boreholes or sufficient borehole information was not available to make adequate rating estimations (e.g., highly broken coal samples which had very little recovery). In these cases, the design GSI ratings were estimated based on the geology.

TUNNEL SUPPORT REHABILITATION

There are two distinctive differences between rehabilitation of existing tunnels and the design of new tunnels:

- Unlike a new tunnel design where the support requirements are typically based on the change in loading conditions created by the tunnel excavation itself, the tunnel rehabilitation support requirements are to be based, in this case, on the future mining and backfill loads on the tunnel.
- Unlike a typical new tunnel where most of the in-situ ground stresses have been released prior to installation of the support system, this tunnel support system would have to be able to handle the full stress changes in the ground from future mining and backfilling, which could be very significant due to the maximum 200 m of overburden anticipated to be removed during future mining, in addition to the blasting vibration impact during mining, and subsequent backfill.

Based on the tunnel geology and proposed future mining sequence, the tunnel rehabilitation works are divided into three primary zones or blocks in the order of descending priorities as shown in Figure 7:

- *Block 1*—comprises the tunnel sections in the upstream end of the alignment. It is located where the tunnel is sufficiently close to the coal seam to be mined, that its behaviour will likely be strongly controlled by that seam. This zone is also closest to the future surface mining operations directly overhead with the greatest impact on the tunnel. This block was therefore considered to be the most critical, generally requiring the most extensive rehabilitation measures compared to Blocks 2 and 3.
- *Block 2*—comprises the tunnel sections in the middle of the alignment. It is located where the tunnel passes through a reverse fault

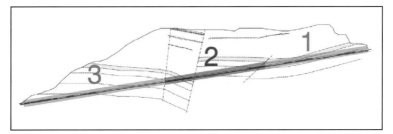

Figure 7. Tunnel rehabilitation blocks

identified during the geotechnical investigation to the western end of the Primary Fault zone identified both during construction in 1969 and subsequent geotechnical investigations. With the exception of the Primary Fault zone with poor ground conditions requiring major strengthening of tunnel supports, the remaining significant sections in this block have good quality Morrissey Formation sandstone in the tunnel roof, requiring minimal additional support.

- *Block 3*—comprises the tunnel sections in the downstream end of the alignment. It is located west of Block 2 and the Primary Fault zone. It is sufficiently far below the coal seam that the tunnel behaviour is not likely to be affected significantly by future mining. Over half of the tunnel length in this block is in the better quality Morrissey Sandstone and Passage Beds which would generally require minimal additional support. However the downstream half of this tunnel block is in the weaker Fernie Shale, and some of the existing tunnel support in this area has deteriorated such that support rehabilitation would still be required regardless of the future mining.

Design Criteria and Approach

Tunnel support design criteria are often based on maximum controlled displacement measured inside the tunnel during excavation to ensure proper support behavior. However in this case, relatively large deformations can be tolerated as long as the tunnel integrity can be maintained with no significant disruption to the conveyor operation. Therefore, the following tunnel rehabilitation design criteria were adopted:

- Greatly minimize the risk of tunnel collapse or damage that could result in unplanned shutdowns of the conveyor system and adjacent utilities within the tunnel, by preventing any failed rock or kinematically unstable rock blocks and wedges from falling into the tunnel.

- Minor damage to the tunnel including cracks and yielding in the tunnel lining system that can be readily repaired may be tolerated, but major spalls that could require detailed repair and maintenance efforts and/or interrupting service of the conveyor or tunnel utilities for extended periods of time will not be desired.

The tunnel rehabilitation design approach consisted of three steps to be described in the following sections: (i) semi-empirical support design assessment, (ii) finite element continuum modelling, and (iii) kinematic stability analyses of rock blocks or wedges.

Semi-Empirical Design Assessment

To check the basic tunnel support requirements for the site geology, closed-form analytical solutions were used to estimate the maximum support capacities for the two general types of systems currently being used in the tunnel, namely steel rib support for poor ground conditions and pattern bolting for the better (fair to good) rock conditions. The results are shown in Figure 8 together with the estimated loading requirements from empirical design charts for comparison purposes. The results show that the steel ribs would have an adequate capacity to support the anticipated ground load, provided that the steel ribs are continuously blocked/grouted against the surrounding ground. Furthermore, the results also indicate that the current pattern bolting would be sufficient to support the anticipated ground load in the better ground conditions along the tunnel.

Numerical Models

In order to estimate and evaluate impacts of future mining and backfilling on the tunnel support, a series of 2-D finite element numerical models using PLAXIS^{TM} software were performed during design for several different cross sections along the tunnel alignment, which represent the different tunnel rehabilitation zones, sections, and geology as described previously. Locations of these model sections are

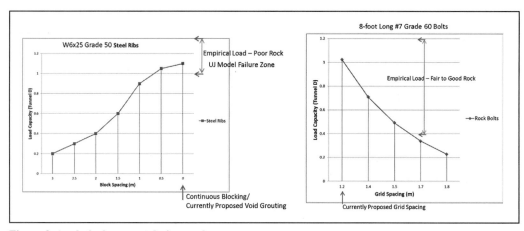

Figure 8. Analytical support design analyses

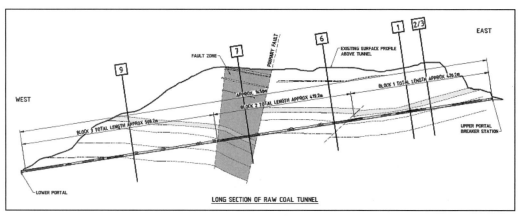

Figure 9. Location of model sections along the tunnel

shown in Figure 9 (indicated by the numbered boxes).

Each of the section models were run using the typical project construction sequence as follows (which is also illustrated in Figure 10):

- **Original Tunnel Construction**—Initial tunnel construction with assumed original ground surface.
- **Historic Mining**—Current condition after past mining/backfilling had taken place.
- **Future Mining**—Planned future mining of the coal seam around the tunnel.
- **Blasting**—Blast impacts during future mining, for both smaller blasts very close to the tunnel and for larger blasts further from the tunnel.
- **Future Backfill**—Planned future backfill after mining is completed.

To simulate the near-site and far-site blasting during future mining, a very short duration stress pulse was applied at different locations in the model (indicated by downward arrow in Figure 10), and the pulse magnitude and time duration were adjusted each time or re-calibrated to achieve the targeted maximum PPV and frequency monitored in the model around the tunnel perimeter. Figure 11 shows typical estimated vibration measurements at the tunnel in the form of time functions of the particle velocities for the different types of near-site and far-site blasting motions. Based on actual measurements during past mine blasts, the lower bound vibration frequency range of 1 Hz to 2 Hz was targeted for the far-site blast and the upper bound vibration frequency of 20 Hz was targeted for the near-site blast.

The results from each modelling sequence were evaluated for maximum support forces (i.e., axial, bending and shear forces for steel ribs and axial forces and skin friction for bolts) and the corresponding

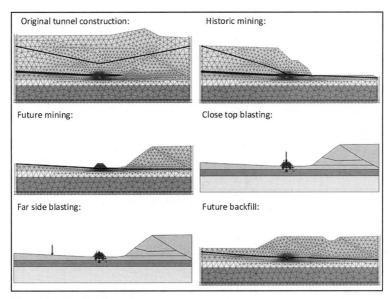

Figure 10. Section 1 model run sequence

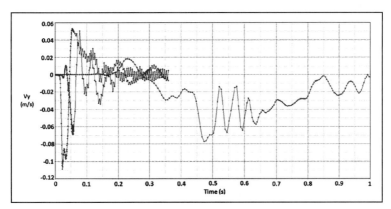

Figure 11. Typical model blast response

maximum stresses (compression and tension) in the support elements. The primary conclusions derived from the modelling efforts are as follows:

- A 5D (where D is the tunnel diameter) no-mining zone should be established around the tunnel to accommodate stress redistribution and provide rock mass stability during future mining and backfilling, especially for the most critical tunnel areas of Block 1 and Block 2 Primary Fault zone.
- The existing steel rib support would be able to support the tunnel for future mining purposes without failing. However some compressive yielding (during backfilling) and high tensile stresses (during mining) are anticipated

in the critical tunnel areas. Therefore due to variability in the steel ribs (i.e., size and shape) along the tunnel, and uncertainties in the steel strength and long-term deterioration (i.e., corrosion), a nominal thickness of shotcrete would be used to provide a margin of safety against undesired tunnel performance and additional steel reinforcements would be required where existing steel ribs show signs of significant deterioration.
- A thick layer of shotcrete would be required within the Block 2 Primary Fault zone to help reduce the expected high stresses in the existing steel ribs during future mining and backfilling. Furthermore, much larger tunnel displacements are anticipated within this

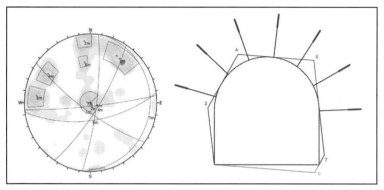

Figure 12. Footwall strata pole plot and typical kinematic model

zone, which may lead to potential cracks in the invert slab and distortions of the conveyor footings and utility brackets due to differential movement within the tunnel. The risks associated with the design have been assessed and those related to potential failure of the design in whole or in part are to be mitigated by comprehensive surveillance which in turn requires a monitoring plan.

- After tunnel rehabilitation, the currently used 50 mm/s maximum PPV vibration criterion during production mine blasting can be increased to 100 mm/s for relatively close-in blasting with frequencies above 10 Hz. However for more distant blasts with frequencies below 10 Hz, the maximum PPV criterion should be 75 mm/s to bring the steel rib stresses in the critical tunnel areas to acceptable levels.

- The pattern rock bolting currently used in some sections of the tunnel was shown to be adequate for the deformation-induced loading, provided that a minimum of 1.5D of fair quality rock cover is present directly above the tunnel. Blasting impacts on axial and frictional forces in the rock bolts were shown to be minimal, assuming the bolts are fully grouted or bonded to the ground.

Kinematic Stability

Kinematic stability analyses were performed over the entire tunnel length using the UNWEDGE™ software. These analyses evaluate the stability and support requirements for the tunnel where the loading is primarily controlled by the weight of kinematically unstable rock blocks or wedges formed by joints and/or discontinuities that are naturally present within the surrounding rock masses. Data for the joint sets existing along the tunnel were obtained from the combination of in-tunnel mapping and geophysical surveys conducted in the underground boreholes.

The collected joints data has been used to develop summary pole plots such as shown on Figure 12 for the critical strata above the tunnel, which is then used to analyse the kinematic stability of all possible joint combinations, one of which is shown in the figure. The study results concluded that pattern rock bolting would likely be necessary to arrest all kinematically possible rock blocks and wedges along the tunnel length. Furthermore using conservative assumptions of joint strength and bolt pullout capacity obtained from field and laboratory testing, and providing a minimum 30% bolt length of end anchorage zone is maintained, the currently used pattern bolting can provide in-tunnel stability with a minimum acceptable Factor of Safety (FS) of 1.6. However, the bolts will be fully grouted for redundancy design and corrosion protection purposes.

Rehabilitation Scheme

Based on the tunnel rehabilitation design approach and analyses described above, the recommended final rehabilitation work will involve three general types of new support, which vary depending on the existing support and location in the tunnel:

- For tunnel sections with existing steel (i.e., yielding) arches and timber cribbing, a new structural shotcrete arch will be formed on the inside face of the steel arches, and all voids behind the arch including timber cribbing will be completely backfilled with cellular grout/contact grout to establish continuous contact with surrounding ground. A nominal shotcrete thickness of 100 mm will be used for all of the sections, except for the Block 2 Primary Fault zone where a 200-mm thick shotcrete liner will be used. Additional lattice girder reinforcement will be added at select locations where the existing steel ribs showed significant deterioration.

- For sections with bolted roof support, new pattern rock bolts will be installed in areas currently supported by spot bolts. In other areas where pattern bolts have already been installed, previous non-destructive tests have indicated that some of these bolts may have significant amounts of voids due to improper previous installation. Therefore, a percentage of the existing pattern bolts will be supplemented with new roof bolts installed in the space between the existing bolts, depending on the suspected condition of the existing bolts. New steel wire mesh covering the full roof surface will be installed with the roof bolts.
- For sections currently with bolted roof support in critical tunnel areas, new lattice girders will be installed to form a new tunnel arch inside the existing profile and a structural shotcrete lining will be formed between the lattice girders, incorporating the girders into the arch. All voids behind the new arch will be filled with cellular grout. This will include bulk filling with grout three passing bays located within the critical zone in Block 1.

Constructability Considerations

A key objective of the project is constructability in order to ensure that the proposed design can be built using the anticipated construction methods while remaining within the operational, schedule, cost, and other constraints of the project. Two particular aspects of constructability were identified and pursued during the project study and design development: firstly, constructability of the proposed underground geotechnical drilling program, with the conveyor belt fully operational; and secondly, constructability of the tunnel rehabilitation design itself.

The initial constructability review for the underground drilling program laid a good foundation for the successful implementation of the program. In turn, the experience and lessons learned of successfully implementing the underground drilling program formed a useful basis for the constructability review for the tunnel rehabilitation design and methodology. A particular challenge during the reviews was the requirement to perform the work around, and without disturbing, the operating conveyor. This led to the need for proactive thinking ahead to prepare and stage materials and equipment for tasks requiring conveyor shutdown and to be able to quickly respond to opportunities presented by unplanned conveyor stoppages.

As a part of this review process, an independent contractor was engaged to provide input to the project team during the design development. The prime intent was to identify any fatal flaws or areas for refinement with the design team's approach, and no such fatal flaws were identified. The review also confirmed that the work could be constructed with the proposed approach. Many useful refinements were noted and incorporated into the final design.

A critical constructability issue identified was the difficult application of shotcrete to the sidewall on the conveyor side due to limited access around the conveyor. In these locations, cast-in-place concrete elements will be constructed in lieu of the shotcrete and formed up to a height above the conveyor that will allow the remainder of the shotcrete arch to be sprayed with access from a work platform. A trial form system for the cast-in-place concrete wall was built in the tunnel as a part of this review to test the suitability of its use inside the tunnel.

The project requires many temporary works in order to facilitate the construction of the main rehabilitation components. These include mobile platforms in the tunnel, materials handling, staging of materials, guarding of the conveyor, and the setup of surface facilities such as offices, mine dry (change house), warehousing, heated storage, and shotcrete and grout handling equipment. It also requires a means to capture and treat free-flowing groundwater that is generated from within the tunnel which will be impacted by contact with the active working zone during the application of shotcrete. A final complexity is that the work must be completed in keeping with Class 1 Division 1 Hazardous Location standard with the corresponding additional safety requirements, especially for electrical equipment.

CONCLUDING REMARKS

The current mine plan for the continued operation of Elkview Mine involves the extraction of coal reserves under Baldy Ridge. An existing tunnel crossing under the ridge coincides with the coal reserves. Therefore preservation of the tunnel and its continued use as the only conveyance for coal transport to the processing plant is critical for future mine development. This paper has presented the design approach, methodology, and modeling results for the rehabilitation of the ground support structures of the Raw Coal Tunnel in order for it to allow for such future mining to within an envelope around the tunnel. There are in fact two such envelopes: the first is driven by the tunnel support requirements to maintain tunnel stability during future mining and backfilling, which is defined as a no-mining zone radius of 5 times the tunnel diameter (D). The second envelope is a wider standoff distance for full-scale production blasting, one of which is driven by production blasting practices and the maximum PPV vibration criterion at the mine site resulting from this study.

A thorough geotechnical investigation program in the tunnel and a test blast study were successfully completed for design purposes. Realistic design criteria, supported by data from the various studies, were developed specifically for this project. This led to an optimum design that is neither too conservative nor too aggressive and addresses the site's specific risk profile. State-of-the-art blasting vibration impact models were performed to allow for maximum future mining production by optimizing the vibration limit currently imposed at the mine site.

A critical feature of the rehabilitation is that most of the work will need to be performed while the conveyor in the tunnel is operating, which is driven by the production targets of the processing plant during the rehabilitation period. For this reason, a comprehensive series constructability reviews were performed to evaluate whether the proposed tunnel rehabilitation program is feasible at the estimated production rates of construction and to consider various ways of accomplishing this work while minimising the disruption to the plant. Nevertheless, the work will require shutdown hours for the installation of wall formwork in the tight space next to the conveyor.

ACKNOWLEDGMENTS

The authors wish to acknowledge Teck Resources, as the Owner of the project described in this paper. This article represents the opinions and conclusions of the authors and not necessarily those of Teck. This article shall not be used as evidence of design intent, design parameters or other conclusions that are contrary to the expressed provisions in the contract documents for the Teck BRE Tunnel Rehabilitation Project.

REFERENCES

AECOM (2017), "*Final Feasibility Study Report,*" Report Number 60444413-10-0000-REP-0010-R2, March 22, 2017.

AECOM (2017), "*Final Design Report,*" Report Number 60546109-10-0000-REP-0005-R2, October 20, 2017.

Poulton, T.P., et al. (1994), "*Geological Atlas of the Western Canada Sedimentary Basin, Chapter 18: Jurassic and Lowermost Cretaceous,*" Canadian Society of Petroleum Geologists.

TRACK 2: DESIGN

Session 1: Resiliency

Dawn Dobson and Michael Brethel, Chairs

In-Situ Fire Test for the Design of Passive Protection for the Lafontaine Tunnel

Jean Habimana and Reza Showbary
Hatch

Jos Bienefelt
Efectis

ABSTRACT: The Ministère des Transports, de la Mobilité durable et de l'Électrification des Transports (Quebec Ministry of Transportation) is planning to undertake major safety upgrade and rehabilitation work on the Louis - H. La Fontaine Tunnel to prolong its lifespan and to comply with current codes, standards and best practices in fire life safety as well as emergency egress in case of a major fire event. The tunnel, which is approximately 1 mile long, is one of the main links between Montreal and the City of the Longueuil and is heavily trafficked by cars and trucks as it serves the nearby Port of Montreal. The paper focuses on in-situ fire tests that were carried out in the tunnel in summer 2016 and on how the results of the tests were incorporated into the requirement for the design of the passive protection to maintain its structural integrity during a major fire event in the tunnel.

PROJECT BACKGROUND

The Louis-Hippolyte-La Fontaine road tunnel (hereafter Lafontaine Tunnel) is an immersed tube tunnel which is located below the St Lawrence River and connects the City of Montréal with Île-Charron and Boucherville along Autoroute 25. It was constructed between 1963 and 1967. During the first phase of construction, the approaches on the north and south riverbanks were dredged and cast-in-place. The section of the tunnel between the approaches consists of seven precast tunnel elements, each approximately 110 m in length, 36.73 m in width, and 7.70 m high, for a total length of 768 m; installation of the elements began from the southern riverbank and progressed toward the northern riverbank, with Section 1 being the first and Section 7 being placed last. The tunnel has two traffic tubes that are separated by a service corridor. Each tube carries three lanes of traffic in unidirectional traffic flow, one tube for northbound and the other for southbound traffic, respectively. The tunnel is one of the five crossings of the main branch of the Saint Lawrence River in the Montreal urban area and is a major transport link to the nearby Port of Montreal. It is heavily used by heavy goods vehicle that is higher than typically experienced by tunnels as it serves the port. However, transport of dangerous goods cargoes is prohibited within the tunnel. Figure 1 shows an aerial view of the tunnel with its north and south towers and Figure 2 shows its profile.

The Ministry wants to undertake major rehabilitation program to do repair and upgrade work to meet current codes and standards and best practices to avoid any major intervention on the tunnel for the next 40 years. This rehabilitation work includes structural systems, ventilation, fire protection, water and drainage, mechanical, pavement structure, electrical and lighting systems, road safety and security for users and workers, communication, traffic signals, management and maintenance of traffic, mitigation measures during construction, as well as public transit systems within the tunnel during and after the rehabilitation work. Part of the project is to improve the resistance of the existing structure to a design fire event, which has a power of 50 megawatts and is based on the modified hydrocarbon curve as described below. The fire tests that were carried out in the tunnel, as described below, were part of the preliminary design study that was completed in winter 2017.

DESIGN FIRE EVENT

Over the last two decades there have been a number of serious fires in road tunnels that have cause extensive loss of life and severe collateral loss to the infrastructure that include structural damage and long-term financial effects to the local infrastructure. This has led to a series of updates in codes and regulations requirements, in particular the American National Fire Protection Association (NFPA) 502, which now includes strict requirements for fire life

Figure 1. Aerial view of the Lafontaine Tunnel

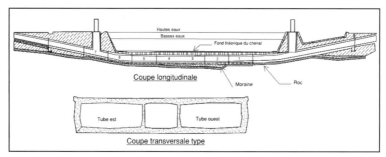

Figure 2. Profile and cross section of the Lafontaine Tunnel

safety infrastructure and protection measure to the structure of the tunnel. There are a number of other codes, standards and best practice recommendations in industry that were developed in Europe and Asia that were also used in this study.

One important aspect of fire life safety design in the tunnel is the design fire event, which characterizes the size of the fire and its duration. Design fire events are represented by time/temperature fire curves (also known as fire curves) that illustrate the increase of temperature in a tunnel over time during its several stages such as ignition, growth, spread and decay. Over recent years, several fire curves have been developed internationally. The magnitude and shape of these curves vary as they were developed under different conditions in laboratory and real-world conditions. The most known fires curves, as shown in Figure 3, are the Hydrocarbon Modified (HCM) curve that simulates petroleum combustion, such as from a truck fire, the Rijkswaterstaat (RWS) curve that was developed in Netherlands and

reconfirmed in fire tests in Norway is based on worst case scenario a 50 m³ oil or petroleum tanker that has a fire load of 300 MW, and the ISO 834 curve that is based on burning of general materials and contents found in buildings. The HCM curve was judged to be more applicable to the Lafontaine Tunnel and time of tenability was assumed to be 2 hours. This was judged to be an extreme scenario given transport of dangerous goods is forbidden within the tunnel. The size of the design fire is 50 MW.

IN-SITU FIRE TEST

Background on Spalling Process

When the concrete is subjected to a rapid increase of temperature such as what happens during a fire event, it can experience what is known as a spalling phenomenon, which can significantly reduce its strength and may lead to structural failure even a collapse of the tunnel. Spalling is mainly caused by two phenomena; rapid build-up of pore pressures

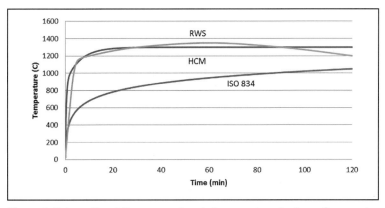

Figure 3. International fire curves for air temperature during a fire event

and thermal stresses. In the first mechanism, the rapid increase of temperature transforms the pore pressures into steam which by expanding causes a rapid increase of tensile stresses and breaks pieces of concrete near the fire surface. This phenomenon is also known as explosive spalling. In the second, a high temperature gradient is caused by a high heating rating leading to thermal stresses close to the surface. It is generally agreed in literature (Klingsch, 2014) that explosive spalling is considered the most violent form of concrete spalling that occurs at high temperatures and has significant influence on structural performance.

Spalling is a rather unpredictable phenomenon, depending upon several criteria such as concrete mixture and strength, permeability, porosity, moisture content, etc. It is theorized that high pore pressure may only serve as a trigger for spalling, rather than the main driving force. If the pore pressure were to open a crack, the available volume for the vapor would immediately increase significantly and pore pressure would quickly drop. This process leads to internal cracks parallel to the surface layer followed by surface buckling and finally spalling. Moisture trapped in the concrete has therefore a significant influence on spalling behavior. Relative humidity (RH) is the ratio of the partial pressure of water vapor to the equilibrium vapor pressure of water (saturation) at the same temperature. The RH of a concrete mass has been found to decreases with age. Beginning at 100 percent RH after casting, RH decreases with time as the concrete cures and water diffuses to the outside environment. The reduction in RH is influenced by the concrete's strength and depth from surface. For the same age, higher RH will occur in normal strength concrete than in high-strength.

The dry-wet cycle, such as from rain and car splashes, is an environmental condition that impacts the durability design of concrete structures. As concrete undergoes wetting, RH rises to 100 percent

in a short time. By contrast, as concrete undergoes drying, RH gradually drops. Zhang et al. (2012) determined that there is a zone near the surface that is influenced by external moisture. For normal and high-strength concrete, this is approximately 60 to 80 mm and 40 to 50 mm deep, respectively.

Spalling will occur more frequently:

- in concretes with lower water/cement ratio since permeability is reduced
- in water saturated concrete because of initial pore saturation
- at high heating rates that create steep thermal gradients and rapid build-up in pore pressures

Description of In-Situ Fire Tests

In order to assess the structural behavior of the in-situ concrete that takes into account its intrinsic properties as well as stress conditions, two in-situ tests were performed in the tunnel. The tests were conducted by Efectis of Netherlands using its Mobifire test apparatus to apply a heat exposure to the surface of the tunnel. The mobile furnace was applied to a surface area of 1 m × 1 m. There are typically two types of tests that can be conducted in the tunnel:

- Test 1 applies the design time-temperature curve for the whole duration of the design event, i.e., 120 minutes and record the results
- Test 2 applies a set of progressively increasing temperatures to the surface to assess the spalling performance of the structure. Thermocouples are installed in the structure to record the temperature experienced for each thermal exposure. As soon as spalling occurs, the test is terminated

It is the Test 2 that was performed in the Lafontaine Tunnel mostly because the Ministry did not want to risk significant damage to the existing structure.

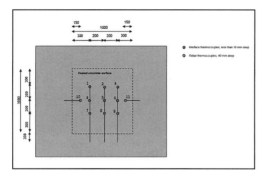

Figure 4. Location of thermocouples in the test area

Figure 5. In-situ fire test set up during the test

Figure 6. View of the inside of the furnace after the fire test

The test goal was twofold, in part to assess if the existing structure is capable to withstand a design fire event without spalling and to assess how long it takes for spalling to occur, if any. The intent of the test was therefore to assess the level of thermal stress the structure can absorb before spalling commences. By applying increasing levels of thermal stress to the structure until the onset of spalling is observed, the actual level of resistance to spalling

can be estimated. The results were used to determine the required level of fire protection that would need to be added to the structure to meet the requirements of NFPA 502-2014 Section 7.3 during the next phase of the project.

On June 18th 2016, two fire tests were performed in the Lafontaine Tunnel. The tests were performed on the upper portion of the interior tunnel wall of caisson 3, which was determined by the preliminary design team as representative of existing conditions in terms of stress conditions as well as concrete properties. To monitor the concrete temperature, twisted type K thermocouples were embedded in the concrete and covered in mortar. Thermocouples logged the furnace (i.e., air) temperature and the concrete temperature near surface (10 mm or less) and at 40 mm depths (see Figure 4).

Temperature loading was applied directly to the concrete without any protective fire-resistant measures using a telehandler, a propane fired mobile furnace. However, to limit the impact area, a high temperature special ceramic wool was applied to the surrounding concrete face.

The furnace had a single premixed burner with a maximum heat release rate of 500 kW. The heat release rate was manually controlled to follow the HCM curve. Figure 5 shows the fire test set up and Figure 6 shows the inside of the furnace after the test is completed, where it is clearly seen that concrete spalling occurred by removing small pieces of the concrete that detached like "chips." Closer observation showed that the detached pieces broke through aggregates and had a thickness of approximately 2 to 5 centimeters.

TEST RESULTS AND INTERPRETATION

Fire Test Results

The concrete temperature results for all the thermocouples are plotted in Figures 7 and 8 for the depth of 10 mm or less and 40 mm. Spalling was reported to begin after 1.5 min and 2 min, and the test was intentionally stopped after 3 min and 9 min for Panel Test 1 and 2 respectively. During Panel Test 1, an issue in the drainage of the furnace propane line led to a rapid temperature increase after the test had stopped at 7 min.

The rapid increase of the thermocouples in Panel Test 2 is interpreted as spalling of the concrete and grout cover. Temperature quickly increases to match the measured air temperature from approximately 4 to 9 minutes. For the near surface thermocouples, this was seen to occur at around 400°C (see Figure 9). This same behavior is not seen for Panel Test 1 as the mortar above the thermocouples did not spall. The variability among recording of thermocouples with depth of 10 mm is due to the fact that

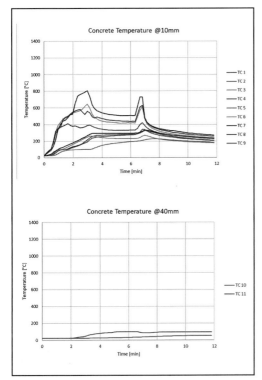

Figure 7. Concrete temperature during in-situ
fire tests for panel 1 at 10 and 40 mm depth

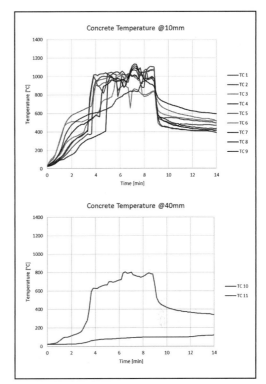

Figure 8. Concrete temperature during in-situ
fire tests for panel 2 at 10 and 40 mm depth

depths of thermocouples are only nominally 10 mm
and their actual depth is 10 mm or less.

Since furnace in Panel Test 2 performed closer
to the design intention, this panel was used to calibrate thermal properties of concrete.

The immediate conclusions from these tests,
which were used in the subsequent model calibration, are:

- Spalling is predicted to occur at temperature
  of around 400°C
- Temperature recorded from thermocouples
  with depth of 10 mm or less are expected to
  provide an upper bound response for temperature at depth of 10 mm under HCM fire.
- Temperature recorded from thermocouples
  with depth of 10 mm or less are expected to
  provide a lower bound response for concrete
  surface temperature under HCM fire.

Calibration of Numerical Modeling

A one-dimensional (1D) thermal model was first
created to calibrate the concrete properties using
the results from the in-situ fire tests. A simple one
dimensional 1000 × 1000 mm FLAC3D thermal

analysis was conducted without including mechanical behavior. A line of 10 mm thick elements was
generated to represent a nominal cross section of
the tunnel at the interior wall. Data from furnace air
temperature recorded by the thermocouples in the
in-situ test showed that furnace temperature closely
matched the theoretical HCM curve. Therefore, the
HCM was applied in the FLAC3D model. However,
a direct application of HCM curve to the concrete
surface will generate a far aggressive rise of temperature as compared to the in-situ test or actual fire
event, where heat convection will occur and delay
the increase of the temperature at the surface of the
concrete. This effect was considered using a convection boundary which requires the air temperature
(i.e., HCM) and the effective heat transfer coefficient
to determine the material surface temperature. The
effective heat transfer coefficient was assumed to be
the sum of the radiative and convective heat transfer
coefficients.

The parameters adjusted during calibration
were the effective heat transfer coefficient, the concrete specific heat and the concrete thermal conductivity. The grid-point temperature at surface, 10 mm
and 40 mm depths were monitored and compared
with the recorded thermocouples measurements.

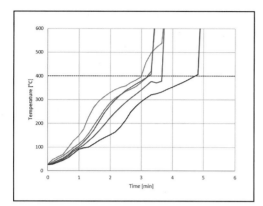

Figure 9. Concrete spalling at approximately 400°C for selected thermocouples

Using the thermocouple data from the in-situ fire test, the model was run using a trial-and-error process. Calibration was considered complete when the following criteria were met:

- The predicted surface temperature was greater than the recorded temperature in any thermocouples with a depth of 10 mm or less
- The predicted temperature at depth of 10 mm was less than or equal to the recorded temperature in any thermocouples with a depth of 10 mm or less
- The predicted temperature at depth of 40 mm was approximately equal to the temperature in thermocouples with a depth of 40 mm without any noticeable spalling or sudden rise in temperature
- Calibration is performed only for the time interval of data where spall did not occur
- Parameters have realistic values. This qualifier was significant as all three parameters directly impacted the result which can lead to infinite solutions with generally unrealistic parameters. The Eurocode 2 Part 1.2, was used as a guide during the calibration process

As shown on Figure 10, a reasonable fit was observed between 0 and 3.5 minutes, which corresponds to the period before spalling occurred for the average thermocouple, using the following parameters:

- Thermal Conductivity $= 2.275e^{-0.0010}$ [W/m °C]
- Specific Heat = Eurocode model for concrete with 3.0% moisture content
- Effective Heat Transfer Coefficient = 250 W/m^2 °C

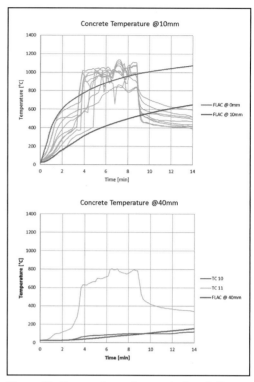

Figure 10. Comparison of numerical modeling and in-situ concrete profile at 10 and 40 mm depth

Prediction of Spalling in the Tunnel for the Design Fire Event

In order to predict the behavior of the existing structure of the Lafontaine Tunnel when subjected to a fire event, constitutive models that appropriately describe the behavior of concrete and steel reinforcement under fire needed to be employed. The complexity quickly increases when concrete is correctly recognized as a multiphase system where the voids of the solid concrete are partially filled with liquid and partly with a gas. To fully capture this behavior, it is required to include constitutive equations for mechanical, thermal and fluid behavior.

The mechanical and thermal models were derived from a combination of experimental results and values from the available codes and standards. One of the standard codes that provide the principles, requirements, and rules for the design of structures exposed to fire is Eurocode 2 Part 1-2. The code comprehensively prescribes mathematical models for the strength, deformation, thermal, and physical properties of concrete and steel reinforcement when subjected to fire. The mechanical properties of concrete, steel rebar and post-tension cables, that could not be

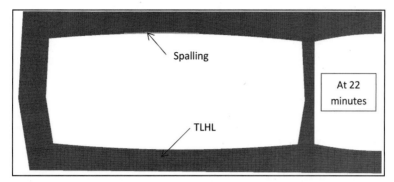

Figure 11. Prediction of the onset of spalling in the crown of the tunnel after 22 minutes

calibrated from the in-situ fire test experimental data, followed the mathematical models established by Eurocode 2 Part 1-2. These properties include stress-strain behavior in compression and tension as well as the coefficient of thermal expansion, all are variable as a function of temperature.

A spall criterion was developed based on the depth recommendations by Eurocode 2 and data obtained from in-situ fire test. First, spalling was observed to occur once the concrete reached approximately 400°C. Once spalled, the temperature of the thermocouples rapidly increased to match the furnace air temperature because the once embedded concrete thermocouple were now directly exposed to the air from the furnace as illustrated in the Figure 9. Second, spalling was observed to each a maximum depth of 40 mm, which corresponds to the depth of the reinforcement rebars. This aligns with general literature observations, that spalling typically does not exceed the concrete cover as deeper concrete is confined by reinforcement. In the model, the depth of spalling is limited to 50 mm which is the thickness of one element. The element exposed to heat spalled once the temperature of its centroid reaches 400°C.

A detailed FLAC3D model was next created to simulate all stages of construction sequence and fire. An elasto-plastic stress-strain constitutive model with the Mohr-Coulomb failure criterion was used to simulate the behavior of the ground.

In order to obtain a more accurate representation of the response of the concrete during a fire event, the concrete was modelled using brick elements; cable elements were used to explicitly represent the reinforcing rebars and post-tensioning tendons in the tunnel elements. FLAC3D strain-softening constitutive model was used to define the stress-strain response of the concrete, and linear elastic-plastic constitutive model was used to define the stress-strain response of the reinforcing rebars and post-tensioning tendons.

Fire was simulated by imposing surface temperature at the tunnel lining intrados in one tube or compartment of the tunnel. Temperature was applied to the crown and interior walls and excluded the invert. It was assumed the road surface and screed layer would insulate the invert during the fire event. The magnitude of the surface temperature at any given period of fire event varied corresponding to the temperature-time curve obtained from the design fire curve.

The numerical analysis predicted spalling will begin after approximately 22 minutes in the ceiling for areas that exposed to the full HCM fire curve, and by 25 minutes it will propagate to the walls. The depth of spalling reached approximately 50 mm. Unsurprisingly, the time estimate differs from the in-situ fire test record of 1.5 minutes. It should be noted that the maximum spall depth was not reached for the in-situ fire tests as they were stopped at the onset of the spalling; however, concrete spall pieces were reported to have a maximum thickness of 20 mm. At these depth of spalling, there is a potential to expose the reinforcement rebars and post-tensioned cables which could lead to severe localized damage of the existing structure. Therefore, passive fire protection such as a protection board is recommended. The model was finally used to determine the performance requirements for passive fire protection. It was predicted a protection board with a thickness of 30 mm, thermal conductivity of 0.3 W/m°C and specific heat of 900 J/kg°C protect the structure to meet the criteria of NFPA 502-2014 Section 7.3. A combination of board thickness and its thermal conductivity can be expressed as a single thermal resistance parameter:

$$\text{Thermal Resistance} = \frac{\text{Thickness}}{\text{Conductivity}}$$
$$= \frac{0.03}{0.3} = 0.1\,\text{m}^2\,°\text{C/W}$$

As thermal resistance and specific heat can vary with the temperature, the above values are considered minimum requirement over the range of possible temperatures which is up to 1300°C.

CONCLUSIONS

As part of a major rehabilitation project that the Transportation Ministry of Quebec is planning to undertake for the Lafontaine Tunnel in Montreal, in-situ fire tests were carried out in the tunnel during summer 2016. The goal of the tests was to characterize the resistance of the existing concrete in its actual conditions to a design fire event and predict when spalling will start to occur. The destructive test consisted of a mobile furnace that applied the design fire curve temperature directly to the concrete with several thermocouples installed at different location and depth of the area that was tested. Given that the Lafontaine Tunnel is heavily trafficked by a large number of trucks but transport of dangerous good is prohibited within the tunnel, the HCM curve was selected for the design fire event for a duration of two hours with a power of 50 MW. The fire tests applied the HCM curve directly to the concrete and spalling occurred within the first 1.5 to 3 minutes, respectively, when the temperature of the concrete reached approximately 400°C. The results of the tests were used to back calculate the properties of the in-situ concrete, i.e., an effective heat transfer coefficient, thermal conductivity and specific heat. The calibrated parameters were then used to predict the behavior of the tunnel structure in the case of fire event in the tunnel. Numerical model considered thermal and mechanical properties of concrete and steel varying with temperature and explicitly included steel rebar and post-tensioning cables. Numerical models predicted spalling will likely occur after 22 minutes of fire ignition, which is more realistic and compares better with observed times in previous fire incidents or full-size fire tests. Finally, this exercise led to determination of performance requirements that will be used during the detail design for the passive fire protection measures to be implemented later on to meet the applicable road tunnel fire life safety standard. Passive protection board will need to have a minimum thermal resistance of 0.1 m² °C/W, which represents a combination of board thickness and its thermal conductivity with a minimum specific heat of 900 J/kg°C. Finally, it should be noted that fire life safety measures will follow a holistic approach that encompasses a series of several measures to be implemented in case of a fire event within the tunnel, passive protection boards being one of them.

REFERENCES

Bazant, Z. P., 2005. "Concrete Creep at High Temperature and its Interaction with Fracture: Recent Progress." *Concreep-7 Conference: Creep, Shrinkage and Durability of Concrete and Concrete Structures.*

Dwaikat, M. and Kodur, V., 2009. "Response of Restrained Concrete Beams under Design Fire Exposure." *Journal of Structural Engineering*, Vol. 135, Issue 11.

European Committee for Standardization, 2004. "Eurocode 2: Design of Concrete Structures, Part 1-2: Structural Fire Design," EN 1992-1-2.

Jansson, R. 2008. "Material Properties Related to Fire Spalling of Concrete." *PhD Thesis*, Lund University.

Klingsch, E., 2014. "Explosive Spalling of Concrete in Fire." *PhD Thesis*, Swiss Federal Institute of Technology, Zurich.

Kodur, V.K.R, Wang, T.C., and Cheng, F.P., 2004. "Predicting the Fire Resistance Behaviour of High Strength Concrete Columns," *Cement and Concrete Composites*, Vol. 26, Issue 2.

Maraveas, C. and Vrakas, A.A., 2014. "Design of Concrete Tunnel Linings for Fire Safety," *Structural Engineering International*, Vol. 24, Issue 3.

Mindeguia, J., et al, 2011. "Influence of Water content on Gas Pore Pressure Concrete At High Temperature" 2nd International RILEM Workshop on Concrete Spalling due to Fire Exposure, RILEM Publications SARL.

Peter, A., et al, 2014. "Numerical study of heat and moisture transport through concrete at elevated temperatures." *Journal of Mechanical Science and Technology*, Vol. 28, Issue 5.

Phan, L. T., 2008. "Pore Pressure and Explosive Spalling in Concrete," *Materials and Structures*, Vol. 41, Issue 10.

Saito, H., 1965. "Explosive Spalling of Prestressed Concrete in Fire." *Building Research Inst.*

Shorter, G. and Harmathy, T.Z., 1965. "Moisture Clog Spalling." *Proceedings of Institution of Civil Engineers.*

Ulm, F., Acker, P., and Lévy, M., 1999. "The Chunnel Fire II: Analysis of Concrete Damage," *Journal of Engineering Mechanics*, Vol. 125, Issue 3.

Vitek, J.L., 2008. "Fire Resistance of Concrete Tunnel Linings," *Tailor Made Concrete Structures: New Solutions for Our Society*, Editor: Walraven, J.C. and Stoelhoerst, D., CRC Press.

Zhang, J., et al, 2012. "Interior Relative Humidity of Normal- and High-Strength Concrete at Early Age." *Journal of Materials in Civil Engineering*, Vol. 24, Issue 6.

Preliminary Assessment—Rehabilitation and Expansion of the Central City Tunnel System in Minneapolis, MN

Greg Sanders, Michael Gilbert, Mahmood Khwaja, and Bill Lueck
CDM Smith

ABSTRACT: The tunnels of the Central City Tunnel System in downtown Minneapolis, MN, were, primarily, excavated between 1935 and 1940. The City's growth and increased paved areas are channeling additional flows to the, now, undersized system, resulting in unintended system pressurization and extensive maintenance.

This paper focuses on the technical aspects of the preliminary design, and the design challenges faced to enhance and expand the tunnel system capacity. One of the key decisions addressed during the preliminary design is the choice between expanding the existing tunnel cross section or constructing new tunnels parallel to the existing storm main tunnels. Preliminary engineering effort focusing on an integrated hydraulic modelling and tunnel design/construction approach are presented. Project geology is also captured, emphasizing the unique features of St. Peter Sandstone, highlighting the challenges it creates for design and construction.

SYSTEM OVERVIEW

The City of Minneapolis (City) contracted with CDM Smith to update and provide a conceptual design for improvements to mitigate surcharge flooding in the Central City Stormwater Tunnel System. As a part of the project, CDM Smith conducted a field survey and condition assessment of the existing tunnel system. Information from the survey was then used to update the existing XPSWMM model and develop system-wide alternatives.

The CCSTS provides stormwater runoff drainage for nearly the entire area of the City downtown commercial district. The system consists of deep stormwater tunnels constructed in the St. Peter Sandstone, approximately 70-feet (21-m) below the street's surface. The primary tunnels comprising the Central City stormwater tunnel system are located below Hennepin Avenue, Nicollet Mall, Lesalle Avenue, Marquette Avenue South, 2nd Avenue South, South 5th Street, Washington Avenue South, Portland Avenue South, 2nd Street South, and Chicago Avenue South, as shown in Figure 1. This network of tunnels conveys the runoff from a 305-acre tributary area that is generally bound by Hennepin Avenue and 1st Avenue North to the east, 12th Street to the south, 4th Avenue South and 7th Avenue South to the west, and 2nd Street South. These tunnels were constructed between 1936 and 1940, except for the Marquette Avenue South tunnel, which was constructed between 1963 and 1964.

The CCSTS operates as a gravity flow system. These tunnels were constructed within the St. Peter Sandstone layer of bedrock and emerge from the bedrock at the Mississippi River below St. Anthony Falls. The Central City and the adjacent Chicago Avenue tunnel system converge into a single outfall at the Mississippi River. The runoff discharges from the converged outfall to a side channel of the Mississippi River, called a tailrace, located near the Guthrie Theater. Minneapolis Division of Surface Water and Sewers provided 32 historic plats detailing the plan and profile of the tunnel system.

The tunnel plats show nine different cross-section configurations. Eight configurations within the overall system generally show the same geometric "cathedral" shape with the inside dimensions varying from 4- to 6-feet (1.2- to 1.8-m) in width and 6- to 8-feet (1.8- to 2.4-meters) in height. For analysis these eight configurations were reduced to three configurations with regards to tunnel liner, cross-sectional area, and support. For simplicity, the three configurations are described according to the three types of tunnel support used during construction: (1) none required; (2) light timber; and (3) heavy timber.

The stormwater tunnels on Hennepin Avenue, LaSalle Avenue, and Marquette Avenue between 4th and 7th Street South, and Nicollet Avenue between 9th and 10th Street South all have sanitary sewers that either cross, or aligned with, the storm tunnels, but are located below the invert of the stormwater tunnels. These sanitary sewers are clay pipes encased in concrete and range from 12- to 24-in (304 to 610 mm) in diameter. The separation between the top of the sanitary tunnel and the Central City stormwater tunnel is minimal, ranging from immediately beneath the stormwater tunnel to 2.75 feet (840 mm).

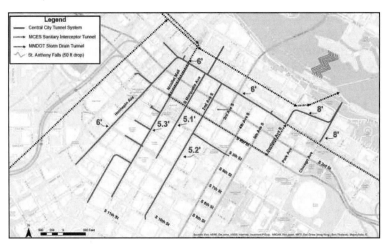

Figure 1. Central City tunnel system

The Central City stormwater tunnel, as it approaches the convergence structure, is a 7.5-foot (2.3 m) wide by 7.9-foot (2.4 m) tall cathedral shape tunnel constructed of block below the springline, and liner above. The Chicago Avenue tunnel, as it approaches the convergence structure, is an 8-foot diameter circular brick structure. The outfall structure, below the convergence of the Central City and Chicago tunnels, has unique cross-section configurations: a mushroom shape at the convergence that transitions to a cathedral shape for approximately 50-ft (15.2 m) immediately upstream of the outfall structure at the Mississippi River.

Pressurization of the Central City stormwater tunnel segments has been an ongoing issue for the City, leading to repeated and expensive maintenance repairs. The current deep tunnel stormwater system, was not designed for the characteristics of drainage inflow that consist of an increase in runoff volumes and shortened time-of-concentration caused by increasing impervious surfaces in the downtown commercial district. Pressurization of the tunnel during large, intensive rainfall events has caused the liner to crack, contributing to liner failure and erosion of the sandstone immediately outside of the tunnel liner at multiple locations. The maintenance and repair process, typically, requires identifying the void locations caused by erosion, filling the voids with grout, and repairing the cracks that led to the creation of these voids.

GEOLOGICAL SETTING

The general subsurface geological profile in the drainage system area is very consistent as shown on the available tunnel plat drawings. This general geological profile consists of:

- Overburden—sand, gravel, boulders and in some locations a thin layer of clay below the granular material.
- Weathered rock—described differently on profiles ranging from hardpan and boulders to broken limestone.
- Rock—predominately a 15 to 40 foot (4.6 to 12.2 m) thick stratum of limestone that serves as a cap rock to a very thin soapstone, overlying St. Peter Sandstone.

Groundwater was not identified on the plat sheets. However, during recent work on the Nicollet Mall project, which is within drainage area of this project area, CDM Smith drilled borings to a depth of 40 feet (12.2 m) without encountering any groundwater.

The CCSTS is located entirely within the St. Peter Sandstone. This rock is unique in that it is composed of a very uniform sand size grains that are 99 percent quartz. The rock strength is developed from compressive loads and it exhibits almost no cohesion. The sandstone becomes harder and denser with depth. The rock is also very friable. Turbulent water in contact with fresh surface of sandstone will cause a rapid disintegration of the rock.

The total unit weight of the rock as reported in existing data is 135±4 pcf (21.2 kg/m^3 ± 0.6 kg/m^3). Gradation, sieve, analyses of the sandstone indicate that approximately 90 percent of the sand grains are between to 140 and 60 sieve sizes. This is indicative of a fine sand. Porosity of the rock averages 0.28. Unconfined compressive strength testing of 11 samples ranged from 680 psi to 2,810 psi (4.7 MPa to 19.4 MPa) with an average strength of 1,570 psi (10.8 MPa). Published friction angles of the St. Peter Sandstone typically range from 54° to 65°.

HYDRAULIC MODELING

The purpose of the hydraulic modeling analysis was to determine the extent of improvements to the Washington Avenue leg of the Central City stormwater tunnel in preparation for a capital improvement project that the City has scheduled for construction starting in 2020. As part of this analysis, the XPSWMM model was used to determine the equivalent hydraulic diameter needed to provide additional hydraulic capacity for the CCSTS to prevent pressurization of tunnel during a design rainfall event.

To determine whether the observed pressure surcharge in each tunnel leg was a product of downstream constraints plus tailwater or an individual tunnel leg is being constrained within a segment of the tunnel, free discharge conditions were created at the points where the Hennepin Avenue, Nicollet Mall, Marquette Avenue, and 2nd Avenue tunnels discharge into the Washington Avenue tunnel. Each leg was analyzed using the 10-year, 100-year, and 500-year design storms to determine the level of service of each tunnel leg. discharge into the Washington Avenue tunnel. The following describes the hydraulic capacity of each of these tunnel segments when not influenced by the hydraulic grade line (HGL) of the Washington Avenue stormwater tunnel:

- Hennepin Avenue: Hennepin Avenue operated without surcharge for a 10-year design rainfall, had negligible surcharge during a 100-year design rainfall, and had 5 feet to 10 feet of surcharge for a 500-year design rainfall.
- Nicollet Mall: Nicollet Avenue, including contributing flows from the LaSalle Avenue tunnel, had negligible surcharge during a 10-year design rainfall, and 20 feet to

50 feet of surcharge during a 100-year design rainfall.
- Marquette Avenue: The Marquette Avenue tunnel conveyed the runoff from all rain events within the crown of the pipe, including a 500-year design rainfall.
- 2nd Avenue South: The 2nd Avenue South tunnel surcharged as much as 30 feet during a 2-year design rainfall and had significantly greater surcharge during the larger design rainfall events.

The tunnel segments were recombined to assess how the hydraulic conditions of Washington Avenue, in combination with the known deficiencies in hydraulic capacity of each tunnel leg, influenced the total flow. The most significant changes occurred in the Hennepin Avenue and Marquette Avenue legs of the system, changing from no surcharge or negligible surcharge to surcharge in all design rainfall events. However, the Chicago Avenue tunnel and the converged Central City/Chicago Avenue outfalls have sufficient capacity for all modeled rainfall events.

The hydraulic analysis indicates the need for hydraulic improvements to Washington Avenue, 2nd Avenue South, and Nicollet Mall. The Marquette Avenue and Hennepin Avenue tunnel legs will have improved hydraulic performance after improvement of the Washington Avenue tunnel segment, and therefore do not need further analysis. The proposed 2020 construction will focus on improvements to the Washington Avenue tunnel segment.

PRELIMINARY DESIGN ALTERNATIVES

Initial increased conveyance capacity alternatives were developed for the Nicollet Avenue, 2nd Avenue South, and Washington Avenue tunnel segments

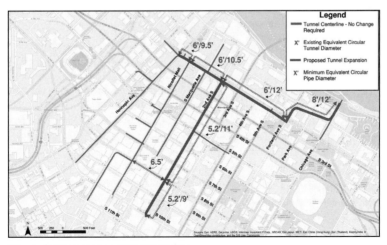

Figure 2. Proposed tunnel expansion

found to have insufficient hydraulic capacity to convey the runoff from a design rain event. The Existing Conditions XPSWMM model was used to compute the equivalent circular cross-sectional area for each hydraulic option. Cross-sectional areas for a 10-year and a 100-year rainfall events were developed to establish the incremental cost differences for mitigating system pressurization risks for the respective design rain events. Based on the XPSWMM model results of these two methods the following alternative were developed.

Expanded Tunnels

This alternative increases the size of the existing tunnel cross-sectional area. The minimum cross-section area of an equivalent circular tunnel was computed for both the 10-year rainfall event (4.27 inches of rainfall in 24 hours) and the 100-year rainfall event (7.47 inches of rainfall in 24 hours), as estimated by NOAA Atlas 14, Volume 8. The actual cross-sectional shape of an expanded tunnel will likely not be circular, given the characteristics of the engineering properties of the St. Peter Sandstone, available headspace between top of tunnel and top of St. Peter Sandstone, and conflicts with the Metropolitan Council Environmental Services (MCES) interceptor. An in-depth description of the shapes and cross-section area considered is presented in the construction alternative section.

Parallel Tunnels

This alternative involves construction of a new parallel tunnel adjacent to the existing tunnel. The minimum cross-sectional area of a circular parallel tunnel was computed for both the 10-year design rain event and the 100-year design rain event. A parallel tunnel could either be circular or it could be another shape if any of the constraints described in the Tunnel Expansion option are encountered. Tunnel sizes for the 10-year and the 100-year rainfall events are discussed in the construction consideration section of this paper.

Final Alternative

After the initial alternatives were developed for the entire system additional refinements were made to the model to specifically address the proposed 2020 project along the Washington Avenue Tunnel alignment. For this section a combined approach was proposed that would construct a new parallel tunnel east and west of Portland Avenue and expand portions of the existing tunnel alignment along Portland Avenue and at the junction with the Chicago Avenue tunnel system.

GEO-STRUTURAL ANALYSES

Concurrent with the hydraulic analysis, CDM Smith completed a geo-structural evaluation of the existing tunnel system with the goal of identifying and evaluating any risks associated with enlarging tunnel cross-sections to increase hydraulic capacity of the system and for repairing the existing tunnel segments.

Existing Tunnel Analyses

To analyze the existing tunnel system, Rocscience Phase 2 software program was utilized. After review of the existing tunnel plats, five existing tunnel configurations were identified for evaluation. The existing tunnel materials were modeled using Mohr-Coulomb failure criteria to account for the lack of reinforcing steel and an inability to resist tensile stress. Therefore, the model assumed that when a very low tensile stress was applied to the liner failure of the tunnel liner would occur. The analysis configurations consisted of the following:

- Profile Analysis of the excavation of the storm drainage tunnel at locations where it crosses directly above an existing sanitary sewer tunnel. This analysis was performed to evaluate the magnitude of change in stress on the existing underlying sewer tunnel due to excavation above it.
- Cross-Sectional Analysis: An analysis of different support types was performed using adjusted rock strength values depending on the existing liner support system, including no timber, light timber and heavy timber support behind the liner. The analysis consisted of applying a cyclical internal pressure to the tunnel representing loads experienced during a the 100-year rainfall event, as predicted by the XPSWMM existing conditions model, as performed by CDM Smith. The frequency of the cyclical loading was based on a review of 5 years of historic pressure data provided by the City. During this 5-year period, there were 6 events that surcharged the tunnel at the pressure meters. These surcharges ranged from 4 feet to 38 feet (1.2 m to 11.6 m) above the tunnel crown. For the modeling, we extrapolated this to 20 surcharge loadings, representing the occurrence of one surcharge event every 5 years for a period of 100 years. The applied internal pressure represented by the 100-year rain-event is predicted to be 35 psi, (0.24 MPa) or 80.7 feet (24.6 m) of water. This represents a factor of slightly greater than twice the measured event. Each

of the three different existing liner conditions and locations, were modeled as follows:

- No Lining Support: There are several locations shown on the City's tunnel plats where the tunnel liner is shown as concrete placed against the sandstone without initial support. Figure 3 represents a typical No Support segment. The average rock strength parameters were used, without strength reduction, since there is no initial liner support. It was assumed that the St. Peter Sandstone at these locations was in good condition, with few joints or loose materials and a strength reduction was not applied to the model.

- Light Timber Support: At locations where, light timber support was identified, the drainage tunnel is approximately 6-feet-high by 6-feet-wide (1.6 m × 1.6 m). A light wood support encompasses the upper portion of the tunnel from springline to crown and back to the springline in a trapezoidal configuration. It was assumed that the ribs were used in locations where the rock quality exhibited some joints or fractures, requiring some additional initial support. To account for this condition, the model used a reduced rock strength of the intact rock. It is assumed that the timber supports provides a seepage path for groundwater outside the tunnel and leakage through the tunnel to cause erosion of the sandstone. This results in a source of sand to migrate through cracks in the lining and creates an ongoing process of deterioration of lining support by creating progressively larger areas of unsupported lining.

- Heavy Timber Support: Heavy timber support locations consist of wood ribs that fully surround the tunnel perimeter with wood lagging. Heavy timber supports were used where the rock quality was significantly poorer than at other segments of the tunnel, requiring this stronger initial support. To account for this condition, the model used the reduced rock strength of the intact rock. The same process of loss of strength of the initial support system was used to model the behavior of the tunnel as a function of time and cyclical loads.

- Reduction in Liner Strength: The purpose of a reduction in strength model was to account for degradation of the underlying timber supporting the tunnel liner related to the environmental cycles of wet and dry conditions. As the wood both shrinks in volume and decreases in strength, deformation of the sandstone would follow with each cyclical loading due to a storm event. This loss

of external support originally provided by the sandstone, causes a tensile loading on the unreinforced segments of the concrete liner. The tensile loading results in liner cracks, which creates a pathway for seepage of groundwater from outside the tunnel liner during non-storm events, and leakage into the sandstone during a pressurized storm event to cause erosion of the sandstone. The resulting sand migration through liner cracks likely results in an ongoing process of cracking of liner by creating increasing areas of unsupported lining over time. To account for

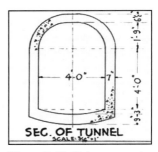

Figure 3. No lining support

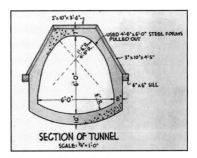

Figure 4. Light timber support

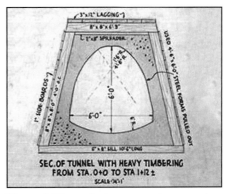

Figure 5. Heavy timber support

this long-term reduction in liner strength, it was assumed that the timber strength reduced by 5 percent between each cyclical loading event.

- Concrete Liner Loading: This model provided an assessment of the liner after each loading event was conducted. Providing there was continuous rock support against the liner, deformations were found to be minimal. However, where joints were formed due to shrinkage of the unreinforced concrete, the measured cracks were of sufficient size to allow passage of sand grains into the tunnel. This loss of ground was modeled by assuming a void behind the tunnel lining at each tunnel crack.
- Combined Effects: To evaluate locations where several factors may increase the loads an analysis was performed taking into account a combined effect of nearby sanitary sewers, the weakened condition of the tunnel lining, and disturbance to the rock.

Expansion of the Tunnel System

In addition to performing an analysis of the existing tunnel, CDM Smith performed a similar analysis on the proposed parallel and expanded tunnel configurations. During the analysis two constraints were identified for the proposed tunnel expansion.

A review of the relationship between the existing tunnel and the caprock above the tunnel, as drawn on the tunnel plats, showed that several tunnel segments are close to the caprock and have limited space available for vertical tunnel expansion without penetrating the caprock. According to the historical data, tunnels excavations, at elevations above the limestone caprock, are significantly more challenging to support and are double to triple the construction cost than if the excavations were below the caprock. Therefore, primarily horizontal tunnel expansion, with limited vertical expansion, was evaluated. To maintain a gravity system, lowering the invert for expansion was eliminated from consideration.

Additionally, there are several adjacent sanitary and storm drain tunnels that either share a wall or are very close to one another. Because of these adjacent tunnels, it was concluded that the storm tunnel cannot be lowered, or substantially re-aligned due to the conflicts created by these nearby, and crossing, sanitary tunnels. Therefore, the adjacent tunnel expansion analysis only evaluated the option to increase the cross-sectional area of selected tunnels along their existing alignment to increase the hydraulic capacity of the stormwater tunnel system.

For expansion of the existing tunnel two possible configurations were identified and evaluated. The configurations consisted of the following:

- **Excavation Within the Existing Tunnel.** The increase in the tunnel cross-section would be constructed using sequential excavation method to reduce excessive stresses on the lining left in place. CDM Smith assumed a sequential excavation on both sides of the existing tunnel would require a minimum width of 8 feet for equipment access. The excavation width would likely result in a flat roof that would not be stable as a function of the sandstone structure. However, the sandstone could be made stable with rock bolts fully anchored into the overlying limestone rock.
- **Excavation Adjacent to the Existing Tunnel.** To evaluate this condition a sequential excavation of tunnel adjacent to the existing tunnel to a width of 8 feet was assumed. Additionally, assuming a relatively flat excavation roof this excavation can be made stable with rock bolts anchored into the limestone.

RESULTS

System Rehabilitation

The Phase2 model results all indicate that the existing tunnel structures are stable where the tunnel liner is in contact with the St. Peter Sandstone. However, accumulations of sand can be an indication of eroded sandstone behind the liner, causing additional stress on the liner. It is assumed that the reason for these deposits is the combination of deterioration of the wood supports, and the natural behavior of the friable sandstone that causes the fine sand to erode as the groundwater moves along the outside of the tunnel liner. The basis for this assumption is that the tunnel liner consists of unreinforced concrete that ranges in thickness from 7 inches to 12 inches (178 mm to 305 mm), based on the details shown on the tunnel plats provided by the City. There is no indication of expansion joints being installed in any of the tunnels. Inspections of the tunnels indicated that spacing of vertical cracks and transverse cracks averaged about 57 feet (17.4 m) apart, measured along the tunnel axis. Considering that 90 percent of the sandstone grain size is fine sand and can pass through about 90 percent of the observed cracks, it is possible for sand grains to migrate and thus create void spaces that, as shown in the analyses, will self-perpetuate the deterioration of the tunnel lining as a function of surcharge loading. Based on the Phase2 analysis results an increase in cracking should be anticipated in locations where poor rock conditions or timber supports are present. This is supported by the locations where cracking was observed in the inspection data.

Some rehabilitation of the existing tunnel system is required in the form of repairing the cracks and filling any void that are present behind the lining to stop further long-term deterioration of the liner. Determination of the locations and approximate volume of voids behind the liner could be conducted by an extensive geophysical survey from inside the tunnels. The results of such a survey could then be used to quantify a program of repairs to the tunnel. This rehabilitation operation would mitigate the risk of failure for the existing tunnel system. However, it would not provide any increase in the hydraulic capacity of the tunnel. Therefore, the tunnel would still be subject to surcharge and street flooding.

System Expansion

The modeling performed indicates that where there is adequate sandstone cover over the tunnel such that the tunnel can be enlarged laterally with a minor amount of vertical expansion needed to create a stable shape. To maintain stability of the existing tunnel, which must remain in use during construction, external braces that support the tunnel liner would be required. Rock anchors and a new shotcrete liner would be required for tunnel support. Depending on the increased size of the tunnel, excavation can be performed either by hydraulic lance or using a roadheader. These excavation procedures will be discussed in greater detail later.

As the tunnels advance closer to the river and maintain their gravity slope the sandstone thickness above the tunnel crown increases and there is adequate sandstone cover to expand the tunnel upward and maintain a cathedral shape for stability purposes. This expansion should be limited in height to maintain about 2 feet of sandstone above the crown of the expanded tunnel cross-section. A first estimate of the cathedral shape can be calculated based on the friction angle of the sandstone. The height above the springline (vertical wall) of the tunnel is about the sum of half the existing tunnel width plus the proposed increase in the width divided by tangent of the friction angle divided by two.

$$\text{Height above springline} = \frac{\dfrac{W}{2} + \Delta W}{\tan\left(\dfrac{\phi}{2}\right)}$$

Based on the available historical data the friction angle for the St Peters Sandstone is about 60° to 62°. The advantage of a vertical expansion in the sandstone is that it eliminates the need for the rock anchors.

There are relatively short segments of the existing tunnels that are shown to have heavy timber support. Our interpretation of using this initial support system is that the rock is in poor condition relative the other sandstone encountered in the Central City stormwater tunnels. These areas may require some additional ground modification such as grouting or using a welded wire mesh to prevent fall out of rock during the excavation for the tunnel expansion.

PROPOSED TUNNEL CONSTRUCTION

As previously stated construction of the Washington Avenue portion of this project is anticipated to begin in 2020. The preliminary alignment consists of both a parallel tunnel and some portions of the alignment where the existing tunnel will be expanded to meet the hydraulic capacity requirements. The required hydraulic capacity of the parallel tunnels would range in size from the equivalent of a 6.5- to 12-foot (2 m to 3.6 m) internal diameter circular tunnel. The changes in the proposed size of the parallel tunnel, need to maintain flow in the existing tunnel and location of several cross passages greatly increase the complexity of the proposed construction. The following tunneling methods were considered for construction of a parallel tunnel:

- Hydraulic Lance. The original tunnel construction used hand-held lances that emit highly pressurized streams of water that cut through the sandstone. The benefits of the approach are the ability to excavate in small spaces and to create non-circular shapes. The disadvantages include slower pace of excavation and limited number of contractors having experience with the hydraulic lance. Hydraulic lances are advantageous as a secondary method used for areas, such as transition structures, which will have a unique shape that cannot be created by a boring machine.
- Tunnel Boring Machine (TBM). Advantages include large boring face and efficient boring speed. Disadvantages include need for large diameter access shaft, longer time to set-up and inability to maneuver machine through tight radius curves.
- Road Header Machine. Advantages include smaller area needed for equipment installation, and ability to maneuver into non-circular shapes and non-straight alignments. Disadvantages include slow rate of advancement and the need for more personnel in the tunnel.

The hydraulic lance method has not been used for several years in the Minneapolis area and finding labor and equipment using this method can be a limitation for this method. Generally, a TBM is a more economical method of tunnel excavation given the

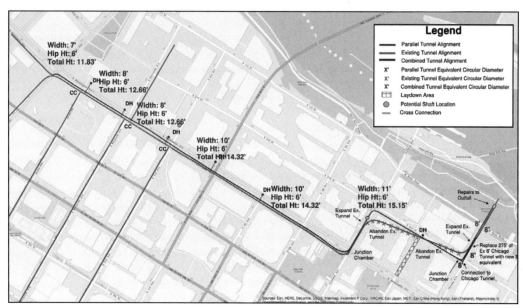

Figure 6. Proposed tunnel construction

proposed length of approximately 3500 ft. However, the alignment requires three 90 degree turns where new shafts would be required. The tunnel alignment also would be required to cross 6 lateral connection tunnels. An additional limitation for excavating the tunnel with a TBM is that the tunnel diameter is set by the machine. The required hydraulic capacity reduces to the west and therefore a TBM would perform unnecessary excavation.

The use of a roadheader also has limitations. These limitations are based on the size of the machine versus the excavation size. Roadheader power and ability to cut rock is a function of the machine size. To excavate a tunnel that is only about 8 feet in height and of the rock strength presented in the modeling report a small machine will be sufficient. A hydraulic roadheader is able to excavate the rock into any cross-sectional shape that has been determined to be the most stable, creating tunnels that are able to obtain the required equivalent hydraulic capacity as it changes along the alignment. The other advantages are: it can make very short radius turns that would eliminate the need for a shaft extending to the street. Use of the road header also allows for construction of a non-circular tunnel and is particularly apt for constructing a tunnel with a non-circular (Cathedral or other) shape that takes advantage of the properties of the St Peter Sandstone. The disadvantages to the roadheader is that the shape not circular, because of the high quartz content tool wear can be expected to lead to higher tool wear/replacement; and, advancement rate is less than that of a TBM.

CONCLUSION

Final design, Construction and rehabilitation of the Central City Tunnel system will face many challenges. The contractor will be required to maintain the existing flow in the tunnel. The surface is densely covered with utilities and high traffic streets that will make the location and construction of shafts difficult. The unique properties of the St. Peter Sandstone also present its own challenges. However, the expanded tunnel system will greatly improve the performance of the tunnel system and greatly reduce annual maintenance cost.

REFERENCES

Central City Tunnel System Hydrologic and Hydraulic Analysis Modeling Using XPSWMM, Central City, Eleventh Ave, and Chicago Ave Tunnel Systems, June 2015.

Central City Tunnel System Feasibility Study, Central City Tunnel System Pressure-Mitigation Options, June 2015.

Engineering aspect of the St. Peter Sandstone in the Minneapolis-St. Paul area of Minnesota, Charles M. Payne, University of Arizona, 1967.

GSI: A Geologically Friendly Tool for Rock Mass Strength Estimation, P. Marinos and E. Hoek. 2000.

National Oceanic and Atmospheric Agency (NOAA) Atlas 14, Volume 8, Version 2 for Minneapolis.

Recent Passive Fire Protection Strategies for US Highway Tunnels

Wern-Ping Nick Chen
Jacobs Engineering

ABSTRACT: Tunnel structure fire protection strategy for the US highway tunnel is inconsistent. The National Fire Protection Association 502 (NFPA-502) technical committee provided its first tunnel structure fire protection clause in 2004. Since then, it has made several updates for this clause. The trend of these updates is to replace prescriptive requirements with performance based requirements. This paper reviews these requirements in chronological order; and examines the use of the requirements in three US highway tunnels constructed in the last decade. From the literature review and the case studies, this paper explains why the tunnel structural fire protection strategy for the US highway tunnels is inconsistent, provides recommended practice, and offers suggested revisions for NFPA-502s consideration.

INTRODUCTION

Tunnel structure fire protection strategy for the US Highway tunnel is inconsistent. One of the reasons can be attributed to the lack of a standard for this practice. The National Fire Protection Association 502 (NFPA-502) technical committee provided its first tunnel structure fire protection clause in 2004. Since then, it has made several updates to this clause. The trend of these updates is to replace prescriptive requirements with performance based requirements.

Through the years, the tunnel industry has mostly adopted the NFPA-502 Standards as project design criteria. However, by observing the US highway tunnels constructed in the last decade, it is evident that a unified approach does not exist. This begs the question do we need a unified approach? The purpose of this paper is to investigate the reason why the US tunnel structure fire protection practice is inconsistent, to provide a recommended practice, and to offer suggestions for NFPA-502s consideration.

In the following sections, this paper provides literature reviews of NFPA-502 requirements in chronological order and examines three US design-build highway tunnels constructed in the last decade. From the literature review and the case studies, this paper explains why the tunnel structure fire protection strategy for the US highway tunnels is inconsistent, provides recommended practice, and offers suggested revisions for NFPA-502s consideration.

LITERATURE REVIEW

This section reviews the tunnel structure fire protection requirements of the following NFPA-502 editions: 2004, 2008, 2011, and 2017. Only relevant and critical clauses are described here. Please note that

NFPA-502 2014 edition was reviewed; however, it is not presented here, since the requirements in the 2014 edition are similar to that of 2017 edition. The Summary section will discuss the relevant revisions of the 2014 edition.

NFPA-502, 2004 Edition

The 2004 edition includes new requirements for the protection of concrete and steel tunnel structures as described below.

General Clause

NFPA-502 paragraph 7.3 Protection of Concrete and Steel Structures states "Regardless of tunnel length, all primary structural concrete and steel elements shall undergo a fire engineering analysis to ensure that the tunnel structure can withstand the anticipated fire severity based on the type of traffic to be permitted."

Fire Curve and Fire Resistance Rating

A.7.3 of this edition requires a 4-hour fire resistance rating in accordance with the time/temperature curve, such as ANSI/UL 1709. Fire resistance rating of 2 hours in accordance with ANSI/UL 1709 is permitted where the anticipated design fire size is 20 Megawatts (MW) or less and trucks carrying flammable liquid are prohibited in the tunnel.

Please note that the ANSI/UL 1709 fire curve is the Hydrocarbon Curve, shown in Figure 1.

Structure Temperature Limits

For tunnel lining with cast-in-place concrete, the temperature on the concrete surface should not

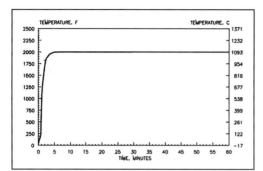

Figure 1. ANSI/UL 1709 Fire time/temperature curve (UL 1709)

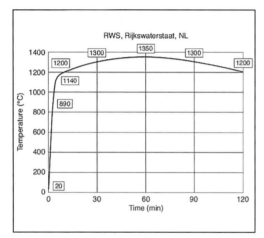

Figure 2. RWS time-temperature curve (NFPA-502, 2008)

exceed 380°C (716°F) after the defined fire exposure conditions.

The temperature of the steel reinforcement in the concrete, assuming a minimum 1 in. cover, should not exceed 250°C (482°F) under the same fire exposure conditions.

Concrete Spalling

Tunnels with precast (high strength) concrete elements should be protected such that *explosive* spalling does not occur under the defined fire exposure conditions.

The following are not defined in this edition: spalling of the cast-in-place concrete, the definition of the "high strength" concrete, and the definition of the "explosive spalling."

Design Fire

A design fire is defined as a fire's Heat Release Rate (HRR), in megawatts (MW), designated

in conjunction with the AHJ (Authority Having Jurisdiction). The selection of the design fire size shall consider the types of vehicles that are expected to be used in the tunnel. The peak HRR of typical vehicle is 100 MW from tankers.

The fire growth rate, rate of change of the fire's HRR, was undefined.

NFPA-502, 2008 Edition

The 2008 edition adds specific requirements for fire tests for tunnel structure elements and fire protection material. The fire time-temperature curve is also revised.

General Clause

Paragraph 7.3.1 requires that all primary structural concrete and steel elements shall be protected (1) to maintain life safety and provide a tenable environment, (2) to mitigate structural damage and prevent progressive structural collapse, and (3) to minimize economic impact.

Fire Curve and Fire Resistance Rating

Paragraph 7.3.2 requires the structure be capable of withstanding the Rijkswaterstaat (RWS) time temperature curve or another curve that is acceptable to the AHJ. Fire resistance rating is based on a 2-hour structural exposure condition under RWS curve.

Structural Element Temperature Limits

Paragraph 7.3.3 requires that after a 2-hour fire exposure, tunnel lining with cast-in-place concrete shall be protected such that:

- The temperature of the concrete surface does not exceed 380°C (716°F).
- The temperature of the steel reinforcement in the concrete, assuming a minimum 1 in. cover, should not exceed 250°C (482°F).

Concrete Spalling

Tunnels with *precast*, high-strength, concrete elements shall be protected such that *explosive spalling* is prevented.

The followings are not defined in this edition: spalling of the cast-in-place concrete, the definition of the "high strength" concrete, and the definition of the "explosive spalling."

Design Fire

The selection of the design fire size shall consider the types of vehicles that are expected to be used in the tunnel. The RWS requirements are adopted as a design fire curve that is representative of typical tunnel fires. Representative fire HRRs that correspond to the various vehicle types are provided. The peak HRR of typical tankers is revised from 100 to 300 MW.

The fire growth rate data are provided based on test fires.

NFPA-502, 2011 Edition

The 2011 edition further develops performance-based design approaches for tunnels. Table 7.2 is updated to provide a more comprehensive review of the required systems for tunnels based on tunnel category.

General Clause

Its general clause is similar to that of 2008 edition.

Fire Curve and Fire Resistance Rating

The fire curve is still based on the RWS curve, while the fire resistance rating of tunnel *concrete* structural elements during a 2-hour of fire exposure shall be designed or protected such that *explosive spalling* is prevented.

Structural Element Temperature Limits

Paragraph 7.3.4 requires that structural fire *protection material*, where provided, shall satisfy the following failure criteria:

Tunnels with *cast-in-place* concrete structural elements shall be protected such that:

- The temperature of the concrete surface does not exceed 380°C (716°F).
- The temperature of the steel reinforcement in the concrete, assuming a minimum 1 in. cover, should not exceed 250°C (482°F).

Concrete Spalling

Tunnels *concrete* structural elements shall be protected such that *explosive spalling* is prevented. Paragraph A.7.3.3.(1) further explains/defines the "explosive spalling" phenomenon.

Design Fire

The design fire is similar to that of the 2008 edition, with the following key commentaries: "Each engineering objective should have an appropriate design fire curve adapted to take into account project-specific factors directly relating to the engineering objective to be achieved, and the design fire is not necessarily the worst fire that may occur. Engineering judgment should be used to establish the probability of occurrence and the ability to achieve practical solutions.

Therefore, different design scenarios are often used for various safety systems."

No revision is made to the fire growth rate clause.

NFPA-502, 2017 Edition

This edition revises the list of considerations to be taken into account during an engineering analysis and adds guidance in the annexes. The tunnel structure fire protection clause is generally un-revised.

General Clause

Paragraph 7.3.1 is revised such that the tunnel design shall prevent progressive collapse of primary structural elements in addition to life safety to achieve the following functional requirements: (1) Support fire fighter accessibility, (2) Minimize economic impact, and (3) Mitigate structural damage. Though the wording of this clause is revised, its principal remains the same as that of previous editions.

Fire Curve and Fire Resistance Rating

The fire curve is still the RWS curve. A new addition is that an engineering analysis is required as described in Chapter 4.

"During a 120-minute period of fire exposure, the following failure criteria shall be satisfied: (1) Regardless of the material the primary structural element is made of, irreversible damage and deformation *leading to progressive structural collapse shall be prevented*. (2) Tunnels with concrete structural elements shall be designed such that fire-induced spalling, which *leads to progressive structural collapse*, is prevented."

In the Annex, further analyses to preventing progressive structural collapse and mitigation of structural damage are discussed. It requires analyses to include the following effects of the fire on the primary structural elements: (1) Loss of strength causing failure, (2) Loss of stiffness causing plastic deformations, (3) Loss of concrete durability due to cracking, which could lead to structural collapse (taking into account that some cracking, both during and after a fire, can occur at the non-visible external perimeter of the structure that cannot be detected or repaired), and (4) Specifically for concrete: fire-induced spalling, which could lead to structural collapse.

Structural Element Temperature Limits

No revision is made to structural element temperature limits.

Concrete Spalling

Paragraph A.7.3.4 adds "The concrete is protected such that fire-induced spalling is prevented."

Explanation of the "spalling" is provided in the Annex.

Design Fire

The design fire and the fire growth rate in this edition are similar to that of the 2011 edition. The term "tanker" that generates a HRR form 200 to 300 MW is revised to "Flammable/combustible liquid tanker."

CASE STUDIES

This section provides three case studies for the US highway tunnels constructed in the last decade, including the Port of Miami Tunnel (POMT), the Ohio River Bridge East End Crossing Tunnel (ORBEE Tunnel), and the Parallel Thimble Shoal Tunnel (CBBT Tunnel).

The Port of Miami Tunnel

The POMT is the first US underground tunnel project constructed by a Public Private Partnership (PPP) contract. The project is located in Miami, FL, connecting two man-made islands, the Watson Island

on the North and the Dodge Island (where the Port of Miami is located) on the South, as shown in Figure 3. The tunnel is a twin-bored tunnel, approximately 3,900-ft long for each bore, under Biscayne Bay with 2-ft thick precast concrete segmental lining and traditional rebar reinforcement; its inside and outside diameters are 37-ft and 41-ft, respectively. Figure 4 shows its schematic tunnel cross section. The owner of this facility is Florida Department of Transportation (FDOT); the Design-Builder is Bouygues Construction Work Florida; and the Engineer of Record is Jacobs Engineering. The tunnel was open to traffic on August, 2014.

Design Criteria

In accordance with the Contract Agreement (CA), tunnels and all "critical anchorages" shall be designed to resist failure due to fire in accordance with NFPA-502 2008 edition and ITA (2004) Guidelines for Structural Fire Resistance for Road Tunnels. Tunnel type is classified as Category 2 (ITA) road tunnels in soft ground, and the entire tunnel structure, inclusive of cross passages and emergency exit stairs, shall

Figure 3. POMT schematic project site plan

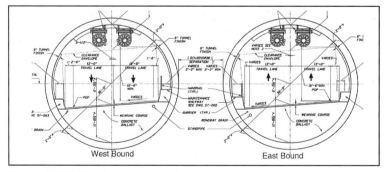

Figure 4. POMT schematic tunnel cross section

be designed to have a fire resistance level N3 (ITA) when subjected to temperature rise as defined by the RWS curve for a period of 120 minutes. Please note the tunnel type category and fire resistance level are defined based on ITA classifications.

The tunnel design fire size is 100 MW, selected due to the high percentage of buses and heavy goods trucks and the potential of a multiple-vehicle fire.

Implementation

The concrete mix of the precast segments consists of 50% Portland cement, 10% fly ash, calcareous aggregate, 40% slag, and has a water cement ratio, w/c, of 0.32. Its 28-day compressive strength, f_c', is 8,000 psi. With this low w/c ratio and low permeability of the mix, explosive spalling during a fire event is likely.

A fire test of the tunnel segments was conducted by Efectis, using the MobiFire mobile furnace (van der Waart van Gulik, 2016) on site. The fire exposed area on a half segment was approximately 10.76 ft² (1 m²). This area is smaller than in traditional tests, where the full internal surface of a segment is exposed to fire in a laboratory furnace; however, it was consider sufficient for this fire spalling test purpose as spalling is not allowed.

In order to obtain a targeted interface temperature between the passive fire protection material and the concrete surface, the test specimens were exposed to different generic time-temperature curves. The generic interface temperature curves represent shapes that may be expected when testing different thicknesses of typical fire protection material. The curves are not specific for one type of material but are based on Efectis' experience with different tunnel fire protection systems. A typical

interface temperature curve includes the following characteristics: a moisture plateau of constant temperature around 100°C, and a temperature increase after the moisture plateau depending upon the properties of the fire protection material. Figure 5 shows typical interface temperature curves of common fire protection materials experiencing RWS fire curve.

Seven fire tests were carried out iteratively, with varying fire growth rate to investigate the potential concrete spalling behavior and to provide decision making for fire protection material selection. Targeted interface temperatures vary from 250 °C (Test 1), 380 °C (Test 2), 380 °C (Test 3), 800 °C (Test 4), 450 °C (Test 5), 420 °C (Test 6), and 400 °C (Test 7). Figure 6 shows the results from these tests.

Based on the fire test results, it was concluded that the fire protection material shall satisfy the following interface time temperature criteria: 246 °C at 60 min., 314 °C at 90 min., and 380 °C at 120 min, corresponding to the results from Test 2 and Test 3. Promat®-T facetted board system of 21 mm (0.83 inches) was selected as the fire protection material, as shown on Figure 7.

The Ohio River Bridges East End Crossing Tunnel

The ORBEE Tunnel is part of a PPP project, which connects the transportation network between Louisville, Kentucky and southern Indiana. It is located about nine miles northeast to downtown Louisville, as shown on Figure 8. The twin-tube highway tunnel provides northbound and southbound lanes of the Kentucky approach to a new bridge across the Ohio River. The tunnel was constructed by drill and blast method with a 16-inch thick cast-in-place concrete final lining. Tunnel length is about 1,700-ft for each bound, with roadway width of 40-ft, for 3-lane traffic, and a vertical clearance of 26.5-ft. Figure 9 shows its typical tunnel cross section

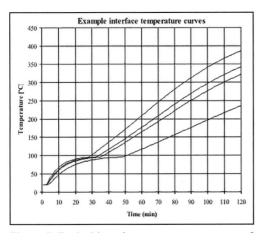

Figure 5. Typical interface temperature curves of common fire protection materials experiencing RWS fire curve (van der Waart van Gulik)

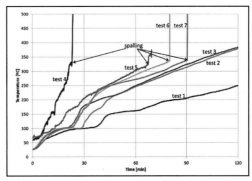

Figure 6. The average recorded interface temperatures during POMT fire tests (van der Waart van Gulik)

Figure 8. ORBEE Tunnel site plan, shown on green dotted line

Figure 7. POMT fire protection board

at cross passage. Indiana Financial Authority (IFA) oversaw the project during the design-build phase, and Kentucky Transpiration Cabinet will maintain the tunnel after construction is completed. Jacobs Engineering is the Engineer of Record. The tunnel was open to traffic in December, 2016.

Design Criteria

The project Technical Provisions (TP) require the fire size to be determined by an assessment of the expected type of vehicles in the tunnel (vehicles with dangerous goods are not allowed), and should not be less than 300 MW (based on tanker trucks in the tunnel, as noted in NFPA-502 2011, Annex A). The TP also allow the consideration of using a fire suppression system as a means of reducing the fire size. The Fire Life Safety Report (Jacobs Engineering, 2013-1) documented the technical design approach to protect the environment and the structure by reducing the fire size and cooling the environment using a water-based fixed fire suppression system and enhanced tunnel drainage system to capture flammable liquid spills and to reduce the fuel pool size.

Full scale fire tests with the proposed fire detection and fixed fire suppression system were performed in Seattle and witnessed by Harrods Creek Fire Department, the AHJ of this project. The tests demonstrated the fast fire detection and suppressing

system is able to constrain a 300 MW flammable liquid fire growth to 50 MW in 2.5 minutes after the fire ignition. Figure 10 shows the design fire growth curve, which corresponds to a growth rate of 20 MW/min. This growth rate is much faster than an ultrafast growth rate fire.

Implementation

The concrete mix of the cast-in-place concrete lining consists of 80% Portland cement, 20% fly ash, and has a water cement ratio, w/c, of 0.41. Its f_c' is 4,400 psi. With the normal w/c ratio and the low strength concrete mix, explosive spalling during a fire event is unlikely.

Computational fluid dynamics (CFD) modeling was conducted (Jacobs Engineering, 2013-2) to simulate the design fire event of this project, which includes six (6) pairs of jet fans in the northbound (downward grade) tube and six (6) in the southbound (upward grade) tube. The tunnel grade is about 0.6% to 4%, in a downward direction from south to north.

From the CFD model results, only three fire events indicate the concrete lining surface will exceed the NFPA-502 temperature limit of 716 °F (380 °C). Figure 11 shows the worst fire time temperature curve from the CFD results. Though it is expected that concrete explosive spalling is unlikely to occur, IFA requested laboratory fire test of the cast-in-place concrete lining with the fire time temperature curves derived from the CFD analyses.

A specific concrete explosive spalling test procedure was developed (Jacobs Engineering, 2014). Fire test was performed on April 2, 2015, using "high humidity" specimens, in CTL Group's laboratory in Chicago. The nominal test specimen geometry is 10-inches thick by 60-inches wide and 60-inches long. The width and length each was extended minimum 12-inches to provide a "furnace overlap."

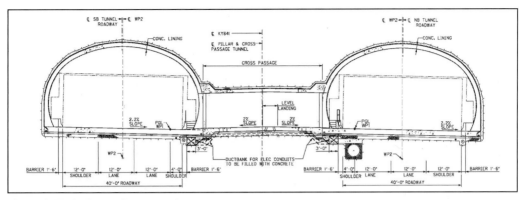

Figure 9. Typical tunnel cross section at cross passage

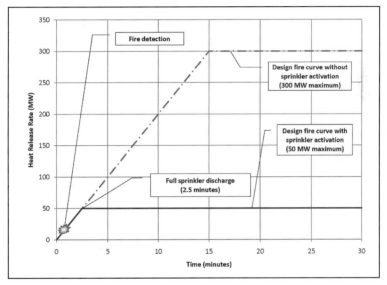

Figure 10. Design fire growth curve

Explosive spalling is defined as spalling involving the projection of chips or pieces of concrete from the concrete mass at a high velocity. Explosive spalling does not include chips or pieces of concrete that drop out from concrete mass under gravity alone. If explosive spalling occurs during the fire test using the proposed fire time temperature curves, the concrete mix design is considered inadequate.

During the tests, no explosive spalling was heard, and no explosive spalling was visually observed or recorded by the video cameras (Chen, 2016). Figure 12 shows a typical post fire test specimen, with no sign of concrete spalling.

Based on the fire test results and in conjunction with the fire suppression system, the tunnel lining is proved adequate for the design fire without the addition of a passive fire protection system.

The Parallel Thimble Shoal Tunnel

The CBBT Tunnel entails the design and construction of a new tunnel crossing, located parallel to the existing tunnel crossing, of the Chesapeake Bay Thimble Shoal Channel between Portal Islands Nos. 1 and 2, through a Design-Build contract. Figure 13 shows a schematic tunnel alignment plan connecting these two islands. The tunnel is a single-bored tunnel, approximately one (1) mile long, with 1.5-ft thick precast steel fiber reinforced concrete segmental lining; its inside and outside diameters are 39-ft and 42-ft, respectively. Figure 14 shows the schematic tunnel cross section. The project Owner is the Chesapeake Bay Bridge and Tunnel District (the "District") and the District's Design Manager is Jacobs Engineering. The Design-Builder is Dragados.

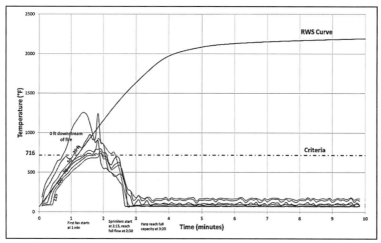

Figure 11. The worst fire temperature curve from the CFD results; the "Criteria" is the NFPA-502 concrete surface temperature limit (1st fan starts at 1 min., sprinklers start at 2:15 min. with full flow at 2:30 min, and fans reach full capacity at 3:20 min.)

Figure 12. Typical post fire test specimen

Figure 13. Proposed CBBT Tunnel schematic (adapted from http://www.cbbt.com/project -description)

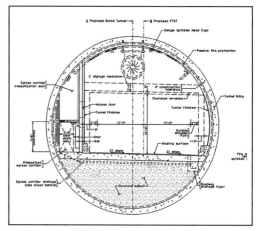

Figure 14. CBBT Tunnel schematic tunnel cross section

Design Criteria

The project Technical Requirement (TR) states the tunnel structure above the roadway traffic barriers, including all critical anchorages, be designed to resist structural failure when subjected to a design fire with peak HRR rate of 150 MW developed in fifteen (15) minutes, without the consideration to any mechanical fire suppressing systems. Passive fire protection board is required and shall extend over the entire exposed traffic lane perimeter above the roadway tunnel traffic barriers, and shall be designed so that the concrete temperature at the interface between concrete and fire protection board is below 250°C and 380°C for concrete with f_c' equal to or greater than 6,000 psi and less than 6,000 psi, respectively. The fire protection board and its anchors shall meet the performance requirements of NFPA-502 2014.

Implementation

At the time of writing this paper, the actual concrete mix of the tunnel lining is not yet finalized; however, the following parameters are preliminary determined: steel fiber will be used; concrete mix will contain pozzolanic material, such as fly ash or slag; water cement ratio is around 0.35; its f_c' is 7,500 psi. The steel fiber-reinforced concrete equivalent flexural strength will be based on the ASTM C1609 test: 700 psi at L/600 beam deflection; and 500 psi at L/150 beam deflection; where the "L" is the length of the test beam specimen.

Based on this preliminary concrete mix and strength information, it is expected that fire protection board will be used because of the fire load and the interface temperature requirements.

SUMMARIES AND RECOMMENDATIONS

This section provides summaries from the NFPA-502 literature reviews and the case studies. It then provides suggested revisions to NFPA-502 and recommended practice for tunnel structure fire protection.

NFPA-502 Literature Review Summary

Table 1 summarizes the review. From Table 1, it is noted that NFPA-502 has reached consensuses on most of the tunnel structure fire protection criteria, including principles on design fire size, fire curve, peak HRR, fire growth rate, temperature limits on

concrete surface and steel surface; however, the concrete spalling criteria is still in progress. Specifically, the following criteria require further considerations:

- The temperature limit of cast-in-place (CIP) concrete at 380°C (716°F) can result in spalling or explosive spalling, since the CIP concrete can be of high strength (say 9,000 psi) and low w/c ratio. The spalling temperature of high strength concrete, irrespective of it being CIP concrete or precast concrete, is about 250°C (482°F) (Chen, 2012), which depends upon its f_c' and mix.
- NFPA-502 does not speculate the temperature limit of precast and high strength concrete. However, it generally describes the explosive spalling phenomenon of precast concrete, even though the precast concrete can be of low strength and not having explosive spalling issue.
- Though tunnel structure fire protection requirements between NFPA-502 2014 edition and 2017 edition are similar, their concrete spalling criteria are different.
  - The 2014 edition requires: in Section 7.3.3, "Tunnels with concrete structural elements shall be designed such that fire-induced *progressive spalling* is prevented"; and in Section 7.3.4, "Precast concrete is protected such that fire-induced *progressive*

Table 1. NFPA-502 literature review summaries

Tunnel Structure Fire Protection Criteria	NFPA-502 Edition			
	2004	2008	2011	2017
Design Fire HRR	Based on the type of vehicles allowed in the tunnel	*-	-	-
Fire Curve	ANSI/UL 1709	RWS or equivalent	-	-
Peak HRR	100 MW	200 to 300 MW	-	-
Fire Growth Rate	undefined	Provided guidelines	Provided data from fire tests	-
Spalling	*Precast concrete* (high strength concrete) *explosive spalling* is not allowed	-	*Concrete explosive spalling* is not allowed	Concrete structure designed to prevent spalling that leads to collapse; crack that leads to collapse or durability issue is not allowed
Concrete Surface Temperature Limit	380°C (716°F) for cast-in-place concrete	-	-	-
Steel Surface Temperature Limit	250°C (482°F)	-	-	-
Structure Fire Protection material Criteria	None	Provided	-	-

*Note: The "-" indicates the clause remains the same as that of the previous edition.

spalling is prevented." The Section 7.3.4 can be interpreted as "progressive spalling is acceptable" for CIP concrete, which contradicts the Section 7.3.3. The "progressive spalling" seems to be the "explosive spalling" based on descriptions in the Annex of the 2011 and 2014 editions. This can lead to the interpretation that non-progressive spalling or non-explosive spalling is acceptable.

– The 2017 edition implies that spalling is acceptable if it does not lead to structure collapse, and cracking is acceptable if it does not lead to collapse or durability issue, irrespective to the type of tunnel (CIP or precast; high or low compressive strength) and the ground condition (above groundwater, underwater, soft ground, or hard rock) of the tunnel.

– However, cracks can lead to safety and economic impacts (without collapse), such as inundation of a subaqueous tunnel is not acceptable. In addition, if the concrete is not protected, cracks will occur when subjected to the RWS fire; however, cracking and spalling will be repaired after a fire event and durability issues can be resolved.

From the performance requirement perspective, specifying concrete surface and steel surface temperatures are unnecessary, if structure performance requirements, (such as safety, stability, or groundwater infiltration criteria), are specified by the designer engineer. The same principle should be applied to the spalling and cracking requirements.

Tunnel Project Case Study Summary

Table 2 summarizes the reviews. From Table 2, it is observed that both the POMT and the ORBEE Tunnel projects strictly follow the NFPA requirements, while the CBBT Tunnel project does not adopt NFPA spalling and temperature requirements and specifically defined the fire growth rate, which is one of the key concrete spalling parameters. The principle behind the CBBT project is that once the spalling performance criteria, no spalling and cracking, are specified, all other temperature criteria become redundant. This inconsistent approach among projects is because the NFPA clauses are progressing and a best practice is not yet finalized. Some other observations include:

• The NFPA-502 2008 spalling clause can lead to an interpretation that spalling of the precast concrete lining is acceptable; however, spalling can lead to groundwater infiltration

into the POMT resulting in unacceptable safety and cost impacts.

• POMT utilized fire tests to justify that the interface temperature of 380°C (716°F) is acceptable for the precast high strength concrete segments, resulting in a cost savings from a thinner fire protection system. ORBEE Tunnel integrated a fire suppressing system and a lining fire test, resulting in a cost savings by proving that passive fire protection is not needed. For the CBBT Tunnel project, it is expected that no lining fire test will be conducted, since the interface temperature is prescribed in its TR.

• In addition, CBBT Tunnel adopts steel fiber reinforced precast concrete segmental lining; however, NFPA does not have the clause for steel fiber reinforced concrete yet. Does it have to?

From these case studies, it is concluded that the inconsistent tunnel structure fire protection design approach can be attributed to the following reasons: 1) NFPA is still developing its standard for this subject; however, its direction seems unclear and 2) It's impossible to prescribe a single methodology to cover all tunnel projects; each project has its own performance requirements depending upon its geological and hydrogeological conditions (hard rock or soft ground; above or below groundwater level, under ship channel or not), tunnel lining concrete properties, fire suppression system, and many other project specific factors.

It is also concluded that the responsibility to adopt/modify NFPA standards for a specific project shall reside on the design engineer of the project; specifically, the owner's engineer who specifies the criteria either for design-bid-build or design-build projects. A project example adopting this concept is the CBBT Tunnel project.

Recommendations

Recommendations for NFPA future considerations include:

• Stay away from prescriptive clauses for concrete surface temperature, spalling, cracking and durability requirements. Avoid the descriptions of "precast" or "cast-in-place" concrete. It is the concrete mix, strength, and permeability that matter. Avoid the "spalling," "explosive spalling," or "progressive spalling" descriptions, it is the geological and hydrogeological condition that matter.

• Provide tunnel structure performance requirements and let the owner's engineer

Table 2. US tunnel case study summaries

Tunnel Structure Fire Protection Criteria or Ground Condition	Project		
	POMT	ORBEE Tunnel	CBBT Tunnel
NFPA-502 Edition	2008	2011	2014
Design Fire Curve	RWS	Implicit RWS (ANSI/UL 1709 was defined in the Structural TP Section)	RWS
Design Fire Peak HRR	100 MW	300 MW	150 MW
Fire Growth Rate	Implicit	Implicit	Explicit—peak HRR in 15 minutes
Spalling	Precast concrete explosive spalling is not allowed	Concrete explosive spalling is not allowed	*Specify interface temperature to control concrete spalling
Fire Test	Unspecified, but was tested in the field	Unspecified, but was requested and conducted in laboratory	Unspecified and unlikely to occur (except the passive fire protection panel)
Integration of the tunnel ventilation system (and CFD modeling) to the tunnel structure protection	No	Allowed and †performed	Not allowed
Tunnel Lining Concrete f_c'	8,000 psi	4,400 psi	7,500 psi
Concrete Mix W/C Ratio	0.32	0.41	0.35
Concrete Surface Temperature Limit	380°C (716°F) for CIP concrete	380°C (716°F) for CIP concrete	*None
Steel Surface Temperature Limit	250°C (482°F)	250°C (482°F)	*None
Interface Temperature between the Concrete and the Fire Protection Material	Undefined	Undefined	*250°C for $f_c' \geq 6,000$ psi, and 380°C for $f_c' < 6,000$ psi
Ground Condition	Soft ground, mixed face, under ship channel	Hard rock	Soft ground, mixed face, under ship channel
Tunnel Lining Type	Precast Concrete Segmental Lining	CIP concrete	Precast Concrete Segmental Lining
Tunnel Lining Steel Reinforcement Type	Conventional rebar	Conventional rebar	Steel fiber
Construction Method	Closed face tunnel boring machine	Drill and blast	Closed face tunnel boring machine

* This requirement is a project specific requirement.
† The first US tunnel to do so.

specify prescriptive requirements to suit specific project situation.
- Performance requirements, after a fire event, including global tunnel stability must be maintained and spalling and cracking that result in unacceptable groundwater infiltration causing life safety and economic impacts is not allowed.

A right practice must come from right requirements. Owners must realize that NFPA is a standard, not a code, and shall not adopt its clause without modifications. It is the responsibility of owner's engineering to specify appropriate requirements to suit a specific project. Until perfect NFPA-502

requirements are developed, recommendations to owner's engineer include:

- Communicate with owner and AHJ that interpretations and modifications to NFPA-502 are needed to suit project need.
- Make judgment on providing prescriptive requirements such that bidders are bidding on the same requirements, resulting in fair competition and risk reduction from potential inferior construction and disputes. These requirements can include: the design fire size, fire design growth rate, interface temperature if spalling is not acceptable and fire protection material is required, whether integrating fire suppression system and tunnel structure

fire protection system is acceptable or not, and fire test requirements. Specific requirements will depend upon tunnel location, ground condition, and tunnel lining concrete properties.

CONCLUSIONS

From literature reviews and case studies, this paper has proved the followings, in regarding to tunnel structure fire protection practice in the US highway tunnels:

- Current NFPA-502 structure temperature and spalling clauses require updates to avoid potential miss-interpretations leading to unacceptable safety and cost impacts.
- The inconsistent practice for this subject may be caused by the unclear NFPA-502 requirements. It is impossible to have a single consistent prescriptive approach for this subject.

This paper also made the following recommendations:

- NFPA-502 should replace all related prescriptive requirements by performance-based requirements.
- Owners' engineer should adopt and modify NFPA requirements to suit specific project needs. Prescriptive project requirements are needed for consistent and fair bidding purpose and to reduce owner's risk.

REFERENCES

ASTM C 1609. *Standard Test Method for Flexural Performance of Fiber-Reinforced Concrete (Using Beam with Third-Point Loading)*.

Chen, W. 2012. *"Passive Fire Protection Design for High Performance Precast Concrete Segmental Lining,"* World Tunneling Congress 2012, Bangkok.

Chen, W. 2016. *"Fire Design for the Concrete Lining of the Ohio River Bridges East End Crossing Tunnel,"* World Tunneling Congress 2016, San Francisco.

ITA International Tunnelling Association. 2004. *"Guidelines for Structural Fire Resistance for Road Tunnels,"* Working Group No. 6 Maintenance and Repair.

Jacobs Engineering. 2013-1. *"Ohio River Bridges—East End Crossing, Section 4 Tunnel Fire Life Safety Report."*

Jacobs Engineering. 2013-2. *"Ohio River Bridges—East End Crossing, Section 4 Tunnel CFD and Egress Analysis Report."*

Jacobs Engineering. 2014. *Ohio River Bridges—East End Crossing, Section 4 Tunnel Concrete Lining Explosive Spalling Testing Procedure.*

NFPA-502 2004, 2008, 2011, and 2017. *"Standard for Road Tunnels, Bridges, and Other Limited Access Highways."*

UL 1709 2017. *"STANDARD FOR SAFETY - Rapid Rise Fire Tests of Protection Materials for Structural Stee*l.*"*

van der Waart van Gulik, T. G., Breunese, A. J., DeGraff, L., and Bourdon, P. 2016. *"Design, Assessment and Application of Passive Fire Protection in the Port of Miami Tunnel According to NFPA 502,"* Seventh International Symposium on Tunnel Safety and Security, Montréal.

The Use of Numerical Analysis in Rehabilitation of Baltimore's Howard Street Tunnel

Saeid Rashidi and Verya Nasri
AECOM

ABSTRACT: This paper discusses the use of numerical modeling for optimization of tunnel rehabilitation scheme to improve the clearance of the Howard Street Tunnel in Baltimore, MD. With a total length of approximately 8,700 feet, the Howard Street Tunnel consists of three sections: concrete box section, original cut-and-cover section, and original mined section. The Howard Street Tunnel currently has a minimum vertical clearance of 19'-4" but a vertical clearance of 21'-0" should be achieved by lowering the track profile and/or structural modifications. Finite element modeling was used to determine the maximum allowable depth of cut in the brick arch at the cut-and-cover and mined sections of the tunnel. The results of these analyses allowed the designers to formulate a cost-effective solution which minimized the risk of compromising the structural integrity of the tunnel.

INTRODUCTION

The objective of the Howard Street Tunnel Clearance Project (Project) is to obtain a double stack vertical clearance envelope throughout the Howard Street Tunnel (HST). The Howard Street Tunnel currently has a minimum vertical clearance of 19'-4" but a vertical clearance of 21'-0" should be achieved by lowering the track profile and/or structural modifications. The additional clearance would allow shipping containers from the port of Baltimore, for the first time, to be stacked two-high atop trains, a far more efficient way to move them.

The clearance envelope is defined by the National Gateway Clearance Template which is shown in Figure 1. The height of the template above the top of rail is 21'-0". A curvature allowance of 1.25" per degree of track curvature along plane of super-elevated top of rails is applied for curved track.

The possible solutions for increasing the clearance inside the tunnel are track lowering, structural modifications, or a combination thereof. Track lowering without structural modifications is preferred due to its lower cost, faster construction, and minimal risk but it will not be sufficient alone to provide the required clearance along most of the tunnel. The structural modifications include cutting the arch and/or invert of the tunnel in order to increase its clear height.

The design is hinged on an understanding of the structural behavior of the tunnel liner and its interaction with the surrounding ground to avoid costly and unnecessary measures such as liner reinforcement and hardening, ground improvement, and added temporary or permanent support to the extent possible.

TUNNEL DESCRIPTION

The Howard Street Tunnel is considered part a major freight link servicing the east coast. The tunnel was originally constructed to provide two tracks but is currently servicing a single track. The east portal of the tunnel is located adjacent to the Mount Royal Train Station just east of the intersection between North Howard and Preston Streets. The tunnel runs under Howard Street to between Camden Street and Conway Street where its original west portal was located. The tunnel was later extended approximately 1,400 feet as part of the I-395 extension with the west portal currently located between Camden Yards and M&T Bank Stadium. The total length of the tunnel is 8,703 feet, including the original 7,340-foot tunnel constructed from 1890 to 1895 and the extension built in 1982.

The HST consists of three different sections:

1. The original mined tunnel section constructed using sequential excavation method (SEM). A typical section of the mined tunnel is shown in Figure 2.
2. The original cut and cover section (Figure 3) including the steel girder coverway and the brick arch coverway.
3. The concrete box section at the west end of the tunnel constructed in 1982 as part of the I-395 extension.

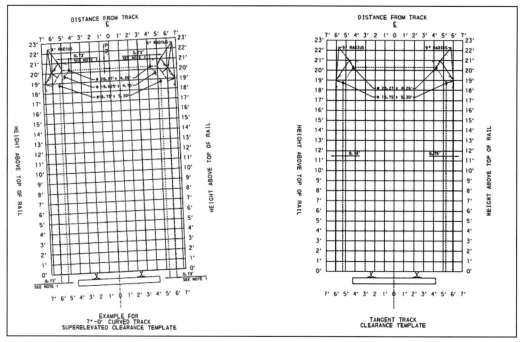

Figure 1. National gateway initiative clearance template

GEOTECHNICAL CONDITIONS

Soil and Rock

Site Geologic History

The tunnel is located east of the Fall Line in the Atlantic Coastal Plain physiographic province. West of the Fall Line is the Piedmont Plateau, which separates the Atlantic Coastal Plain from the Appalachian Mountains. The Fall Line separates the steeper bedrock topography to the west from the flatter Coastal Plain sediments to the east. The tunnel lies in the Western Shore Uplands Region near the Uplands boundary with the Western Shore Lowlands Region to the north.

Surficial deposits typically include fills in developed areas that are underlain by the Coastal Plain Cretaceous age Potomac Group sediments. The Potomac Group consists of interbedded alluvial and deltaic gravels, argillaceous sands, and white, dark gray, gray and multicolored silts and clays.

Beneath the Coastal Plain deposits is Precambrian age bedrock, as seen in several I-395 boring logs which encountered Baltimore Gneiss, and the Red Line borings which encountered granite and amphibolite. These dense crystalline rock formations are an extension of the eastern edge of the Piedmont Plateau. The bedrock is often weathered to decomposed at the contact with the overlying Coastal Plain sediments.

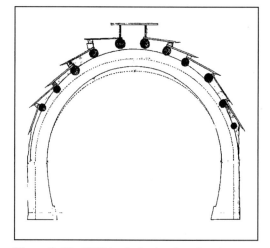

Figure 2. The HST typical mined section

Historical boring logs from several projects and recent boring logs were reviewed to examine the subsurface geology along the Howard Street Tunnel. The tunnel is generally contained in fine-grained silt and decomposed rock down station of 23+50. Up station of 23+50, the silt and decomposed rock layer generally drops off and the tunnel is surrounded by more permeable natural sands and gravels that extend 10 to 20 feet beneath the tunnel invert.

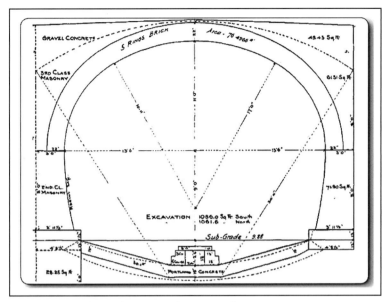

Figure 3. The HST typical cut-and-cover section

Subsurface Profile

An interpretation of subsurface conditions along the tunnel alignment was made from available boring data. The subsurface strata are described and summarized below in order of increasing depth.

Fill (F). A layer of fill blankets the tunnel alignment and generally comprises loose to compact fine to coarse sand, silt, clay, gravel, brick fragments, concrete and asphalt and variable mixtures of these soils and materials.

Sand (S). This soil stratum underlies the fill and generally extends below the tunnel between approximately STA. 36+00 and the west portal. This stratum typically comprises loose to very compact sand layers with varying amounts of silt interbedded with gravel, clayey silt, clayey sand and silty clay.

Organic Silt (O). Organic silt was encountered in two borings, P2 and C-215, near the west portal.

Silt and Clay (M). Underlying stratum S is a compact to very hard deposit of silt and clay.

Decomposed Rock (DR). Very compact to hard micaceous fine to coarse sand with varying amounts of silt and clay to silts and clays. This stratum underlies Strata S and M and generally lies at or within a few feet below the tunnel invert between STA. 30+00 and the east portal.

Rock (R). Rock was encountered in a few borings that extended below the tunnel and is generally described near the surface as light grey to grey weathered to slightly weathered gneiss and granite, broken to jointed, weakly foliated. Boring No. 4-T-B-209w made for the Baltimore Red Line

near the intersection of Lombard Street shows the rock quality increases with depth with core recovery ranging from 95 to 100 percent and RQD ranging from 53 to 100 percent.

GROUNDWATER

Groundwater in the soils around the tunnel (and within the tunnel) is generally within a few feet of the invert of the tunnel between STA. 40+00 and 87+00. The groundwater level in the soils surrounding the tunnel rises (relative to the tunnel invert) from STA. 40+00 to approximately 20 feet above the tunnel invert, at STA. 27+70. Water levels observed in historical borings made a block or more perpendicular from Howard Street indicate that groundwater at that time was 8 to 15 feet higher than that observed adjacent to the tunnel.

Water was observed to be weeping through the tunnel walls and roof throughout the alignment. Groundwater was observed to collect and flow within the ballast on either side of the tracks. In addition, a box drainage culvert is located beneath the tracks and is approximately 0.96' × 2' in section. Both the culvert and the ballast drain into a 48-inch sewer pipe at the low point of the tunnel at STA. 71+50.

STRUCTURAL MODIFICATIONS

The existing tunnel cross section cannot accommodate the required clearance without some level of structural modification in the arch and invert at several locations. During the early stages of this study, the design team recommended avoiding any structural alteration of the tunnel arch, especially in the shallow cut-and-cover section, in order to minimize the risk of structural failures. The proposed approach was to obtain the required clearance by replacing or modifying the tunnel invert which is not a structural (load bearing) element. Numerical analysis was performed to evaluate whether notching of the arch in the cut-and-cover section would be an option.

Numerical Analysis

Finite Element analysis was used to evaluate the overall integrity of the archway section of the Howard Street Tunnel after removing a portion of the tunnel liner, which was proposed as an option to accommodate a larger train envelope. This section delineates underlying methodology and assumptions used in evaluating existing state of stress in the liner, and assessing potential level of distress due to cutting the liner.

Background

The arch portion of the liner consists of five rings of brick masonry with total thickness of 22 inches. The liner arches over a clear span of 27 feet measured on the springline and rises 11 feet above the springline. For a sketch of the liner and soil strata refer to Figure 4.

The liner supports overlying backfill and existing traffic load through a combination of axial (hoop), and bending moments in its existing form. Any alteration in the liner causes a reduction in the strength and a redistribution of stresses, which in turn, can potentially overstress the liner beyond its capacity. It is presumed that no initial defects—such as loose bricks, or deteriorated mortar—exist that may further impede the existing load bearing capacity of the liner.

Modeling

Finite Element analysis package Plaxis was used to simulate the liner response to 50% thickness reduction. To arrive at the existing state of stresses in the liner as well as in-situ soil stresses, construction sequence of the tunnel (such as support of excavation at the time of initial construction) needs to be modelled. In lieu of modelling the actual construction sequence at the time of tunnel construction which is not exactly documented, it was assumed that the tunnel was built in an open-cut as wide as the span of the tunnel, and with side slopes of 3H to 2V; and the open-cut was then backfilled upon completion of liner installation.

For the complete construction sequence used in the model, refer to Figure 5. Traffic loads and adjacent building loads were applied as uniform loads of 300 psf and 1500 psf, respectively; in both symmetric and asymmetric load patterns.

· The liner was modeled with continuum elements for two reasons; namely, a) obtaining detailed stress distribution in the liner, and b) adopting nonlinear material model (Mohr-Coulomb) for brick/mortar. Mohr-Coulomb model allows using a plastic failure criterion under shear or tensile stresses. Mechanical properties of materials used in the model are reported in Table 1. The tensile strength of the brick/mortar material was set to the lowest value which would not cause solution convergence problems.

The soil cover above the crown is approximately 5 feet at the archway stretch of the alignment. Different soil strata are included in the model based on the geotechnical profile. No water pore pressure is considered in the model, since water table is reported below the tunnel invert in geotechnical profile. Tunnel invert was not considered a structural element and therefore was not included in the model.

Analysis Results

The state of the stress in the tunnel liner in the existing condition (before cutting) is compared with that of the future condition (after cutting) based on the results obtained from the analysis of the following four (4) load cases:

1. Load Case 1—symmetric traffic loading
2. Load Case 2—asymmetric traffic loading

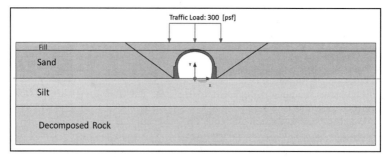

Figure 4. Geometry sketch—load case 1 shown

Active Parts in the Model—Load Case 1 shown	Construction Stage
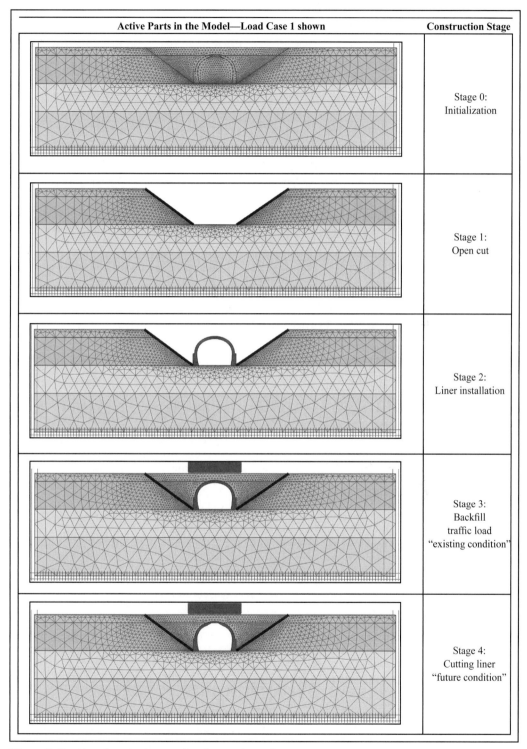	Stage 0: Initialization
	Stage 1: Open cut
	Stage 2: Liner installation
	Stage 3: Backfill traffic load "existing condition"
	Stage 4: Cutting liner "future condition"

Figure 5. Construction staging—cut and cover tunnel

Table 1. Mechanical properties of materials—cut and cover tunnel

Soil	Thickness [ft]	E [kip/ft²]	c [kip/ft²]	φ	K₀	Material Model
Fill	6	150	0.00	30	0.5	Mohr-Coulomb
Sand	21	300	0.05	36	0.6	Mohr-Coulomb
Silt	21	200	1.00	25	0.6	Mohr-Coulomb
Decomposed Rock	29	300	2.00	30	1.0	Mohr-Coulomb
Brick/Mortar Liner	1.83	200,000	16.00	56	-	Mohr-Coulomb

3. Load Case 3—symmetric traffic loading + asymmetric adjacent building loading
4. Load Case 4—asymmetric traffic loading + asymmetric adjacent building loading

The maximum compressive principal stresses in the tunnel liner before and after cutting are presented in Table 2. The plot of liner principal stress is shown in Figure 6 for Load Case 4 which results in the largest stresses after cutting. Cutting the arch to its half-depth will result in 65% increase in the maximum compressive principal stress.

ACI 530, Building Code Requirements for Masonry Structures Section 2.2 provides requirements for unreinforced masonry. Masonry members that are subjected to axial compression, flexure, or to combined axial compression and flexure shall be designed to satisfy the following equation:

$$\frac{f_a}{F_a} + \frac{f_b}{F_b} << 1 \tag{1}$$

where,

F_a = allowable compressive stress due to axial load only, psi

F_b = allowable compressive stress due to flexure only, psi

f_a = calculated compressive stress in masonry due to axial load only, psi

f_b = calculated compressive stress in masonry due to flexure only, psi

Neglecting slenderness effects,

$$F_a = \frac{f'_m}{4} \tag{2}$$

$$F_b = \frac{f'_m}{3} \tag{3}$$

where,

f'_m = specified compressive strength of masonry, psi

An estimate of f'_m was obtained by averaging the core sample compressive strengths. The brick cores obtained from tunnel arch exhibited varying compressive strengths ranging from 1252 psi to 4585 psi with an average of 2775 psi (400 ksf). Therefore,

Table 2. Maximum compressive principal stress in tunnel liner—cut and cover tunnel

Load Case	Compressive Principal Stress (Before Cutting) [ksf]	Compressive Principal Stress (After Cutting) [ksf]	Change [%]
Case 1	38	40	5%
Case 2	41	50	22%
Case 3	94	121	29%
Case 4	97	160	65%

F_a = 694 psi = 100 ksf
F_b = 925 psi = 133 ksf

Since the Finite Element analysis results do not differentiate the axial and flexural stresses, the stress interaction sum cannot be readily calculated; however,

$$\frac{f_a}{F_a} + \frac{f_b}{F_b} > \frac{f_a}{F_b} + \frac{f_b}{F_b} = \frac{f_a + f_b}{F_b} = \frac{160}{133} = 1.2$$

This leads to the conclusion that a 50% cut in the arch will result in overstressing the arch. Since smaller cut depths do not provide sufficient clearance, the results of the analysis indicate that cutting the arch of the cut and cover tunnel section is not a viable option for achieving additional clearance.

CONCLUSIONS

Finite Element analysis was used to evaluate the arch cutting option in order to improve the clearance of the Howard Street Tunnel. The complete construction sequence of the cut-and-cover section of the HST was simulated in the analysis. The liner was modeled with continuum elements to obtain detailed stress distribution in the liner, adopting a nonlinear material model (Mohr-Coulomb) for bricks/mortar. The results of the analysis indicate that cutting the arch of the cut and cover tunnel section is not a viable option for achieving additional clearance.

AKNOWLEDGMENT

The authors thank Michael Hoey and Robert Humbert for their contributions to this paper.

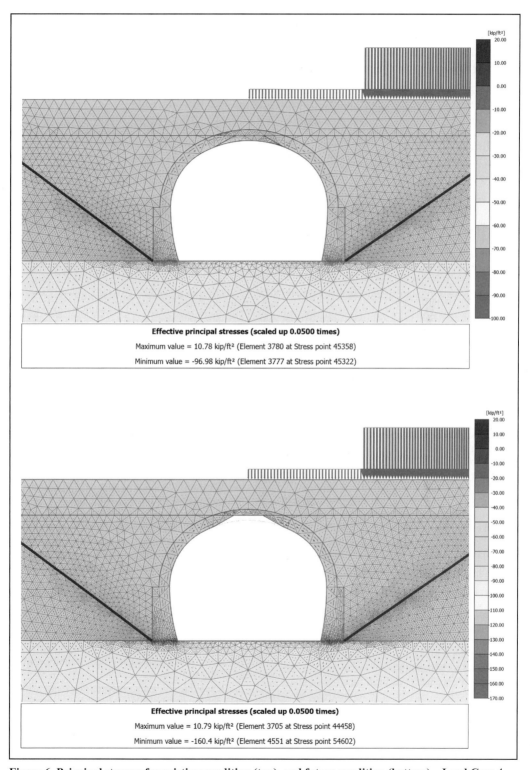

Figure 6. Principal stresses for existing condition (top), and future condition (bottom)—Load Case 4

Evaluation of a Segmental Tunnel Lining Response to a Strike Slip Fault Rupture

Angel Del Amo, Robert Goodfellow, Moustafa Awad, and Bingzhi Yang
Aldea Services

Kurt Braun
L-7 Services

ABSTRACT: Several underground projects in seismic areas are currently dealing with active fault crossings, in such locations the tunnel lining performance is a key element for the project's success. This paper describes various special tunnel lining systems that could be designed to accommodate the imposed displacements on the tunnels due to the risk of a strike slip fault rupture when moderate movements are anticipated. It also emphasizes the lining response in oblique crossings compared to perpendicular crossings. Numerical analyses included advanced three-dimensional numerical models and special structural details. Finally, it provides recommendations to improve the tunnel performance during a fault movement.

INTRODUCTION

Seismic design for tunnels varies from typical analyses of above-ground structures. In general terms, underground structures are known to perform relatively well during seismic events compared to above-ground structures since, among other reasons, they are not subjected to inertial forces in the same way. During extreme ground failure events such as a fault rupture, tunnels can be subjected to moderate to severe damage associated to large ground displacements when compare to seismic ground shaking. This paper presents a number of significant parameters to be considered when evaluating the behavior of a segmental lining during a fault rupture. Recommendations to improve the lining performance during seismic extreme events are also provided.

PAST CASE HISTORIES

Although there are several publications available regarding the evaluation of tunnels subjected to fault displacements, most of them have been typically focused on the analysis of buried steel pipelines. The behavior of a segmental tunnel lining under a fault movement is expected to differ from a buried pipe.

To date there has been a limited number of previous case histories that have evaluated for the potential use of a Precast Concrete Tunnel Linings (PCTL) to accommodate an active fault movement. As a reference, table 1 shows similar experiences and studies in California, including also the use of steel liners.

It is noted that others have investigated fault ruptures for segmental linings, but the majority of those evaluations have focused on a vertical or lateral fault acting perpendicular to the tunnel. The direction of the slip relative to the tunnel axis will involve a significant number rings and introduce soil displacements not parallel to the circumferential joints.

TUNNEL LINING SYSTEMS

Although there is a large variety of lining systems, this paper is based on the behavior of a Precast Concrete Tunnel Lining (PCTL) during a fault rupture. Currently, this type of tunnel lining has been implemented in a variety of geotechnical environments ranging from soft soils to hard rock involving the use of mechanized shield tunneling.

An internal diameter (ID) consisting of 18'-10" and a 12-inch-thick double gasketed precast concrete segments bolted (radial joints) and dowelled (circumferential joints) was selected for evaluation purposes since it is believed to represent typical dimensions currently used in the tunnel industry for different purposes (e.g., metro lines, sanitary sewers, etc.) (Figure 1)

Depending on magnitude to the fault movement, more ductile solutions as steel liners are typically preferred when large fault movements or stringent performance criteria are anticipated for the following reasons:

Table 1. Similar examples in California

Project	Material Type	Fault (s)	Fault Movement	Tunnel ID	Tunnel Thickness	Singular Details
The South Bay Tunnel Outfall Project (San Diego, CA)	Soil	Coronado Banks Fault & Rose Canyon Fault	3"	11'-2"	9"	PCTL. Continuous longitudinal and hoop reinforcement
State Route 75/282 Transportation Corridor (San Diego, CA)	Soil	Coronado Fault	Right lateral (22") Left Lateral (14")	33'-6"	18"	PCTL. Requires mitigation measures to avoid soil/water entering circumferential joints
New Irvington Tunnel, San Francisco (CA)	Rock	Four Secondary Faults	6"	8'-6"	Unknown	Two-pass lining system. Use of special backfill
JWPCP Effluent Outfall Tunnel (Carson, CA)	Rock	Palos Verdes (PVF) & Los Cabrillos (CF)	20" (ODE) & 4" (ODE)	18'-10"	12"	Two-pass Lining (PVF) PCTL. Post-tensioning system (CF)
SR 710 North Study (Los Angeles, CA)	Rock	Raymond, Eagle Rock and San Rafael Faults	Vertical disp.: Raymond (4" to 40") Eagle Rock (20") San Rafael (10")	22'-0" 55'-2"	20"	Prefabricated steel segments Vault concept

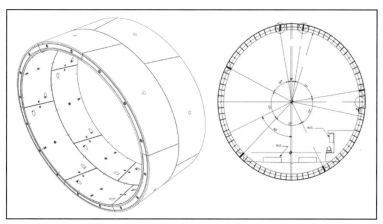

Figure 1. Tunnel lining systems: precast concrete tunnel lining (left) and steel lining (right)

- Ability to sustain high deformations prior to its fracture (ductility).
- Able to remain mainly elastic (with minimal inelastic deformations).
- Limited structural repairs after the fault movement.
- Likely bending of the flanges may help reducing the gaps.

In our case, we have assumed a fault displacement of a few inches where both systems may be applicable.

FAULT CROSSING

Type of Fault

Generally speaking, there are four different types of faults: (1) normal, (2) reverse, (3) strike slip and (4) oblique.

Depending on the type of fault and the relative angle between fault and tunnel the lining can be subjected to different loading conditions. Normal faults induce tensile stresses on the lining whereas reverse faults cause compressive stresses. A strike slip fault rupture motion causes horizontal ground displacements which also tends to increase the pressures on the lining and may induce longitudinal deformation depending on the interception angle (Figure 2).

Although the next sections describe a strike slip fault motion, since the cross-sectional shape of the tunnel is circular, some of the conclusions and findings are also applicable to normal and reverse faults.

Relative Movement

A tunnel crossing an active fault is subjected to shearing displacements than can range from a few inches to several feet. Depending on the ground conditions and the fault characteristics, those displacements can be concentrated in a very narrow zone or spread-out a certain distance.

In our particular case, the following lateral displacement along the fault strike were considered:

- Ordinary Design Earthquake (ODE): 0.5 inches.
- Maximum Design Earthquake (MDE): 5 inches.

Orientation

The orientation between the fault trace and the tunnel plays a key role in the lining behavior. For Precast Concrete Tunnel Linings (PCTL) relative shear motion placing the tunnel in compression (bursting and spalling) tends to be more harmful than a shear motion that elongates the tunnel (joint openings). For both cases, the potential flooding of the tunnel needs to be considered although it is anticipated to be more problematic when the tunnel is elongated.

Different relative angles between the tunnel and the fault slip to evaluate the lining response under two different situations (Figure 3):

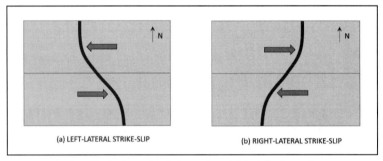

(a) LEFT-LATERAL STRIKE-SLIP

(b) RIGHT-LATERAL STRIKE-SLIP

Figure 2. Strike slip faults: (a) left-lateral and (b) right-lateral

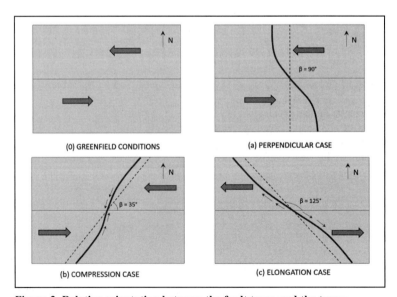

(0) GREENFIELD CONDITIONS

(a) PERPENDICULAR CASE

$\beta = 90°$

(b) COMPRESSION CASE

$\beta = 35°$

(c) ELONGATION CASE

$\beta = 125°$

Figure 3. Relative orientation between the fault trace and the tunne

- Perpendicular case (90 degrees)
- Compression case (35 degrees)
- Elongation case (125 degrees)

Ground Conditions

In general terms, if an abrupt fault displacement is expected in stiff ground (rock conditions) the tunnel lining may be subjected to severe localized damage following the fault trace. On the other hand, in soil conditions the reduction on the soil modulus may induce a snaking motion of the tunnel lining.

The evaluation performed assumed the fault to be located in alluvial soils and is expected that the tunnel will follow a S-shape curve.

SEISMIC PERFORMANCE CRITERIA

When dealing with a fault rupture during a seismic event, it is important to precisely define the performance criteria during the design earthquake. The current practice in the tunneling industry is to define two levels of design earthquakes: Ordinary Design Earthquake (ODE) and Maximum Design Earthquake (MDE).

Ordinary Design Earthquake (ODE)

For the ODE, the basic criteria are that the tunnel lining will not be damaged allowing tunnel operations to remain in service. Given the ODE maximum anticipated fault rupture movement of less than 0.5", the tunnel lining will not be subjected to significant ground displacements and the ODE fault rupture is not currently anticipated to control design of the special seismic tunnel lining. The post-ODE tunnel lining may have increased stepping between the segments and some additional flexural cracking, but nothing that would require repair.

Maximum Design Earthquake (MDE)

Due to the extreme nature of the MDE fault rupture, the primary criteria is that there shall be no collapse and no catastrophic inundation with danger to life, and any structural damage shall be controlled. Accordingly, a PCTL could be designed to withstand the MDE fault rupture of 5 inches without collapse, but with some inevitable cracking and spalling of the concrete in the vicinity of the fault rupture which will require some post-MDE repair effort.

Materials Performance

Some administrations, line LACTMA or Sound Transit, require to guarantee that the following strains must not be exceeded during the MDE:

- Pure compression concrete strain <0.002
- Flexural concrete strain <0.004

- Rebars tensile strain < 0.006

For reparability, the above-mentioned limits can be increased up to 0.0033 for concrete and 0.02 for steel rebars. It is worth mentioning that all these limits were likely developed for transient ground motions rather than for fault rupture events.

Regarding the shear stresses, in order to enable post-earthquake repair works, it is recommended to limit the maximum allowable shear to $10 \cdot \sqrt{fc'}$; giving a maximum allowable shear strain of around 0.0004.

SPECIAL PRECAST CONCRETE TUNNEL LINING

The primary focus of the design is to make the lining as robust as practical to ensure that the tunnel rings remain structurally functioning as individual rings and to minimize the subsequent damage to the concrete.

The design of the special PCTL requires the evaluation of some specific details to enhance the ductility of the lining and guarantee that the ring survives the fault rupture (Figure 4):

- Hoop reinforcement by means of post-tensioned strands or continuous threaded bars.
- "Fuses" by means casted rolled tubular steel structural element into the segments to activate buckling under high compressive forces.
- Dense rebar mats ("grillage") incorporated in the segments with the ability to deform significantly but act as a mesh to keep the broken concrete in place.
- Addition of steel fibers to enhance the concrete ductility.

3D MODELS FOR THE FAULT RUPTURE

In order to determine the effects of a fault rupture on a tunnel lining, 3D models were developed. The purpose of the models was to simulated the aforementioned displacements on the tunnel lining and to determine whether the proposed tunnel lining would be suitable.

Methodology Approach

The evaluation of the fault rupture starts defining the model boundaries and sheared zone characteristics, free-field behavior and tunnel lining response.

Model Boundaries

Different model boundaries were evaluated and it was finally decided to divide the model into two different blocks: a fixed block with standard boundary conditions and a second moving block with the fault movements imposed.

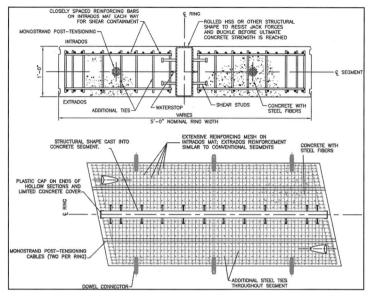

Figure 4. Special precast concrete tunnel lining details (post-tensioning, grillage and fuses)

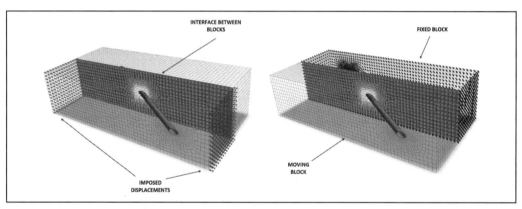

Figure 5. Boundary conditions, imposed displacements and interface between blocks

The contact between blocks was simulated by an interface element in order to reflect an abrupt displacement between blocks (Figure 5).

Greenfield Conditions

Greenfield conditions is referred to the model response in the absence of the tunnel and it is the first step prior to evaluating the tunnel lining response.

Various greenfield models were run to get the required imposed movements at the boundaries, in order to make sure that 5 inches of displacement were exactly occurring at the tunnels depth (Figure 6). From these analyses, it was concluded that it was necessary to increase the boundary displacements.

The ground response was simulated based on a Mohr-Coulomb constitutive model. Using a more advanced constitutive model, such as hardening soil small strain (HS-SS), was understood to simulate the ground response more accurately but hindered the models' convergence. For this reason, a strain-compatible Young's modulus that matched the HS-SS constitutive model was selected (Figure 7).

Refined FEA Models

The purpose of these models was to simulate the fault crossing for different relative angles between the fault and the tunnel alignment. The models incorporated the circumferential joints between rings

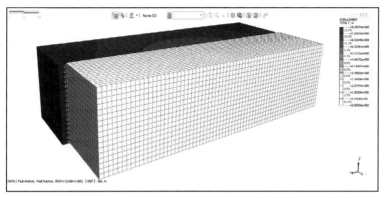

Figure 6. Horizontal displacements between blocks

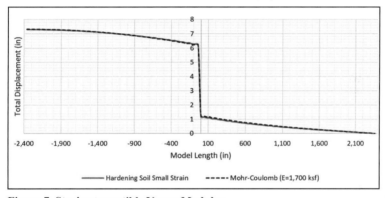

Figure 7. Strain compatible Young Modulus

as interface elements and for the relative angle of 125 degrees (elongation case) bolts located on the circumferential joints to accurately simulate the joints opening. Models' were developed to evaluate:

- Longitudinal and hoop forces, bending moments and shear forces in the rings during the fault rupture.
- Obtain more accurate results in terms of stresses and strains.
- Evaluate the movements and rotation between rings (i.e., stepping & rotation).
- Effect of the "fuses" decreasing the shear stresses for a relative angle of 35 degrees (compression case).

Several discrete rings were also modeled within the fault trace to simulate their interaction between the fault rupture. Additionally, the radial joints between rings were tried to be simulated via interface elements (Figure 8).

Regarding the concrete stiffness, two different values of 100 percent and 35 percent of the original concrete Young modulus were analyzed. The

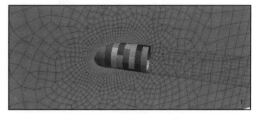

Figure 8. 3D view—Discrete segments at fault crossing

latter value tries to simulate the inelastic response and cracking during the MDE fault rupture event. Based on the models results, it is anticipated that the rings directly impacted by the fault trace (2–12 rings depending on the strike/dip angles) will be severely distorted and that the 35 percent stiffness could be more representative of segment's inelastic behavior directly impacted by the fault crossing.

Results

The first outcome of the models is the expected snaking action of the tunnel in soil conditions (Figure 9):

Regarding the perpendicular crossing, the extent of damage predicted is quite limited because of the presence of circumferential slip joints without any type of bolting systems. In this particular case, the tunnel behaves as individual rings which can move relative to each other (Figure 10).

In general, joint rotation and opening for β = 125 degrees was anticipated (Figure 11) while, due to the high compressive stresses, joints for β = 35 are expected to remain closed with a small stepping in some cases (Figure 12).

In order to evaluate the length of damage at the fault crossings, the tunnel lining was modeled as an elasto-perfectly plastic material governed by equivalent Mohr-Coulomb properties. Figure 13 shows an anticipated length of damage.

The fault sheared zone imposes stress concentration and differential displacements on the tunnel lining. The modeling indicated that damage may be limited to the tunnel lining directly contacting the fault trace with additional localized damage along

portions of the circumferential joints for these rings in contact with the fault.

The post-MDE tunnel lining is anticipated to be stable but exhibit intense localized cracking along the fault trace, increased flexural cracking due to ring distortion, increased joint stepping, and for the elongation case limited opening of the circumferential joints. For these reasons, it is believed that the localized damage may be repairable and does not represent a mechanism for collapse during the MDE.

Secondary effects associated with the MDE damage could include groundwater infiltration (accompanied by potential gas infiltration) and minor soil infiltration. All of these items are anticipated to occur for either the steel tunnel lining or for the VECP precast concrete tunnel lining due to the extreme nature of the MDE fault rupture event.

SEISMIC DETAILS

The analysis revealed the need for special engineering considerations, required to resist the design

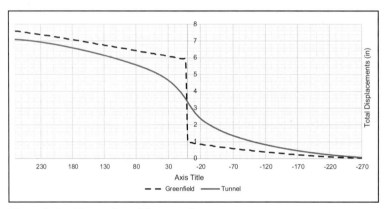

Figure 9. Snaking action of the tunnel

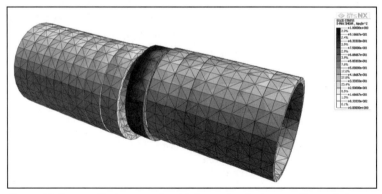

Figure 10. Perpendicular case (β = 90 degrees)—Deformed Shape (joints opening)

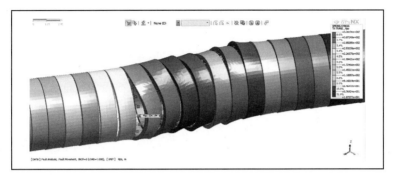

Figure 11. Elongation case (β = 125 degrees)—deformed shape (joints opening)

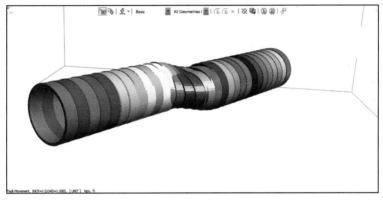

Figure 12. Compression case (β = 35 degrees)—deformed shape (joints stepping)

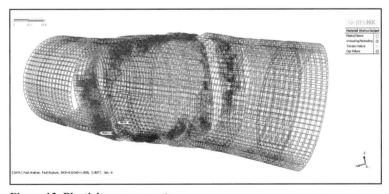

Figure 13. Plasticity on segments

loading case of the MDE. Special measures are described in the following sections for each of the compressive forces, tensile forces, shear stresses and joint openings anticipated in our analysis. The following recommendations are understood to be valid if fault movements are small (i.e., few inches) or when the displacements are distributed over a relative wide zone.

Longitudinal Forces

For the anticipated longitudinal forces for the compression case (β = 35 degrees), one concept is to

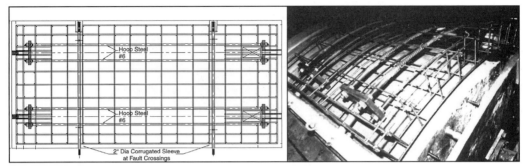

Figure 14. Continuous hoop reinforcement—San Diego Effluent Outfall (1995)

create a collapsible section utilizing a hollow rolled rectangular tubing section (or other standard or built-up section) with shear anchors and integral waterstop embedded into the concrete. The tubing wall thicknesses is designed to buckle when subjected to longitudinal forces significantly greater than the jacking force (but less than the concrete compression capacity; 150 Kips/ft $< P_{buckling} < 500$ Kips/ft).

Another option is to use ¼" packers instead of the ⅛" typically used in the tunneling industry.

Shear Stresses

The numerical models showed that the fault rupture induces considerable in-plane shear stresses in the tunnel lining, making it necessary to improve the compressibility and stepping between rings to reduce these stresses.

One potential remedy is to recognize that the concrete could fracture significantly and to provide a ductile restraint to ensure that the broken lining remains in place. The concept was to install rebar mats that had the ability to deform significantly but act as a mesh to keep the broken concrete in place. The grillage could be a simple heavy-welded wire mesh with a close wire-to-wire spacing. In addition, the use of steel fiber reinforced concrete for the segments could maximize the retention of the damaged concrete in the matrix. Multiple post-tensioning strands or continuous hoop reinforcement may further assist in increasing the shear capacity of the concrete and keeping any damaged concrete in place.

From the modeling it was also clear that rolled tubular steel structural element casted into the segments could buckle under high compressive forces ($\beta = 35$ degrees) reducing the shear stresses in the lining.

Tensile Hoop Forces

Tensile hoop forces were expected due to the distortion of the rings directly affected by the fault rupture. A post-tensioning system could be designed similarly to that used on LA Effluent Outfall Project and to provide about 90–150 Kips/ft of compression uniformly distributed within each ring. The post-tensioning strands could benefit the lining by maintaining the circular ring compression and thereby, maximize the ability of the segmental lining to act as independent structural rings.

Another alternative is to consider the San Diego Outfall concept where continuous hoop bars were used to withstand the tensile hoop forces (Figure 14) and enhance the tunnel ductility. These feasible solutions could allow the lining to survive the specified MDE.

Relative Movement Between Circumferential Joints

From the numerical modeling it was concluded that the maximum fault rupture displacement of 5 inches may not occur at any single joint. According to our analysis, the joints immediately adjacent to the fault trace could experience a maximum relative joint movement of 2 in, for the worst-case scenario. The circumferential joints should be designed for specified leakage criteria during the MDE so that adverse consequences should be mitigated during the design stage.

One alternative approach proposed was a continuous longitudinal steel rod connection, similar to the one used in the San Diego Outfall Tunnel for the faults crossings, or the use of seismic washers. These elements work together to absorb displacement energy and reduce the relative movements across joints.

Current concepts identified to prevent soil ingress include utilizing custom oversized EPDM gaskets to maintain gasket contact (Figure 15). If additional measures were deemed necessary, expansion steel bands (or a similar sealing/restrain systems) would also be used to minimize the soil and water ingress after the MDE.

Backfilling

In some cases, TBM over-excavation and the use of frangible material as cellular concrete between

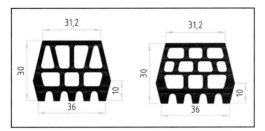

Figure 15. Details of oversized gaskets to minimize soil and water entering the tunnel (Datwyler)

tunnel lining and the ground could help the tunnel to absorb or accommodate some of the fault movements. Two key elements to be considered are the relative stiffness with the surrounding ground and its durability during the design life of the infrastructure. A sufficiently large over-excavation between the ground and the segmental lining may be a challenge in urban environments where settlements need to be minimized.

CONCLUSIONS

When possible, any tunnel alignment should be selected avoiding active fault crossings. Since fault crossings are in some cases unavoidable, a good characterization of the fault is highly recommended to select a tunnel alignment as perpendicular as possible. In any case, tunnel cross section needs to account for the accommodation of the fault offset (i.e., widened tunnel section).

From all the evaluations, it is apparent that the orientation of the fault movement relative to the tunnel axis is an important parameter in the stresses ultimately induced in the lining. The fault strike skew angles considered introduced significant longitudinal forces into the lining. In addition, the skew angle combined with the fault dip angle results in a shear action involving numerous rings which cannot be relieved by circumferential joints alone. Initial evaluation of a perpendicular fault crossing with a vertical dip confirms that the longitudinal forces are significantly diminished and this design scenario could be unconservative.

The continuous reinforcement concept (e.g., conventional rebars or post-tensioning system) supplemented by a "fuse" lining concept along with an extensive intrados reinforcing mesh and fiber concrete appears to have merit for addressing the lining stress issues associated with the fault crossings.

For elongation case ($\beta = 125$ degrees) it is crucial to minimize water/soil ingress into the tunnel making it necessary to guarantee that the gaskets remain in contact after the fault rupture.

With regards to overall stability, the design should maximize the ability of the rings to maintain circular ring action after the fault rupture; continuity of ring structural action is critical to ensure that no collapse mechanism develops. This requires that damaged sections of lining remain in place and continue to transfer hoop compressive forces after the MDE fault rupture.

Finally, contingency and repair plans after the fault rupture event need to be considered during the design stage.

REFERENCES

Boyce G., Klein S., Sun Y., Feldsher T. and Tsztoo. D (2010), *New Irvington Tunnel Design Challenges*, North American Tunneling Conference 2010, pp. 325–336.

Federal Highway Administration (2004*), Seismic Retrofitting Manual for Highway Structures: Part 2—Retaining Structures, Slopes, Tunnels, Culverts and Roadways*, pp. 145–242.

Gregor T., Garrod B. and Young D. (2007), *Analyses of underground structures crossing an active fault in Coronado, California*. Underground Space—the 4th Dimension of Metropolises, pp. 445–450.

Jacobs Associates/CH2M Hill (2014), SR 710 North Study. Preliminary Design Concepts for Fault Crossings. Technical Memorandum 6.

Metro Rail Design Criteria (2017), *Section 5. Structural/Geotechnical*. Los Angeles County Metropolitan Transportation Authority (LACTMA).

Navin. S., Kaneshiro. J., Stout L. and Korbin G. (1995), *The South Bay Outfall Project San Diego, California*, Rapid Excavation Tunneling Conference 1995, Chapter 40.

Van Greuen J., Sun Y., Hughes, Geraili R., Kaneshiro. J., Haug. D. and Vanderzee. M. (2016), *Technical Approach to Lining Design for Internal Pressure and Fault Offsets on the JWPCP Effluent Outfall Tunne*l, World Tunneling Congress 2016, San Francisco.

Young D., Dean A., Warren S., Haldin A. and Gregor T. (2008), *Preliminary Seismic Design Considerations for a Highway Tunnel in Coronado, California*, pp. 405–416.

Session 2: Sequential Excavation Methods

Bill Dean and Mun Wei Leong, Chairs

Design and Construction Challenges in Urban Settings for the NATM Tunnels' Line 2 of the Riyadh Metro

Sandeep Pyakurel, Kurt Zeidler, and Andreas Gerstgrasser
Gall Zeidler Consultants

Iain Thomson
Bechtel

Abstract: The Riyadh Metro Project entails the construction of six new lines. ArRiyadh Development Authority, the executive arm of the High Commission for the Development of Arriyadh contracted BACS consortium to design and build Lines 1 and 2; Gall Zeidler Consultants designed mined tunnels which are sections of the tunnels for the Lines 1 and 2 including emergency egress shafts and connection adits.

This paper describes design and construction of Line 2 mined tunnels in an urban settings including challenges encountered during excavation in close proximity to the building foundations, piers and methods to control inflow of groundwater in ground conditions varying from disturbed brecciated to fresh intact limestone.

INTRODUCTION

The Riyadh Metro is a rapid transit system which will be serving the city of Riyadh that consists of 6 metro lines with an overall length of 176 kilometers including 85 stations. The project is envisioned to meet the transportation needs of Riyadh's growing population which is estimated to increase by more than 8.3 million by 2030. The network will connect the King Khalid International Airport and King Abdullah Financial District with the main universities, downtown and the public transport center. The project will also reduce traffic congestion and improve air quality. The total estimated project cost is $23 billion. The metro project is owned and operated by ArRiyadh Development Authority (ADA). ADA awarded the design and construction contracts of the Riyadh metro to the ANM, FAST and BACS consortiums in October 2013. The BACS consortium for the design and construction of the Metro Lines 1 and 2 with an overall contract value of $10 billion includes Bechtel, Almabani General Contractors, Consolidated Contractors Company and Siemens and is led by Bechtel.

Contracted by the BACS consortium, Gall Zeidler Consultants (GZ) provided technical guidelines and requirements for the design of all structures associated with the permanent and temporary works for the construction of the Line 2 running tunnels and Lines 1 and 2 emergency egress shafts. This paper describes the design and construction of the Line 2 running tunnels that were excavated using the sequential excavation method (SEM).

Project Description

The Metro Line 2 runs mostly at grade from east to west along King Abdullah Road, and will extend more than 25km with 14 stations including 2 transfer stations. The running tunnels, shown in Figure 1, are part of the Line 2 tunnel. The total length of the running tunnels is approximately 2km, with a 1-km section driven westbound of the Olaya Station that connects to the cut-and-cover tunnels adjacent to the 2B1 Station, and a 1-km section driven eastbound of Olaya Station that ends at the 2B4 Station. The alignment is generally straight in a SW-NE direction and has a curvature around the Olaya Station to accommodate the transitions in and out of the station. The running tunnels were excavated from a temporary shaft built within the footprint of the 2B2 Olaya transfer Station. The tunnel is located at depths varying from 5m to 17m below ground surface. The vertical alignment starts at an approximate elevation of 612 m to the east of station 2B1, deepens to an elevation of 605m and then slightly rises to a level of 607m at the Olaya Station. East of the Olaya Station to the end of the mined section, the alignment ascends to an approximate elevation of 612m.

GEOLOGY AND GROUND CONDITIONS

The regional geology within the project area comprises thick sedimentary successions of Jurassic-Cretaceous limestone and anhydrite which follow the general structural setting of the region and gently dip from SW to NE. The general stratigraphy along the Line 2 consists of a thin layer of fill material and

Quaternary alluvial deposits comprising interbedded layers of silty sand and gravel overlying anhydrite and limestone rock members of Cretaceous and Jurassic age. Three different rock formations namely the Sulaiy Formation, Arab Formation and Jubaila Formation are encountered from top to bottom respectively. The Cretacious Sulaiy Formation consists predominantly of limestone with calcarenite beds and outcrops along East Riyadh. The Jurassic Arab Formation comprises limestone and brecciated limestone. The Jurassic Jubaila Formation comprises limestone with some calcrenite beds and mostly outcrops along West Riyadh.

The running tunnel alignment cuts through the carbonate members of the Arab Formation and encroaches locally into the Sulaiy Formation at the east close to Station 2B4. Thin layers of fill and Quaternary deposits comprising mostly silty sand and silty clay with limestone fragments and cemented granular materials overlie the Arab Formation. The Arab Formation is a contorted and brecciated alternation of limestone and thin beds of anhydrite consisting of four stacked carbonate-evaporite cycles, named Upper Limestone Breccia, Arab-C Disturbed Bedded Limestone, Lower Limestone Breccia and Arab-D Undisturbed Bedded Limestone. The mined tunnel plan and profile view with geology is shown in Figure 2.

The Upper Limestone Breccia member represents cemented breccia of collapsed blocks of the overlying Sulaiy formation mixed with limestone blocks from Arab formation. This member originates from the dissolution of the Hith Anhydrite member initially present between the Sulaiy and Arab Formations. The breccia is generally matrix-supported and comprises clasts ranging from coarse

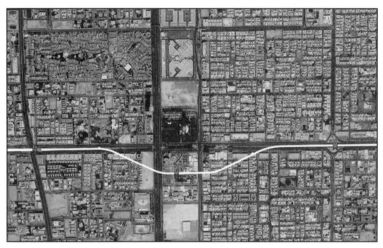

Figure 1. Line 2 Tunnel alignment (red) Running Tunnel and its proximity to the other structures

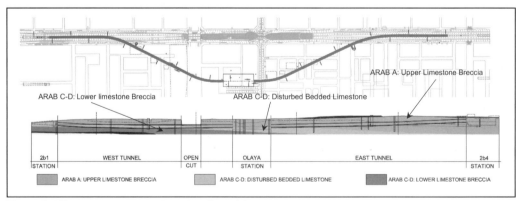

Figure 2. Mined tunnel plan and profile view with geology

gravel to very large sized boulders. The Arab-C member comprises very thin to thinly bedded calcareous claystones, calcareous siltstones and reddish brown partially dolomitised calcarenite and fine-grained siliceous limestone. This member is similar to the one encountered in the Upper Limestone Breccia originating from the dissolution of breccia beds. The Lower Limestone Breccia comprises moderately weathered, moderately to well-cemented, moderately strong, creamy white to yellowish brown, matrix-supported clasts of limestone breccia. Some larger clasts have been dolomitised and are strong to very strong. The matrix comprises yellowish brown, mottled white silt, some clay, and minor quantity of fine sand. The Arab-D member comprises grey to bluish limestone and some calcarenite beds. The limestone is slightly weathered, thick to very thin sub-horizontally bedded and undisturbed.

The west tunnel is mostly within the Disturbed Bedded Limestone with tunnel roof reaching the Upper Limestone Breccia at higher elevations. The east tunnel is mostly within the Upper Limestone Breccia while the invert reaches the lower Disturbed Bedded Limestone near the Olaya Station. Both of these members are highly heterogeneous and display varying behavior based on the intactness of the rock. Weathered profiles characterize the shallow strata especially at the eastern part of the alignment, however localized seams of deteriorated material were expected along deeper level. Sub-horizontal and sub-vertical discontinuities and vugs characterize the entire Arab-C Formation and local failures were expected during excavation.

Dissolution cavities were expected in all limestone formations. These cavities originated from dissolution of limestone and anhydrite beds by acidic groundwater along fractures and discontinuities, which is a typical karstification process in carbonate rocks. The upper members of the Arab formation are most prone to such cavities.

The carbonate rocks encountered along the alignment of the Line 2 mined tunnels comprise the main regional aquifer in the Riyadh area. The groundwater level is highly variable in the city of Riyadh as it is influenced by infiltration from sewage and is hydraulically connected to the shallower aquifers developed in the gravel in surrounding areas. Therefore, groundwater levels displayed a significant seasonal variation along with very high permeability of the limestone formation led to significant groundwater inflow in the tunnel.

TUNNEL DESIGN AND ANALYSIS

The cross section has maximum height of 9m and width of 10.9m at the tunnel springline (Figure 3). The excavation and support design followed the principles of SEM which is also referred to as New Austrian Tunneling Method (NATM). The excavation and support method was selected to preserve the load-bearing capacity of the surrounding rock mass which mobilizes the strength of the ground and allows implementation of lighter support. The selected excavation and support method also considered the timing of the support installation to optimize support requirements. The design also provided toolbox item such as rebar spiles as local measure to provide additional support when soft, highly fractured rock or soil was encountered in the crown.

The primary lining comprised 250mm of steel fiber-reinforced shotcrete lining and systematic rock dowels. The rock dowels were designed with a length of 3m spaced in a 2m × 1.5m staggered pattern. The dashed line on Figure 3 indicate those rock dowels within which were omitted during construction due to the prevailing good ground conditions. The excavation and support sequence typically included a 1.5m long top heading round followed by installation of rock dowels and shotcrete lining. During construction, the 1.5m round length was gradually increased to a maximum of 3m upon encountering of ground with relatively good rock mass quality. The cycle was repeated twice followed by staggered excavation and support of bench/invert with a round length of 3m. The secondary lining comprised 300mm of cast-in-place steel fiber-reinforced concrete. A full-round waterproofing membrane was placed between the primary and secondary linings to prevent water inflow into the tunnel.

The running tunnels were also connected to the two Emergency Egress Shafts by means of smaller connecting adit tunnels. The adits were constructed using SEM by breaking out from the Line 2 SEM running tunnels. The tunnel design was in accordance with BS EN 1990, BS EN 1991, BS EN 1992, and BS EN 1997. Structural loading predicted in the numerical models was in accordance with BS EN 1997, Design Approach 2. The primary lining was assumed to degrade over the design life, and was designed to provide temporary support prior to installation of the secondary lining.

The sprayed primary lining concrete mix was designed to a compressive strength of class C20/25 and early age strength gain per J1 curve. A performance-based requirement was followed for the steel fibers which were specified to provide a strength of class D3/S2 as defined by BS EN 14487-1: 2005. The secondary lining mix was specified to have a compressive strength of C30/37 class along with residual flexural strength of 2Mpa at mid-span deflection corresponding to a crack mouth opening displacement (CMOD) of 3.5mm. The forces imposed on the lining were determined using a 2D and 3D finite element model (Figure 4). The three-dimensional finite element analyses were performed

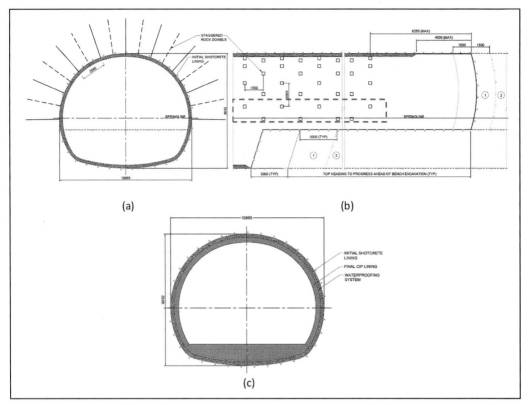

Figure 3. Typical tunnel geometry: (a) primary lining, (b) typical excavated sequence with rock dowels, and (c) secondary lining

with the software Rocscience Phase 2 for 2D analysis and Midas GTS NX for 3D analysis. Ground elements were modelled with 4-noded tetrahedral elements. Sprayed concrete lining elements were simulated with two-dimensional three-noded plate elements, and rock dowels were modelled with one-dimensional embedded truss elements.

Green (fresh) sprayed concrete is subjected to creep during installation due to deformations caused by ground loadings. Based on experience, a stiffness of 7.5GPa was used for green shotcrete to account for creep during installation of the first layers of sprayed concrete. A stiffness 15GPa was used to model the stiffness of the hardened sprayed concrete for the primary lining. The designed ground parameters are shown in Table 1.

A waterproofing system was required to provide long-term protection of the running tunnels against moisture and water inflow. In order to ensure durability of the tunnels, the design specified installation of fully tanked sheet PVC waterproofing membrane between the primary and secondary linings to act as a permanent water barrier all around the mined section.

The tunnel lining was designed for fire exposure in accordance with the Riyadh Metro project requirements. The concrete mix was designed for a 3 hours of minimum fire resistance (Figure 5). The fire design curve ISO-834-1 with temperature curved extended to a minimum of 180 minutes was considered in the design. The ISO 834-1 for 180 minutes (HC and RWS curves are commonly used for road tunnels due to the higher fire HRR). For improved fire resistance of the secondary lining a microfilament Polypropylene (PPE) fibers were added to the concrete mix. Fire tests were conducted to determine PPE fiber dosage and the test results indicated that a minimum of 1kg/m^3 of the PPE fibers is sufficient to achieve the required fire resistance. Further, the PPE fibers were only added to the secondary lining mix for the arch portion of the tunnel and no PPE fibers were added for the invert mix since there was a second stage concrete to cover the entire invert.

Ground Movement

During design development, calculations were performed to estimate the surface settlements induced by the tunnel excavation. Representative cross

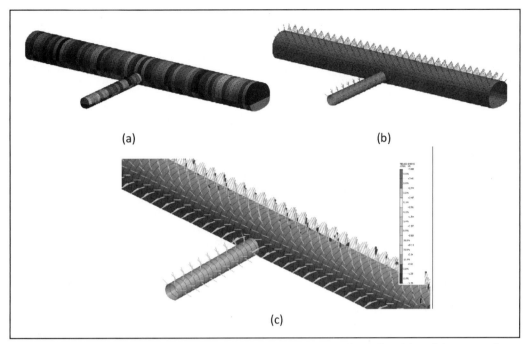

Figure 4. 3D Finite element model for the mined tunnel with connecting Adits: (a) model construction stages, (b) stage after support installation, and (c) axial forces developed in the rock dowels

Table 1. Ground parameters used in the design

Strata	Unit Weight (kN/m³)	Young's Modulus of Elasticity (MPa)	Poisson's Ratio	Friction Angle	Cohesion (MPa)	Coefficient of Lateral Earth Pressure at Rest
Fill/Soil	18	35	0.30	35	10	0.40
Arab Formation—Upper Limestone Breccia (Highly Weathered)	24	1100	0.34	42	110	0.52
Arab Formation—Upper Limestone Breccia (Moderate Weathered to Fresh)	24	4800	0.34	49	190	0.52
Arab Formation (Arab C)—Disturbed Bedded Limestone	25	8000	0.34	56	470	0.52
Arab Formation—Lower Limestone Breccia	25	7000	0.30	49	220	0.52

sections in both 2D and 3D models were developed taking into account construction sequence, geology, and proximity to existing structures, tunnel geometry, and the construction of break-outs. Based on the results, settlement contour lines were derived along the entire Line 2 alignment. The maximum surface settlement above the tunnel centerline was predicted to be slightly over 2mm with virtually no adverse impacts on buildings and infrastructure during construction of the running tunnels. To measure ground movement around the tunnel opening during excavation, the design required installation of in-

tunnel convergence monitoring cross-sections at predetermined tunnel sections and surface settlement monitoring points. Each in-tunnel monitoring section consisted of 5 convergence monitoring points (3D optical targets) around the tunnel periphery, while surface settlement monitoring was carried out using automated total stations and optical targets arranged along the alignment.

CONSTRUCTION

The Line 2 running tunnels were constructed following the SEM principles. Two Alpine roadheaders were used for the excavation (Figure 6). The tunnel excavation commenced from temporary shafts located within the footprint of the Olaya station (east and west sides), which was still to be built. The running tunnel excavated westbound transitioned to an open-cut excavation while the eastbound tunnel transitioned into the 2B4 Station.

The tunnel construction started in November 2014 and was completed in December 2015. Excavation and support was a 24-hour operation of two 12-hour shifts for 6 days per week. During construction, the excavation sequence was adjusted depending on the ground conditions. In better ground, the top heading round length was increased to a maximum of 3m. Similarly, the timing and round length of bench/invert excavation was modified accordingly since no immediate ring closure was required in good competent rock. The bench/invert excavation and support installation was usually

implemented after excavation and full support of 10-15 rounds of top heading. Systematic probing was performed to allow investigation of deteriorated ground and/or groundwater ahead of the advancing tunnel face. Typical probing comprised 17-meter long probe holes with minimum overlapping of 5m. The tunnel geometry was locally enlarged when soft, highly fractured rock or soil was encountered in the tunnel crown to facilitate installation of the grouted steel pipes to provide additional support and minimize ground instability.

Throughout the project, two Senior SEM Engineers were available at all times on site to ensure that the design was adhered to throughout the construction process. This was achieved by leading the daily Shift Review Group meeting (SRG) and preparing the Required Excavation Support Sheet (RESS) in agreement with the tunnel construction team which was represented by the BACS Construction

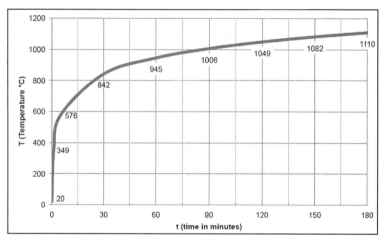

Figure 5. Standard ISO 834 temperature-time curve for a 3-hour fire duration

Figure 6. Tunnel excavation with roadheader (left) and secondary lining formwork (right)

Figure 7. Running tunnel (dashed red line) and its proximity to above ground buildings and viaduct

Manager, Head Geologist, Head Surveyor, Health and Safety Officer and Quality Manager. Adaptations to cater for geological and hydrological conditions as well as to construction and logistic requirements were made in agreement between all SRG members during the daily meetings. During the daily meetings, monitoring data, observations and any unexpected occurrence as well as specific issues such as quality of workmanship were reviewed and adaptations or corrective measures agreed upon. No tunnel excavation was permitted without a RESS being in effect and signed by all involved.

Challenging Urban Settings

The tunnel passed under building foundations and underneath the viaduct pier and therefore it was crucial to minimize ground movement during excavation to avoid impact to the overlying structures. Figures 1 and 7 also illustrates the critical structures along the tunnel alignment. The round length of the tunnel heading advance was reduced at these locations. Similarly, attention was given during construction for any possible future construction and tunnel support design was modified accordingly to make such future construction easier. For instance, GRP rock dowels were used in lieu of steel dowels at locations of future breakouts and at the end of the tunnels at 2B4 stations where new structures had to be constructed directly above the tunnel.

No discernible settlement or ground movement was observed and none of the above ground structures was impacted during construction.

A section of the west tunnel passed below an existing open cut excavation with the tunnel crown daylighting into the open cut. This open cut had been excavated prior to the tunnel construction, for a large commercial building foundation and was later backfilled.

Tunneling in such setting of minimal cover was very challenging and required a careful and well-thought out design. Mined tunnel construction under the open cut area was made possible by the installation of a temporary, roof-shaped slab ('turtle back') alorng the open cut section above the tunnel roof. The existing backfill material was used as earthform. A plastic sheet on top of the fill material was used as separation layer. The 'turtle back' was nominally reinforced with wire mesh. The concrete roof slab was finally cast with the finished structure forming an arched-type lid above the tunnel section. The mined tunnel was excavated under this canopy in 3-meter top heading rounds. Shotcrete primary lining was placed against all exposed rock surfaces in the sidewalls. The two ends of the concrete canopy lid at the interface with the existing rock were reinforced with spiles to avoid overbreak at the interfaces, prevent movement and maintain tunnel profile. Figure 8 illustrates a section of the concrete canopy lid during construction. In the same area, Figures 8c and 8d also illustrates the close proximity of the tunnel crown with respect to the foundations of residential buildings and close proximity to the above ground bridge viaduct highlights the tunneling complexity in such urban settings.

Challenging Geology

One of the anticipated primary challenges during construction was the presence of fractures and dissolution features within the limestone. The aquifers typically developed in carbonate rocks have high permeability due to increased secondary porosity associated with karstic or other dissolution features such as vugs. These vugs were observed in borehole cores recovered during the geotechnical exploration program. These dissolution features developed along the sub-horizontal and sub-vertical joints and discontinuities, encountered in the Arab Formation. Therefore, the presence of undetected voids and cavities presented significant construction risk. Very thin interbeds of dark brown very weak laminated marls were also identified which represented weak seams within the rock. The fractured and less cemented layers within the limestone and breccia were also prone to develop rock slabs for fall out. In general, though, the actual ground conditions encountered during excavation were mostly favorable with competent rock except for few instances, mostly along the west alignment, where cavities and rock slabs were noted within the disturbed bedded limestone.

As part of the risk mitigation, systematic probing was performed from within the tunnel to identify geology ahead of the advancing tunnel face. The probing provided information regarding the presence of weak or fractured rock as well as the presence of any cavities or voids. In case that such adverse

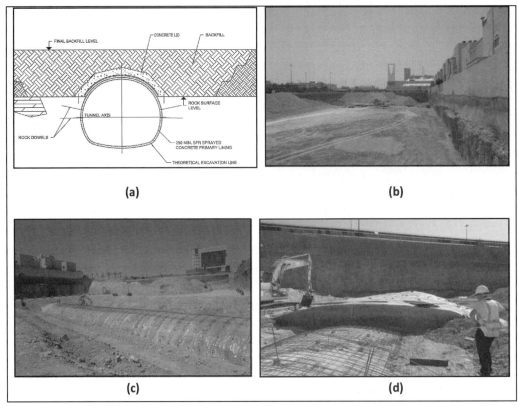

Figure 8. Construction of west tunnel section within the open cut area: (a) tunnel configuration with respect to backfill and canopy lid, (b) open cut area before tunnel excavation, (c) backfill with PVC above the tunnel section, and (d) tunnel concrete canopy

ground conditions were anticipated, mitigation measures such as reduction in round length and filling of the voids were adopted along with other local support measures (tool box items such as spiling) prescribed by the design. Rebar spiles were particularly required to be installed at the break-in and break-out locations and under shallow cover close to the tunnel portals to provide additional support, limit overbreak and minimize ground settlement. Overall, the carefully designed robust systematic support system allowed completion of the tunnel excavation and support on schedule avoiding delays despite the challenging ground conditions at due to dissolution features and fractured rock.

Groundwater Management

Groundwater control was an additional challenge during construction. The presence of dissolution features and pores increases the permeability in bedded limestone and breccia and provides preferred groundwater flow path which increases groundwater inflow. The groundwater level was varying over the

length of the tunnel, but significantly above crown level along most part of the westbound tunnel and at invert level or below along the eastbound tunnel and therefore significant groundwater inflows were expected during excavation at locations of the westbound tunnel. Furthermore, probe holes were converted into dewatering holes when required to allow gravity drainage. The groundwater discharge rate through the probe holes was noted. It was planned install drain holes around the excavation periphery, if pre-determined values were exceeded, to inject grout and control the groundwater inflow. Such measures were not required to be implemented during the construction. Drain mats and flexible drain hoses encased in shotcrete were used to collect localized water inflow and allow shotcrete installation while channeling the water into small temporary sumps located near the invert. As the tunnel excavation progressed, a trench was excavated along the invert centerline and a 12-inch perforated pipe was installed along the trench to divert all collected groundwater along the tunnel to a temporary pump sump (Figure 9). The temporary sump was located

Figure 9. In-tunnel groundwater control: (a) trench with inlet and outlet piles, (b) temporary sump

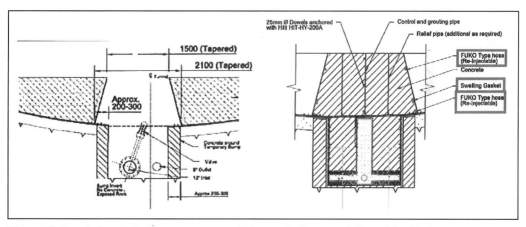

Figure 10. Front view of the temporary sump: before and after completion of final lining

in the middle of the west tunnel at its lowest point and contained three 6-inch dewatering pumps. In an urban environment, discharge of groundwater poses an additional challenge on the construction team. The invert-trench between the temporary sump and the open excavation for the Olaya Station also accommodated two solid 8' pipes, which were used for groundwater discharge back into a pump sump at Olaya Station. From there the water was pumped to the surface and discharged. The pumps were in continuous operation until completion of the final lining due to the high groundwater recharge rate.

The high inflows resulted in increased potential for flooding of the tunnels in case of failure of the dewatering pumps and pumping redundancy was in place to mitigate the risk. At one instance, groundwater discharge of as much as 1500cm³/min had to be pumped from the west tunnel to provide safe working conditions. The east tunnel was relatively dry, not requiring major groundwater management measures.

Closure of the sump for installation of the invert final lining was a significant construction challenge as the pump had to be in operation at all times. A temporary closure of the pump could easily flood the tunnel due to the large water inflows and would cause invert heave in the tunnel section where the arch was not completed. Therefore, a carefully designed plan was executed to finally close the temporary sump pit and complete the secondary lining installation. Before closing of the sump pit, the secondary lining including the invert was completed along the tunnel alignment. For the closure of the sump pit, at first, the feeding pipes into the sump pit were cut flush to the sump walls to provide a flat substrate for installation of PVC membrane and then ground anchors along with blinding and gravel layer were installed on the base of the sump. A PVC membrane along with drainage mats were installed over the sump faces. A perforated T-shaped drainage pipe was then installed to keep the dewatering system running and prevent build-up of groundwater pressure during concrete curing.

A 4-inch drainage layer comprising of clean crushed stone was placed above the sump pit to allow the filtered water reach into the perforated T-pipe. A PVC membrane was installed on top of the drainage layer and connected to the PVC membrane of the sump walls. The sump pit was then filled with concrete under continued pumping to avoid groundwater pressure building during curing. Additional relief pipes were attached to the top of the finished concrete extending all the way to the top surface of the finished invert secondary lining (Figure 10). If water discharge was observed through the pressure relief tubes, a pumping was implemented to reduce the water pressure and ensure depressurized conditions for the invert pour above the sump.

CONCLUSION

The Line 2 running tunnels were successfully constructed without any delays or impact to the utilities and structures located above the tunnel alignment. A systematic and robust design along with ground probing and presence of experienced on-site SEM personnel were key for the successful execution of the project. The presence of dissolution features, including cavities, groundwater, and fractured rock presented significant construction challenges which were addressed with consideration of the geology and careful interpretation of ground probing ahead of the tunnel face. Identification of adverse ground conditions ahead of the tunnel face including the presence of groundwater allowed mitigation of such risks by either reducing the round length or implementing grouting program. Although the running tunnels were excavated in an urban setting in close proximity to building foundations and viaduct piers, no adverse impact to the existing structures was observed as a result of a carefully designed and executed tunnel excavation and support. Adherence to the design was ensured by experienced SEM Engineers through the daily SRG meeting where the excavation and support requirements were discussed and agreed for implementation.

Downtown Bellevue Tunnel—Analysis and Design of SEM Optimization

Christoffer Brodbaek and Derek Penrice
Mott MacDonald

Jake Coibion
Guy F. Atkinson Construction

Chad Frederick
Sound Transit

ABSTRACT: The Downtown Bellevue Tunnel (DBT) is part of Sound Transit's $3.7 billion light rail extension connecting Seattle with the cities of Bellevue and Redmond. In December of 2015 Atkinson Construction was awarded the DBT construction contract, which includes a 1,985 foot long, 38 foot wide Sequentially Excavated (SEM) tunnel. As part of a subsequent construction schedule reduction effort, in areas with favorable ground conditions Atkinson proposed that the excavation of the SEM tunnel be revised from six to three headings. To validate the optimized excavation sequence, three-dimensional numerical modeling was required to demonstrate its safety and reliability. This paper describes the numerical modeling—addressing initial liner design, ground deformations and three-dimensional excavation stability—and compares analysis results with construction field measurements and observations.

INTRODUCTION

The Sound Transit East Link light rail extension from Seattle to Bellevue and Redmond in Washington State includes a 2,500-foot long Tunnel in Downtown Bellevue. The E330 Contract, awarded to Atkinson Construction includes approximately 2,235 feet of tunnel construction. The DBT corridor—shown in Figure 1—comprises a short section of cut and cover structure, required to accommodate sidewall mounted ventilation fans at the south end of the alignment and a 1,985-foot long SEM tunnel. The north cut and cover structure and future Bellevue Downtown Station are part of an adjacent construction contract.

A typical section of the final configuration of the new tunnel is shown in Figure 2. It consists of two tracks separated by a center concrete wall, with a 12-inch permanent conventional reinforced concrete liner cast upon completion of the SEM excavation and placement of the initial support of excavation. The initial support of excavation consists of a 10-inch thick fiber reinforced shotcrete liner. The focus of this paper is the design of the tunnel excavation and initial support of excavation utilizing the Finite Element Method (FEM) at the southern part of the SEM tunnel, where a revised, three-heading advance was deployed. In the northern part

of the SEM tunnel a six-heading excavation will be maintained due to shallower cover, poorer ground conditions and a shallower groundwater table. A six-heading excavation was originally planned for in the southern part of the tunnel, but through construction optimization assessment, described in Penrice et al (2017) and in this paper, it was concluded that it was feasible to perform the southern part of the SEM tunnel as a three-heading excavation. Definitions of the south cut-and-cover, three-heading and six-heading zones part of the E330 Contract are shown on the tunneling profile in Figure 3.

GROUND CONDITIONS AND MATERIALS

The ground conditions along the southern part of DBT are shown on geological profile in Figure 4. They consist of glacially over-consolidated soils with mining of the SEM tunnel predominately performed in the Vashon Lodgement Till (Till, Qvt), which is a mixture of clays, silts, sand and gravel, with a significant amount of fine-grained materials. The Till is overlain by 3 to 10 feet of sandy Fill (Hf) and underlain by Vashon Advance Outwash (Outwash, Qva) and Pre-Vashon Lacustrine Deposits (Qpnl). The two latter ground types are also glacially over-consolidated deposits, although weaker and softer than the Till.

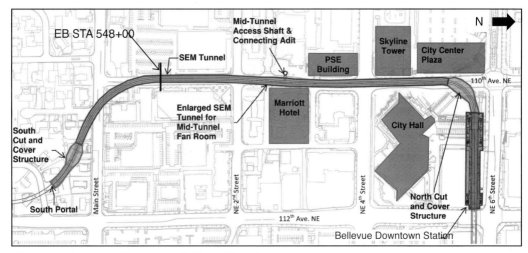

Figure 1. Alignment of Downtown Bellevue Tunnel

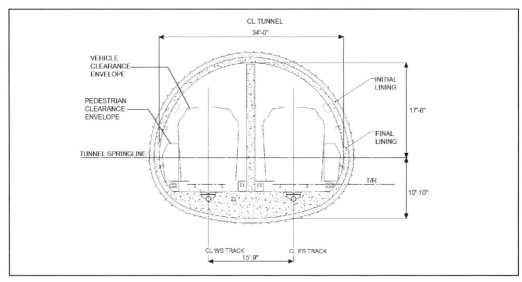

Figure 2. Final configuration of light rail transit in SEM tunnel

Mean geotechnical ground parameters used in the design of the 6-heading excavation sequence and subsequently used in the redesign of the 3-heading sequence are provided in Table 1. The regional water table is generally located at the contact of the Till and the Outwash, but local perched water tables are present within sand lenses within the Till formation, requiring local drainage during excavation.

Typical grain size distributions of Till are shown in Figure 5, indicating that the fines content of the Till typically varies between 25 and 55 percent. The Till is described as firm to slow raveling above

the water table and generally fast raveling when saturated or below the water table.

The specified unconfined compressive strength for the shotcrete initial liner is shown in Table 2, including specified minimum early age strength, which is applied in the finite element analysis to establish time dependent liner stiffness based on excavation rates and productivities. The 3-day and 28-day shotcrete strengths are consistent with the as-bid specifications. The early age shotcrete strengths were agreed with Atkinson based on what was achievable based upon the proposed mix designs,

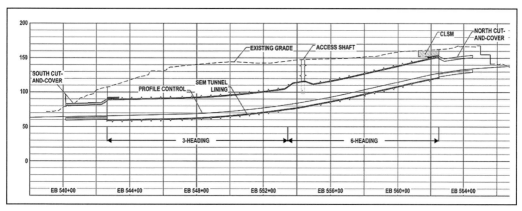

Figure 3. Tunneling profile

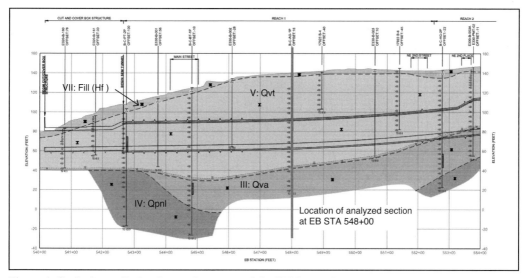

Figure 4. Geologic profile for the southern part of the SEM tunnel

and what was necessary to achieve desired rates of production.

CONSTRUCTION SEQUENCES AND GROUND SUPPORT

Originally it was planned that the entire tunnel excavation be performed with the six-heading sequence comprising left and right drifts, each of which is subdivided into top heading bench and invert excavations, as shown in Figure 6. The six-heading excavation sequence met all safety, stability and deformation requirements for the project along the tunnel alignment. Due to a delayed notice-to-proceed for the construction of the DBT caused by delays in right-of-way acquisition on the project; Sound Transit and Atkinson met shortly after contract

award to find ways to recover the project schedule, as completion of DBT is on the critical path for completion of the East Link extension project. As such expediting the construction of the SEM tunnel was an important element in the construction schedule improvement, and it was estimated that a two to three months schedule reduction could be achieved by simplifying the excavation and initial support in the southern half of the SEM tunnel, due to the more favorable ground conditions in this part of the tunnel.

As a result, it was agreed that Mott MacDonald, as Engineer of Record for DBT would evaluate if it would be safe to perform the southern half as a three-heading excavation comprising a single top heading, bench and invert excavation. The evaluation was to be performed as part of a no-cost change order executed between Sound Transit and Atkinson. To limit

Table 1. Mean geotechnical design parameters for SEM tunneling

Soil Unit/Name	Thickness (ft)	γ (pcf)	E' (ksf)	v (-)	φ (deg)	c' (ksf)	K₀ (-)
VII Fill (Hf)	3	130	1,000	0.35	37	0	0.40
V Vashon Lodgement Till (Qvt)	95	140	10,800–37,800	0.35	42	0.5	1.22
III Vashon Advance Outwash (Qva)	38	135	11,880	0.35	40	0.05	0.80
IV Pre-Vashon Lacustrine Deposits (Qpnl)	>44	125	8,100	0.35	32	0.3	1.22

g = unit weight, E'= Young's modulus; v = Poisson's ratio, φ = internal friction angle, c' = effective cohesion, K₀= at rest earth pressure coefficient.

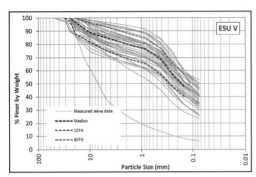

Figure 5. Grain size distribution of Vashon Lodgement Till

Table 2. Specified shotcrete strength

Age of Shotcrete	Shotcrete Compressive Strength, $f'c$ (psi)
2 hours	200
8 hours	800
16 hours	1525
1 day	2000
3 days	3000
28 days	5000

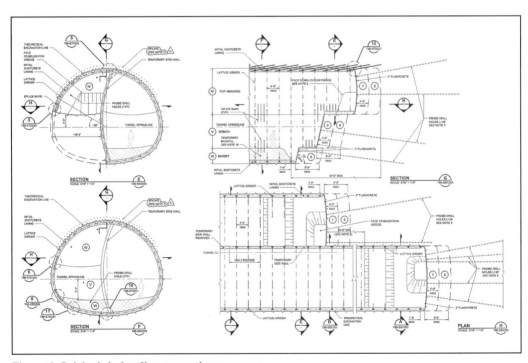

Figure 6. Original six-heading excavation sequence

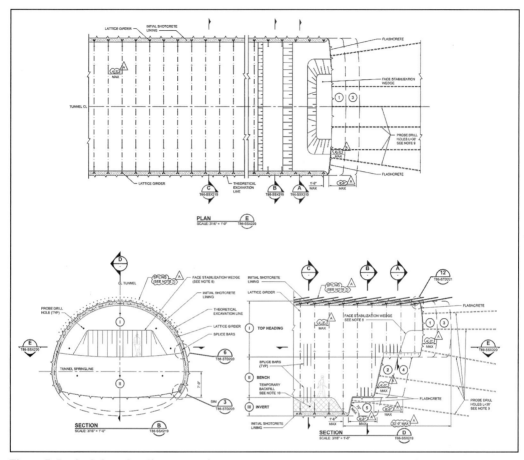

Figure 7. Revised three-heading excavation sequence

expenditures, the evaluation would be performed in two stages. The first stage would comprise sufficient numerical modeling to demonstrate the feasibility of the 3-heading concept and provide Sound Transit and Atkinson with sufficient information to make a go/no go decision. Stage 2, comprising modifications to the Contract Documents to depict the 3-heading sequence would only be implemented in the event of a Stage 1 'go' decision. The revised excavation sequence, shown in Figure 7, eliminates the temporary center wall and simplifies the overall excavation process. The excavation optimization of the southern part of the SEM tunnel required evaluation of ground volume loss, anticipated ground deformations, initial liner design forces and stability of the face during excavation. Each of these construction stages may conveniently be evaluated with 3D finite element modeling incorporating the construction sequence, excavation productivity and shotcrete strength gain.

Standard support measures required to construct the tunnel under stable ground conditions include flashcrete, fiber reinforced shotcrete lining and lattice girders. Supplemental toolbox items deployed as required during construction to pre-support, support or improve the ground around or within the tunnel include rebar spiles, grouted pipe spiles, metal sheets, face stabilization wedge, face bolts, grouting, additional shotcrete and vacuum dewatering. The three-heading SEM tunnel evaluation was based on deployment of standard support measures only.

Both the six-heading and the three-heading excavation sequences were based on 4-foot advances of the top heading and bench and an 8-foot advance of the invert with lattice girders every 4 feet as well. The 6-heading excavation comprised a top-top-bench-bench-invert sequence. At the request of Atkinson, this was modified slightly for the 3-heading excavation to a top-bench-top-bench-invert sequence to suit construction logistics. One full 8-foot advance cycle of the three-heading excavation progress as follows: (1) 4-ft top heading, (2) 4-ft bench, (3) 4-ft top heading, (4) 4-ft bench and

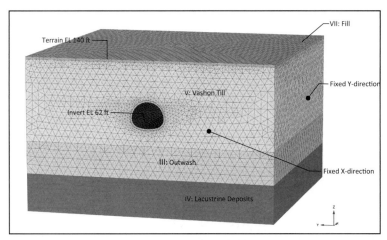

Figure 8. 3D finite element model at STA 548+00

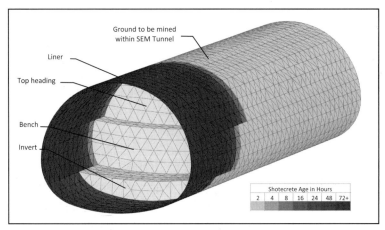

Figure 9. Detailed finite element model of tunnel

(5) 8-ft invert. During construction the maximum advance length was increased slightly to 4.25 feet and 8.5 feet top/bench and invert, respectively, since observed deformations were small and well below project defined ground deformation limits.

3D FINITE ELEMENT ANALYSIS

The 3D analysis was performed at the area with the largest overburden along the corridor which is around STA 548+00 and was considered one of the more critical areas in the southern half of the SEM tunnel. The overall dimensions of the finite element model at STA 548+00 are shown in Figure 8. The detailed elements at the tunnel are shown in Figure 9. Face wedge or other ground stabilizing toolbox items deployed on an as required basis during excavations were not included in the finite element model as each of these items are optional and are planned only to

be used for local mitigation of adverse mining conditions. The construction sequence for the three-heading concept was incorporated into the finite element analysis evaluating each stage of excavation and shotcrete application in a stepwise fashion as shown in Figure 10.

A typical bench excavation stage from construction at STA 548+00 is shown Figure 11. It is noted that no stabilizing face wedge or other toolbox items were deployed during construction at this station, as the Till was dry and stable and exhibited excellent stand-up time. As such, the actual excavation conditions were close to the assumptions applied in the finite element analysis.

The three headings are mined in a stair-step fashion to replicate the excavation conditions. In the finite element analysis liner properties for each advance are updated according to applied shotcrete

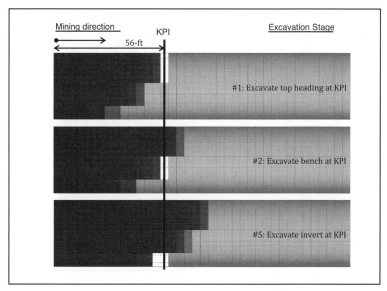

Figure 10. Detailed excavation sequence applied in analysis

thickness and estimated shotcrete strength gain based on anticipated construction cycle times provided by Atkinson Construction and listed in Table 3. This is done to replicate the liner deformations occurring during mining and assess the liner design forces for the interim conditions, without ring closure and early shotcrete strength. Additional information on liner capacity verification may be found in Penrice et al (2017).

It is noted that a total cycle time for an 8-foot advance in the analysis was assumed to be 38 hours. The average cycle time experienced during construction in and around STA 548+00 was 36 to 38 hours.

One of the key outputs of the analysis is accumulated ground displacements which are shown in Figure 12 for a longitudinal section along the centerline of tunnel. The shape of the ground deformation field is typical of an advancing tunnel with small deformations at the face of excavation and increasing deformations farther away from the face of excavation.

The maximum ground deformations obtained via FEM are of the order 0.2 inches along the centerline of the tunnel. Some elastic invert heave is noticed, which is primarily a result of the simplified isotropic linear elastic perfectly plastic Mohr-Coulomb ground material model utilized in the analysis.

The theoretical volume loss from mining of the SEM tunnel was calculated to be 0.11% for the three-heading excavation. This is comparable to computed volume losses for the six-heading excavation scheme which varied between 0.09% and 0.18% at different

Figure 11. Bench excavation with exposed side walls prior to shotcrete application

Table 3. Design excavation and shotcrete cycle times for 8-foot advance

Advance	Round Length	Duration	Description
1)	4-ft	10 hrs	First top heading advance.
2)	4-ft	5 hrs	First bench advance.
3)	4-ft	10 hrs	Second top heading advance.
4)	4-ft	5 hrs	Second bench advance.
5)	8-ft	8 hrs	Invert excavation and closure of ring.
1)-5)	**8-ft**	**38 hrs**	**Total cycle time for 8-ft advance**

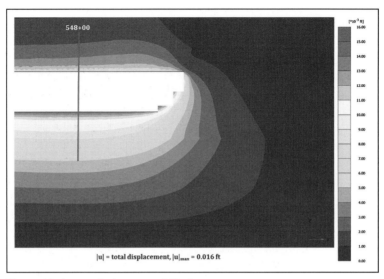

Figure 12. Accumulative ground displacements

locations along the alignment. This was a very favorable outcome of the three-dimensional finite element analysis performed, as it documented that switching to a three-heading excavation should not increase tunneling induced deformations nor adversely impact adjacent properties, structures and utilities. The main reason for the similar ground deformation response from three- and six-heading excavation schemes is attributed to favorable engineering properties of the Till, where the ground strength facilitates excavation be performed in stable ground with small plastic deformations. Additional discussion on face stability is included in section Tunnel Face Stability of this paper.

The ground deformation analysis concluded that the anticipated ground deformations were small, indicating stable ground conditions during mining of the SEM tunnel and generally supporting a 'go' decision for implementation of the 3-heading sequence.

FIELD OBSERVATIONS

Monitoring during mining of the SEM tunnel is important to understand the actual ground response to tunneling and to provide early indications of softer than anticipated ground, shotcrete anomalies or other adverse conditions. The monitoring results also serve as an indicator of potential deployment of toolbox items required for continued safe excavation of the SEM tunnel. The project includes prescribed instrumentation requirements including surface and building mounted monitoring points, borehole installed inclinometers and extensometers, in-tunnel convergence monitoring and strain gauges applied

on lattice girders. Convergence monitoring and lattice girder strain gauge arrays are generally located at 50-foot spacings along the alignment, with tighter spacings at the outset of the SEM tunneling.

Limits on acceptable ground deformations were established through a project-wide building vulnerability assessment. Due to different structures, foundations, utilities and varying ground conditions along the alignment different specification limits were developed at different stations. In the area around STA 548+00 the surface ground movement limits were 0.3-inch trigger (amber) level and 0.45-inch maximum level. Maximum deformations of the tunnel during construction measured at the convergence monitoring points were also limited by 0.3 inch as a threshold (amber) level and 0.45 inch as limiting (maximum) level.

With the excavation sequence, explicitly modelled tunnel convergence may be evaluated as a function of tunnel face distance at STA 548+00—similar to a longitudinal displacement profile conventionally used in tunnel engineering. The tunnel side wall displacements obtained from the finite element analysis are shown as solid black and blue lines in Figure 13 at the tunnel crown and springline, respectively. The crown displacement shown in Figure 13 is the total vertical displacement of crown and invert to adjust for the global elastic heave at the invert from the finite element model. The 3D analysis captured increasing displacements with distance from the excavation face, confirming the stabilizing 3D effect of the face. At approximately 28 feet from the excavation face—which approximately corresponds with the distance to ring closure, the deformations

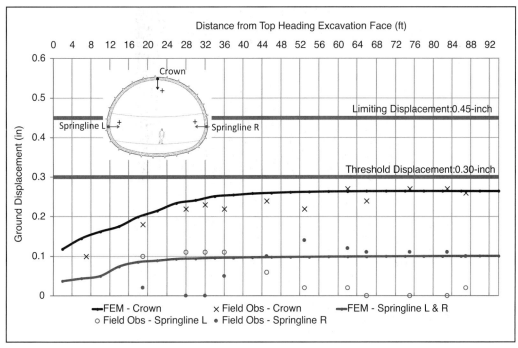

Figure 13. Tunnel displacements vs. distance to face of tunnel—comparison of FEM with field observations

at springline become constant with 0.1-inch convergence. The crown deformation would become constant at approximately 40-foot distance to the face of the excavation with a maximum convergence of 0.26-inch. The maximum displacements computed via FEM were less than the specified threshold (amber) and limiting (maximum acceptable) displacements.

To compare the results from FEM, the observed convergence monitoring results obtained during construction were analyzed and plotted as discrete points against FEM results in Figure 13. It can be seen that there is generally good agreement between results from FEM and the field observations especially considering the small deformations at hand. The field observations confirmed maximum convergence is less than the threshold and limiting displacements and the convergence would increase with increasing distance to the face of excavation, most pronounced at the crown. Due to the small 0.1-inch to 0.3-inch displacements there is significant scatter in the field measurements. At the crown the maximum ground displacement was 0.26-inch by FEM versus 0.27-inch observed. The springline observations deviate a little more with 0.14-inch maximum deformation observed versus 0.1-inch by FEM. Additional information on construction of the Downtown Bellevue Tunnel in general may be found in Wongkaew et al. (2018).

TUNNEL FACE STABILITY

Maintaining adequate face stability during mining is critical for safety and ground movement control. A detailed face stability analysis was required to quantify the safety margin against collapse at the face during mining—the three-dimensional FEM model created for the project is suitable for such safety evaluations through a special ground strength-reduction procedure implemented in PLAXIS 3D, see PLAXIS 3D (2015) for more details.

Multiple face stability features were included as toolbox items for SEM construction such as rebar spilling at the crown, face stabilization wedge left in place at the center of the excavation and immediate application of 2-inch flashcrete on exposed ground. Each of these items increases the face stability and prevents local wedging and raveling of ground during the excavation process. In addition, probe drilling at the face was specified and is performed on a bi-weekly basis to identify imminent ground and groundwater conditions.

Going from a six-heading excavation to a three-heading excavation required additional confirmation of global face stability due to the increased area of exposed ground and time of exposure in each excavation stage of the typical 8-foot advance cycle. A face stability safety evaluation was performed in PLAXIS 3D, utilizing the $\varphi'\text{-}c'$ strength reduction method,

Table 4. Summary of face stability analysis

	FS - 3D FEM		FS - Empirical	
	Mean Design Strength	**Lower Bound Strength**	**Mean Design Strength**	**Lower Bound Strength**
Excavation	$c' = 0.5$ **ksf**	$c' = 0.2$ **ksf**	$c' = 0.5$ **ksf**	$c' = 0.2$ **ksf**
Top heading	2.0	1.3	1.8	1.0
Bench	2.1	1.3	1.8	1.0
Invert	2.1	1.2	2.0	1.2

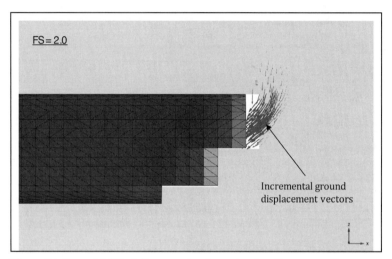

Figure 14. Face stability failure mechanism—top heading excavation

where the soil strength is incrementally reduced until the critical ultimate limit state of the ground is reached according to the following relationship:

$$FS = \frac{c' + \tan \varphi'}{c'_{red} + \tan \varphi'_{red}}$$

c' and φ' are the actual effective cohesion and friction angle of the ground, and c'_{red} and φ'_{red} are the reduced effective cohesion and friction angle at the ultimate limit state of the ground. Consequently, FS expresses a safety margin against at large face collapse during mining.

As discussed under Ground Conditions and Materials the Till stand-up time facilitates the SEM construction and is expressed by the mean design effective strength parameters, $\varphi' = 42$ degrees and $c' = 0.5$ ksf. Due to presence of local sand pockets and potential degradation of exposed ground during mining a lower bound cohesion assessment was performed to evaluate stability under degraded conditions. The lower bound effective strength parameters of the Vashon Till applied is $\varphi' = 42$ degrees and $c' = 0.2$ ksf.

Vermeer et al (2002) developed empirical face stability expressions based on traditional soil mechanics principles. Factors of safety were computed using these empirical expressions and compared with results from the 3D finite element analysis as shown in Table 4. A factor of safety equal to or greater than 1.5 was considered an acceptable safety level on the project. It is seen that under the anticipated ground conditions, applying the mean design ground strength, both the 3D FEM and the empirical expressions result in factors of safety exceeding the minimum desirable value of 1.5. Hence, under normal mining conditions it was anticipated that the three-heading excavation could proceed with minimal use of face stabilizing toolbox items.

For areas with potentially sandier soil layers the lower bound strength of c' of 0.2 ksf yields factors of safety 1.2 to 1.3 using the 3D finite element method and factors of safety between 1.0 and 1.2 using empirical methods. This means that with lower bound soil strength the excavated face would be stable, albeit with a factor of safety less than the minimum desired factor of safety of 1.5. This was agreed to be an acceptable risk level among the project team since: a) It is unlikely to have the entire excavated face dominated by materials characterized by the lower bound ground strength, b) There would typically be a transition to weaker ground that should be identified from face mapping performed

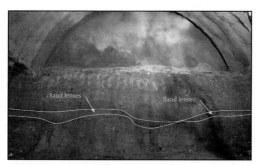

Figure 15. Bench excavation. Lower cohesion sand lenses indicated.

during excavation and probe hole explorations, and c) Additional face stability can be provided during tunnel excavation by toolbox measures including face wedge left in place, placement of additional shotcrete, spiles and grouting.

Figure 14 shows the face failure mechanism for the top heading excavation with exposed ground/unlined. The stability results confirm the traditional chimney shaped soil wedge often assumed in face stability evaluation for cohesive and friction grounds due to soil arching around the opening.

During construction it has become clear that the anticipated sand lenses encountered in the geotechnical explorations are present but their extent is generally quite limited, as shown in Figure 15. Since the sand lenses are local, their reduced cohesive strength has not adversely compromised the stability of the face and the excavation of the three-heading SEM tunnel has generally taken place without deployment of face stabilizing toolbox items, for sections with sand lenses like those shown in Figure 15. In some locations where sand lenses have been encountered in the crown, rebar spiles were locally installed for additional ground support.

CONCLUSIONS

The excavation optimization of the SEM tunnel was driven by necessity, to generate a two to three month construction schedule reduction for the DBT, whose completion lies on the critical path of the larger East Link project. The proposed optimization of the excavation sequence for the DBT comprised eliminating the six-heading excavation and replacing it with a more efficient three-heading excavation scheme in the southern half of the SEM tunnel. The results from the detailed 3D finite element analysis performed were instrumental in informed decision making to proceed with the proposed optimized excavation scheme.

The finite element analysis demonstrated that ground deformations were below the specified deformation limits as it relates to tunnel convergence and did not significantly increase the theoretical volume loss due to tunneling when compared to the originally planned six-heading excavation scheme. The theoretical results obtained with FEM have since been corroborated by field observations, confirming that the three-heading excavation could be performed with ground movements less than the safe deformation limits defined for the project through a settlement vulnerability assessment of adjacent buildings and properties. The small deformations computed with 3D finite element analysis are also an indication of relatively stable excavation conditions.

To quantify the margin against face instabilities, a face stability safety evaluation was performed—which showed factors of safety of 2.0 or greater for mean design ground strength values—exceeding the minimum desired safety factor of 1.5. Again, the results of the FEM and empirical analyses have been borne out in the field, through observation of Till stand up time performance, and from the lack of face stabilizing toolbox measures applied at the analysis location.

The field corroboration of the FEM outcomes provides confidence in the value of FEM as a tool to facilitate sound design decision making.

REFERENCES

Penrice, D., Yang, H., Frederick, C. and Coibion, J. (2017). Downtown Bellevue Tunnel—Concept Optimization Through Team Collaboration. Rapid Excavation and Tunneling Conference, San Diego 2017.

PLAXIS 3D (2015). PLAXIS 3D Reference Manual, Delft, Netherlands.

Vermeer, P., Ruse, N., and Marcher, T. (2002). Tunnel Heading Stability in Drained Ground, Felsbau Journal for Engineering Geology, Geomechanics and Tunnelling, 20 no. 6, 2002.

Wongkaew, M., Murray, M., Coibion, J. and Frederick, C. (2018). Design and Construction of the Downtown Bellevue Tunnel. North American Tunneling, Washington D.C. 2018.

Elasto-Plastic Design of Fiber-Reinforced Concrete Tunnel Linings

Axel G. Nitschke
WSP USA

ABSTRACT: The most advantageous property of Fiber Reinforced Concrete (FRC) is its toughness, changing the failure mode of concrete from brittle to elasto-plastic.

In the plastic stage, the FRC tunnel lining effectively softens, which allows load re-distribution within the structure and consequentially increasing its ultimate load-bearing-capacity. However, to take full structural and economic advantage of this effect, the lining design's focus should shift from a "strength" perspective to a "deformation" perspective respectively from a design concept based on "cross-section"-failure to design concept based on "system"-failure.

The paper discusses the load bearing behavior of FRC in tunnel linings under combined moment-normalforce loading. Based on the findings, options for elasto-plastic failure-mode design of FRC tunnel linings, plastic hinges, and the paradigm shift that is required to take full structural and economic advantages of FRC for tunnel linings.

INTRODUCTION

The use of Fiber Reinforced Concrete (FRC)—with steel or macro-synthetic fibers—has technical and economic advantages which stem from the fact that fibers transform the post-cracking behavior from a brittle failure mode typical of unreinforced concrete into an elasto-plastic behavior. Numerous codes and guidelines provide qualitative or quantitative design approaches [9,10,11,12,13,14]. However, current modeling of the load-bearing behavior based on a Stress-Strain-Relationship (SSR) for tunneling applications appear to underutilize the structural and economic potential of FRC.

A previous article of the author [1] discussed the load bearing capacity of FRC tunneling linings under combined moment-normal force loading conditions and typical results based on a parametric study. The article also identified the weakness of the current concept and suggested a path to more fully utilize the structural and economic potential of FRC. This article complements the previous paper and focusses on a design concept that takes full advantage of the elasto-plastic failure behavior of FRC in order to utilize the full economic benefits of the material.

The concept discussed herein is theoretical in nature. To limit the scope of this paper, the discussion is focused on the load bearing capacity under cracked conditions. Therefore, design concepts that do not utilize the toughness potential of FRC (i.e., by limiting it to un-cracked conditions) are not discussed herein. Furthermore, the focus lays solely on combined moment-thrust or moment-normalforce (M/N) loading of tunnel linings in which bearing

capacity relies on a tunnel arch. This is typical for soft ground tunnel linings and rock tunnels with soft-ground-like behavior. They are not useful in tunnels with no arching effect, which is typical for tunnels in rock with relatively thin linings or with an irregular shape.

Different international codes and guidelines for FRC provide testing procedures based on simply-supported beam tests that are utilized to define a SSR by basically amending the known trapezoidal or parabolic SSR for concrete on the compression side with assumptions for a SSR on the tension side. The SSR models a homogeneous, composite material behavior and not discrete fibers. It is therefore irrelevant which types of macro fibers—steel or synthetic—are used.

The use of a SSR is typically evaluated based on beam test data. Under elastic (un-cracked) conditions the beam theory and the classical mechanics for materials apply. However, after the initial cracking of the FRC the material is no longer homogeneous and the theoretical conditions for beam-theory no longer apply. The bearing behavior of FRC in beam tests in a cracked state are better described using a stress-crackwidth relationship rather than a stress-strain relationship. It is important to understand that for the above reason a SSR cannot be measured directly in standard FRC beam tests. The codes and guidelines are therefore describing testing procedures that measure external forces and deformations, which are then transformed into stresses and "equivalent" strains via an equivalence model which

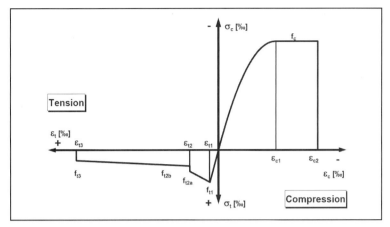

Figure 1. Generic stress-strain relationship for fiber-reinforced concrete

implies several assumptions. For further discussion refer to [1] and [2].

STRUCTURAL BASICS OF FRC DESIGN

The biggest difference between the sectional strength of unreinforced or rebar reinforced concrete and FRC is that the concrete in unreinforced or rebar reinforced concrete has theoretically no bearing capacity in tension. In the modeling of conventionally reinforced concrete sections, all tension is supported by the rebar. Since the location of the rebar is known, the location of the resulting tensile force is also known, and this simplifies the calculation of the equilibrium compared to FRC sections. The computation of axial equilibrium in FRC sections is much more challenging because the location of the resulting tension force is an unknown during the computation and moves if the external load and the distribution of the strain over the cross-section changes. The design assumptions for the calculation of the sectional strength for FRC based on a SSR are presented and further discussed in detail in [1].

It is, however, crucial to realize, that with this approach FRC is assumed to be a macroscopically homogeneous and isotropic material [4]. The material properties of a single fiber in the model becomes irrelevant. Therefore, the fibers and the concrete are modelled using a single SSR relationship and not two, (i.e., as for steel rebar and concrete.)

After the cracking of the material under tension, the material properties in the model are based on strains rather than a discrete crack. In the model the cracked material is also viewed as homogeneous and isotropic. Since this is in the area around the crack, it is obviously not the case. This circumstance is very important to realize and understand when evaluating the sectional strength of FRC using a SSR. During the evaluation of material testing data based on beam

tests (and subsequently the design of the structure), it is assumed that the crack is "smeared" over a certain length into an "equivalent strain," which is also referred to as "integral approach" [5].

Fibers influence the bearing behavior in multiple ways. However, three properties are most relevant for application in tunnels [4]. (1) They slightly increase the flexural tensile strength, which is mostly needed if improved properties under un-cracked conditions are desired, (i.e., to design for serviceability.) However, for the case of ultimate bearing capacity of tunnel linings, the residual flexural tensile strength under cracked conditions (2) and the increase of the toughness (3), are the major benefits. Hence, the focus of this paper is on the performance improvements attributable to (2) and (3).

The provision of a reliable and usable post cracking tensile strength transforms the brittle failure mechanism of plain concrete into a ductile failure mode. This is a material property that provides major engineering and economic advantages, especially if utilized to facilitate system failure of a tunnel lining rather than a cross section failure at one, presumably most critical location. Providing a concept for the design of a system failure is the main subject of this paper.

STRESS-STRAIN RELATIONSHIP OF FRC

A generic SSR and nomenclature of the variables used throughout this paper is shown in Figure 1. The tension side is represented by three sections, which model and control the bearing behavior in the three different phases (elastic, micro-cracking, macro-cracking). A detailed discussion about the three different phases is provided in [1].

The compression side uses a classical parabolic-constant shape.

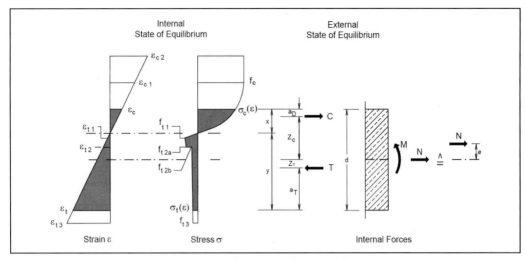

Figure 2. Calculation of equilibrium between internal and external forces

Table 1. Stress-Strain-Relationship used in the Parametric Study

Stress	Tension				Compression	
	f_{t3}	f_{t2b}	f_{t2a}	f_{t1}	f_{c1}	f_{c2}
[MN/m²]	0.5	1.0	4.0	4.0	−40.0	−40.0
[% of f_c]	1.25%	2.5%	10%	10%	100%	100%
Strain	ε_{t3}	ε_{t2b}	ε_{t2a}	ε_{t1}	ε_{c1}	ε_{c2}
[‰]	10.0	0.16	0.16	0.12	−2.0	−3.5

Cross section: thickness d = 0.25 m ; width b = 1.0

The load bearing capacity of a cross section based on the SSR is calculated by finding the equilibrium between internal and external forces as shown in Figure 2 and discussed in detail [1], with the curvature κ of a cross section defined by Equation 1.

Equation 1: Curvature

$$\kappa = \frac{|\varepsilon_c| + |\varepsilon_t|}{d} \left[\frac{1}{m}\right] \tag{1}$$

LOAD-BEARING BEHAVIOR OF FRC UNDER COMBINED M/N-LOADING

Moment-Normalforce Interaction Diagrams (MNID) are typically used during the design of tunnel linings (and columns under combined moment-normalforce (M/N) loading in general) for rebar reinforced linings as well as FRC. However, while generic MNIDs are available for rebar reinforced members, a SSR-specific MNID must be developed for FRC. Generic MNIDs for FRC can be developed in a similar fashion to rebar reinforced members if the diagrams are

dimensionless and all strength values are provided. The dimensionless normalforce n can hereby be interpreted as the utilization towards the maximum thrust under pure compression (see n = 1.0 (or 100%) for m = 0 in Figure 3 and Equation 2).

The SSR used for the following parametric study and graphs in this paper is defined in Table 1 and represents typical values for FRC (i.e., used for an initial tunnel lining.) For the nomenclature and shape refer to the SSR in Figure 1.

Figure 3 shows the MNID for the SSR used. Figure 4, Figure 5, and Figure 6 show the results of a parametric study for the bearable moment under varying normalforces for the same SSR. Figure 4 displays the results over the strain on the tension side ε_t, while Figure 5 shows it over the strain on the compression side ε_c. Figure 6 presents the same results over the curvature κ.

All results shown can also be easily transferred into non-dimensional plots by using Equation 2 and Equation 3 for the non-dimensional normalforce n and the non-dimensional moment m and as shown with the secondary axes in Figure 3 and Figure 6.

Equation 2: Non-Dimensional Normalforce *n*

$$n\,[-] = \frac{N}{f_c * b * d}$$

$$= \frac{N\,[MN]}{40 * 1.0 * 0.25 \left[\frac{MN}{m^2} * m * m\right]} \tag{2}$$

$$= \frac{N\,[MN]}{10\,[MN]} = \frac{N\,[kN]}{10{,}000\,[kN]}$$

Equation 3: Non-Dimensional Moment _m_

$$m\,[-] = \frac{M}{f_c \times b \times d^2}$$

$$= \frac{M\,[MNm]}{40 * 1.0 * 0.25^2 \left[\dfrac{MN}{m^2} * m * m^2\right]} \quad (3)$$

$$= \frac{M\,[MNm]}{2.5\,[MNm]} = \frac{M\,[kNm]}{2,500\,[kNm]}$$

The dimensionless plots allow to easily investigate i.e., the effect of different strength SSRs or different lining dimensions. The plots can be basically just linearly scaled up or down using the formula above. Hence, the characteristic of the SSR is defined by the different strength values for the tensile strengths f_{ti} at defined strains ε_{ti} relative to the compressive strength f_c. These can also be expressed as a relative value i.e., in percent as shown in Table 1, which highlights the flexibility of the dimensionless graphs provided and a dimensionless approach in general.

A good rule of thumb is that tunnel linings are typically using between 5% to 30% of the compression capacity of a member [2]. That means, that typically only the lower third of an MNID is relevant for most designs. So, for example, in the MNID in Figure 3, a typical utilization in a tunnel lining design would be between n = 0.05 to 0.3 or N = 500 to 3,000 kN.

The different lines in Figure 3 in the MNID represent specific strains. Left of the line marked with "$\varepsilon_t = 0‰$"—tensile strain zero—all members are under full compression. The "$\varepsilon_t = 0.12‰$" respectively "$\varepsilon_t = 0.16‰$"-lines represent the end of the elastic (un-cracked) behavior respectively the beginning of macro-cracking. Micro-cracking occurs between the two lines. The outermost line represents equilibriums at failure, where either the maximum tensile strength at $\varepsilon_{t3} = 10.0‰$ or the maximum compressive strength at $\varepsilon_{c2} = -3.5‰$ is reached.

The triangle in Figure 3 marks the spot, where the cross section fails by reaching both criteria simultaneously, the maximum allowable tensile strength, here 10‰, and the maximum allowable compressive strength, here −3.5‰. In the area below the triangle, in this example at a normalforce of roughly N = 2,000 kN (n = 0.2), the lining will fail on tension (see also Figure 4), while cross sections above a normalforce of N = 2,000 kN (n = 0.2) fail on compression (see also Figure 5).

The circle in Figure 3 marks the spot, where the material behavior is "ideal-elasto-plastic"—the maximum moment reached at the end of the micro-cracking phase is identical with the moment borne in the macro-cracking phase until failure is reached. The lining is basically capable to "hold" the moment constant while plasticly rotating. For SSR of the parametric study this behavior is achieved at a thrust of N = 500 kN (n = 0.05) (see Figure 4, Figure 5, and Figure 6). Higher normalforce induces a strain-hardening behavior, while lesser thrust will lead to strain softening. The latter would also be observed for pure bending (n = 0) or would be measured in a beam test (compare Figure 3 with Figure 4, Figure 5, and Figure 6).

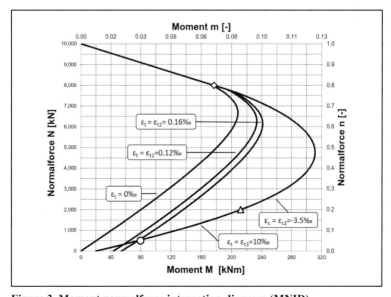

Figure 3. Moment normalforce interaction diagram (MNID)

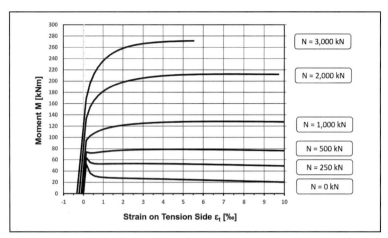

Figure 4. Moment-tension strain—Curves with different normalforces

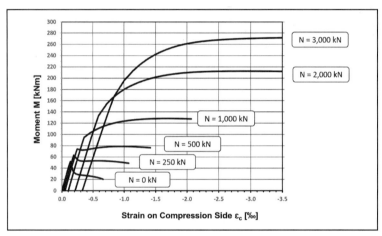

Figure 5. Moment-compression strain—Curves with different normalforces

Therefore, it should be highlighted and emphasized that FRC that shows "strain-softening" behavior in a beam test under pure bending can nonetheless provide "elasto-plastic" or "strain hardening" behavior" in a tunnel lining simply due to the presence and influence of a normalforce. At which normalforce the specific FRC tunnel lining performs "strain-hardening" can easily be identified with the characteristic strain curves discussed above and shown in Figure 3. For the SSR utilized for the parametric study strain-hardening can be observed already at relatively small normalforce influence of 5% of the compressive bearing capacity (see n = 0.05 in Figure 3), which is in a typical range for FRC used in tunnel linings.

The diamond in Figure 3, at a normal-force of ca. 8,000 kN (n = 0.8), marks the point and the

area above, where the lining would fail on compression, while being under full compression. The lining would fail as a fully compressed arch, which is untypical for modern tunnel lining designs. Since FRC improves primarily the material properties under tension, a FRC design governed by compression failure does not make sense.

The results of the above parametric study, which are representative of FRC behavior observed in tests [2, 5, 6], also show that the moment bearing capacity in the elastic, the microcracking, as well as the cracked phase, are all increased under the influence of an increasing normal compressive force. While typical FRC simply supported beam tests under pure bending (N=0) show a strain-softening behavior, it is emphasized that an increased normal force leads to a quasi elasto-plastic and a quasi-strain-hardening

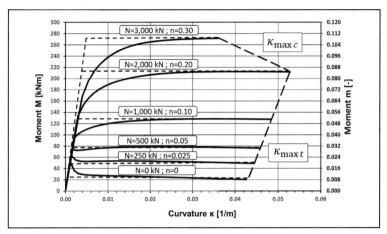

Figure 6. Moment-curvature—Curves with different normalforces

effect. The term "quasi" is used because the bending behavior is a characteristic of the structural system; material properties do not actually change. For the same material, the bearing capacity increases with an increased acting normal force (compare Figure 3, Figure 4, Figure 5, and Figure 6).

Characteristic—and more pronounced with an increased amount of normalforce—is the fact that a FRC tunnel lining can "hold" a certain level of moment during the macrocracking phase. This is the well-known increase of "toughness" and major benefit of FRC. The results can be simplified with good approximation as an ideal-elasto-plastic behavior (see dashed lines in Figure 6), which moment level for a given SSR is primarily controlled by the acting normalforce.

Figure 4 displays the moment over the strain on the tension side ε_t and Figure 5 displays the moment over the strain on the compression side ε_c, both under the influence of different normalforces. Only for pure bending (N = 0) the curve starts at $\varepsilon_t = 0$ respectively $\varepsilon_c = 0$. For all other cases the cross section is under full compression until the moment reaches a certain level (see Figure 4).

Up to a normalforce of approximately N = 2,000 kN the cross sections fail on tensions by reaching the maximum tension strain ($\varepsilon_{t3} = 10.0‰$). Above N = 2,000 kN the cross sections fail on compression by reaching the maximum allowable compression strain ($\varepsilon_{c2} = -3.5‰$) (compare Figure 3, Figure 4, and Figure 5). Consequentially, the strain on the tension side does not reach the maximum tension strain ($\varepsilon_{t3} = 10.0‰$) at failure for normalforces above N = 2,000 kN (see Figure 4) and the strain on the compression side does not reach the maximum compression strain ($\varepsilon_{c2} = -3.5‰$) at failure for normalforces below N = 2,000 kN (see Figure 5).

Since the curvature is dependent on the strains on both sides of the cross section, tension as well as compression, the maximum curvature at failure is reached, when both strains are maximized ($\varepsilon_{t3} = 10.0‰$ and $\varepsilon_{c2} = -3.5‰$), which is the case at a normalforce of about N = 2,000 kN for the FRC chosen for the parametric study. For normalforces either above N = 2,000 kN ($\varepsilon_{c2} = -3.5‰$) or below N = 2,000 kN ($\varepsilon_{t3} = 10.0‰$) the curvature at failure is smaller than the maximum curvature (see Figure 6).

Based on the curvature at failure at different normalforces it is also possible to define a curvature failure criteria for tensile failure and compression failure based on the normalforce by simply evaluating the corresponding linear equation (see Figure 6):

Equation 4: Maximum Curvature for Tension Failure

$$\kappa_{max\,t}\left[\frac{1}{m}\right] = 0.0426 + \frac{N\,[kN]}{196,078} \tag{4}$$

Equation 5: Maximum Curvature for Compression Failure

$$\kappa_{max\,c}\left[\frac{1}{m}\right] = 0.0862 - \frac{N\,[kN]}{59,880} \tag{5}$$

These failure criteria can be use i.e., in a design concept.

PROPOSED ELASTO-PLASTIC DESIGN APPROACH FOR FRC TUNNEL LININGS

What do these results mean for a FRC tunnel lining design and how can they be utilized? The general desire from a structural perspective for a cracked tunnel lining requires that the bearing capacity under

Table 2. Iterative elasto-plastic design procedure for FRC tunnel linings

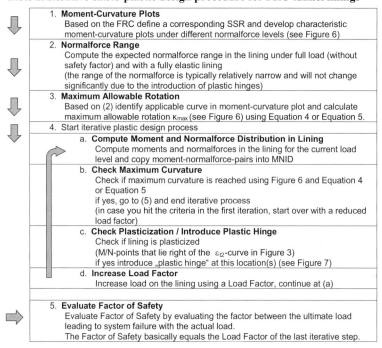

1. **Moment-Curvature Plots** Based on the FRC define a corresponding SSR and develop characteristic moment-curvature plots under different normalforce levels (see Figure 6)
2. **Normalforce Range** Compute the expected normalforce range in the lining under full load (without safety factor) and with a fully elastic lining (the range of the normalforce is typically relatively narrow and will not change significantly due to the introduction of plastic hinges)
3. **Maximum Allowable Rotation** Based on (2) identify applicable curve in moment-curvature plot and calculate maximum allowable rotation κ_{max} (see Figure 6) using Equation 4 or Equation 5.
4. Start iterative plastic design process
a. **Compute Moment and Normalforce Distribution in Lining**   Compute moments and normalforces in the lining for the current load level and copy moment-normalforce-pairs into MNID

b. **Check Maximum Curvature**
    Check if maximum curvature is reached using Figure 6 and Equation 4 or Equation 5
    if yes, go to (5) and end iterative process
    (in case you hit the criteria in the first iteration, start over with a reduced load factor)

c. **Check Plasticization / Introduce Plastic Hinge**
    Check if lining is plasticized
    (M/N-points that lie right of the ε_{t2}-curve in Figure 3)
    if yes introduce „plastic hinge" at this location(s) (see Figure 7)

d. **Increase Load Factor**   Increase load on the lining using a Load Factor, continue at (a)
5. **Evaluate Factor of Safety** Evaluate Factor of Safety by evaluating the factor between the ultimate load leading to system failure with the actual load. The Factor of Safety basically equals the Load Factor of the last iterative step.

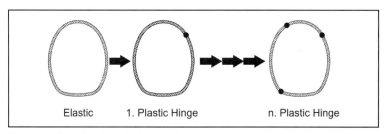

Elastic   1. Plastic Hinge   n. Plastic Hinge

Figure 7. Introduction of plastic hinges in FRC tunnel linings during iterative design process

cracked conditions should be equal to or higher than the bearing capacity in the elastic state. The behavior shall be at least "elasto-plastic" or show "strain-hardening." In the previous section, it was shown that these conditions are highly dependent on the amount of normalforce in the system. However, current tunnel designers typically do not take the expected normalforce into consideration when designing FRC tunnel linings nor do their designs benefit from it. Therefore, a lot of structural as well as economic potential of FRC remains under-utilized.

In a classic tunnel design the forces of the lining are determined in a linear-elastic model. Representative pairs of moments and normalforces from this analysis are then transferred into a MNID to ensure that the load combinations can be borne by the FRC lining (lie left of the failure line in Figure 3).

As discussed above, the inclusion of fibers increases the moment bearing capacity compared to unreinforced concrete when a section is subject to a large compressive normal force. The benefit of FRC lies in the added toughness of the material, which allows—under elasto-plastic or strain-hardening conditions—to "hold" a moment in a lining even under plastic deformation of the lining. However, these benefits are currently not utilized in a standard linear-elastic structural analysis. The structural and economic potential can be activated in a non-elastic structural analysis using, for example, a concept typically used for a simplified method for a non-elastic design of steel frames. Structurally a cracked FRC lining under normalforce acts like a "plastic hinge," which still transfers—"holds"—a moment while rotating (see Figure 6 and Figure 7).

The plastic hinges soften the lining locally, while provoking a load re-distribution in the static system, which increases the bearing capacity of the system. To utilize this effect, the design has to focus on a system failure rather than the failure at a specific cross section that reaches the maximum bearable moment first. A design with plastic hinges would follow the procedure introduced in Table 2.

The governing failure criteria with the procedure introduced above is not the maximum bearable moment at a critical cross section like in an elastic design approach, but the maximum rotation controlled by either reaching the maximum allowable tensile strain ε_{f3} or maximum allowable compressive strain ε_{c2}. As an additional failure criteria, a maximum allowable number of plastic hinges in the system could be introduced or a minimum distance between adjacent plastic hinges could be defined.

The "plastic hinges" have the elastic lining stiffness before they are identified as plastic hinges and "activated" by the above iterative process. "Plastic hinges" after activation act like free hinges with no remaining bending stiffness after activation. However, it should be noted that this approach is theoretically only valid for FRC that acts "ideal elasto-plastic" under the present normalforce; meaning that the cracked cross section still transfers or "holds" the maximum elastic moment. As shown above this assumption meets the actual results reasonably well (see Figure 6). Alternatively, the strain-hardening behavior of FRC under the influence of normalforce can also be utilized by introducing a rotational spring rather than a free hinge, which allows to further utilize the structural benefits.

If the used software does not allow such an iterative process and the modelling of elasto-plastic material behavior, each step may be conducted elastic with classical free hinges, but with a slightly modified static system, with an increasing number of hinges, in each step. Eventually all results will then be superimposed to evaluate the ultimate bearing capacity.

The iterative procedure above uses the increased load as variable to gradually introduce the plastic hinges. The introduction of the plastic hinges makes the system locally softer and provokes a load re-distribution in the system when the load is further increased. The failure mode is hereby moved from a classical failure of the most critical cross section to a system failure by utilizing the full structural potential and providing a more economic design approach.

CONCLUSION

The paper presented and discussed the basics of FRC tunnel lining design using a Stress-Strain Relationship. The impact of normal force within the lining and the impact of a change of post-cracking behavior from strain-softening to strain-hardening was discussed in detail by means of a parametric study.

Typical design tools, like a moment-normal-force interaction diagram for FRC was discussed and a curvature based failure criteria, based on the maximum allowable strains for tension and compression, was introduced and developed in an example.

An iterative process of an elasto-plastic design approach for tunnel linings based on the idea of a system failure rather than a cross section failure was introduced. The concept allows to utilize the full structural potential and with that the full economic potential of FRC for tunnel linings.

Current tunnel lining design does not fully utilize the potential of FRC because it (1) disregards the positive benefits of the compressive force, which are not related to the material properties itself and (2) designs the tunnel lining in an elastic state. The introduced new design approach, utilizes the full structural potential of FRC by introducing plastic hinges in an iterative elasto-plastic design process. The benefits of the new design approach were presented and discussed and are ready for practitioners to use the full structural and economic potential of FRC in tunnel lining designs.

LITERATURE

[1] Nitschke, A.; Bernard, E.: Load-Bearing Capacity of Fiber-Reinforced Concrete Tunnel Linings. 23rd Rapid Excavation & Tunneling Conference (RETC 2017) & Exhibit Proceedings, 4–7th June 2017, Sand Diego, CA, 2017.

[2] Nitschke, A.: Tragverhalten von Stahlfaserbeton für den Tunnelbau. Dissertation. (in German. Load Bearing Behavior of Steel Fiber Reinforced Concrete for Tunneling. Doctor Thesis.) Technisch-wissenschaftliche Mitteilungen des Instituts für konstruktiven Ingenieurbau der Ruhr-Universität Bochum, TWM 98-5, 1998.

[3] Ruhr-Universität Bochum, Lehrstuhl Prof. Maidl, Nitschke, A., Ortu, M.: Bemessung von Stahlfaserbeton im Tunnelbau. Abschlußbericht. (in German: Design of Steel Fiber Reinforced Concrete for Tunneling. Final Report) Research Project funded by the Deutscher Beton-Verein E.V. (DBV-Nr. 211) and the Arbeitsgemeinschaft industrieller Forschungsvereinigungen (AiF-Nr. 11427 N). Fraunhofer IRB Verlag, 1999. ISBN 3-8167-5455-4.

[4] Maidl, B.: Steel Fibre Reinforced Concrete. Ernst & Sohn, Berlin, 1995. ISBN 3-433-01288-1.

[5] Dietrich, Jörg: Zur Qualitätsprüfung von Stahlfaserbeton für Tunnelschalen mit Biegezugbeanspruchung. Dissertation. (in German: About Quality Testing of Steel Fiber Reinforced Concrete for Tunnel Lining under Flexural Tension. Doctor Thesis) Technisch-wissenschaftliche Mitteilungen des Instituts für konstruktiven Ingenieurbau der Ruhr-Universität Bochum, TWM 92-4, 1992.

[6] Maidl, B.; Nitschke, A. ; Ortu, M.: Bemessung von Tunnelschalen mit dem M/N-Prüfkonzept. (in German: Design of Tunnel Linings with the M/N Testing concept) In: Taschenbuch für den Tunnelbau 1999. Glückauf Verlag, Essen, 1998.

[7] American Concrete Institute Committee 318: Building Code Requirements for Structural Concrete (ACI 318-14). 2015.

[8] American Concrete Institute: The Reinforced Concrete Design Handbook. Volume 1: Member Design SP-17(14). Building Code Requirements for Structural Concrete (ACI 318-14). 2015.

[9] American Concrete Institute: Report on Indirect Methods to Obtain Stress-Strain Response of Fiber-Reinforced Concrete (FRC) (ACI 544.8R-16). 2016.

[10] American Concrete Institute: Design Considerations for Steel Fiber Reinforced Concrete (ACI 544.8R-16)). Reapproved. 2009.

[11] American Concrete Institute: Report on Design and Construction of Fiber Reinforced Precast Concrete Tunnel Segments (ACI 544.7R-16). 2016.

[12] American Concrete Institute: Guide to Fiber-Reinforced Shotcrete (ACI 506.1R-08) First Printing. 2008.

[13] German Society for Concrete and Construction Technology (DBV): Guide to Good Practice—Steel Fibre Concrete. 2001.

[14] Deutscher Ausschuss fuer Stahlbeton (German Committee for Reinforced Concrete) (DAfStb): DAfStb Richtlinie fuer Stahfaserbeton (DAfStb Guideline for Steel Fiber Reinforced Concrete). 2010.

Ground-Liner Interaction During Seattle Northgate Link Cross-Passage Construction

Tamir Epel and Mike Mooney
Colorado School of Mines

Kurt Braun
L-7 Services LLC

Mike DiPonio and Nate Long
Jay Dee Contractors, Inc.

ABSTRACT: This paper presents the results and analysis of field measurements to characterize ground-liner interaction and the development of thrust force and bending moment in segments adjacent to openings during cross passage construction. The instrumentation and data collection of segment strain measurements during cross passage construction made possible by embedded wireless sensors enables a complete picture of thrust-moment evolution and ground-liner interaction. The results show how different excavation and construction activates load the cross passage opening and the contribution of different supporting elements.

INTRODUCTION

Cross passages (CP) are an essential requirement in modern twin transit tunnels for emergency egress. In most countries, the minimum requirements are outlined in fire protection regulations, which usually require a maximum distance of 200–250 m between cross passages, as in the National Fire Protection Association Standard (NFPA 130) required in the US. In current US practice, cross passages are constructed via the sequential excavation method (SEM). Prior to the opening in the existing support of the completed main tunnel, ground improvement is often needed, followed by temporary support using structural elements around the opening, then opening of the existing lining and careful excavation and support towards the opposite tunnel.

The three-dimensional nature of a cross passage opening and excavation perpendicular to the main tunnel is complex and has not been extensively addressed in the literature. Most of the scarce research that has been done, was done on stress concentration around openings in circular tunnels with monolithic lining (Jones 2007, Spyridis and Bergmeister 2015). Jones (2007) studied stresses in sprayed concrete tunnel intersections by 3D computational modeling. He found that the maximum axial stress concentration factor was about five and the maximum bending stress concentration factor was just over two for the base case model. In addition, he concluded that construction sequence and the explicit modelling of

the ground structure interaction controlled the stress concentration at the tunnels junction. Kuyt et al. (2016) studied field data collected from cross passage construction between twin segmental lining supported tunnels in mixed ground conditions with CP opening supported by steel segments. They found that a relatively minor (10%) transfer of loads onto the segments surrounding the opening, as the ground and installed steel jacking props took most of the hoop forces from the opening ring.

Due to limited available knowledge, designers approach CPs with a high degree of conservativism. Either a full ring beam (so called 'hamster cage') or vertical steel propping with or without a header beam is the most commonly used approache (see Figure 1), though steel segments can also be found in some projects. With the advances made in 3D computational modeling, less conservative solutions are being used in recent years. Shear bicone dowels are common, although, in most cases, additional elements are used as "belts and suspenders." In this paper, data from strain gauges installed in the segmental lining at two cross passage opening locations on the Seattle Northlink tunnel project are analyzed during the excavation process.

PROJECT BACKGROUND

Northgate Link Project

The Northgate link Extension project includes 5.6 km (3.5 mi) of twin bored tunnels and 23 cross passages that is expected to be completed by 2018. The tunnels run from the University of Washington to Maple Leaf Portal in north Seattle. The twin bored tunnels were constructed using two EPB TBMs with excavation diameter of 6.64 m and supported in a single pass, gasketed segmental lining, 25 cm (10 inch) thick, and outer diameter of 6.25 m (20.5 ft). The geology of the area through which the tunnels were constructed, consists of complex and highly variable interlayered glacial and non-glacial soil deposits (Figure 2). As part of the geotechnical baseline work, the geological units were grouped into engineering soil units (ESU) based on their behavioral characteristics. All tunnel excavation was done in the following glacially over ridden ESUs; till and till-like deposits (TLD), cohesionless sand and

gravel (CSG), cohesionless silt and fine sand (CSF) and cohesive clays and silts (CCS). Most cross passages were excavated under the groundwater table with a water pressure at the invert of up to 300 kPa (3 bars), and highly variable interlaying soil units with permeability ranging from 10^{-2} to 10^{-7} cm/sec.

Geology and Dewatering

In this paper, two of the 17 monitored cross passages (CP) will be discussed, CP-39 and CP-41. The two cross passages were excavated according to support category 2 (Pyakurel et al.) and have cross sections with excavation width of 5 m (16.33 ft) and excavation height of 5.5 m (18 ft). The support system of support category 2 included 20 cm (8 in of shotcrete (including 5 cm of flashcrete), 3 bar lattice girders spaced at 90 cm (3 ft), and systematic 3 m (10 ft) long spiling. The excavation of both cross passages was performed from the NB tunnel towards the SB tunnel, where the instrumented rings are located. The excavation was performed by top heading and bench with round lengths of 90 cm (3 ft).

The segmental lining of the main tunnel was saw-cut, creating a 3×3 m² opening (the width equaling exactly 2 segments). The opening was supported by 6 shear bicone dowels that were pre-installed on each segment's circumferential joint and 4 shear bicones on each key or counter key circumferential joint (32 in total per each circumferential joint). Three circumferential joints are connected at each opening by the shear bicones, the two rings that were cut and the adjacent ring on each side (see Figure 3). The shear bicones were used to transfer the load from the opened segmental rings to the adjacent rings. In

Figure 1. Top: Ring beam installed in Northgate link CP. Bottom: Vertical steel propping.

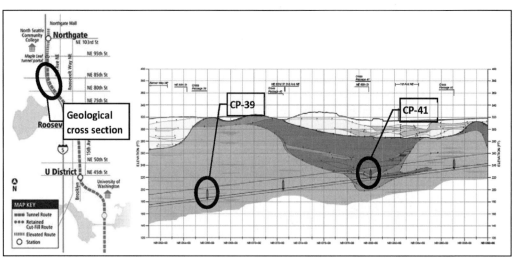

Figure 2. Left: Northgate link alignment in plan view (source Sound Transit). Right: Geological cross section between station 1362+00 to 1390+00.

addition, vertical steel propping was installed at the break-outs and break-ins at the cross passages discussed in this paper.

The excavation of CP-39 was expected to be performed through cohesionless sand and gravel with permeability in the order of 10^{-2} to 10^{-5} cm/sec. With the groundwater table before excavation of CP-39 approximately 1.5 m (5 ft) above the CP invert, systematic depressurization prior to break out was necessary to improve short term stability of the cross-passage excavation. Surface dewatering started 30 days prior to the break-out of CP-39 and was lowered to about 3 m beneath the CP invert. No water was encountered during the excavation of the CP. During the excavation of CP-39, cohesionless sand and gravel was encountered, and exhibited slow raveling behavior during the excavation of the top heading that required pocket excavation. The geological cross section of CP-39 consists of a top layer of about 4 m of TLD over laying CSG continuing down under the tunnel invert (Figure 4a).

For CP-41, although the groundwater table was expected to be 13 m above the CP invert, probe drilling performed before cross passage excavation found dry conditions, and no dewatering was needed. During the excavation of CP-41, stiff to very stiff silty clay (CCS) was encountered at the top heading with a layer of CSF at the SB tunnel intersection CP crown, and very dense peat in the bench/invert that exhibited very stiff behavior, similar to the CCS (Figure 4b). Some raveling behavior of the clay occurred in the crown due to very thin sand layers (< 5mm). The geological cross section of CP-41 consists of non-glacial deposits from the surface to a depth of about 8 m over-laying CCS down to about 3 m under the tunnel invert over-laying CSG and a thin layer of CSF just above the tunnel crown. No ground improvement was performed at either CP.

INSTRUMENTATION AND MONITORING

Segment Strain Gauges

Vibrating wire strain gauges were installed in select precast concrete segments. In total, 102 segments in 34 rings at 17 cross passage locations were instrumented. Two out of the 17 cross passages that were monitored are discussed in this paper. At each cross passage, two rings were instrumented. One instrumented ring that was eventually cut for the cross-passage entrance will be referred to as opening ring, and a second instrumented ring adjacent to the opening will be referred to as adjacent ring (see Figure 5). Each ring is comprised of four segments plus geometrically similar key and counter key. Three segments of each selected ring were instrumented—two full segments plus a key or counter key. Each instrumented segment was outfitted with a set of two vibrating wire strain gauges welded to the reinforcement cage, one at the intrados and one at the extrados as shown in Figure 6.

The strain gauges were installed on supplemental steel rebar (sister bar) welded to the longitudinal reinforcement of the segment, in the direction of the primary reinforcement (Figure 6). The strain gauges were installed in a slightly asymmetrical configuration due to limitations in the casting process (Figure 6b). The intrados strain gauge center depth from the segment extrados is 58 mm (2.3 in) as the strain gauge is installed on #3 rebar (9.5 mm or 0.37 in in diameter) welded to the underside of the longitudinal 14 mm (0.55 in) diameter (i.e., away from the intrados concrete face) rebar welded to the primary reinforcement (14 mm diameter), and minimum clearance is 25 mm (1 in). The extrados sister bar was installed on the exterior of the longitudinal reinforcement resulting in a depth of approximately 37 mm (1.46 in) for the extrados stain gauge center axis.

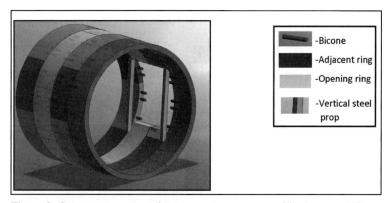

Figure 3. Cross passages opening support measures and insturamented rings with 32 bicones and vertical steel props

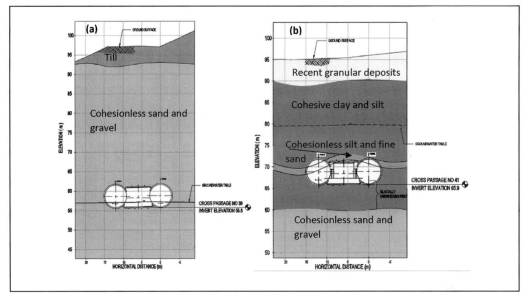

Figure 4. (a) Geological cross section of CP-39. (b) Geological cross section of CP-41.

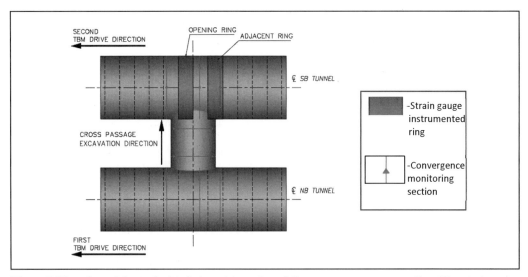

Figure 5. Plan view of the typical twin bored tunnels and the cross-passage connection. The strain gauge instrumented rings are marked in green and the convergence monitoring cross sections are marked in red.

With two strain gauges near opposite faces of the concrete segment, the thrust force and bending moment can be calculated. The sign convention of the thrust forces and bending moments presented in this paper are positive for compression thrust force, and positive bending moments when the segment intrados is in tension see Figure 6-d. The monitoring schedule included an initial zero strain reading after

the welding of the supplemental rebar on which the strain gauge is installed, prior to casting. The strain gauge readings began about 110 days after the ring installation for CP-39 and 150 days for CP-41.

The strain gauges were part of a wireless sensing system developed by Phase IV Engineering. The layout, implementation and recording program was developed in cooperation between the

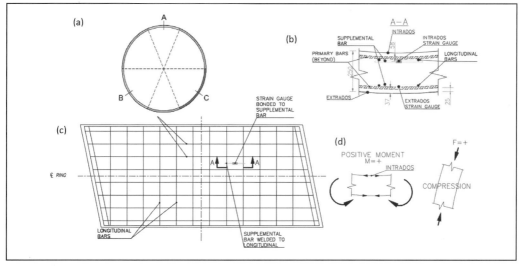

Figure 6. (a) Cross section of a typical instrumented ring, with the instrumented segments marked A, B & C. (b) Segment cross section at strain gauge location (dimensions in mm). (c) Strain gauge location on a typical segment plan view (d) sign convention for bending moments and thrust force.

tunnel contractor JCM Northlink and designer L-7 Services. Collection of the strain gauges readings required passing flat panel reader antennae of the reading unit within 30 cm (12 inches) of the concrete surface in the vicinity of the embedded sensor at a frequency of every two weeks. Sufficient continuous data during cross passage excavation was collected only from CP-39 and CP-41. Due to obstructions, as utilities and invert concrete slab not all strain sensors data were collected continuously and some sensors malfunctioned and no reading were collected or no readings were collected at the time of the cross-passage construction. In other cases, only one strain gauge of a set of two worked and the development of thrust-moment loading could not be analyzed. Unfortunately, the vertical steel props installed at the cross-passage openings were not monitored despite their important role in load redistribution.

Convergence and Geotechnical Monitoring

Convergence monitoring was performed in both running tunnels at the CP openings, and inside the CPs. Three cross sections in each running tunnel at each CP opening were instrumented (see Figure 5). In each cross section, five optical reflective targets were installed according to the configuration shown in Figure 7. A manual survey was performed on a daily basis, beginning 20 days prior to break-out. Inside each cross passage, two cross sections were instrumented in the first one-third of the excavation length and in the second one-third of the excavation length. Three targets were installed in each cross section, one at the crown and two at the spring line of the CP as

shown in Figure 7. The targets were installed in the full initial lining (20 cm of shotcrete) of the top heading. The first measurement ("zero" measured convergence) was taken before advancing to the next round of excavation. With maximum displacement measurements of up to 5 mm only and a measurement accuracy of ±1 mm, a connection between specific activities and the convergence monitoring could not be determined. Further, to establish a trend of behavior, the strain gauge data was analyzed using a moving average filter. Surface settlement was also measured on a daily basis, at three points above each CP; over the axis of the NB main tunnel, over the mid-length of the CP and over the axis of the SB tunnel.

EXPERIMENTAL RESULTS

Thrust-moment results and convergence monitoring data are presented against a timeline of the construction sequence for each CP in Figures 8a and 9a. Displacement measurements are very small; recall due to manual survey, these measurements have an accuracy of ±1 mm. Therefore, much of the day to day 0.5 mm fluctuations in these displacements are likely not reliable. It is also worth noting that strain gauge data were recorded continuously, so their changing response is temporally accurate. Convergence monitoring data, however, was recorded manually and is reported to the nearest day. Therefore, temporal comparison of strain and convergence events is challenging.

The typical round of excavation of the top heading (TH) took between 1–2 days (with the exception of TH2 in CP-39 that took 5 days), and the typical

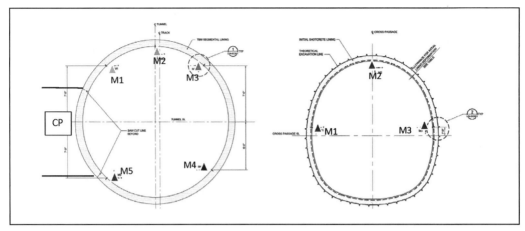

Figure 7. Convergence monitoring optical target locations. On the left are running tunnel target locations, and on the right CP target location.

excavation of the bench (B) took one day (with the exception of B2 in CP-41 that took 4 days), excluding weekends and Christmas that took place during the excavation of CP-39. For both CPs, the excavation stages were grouped into four construction sequences according to their influence on the measured stresses.

Figure 8b shows thrust forces and moments development in the ring adjacent to the CP-39 opening. The pre-CP positive bending moment indicates the ring is egging. During construction of CP-39, only one set of strain gauges was functional in each instrumented ring, denoted segment B in each ring. No significant change in SB ring segment loading occurs during sequence #1. At this point, the distance between the face of excavation to the SB tunnel side wall greater than 3 m (0.5D).

During sequence #2, a 7% increase in thrust force is observed after TH4 excavation when the CP face is 0.33D from the SB tunnel. No noticeable change in bending moment occurs. At the beginning of sequence #3, another 7% increase in thrust force is observed after the excavation of TH5, together with a 40% decrease in bending moment. These load changes suggest some loss of lateral soil support at the springline of the SB tunnel that would manifest as a net squatting behavior, i.e., the intrados is compressing and extrados extending. Following the excavation of TH5, no work was performed for a week (Christmas). Another significant decrease in bending moment is observed during this time. The bending moment is near zero at this point; however, the thrust force is at its greatest.

In sequence #4, a sharp change in bending moment occurs as the last CP rounds are excavated. A significant negative moment develops, consistent with ring squatting, as the lateral ground support is removed. The squatting behavior is clear in the SB tunnel convergence monitoring shown in Figure 8c. Here, the crown monitoring prisms M1 and M2 move downward (positive displacement is upward) while the invert monitoring prism M5 moves upward.

During the break-in cutting of the opening ring, a sharp decrease in adjacent ring thrust force and bending moment is observed. This maybe a result of releasing the moment transferred from the opening rings through the circumferential joints by coupled forces transferred through the high shear capacity of the bicone shear dowels. However, this releasing of moments is not seen in the convergence monitoring as a result of the break-in cutting of the opening ring. Following the opening both data sets show a stable condition with no additional loading and displacements. Following the break-in to the SB tunnel, resultant thrust force in the adjacent ring is only 10% less (1600 kN/m) than it was prior to CP construction (1750 kN/m). The bending moment changed significantly from +22 kN-m/m to –8 kN-m/m.

The opening ring analysis is not shown here as no significant change in loading was observed throughout the excavation sequence. However, the opening ring shows a steady to slight decrease of about 20% thrust force in the tunnel crown and an increase in positive bending moments (inner fiber in tension) during the first sequence. This may also be a result of the ring deforming to a more squatting position as is seen in the adjacent ring due to the relaxation of the ground behind the side wall allowing the ring to deform and transferring some of the thrust forces to the vertical steel prop.

The loads on the opening ring crown remain stable during sequences #2 & #3, and during sequence #4 the moments drop gradually down to zero at the time of saw-cutting of the ring. The

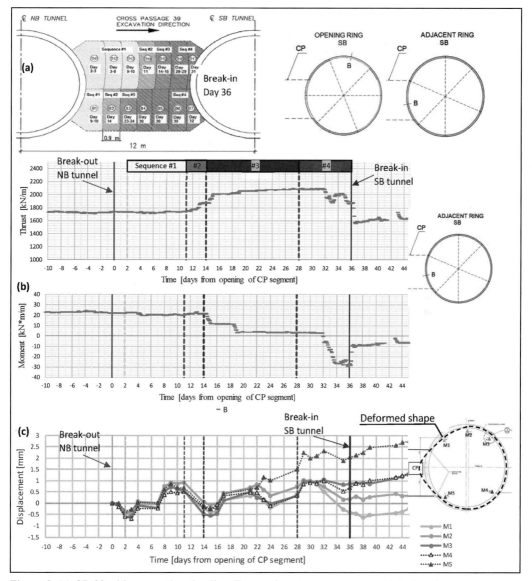

Figure 8. (a) CP-39 with excavation timeline. Excavation stages are grouped into 4 major sequences. (b) Adjacent ring thrust and moment development vs CP construction timeline (day zero is the opening of NB segments at the CP entrance). During the time of CP-39 construction, only segment B at the side wall was working. (c) CP-39 SB adjacent ring vertical displacement vs time (day "zero" is NB break-out), on the right target locations and deformed ring in dashed line. Positive displacement is upward.

gradual reduction in bending moment as the soil behind the ring is excavated is due to the segment unconfined by the soil. From the CP convergence, a similar displacement is observed at the crown and side walls that may indicate a lateral stress ratio close or smaller to one ($K_0 \leq 1$) as the tunnel geometry has a small radius crown and large radius walls (i.e., more deformation is accepted at the large radius).

During the construction of CP-41 no strain gauges in the opening ring functioned correctly. In the adjacent ring, two sets of strain gauges worked from the time of break-out. The sensors in segment A located between springline and invert worked continuously throughout the CP construction. The sensor in segment B, located at the tunnel crown, stopped working 7 days after break out.

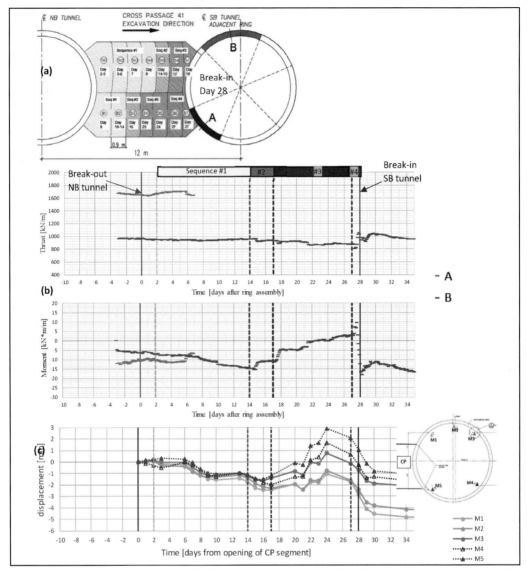

Figure 9. (a) Longitudinal section of CP-41 with construction timeline. Excavation stages are grouped into 4 major sequences. (b) Adjacent ring thrust and moment development over CP construction timeline (day zero is the opening of NB segments at the CP entrance). During the time of CP-41 construction, segment B worked up to day 7 after Break-out and segment A worked throughout the CP construction. (c) CP-41 SB adjacent ring vertical displacement vs time (day "zero" is NB break-out), on the right target locations.

The thrust-moment loading prior to CP construction indicates higher lateral than vertical stress conditions, as the crown moment is negative. It is difficult to interpret the ring flexural state from segment A data due to its location at the 7:20 clock position. In this clock position vicinity, and assuming an egging shape, moment would be transitioning from negative at the invert to positive at springline.

In sequence #1, no significant change in thrust force is observed in either segment (segment B data suggests some change but then goes off-line on day 7). The bending moment in segment B experiences a sharp change around day 6, decreasing in negative magnitude. This would be consistent with a net squatting behavior due to a reduction in lateral ground support. Segment A gradually grows more negative throughout sequence #1, from day 6 to day 14 when the TH4 face is about 2 m (0.33D) away and the bench is about 4 m (more than 0.5D). This results in a greater relaxation above the spring line than below, and may be the reason for this increase in negative moment.

As the excavation of the bench advances in sequence #2, significant overbreak occurs at TH5 and a 21% decrease in negative moment is observed in segment A. In sequence #3, two sharp decreases in negative bending moment occur in segment A as TH excavation reaches the SB tunnel segmental lining, and as the bench excavation advances to less than 2 m (0.33D) from the SB tunnel. The bending moment changes from negative to positive as the ring is no longer constrained by the soil at the CP top heading.

In sequence #4, a sharp change from a positive bending moment back to negative bending moment happens as the last bench is excavated and the break-in segments are cut. Following the break-in of the SB opening, the final value of thrust force returns to the about the same value as before CP construction and the bending moment decreases by 150%.

Figure 9c shows the convergence monitoring data with a relatively high uncertainty in measurement of ±1 mm for maximum measured displacement of 4 mm. Nevertheless, a general trend of squatting is observed beginning in sequence #2. While the cross-passage convergence is not shown here due to space, the side walls converge inwards 3–4 mm the crown deformed outwards 3–6 mm, indicating high horizontal stresses as expected in the CCS. Stability of the deformation was achieved only after break-in, and at the cross section at the 1st third of CP length, 75% of the deformation was achieved after the bench of the monitored section was excavated (sequence #3).

ACKNOWLEDGMENT

The authors thank Marte Guitierrez for his contribution to this paper.

CONCLUSIONS

The strain measurements collected during excavation of the Northgate Link CP construction show the change in internal thrust and moment in the segmental lining at CP openings. The development of the loads is linked to the advance of the excavation towards the break-in. While the measurements collected and presented here only address behavior at the break-in side, a number of important observations and conclusions were made. Significant relaxation of the ground behind the break-in support was observed when the distance from the face of excavation was smaller than 0.5–0.7D. This relaxation behind the CP opening results in unsymmetrical squatting of the running tunnel and increased bending moments.

While the bicone shear dowels are designed to transfer the thrust force from the opening rings to the adjacent rings, the bicones on the break-in side appear to transfer the ring deformation to the adjacent rings increasing the moment by up to 300% from geostatic pre-CP conditions before saw-cutting the opening. After the opening is cut, the transfer of moments is released. Unlike a common design assumption that a stress concentration of 3–5 times exists around an opening, when using additional opening support elements such as the ring beam or vertical steel props, the concrete segmental lining around the opening experienced no increase in thrust force and even a decrease.

REFERENCES

Agus S.S., Enferadi N., Amon A. 2016. Aspects on Design of Tunnel Cross Passage, Conference: Underground Singapore 2016. https://www.researchgate.net/publication/310352864_Aspects_on_Design_of_Tunnel_Cross_Passage.

Akai, S. Yamauchi, K. Kawasaki, H. Liebno, D. Venturini. G. 2017. Cross-Passage Mining Using Different Supports in Different Grounds, Rapid Excavation and Tunneling Conference 2017.

Pyakurel S. Klary, W. Gall, V. Long, N. Pooley, A. 2017. Risk Reduction, Management, and Mitigation from Experience-Based Learning During Construction of Cross Passages, Seattle, Washington, Rapid Excavation and Tunneling Conference 2017.

Jones, B.D., 2007. Stresses in sprayed concrete tunnel junctions. Doctoral thesis, University of Southampton.

Kuyt, J. Mooney, M. Mangione, M. Li, Z. 2016 Observed Loading Behavior During Cross Passage Construction, World Tunneling Conference 2016.

Design and Construction of the Downtown Bellevue Tunnel

Mike Wongkaew and Mike Murray
Mott MacDonald

Jake Coibion
Guy F. Atkinson Construction

Chad Frederick
Sound Transit

Mun-Wei Leong
McMillen Jacobs Associates

ABSTRACT: Construction of Sound Transit's East Link light rail tunnel using the Sequential Excavation Method (SEM) is well underway in downtown Bellevue, Washington. This paper describes the design and construction to date of the 2,000-foot long, 38-foot wide soft ground tunnel including refinements adopted in a collaborative team environment. These include the change from a six-heading to a three-heading excavation sequence, reduction of prescribed pre-support and face support, and changes in shotcrete supply and equipment. Productivity analyses are provided to illustrate these benefits. The paper also discusses the instrumentation program and the testing program to practically confirm shotcrete early strengths in the field.

DOWNTOWN BELLEVUE TUNNEL

East Link is a 14-mile extension of the existing Sound Transit (ST) light rail transit system from downtown Seattle, across Lake Washington, to the cities of Mercer Island, Bellevue and Redmond. The East Link includes 10 new stations. The Downtown Bellevue Tunnel (DBT) is the only tunnel and extends through downtown Bellevue with relatively low cover between the at-grade East Main and Bellevue Downtown stations (Figure 1). The DBT consists of the 250-foot-long south cut-and-cover portal structure, the 1,983-foot-long SEM tunnel, the mid-tunnel access shaft and connecting adit, and the 200-foot-long north cut-and-cover structure. The DBT was originally planned as a cut and cover tunnel but this was changed to reduce surface disruption and impact on the community.

The typical SEM tunnel excavated cross-section is a large 37.7-foot-wide by 30.5-foot-high ovoid sized for twin tracks (see Figure 2). The cross section is enlarged to 42.3 feet wide by 37.7 feet high (Figure 3) near the tunnel's mid length to provide the space for an emergency ventilation fan room above the tracks. The 12.8-foot-wide by 13.2-foot-high adit and the 20.7-foot-diameter by 50-foot-deep shaft provide maintenance access to the fan room from the

surface (see Figure 4). The shaft and adit will also be constructed using SEM.

Geology

The design geologic profile along the DBT (see Figure 5) indicates glacial deposits typical of the Seattle area and consists of a thin layer of fill overlying a glacially over-consolidated stratigraphic sequence that includes Vashon till, Vashon advance outwash deposits, and pre-Vashon glaciolacustrine deposits. The design groundwater table generally follows the top of the advance outwash which is expected to be encountered in the tunnel face during the second half of the tunnel.

DESIGN OF THE SEM TUNNEL, MID-TUNNEL ACCESS SHAFT AND CONNECTING ADIT

Excavation Sequence and Initial Supports

H-J-H, a design joint venture of HNTB, Jacobs Engineering and Mott MacDonald, was awarded the East Link detailed design in 2012 and the final design was completed in 2015. The design was in accordance with Sound Transit Design Criteria Manual and utilized two- and three-dimensional numerical modeling as a primary tool to estimate the ground

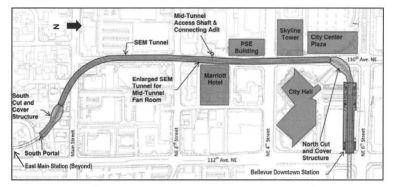

Figure 1. Location and extent of the Downtown Bellevue Tunnel

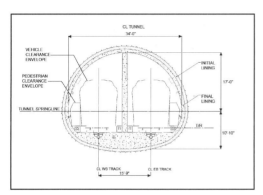

Figure 2. Typical SEM tunnel cross section

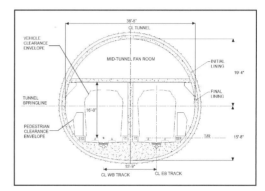

Figure 3. Enlarged SEM tunnel cross-section

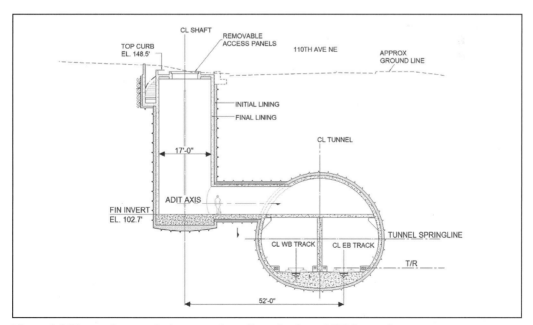

Figure 4. Mid-tunnel access shaft, connecting adit, and enlarged SEM tunnel

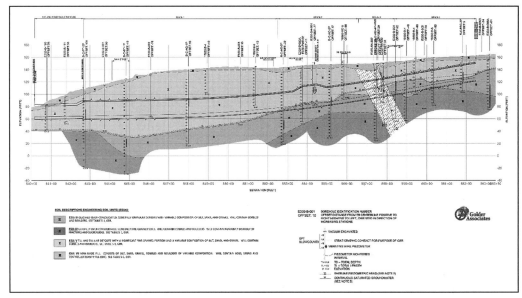

Figure 5. Geologic profile of Downtown Bellevue Tunnel (taken from the GBR)

response to the excavation sequence and the load effects in the initial and the final lining (Wongkaew et al., 2015).

The design of the SEM tunnel initial lining is based on the single side wall drift method successfully used at Sound Transit's similarly sized Beacon Hill station platform tunnels (Murray et al., 2006). Lattice girders are installed in the 10-inch-thick lining and generally spaced at 4-foot advances with a top heading, top heading, bench, bench and double invert excavation sequence. The enlargement is excavated in the same fashion following the completed side wall drift with a minimum lagging distance of 24 feet and later removal of the temporary sidewall. Six excavation sequence and initial support types are summarized in Table 1. Figure 6 shows the excavation sequence and initial support Type 2.

The final design of the connecting adit is based on full width top heading, top heading and double invert excavation sequence with 3'-4" maximum round length. The final design of the shaft is based on 4'-9" maximum round length. The SEM tunnel, adit and shaft are initially lined with 10-inch-thick fiber reinforced shotcrete, inclusive of 2-inch flashcrete.

Toolbox Items

As typical with soft ground SEM, the base design includes toolbox items with separate payment items and includes rebar and grouted pipe spiling in addition to the prescriptive presupport requirements, metal sheeting, additional shotcrete, welded wire fabric, chemical and cementitious grouting, face

Table 1. SEM excavation sequence and initial support types in the final design

Type	Cross Section	Feature	Presupport	Round length
1	Typical	Portals	Pipe arch canopy	4'-0"
2	Typical	—	Rebar spiling	4'-0"
3	Tapered	—	Rebar spiling	4'-0"
4	Enlarged	—	Rebar spiling	4'-0"
5	Typical	Building basements	Grouted pipe spiling, extended into the bench on the basement side	4'-0"
6	Typical	Anomaly zone	Grouted pipe spiling	3'-0"

bolts, and vacuum dewatering. Face wedge, probing, and dewatering from the surface are not separately measured.

Waterproofing

The initial fiber reinforced shotcrete lining will receive an application of 1.5-inch-thick smoothing shotcrete. Waterproofing membrane system will be applied against the smoothing shotcrete and fully encapsulate the final lining. The specification allows the contractor to select between a compartmentalized PVC sheet waterproofing membrane system and a self-adhered spray-applied waterproofing membrane system.

To date, the contractor has been exploring the Mapei Mapelastic TU spray-applied waterproofing

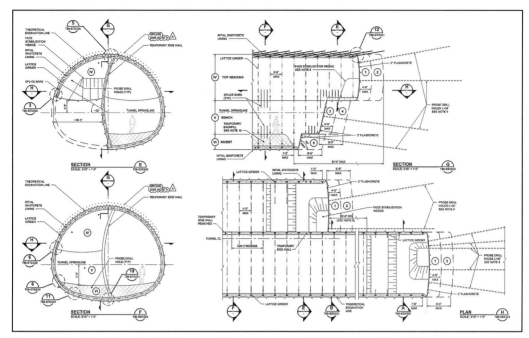

Figure 6. Typical six-heading SEM excavation sequence and initial support Type 2

membrane system and performed limited field trials consisting of applying a smoothing shotcrete mix with crystalline admixture on the completed initial lining to evaluate its effectiveness for sealing damp patches and minor seepage. Preconstruction trial with spray applied membrane has not yet started.

Final Lining and Interior Structures

The final linings of the SEM tunnel, shaft and adit have been designed as cast-in-place bar-reinforced concrete. The SEM tunnel and shaft final lining are 12 inches thick. The adit lining is 10 inches thick. The inverts have variable thickness because of the excavation geometry. The center dividing wall within the SEM tunnel is 1'-6" thick with no structural connection to the tunnel crown. At the mid-tunnel fan room, the center wall is 17'-3" high and supports the 1-foot-thick fan room slab. The concrete for the final lining and interior structures will have a minimum compressive strength of 5,000 psi and include a minimum of 1.7 lb/CY of micro-synthetic fibers for fireproofing and shrinkage crack mitigation.

The contractor is in the process of preparing the final lining concrete placement drawings as of this writing. The contractor has proposed to increase the construction joint spacing of the SEM tunnel final lining and the center dividing wall from the specified 30 feet to 40 feet to increase productivity. Sound Transit has agreed to the increase, which will enable the contractor to start procuring the steel traveling

form for the arch. The laboratory testing of the early age thermal and shrinkage characteristic of the proposed concrete mixes is in progress. The data will be used to determine if the shrinkage and temperature reinforcing that were designed on the basis of 30-foot construction joint spacing requires modifications.

The contractor is also exploring a shotcrete alternative for constructing the center dividing wall and the final lining of the enlarged SEM tunnel, shaft and adit. The project team is in the process of developing a preconstruction trial program, including mock-up requirements, to prequalify the means and method and the crew. The main challenges will be achieving satisfactory shotcrete encasement of the relatively congested rebar cages and surface finish of overhead shotcrete final lining.

CONSTRUCTION PLANNING AND MANAGEMENT

Contract Interfaces

The East Link $3.7 billion program is divided into various Contract Packages. For the Downtown Bellevue Tunnel and vicinity, Contract E330 provides excavation, initial support and final lining of the SEM tunnel, shaft and adit, and the south cut-and-cover structure and portal.

Contract E335 will construct the East Main Station just south of the south portal, the north cut-and-cover structure, and the Bellevue Downtown

Station. The E335 contract also includes construction of ballasted and direct fixation tracks; and installs all the ventilation, lighting, fire and life safety elements, and all necessary wiring for those systems in the Downtown Bellevue Tunnel.

Once E335 is complete, the E750 Systems Contract will install the signaling, traction power stations, overhead catenary (OCS), Supervisory Control and Data Acquisition (SCADA), and communications systems for the East Link Project. The E750 contractor will also be responsible for the integration, testing, and commissioning of all systems elements (hardware and software) of the East Link Segment.

The E330 Construction Management (CMC) team is working with the follow-on E335 and E750 contracts to ensure smooth handover of work elements and coordination between the contracts. The E330 CMC coordinated with ST's Real Estate and ST's Utility Coordinator to ensure all the properties, including construction easements and rights of entry, were received without delay and coordination with individual utility owners was completed.

Contract E330 Procurement

The E330 Contract was a traditional design-bid-build contract. The procurement was handled by Sound Transit with assistance by the E330 CMC and the designers during the bid phase. There were seven addenda issued during the bid period. The E330 CMC team and designers worked with ST in responding to questions from the bidders. Seven bids were received on October 27, 2015, with the winning bid by Guy F. Atkinson at $121,446,551. The second bidder was Downtown Tunnel Partners at $142,751,000, with Salini-Impregilo Healy closely behind at $142,809,000. The engineer's estimate at the time was $156,929,508. The overall bid results showed a great competitive bidding environment at the time, and Sound Transit was pleased with the bid results. The CMC team assisted Sound Transit in evaluating Atkinson's bid and issuance of the Contract.

The single lump sum bid item for the tunnel work was $117,025,000. SEM-related toolbox items with separate payments were: rebar and grouted pipe spiling, metal sheeting, additional shotcrete, welded wire fabric, chemical and cementitious grouting, face bolts, and vacuum dewatering. Additional bid items included a Trench Safety Item required by Washington State law, and eight provisional sums for unknown utility conflicts, archaeological investigation, art work, community construction impact mitigation, unknown hazardous and contaminated substances, partnering, Dispute Resolution Board (DRB), and other small items totaling $3,381,301. A separate bid item for an additional 6 inches of

shotcrete and reinforcing in the Skyline building parking basement was also included for $1,040,250.

Construction Schedule

Sound Transit issued a Limited Notice to Proceed (LNTP) to Atkinson on December 15, 2015. The issuance of the LNTP allowed Atkinson to begin providing submittals for the work, allowing Sound Transit's early approval and construction to begin as close to the planned NTP of February 5, 2016 as possible. NTP was issued to Atkinson on February 8, 2016.

Atkinson needed to complete four major milestones on the project:

- Milestone 1—CLSM Installation on 110th Avenue NE at the North Portal (300 Calendar Days after NTP)
- Milestone 2—Complete Pipe Canopy at the North Portal in 50 Calendar Days
- Milestone 3—Completion of Tunnel (1,445 Calendar Days after NTP)
- Milestone 4—Substantial Completion (1,582 Calendar Days after NTP)

The Critical Path of the project was to develop the site, complete the soil nail walls for the South Portal, install the pipe arch canopy from the South Portal, excavate 1,983 feet of SEM tunnel, install the majority of the permanent concrete tunnel final lining, and install finishes through the south cut-and-cover structure. Milestone 2 was planned to be 50 days duration for the use of the E335s north cut-and-cover structure site to install a 70-foot-long pipe arch canopy at the North Portal.

Atkinson's baseline schedule indicated the overall tunneling activity would begin on December 15, 2016 and complete 2 years later with an average production rate of 3.2 feet/day. To create opportunities for this and the follow-on contracts, the E330 project team worked collaboratively to develop an alternative three-heading excavation sequence. Other ideas included working with the City of Bellevue to implement a short full road closure of 110th Avenue NE for additional ground improvement and allowing the project team to replace the North Portal pipe canopy and its associated dependency with the E335 construction sequence.

Because of issues associated with the wettest winter on record in the Pacific Northwest, which interfered with the construction of the soil nail wall, Atkinson started tunneling more than two months later than expected, but with the alternative three-heading excavation sequence and better than expected ground behavior, Atkinson was able to accelerate construction of the tunnel and achieve production rates as high as 6 feet/day. They are currently on schedule to complete tunneling by December 2018, which is in line with the baseline schedule.

Construction Management

The E330 project team involves many parties: Sound Transit Construction Manager, Resident Engineer and their construction management team, the design team, the Contractor, and various third parties including the City of Bellevue. The Sound Transit Construction Manager oversees the Resident Engineer, acts as the liaison between the Sound Transit and the CMC and also ensures the work performed by the Resident Engineer's CMC team is in compliance with Sound Transit's procedures and processes. The Resident Engineer is the contractual point of contact between Atkinson and Sound Transit.

The Resident Engineer is responsible for administering the Contract and verifying that Atkinson completes the work in compliance with the Contract Documents. The CMC team is responsible for processing of submittals and RFIs, and review of progress payments and schedules. The CMC is also responsible for administering changes in the field, negotiating the change orders with Atkinson, and coordinating design inquiries and changes with the design team.

The design team provides technical support advice to the Resident Engineer. Because of the nature of SEM, the design is not final until all the support needed is installed so the design team also provides a SEM Resident Engineer and SEM Inspectors working shifts to verify support requirements and monitor the work occurring at the face of the tunnel. The three parties (CMC Resident Engineer, Atkinson, and SEM Resident Engineer) all agree at the daily SEM meetings what support will be required for that day's activities. This follows a joint inspection of the tunnel face every morning 6 days/week along with a review in the office of face maps, probing, instrumentation, lining scans and quality control including shotcrete test results. The output is a jointly signed Required Excavation and Support Sheet (RESS) which is distributed to all supervisory staff.

CONSTRUCTION OF ENABLING WORK

South Portal and Head Wall

The initial portion of the critical path of the project was to develop the South Portal site. After NTP was issued, Atkinson immediately began hazardous material abatement and demolition of the 14 existing structures within the site, abandoned the existing utilities, installed new drainage and permanently closed a road. The site was then graded and a 20-foot-high temporary sound wall was installed around the site.

Atkinson's mass-excavation and earth retention subcontractors mobilized in August 2016 to install the soil nail walls for the south portal. The shotcrete walls were typically 6 inches thick, with nails in a 6'×6' pattern of #8 to #10 bar varying from 20 to 30 feet long. A Klemm KR806 was used to drill the holes for the soil nails. A total of 809 soil nails were installed, 42 of which were fiber-glass nails for where the tunnel was going to be excavated through the wall. The mass-excavation subcontractor excavated the soil from the area in 6-foot lifts while the earth retentions subcontractor installed the soil nails and shotcrete. The portal structure was 525 feet long and, at the headwall, the portal wall was approximately 52 feet deep. The soil nail wall was completed two months behind schedule because of weather.

South Portal Pipe Canopy

The same earth retention subcontractor used its Klemm 806 drill rig to install the pipe canopy as the soil nail portal wall was being excavated. A total of 55 canopy pipes were installed in two staggered rows, leaving the pipes 18 inches center-to-center in an arch spanning 120 degrees centered on the tunnel. The 70-foot-long canopy pipes were 6–5/8-inch-diameter drill casing. There were one-way port valves every 12 inches. The location of each pipe installed was surveyed using an Inertial Sensing Gyro. The canopy pipes were then stage-grouted using a double packer system up to a maximum pressure of 70 psi. Grout take in till was generally limited. All canopy pipes were found to be within the specified 1% alignment tolerance and did not encroach into the tunnel excavation envelope. During SEM tunneling, the grouted canopy pipes helped reduce overbreak above the pipes where a sand layer was present in the tunnel crown. The soil nail wall portal and pipe canopy were completed in late January 2017 after the team battled the wettest winter on record for the Seattle area. Subsequently, a horizontal core was advanced through the center of the tunnel to investigate the ground prior to tunneling. Difficulties in advancing the core from the south portal beyond the first 182 feet in one drilling operation led to questioning the benefit vs. scheduling impact, eventually leading to the deletion of the core hole for the remainder of the project. The function of the horizontal core investigation was subsequently changed, and additional probe holes were drilled on approximately 32-foot intervals through the length of the tunnel. The change from the horizontal core to probe hole without core recovery provided for cost savings to Sound Transit, and the overall project benefited from the time savings.

North Portal Ground Replacement

One of the first work activities to be completed was the Controlled Low Strength Material (CLSM) work at the North Portal. The weak in situ material above the tunnel would be excavated and replaced with CLSM. This work needed to be completed prior to the E335 Contract mobilizing on site, 300 days after NTP. As the CLSM work was starting, an electrical vault in the CLSM area was found to be deeper than anticipated and intruding into the tunnel excavation and pipe canopy zone. The team worked out a design to install a shallower replacement vault with raised invert elevation and to support it on a 2,000 psi concrete arch constructed concurrently with the CLSM work. This arch construction was made easier to install because Atkinson's proposal to extend the CLSM zone deeper (Figure 7). This deepening would also allow Atkinson to eliminate the pipe canopy from the North Portal, thereby eliminating the need to stop the mass excavation of the E335 contractor to allow E330 contractor access to install the canopy.

The City of Bellevue allowed the team to fully close 110th Avenue NE for a short period, allowing the Contractor full access to the CLSM site. Atkinson completed the CLSM work at the North Portal in two and a half months, completing both Milestone 1 and Milestone 2 activities one and a half months ahead of schedule.

CONSTRUCTION OF THE SEM TUNNEL

Site Layout

The Project is located at the southwest corner of 112th Avenue SE and Main Street in downtown Bellevue, WA. Almost the entire site location is on a severe slope that needed to be developed to support the tunnel operation. A 52-foot-high soil nail wall was cut into the slope to facilitate the portal opening and ultimately serve as the support of excavation for the south cut-and-cover structure and retained cut section to the south. Because of the severe slope and footprint of the soil nail wall cut, usable space on the jobsite was at a premium. All outside haul traffic traveled through the site on a one-way road that included an automatic wheel-wash to prevent tracking on the city streets. All deliveries had to be just-in-time as no queuing disruption to traffic outside the portal was allowed at any time without a permitted stationary traffic control plan. The 20-foot-high sound barrier wall that was constructed around the entire perimeter of the site as part of Sound Transit's "good neighbor" initiative allowed the tunnel construction operation to carry on 24 hours per day.

Three-Heading SEM Alternate Design

Concurrent with the site development and construction of enabling work, the E330 project team worked collaboratively to plan and design an alternative SEM excavation sequence that will increase the opportunity for overall time savings on the project. As illustrated in Figure 8, the alternative three-heading SEM includes full width top heading, bench, top heading, bench and double inverts excavation sequence. For a detailed discussion of the design development and numerical modeling of the three-heading sequence, see Penrice et al. (2017) and the companion paper Brodbaek et al. (2018), respectively. The three-heading SEM was incorporated into Contract E330 by a change order as excavation sequence and initial support Type 2A. Its implementation is permitted in the first 50 percent of the tunnel length, where full face of Vashon till is anticipated, subject to additional

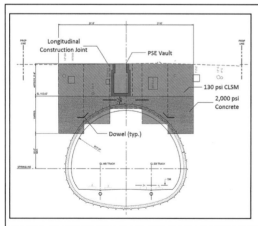
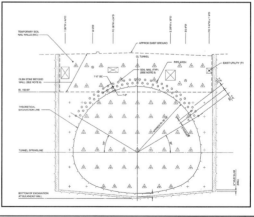

Figure 7. As-bid north portal pipe canopy and ground replacement (left) and the contractor's proposal to increase the ground replacement depth and eliminate the pipe canopy (right)

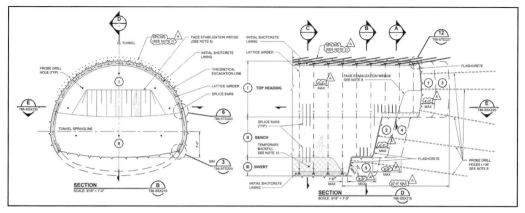

Figure 8. Three-heading SEM excavation sequence and initial support Type 2A

monitoring requirements and criteria for reverting to the as-bid six-heading design.

Tunnel Utilities

The Downtown Bellevue Tunnel will be utilizing both a full-width excavation heading and a side drift excavation sequence. The utility arrangement must take both excavation sequences into account. Tunnel lights, tunnel ventilation, and the drainage system are advanced in parallel lines to support either excavation sequence. Tunnel power, water, and air lines are carried up to the heading on only one side of the tunnel and can be split into the side drifts using headers and umbilical cables at the heading. Ventilation, air and water are advanced simultaneously with the excavation. The lights are advanced every 35 feet. The power center and personnel walkways are advanced every 150 to 200 feet.

Turn Under Sequence

In early February 2017, three-heading SEM tunneling commenced after the completion of the pipe arch canopy. The 6-inch thick soil nail wall headwall was line drilled along the excavation profile of the top heading. The first 4-foot top heading advance took approximately 72 hours to complete. Fortunately, the ground conditions were known to be stable from the experience with the soil nail wall construction, and there were no tunnel stability issues. Two more top headings were removed prior to commencing the bench excavation. Invert closure occurred several advances later. The invert closure is kept within 26 feet of the advancing top heading. See Figure 9 for the construction photo of the south portal.

Two important lessons were learned during the first 72 hours of tunneling that resulted in the Contractor temporarily halting further excavation while these issues were resolved. First, the ambient

air temperature when tunneling commenced was just above freezing. The shotcrete operations could heat the mix but not at the volume required for production tunneling. Accelerator temperatures also fell below 50 °F, which resulted in excessive initial and final set times—up to three hours for the final set. Second, even though the shotcrete nozzlemen on the Contractor's crew were highly experienced and all ACI certified, they did not have much time prior to tunneling to work with the proposed shotcrete robot arrangement. Each nozzleman's experience is different, but a 40-hour requisite training on the specific robot will help produce a quality finished product from the onset of tunneling.

SEM Tunnel Construction Cycle and Productivity Analysis

Excavation and support in SEM tunneling starts with the installation of pre-support. Excavation itself can then commence. The general sequence for the Downtown Bellevue Tunnel is to excavate, flash the exposed ground, install the lattice girder, shotcrete in the legs of the girder, install presupport (spiles) and then install the remainder of the fiber reinforced shotcrete lining.

The initial lining system consists of a 2-inch fiber reinforced flashcoat layer of shotcrete, lattice girders, and an 8-inch fiber reinforced development layer of shotcrete. The strength development of the initial lining shotcrete ground support was prescribed based on a design that modeled the proposed excavation cycle with the most optimistic cycle times. For the six-heading SEM, the perimeter lattice girders are generally used for profile control. However, for the three-heading SEM the model also required structural contribution of the lattice girders during the early strength development stage of the shotcrete, which took up to 12 hours.

Figure 9. Construction photo of the south tunnel portal

Each 8-foot production "round" is made up of two 4-foot top heading advances, two 4-foot bench advances and one 8-foot invert closure advance. Probe drilling occurs every other round, with a minimum of 16 feet of probe data being kept in front of the excavation at all times. The top heading advance (Figure 10) is the most time consuming—it is where the presupport is installed, there are more shotcrete stages in the top heading, and the top heading is typically where the majority of any pocket excavation will take place. Depending on the ground conditions, a top heading advance can take from one shift to as long as three shifts from start to finish. The bench typically takes a little longer than half a shift, and an invert will usually take one shift to complete. A typical probe drill cycle will take about half a shift to complete and is highly dependent on ground conditions. Loose gravelly ground is very difficult to drill.

Tunneling operations are conducted six days per week and 24 hours per day. For each of the three 8-hour shifts, the typical tunnel heading crew consists of four operators, six laborers, a foreman, a SEM superintendent and a field engineer. A surveyor and a shotcrete testing technician are at the tunnel heading as required. The major equipment used in the tunnel include two Liebherr 950 excavators, a Deere 135 excavator, a loader, a haul truck, a drill jumbo, a telehandler, a manlift, a shotcrete robot, a shotcrete pump and a concrete mixer truck.

In SEM tunneling the production rates are highly dependent on ground conditions and ground behavior as these factors determine the presupport, the length of the cut, and the number of pockets.

Because the ground conditions within the E330 tunnel proved to be more favorable than expected, several modifications to the initial excavation and support sequence were allowed. These modifications included reducing the prescriptive spiling to an as-needed quantity based on the ground behavior, elimination of the face wedge requirement, increasing the advance length from 4'-0" to 4'-3", and eliminating the immediate flash shotcrete for the bench and invert advances. These modifications have allowed tunnel production rates to easily beat the original estimate.

Another significant factor to SEM tunnel productivity is the shotcrete delivery. Because of the aggressive schedule requirements, the tunnel operation began while the on-site shotcrete batch plant had yet to be fully operational. The tunnel excavation began using the backup dry super sacs that were mixed in a mixer truck and then applied using the wet-mix process. This was slow and tedious. The robot that was initially supplied was also underperforming and undersized for the shotcrete demand. Approximately three months into tunneling, the batch plant was fully operational and a Normet Spraymec had been procured, which helped improve the shotcrete placement productivity by a factor of 3.

With SEM tunneling, the learning curve for the crews is lengthy. This is a process of many steps going into each individual advance and five different advances to achieve one round. Three crews working 24 hours per day means it could be several days or even weeks until one crew repeats an identical step within the 8-foot round. Less repetition plus

Figure 10. Excavation of a full-width top heading advance

slow production (i.e., SEM vs. TBM) equates to a longer learning curve. For the three-heading SEM tunnel, production averaged 15 feet per week for the first two months of mining. By the end of the fourth month the tunnel was averaging 25 feet per week, and by month six the weekly tunnel production was consistently over 30 feet.

Exclusion Zone

For safety reasons, no personnel are allowed under or in front of unsupported ground at any time. Also delineated as a personnel exclusion zone is the 10-foot zone around an area where shotcrete is being sprayed overhead and during the initial curing period of that overhead shotcrete. A shotcrete compressive strength of 75 psi is required to clear the exclusion zone.

MACRO-SYNTHETIC FIBER REINFORCED SHOTCRETE

Mix Design

Four wet-process fiber reinforced shotcrete mixes have been used to date for the flashcrete and the initial lining. All mixes include 3/8-inch pea gravel, sand, water reducing admixture, hydration controlling admixture, and accelerating admixture with a typical dosage of 5 and 8 percent. BarChip 54 macro-synthetic fibers are used as an approved substitute for the specified steel fiber reinforcement. The polyolefin fibers are 54 mm long and have a rectangular, fully embossed cross section with an equivalent

diameter of 0.90 mm. Table 2 lists the four mixes in the chronological order of their development. The combined dry ingredients of RAM 080 and RAM 092 mixes are delivered to the site in super sacks, which are then mixed with water, fibers, and liquid admixtures in a mixer truck. The use of RAM 080 mix was discontinued after its variant with lower fiber dosage (RAM 092) had demonstrated satisfactory performance. After the onsite batch plant became operational, RAM 097 mix became the primary mix batched from the material stored on site. RAM 108 mix is the slag-free variant of RAM 097 mix; its performance data are limited as of this writing.

Performance Requirements and Testing Methods

Table 3 summarizes the design performance and testing requirements of the fiber reinforced shotcrete. In addition, 75 psi minimum compressive strength is adopted by Atkinson for lifting the exclusion zone. McMesin penetrometer is used for the ASTM C803 test up to 150 psi. The 3- and 6-hour compressive strengths are tested by crushing 3"×3"×14" long specimens with a calibrated 10-ton Enerpac hydraulic arbor press between 3"×3" steel bearing plates, as shown in Figure 13. Specimens for ASTM C39 core tests and ASTM C1550 round determinate panel tests are transported to an off-site laboratory for testing.

Field Performance

Figure 11 shows the compressive strength development of the shotcrete mixes RAM 080, 092, and 097. The early age strength development for the first

Table 2. Wet-process macro-synthetic fiber reinforced shotcrete mixes approved to date

Mix ID	Cement, Type I/II	Fly Ash	Slag	Silica Fume	BarChip 54 fibers	W/cm Ratio
			Lb/CY			
RAM 080	825	115	-	95	15	0.35
RAM 092	825	115	-	95	13	0.35
RAM 097	705	-	40	50	13	0.38
RAM 108	705	-	-	50	13	0.40

Table 3. Performance and testing requirements for macro-synthetic fiber reinforced shotcrete

Test	Compressive Strength (psi)						Energy Absorption (J)	
	ASTM C803	ASTM C116		ASTM C39			ASTM C1550	
Specimen	2'×2' panel	6 beams		9 cores from 3'×3' panel			6 round panels	
Age	1 h	3 h	6 h	24 h	7 d	28 d	7 d	28 d
Passing criteria	100	300	600	2,000	4,000	5,000	350	450
	Avg. of 10 penetrations	Avg. of 3 beams each		Avg. of 3 cores each			Best 2 out of 3 panels	
Frequency	1 set per advance			1 set daily, reducing to 1 set per 50 CY at 3rd month, reducing to 1 set per 100 CY at 6th month				

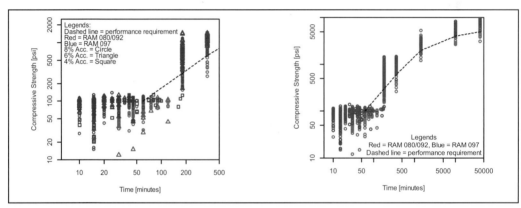

Figure 11. Shotcrete compressive strength development: first 6 hours (left) and first 28 days (right)

six hours is plotted at a larger scale on the left. The wider scatter and lower strengths for a given curing age generally occurred during the first four months of tunneling. The shotcrete strength development has since been improved after the use of the on-site batch plant, Normet shotcrete robot, and increased experience of the field staff with the means and method. Figure 12 shows the energy absorptions from the round determinate panels tested during construction. An upward trend in the energy absorption corresponding to the increase in compressive strength and the fiber dosage can be observed.

INSTRUMENTATION

Instrumentation and Monitoring Program

SEM is a construction method where the design of the excavation sequence and amount of the initial

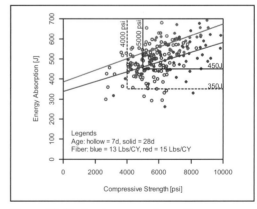

Figure 12. Shotcrete energy absorption vs. compressive strength

support are based on the anticipated subsurface conditions. The assumptions are confirmed or refined by field observation, including in situ measurements and monitoring. The instrumentation provides early information on the interaction of the SEM construction process and its effect on ground, groundwater and the initial support. This permits timely adjustment of the excavation sequence and additional SEM support measures, as and when required to maintain the stability and safety of the excavation and prevent damage to structures and utilities. For the DBT, two action levels for instrumentation measurements have been established and contingency measures planned for each level.

Figure 13. ASTM C116-based device for testing 3- and 6-hour compressive strengths

Tunnel Instrumentation and Monitoring

The typical layout of the instruments includes a transverse array of tunnel convergence monitoring points, lattice girder strain gauges and surface monitoring points at 50 foot longitudinal spacing, and a multi-point borehole extensometer and a pair of inclinometers at 200 foot longitudinal spacing. As part of the design change from six-heading to three-heading SEM, three additional arrays of tunnel lining strain gauges and convergence points are installed within the first 200 feet of tunneling and five additional multi-point bore hole extensometers are installed within the first 50 percent of the tunnel length.

Utilities and Building Instrumentation and Monitoring

Shallow buried utilities are monitored indirectly with the surface monitoring points and upper anchor of multi-point extensometers. Some high-risk utilities are directly monitored with utility monitoring points. Buildings partially or fully within the zone of influence are monitored with structural monitoring points around their perimeters and crack gauges. The basement wall of the Skyline Building, which is within 4 feet of the tunnel excavation, also receive strain gauges and tiltmeters.

Summary of Monitoring Results

Table 4 lists the instruments, action levels, and monitoring results for the first 900 feet of three-heading SEM tunneling. The results confirm satisfactory performance of the excavation sequence and initial lining. Back analysis of the vertical (crown and invert)

Table 4. Instrumentation monitoring results for the first 900 feet (45 percent) of SEM tunneling

		Action Levels		
Instruments	Unit	Trigger (Threshold)	Maximum (Limiting)	Range of Values to Date
Tunnel Convergence Points	Inch	0.30	0.45	0.01—0.35 (Crown) 0.05—0.31 (Spring line) Nil—0.15 (Invert)
Tunnel Lattice Girder Strain Gauges	Microstrain	1000	1400	241—1218 (Crown) 71—668 (Spring line) 100—708 (Invert)
Surface & Utility Monitoring Points	Inch	0.30	0.45	Nil—0.29
Inclinometers	Inch	0.20	0.30	Nil—0.20
Extensometer	Inch	0.75	1.10	Nil—0.18
Building Monitoring Points	Inch	0.30	0.45	Nil—0.12

Figure 14. Commencement of the six-heading excavation sequence for the second half of the DBT

and horizontal (spring line) convergence readings imply an equivalent volume loss of 0.14 percent or less, which is in line with the predicted volume loss of 0.08 to 0.17 percent.

CURRENT PROJECT STATUS

The first 50 percent of the SEM tunnel with three-heading excavation sequence has just been completed, ahead of schedule. This was attributed to the favorable ground conditions and collaborative working relationship of Sound Transit, CMC, Contractor, and the designers to incorporate productivity improvement measures. The Vashon till exhibited firm behavior and good standup time, except for occasional slow raveling behavior where a thin layer

of sand or gravel was in the crown. Minor seepage from some excavated sand lenses and probe holes was observed but presented negligible impact on the face stability. The six-heading excavation sequence has recently begun (see Figure 14).

In comparison to the preliminary design of the Downtown Bellevue Tunnel as a cut-and-cover tunnel, the SEM tunnel has been proven to significantly reduce surface disruption and impact on the community.

REFERENCES

Brodbaek, C., Penrice, D., Coibion, J. and Frederick, C. 2018. Downtown Bellevue Tunnel— Analysis and design of SEM optimization. In *Proceedings of the North American Tunneling Conference 2018.*

Murray, M., Redmond, S., Sage, R., Langer, F. and Phelps, D. 2006. SEM in Seattle—Design and construction of the C710 Beacon Hill station tunnels. In *Proceedings of the Tunnelling Association of Canada 19*th *National Conference, Vancouver, B.C.*

Penrice, D., Yang, H., Frederick, C. and Coibion, J. 2017. Downtown Bellevue Tunnel—Concept optimization through team collaboration. In *Proceedings of the Rapid Tunneling and Excavation Conference 2017.*

Wongkaew, M., Penrice, D., Theodore, J. and Patton, B. 2015. East Link—Final design of the Downtown Bellevue Tunnel. In *Proceedings of the Rapid Tunneling and Excavation Conference 2015.*

The Influence of Material Models for the Effective Design of the Primary Lining

Martin Bakoš, Juraj Ortuta, and Peter Paločko
Amberg Engineering Slovakia

ABSTRACT: The finite element method made a significant impact of static assessment of tunnel primary lining. The Mohr–Coulomb model is a widely used model. However, this model is not economically optimal. Designer can use other material models which are more suitable for rock environment resulting in more economical design.

Tunnel Soroška is designed as 4.2 km long road tunnel in karst rock. One tunnel tube for both directions will be built in the first phase. The construction had an interesting result during the calculation of primary lining.

The article presents a summary of geotechnical inputs for the efficient selection of the material models and a method of optimizing the calculation that takes the economic and environmental aspects of the design work into account.

INTRODUCTION

One of the main goals for the Slovak Republic is to connect the local transport network to European transport network. The proposed motorway R2 lot Rožňava—Jablonov and Turňou is in accordance with the development program of the Slovak national motorway network.

Proposed tunnel, Soroška, is located between the villages Lipovník and Jablonov nad Turňou.

This area is a part of Slovenský kras national park and it is considered as a protected area. The terrain creates steep slopes up to 15%.

Mountain of Slovenský kras consists of limestone. Tableland is divided by deep valleys with many surface and underground karst phenomena. About 1300 caves are located in this area.

The route of the tunnel Soroška crosses several protected areas.

The areas of importance are:

- Slovenský kras national park and associated protected areas
- Protected area NATURA 2000—protected bird area SKCHVU027 Slovenský kras
- Area NATURA 2000—habitat of European importance Hrušovská forrest-steppe SKUEV0352
- UNESCO objects—national nature monument Hrušovská cave and Krásnohorská cave—listed in UNESCO heritage

ENGINEERING GEOLOGICAL AND HYDROGEOLOGICAL DESCRIPTION OF THE SITE

Site of the planned tunnel belongs to the most interesting members of table karst in the Slovakia. All possible karst phenomena, such as caves, abyss, sinkholes, karst ponds, karst springs and others can be found.

According to geological investigation the geological structure of the area is simple. Young Triassic limestone covers the main area in the center. The bedrock consists of lower Triassic layers mainly slate, limestone and dolomite.

From geotechnical point of view, the structure is more complicated. Rocks of Silická plane are highly tectonic disrupted by steep faults of NW-SE orientation. Kars phenomena are also present. In the Figure 1, there is visible contrast between Sinské layers consisting mainly of marl slates with layers of gray mainly marly limestone.

The main interface of two geological structures is in km 2,1. In this location route comes from Sinský limestone layer to Wetterstein slates.

The next problem are karst phenomena. Those were detected in km 3,000–3,700. The measurement shows deep low resistance zone. This zone crosses the tunnel in 250 m section.

Detailed engineering geological investigation was performed in 2017. This investigation was found several new caverns 8–9 m high and several large hollow volumes (height about 23–25 m).

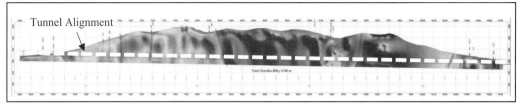

Figure 1. Geophysical measure on tunnel alignment

Figure 2. Hrušovská cave is UNESCO listed heritage

These features resulted in the need of use several auxiliary measures to protect the tunnel during excavation.

Slovenský kras is a unique nature complex due to the presence of groundwater. It is characterized as extreme rich groundwater source and it is important source for water management.

Because of this, a part of national park is declared as protected zone of natural accumulation of the water.

The flow of underground water in rock mass is quite complex. Infiltrated water keeps vertical direction and later it is changed to horizontal direction, called siphon flow.

Flow velocity of karst water is important hydrogeological characteristics. Due to high permeability on the surface, water is accumulated in deeper carbonate rock mass.

CONSTRUCTION OF THE TUNNEL LINING NUMERICAL MODEL

The problem of mathematical modelling is the assumption including a wide spectrum of geotechnical problems.

While designing the tunnel designer must keep in mind whole context and possible complication of future work and must deliver technically suitable and economically feasible solution.

Manmade constructions are defined more or less exact and it is a good base for static calculation.

However, rock environment is for geotechnical engineer a construction material and it creates a basic element of bearing system lining-rock. Parameters of this rock environment are very limited and geotechnical engineer must carefully decide on input parameters for static calculation.

The results of this static calculation directly influence the safety and price of the construction. The second is quite high in case of infrastructure projects.

While gravitational stress can be calculated, other parts of geotechnical state—residual and tectonic stress, can be only estimated. The wrong interpretation will lead to bad assumptions and inaccurate calculations. This leads into two options: economically unfavorable design, or worse, to undersized structure.

DEFINITION OF MODEL BOUNDARY CONDITION

To define behavioral model of rock environment, linear model and non-linear models were used: Mohr Coulomb, Ducker Prager and Hoek Brown. Geotechnical calculations aimed to calculate the internal forces and deformations in primary lining. Staged construction was also taken into account.

Figure 3. Intact drill core form tunnel route

Material properties in the calculation are based on engineering geological calculation. Cores were taken directly along tunnel route (Figure 3).

The interaction between primary lining and rock environment is secured in the upper part of arch by rock bolts. Rock bolts, 4 m long, are in the model substituted with anchored area. Anchored area located in sandstone is modelled with higher cohesion. In other areas (without anchoring) interface between lining and cohesion is described by fictitious contact layer by shear and normal stiffness.

INFLUENCE OF FAULT ZONES TO STRESS CALCULATION

To evaluate the rock mass faults, influence was applied the assumption that in place of fault/crack lowers contact stress. This stress can vary according to size and filling (water or air) of fault/crack.

There is however a question how to evaluate vertical deformations according to different compressibility of water and air. In this case, we can work with the assumption that the geology age is reached equilibrium stage and the deformation is zero.

This way, we can limit the number of variables influencing primary stress state (Figure 4 and 5).

INTERNAL FORCES ON PRIMARY LINING

The calculation was performed in eight construction phases. Primary lining is constructed during the third phase of calculation. Therefore phase 1 and 2 are not taken in account. From comparison of maximum forces in individual phases results that the value of those forces and also position of monitoring points for individual calculation stages.

The results are compared to Mohr-Coulomb material model because this is the most widely used model and it is relatively light from view of input parameters (Table 1).

According to the evaluation of the most important combinations we can compare the difference between Mohr-Coulomb model and Hoek-Brown model. At extreme compression of forces this difference is about 45%. At maximum tensile forces this difference is between Mohr-Coulomb and Hoek-Brown 27% however between Mohr-Coulomb and Drucker-Prager is difference up to 63%. The comparison of bending moment is as following:

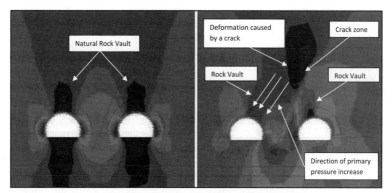

Figure 4. Equivalent relative deformation around the tunnel (left: without influence of cracks, right: with influence of cracks)

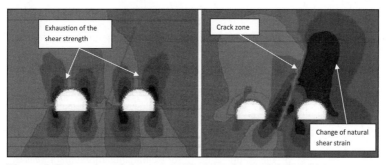

Figure 5. Exhausting of shear strength around tunnel (left: without influences of cracks, right: with influence of cracks)

Table 1. The comparison of the internal forces in representative points

			Material Model			
			Elastic	**Mohr-Coulomb**	**Drucker-Prager**	**Hoek-Brown**
Values in the Representative Points	Maximum compression forces	N (kN/m)	**–1396,8**	**–1309,5**	**–1296,9**	**–742,7**
		M (kNm/m)	0	0,0	5,2	4,0
		Q (kN/m)	–41,2	–50,8	9,0	–8,5
	Comparison to the Mohr-Coulomb model		106,7%	100,0%	99,0%	56,7%
	Maximum tension forces	N (kN/m)	**406,2**	**352,8**	**129,6**	**258,2**
		M (kNm/m)	0	0,0	0	0,0
		Q (kN/m)	32,8	25,4	28,5	9,1
	Comparison to the Mohr-Coulomb model		115,1%	100,0%	36,7%	73,2%
	Maximum bending moment	N (kN/m)	–1111,6	–1093,6	–1280	–138,6
		M (kNm/m)	**–13,5**	**–13,5**	**–17,2**	**–17,9**
		Q (kN/m)	0,5	0,9	1,2	–5,3
	Comparison to the Mohr-Coulomb model		100,0%	100,0%	127,4%	132,6%
	Maximum transverse forces	N (kN/m)	–1396,8	–1309,5	–1117,1	–566,8
		M (kNm/m)	0	0,0	0,0	0,0
		Q (kN/m)	**–41,2**	**–50,8**	**–40,2**	**–32,2**
	Comparison to the Mohr-Coulomb model		81,1%	100,0%	79,1%	63,4%
	Maximum deformations	d (mm)	**4,2**	**4,2**	**4,3**	**3,1**
	Comparison to the Mohr-Coulomb model		100,0%	100,0%	102,4%	73,8%

the difference between Mohr-Coulomb and Hoek-Brown is 33%.

Maximum difference in transverse forces is 37% and in deformation about 26%.

THE EVALUATION OF REINFORCED CONCRETE CROSS SECTION

Internal forces for individual material models were evaluated for design of lining. Primary calculation was performed at first. This calculation was later modified (in the view of input parameters: class of concrete, amount of reinforcement) to reach an effective design.

The result is an evaluation of usage ratio of the cross-section. This evaluation is shown in Table 2 and 3.

Based on the results we can say, at concrete thickness 250 mm and class C20/50 (according EN 206 [5]) the compression force in the three material models is crucial and results in crushing of concrete. Therefore, the change of reinforcement has no influence to bearing capacity.

However, for Hoek-Brown model tension force is important and the change of reinforcement influences the usage ratio of the cross-section over 10%.

When we retain thickness of the lining and increase concrete class to C25/30 concrete crushing is not observed and the influencing parameter is tension force. Because of this the change of reinforcement has large impact on the evaluation of the cross section.

When we retain concrete class C20/25 and increase thickness of the primary lining to 300 mm the usage ratio of the cross-section is the same. Cross section bears compression forces and the decisive parameter is tensile forces. The differences in the different material models of the rock mass are resulted by extreme tensile forces during section evaluation.

CONCLUSIONS

At the beginning of every geotechnical calculation engineer needs to answer two basic questions connected with modeling process: What will be modelled and what kind of results are expected from the calculation?

The first question is connected with the model. There is no need to build a complex model without feedback at the beginning. The second problem is connected with mathematical analysis. We need to know what to expect from the calculation. The different approach is used for evaluation of deformations caused by anthropogenic impact and different for examination of internal forces in structure.

Table 2. Usage ratio of the cross-section: primary design

Lining Thickness 250 mm, Concrete C20/25, Reinforcement Mesh 9×9×150×150 mm			1st Stage	2nd Stage	3rd Stage	4th Stage	Usage Ratio of the Cross-Section	
Material Model	Elastic	N (kN/m)	−1396,8	406,2	−1111,6	−1396,8	—	Crushing of concrete, normal force breaks ultimate limit state, insufficient degree of reinforcement
		M (kNm/m)	0,0	0,0	−13,5	0,0		
		Q (kN/m)	−41,2	32,8	0,5	−41,2		
	Mohr-Coulomb	N (kN/m)	−1309,5	352,8	−1093,6	−1309,5	—	Crushing of concrete insufficient degree of reinforcement
		M (kNm/m)	0,0	0,0	−13,5	0,0		
		Q (kN/m)	−50,8	25,4	0,9	−50,8		
	Drucker-Prager	N (kN/m)	−1296,9	129,6	−1280,0	−1117,1	—	Crushing of concrete
		M (kNm/m)	5,2	0,0	−17,2	0,0		
		Q (kN/m)	9,0	28,5	1,2	−40,2		
	Hoek-Brown	N (kN/m)	−742,7	258,2	−138,6	−566,8	65,30%	
		M (kNm/m)	4,0	0,0	−17,9	0,0		
		Q (kN/m)	−8,5	9,1	−5,3	−32,2		

Table 3. Usage ratio of the cross-section: after modification

Lining Thickness 250 mm, Concrete C25/30, Reinforcement Mesh 9×9×150×150 mm			1st Stage	2nd Stage	3rd Stage	4th Stage	Usage Ratio of the Cross-Section	
Material Model	Elastic	N (kN/m)	−1396,8	406,2	−1111,6	−1396,8	—	Breaking ultimate limit state
		M (kNm/m)	0,0	0,0	−13,5	0,0		
		Q (kN/m)	−41,2	32,8	0,5	−41,2		
	Mohr-Coulomb	N (kN/m)	−1309,5	352,8	−1093,6	−1309,5	89,30%	
		M (kNm/m)	0,0	0,0	−13,5	0,0		
		Q (kN/m)	−50,8	25,4	0,9	−50,8		
	Drucker-Prager	N (kN/m)	−1296,9	129,6	−1280,0	−1117,1	32,80%	
		M (kNm/m)	5,2	0,0	−17,2	0,0		
		Q (kN/m)	9,0	28,5	1,2	−40,2		
	Hoek-Brown	N (kN/m)	−742,7	258,2	−138,6	−566,8	65,30%	
		M (kNm/m)	4,0	0,0	−17,9	0,0		
		Q (kN/m)	−8,5	9,1	−5,3	−32,2		

To minimize associated errors, the decision about a material model must be based on the real geological information about modelled rock. This fact emphasizes the importance of geological investigation and mapping before start of the design process.

Based on the results of this investigation geotechnical engineer can choose the most suitable numerical model for the current conditions to provide economically efficient and structurally safe design. This plays a decisive role for tunnel design.

REFERENCES

[1] Ortuta, Paločko: Vplyv seizmického zaťaženia na geotechnické konštrukcie /THE EFFECT OF SEISMIC LOADING ON A GEOTECHNICAL CONSTRUCTION/: Aktuálne geotechnické riešenia a ich verifikácia, Bratislava 05.–06. júna 2017(in Slovak)

[2] Hlaváč, Ortuta, Chabroňová: Materiálové modely a ich vplyv na výpočet primárneho ostenia seizmického zaťaženie na geotechnické konštrukcie/The Material Model and Their Influence for the Design Primary Linning/: Aktuálne geotechnické riešenia a ich verifikácia, Bratislava 05.- 06. júna 2017(in Slovak)

[3] Chabroňová, Ortuta, Paločko: The Seismic Loading and his Impact on the Geotechnical Structure: 17th International Multidisciplinary Scientific GeoConference SGEM 2017, 27 June–6 July, 2017, Bulgaria

[4] Eurocode 8: Design of structures for earthquake resistance

[5] European standard EN 206—Concrete—specification, performance, production and conformity

Shallow Cover SEM Tunneling for the Purple Line Project

David Watson, Philip Lloyd, and Rafael Villarreal
Mott MacDonald

Richard Taylor
Traylor Brothers

ABSTRACT: The Plymouth Tunnel is part of the 16.2-mile, 21-station, east-west, Light Rail Transit (LRT) Purple Line Project undertaken by the Maryland Transit Administration (MTA). With an overall length of 1220-feet, the tunnel comprises only a short section of the line. The tunnel follows a relatively shallow vertical alignment with a maximum depth of cover above the tunnel crown of approximately 40-feet, and to a minimum of approximately 15-feet at the portals. The overburden material consists of soils, disintegrated and decomposed highly weathered rock traversing under a residential area. The groundwater table is located above the tunnel crown along most of the tunnel alignment. The ground surface topography above the tunnel alignment forms a hill like promontory that drains the groundwater towards two creeks to the north and south of the alignment. This paper will discuss present the ground conditions, the design approach and constructability considerations for the tunnel.

INTRODUCTION

The Purple Line is a 16.2-mile Light Rail Transit (LRT) line that will extend west to east linking the Red, Green, and Orange lines of the Washington Metro transportation system, providing transportation connection of the Maryland and Virginia suburbs to Washington, DC. Maryland Transit Administration (MTA) has undertaken this significant rail infrastructure project, planned along a 16.2-mile LRT alignment including 21 stations, of which 19 are above-grade and two are underground, see Figure 1. Manchester Place Station is one of the underground stations and is located near the center of the Purple Line alignment in Silver Spring, Maryland. The Plymouth Tunnel connects into the east end of Manchester Place Station and extends west with an overall length of 1,220 ft. The tunnel comprises a short, but critical, section of the Purple Line alignment as it traverses under a residential area with low cover between the tunnel crown and ground surface see Figure 2.

The Plymouth Tunnel mined alignment traverses from Manchester Place Station, beneath Plymouth Street and Flower Street before connecting with the East Cut & Cover tunnel under Arliss street at its eastern portal. The East Cut & Cover tunnel transitions into a retained cut located along Arliss Street that brings the alignment back up to grade.

Starting from the West end of the tunnel, the vertical alignment continues the 0.35% grade from Manchester Place Station through a vertical curve increasing the gradient to approximately 1.66% through the remainder of the SEM tunnel, East C&C tunnel and retained cut, to grade. The Plymouth Tunnel follows a relatively shallow vertical alignment with a maximum depth of cover above the tunnel crown of approximately 40-feet, reducing to a minimum of approximately 15-feet at the portals. The overburden material consists of soils, disintegrated and decomposed highly weathered rock traversing under a residential area.

Considering previous tunneling experience in the Plymouth Tunnel area, as well as tunnel length, anticipated ground conditions and site constraints, the Sequential Excavation Method (SEM) was selected as the most favorable tunneling method to be utilized for the project. In addition, the SEM tunneling method was considered to be most applicable in the expected mixed face conditions, since excavation method and support elements can be adjusted to meet the actual encountered ground conditions.

The mined section of the Plymouth Tunnel is scheduled to be excavated from East to West. The tunnel cross section is divided into two headings: a top heading and a bench, see Figure 3. The SEM tunnel will be excavated first by the top heading through the entire tunnel, followed by the bench excavation. The standard round length for each heading will be 4-feet for the top heading and 8-feet for the bench excavation, respectively. Due to the tunnel vertical alignment, the tunnel will be excavated on a downgrade for the entire length of the tunnel. Groundwater

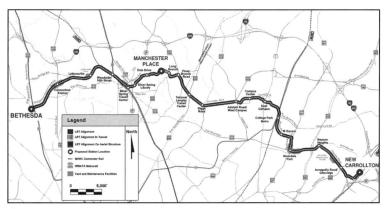

Figure 1. General location of the Plymouth Tunnel

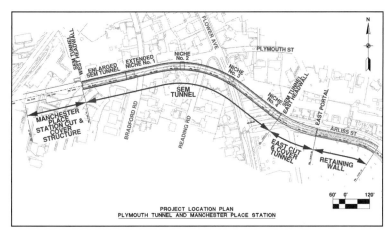

Figure 2. Plymouth Tunnel general layout

control will be required throughout construction to maintain the heading in a suitable condition for construction operations and to ensure that the tunnel dewatering drains drilled through the face and the sidewall remain operational.

GROUND CHARACTERIZATION

The overburden material consists of soils, disintegrated and decomposed highly weathered rock traversing under a residential area. The ground along the tunnel alignment is divided into five Ground Classification zones (GC's), designated GC-1 through GC-5. The locations of the GC zones and the anticipated static groundwater level are shown in the geological profile depicted in Figure 4. The GC's were delineated based on grouping materials with generally similar geotechnical characteristics as shown in Figure 5. These five GC's are the basis for

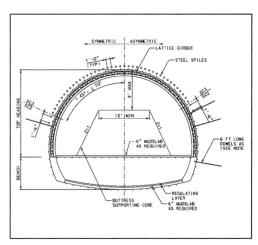

Figure 3. Plymouth Tunnel SEM sequence

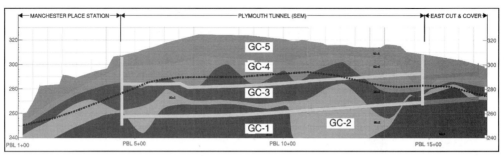

Figure 4. Geological profile for the Plymouth Tunnel. Tunnel shown in green lines, West (left) and East (right) headwalls shown in cyan. Groundwater table shown as dashed blue line.

Stratum	Ground Classification	Description
Surficial Soil	GC5	Predominant soil type is loose to medium dense silty sand, soft to firm sandy silt and sandy lean clay
		Various composition, potentially gravel fills and construction debris of various origin
		Fragments of Saprolite - ubiquitous discontinuities with negligible cohesion
Saprolite	GC4	Predominant soil type Saprolite with numerous fractures and almost cohesionless
		Highly weathered Saprolite could be classified as silty sand or sandy silt
Highly Weathered Rock	GC3	Primarily blocky to seamy decomposed rock
		Quartzite fragments (olistoliths), ranging from gravel to boulder size, randomly distributed
Weathered, slightly weathered to fresh rock	GC2	Weathered, slightly weathered to fresh bedrock that is still blocky and seamy
		Seams of Saprolite and decomposed rock, and zones of shattered rock
Slightly weathered to fresh rock	GC1	Consist of bedrock with rock mass conditions typically ranging from massive to very blocky, depending on local joint orientations.
		Seams of Saprolite and decomposed rock, and zones of highly fractured rock with fractures

Figure 5. Ground classification

describing anticipated ground conditions and behavior along the tunnel alignment.

Actual GC boundaries encountered in the excavations will typically be indistinct, and changes in ground conditions across the boundaries will be gradational. Distribution of the actual GC boundaries will be irregular, due to the variable extent of weathering within the weathering profile. Localized seams of more highly weathered material will extend down along joints and foliation from the more weathered zones above into the less weathered zones below; and, pinnacles of less weathered rock will protrude upward into the more highly weathered material above.

Actual positions of the GC boundaries observed in the SEM tunnel will also vary laterally across the tunnel face. Figure 6 shows a cross section at a selected tunnel station, showing the interpreted GC boundaries at the tunnel face, that demonstrate the variation across the face.

The anticipated groundwater conditions are shown within the geotechnical profile. Lowering the groundwater by dewatering, whether the crown of the tunnel is within GC3 highly weathered rock, Saprolite GC4 or soil/fill GC5, prior to tunnel excavation is an important factor contributing to ground stability during SEM tunnel excavation and support.

The MTA drilled 29 borings and excavated and supported one test shaft in a prior phase of geotechnical investigations. Between August to September 2016, an additional 15 borings were drilled, with locations dispersed among the existing 29 boring

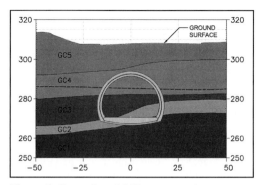

Figure 6. Ground variability at tunnel cross section

locations and test shaft. All borings were drilled with the hollow stem auger (HSA) and NQ rock core method. The borings extended to depths of approximately 15 to 60 feet below ground surface in soil and decomposed rock, where auger refusal was encountered. Approximately 10 to 40 feet of rock coring was performed.

GROUNDWATER DRAWDOWN MODELING

The groundwater table within the ground along Plymouth Tunnel alignment is generally located above the tunnel crown along most of the tunnel alignment. The ground surface topography above the tunnel alignment forms a hill like promontory parallel/sub parallel to the drivage direction of the tunnel during construction that drains the groundwater towards two creeks to the north and south of the tunnel. Groundwater will have significant bearing on excavation stability for some specific ground classes in the Plymouth Tunnel. Groundwater levels vary from a peak of about 12-feet above the tunnel crown at the deepest point to below the tunnel crown within the alignment approaching each portal and within the tunnel shoulders at the portals, between 5 and 8-feet below the tunnel crown.

An analytic approach to modeling the dewatering of the Plymouth Tunnel was considered to be more labor intensive, and less able to represent the dynamics of the system, than an approach using USGS MODFLOW-96 (Harbaugh and McDonald 1996). The use of MODFLOW also allowed flexibility, such that a quickly created, initial numerical model, could be easily refined to accommodate a variety of possible construction-dewatering scenarios.

Due to the location of the Plymouth Tunnel in an urban residential area, and due to the orientation of geological features with respect to the tunnel alignment and the expected dewatering mechanism,

it was deemed preferable to limit dewatering borings to those that would be drilled from within the advancing tunnel excavation itself, thereby optimizing dewatering efficiencies and minimizing disruption to the local neighborhood and streets. Therefore, the initial purpose of the MODFLOW model was to evaluate the efficacy of face drains in the dewatering of the tunnel. Subsequently, the MODFLOW model was refined to determine if it was possible to reduce the dewatering drainage yields to an average annual yield of less than 10,000 gallons per day (gpd). Obtaining this objective would allow construction on the tunnel to begin promptly, whereas the expected turnaround time to acquire the water-appropriation permit, required by the State of Maryland for diversions exceeding 10,000 gpd, is 18 months.

The model domain consisted of a rectangle that extended 5,000 feet from east to west and 4,200 feet from north to south that encompasses the main stems of Sligo Creek and Long Branch Creek, which are modeled using MODFLOW's River Module. There were initially 100 columns and 84 rows of 50×50 ft cells. However, the excavation of the tunnel is expected advance at a rate of 8 to 12 feet per day, so the MODFLOW model was to represent the progressive opening of the tunnel on each day subsequent to the last. Therefore, after calibration and in preparation for the simulation of dewatering in the vicinity of the tunnel, the MODFLOW grid was refined from 50×50 ft cells to 10×10 ft cells and another layer added. This adequately represented the tunnel and the daily advance when the tunnel was oriented more or less parallel to the grid in the east-west direction.

The MODFLOW results show that in a dewatering configuration where the top heading is driven a week or more ahead of the bench excavation, and the tunnel secondary lining and waterproofing to complete the permanent lining system is not installed until an indefinite time after the bench is completed, face drains appear to effect a significant (10 to 30 percent) reduction of the maximum heading inflows in the middle of the tunnel. The effect at the ends is less significant. The objective of lowering the water levels in the aquifer surrounding the tunnel excavation to below the saprolite during excavation which is to within a few feet above the crown where the ground water is highest would be met. However, the simulated peak inflows come with the excavation of the bench. The total amount of the dewatering discharge gradually increases as the excavation proceeds. By the end of tunnel excavation, the overall discharge rate is predicted to exceed 50,000 gpd. Peak top heading inflows are expected to average less than 200 gpd in the top heading and less than 300 gpd when the bench is excavated.

GEOTECHNICAL INSTRUMENTATION AND MONITORING

A Tunneling Monitoring Plan (TMP) was developed for the Plymouth Tunnel and required Support of Excavation (SOE) for the open cut excavation areas of the project. The TMP provides a description of the geotechnical monitoring program that will be used to ensure the safety and stability of the tunnel and open cut excavations and the integrity of any overlying or adjacent structures or utilities. The TMP defines how to monitor the performance of the underground construction as one measure to mitigate the potential impacts of ground movements; ground movements are detected and construction procedures can be modified if the displacements reach response limit values.

The instrumentation for the Plymouth Tunnel and SOE are targeted to monitor ground movements, groundwater elevations, building movements, and convergence/confinement movements of the works during construction. Instrument types will include shallow near-surface monitoring points (e.g., PK nail, prisms, etc.), building monitoring devices (e.g., crack gauges, prisms, etc.), subsurface ground movement monitoring devices (e.g., multiple point borehole extensometers, inclinometers, etc.), convergence monitoring arrays (e.g., survey targets), and piezometers. Seismographs will be utilized to monitor vibration and air overpressure levels during blasting activities.

Instruments will generally be located within the Zone Of Influence (ZOI) to register ground

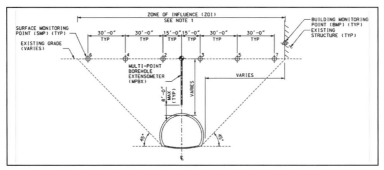

Figure 7. Instrumentation within tunnel cross section during SEM tunnel excavation

Table 1. Threshold and limiting values for different instruments within the Plymouth Tunnel

Instrument	Structure To Be Monitored	Measurements	Section, Station, or Instrument	Threshold Value	Limiting Value
SMP—Surface Monitoring Point	Ground Surface	Vertical Displacement	Plymouth Tunnel PBL 4+50 - 5+00	0.5"	0.75"
			Plymouth Tunnel PBL 12+50 - 14+75	1.0"	1.25"
MPBX— Multipoint Borehole Extensometer	Ground at various depths	Vertical Displacement	Plymouth Tunnel PBL 4+50 - 12+50	0.4"	0.6"
			Plymouth Tunnel PBL 12+50 - 14+75	1.0"	1.5"
CMP— Convergence Monitoring Point	SEM Initial Lining	Total Movement	Plymouth Tunnel PBL 4+50 - 12+50	0.4" Crown 0.15" Walls 0.1" Invert	0.6" Crown 0.3" Walls 0.2" Invert
			Plymouth Tunnel PBL 12+50 - 14+75	1.0" Crown 0.8" Walls 0.6" Invert	1.5" Crown 1.2" Walls 0.8" Invert

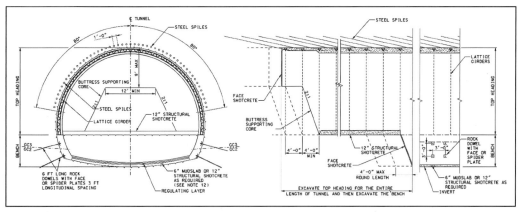

Figure 8. SEM tunnel ground support: cross section (left) and elevation (right)

movements and changes in groundwater elevations, see Figure 7. The instrument readings will be used to confirm the design through checking against predicted ground movements and changes in groundwater elevations. Instruments will also be installed on select properties and utilities within the ZOI to provide early warning of any potential damage. Similarly, convergence monitoring arrays within the SEM tunnel will provide assurance that the tunnel lining is performing as intended.

A series of trigger levels were developed for each type of instrument depending on the type of instrument, instrument accuracy, location of instrument, sensitivity of structures beings monitored and predicted ground movements, see Table 1. Response action plans will be developed that provide a general plan of action to be taken in the event of trigger levels being reached or exceeded. Each specific instance of a trigger level being reached will require a specific assessment of the particular conditions surrounding the exceedance in order to develop a meaningful course of action. Typical actions will be identified ahead of time, and contingency measures will be made available for rapid deployment.

For SEM construction, instrumentation and monitoring is also an integral part of construction, as it is the verification of the design assumptions made regarding the interaction between the ground and initial support during the excavation process.

GEOTECHNICAL/STRUCTURAL DESIGN FOR SHALLOW COVER SEM TUNNEL IN MIXED GROUND

The typical SEM tunnel cross section is shown in Figure 8, it consists of a horse shoe shape with a width of 35-feet and a height of 27-feet, approximately to outer face of the initial lining. In addition to the typical SEM cross section, the Plymouth Tunnel included one enlarged cross section immediately east

of the west portal, which houses the rail turnout, and four niches with slightly larger section than the typical section primarily for jet fans and utilities.

The initial lining consists of a 15-inch Fiber Reinforced Shotcrete (FRS) lining with a 6-inch thick mud slab (as needed), constructed in stages as the excavation of the tunnel progresses. The technical provisions for the project required any initial support to be 6-inch thick minimum to be considered part of the permanent lining. The 15-inch initial lining satisfied this requirement, and was considered part of the permanent lining, however the design assumed that the first 3 inches in contact with the ground will be sacrificial for the permanent condition. The permanent tunnel lining of the Plymouth Tunnel consists of a double shell lining composed of a 12-inch initial FRS lining, a PVC water proofing membrane, and a 12-inch secondary FRS lining, see Figure 9. The tunnel invert consists of a reinforced concrete invert slab, with a variable thickness in cross section.

The design of the tunnel lining incorporated several methods including empirical methods, ground stress strain methods, force equilibrium methods and structural methods. To capture the structural demands on the tunnel lining during all stages of excavation of the SEM tunnel, ground stress strain methods were developed by means of soil structure interaction models using the software packages Plaxis 2D and MIDAS GTS NX. To comply with the temporary and permanent load combinations required by the Technical Provisions, 2D structural models were developed using the software SAP2000 Ultimate.

Soil Structure Interaction Models

The sequential excavation method, including ground support elements and initial lining design, was analyzed using stress-strain methods by means of Plaxis 2D and MIDAS GTS NX. This analysis included the ground stresses generated due to the sequential excavation of the tunnel. A series of cross sections identified as critical along the tunnel alignment were selected for analysis in Plaxis 2D. To investigate the benefits of the three-dimensional effect of the excavation, two 3D models using MIDAS GTS NX for selected sections of the tunnel were included in the design of the sequential excavation method of the tunnel. In addition, a FLAC2D model was developed for the steady state seepage flow analysis to determine the groundwater drawdown curve, which was used as input in the Plaxis 2D and MIDAS models. All 2D models included both the initial and the secondary tunnel lining installed as per anticipated construction sequence. 3D modelling covered only the tunnel excavation and the installation of the initial

lining and not the permanent conditions with secondary lining in place.

Due to the different ground characterization present within the Plymouth Tunnel area, two different soil models were utilized for the soil structure interaction analysis. Ground classes GC-1 and GC-2 were modeled using the Hoek Brown material model, and ground classes GC-3 through GC-5 were modeled using the hardening soil material model.

The Plaxis 2D models included five tunnel cross sections selected to cover each principal setup of ground condition and each principal cross section of the Plymouth Tunnel, see Figure 10 for a typical cross section. Four cross sections with typical tunnel geometry and two cross sections with enlarged tunnel geometry were investigated. The 2D analysis included a parametric study to investigate the influence of variability in the ground relaxation parameters, tunnel shape and the construction sequence on structural forces, displacements and ground settlement. Sensitivity analysis of the initial lining was performed to assess variability in the ground parameters and the influence of implementing elephant feet in the initial tunnel lining. The 2D cross sections with the highest demands were checked for seismic loading.

The 3D analysis included two models as described below:

- Model 1—East portal through to PBL 12+00—incorporates modelling of canopy tubes, spiles and the lowest ground cover to determine ground relaxation ratio.
- Model 2 –PBL 6+50 to 8+00—at the location of the maximum overburden, incorporates

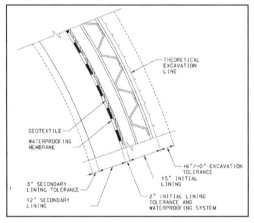

Figure 9. Double shell permanent lining

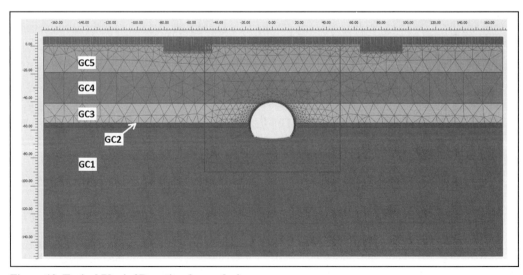

Figure 10. Typical Plaxis 2D section for analysis

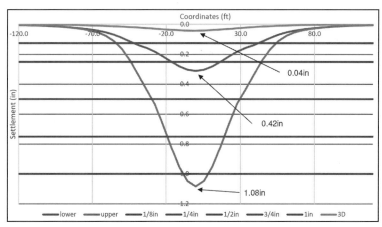

Figure 11. Comparison 3D vs. 2D settlement troughs

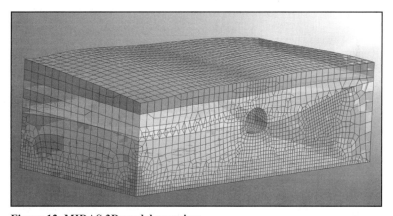

Figure 12. MIDAS 3D model overview

typical and enlarged cross sections as well as transition zone between them and provides the baseline for determination of ground relaxation ratio used in 2D analyses with invert embedded into GC2 or GC1

The soil structure interaction models provided prediction of the settlement trough caused by tunnel excavation and groundwater drawdown, and the results of the 2D and 3D models were compared. The results show that the induced settlement from the 3D models is smaller than the settlement obtained with the 2D models, see Figure 11. This is reasonable considering that the 3D models incorporate the temporary ground support (spiles, canopy), as well as the three-dimensional effects of the ground.

The stability of the tunnel face was checked by empirical methods using the procedure described by Jancsecz & Steiner. This method is based on three-dimensional failure scheme that consists of a soil wedge and a soil silo mode of failure and incorporates

the theorem of influence of excess pore pressures on the face stability. Additionally, the tunnel face stability was also checked by 3D analysis using MIDAS with influence of all supporting elements including face dowels and spiles, see Figure 13. Other temporary ground support elements including canopy pipes and spiles were designed using empirical models and compared with the outputs from the 3D model.

Structural Models

To comply with the load combinations for the temporary and permanent load cases, the tunnel lining, including the tunnel invert slab were modeled and checked using structural methods by employing the software SAP2000 Ultimate.

All sections were modeled in 2D using 1-foot wide beam elements. To simulate the soil structure interaction, radial and tangential spring elements were applied to the initial lining using the procedure described in the US Army Corps of Engineers

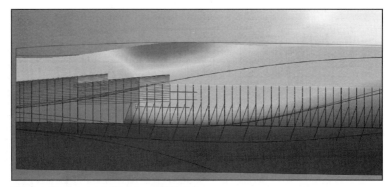

Figure 13. Face stability from MIDAS

Manual EM 1110-2-2901 Engineering and Design Tunnels and Shafts in Rock. The radial springs were modeled as compression only elements. Tangential springs were model to resist shear in both directions along the length of each beam element. Additional compression only springs were added at the toe of the initial lining to simulate the bearing resistance of the soil at these locations. The waterproofing membrane was modeled using compression only springs located between the initial and secondary linings, allowing a slip interface at the membrane. The structural analysis assumed that all ground loads are applied at the outer face of the initial lining, and the groundwater pressures applied at the interface between initial and secondary lining. This is based on the location of the waterproofing membrane and the assumption that the groundwater table would build up after the secondary lining has been constructed.

Load Sharing and Overburden Pressure

The double shell lining design of the Plymouth Tunnel assumes that the initial and secondary lining act together to carry the ground loads, and the groundwater pressure is carried by the secondary lining. Due to the presence of the waterproofing membrane between the initial and secondary linings, and the excavation sequence of the tunnel, the percentage of the ground load carried out by each lining was unknown and needed to be determined. A series of sensitivity analysis were performed in the structural models by varying the normal stiffness of the waterproofing membrane, thickness of the tunnel lining, and accounting for ground arching. The result showed that in the long term, the percentage of the ground load share was on the order of 70% carried by the initial lining, and 30% by the secondary lining.

The demands of the soil structure interaction and structural models were in good agreement, considering the simplifications of the structural model as it relates to the soil structure interaction, load combinations, and ground loads.

INTERDISCIPLINARY DESIGN FOR SEM TUNNEL FOR LRV

An extensive study and program of coordination between multiple disciplines was undertaken to verify that the interior dimension of the Plymouth Tunnel, including the SEM Tunnel, East Cut and Cover Tunnel and Retained Cut, could accommodate all the necessary items of infrastructure to safely service and support the operation of Light Rail Vehicles through the underground and below grade portions of the alignment in Segment 4 for the Plymouth Tunnel. Key items that needed consideration included:

- Construction tolerances
  - Excavation
  - Initial
  - Secondary
- Track alignment and variation between tracks—a best-fit alignment for the tunnel was created with a unique Project Baseline (PBL) alignment developed to minimize elevation differences between each track and the tunnel.
- Trackwork including details for fixing track and vibration mitigation measures.
- Egress walkway including handrail
- Light Rail Vehicle dynamic and static clearance envelopes
- Train control systems
- Overhead contact system (OCS)
- Communication systems
- Tunnel Ventilation (Jet fans)
- Dry fire Standpipe and fire hose valves

CONSTRUCTION AND DESIGN INTEGRATION

The Purple Line project is being delivered though a P3 model. With respect to the design and construction of the Purple Line, the nature of a P3 contract with the owner results in what is essentially a design-build arrangement between the construction team and the design team.

In contrast to the more traditional design-build-build delivery method for infrastructure projects, the design-build method allows for contractor involvement in the design process from bid-time through to release for construction. This level of involvement presents great opportunities. Having input on alternative designs, materials, means and method, and constructability issues during the design process allows the contractor to influence the design in such a way as to reduce construction time and cost, and benefit the project as a whole.

During the development of the Plymouth Tunnel design, Purple Line Transit Constructors (PLTC), was involved in the design process via regularly scheduled coordination meetings, members of the design team being co-located with the PLTC staff, and formal constructability reviews scheduled at the completion of the various levels of design (preliminary, intermediate etc.). At the initiation of detailed design, PLTC identified several areas in the construction of Plymouth Tunnel where reductions in construction time and cost could be achieved. Working with the design team several updates to the design were determined to be feasible and as the design was developed, it was able to be changed to optimize constructability. Several of these items are discussed in greater detail below.

Excavation Round Length

The initial design of the Plymouth Tunnel was based on progressing excavation in 3-foot rounds. Increasing the round length in the heading would allow PLTC to progress excavation activities at a faster rate and reduce the time needed for this activity. This was identified as a key item with the design team and achieving this would be dependent on the ground conditions anticipated and the ground support classes developed. Additional geotechnical investigations were performed and a more in-depth understanding of the ground was gained and models refined. This resulted in the round length being able to increase to 4 foot in the final design.

Top Heading Excavation Profile

The initial design of Plymouth Tunnel had three features identified by PLTC that if revised would benefit construction activates.

1. The height of the top heading was originally 18'-0" and limited the size of the excavation equipment that could be utilized.
2. The curved invert for the heading excavation and resulting driving surface would cause

inefficiencies with excavation activities and limit the size and type of equipment utilized.
3. The initial design of the heading support utilized elephant's feet at the base of the initial lining. Constructing the elephant's feet required additional excavation and materials as well as adding to the excavation schedule. If eliminating the need for these was deemed feasible this would reduce construction time and cost.

During design development, the design team was able to increase the heading excavation height to 20'-0", allowing PLTC to utilize the preferred equipment for tunnel excavation, a Liebherr 950 tunnel excavator. In addition to this, the heading invert was able to be flattened out and the elephants foot eliminated from the design.

Spile Length

The drill jumbo being used for the excavation of the Plymouth Tunnel is an Atlas Copco E2C. The drill steel length on the E2C is just over 15 ft. The construction team identifies that if the spile length could be limited to a maximum 15 ft in length, and the weight of the spiles limited to 50 lb each, that installation time would be greatly reduced. The reduction in installation time is a result of not having to add additional drill steel and spiles that are less than 50 lb can be handled by one individual. After utilizing the supplemental geotechnical investigation data and refining models, the design was able to be updated and it was confirmed that spiles less than 15 ft and 50 lbs in weight could be used with a 4 ft round length in the heading in conjunction with spiling every round.

CONSTRUCTION ACTIVITIES

All construction and mucking activities will be conducted from a staging area located at the eastern portal on Arliss Street. The Plymouth Tunnel will be constructed using Sequential Excavation Method (SEM) and cut-and-cover tunneling at the eastern portal.

The east cut and cover section is to be excavated and supported before SEM tunnel excavation can begin. The support of excavation will consist of soldier piles and lagging braced by a combination of walers, struts, rock bolts, and tie-backs. The east tunnel portal will be supported with braced soldier piles and lagging above the tunnel eye and below by a soil nail wall reinforced with fiber glass dowels within the tunnel eye itself. The cut and cover section will be excavated leaving behind a ramp to allow equipment to access the tunnel from Arilss Street.

The mined tunnel section will be constructed as a single tube, driven from the eastern portal toward the West Portal located at the Manchester Place Station site. The top heading will primarily be excavated through weak to very weak rock using a tunnel excavator equipped with interchangeable digging tools such as ripper bucket, hydraulic hammer, and milling head. The bench is expected to be driven through medium strong to strong rock using drill-and-blast methods as mechanical means of excavation are inefficient in these conditions. At the two portals, a pipe canopy will be installed as pre-support due to the shallow cover and the proximity to the fill material above the tunnel alignment.

The top heading will be driven all the way to the western portal, prior to excavation of the bench, to optimize equipment utilization and production. The heading excavation will be supported with fiber reinforced shotcrete, and lattice girders. Where required, spiling will be done at the excavation face to provide additional ground support along with other SEM 'toolbox' items required by the face conditions encountered. The bench will be supported with fiber reinforced shotcrete and rock bolts are required. To allow tunnel excavation to proceed without losing face stability and to facilitate placement of the initial lining under reasonably dry conditions, the groundwater level will be lowered in the vicinity of the heading. Dewatering will be accomplished by a systematic combination of underground drainage methods including probe holes and face drains in the heading.

Upon completion of tunnel excavation, sheet membrane will be placed on the inside face of the initial liner to provide waterproofing prior to placing the 12-inch thick secondary lining. Once the tunnel is complete the cut and cover section and retained cut section at the east end will be completed. The cast-in-place box will be constructed in three lifts (base slab, walls, and roof) of cast-in-place reinforced concrete with a sheet waterproofing membrane. The support of excavation for the retained cut section will consist of soldier piles and lagging supported by a combination of tie-backs and rockbolts. After excavation, the retained section will be constructed in two lifts (base slab and side walls) of cast-in-place reinforced concrete with a sheet waterproofing membrane.

CONCLUSIONS

Detailed design and close collaboration between designer and contractor throughout the design process has resulted in an optimized design for the Plymouth Tunnel. The sprayed concrete lining has been designed as a double shell, including long-term load carrying capacity of the initial lining. Optimizations of the design included increasing the round length and optimizing spile selection to improve cycle times.

ACKNOWLEDGMENTS

The authors would like to thank the MTA and PLTC for the opportunity of working on the design of the Purple Line Project.

REFERENCE

Harbaugh, A.W., and M.G. McDonald. 1996. User's documentation for MODFLOW-96, an update to the U.S. Geological Survey modular finite-difference ground-water flow model: U.S. Geological Survey Open-File Report 96-485. vi + 56 p.

Session 3: Rock Tunneling

Desiree Willis and Carmen Scalise, Chairs

Design and Construction of South Hartford Conveyance and Storage Tunnel

Verya Nasri and James Sullivan
AECOM

Vincent Prestia
Kenny Construction Company

Andrew Perham
Metropolitan District Commission

ABSTRACT: The South Hartford Conveyance and Storage Tunnel (SHCST) is a major component of the Hartford Metropolitan District's Clean Water Project (CWP). This tunnel will capture and store Combined Sewer Overflows (CSO) from the southern portion of Hartford, CT and Sanitary Sewer Overflows (SSO) from West Hartford and Newington, CT. The project components include a deep rock tunnel 21,800 feet in length with a 21'-1" excavated diameter, several miles of consolidation sewers, multiple hydraulic drop shafts with deaeration chambers and a 27 MGD tunnel dewatering pump station. AECOM and Black & Veatch JV led the design and Kenny and Obayashi JV is currently building the tunnel. This paper discusses the major aspects of the design and construction of the SHCST.

INTRODUCTION

The South Hartford Conveyance and Storage Tunnel (SHCST) project is a significant component of the Hartford Metropolitan District's (MDC) Long Term Control Plan (LTCP) which is overseen by the Connecticut Department of Energy and Environmental Protection (CTDEEP). This project will address a portion of the MDC's Clean Water Project (CWP), which will reduce combined sewer overflows (CSOs); eliminate sanitary sewer overflows (SSOs); and reduce nitrogen released into the Connecticut River.

The purpose of the SHCST project is to eliminate West Hartford and Newington SSOs, eliminate Franklin Area CSOs discharging to Wethersfield Cove and to minimize CSO discharges to the South Branch Park River.

During 2012, the MDC conducted an evaluation of the potential of connecting the proposed North Tunnel (originally proposed as an independent tunnel with its own pump station) into the South Hartford Conveyance and Storage Tunnel. This evaluation concluded that the two proposed tunnels could reasonably be connected together and operated as a single tunnel system utilizing the tunnel pump station at the eastern terminus of the South Hartford Tunnel (Figure 1). It also was concluded that this alternative was less costly than two independent tunnel systems.

During dry weather, the SHCST will not receive flow as the existing MDC collection system can adequately convey flow to the Hartford Water Pollution Control Facility (HWPCF). During wet weather, when the capacity of the existing collection system is exceeded, the SHCST will receive overflows that would have previously discharged directly to receiving waters.

This paper summarizes the design and construction aspects of the SHCST project. New diversion structures will be constructed at each CSO/SSO relief point to divert overflows to new consolidation sewers (near surface). These, in-turn, will discharge flow to hydraulic drop shafts which will convey the flow in a controlled manner to the deep rock storage tunnel. Once in the tunnel, flow will be pumped to the new headworks at the HWPCF. The components of the SHCST project described in this paper are as follows:

- Deep rock tunnel (18' ID @ 21,800 LF) with a TBM launch shaft near the HWPCF in Hartford and a TBM retrieval shaft in West Hartford
- 12,200 LF of near-surface consolidation sewers (24" to 66" in diameter)

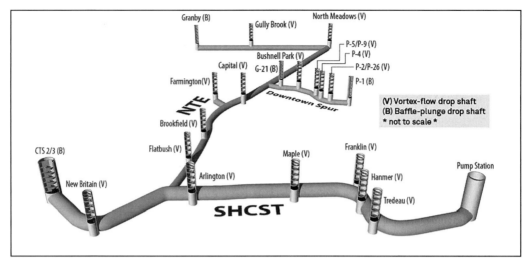

Figure 1. SHCST and North Tunnel integration

- Seven hydraulic drop shafts
- 27 MGD tunnel dewatering pump station
- Odor control at all potential air release points.

DESIGN CRITERIA

The sizing of the tunnel was based on the volumes from the 1-year, 18-year and 25-year design storm per the LTCP and updated collection system modeling from the MDC's Program Management Consultant. The LTCP specified a different level of control for each tributary area. Table 1 shows the peak flows and volumes to be stored in the SHCST for each major source and respective design storm.

Surge, air entrainment and pressure waves can occur in CSO tunnels filling rapidly, with detrimental results such as geysering, blowback and flow instabilities. Based on the hydraulic analysis, it appears that surge in the SHCST is unlikely, due to the relatively large tunnel diameter in comparison to the incoming peak flows.

Sediment deposition can present an ongoing maintenance burden if not controlled. Based upon the initial sediment deposition analysis and modeling, a slope of 0.1% appears adequate for the deep rock tunnel to cost-effectively minimize sediment deposition issues. This slope is consistent with the state of practice for other large diameter CSO tunnels as steeper slopes will increase project cost. The tunnel will still require periodic maintenance to remove sediment build-up over the life of the facility.

The capacity of the tunnel dewatering pump station has been established by the MDC as 27 MGD. At this rate, the 25-yr design volume for both the North and South tunnels will be dewatered in 4.2 days. The 1-yr design volume would be dewatered in

Table 1. Tributary overflows to the SHCST

Contribution	Design Storm	Peak Flow (MGD)	Volume (Mgal)
West Hartford/ Newington SSO	25-yr	27	17
South Branch Park River CSO	1-yr	68	6
Franklin Area Relief	18-yr	313	39
Total			62

approximately 2.5 days. A typical CSO tunnel is dewatered in 24 to 48 hours.

The operation of the tunnel must not result in odor complaints. As such, odor control has been assumed at each drop shaft location.

An alignment study was conducted to evaluate various configurations of rock tunnels and consolidation conduits. Several conceptual rock tunnel alignments and associated consolidation conduit options were developed and evaluated. The purpose of this alignment study was to identify a cost effective and acceptable tunnel alignment that balances the expectations of the many stakeholders impacted by the project. All the alignments began in property owned by the District adjacent to the HWPCF. The selected alignment significantly reduced the length of consolidation conduits and their surface impact.

GEOTECHNICAL SETTINGS

The site area lies in the Central Lowlands physiographic province that extends in a north-south direction in the middle of the state. The central lowland area consists mainly of the sedimentary rocks and the associated igneous basalts of Triassic and Jurassic age. The Hartford Basin of Connecticut and southern Massachusetts is a half graben in structure, 90 miles long, and filled with approximately 13,000 ft of sedimentary rocks, and basaltic lavas and intrusions (Hubert et al, 1978). The source area for the sedimentary rocks was mainly the metamorphic rocks of the Eastern Highlands. Volcanic flows separated the deposition of the lacustrine and fluvial deposits, which were derived from the erosion of the highlands to the east. Displacements along the faults continued throughout the depositional period. The depositional sequence resulted in a series of features including the alluvial fan, lake, alluvial mudflats and floodplain deposits separated by basaltic flows.

Following the deposition of most of the sediments, the tectonic activity continued along the east edge of the basin. Displacements along the eastern border fault rotated the basin downward to the east that resulted in the easterly dipping beds. The Jurassic extensional tectonics is associated with the separation of the continents. That was the last major tectonic episode affecting the geology of the region. Age dating of the Triassic/Jurassic faulting in southern Connecticut has indicated that the last activity along the faults is approximately 175 million years ago (NNEC, 1975). All faults in the project area are therefore considered to be inactive.

The region has undergone a period of glaciations that has reshaped the terrain. Glaciers ground down the area's peaks, scraping away any weak or weathered rock and laying down a heterogeneous layer of ground-up rock. This till layer is present over much of the lower lying bedrock surfaces. The sediments of Glacial Lake Hitchcock filled in the deeply-incised Connecticut River Valley. The lake deposits are present in varying forms from Rocky Hill, Connecticut to Northern Vermont. Glaciers shaped the topography and left the area with much of the topographic relief present today. More recent alluvial deposits are common along the Connecticut and Park Rivers and their tributaries.

In the site area, the following soils are present overlying the bedrock, in general order of sequence from ground surface downwards: Artificial Fill, Alluvium, Beach Deposits of Lake Hitchcock, Glaciolacustrine Deposits, Glaciofluvial Deposits, and Glacial Till. Bedrock is not widely exposed in the project area. The formations that are in the general vicinity of the project and potentially could be encountered along the proposed tunnel are the Portland Arkose, the Hampden Basalt, and the East Berlin Formation. These units consist of shale and basalt with fractured and fault zones (Figure 2).

The final geotechnical investigation program consisted of 55 deep rock borings, 50 shallow borings, and 5 geophysical survey lines. The program included geophysical logging (acoustic televiewer) performed in 21 of the deep borings and 5 of the shallow rock borings, water pressure (packer) testing performed in 30 of the deep borings and 8 of the shallow rock borings, 6 in-situ stress determinations in two deep borehole, falling head tests completed in the soil profile in selected borings, observation wells installed in 13 of the deep borings as well as 13 of the shallow borings, and 22 vibrating wire piezometers installed in 16 borings. The program also included the monitoring of groundwater levels, and the completion of soil and rock laboratory testing. Several phases of comprehensive geotechnical investigations were planned and implemented in order to obtain sufficient subsurface information for design and address the geotechnical challenges discussed in the following section.

GEOTECHNICAL CHALLENGES

There were several geotechnical challenges which needed to be investigated and addressed during the design. The main geotechnical challenges were:

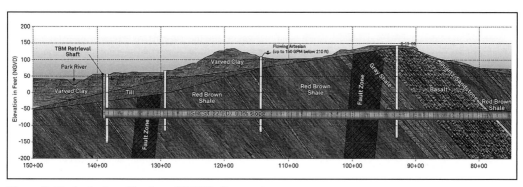

Figure 2. Geological profile along SHCST alignment

- Consolidation settlement of varved clay as a result of groundwater lowering due to shaft and tunnel construction and the settlement impact to the existing facilities
- Faults and fracture zones along the tunnel drive and tunnel construction impact
- Artesian groundwater inflow for a portion of the tunnel drive and its design and construction impacts

In order to evaluate the impact of consolidation settlement on the facilities, the extent, thickness, and physical and mechanical properties of varved clay had to be established first. Test borings were drilled along the tunnel alignment as well as the shafts to establish the subsurface profile and varved clay thickness. In situ tests such as falling head and packer tests were performed to establish the permeability of the soils and rock. Geotechnical laboratory tests were performed to establish the properties of soil and bedrock as well as consolidation characteristics of the varved clay.

Consolidation settlement may impact several existing facilities in the vicinity of the launch shaft and starter tunnel. These facilities include a flood control embankment dike, transmission towers, railroad tracks, wastewater facility structures and a pump station. Other facilities that may have potential impact are the Amtrak Rail and residential and industrial buildings along the tunnel alignment.

In order to evaluate the impact of consolidation settlement of the varved clay as a result of the tunnel and shaft construction, a three-dimensional model was utilized. The effect of groundwater lowering at the bottom of varved clay layer increases the effective stress in varved clay and thereby results in consolidation settlement. The time rate of change in effective stress (i.e., pore water pressure reduction) was calculated by MIDAS GTS finite element program based on transient flow analysis and the consolidation settlement was calculated in accordance with one dimensional consolidation theory.

Three-dimensional modeling provided sufficient results to compute the total settlement and angular distortion of the facilities with time. The settlement associated with tunnel ground relaxation was also evaluated using finite element model and is found to be negligible due to the deep rock nature of tunnel alignment. The finite element modeling results were used to address and mitigate the settlement of the facilities to an acceptable level.

Faults and fracture zones were evaluated based on the borings drilled along the tunnel alignment. A desktop study was performed first to identify the potential fault and fracture locations based on the regional geological maps, previous tunnel projects in the vicinity and published papers. The field investigation programs were performed in four sequential phases to insure cost effective, targeted and focused investigations in the fault and fracture zones.

Artesian groundwater pressure was identified to be present in bedrock in the central portion of the tunnel alignment. The artesian pressures were measured in the boreholes and pressures as high as 30 ft above the ground surface were recorded. The effect of artesian pressure on the tunnel construction as well as the tunnel lining has been accounted for.

Based on the above geotechnical challenges and evaluation of cost estimate and risk mitigation, it was concluded that shielded TBM with one-pass lining system is the preferred alternative for the deep rock tunnel.

MAIN TUNNEL

The deep rock tunnel is approximately 21,800 feet in length and has a finished internal diameter of 18 feet. The tunnel is excavated by a Tunnel Boring Machine (TBM) which is suitable for tunneling in hard rock conditions. The tunnel profile is entirely within bedrock at a depth low enough to accommodate the North Tunnel system (part of a separate contract). The tunnel is bored by a 21'-1" diameter Herrenknecht single shield hard rock TBM (Figure 3). The TBM will dump muck onto a horizontal tunnel conveyor which transports muck to a vertical conveyor belt in the launch shaft. The running tunnel belt is positioned at 12-o'clock in the tunnel to allow for adit construction to begin while mining is ongoing. The vertical conveyor belt moves the muck to the surface and dump onto a transfer conveyor, which dumps onto a stacker conveyor to create piles of muck which is segregated daily, then sampled and characterized prior to being cleared to be disposed of from site.

It is anticipated that the rock mass along the tunnel alignment primarily consists of competent shale, sandstone, and basalt bedding dipping 10 to 20 degrees with occasional known fault zones.

Case Foundation Company completed the 39 ea. 60" diameter secant piles 75'-0" deep, which make up the support of excavation (SOE) of the 39'-0" inside diameter overburden portion of the launch shaft. The shaft was excavated to top of rock and a ring beam poured at the interface as part of the structural liner section within the overburden. After the overburden concrete operation was complete, drill and shoot operations began and continued until rock excavation reached just above the arch of the starter tunnel. At that time the shaft was lined bottom up to meet the overburden stage. The balance of the drill and shoot of the shaft was completed next, and then concrete-lined below starter tunnel invert elevation, the lower 19' of shaft is backfilled with sand to

Figure 3. Herrenknecht single shield hard rock TBM

starter tunnel elevation. The work was sequenced in this manner so drill and blast work is not required to be performed at the end of the contract after TBM mining is completed. After a working slab is poured at starter tunnel elevation, drill and shoot of the starter/tail tunnels is commenced. For the retrieval shaft, Case Foundation Company installed 46 ea. 48" diameter secant piles which make up the support of excavation (SOE) of the 32'-0" inside diameter overburden portion of the shaft. Then the shaft is excavated and concrete lined the same as the launch shaft.

After the rock excavation and the completion of the concrete lining of the launch shaft, drill and shoot operations of the starter/tail tunnel commences. The starter tunnel from center of the launch shaft is approximately 480'-0" and the tail tunnel from the center of the launch shaft approximately 80'-0"±. The interface of the starter tunnel, tail tunnel and launch shaft is shaped in a manner to accept all equipment associated with TBM operations. Three rail spurs are present within the starter tunnel, with two of them continuing through the launch shaft into the tail tunnel. One rail spur within the launch shaft is utilized for the loading of precast segment annulus grout mix to be batched and loaded directly into agitator rail cars. Since the TBM was manufactured and arrived prior to completion of the starter/tail Tunnels, its assembly is completed as much as possible above ground, and then it is lowered into the shaft by a strand jack gantry system straddling the top of the launch shaft.

One-pass lining system was selected based on the geotechnical challenges discussed in the previous section and as a means of risk mitigation. The 5+1 rhomboidal system assembled ring by ring is selected for SHCST tunnel lining (Figure 4). It consists of five full-size parallelogrammic and trapezoidal segments and one small key segment in a ring. Advantages include staggered longitudinal joints, continuous

ring building and compatibility with a dowel type connection in circumferential joints, which results in a faster ring assembly process comparing to rectangular systems. Universal rings are selected for this project assembled from rings with circumferential joints inclined to the tunnel axis on both sides. One of the main advantages of this ring system is using only one set of forms for segment production. With 18' as the internal diameter of the tunnel, 12" was selected as the thickness of segments which is the common value used in practice for this size of tunnel. Lining thickness was verified during the design procedure. Longitudinal and circumferential joints were designed as completely flat joints which are advantageous to other types of joints for transfer of loads between segments and rings, and also have proven to have a superior sealing performance. Bolt connection was designed for longitudinal joints and dowels were chosen for connecting rings in circumferential joints as they require less work for the construction of the segment form and less manpower in the tunnel as the insertion is automatically performed by the erector when the segment is positioned.

The gasket profile was designed as fiber anchored which was the first application of this type in the world (DATWYLER M80207 "type South Hartford") providing watertightness under the maximum expected groundwater pressure of 140 psi. This gasket profile guarantees watertightness for 1.5 times maximum working water pressure considering a combination of gasket differential gap of 0.2" and bearing surface offset of 0.4". The analysis and design of the segmental lining complies with the latest guideline (ACI 544.7R, 2016) in order to satisfy the intended objectives of the project during construction and service life of the tunnel. Cement type was Portland cement type II per ASTM C150. Type II Portland cement is sulfate-resistant cement, recommended for concrete linings exposed to the sulfate attack. In wastewater tunnels, the lining must also be designed to withstand corrosion attack from the influents which contains sulfate, as well as the groundwater that may contain corrosive chemicals. A quantitative approach, adopted by EPA and ASCE, was used to assess the corrosion of the final lining. Loss of concrete material as a function of time, concrete properties and CSO characteristics was estimated and considered in this project. For the first time in North America, double hooked-end steel fibers (Dramix® 4D) were designed for reinforcement. This new type of steel fiber satisfies the serviceability requirements by limiting time-dependent effects of creep on crack opening and more significantly guarantees ductility requirements in conventional fiber dosage rates by providing an ultimate bending moment higher than the cracking bending moment.

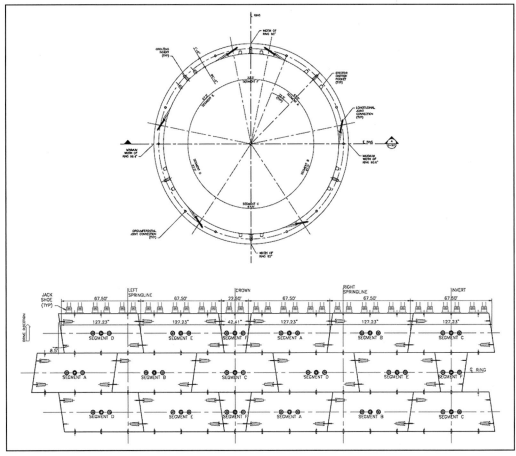

Figure 4. SHCST segmental ring geometry

The cost estimate for the entire SHCST Project is approximately $500 M. The project construction duration is estimated at approximately 6 years. The recommended contract packaging was to release six construction contracts: 1) Preliminary Utility Relocation, 2) Tunnel, 3) Pump Station, 4) Franklin/Maple Consolidation Conduits, 5) Flatbush/Arlington/Newington/New Britain Consolidation Conduits, and 6) West Hartford Consolidation Conduit. The contracts were grouped to align construction skill sets but allow for the phased release of the bid packages. The overall construction schedule is to be coordinated such that the tunnel, pump station and consolidation conduit contracts are constructed independently but conclude coincidently.

MDC management has stated that a goal for the project is that odor complaints must not occur. Therefore, the odor control strategy for the SHSCT system is focused on minimizing odors from the two main shafts at the tunnel ends and at the six intermediate drop shaft sites.

DROP SHAFTS

Seven hydraulic drop shafts are used to convey flow in a controlled manner from the shallower consolidation conduits to the deep rock tunnel. A two-level screening process was used to assess the characteristics of each site and to recommend either a baffle-plunge or tangential vortex based upon cost effectiveness, hydraulic performance, and operation and maintenance considerations (Figure 5).

The tangential vortex drop structure type was selected for all of the sites along the tunnel alignment (with the exception of the TBM retrieval site) due to its widely accepted use for deep rock CSO storage tunnels, history of acceptable performance, and cost effectiveness when compared to the baffle-plunge drop structure. The baffle-plunge drop structure type was selected for the deep rock tunnel TBM retrieval site because of the existence of the larger diameter shaft being constructed at this site for the TBM retrieval. Once such a large shaft is present, the

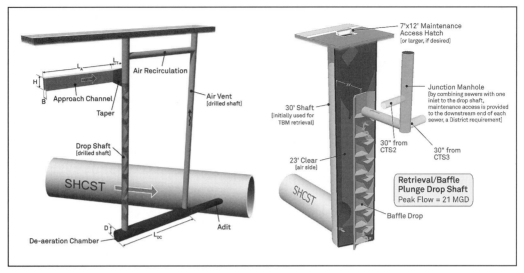

Figure 5. Vortex and baffle drop shaft alternatives

baffle-plunge becomes ideally suited for such applications because of its compact surface area impact.

The drop shafts at the TBM launch and retrieval areas are located within larger sites, whereas, others along the alignment are located in small independent sites. All shafts are drilled and cased by Case Foundation Company utilizing a Bauer drill rig for the overburden sections. Case also downs bore the rock excavation utilizing a Wirth drill rig. The inside diameters of the drop shafts/vent shafts vary from 30" to 72". Vertical mortar-lined carrier pipes are installed by floating them into place from the surface, then grouting the annulus utilizing a tremie placement method. The pipes are capped with a steel plate and abandoned for a future contract to tie into surface sewers.

All adit and deaeration chamber work is completed through the main tunnel since there is no access through the drop shaft. Adits and chambers for drop shafts 3 to 7 are serviced from the launch shaft and mined/lined as the TBM bores ahead. By nature of where they land in the tunnel alignment, adits and chambers for drop shafts 1 and 2, and the North Tunnel stub horseshoe excavation are serviced from the retrieval shaft after mining is completed.

At each adit the existing segment liner is supported by pinning them back to sound rock through each segment's grout port. Rock excavation is performed after an opening is saw cut via drill and shoot. The drilling aspect of the operation is accomplished utilizing jack legs, due to the small heading size.

The deaeration chamber at the bottom of each drop shaft is slightly larger in size due to the fact that it is formed and poured with cast in place concrete which is pumped from the tunnel. The adits are lined

with mortar-lined steel carrier pipe which is transported into the tunnel, welded, blocked in place, and then grouted. The last operation is the patch of the area where the segments were removed at the tunnel/adit intersection.

CSO/SSO CONSOLIDATION CONDUITS

New diversion structures constructed near existing CSO/SSO locations will utilize transverse or side flow weirs to direct the design overflows from existing pipes into the consolidation conduits. These conduits then convey flows to the deep rock tunnel through either vortex or baffle drop shafts.

The consolidation conduits will be installed using a combination of microtunneling, guided boring, shallow rock tunneling, and open cut construction techniques. It is anticipated that three consolidation pipe branches along the selected alignment will be installed using microtunneling methods. This includes a 24-inch guided bore of the Newington Consolidation Pipe (NCP), a 42-inch microtunnel installation of the New Britain Consolidation Pipe (NBCP), and a 48-inch microtunnel installation of the Flatbush Consolidation Pipe (FCP). When considering microtunneling as the likely means of installation, effort has been made to locate conduits within soil; however, there is the potential for mixed-face microtunneling in areas of till.

The open cut method of pipe installation will be utilized for installation of the 30-inch West Hartford Consolidation Pipe (WHCP), the southern section of the 24-inch NCP, and the 27-inch Arlington Consolidation Pipe (ACP). The open cut method creates more temporary disturbance to traffic, business and residences as this work is performed primarily

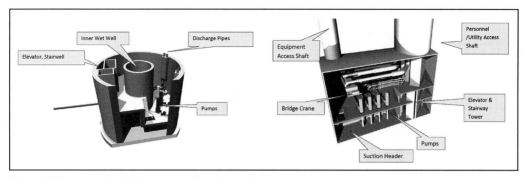

Figure 6. Cavern and shaft pump station alternatives

within the roadways; however, it may be the preferred installation method due to the depth of the pipe, geotechnical conditions, and cost considerations.

Based on existing geotechnical information, it is anticipated that the 66-inch Franklin Avenue Consolidation Pipe (FACP) and the 60-inch Maple Avenue Consolidation Pipe (MACP) will be constructed using an open face rock tunneling machine. Consideration is given to standardizing the diameters of these tunnel consolidation sewers to potentially reduce costs.

PUMP STATION

The Tunnel Pump Station (TPS) is designed to pump out the SHCST following storm events so that the flow can be treated at the HWPCF. At this point, stored flows will receive adequate treatment prior to discharge to the Connecticut River. The proposed TPS will be located within the HWPCF complex.

The TPS will be designed to pump out at a maximum 27 MGD capacity. This rate will allow the 62 MG SHCST to be pumped out within 55 hours (2.3 days). The proposed tunnel invert elevation at the TPS site is –170 feet and the discharge elevation at the plant is +6 feet. Therefore, the total maximum static head is 176 feet.

The recommended pump equipment consists of four 9 MGD vertical non-clog centrifugal pumps. This will provide a firm pumping capacity of 27 MGD with one unit out of service. Variable frequency drives (VFDs) are recommended for the pumps as turn-down capability to approximately 4 to 5 MGD can be achieved.

The TPS will discharge directly to the new Headworks facility at the HWPCF. The force main is currently sized to be 36-inches in diameter. The recommended connection point at the discharge end is at a new junction structure just upstream of the new influent pumping station. A surge tank will be provided on the discharge force main to minimize surges in the system. The surge tank will be situated at the TPS site.

Two pump station configurations were initially presented as the finalist options. One of these is a cavern pump station and the other is a circular pump station with a suction header pipe system (Figure 6). A comparative assessment of the capital costs of these two configurations was then prepared and it was concluded that the cost of the circular pump station is less than that of the cavern pump station. On this basis, the circular pump station was chosen for the project.

A new 9,800 kW overhead electrical power service from CL&P is required for the tunnel boring machine. This power feed will be converted to a permanent power feed for the TPS, once the TPS is completed and made operational. Current power requirements for the TPS and related facilities are on the order of 3,055 kW.

Screenings and grit capture is accomplished in a separate 35-foot diameter dedicated shaft. The shaft which is used as the launch shaft for the TBM tunnel is converted to the grit/screenings shaft. Bar screens are provided to protect the TPS pumps from solids and debris which would either clog or damage the pumps. A rake lowered by crane either pushes or pulls the screenings up from the shaft. Grit and other heavier debris are removed from the pit by a clamshell bucket. The screenings shaft is used for tunnel construction, allowing construction of the TPS to proceed in parallel with tunnel construction.

The TPS and the Grit/Screenings facility are roughly 150 feet apart and are connected with a 48-inch diameter suction header. An at-grade building is constructed to house support facilities critical to the operation of the pump station and to allow for pump station access and egress. Personnel access/egress will be by elevator. A separate stair tower is provided for emergency situations. The grit/screenings facility is also enclosed in a building to better contain odors and to promote a more visually appealing facility to neighboring businesses.

Case Foundation Company completed the 70 ea. 60" diameter secant piles, 75'-0" deep that

Figure 7. Completed shaft in the overburden

make up the SOE of the 74'-0" inside diameter overburden portion of the pump station shaft. This followed by the construction of the liner plate/shaft ring floodwall. The construction of the pump station shaft is similar to the launch shaft: overburden excavation to rock, pour ring beam, form/pour to surface (Figure 7), drill and shoot to the bottom of shaft, and concrete line up to the overburden concrete. As part of the work, the suction tunnel between the launch shaft and pump station shaft is drilled and shot and pipe carried in and grouted.

CONCLUSIONS

This paper presents the design and construction aspects of a deep rock conveyance and storage tunnel, drop shafts, consolidation conduits, and a pump station in Hartford, CT. The geological settings, subsurface investigation program, and geotechnical challenges are discussed. The criteria for the selection of preferred alignment, type of the TBM, details of the liner for tunnel and shafts are described. Relevant alternatives for the drop shafts and the pump station are explained and recommended options are presented.

REFERENCES

Hubert, J.; Reed, F.; Dowdall, W. and Gilchrist, J. 1978. *Guide to the Mesozoic Redbeds of Central Connecticut*, Guidebook No. 4, Connecticut Geology and Natural History Survey, Hartford, CT.

Northeast Nuclear Energy Company 1975. Geologic Mapping of the Bedrock Surface, Millstone Nuclear Power Station Unit 3, Docket No. 50–423.

Nasri, V.; Sullivan, J.; Pelletier, A. and McCarthy, B. 2013. Design of South Hartford *Conveyance and Storage Tunnel*, Rapid Excavation Tunneling Conference 2013, Washington, DC, June 23–26, 2013, pp. 546–556.

Bakhshi, M., and Nasri, V. (2017), Design of Steel Fiber-Reinforced Concrete Segmental Lining for the South Hartford CSO Tunnel. Rapid Excavation Tunneling Conference 2017, San Diego, CA, June 4–7, 2017, pp. 706–717.

Approach to Characterizing BIM Rocks for Tunnel Design Using As-Built Tunnel Data

Shawna Von Stockhausen and David J. Young
Mott MacDonald

Christopher Slack
GEI Consultants

ABSTRACT: Characterization of block in matrix (BIM) rock is challenging due to strength differences of up to two orders of magnitude between blocks and matrix. The Franciscan Complex is a classic BIM rock widespread throughout Northern California. Several tunnels are planned to be constructed in this material by the Santa Clara Valley Water District and others over the next five years. This paper describes a geotechnical characterization process involving laboratory testing, back-analysis using the convergence-confinement method, and as-built tunnel records for a tunnel that encountered squeezing conditions to develop geotechnical design parameters for one of these new tunnels. The likelihood, extent and distance from the face for squeezing behavior in matrix dominated ground is explored in this paper.

INTRODUCTION

A mélange body is a mixture of relatively strong blocks of rock, ranging in size from cobble size fragments to house size blocks and larger, embedded within a weaker matrix of sheared argillite, shale, siltstone or serpentinite. Mélanges are represented in Northern California within the Franciscan Complex. Mélanges are part of the large family of BIM rock, which are mixtures of rocks composed of geotechnically significant blocks, within bonded matrices of finer texture (Medley 1997).

The Franciscan Complex is known to be a chaotic geologic mixture of relatively strong clasts or blocks of rock, consisting of a variety of lithologies and sizes, embedded within matrices of weaker shale, sandstone, serpentinite, and clay. A consequence of this geologic mixture, especially regarding the prediction of ground conditions, is that the Mélange typically lacks spatial continuity and therefore exhibits frequent and abrupt variations in geotechnical characteristics.

Of particular interest to tunnel designers, contractors, and owners is the ability to characterize the engineering properties and behavior prior to and during construction. Medley has presented the concept of volumetric block proportion (VBP)—the percentage of the face that contains blocks—as a means of understanding and characterizing the properties and behavior of BIM rocks during excavation since the mid-1990s. Squeezing ground is one such example— even under low cover, significant squeezing can occur when the excavated face is dominated by matrix material (low VBP). By taking the concepts presented by Medley and adapting to tunnels, it is possible to predict squeezing behavior in BIM rock tunnels.

REGIONAL GEOLOGY OF NORTHERN CALIFORNIA

The Franciscan complex or Franciscan Assemblage has been mapped in Northern California by Andrew Lawson (Byerly, P., and G. D. Louderback. 1964). It is typically described as a Jurassic to Cretaceous-age (205 to 65 Million years ago (Ma)) marine sedimentary and volcanic rock unit, deformed (i.e., sheared, faulted, and folded) and locally characterized by a chaotic mixture of rock blocks of different sizes and lithologies that are encased in a sheared, soil-like matrix.

The highly fractured Franciscan Complex bedrock is not easily characterized by standard in-situ and laboratory test methods for rock or soil. Much of the core is not durable enough for ASTM sample preparation for unconfined or triaxial compressive strength testing. As such, laboratory strength test results tend to be biased toward the more intact and

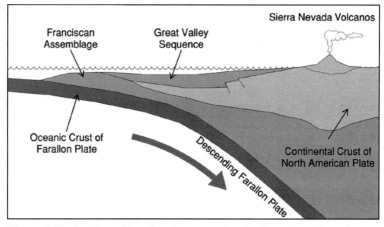

Figure 1. Typical depositional environment for the Franciscan Complex (assemblage)

Figure 2. Core photo from Boring O-5 showing highly fractured rock

Table 1. Case history block properties for tunnel projects in Franciscan Complex

	Uniaxial Compressive Strength Average*	
	(psi)	No. of Tests
Guadalupe (all)	3,693	32
Sandstone	3,729	26
Greywacke	3,139	4
Greenstone	1,857	1
Serpentinite	2,729	1

*From combination of point load and laboratory uniaxial compression tests for valid test results only

therefore stronger sample specimens, which are durable enough to survive coring and sample preparation. When cored, extremely poor-quality rock (similar to that shown in Figure 2) can be reduced entirely to pieces smaller than 4 inches by hand or light hammer blow. The generally poor condition of the rock means that it is highly susceptible to mechanical breakage during coring. Rock quality designation (RQD) is influenced by the soundness criteria (Deere and Deere, 1989), which results in zero RQD for intact core pieces. Further, packers are not easily seated for leakage testing particularly in lower permeability test intervals. Some of the best data can be obtained from borehole televiewer and caliper measurements; however, borehole stability issues can limit the availability of televiewer data particularly in angle holes and poor rock quality zones.

The Blocks

Typical rock types represented by blocks in the Franciscan Complex include greywacke, basalt, gabbro, greenstone, serpentinite, silica-carbonate, sandstone and siltstone.

Rock block strength data is typically evaluated using point load and unconfined compressive strength tests. What is often found is that there is not a significant variation in strength apparent between lithologies as illustrated in Table 1.

The Matrix

Samples of matrix need to be carefully handled from the split core barrel to the laboratory. Finding testable samples is a challenge. For example, on the Guadalupe tunnel project, of 590 feet of core along the outlet tunnel alignment, only two triaxial test samples were suitable for testing. The focus is on finding good quality representative samples to test rather than having a large quantity of lower quality test results. Tri-axial tests are considered more representative than the point load and unconfined data which is influenced by defects, mechanical breaks and sample selection bias. The three-point tri-axial test is recommended so that each sample yields

Table 2. Test results on mélange matrix samples

Project	Laboratory UCS (psi)	LL%	PI %	Phi (deg)	C (psi)
Richmond Transport	40–70 (2 tests)	28–47	9–27	12–15.5	5–6
Lenihan	135 (average)				
Guadalupe	20–26 (2 tests)	29	13–14	19–28 (2 tests)	7–8 (2 tests)

Table 3. Triaxial test results on mélange matrix samples from Guadalupe Outlet Tunnel

Sample Information		Triaxial Testing Total Strength			Equivalent UCS*
Exploration ID	Depth (ft)	c (psf)	φ (degrees)	E50 (psi)	psf/psi
O-5	141	1,000	19.4	18,586	2,825/20
O-4	114	1,100	28.3	17,134	3,683/26

*Equivalent UCS is derived from the total strength parameters by calculation assuming linear strength envelope.

Mohr-Coulomb strength parameters. Test results for mélange samples from local projects are shown in Table 2 with the triaxial test results from Guadalupe shown in Table 3.

ROCK MASS CHARACTERIZATION METHODS

In order to design initial support for a tunnel, the rock mass must first be characterized. Several rock mass characterization methods have been developed and are used in industry: Terzaghi in the 1940s, the Geological Strength Index (GSI) system in the 1990s, and volumetric block proportion (VBP) in the 1990s, as well. Each of these methods are discussed in the following sections as they relate to the design of the initial support of a tunnel in BIM rock, the Guadalupe tunnel.

Terzaghi

The Terzaghi method first published in the 1940s is still used today for steel supported tunnels (Proctor and White 1988). The basic rock mass conditions and range of rock loads on steel supports developed for each of these rock conditions is summarized in Figure 3. This system provides a range of support loads for rock conditions based on construction histories of drill and blast tunnels and does not take into account specific design parameters of the rock mass. In the case of the Guadalupe outlet tunnel, the Franciscan Complex with its chaotic structure would be characterized as blocky and seamy to completely crushed but chemically intact, and possibly even squeezing and swelling, which results in a wide range of possible design loads on steel supports. The squeezing conditions consider depth but not rock

mass strength, so it is not particularly sensitive to the ratio of matrix to blocks in the rock mass. Stand-up time and convergence are not outcomes of this type of analysis; therefore this approach has been used only for checking reasonableness in design outcomes.

Geological Strength Index (GSI) System

The Geological Strength Index (GSI) introduced by Hoek (1994) and Hoek et al. (1995) is a system of rock mass characterization that has been developed in engineering rock mechanics to meet the need for reliable rock mass properties required as inputs into numerical analysis or closed form solutions for designing tunnels, slopes or foundations in rocks (Marinos and Hoek, 2001). The system is qualitative in nature and is not always useful for lower quality rock. It is also intended to quantify rock mass strength and not to be used as a means of determining tunnel reinforcement or support and for that reason should only be considered when used in concert with other methods. A GSI system applicable for BIM rock is presented in Figure 4 (GC1, GC2, and GC3 will be defined in a later section). This GSI is most useful in the design of support for BIM rock tunnels by using it to establish engineering parameters for matrix dominated and block dominated conditions (GC3 and GC1, respectively) and then using other published relationships to calculate the parameters for conditions ranging between (GC2).

Volumetric Block Proportion

As was seen earlier, the difference in strength between blocks and matrix can be up to two orders of magnitude and the range in block strength from Table 1 is much less so it is evident that the VBP should control

Rock condition	Rock Load, H_p (ft)	Remarks
Hard and intact	Zero	Light lining, required only if spalling or popping occurs
Hard stratified or schistose	0 to 0.5 B	Light support. Load may change erratically from point to point
Massive, moderately jointed	0 to 0.25 B	
Moderately blocky and seamy	0.25B to 0.35 $(B + H_t)$	No side pressure
Very blocky and seamy	(0.35 to 1.10) $(B + H_t)$	Little or no side pressure
Completely crushed but chemically intact	1.10 $(B + H_t)$	Considerable side pressure. Softening effect of seepage towards bottom of tunnel requires either continuous support for lower ends of ribs or circular ribs
Squeezing rock, moderate depth	(1.10 to 2.10) $(B + H_t)$	Heavy side pressure, invert struts required. Circular ribs are recommended
Squeezing rock, great depth	(2.10 to 4.50) $(B + H_t)$	
Swelling rock	Up to 250ft. irrespective of value of $(B + H_t)$	Circular ribs required. In extreme cases use yielding support

Figure 3. Terzaghi rock loads on steel supports (FHWA 2009)

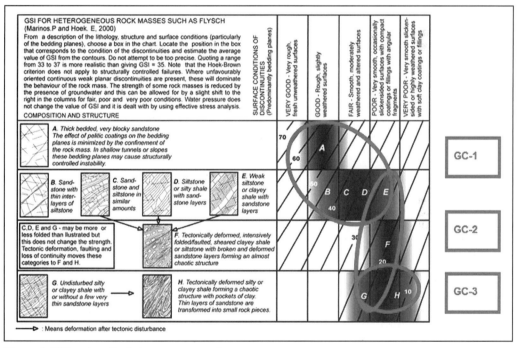

Figure 4. GSI system for heterogeneous rock masses

rock mass parameters more so than the lithology of the rock blocks. This is the same conclusion documented by Medley and others (Sonmez et al., 2006) and observed during excavation and support of tunnels in the Franciscan Complex. Therefore, ground characterization for tunneling in the Franciscan Complex should use a BIM model based on VBP. Generally speaking, this method of characterization involves establishing generalized design parameters for the matrix and for the blocks and then assigning rock mass parameters based on the VBP. Table 4 summarizes such a classification system recently developed for the Santa Clara Valley Water District's (SCVWD) Guadalupe Outlet Tunnel. The anticipated ground behavior conditions were separated into three ground classes. Each ground class has a corresponding range of VBP. The VBP ranges for each ground class, and the correlations with other classification

Table 4. Site-specific rock mass classification based on VBP

Ground Classes (GC)	Volumetric Block Proportion, VBP (%)	Terzaghi Ground Condition	GSI Ground Condition*	Description
GC1	71–100%	Massive Moderately Jointed to Moderately Blocky and Seamy	A to E	Consists mainly of blocks: • either poor quality rock mass with thick clay filled discontinuities; • or blocks occupying more than 70% of the tunnel face.
GC2	31–70%	Very Blocky and Seamy	E to G	Mixture of block and matrix.
GC3	0–30%	Crushed to Squeezing	G to H	Mainly matrix with presence of small fragment of blocks limited to a maximum of VBP = 30%

*GSI system is not directly applicable for all BIMROCK. GSI system here serves only as a standardized tool to help identify these heterogeneous rocks in the field.

Table 5. Classification of face regions

Face Region Classification	Face Region Block Proportion, FRBP (%)	Description Based on Face Mapping
Block Dominated	100%	Predominantly blocky and seamy rocks with moderately hard to very strong Sandstone or serpentinite blocks are hard - very hard; strong to very strong Massive matrix/sandstone—mostly hard and strong Chert is strong, very hard
Mixture of Block and Matrix	50%	MX/SP/SS: Mixture of Matrix / Serpentinite / Sandstone MX/SS/BS: Mixture of Matrix / Sandstone / Blocky and Seamy ground Interbedded shale matrix and sandstone
Matrix Dominated	0%	Black shale matrix—moderately hard; moderately strong Clayey matrix - dark brown, very stiff SSMX-C: Sandstone Matrix, crushed C MX (Crushed Matrix) with blocks of greywacke Matrix Shear Zone Crushed, thinly bedded chert Soft and friable chert/serp mixture

systems, are carefully evaluated to be consistent with documented literature, rock mechanics theory and calibrated against observed conditions and observed behavior from the SCVWD Lenihan Outlet Tunnel.

USE OF VBP AND THE CONVERGENCE-CONFINEMENT DESIGN TOOL

Data Collection

During and after construction, VBP can be assessed using simplified face maps focused on extent of blocks and matrix as shown in Figure 5 from the Lenihan Outlet Tunnel project. Three categories are used to classify the tunnel face region: "Block dominated," "Matrix dominated," or "Mixture of Block and Matrix." Face Region Block Proportion, FRBP (%) is assigned to each face region class (i.e., it is assumed that 100% of the "Block dominated" face area is block, 50% of the "Mixture of Block and Matrix" is covered by blocks and none of the

"Matrix dominated" area is covered by blocks). Table 5 summarizes the principals for classification for the Lenihan data.

The sum area of each face region is multiplied with the FRBP% corresponding to the Face Region Classification. Then the sum of each region's block proportion is summarized, and divided by the total face area. The result is considered to be the Volumetric Block Proportion (VBP%) of the face (i.e., weighted average of the FRBP is considered the VBP of the face). An example calculation is shown in Figure 5.

Analysis Method

The convergence-confinement method is recommended as the method to develop initial support design concepts based on VBP in Franciscan Complex rock. The convergence-confinement method combines concepts of ground relaxation and support stiffness to analyze the interaction between

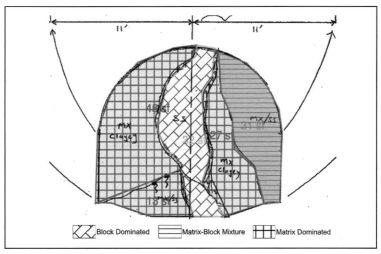

Figure 5. Face regions for T6+92 of Lenihan Tunnel

the ground and ground support (USACE, 1997) and it provides corresponding anticipated radial convergence.

The basic steps utilized in the analysis were:

1. Calculate the maximum stiffness and capacity of the rock support scheme.
2. Calculate the ground properties for design.
3. Using the support schemes from Step 1 and the ground properties from Step 2, use the convergence-confinement method to predict the convergence. If support schemes from Step 1 are insufficient, new schemes are developed and Step 3 is repeated.
4. *If construction data is available,* compare convergence data from construction against that calculated in Step 3 in order to validate predicted ground properties.

The Lenihan Outlet Tunnel provided construction data to validate the model as discussed below.

Case History Example from Lenihan Outlet Tunnel

The Lenihan Dam Outlet Tunnel Modification Project is in Los Gatos, California approximately 6 miles west of Guadalupe Dam. The 54-inch outlet at Lenihan Dam was constructed inside a new access tunnel and connected to a new sloping intake riser in the reservoir by an approximately 35-foot-deep shaft. The ground consisted of Franciscan Complex, and cover ranged from 10 to 320 feet. The tunnel was a 12.5-foot by 14-foot (H by W) D-shaped tunnel excavated full face over a 2,030 foot length by a combination of roadheader and drill and blast methods. Initial support consisted of W6×25 steel supports spaced at

4 and 5 feet centers with 3 to 9 inches in thickness of shotcrete as shown in Figure 6. The shotcrete thickness was completed within approximately 28 feet of the face (Von Stockhausen, 2008).

When necessary, in areas of low stand-up time to limit and control overbreak, No. 9 rebar spiles were used for pre-excavation ground support in the crown of the tunnel. Spiles were recorded for approximately 9.5 percent (47/499) of the steel sets installed.

A mudslab was placed approximately once every two weeks prior to weekend closure. In areas where squeezing ground was experienced the mudslab was placed more frequently. This served not only to protect the invert against degradation but also to control convergence by acting as an invert strut.

Table 6 provides average rock properties for the Franciscan Complex encountered in the Lenihan tunnel.

A series of convergence monitoring stations were analyzed as summarized in Table 7.

The maximum average convergence measured in the tunnel was approximately 1% at STA 12+00. This occurred at approximately the maximum cover point which coincided with a zone of matrix rich ground (VBP between 27 and 57 percent based on nearby face maps). The majority of convergence readings were less than 0.7%. For the convergence measurements at Lenihan, the initial readings were taken between 5 and 91 feet back from the face, though the majority were between 5 and 16 feet back from the face. Because monitoring systems were not installed immediately at the face the actual radial convergence was higher that what was able to be measured. Using the convergence-confinement method for the estimated VBP, the calculated convergence

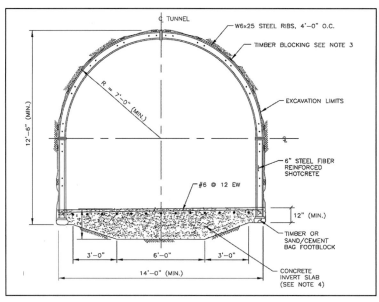

Figure 6. Lenihan Outlet Tunnel initial support (Courtesy of Santa Clara Valley Water District)

Table 6. Average block and matrix properties for Lenihan Tunnel

	UCS (psi)	Is50 (psi)	E (psi)
Block	7,234	538	2,259,578
Matrix	341	28	89,430

(occurring after monitoring system installation) was compared to the measured readings at 8 locations in order to validate the method. These eight locations were selected because they were the locations where monitoring was installed within 20 feet of the face and the mudslab was installed at least 10 feet behind the monitoring. Calculated and measured convergence for these points are summarized in Figure 7.

Allowing for the radial convergence that is predicted to have occurred prior to the monitoring system installation, Figure 8 summarizes the estimated total radial convergence that occurred at the Lenihan tunnel for the monitoring stations summarized in Table 7.

Allowable Radial Convergence

Guidelines have been published by Hoek and Marinos (2000) that approximate the relationship between strain and the degree of difficulty associated with tunneling through squeezing rock (Figure 9). These guidelines are consistent with the findings from the Lenihan tunnel. Twenty-five pound per foot steel sets and shotcrete with a concrete invert strut are indeed heavy support for a 14 foot wide tunnel.

However, another finding of the analysis is that while this support was used throughout the tunnel, it was probably more than was needed for a significant proportion of the tunnel. This is not uncommon for short tunnels where a uniform steel set size and spacing is used to simplify the work.

Using the behavior of the support at the Lenihan tunnel, it is considered reasonable and manageable during construction to allow the total allowable convergence (including movement prior to monitoring system installation) to enter the category defined by severe squeezing in Figure 9.

DESIGN APPROACH FOR GUADALUPE OUTLET TUNNEL

The rock mass properties for initial support design were initially developed for the range of possible VBP (see Table 8) using the geotechnical properties for matrix, blocks and the VBP as described earlier. These were then simplified to three ground classes for design as shown in Table 9. Previous work and calibration analysis from Lenihan has shown that the properties of rock mass with VBP of 15% or less were found to be similar to matrix and similarly, the rock mass properties for VBP of 85% or greater were found to be similar to rock block properties.

Initial support designs were developed for each ground class based on analysis of expected value ground properties for each ground class.

The percentage of tunnel anticipated to encounter each ground class can be predicted by assessing

Table 7. Lenihan Tunnel convergence monitoring stations

Station	Support Class	VBP (%)	Rock cover, H (ft)	Steel Sets	Shotcrete	Mudslab	Monitoring
				Installed Distance from Tunnel Face (ft)			
3+86 (1)	1	34%	220	1	9	67.5	3.5
5+82 (2)	1	23%	220	1	9	24	12
6+98 (3)	1	30%	250	1	9	56	12
8+32 (4)	2	37%	300	1	9	27	6
9+14 (5)	1	56%	300	1	9	63	39
9+62 (6)	2	25%	300	1	9	16	11
12+00 (7)	1	37%	300	1	9	65	1
12+26 (8)	1	42%	290	1	9	39	87
14+50 (9)	1	19%	200	1	9	31	27
16+00 (10)	3	16%	140	1	9	16	5
16+76 (11)	2	28%	130	1	9	16	5
17+84 (12)	3	15%	100	1	9	33	5
18+80 (13)	2	19%	80	1	9	45	1
19+52 (14)	2	39%	70	1	9	69	61

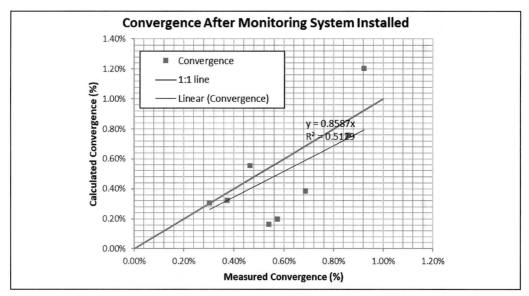

Figure 7. Measured versus calculated average radial convergence (occurring after installation of monitoring system) for the Lenihan Outlet Tunnel case history

the distribution of VBP in geomechanically consistent regions on the anticipated geologic profile for the tunnel alignment using borehole data if block and matrix data are captured on the boring logs. after Medley (1997) that found that the threshold between blocks and matrix is approximately equal to 5% of the critical dimension, in this case the tunnel size. For the 12 foot by 12 foot Guadalupe tunnel, 6 inches was selected to define a block for logging purposes. It was found that the upper 20 feet of bedrock in most of the boring logs is less "block rich," possibly due to weathering and/or stress relief, and is therefore not considered in assessing the distribution

of ground classes at tunnel depth. Data should be analyzed in bins about the size of the tunnel heading which means that ground class could change on a round by round basis.

CONCLUSIONS AND RECOMMENDATIONS FOR FUTURE USE

For characterizing matrix strength, consider the three-point tri-axial test and focus on getting a few high-quality test results rather than a lot of poor quality tests. Core should be obtained by careful triple tube coring using a split tube core barrel. Special

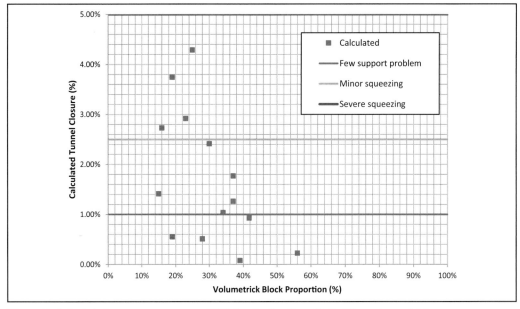

Figure 8. Calculated convergence versus VBP for the Lenihan Outlet Tunnel case history

Table 8. Estimated rock mass properties for Guadalupe Tunnel

Volumetric Block Proportion (%)	Uniaxial Compressive Strength, UCS (psi)	Young Modulus, E (psi)	Equivalent Friction Angle, φ (deg)
0 to 15%	27	17,991	23
20%	33	23,126	23
25%	40	29,726	23
30%	50	38,210	23
35%	61	49,116	27
40%	75	63,134	30
45%	93	81,153	33
50%	114	104,315	36
55%	140	134,087	39
60%	172	172,357	43
65%	212	221,550	46
70%	261	284,782	49
75%	321	366,062	49
80%	395	470,539	49
85% to 100%	486	604,836	49

care is needed when handling the samples from the split tube core barrel to the lab to avoid damage and maintain the in-situ moisture content.

Matrix strength properties can be derived from case history information using the convergence-confinement method, as was presented in this paper for the Lenihan Outlet Tunnel, if sufficient tunnel records are kept on the timing of installation of different support elements, convergence measurements and face mapping focused on VBP. This case history information appears to be broadly applicable

for tunnels in Franciscan Complex rock in Northern California.

The convergence confinement method is useful for initial support sizing and timing of installation. The method also provides anticipated convergence values to compare with in-tunnel convergence measurement to confirm the base design or to identify when supplementary measures are needed. Some simplifying assumptions are needed regarding the calculation of support stiffness for elements not installed at the face. Analysis of tunnel as-built data

	Strain ε %	Geotechnical issues	Support types
A	Less than 1	Few stability problems and very simple tunnel support design methods can be used. Tunnel support recommendations based upon rock mass classifications provide an adequate basis for design.	Very simple tunnelling conditions, with rockbolts and shotcrete typically used for support.
B	1 to 2.5	Convergence confinement methods are used to predict the formation of a 'plastic' zone in the rock mass surrounding a tunnel and of the interaction between the progressive development of this zone and different types of support.	Minor squeezing problems which are generally dealt with by rockbolts and shotcrete; sometimes with light steel sets or lattice girders are added for additional security.
C	2.5 to 5	Two-dimensional finite element analysis, incorporating support elements and excavation sequence, are normally used for this type of problem. Face stability is generally not a major problem.	Severe squeezing problems requiring rapid installation of support and careful control of construction quality. Heavy steel sets embedded in shotcrete are generally required.
D	5 to 10	The design of the tunnel is dominated by face stability issues and, while two-dimensional finite analyses are generally carried out, some estimates of the effects of forepoling and face reinforcement are required.	Very severe squeezing and face stability problems. Forepoling and face reinforcement with steel sets embedded in shotcrete are usually necessary.
E	More than 10	Severe face instability as well as squeezing of the tunnel make this an extremely difficult three-dimensional problem for which no effective design methods are currently available. Most solutions are based on experience.	Extreme squeezing problems. Forepoling and face reinforcement are usually applied and yielding support may be required in extreme cases.

Figure 9. Approximate relationship between strain and the degree of difficulty associated with tunneling through squeezing rock

Table 9. Rock mass properties for Guadalupe Tunnel

Ground Class (See Table 4)		Characteristic Volumetric Block Proportion, VBP (%)	Uniaxial Compressive Strength, UCS (psi)	Young Modulus, E (psi)	Equivalent Friction Angle, φ (deg)
GC1	Expected Value	77.5%	356	415,026	49
GC2	Expected Value	50%	114	104,315	36
GC3	Expected Value	22.5%	36	26,219	23

such as the Lenihan Outlet Tunnel data helps constraint the number of variables in the approach.

The characterization and design approaches discussed in this paper are applicable to tunnels in Franciscan Complex including the upcoming outlet tunnels for the Santa Clara Valley Water District and the high-speed rail tunnels in Northern California.

ACKNOWLEDGMENTS

Special thanks to Santa Clara Valley Water District, particularly Hemang Desai and Bal Ganjoo for allowing their project information to be published.

REFERENCES

Byerly, P., and G. D. Louderback, 1964. Andrew Cowper Lawson, July 25, 1861-June 16, 1952. Biographical Memoirs, 37:185–204.

Deere, D. U., and Deere, D.W, 1989. "Rock Quality Designation (RQD) After 20 Years," US Army Corps of Engineers Waterways Experiment Station, February, Contract Report GL-89-1.

Federal Highway Administration (FHWA) (2009) Technical Manual for Design and Construction of Road Tunnels—Civil Elements, FHWA-NHI-10-034.

Hoek E, and Marinos P 2000. Predicting Tunnel Squeezing Problems in Weak Heterogeneous Rock Masses. Tunnels and Tunnelling International, Part 1 November pp45–51, Part 2 December pp 33–36, 2000.

Hoek E, Carranza-Torres CT and Corkum B, 2002. Hoek-Brown failure criterion—2002 edition. Proc. North American Rock Mechanics Society.

Hoek E. (1994) Strength of rock and rock masses, ISRM News Journal, 2(2).

Hoek E., Kaiser P.K. and Bawden W.F. (1995) Support of underground excavations in hard rock. Rotterdam, Balkema.

Marinos, P., & Hoek, E. (2001). Estimating the geotechnical properties of heterogeneous rock masses such as flysch. Bulletin of engineering geology and the environment, 60(2), 85–92.

Medley, Edmund. (1997). Uncertainty in estimates of block volumetric proportions in mélange bimrocks. Engineering Geology and the Environment.

Proctor and White, 1988, Rock Tunneling with Steel Supports, Commercial Shearing Company.

Sonmez, H., Gokceoglu, C., Medley, E. W., Tuncay, E., & Nefeslioglu, H. A. (2006). Estimating the uniaxial compressive strength of a volcanic Bimrocks. International journal of rock mechanics and mining sciences, 43(4), 554–561.

United States Army Corps of Engineers (USACE) (1997) Engineering and Design of Tunnels and Shafts in Rock, EM-1110-2-2901.

Von Stockhausen, S (2008). Tunnelling for the new Lenihan Outlet, Tunnels and Tunnelling International, November, 29–32.

High In Situ Stress and Its Effects in Tunnel Design: An Update Based on Recent Project Experience from WestConnex M4 East and New M5 Tunnels, Sydney, Australia

D. Tepavac, P. Mok, and Y. Sun
McMillen Jacobs Associates

D. Oliveira
Jacobs Engineering Group

H. Asche
Aurecon Group

S. Simmonds
CPB Samsung John Holland Joint Venture

ABSTRACT: The WestConnex project is part of an integrated transport plan for Sydney. It includes three underground multilane tunnels: M4 East, New M5, and M4-M5 Link. Each of these tunnels is a project of its own. This paper focuses on two of the three tunnel projects (M4 East and New M5). The results of design analyses as well as tunnel support solutions that have been developed to address the anticipated adverse effects of high in situ stress in these two projects are presented. Relevant data were obtained from ongoing tunnel excavations and construction monitoring program of both projects. The developed tunnel support solution and corresponding available construction data are compared.

INTRODUCTION

The high virgin horizontal in situ stress field in the Sydney Basin and its impact on civil engineering projects is a well-known and accepted phenomenon with significant literature (e.g., Pells 2013). The prevailing high stress effects depend on many factors—rock quality, tunnel orientation, proximity of geological features to tunnel crown, size and shape of opening, depth of excavation, and stress magnitude. The behavior of the rock mass under high in situ stress conditions can cause stress fracturing and consequent dilation of the tunnel periphery, resulting in rock spalling at the tunnel crown/invert or raveling of rock blocks on the tunnel sidewall. This type of failure is of a brittle nature and may create construction and safety risks during tunnel excavation if it occurs behind the excavation face where ground support has already been installed. Therefore, the associated risks need to be managed during construction. This paper presents the design strategy adopted to mitigate these adverse tunneling conditions for the WestConnex Project to-date.

OVERVIEW OF THE WESTCONNEX M4 EAST AND NEW M5 TUNNEL PROJECTS

WestConnex is one of the NSW Government's key infrastructure projects. This 33 km (20.5 mi) project aims to ease congestion and connect communities, and is the largest integrated transport and urban revitalization project in Australia. It was a key recommendation of the State Infrastructure Strategy released in October 2012. It brings together a number of important road projects, which together form a vital link in Sydney's Orbital Network. These road projects include a widening of the M4 east of Parramatta, a duplication of the M5 East, and new sections of motorway to provide a connection between these two key corridors. The WestConnex project includes a number of stages: Stage 1a: M4 Widening; Stage 1b: M4 East; Stage 2: New M5; and Stage 3: M4–M5 Link. The tunnel design referenced in this paper relates to Stage 1b and Stage 2 of the WestConnex Project (Figure 1).

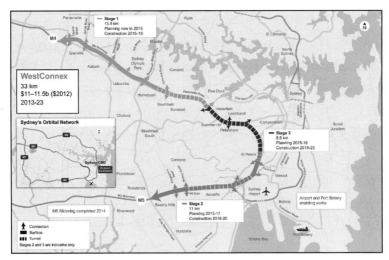

Figure 1. WestConnex corridor (Transport Sydney 2013)

WHAT ARE THE UNDERLYING CAUSES OF HIGH HORIZONTAL STRESS?

The effects of high horizontal stress in tunnels and underground excavations have been well documented in the Australian literature (Pells 1993; Oliveira and Diederichs 2017). As discussed by Oliveira and Diederichs (2017), two rock failure mechanisms may be observed at stress levels lower than the rock unconfined compressive strength (UCS):

- Brittle failures involving crushing, spalling, and or slabbing of intact rock blocks, more often associated with buckling of thin sandstone beds in Sydney
- Shear failures associated with planes of weakness, either pre-existing or induced by the excavation process, such as faults, cross bedding partings, and bedding shears

The primary cause of such failures is associated with removal of confining rock during tunnel excavation, which causes stress redistribution and results in stress concentration around the excavated periphery. This induced stress condition can be estimated using classical solutions for stress distribution around an elastic circular opening such as that given in Figure 2. As shown, the horizontal stresses within the tunnel crown increase proportionally with the horizontal-to-vertical stress ratio (λ). Therefore, the effects of induced high horizontal stresses can be expected to be more pronounced with increasing depth.

Considering that rocks like sandstone may fail at stress levels of approximately 50% of the UCS (Oliveira and Diederichs, 2017), it can be easily demonstrated (using the classical solution of Figure 2) that the risk of stress-induced failure increases

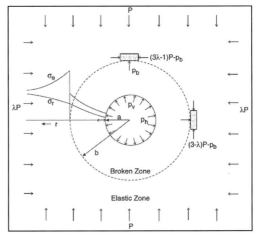

Figure 2. Stress distribution around an elastic opening

Table 1. Approximations of in situ stress magnitude in the Sydney Basin

	Bertuzzi (2014)	Oliveira and Parker (2014)
Major Horizontal Stress, σ_H	$2\,\sigma_V\,+$ (1.5–2.5) MPa	(3.0–3.8) $\sigma_V\,+$ (1.0–1.2) MPa
Minor Horizontal Stress, σ_h	$0.7\,\sigma_H$	$0.61\,\sigma_H$

Note: σ_V denotes vertical stress.

beyond depths of 40 to 50 m (131–164 ft) in Sydney. Therefore, the excavation induced stresses estimated with recent approximations of in situ stress magnitude (Table 1) approach values of approximately

Table 2. Tunnel depth assessment for susceptibility to elevated stress condition and its effects

In Situ Stress Approximation	Depth of Cover (m)	Vertical Stress Before Excavation, σ_V (MPa)	Horizontal Stress Before Excavation, σ_H (MPa)	Horizontal-to-Vertical Stress Ratio, λ	Horizontal Stress After Excavation, σ_1 (MPa)
Bertuzzi (2014)	40	0.96	$2\,\sigma_V + 2.5 = 4.42$	4.60	12.29
Oliveira and Parker (2014)	50	1.20	$3\,\sigma_V + 1.0 = 4.6*$	3.83	12.59

* Note the lower range of Oliveria and Parker's (2014) approximation was used since a UCS of 25 MPa was adopted in this example. The upper range of Oliveria and Parker's approximation corresponds to Sandstone with a UCS of 30 MPa.

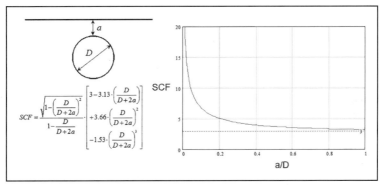

Figure 3. Stress concentration factor of a circle on an edge (after Asche and Cooper 2002)

12.5 MPa, which are equivalent to 50 percent for an average Hawkesbury Sandstone UCS = 25MPa (Table 2).

The effect of such high horizontal stresses is often altered locally by the presence of major geological features such as valleys, fault zones, and dykes, but also varies with orientation and with respect to stronger and/or weaker bands of rock—all making for variability and unpredictability. In addition, another important factor is the presence of planes of weakness, particularly bedding partings near an excavated tunnel, which magnify the induced stresses. For example, the simplified solution provided in Figure 3 (Asche and Cooper 2002), indicates that the stress concentration factor of SCF = 3 given in Figure 2 may increase to approximately SCF = 7 for a low friction bedding parting located at about 0.5 m (1.6 ft) above a 7 m diameter (23 ft) tunnel.

WHAT ARE THE ISSUES ASSOCIATED WITH HIGH HORIZONTAL STRESS?

The immediate effect of elevated excavation induced horizontal stresses near a tunnel is the occurrence of significant shear displacements on subhorizontal discontinuities. Such an effect may cause damage to

rock bolt corrosion protection, thus affecting durability, and in more severe cases tensile rupture of the bolts (Oliveira and Diederichs 2017). The shear displacements may also induce local loosening of rock wedges or blocks near excavation shoulders. Because of the way the excavation stress release occurs, this can be a nuisance behind the face. In addition, such stress concentrations cause localized rock mass failures in the crown (as previously discussed), which require appropriate ground support with rock bolts and shotcrete.

An important aspect for design of such a ground support is that the staged release of excavation induced stress means that the effect does not necessarily occur entirely at the excavation face. This becomes more pronounced when taking into account 3D effects such as the orientation of the excavation in relation to subhorizontal discontinuities that affect stress concentration, as discussed above. For instance, driving downhill means that bedding partings would typically rise into the roof, transitioning from a low stress concentration factor (SCF) to an extreme or high SCF environment where brittle failure is more likely, thus giving rise to the feeling that this is "unpredictable." This condition is shown in

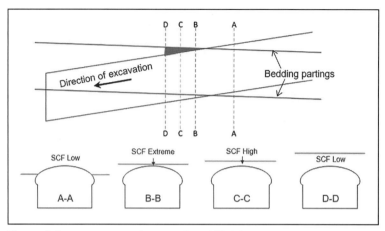

Figure 4. Stress concentration on a downhill drive

Figure 4, where the SCF transitions from low at section A-A to extreme at section B-B. A real example of this case is presented in Figure 5, where the roadheader marks indicate failure occurred close to the face. On the other hand, driving uphill can mean that failure is triggered backwards from a failure initiating closer to the face when the stresses exceed the applicable spalling limit—for example, at section B-B and propagating to C-C (Figure 6). This also gives rise to the feeling that it is "unpredictable." Oliveira and Diederichs (2017) presented a simplified numerical to illustrate initiation of a brittle failure at a distance of 2 to 2.5 times the height of the excavation (Figure 7).

The major implication of potential stress induced failure postexcavation relates to safety risks for construction personnel who require access within the tunnels. Current tunnel construction practice is to have no personnel entry under unsupported ground because of legislative requirements. Supported ground is currently defined, within competent Sandstone on the New M5 and M4 East, as when both rock bolts and shotcrete have been installed, with the shotcrete having gained a certain minimum strength. Stress-induced spalling is therefore an issue if it occurs within supported ground conditions.

Given that rock bolts and shotcrete are installed soon after excavation, confinement is provided to the tunnel periphery. The level of confinement provided by the rock bolts and shotcrete is low. However, rock bolts provide sufficient retaining capacity should the spalled rock be wide enough to be captured by the rock bolts. Analysis results presented herein suggest that the plausible stress-induced spall rock size rarely exceeds the extent of one rock bolt spacing. Any residual spalled rock not retained by rock bolts must, therefore, be accounted for in the design of the applied shotcrete.

Figure 5. Stress-induced spalling in the M5 East Tunnel (after McQueen et al., 2017)

However, the process of stress fracturing is complex and dependent on multiple factors. Therefore, the ground-support interaction is also complex and so is the design, particularly with shotcrete applied early to the excavated rock (because of supported ground requirements). Such complexity can only be addressed by an observational approach during construction in an attempt to manage such risks.

DESIGN STRATEGY

There is no practical way to avoid stress-induced spalling and other related consequences, and the associated risk of such an event occurring after support installation (i.e., under supported ground) cannot be ignored. The consequences of rock spalling on permanent shotcrete lining (particularly postinstallation) are that relatively large volumes of broken rock may build up behind the permanent shotcrete lining if such failure occurs some distance behind the excavation face. This buildup of broken or spalled rock could induce fallout of shotcrete and rock. This

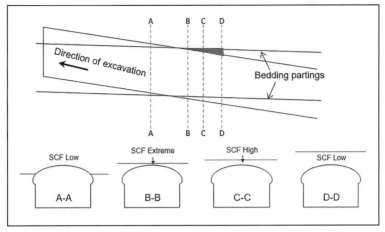

Figure 6. Stress-concentration on an uphill drive

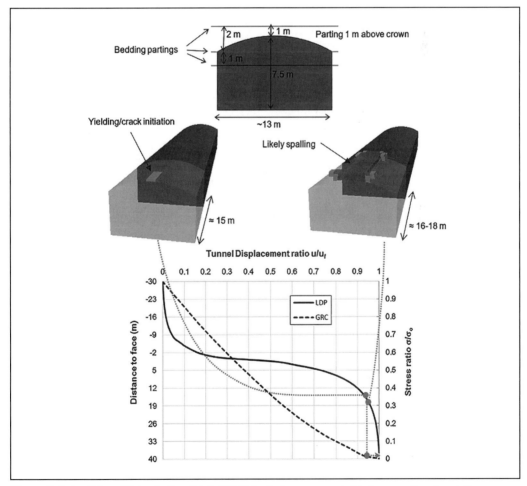

Figure 7. Potential spalling of a 1 m thick (3.3 ft) bed above crown behind excavation face (after Oliveira and Diederichs 2017)

raises a safety risk for the construction personnel as well as for the long-term end users of tunnels.

The large shear movement associated with stress-induced spalling may also impact the longevity of installed rock bolt reinforcement. The large shear movement may exceed the allowable shear limit of the rock bolts' corrosion protection (plastic sheath). The damage of the corrosion protection means that the residual design life of the rock bolts is reduced and does not meet the design durability requirement. Rebolting of sheared rock bolts will be necessary to maintain the design life requirement in this case.

The design aim is then to implement controls to reduce the probability of ground support damage should stress-induced spalling occur. Given that stress-induced spalling cannot be fully avoided with reasonably practical means, the consequences are unlikely to change. The associated risks detailed above therefore remain. Based on practical risk analysis, control measures must be applied such that the likelihood of ground support damage is reduced, and the overall risk level of a stress-induced spalling event of shotcrete and failure bolts is thereby reduced to an acceptable level.

The viability of any proposed control measure depends on its ability to meet the construction constraints where applicable/possible such that it must perform the following:

- Integrate and be compatible as much as possible with the already developed typical tunnel excavation and support installation sequence for tunnels without a stress-induced spalling problem.
- Incorporate either rock bolts or shotcrete as tunnel support. Support elements other than rock bolts and shotcrete may yield procurement difficulty and increase construction complexity.
- Conform to the protocols of the project-wide instrumentation and monitoring plan in the context of identification and transition in and out of elevated stress conditions.
- Yield tangible triggers to facilitate site-based observations. This follows, in principle, the observational approach of conventional tunneling with sequential excavation.

The site feedbacks, guided by in situ observational triggers, are adopted to make adjustment(s) to the already developed tunnel support design. Control measures for elevated stress condition are thus developed as adjustments to already developed typical tunnel support with associated monitoring triggers.

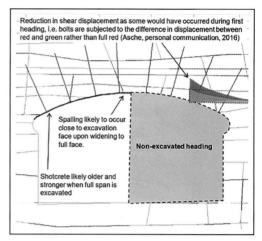

Figure 8. Benefits of split heading on tunnel support (after Oliveira and Diederichs 2017)

Selected Control Measures for Elevated Stress Condition

The selected risk control measures adopted for elevated stress conditions are:

- *Mandatory split headings with minimum lag distance between headings*: Typical tunnel cross sections applicable to these WestConnex projects are excavated using multiple headings. Each heading is approximately 5.5 m to 7m wide by 6 m high. A 10 m minimum lag distance (approximately 1.5 times the single heading span of 7 m) between headings has been adopted. The main objective of the split heading (Figure 8) is to induce the brittle failure near the face of the second heading and reduce the shear displacements within the second heading.
- *Staggered rock bolt pattern*: For the New M5 tunnels, the typical rock bolt pattern was adopted as square. However, for ground support against an elevated stress condition, the rock bolt pattern is altered to staggered. For the M4 East tunnels, the staggered rock bolt pattern is typical, and thus this control is less sensitive. Rock bolt spacing is also tightened where applicable. The benefits of a staggered pattern is its increased ability to contain fracture propagation within a bolt spacing. This prevents fractures from extending over multiple bolt spacings, with associated increased displacements, as would be observed for a square pattern (Figure 9).
- *Increased shotcrete thickness*: The primary shotcrete developed for these projects is generally quite thin and relies on adhesion

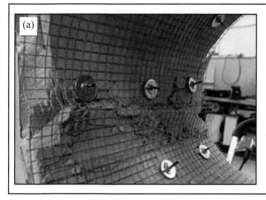

Figure 9. Square (a) vs. staggered (b) reinforcement pattern performance during brittle failure (after Villaescusa et al., 2016)

to the substrate. The thinnest crown primary shotcrete thickness is 55 mm (2.2 in.) for Sandstone Class I in the New M5 tunnels and 60 mm (2.4 in.) in the M4E tunnels. The minimum shotcrete thickness adopted for elevated stress condition is 125 mm (4.9 in.) targeting flexural capacity.

Numerical Analysis

Numerical analyses were undertaken to assess responses of the rock mass and installed tunnel support subjected to elevated stress conditions. These numerical analyses include continuum (utilizing FLAC2D and FLAC3D) and discontinuum analyses (utilizing UDEC and RS2). The studies undertaken utilizing these analyses aided development of the adopted control measures for elevated stress conditions.

To assess the tunnel support performance when the support is subjected to elevated stress condition, additional rock material modelling was undertaken. Recent research development has shown that to better capture the stress-induced spalling zone and extent around a tunnel, a modified failure criterion for the rock mass should be used (Diederichs et al. 2010; Oliveira and Diederichs 2017). Table 3 presents a set of modified Hoek-Brown rock mass material parameters adopted for these projects.

Figure 10 presents the analysis results as part of the evaluation of the stress-induced spalling effects. Two heading excavations followed by bench excavation have been adopted for the analysis results shown. A single bedding parting set at 1 m (3.3 ft) from the tunnel crown has been analyzed, with and without support installed. Rock spalling is likely to extend towards the full depth of rock bounded by the tunnel excavated periphery and bedding parting. With support installed, the rock spall size is much

Table 3. Adopted rock mass properties for rock spalling analysis

Rock Mass Properties		Ground Types		
		Type 1	Type 2	Type 3
Unconfined compressive strength, MPa	UCS	30	25	20
Unit weight, kN/m³	γ	24	24	24
Rock mass modulus, MPa	E_m	6,000	4,000	3,500
Poisson's ratio	ν	0.2	0.2	0.2
Peak parameters	m	0.750	0.625	1.250
	s	0.063	0.063	0.063
	a	0.25	0.25	0.25
Residual parameters	m	9	9	9
	s	0.0001	0.0001	0.0001
	a	0.75	0.75	0.75

reduced. This confirmed the beneficial confinement effects provided by the rock bolts and shotcrete.

By adopting the same excavation sequence and material model, the effect of rock bolt shearing was assessed. Twin mainline tunnels were analyzed within the same cross section to capture the effects on the adjacent tunnel (Figure 11). The results showed that potential rock bolt shearing is only likely to be confined to the lead heading of the first excavated tunnel. The magnitude of shear displacements is likely to exceed the shearing limit for corrosion protection, but does not exceed the ultimate structural capacity of rock bolts. That is, if elevated stress conditions are observed, rebolting for long-term reinstatement will be required and is likely to be limited to the rock bolts installed within the lead heading of the first excavated tunnel. This concludes that if there is sufficient lag distance between the lead and trailing headings of the first tunnel, the extent of

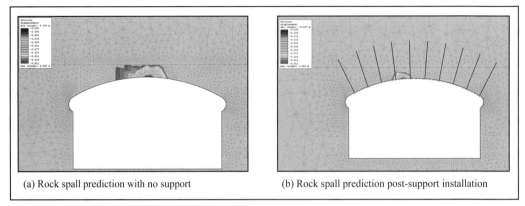

| (a) Rock spall prediction with no support | (b) Rock spall prediction post-support installation |

Figure 10. Stress-induced spalling analysis results, with and without support installed

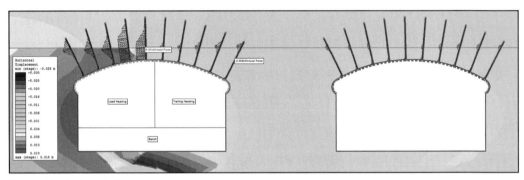

Figure 11. Rock bolt shear assessment

rebolting required will likely decrease. Confirmation of the extent of rebolting typically is done using endoscope observations.

Based on the spalling analysis results, it was assessed that the most likely rock spall depth ranges from 0.5 m to 0.7 m (1.6–2.3 ft; subject to different ground conditions). Rock spall extent is slightly greater than a one-bolt spacing (Figure 10). Staggered rock bolt patterns were then adopted/confirmed to better arrest and contain spalling rock as discussed above. With the rock bolts arranged in a staggered manner, the longitudinal strip of spall rock is arrested and contained. The staggered rock bolt pattern is therefore considered more suitable for elevated stress conditions.

Additional analyses were undertaken to evaluate the performance of the primary shotcrete lining. The prevalence of rock spall within a one-rock-bolt spacing suggests that the shotcrete needs to provide retaining capacity to retain the rock spall. Stress-induced spalling is a fracture process that likely induces multiple fractures within the rock spall, especially when the spall rock size is relatively small (Figure 12). The mechanism involves a load transfer

of the spall rock weight to the shotcrete, which in turn transfers to the rock bolts through the connection between the shotcrete and the rock bolt. A typical rock bolt–shotcrete connection is facilitated using handle bar plates. The primary shotcrete lining is therefore critical for containment of the rock spall.

Deterministic structural analysis based on Barret and McCreath (Oliveira and Diederichs 2017) was then undertaken to assess the mechanical response of the shotcrete. This analysis assumes that the process of stress fracturing is associated with ground stress dissipation (i.e., stress transfer from the rock mass movement to the shotcrete is reduced to a negligible magnitude). However, it was also realized that some level of in situ stress remains within the spall rock boundary, providing frictional restraint (Figure 12). The overall rock spall weight was then adjusted accordingly when applied to the deterministic analysis to determine an appropriate shotcrete thickness.

The assumption of stress dissipation used in the deterministic analysis is not definite. There is limited evidence or in situ experience available to confirm this assumption. The complexity of the stress fracturing process and the associated ground stress transfer/

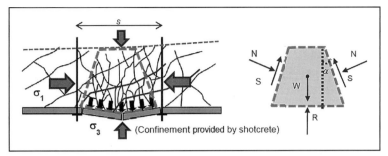

Figure 12. Assumed shotcrete mechanical response to support stress-induced spalling rock (after Oliveira and Diederichs 2017)

dissipation to the shotcrete is not completely understood. The implication is that the risk of shotcrete failure is further increased, regardless of shotcrete thickness (reasonably practical thickness), because of the magnitude of plastic strain experienced during stress fracture. The primary shotcrete was analyzed using numerical analyses to confirm adequacy of the adopted shotcrete thickness.

Design Outcomes

The different analyses undertaken indicate the adequacy of the control measures adopted for the elevated stress conditions in these projects. To facilitate implementation of these control measures, a set of observational criteria was developed:

Elevated Stress Condition Susceptibility Definition

Tunnel sections defined below shall be classified as areas subject to potential rock spalling risk:

- First bedding ≤1.5 m (4.9 ft) above crown
- Presence of any seams or shears ≤1.5 m above crown
- Presence of shale lens or mudstone facies ≤1.5 m above crown

Note the first bedding distance of 1.5 m from tunnel crown was set to provide an early alarm to initiate in situ observations earlier. This was to better react and implement control measures in advance.

Elevated Stress Condition Observational Triggers/ Identifiers

High stress conditions are evidenced by a combination of (but not limited to):

- Ground cover greater than 45 m (148 ft).
- Higher horizontal ground movement measurements (above 75% of the amber warning level of the typical conditions).

- Spalling or cracking of shotcrete within 2 weeks of application.
- Higher horizontal movements detected in endoscopes (more than 9 mm [0.4 in.]).
- Presence (or predicted presence) of a subhorizontal discontinuity or adverse feature(s) within 1.5 m (4.9 ft) of the tunnel crown. This can be predicted:
  – If the bedding is rising in the face: From geological face mapping.
  – If the bedding is falling towards the crown: From geotechnical endoscope mapping.

The above observational triggers/identifiers were also adopted to aid removal of elevated stress control measures.

The strategy of the elevated stress control measures is:

- To promote stress-induced spalling to occur at the lead and trailing heading interface. This is to permit/aid removal of spall rock during trailing heading excavation, and constrain spall rock to unsupported or less trafficked supported ground.
- To reduce the uncertainty related to rock bolt shearing. Endoscope installation density is also increased (doubled) to better define the affected extent of rock bolts.
- To give time for shotcrete strength gain to provide support to spall rock.
- To control spall rock failure after support installation with increased shotcrete thickness and tightening of rock bolt spacing such that tangible damage (i.e., not sudden brittle failure fallout) can be observed and minimized.

It should also be noted that, although the M4 East project incorporated a few of these key control measures, the majority were developed as part of the New M5 project given that on average the tunnels are located at greater depths.

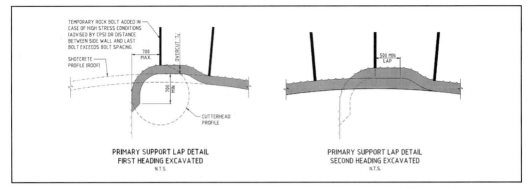

Figure 13. Lap detail adopted for addressing spalling risks between lead and trailing headings

CONSTRUCTION OBSERVATIONS TO-DATE

Limited construction data are available to-date from these projects that may indicate stress concertation. However, anecdotal observations to date across both projects indicate larger overbreak in sections that are deeper than 45 m (148 ft). Additionally, lead heading shotcrete near the central temporary haunch has exhibited a slightly higher frequency of cracking than in other parts of the tunnel. Although this correlates well with spalling analysis results, a new lap detail at the lead and trailing heading interface was developed to mitigate risks associated with overstress of the primary shotcrete lining (Figure 13).

CONCLUSIONS

Limited construction data are available to-date to provide sufficient validation to the design implemented for high stress condition for the WestConnex Project. Construction observations are ongoing to provide feedback to the performance of the selection system. However, the construction data collected to-date show that the expectation for elevated stress conditions perceived from the design analysis presented in this paper holds. This concludes that an observational based approach allows for selection of adequate tunnel support design to manage the risks associated with elevated stress conditions. Coupled with engineering judgment, the design principles discussed in this paper although developed for Sydney Sandstone, are applicable to other conventionally excavated tunnels in horizontally bedded strata of similar stress states.

REFERENCES

Asche H.R. and Cooper D.N. 2002. Estimation of tunnel support requirements for TBM driven rock tunnels. *ITA World Conference, Sydney.*

Barrett, S.V.L. and McCreath, D.R. 1995. Shotcrete Support design in Blocky Ground: Towards A Deterministic Approach. *Tunneling and Underground Space Technology.* 10(1): 79–89.

Bertuzzi, R. 2014. Sydney Sandstone and shale parameters for tunnel design. *Australian Geomechanics.* 49(2): 95–104.

Diederichs M.S., Carter T., and Martin C.D. 2010. Practical rock spall prediction in tunnel. *Proceedings of World Tunnelling Congress '10—Vancouver.*

McQueen L.B., Bewick R.P., Sutton J., and Morrow A. 2017. Stress-induced brittle failure of the Hawkesbury Sandstone—Case study from crack initiation to tunnel support. *16th Australasian Tunnelling Conference 2017.*

Oliveira D., and Diederichs, M. 2017. Tunnel support for stress induced failures in Hawkesbury Sandstone. *Tunnelling and Underground Space Technology.* 64: 10–23.

Oliveira, D.A.F., and Parker, C.J. 2014. An alternative approach for assessing in-situ stresses in Sydney. *15th Australasian Tunnelling Conference 2014.*

Pells, P.J.N. 1993. Rock mechanics and engineering geology in the design of underground works. 1993 E.H. Davis Lecture, Australian Geomechanics Society.

Transport Sydney. 2013. WestConnex— Building for the Future https://transportsydney.files.wordpress.com/2013/09/2013-09-19-westconnex-map.png.

Villaescusa E., Kusui A., and Drover C. 2016. Ground support design for sudden and violent failures in hard rock tunnels. *9th Asian Rock Mechanics Symposium,* 18–20 October 2016, Bali, Indonesia.

Design and Construction Considerations for Atwater Water Intake and Conveyance Tunnel in Montreal

Jean Habimana and Benoit Rioux
Hatch

Stephane Dusablon
CRT

ABSTRACT: The City of Montreal plans to secure the existing water intake of the Atwater Treatment Plant by permanently isolating the section of the channel adjacent to nearby highway that is at risk from contamination and spills and constructing a new intake upstream. The new intake will be linked to the existing intake by a 877 m (0.5 mile) long 6 m (20 ft) diameter 30 m (100 ft) deep tunnel, two 6 m (20 ft) diameter vertical shafts, and an cut and cover connecting structure to the existing valve chamber. The geology consists of shale that is known to be fissile and slaking and is being excavated by drill-and-blast. The paper provides design considerations of this siphon tunnel as well as construction updates.

PROJECT BACKGROUND

The Atwater Potable Water Treatment Plant (Usine de filtration d'eau d'Atwater) was first commissioned in 1918 and has over the years undergone a series of significant upgrades with the last one being in 2008. The plant is the second largest potable water treatment plant in Canada and it currently provides potable water to approximately 42 percent of the Montreal island population, which is around 1,8 M of citizens. It has a capacity to treat 1,500,000 m³ of water per day (396 MGD). Its average daily treatment is approximately 650,000 m³ of water per day (172 MGD). The plant takes its water from a nearby Aqueduct Channel that was originally built in 1854 to provide clean water to the southeast area of Montreal. The channel has also undergone several upgrades and reconstruction to its current configuration of approximately 40 m (131 ft) wide and approximately 8 m (26 ft) deep. The channel takes its water from the Saint Lawrence River at approximately 8 km (5 miles) upstreamand at 2.4 km upstream of Lachine Rapids. The channel ends approximately 400 m (0.25 mile) downstream the current intake structure of the treatment plant.

Near the plant, the channel is bounded to the north by a very busy section of Highway 15 (see Figure 1), which links the Champlain Bridge to the Turcot Interchange, that are both having major reconstruction and replacement works as they have reached the end of their design life. It should be noted that the Champlain Bridge is the most traveled bridge in Canada. The portion of Highway 15 that boarders

the channel is set to undergo major works starting mid-2018. Following a Quebec provincial regulation on water withdrawals and their protection that went into effect in August 2014 and requires to evaluate risks around all water withdrawals within the province, the Montreal Potable Water Department has undertaken a risk analysis exercise for the Atwater plant. The Highway 15 scored the highest among all risks that were identified. In addition, the planned works on that section of Highway 15 may further impact the quality of the water in the channel and lead to the plant closure for a long period of time. Furthermore, with the expected increase of the traffic on that stretch of the highway there is an increasing risk of a potential chemical spill into the channel from the highway as it is travelled by trucks carrying dangerous goods and chemical materials. The Potable Water Department therefore recommended to secure the current intake from potential risks of contamination from the surrounding environment.

To alleviate these risks, the City of Montreal is currently undertaking the construction of a new intake structure to permanently secure the water source of the Atwater plant by isolating the section of the channel most at risk from sources of contamination. This will be achieved by building a new water intake within the channel approximately 1 km (0.6 mile) away from the current intake location and conveying the water through a siphon system that will carry water to the existing facilities.

The project is founded by the provincial government as well as the City of Montreal. At the time of the writing the paper, the project is

Figure 1. Aerial photo of Atwater Tunnel Treatment Plant and its vicinity to Highway 15

under-construction. Shafts excavation is completed and the tunnel excavation is beginning.

The new intake structure that is being built will connect to the existing plant facilities by two deep shafts and a conveyance tunnel as well as a downstream connecting underground channel to the existing facilities at the plant. The main components of the project are:

- An intake structure that is approximately 21 m (69 ft) wide, 55 m (180 ft) long of which 10 m (33 ft) are in the channel that is approximately 6 m (20 ft) deep. The intake is located in a park area slightly to the west of Galt street to the north of Champlain Boulevard. The intake structure is being built near a high voltage electric tower. The selection of the intake location was based on the need to minimize the project cost by optimizing the tunnel length as well as to reduce traffic disruption, noise and vibration during construction to the nearby residential neighborhood. Architectural and landscaping considerations are being planned by the City for the service building that will house operational equipment.
- An upstream shaft that is 6 m (20 ft) diameter and 35 m (115 ft) deep. The shaft is being sunken in the park area near the Galt Street. Work is being carried out near a high voltage tower and the contractor had to take appropriate steps to comply with safety measures. For instance, some of the equipment had to be adjusted to ensure that their swing meets

maximum height requirements. At this location, there is major underground power utility, that could not be relocated within the project schedule, it was suspended and is being supported by a temporary bride structure.
- A conveyance tunnel that is approximately 877 m (0.5 mile) long and 35 m (115 ft) deep that slopes at 0.2 percent towards the treatment plant. Its inside diameter is 6 m (20 ft) and final lining is a cast in place reinforced concrete lining. The tunnel runs underneath a mostly residential area with some commercial and industrial facilities. It terminates near a Siemens facility that houses jet engines testing equipment.
- A downstream shaft that is also 6 m (20 ft) diameter and 35 m (115 ft) deep. The shaft is being sunken past the Siemens facility on the Atwater treatment plant site. There is some known soil contamination at this site that was properly disposed based on environmental regulations.
- An underground cut and cover portion to connect to the existing valve chamber, which is approximately 9.6 m (130 ft) wide and 115 m (377 ft) long.

GEOTECHNICAL SETTING

Overburden Geology

The stratigraphy of the Montreal area includes a variety of deposits dating back to the Pleistocene or recent times. At the project site, the overburden can be described as consisting of a thin 1 m layer of fill material covering the natural soil, which is composed of till varying from 9 to 10 m thick. The composition varies from silt with some sand, gravel and clay to silty and gravelly sand. Occurrence of cobbles and boulders is typical in the till, particularly near the bedrock surface. Compactness varies with depth from loose to compact over the first 1.5 to 4.0 m and becoming dense to very dense thereafter. To perform the majority of the exploration boreholes, the coring was necessary to sample all the till deposit layer and reach the bedrock. Based on previous experience in the area, the size of boulders can reach 5 meters or more, but the bigger size measured on both sites was 2.3 m and with a mean of 1.6 m.

The bottom of the Aqueduct canal, from where the new intake will divert the water, is covered with a three layer of non-consolidated sediment of thin particles coming from the deposit of fines from the water pumped from the Saint-Lawrence River. The contract documents warned that any disturbance of this layer would result in increasing suspended solids concentration that can the water quality at the existing intake.

Regional Geology of Montreal Island

The geology of Montreal Island has been presented in detail by T.H.Clark and Grice (1972) and later by Durand (1978). Briefly, the geological setting consists of a variety of Pleistocene and Recent deposits that overlie early Paleozoic sedimentary rocks and Precambrian rocks, both of which are cut by Mesozoic intrusions. The principal events affecting the near surface rock assemblage were faulting, gentle folding, and minor metamorphism in the Mesozoic era and multiple glaciations and isostatic movements during the Pleistocene epoch (Grice 1972).

Bedrock geology in the region consists of Ordovician dolomite, limestone, and shales that are either exposed at the surface or underlie the Pleistocene and Recent deposits, except where they have been cut through by small bodies of dikes and sills associated with Monteregian Hills Mesozoic intrusions, one of which created Mont Royal. During the Mesozoic, the sedimentary rocks were faulted and gently folded and display a predominant bedding structure that dips a few degrees to the northeast. Except at Mont Royal, which is an 83 m high remnant of an intrusive plug immediately to the west of the downtown area, the bedrock surface has been reduced to low relief by glacial erosion.

There are five (5) major rock formations in the Montreal area: (1) pure crystalline limestones or dolomites with negligible shale, (2) fossiliferous and shaley limestones, (3) shales and (4) a massive intrusive rock which forms Mont Royal, and (5) the dikes and sills that are present to a lesser extent.

All these formations can be expected to be weathered and fractured close to the surface and some are locally cut by dikes and sills. Altered and fractured zones do not ordinarily exceed 2 m in width, however, when adjacent to the principal faults, they have been observed to be wider and deeper.

Solution cavities or glacio-tectonic surface features are encountered occasionally in the carbonate rocks; the principal lithological weakness is due to the softening of shale beds within limestone units and along joints and fractures zones in the shale.

Local Geology and Ground Characterization

The local geology indicates that the Atwater tunnel is located entirely in the Shale of the Utica Group, which is part of the carbonate sedimentary sequence of rocks that were deposited during the Ordovician age. The shale is fissile and slaking, but not swelling. The shale is known to rapidly lose its properties on contact with air and water. From Thériault 2012, the Utica Shale is formed by a clay size particle, essentially composed of 60 percent of calcite and dolomite; 20 percent of quartz and feldspath; and 20 to 25 percent clay minerals (illite, chlorite or kaolinite).

Figure 2 shows the location of the project with regard to major known faults and anticlines. No major fault is anticipated along the tunnel axis and no provision is planned in the GBR. The alignment of tunnel is located parallel to the Villeray anticline axis which gives the bedding a shallow 2–4° dip. The Shale rocks in the area are cut by an orthogonal system of vertical joints, a sub-horizontal joint set that follows along the bedding planes (±142N/02) and 2 families inclined joints (093N/30 and 281/27) and 1 subvertical (271/88) (Figure 3).

It has also to be noted that the shale of Utica is known to have a potential of explosive gas. Some gas bubbles were visible from the bedrock surface and after each blast, but the concentration of methane or

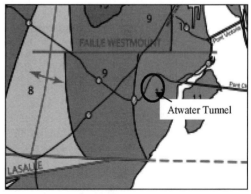

Figure 2. Local geology in the vicinity of the tunnel alignment (after Clark, 1972)

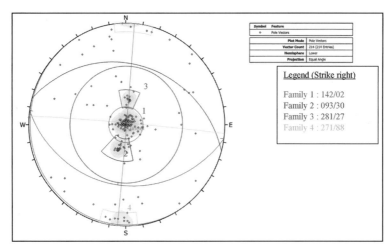

Figure 3. Stereonet representation of structural discontinuities along the alignment

Table 1. Measured properties of shale, sills and dykes

	Shale			Sills and Dykes		
	UCS (MPa)	Young Modulus (GPa)	BTS Modulus (MPa)	UCS (MPa)	Young Modulus (GPa)	BTS Modulus (MPa)
Minimum	40	13	0.24	113		
Maximum	131	20	1.13	333	76	1.4
Average	94	17.5	0.90	201		

other gas were still negligible, well below the lower explosive limit.

The measured properties of shale, dykes and sills are given in table 1. The shale is relatively low strength with an average unconfined compressive strength of approximatively 94 MPa. There are igneous intrusions in form of many dikes and sills that were associated with the formation of Mount Royal or/and the Appalachia's Orogeny that are irregularly scattered. They exhibit a much higher strength than the surrounding shaley rocks that averages approximately 201 MPa. The thickness of the sills can be as much as 3 m. It should be noted that in contact of intrusions, the shale was generally recooked on few millimeters to meters when enclaved in between two intrusions, and has therefore higher strength.

DESIGN CONSIDERATIONS

Alignment Selection and Description

The selection of horizontal alignment was heavily tied to the selection of the new intake location. As indicated above the new intake is located in a park area outside the portion of the Highway 15 that presents a risk of spill into the canal. Considerations associated with visual impact of the service building after commissioning, availability of laydown area and minimization of the impact to the community during construction as well as the size of the intake itself were used to select that location. The depth of the intake and its configuration were selected to avoid bringing sediments into the tunnel. Hydraulic design governed the size of the tunnel and the shaft and was based on an operating volumetric flow rate of 20.8 m^3/sec. A siphon system was therefore designed to reach a constant velocity that is above 0.6 m/sec to prevent sedimentation on the bottom of the tunnel. Screens and gates within the intake will prevent organic debris such as tree leaves and seaweeds to enter the system. A combination of hydraulic and cost analyses resulted in selection of an optimum inside diameter of 6 m for shafts and tunnel.

Vertical alignment was selected based on geological considerations to allow enough rock cover for a full face excavation of the tunnel while minimizing ground support as discussed below, which lead to a tunnel depth of 30 m (100 ft).

Selecting the Retaining Walls and Intake Cofferdam

The new intake and the buried conduit are founded on bedrock, which is approximately 10 meter deep, and therefore require deep excavations. Considering that the existing Aqueduct Canal is at less than

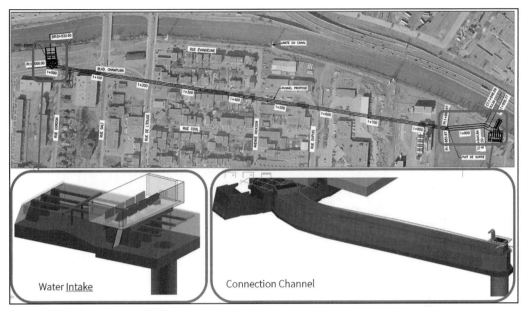

Water Intake

Connection Channel

Figure 4. The Atwater tunnel alignment

100 m away, it governs the groundwater regime and provides a continuous recharge for the aquifer. The choice and design of the support of excavation system was left to the contractor and it was specified to be watertight to limit the water inflows inside the excavation, especially towards each of the shafts and in the tunnel. The contractor elected to use a combination of sheetpiles and drilled piles. Figure 4 shows the alignment with a closeup view of the new intake and the connecting channel to existing valve chamber.

To seal off the soil-rock interface of the sheet piles and to ensure they are embedded in the sound rock level, the contractor chose to use a pre-excavation, which allowed the removal of any boulders. The excavation was filled by a soil-bentonite mix and sheet piles were inserted in this bentonite-fill material to the rock.

To erect the cofferdam in the Aqueduct Canal, the contractor elected to build a jetty by using compacted coarse grained material from which the sheet piles were driven to the top of the rock. A double screen system was used to prevent suspended solids to get into the downstream water and reach the existing intake. This allowed the installation of the jetty in a controlled manner under tight monitoring of suspended solids in the water. Figure 5 shows an aerial photo of the jetty and screen system as well as the sheet piles cofferdam being built in the canal.

Selecting the Rock Excavation Method

Alternative construction methods evaluated for the project included an open face Tunnel Boring Machine (TBM), roadheader, and drill and blast. Given the short length of the tunnel, the use of a TBM was considered not cost effective. Roadheader excavation of a medium weight class was judged applicable to this type of rock with compressive strengths varying from 40 to 131 MPa and an average of 94 MPa, as well as an average Cerchar Abrasivity Index of 0,8. The use of the roadheader would have however required widening each shaft to insert the roadheader. Two machines would have been required to guarantee the schedule, which requires excavating from both sides simultaneously. This method was not however precluded. Drill and blast was therefore the chosen excavation method.

The main challenge of drill and blast is the need to limit vibration through the highly populated and urban areas and to comply with the City of Montreal maximum peak particle velocity (PPV) vibration criteria of 25 mm/sec. In addition, at the new intake site, a 120 kV underground power utility presents a more restrictive operation because the shaft is at approximately 10 m above the first blast. To meet these requirements controlled blasting was used by optimizing round length, powder factor, drilling diameter and spacing of holes as well as charge per delay based on previous experience in the area (Habimana et al., 2015).

Figure 5. Aerial photo of the construction of the support of the excavation in overburden system at the intake structure

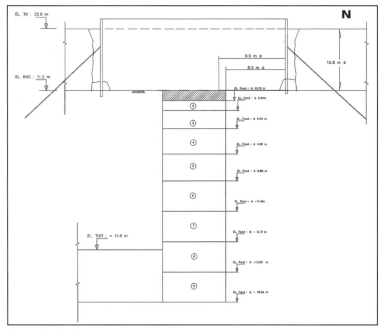

Figure 6. Shafts blasts design

To reduce to groundwater flow from the top of bedrock during the shaft excavation, a 10 m (33 ft) deep curtain of grouting has been completed at 3 m (10 ft) around the side of shafts wall. Grouting quantity was minor for the first holes where sub horizontal connectivity was found. A total of 3.7 m³ (131 ft³) of cement was injected.

For the drill and blast excavation of the shaft, bench height increased from the top to the bottom, as showed in Figure 7, from 1 to 5 m (3 to 16 ft). The first bench of the new intake shafts was excavated mechanically due to poor quality of the rock in surface.

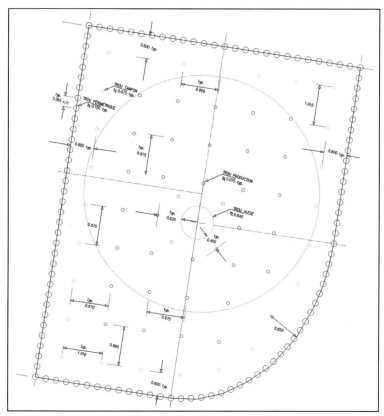

Figure 7. Typical shafts blasts design and pattern

The shaft size has been oversize for the entry of an underground drilling Jumbo and loader. A typical shaft's blast design and pattern are given by Figure 6. A 150 mm diameter line drilling was spaced by 265 mm center-to center. An 840 mm diameter destructive borehole was drilled in the middle of the shaft to limit the effect of sinking shut. The powder factor was varied from 1.0 to 1.95, depending on the size/proportion of sills and dykes observed in the 840 mm diameter borehole. Their position and thickness in each bench are critical to know because the shale will have the ability to keep the energy without breaking important sill layers.

Characterization of Expected Ground Conditions and Design of Initial Support System

Rock mass classification was performed using the RMR method. Ground conditions are expected to consist of a generally good quality rock mass. Table 2 provides a summary of the four RMR classes of excavation that were considered and the corresponding initial support categories for a drill and blast operation.

The design of the support requirements and sequence of excavation was done by a combination of empirical RMR, analytical, and numerical methods.

The contractor has introduced a change to replace the steel fiber by polysynthetic fiber shotcrete, which was accepted after a series of lab test results that showed a mix that comply with specification requirements of tenacity and UCS. Given the Utica shale is known to exhibit fissile and slaking behavior a protecting layer of shotcrete was required to be applied less than 5 hours after blasting for each cycle before the installation of the rest of the initial support system.

Final Lining Design and Construction Considerations

For ease of the construction, a horse-shoe shape with a cross-section of approximately 6.6-m-wide and 6.6 m high (Figure 8) was used for the excavation but a circular shaped cross section with an inside diameter of 6 m will be used for the cast-in-place concrete for the final lining. Collapsing forms will be used and

Table 2. Summary of expected ground conditions and ground support systems

Class of Excavation	I	II	III	IV
RQD	Excellent	Good	Fair	Very poor to poor
RMR class	81 to 100	61 to 80	41 to 60	21 to 40
Percentage of the length	75%		23%	2%
Advance per round	3.5 m		2.5 m	1.2 m
Initial Support Category	Pattern bolting at 2.0 m spacing with resin anchored 2.5 m long bolts (25 mm) and steel fiber reinforced shotcrete (100 mm)		Pattern bolting at 1.5 m spacing with resin anchored 2,5 m long bolts (25 mm) and steel fiber reinforced shotcrete (100 mm)	Pattern bolting at 1.2 m spacing with resin anchored 2,5 m long bolts (25 mm) and steel fiber reinforced shotcrete (100 mm)

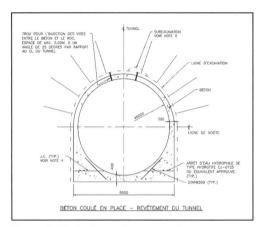

Figure 8. Typical tunnel cross-section

installation will be done from both ends to meet the tight schedule.

CONCLUSIONS

The Atwater Conveyance Tunnel Project which is currently under construction, will mitigate the risk of potential contamination and spills in the Aqueduct Channel by providing a new intake that is outside the zone at risk. Its construction is currently using drill and blast method for the rock excavation and a combination of sheet piles and drilled piles for the support of the excavation in overburden. The choice of drill and blast was based on the expected geological conditions as well as the need to meet the tight schedule, which requires excavation and installation of the cast-in-place concrete lining from both ends. Appropriate measures are being taken to minimize the impact on surrounding community as well as on the existing treatment plant. A double screen system

was used to limit the amount of suspended solids in the water during the construction of the jetty in the channel from which the support of the excavation in overburden was installed successfully.

The ground characterizations resulted in three main classes of excavation that can be excavated by a regular full-face excavation with 1.2 to 3.5 m advances and rock support consisting of 2.5 m long resin anchored bolts with reinforced fiber. To address the fissile and slaking behavior of the Utica Shale a protective shotcrete layer is being applied immediately after the excavation.

REFERENCES

Clark, T.H. (1972), Montreal Region, (Publication no. 152). Québec, Ministère des Richesses Naturelles.

Durand M. (1978). Classification des phénomènes et cartographie géotechnique des roches rencontrées dans les grands travaux urbains à Montréal, Canada. International Congress of Engineering Geology, Madrid, Spain, Vol. 1, 45–55.

Grice, R.H. (1972). Engineering Geology of Montreal, *International Geological Congress*, Canada, B-18, 15pages.

Habimana, J., Kramer, G.E., Revey, G. (2015). Planning, Design and Construction Considerations for a Drill and Blast Utility Tunnel underneath the Montreal Neurologic Institute with Highly Sensitive Equipment at McGill University Campus, Rapid Excavation Conference, New Orleans, 2015, 784–796.

Thériault, R., 2012. Caractérisation du Shale d'Utica et du Groupe de Lorraine, basses-terres du Saint-Laurent - Partie 1 et 2. Ministère des Ressources naturelles et de la Faune, Québec; DV 2012–03.

Final Design of the River Des Peres Tunnel in St. Louis

Patricia Pride
Metropolitan St. Louis Sewer District

Mike Robison, Jon Bergenthal, and Wojciech Klecan
Jacobs Engineering

ABSTRACT: The $585 million Lower and Middle River Des Peres (LMRDP) CSO Storage Tunnel is the largest component of Metropolitan St. Louis Sewer District's (MSD) consent decree-driven, multibillion dollar, 23 year, Project Clear. Final design of the 8.5 mile long, 30-ft diameter tunnel, which will be excavated through limestone and dolomite at depths in excess of 200 feet, was completed in 2017. This CSO conveyance and storage project has 34 intake structures, five construction shafts, two long connecting tunnels and excavation of a large, cavern-style pump station. This paper describes the planned construction, including requirements for initial support for shafts, tunnel and cavern, and initial support of excavation for a large intake structure.

INTRODUCTION

The LMRDP CSO Storage Tunnel is the largest single component of MSD's CSO control system improvements planned to meet consent decree requirements for MSD's Lemay Service Area. The 13.7 km (8.5 mi) long, deep, hard rock tunnel will have a finished diameter of 9.1 m (30 ft). Beginning at the Lemay Wastewater Treatment Plant (WWTP), the alignment follows the channelized River Des Peres to the outfall of the existing Forest Park tubes, as indicated in Figure 1.

The project includes three construction shafts, 34 intakes and a deep, cavern-style pump station located adjacent to the WWTP. The tunnel will be driven from the Defense Mapping shaft to the Macklind shaft. Another shaft is located near the midpoint at the tunnel in River Des Peres Park.

The Forest Park intake, located at the upstream end of the tunnel just beyond the Macklind shaft, is a large, tangential vortex intake structure. With a peak flow rate of 170 m³/s (6,000 cfs), it will have a 6.1 m (20 ft) diameter dropshaft to transfer flow into the tunnel.

The project includes another three large tangential vortex-flow intake structures with peak flow rates ranging from 28 to 44.5 m³/s (990 to 1570 cfs). These large intakes are located at points where three large sewers (formerly creeks) enter the River Des Peres. Dropshaft diameters for these intakes range from 3.0 to 3.7 m (10 to 12 ft) in diameter. The 30 remaining intakes are smaller tangential vortex-style, with dropshafts in the range of 0.8 to 1.7 m (30 to 66 inch) in diameter.

Each intake structure drops flow into a deaeration chamber constructed adjacent to the tunnel at tunnel depth. Intake tunnels transfer flow from the deaeration chambers to the main tunnel. Most of the intake tunnels are short, however, two are long. The Glaise Creek tunnel will extend 762 m (2,500 ft) from the main tunnel to the Alaska Park intake, while the Rock Creek tunnel will extend 823 m (2,700 ft) from the main tunnel to the Rock Creek intake.

The dewatering pump station, located at the downstream end of the tunnel, will be built inside a large, horseshoe-shaped cavern with two access shafts. Excavation of the cavern and its access shafts will be performed by the tunnel contractor. Build-out of the dewatering pump station will be performed under separate contract.

GEOLOGIC CONDITIONS

St. Louis lies at the northeastern tip of the Ozark Uplift and is bordered to the east by the Mississippi River and the north by the Missouri River. Physiographic regions in the St. Louis area include alluvial plains along the rivers, hilly uplands located in the southern portion of St. Louis County, and the rolling upland located in the central and northern portions of the county. The uplands are characterized by a relatively thin veneer of silty and clayey soils overlying bedrock.

Bedrock in the region is comprised of horizontally stratified limestone with some dolomite and a few thin shale layers. The rock is generally intact but at tunnel depth can be slabby in the crown, where it breaks along bedding planes. The stratified nature of the rock will inhibit the downward movement

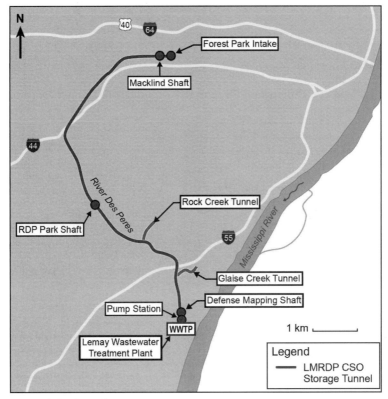

Figure 1. Project alignment

of groundwater, except along faults. Some of the limestone layers are very strong, with unconfined compressive strength (UCS) in excess of 200 MPa (30 ksi).

A large fault system crosses the tunnel alignment, as shown on Figure 2. This fault zone contains a number of high-angle faults with both normal and reverse movement and up to 40 feet of vertical offset. Some of the faults contain several feet of gouge, along with zones of highly fractured rock and rotated blocks. Deep weathering and hydrothermal mineralization are present in some areas.

SHAFTS

Each of the three construction shafts are designed for an internal diameter of 13.7 m (45 ft).

At the Defense Mapping site, the 22 m (73 ft) thick soil zone consists of fill overlying alluvium. The alluvium is potentially contaminated due to prior usage of the site. Concrete diaphragm walls are required for support of excavation, in order to restrict the flow of potentially contaminated groundwater, in compliance with restrictions imposed by an environmental covenant. Additionally, pre-excavation grouting of the upper section of bedrock is required

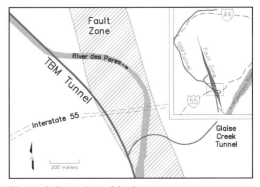

Figure 2. Location of fault zone

prior to concrete diaphragm wall construction. This approach was determined to be the best method for limiting the movement of contaminated groundwater into the shaft.

At the River Des Peres Park shaft, the 7.3 m (24 ft) thick soil zone consists of fill overlying alluvium, which is required to be pre-supported with sheet piles. At the Macklind shaft, the 11 m (36 ft) thick soil zone consists of fill and weathered shale, which can be supported concurrently with excavation using liner plates or ring beams with timber lagging.

At each shaft, the underlying rock will be supported with rock dowels, welded wire mesh and shotcrete.

TBM TUNNEL

The TBM Tunnel will be excavated with a single tunnel boring machine (TBM) beginning at the Defense Mapping shaft and terminating at the Macklind shaft. Near its mid-point, the River Des Peres Park shaft can be used for major TBM maintenance, if needed.

All muck from the TBM Tunnel will be removed through the Defense Mapping shaft. The first 230,000 m³ (300,000 cubic yards) of muck will be used to fill the Defense Mapping site for future construction of the Dewatering Pump Station and Enhanced High Rate Treatment Facility. Prior to construction of the LMRDP Tunnel, the general area around the Defense Mapping shaft and the Dewatering Pump Station will be raised above current grade with processed shot rock and tunnel muck from the adjacent Jefferson Barracks Tunnel. Construction of the Jefferson Barracks Tunnel began in March 2017.

The rock in tunnel has been grouped into three types according to expected behavior related both to support needs and potential difficulty of advancing the excavation due to stability issues, as follows:

- Type A ground needs light to minimal support with the focus on controlling overstress failure in the crown area and isolated wedges. Type A ground typically correlates to Rock Mass Rating (RMR) values greater than 60.
- Type B ground needs substantial support but can hold pattern rock dowels. The ground tends to be blocky such that progressive raveling can occur if the ground is inadequately supported. Thinly bedded strata under high horizontal stress conditions can produce Type B ground. Another cause of Type B ground is the presence of multiple intersecting joint sets. Type B ground typically correlates to RMR values between 60 and 40.
- Type C ground does not hold rock dowels well because of gouge, weathered seams, voids or tendency of the rock to deteriorate. Type C ground can often be described as blocky and seamy. Type C ground typically correlates to RMR values less than 40.

Figure 3 shows the initial support requirements for each of the three ground types in the TBM Tunnel. Based on analysis, 80 percent of the tunnel is expected to require Type A support, while 18 percent will require Type B support, and 2 percent will require Type C support.

The large fault system crosses the TBM Tunnel in the vicinity of Interstate 55, as indicated in Figure 2. The tunnel has been routed around the worst part of

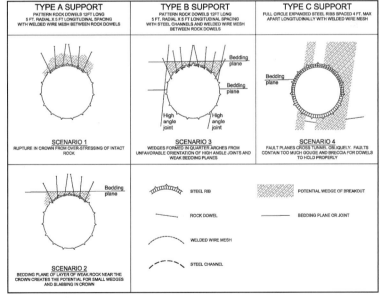

Figure 3. Initial ground support types for TBM tunnel

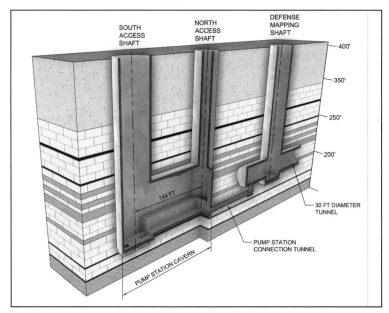

Figure 4. Profile of Pump Station cavern

this area, but still runs along the edge of it. Probing and pre-excavation grouting will be used during excavation to reduce the potential for ground and groundwater inflow problems in the I-55 fault zone.

Groundwater inflow into the excavated tunnel is expected to be around 900 Lpm (500 gpm), assuming the pre-excavation grouting program in the fault zone is properly completed.

PUMP STATION CAVERN

The Pump Station cavern consists of a 9.1 m (30 ft) diameter North access shaft, a 15 m (50 ft) diameter South access shaft and a 46 m (150 ft) long by 18 m (58 ft) wide by 26 m (84 ft) high cavern, as illustrated in Figure 4. The cavern and access shafts are located on the Defense Mapping site, adjacent to the Defense Mapping shaft.

Similar to the Defense Mapping shaft and because of the environmental covenant, the soil zone at the North and South access shafts will require concrete diaphragm walls for initial support. Support of the underlying rock will be through the use of rock dowels, welded wire mesh and shotcrete.

At cavern depth, ground conditions consist of alternating limestone and dolomite layers with occasional thin shale layers. A thinly bedded, relatively weak limestone layer, several feet thick, is present about 3 m (10 ft) above crown of the tunnel.

The cavern will be excavated by drilling and blasting a top heading from the shafts towards the center of the cavern. The remainder of the cavern and shafts will be excavated in benches to bottom. Initial

ground support in the cavern crown consists of 5.8 m (19 ft) long, high-capacity, tensioned rock bolts at 1.5 m (5 ft) spacing and split-spaced 3.7 m (12 ft) long rock bolts around the center section of the cavern crown and welded wire mesh, as shown on Figure 5. High capacity, 5.5 m (18 ft) long, grouted rock dowels and welded wire mesh will be used for supporting the cavern sidewalls.

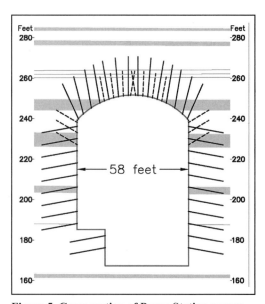

Figure 5. Cross section of Pump Station cavern

The cavern sidewalls are anticipated to move inward up to about 25 mm (1 inch) on each side due to high horizontal stresses, based on FLAC 3D/UDEC numerical modeling. Over-stressing is expected in the ground immediately above the cavern crown. Ground displacements will be monitored during excavation and will include several multi-point borehole extensometers installed from the surface and terminating just above the cavern crown.

FOREST PARK INTAKE

The Forest Park intake is one of the largest tangential vortex intake structures ever designed. The intake will transfer flow from the existing Forest Park tubes down to tunnel elevation. The existing tubes are a pair of 8.8 m (29 ft) horseshoe-shaped tunnels that were constructed by cut-and-cover in the early 1900s, as well as a 4.9 m (16 ft) horseshoe-shaped tunnel. Downstream of the tubes, the River Des Peres has been channelized.

The Forest Park intake includes a diversion structure located in the channel of the River Des Peres. From the diversion structure, flow drops and enters an approach channel and inlet structure. The inlet structure is a tangential vortex that directs flow to a dropshaft. At the base of the dropshaft is a large deaeration chamber, located immediately upstream of the Macklind Shaft. Figure 6 shows the general arrangement of the Forest Park intake.

The hydraulic performance of the Forest Park intake was initially modeled using Computational Fluid Dynamics (CFD) to confirm that it could convey the peak design inflow. CFD modeling was followed by construction of a 1/16 scale physical model to further evaluate the hydraulic characteristics and determine final dimensions and configuration. Construction of the inlet structure and approach channel will require excavating a deep slot, up to about 30 m (100 ft) below grade in the south slope of the River Des Peres channel. Initial ground support will consist of secant piles in the soil zone and a combination of rock dowels, welded wire mesh and shotcrete in the bedrock.

OTHER FACILITIES

The project includes almost 3,400 m (11,000 ft) of drill and blast tunnels and deaeration chambers to connect dropshafts to the TBM Tunnel. About half of this length will be intake tunnels having a minimum excavated diameter sufficient to install 1.8 m (6 ft) diameter fiberglass pipe. The remainder of the deaeration chambers and larger diameter Glaise Creek, Rock Creek and Wherry Creek tunnels will be lined with cast-in-place concrete. Initial support systems for the drill and blast tunnels and deaeration chambers consist of rock dowels, welded wire mesh

and shotcrete in Type A and B ground and steel ribs in Type C ground.

A 6.1 m (20 ft) diameter emergency overflow shaft will be located near the downstream end of the TBM Tunnel. Although each of the intakes is equipped with slide or roller gates that can be closed when complete filling of the tunnel is expected, an emergency overflow is included in the system in case of gate failure. Of particular concern are the gates at the Forest Park intake and the other three large intakes, which contribute the majority of the inflow.

The dropshafts and vent shafts for all except the largest intakes will be excavated using either raise boring or blind boring techniques. The choice of construction technique will be left to the contractor. The dropshaft for the Forest Park intake is expected to be blasted before lining with concrete. The main ventilation shafts on the TBM Tunnel alignment, located at the River Des Peres Park shaft and Macklind shaft, will also be blasted and lined with concrete.

CONSTRUCTION SCHEDULE AND COST ESTIMATE

Construction of the LMRDP Tunnel is expected to take about 6-½ years. The contractor will begin construction at the Defense Mapping site, sinking not only the Defense Mapping shaft, but also the North and South access shafts for the pump station cavern. Shot rock and tunnel muck will be processed and placed on the Defense Mapping site to complete its filling.

Excavation of the TBM Tunnel and the cavern can take place simultaneously. At the three-year mark, the tunnel contractor will demobilize from a portion of the Defense Mapping site to allow the pump station contractor to begin work. By this point, TBM excavation is expected to be complete and the tunnel contractor will have demobilized his muck processing equipment from the Defense Mapping site. The pump station contractor will begin with lining the cavern and its two access shafts.

After 54 months, the tunnel contractor must completely vacate the Defense Mapping site to allow the pump station contractor to install screening equipment in the Defense Mapping shaft and to complete build-out. By this point, concrete lining of the section of tunnel extending from the Defense Mapping shaft to the River Des Peres Park shaft is expected to be complete. The tunnel contractor will move his tunnel construction water pumping station up to the emergency overflow shaft, and lining of the section of tunnel upstream of the River Des Peres Park shaft can take place.

The final six months of the schedule is used for intake activation. This work must be coordinated with the pump station contractor and the wastewater treatment plant operator. In cooperation with and at the direction of both the pump station contractor and

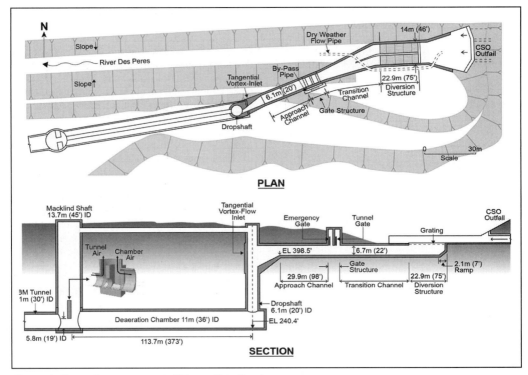

Figure 6. General arrangement of Forest Part Intake

MSD, the tunnel contractor will activate each of the individual intakes.

The Opinion of Probable Construction Cost prepared at the 95% design stage in mid-2017 totaled $585 million in current dollars. Of that cost, roughly $48 million was carried as contingency.

SUMMARY

The Metropolitan St. Louis Sewer District has completed design of the LMRDP CSO Storage Tunnel, which is the largest component of its consent decree-driven Project Clear. The work included modeling and design of one of the largest tangential vortex intakes ever, design of a deep, large diameter hard rock tunnel, and design of excavation and initial ground support for a large cavern-style dewatering pump station. At an estimated cost of $585 million, the tunneling project is expected to require 6-½ years for construction, including activation.

ACKNOWLEDGMENTS

The authors would like to acknowledge Jack Raymer, the primary author of the project's Geotechnical Baseline Report, for providing input and figures related to ground conditions and initial support.

REFERENCES

Pride, P., Robison, M., Bergenthal, J., and Klecan, W., 2015. St. Louis' Sewer Tunnel Vision. In *Rapid Excavation and Tunneling Conference, 2015 Proceedings,* Edited by Mark C. Johnson and Shemek Oginski, Littleton, CO: SME.

Klecan, W., Pride, P., Hallsten, J., Craig, A., Lyons, T. and Kharazi, A., 2017. Diverting 4400 MGD into a Deep CSO Tunnel: Forest Park Intake Design. In *WEFTEC 2017 Proceedings of the Water Environment Federation,* Chicago, IL: WEF.

Successful Implementation of "SEM–Single Shell Lining Concept" for the Emergency Tunnel at the Brenner Base Tunnel Project—Theory and Praxis of an Innovative Shotcrete Model

Thomas Marcher
SKAVA Consulting ZT GmbH

ABSTRACT: The Brenner Base Tunnel is a railway tunnel between Austria and Italy through the Alps with a length of 64 km. The construction lot Tulfes-Pfons was awarded to the Strabag/Salini-Impregilo consortium in 2014. The construction lot includes 38 km of tunnel excavation work and consists of several structures such as the 9 km long Tulfes emergency tunnel. Such a service (non-public) tunnel does not necessarily require a tunnel lining system with two shells, but under certain boundary conditions can be achieved by a single shell lining approach. The required conditions and limitations for the single lining approach are reflected and a proposal for structural verification is provided. For verification approach a novel constitutive model for the shotcrete design is used.

INTRODUCTION

The Brenner Base Tunnel (BBT) is a flat railway tunnel between Austria and Italy. It runs from Innsbruck to Fortezza. Including the Innsbruck railway bypass the entire tunnel system through the Alps is 64 km long. It is the longest underground rail link in the world.

The BBT consists of a system with two single-track main tunnel tubes 70 meters apart that are connected by crosscuts every 333 metres (see Figure 1). A service and drainage gallery lies about 10–12 metres deeper and between the main tunnel tubes. It is constructed ahead of the main tunnels and will be used as an exploratory tunnel for them. Four connection tunnels in the north and south link to the existing lines and also belong to the tunnel system, with a total length of approx. 230 km. Three emergency stops, each about 20 km apart, are planned in Ahrental, St. Jodok and Trens (see Figure 2). The emergency stops serve to rescue passengers from trains with technical difficulties. In addition, all emergency stops are accessible through driveable approach tunnels. A detailed project description is provided e.g., in [Eckbauer et al., 2014].

The two-track Inn Valley Tunnel has been already completed in 1994 and will be integrated into the overall tunnel system. This will reduce the travel time from Germany greatly. The 8 km long section of the Inn Valley Tunnel between the Tulfes portal and the link with main Brenner tubes is a part of the BBT system and will, for reasons of safety, be retrofitted with a rescue tunnel.

This construction lot "Tulfes-Pfons" was awarded to the Strabag/Salini-Impregilo bidding consortium in 2014. Excavation works will last until Spring of 2019. The construction lot includes 38 km of tunnel excavation work. It consists of several structures such as the Tulfes emergency tunnel, the Connection Ahrental access tunnel—Innsbruck emergency stop, the Innsbruck emergency stop with central tunnel and ventilation structures, the Main tunnel tubes, various Connecting tunnels and Ahrental-Pfons exploratory tunnel (see Figure 3).

The paper focuses on the experiences with the Tulfes emergency tunnel. The emergency tunnel is being driven parallel to the existing Innsbruck railway bypass; it will be 9 km long and the excavation cross-section is 35 m^2 (see Figure 3). The drill-and-blast excavation work (SEM) on this tunnel starts from three points at the same time: from Tulfes westwards (already completed), from the Ampass access tunnel eastwards and again westwards. The excavation works for this emergency tunnel were completed in Summer of 2017.

OUTLINE OF GEOLOGY

The BBT is geographically positioned in the Central Eastern Alps. Geologically, the tunnel is crossing the collision zone of the European and the Adriatic (African) plate. The main lithological units are the

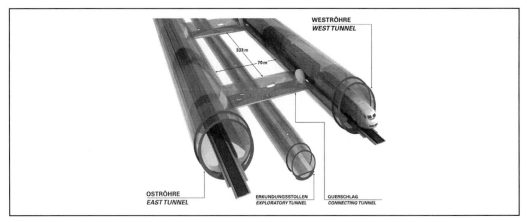

Figure 1. Brenner Base Tunnel System with two main tunnels and the exploratory tunnel (Eckbauer et al., 2014)

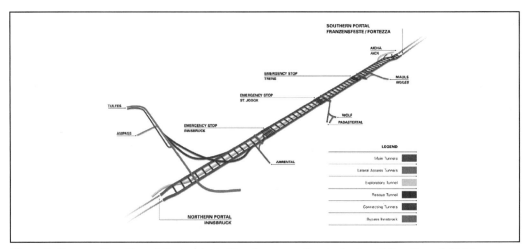

Figure 2. Brenner Base Tunnel System including the connection to the Inn Valley Tunnel (Eckbauer et al., 2014)

Figure 3. Construction lot Tulfes-Pfons of the Brenner Base Tunnel project

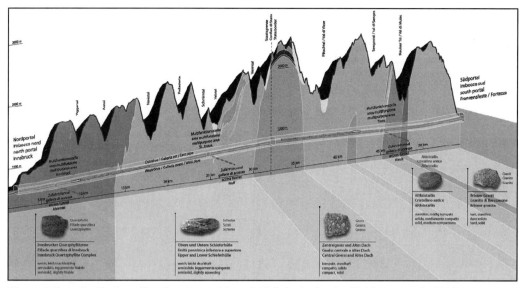

Figure 4. Longitudinal profile of geology with main lithological units (Eckbauer et al., 2014)

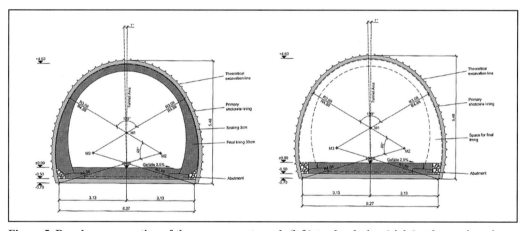

Figure 5. Regular cross section of the emergency tunnel: (left) tender design (right) value engineering

Innsbrucker Quarzphyllites, the Bündnershists, the central gneisses and the Brixener granites (see Figure 4).

From the hydrogeological point of view the water ingress is very limited in the homogenous sections of the phyllites, schist, gneisses and granites. The amount of water is expected to increase only with advance through some fault zones. A lowering of the water table at the surface is prohibited in these fault zones because of environmental aspects.

The overburden varies between 1,000 and 1,500 m over most of the tunnel. The maximum of 1700 m will be reached in the central gneisses.

PROJECT OPTIMISATIONS

Such an emergency tunnel does not necessarily require a tunnel lining system with two shells (double shell lining system), but under certain boundary conditions can be achieved by a single shell lining approach. This is especially valid for non-public tunnels where lower levels of water tightness are acceptable. Single shell lining systems offer the most efficient lining design as they take both the temporary and long-term loads. Additionally, the construction is very fast compared to a double shell (or even composite lining systems) where multiple stages of construction are required.

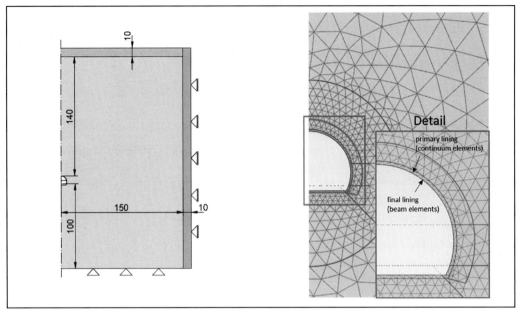

Figure 6. Boundary conditions (left) and FE mesh (right)

The tunnel cross-sectional area remains unchanged compared to the tender design (see Figure 5). The permanent shotcrete lining shall have a minimum thickness of d = 25 cm with two layers of reinforcement. The effectively applied shotcrete thickness shall be determined based on back-calculations taking into account all available on-site information. Additionally the following requirements have been defined:

- increased shotcrete quality
- design for service life of 200 years (same as for main tunnels)
- increased concrete cover of 55 mm
- maximum crack widths ≤0,3 mm
- alkali-free accelerator to be used

In addition the applied monitoring concept has been adopted:

- five measuring bolts per cross section with installation interval 5–10 m in weak rock/fault zones and 10–30 m in competent rock formations
- timely implementation of the first convergence measurements
- measurement of the shotcrete thickness by means of tunnel scan
- continuous tests for stiffness and strength development of shotcrete

For the permanent shotcrete lining the verification analyses have to be carried out for the ultimate limit state (ULS) and the serviceability limit state (SLS) as well as for the durability of the shotcrete. This verification assessment is done in two steps:

1. Design of the shotcrete lining on the basis of hydrogeological, geological and geotechnical forecasts.
2. Verification on the basis of the measured (monitored) deformations, the actual shotcrete thicknesses and the determined shotcrete properties (modulus of elasticity, strengths).

The proof of suitability for the permanent use is provided by the proof that the analysed crack widths do not exceed the maximum crack width as defined for the project.

TUNNEL BOUNDARY CONDITIONS

The reference calculation (see Figure 6) considers a service tunnel with a width and height of W/H = 14,4/10,9 m and an overburden of 650 m above the crown. The rock mass has been modelled as Mohr-Coulomb material with a rock mass stiffness of E_{rm} = 12000 MPa, Poisson ratio v = 0,2 and a rock mass strength of c_{rm} = 2,0 MPa and φ_{rm} = 34,3°.

The shotcrete itself has been modelled using two different approaches: (1) using a Mohr Coulomb model based on the concept of "hypothetic stiffness" which is in fact based on experience and has been introduced e.g., (John et al., 2003), which considers a reduced stiffness of young shotcrete of 5 GPa and 15 GPa for hardened shotcrete, and (2) the shotcrete model mentioned above which has been implemented in the model code PLAXIS as a user defined shotcrete model (Schädlich et al., 2014 and Saurer et al., 2014).

CONSTITUTIVE MODEL OF SHOTCRETE

The theoretical background to the novel constitutive model has already been presented in (Schädlich et al., 2014) and validated against a practical engineering task, which is an executed high-speed railway tunnel project in Germany, built according to the SEM design philosophy (Saurer et al., 2014). Here, only a short summary of the constitutive model is provided. Fundamental equations are provided in (Schädlich et al., 2014). The model has been implemented in the finite element software PLAXIS 2D 2012 (Brinkgreve et al., 2012). Note that compression is defined as a negative number.

Model Parameters

The model parameters for the shotcrete model are presented in Table 1.

Yield Surfaces and Strain Hardening/Softening

Plastic strains are calculated according to strain hardening/softening elastoplasticity. The model employs a Mohr-Coulomb yield surface F_c for deviatoric loading and a Rankine yield surface F_t in the tensile regime (see Figure 7).

Strain hardening in compression follows a quadratic function up to the peak strength, with subsequent bi-linear softening (Figure 8). Due to the time dependency of the involved material parameters, a normalised hardening/softening parameter $H_c = \varepsilon_3^p / \varepsilon_{cp}^p$ is used, with ε_3^p = minor plastic strain. While hardening yields a rotation of F_C, softening is modelled by a parallel shift of the failure envelope (cohesion softening). The softening rate is governed by the fracture energy G_c, which is used within a smeared crack approach to ensure mesh independent results.

The model behaviour in tension is linear elastic until the tensile strength f_t is reached. Linear strain softening follows, governed by the major principal plastic strain and the fracture energy G_t.

Time Dependent Stiffness and Strength

Two types of functions to model the time dependent strength have been implemented in the model: (1) according to the recommendations by CEB-FIB model code; (2) according to the early strength

Table 1. Parameters of the shotcrete model and values considered in the example calculation

Name	Unit	Value	Remarks
E_{28}	GPa	30	Young's modulus after 28 days
ν	-	0,2	Poisson's ratio
$f_{c,28}$	MPa	33	Uniaxial compressive strength after 28 days
$f_{t,28}$	MPa	3,3	Uniaxial tensile strength after 28 days
ψ	°	0	Angle of dilatancy
E_1/E_{28}	—	0,6	Ratio of Young's modulus after 1 day and 28 days
$f_{c,1}/f_{c,28}$	—	−2	Ratio of f_c after 1day and 28 days
f_{c0n}	—	0,15	Normalized initial yield stress (compr.)
f_{cfn}	—	1	Normalized failure strength (compr.)
f_{cun}	—	1	Normalized residual strength
ε_{cp}^p	—	−0,03, −0,002, −0,001	Plastic peak strain in uniaxial compression at shotcrete ages of 1 hour, 8 hours and 24 hours
$G_{c,28}$	kN/m	6,1	Fracture energy in compression after 28 days
f_{tun}	—	1	Normalized residual tensile strength
$G_{t,28}$	kN/m	0,03	Fracture energy in tension after 28 days
φ^{cr}	—	2,0	Ratio of creep vs. elastic strains
t_{50}^{cr}	days	3,0	Time at 50% of creep
ε_∞^{shr}	—	0,0005	Final shrinkage strain
t_{50}^{shr}	days	70	Time at 50% of shrinkage

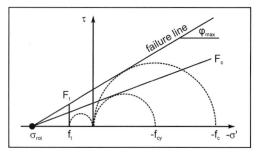

Figure 7. Yield surface and failure envelope of the shotcrete model

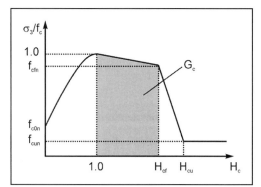

Figure 8. Normalized stress strain curve in compression

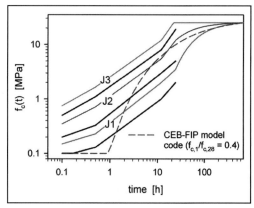

Figure 9. Time dependency of shotcrete strength

Table 2. Calculation phases

Phase	Name/Description
1	In-situ stress state
2a	Stress release of the Top Heading
2b	Excavation of the Top Heading, installation of the Shotcrete in the Top Heading
3a	Stress release of the bench
3b	Excavation of the Bench, installation of the Shotcrete in the Bench

with ε_∞^{shr} being the final shrinkage strain and t_{50}^{shr} the time when 50% of shrinkage has occurred.

TUNNEL LINING CALCULATIONS

Initially, calculations using a linear elastic model, a Mohr-Coulomb Model and the shotcrete model without applying the time dependency, which means that using the common approach with reduced stiffness have been performed to benchmark the result. The results are marked as 'calculations with hypostatic stiffness'.

The calculation phases in Table 2 have been considered in the model.

The shotcrete strength has been considered according to the J2 curve described above. In order to consider the effect of time in the model, two additional analyses have been performed marked with 'calculations with time-dependent stiffness'.

Initially, a benchmark model using linear elastic material, Mohr-Coulomb material and the UDM shotcrete model with the hypothetic stiffness approach (i.e., with E = 5/15 GPa) has been calculated to compare the resulting effects of actions in the elastic mode. (i.e., ignoring implicit time dependency).

As can be observed in Figure 10 the effects of actions remain within the elastic limits and results from all three models are comparable. Hence the

classes of EN 14487-1. Here, the latter approach has been considered; mean values of the class J2 have been assumed in the model, as shown in Figure 9.

The change in shotcrete ductility is represented by a time dependent plastic peak strain ε_{cp}^P. Input values are the plastic peak strains at t = 1 h, 8 h and 24 h. Beyond 24 h, ε_{cp}^P is assumed to be constant.

Creep and Shrinkage

Creep strains ε^{cr} increase linearly with stress σ and are related to elastic strains via the creep factor ϕ^{cr}.

$$\varepsilon^{cr}(t) = \phi^{cr} \cdot \sigma \cdot D^{-1} \cdot \frac{t - t_0}{t + t_{50}^{cr}} \quad (1)$$

where D is the linear elastic stiffness matrix. The evolution of creep with time t is governed by the start of loading t_0 and the parameter t_{50}^{cr}. For shotcrete utilization higher than 45% of f_c, non linear creep effects are accounted for according to (Eurocode 2, 2004).

Shrinkage strains are calculated as

$$\varepsilon^{shr}(t) = \varepsilon_\infty^{shr} \frac{t}{t + t_{50}^{shr}} \quad (2)$$

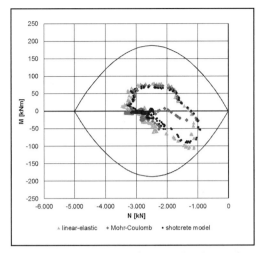

Figure 10. Comparison of effects of actions using the hypothetic stiffness approach

model allows comparing the results in terms of time effects shown hereafter as a next step.

As a next step, the time dependency of the shot-crete strength and stiffness have been included in the model, while still considering E = 15 GPa as maximum stiffness instead of consideration of implicit creep and shrinkage. Here, the boundary conditions as shown in the first chapters have been considered.

As a last step, the effects of shrinkage and creep have been modelled using the above expressed equations, considering realistic parameters for both, material model parameters and construction time. For this model, the parameters as stated in Table 1 have been considered.

However, note that for fully time dependent calculations, boundary conditions in terms of construction time and the time dependent behaviour of shotcrete are important factors that have a major influence on the calculation results.

INTERPRETATION OF RESULTS

Shotcrete, a key support element in SEM tunnelling, exhibits a significant time dependent behaviour, in particular during the initial hours of curing. This is important, because once applied as primary lining, the shotcrete is immediately loaded due to the excavation process.

In practical tunnel-engineering crude simplifications are usually adopted with respect to modelling the mechanical behaviour of shotcrete in numerical analysis, The paper at hand presents the application of a novel constitutive shotcrete model using realistic boundary conditions for a shotcrete tunnel

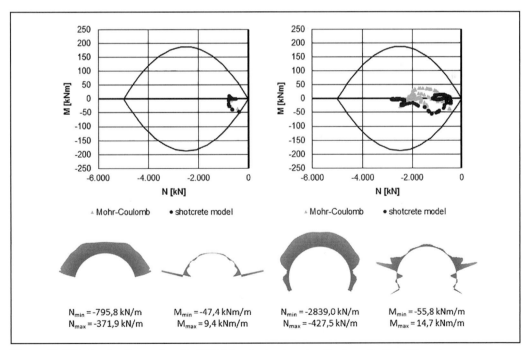

Figure 11. Effects of actions using the UDM for shotcrete using the hypothetic stiffness approach (lower part) and comparison with MC results as an interaction plot (upper part) for both excavation of crown and crown + bench

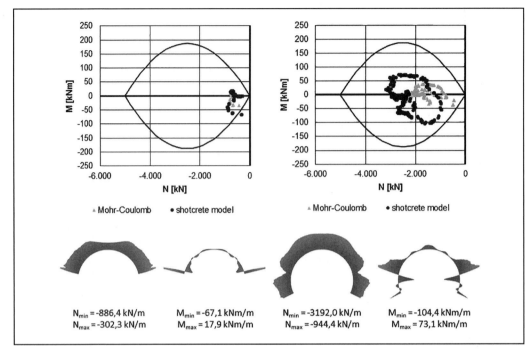

Figure 12. Effects of actions using the UDM for shotcrete using full stiffness approach with creep and shrinkage (lower part) and comparison with MC results as an interaction plot (upper part) for both excavation of crown and crown + bench

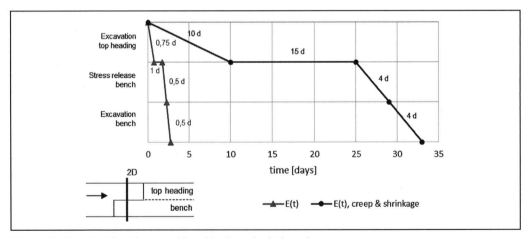

Figure 13. Construction time considered in the calculation phases

lining excavation. The model has first been verified by applying a classical application of a hypothetic E-modulus for shotcrete, which has been manually changed during calculation (see Figure 10). Only the strength has been calculated implicitly based on elastoplastic strain hardening/softening plasticity.

After this successful validation, the shotcrete stiffness has been determined automatically as a function of time. An automatic increase of shotcrete stiffness with progressing excavation has been applied (see Figure 13). The resulting stresses and strains show a realistic range of values (see Figure 11 and 12).

The benefit of such calculation is that there is no longer need for manual adaptation of strength and stiffness with time. Of course, such an advanced constitutive model requires both a higher number of parameters and more detailed knowledge about construction time.

CONCLUSION

The present paper illustrates the required boundary conditions and limitations for the single lining approach. A proposal for geomechanical and structural verification is provided. For the verification approach both, the use of a novel constitutive model with elastoplastic strain hardening/softening plasticity and time dependent strength and stiffness, creep and shrinkage are considered.

The paper illustrates a successful implementation of such a "SEM–Single Shell Lining Concept" for the Emergency Tunnel of the Brenner Base Tunnel Project.

ACKNOWLEDGMENT

The author thanks Cristian Kaiser for his contribution to this paper.

REFERENCES

Eckbauer, W., Insam, R. and Zierl, D. 2014. Planning optimisation for the Brenner Base Tunnel considering both maintenance and sustainability. Geomechanics and Tunnelling 7 (2014), No. 5.

Bergmeister, K. 2015. Life cycle design and innovative construction technology. Swiss Tunnel Congress 2015.

Insam, R. and Rehbock-Sander, M. 2016. The Brenner Base Tunnel. RETC conference proceedings, San Francisco, 2016.

B. Schädlich, and H.F. Schweiger, *A new constitutive model for shotcrete*, Numerical methods in Geotechnical Engineering, Hicks, Brinkgreve, Rohe (eds.) (2014) 103–108.

E. Saurer, T. Marcher, B. Schädlich, and H.F. Schweiger, *Validation of a novel constitutive model for shotcrete using data from an executed tunnel*, Geomechanics and Tunnelling 7, No. 4 (2014), 353–361.

R.B.J. Brinkgreve, E. Engin, and W.M. Swolfs, *Finite element code for soil and rock analyses. User's Manual.* Plaxis bv, The Netherlands, 2012.

M. John, B. Mattle, T. Zoidl: Berücksichtigung des Materialverhaltens des jungen Spritzbetons bei Standsicherheitsuntersuchungen für Verkehrstunnel, in: DGGT (Hsg.): Tunnelbau 2003, S 149–188.

Eurocode 2: EN 1992-1-1: Design of concrete structures—Part 1-1: General rules and rules for buildings (Authority: The European Union Per Regulation 305/2011, Directive 98/34/EC, Directive 2004/18/EC), 2004.

TRACK 2: DESIGN

Session 4: Challenging Design Issues

Peter Chou and Hamed Nejad, Chairs

Design and Monitoring of an Effective Deep Excavation Cut-off

Eric Sekulski, Jon Leech, and Michael Chendorain
Arup

John Yao
LA Metro

Mark Ramsey
HNTB

ABSTRACT: Exposition Station is a 70-ft-deep cut-and-cover structure and a TBM launch portal for the Crenshaw/LAX Transit (light rail) Project. Ground conditions comprise sequences of clays and sands with two aquifers separated by an aquitard layer. Station excavations extend near or into the aquitard. A pressure-grouted base was initially considered necessary for a groundwater cutoff. With additional investigation and analysis, the project team demonstrated adequate cutoff could be achieved by keying the CSM wall into the aquitard, providing considerable savings to the project. This paper presents the work and challenges to demonstrate and successfully implement the alternative solution, including additional subsurface investigations, multiple aquifer tests, analyses, and a monitoring and mitigation program.

BACKGROUND AND SETTING

The Crenshaw/LAX Transit Corridor Project is a Light Rail Transit (LRT) that extends 8.5 miles between the existing Metro Green Line near Aviation/LAX Station, and the Exposition Line at Exposition Boulevard in Los Angeles. The extension comprises eight stations at or below ground (Figure 1). Construction on the corridor project began in January 2014 and train services are planned to begin by 2019. The Los Angeles County Metropolitan Transportation Authority (LA Metro) is the owner of the project. This design-build project is being constructed by Walsh-Shea Corridor Constructors, and is supported by a design team led by HNTB Corp, and Arup as the lead engineer of underground guideways, Expo Station, and tunnels.

Exposition (Expo) Station is one of three underground stations of the project and is at the northern terminus of the LRT alignment, located along Crenshaw Boulevard between Exposition Boulevard and Rodeo Place (Figure 2). The station is approximately 820 ft in length and minimum 65 ft wide. The cut-and-cover station was excavated to a depth of 67 ft below ground surface (ft bgs) following construction of cutter soil mix walls (CSM) surrounding the excavation. The CSM wall comprised 2.5-ft-thick soil-cement mix (minimum 300 psi) with steel beams spaced at 5.5-ft centers, and was selected to provide lateral support of excavation and limit groundwater ingress during construction.

Pre-bid geotechnical investigations at Expo Station indicated groundwater about 20 ft above the anticipated base of excavation. Technical requirements for the project based on this data included a geotechnical baseline report (GBR) that defined several key assumptions of ground behavior. In particular, lowering of groundwater levels outside of the excavation was not to be permitted in order to reduce potential for excessive settlement of adjacent roads, buildings, and utilities due to presence of peat near the surface, as well as to reduce the potential migration of known contamination at nearby sites. In addition, the GBR stipulated that a clay layer below the proposed station that was encountered in the preliminary borings shall not be assumed continuous for purposes of groundwater cut-off.

The design-builder proposed to demonstrate through additional testing and analysis that the deep clay layer could perform as an effective groundwater cut-off during excavation. As a result, the CSM wall could be keyed into the clay layer to prevent loss of groundwater above, and form a barrier capable of resisting hydrostatic uplift forces from the deeper aquifer below. This would provide cost savings and avoid disturbance of the natural aquitard expected to exist just below the proposed base of excavation.

Figure 1. Map of Crenshaw/LAX transit project, red box indicating Expo Station

GROUND CONDITIONS AND INVESTIGATIONS

The northern part of the project alignment, which includes Expo Station, lies within the Downey-Tustin Plain of the Los Angeles Basin. Subsurface conditions comprise unconsolidated Holocene-age stream channels and floodplain deposits sediments referred to as Young Alluvium which overlie late-Pleistocene Old Alluvium. Subsurface materials as categorized in the GBR include Type 1 (fine-grained silt, clay, or organic soils) and Type 2 (fine- to coarse-grained sands and gravels with cobbles) soils. Young and Old Alluvium can include both Type 1 and 2 soils. The primary water bearing units included a shallow aquifer separated from a deep aquifer by an aquitard.

The pre-bid geotechnical investigation included several borings at Expo Station. Supplemental ground investigations were undertaken in 2014 and 2015 by the design-builder for purposes ranging from geotechnical design to installation of instrumentation for construction monitoring. A summary of geotechnical and hydrogeologic subsurface conditions is provided in Table 1.

Hydrogeologic conditions comprise a "shallow" unconfined aquifer within the upper Type 2 old alluvium and a "deep" confined aquifer within the lower Type 2 old alluvium, separated by a potential aquitard of Type 1 (cohesive) old alluvium. Piezometric readings in the "deep" aquifer indicated groundwater levels that were 2 to 3 ft lower than in the "shallow" aquifer.

Geological cross sections developed following the supplemental investigations along the west and east sides of the station are presented in Figures 3 and 4, respectively. The cross-sections (locations as indicated in Figure 2) illustrate the extent, thickness, and composition of the aquitard underlying the station.

The aquitard was encountered in all borings and the inferred thickness varied from 6 to 21 ft. Pre-bid boring R-11-135 indicated the thinnest area of the aquitard (6 ft); however, this is possibly due to the sampling interval (samples were collected at 5-ft intervals or greater). Later borings advanced specifically to demonstrate aquitard thickness, including continuous soil sampling, which provided data indicating a thicker aquitard consistently on the order of

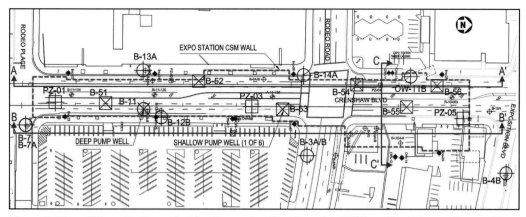

Figure 2. Plan of Expo Station; select borings and wells relevant to 2015 aquifer tests are shown with emphasis; note that soil profiles A–A' and B–B' are presented in Figures 3 and 4, respectively, and lab data from samples along section C–C' is presented in Figure 6.

Table 1. Summary of stratigraphy at Expo Station

Geologic Unit (Hydrogeologic Unit)	Typical Depth (EL)	Thickness (ft)	Description	Groundwater Depth (EL)
Young Alluvium – Type 1 (Unsaturated)	0 to 30 ft bgs (EL +112 to +82)	30	Medium-stiff to very stiff lean clay and sandy silt, layers of medium-dense to dense silty sand.	Dry, variable perched conditions
Old Alluvium – Type 2 (Shallow Aquifer)	30 to between 70 and 90 ft bgs (EL +82 to between +42 and +22)	40 to 60	Dense to very dense coarse to fine sand with gravel; occasional cobbles and boulders.	48 to 49 ft bgs (EL +63 to +64)
Old Alluvium – Type 1 (Aquitard)	Varies, between 70 and 100 ft bgs (EL +42 and +10)	6 to 21	Stiff to very stiff lean clay, sandy clay, and silts; some discontinuous lenses of silty sand.	49 to 51 ft bgs (EL +61 to +63)
Old Alluvium – Type 2 (Deep Aquifer)	Confirmed up to 152 ft bgs (EL –40)	> 50	Very dense sand with silt and gravel.	51 to 52 ft bgs (EL +60 to +61)

Elevation is relative to NAVD88.
Groundwater Depths cited are pre-construction.

10 to 20 ft thick. However, due to the variability in thickness and composition of the aquitard, aquifer testing was considered prudent to directly test for hydraulic connectivity across the aquitard.

Laboratory testing (which included 89 Atterberg Limits, 82 grain-size distributions, and 4 falling head hydraulic conductivity tests) were performed on soil samples collected between depths of 67 and 100 ft (i.e., within the expected depth range of the aquitard). Atterberg limit testing results from samples associated with the aquitard are presented in Figure 5. The majority of samples were classified as USCS 'CL' with a plasticity index ranging from 7% to 29% with an average of 15%. Figure 6 shows the results of a subset of grain size distribution plots for an east west cross-section (See Figure 2 for location). The falling

head tests on samples of aquitard clay soils from borings B-11, B-13B, B-51, and PZ-02 from depths between 91 and 96 ft indicated hydraulic conductivities from 2×10^{-10} to 1×10^{-9} m/s, indicative of very low to essentially impermeable material. The data illustrates variability within the aquitard but clearly suggests a predominantly fine-grained material with low to moderate plasticity and low permeability.

AQUIFER TESTING AND MONITORING

A total of 19 observation wells (mix of dual and single level installations), 5 nested vibrating wire piezometers, 3 deep aquifer pumping wells, and 6 shallow aquifer pumping wells were installed for investigation, testing, and construction monitoring and mitigation.

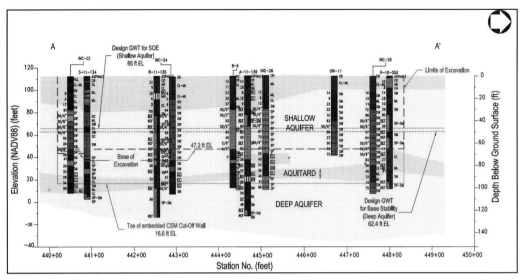

Figure 3. Soil Profile A-A' showing western side of Expo Station following supplemental investigations

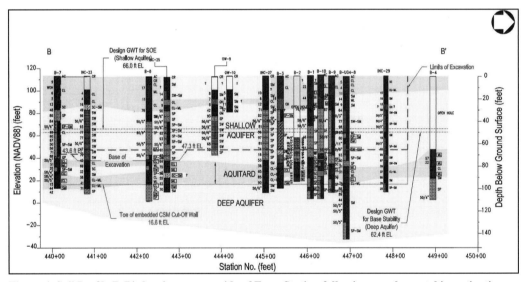

Figure 4. Soil Profile B-B' showing eastern side of Expo Station following supplemental investigations

The data collected and instrumentation installed were used for engineering evaluations and subsequent construction monitoring and mitigation. Aquifer testing was performed prior to excavation below the groundwater table in order to demonstrate a lack of hydraulic connectivity between the shallow and deep aquifers (i.e., the aquitard behaves as a cut-off). Aquifer testing included two phases of testing. The first phase, in 2014, included a series of slug tests (variable head tests), a step drawdown

test, and a constant rate test (CRT) at the north end of the station. The second phase occurred in 2015 after CSM walls had been installed, and consisted of a second-deep aquifer CRT as well as a shallow aquifer test. The 2015 program included five nested vibrating wire piezometers (VWPs) within the station footprint, each with three response zones (shallow aquifer, aquitard, and deep aquifer), as discussed further below.

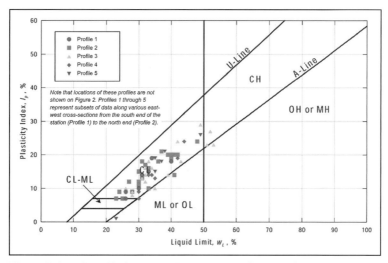

Figure 5. Atterberg limit test results on samples from aquitard

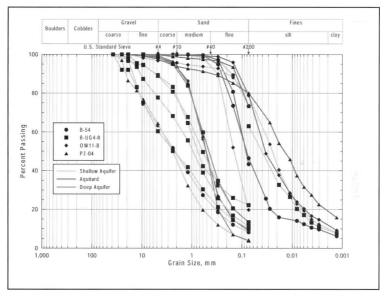

Figure 6. Grain size distributions along cross section C-C' at Expo Station

2014 Testing

The first deep aquifer CRT, conducted in 2014, was pumped at a rate of 15 gallons per minute (gpm) for 72 hours, with an estimated radius of influence (ROI) of 600 ft, and a maximum drawdown of 16 ft. Drawdown within deep aquifer observation wells ranged between 0.1 and 1.3 ft. The results of both time drawdown and distance drawdown analysis (using Aqtesolv version 4.5, Hydrosolv, 2017) indicated a hydraulic conductivity of the deep aquifer of

around 2×10^{-3} m/s. There was no observable drawdown in any of the shallow monitoring wells during the test, indicating a lack of hydraulic connectivity between the shallow and deep aquifer. Construction of the CSM wall and excavations above the groundwater table were allowed to proceed while the need for and details of further evaluations were being considered by the project team. Further aquifer testing was performed after installation of the CSM wall was complete and while excavation above the groundwater table was in progress.

2015 Testing

The second-deep aquifer CRT, conducted in 2015, was performed using a new and larger well (B-11) that was installed to 152-ft depth within the deep aquifer and located near the south end of the station, in order to increase drawdown and expand coverage of the testing to a greater portion of the station footprint. The arrangement of wells relative to the station footprint is presented in Figure 2; note that only the subset of wells whose data is presented in Figures 7 and 8 are shown with emphasis. Following a step drawdown test, well B-11 pumped at a rate of 99 gpm for 72 hours, with a maximum drawdown at the well of 21 ft and an ROI which extended well beyond the CSM limits. Maximum drawdowns of between 3.4 and 6.8 ft were observed at all deep aquifer observation wells and VWPs during testing (see Figure 7).

Reductions in pressure were observed at all aquitard VWPs during the CRT. The rate of response and total pressure reduction varied by location. VWPs PZ-04 and PZ-05 experienced approximately 95% of the pressure reduction that was observed in the deep aquifer nested at the same location. PZ-01 and PZ-02 showed less response in the aquitard, between 15 and 25% of the observed drawdown in the deep aquifer. The results indicated that within the aquitard the pressure response (and thus the inferred connectivity) generally dissipated with distance away from the top of the deep aquifer.

Minor but measurable changes in water level ranging from a reduction of 0.07 ft to an increase of 0.08 ft were observed within the shallow wells during the CRT. The principle mechanism proposed for the minor changes in the shallow aquifer was mechanical unloading associated with the ongoing excavation of material within the site limits during the CRT, because the fluctuations in water levels continued even after pumping had stopped.

Analysis of the pumping test data (using similar methods employed for the previous tests) indicated that the effective hydraulic conductivity of the deep aquifer was approximately 3×10^{-4} m/s (an order of magnitude lower than the first test). The difference is likely attributed to the difference in ROI between the two tests. Estimates of the effective aquitard hydraulic conductivity ranged from 7×10^{-9} to 4×10^{-8} m/s.

The shallow aquifer test, also conducted in 2015, involved a three-stage aquifer pumping test performed simultaneously from 6 shallow aquifer wells (B-51 to B-56), all located within the completed CSM cut-off wall. The shallow aquifer test was included to evaluate the potential effect of unloading and reversal of hydraulic gradients across the aquitard, as well as to test the ability of the CSM to perform as a lateral cut-off. The test comprised three rounds of pumping (80 or 130 hours each) and recovery periods with durations between 40 and 43 hours (except the final stage which was left to recover with monitoring continuing for 250 hours). The periods of pumping and recovery varied due to site constraints and construction activities. The total pumping rates during testing ranged between 45 and 94 gpm, and an estimated 1.5 million gallons of water was pumped. Groundwater levels observed within the shallow aquifer inside the CSM wall reduced between 1.6 and 3.2 ft by the end of the final pumping stage. A slight decrease (up to 0.2 ft) in groundwater head was observed in deep aquifer observation wells, but only those within the limits of the station footprint; and it was noted that the change in water levels did not correlate to the pumping and recovery periods. Thus, this slight change in the deep aquifer was similarly assumed to be related to excavation of soil within the overlying site area during the shallow aquifer test. The groundwater level response in the shallow and deep aquifers and aquitard piezometers during testing of the shallow aquifers is presented in Figure 8.

All five aquitard piezometers responded to pumping in the shallow aquifer, with the greatest response evident in the three southernmost instruments (PZ-01 to PZ-03). The maximum drawdown observed at all aquitard piezometers ranged from between 0.3 and 1.3 ft by the end of the test. The extent of pressure reductions observed in the aquitard VWPs generally followed the opposite trend to the reductions observed during the deep aquifer CRT.

During periods of recovery, it was noted that the shallow aquifer wells within the station footprint showed an increase in water levels. These results indicated that limited recharge was occurring, which were found to be between the gaps of the CSM walls for utility crossings where jet grouting had been performed to close the gaps. Additional localized grouting was subsequently performed to improve the water tightness of these areas.

The aquitard's response to pumping from both the deep and shallow aquifers indicated heterogeneity across the site. VWPs PZ-01 and PZ-02 located to the south responded more closely to pumping of the shallow aquifer whereas PZ-04 and PZ-05 responded more closely to pumping of the deep aquifer. Despite the slight variation in responses within the aquitard, the overall conclusion of the testing programs was that the shallow and deep aquifer units behaved hydraulically independent of each other. The aquitard separating these units were considered sufficiently continuous across the site and was expected to perform as an effective cut-off during construction.

BASE STABILITY

Further engineering evaluations were undertaken to assess base stability of the aquitard during

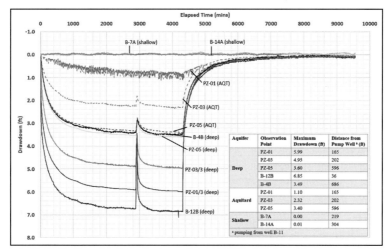

Figure 7. Drawdown observations during the 2015 deep aquifer CRT (B-11)

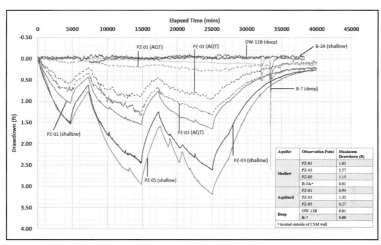

Figure 8. Drawdown observations during the 2015 shallow aquifer CRT (B-51 to B-56)

excavation. These studies included base heave stability, tension cracking of the aquitard, and a filter rules check to evaluate the potential for deterioration of the aquitard leading to quick conditions and/or piping.

Base Heave

Base heave is related to an upward movement of soil at the excavation base due to the upward force of water pressures from within the underlying deep aquifer. The risk increases as the overlying weight of soil and groundwater in the shallow aquifer decreases during excavation. To prevent base heave during excavation, the remaining weight of soil the above aquitard, and the weight of the aquitard itself,

must sufficiently offset the hydrostatic uplift from the deep aquifer. Since the underlying principles are well established, relatively low factors on safety are generally acceptable where supported by significant site data and/or conservative assumptions.

At Expo Station, the critical location occurs where the base of the aquitard is the shallowest, corresponding to the northwest corner of Expo Station in the vicinity of boring R-10-053 (See Figure 3) where the base of the clay layer is EL +24.9 ft (about 22.4 ft below the base of excavation). At the completion of excavation, the factor of safety at this location would be at its lowest and was estimated as 1.2. After the 3-ft-thick station invert slab is poured, the factor of safety would increase to 1.4. This calculation assumed the highest measured groundwater level of EL +62.4, recorded prior to construction.

Tension Cracking

The potential for tension cracking within the aquitard occurs when hydrostatic uplift pressures exceed the weight of overlying soil in a given location. In this situation, upward movement would occur as internal shear strength becomes mobilized to resist the imbalance of forces. Where sufficient downward resistance exists from gravity alone, such as was demonstrated for typical base of excavation levels at Expo Station, the aquitard would not be subject to bending and tension cracking.

Methods used to evaluate the stability of localized 'plugs' against buckling and tension cracking were presented by Hunt and Gill (1997) and rely primarily on perimeter shear, plug geometry, and minimum tensile strength of the plug. These methods were used to assess the sensitivity of the aquitard to these failure modes under several extreme hypothetical scenarios, which included presumption of hydraulic connections within the aquitard (reduced thickness) over a limited horizontal extent. Analysis of these localized 'weak spots' in the aquitard resulted in relatively high factors of safety (>1.6) or required very large increases in groundwater level (> 6 ft) to cause instability. Thus, the risk of tension cracking was evaluated to be low.

Aquitard Deterioration

Although no seepage paths were identified from the aquifer testing, there remained a risk that partial pathways and/or relatively thin regions of the aquitard could deteriorate during excavation leading to quick conditions (boiling) and/or piping. At least in theory, hydraulic connectivity across the aquitard could be established when the unloading of soil weight and dewatering of the excavation, as the station excavation progressed, allowed fines from the top of thin areas of the aquitard to disperse into the overlying shallow aquifer material. Once initiated, continual unloading of weight could then lead to increased material deterioration to the point where hydraulic connectivity would be established leading to piping and/or boiling. A filter rules check was performed (Terzaghi and Peck, 1948) to document the low risk that the aquitard soils would be subjected to this mechanism of deterioration during excavation. The method, as adapted from standard practice of well filter material design, is widely used to evaluate piping risks in dams and similar structures (Ciria 2013, USACE, 1993, Simpson and Katsigiannis, 2015). The check included the primary criteria $D_{15 \,(filter)} \leq 5 \times D_{85 \,(soil)}$, where: $D_{15 \,(filter)}$ is the particle diameter with 15% retention in the overlying shallow aquifer;

and $D_{85 \,(soil)}$ is the particle diameter with 85% retention in the aquitard.

The filter rules check was performed where data was available to compare the upper aquifer with aquitard grain size distributions. In total, 136 grain size distribution curves were reviewed. Checks using the lower limits, upper limits and the extreme case limits were used to assess likely and worst case scenarios based on the upper and lower bounds of the grading for two hydrogeologic units. The lower 5 to 10 ft of the shallow aquifer typically comprised finer grained materials such as silty clay and clayey sand. The transition from coarse to finer soils in the shallow aquifer lowered the risk of deterioration of the aquitard. The evaluation concluded there was a low risk of instability arising from deterioration of possible thin layers of the aquitard.

MITIGATION AND MONITORING PLAN

A significant amount of monitoring instrumentation was installed at the site as a result of the supplemental investigations and testing. This monitoring instrumentation included extraction wells, observation wells, nested vibrating wire piezometers, and borehole extensometers. To manage the residual risk of drawdown of shallow groundwater levels outside the CSM wall, and potential base instability, mitigation and action plans were developed for construction that included establishing trigger criteria specific to the various instrumentation. The overall construction monitoring program for excavation and tunneling had already implemented a real-time web-based system with automated alerts, and specific monitoring for Expo Station associated with groundwater related concerns was integrated into this system.

A traffic light system was set up using green, amber, and red trigger levels at each monitoring location across the site. Principally this included: increases in groundwater level in the deep aquifer, relating to reduced factors of safety for base instability, and reduction in shallow aquifer water levels outside of the CSM. A flowchart was developed outlining the procedure which was required following an amber or red breach (See Figure 9). As part of the mitigation plan, a second-deep aquifer depressurization well was installed to the north of the station so that a rapid implementation of depressurization could take place should there be signs of base instability or if trigger levels are breached.

A weekly monitoring report was produced during excavation below the water table to summarize manually downloaded data and observations. The reporting frequency was reduced to monthly and then bi-monthly periods as construction of the permanent station structure progressed adding weight to resist hydrostatic uplift.

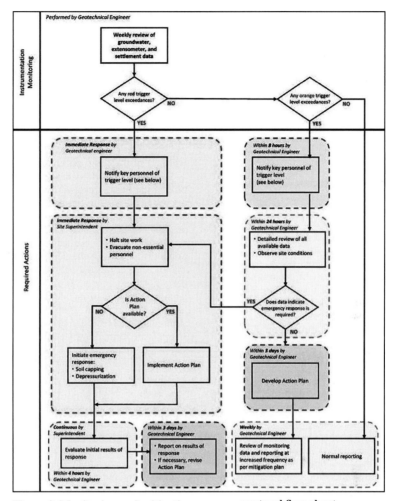

Figure 9. Monitoring and mitigation response protocol flow chart

PERFORMANCE DURING CONSTRUCTION

Construction proceeded as intended with station excavation occurring within the CSM walls. Internal dewatering occurred as the excavation advanced, with the CSM walls cutting off groundwater recharge both laterally from the shallow aquifer flow (through the wall itself) and upwardly from the deep aquifer flow (through the aquitard). Depressurization, by active pumping of the deep aquifer, was only implemented briefly during localized deeper excavations below the general base of excavation level.

Groundwater Levels During Construction

Construction at Expo Station began in 2014; however, construction monitoring of groundwater levels began immediately after completion of the 2015 ground investigation in July 2015 (when the excavation went below groundwater level in the shallow aquifer). Dewatering of the shallow aquifer within the CSM walls was initially undertaken by pumping from the shallow aquifer pumping wells and later by sump pumping during excavation. By September 2015 water levels within the excavation had been lowered to between EL +46 and +49 ft. Shortly after, shallow aquifer vibrating wire piezometers were excavated and water levels within the shallow aquifer within the CSM could no longer be monitored.

Between July 2015 and January 2017, groundwater levels in the shallow aquifer outside of the CSM wall declined by between 2.5 and 2.7 ft. During this period, groundwater levels in the deep aquifer had also declined by a similar amount (2.4 to 2.7 ft). This global decline in groundwater levels at the site indicated that other mechanisms (such as below average rainfall/aquifer recharge from 2012 to 2016) may have been responsible for the declining groundwater levels in the shallow aquifer, rather than dewatering within the station excavation. In addition, groundwater quality sampling and testing confirmed no mobilization of contamination due to the groundwater removal from within the station excavation. Since early 2017, groundwater levels in both the deep and shallow aquifers at the site have started to stabilize and in some cases recovered.

Sump Pit Excavation

The permanent station structure includes a discharge system with two sump pits for collection and removal of water from the station. Construction of one of these sump pits required deeper localized excavations of 13 ft below the typical base of excavation level.

The nearest available geotechnical data to the sump locations indicated that the deepest pit would extend 7 ft below the top of the aquitard in a location where the total aquitard thickness was estimated to be 21 ft. Laboratory tests indicated that the uppermost portion of aquitard within the zone of excavation was likely to comprise more permeable interbedded soils than the lower portions of the aquitard that would remain intact. Stability against base heave now required that internal shear resistance along the perimeter of excavations be included in the calculation. Using methods previously discussed (Hunt and Gill, 1997), factors of safety between 1.2 and 1.3 were estimated when assuming one-third of the estimated undrained shear strength (3,600 psf) of the aquitard, and factors of safety exceeding 3 were predicted with full mobilization of shear strength. Tension crack checks indicated factors of safety above 3.

The recommendation of the geotechnical engineer was that these assumptions would be acceptable for the short-term excavated condition prior to placement of the invert slab, provided that the geotechnical engineer was present to observe the excavation and that contingency depressurization systems were available. Following subsequent review and discussion with LA Metro it was decided to initiate depressurization in advance of (and during) sump construction to lower groundwater levels within the deep aquifer by about 5 ft at the location of the sump pit excavations. This was projected to increase factors of safety from between 1.2 and 1.3 to between 1.4 and 1.5.

Excavation and construction of the sump pits was undertaken between mid-December 2015 and mid-January 2016. Depressurization from well B-11 commenced at a rate of about 100 gpm and achieved a pressure reduction of between 4 and 5 ft of H_2O (250 to 310 psf) at nearby observation wells. Excavation began in 4-ft vertical increments under full-time geotechnical supervision. At a depth of about 9.5 ft in the sump pit, several feet into the aquitard, a saturated silty layer was encountered and the ground was observed to undulate under the weight of the excavation equipment. This behavior was assessed to be due to the phenomena of pumping of saturated non-plastic fines, and not necessarily an indication of instability of the aquitard across the site. Shallow sumps were used to dewater the footprint of the sump pit in advance of the excavation and as a precaution, the second depressurization well in the north was switched on to further lower the water pressure at the base of the aquitard for a brief amount of time while conditions of the sump pit were evaluated.

After further evaluation, the project team decided to re-design the deeper sump pit to be 3 ft shallower as a precautionary measure to avoid the need for further excavation. The footprint of the pit was increased to maintain the necessary capacity. Depressurization from well B-11 continued until completion of pit construction. The permanent sumps were formed by placement of a pre-fabricated steel block-out box into the excavated pit and backfilling the sides of excavation with a lean clay mix. The permanent reinforced concrete structure was then constructed inside of the box.

REMARKS AND CONCLUSIONS

The initial support of excavation system for Expo Station comprised a CSM wall embedded 15 ft below the base of excavation. Uncertainty regarding the continuity and permeability of a deep cohesive layer resulted in the potential need to install a jet grouted plug as thick as 15 ft under the station excavation.

Due to the high cost of a jet grout plug, the design-builder pursued an alternative that utilized a deep CSM wall keyed into an aquitard layer. Extensive geotechnical and hydrogeological investigations and testing confirmed the presence and hydraulic behavior of the aquitard and demonstrated that base grouting could be avoided.

The residual risks of base instability were evaluated and mitigated with effective construction monitoring and contingency planning. Overall, the design resulted in a net savings to the project and avoided unnecessary disturbance of natural conditions.

Currently, construction of Expo Station is substantially advanced with the majority of structure complete and final backfill operations schedule to begin in early 2018, at which point the station will have sufficient dead load to resist long-term uplift forces and temporary under slab drainage will cease.

REFERENCES

Various internal documents not publicly available.

Ciria, 2013. The International Levee Handbook. C731.

Hunt, Steven W., and Gill, S.A.1997. *Bottom stability of shafts with cohesive soil plugs overlying an artesian aquifer*. Proceedings: 1997 Rapid Excavation and Tunneling Conference. 379–404.

HydroSolve. 2017. Aqtesolv aquifer testing analysis software, Version 4.5.

Terzaghi, K. and Peck, R.B. 1948. Soil Mechanics in Engineering Practice.

U.S. Army Corps of Engineers (USACE). 1993. *Engineering and Design, Seepage Analysis and Control for Dams*, Engineer Manual 1110-2-1901. April 30.

Simpson, B., and Katsigiannis, G. 2015. *Safety considerations for the HYD limit state*. Proceedings of the 2015 European Conference on Soil Mechanics and Geotechnical Engineering. September 13–17, Edinburgh, Scotland.

A New Approach to Hydraulics in Baffle Drop Shafts to Address Dry and Wet Weather Flow in Combined Sewer Tunnels

Michael J. Seluga and Frederick (Rick) Vincent
Northeast Ohio Regional Sewer District

Samuel Glovick
Wade Trim

Brad Murray
McMillen Jacobs Associates

ABSTRACT: Sewer authorities in the US are constructing tunnels to control Combined Sewer Overflow (CSO) to meet regulatory requirements. This paper focuses on the design solutions for drop structures that convey an extreme variation of flow. The Doan Valley Storage Tunnel in Cleveland Ohio conveys dry and wet weather flows with a series of dual purpose conveyance and CSO storage tunnels. Baffled drop structures with dry weather drop pipes were designed to address this uncommon flow variation. This paper summarizes the design approach including numerical and physical modeling, air entrainment, energy dissipation, and structural systems for these unique drop shafts.

INTRODUCTION

Background

The Northeast Ohio Regional Sewer District (NEORSD or "District") was established on July 15, 1972 by Court Order. The District is a regional authority in northeast Ohio which has 62 member communities, including the City of Cleveland, as shown in Figure 1. The District is charged with the responsibility for planning, financing, constructing, operating, and controlling three wastewater treatment and disposal facilities, the major interceptor sewers and other water pollution control facilities within its service area.

The NEORSD entered a Consent Decree with the Ohio and United States Environmental Protection Agencies (OEPA/USEPA) and the United States Department of Justice (USDOJ) in 2011 to implement, within a 25-year time limit, a long-term control plan (LTCP) to control CSO currently impacting Lake Erie and its tributary waterways. The plan includes a series of control measures comprised of wet weather plant upgrades, seven CSO storage tunnels, collection system improvements, remote storage tanks, and implementation of green infrastructure technologies limiting storm inflow. The District refers to this plan as Project Clean Lake.

Project Description

The Doan Valley Storage Tunnel (DVT), shown on Figure 1, is part of the NEORSD's LTCP and is the third of seven storage tunnels in Project Clean Lake addressing the requirements of the Consent Decree. The DVT is in an urbanized environment just east of downtown Cleveland and includes nearly four miles of rock tunnels divided into three segments: the DVT, the Woodhill Conveyance Tunnel (WCT), and the Martin Luther King Conveyance Tunnel (MLKCT), shown on Figure 2 and Table 1. The project includes 6 shafts, near surface structures with consolidation sewers, and an emergency overflow.

The DVT system is unique as it will function as both a deep interceptor conveyance sewer and a CSO storage facility. The tunnels will convey dry weather flow (DWF) to relieve the existing Doan Valley Interceptor (DVI) sewer system and improve DVI level of service. The project will also reduce CSO by conveyance and storage during wet weather flows (WWF). WWF will be conveyed and discharge at a maximum rate of 50 million gallons per day (MGD) and excess flow stored within the project's 18 million gallons (MG) capacity. WWF to the tunnel will be controlled by automated gates at Shafts DVT-2 and MLK-2 to manage surge and tunnel

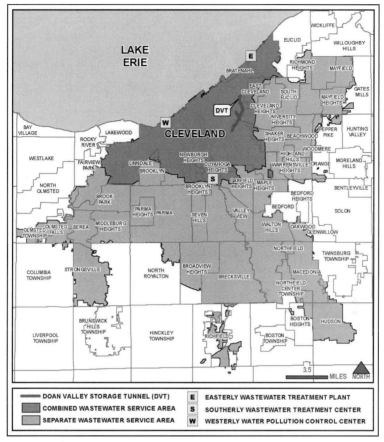

Figure 1. Northeast Ohio Regional Sewer District service area

filling conditions; once the gates close for large wet weather events, the system will overflow to existing permitted outfalls CSO-073 and CSO-223/224 respectively. An emergency overflow at Shaft DVT-1 will address upstream gate failure. Unlike many CSO storage tunnels, the DVT will not have a dewatering pump station and will gravity discharge and dewater up to 50 MGD through a 48-inch dewatering sewer to NEORSD's downstream interceptor sewer system. The discharge rate will be controlled by gates located at Shaft DVT-1. Due to the quantity of WWF and hydraulic head developed in the storage tunnel, these gates will generate extreme discharge velocities. A subsurface energy dissipation chamber at Shaft DVT-1 will mitigate these velocities prior to entering the dewatering sewer. Flow discharged from the DVT will be conveyed to the Easterly Wastewater Treatment Plant before discharge to Lake Erie. Additional information for the DVT tunnel and shaft components are summarized in Table 1 and Table 2, respectively.

This paper focuses on the design solution to address the conveyance of the extreme variation of dry and wet weather flows at Shafts DVT-2, MLK-2, and WCT-3. Shaft DVT-2, shown in Figure 3, is the largest structure and will convey flows from an existing 84-inch CSO culvert and the 8.5-foot diameter WCT into the tunnel system for a range of flow conditions. This structure will capture DWF from the WCT (average of 6.5 MGD), and WWF regulated by the gate in the WCT-1 Flow Control Structure to capture flows from the WCT and the existing CSO culvert (395 MGD for a 5-year storm operating condition).

This paper summarizes the design approach for the DVT-2 drop structure to address the large flow variation, which resulted in a baffled drop structure with dry weather drop pipe; explains how those results were applied to Shaft MLK-2 and Shaft WCT-3; and describes the resulting structural systems coordination and long-term operation and maintenance considerations for these unique drop structures.

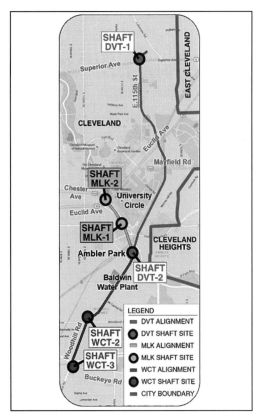

Figure 2. Doan Valley Tunnel System overview

Table 1. DVT tunnel summary

Tunnel	Finished Inside Diameter (ft)	Length (ft)	Approx. Depth Below Ground Surface to Invert (ft)
Doan Valley Storage Tunnel (DVT)	18	10,026	50–115
MLK Conveyance Tunnel (MLKCT)	8.5	2,967	50–90
Woodhill Conveyance Tunnel (WCT)	8.5	6,281	30–80

later detailed design and production phases, the drop structure sizing, baffle spacing and pipe orientations were initially configured using previous numerical and physical modeling studies, closely followed by the CFD modeling results for the DVT project. Various considerations, including site/civil, structural, constructability, geotechnical, and initial CFD model hydraulics, provided additional insight to refine the design prior to constructing the physical model. Then, thorough, regular coordination between the shaft hydraulic modeling team, physical modeling team, and the rest of the project team produced consistent updates that the structural shaft designers could rely upon. Ultimately, design schedule objectives were met without re-work.

DROP SHAFT HYDRAULICS

Baffle Drop Structure Background

Baffle (cascade) drop structures consist of a series of alternating shelves that dissipate energy at each drop and minimize the drop distance. The overall drop is divided into smaller individual drops to manage velocity, erosion, and air entrainment at the invert of the structure.

The earliest known installation of the baffle drop structure on a sewer in the United States was in Cleveland, Ohio, USA in 1914, as illustrated in Figure 5. The structure is also referenced in a contemporary text book at the time of the design (Metcalf & Eddy, 1914). As today, these early structures included circular shafts, semi-circular baffles or shelves, small openings under each baffle for air flow and pressure balancing, and man access on the "dry side" of the shaft. However, unlike current installations, access to the flow side was only through an iron door near the invert and didn't allow for cleaning or inspection of the baffle shelves.

The District's newer tunnel and interceptor systems have incorporated reconfigured baffle drop structures (Lyons, Odgaard, & Jain, 2007; Margevicius, Schreiber, Switalski, Lyons, Benton, Glovick, 2009). The modifications have included a

DESIGN SCHEDULE

It was determined early in the design planning that robust design tools would be needed to address the tunnel system's unique characteristics of CSO conveyance and storage over a wide range of flows and volumes. From the hydraulic design standpoint, both numerical methods using Computational Fluid Dynamics (CFD) and physical modeling at the Iowa Institute of Hydraulic Research (IIHR) were scoped into the project for drop structure design. These additional tools and steps also presented a project scheduling challenge in terms of how to marry the hydraulic design results to the structural design and production of contract drawings under a tight, consent decree driven schedule.

The key to this effort was to start with realistic durations for obtaining reliable hydraulic field data to support model development and flow calibration for the progression of the design. Due to the aggressive overall project deliverables schedule, it became apparent that the solution to the schedule challenge was to employ parallel, concurrent efforts between the hydraulic and structural teams, as shown in Figure 4. To avoid the risk of re-work during the

Table 2. DVT shaft summary

Shaft	Average Dry Weather Flow Input (MGD)	Wet Weather Design Flow* Input (MGD)	Drop Structure Type	Finished Inside Diameter†	Hydraulic Drop Distance	Depth Below Ground Surface
DVT-1	N/A	N/A	N/A	30 feet	N/A	52 feet
DVT-2	6.5	395	Baffle with dry weather drop pipe	50 feet	61 feet	98 feet
MLK-1	N/A	40	Helicoidal	6.5 feet	36 feet	56 feet
MLK-2	2.4	88	Baffle with dry weather drop pipe	18 feet	24 feet	54 feet
WCT-2	3.6	93	Plunge pool with dry weather drop pipe	20 feet	18 feet	49 feet
WCT-3	1.7	93	Baffle with dry weather drop pipe	18 feet	64 feet	82 feet

*Wet Weather Design Flow = 5-year, 6-hour storm event.
†Diameters for MLK-2, WCT-3, DVT-1 & DVT-2 are sized for constructability reasons related to tunneling.

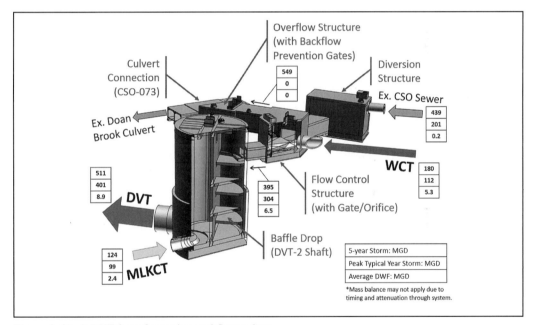

Figure 3. Shaft DVT-2 configuration and flow values

DVT Dropshaft Design Schedule																								
	Year																							
	2015												2016											
Project Design Element	Basis of Design												30%		60%		90%		100%					
	J	F	M	A	M	J	J	A	S	O	N	D	J	F	M	A	M	J	J	A	S	O	N	D
Overall Project Milestones									◆				◆					◆				◆◆		
Hydraulic Modeling/Calibration																								
CFD Modelling																								
Physical Modeling																								
Structural Shaft Design																								

Figure 4. Design schedule for drop shaft hydraulic and structural designs vs. overall project milestones

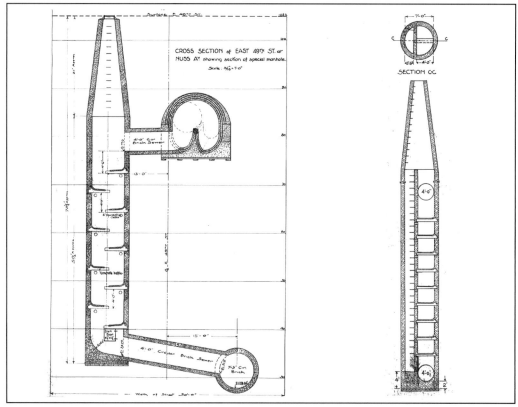

Figure 5. Drop structure to southerly interceptor at East 49th Street and Morgana Avenue, Cleveland, Ohio, circa 1914

reorientation of the "dry side" and "wet side" and an enlarged dry side for improved tunnel ventilation, personnel and equipment access, and energy dissipation and air release prior to tunnel entry. The configuration utilizes the wall openings for air flow at the intermediate steps and the dry side is utilized for additional deaeration and stabilization of flows prior to the connection to the tunnel. Many of the configurations have also used a restricted outlet or adit to further promote the release of entrained air. The additional free surface on the dry side also provides additional surge mitigation and ventilation from the tunnel or interceptor system.

The baffle drop provides other distinct advantages. Unlike many of the alternatives designed to manage significant wet weather flows, the baffle drop structure does not require any devices or structures upstream of the drop to pre-condition the flow. The inlet to a baffle drop consists of a standard sewer pipe without any modifications. Likewise, the baffle drop does not require any device or structure downstream such as a deaeration chamber. In addition, baffle drop structures have been implemented with

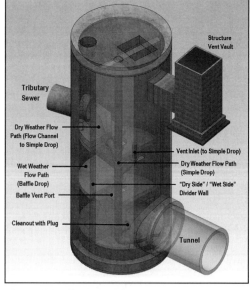

Figure 6. Baffle drop structure configuration with drop pipe (MLK-2)

two sewers connecting at different baffle elevations, which has limited the need for external manholes or junction chambers.

Additional evaluations, sizing, and physical modeling has been performed for other tunnel systems since the development of the initial revised configuration (Odgaard, Lyons, & Craig, 2013). The evaluations have considered variation of the vertical spacings of the baffle shelves, baffle widths, and various wall placements rather than the exact middle of the shaft.

Management of Wastewater Flows Within Baffle Drop Structure

All DVT drop structure locations will convey dry weather flows to the tunnel in addition to the wet weather flows except for MLK-1. To manage the range of flow rates, a combination of a simple vertical drop for the smaller dry weather flows and a baffle system for the larger wet weather flows was developed. The vertical drop pipe avoids sediment deposition on the baffle shelves and minimizes odor release within the drop structure during dry weather periods. The simple vertical drop will be extended beyond the baffle wall and separated from the cascading effect of the baffle drop side. The configuration is illustrated schematically in Figure 6 for the MLK-2 structure.

The inlet configuration has a significant influence on the capacity of the drop pipe. A side connection through the baffle wall was implemented to decrease the potential injury risk of personnel. Straight tee configurations and elbow inlet configurations were considered. Although straight tee configurations are more compact, the capacity is significantly diminished compared to elbow configurations (Rajaratnam, Mainali, & Hsung, 1997).

Drop Structure Screening and Evaluation

An evaluation was performed for each of the drop structure locations for the DVT system to determine the preferred configurations based on siting, tunnel access, hydraulic effectiveness, and anticipated performance. Due to the flow rates, simple drops were considered for the wastewater flows, while the drop structure configurations evaluated for wet weather conveyance were plunge pools, tangential inlet vortex drop structures with a deaeration chamber, plunge inlet drop structures with a deaeration chamber, baffle (cascade) drop structures, and helicoidal (helical ramp) drop structures with continuous guide vanes or with component vanes. The plunge inlet and helicoidal structures had not been previously implemented by the District, but had been implemented by others.

Sizes were developed for each of the drop structures for these configurations to determine siting and constructability requirements. The sizes were developed based on scaling design flow rates relative to previous designs, physical models, and empirical relationships.

Selection of Preferred Drop Structure Configurations

Various drop structure configurations were evaluated for each of the DVT drop structure locations. The District has previously implemented drill, simple, free-jet, plunge pool, tangential inlet vortex with deaeration chamber, and baffle (cascade) drop structures. Plunge inlet drop structures with deaeration chambers and helicoidal (helical) ramp drop structures were also considered for the DVT.

In addition to the primary design criteria of receiving the peak design flows, dissipating the flow energy associated with each drop, minimizing air entrainment to the tunnel, and providing necessary air release from the tunnel, additional criteria were considered including cost, performance, durability, constructability, easement / future land use, operation and maintenance, and access requirements.

Due to siting constraints and to minimize deep construction, the vortex and plunge inlet configurations were less desirable than the helicoidal and baffle configurations. The depths of drop are relatively shallow for the DVT system and the smaller shaft sizes for the vortex and plunge shafts with deaeration chambers were not enough to offset the additional appurtenances necessary for those drop types.

The screening of alternatives led to a preferred configuration of baffle drop structures at the upstream ends of the tunnel segments (MLK-2 and WCT-3) as well as the middle confluence of the tunnel segments (DVT-2). The selection of a baffle drop structure presented numerous advantages at these locations:

- These were tunneling locations with large shaft excavations already necessary for machine extraction (MLK-2 and WCT-3) or mining (DVT-2)
- Tunnel access can be provided through the structure
- The larger surface area was beneficial for surge dissipation
- Tunnel ventilation associated with tunnel filling and surge conditions can be managed with the shaft
- With the baffle drop structures at the upstream locations, deaeration of entrained air can be accomplished with the drop structure and the tunnel free surface without the need of a smaller adit to tunnel connection

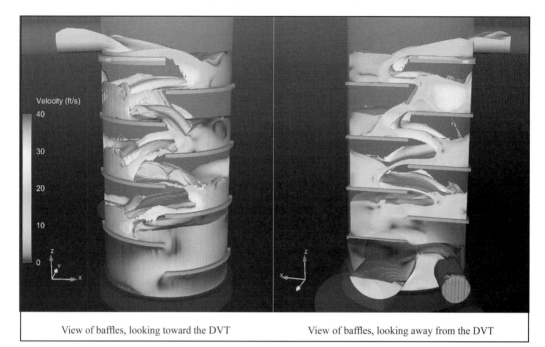

| View of baffles, looking toward the DVT | View of baffles, looking away from the DVT |

Figure 7. CFD modeling of baffle drop structure

- The DVT-2 Shaft is located where the MLKCT connects to the DVT, and the configuration provides a confluence and access to both tunnels without requiring an additional shaft
- Utilization of an internal drop pipe can accommodate the wastewater flows without debris accumulation or the need for an additional shaft

Alternative drop structure configurations were selected for the intermediate shaft locations where the benefits of the baffle drop structure were not as significant. A helicoidal drop structure was selected for the MLK-1 location for significantly reduced footprint benefits and a plunge pool structure with an internal drop pipe was selected for the WCT-2 location mainly due to constructability concerns.

Initial Design of Drop Structures

Initial sizing of the drop structure alternatives was developed for each location based on scaling design flow rates relative to previous designs, physical models, and empirical relationships. In the case of the baffle drop structures, the configurations were determined based on prior physical modeling performed for the District and additional modeling performed by the IIHR.

Computational Fluid Dynamics Modeling (DVT-2)

Computational Fluid Dynamics (CFD) modeling was utilized to better understand the hydraulic performance of the DVT-2 Shaft during the preliminary design.

The CFD analysis was performed using the Flow Science FLOW-3D software. The primary purpose of the analysis was verification of the baffle configuration and the confluence of the MLKCT and the DVT. The dry weather drop pipe was not included in the CFD model due to the small diameter relative to the size of the baffle shelves. Air entrainment was not included in the initial CFD model.

The CFD modeling reflected an initial configuration of DVT-2 consisting of a 40-foot-diameter shaft and a different orientation than the final configuration. After the initial CFD analysis, the overall configuration was changed at DVT-2 due to a reorientation of the WCT, which resulted in a reorientation of the divider wall. The shaft diameter was also increased to 50 feet to better accommodate the connection of the MLKCT and DVT segments.

The depths and velocities are illustrated in Figure 7 for the CFD model. The flows are very dynamic and rapidly-varied on each baffle shelf. The flow characteristics provided good agreement to the subsequent physical modeling.

The model showed that the original baffle outlet opening restricted discharge and caused heads to build up on the lower portion of the shaft. The shaft

Figure 8. Photograph of DVT-2 physical model

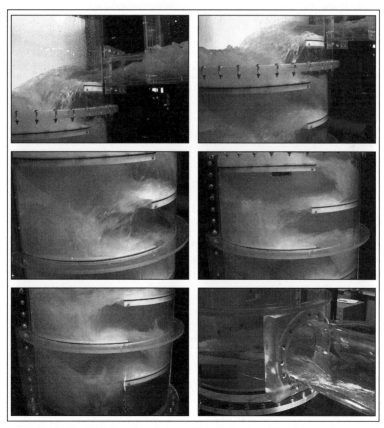

Figure 9. Wet weather design flow of 400 MGD from WCT-1 and 125 MGD from MLKCT, free discharge in DVT

Figure 10. Effect of upstream control gate (400 MGD flow, 2.25 ft weir height, partial gate closure)

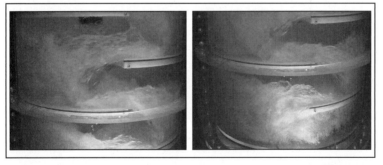

Figure 11. Baffle performance for maximum capacity test (712 MGD)

size also resulted in a sharp turn of the MLKCT flows. A larger opening and larger shaft size were incorporated into subsequent design.

Physical Modeling (DVT-2)

A physical model was constructed for the DVT project by the University of Iowa IIHR – Hydroscience & Engineering Laboratory for the refined DVT-2 Shaft (Craig & Lyons, 2016). The physical model was used to confirm energy dissipation and air entrainment performance, to identify hydraulic performance deficiencies, to assess the performance of the dry weather drop pipe and the diversion to the drop pipe,

the influence of the upstream inflow control gate and corresponding high hydraulic grade line (HGL) on the baffle performance, the effect of downstream tunnel tailwater on the baffle drop performance, and the hydraulic performance of the confluence of the MLKCT and the DVT.

The model was a 1:16 scale model and is shown in Figure 8. The scaling was performed for Froude similitude, which represents the ratio of inertial to gravitational forces and represents the dominant parameter in free-surface flows.

The performance of the structure to manage dry and wet weather flows was verified and validated.

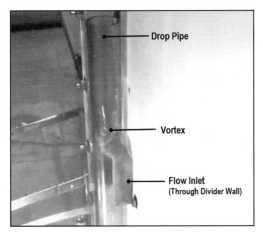

Figure 12. Vortex formation in drop pipe during large influent flows (ventilation was subsequently provided below inlet)

Figure 13. Air entrainment at drop pipe outlet (outlet was subsequently removed)

The wet weather conveyance was satisfactory and no changes were recommended to the baffle configurations including the width and the vertical spacing. Examples of the design flow are shown in Figure 9 for an open upstream control gate and Figure 10 for a partially closed upstream control gate, which reflects the normal operating condition to allow relief.

The structure was found to have a "stress test" capacity of approximately 80% more flow than the design flow if the flow depths were increased to cause splashing through the vent ports in the divider wall approximately 50% of the time. The performance for this condition is shown in Figure 11. Additional flow could potentially lead to increased splashing and intermittent blockages of the vent ports resulting in undesirable pressure fluctuations within the structure.

The model showed that there was a tendency for sub-atmospheric pressure and vortex development in the dry weather flow drop pipe when the depth was increased on the baffle during wet weather flow conditions. Due to the higher head forcing more flow into the drop pipe, pressurization occurred at the change in direction in the vertical pipe. The pressurization caused sub-atmospheric pressures to develop in the vertical pipe below the pressurization. The lower pressures were relieved when an intermittent vortex, shown in Figure 12, developed in the drop pipe, which caused significant noise and intermittent release of air at the invert of the shaft. The condition was addressed with ventilation of the drop pipe below the influent connection.

The outlet of the drop pipe was also examined. A larger outlet was initially considered, as shown in Figure 13, but the benefits of the larger outlet were negligible and it was removed to allow more distance and time for bubble rise prior to entering the DVT.

There was a tendency for sediment deposits between the MLKCT and DVT and this was addressed with a flow bench in the final design.

Application of Physical Model Findings to Overall Project Design

In addition to the refinement of the DVT-2 design due to the CFD and physical modeling, the results were used to develop the design of the other two baffle drop structures located at MLK-2 and WCT-3. Due to the smaller wet weather design flows at these locations, the baffle drop structures consisted of 9.75 and 10-feet wide baffles, respectively, located in shafts with an I.D. of 18 feet. Dry weather drop pipes were also incorporated into the shaft at these locations.

The baffle spacing at WCT-3 was utilized to connect two different sewers on alternating baffle shelves. The direct connections to the shaft eliminated a large external junction chamber.

Shaft Structural Design

The structural design of the baffle drop structures, like all structures, required good communication between all design disciplines (hydraulics, civil, geotechnical, and structural) during preliminary and final design stages. It is rarely possible to complete all hydraulic studies in advance of the structural

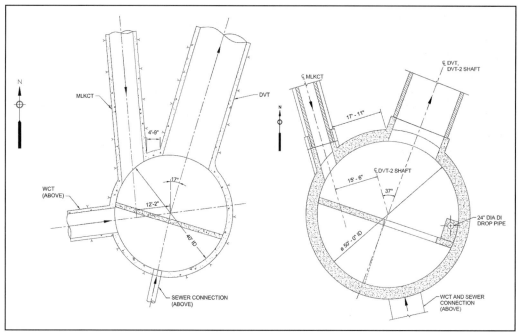

Figure 14. Comparison of the DVT-2 shaft configuration early in preliminary design (left) and final design (right)

design work, and in the case of the DVT drop structures this was amplified by the additional physical modeling requirements and some iterations of the drop shaft design were expected. A primary objective during preliminary design of the shafts was to develop the shaft sizes (diameter and depth) and general configuration of their various components based on conservative estimations of hydraulics, influent sewer and tunnel alignments, and site constraints. The shaft configurations were optimized further during final design and after all physical modeling was complete, but significant revisions were avoided to prevent delays to structural design completion. The hydraulic and structural designers communicated to ensure that each other's assumptions were known and each was aware of how the other's design may affect the final design, e.g., uncertainties about the spacing and width of the baffle slabs which influences the loading on the slab.

Baffle drops have stricter vertical and horizontal layout constraints than some other drop structures. This means that modification to the layout of near surface structures or tunnel and consolidation sewer alignments can lead to significant reconfiguration of the drop structure components. Significant reconfiguration of the drop structure components was avoided in spite of necessary changes to the structures that connect to the DVT-2 Shaft as depicted in Figure 3. The initial alignment of the WCT was rotated

approximately 90 degrees, requiring revisions to the divider wall, dry weather drop pipe, and stagger of the baffle slabs. Other revisions included increasing the diameter to 50 feet and slight alignment changes to the MLKCT and DVT, which improved the efficiency of turning the MLKCT flows within the shaft and improved the stability of the rock pillar between the two tunnels. The evolution of the DVT-2 shaft internal configuration is depicted in Figure 14.

The structural design of the shafts for the DVT project utilized various design worksheets and finite element modeling to evaluate demands on the shaft linings, base slabs, divider wall, baffle slabs, various openings in the lining for connection to sewers and tunnels, and the roof cover. Many of these components are common to other shaft types, but the divider wall, baffles slabs, and dry weather drop pipes have some special considerations.

The divider wall and baffle slabs are pin-connected to the shaft lining and must be significantly thick to resist static and dynamic fluid loads associated with cascading wet weather flows. As a result, they create stress concentrations in the shaft lining near their connections that must be carefully evaluated and reinforced. Hydraulic loads on the baffle slabs and divider wall were calculated using analytical approaches considering conservation of momentum, and insights from CFD and physical modeling were also used to refine these loads. When detailing

Figure 15. Example of shaft lining and divider wall form work from the NEORSD Dugway Storage Tunnel project

the reinforcement in the baffle slabs, consideration was given to whether pinned or full moment connections between the slabs and the divider wall and shaft lining are necessary. A pinned connection is favorable for constructability and was assumed for DVT shafts.

High quality concrete is necessary to achieve acceptable design life of the baffle slabs, divider wall and flow channels. The flow and impact of water on these components can lead to pitting and erosion of concrete surfaces. Cement replacement, such as fly ash, ground granulated blast-furnace slag, or silica fume, was specified to improve the durability of the baffle slabs, which are subject to the flow impact. Also, use of steel fiber reinforcing was required to help improve crack control in areas with turbulent flows. Increased pre-production testing was also specified to confirm that a dense, low permeability concrete will be delivered. Some consideration was given to the practice and capability of local ready mix concrete producers when specifying concrete mixes of exceptionally high strength. The dry weather drop pipes for the DVT project will be ductile iron encased in concrete to reduce long term durability concerns.

Construction Considerations

Subsurface conditions in the DVT project area generally consist of relatively shallow shale bedrock formations overlain by fill and alluvium soil types. Shafts will initially be supported during construction with liner plate and steel ribs within the soil and weathered rock. Rock dowels and shotcrete will be required below the weathered rock elevation. Only the MLK-2 Shaft site is anticipated to require an active dewatering system for lowering the groundwater table within granular alluvium soil types to permit construction.

The sequence of constructing the shafts involves excavating and supporting exposed soil and rock, excavating the tunnels, and constructing the base slabs, linings, dividing walls and baffle slabs. An example of CIP concrete construction for a baffle shaft from another NEORSD project is shown in Figure 15. The DVT shaft design detailing does not require monolithic pours between the shaft lining, divider walls and baffle slabs. Although the shafts will be constructed primarily of cast-in-place concrete, precast concrete elements were incorporated in the roof covers, including beams and removable planks, for accommodating larger equipment to be used for cleaning and maintaining the DVT-2 Shaft.

Tunnel contractors often excavate tail tunnels to provide additional staging area. It can be difficult to accommodate tail tunnels into the permanent baffle shaft structure, even though the prospect of gaining additional volume for a CSO storage project is attractive, because it would generally be located on the wet side of the shaft and would interfere with the baffle slabs or possibly the divider wall. The DVT Project requires that the tail tunnel (if used) be backfilled prior to the construction of the shaft lining.

Long-Term Operation and Maintenance

Long-term operation and maintenance (O&M) for the DVT project were considered since the project's inception and carried as a design consideration from planning through the final design phase. Due to the project's unique function as both a deep interceptor sewer and a storage facility, there were specific design impacts and considerations, as outlined in Figure 16.

Due to the dual-purpose operation of the tunnels providing conveyance and CSO storage, the system will provide sufficient self-cleansing velocities for debris. Therefore, surface structures will not

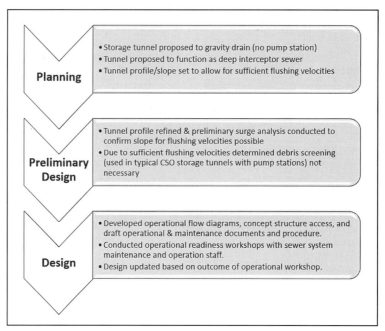

Figure 16. O&M design considerations process

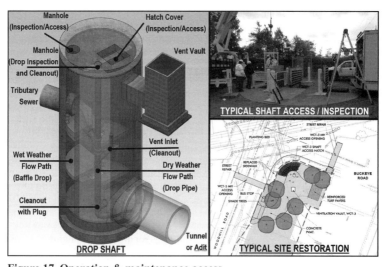

Figure 17. Operation & maintenance access

be equipped with screening typically provided for CSO storage tunnels dewatered through pump stations. However, since typical combined sewage will be able to enter the tunnel system, considerations for frequent preventative maintenance at the structures were discussed with the District's Sewer System Maintenance and Operation Division to develop final surface restoration, structure / shaft access, and specifically cleanouts for the dry weather drop pipes

incorporated in the baffle drop structures, as shown in Figure 17.

CONCLUSION

The development of drop structure designs on a CSO project are usually very site-specific and unique, and the scope of that design effort needs to be considered and properly planned for early in the project timeline to avoid rework. In the case of DVT, the combination of a widely variable rate of DWF and WWF to be dropped to the tunnel level as well as the physical site constraints in an urban environment led to an even more unique drop structure solution than usual. In addition to some of the typical baffle drop structure benefits including: permanent tunnel access, reduced number of drop system components compared to other systems, acceptance of a large flow rates and multiple inputs at different elevations, as well as air management, the selection of a baffle drop structure with integral dry weather simple drop pipe on the DVT project allowed for other additional benefits:

- Consolidating numerous important features into one footprint at the DVT-2 shaft, including: A junction structure for two tunnels (DVT & MLKCT); dropping an extremely wide range of flows (7.1 to 395 MGD); and providing a major permanent tunnel access point without requiring an additional shaft
- Consolidation of dry weather flows with an internal drop pipe without debris accumulation or the need for an additional shaft
- When baffle drop structures are aligned on the main tunnels at upstream locations, deaeration of entrained air was accomplished with the drop structure and utilized the tunnel free surface without the need of a smaller adit to tunnel connection

To realize these benefits, due to its uncommon use, additional design effort in the form of numerical (CFD) modeling and physical modeling was employed. The detailed structural design of the shaft had to be closely monitored against these other parallel efforts but was accomplished with the use of historical precedence at the outset and close coordination of the hydraulic and structural design teams.

REFERENCES

Metcalf, L., Eddy, H. (1914). American Sewerage Practice, Vol. 1: Design of Sewers, 1st edition (New York: McGraw-Hill), p. 544.

Lyons, T. C., Odgaard, A. J., and Jain, S. C. (2007). "Hydraulic Model Study for the Northeast Ohio Regional Sewer District Euclid Creek Tunnel/ Baffle Drop Structure." IIHR LDR Rep. 352, Univ. of Iowa, Iowa City, IA.

Margevicius, A., Schreiber, A., Switalski, R., Lyons, T., Benton, S., Glovick, S. (2009). "A Baffling Solution to a Complex Problem Involving Sewage Drop Structures," 33rd IAHR Congress: Water Engineering for a Sustainable Environment, p. 2,708–2,715.

Odgaard, A. J., Lyons, T., Craig, A. (2013). "Baffle-Drop Structure Design Relationships." J. Hydraul. Eng., 139(9), 995–1,002.

Rajaratnam, N., Mainali, A., Hsung, C. Y., (1997). "Observations of Flow in Vertical Dropshafts in Urban Drainage Systems" ASCE Journal of Environmental Engineering, p. 486–491.

Craig, A., Lyons, T. (2016). "Baffle Drop Structure Design and Hydraulic Model Studies for the Northeast Ohio Regional Sewer District's Doan Valley Tunnel." IIHR LDR Rep. 410, Univ. of Iowa, Iowa City, IA.

Challenges of Design and Construction of Slurry Walls in a Congested Site

Ravi Jain
Parsons Brinckerhoff

Mina Shinouda
Jay-Dee Contractors

Angelo Colasante
Treviicos

ABSTRACT: The First Street Tunnel (FST) is part of DC Water's Clean Rivers Project constructed under a residential area in Washington DC. As a design-Build project, The FST project encompassed rapid design progression and fast-track delivery. Reinforced concrete slurry walls were utilized as the support of excavation for the TBM launching shaft. As the design evolved towards construction stage, additional geotechnical borings revealed differences from provided baseline data. Coupled with permitting considerations, the design needed to be optimized to maintain structural integrity without compromising economy and efficiency. Furthermore, the execution of slurry walls having a depth of 175-ft required a controlled excavation plan which included sophisticated excavation equipment, highly trained personnel and a quality control plan that aimed to ensure the shaft wall was installed to tight tolerances.

INTRODUCTION

First Steel Tunnel, constructed in the Bloomingdale district of Washington, DC, is part of the District of Columbia Water and Sewer Authority's (DC Water) long term combined sewer overflow control plan. The project scope included a precast segmental tunnel, near surface structures, four shafts and three adits connecting them to the main tunnel. The project scope was accelerated to alleviate flooding that has been historically prevalent in the District's historic and densely populated Bloomingdale neighborhood. The focal point of the project and primary construction staging area was located at the intersection of Channing Street and First Street Northwest where a slurry wall support of excavation, permanent base slab and shaft lining, internal structures and permanent cover were constructed in addition to the launching of the Tunnel Boring Machine (TBM). The dimensions of the Channing Street CSA were 300-ft by 350-ft.

The Channing Street Mining Shaft slurry wall was to be constructed within the footprint of McMillan Sand Filtration cells 25 and 26, which were to be demolished as part of the project scope. Figure 1 is taken inside the sand filtration cells prior to demolition. Adjacent cells are similar. Prior to mobilizing within the construction staging area

Figure 1. McMillan sand filtration plant

(CSA), the historic structure had to be tactfully taken apart. The exterior façade—walls and portal entrances around the perimeter of the CSA—were to remain intact and protected during all construction activities. Protection measures on site included earth walls and tension tie rods. Figure 2 is an overview of the site pre-demolition while Figure 3 provides a post-demolition overview.

Once the historical structural system, load paths and distributions of the Sand Filter were realized, analyzed and protection measures designed, the demolition commenced. A combination of the

Figure 2. McMillan Sand Filtration Plant – predemolition (looking north)—Google Maps ©

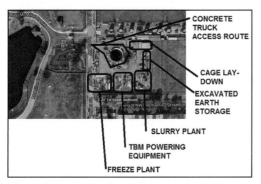

Figure 3. McMillan Sand Filtration Plant—post demolition and slurry wall construction (looking north)—Google Maps ©

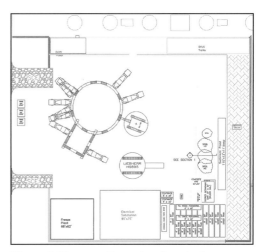

Figure 4. Slurry wall construction staging area

implemented protection measures and ultimate boundaries of the site left a constricted work area.

The Channing Street Mining Shaft (CS-MS) outside diameter of the slurry wall system was approximately 81'-6". This was controlled, primarily, by the required finished inside diameter of the final lining, which was 65'-0". The additional 8'-3" on radius accounted for slurry wall thickness (3'-6"), final lining thickness (4'-0") and slurry wall construction tolerance (9"). The project-specific Mandatory Requirements disallowed the use of a composite linin. The depth of the slurry wall panels was approximately 174-feet. The shaft had a 10-ft toe depth below the bottom of the permanent base slab. The construction tolerance stemmed from contract documents, which required a tolerance of 0.4% relative to depth. The culmination of all dimensions created a large footprint within the CSA and limited the working areas (Figure 4).

SUPPORT OF EXCAVATION SELECTION

Ground Freezing was utilized as the primary support of excavation method (SOE) for most subsurface construction on this project. Figure 5 shows the ground freeze circuit associated with the project. The freeze trench was taken through the back alley of houses to minimize disruption on First Street. The trench was a closed system with supply and return pipes. However, slurry wall shaft system was selected for the CS-MS for two primary reasons—schedule and community impacts, which could include settlements, noise or vibration. Due to the tight schedule, depth of SOE required and critical path nature of the shaft construction readiness, which had a direct correlation to the commencement of mining operations for the tunnel, slurry wall facilitated a quick-start and reduced risk to construction versus the alternatives. Additionally, excavating in slurry imposes little to no vibration issues to surrounding homes and dampens associated noise.

The shaft needed to be constructed in an accelerated manner prompting the need to factor durations of all associated activities when considering alternative construction methods. Options consisted of slurry wall construction, frozen ground and sunken caisson. Ground freeze was viable since it was already utilized at other sites on this contract. Furthermore, the freeze plant itself was located within the Channing Street construction staging area. However, mobilizing the equipment and achieving sufficient ground strength to the required depth would have delayed the project. Ground freeze requires toeing into rock for groundwater cutoff. The slurry wall constructed did not extend into rock. However, a ten-foot toe inconjunction with an external deep well with partial dewatering provided adequate invert stability. Use of a caisson was discarded due to the high risk it presented in this geologic stratum. The caisson, whether cast-in-place or precast concrete, could have had difficulty penetrating the 100-ft thick upper layers of stiff cohesive soils. These considerations, coupled with the site availability and schedule, led to the

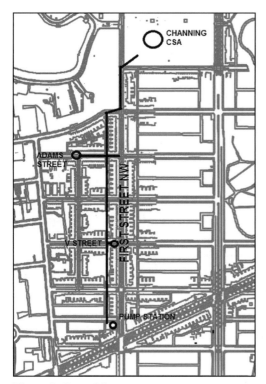

Figure 5. Ground freeze system

selection of a slurry wall with temporary external partial dewatering.

GEOLOGICAL CONDITIONS

The geologic conditions under the McMillan Sand Filtration Site consisted of recent fill, Quaternary alluvium, undifferentiated Cretaceous Patapsco/ Arundel Formation of the Potomac Group (P/A), Cretaceous Patuxent Formation of the Potomac Group (PTX) and bedrock. Fill deposits range from fine-grained to course-grained material containing fragments of wood, metal, concrete and other debris. The alluvium consists of interlayered soft clay, variably dense silt and sand and organic material. Gravel deposits and boulders are present beneath fine-grained material and at the boundary with the underlying Potomac Group soils. The Potomac soils encountered are the previously mentioned P/A and PTX. Table 1 lists the soil groups encountered in this project. Groups G1 and G2 relate to P/A while groups G3, G4 and G5 relate to PTX. G5 was not encountered within the conducted borings. Table 1 presents a descriptive summary of the Cretaceous groups below. The upper 15 feet consisted of fill, followed by a 40-ft band of alluvial material. As subsequently explained, these conditions made polymer slurry ideal over bentonite.

Table 1. Soil groups

Soil Group	Description		Stratum Depth
G1	Highly plastic and fine grained		20 feet
G2	Plastic (lower than G1), fine grained and typically a transition between G1 and coarser-grained soils of G3.		
G3	G3A	Nonplastic silty or clayey sand or mixtures of sand, silt and low-plasticity clay.	95 feet
	G3B	Nonplastic silty or clayey gravel or mixtures of gravel, silt and low-plasticity clay.	
G4	Fine to coarse sand with trace amounts of gravel and fines.		

Initial geotechnical parameters utilized for shaft design were based on borings (designated NEBBE in Figure 6) outside of the CSA as provided by the Request for Proposal (RFP). The contract documents required additional borings to be taken at the Adams Street construction site, roughly 1,000-ft from the Channing Street CSA. However, due to the limited information where the slurry wall was to be constructed, it was recommended by the design-team to shift one of the borings to the Channing Street site. Figure 5 shows locations of existing borings used to develop preliminary design and construction criteria. Also shown are additional borings (designated PB) conducted for final design and determination of parameters. Differences between the existing borings and additional borings included the location of the groundwater table and thickness of Potomac PTX layer. There was a direct effect on design as the design relied on compressive hoop forces and a lower lateral earth pressure coefficient would reduce the axial force. Construction would have also been affected due to soil behavior relative to settlements and movement when dewatering commenced. Due to the small CSA, various construction equipment,

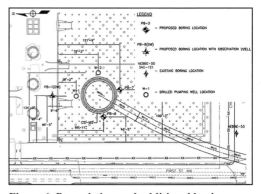

Figure 6. Pre-existing and additional borings

the ground freeze plant, slurry operations and remnants of the McMillan Sand Filter walls, which required space-consuming protection, minimizing settlements and movements was essential. Even the smallest of movement would crack the aged concrete of the historic structure. Strengthening works included placing backfill against the walls to counter lateral movement caused by the arch-geometry of the cells. Concrete bulkheads were constructed within the entrances on the North Sand Filter Wall to constrain the backfilled soil. After demolition of the roof arch, the North Sand Filter Wall behaved as a gravity retaining wall. A berm was placed along the wall, ahead of demolition. The lateral pressure exerted by the berm increased the resistance to overturning, resulting in a capacity/demand ratio of over 2.0. Resistance to sliding was provided by the base slab of the structure, into which the retaining wall was keyed. After demolition of the roof structure, the South and West Sand Filter Walls would not be able to resist the external horizontal soil pressures without the addition of compacted soil berms placed against the perimeter walls. Berms were placed prior to demolition of the adjacent section of structure. After demolition of the roof, the East Sand Filter Wall would not be able to resist the external vertical soil pressure and vehicle live load without the addition of a compacted soil berm placed against the wall and the subsequent installation of 2 tension rods spaced at every 14 feet. The berms were placed prior to adjacent demolition work. The East Wall berm was removed after the installation of the tension rods.

ANALYSIS AND DESIGN

The slurry wall was analyzed using PLAXIS 3D finite element analysis software (Figure 7). PLAXIS accounts for soil-structure interaction and directly considers the mechanical properties of soil materials

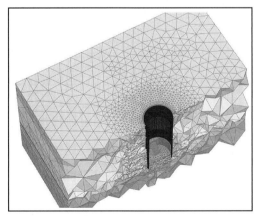

Figure 7. PLAXIS 3D

based on subsurface investigations and testing. Further, a staged construction methodology was utilized within PLAXIS 3D to depict the actual conditions. The governing design condition of a fully excavated shaft relied on partial dewatering until a permanent final liner was constructed. This required a careful balance as lowering the groundwater table limited the compressive hoop forces carried by the shaft, which may increase reinforcement requirements. However, constructing sufficient structure to counter buoyancy was also required at each stage. The groundwater level was at approximately EL 145.00' while the bottom of the slurry wall was at EL −13.00' and the bottom of the slab at EL −3.00'. Dewatering to EL 40.00' was performed until the permanent shaft base slab and lining were fully constructed.

The shaft was analyzed as a "round" structure, but not designed as one. To account for panel offsets and associated eccentric forces, the verticality requirement was multiplied by the depth to the top of the permanent base slab. This eccentricity was then multiplied by the hoop force at the designed elevations to obtain an additional horizontal moment, added to the analyzed moment. For the slurry wall panels, it is critical to keep the required reinforcement to a minimum, utilizing larger diameter bars at maximum allowable spacing, by design. The criteria was utilized to enhance the cage stability and achieve optimal weight for picking and placement. Due to the large compression forces combined with discontinuous reinforcement around the circumference, there was a concern of potential buckling of the reinforcement. To alleviate this risk, the cages were assembled using ties with alternating 135-degree hooks. The circumferential hoop force was correlated with the shear friction capacity provided at panel joints to account for joint closure with increasing depth. A capping beam was utilized near the top of the shaft to provide panel continuity where lateral forces are low. The permanent shaft lining, tunnel opening and collar beam were analyzed, three-dimensionally, separately using structural finite element analysis software SAP2000.

Due to the various site activities ongoing at the Channing Street CSA, an unbalanced "squeeze" effect surcharge was accounted for in the slurry wall design. This consisted of a conservative 500 psf for the first 50-ft of depth and 250 psf for the subsequent 50-ft of depth. This was included in attempt to ensure all concurrent activities are accounted for as opposed to only the surcharge induced by the slurry wall equipment. Typically, a hydromill or clamshell and a picking crane for reinforcement. The reinforcement requirements were subsequently applied around the entire perimeter for that specific depth to minimize impacts due to any unforeseen activities.

The congested CSA coupled with the adjacent McMillan Reservoir posed concerns relative to dewatering. Among the challenges, the dewatering wells further limited the working zone in and around the shaft (Figure 8). To obtain a dewatering permit, a zone of influence and extent of drawdown was developed. This was done to determine if the adjacent reservoir was at risk from drawdown. As the slurry wall construction and subsequent excavation was on the critical path, a contingency plan was developed in case a permit could not be obtained. All contingency measures needed to account for the limited area within the CSA. Among the alternatives considered was excavating on the outside of the shaft and constructing a collar providing the required weight to counter buoyancy effects. However, this was quickly dismissed due to the width and depth required within such a confined area. Additional alternatives consisted of casting a base slab in the wet, tremie slab, and using dead weight until sufficient lining has been cast. This required detailed analyses beyond the already released for construction documents. The slurry wall was analyzed for higher than anticipated lateral forces and hoop loads. Increased hoop loads would require greater joint contact and ultimately, enhanced quality assurance and control.

Additional analysis was performed on the CSA to ensure construction impacts were minimal, not just relative to slurry wall construction, but all adjacent work within the area. This included assessing impacts due to shaft construction and excavation, tunneling and dewatering on McMillan Sand Filter walls, South Clearwell Basin (in close proximity to the west of the CSA), waterlines, sewers and structures within the zone of influence (along First Street NW). This was required towards obtaining a dewatering permit from the governing agency. Since the CSA was congested, unconventional contingency plans were developed documenting the analyses.

POLYMER VERSUS BENTONITE

Upon construction notice to proceed, the slurry wall subcontractor, TREVIICOS, turned their attention to what they believed would be the two most critical factors for a successful slurry wall project:

1. How may the verticality tolerances be achieved and verified.
2. What excavation fluid or "slurry" should be utilized on the project with the aforementioned geology.

These factors are interrelated. Decision on the excavation fluid came late, just before construction began and was first met with some concern. Analysis of the borings showed that cohesive soils found in this region of the Potomac are primarily highly

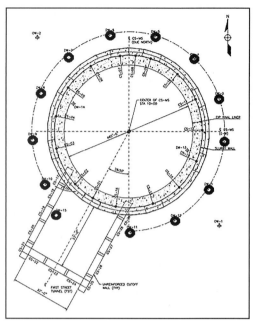

Figure 8. Dewatering wells layout

reactive or "swelling" clays with sand and gravel seams. The clays in the G1 and G2 stratum are very sticky clays with a high swelling capacity. In the G3 strata the clays are less sticky but have the ability to disintegrate into fines that will incorporate themselves into a fluid system very easily. It is known that these soils will lead to Hydromill problems causing the cutting wheel to gum up, high mud densities and excessive disposal volumes. However, given the verticality tolerance required on this shaft, a Hydromill was the only way excavation could occur with adequate monitoring.

Bentonite slurry was an option considered. It is traditionally used in slurry wall construction, is inexpensive and is easily supplied from a variety of mines in the United States. A notion against using bentonite slurry was that when used in these soils, with a Hydromill, it could have a reuse factor of less than two which would require excessive storage capacity to allow for treatment and reuse or would need to be disposed in high quantities. This would drive costs up and potentially affect schedule on which this activity was on the critical path.

Polymer is a well-utilized fluid in the deep foundation industry. However, it is not widely used in slurry wall construction and traditional polymers are rarely used in Hydromill excavation. Additionally, polymer is not easily procured. The polymer TREVIICOS considered is manufactured in Europe. It is pound for pound, ten times the cost of bentonite and had the potential to be very difficult to dispose. The one advantage polymer did have was its lack of blending with the reactive clays and its ability to be separated from the fines found in the Potomac soils through chemical treatment. With this, the fluid could easily be reused multiple times cutting down on disposal, storage capacity and ultimately cost and schedule. Furthermore, polymer retains its filtrate abilities while maintaining both low unit weight and viscosity.

After careful consideration, it was determined that a polymer-based system would be the best solution despite the lack of a prior record of use on DC Water projects. Ultimately, using polymer was extremely successful relative to the cost of supply and disposal, below anticipated levels. As observed during excavation, the shaft wall was of excellent quality with little to no leakage and no over-pours or "bulges."

The slurry mix consisted of a stable suspension of powdered natural bentonite in potable water. Slurry testing was performed twice per eight-hour shift in each panel during excavation, prior to concrete placement, after rainfall and as requested by the Construction Manager. Slurry had to be maintained constant through the excavation and samples were taken from the top and bottom of a panel. The specific quality requirements that had to be met included viscosity, specific gravity, sand content, pH and a filter stress test. Slurry levels were maintained 5-ft above the groundwater table.

SLURRY WALL CONSTRUCTION

Slurry wall construction allows for the installation of a complete foundation or wall system through individual panel construction. The layout of panels is typically driven by structural constraints, equipment availability and other mitigating factors. Also, joints between panels play an important role in panel layout and must be considered as they create a watertight bond. The panel joints are especially critical at increased depths since they become structural connections as panel end bearing is required to transfer highly compressive hoop forces. The Channing Street Mining Shaft (CS-MS) panel layout was primarily driven by the size of the Hydromill. The Hydromill head used on this project had a 9.17-ft (2800 mm) cutting length. Therefore, the circular shaft was broken down into thirty-two 9.17-ft bites. Subsequently, the bites were combined into 16 panels, 8 primary and 8 secondary (closing panels), alternating around the circumference of the shaft. The primary panels were made up of 3 bites and the secondary panels comprised of a single 9.17-ft bites.

The panel installation sequence was driven by layout and the logistics of maneuvering the various equipment around the perimeter of the shaft. The Hydromill (Figure 9) works similar to a vertically suspended Tunnel Boring Machine (TBM). Excavation occurs at the drilling head and proceeds downward as cuttings suspended in the excavation fluid are removed using a hydraulic pump through the cutting head to a separation unit. The cuttings on the end of the Hydromill, like a TBM, are determined by the soil types anticipated to be encountered during excavation. However, at the Channing Street shaft there were not only soils to consider, but also the concrete overbites when constructing closing panels.

Panel joints between adjacent panels were created by over-excavating the width of the primary panels and casting them in a concrete profile wider than required. The Hydromill would then "mill" through the excess concrete while excavating the secondary panels creating a watertight toothed joint – roughed surface bond. At CS-MS the required concrete strength was 6,000 psi which is much higher than a typical slurry wall. The use of this higher strength concrete mix presented a challenge that it will also have a higher early strength gain with the concern that the 7-day strength might reach design capacity. This meant that the Hydromill cutting tools needed the ability to excavate through the Potomac soils, but also the 6,000 psi concrete. A proposed solution was to fit the Hydromill with cutting tools capable of excavating soils as well as concrete. However, there was no one tool that could do both. The decision was made to install soil-cutting tools first and install all the primary panels. Once complete, the Hydromill would be re-tooled and fitted for concrete cutting, which are typically used in rock. However, given the relatively high strength of the concrete they were very effective in removing the concrete at the joints and although production was slightly reduced, the project schedule was met. This had the added benefit of all the secondary panels being single bite so TREVIICOS was able to reduce the equipment necessary, which led to less tremie

Figure 9. Hydromill

pipe and smaller clamshells on the project saving time on demobilization.

Site logistics were heavily considered both during the bid stage and pre-construction process which is typical for construction in urban environments since space is always an issue. Prior to contract award, a plan was developed to allow TREVIICOS to fit a slurry plant and three cage beds while maintaining access into and out of the site via two separate gates. However, site constrains changed significantly just prior to slurry wall mobilization. Due to a construction delay at another contract that is adjacent to the Channing Street Site, one of the two gates to the project site was eliminated. This required all trucking into and out of the site to go through one gate. This made placing concrete a challenge since it required 60 to 70 concrete truck deliveries to place each of the primary slurry wall panels.

Another major hurdle was the work required to support a part of the old filtration plant which was contractually required to be left in place. A concrete retaining wall had to be temporarily supported and repaired during the slurry wall construction phase. The temporary supporting system required impeded on the area allocated for the slurry plant. This required changing the location of the plant to the other side of the work zone. While the problem was solved, it significantly reduced the work zone and slurry storage, cage fabrication and spoil handling areas. Although this reduced work zone impacted the day to day logistics of the slurry wall operation the diligent efforts of the construction team did not allow it to delay the project schedule.

The CS-MS slurry wall was designed to act as a compression ring allowing for excavation inside the shaft without requiring internal bracing. The design called for a minimum contact area between the panel joints of 34" while maintaining a contractually required verticality tolerance of 0.4%. With the ability to monitor and adjust verticality in real time, only the Hydromill could realistically achieve this tolerance. This was a major concern at the start of the project, but through careful operator monitoring of the Hydromill excavation all the panels were cast within specified tolerance. As an additional quality control measure against the Hydromill instrumentation, a Koden was used to verify panel verticality prior to casting concrete. The Koden operates by lowering a cable suspended sonar device into the excavated panel. As the device is lowered, it emits a sonar signal that profiles the excavation. The output of the Koden is plotted in real time on a monitor located at the site. Once the process is completed, the plot was submitted to the design team for record and review. Subsequently, the panel was approved for casting concrete.

The reinforcement cage built for each panel was relatively light relative to their height. Weighing approximately twenty-five tons for the primary panels and eight tons for the secondary panels. All cages were built in two approximately ninety feet long sections. When panel excavation was completed and all slurry was cleaned, the process of installing the steel cage started. Once the bottom cage was placed, the top cage was picked and brought to the hole where it was spliced to the bottom section. The whole assembly was subsequently lowered to its final depth. Out of the sixteen cages placed, only one had an issue during setting. On one of the panels, the cage stopped sinking about twenty feet from the bottom. After several attempts to lower it into place, the cage ultimately reached final depth. The panel was placed without further complications, however during shaft excavation it was discovered that several of the reinforcing bars were displaced during the cage placement and exposed. Although it was unclear what caused the issue it was determined that the misaligned bars would not affect the stability of the shaft and excavation proceeded without delay.

CONSTRUCTION STAGING AREA AND ACCESS ROUTE

As indicated before, at the time of the slurry wall construction, another project was underway on First Street right outside of the project fence line. This project was under a different contract and was supposed to be completed before our work started. The construction work area of this adjacent project blocked one of the two jobsite gate locations. Having a single access point to the jobsite dictated that delivery trucks turnaround on the jobsite to exit from the same gate. This consumed a significant portion of the jobsite that would otherwise be utilized for other construction activities. Furthermore, both projects shared a single access road which complicated deliveries and required additional mitigation measures. While the impacts from the adjacent contract work was relatively mitigated through onsite coordination and owner support, the days with concrete placements for either contract were very challenging.

In addition to the Channing Street shaft, the project comprised of three drop shafts with adjacent near surface structures along the alignment of the tunnel. These additional shafts were located in much congested in residential areas which mandated minimizing the footprint of the work area around each shaft. These small work areas did not allow storing construction materials needed for the work, materials had to be delivered to the work area, as needed, for immediate use. Recognizing the risk of depending on commercial daily deliveries, a decision was made to utilize a portion of the Channing Street site as a staging area for these remote work zones. In addition to reducing staging areas within the site, this further complicated the logistics of the site traffic.

As indicated before, ground freezing was utilized as a support of excavation for the majority of excavation at the drop shafts locations. The constricted work areas for these shafts prohibited locating the freeze plants close to the jobsite. A central freeze plant was built in the Channing Street site and provided service the other three sites. With the long freeze pipe distances the brine had to be pumped, an enlarged freeze plant was needed. Furthermore, commercial power was not available and schedules for power drops from the local power companies indicated long lead times. A decision was made to generate our own power using a natural gas generation plant located at the Channing street site. These two plants consumed a significant area of the jobsite.

The slurry wall panels were excavated using two cranes. One equipped with hydraulic clamshell buckets for the TBM launch cutoff cell (Figure 8) and the other a Hydromill for the shaft. Additionally, a 200-ton crane was utilized to handle the steel reinforcing cages. Maneuvering the sizeable equipment within the small footprint of the site required diligent planning and tactful execution. The presence of these cranes together with the slurry plant, spoil disposal pit, power plant, and freeze plant rendered a very congested site and a challenging operation. An amalgamation of these factors coupled with others led to postponing some of the activities that were scheduled to be performed concurrent to the slurry wall construction, which essentially put the slurry wall on critical path. Nevertheless, the cooperation of the involved parties led to on-schedule operation and successful execution. Figure 10 shows the site during the slurry wall operation.

CONCLUSION

Construction of the Channing Street Mining Shaft slurry wall required close coordination between all involved parties during the design and planning phase. Furthermore, extra caution was exerted during the construction phase due to a wide range of concurrent activities. Unforeseen circumstances,

Figure 10. Channing Street site during slurry wall operation

such as government agency-required analyses for permit approval and delayed work by an adjacent contractor, required the development of unique contingency plans. While local dewatering is common, the adjacent McMillan Reservoir required extensive analysis to determine the zone of influence would not drain it. Communication was a key element towards successful completion of the slurry walls from the planning phase to demobilization.

The authors would like to acknowledge the owner, DC Water, owner's engineers, McMillan Jacobs and the design-build team of Skanska Jay-Dee and Parsons Brinckerhoff in addition to TREVIICOS, the slurry wall subcontractor.

Support of Excavation Structural Challenges and Steel Design for the Crenshaw/LAX Transit Project

Bradley Hoffman, Bingzhi Yang, and Paul Leduc
Aldea Services

ABSTRACT: The Crenshaw/LAX Transit Project is an 8.5-mile extension of the Los Angeles Metro Rail System. Three of the stations along the underground portion of the extension have been constructed using top-down construction methods. The stations were constructed below six lane major arterial streets through South Los Angeles and traffic had to be supported on a decking system above large open pits to maintain traffic flow. The size of the excavations and placement of pin piles within the station excavations created challenges and a need for innovative steel design to achieve the acceptable unbraced lengths (up to 118 feet full length, 60 feet unbraced length) of strut members and make installation feasible. This paper summarizes the solutions that were developed to overcome various design obstacles for the support of excavation systems.

INTRODUCTION

The Crenshaw/LAX Transit Project consists of approximately 8.5 miles of rail from the existing Metro Expo Line at Crenshaw and Exposition Boulevards to the Metro Green Line to service Los Angeles, Inglewood, and El Segundo. This paper discusses the temporary support of excavation structures used in the construction of three underground rail stations, Expo/Crenshaw Station, Martin Luther King Jr. Station, and Leimert Park Station. Expo/Crenshaw and Martin Luther King Jr. stations were constructed using cutter soil mixing (CSM) with steel soldier piles. Leimert Park Station was constructed using soldier piles and wood lagging. All stations were braced using a combination of steel spiral-weld pipe for struts and W-shape steel beams for bracing at corners. All three stations also included a traffic deck that supported traffic and construction loads and served as a strut system to resist earth pressures. Additional "pin piles" were driven in the center of the excavation to serve as columns for supporting the traffic deck and to provide a bracing point for the pipe struts to reduce slenderness. Figure 1 provides the general shape for each of the three stations.

DECK SYSTEM DESIGN

The Los Angeles County Metropolitan Transportation Authority provided a set of drawings detailing loading requirements and design criteria for construction structures. These criteria were based on a combination of ACI 318-11 for structural concrete, AISC 360 for steel using Allowable Stress Design (ASD), "Bridge Design Specifications" by Caltrans, 2004, AASHTO LRFD Bridge Design Specifications, and the Caltrans amendments to the AASHTO LRFD Bridge Design Specifications. In addition to earth pressure loads, the deck structure was required to support dead load comprised of 10-inch thick precast, prestressed concrete deck panels, beam self-weight, k-rail, and the weight of utilities which were suspended below the deck via hangers attached to the underside of the main deck beams. Live load consisted of 85 psf pedestrian loading in accordance with Caltrans BDS, 2004, construction loading, and HL-93 truck loading including impact loads as described in AASHTO, but without the presence factor which would allow for a reduction when considering multiple lanes loaded simultaneously. Longitudinal loads in accordance with Caltrans BDS

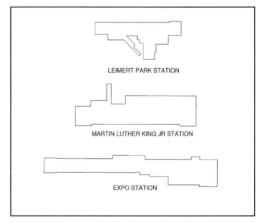

Figure 1. Outlines of the main station structures (NTS)

LEIMERT PARK STATION

MARTIN LUTHER KING JR STATION

EXPO STATION

470

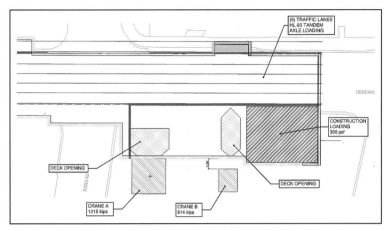

Figure 2. Example of design loads at the north end of Expo Station

2004, Article 3.9, wind loads and a simplified seismic loading were also required; however, these loads generally did not contribute to the governing load combinations for primary deck members. A wind pressure of 20 psf was applied in accordance with Los Angeles County Metropolitan Transportation Authority Construction Structures Loads and Design Criteria, Rev. 6 2013. Selection of the primary deck beam shapes was typically controlled by the truck live loads which were based on the tandem axle configuration, assuming a scenario in which HL-93 tandem axles were present in each lane and centered on any individual beam simultaneously (primary deck beams were oriented perpendicular to the direction of traffic). Tandem axle loads were modeled as two-point loads in each lane of traffic. The magnitude of individual point loads were calculated based on a 12 foot beam spacing, assuming that the tandem axle is centered above the beam. This resulted in mostly W40 sized steel beams. Plate girders were considered for the deck beams and may have been more efficient in terms of weight and serviceability; however, too much time and effort would have been required to fabricate such a large number of plate girders. A few sections were also designed to support construction loading and an occasional crane; however, the most significant crane lifts were performed with the crane on the ground just to the side of the deck. This impacted SOE design through ground surcharge loading, not through load applied directly to the deck.

Thermal loads were applied to the main deck support beams and the struts in accordance with Boone and Crawford (2000). This method accounts for the elastic properties of the soil to provide a more realistic determination of thermal strut loads. The loads were calculated using the following equation:

$$\frac{P_T}{\Delta T} = \frac{-\propto_s LsE_{s(m)}A_sE_s}{2IA_sE_s + LsE_{s(m)}}$$

- Es = Elastic modulus of steel = 29,000 ksi
- Es(m) = Secant elastic modulus of soil (mobilized), determined using Figure 13 of Boone and Crawford (2000)
- As = Cross sectional area of steel strut (deck beam)
- I = Shape factor per NAVFAC 7.01 = 1.0
- s = vertical distance between struts (influence area)
- L = Span of strut (deck beam)

Because of earth pressures being transferred to the deck from all sides, the secondary deck beams were required to transfer earth pressure loading in addition to serving as bracing against lateral torsional buckling of the deck beams. As a result, traditional X-bracing or K-down bracing could not be used between deck beams. Channels were considered first for the secondary bracing members, however the axial load resulting from earth pressures and seismic loads exceeded the capacity of the available channel sizes. Instead, W24 beams were bolted or welded to full depth stiffeners attached to the W40 main deck beams as can be seen in Figure 4. To ensure that these secondary beams provided sufficient torsional restraint, the system was checked using The Steel Bridge Design Handbook: Bracing System Design; Publication No. FHWA-IF-12-052—Vol. 13, published November 2012. This method accounts for the stiffness of individual members and their connection types and ensures that the braces provide proper stability to improve the lateral or torsional stiffness of the main deck beams.

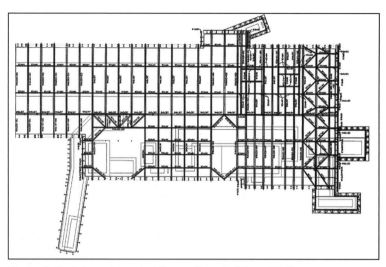

Figure 3. Deck beam layout at the north end of Expo Station

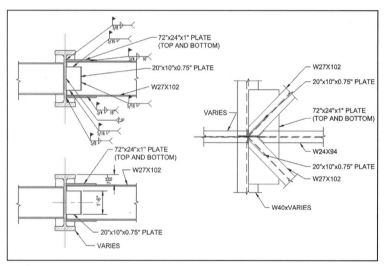

Figure 4. Example connection design at intersection of primary deck beams, secondary beams (W24), and diagonal beams (W27)

As can be seen in Figure 3, diagonal members were also used as part of the deck system to transfer earth loads from the end wall into the side walls on the east and west sides of the excavation. Diagonals were also necessary to transfer loads around deck openings. Two of these openings can be seen in Figure 3 as well. The larger opening was needed for removing muck during excavation and throughout the tunneling process. The smaller opening to the north was used for moving equipment and construction materials in and out of the excavation. The diagonal beams combine with the W24 secondary beams and the W40 deck beams to form horizontal truss systems in the deck. The orientation of the beams in these trusses created challenging connections, especially at nodes where both diagonal and secondary beams converged. To address this, W27 beams were selected for the diagonal members. This allowed for the use of "sandwich plate" connections, as the top and bottom plates would not interfere with the W24 beams at the same location. This configuration can be seen in Figure 4.

Additional challenges were encountered when it became necessary to change the elevation of a section of decking at Expo Station. The elevation change occurred along a header beam that was installed at

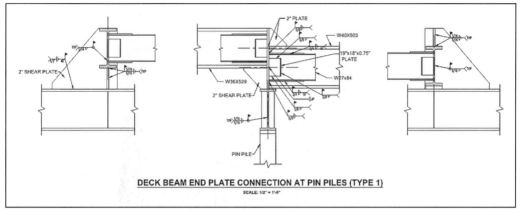

Figure 5. Deck beam connection at pin pile column header beam where elevation change was needed. Also includes the connections of diagonal beams.

Figure 6. Deck beam connections at pin pile column header beam where elevation change was needed

the top of the pin pile columns. The deck beams were originally designed to rest on the top flange of the header through a triangular shear plate that could transfer lateral forces from the deck beams into the header. The elevation change created eccentricity in the connection that produced bending moment in the deck beams that had to be transferred through the connection. This also necessitated the use of a much larger shear plate. One of the connection designs that resulted from this change is shown in Figures 5 and 6. Every member at the connection shown in these figures is subjected to a substantial amount of axial load.

DESIGN AND PRELOAD OF BRACED PIPE STRUTS

Large spiral-weld steel pipe sections were used for the primary struts throughout all three stations. The only exceptions were at corners and headwalls where W shapes were used since they were simpler to fabricate for more extreme angles. All pipe struts that were oriented at 90 degrees to the wall were preloaded to 25 percent of their design loads (per project

specifications) using pneumatic jacks. Angled struts were generally not preloaded for this project. The largest available pipes were 36 inches diameter with a 1-inch thickness. The distance from waler to waler on opposite sides of the excavations could be over 120 feet in certain cases, creating issues with slenderness of the pipe struts. At 120 feet, not only would the recommended slenderness limitations provided by AISC be exceeded, but the bending stresses produced by the self-weight of the pipe would limit the axial capacity of the struts. In order to solve this problem, the struts had to be braced at the center using the pin pile columns. This in itself produced additional design challenges.

There were a few cases, mainly at Leimert Park, where the struts could simply be placed next to the pin piles with W-beams framing around the pipes to brace them to the pin piles. In most instances, however, this was not an ideal solution for bracing struts. Deck openings for moving equipment and material often affected strut placement and certain strut configurations might require the installation of an additional strut in order to maintain a maximum spacing limit. Based on these limitations, it was determined that aligning the struts with the pin piles would be the most effective way to design the bracing system. This also had the added benefit of reducing the length of the pipe struts that had to be lowered into the excavation and welded together at the bottom of the pit.

In order to preload a strut that was braced against a pin pile column in this way, a unique connection design was created to protect the column and the traffic deck that supported it. Most struts that did not need center bracing were preloaded by jacking from one end of the strut and pushing directly against the waler. This was not an acceptable method for these braced struts since preloading from the end would

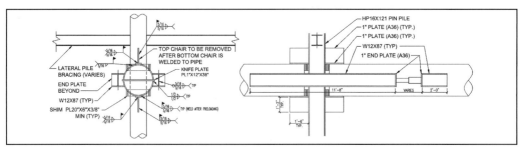

Figure 7. Sliding connection for strut jacking at center pin piles

cause deflection and bending in the pin pile column, as the jacks create a gap between the waler and strut on one side of the excavation only. This deflection is illustrated in Figure 7. The induced deflection in the pin pile would create bending stresses and eccentricity, thereby significantly reducing the axial capacity of the column. It was determined that in order to prevent this, the strut, separated into two pieces, would have to be jacked at the pin pile column as shown in Figure 8. Two long W-beams were weld to both sides of one strut segment and extended a distance past the end of the other strut segment to serve as a guide or sliding rail that prevented horizontal buckling of the system as the two strut segments were jacked away from each other. The strut segment that was not welded to these W-beams could slide freely between them during preloading. Knife plates were welded to the pin piles above and below the strut segments on each side that allowed the struts to move in their axial direction but not buckle vertically as they were preloaded. Once the target preload was reached, the pressure in the jacks was held while the struts were shimmed against the pin pile and the knife plates and sliding rail beams were welded to the struts. This locked the preload into the system so the jacks could be removed. This method was used at most of the braced struts, with a few exceptions where it was not needed, with no issues reported by the contractor. A picture of the connection in its final state is shown in Figure 9.

DESIGN OF STRUT TO WALER CONNECTION

Special attention was paid to the connections between walers and struts given the magnitude of the forces and the fact that the edges of the pipes could create concentrated forces on the flange of the waler causing unwanted deformation. This deformation would not only be detrimental to the strength of the waler, but would allow for the release of stresses in the pipe strut. This could lead to movement of the excavation wall and surface settlement outside the excavation. A 3-dimensional finite element model of the connection was developed to study where local

Figure 8. Sliding connection photo for strut jacking at center pin pile

deformations were likely to occur and how to prevent them. For this model, it was assumed that the strut was located equidistant from two adjacent piles. Walers were typically shimmed and welded to the piles which had been exposed by chipping away the soil-cement mix. The first model contained no modifications to either the pipe or the waler. Deformations were observed most significantly in the waler at the two points of contact between the waler and the strut and also in a bell-out effect in the pipe close to its end. To prevent local deformation of the waler, full depth stiffeners were welded to the web and flanges, centered at roughly the centroid of the contact area between each side of the pipe and the waler's flange. Horizontal and vertical knife plates were also added to the ends of most struts. This served two purposes: it allowed for better distribution of the strut load along the web of the waler, and prevented the bell-out effect in the strut. An additional stiffener was typically added to the waler at the center of the connection, aligned with the vertical knife plate. Samples from this modeling can be seen in Figure 9.

CONCLUSION AND LESSONS LEARNED

Throughout the design-build process, a variety of design challenges presented themselves, especially as modifications were made to the station layout, utilities were uncovered, and changes were made to accommodate construction tasks. There were an

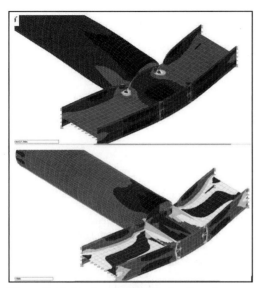

Figure 9. Finite element models developed for studying the deformation of strut-waler connections (deformation magnified ×80). No knife plates or stiffener plates (top). Vertical and horizontal knife plates added to pipe strut and three stiffeners welded to the waler (bottom).

extraordinary number of atypical connection designs that required a strong understanding of the forces involved, the numerous load paths, and the resulting eccentricities. Because design and construction were occurring simultaneously, it was necessary to design the support of excavation systems in such a way that modifications could be made, even after steel had

been delivered and piles were in the ground. For this project, the team of SOE engineers was prepared to make quick changes and produce creative but sound solutions for the tasks encountered.

REFERENCES

LACMTA—Metro Rail Structural Standard (MRSS), Cut and Cover Underground Structures Drawings, Revision 6, Dated October 14, 2013 (Revised Per SBCN 2013-39).

American Institute of Steel Construction (AISC), Steel Construction Manual—Fourteenth Edition, 2011.

American Welding Society (AWS) D1.1/ D1.1M:2006, Structural Welding Code—Steel.

American Concrete Institute (ACI)—ACI 318-11.

Caltrans Bridge Design Specifications (BDS), 2004.

American Association of State Highway and Transportation Officials (AASHTO) Load-and-Resistance Factor Design (LRFD), Bridge Design Specifications, 2012.

2013 California Building Code, Title 24, Part 2, Volume 2 of 2.

ASCE 37-02, Design Load on Structures During Construction.

ASCE 7-98, Minimum Design Loads for Buildings and Other Structures (Revision of ANSI/ASCE 7-95).

ASCE/SEI 7-10, Minimum Design Loads for Buildings and Other Structures.

Boone, S.J. and Crawford, A.M. (2000). "Braced Excavations: Temperature, Elastic Modulus, and Strut Loads." Journal of Geotechnical and Geoenvironmental Engineering. ASCE, October, pp 870–881.

The Steel Bridge Design Handbook: Bracing System Design; Publication No. FHWA-IF-12-052— Vol. 13, published November 2012.

Tunnel Ventilation Systems Harmonization for Optimization of Tunnel Construction

Sean Cassady and David Parker
HNTB Corporation

ABSTRACT: The SR 99 Tunnel Ventilation System (TVS) manages pollutants, providing a safe and tenable environment for motorists in the tunnel during flowing traffic, stopped traffic, and congested traffic. The TVS also mitigates the effects of smoke and heat during a fire incident to facilitate evacuation of tunnel occupants and implementation of firefighting operations.

Three subsystems key to the ability of the TVS are: the extraction ventilation system; the maintenance air ventilation system; and the jet fan ventilation system. The extraction ventilation system is made up of four centrifugal extraction fans in each of the two Tunnel Operations Buildings. The fans connect to a continuous extraction duct running along the east side of the tunnel. The fans pull smoke and pollutants out of the tunnel roadway through the extraction duct and exhaust it out of stacks on top of the buildings. All eight fans are available for smoke extraction. The maintenance air ventilation system is made up of one centrifugal fan and one backup fan in each of the two Tunnel Operations Buildings. These fans connect to a maintenance air duct running along the west side of the tunnel and a utilidor under the roadways. The maintenance air system pressurizes the egress passage, preventing smoke and pollutant infiltration from the roadways. The jet fan ventilation system consists of several jet fans mounted above and alongside the roadways near the tunnel portals. These fans manage the longitudinal airflow drawing fresh air into the roadway tunnel, and pushing smoke and heat out of the tunnel away from motorists. Design development of these three ventilation subsystems establish harmonized operation modes for managing the tunnel environment such that tunnel bore and approach cut and cover size was optimized.

INTRODUCTION

After 10 years of study, the Federal Highway Administration signed a record of decision approving construction of a bored tunnel to replace Seattle's above-street Alaskan Way Viaduct. The central waterfront section of the viaduct will be replaced with a tunnel beneath the city's downtown. The viaduct, built in 1953, is at risk of failure from earthquakes and irreversible loss of use from age and deterioration. It was weakened in the 2001 Nisqually earthquake, and the state says it would fail in the event of another strong earthquake. The tunnel will connect to the new SR 99 roadway under construction south of downtown, and to Aurora Avenue to the north of downtown. The viaduct plays a major role in sustaining the local economy and its citizens' ability to travel to and through Seattle.

As part of the design-build team, HNTB is the lead designer and engineer of record for this project that includes structural, civil, mechanical, electrical, architectural, and geotechnical final design for the tunnel and the tunnel liner, approaches, portals, interior structures. The design-build team charged with completing this 1.7-mile-long double-deck tunnel is using an earth pressure balance tunnel boring machine (TBM). The 57.5-foot-diameter TBM will drill through a mix of fill soil, clay, glacial till and boulders, at depths of more than 200 feet near the stadium area from SODO (approximately Dearborn Street) to South Lake Union (approximately Thomas Street). The tunnel will be the largest diameter ever constructed by TBM technology, and will have state-of-the-art fire/life safety systems, including a single-point extraction ventilation system and a fire suppression system capable of managing a 100-MW fire.

The south portal area served as the primary staging area for construction materials and operational support of the TBM. This area is bounded by the historical Pioneer Square District and Port of Seattle's working waterfront. The north portal area is near the Seattle Center and Gates Foundation World Headquarters.

TUNNEL VENTILATION SYSTEMS OVERVIEW

The Tunnel Ventilation System (TVS) components are sized to maintain acceptable air quality in the tunnel during normal, congested, and standstill operations, and to remove heat and smoke from the

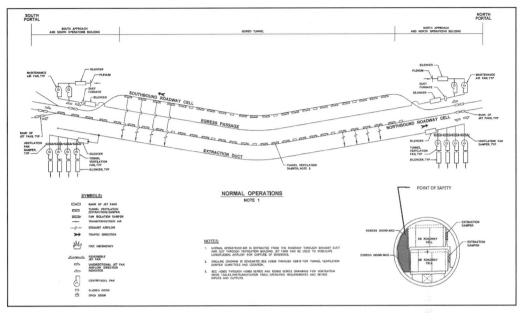

Figure 1. Tunnel ventilation system schematic—smoke extraction, longitudinal tunnel velocity control and maintenance ventilation system

tunnel in the event of a fire. In a fire emergency condition, TVS operations provide tenable conditions beyond 100 feet upstream (towards oncoming traffic) and 600 feet downstream of the fire, and for a sufficient time to allow tunnel occupants to exit to a point of safety. These design and operational conditions were developed to comply with NFPA 502 standard criteria in coordination with Washington State Department of Transportation, Seattle Fire Department, and other stakeholders. The points of safety are provided in the egress rooms and areas behind the egress doors spaced 650 feet apart along the west sides of the northbound and southbound roadways. Tunnel occupant evacuation conditions were evaluated with computer based people movement simulation program called STEPS to confirm door spacing was adequate for mitigating exposure to fire hazards. These doors have a two-hour fire resistance rating. Egress rooms are connected by stairs to an egress passage leading to ground level exits near the tunnel portals.

The TVS consists of three subsystems designed to work together to manage the tunnel environment: the Fire Hazard/Emissions Extraction Ventilation System, the Maintenance Air Ventilation System; and the Jet Fan Longitudinal Ventilation System. The basic configuration shown in Figure 1 consists of a continuous extraction duct with evenly-spaced dampers along the east sides of the northbound and southbound roadways. Jet fans located near the tunnel portals will manage the longitudinal air flow and

Table 1. Tunnel fire life safety systems features summary

Fire and Life Safety System Features Summary Listed in Order of Operational Significance

1. Develop system design that supports concept of operations applicable to the tunnel facility
2. Tunnel features that support self-evacuation
3. Mitigation of fire hazard impacts
4. Prevent progressive structural collapse and limit potentially harmful concrete spalling
5. Mitigation of life safety hazards developed by traffic operations generation of high pollutant concentrations
6. Provide responder access in the event of traffic accident and/or fire incident

move smoke and heat away from motorists stopped upstream of a fire and unable to drive out of the tunnel.

Emergency operation capabilities of the TVS need to incorporate critical aspects of fire life safety system operations. Table 1 provides a list of features important to managing fire life safety conditions within tunnel facility.

The ventilation system is coordinated with the tunnel configuration to optimize efficiency and establish operable volume that is harmonized with the tunnel configuration. Highway vehicle operations establish a dynamic envelope for operating vehicles with a shape that resembles a rectangle. Two lane highway configuration establishes an aggregate

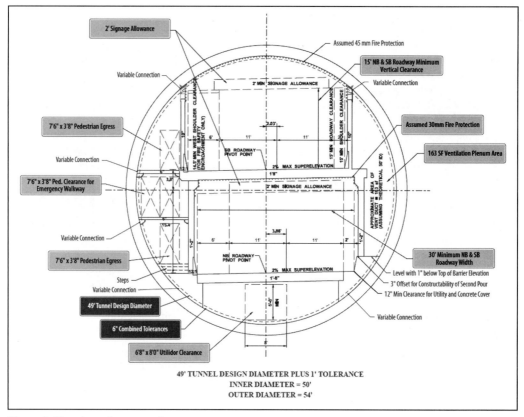

Figure 2. Bore tunnel cross section

traffic operating envelope with a high aspect ratio capable of accommodating roadway shoulder clearances that are necessary for safety and efficient traffic operations. Figure 2 indicates the aggregate traffic operating envelope for SR 99 tunnel supporting bi-directional two lane traffic with segregated traffic enclosures. Note additional volumes created by tunnel liner structure circular cords that are at either side of the traffic enclosures. This paper outlines the synergistic space proofing process utilized to establish these volumes as critical fire life safety features for the SR 99 tunnel facility.

The South Cut-and-Cover section of the tunnel includes 950-feet of northbound-only roadway and 300-feet of a combined lower northbound roadway and upper southbound roadway tunnel. The northbound on-ramp from South Royal Brougham Way enters a tunnel portal immediately to the east of the northbound mainline portal. The on-ramp merge with the mainline roadway ends about 100 feet from the bored tunnel section. Figure 5 shows a cross section of the South Cut-and-Cover Tunnel.

The North Cut-and-Cover section of the tunnel is approximately 450 feet long from the end of the Bored Tunnel to the northbound and southbound roadway portal.

TUNNEL VENTILATION SYSTEM SUBSYSTEMS

Fire Hazard/Emissions Extraction Ventilation System

The tunnel roadway extraction ventilation is provided by eight ventilation fans with four in the South Operations Building and four in the North Operations Building. Flow from the roadways is through tunnel ventilation dampers typically 108.33 feet apart on the east side each roadway. The dampers are typically closed under normal traffic operation. Emissions concentrations may be controlled to levels below criteria limits by adjusted flow from the roadway through the extraction damper and inflow of air from the maintenance air ventilation system and the jet fan banks near the tunnel portals.

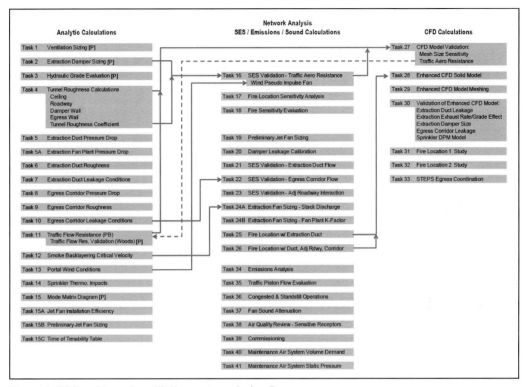

Figure 3. FLS and tunnel ventilation systems design flow process

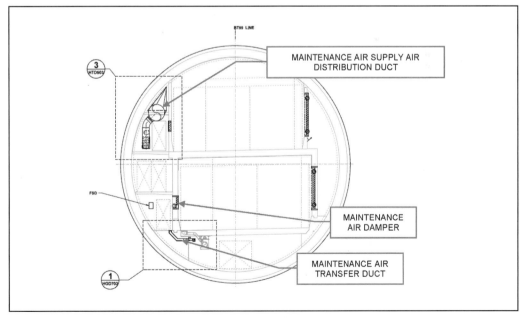

Figure 4. Maintenance air system bore tunnel cross section

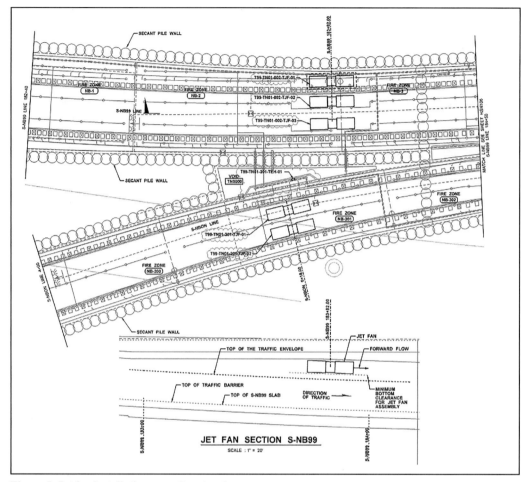

Figure 5. Jet fan installation—south cut and cover

Maintenance Air Ventilation System

The Maintenance Air Ventilation System supplies ventilation to the egress stairways, egress passage, electrical rooms and other equipment spaces, and the Invert Utilidor level. Maintenance air supply fans are located in the North Operations Building and the South Operations Building. Figure 4 indicates the location of the maintenance ventilation supply air distribution system that pressurizes the egress corridor creating a point of safety adjacent to the roadway. Critical pressure relationship between roadway and egress corridor is managed with coordinated operations of the TVS and Maintenance Air System under normal and emergency operation. The maintenance air system functions as a passive fire protection element (does not rely on fire detection to provide fire hazard mitigation) due to the fact that it is on continuously during facility normal operation

periods to provide space conditioning and safe working environment to equipment ancillary spaces.

Jet Fan Longitudinal Ventilation System

Jet fans are located in the tunnel near the roadway portals: three fans about 180 feet in from the northbound roadway south portal; two fans about 160 feet from the northbound on-ramp portal; four fans 200 feet in from the southbound roadway south portal; four fans about 180 feet in from the southbound roadway north portal; and four fans about 220 feet in from the northbound roadway north portal. All fans are located in banks over the roadways except for the northbound roadway north portal where two fans are located on each side of roadway as shown in Figure 5.

Under normal roadway operating conditions, jet fans are available, along with the extraction system, to control CO concentrations if CO sensors indicate exposure criteria will be exceeded. In emergency fire conditions, jet fan ventilation and the extraction system are operated to maintain tenable conditions while tunnel occupant evacuation is in progress and until the fire is controlled.

DESIGN ACTIVITIES SUPPORTING TUNNEL INFRASTRUCTURE

An inter-discipline design flow process is key to maximizing value in a facilities design. Space proofing development along with system evaluation for solution robustness are key steps to validating an approach. It is important to invest time from subject matter experts who can employ simplified engineering analysis methods with confidence for adequate levels of conservatism that will not jeopardize the viability of the design. Figure 3 indicates mechanical engineering activities employed to design FLS

and TVS systems for SR 99. The flow chart indicates where calculation processes where developed at initial preliminary levels of accuracy and then refined for final equipment sizing determination.

CONCLUSION

Design of underground facilities such as tunnels have significant engineering challenges to establish structurally stable solutions within limited right of way space. Ventilation and fire life safety infrastructure are required to establish adequate levels of safety for underground facilities. When methodical processes of inter-discipline coordination are undertaken with activities such as space proofing and concept level ventilation calculations, development of underground structure design can be optimized by harmonization of structural and fire life safety system features.

Vibration Study on California High-Speed Rail Tunnels Based on Earthquake Data and Numerical Analysis

Xiaomin You and Douglas Anderson
WSP USA

ABSTRACT: The paper presents a vibration study for a planned California High Speed Rail (CHSR) tunnel. Proposed tunnel alignment alternatives cross properties near a limestone quarry. There is concern regarding the potential impact of the quarry's blast vibration on the tunnel structure. As it is in an unpopulated area, seismograph monitoring data was unavailable. A numerical simulation was done to simulate quarry blasting impact on the proposed tunnel. Quarry blast events were identified among earthquake seismic data to calibrate the numerical results. The study results can help to identify setback distance to secure integrity of the proposed tunnel.

INTRODUCTION

The two proposed alignment alternatives for the California High Speed Rail (CHSR) in the stretch between Bakersfield and Palmdale both cross property owned by a company that operates a limestone quarry in the neighborhood of the alignments. The quarry conducts blasting operations that generate ground vibration; therefore, there is concern regarding the potential impact of such blast vibration on the proposed CHSR tunnel near the quarry.

WSP carried out a preliminary study to understand the potential impacts of the quarry operations. As the study area is in an unpopulated area, usually collected seismograph monitoring data for regulatory compliance was unavailable to evaluate the vibration induced by quarry blasting. Fortunately, the Southern California Earthquake Data Center (SCEDC) of California Institute of Technology continually collects vibration data from hundreds of seismic instruments in southern California for earthquake study. They identify non-earthquake events (such as quarry blasts), and a catalog of such events in the neighborhood of CalPortland Cement over a fifteen-month period was obtained. Events identified by SCEDC as quarry blasts were analyzed, recorded at a location near the proposed alignments. From these events, a range of likely vibration levels at the alignments was determined.

As a cross check to estimate the magnitude of potential vibration level at the proposed location of the CHSR tunnel, a numerical simulation was done to simulate the quarry blasting and its impact to the proposed tunnel. The tunnel structure and its surrounding ground were included in the model. In absence of blasting details from the quarry, an assumed representative blast was considered in this numerical study and the SCEDC data was used to calibrate the numerical model.

Furthermore, the results obtained from this study were compared to published international guidelines on vibration protection for railway tunnels. This study can help to identify a required setback distance to secure the safety and security of the proposed tunnel.

ALIGNMENT AND GEOMETRY

The two proposed CHSR alignments and the two current active quarry pits (Section 15 pit and Main pit) are shown in Figure 1. Section 15 pit is quite close to the eastern alignment with a shortest approximate distance of 800 ft. The Main pit is about 6800 ft to the closest approach of the tunnel. The proposed tunnel alignments and the approximate locations of the two quarry pits superimposed on the Dibblee Geologic Map are shown in Figure 2.

BLAST-INDUCED VIBRATION

This paper focuses on blast-induced vibration only. Ground vibration generated by trains is not included in the discussion of this paper. Extensive research on the effects of blast-induced vibration has been performed in the US. During the 1940s through the 1980s, US Bureau of Mines (USBM) gathered and analyzed data from residential structures near surface mine blasting. Their findings were summarized in USBM RI 8507 (Siskind, et al., 1980). USBM proposed using Peak Particle Velocity (PPV) as the most practical descriptor of vibration and established a general rule of 2 inch per second (ips) for residential structures at high frequency, with lower PPV limits at lower frequency. Nowadays, similar rules

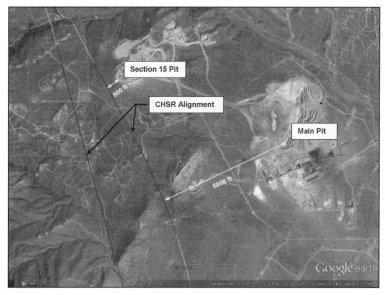

Figure 1. CHSR tunnel alignment and location of the quarry

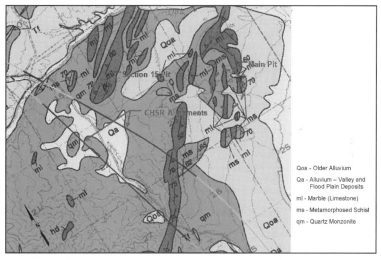

Qoa - Older Alluvium

Qa - Alluvium – Valley and Flood Plain Deposits

ml - Marble (Limestone)

ms - Metamorphosed Schist

qm - Quartz Monzonite

Figure 2. Geologic map (Adapted from Dibblee Geological Map, 2008)

are widely used and accepted for both residential and concrete structures.

Table 1 provides a summary of effect of blast vibration on materials and structures from various researches and tests. Massive concrete structures are far more capable of withstanding blast vibration intensities than residential structures. Table 1 shows in some cases concrete structures were capable to withstand blast vibration as high as hundreds of ips and minimum damage in concrete was observed when PPV is less than 8 ips.

Buried structures and facilities can withstand high vibration because they are constrained by the surrounding ground. Kendorski et al. (1973) conducted a study on the potential effect of blast vibration on a 6-foot by 8-foot tunnel lined with 2 to 11 inches of shotcrete. Hairline cracks was observed when the vibration level exceeded 30 ips.

A more direct study has been conducted determining the vibration levels necessary to cause failure to an abandoned brick-lined railway tunnel by nearby blasting at a quarry in South Wales (Kaslik, et al., 2001). In this case, blasting was taken to the

Table 1. Effect of blast vibration on materials and structures, data from Caltrans (2013)

PPV (in/sec)	Application	Effect	Reference
600	Explosives inside concrete	Mass blowout of concrete	Tart et al. 1980
375	Explosives inside concrete	Radial cracks develop in concrete	Tart et al. 1980
200	Explosives inside concrete	Spalling of loose/weathered concrete skin	Tart et al. 1980
>100	Rock	Complete breakup of rock masses	Bauer and Calder 1978
100	Explosives inside concrete	Spalling of fresh grout	Tart et al. 1980
100	Explosives near concrete	No damage	Oriard and Coulson 1980
50–150	Explosive near buried pipe	No damage	Oriard 1994
25–100	Rock	Tensile and some radial cracking	Bauer and Calder 1978
40	Mechanical equipment	Shafts misaligned	Bauer and Calder 1977
25	Explosive near buried pipe	No damage	Siskind and Stagg 1993
25	Rock	Damage can occur in rock masses	Oriard 1970
10–25	Rock	Minor tensile slabbing	Bauer and Calder 1978
24	Rock	Rock fracturing	Langefors et al. 1948
15	Cased drill holes	Horizontal offset	Bauer and Calder 1977
>12	Rock	Rock falls in underground tunnels	Langefors et al. 1948
12	Rock	Rock falls in unlined tunnels	E. I. du Pont de Nemours & Co. 1977
<10	Rock	No fracturing of intact rock	Bauer and Calder 1978
9.1	Residential structure	Serious cracking	Langefors et al. 1948
8.0	Concrete blocks	Cracking in blocks	Bauer and Calder 1977
8.0	Plaster	Major cracking	Northwood et al. 1963
7.6	Plaster	50% probability of major damage	E. I. du Pont de Nemours & Co. 1977
7.0–8.0	Cased water wells	No adverse effect on well	Rose et al. 1991
>7.0	Residential structure	Major damage possible	Nichols et al. 1971
4.0–7.0	Residential structure	Minor damage possible	Nichols et al. 1971
<6.9	Residential structure	No damage observed	Wiss and Nichols 1974
6.3	Residential structure	Plaster and masonry walls crack	Langefors et al. 1948
5.44	Water wells	No change in well performance	Robertson et al. 1980
5.4	Plaster	50% probability of minor damage	E. I. du Pont de Nemours & Co. 1977
4.5	Plaster	Minor cracking	Northwood et al. 1963
4.3	Residential structure	Fine cracks in plaster	Langefors et al. 1948
>4.0	Residential structure	Probable damage	Edwards and Northwood 1960
2.0–4.0	Residential structure	Plaster cracking (cosmetic)	Nichols et al. 1971
2.0–4.0	Residential structure	Caution range	Edwards and Northwood 1960
2.8–3.3	Plaster	Threshold of damage (from close-in blasts)	E. I. du Pont de Nemours & Co. 1977
3.0	Plaster	Threshold of cosmetic cracking	Northwood et al. 1963
1.2–3.0	Residential structure	Equates to daily environmental changes	Stagg et al. 1980
2.8	Residential structure	No damage	Langefors et al. 1948
2.0	Residential structure	Plaster can start to crack	Bauer and Calder 1977
2.0	Plaster	Safe level of vibration	E. I. du Pont de Nemours & Co. 1977
<2.0	Residential structure	No damage	Nichols et al. 1971
<2.0	Residential structure	No damage	Edwards and Northwood 1960
0.9	Residential structure	Equivalent to nail driving	Stagg et al. 1980
0.5	Mercury switch	Trips switch	Bauer and Calder 1977
0.5	Residential structure	Equivalent to door slam	Stagg et al. 1980
0.1–0.5	Residential structure	Equates to normal daily family activity	Stagg et al. 1980
0.3	Residential structure	Equivalent to jumping on floor	Stagg et al. 1980
0.03	Residential structure	Equivalent to walking on floor	Stagg et al. 1980

point of failure, since the quarry was to expand to the area of the tunnel. It was not until vibration levels substantially exceeded 20 ips that a displacement of the unmaintained lining was observed.

TUNNEL PROTECTION

For the safety of the railway tunnels and running trains, agencies develop railway protection measures to regulate activities, including blasting, which may expose the railway to danger. For example, Singapore Land Transport Authority generally does not permit blasting using explosives within 60 m (200 ft) of the underground rapid transit system structure; however, blasting could be allowed within the railway protection zone if some special conditions can be met, including controlling of vibration less than 15 mm/s (0.6 ips) PPV.

Hong Kong Mass Transit Railway (MTR) adopts a similar strategy to protect its railway structures and operation safety. MTR defines the boundary of the railway protection areas as 30 m (100 ft) outside the railway structures. Construction works carried out within this protection areas are subject to approval before the works can commence. The vibration limit induced to railway structures is set to 25 mm/s (1.0 ips) for blasting.

This study was aimed to estimate an upper limit for potential blast-induced vibration at the CHSR tunnels. Such information helps the designer and owner to develop mitigation measures or identify a set-back distance for future quarry expansion.

SEISMIC DATA AND VIBRATION ANALYSIS

As noted earlier, because of the quarries' remote location, no routine seismograph monitoring is done to demonstrate compliance with regulations. Fortunately, California Institute of Technology's Southern California Earthquake Data Center (SCEDC) continually collects vibration data from hundreds of seismic instruments in southern California for earthquake study. They identify non-earthquake events (such as quarry blasts), and a catalog of such events in the neighborhood of the quarries over a fifteen-month period were obtained.

A total of 99 events identified by SCEDC as quarry blasts were analyzed for this study, and it was found that all but about 5 of them have the characteristics such as waveforms and dominant frequencies normally associated with quarry blast vibration. This information was very useful because, for events recorded at a SCEDC instrument location fairly close to the quarries (about 16,500 feet to the closest approach), the peak vibration levels for each of the events were able to be determined, as well as the frequency content.

The measured peak vibration levels, termed "peak particle velocity" or "PPV," ranged from a maximum of about 0.04 in/s to a minimum of about 0.004 in/s, a range of about an order of magnitude. However, locations of events identified by SCEDC have an uncertainty of ±5 km (±16,400 feet), so it is not practical from the SCEDC locations to determine where in the quarries the blasts occurred, since the quarries range in size from 0.25 km (820 feet) to 1 km (3,280 feet). Though we knew the time and date of events, we were not provided explosive charge weights to associate with specific blasts. Dominant frequencies determined from Fourier analysis are typically in the range from 1 to 10 cycles/second (Hz), often around 3 Hz.

FINDINGS FROM SEISMIC DATA ANALYSIS

Based upon these data and industry-standard regression methodology, the preliminary analysis indicated that maximum likely PPV values would be 0.24 in/s

at the closest approach of current quarry blasting to either proposed alignment, 800 feet. Scatter in PPV values with similar blast design and at equal distances can be up to an order of magnitude. Therefore, it is reasonable to conclude that a likely lower bound for vibration would be 0.024 in/s.

Because the proposed CHSR alignments in this area are substantially underground, these PPV values were compared with other studies documenting blast damage on underground structures. It is noted that in these studies vibration levels much greater than those calculated here were needed to damage engineered underground structures.

These quarry blast data and analyses serve several other purposes:

- First, they served as constraints on numerical modeling studies performed concurrently for this project, discussed below.
- Second, they provide context and bounds for a proposed quarry blast monitoring program.
- Third, they serve as a template for ongoing vibration monitoring during CHSR operation.

In summary, this part of the study indicated that blasting at current pit locations would be unlikely to affect tunnel structure at either alternative alignment. Expansion of quarry operations, as is normal procedure, would require further evaluation.

QUARRY BLASTING

At the time this study was performed, very limited information was provided by the Quarry. No detailed information on blasting operation was provided. Hence, design elements for a typical multi-row quarry blast with vertical holes (Figure 3) were assumed. Such assumptions were based on aerial maps from Google Earth and WSP understanding of typical quarry blast practices. Based upon that information, the charge weight was assumed to be 300 pounds per hole at the Section 15 pit, and 1000 pounds per hole at the Main Pit.

It should be noted that assumed delay times can affect the vibration substantially. Also, there can be constructive (increased PPV) or destructive (decreased) vibration depending on the frequencies amplified by the geological path (Anderson, 2008). Such analysis requires information on wave propagation characteristics that was not available, so we did not correct for such effects.

MODEL SETUP

This blast analysis is performed via 3D finite element analysis by using ANSYS Autodyn in two steps: (1) set up a model with a single blast hole to obtain the blast pressure within the hole; and then (2) apply the blast pressure to the walls of 32 blast holes sequentially in the full 3D model to investigate the ground vibration due to the quarry blast.

Considering that the blast hole is a long cylinder, in the first step an axisymmetric model (Figure 4) was set up to model the single blast hole to improve the calculation efficiency. In the second step, a full 3D model (Figure 5) was set up to include 32 blast holes, the CHSR tunnels and the surrounding ground. External surfaces (bottom, sides and end surfaces) were assigned an infinite boundary condition, which allows the pressure wave to propagate; at the top surface of the 3D model, a free boundary was used.

NUMERICAL SIMULATION RESULTS

Blast pressure within a single hole was obtained using an axisymmetric model. Figure 6 presents nine pressure history curves measured (as calculated in the numerical model) at the wall of the blast hole. The charge length is 27 ft, and therefore the charge weight is 300 pounds, representative of Pit 15. The figure shows the travel time from the bottom to the

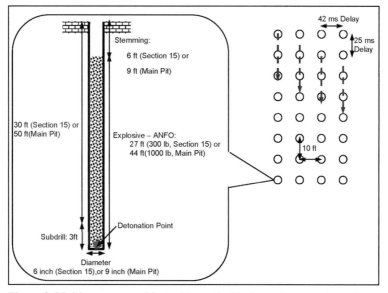

Figure 3. Multi-row quarry blast with vertical holes and assumed blast delay timing and sequence

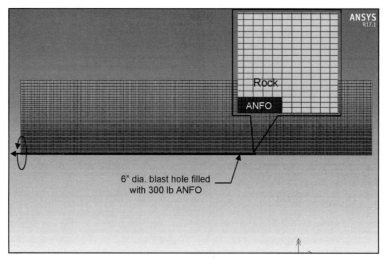

Figure 4. Single blast hole, shown as horizontal in axisymmetric model

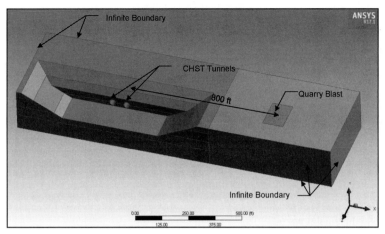

Figure 5. Full 3D model showing location of blast relative to CHSR tunnel portals

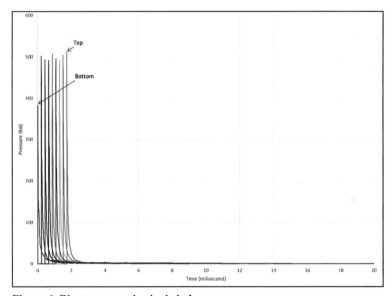

Figure 6. Blast pressure in single hole

top is around 2 milliseconds and the peak pressure over the wall is 380 to 510 ksi.

Typical detonation velocity for the most commonly used commercial explosive, ANFO, is between 10,000 to 19,000 ft/s and detonation pressure between 100 to 1300 ksi (FHWA, 1991). For a 27-foot blast hole charged with explosive, it is estimated that the shock wave would take about 1.4 to 2.7 millisecond to travel from the bottom to the top. The numerical simulation results indicated 2 ms of travel time, which fits in the estimated range based on only detonation velocity and hole depth.

The blast pressure at the wall can be roughly calculated as one half of the detonation pressure and thus is estimated as 50 to 650 ksi. Again the blast pressure of 380 to 510 ksi from the numerical simulation is within the estimated range.

Four virtual gauge points were evenly distributed along the tunnel circumference to record the blast-induced vibration in the numerical analysis. Figure 7 shows the velocity histories along x-, y-, and z- (transverse, vertical, and longitudinal) directions, respectively. The full 3D model runs from blast initiation to 1 second (1000 millisecond).

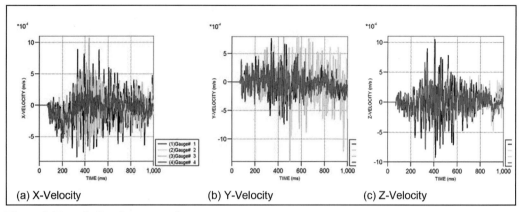

(a) X-Velocity (b) Y-Velocity (c) Z-Velocity

Figure 7. Tunnel vibration at portal

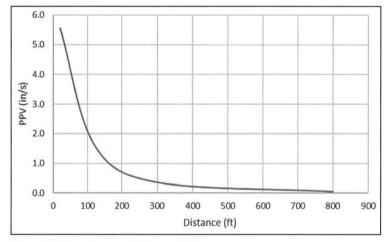

Figure 8. Vibration amplitude vs. distance

As shown in Figure 7, the calculated PPV at the tunnel was 0.0014 m/s (0.06 ips), in the case that the Section 15 pit is located about 800 ft from the tunnel. Comparing to the results from the SCEDC Seismic Data Study, a PPV of 0.06 ips from the numerical analysis is consistent with the estimated range of 0.024 ips to 0.24 ips. However, this vibration level is based upon assumed blast parameters such as delay times, which can affect the vibration substantially. To account the uncertainty in delay times, an amplification of 5 is reasonable and thus the maximum PPV is 0.30 ips, which is still well below the vibration level needed to damage engineered underground structures.

To further evaluate the attenuation of vibration with distance to the quarry blasting, additional cases with shorter distances were considered in the numerical study. Figure 8 presents a summary of the

PPVs from various cases and shows the attenuation of vibration with distance to the quarry blasting:

- The PPV at the tunnel decreases as moving further away from quarry blasting.
- When the distance is 100 ft or further, the PPV is expected to be below 2 ips.
- When the distance is 160 ft or further, the PPV is expected to be below 1 ips.

This graph shows the change in vibration with distance for a charge weight of 300 pounds. At different charge weights, the graph would need to be modified appropriately. A commonly used weighting factor is that vibration scales directly as the square-root of the charge weight per delay. Therefore, vibration at all distances for the Main Pit would be approximately 1.8 times that shown in this graph. However, it is recommended performing additional analysis if there is any change to the blasting pattern, ground condition, and/or charge weight. In addition to the impacts to the structure, train operation, which was not within the scope of this study and would also be impacted by ground vibration, is another important factor that should be considered in considering the guideline value of set-back distance.

CONCLUSION

Quarry blast seismic records obtained from the SCEDC have been analyzed to determine the range of vibration parameters generated at approximately 16,500 feet from the quarries. Analysis of the data indicates maximum likely vibration levels at the current closest approach of the quarry blasting to the proposed tunnel alignment (800 feet) would be, for a 300 pound charge, within a PPV range of 0.024–0.24 in/s. Based on the numerical simulation, the PPV at the tunnels due to typical blasting in the closest quarry was estimated to be around 0.06 ips, which fit in the range of 0.024 ips to 0.24 ips as estimated based on SCEDC Seismic Data Study. However, this vibration level is based upon assumed blast parameters such as delay times, which can affect the vibration substantially. To account the uncertainty in delay times, an amplification of 5 is reasonable and thus the maximum PPV is 0.30 ips, which is still well below the vibration level needed to damage engineered underground structures.

The results and conclusions are based on the analysis for blast-induced vibration only. Mass ground movement and flying debris/rocks may be of a concern at short distance if the CHSR is not in a tunnel. Furthermore, it would be useful to have a quarry blast monitoring program with sensors at locations coordinated with the proposed alignments. The results of this study provide bounds on likely upper and lower levels of vibration found in a limited monitoring study. The results of this study can help in developing appropriate mitigation measures to assure safe and secure operation of CHSR, consistent with the needs of all stakeholders.

ACKNOWLEDGMENT

The authors would like to thank California High Speed Rail Authority for the permission to publish this paper.

REFERENCES

Anderson, D.A., "Signature Hole Blast Vibration Control—Twenty Years Hence and Beyond," Proceedings of the 33rd Annual Conference on Explosives and Blasting Technique, New Orleans, LA, 2008.

ANSYS Mechanical User Guide R17.1. ANSYS, 2016.

Caltrans, "Transportation and Construction Vibration Guidance Manual," September 2013, available at http://www.dot.ca.gov/hq/env/noise/pub/TCVGM_Sep13_FINAL.pdf.

Dibblee, T.W, "Geologic Map of the Cummings Mountain and Tehachapi 15 Minute Quadrangles, Kern County, California," Dibblee Foundation Map DF-397, 2008.

Dowding, C.H., "Blast Vibration Monitoring and Control," Prentice-Hall, INC. Englewood Cliffs, NJ, 1985.

FHWA, "Rock Blasting and Overbreak Control," Retrieved from http://www.fhwa.dot.gov/engineering/geotech/pubs/012844.pdf, pages 38–39, 1991.

Hongkong MTR, Control Criteria of Effects on Railway from Adjacent Construction Works. Retrieved from https://www.mtr.com.hk/en/corporate/operations/protection_criteria.html, 08/01/2016.

Kaslik, M., W.J Birch, and A. Cobb, 2001. "The effects of quarry blasting on the structural integrity of a disused railway tunnel," Proceedings of the 27th Annual Conference on Explosives and Blasting Technique, Orlando, FL.

Kendorski, F.S., C.V. Jude and W.M. Duncan, "Effect of Blasting on Shotcrete Drift Linings," Mining Engineering, v 25, pp. 38–41, 1973.

Land Transport Authority, "Guide to Carrying out Restricted Activities within Rail Protection and Safety Zones," Singapore, 2009.

Siskind, D.E., M.S. Stagg, J.W. Kopp, and C.H. Dowding, "Structure Response and Damage Produced by Ground Vibration From Surface Mine Blasting," USBM RI 8507, 1980.

Tunnel Cross Passage Seismic Analysis Considering 3D Wave Propagation

Yue Shi, Peter Chou, and Danny Lin
Parsons Corporation

ABSTRACT: Cross passage is a rigid connecting structure between twin bored tunnels and serves as a vital means of egress in the event of an emergency. For its seismic analysis, strains induced from the propagation of seismic waves are important parameters as well as dynamic forces transferred from adjacent tunnels. Closed-form solutions based on plane formulations have been provided in classic literatures. However, the behavior of a rigid structure between two relatively flexible tunnels during a seismic event is rather complex, and closed-form solution cannot fully capture the 3D effects of the interacting structures. An expansion of the solutions to 3D space is therefore of significance for the cross passage design. This paper presents a detailed expansion and an application to a finite element model for a transit project in Southern California to show how design can be facilitated in modeling this complex behavior.

INTRODUCTION

Earthquakes pose a unique threat to underground structures. Serving as part of the emergency egress, cross passages are especially important when it comes to seismic design. Generally, as they are surrounded by the media transmitting the seismic energy and tend to move along as seismic event occurs, seismic behavior of cross passages is directly related to the seismic waves propagating through the soil. So far two general seismic design approaches have been developed: Closed-form solution and 3D soil-structure interaction numerical modeling.

Closed-form solution utilizes the free-field ground deformation (Newmark, 1967) responding to varying directions of wave propagation, and modifies this deformation by incorporating the soil-structural interaction. Seismic forces can then be calculated from the modified deformation field based on classical beam theory. Consequently, these forces can be applied to a 3D numerical model to perform analyses and design.

Another approach is 3D soil-structure interaction numerical modeling. A 1D site response analysis is first performed. Free-field ground displacements can then be obtained and applied as boundary conditions in a 3D numerical model accounting for soil-structure interaction. The governing structural responses can then be captured from so called "pseudo-static" 3D finite element analyses.

These two approaches will be discussed first in this paper. Then an example project will be presented using these two approaches.

CLOSED-FORM SOLUTION

To facilitate the formulations presented in the later section, a brief review of some basics regarding seismic waves can be helpful. As well known by the most, seismic waves consist of body waves and surface waves. Particularly, located away from the ground surface, tunnels are much more influenced by the body waves. Two types of body waves, primary and secondary waves, have different effects on tunnels due to their respective characteristics.

Primary waves (P-waves) are compressional waves. If they propagate along the longitudinal direction of the tunnel, local compression and extension would be induced within the tunnel. This effect can be dealt with more easily with an appropriate design of longitudinal joints between tunnel liners. Therefore P-waves are generally not the primary concern for the seismic design of tunnels and its cross passages.

Secondary waves (S-waves), however, can become quite tricky to analyze when the direction of their propagation is not determined. As typical in shear waves, particles move perpendicular to the direction of the wave propagation. In this way, if a S-wave propagates along the longitudinal direction of the tunnel, the structure would mimic a bending behavior of a long and slender beam (Figure 2 left). In another case, when propagated transversely across the tunnel, S-wave would cause a unique tilting effect, known as "Ovaling" (Figure 2 right), to the cross section, which requires separate analysis and design for the structure.

Comparing the effects of these two types of seismic waves, one can readily recognize that S-wave is

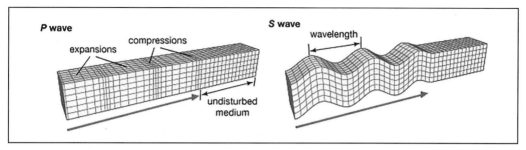

Figure 1. Seismic body waves (from Britannica, 2012)

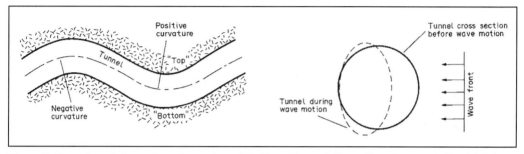

Figure 2. Curvature deformation along a tunnel (left) and hoop deformation of cross-section (right) (from Owen and Scholl 1981)

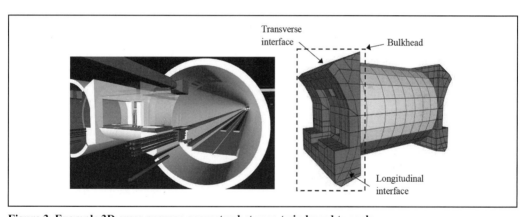

Figure 3. Example 3D cross passage geometry between twin bored tunnels

more detrimental to the structural integrity of tunnels. Hence S-wave as shown in Figure 1 will be the focus of the discussions to be followed.

On the other hand, serving as the connecting structure of tunnels for emergency, cross passages have their own characteristics when it comes to seismic response. Because they are located between twin tunnels and increase the local rigidity of the structure, more complex seismic behaviors can be expected. Seismic design of a transit tunnel in Southern California is used herein as an example. The project is located in a high seismic area. A basic

layout of a cross passage connecting twin tunnels is illustrated in Figure 3.

Design Approach

At seismic design phase, the interfaces between the cross passage and bored tunnels required special attention as to how design forces could be transferred. Cross passage bulkheads, the structure located at the interface as load transfer elements, were identified as the most important for the design. Assumption was that seismic forces would be transferred in two orthogonal directions. One along the

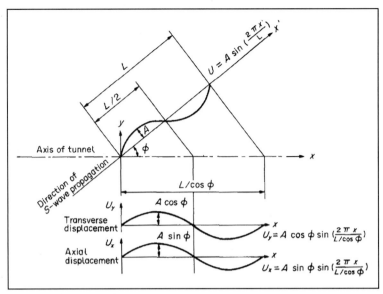

Figure 4. 2D tunnel displacement due to S-wave propagation in the plane of tunnel axis t (from St. John and Zahrah 1987)

axis of the twin tunnels, the other the circumferential (transverse) direction of the ring segment. Based on this assumption, with possible seismic effects of shear waves considered, three scenarios could exist at the interfaces:

1. Longitudinal loads only: This applies when shear wave is propagated on the plane where axis of the tunnels is located (Figure 2 left). The presence of longitudinal joints makes the transverse deformation negligible in this case.
2. Transverse loads only: This applies when shear wave is propagated transversely across the tunnels, i.e., perpendicular to the plane of the axis of tunnels (Figure 2 right).
3. A combination of longitudinal and transverse loads: This is a more general scenario when shear wave is propagated in a direction that could induce seismic loads in both directions (Figure 5).

To address these cases one by one, closed-form solutions for the first two scenarios will be presented at first. Then they will be combined and the problem will be expanded to a three-dimensional space applicable to Case 3.

Longitudinal Seismic Effect

To discuss the longitudinal seismic behavior of tunnels more rigorously, a very meticulous presentation can be found by St. John and Zahrah (1987). This

classic formulation starts with a simple case where soil-structure interaction is neglected, i.e., free-field solution.

Equation of motion of a plane wave propagation is first introduced: Assume a simple sinusoidal displacement function

$$u = A \sin\left[\frac{2\pi}{L}(x - ct)\right] \tag{1}$$

where L is the wavelength; c is the apparent wave propagation velocity; u stands for particle displacement and A its amplitude; x and t are spatial and temporal coordinates.

For a shear wave, particles move perpendicular to the direction of the propagation. Therefore, u describes pure transverse displacement. In a more general case, when there is an angle ϕ between the direction of wave propagation and the axis of the tunnel in the plane, a projected longitudinal displacement would occur along the tunnel, i.e.,

$$u_x = u \sin\phi \tag{2}$$

and the transverse displacement

$$u_y = u \cos\phi \tag{3}$$

The corresponding wave length and apparent wave propagation velocity along the tunnel should be

$$L_x = \frac{L}{\cos\phi} \tag{4}$$

$$c_x = \frac{c}{\cos \phi} \tag{5}$$

Therefore along the tunnel axis we have

$$u_x = A \sin \phi \sin \left[\frac{2\pi}{L/\cos \phi}(x-) \right] \tag{6}$$

The axial strain is then given by

$$\varepsilon_x = \frac{\partial u_x}{\partial x}$$
$$= A \sin \phi \cos \left[\frac{2\pi}{L/\cos \phi}(x-) \right] \cdot \frac{2\pi}{L/\cos \phi} \tag{7}$$

Thus at a particular location and moment, the maximum axial strain should be

$$\varepsilon_{x,\max_{free}} = \frac{2\pi}{L} A \sin \phi \cos \phi \tag{8}$$

Formulations above are based on the free-field assumption. Realistically, certain amount of soil-structure interaction exists between the tunnel and surrounding soil. To account for this interaction, simple beam theory is introduced with respect to axial force:

$$E'A_c \frac{d^2 u_a}{dx^2} = K_a(u_a - u_x) \tag{9}$$

where E' is the Young's modulus of the tunnel, A_c the cross-sectional area of the tunnel, u_a the axial displacement of the tunnel, K_a the assumed stiffness coefficient of the linear elastic springs in the longitudinal direction. Assume a reduction factor R_a for the axial displacement of the tunnel due to soil-structure interaction, i.e.,

$$u_a = R_a u_x \tag{10}$$

Substitute it in, one can easily obtain

$$R_a = \frac{1}{1 + \dfrac{E'A_c}{K_a}\left(\dfrac{2\pi}{L}\right)^2 \cos^2 \phi} \tag{11}$$

Therefore the maximum axial strain of the tunnel with SSI considered should be

$$\varepsilon_{x,\max_{SSI}} = \frac{\dfrac{2\pi}{L} A \sin \phi \cos \phi}{1 + \dfrac{E'A_c}{K_a}\left(\dfrac{2\pi}{L}\right)^2 \cos^2 \phi} \tag{12}$$

Transverse Seismic Effect

As introduced earlier, when shear wave is propagating in a plane perpendicular to the tunnel axis, it leads to a shear deformation mode termed "Ovaling." The wave equation should consist of pure transverse

component, i.e., (l stands for longitudinal coordinate and n transverse coordinate)

$$u_n = A \sin \left[\frac{2\pi}{L}(l - ct) \right] \tag{13}$$

Thus

$$\gamma = \frac{\partial u_n}{\partial l} = A \cos \left[\frac{2\pi}{L}(l - ct) \right] \frac{2\pi}{L} \tag{14}$$

since

$$\dot{u}_n = \frac{\partial u_n}{\partial t} = -A \cos \left[\frac{2\pi}{L}(l - ct) \right] \frac{2\pi}{L} c \tag{15}$$

Compare the above two equations, we have

$$\gamma = -\frac{1}{c} \dot{u}_n \tag{16}$$

u_n stands for the transverse particle displacement, c the wave velocity. Neglect the sign of shear strain and substitute $\dot{u}_n$ and c with known site seismic parameters V_s (peak particle velocity) and c_s (apparent shear wave velocity), the maximum shear strain can be obtained by

$$\gamma_{\max} = \frac{V_s}{c_s} \tag{17}$$

Combined Seismic Effect

The two scenarios we have discussed so far deal with the seismic strains in two orthogonal directions. One has shear wave propagation direction and axis of the tunnel coplanar, and the other perpendicular to each other. To cover a more general case in the three-dimensional space. Another orientation needs to be introduced: The angle between the plane of the shear wave and the plane of the tunnel axis, denoted as β (See Figure 4).

Apply basic geometry concept here. The shear wave can be resolved into two components corresponding to the two planes the previous formulations have been based on. The one on the plane of tunnel axis has an amplitude

$$A_l = A \cos \beta \tag{18}$$

Then the maximum axial strain should be

$$\varepsilon_{l,m} = \frac{\dfrac{2\pi}{L} A \cos \beta \sin \phi \cos \phi}{1 + \dfrac{E'A_c}{K_a}\left(\dfrac{2\pi}{L}\right)^2 \cos^2 \phi} \tag{19}$$

On the plane across the tunnel axis, the shear wave velocity should be modified to

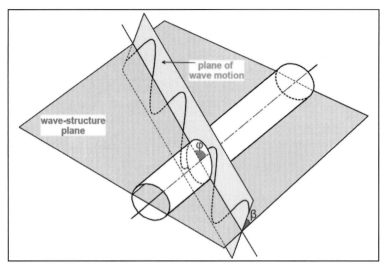

Figure 5. General 3D shear wave propagtion scenario (G. Boukovalas, 2010)

$$c_{s,n} = \frac{c_s}{\sin\phi} \qquad (20)$$

And the corresponding peak particle velocity is given by

$$V_{s,n} = V_s \sin\beta \qquad (21)$$

Thus the maximum shear strain

$$\gamma_m = \frac{V_{s,n}}{c_{s,n}} = \frac{V_s}{c_s}\sin\beta\sin\phi \qquad (22)$$

Figure 4 illustrates this more general case. In a case of 3D wave propagation, angles β and ϕ may vary. Thus tunnel seismic design should consider all possible combinations and select values of β and ϕ that would yield the most critical design forces.

Convert Seismic Strains into Forces

To be applied to a structural analysis model, seismic strains obtained so far need to be converted to forces acting on the interface structure (bulkhead). In addition, certain realistic restraints exist on the range of the seismic forces as closed-form solutions are based on ideal theoretical scenarios.

Longitudinal Force

Axial strain has been calculated using the Equation 12. However, realistically this force is bounded by the ultimate frictional force that can be developed between the tunnel and the surrounding soil. An equation proposed by Sakurai and Takahashi (1969) takes the following form to estimate this upper bound.

$$\varepsilon_{l,m_f} = \frac{fL}{4E'A_c} \qquad (23)$$

where f is the ultimate friction force per unit length and L the wave length of the shear wave.

To estimate f, provisions regarding sliding failure between concrete and soil in AASHTO Article 10.6.3.4 were used.

$$f = R_\tau = V\tan\delta \qquad (24)$$

where V is the total vertical force and can be estimated by the soil weight within the area of soil arching (approximately twice the tunnel diameter)

$$V = 2\gamma_s D \qquad (25)$$

δ is the friction angle between foundation and soil and

$$\tan\delta = 0.8\ \tan\phi_f \qquad (26)$$

for precast concrete where ϕ_f is the internal friction angle of the soil.

Combine the above parameters, the ultimate friction force per unit length can be expressed as

$$f = 2\gamma_s D 0.8\ \tan\phi_f \pi D = 1.6\pi\gamma_s D^2\tan\phi_f \qquad (27)$$

In the end, the design axial strain should be the smaller value

$$\varepsilon_{l,des} = \min(\varepsilon_{l,m}, \varepsilon_{l,m_f}) \qquad (28)$$

Thus the design longitudinal load

$$N = E'A_c\varepsilon_{l,des} \qquad (29)$$

Transverse Force

The transverse force that can be induced from the ovaling effect is directly related to two parameters that describe the interaction between the tunnel and surrounding soil, the flexibility ratio F and compressibility ratio C (Merritt et al., 1985), defined as

$$F = \frac{E_m(1 - v_l^2)R_m^3}{6E_l I_e(1 + v_m)} \tag{30}$$

and

$$C = \frac{E_m(1 - v_l^2)R_m}{E_l t_e(1 + v_m)(1 - 2v_m)} \tag{31}$$

where

E_m = strain compatible elastic modulus of the surrounding soil
E_l = elastic modulus of the concrete lining
R_m = nominal radius of the concrete lining
I_e = effective moment of inertia of the concrete lining (stiffness reduction due to circumferential joints considered)
v_l = Poisson's ratio of the concrete lining
v_m = Poisson's ratio of the surrounding ground
t_e = effective thickness of the concrete lining

To determine transverse thrust (assume no moment transfer across the interface for this design), two types of interaction condition have been identified. Full-slip condition is when no friction or shear occurs between the ground and the lining, whereas no-slip condition assumes full friction between them.

In full-slip condition (Wang, 1993),

$$T_{max\_fs} = \frac{K_1 \gamma_m E_m R_m}{6(1 + v_m)} \tag{32}$$

where

$$K_1 = \frac{12(1 - v_m)}{2F + 5 - 6v_m} \tag{33}$$

In non-slip condition (Hoeg, 1968; Schwartz and Einstein, 1980),

$$T_{max\_ns} = \frac{K_2 \gamma_m E_m R_m}{2(1 + v_m)} \tag{34}$$

where

$$K_2 = 1 + \frac{F(1 - 2v_m)(1 - C)}{F[(3 - 2v_m) + (1 - 2v_m)C]} \atop {+ C(2.5 - 8v_m + 6v_m^2) + 6 - 8v_m}} \tag{35}$$

In reality, the interaction between the ground and the tunnel should be somewhere between full-slip and non-slip condition. Considering this, an interface stiffness coefficient η was introduced to represent the real ground condition. The minimum value of this coefficient was estimated to be 0.01 (a value close to 0) for full-slip condition and the maximum 1.0 for non-slip condition. To simplify the problem, linear interpolation was used to determine the thrust corresponding to η:

$$T_{des} = T_{max\_fs}\left(1 - \frac{\eta - 0.01}{0.99}\right) \atop {+ T_{max\_ns} \frac{\eta - 0.01}{0.99}}} \tag{36}$$

Determine the Governing Design Forces

It can be concluded that once the properties of ground and structure are known, the only variables that would determine the design forces are the two angles β and φ as shown in Figure 4. Due to the complexity of the formula leading to the final design forces, it is not straightforward to find the critical combinations of these two angles for the cross passage design. Therefore, a series of combinations was investigated with an increment of 15 degrees. Both axial and shear strains for a typical design section are tabulated in Table 1 and Table 2.

Seismic stains above then needed to be applied into a 3D finite element model. This required a conversion into design forces, or more specifically stresses since either plate or solid element is more appropriate in modelling the bulkheads, the conjunction of bored tunnel and cross passage. Furthermore, since the longitudinal and transverse stresses are perpendicular to each other, a scalar value needs to be used to help determine the most critical design combination. Following the convention of seismic design, the Square Root of Sum of Squares (SRSS) was used. In addition, maximum friction force discussed earlier was calculated to restrict the longitudinal stress to be applied. The results for the same design section are tabulated in Table 3, Table 4 and Table 5.

From Table 5, the critical combination can be easily identified as the one with β=60° and φ=75°. In this case axial stress is equal to 110.94 ksf and hoop stress 9.14 ksf. Repeat the same procedure to find design stress combinations for other design sections.

Once the critical design force combinations have been determined among all design sections, they can then be applied to finite element models as shown in Figure 5 and included into design load combinations. Typical structural analysis and design can then be performed to determine the member dimensions as well as the reinforcements.

Table 1. Axial strain ε_l for different combinations of β and ϕ in a typical design section

$\beta \setminus \phi(°)$	0	15	30	45	60	75	90
0	0	0.000151	0.000304	0.000451	0.000546	0.000446	0
15	0	0.000146	0.000293	0.000435	0.000528	0.000430	0
30	0	0.000131	0.000263	0.000390	0.000473	0.000386	0
45	0	0.000107	0.000215	0.000319	0.000386	0.000315	0
60	0	0.000075	0.000152	0.000225	0.000273	0.000223	0
75	0	0.000039	0.000079	0.000117	0.000141	0.000115	0
90	0	0	0	0	0	0	0

Table 2. Shear strain γ_m for different combinations of β and ϕ in a typical design section

$\beta \setminus \phi(°)$	0	15	30	45	60	75	90
0	0	0	0	0	0	0	0
15	0	0.000141	0.000272	0.000384	0.000471	0.000525	0.000544
30	0	0.000272	0.000525	0.000742	0.000909	0.001014	0.001050
45	0	0.000384	0.000742	0.001050	0.001286	0.001434	0.001485
60	0	0.000471	0.000909	0.001286	0.001575	0.001757	0.001819
75	0	0.000525	0.001014	0.001434	0.001757	0.001959	0.002028
90	0	0.000544	0.001050	0.001485	0.001819	0.002028	0.002100

Table 3. Axial stress σ_l (ksf) for different combinations of β and φ in a typical design section

$\beta \setminus \phi(°)$	0	15	30	45	60	75	90
0	0	106.28	110.94	110.94	110.94	110.94	0
15	0	102.76	110.94	110.94	110.94	110.94	0
30	0	92.20	110.94	110.94	110.94	110.94	0
45	0	75.31	110.94	110.94	110.94	110.94	0
60	0	52.79	106.98	110.94	110.94	110.94	0
75	0	27.45	55.60	82.35	99.24	80.94	0
90	0	0	0	0	0	0	0

Table 4. Hoop stress σ_h (ksf) for different combinations of β and φ in a typical design section

$\beta \setminus \phi(°)$	0	15	30	45	60	75	90
0	0	0	0	0	0	0	0
15	0	0.73	1.41	1.99	2.45	2.73	2.83
30	0	1.41	2.73	3.86	4.73	5.27	5.46
45	0	1.99	3.86	5.46	6.69	7.46	7.73
60	0	2.45	4.73	6.69	8.19	9.14	9.46
75	0	2.73	5.27	7.46	9.14	10.19	10.55
90	0	2.83	5.46	7.73	9.46	10.55	10.93

Table 5. Combined (SRSS) stress σ_{com} (ksf) for different combinations of β and φ in a typical design section

$\beta \setminus \phi(°)$	0	15	30	45	60	75	90
0	0	106.28	110.94	110.94	110.94	110.94	0
15	0	102.76	110.95	110.96	110.97	110.97	2.83
30	0	92.21	110.97	111.01	111.04	111.07	5.46
45	0	75.34	111.01	111.07	111.14	111.19	7.73
60	0	52.85	107.08	111.14	111.24	111.32	9.46
75	0	27.59	55.85	82.69	99.66	81.58	10.55
90	0	2.83	5.46	7.73	9.46	10.55	10.93

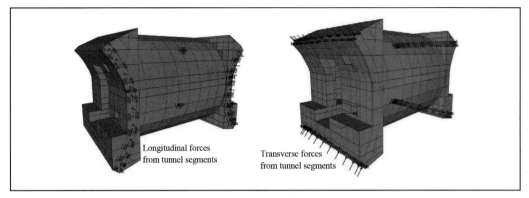

Figure 6. Longitudinal and transverse loads applied to 3D structural FE models

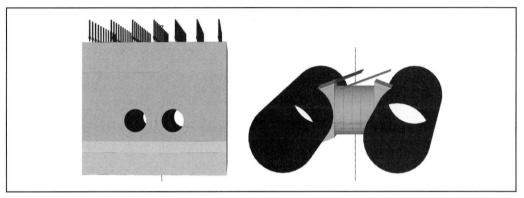

Figure 7. Example of 3D soil-structure interaction models

A summary of this design procedure is as follows:

1. Calculate seismic strains with varying combinations of wave propagations angles: longitudinal strain from Eq. (19), Eq. (23) and Eq. (28) and transverse strain from Eq. (22).
2. Convert strains into structural forces (stresses): longitudinal from Eq. (29) and transverse from Eq. (36).
3. Determine the most critical design case following the SRSS rule.
4. Apply the critical design forces into structural analysis model and perform the design.

3D SOIL-STRUCTURE INTERACTION NUMERICAL MODELING APPROACH

As discussed above, closed-form solutions can estimate the seismic forces acting on cross passage bulkhead walls transferred from adjacent tunnel liners. However, they cannot accurately calculate cross passages' own ovaling effects (due to short length) and shadowing effects from interacting with adjacent

tunnels (closed-form solutions consider green field only). This is particularly true because cross passage is significantly more rigid than the surrounding soils and adjacent structures, which may attract more loads onto itself. Therefore, complex 3D SSI numerical models consisting of all structure elements (twin tunnels and one cross passage) were also introduced to the design to address above issues.

In our case, 3D numerical soil-structure interaction analyses were performed using FE package PLAXIS 3D. To minimize the boundary effect, sufficient dimensions of the model were required. On the other hand, 3D finite element analysis is rather time-consuming when model size becomes too large. After preliminary analyses and sensitivity test, the dimensions of the models were determined to be around 200-ft long, 135-ft wide and 120-ft deep. The length of the model should cover at least 5 times the cross passage diameter at each side of the cross passage. The width is less critical because the cross passage is blocked by adjacent tunnels but sufficient space should still be ensured beyond the tunnels. The depth (height) of the model was determined based on

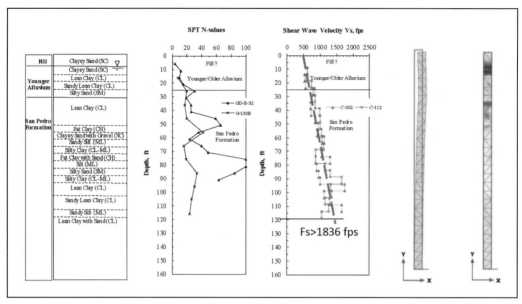

Figure 8. Soil profiles (left) deformed mesh of 1D dynamic analysis (right) and shear strain plot (from Plaxis)

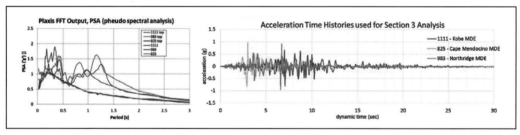

Figure 9. Pseudo response spectral analyses from Plaxis and example time-history ground motions used

the depth where the ground stiffness is sufficiently high (shear wave velocity greater than 1836 fps.).

Design Approach

For this design approach, soil stratigraphy, dynamic soil parameters and their shear wave velocities were determined first as shown in Figure 8. Then a full dynamic time-history analysis of an 1D or 2D soil model subject to ground motions was performed to estimate racking displacement of the model. Example results are also shown in Figure 7 based on an estimated soil shear wave velocity profile. This step can be replaced by the conventional SHAKE analysis of a 1D soil column. However, since SSI model is used in the final step, dynamic analyses of a 1D or 2D SSI model are recommended. In addition, the ground motions applied to the model should be preprocessed by special pseudo spectral analyses (see Figure 9 as an example), and the goal was to obtain

the maximum free-field displacement response of the ground to be applied to the 3D SSI Model.

Benchmark Plaxis Spectra of Free Field Motion

Though the motions were matched to a target response acceleration, all three records provided were applied to the model due to variability of the time histories, which could be significant for design. See Figure 8 for the 1-D output of the pseudo response spectral analysis (Plaxis Fast Fourier Transform) for the three time histories at the base of the model the ground surface (see Figure 8).

Racking Analysis on 3D SSI Model

Racking (or ovaling) analysis is very commonly used in 2D tunnel seismic design (see Figure 10). Once maximum free-field displacements are obtained from the dynamic analyses, the same racking analysis

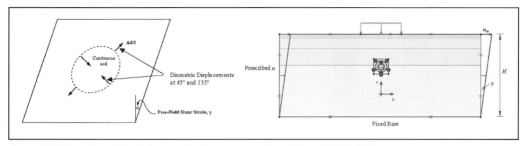

Figure 10. Tunnel ovaling deformation due to ground racking (2D Model)

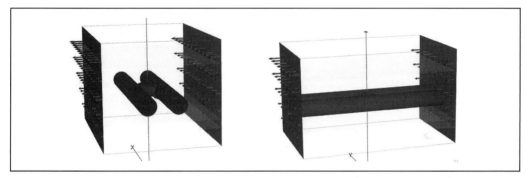

Figure 11. Maximum free-field displacement on transverse faces (left) and longitudinal faces (right)

approach can be applied to a 3D SSI model to simulate the 3D effects from the tunnels.

Once the shear strains along the base and top of the models are obtained, the envelope of maximum free-field shear strain responses at elevations of bored tunnels and cross passage can be obtained and applied as a linearly distributed pseudoptotic displacement at the vertical boundaries of the model. The 3D SSI model consists of twin bored tunnels and a cross passage (See Figure 11), and the maximum racking displacements were applied along the transverse direction and longitudinal direction respectively as shown in Figure 10. Because these displacements are applied as free-field displacement, the model should be sufficiently large to eliminate boundary effects as described earlier.

Result

Theoretically, the maximum diametric displacements should occur at 45° and 135°. Node displacements at these locations are therefore extracted from the results to compute the diametric displacements as a measure of the ovaling deformation These diametric displacements will be used in cross passage 3D structural model for seismic design. The example cross passage deformation under these two racking scenarios are shown in Figure 12.

In addition to displacements, the stresses (axial, shear and bending moments) from 3D SSI models

can also be used to compare with the 3D structure model.

CONCLUSION

The goal of this paper is to present two approaches for estimating seismic responses of the twin tunnels with cross passages in a 3D wave propagation scenario. During earthquakes, deformations of tunnels are assumed to be primarily related to the ground deformation. The structures, therefore, are designed to accommodate these deformations. Inhereantly more accurate in simulating ground deformations, SSI 3D numerical modeling approach is more favorable in this respect.

However, the described 3D SSI numeical approach considers only the seismic waves propagating perpendicular to the tunnel longitudinal axis (so that cross passage ovaling occurs). The longitudinal (curvature and compression) effects, induced by components of seismic waves that propagate along the longitudinal axis, cannot be estimated in this way. Due to the relatively short length of cross passage, curvature and compression effects are expected to be small. Conversely, closed-form solutions, following classic structural mechanics theory, are more inclusive in dealing with various wave propagation scenarios in terms of the orientation. The whole design procedure is also much more convenient to be implemented. However, closed-form solutions

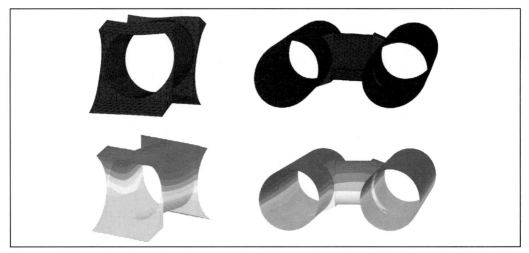

Figure 12. Cross passage deformation on transverse faces (left) and longitudinal faces (right)

usually make conservative assumptions for soil-structure interaction (and in this case, the interaction with adjacent tunnels), which may not predict the actual interface forces accurately.

In addition, seismic strains tend to be most pronounced at the interfaces of bored tunnels and cross passages. For tunnel structures with considerable structural discontinuities (such as joints between tunnel segments and cross passages), closed-form solutions are more viable to calculate the forces transferred across the interface (as for the bulkhead design in the example), and account for discontinuities in the 3D structural models.

In the end, each approach has its own merits. Therefore, it is recommended that both be performed when designing cross passages in a high seismicity area, if applicable.

ACKNOWLEDGMENT

The authors wish to thank the Los Angeles Metropolitan Transportation Authority Engineering team for their valuable comments from the previous project, and also acknowledge the contributions of the tunnel contractor Skanska, Traylor, and Shea from the example project.

REFERENCES

AASHTO LRFD Bridge Design Specification 6th Edition, 2012.

Bouckovalas, G., Kouretzis, G. (2010) "Seismic Design of Underground Structures" Lecture.

Encyclopedia Britannica, "Seismic wave," 2012.

Hoeg, K. (1968), Stresses against underground structural cylinders. J. Soil Mech. Found. Div., ASCE 94 (SM4) 833–858.

Merriitt, J.L., Monsees, J.E., Hendron, A.J., Jr. (1985), Seismic design of underground structures. Proceedings of the 1985 Rapid Excavation Tunneling Conference, vol. 1, pp. 104–131.

Newmark, N. M. (1967). Problems in wave propagation in soil and rock. Proc. Int. Symp. Wave Propagation and Dynamic Properties of Earth Materials, New Mexico: Univ. of New Mexico Press.

Owen, G.N. and Scholl, R.E. (1981), Earthquake Engineering of Large Undrground Sructures. JAB-7821. San Francisco: URS/John A. Blume.

PLAXIS 3D reference manual, 2016.

Schwartz, C.W., Einstein, H.H. (1980). Improved design of tunnel supports: vol.1—simplified analysis fro ground-sturcture interaction in tunneling. Report no. UMTA-MA-06-0100-80-4. USDOT, Urban Mass Transportation Administration.

St. John, C.M., Zahrah, T.F. (1987). Aseismic design of underground structures. Tunneling Underground Space Technol. 2 (2), 165–197.

Sakurai, A., Takahashi, T. (1969). Dynamic stresses of underground pipeline during earthquakes. Proceedings of the Fourth World Conference on Earthquake Engineering.

Schnabel, P.B., Lysmer, J., Seed, B.H. 1972. SHAKE—a computer program for earthquake response analysis of horizontally layered sites. Report no. EERC 72-12. University of California, Berkerley, CA, USA.

Wang, J.-N. (1993). Seismic Design of Tunnels: A State-of-the-Art Approach, Monograph, monograph 7. Parsons Brinckerhoff, Quade and Douglas Inc, New York.

TRACK 2: DESIGN

Session 5: Precast Concrete Tunnel Linings

Andre Solecki and John McCluskey, Chairs

Design of the San Francisco Public Utilities Commission's Channel Tunnel

Michael Deutscher and Samer Sadek
Jacobs Engineering

Manfred Wong
San Francisco Public Utilities Commission

ABSTRACT: The San Francisco Public Utilities Commission's (SFPUC's) Central Bayside System Improvement Project (CBSIP) includes the proposed 24 foot internal diameter, approximately 1.7 mile long Channel Tunnel (CHTL) for gravity conveyance and storage of wet and dry weather flows. The tunnel profile is approximately 110 to 150 feet below ground surface, located primarily within the Franciscan Complex. The shafts at either end of the tunnel require slurry diaphragm walls to achieve excavated diameters and depths of up to approximately 130 feet and 180 feet, respectively.

This paper presents the design progress to date, including a discussion of tunnel design challenges in complex geologic conditions and the design process of a large diameter slurry wall shaft in an active seismic zone.

INTRODUCTION

The Central Bayside System Improvement Project (CBSIP) is a major component of the San Francisco Public Utility Commission's (SFPUC's) city wide $6.9 billion Sewer System Improvement Program (SSIP). The primary objective of the CBSIP is to provide seismic resiliency, system reliability and operational flexibility that are not currently provided by the existing 66 inch diameter Channel Force Main (CHFM), a vital link in the Bayside wastewater system that has ruptured multiple times since the 1980s.

The proposed Channel Tunnel (CHTL) is the centerpiece of the CBSIP. At a proposed 24 foot internal diameter, the CHTL would be the largest diameter bored tunnel in the Bay Area to date. The proposed CHTL alignment is approximately 1.7 miles long and includes seven curves with radii ranging from 700 to 1,000 feet to stay more or less beneath the street right-of-way as shown in Figure 1. The CHTL will provide gravity conveyance and storage of wet weather (WW) flows and gravity conveyance of dry weather (DW) flows at a 0.2% downgrade from the existing Channel Pump Station (CHS) near the north end of the tunnel to the new Central Bayside Pump Station (CBS) at the south end. From there the flows will be delivered to the existing Southeast Water Pollution Control Plant (SEP; a.k.a., Southeast Plant) via two 48 inch diameter force mains.

The CHTL requires deep, large diameter shafts at either end to facilitate bored tunnel construction and to house permanent structures. The CHTL Launching Shaft is located at the south end of the CHTL, just south of Islais Creek, and will house the permanent CBS structure upon completion of tunneling. With a proposed excavated diameter of approximately 131 feet and an anticipated excavated depth of 183 feet, the Launching Shaft would be one of the largest, deepest shaft excavations in the Bay Area to date. The CHTL Retrieval Shaft is located at the north end of the CHTL, just south of Mission Creek, and will house a permanent drop structure to drop flows from the CHS connecting pipeline into the tunnel. The Retrieval Shaft is also a significant excavation with an excavated diameter of approximately 52 feet and an anticipated excavated depth of 140 feet.

The project also includes three major connections:

1. The Northern Connection, a connection between the CHS and CHTL Retrieval Shaft at the Parcel P7 site near the intersection of Owens Street and Mission Bay Drive, which will likely include a 9.5 foot (±) inside diameter microtunnel, approximately 600 feet long, along with an associated microtunnel retrieval shaft in Berry Street;
2. The Southern Connection, consisting of twin 48 inch diameter force mains between the CBS (CHTL Launching Shaft) and the SSIP Headworks at the SEP, which also will likely include trenchless installations; and
3. The Intertie Connection, a connection between the existing Islais Creek Transport/

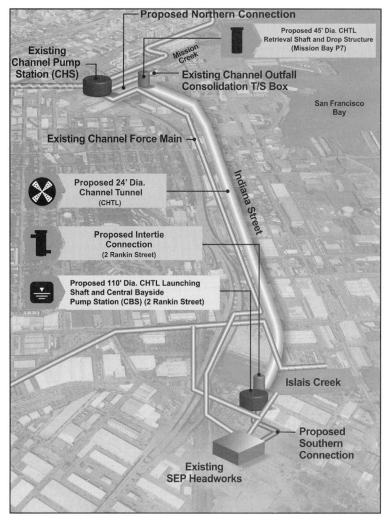

Figure 1. CHTL alignment

Storage Box (ICT), located near the south end of the CHTL, and the CHTL via a drop structure and adit connection at tunnel level.

Project Procurement and Schedule

The project will be delivered using traditional design-bid-build procurement methods. The project design team concluded 35% Design in 2017 and are anticipated to conclude final (100%) design of the CHTL and associated shafts by the end of 2018 (design of the CBS permanent structure will continue beyond 2018). Bid and award of the CHTL construction contract, including temporary support and excavation of the Launching Shaft and the Retrieval Shaft, are presently envisioned to occur in 2020 with construction occurring between mid-2020 and mid-2022.

Construction of CBS structures and force mains will likely be in separate later contracts.

Project Team

The project design team is led by SFPUC and their prime consultant Stantec. Jacobs Engineering Group, Inc. (Jacobs) is a major subconsultant, responsible for the design of the CHTL tunnel and the temporary support of excavation for the Launching Shaft and Retrieval Shaft. Jacobs is also responsible for geotechnical engineering for the project. The permanent CBS structure, to be constructed in a separate contract at the conclusion of tunneling operations, is being designed by the San Francisco Department of Public Works (SFPW).

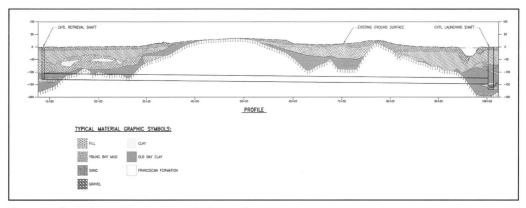

Figure 2. Geologic profile along tunnel alignment

Scope

The focus of this paper is the constructability and design of the CHTL and associated Launching Shaft and Retrieval Shaft as of the conclusion of preliminary design (35% Design).

PROJECT GEOLOGY

The project is located along the eastern side of the San Francisco peninsula, about 2,000 feet west of San Francisco Bay. The ground surface along the alignment is typically 10 to 40 feet above bay level. Mission Creek lies at the northern end of the alignment and Islais Creek lies at the southern end; both are tidal, as is the bay. Potrero Hill lies further to the west and rises about 300 feet above the bay.

The geology along the CHTL alignment consists mostly of the Franciscan Complex overlain by the soils of the Bay Sequence. A transitional interval of colluvium and alluvium commonly occurs at the top of the Franciscan Complex.

The Franciscan Complex is a melange, or chaotic mixture, of metamorphic, sedimentary and igneous rocks. The most common rocks along the project alignment are serpentinite, graywacke sandstone and shale-matrix melange (commonly abbreviated to shale or melange on the boring logs); minor lithologies include chert, quartzite and greenstone. Unconfined compressive strength testing of intact specimens of greywacke sandstone yielded strengths as high as 16,791 psi (115.77 MPa) but were typically under 10,000 psi (68.95 MPa) while unconfined compressive strength testing of intact specimens of serpentinite returned strengths as high as 11,063 psi (76.27 MPa) but were typically under 3,000 psi (20.68 MPa). Unconfined compressive strength testing of the mélange (shale mélange, serpentinite mélange) yielded highly variable strength results, from as high as 8,988 psi (61.97 MPa) to as low as 44 psi (0.31 MPa). Much of the rock is highly sheared

and fractured, with secondary talc and epidote along the fractures. Naturally occurring asbestos is present in places within the Franciscan Complex while other areas contain elevated concentrations of heavy metals, such as chromium and nickel.

The five main soil units of the Bay Sequence at the project site, in descending order, are: artificial fill, Young Bay Mud; Old Bay Clay; Bayside Sands and Lower Sands. Much of the 10 to 25 feet thick fill layer along the CHTL alignment consists of demolition debris from the 1906 San Francisco Earthquake and accordingly is uncontrolled and contains hazardous substances in places. The Young Bay Mud is up to 70 feet thick at shaft locations and consists of mainly soft to very soft, normally consolidated to slightly overconsolidated, silty fat clay (CH) with organics and layers of peat. The Old Bay Clay is a stiff to hard, overconsolidated, silty to sandy clay (CL-CH) with some layers of clayey sand and minor layers of sand and gravel. The Bayside Sands are a series of fine-grained, uniform sand lenses that tend to be inter-fingered with the Young Bay Mud and the Old Bay Clay; the Lower Sands below tend to be coarser and less uniform. Colluvial and alluvial clay deposits above the Franciscan Complex can be difficult to distinguish from Old Bay Clay and are sometimes combined together with it.

Geology Along the Tunnel Alignment and at Shaft Locations

The geologic profile along the tunnel alignment, including shaft locations, is presented in Figure 2. The proposed tunnel profile is approximately 110 to 150 feet below ground elevation (measured to the crown) and is located mostly within the Franciscan Complex (rock and mélange) with short lengths of mixed face and soft ground tunneling conditions encountered within basins at either end. The mixed face conditions consist of Old Bay Clay over Franciscan Complex. The soft ground tunneling

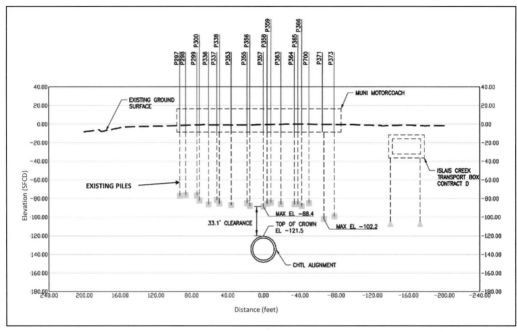

Figure 3. Typical tunnel cross section of proposed CHTL relative to the existing Muni Motor Coach Facility and the existing Islais Creek Transport/Storage Box (ICT), taken near Tulare Street

conditions consist primarily of Old Bay Clay but will also encounter sandy layers. The groundwater table is typically within a few feet of the ground surface.

The shaft excavations will be entirely within the soils of the Bay Fill Sequence. Both the launching and retrieval shafts will encounter fill, Young Bay Mud and Old Bay Clay as well as the Bay Side Sands and possibly the Lower Sands. The groundwater table at both shaft locations is within a few feet of the ground surface.

The relatively deep tunnel profile presented in Figure 2 was selected over shallower profiles in order to avoid existing piles along the alignment, which have historically been used to provide foundation support below the softer soils of the Bay Fill Sequence (i.e., Young Bay Mud) and toed into stiffer/denser soils (e.g., Old Bay Clay, sandy layers). For example, the tunnel profile was set to avoid piled foundations for both an existing SFMTA Motor Coach Facility and the Islais Creek Transport/Storage Box (ICT) just north of Islais Creek; a cross section of the tunnel location relative to these structures is presented in Figure 3. The deeper tunnel profile was also selected to avoid tunneling in the Young Bay Mud and to mitigate near-surface ground movements.

TUNNEL CONSTRUCTABILITY AND DESIGN

The proposed tunnel will be constructed using a pressurized face Tunneling Boring Machine (TBM) along with a precast concrete liner that will serve as both initial and final lining (i.e., a one-pass lining system). A pressurized face TBM with one-pass lining system was selected primarily due to the extremely variable and difficult geologic conditions that will be encountered along the tunnel alignment, including sheared shale with embedded blocks and slabs of stronger greywacke; soft ground tunneling conditions that include saturated sand (i.e., potentially flowing ground); and mixed face tunneling conditions.

Selection of Geotechnical Parameters and Critical Cross Sections for Analysis

The variable geologic conditions along the alignment posed a challenge both to the selection of geotechnical parameters and to the identification of potentially critical cross sections for the structural analysis of the liner. Given the geologic history of the site, it is possible for relatively stiff geologic materials to be juxtaposed with relatively soft geologic materials in virtually any configuration at virtually any location along the tunnel alignment. For example, in the full face Franciscan Complex reaches of the tunnel alignment, strong greywacke sandstone can be located immediately above or below weak mélange. Similarly, in the soft ground reaches, Old Bay Clay can be located above or below very dense sands.

To address this challenge, the general approach used to select geotechnical parameters for subsequent structural analysis of the liner was as follows:

- For reaches of the tunnel in soil, a statistical analysis was performed for each soil parameter required for structural analysis (i.e., unit weight, lateral coefficient of earth pressure at rest, stiffness, small strain stiffness and Poisson's ratio) following the recommendations by Raymer (2010), with consideration to test data from soil specimens originating near the elevation of the tunnel only. Because the various soil layers at tunnel level can occur adjacent to each other in any sequence, the data for all soil layers at tunnel level was grouped together for the statistical analyses to determine reasonable upper and lower bound parameters. In future design stages (as more laboratory testing results become available), the data for the soil at the opposite ends of the tunnel will be considered separately; this should reduce the contrast between the upper and lower bound soil parameters within individual cross sections, which could in turn improve the efficiency of the design.
- For reaches of the tunnel in Franciscan Complex, a similar statistical approach was used. Unlike the statistical analyses for the soil parameters, however, the statistical analyses for the Franciscan Complex considered test data from all elevations (as opposed to test data from specimens originating near the elevation of the tunnel only); this is considered to be a reasonable approach given the lack of correlation of geotechnical parameters with depth within the Franciscan Complex.

Structural Analysis

Potentially critical cross sections were identified for structural analysis, including sections within full face soft ground, full face Franciscan Complex and mixed face conditions. The upper and lower bound geotechnical parameters, as determined by the statistical analyses described above, were considered both individually and in combination for the analysis of the potentially critical cross sections.

Structural analysis performed for 35% Design addressed initial static loading conditions, permanent (long term) static loading conditions and seismic loading conditions (using small strain stiffness parameters). Construction loading conditions (e.g.,

mold stripping, stacking, erection, jacking loads, etc.) will be addressed during subsequent design stages.

For sections in full face Franciscan Complex, the vertical effective earth pressure was determined based on a height of rock column equal to twice the tunnel diameter, as opposed to full overburden, in order to account for ground arching. For sections in soft ground, or mixed face conditions, the vertical effective earth pressure was determined based on the full height of overburden (i.e., no ground arching). Horizontal effective earth pressures were calculated using the coefficient of earth pressure at rest, as determined by the statistical methods described above. For reaches in a full face Franciscan Complex, two horizontal earth pressures cases were considered:

- Assuming ground arching, in which case the horizontal effective earth pressure is calculated as the coefficient of lateral earth pressure at rest multiplied by the vertical effective earth pressure based on a height of overburden equal to twice the tunnel diameter; and
- Assuming squeezing ground conditions, in which case the horizontal effective earth pressure is calculated as the coefficient of lateral earth pressure at rest multiplied by the vertical effective earth pressure based on the full height of overburden.

The structural analyses performed for potentially critical cross sections included:

- Closed form calculations (Muir Wood, 1975; Curtis, 1976) for all critical locations, with different loading scenarios and geotechnical parameters, considering both initial and permanent conditions;
- Numerical beam spring modelling for selected critical sections with different loading scenarios and geotechnical parameters to verify the closed form calculations and to evaluate the effect of varying ground stiffness within the same cross section; and
- Closed form calculations for initial assessment of seismic loads (Hashash, 2001) for a return period of 975 years as required by SFPUC design guidelines. The analysis used a Peak Ground Acceleration (PGA) of 0.55 g as recommended by the project Seismic Hazard Assessment (Stantec, 2017). The analysis for 35% Design included determination of seismic loads due to ovaling of the tunnel section only; determination of seismic loads for compression-extension and for longitudinal bending of the tunnel will be performed in subsequent design stages.

Table 1. Precast concrete segmental lining design base data (35% design)

Parameter	Assumptions Used in the Analysis
Material	Reinforced Concrete or Steel Fiber Reinforced Concrete (as two alternatives)
Lining thickness	12-inches (for conventional reinforcement) 15-inches (for fiber reinforcement)
Setting out internal diameter	24-feet
Nominal excavated diameter	27-feet (12-in Liner) 27.5-feet (15-in Liner)
Design Life	100 years
Ring length	5 feet (60-inches)
Ring configuration	5+1 (4 standard segments, one counter key segment plus 1 key segment)
Ring geometry	Universal taper
Segment geometry	Rhombodial segments

Structural Design

The axial forces, shear forces and bending moments in the tunnel liner, as determined from the structural analysis above, were used for the tunnel liner structural design. The tunnel liner was designed as a short beam-column. The results of the structural analyses were plotted on a moment-axial load interaction diagram along with the failure envelopes for: (a) plain concrete sections; (b) conventionally reinforced concrete sections; and (c) steel fiber reinforced concrete (SFRC) sections. The structural analysis and design performed to date demonstrate that either a 15-inch thick SFRC segmental lining or a 12-inch thick conventionally reinforced concrete segmental lining are adequate for the load case combinations as set in the project Design Criteria (Jacobs et al., 2017). Information on the precast segmental lining design at the 35% Design stage is presented in Table 1.

The use of steel fiber reinforced concrete is considered as an alternative to conventional reinforced concrete despite the greater thickness because the use of steel fibers offers advantages with respect to segment casting and with respect to durability. The ring length and configuration have been selected with consideration to the opposing objectives of minimizing the number of joints while maintaining a reasonable size of individual segments for handling considerations. The use of a universal taper affords the tunnel liner the ability to navigate the planned curvature of the tunnel alignment (along with minor deviations from the design line and grade) without the need for multiple types of ring sets. Rhombodial segments have been selected for

ring build considerations, particularly with respect to minimize the possibility of dislodging the gaskets while erecting the ring.

SHAFT CONSTRUCTABILITY AND INITIAL SUPPORT OF EXCAVATION DESIGN

Plan and profile views of the proposed CHTL Launching Shaft are presented in Figures 4 and 5, respectively, while plan and profile views of the CHTL Retrieval Shaft are presented in Figures 6 and 7, respectively. The Launching Shaft support of excavation (SOE) has a proposed internal diameter of 121'-2", which includes a 7 inch construction tolerance on radius to provide room for a 110 foot internal diameter permanent structure with 5 foot thick concrete walls. The Retrieval Shaft SOE has a proposed internal diameter of 46'-2", which includes a 7 inch construction tolerance on radius to provide room for a 39 foot internal diameter permanent structure with 3 foot thick concrete walls. The required depths of excavation for the Launching Shaft and Retrieval Shaft are approximately 183 feet and 140 feet, respectively.

Both the CHTL Launching Shaft and the CHTL Retrieval Shaft will be constructed using slurry walls for initial SOE. Slurry walls are considered appropriate given the high water tables and the significant depths of potentially unstable soil conditions at the shaft sites, including squeezing clays and flowing sands. Other SOE systems that were considered (and the main reason why each was not selected) include: secant pile walls (not typically used for the excavation depths considered); cutter soil mixing (strength of soil-cement mixture insufficient to withstand forces at depth with a reasonable wall thickness); and ground freezing (difficulty in freezing highly plastic clays).

The slurry wall panels for each shaft will be arranged in a circular configuration in plan; the circular configuration will result in lower lateral loads and reduced bending moments on the shaft walls (as compared to a rectangular or square configuration). The slurry wall panels will be keyed a minimum of ten feet into the Franciscan Complex to provide a groundwater cutoff, which will allow the shafts to be excavated in the dry with reduced dewatering efforts (dewatering of the soils within the confines of the shaft only). Packer tests performed in the underlying bedrock during the final geotechnical investigation program indicate low permeability at the proposed shaft locations (Jacobs, 2017). Supplemental grouting may be required at the toe of the slurry wall panels to improve the groundwater cutoff where zones of permeable features in the rock may exist.

The key constructability issues for the shafts include space constraints at the Launching Shaft site, which is only 1.72 acres; the installation tolerance of the slurry panels, which are over 200 feet

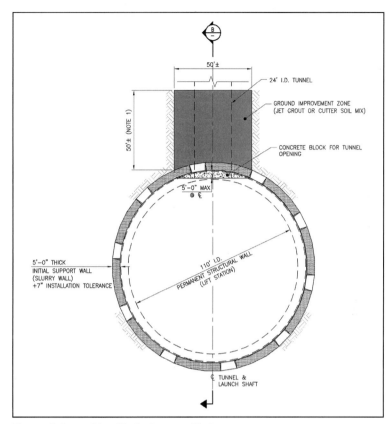

Figure 4. Launching Shaft slurry wall plan

deep; the presence of deep foundation obstructions from previous buildings at the sites (these have been investigated with ground penetrating radar scans and test pits); and the stability of slurry trenches through very soft soils (Young Bay Mud), which may require ground improvement in advance of slurry panel excavation.

As of the conclusion of 35% Design, the slurry wall is designed as a temporary SOE only; the reinforced concrete permanent shaft structures are designed assuming no contribution from the slurry wall with the exception that the permanent base slab will be tied into to the slurry wall to provide resistance to hydrostatic uplift forces. The potential for incorporating the slurry wall into the permanent structures as a means to reduce the thickness of the final lining will be investigated during 65% Design.

Selection of Geotechnical Parameters

Geotechnical parameters for shaft SOE design were determined for each soil layer with consideration to statistical evaluation of test data following the recommendations of Raymer (2010). The upper and lower bounds were established based on judgement with

heavy consideration given to the statistical model, but also considering the number of data points, the sensitivity of the parameter being assessed, typical ranges of values and the quality of the data collected. In cases where the geotechnical parameter in question is known to be heavily dependent on depth (e.g., undrained shear strength of the Young Bay Mud), statistical methods were supplanted by depth based functions. Statistical analysis of unit weight, plasticity index, and fines content considered data collected from all available borings due to the relatively low variability of these properties. Conversely, analysis of Standard Penetration Test "N"-Values, stiffness, strength, and permeability considered only data from borings at the shaft location in question.

Structural Analysis Overview

The analysis of the slurry walls for the shafts included both closed form (hand) calculations and numerical modeling methods using finite element software (i.e., Plaxis 3D), as discussed in the sections that follow. The closed form analyses, which were performed for both the launching and retrieval shafts, considered static loading conditions only. The

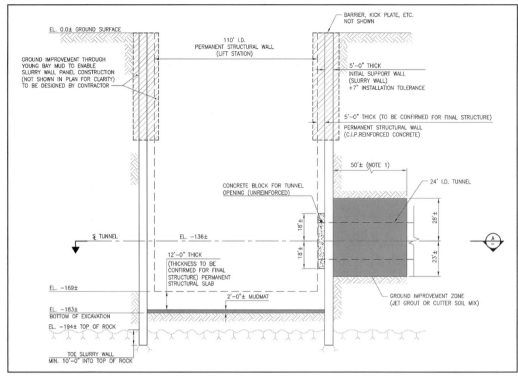

Figure 5. Launching Shaft slurry wall elevation

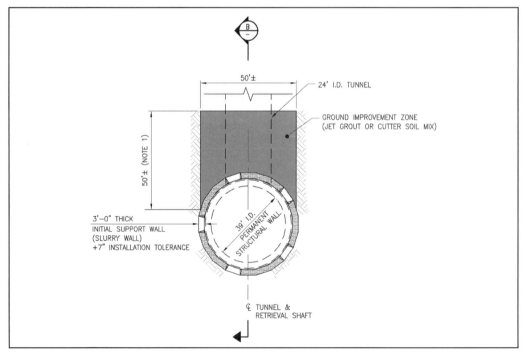

Figure 6. Retrieval Shaft slurry wall plan

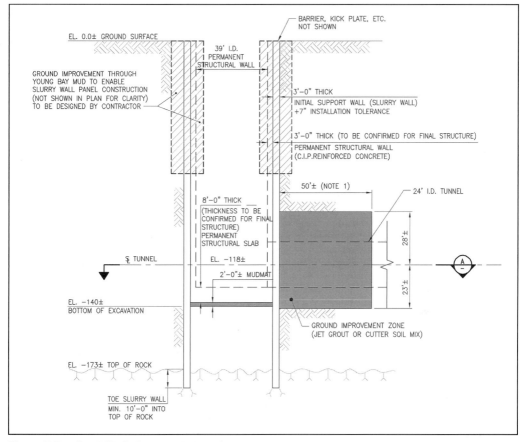

Figure 7. Retrieval Shaft slurry wall elevation

finite element analyses, which were performed for the launching shaft only (i.e., the larger of the two shafts), considered static loading conditions for 35% Design and will include dynamic loading conditions in subsequent design stages (dynamic finite element analysis was on-going at the time of writing). In lieu of full dynamic finite element results, the outcome of a simplified seismic analysis is presented below.

Closed Form Analysis

Closed form (hand) calculations were performed to calculate the maximum compressive stress in the slurry wall concrete for initial sizing. Slurry wall effective thickness reduction due to construction tolerance was considered in the calculations. Stress concentration at the tunnel opening was estimated based on closed form solutions for circular holes in strained plates.

The results of the closed form analysis for the Launching Shaft show stress concentrations on the proposed five foot thick slurry wall at the tunnel opening will be close to 5,000 psi, the compressive

strength of the concrete. To address the stress concentration issue, localized reinforcement of the slurry wall in the form of a reinforced concrete block was considered at the tunnel opening in the subsequent finite element analysis below.

The results of the closed form analysis for the Retrieval Shaft indicate the proposed three foot thick, 5,000 psi slurry wall panels can accommodate the anticipated design loads, including at the CHTL opening location (i.e., a concrete block is not required at the opening location).

Finite Element Analysis

The Hardening Soil (HS) Model was used in the Plaxis 3D static finite element analysis to simulate the overburden soil layers (i.e., Bay Sequence Strata), while the Jointed Rock Model was used to simulate the Franciscan Complex. The Launching Shaft slurry wall was modelled as individual concrete panels with interface elements between the panels. The analysis considered the sequence of excavation including the potential for unbalanced loading conditions during shaft excavation due to surcharge loads (up to 600 psf) and uneven excavation from one side of the shaft to the other (up to a 30 foot difference). The static finite element analysis also considered the stress concentrations at the opening for the tunnel break out.

The results of the static finite element analysis are presented in Figure 8. The results indicate the slurry wall can adequately accommodate the analyzed unbalanced loading conditions in compression. Conversely, similar to the closed form analysis, the static finite element analysis indicated an overstressed condition (compressive failure) in the five foot, 5,000 psi slurry wall panels at the location of the opening for the CHTL. Accordingly, a reinforced concrete block was added at the opening (maximum 4.5 feet thick at the tunnel centerline) to reduce the stress in the slurry wall concrete to acceptable levels, as demonstrated by additional finite element analysis, as shown in Figure 9.

It should be noted that the above stress evaluations are preliminary and for the purpose of validating the numerical model and closed-form hand calculations. The shaft design and its corresponding calculations are expected to be updated during the final design stage, including a more rigorous accounting of both load and resistance factors.

Dynamic finite element analysis of the shaft was on-going as of the time of writing. The Hardening Soil Model with Small-Strain Stiffness (HS small model) is used to simulate the overburden layers for the dynamic analysis.

Seismic Analysis to Date

Seismic design is considered necessary for the slurry wall structures despite their being temporary SOE only (as of 35% Design) since the structures are significant in size and will be open for a considerable time (i.e., three to five years). The seismic analysis considers acceleration time histories from an 80 year return earthquake, as per the project design criteria (Jacobs et al., 2017), which is considered sufficient for the given timeframe.

Free-field ground deformation was modeled using ProShake Version 2.0, which is a one-dimensional, equivalent linear ground response analysis program commonly used to simulate free-field responses in soil layers given earthquake acceleration time histories at top of Franciscan Complex. A simplified soil profile was used by assuming only three soil layers in the model. Six earthquake acceleration time histories (i.e., two time histories from each seismic event, one in the direction of the strong motion and the other in the direction perpendicular to the strong motion), as provided in the project Seismic Hazard Assessment (Stantec, 2017) were input as Franciscan Complex accelerations. The

largest relative ground deformation (i.e., the differential displacement between ground surface and top of Franciscan Complex) generated by the analysis, in the direction of the strong motion, is about 1.6 inches. Conservatively assuming the maximum relative deformation is also 1.6 inches in the direction perpendicular to the strong motion, the maximum resultant relative deflection between the ground surface and Franciscan Complex is calculated to be about 2.3 inches (i.e, $\sqrt{2} *1.6=2.3$ inches).

Free-field ground deformation was also modelled using Plaxis 3D dynamic analyses (i.e., a simplified model that does not consider the shaft structure). A simplified soil profile (i.e., assuming only four different soil layers in the model) was used to reduce the number of elements and hence reduce the calculation time. The acceleration time histories from the governing seismic event (as determined from the ProShake analysis above) were used as dynamic input at the top of the rock elevation of the model. The earthquake accelerations were applied in both the horizontal X and Y directions (i.e., the acceleration in the vertical Z direction was neglected). The dynamic boundary conditions in the two horizontal directions (i.e., at X_{min}, X_{max}, Y_{min} and Y_{max} locations) were set as a free-field boundary condition. The free-field boundary simulates the propagation of waves into the far field with minimum reflection at the boundary. According to Plaxis 3D manual, this boundary condition is generally preferred for earthquake analysis. The maximum relative deformation between the ground surface and Franciscan Complex from the Plaxis 3D free-field dynamic analyses is about 0.26 foot (3.1 inches). Note this deformation is the resultant ground vibration of both the X and Y directions.

Ground surface displacement response time histories calculated from both the ProShake and the Plaxis 3D dynamic analyses were plotted together for comparison in Figures 10 and 11 for the X and Y directions, respectively. It can be seen that good agreement was achieved between the two models; the minor differences in results can be attributed to differences in the basic calculation assumptions and boundary conditions used in the two programs. Note that in the Y direction the result from the Plaxis 3D dynamic analysis indicates a minor permanent ground displacement of about 0.05 feet (0.6 inches) at the end of ground shaking.

In order to estimate the stress imposed on the shaft SOE by the design seismic event, the maximum free-field ground deformation for the 80-year return earthquakes analyzed, as determined by the Plaxis 3D dynamic analysis described above, was imposed on a Plaxis 3D finite element model of the shaft by applying a uniform lateral load to the shaft to produce 3.1 inches of deflection at the top of the shaft

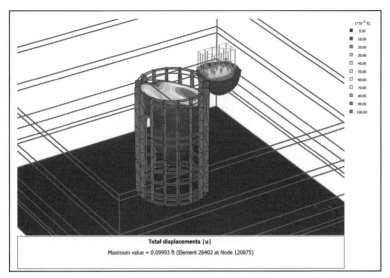

Figure 8. Uneven excavation with live load surcharge—total displacement (3D plot)

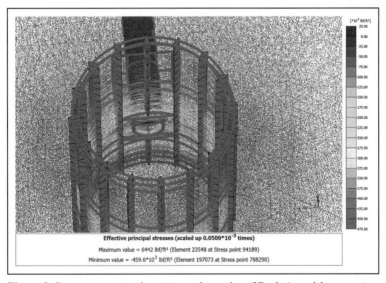

Figure 9. Stress concentration at tunnel opening (3D plot)—with concrete block

relative to the Franciscan Complex. This approach is assumed to be conservative as the shaft deformation will likely be less than the free-field ground deformation. The net increase in the shaft concrete stress is about 300 psi (0.3 ksi) which is considered minor. Accordingly, it is concluded that the seismic load case is unlikely to govern design of the slurry wall considering earthquakes with an 80-year return period.

SUMMARY

When completed, the CBSIP will provide much needed seismic resiliency, system reliability and operational flexibility to San Francisco's sewer system for the twenty first century. The project will feature the largest bored tunnel, as well as some of the largest, deepest shaft excavations, constructed in the Bay Area to date. The main design challenges include: tunnel liner design in a highly variable,

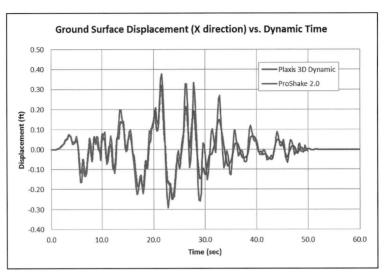

Figure 10. Comparison of ProShake and Plaxis 3D results—X direction ground displacement

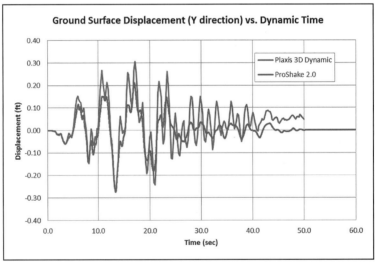

Figure 11. Comparison of ProShake and Plaxis 3D results—Y direction ground displacement

complex geology; design of deep, large diameter shafts in soft, potentially unstable soils under high head of water with large openings for tunnels and other connections; and adequately addressing seismic design requirements. The main construction challenges include: bored tunneling in highly variable, complex geology requiring mechanized tunneling using a pressurized face TBM with one pass segmental lining; and installation of slurry walls through potentially unstable soils to depths exceeding 200 feet. The project team has recently completed 35% Design, which features a CHTL lining consisting of either 12 inch thick conventionally reinforced segments or alternatively 15 inch thick SFRC segments. The design of the initial SOE for the Launching Shaft and Retrieval Shaft includes five foot thick and three foot thick slurry wall panels, respectively. The design is currently scheduled to advance though 2018 with bid and award of the CHTL construction contract in 2020.

REFERENCES

Curtis, D.J. 1976. Discussions on Muir Wood: The circular tunnel in elastic ground. Geotechnique 26, Issue 1. Thomas Telford, London, pp. 231–237.

Goldman, H.B. 1969. Geologic and engineering aspects of San Francisco Bay Fill. California Bureau of Mines and Geology, Special Report 97, San Francisco.

Hashash, Y.M.A., Hook, J.J., Schmidt, B. and Yao, J.I.C. 2001. Seismic design and analysis of underground structures. Tunneling and Underground Space Technology 16 (2001), pp. 247–293.

Jacobs and Stantec. 2017. Channel Tunnel, Temporary Shafts and Connections Draft Design Criteria Technical Memorandum dated August 2017. Prepared for the San Francisco Public Utilities Commission for the Central Bayside System Improvement Project.

Jacobs. 2017. Phase 3 Geotechnical Data Report, Preliminary Draft dated September 2017. Prepared for the San Francisco Public Utilities Commission for the Central Bayside System Improvement Project.

Muir Wood, A.M. 1975. The circular tunnel in elastic ground. Geotechnique 25, Issue 1. Thomas Telford, London pp. 115–127.

Raymer, J. 2010. Geotechnical variability and uncertainty in long tunnels. Proceedings of the 2010 North American Tunneling Conference, Society for Mining, Metallurgy & Exploration, Inc., pp. 16–322.

Stantec. 2017. Draft Final Task 9.6 Seismic Hazard Assessment dated August 10, 2017. Prepared for the San Francisco Public Utilities Commission for the Central Bayside System Improvement Project.

Wakabayashi, J. 2004. Contrasting settings of serpentinite bodies, San Francisco Bay Area, California: derivation from the subducting plate vs. mantle hanging wall. International Geology Review, Vol. 46, pp. 1103–1118.

Designing a Bored Tunnel for the Unthinkable—A Marine Vessel Collision Assessment

David Watson, Chris Pound, Christoph Eberle, and Botond Beno
Mott MacDonald

ABSTRACT: Bored tunnels are generally constructed underground to avoid surface features, whether manmade or naturally occurring. Designing a bored tunnel to avoid potential marine vessel collisions can be viewed as somewhat exceptional. The design of the Parallel Thimble Shoal Tunnel was one such case where engineers needed to find innovative, yet code compliant, design solutions for this unusual scenario. A codified probabilistic approach, intended to assess the vessel impact on bridges, in conjunction with empirical vessel collision test data, was used to calculate the vessel collision force. The integrity of the precast concrete segmental tunnel liner was then assessed using complex ground structure finite element modelling. The results of the analyses were compared against the owner's serviceability design state criteria and successfully verified.

INTRODUCTION

The Chesapeake Bay is an estuary that is surrounded by the North American mainland to the West, and the Delmarva Peninsula to the east, with more than one hundred major rivers and streams flowing into the drainage basin. The geographic setting of Chesapeake Bay is shown in Figure 1.

Chesapeake Bay hosts a number of naval centers which specify the requirement for a trestle free section of the bay crossing that allows access for large cargo and military vessels. The Parallel Thimble Shoal Tunnel Project includes the construction of a new 1.1-mile-long tunnel under Thimble Shoal Channel of the Chesapeake Bay, which will be located west of the existing tunnel crossing. When complete, the new tunnel will carry two lanes of traffic southbound and the existing tunnel will carry two lanes of traffic northbound. An aerial picture the existing crossing facility is shown in Figure 2.

Due to the heavy vessel traffic, the tunnel lining was assessed against the unlikely events of a sunken ship resting above the tunnel and a ship colliding against the tunnel protection layer, the latter being the focus of this paper.

The proposed tunnel alignment, cover and anticipated ground conditions are shown in Figure 3.

TECHNICAL REQUIREMENTS

The project technical requirements called for calculations to verify the adequacy of the engineered fill berm (tunnel protection layer) acting in combination with the tunnel lining to resist a potential ship collision scenario. This requirement called for a ground structure interaction analysis followed by specific structural checks of the tunnel lining itself. To satisfy this criterion, the design approach presented in Figure 4 was used.

The technical requirements define the performance criteria expected for various load combinations as defined by American Association of State Highway and Transportation Officials, Load and Resistance Factor Design (AASHTO LRFD) specifications. Ship collision was categorized as Extreme Event II meaning that "This event can produce light damage to the tunnel, minor leaks, and some loss of service."

AASHTO LRFD only considers a vessel collision scenario applicable to immersed tube tunnels and in Chapter 10 Tunnel Lining, rules out the Vessel Collision Force (CV) for bored tunnels by stating that "This force is not applicable since it would only be applied to immersed tunnels." Notwithstanding the above, the unique setting of this bored tunnel, at a shallow depth in a busy shipping channel with an engineered berm emerging above the water level, the design team evaluated potential ship impact loading for the bored tunnel lining. Locations of potential ship impact on the engineer filled berm tunnel protection at both islands is indicated in Figure 5.

SHIP COLLISION FORCE—PROBABILISTIC ANALYSIS

Given that there are no established codes or standards for deriving ship impact loading on bored tunnels, a codified probabilistic approach intended to be used on for the design of bridge piers was proposed by the design team. After consultation with the project owner and owner's engineer, this approach was

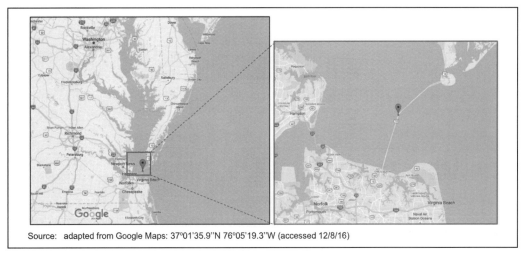

Source: adapted from Google Maps: 37°01'35.9''N 76°05'19.3''W (accessed 12/8/16)

Figure 1. Geographic setting of Chesapeake Bay

Figure 2. Existing Thimble Shoal crossing, surface structures

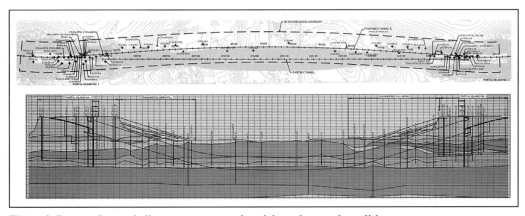

Figure 3. Proposed tunnel alignment, cover and anticipated ground conditions

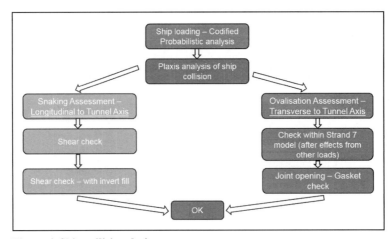

Figure 4. Ship collision design process

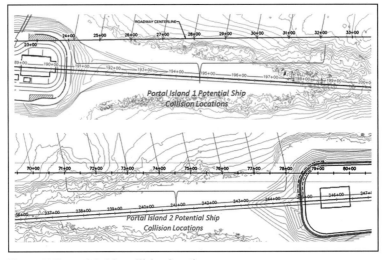

Figure 5. Potential ship collision locations

mutually agreed upon to be a reasonable approach to quantifying the ship impact load.

The details of this method are set out in AASHTO (2009, 2014) Section 4.8—Method II for Design Vessel Selection—"Probability-based analysis procedure for determining the design vessel.," but in essence, the aim is to prove that the Annual Frequency of Collapse (AF), is less than 0.1 in 100 years for *Typical Bridges* and less than 0.01 in 100 years for *Critical/Essential Bridges* (in this case tunnel). For this project, the proposed tunnel is categorized as Critical/Essential therefore a return period of less than 0.01 in 100 needed to be achieved.

This method is used to allow engineers to consider the unlikelihood of a marine vessel striking an offshore structure when determining the design impact load applicable to the Annual Frequency of Collapse. The approach considers a reduction in ship impact forces that would be predicted by other means, (e.g., small scale physical testing or ship collision modelling modeling) to satisfy the Annual Frequency of Collapse criteria. In other words, the method allows the engineer not to design for an absolute worst case combination of the largest possible vessel travelling at full speed hitting an offshore structure that is some distance away from the navigable channel.

The Annual Frequency of Collapse is a combination of multiple probabilities that consider the likelihood of Aberrancy (an incident), coastal conditions, waterway geometry, structure capacity, and the level/type of vessel traffic. First, we need to establish the probability that a vessel becomes aberrant, then, the probability that a vessel will strike the structure given that it becomes aberrant, and finally the probability that the structure will collapse given that a vessel is aberrant and strikes the structure. Once these variables are determined the annual frequency of collapse can be computed by: AF = (N) × (PA) × (PG) × (PC) × (PF). A graphical representation of this design process is illustrated in Figure 6.

To apply this methodology to a tunnel, the length of tunnel protection layer that could feasibly be impacted by a ship was discretized into 100-ft sections. Each section was evaluated using the probabilistic method, with the final Annual Frequency of Collapse being calculated by a summation of the results for all 100-ft lengths.

GROUND STRUCTURE INTERACTION ANALYSIS

Once the design loads were established for every section of the tunnel (at 100-ft intervals), the most critical sections in terms of impact on the tunnel lining were evaluated. The most critical sections were

selected not only by looking at the largest possible design impact loads but also by considering the geometry of the engineered fill berm protecting the tunnel from such loads and the geological conditions around the tunnel. The critical sections selected at each portal island are Station 30+00 on Portal Island No. 1 and Station 71+00 on Portal Island No. 2.

To determine the displacements for the longitudinal checks and to determine forces within the lining for the cross-section checks, a PLAXIS (geotechnical analysis & design software applied for deep foundations, tunnels and dams) analysis was undertaken for both sections. A typical PLAXIS geometry is illustrated in Figure 7. The ship collision was assessed for both short term and long terms conditions in two locations: tunnel centerline and sheet pile location. Two long term assessments were undertaken at Station 71+00 to account for degradation of the sheet piles.

The basis of the assessment is as follows:

- The ship collision loading was assumed to occur at the ship keel path

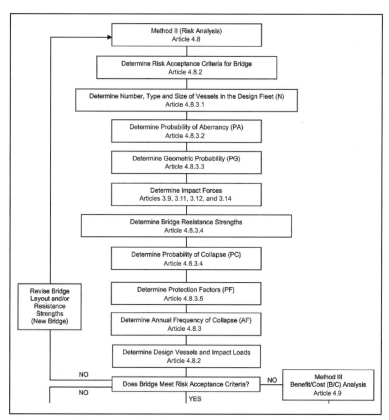

Figure 6. Design process based on AASHTO (2009, 2014) Section 4.8, Method II for Design Vessel Selection

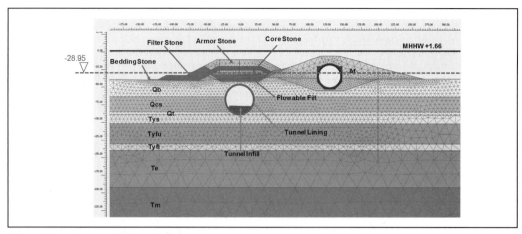

Figure 7. Typical PLAXIS model geometry

- Ship collision load increased until berm starts to fail. This stage matches the highest load on the tunnel
- Once the failure load is identified, movements and forces/moments in the segmental lining are predicted for a given impact load level
- These are then compared to the maximum collision load back-calculated to the AASHTO methodology to give a maximum ship beam length
- Tunnel lining is then checked longitudinally and in cross section

The most onerous results from the analyses at both critical sections was obtained in a format presented below:

Station 30+00 (Short Term)—Displacements

The maximum applied point load per foot (kip/ft); Horizontal Displacement at Axis Level (in); and Maximum Diametral Change (in) from the following two runs:

- In Run # 1, maximum applied point load per foot run was calculated by obtaining limiting passive resistance using PLAXIS analyses. Concluding that the maximum allowable ship beam for failure to occur is 82 ft.
- In Run # 2, maximum applied point load per foot was calculated by normalizing maximum collision load back-calculated using AASHTO method by assumed original ship beam of 141 ft.

Station 71+00 (Short Term)—Displacements

The maximum applied point load per foot (kip/ft); Horizontal Displacement at Axis Level (in); and Maximum Diametral Change (in) from the following two runs:

- In Run # 1, maximum applied point load per foot run was calculated by obtaining limiting passive resistance using PLAXIS analyses. Concluding that the maximum allowable ship beam for failure to occur is 53 ft.
- In Run # 2, maximum applied point load per foot was calculated by normalizing maximum collision load back-calculated using AASHTO method by assumed original ship beam of 141 ft.

The results of the analyses indicated that the most significant displacements occurred at Station 30+00 when the impact load was assumed to applied over a 82 ft ship beam impact area, which is appropriate for the passive case i.e., the impact area over which the impact load is needed to be applied to initiate berm failure.

SNAKING—LONGITUDINAL ASSESSMENT

To check for bending moments due to the longitudinal snaking effect, the two cases described above were considered for the worst case (Station 30+00 short term) and checked in a MIDAS (Structural analysis & design software for advanced nonlinear analysis) beam spring model where the tunnel was modelled as a linear 1D element with properties assigned as a 42 ft diameter 1.5 ft thick tube; meaning the tunnel was treated as a monolithic continuous thick shelled cylinder.

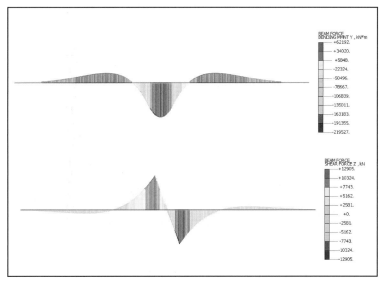

Figure 8. Exemplary longitudinal bending moment and shear force in the liner

This model was then calibrated to give the calculated PLAXIS deflection for a uniformly distributed load (UDL) applied over the full length of tunnel. The maximum impact load was then applied over 82 ft or 141 ft as appropriate and the resulting deflections were calculated. Comparing the MIDAS and PLAXIS deflection results confirms that the longitudinal stiffness of the tunnel has a significant impact on the longitudinal tunnel deflection.

Section 30+00 (Short Term)—Longitudinal Assessment

For both Run #1 and Run #2 described above, a longitudinal assessment was made in MIDAS to provide Maximum displacement (in); Length of deflection (ft.); Maximum Bending moment (ft. kip); Maximum fiber stress (psi); Maximum Shear force (kip)

A graphical representation of an exemplary set of results is illustrated in Figure 8.

Fiber Stress Check

For the worst load case this check resulted in a bending moment and extreme fiber stress that were well within the capacity of the liner.

Shear Check

The analysis indicated significant shear force on the circumferential joints which on first assessment was not within the capacity of the shear cones and consequently required further assessment by taking into consideration the additional shear resistance provided by the permanent ballast concrete within the tunnel.

To check the effect of the shear force generated at the circumferential joints the tunnel was further modelled as 2D plate elements with the individual tunnel rings modelled. This enabled shear interface between the rings with the same spring stiffness as used in the beam spring model for standard load combination structural checks of the segments. The results were then checked against shear cones and the capacity of the surrounding concrete.

The assessment was made against International Federation for Structural Concrete fib Model Code 2010 which showed that the ballast concrete activates adequate resistance at very small deformations. Therefore, shear transfer through the invert is feasible and within capacity for the shear forces generated by a ship collision.

OVALISATION—TRANSVERSE ASSESSMENT

The key effects of the ovalisation of the tunnel resulting from the ship collision are twofold:

- Increased bending moments in the segments; and
- Increased rotation of the radial joints, leading to increased utilization in bursting

Just as with the longitudinal check, due to the higher ovalisation and higher load eccentricity (ratio of M/N) Station 30+00 short term Run #1 was defined as the governing case.

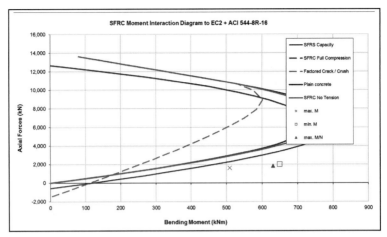

Figure 9. Tunnel lining capacity curve at Station, 30+00 (SRF = 0.9)
(SI units are based on validated fib Model Code spreadsheets)

M/N Capacity Check

Due to the nature of the design case under consideration the M/N capacity check was carried out considering the accidental limit state. Figure 9 represents the capacity of the liner considering a strength reduction factor (SRF) of 0.9 meaning a reduced level of safety corresponding to an accidental limit state.

Station 30+00 (Short Term)—Transverse Assessment

The transverse assessment was made to provide bending moment (lbf ft/ft); compressive force (lbf/ft); position (deg); design bending moment (accidental); and design compressive force (accidental) for Maximum Moment; Minimum Moment and Maximum Moment/Normal Force combinations and plotted against the capacity of the liner.

It was evident that the capacity requirements were locally exceeded. The local exceedance of bending moments is not equivalent with failure of the structure, as the bending moments in the cracked section will redistribute to adjacent sections. This effect was then modelled in an additional run of the PLAXIS analysis considering plastic deformation of the lining.

Plastic Hinge Check

The plastic deformation is represented by a "cap" to the bending moment of the beam elements representing the tunnel lining in the PLAXIS model. This cap has been derived from equilibrium conditions and marks the point where resistance of the concrete section is just not sufficient to establish equilibrium. The lining deforms fully elastic up to this point, and fully plastic thereafter. The plastic moment capacity

was assessed with a strength reduction factor of 0.9, representative for the accidental limit state.

As the plastic moment is depending on the compressive force in the liner which is also subject to redistribution, the model has been run for a number of bands of compressive force and corresponding plastic moment.

Plastic Bending Moment Caps

For each Compressive force band (lbf/ft) a corresponding reference compressive force (lbf/ft) and bending moment cap (lbf ft/ft) was established.

The result of this additional analysis concluded that the plastic cap of the bending moment, and in consequence the corresponding plastification and local cracking of the lining, lead only to the formation of local plastic hinges, and that the lining system is adequately robust to carry the loads in the considered ship impact case.

Shear Check

The shear forces generated due to transverse deformations in the ring were also analyzed. Results for the extracted shear cone forces showed that those are exceeding the limit established from strut-and-tie calculations and validated through published tests. This is assuming that, in the transverse direction, no shear resistance can be gained from the ballast or any of the internal structures. It is therefore expected that the concrete around the dowels could fail locally, and that repairs after the collision may be required. This local damage will not lead to critical system failure, as the cross section of the concrete tunnel is maintained and therefore able to accommodate the compressive loads. This failure mechanism will provide visible evidence from within the tunnel allowing the reinstatement of damaged zones after a possible collision event, in line with the performance criteria established by the project Technical Requirements.

Bursting Check

The combination of maximum radial deflection from ship impact and build tolerances and the maximum compressive force in the lining was checked for tensile splitting, and demonstrated to work to the full ULS (ultimate limit state) level of safety for material resistance. The bursting checks resulted in satisfactory utilization rates using design material parameters as well as peak stresses just in excess of 9,000 psi for the check of the radial joints. It is therefore expected that the ship collision situation would lead only to localized, repairable damage at the radial joints. In practice, the damage may be less than anticipated or even non-existent at the joints, considering that the 28-day characteristic concrete properties (7,500 psi) are likely to be far exceeded.

Gasket Performance Check

The maximum ovalisation of the tunnel lining, defined as change of the diameter, was also extracted from the models. The gasket performance was assessed by calculating the delta deformation of the compressed gasket resulting from the total maximum rotation from build tolerances and ship impact. This rotation is assumed to occur around the closest point of the contact area to the segment intrados, resulting in the most adverse opening possible for the given geometry.

For a maximum radial deformation of the lining and a build tolerance of 0.4% the total delta increase of the groove bottom distance (GBD) in the centerline of the gasket was established. The gaskets have been tested against 40 bar water pressure with up to 0.78 in (20 mm) lateral offset between the gaskets at a delta GBD of 0.20 in (5 mm). Due to the shear cones and guide rods a much reduced stepping and lipping tolerance of less than 1/6 in will be achieved in the design, adding considerable reserves to the water tightness performance of the gasket. It was therefore concluded that the gasket performance would not to be impaired due to a ship impact.

CONCLUSION

Ship collision force:

- Codified probabilistic approach used to determine required tunnel capacity (i.e., max. ship collision force the tunnel is required to resist).
- Maximum ship collision forces occur near to the middle of the berm section

- Ship collision forces on the berm section nearest to Portal Island 1 are greater than those for Portal Island 2
- Two-dimensional analyses using PLAXIS show tunnel lateral deflections are tolerable under the maximum ship collision loading

Snaking:

- Analysis of the longitudinal tunnel deflection show that the tunnel stiffness significantly reduces these maximum lateral deflections
- This reduces the extreme fiber compressive stresses to acceptable values
- The consequence is that significant shear force develops on the circumferential joints unless this force is relieved by slip
- Calculations indicate that the ballast concrete has sufficient shear capacity to resist the maximum shear forces

Ovalization:

- The PLAXIS analysis showed that the maximum ovalization on diameter of the tunnel is tolerable.
- Closed form analysis predicts overstress from the bending moment combined with low axial force
- In the event of a ship collision, flexural cracks may form that are repairable and will not cause a structural failure of the tunnel lining
- Joint rotation in combination with build tolerances does not result in deformation of the gasket incompatible with the water proofing requirement.
- The ship collision may lead to local cracks or damage in the vicinity of the shear cone recesses due to local overstress, which can be easily identified and repaired if vessel impacts occur.

All expected deflections and cracking are within acceptable limit based on the serviceability limit state described in AASHTO LRFD and Technical Requirements Basic Strength and Service A. 4. "Extreme Event II: Combination used to design for any one of these three loads: Design earthquake, Ship sinking, Anchor collision, plus other loads. This event can produce light damage to the tunnel, minor leaks, and some loss of service."

In conclusion, this unique problem was solved with a rational, risk based approach to the loading conditions and sensible application of serviceability criteria. The engineering approach to solving this problem was based on an established codified method but modified to suit the specific requirements of the project in a rational manner.

Designing Reinforcement for Precast Concrete Tunnel Lining Segments

Jimmy Susetyo, Michael Dutton, and Tomas Gregor
Hatch Corporation

Michel Mongeau
Numesh Inc.

ABSTRACT: The paper outlines the factors that need to be considered by designers to make appropriate choice for reinforcement in precast concrete tunnel lining for various project applications. The choices considered in the paper are traditional grade 60 rebar cages, the state-of-the-art grade 80 welded wire reinforcement (WWR), and steel fiber reinforcement. The discussed factors will include consideration for ground condition (including seismic and hydrostatic conditions), long-term safety and serviceability requirements, and the type of application (such as sewer or transit tunnels). The benefits and disadvantages of each option are addressed, and relative costs of the solutions are compared through a case study analysis.

INTRODUCTION

The use of tunnels has increasingly become a common necessity, particularly in developed urban settings. Tunnels have been built for many purposes, from small utility tunnels for linear infrastructure (wastewater or clean water tunnels) to large diameter tunnels for road and transit systems. By utilizing tunnels, infrastructure projects can be realized with minimal interference to the existing surface structures and facilities.

Depending on the ground conditions and the project requirements, there are many means and methods to construct a tunnel. Some examples of the methods include cut-and-cover, drill and blast, mining using mechanized excavator, sequential excavation method, or by tunnel boring machine (TBM). Various support system for the tunnels also exist, ranging from no lining in self-supporting ground, rock bolts, steel ribs and lagging, cast-in-place lining, shotcrete lining, to precast concrete tunnel lining (PCTL).

PCTL is normally used in conjunction with TBM, particularly to support tunnel in soft ground, although it can and has been used for hard ground support as well. The lining system consists of a number of curved precast concrete segments, connected by mechanical connections to form a circular lining ring. Rings are connected along their circumferential faces to create tunnel. The mechanical connections between segments are typically provided using bolts or dowels along the radial and circumferential joints.

Gaskets are also typically provided along the joints to ensure watertightness of the lining system.

Due to the discrete nature of the segments, PCTL reinforcement cannot normally continue from one segment to another. The most common PCTL reinforcement consists of conventional steel rebar or welded wire reinforcement (WWR). More recently, steel or synthetic fibers have also been used.

PURPOSE OF PAPER

This paper outlines factors that need to be considered by tunnel designers to make an appropriate choice for reinforcement in PCTL. Three types of reinforcement will be discussed: conventional steel rebar, high strength WWR, and steel fibers. The advantage and disadvantage of each reinforcement type, as well as its applicability for a tunnel project, will be discussed. In addition, three case studies are presented to compare the different approaches to reinforce PCTL and the subsequent material cost.

DESIGN OF PRECAST CONCRETE TUNNEL LINING

The design of the PCTL depends on the tunnel geometrical requirements, ground/exposure conditions, and the load demand, as well as the TBM arrangement. In order to optimize the design, it is desirable to have a lining that is as thin as possible. However, there are practical limits and code requirements that may govern the thickness of the PCTL, including requirement for concrete cover, for minimum reinforcement, and for reinforcement placement. The type and size of the tunnel gasket and segment connectors also affect the minimum lining thickness.

Requirement for Cover

Adequate concrete cover must be provided to protect the reinforcement from corrosion and other detrimental factors. The minimum required concrete cover depends on several factors, including the design life, exposure conditions, requirement for fire resistance, protective systems (e.g., cathodic protection, coating), and the consequence of corrosion.

Increasing concrete cover is an effective means of protecting reinforcement from corrosion. However, increasing concrete cover will also increase the thickness of the lining, thereby increasing the concrete material cost and segment weight. The thicker lining also leads to a larger excavation. As a result, the overall cost of a tunnel project is increased with the use of thicker lining.

Depending on the codes and standards being employed, different cover requirement exists. ACI 318-14 (ACI Committee 318, 2014) specifies that precast concrete member in contact with ground shall have a minimum cover of 2 inch [51 mm] when using No. 14 and No. 18 bars, 1.5 inch [38 mm] when using No. 6 through No. 11 bars, and 1.25 inch [32 mm] when using No. 5 bars and smaller. Meanwhile, CSA 23.4-16 (CSA Group, 2016) specifies a minimum cover for precast concrete of 1 inch [25 mm] to 2 inch [50 mm], depending on the exposure condition. Additional concrete cover may also be needed depending on project specific requirements (e.g., for fire resistance, etc.).

Requirement for Minimum Flexural Reinforcement

Code requirements also exist for the minimum concrete flexural reinforcement. The purpose of the minimum flexural reinforcement requirements is to ensure good crack control and ductility upon cracking, including cracking caused by temperature change and shrinkage.

The behavior of PCTL can be associated with that of a slab. For non-prestressed slab, ACI 318-14 recommends a minimum reinforcement for shrinkage and temperature of $0.002 \times A_g$ (with A_g being the gross cross-sectional area of the member) when steel reinforcement with a yield strength (f_y) smaller than 60 ksi [414 MPa] is used; the minimum shrinkage and temperature reinforcement is reduced to the greater of $(0.0018 \times A_g \times 60,000/f_y)$ or $0.0014 \times A_g$ when steel reinforcement has a yield strength greater than 60 ksi [414 MPa]. In CSA A23.3-14 (CSA Group, 2014), the minimum shrinkage and temperature reinforcement is $0.002 \times A_g$, regardless of the yield strength. CSA A23.3-14 also requires that the flexural reinforcement be proportioned to provide a flexural moment resistance (M_r) of at least 1.2 times cracking moment (M_{cr}).

In addition to the minimum reinforcement requirement, the maximum spacing of reinforcement is also prescribed to ensure good distribution of the reinforcement. ACI 318-14 prescribes a maximum spacing of deformed reinforcement of the lesser of $3h$ or 18 inch [460 mm], whereas CSA A23.3-14 prescribes the maximum reinforcement spacing to be the lesser of $3h$ or 20 inch [510 mm]. In both codes, h denotes the thickness of the member.

Requirement for Reinforcement Placement

The use of reinforcing steel in a thin section can be problematic as its presence inhibits the flow of concrete, making it more difficult to consolidate. In order to ensure good flow of concrete, codes and standards generally specify a minimum spacing requirement for reinforcement. This requirement may provide constraint to the PCTL dimensions.

ACI 318-14 prescribes that the clear spacing between horizontal non-prestressed reinforcement shall be at least the greatest of 1 inch [25.4 mm], bar diameter, or 4/3 times the maximum size of the coarse aggregate. Meanwhile, CSA A23.1-09 (CSA Group, 2009) prescribes that the clear spacing between parallel bars shall be at least the greatest of 1.4 times bar diameter, 1.4 times the maximum size of the coarse aggregate, and 1.2 inch [30 mm].

REINFORCEMENT TYPES

Traditionally, PCTL reinforcement typically consists of steel cage built from plain or deformed steel rebars, such as the ASTM A615 carbon steel bar, ASTM 706 low-alloy steel bars, or CSA G30.18 bars. However, recent years have seen the increasing use of ASTM A1064 WWR replacing the plain or deformed steel rebars. In addition, fibers, particularly steel fibers, have also been used in some applications as a complete replacement of the rebars.

In order to select the appropriate reinforcement type for an optimal PCTL design, an understanding of the characteristics of each reinforcement type is required. Below are some discussions on the advantages and disadvantages of each reinforcement type.

Conventional Steel Rebar

Conventional steel rebars are the most common type of concrete reinforcement. They are widely available with different sizes. They are hot-rolled and manufactured in accordance to ASTM A615 (for carbon steel bars), ASTM A706 (for low-alloy steel bars), or CSA G30.18. Both plain and deformed rebars can be used, although deformed rebars offer superior performance and greater bond with the concrete

Table 1. Equivalent resistance of welded wire reinforcement to conventional steel rebar

Rebar ASTM A615		Cold-Drawn Wire ASTM A1064					
		F_y = 70 ksi [485 MPa] Steel Reduction = 14.3%		F_y = 75 ksi [515 MPa] Steel Reduction = 20%		F_y = 80 ksi [550 MPa] Steel Reduction = 25%	
Yield Strength (F_y) = 60 ksi [414 MPa]							
Designation	Diameter (in [mm])	Area (in^2 [mm^2])	Size D	Area (in^2 [mm^2])	Size D	Area (in^2 [mm^2])	Size D	Area (in^2 [mm^2])
No. 4	0.5 [12.7]	0.2 [129]	D17.2	0.172 [111]	D16	0.160 [103]	D15	0.150 [97]
No. 5	0.625 [15.9]	0.31 [199]	D26.6	0.266 [172]	D24.8	0.248 [160]	D23.3	0.233 [150]
No. 6	0.75 [19.1]	0.44 [284]	2×D18.9	0.378 [244]	2×D17.6	0.352 [227]	2×D16.5	0.330 [213]
No. 8	1.0 [25.4]	0.79 [510]	N/A	N/A	N/A	N/A	2×D29.7	0.594 [383]

due to the indentation. The typical yield strength for the rebars are Grade 60 for ASTM A615 and A706 bars or Grade 400 for CSA G30.18 bars, although higher yield strengths are also available (e.g., Grade 75, 80, and 100 for ASTM A615 bars, Grade 80 for ASTM A706, and Grade 500 for CSA G30.18 bars). The size of the bars ranging from No. 3 [10 mm] to No. 20 [64 mm].

Due to the wide availability, steel rebars are typically cost-effective in terms of material cost. However, additional labor cost and assembly time will be required to lay the rebars in the PCTL mold.

Welded Wire Reinforcement (WWR)

In contrast to the hot-rolled conventional steel bars, WWR is constructed from cold-drawn steel wires that are arranged at desired angles to each other and electrically resistance welded at the intersection. WWR is manufactured in accordance to ASTM A1064 and can be made of either plain or deformed wires. WWR can be customized to suit the design requirements, with different wire sizes up to D31 (area = 0.31 in^2 [200 mm^2]) and variable wire spacing in the transversal direction and 2 inch increment spacing in the longitudinal direction. The WWR can be manufactured bent or curved, and are available galvanized.

An advantage of WWR over conventional steel rebars is that the WWR is widely available at grades higher than Grade 60 ASTM A615/A706 bars or Grade 400 CSA G30.18 bars. Although conventional steel rebars are also available at higher grades, they are not commonly available and must be specially ordered. WWR is widely available as Grade 70 (70 ksi [485 MPa]), Grade 72.5 (72.5 ksi [500 MPa]), Grade 75 (75 ksi [515 MPa]), and Grade 80 (80 ksi [550 MPa]). The higher steel grade of the WWR enables a reduction of the steel area for the same resistance as the conventional steel rebars. As much as 25% reduction of steel area can be achieved with the use of WWR, as indicated in Table 1. This steel area reduction will result in lighter weight and cost of reinforcement cage.

Another advantage of WWR over conventional steel rebars is the faster installation time and reduction in labor cost. Since the WWR production is highly mechanized, less time and labor are required to assemble the reinforcement cage for the PCTL during the production of the concrete lining. WWR can also be designed using smaller wire at closer spacing to reduce crack widths.

Despite the definite advantages of the WWR over the steel rebars, there are a couple of considerations that need to be made when choosing the WWR. One consideration is material cost. The material cost of WWR is relatively more expensive than conventional steel rebars due to the required production process (cold drawing, cutting, and welding). However, the extra cost may be more than offset by the steel weight reduction or with savings from labor cost and assembly time. Another consideration is the available wire size. The WWR is available from size W1.4/D1.4 (area = 0.014 in^2 [9.2 mm^2]) to W31/D31 (area = 0.31 in^2 [200 mm^2]). Although this size range should cover most, if not all, wire sizes typically used in a PCTL, certain load condition may require larger bar size.

Steel Fibers

Steel fibers have been used more often in recent years as an alternative way to reinforce concrete. Research has shown that the addition of discontinuous discrete steel fibers significantly improves the behavior of otherwise brittle concrete. Enhancement in the concrete behavior includes increased tensile and flexural strength of the concrete (Shah and Rangan, 1971), increased post-cracking ductility of the concrete (Vandewalle, 1999), increased energy absorption capacity of the concrete (Balaguru et al., 1992), and reduced crack width and crack spacing (Grzybowski and Shah, 1990; Banthia et al., 1993).

When used as reinforcement in PCTL, the use of steel fibers may allow for a reduction of the thickness of the segments since the requirements for concrete cover and reinforcement spacing discussed in previous section do not apply. Corrosion of individual steel fibers may still occur. Nevertheless, due to the discrete nature of the fibers, the corrosion is limited to the affected fibers and will not propagate, unlike corrosion of steel rebar or WWR. As a result, corrosion of individual steel fibers does not necessarily compromise the structural integrity and strength of the segments. Moreover, since steel fibers are dispersed in the concrete, they have a higher chance than steel rebars/WWR to intersect cracks, leading to improved durability and reduced crack widths. The use of steel fibers does not require labor and time associated with cage assembly during PCTL manufacturing, since the fibers are directly added to concrete mix during production, resulting in savings of cost and time.

The effectiveness of steel fibers in improving the concrete behavior depends on several factors, including the fiber content, the physical properties (length, diameter, tensile strength) of the fibers, and the orientation of the fibers.

Fiber content significantly influences the properties of hardened concrete and the workability of freshly mixed concrete. An increase in fiber content will result in an increased improvement of concrete behavior. Shah and Rangan (1971) found that doubling the fiber content from 0.5% to 1.0% increased the concrete flexural toughness from 5 times to 15 times that of the plain concrete and increased the flexural strength from 1.1 times to 1.7 times that of plain concrete. However, fiber addition also leads to the reduction of concrete workability since the presence of fibers reduces the amount of concrete paste volume fraction in which the fibers and the coarse aggregate can freely move. Increasing fiber content will also lead to an increase of tendency of fiber balling. The maximum amount of fiber that can be added before the concrete mixture becomes unworkable depends on the placement condition, the amount of the conventional steel reinforcement present, the composition of the concrete mixture, the shape and aspect ratio (ratio of length to diameter) of the fibers, and the type and amount of the water-reducing admixtures used. ACI 544.3R-08 suggested that the typical amount of steel fiber volume content range from 0.2% to 1.0% (26 to 132 lb/yd^3 [15 to 78 kg/m^3]). This limitation of maximum fiber content constraints the level of improvement offered by the steel fibers. As a result, steel fibers may not be the best solution for situation with high load demand. The use of steel fibers has thus limitations.

A high fiber aspect ratio leads to better bond between the fibers and the concrete due the high surface area of the fibers. Shah and Rangan (1971) indicated that substantial improvements in the residual post-cracking tensile strength and the toughness of the concrete were observed when high aspect ratio fibers were used. However, an increase in the fiber aspect ratio also leads to a reduction in the concrete workability (Johnston, 2001) and an increase in tendency of fiber balling (ACI Committee 544, 2008).

Fiber orientation is perhaps one of the influencing factors that is the most difficult to measure. The most efficient condition is when all fibers are aligned parallel to the direction of the tensile stress (Shah and Rangan, 1971). As the fibers become more oblique to the loading direction, their actual embedment length and the number of fibers crossing the cracks will decrease, reducing in their effectiveness (Shah and Rangan, 1971; Johnston, 2001). It is common to assume that the fibers are randomly oriented in two- or three-dimensional space. However, it is hard to ascertain the fiber orientation or the degree of randomness in the orientation. This increases the variability of the level of improvement offered by the steel fibers.

Nevertheless, steel fibers have been used successfully to replace or used in conjunction with steel rebars/WWR, and remain a viable option for PCTL reinforcement.

FACTORS INFLUENCING SELECTION OF REINFORCEMENT TYPE

Diameter of Tunnel

The diameter of the tunnel dictates the size of the excavation, which largely governs the amount of load that needs to be resisted by the tunnel lining. Normally, the larger diameter of the tunnel, the thicker the tunnel lining needs to be. Figure 1 illustrates the relationship between the tunnel diameter and the lining thickness. The data for the figure were compiled from 44 past tunnel projects that cover a variety of diameters, ground conditions, reinforcement type, and loading cases. As indicated by the figure, a linear correlation between the tunnel diameter and the lining thickness can be drawn.

The diameter of the tunnel may influence the selection of reinforcement type to be used. A small diameter tunnel, which typically does not require thick lining, may need to use a thick lining when steel rebar or WWR is used in order to satisfy the concrete cover requirement. The use of steel fibers may allow the use of a thinner lining while still providing the same structural capacity. On the other hand, steel rebar or WWR is likely to result in a more efficient reinforcement design in a large diameter tunnel than steel fibers since the thick lining required by such large diameter tunnel will be able to easily accommodate the concrete cover requirement. Since the performance of the steel fibers are tightly related to the volume fraction, a large mass quantity of steel fibers will be required to provide the same level of structural capacity due to the large volume of concrete in the thick lining. In other words, in thick linings, steel rebars or WWR that are placed at the locations where they are actually structurally required (i.e., near the faces) will be more efficient

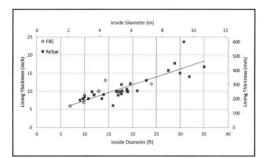

Figure 1. Historical relationship between tunnel diameter and lining thickness

than steel fibers that are distributed evenly through the lining thickness; the sheer cost of the required steel fiber volume offsets the labor savings offered by the implementation of the fibers when compared to steel rebars or WWR.

Load Demand

Loading that needs to be resisted by the PCTL may come from the surrounding ground stresses, hydrostatic pressure, TBM grout pressure, lining handling and installation loads, ground motion (e.g., due to seismic activity), or any other types of project-specific loading (e.g., train and traffic loads, loads exerted by adjacent structures, rock wedge load, etc.). The reinforcement of the PCTL must be designed to accommodate all loads to ensure the structural integrity of the lining.

As previously discussed, there is a limit on the maximum amount of steel fibers that can be added to a concrete mixture. Due to this limit, steel fibers are not suitable when high load capacity is demanded; steel rebar or WWR will be able to provide the capacity required by the high load demand more efficiently and effectively.

Ground Condition (Exposure Condition)

In most, if not all, cases, the PCTL will be in direct contact with the ground. As such, the concrete will be subject to any chemical substances that may exist in the ground, including chloride (e.g., from deicing agent, naturally-occurring salt deposit, seawater) and sulfate (e.g., from seawater, naturally-occurring sulfate minerals, fertilizer). Sometimes, the concrete may also be exposed to corrosive gas from municipal sewage or industrial effluent. The PCTL may also be subject to freezing and thawing. These ground conditions will affect the durability of the concrete.

When exposed to these severe ground conditions, most codes and standards will mandate a minimum concrete cover requirement. This requirement will affect the required dimension of the tunnel lining

and may increase the thickness of the lining beyond what is demanded by the load. Due to the improved durability, the use of steel fibers may allow the use of a thin lining while maintaining the required durability level.

CASE STUDIES

To showcase the difference in choice between conventional steel rebars, WWR, and steel fibers, three case studies are presented. Each study case is based upon a real tunnel scenario, in which the reinforcement specified provided sufficient structural capacity for the anticipated load cases. The tunnel dimensions and reinforcement design are presented in Table 2. The selected cases cover a wide range of project experience; Case 1 is a large diameter transit tunnel, Case 2 is a tunnel in a seismic-prone region, and Case 3 is a typical water conveyance tunnel.

In order to fairly evaluate the relative production cost differences between the three reinforcement options, an equivalent design for each type of reinforcement was determined as presented in Table 3. The design was considered equivalent when the PCTL's flexural capacity was effectively equal at the required level of axial force.

The equivalent conventional rebar design was established for all three cases assuming typical Grade 60 steel, while the equivalent WWR design was calculated for Case 3 assuming Grade 70 steel. The hoop and longitudinal bar size and quantity were adjusted to find a comparable capacity. An equivalent steel fiber design was found for Case 1 and Case 2 by determining an equivalent flexural strength. A fiber dosage was then estimated using a manufacture's test data curve to relate flexural strength to dosage. For the study purpose, collated hooked-end steel fibers with a length of 2.4 inch [60 mm], aspect ratio of 80, and tensile strength of 178 ksi [1,225 MPa] were used.

The moment-axial force interaction curves illustrated in Figure 2 show the sectional capacity of the PCTL. The discrete data points in the figure indicate the predicted internal forces in the lining due to the design loads at various locations along the lining circumference. Data points inside the curves indicate that the predicted internal forces are within the PCTL design capacity. The solid capacity line indicates the original design for that case. As was the goal of the design, the PCTL has similar capacity in the tensile controlled region for each reinforcement option.

As illustrated by Figure 2, it is possible to have the same capacity curve with the steel rebars and WWR, since they both are available at various sizes. The only difference between the design using steel rebars and WWR are that WWR can use smaller bar diameter than rebar due to the higher yield strength.

Table 2. Case study tunnel dimensions and reinforcement

Case		1 (Transit Tunnel)	2 (Seismic Prone Region)	3 (Water Tunnel)
Inside Diameter		35.1 ft [10.7 m]	18.8 ft [5.7 m]	9.8 ft [3 m]
Segment Length		6.6 ft [2 m]	5 ft [1.5 m]	3.9 ft [1.2 m]
Segment Thickness		16.7 in [425 mm]	10.5 in [267 mm]	6.9 in [175 mm]
Original Design	Concrete Compressive Strength	8700 psi [60 MPa]	6500 psi [45 MPa]	5800 psi [40 MPa]
	Reinforcement Type	WWR	WWR	Steel Fiber
	Steel Grade	70 ksi [485 MPa]	80 ksi [550 MPa]	N/A
	Hoop Reinforcement (Per Ring)	12 - D29 [MD187.1] (Top and Bottom)	14 - D20 [MD129] (Top and Bottom)	42 lb/yd³ [25kg/m³] of steel fibers
	Longitudinal Reinforcement (Per Ring)	D15.5 [MD100] at 2.1° (Top and Bottom)	D11 [MD70.97] at 6° (Top and Bottom)	

Table 3. Reinforcement design for analyzed case studies

Case	Reinforcement Type	Segment Thickness	Steel Grade	Hoop Direction (Per Ring—Top & Bottom)	Longitudinal Direction (Per Ring—Top & Bottom)	Equivalent Flexural Strength	Fiber Dosage	Total Steel Weight per Ring
1	WWR*	16.7 in [425 mm]	70 ksi [485 MPa]	12—D29 [MD187.1]	D15.5 [MD100] at 2.1°	N/A	N/A	3,910 lb [1,775 kg]
	Rebar		60 ksi [410 MPa]	10—#6 [20M]	#4 [15M] at 2.3°	N/A	N/A	4,840 lb [2,195 kg]
	Fiber		N/A	N/A	N/A	540 psi [3.72 MPa]	67 lb/yd³ [40 kg/m³]	2,620 lb [1,188 kg]
2	WWR*	10.5 in [267 mm]	80 ksi [550 MPa]	14—D20 [MD129]	D15.5 [MD100] at 6°	N/A	N/A	1,400 lb [635 kg]
	Rebar		60 ksi [410 MPa]	13—#5 [15M]	#4 [15M] at 8.2°	N/A	N/A	1,865 lb [845 kg]
	Fiber-1		N/A	N/A	N/A	1,000 psi [6.89 MPa]	267 lb/yd³ [158 kg/m³]	2,615 lb [1,186 kg]
	Fiber-2	16 in [406 mm]	N/A	N/A	N/A	500 psi [3.45 MPa]	76 lb/yd³ [45 kg/m³]	1,159 lb [526 kg]
3	WWR	6.9 in [175 mm]	70 ksi [485 MPa]	4—D14 [MD90.3]	D15.5 [MD100] at 14.1°	N/A	N/A	230 lb [105 kg]
	Rebar		60 ksi [410 MPa]	4—#4 [15M]	#3 [10M] at 12.1°	N/A	N/A	300 lb [135 kg]
	Fiber*		N/A	N/A	N/A	334 psi [2.3 MPa]	42 lb/yd³ [25 kg/m³]	115 lb [52 kg]

* Original design

Figure 2 also shows the capacity curves provided by steel fibers. As indicated in the figure, steel fibers can satisfy the flexural capacity demand required in Case 1 and 3. An exception was Case 2. Due to the high flexural demand in Case 2 (seismic prone case), an unrealistic high equivalent flexural strength of 1,000 psi [6.89 MPa] is required to achieve a comparable capacity. This demand requires a steel fibers dosage of 255 lb/yd³ [151 kg/m³], which is very high. This highlights the limitation of the use of steel fibers in high load demand situation. Increasing the lining thickness from 10.5 inch [267 mm] to 16 inch [406 mm] will result in a more feasible reinforcement solution of 72 lb/yd³ [42 kg/m³] of steel fibers.

A cost and schedule estimate for each reinforcement option is presented in Table 4. The material cost includes the cost to produce the members to the desired length and shape. It includes production of the curved and straight ladders for WWR, and cutting and bending of reinforcement for steel rebars. The material cost for WWR was assumed at US$ 60 per cwt (100 lb [45.4 kg]); for steel rebars, it was assumed at US$ 43/cwt. The material cost for the steel fibers was taken as US$ 1/lb. The labor time and cost represents the time and cost to assemble the cage in the PCTL manufacturing plant. Labor cost in the plant was assumed at UD$ 40/hour. For WWR, a labor time of 4 hours per ton (2,000 lb [907 kg])

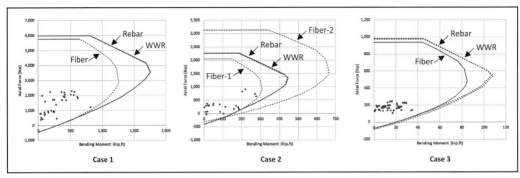

Figure 2. PCTL capacity curves for Case 1, Case 2, and Case 3

Table 4. Production cost and time estimate per PCTL ring

	Case	1 (Transit Tunnel)	2 (Seismic Prone Region)	3 (Water Tunnel)
Rebar	Material Cost (US$)	$ 2,081	$ 802	$ 129
	Labor Time (hours)	17.0	6.5	1.0
	Labor Cost (US$)	$ 678	$ 261	$ 42
	Total Cost (US$)	$ 2,759	$ 1,063	$ 171
WWR	Material Cost (US$)	$ 2,346	$ 840	$ 138
	Labor Time (hours)	8.0	3.0	0.5
	Labor Cost (US$)	$ 313	$ 112	$ 18
	Total Cost (US$)	$ 2,659	$ 952	$ 156
Fibers*	Material Cost (US$)	$ 2,620	$ 1,159	$ 115
	Labor Time (hours)	0.0	0.0	0.0
	Labor Cost (US$)	0.0	0.0	0.0
	Total Cost (US$)	$ 2,620	$ 1,159	$ 115

* Fiber-2 was considered for Case 2.

of steel was assumed. For rebar, the labor time was assumed as 7 hours/ton. Steel fibers do not have labor time and cost associated with it since no cage assemble is required.

As indicated in Table 4, WWR has 5–13% higher associated material cost than steel rebars, but 54–57% lower labor cost and assembly time, resulting in 4–10% lower total cost and 54–57% lower assembly time. This indicates that the higher material cost can be recovered with savings from labor cost and assembly time. The material cost for steel fiber is 11% less to 45% more than steel rebars. However, due to absence of labor needed for cage assembly, the total cost is 33% less to 9% more than steel rebar. Note that the use of steel fibers in Case 2 requires a thicker lining; the use of original thickness will require unreasonable high amount of steel fibers to make it work.

It should be noted that the cost and schedule estimate is presented for the purpose of illustration only, and will vary from one market to another. The cost also does not consider other factors such as

profit margins, delivery charges, or cost associated with a larger excavation in Case 2 with steel fibers.

SUMMARY AND CONCLUSIONS

Discussions have been made on three types of reinforcement: conventional steel rebar, high strength WWR, and steel fibers.

The steel rebar and high strength WWR offer the same level of reinforcement for the concrete. However, the higher steel grade of the WWR enables a reduction of the steel area as much as 25% for the same resistance as the conventional steel rebars, resulting in lighter weight of reinforcement cage. The use of WWR also allows for faster installation time and reduction in labor cost. It also allows a reinforcement design using smaller wire at closer spacing to reduce crack width. However, WWR costs more $/lb than conventional steel rebar and it has a more limited range of available size. Note that the extra cost may be more than offset by the steel weight reduction or with savings from labor cost and construction time, and that the available sizes cover the typical rebar sizes for most project.

The addition of steel fibers into concrete has been shown to increase the tensile and flexural strength, post-cracking ductility, and energy absorption capacity of the concrete. Reduced crack width and crack spacing has also been observed. When used as reinforcement in PCTL, the use of steel fibers will allow for a reduction of the thickness of the segments since the requirements for concrete cover and reinforcement spacing do not apply. It also allows for savings in PCTL production cost and time, since the steel fibers will be added directly into the concrete mix, negating the need for reinforcement cage assembly. However, the addition of fibers leads to a reduction in concrete workability, and there is a limit on the maximum amount of fibers that can be added to the concrete. This limitation of maximum fiber content constraints the level of improvement offered by the steel fibers. As a result, steel fibers may not be the best solution for situation with high load demand.

Three case studies are presented to compare the different approaches to reinforce PCTL. Case 1 is a large diameter transit tunnel, Case 2 is a tunnel in a seismic-prone region, and Case 3 is a typical water conveyance tunnel.

As indicated by the studies, the same concrete capacity can be had with steel rebars and WWR since they both are available at various size. However, smaller bar diameter can be used with WWR due to the higher yield strength. Steel fibers can also satisfy the flexural capacity demand required in Case 1 and 3. Due to the high flexural demand in Case 2 (seismic prone case), an unrealistic high equivalent flexural strength is required to achieve a comparable capacity, resulting in unreasonably high dosage of fibers. This highlights the limitation of the use of steel fibers in high load demand situation. An increase of the lining thickness was required to achieve a more feasible solution.

A cost and schedule estimate for each reinforcement option is presented. Note that the cost and schedule estimate is presented for the purpose of illustration only, and will vary from one market to another. The cost only includes material production and cage assembly cost, and does not consider other factors such as profit margins, delivery charges, or cost associated with a larger excavation in Case 2 with steel fibers.

Depending on the assumed reinforcement material cost, labor rate and the particular Study Case, WWR has 5–13% higher associated material cost than steel rebars, but 54–57% lower labor cost and assembly time, resulting in 4–10% lower total cost and 54–57% lower assembly time. The material cost

for steel fiber is 11% less to 45% more than steel rebars. However, due to absence of labor needed for cage assembly, the total cost is 33% less to 9% more than steel rebar. Note that the use of steel fibers in Case 2 requires a thicker lining, whose associated cost is not considered in the study.

REFERENCES

ACI Committee 318, 2014. "Building Code Requirements for Structural Concrete (ACI 318-14)," 524pp.

ACI Committee 544, 2008. "ACI 544.3R-08 Guide for Specifying, Proportioning, and Production of Fiber-Reinforced Concrete," 16pp.

Balaguru, P., Narahari, R., and Patel, M., 1992. "Flexural Toughness of Steel Fiber Reinforced Concrete," ACI Materials Journal, Vol. 89, No. 6, November-December 1992, pp. 541–546.

Banthia, N., Azabi, M., and Pigeon, M., 1993. "Restrained Shrinkage Cracking in Fibre-Reinforced Cementitious Composites," Materials and Structures, Vol. 26, No. 7, August 1993, pp. 405–413.

CSA Group, 2009. "A23.1-14/A23.2-14 Concrete Materials and Methods of Concrete Constructions/Test Methods and Standard Practices for Concrete," 690pp.

CSA Group, 2014. "A23.3-14 Design of Concrete Structures," 297pp.

CSA Group, 2016. "A23.4-16 Precast Concrete—Materials and Construction," 91pp.

Grzybowski, M. and Shah, S.P., 1990. "Shrinkage Cracking on Fiber Reinforced Concrete," ACI Material Journal, Vol. 87, No. 2, March-April 1990, pp. 138–148.

Johnston, C.D., 2001. "Fibre-Reinforced Cements and Concretes," Gordon and Breach Science Publishers, Ottawa, Canada, 372 p.

Shah, S.P. and Rangan, B.V., 1971. "Fiber Reinforced Concrete Properties," ACI Journal, Vol. 68, No. 2, February 1971, pp. 126–137.

Shah, S.P., Weiss, J., Yang, W., 1998. "Shrinkage Cracking—Can It Be Prevented?," Concrete International, Vol. 20, No. 4, April 1998, pp. 51–55.

Vandewalle, L., 1999. "Influence of Tensile Strength of Steel Fibre on Toughness of High Strength Concrete," Proceedings of Third International Workshop on High-Performance Cement Composites (Mainz, Germany), H. W. Reinhardt and A. E. Naaman, eds., RILEM Publications, Bagneux, France, 1999, pp. 331–337.

Evolution and Challenges of Segmental Liner Design and Construction for SR 99 Tunnel

Yang Jiang, Gordon Clark, and Jerry Wu
HNTB Corporation

ABSTRACT: The breakthrough of the SR 99 Tunnel Boring Machine (named "Bertha") on April 4, 2017, in Seattle, USA, signifies the completion of tunnel lining construction for one of the world's largest bored tunnels. The 57.5-foot diameter, 9,300-foot-long-tunnel under Seattle reached depths of 215 feet in an active seismic region. This paper focuses on the evolution of the SR 99 tunnel one-pass liner and interior structures from conceptual through final designs, including incorporating evolving requirements in roadway alignment, traffic flow, fire life safety, seismic safety, constructability, impact of interior structures, and other geo-structural challenges.

INTRODUCTION

The Alaskan Way Viaduct and Seawall Replacement Program (AWVSRP) is a joint effort of the Washington State Department of Transportation (WSDOT), the Federal Highway Administration (FHWA), King County, and the City of Seattle to replace the Alaskan Way Viaduct. The viaduct was constructed in the 1950s and has reached the end of its useful life. In 2001, after the Nisqually Earthquake, the viaduct was forced to temporarily close for inspection and repairs. In 2018, after 6 years of construction, the viaduct will be replaced by a bored tunnel with cut-and-cover approach structures. The evolution and challenges of the segmental liner design are the focus of this paper.

In December 2010, WSDOT awarded the SR 99 Bored Tunnel Design-Build Project to Seattle Tunnel Partners (STP) based on best technical solution and cost. STP is a joint venture of Dragados USA and Tutor Perini Corp. The design team includes HNTB Corporation, Intecsa of Spain, Hart Crowser, Inc., and Earth Mechanics, Inc. In particular, HNTB was responsible for the design of the lining and approach structures. The SR 99 Tunnel Project consists primarily of a 1500-foot cut-and-cover structure at the south end, a 9300-foot long bored tunnel, a 460-foot cut-and-cover structure at the north end, and two operations buildings.

EVOLUTION OF THE BORED TUNNEL CROSS SECTION

The SR 99 Bored Tunnel was the largest diameter tunnel in the world at the time of design and the beginning of construction. The cross section of the tunnel went through a natural evolution from conceptual through final design. Many studies were conducted by WSDOT, its consultant Parsons Brinckerhoff (now WSP but referred to throughout this paper as PB) during preliminary design, and by the design build (DB) team during final design and construction.

Minimum Tunnel Cross Section

Prior to Request for Proposal (RFP), PB performed planning and preliminary engineering studies for the tunnel cross section. The preliminary engineering effort brought together the studies, trial designs, and evaluations of the tunnel cross-section to provide a "verification" that the proposed tunnel cross-section complied with engineering, operations, safety, cost, and WSDOT requirements. Key considerations for establishing the tunnel cross-section were highway geometrics, fire life safety requirements, constructability, and project budget. The tunnel cross-section was a critical item that was fixed for Reference Drawings and many other documents to be prepared in response to the Final RFP for the proposed SR99 Bored Tunnel. The preliminary design of the SR 99 Bored Tunnel concluded that a 49-foot interior diameter tunnel would accommodate two stacked roadways with required egress, signage, systems, and ventilation. This conclusion was based on 10 assumptions. The facility would have: 1) two 11-foot-wide lanes; 2) a stacked continuous egress corridor along the east side of the tunnel; 3) 2-foot-wide shoulders on the west side of the tunnel and 6-foot-wide shoulders on the east side of the tunnel; 4) southbound lanes stacked on top of northbound lanes; 5) minimum vertical vehicle clearance of 15 feet; 6) 2-foot signage envelope above the vehicle clearance; 7) a continuous egress walkway;

8) a common ventilation duct supporting a single point extraction (SPE) ventilation system; 9) space for all required conduits and piping for lighting and sprinklers, systems; and 10) adequate space for other egress and tunnel systems.

Geometry Deviations

The above 10 assumptions included several approved deviations from the WSDOT Design Manual:

- Lane width: The travel lanes are typically 12 feet wide for a design speed of 50 mph or 11 feet wide with justification. Lane widths of 11 feet were justified and approved in the February 2010 Corridor Analysis for the SR 99 Tunnel.
- Shoulders: Shoulders enhance roadway safety and capacity by extending sight distance, providing emergency vehicle access, and allowing stalled vehicles to stop outside of the travel way. Reduction of shoulder width reduces the ability to provide these safety features. However, it is common to consider reducing shoulder width to maintain lane width. Reduced lane widths can result in reduced capacity. Typically, the outside shoulder (right side when facing the direction of travel) is wider than the inside shoulder. In order to provide access to the continuous egress corridor along the west side of the tunnel, a 6-foot-wide shoulder on the east side of the tunnel and a 2-foot-wide shoulder on the west side were approved by WSDOT Headquarters Design and FHWA and are documented in the Corridor Analysis.
- Vertical clearance: Minimum required vertical clearance for vehicular traffic varies from 14 feet to 16.5 feet. A higher vertical clearance will accommodate taller trucks and special loads and require fewer of these to detour around the tunnel on surface streets. A 15-foot minimum clearance over the roadway with specific exceptions in the southbound shoulders was approved by WSDOT Headquarters Design and FHWA.

Interior Structures

The preliminary design proposed the interior cross-section, shown in Figure 1, to be constructed using a combination of cast-in-place (CIP) and precast concrete members. The sequence for constructing the interior structure was considered in detail as a necessary step in determining the impact on overall diameter. Several construction stages would be required to construct the interior structure. Where sensible from the schedule and practical perspectives, precast

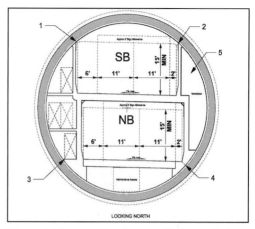

Figure 1. Cross section from preliminary design

concrete was proposed. For the lower roadway, the basic roadway deck would be precast beams with a 6-inch CIP concrete deck, rotated as required for super-elevation and drainage. Lower roadway walls would be vertical CIP concrete. CIP concrete for the upper roadway would be placed to the required cross-slope for super-elevation and drainage. The upper walls were initially considered to be built as precast concrete panels. Several constructability issues were identified: practical ability to fit precast panels between the CIP roadway barrier and the as-built tunnel lining, sealing for duct air pressure, and accommodation of the SPE louvers. After evaluation, the upper roadway walls were changed to CIP concrete in the Final Design.

The liner and final tunnel diameter were dependent on the required space for interior structures, equipment, signage, and minimum clearances. The critical points defining the interior are shown in Figure 1 and numbered 1–4. During the study, the area of the ventilation duct was kept dependent upon tunnel diameter and the interior structure configuration, which reduced the complexity of the study. It was thought that if required, the ventilation area could be increased by enlarging the tunnel diameter while the interior configuration remained constant. This would be necessary if the available area was below an acceptable limit as determined by ventilation analysis. Preliminary sizing suggested that 135 square feet of duct area without obstructions would be adequate for ventilation. This cross-sectional area is the net free space after fire proofing has been applied to the duct surface. The thickness of the fire proofing material was assumed to be 1 inch for preliminary design purposes.

The thicknesses of structural members within the bored tunnel were kept constant throughout the study. Both roadway decks were proposed to be 2 feet thick. The top walls were set at 10 inches thick and the lower walls at 18 inches thick. These preliminary sizes changed as the design progressed. Changes in these thicknesses affected tunnel diameter in a comparable manner as the roadway width and clearance dimensions. The interior structure dimensions and the systems established the inside diameter (ID) of the tunnel lining, to which the liner thickness was added to establish the final tunnel lining diameter. For the evaluations, a fixed 2-foot-thick tunnel lining was selected. This was based on as-built plans of other large diameter tunnels to select a starting place for conceptual design. Preliminary structural analysis showed this dimension to be reasonable. Thus, 2 feet were added to the radius, or 4 feet to the diameter, such that the 50-foot ID required a 54-foot outside diameter (OD).

Baseline RFP Cross Section

Later in the Design-Build RFP process, WSDOT required a minimum ID of 50 feet that included construction tolerances in the tunnel. Assuming a 24-inch-thick tunnel lining, the 54-foot OD would require an approximate 56-foot diameter tunnel boring machine (TBM). With some refinements, it was determined that the necessary components of the tunnel including roadway clearances, systems, egress, and ventilation would fit within a 49-foot circle. This was achieved while maintaining the approved lane configuration of (2'-11'-11'-6'), a 30-foot-wide roadway, spaces for the tunnel systems, an egress passageway, and the required 135 square feet for the ventilation duct. Due to the construction tolerance, the outside diameter of the tunnel lining was shown to be 54 feet, which meant that a TBM of about 56 feet would be necessary to construct the tunnel. A TBM of this size had never been constructed; however, a review of current and anticipated tunneling technology and conversations with two large diameter TBM manufacturers provided sufficient assurance that it could be done. The preliminary design concluded that a pressurized-face TBM of this diameter was unprecedented but was within the capabilities of the global tunneling industry.

DB Proposed and Final Cross Section

While some questioned the technical feasibility of the WSDOT's proposed 56-foot minimum diameter for the TBM, HNTB studied several recent successfully bored tunnels around the world, and recognized that WSDOT´s minimum requirements were feasible and achievable. HNTB's research showed that TBM technology had steadily improved and diameters of bored tunnels had increased over the years. For example, in 1990 an Earth Pressure Balance (EPB) TBM had a maximum excavation diameter of 21 feet; by 1994 that had increased to 31 feet; by 2002 the diameter achievable was 39.5 feet; and by 2006 it had reached 49.25 feet. This steady increase in diameter assured the engineers that the 56-ft diameter machine proposed by WSDOT would be possible by 2012 when the tunnel was to be excavated. The STP team had recent experience with large diameter tunnels bored using a TBM, including the M-30 South Bypass South Tunnel Project through downtown Madrid completed by Dragados in 2007.

To accommodate WSDOT's needs for the Project, STP selected a 57.5-foot EPB TBM, a diameter slightly greater than the minimum specified by WSDOT. By doing this STP offered WSDOT a tunnel configuration that would enhance vehicular safety and traffic operations and exceeded WSDOT's baseline tunnel requirements. The STP-proposed tunnel increased the curb-to-curb roadway width within the bored tunnel from 30 feet to 32 feet and increased the vertical clearance from 15 feet to 15 feet 6 inches (Figure 2). The increased overall diameter produced the largest bored tunnel in the world at the time of design. This proposal was a result of the STP team members' combined design and construction experience.

In developing the final tunnel configuration, HNTB started with WSDOT's basic configuration and then carefully evaluated many options with the aim to improve the safety of the traveling public. The team sought to maximize the horizontal roadway envelope to improve the drivers' sight distance, provide a west shoulder that is wide enough for first-responder access or parking of a disabled vehicle, and allow routine maintenance and repairs without closing traffic lanes.

HNTB's tunnel configuration enlarged the tunnel diameter to accommodate two 11-foot-wide travel lanes and 8-foot-wide west and 2-foot-wide east shoulders in each direction, significantly improving tunnel safety and traffic operations. The increased diameter specifically:

- Reduced the severity of the pre-approved vertical clearance and shoulder deviations
- Maintained two lanes of traffic while allowing emergency access for disabled vehicles and first responder vehicles in the shoulders
- Provided a uniform shoulder travel way
- Provided a greater horizontal stopping sight distance
- Facilitated tunnel operations and maintenance activities without closing the number of active traffic lanes
- Provided potential of accommodating larger vehicles for transit to move goods and services through the tunnel
- Widened egress pathways for individuals to evacuate the tunnel

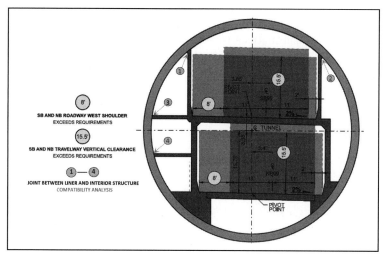

Figure 2. HNTB proposed cross section

In addition to the horizontal width increase, HNTB evaluated impacts to the roadway, systems, and egress using 3-inch increments in the vertical clearance. At each 3-inch increment, the team re-examined the design parameters for the roadway travel way, shoulder vertical clearance, and roadway shoulder widths and ensured that the egress pathways and protective enclosures, overhead signing, mechanical service systems, fire protection, exhaust air duct, utilidor, and maintenance access travel passages and clearance requirements met or exceeded WSDOT's requirements. The final configuration was developed after conducting a Tunnel and Roadway Systems Space Allocation, Coordination, and Verification Study, which included an extensive interdisciplinary analysis of the space requirements inside the finished bored tunnel and a coordinated constructability review of construction space requirements. This study concluded that the HNTB-proposed tunnel configuration would accommodate the proposed roadway lanes, shoulders, and vertical clearances; fireproofing and interior structural element thickness; wider egress pathway requirements; and space for signage, tunnel ventilation, and the tunnel mechanical and electrical systems.

HNTB's tunnel clearance envelope featured a consistent vertical and horizontal cross section from the cut-and-cover sections through the bored tunnel so constant lane and shoulder widths from the portals through the bored tunnel could be maintained. The key benefits of this feature would be improved driver safety and access in the event of emergency situations, including pedestrian egress, as well as the capability of standardizing the interior structure elements.

LINER DESIGN CHALLENGES

LRFD Design

The tunnel liner was analyzed and designed for two stages of structural configurations. The first was the tunnel ring only, which represented the condition of tunneling operations. The second was the tunnel ring with interior structures, systems and associated loads, which represented the in-service condition.

A two-step approach was adopted for the analysis of the tunnel liner for static and seismic load conditions. The strength design of tunnel liners was in accordance with the AASHTO Load and Resistance Factor (LRFD) method, which takes into account statistical variation of member strength and the magnitude of the applied loads. One of the advantages of the two-step approach was to allow the use of different load factors for different loads in the LRFD load combinations. The other advantage was to separate the analytical efforts done respectively by the geotechnical and structural engineers, and to bring the computing effort for the time history analysis to a manageable level.

In the two-step analysis, as described in Section 13.5.1.4 of the FHWA Tunnel Manual [1], the tunnel structure is analyzed separately from the surrounding geological medium by using non-linear springs to simulate the effects of surrounding soils. First, the non-linear soil springs, external forces imposed on the tunnel lining due to soil, water, and surcharge from building weight were generated by the geotechnical engineers through 2D FLAC continuum models. Second, the structural engineers performed separate analyses through 2D CSiBridge beam-spring models by using the external forces and simplified bi-linear soil springs provided by the geotechnical engineers. In addition, the forces and deformation of the liner from the geotechnical models and structural models were compared on an unfactored basis for soil and hydrostatic pressures to ensure the soil-structure interactions were captured correctly in both models.

The force comparison was performed for all six design sections, and in all cases, there was good agreement between the geotechnical and structural models. Figure 3 shows a sample comparison of the axial forces in the liner. Shear, bending moment, and deformation comparisons were also performed, and there was an acceptable agreement between two models, with the structural models producing more conservative results.

The discrepancy may be attributed to the fact that 2D FLAC is a continuum model with more refined mesh, whereas the structural models had a coarser mesh. After verification of the two models, the results from the structural model were used to design the liner, with and without interior structures, using the LRFD method.

Seismic Design

Seismic analysis was performed by using non-linear time history analysis to simulate the ground motions and soil-structure interaction to meet WSDOT's technical requirements.

Two levels of design earthquakes were considered in the design, expected earthquakes and rare earthquakes. Under the expected earthquakes, with ground motions representing an earthquake with a 108-year return period, the liner was designed to sustain only minimal damage to the segments and joints; there would be no loss of water tightness; and the lining would respond in an elastic manner. Under the rare earthquakes, with ground motions representing an earthquake with a 2,500-year return period, the

lining was designed to not collapse, however inelastic deformations would be expected but would be kept to acceptable levels in order to allow repair. The final design provided for a liner that will maintain water tightness for both design earthquakes.

To ensure the liner designs met technical requirements, three different types of models were developed to perform comprehensive analyses including:

- FLAC 2D continuum model and CSiBridge 2D beam-spring model for sectional analysis
- CSiBridge beam-spring model for longitudinal global analysis
- CSiBridge 3D finite element shell model for local analysis and gasket design

Sectional analysis was performed using a two-step approach to determine sectional design force and ovaling. First, deformations of the soil surrounding the liner due to the seismic waves propagating from bedrock through soil media and in absence of the liner were computed with a continuum FLAC model by geotechnical engineers. Second, the resulting ground deformations were imposed on the liner through supporting elements (non-linear springs) with a CSiBridge 2D beam-spring model to analyze the soil-structure interaction. This was done by structural engineers to determine the axial, bending, and shear forces and ovaling deformation. In addition, the predicted ground motion surrounding the liner calculated by the FLAC model was closely compared with the liner ovaling in a CSiBridge model to ensure consistency between the two models. A total of 96 CSiBridge 2D models (eight representative sections, six sets of time history records, with and without interior structures) were used to investigate the transverse response along the entire bored tunnel.

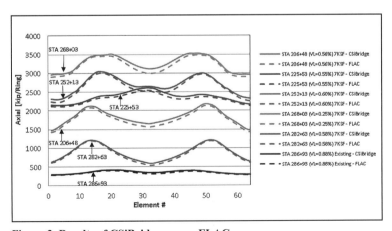

Figure 3. Results of CSiBridge versus FLAC

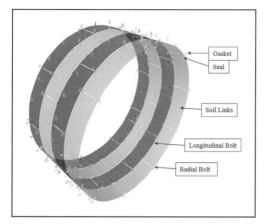

Figure 4. Local model for gasket design

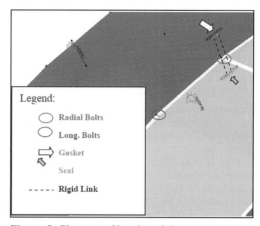

Figure 5. Close-up of local model

Global analysis was performed by using a CSiBridge 3D beam-spring model to determine longitudinal design forces, curvature deformation of the tunnel, expansion and the contraction at the interface between the liner and headwall at each end of the bored tunnel. The primary purpose of the global analysis was to capture the longitudinal response of the bored tunnel at locations where structural stiffness changed or soil conditions changed. Similar to the sectional analysis, a time history analysis was performed by imposing a free-field displacement time history to all soil springs connected to the bored tunnel and approach structures to determine axial, bending, and shear forces as well as deformations. A total of 24 CSiBridge 3D models (compression and tension models, 6 sets of time history records, with and without interior structures) were used to investigate the longitudinal response along the entire bored tunnel.

Lastly, a local 3D finite element shell model of the tunnel was used to predict local behavior and select the appropriate type and size of liner gaskets to ensure water tightness during seismic events, as described in next section.

Serviceability and Water Tightness

A 3D finite element model of the tunnel was built to predict the local behavior of gaskets at the circumferential and longitudinal joints under seismic loading. The performance of the gasket is related to the opening of the joints, which can result from the deformations of the liner under seismic loading. To study the joint opening, the expected seismic deformation from ovaling effects and longitudinal curvature were applied to the models. The ovaling and curvature deformations were predicted from the CSiBridge 2D beam-spring models and CSiBridge 3D beam-spring models, respectively.

Figure 4 shows a total of four rings with shell elements in the model. Each ring is rotated 12.86 degrees relative to the previous ring to avoid cross joints. Figure 5 is a close-up view of the local model. There is an arbitrary 2-inch gap at the edge of the segments. These gaps are included to create a visible connection between the segments and enable link elements representing the gasket and concrete contact to be provided at the gap. Segments are connected in the radial direction by two bolts and in the longitudinal direction by two alignment/shear cones and three bolts per segment. The primary structural elements in the model are summarized below.

- Gasket: visible link with offset at extrados with proper constraint
- Seal: visible link with offset at intrados with proper constraint
- Soil links: soil-structure interaction spring, compression only
- Longitudinal bolts: axial stiffness of bolt plus shear stiffness of shear cone
- Transverse bolts: axial stiffness of bolt

Two models were used in the analysis. The first model was configured to study the joint opening at the radial joint due to transverse response of the liner. The second model was used to study the joint opening at the circumferential joint due to the longitudinal response. The transverse results show the largest joint opening from seismic displacement is 0.12 inch, whereas the longitudinal results show the largest joint opening to be 0.05 inch. Together with the openings from static load, non-uniform grouting and construction, HNTB engineers selected the appropriate gasket type and size to ensure water tightness under expected water pressure and design earthquakes. The selected gasket was capable of maintaining a minimum working pressure of

7.5 bats, or a minimum design pressure (test pressure) of 15 bars with a differential gap of 0.32 inch and a bearing offset of 0.79 inch.

LINER CONSTRUCTION CHALLENGES

Geometry Control

Tunneling in an urban area is always challenging due to existing foundations of buildings and bridges, complex geology due to reclaimed ground, and also due to other existing tunnels. This was especially true for "Bertha," the SR99 Tunnel TBM. Engineers needed to design a tunnel not only strong enough to resist earth and hydrostatic pressures in a complex soil condition in an active seismic zone underneath Seattle, but also to accommodate the need for providing horizontal curvature, vertical profile, and accidental deviations caused by the TBM during tunneling. This demanded a lining that could be continuously adjusted to match the curvature of the tunnel. HNTB engineers selected a 6'-6" wide universal ring made up of 10 tapered segments that included a key. This proved to be an excellent choice that is believed to be a factor in the successful lining of the tunnel.

Each ring consisted of seven rectangular A-segments, two right trapezoid B/C segments, and one isosceles trapezoid key segment as shown in Figure 6. The key segment is the narrowest segment at 6.4 feet, while the A4 segment is the widest segment at 6.6 feet.

The first geometry check was to ensure that the theoretical minimum radius achievable by the adjacent universal rings is less than the instantaneous radius resulting from the combined effects of horizontal curvature and vertical profile. Based on the calculated radius from 28 possible configurations of adjacent rings connecting by shear cones, it was found that the theoretical minimum radius was 1,721 feet, and the instantaneous radius from the horizontal curvature and vertical profile was 5,571 feet. Therefore, the taper of universal ring was appropriate to achieve the project horizontal and vertical alignment.

The second geometry check was to ensure that the longitudinal joints of adjacent rings would not line up to form cross joints and cause potential water leakage. Seventeen of 28 possible configurations were eliminated due to the formation of cross joints. The HNTB engineers provided the TBM operators a positioning matrix of 11 possible configurations, which specified the required rotation angle for the subsequent ring to avoid cross joint formation.

Gasket Performance in Exposure Condition

In December 2013, the TBM was stopped at approximately 1,000 feet into the planned 9,300-foot-long route, after measuring increased temperature in the TBM bearing. While investigating the cause of the

Figure 6. Ring segments

elevated temperatures, STP discovered damage to the machine's main bearing seal system and contamination within the main bearing. It was decided that repairs were required before the machine could continue tunneling. STP and the TBM manufacturer, Hitachi Zosen, completed repairs to the machines in December 2015 [2]. As a result, thousands of fabricated precast tunnel segments had to be stored outdoors near the south launch pit under exterior weather conditions without protection for up to three years, waiting to be assembled in the tunnel.

The elastomeric gasket is a key component to ensure water tightness along the entire tunnel. STP had some concerns of the gasket performance during the design life of the tunnel due to the longer than anticipated outdoor storage time, extended from one year to up to three years without protection. To demonstrate that the sealing gasket's properties would not degrade from the extended outdoor storage, test specimens of gaskets were cut off from the gaskets that had been installed on the precast tunnel segments and sent to the gasket manufacturer. The specimens represented 2, 8, 13, 20, 26, and 30 months of outdoor exposure. The tests included shore hardness, tensile strength, elongation at break, and compression set.

Based on the testing results from outdoor storage shown in Figure 7, the elastomeric gasket properties were not affected and were well within the technical requirements. It was confirmed that the gasket had not degraded, and the water tightness along the entire tunnel was assured.

INTERIOR STRUCTURES

Seismic Design and Compatibility Study

A double-deck highway structure was constructed inside the tunnel lining. For seismic design, a 2D

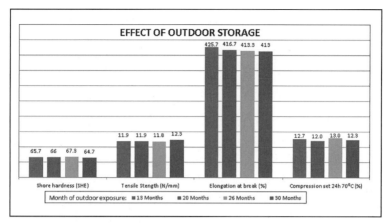

Figure 7. Effect of outdoor storage

frame model representing a typical cross section of the interior structures was created to perform response spectrum analysis assuming a Site Class C. The seismic demands in horizontal and vertical directions were calculated and then combined by using the 100% + 30% rule.

In addition, a compatibility study was performed to investigate the relative displacement between the bored tunnel and interior structures to avoid contraction during an earthquake. Several joints were provided and they were designed to allow relative movement between the tunnel and interior structures as shown in Figure 2, so there would not be any load transfer between the tunnel liner and the interior structures during a seismic event. Therefore, it was also critical to determine the required size of gap between the tunnel and interior structures to allow independent movement. Two representative sections at different stations were selected to perform time history analysis to quantify the relative displacements, and the analysis showed that a 1-inch gap was sufficient to prevent impact between the tunnel liner and the interior structures under a 2500-year return period earthquake.

Construction Sequence

It is common to bore a tunnel first and then construct the interior structures as a separate activity, but STP decided to start the interior structure construction simultaneously with tunneling to accelerate the schedule and save overall construction time. The building of interior structures started when the TBM was approximately 2,000 feet into the 9,300-foot bore. The adopted construction sequence was divided into six stages after many refinement and iterations, as shown below and in Figure 8.

1. Build west and east corbels with post drilled dowels in tunnel liners
2. Build NB roadway walls on top of corbels with cast-in-place dowels in corbels
3. Build SB roadway
4. Build SB roadway west wall, egress floor, utility floor and ceiling slab
5. Build SB roadway east wall
6. Build NB roadway with precast panel option

There were several discussions among the construction team, design team, and WSDOT to finalize the structural element type and construction sequence. For instance, the NB roadway design was changed from cast-in-place to precast panel option. The precast panel option allowed STP to drive vehicles along the tunnel invert to transport liners and tunneling supplies from the south portal to the TBM, and hence the NB roadway construction could not start until the completion of tunneling. Therefore, the precast panel option was more suitable to meet the construction schedule, and the design team changed the design immediately to fulfill the project's needs.

For this design-build project, Engineers, Contractors and the Owner worked closely under one roof to solve the Project's challenges. As of writing this paper, there were nearly 6,000 submittals and RFIs generated to review construction documents and address field conditions. In addition, each document required at least two engineers reach an agreement before execution.

CONCLUSIONS

SR 99 Tunnel is a significant undertaking due to its size, geologic location, and the site conditions under which the tunnel was built. Being one of the largest

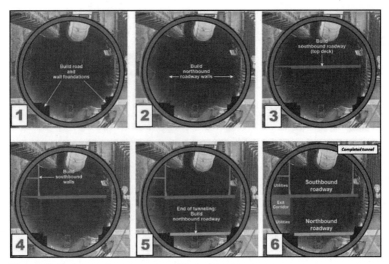

Figure 8. Construction sequence of interior structure

bored tunnels in the world, the tunnel cross section evolved from the conceptual section, to the RFP baseline section, to the final HNTB adopted section, with increasing enhancements. The fact that the tunnel is in a highly active seismic region and needed to remain fully operational during the construction of the tunnel made the design a challenge. The successful design of SR 99 Tunnel shows that 1) preliminary design is critical in setting the parameters for the tunnel section; 2) stringent seismic performance criteria can be satisfied with advanced analysis tools; and 3) Final Designers need to be flexible and adjust the design to changes that occur during construction.

ACKNOWLEDGMENTS

Authors would like to acknowledge Mike Swenson of Hart Crowser for his work on FLAC continuum models, and Tim Moore of WSDOT for his oversight during the design of the SR 99 Tunnel.

REFERENCES

[1] FHWA "Technical Manual for Design and Construction of Road Tunnels - Civil Elements," 2009.

[2] WSDOT "Alaska Way Viaduct Replacement Program—Reparing Bertha" 2007. http://www.wsdot.wa.gov/Projects/Viaduct/About/Tunneling

Guide for Optimized Design of Tunnel Segmental Ring Geometry

Mehdi Bakhshi and Verya Nasri
AECOM

ABSTRACT: Size of circular segmental tunnel linings installed in the rear of the TBM shield is defined by the internal diameter, thickness, and length of the ring. This paper provides guidance to size tunnel inner section considering the internal space required during the service; segmental ring thickness according to different major parameters such as internal diameter, minimum required reinforcement, durability; and length of the ring. In addition, segmental ring systems are presented and ring segmentation and configurations are discussed. Segment systems from the perspective of individual segment geometry are presented with a special focus on key segment geometry. Governing load cases and load combinations that may impact the initially-assumed size of rings and segments are summarized. Based on provided advantages and disadvantages, best practice is recommended for an optimized design of tunnel segmental ring geometry.

INTRODUCTION

A common design approach for concrete tunnel segments starts with selecting an appropriate geometry including thickness, width and length of segments with respect to the size and loadings of the tunnel. Specified compressive strength (f'_c) and type and amount of reinforcement are pre-selected as the next part of iterative design procedure. Next step is to study governing load cases and load combinations that may impact the initially-selected size of rings and segments (Bakhshi and Nasri, 2017a; 2016). Methods of calculation for required strength against these load cases can be found elsewhere (Bakhshi and Nasri, 2014). Using strength reduction factors specified by structural codes, the design strength of segments is compared with required strength for factored load cases (Bakhshi and Nasri, 2013). The geometry, compressive strength, and reinforcement of segments should be checked against the demand forces under all load cases as well as satisfying all service conditions (Bakhshi and Nasri, 2015). If results of analysis show that provided strength is not adequate, segment cross section, concrete strength and reinforcement will be modified until required strength is determined. The design procedure starts with initial considerations for segmental ring system and geometry that is discussed in this paper and further checked against the demand of forces under loading cases occurring from the time of casting to the final service condition (Bakhshi and Nasri, 2017b).

SEGMENTAL RING GEOMETRY AND SYSTEMS

Segmental tunnel linings installed in the rear of the TBM shield are generally in the shape of circular rings. Size of the ring is defined by the internal diameter, thickness, and length of the ring.

Internal Diameter of the Bored Tunnel

The dimensions of the tunnel inner section should be determined considering the internal space required during the service, which depends on the intended use of the tunnel.

For the railroad and subway tunnels, the inner dimensions of tunnels in a single track case are generally governed by the train clearance envelope (clearance gauge), track structure, drainage trough, structure of the overhead catenary and emergency evacuation corridor (egress space). In a double track and twin tunnel cases, tunnel inner dimensions are additionally governed by distance between the centers of tracks, and the cross passageway. The internal diameter of the tunnel is first set by obtaining a circle that satisfies these conditions. Then, the electrical equipment, water pipes, and other equipment are installed in the unoccupied space inside this circle. In general, ventilation space does not need to be considered if egress space and cross passageways are allocated unless is specifically directed by the project's technical requirements (RTRI, 2008).

For the road tunnels, the geometrical configuration of the tunnel cross section must satisfy the

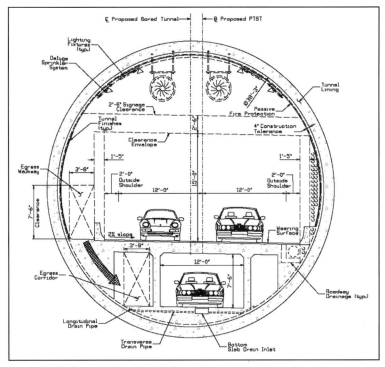

Figure 1. Typical spaceproofing section for TBM-bored road tunnel

required horizontal and vertical traffic clearances, shoulders or sidewalks/curbs, barriers, fans and suitable spaces for ventilation, lights, traffic control system, and fire life safety systems including water supply pipes for firefighting, cabinets for hose reels and fire extinguishers and emergency telephones. As shown in Figure 1, smallest tunnel encircling these clearances and elements are considered as the minimum internal tunnel diameter. The available spaces in a circular cross section can be used to house other required elements for road tunnels including tunnel drainage, tunnel utilities and power, signals and signs above roadway lanes, CCTV surveillance cameras, communication antenna and equipment, and monitoring equipment of noxious emissions and visibility (AASHTO-DCRT-1, 2010). Note that for the spaceproofing of road and railway tunnels, it is crucial to consider impact of maximum superelevation of the tightest curve on the alignment on the rotation of clearance envelopes.

The internal size of the water and wastewater or Combined Sewer Overflow (CSO) tunnels is based on the volumes of design conveyed water or design storm (e.g., 1-year, 18-year and 25-year) specified by local authorities and updated collection system modeling. Often for CSO tunnels, tunnel cross sectional area is determined by dividing the required storage volume for designed capture rate (e.g., 75%) over

total length of tunnel alignment. Other tunnels such as pedestrian, utility and pipeline tunnels need to be reviewed on a case by case basis based on the design criteria established by the owner or designer.

In addition, in determining the ring internal diameter, sufficient construction tolerance should be provided. DAUB (2013) recommends a radius tolerance of R = ±100 mm for TBM-bored tunnels. Therefore, the internal diameter of the tunnel needs to be made 200 mm greater than the required internal structural boundary.

Thickness of the Ring and Outside Diameter

The thickness of segmental lining ring as a structural member is determined in accordance with the result of design calculation. The segment design is an iterative procedure which starts with assumption of a reasonable thickness and later optimized during detailed design calculation. Therefore, it is crucial to consider a reasonable thickness for segmental rings in the beginning of the design process. A review of more than 100 projects published in ACI 544.7R (2016), AFTES (2005), Groeneweg (2007) and Blom (2002) shows that the ratio of internal tunnel diameter (ID) to the lining thickness falls in a specific range of 18–25 for tunnels with ID of more than 5.5 m, and 15–25 for tunnels with ID of 4–5.5 m. JSCE (2007) recommends that as a starter, the ring thickness to be less than 4% of the outer diameter of segmental ring, which translates into an ID to thickness ratio of 23. Most of tunnels are larger than 4m in internal diameter and therefore, it is suggested to consider 1/23rd of ID as the initial lining thickness. For tunnels under 4 m diameter, the lining thickness is not a function of tunnel diameter and ranges between 150–280 mm. Note that reviewing these projects reveals that there is little if no difference between designed thickness of segmental ring and type of reinforcement, i.e., rebar or fiber.

During the analysis and design stage, capacity of the lining section with selected thickness must be sufficient when transverse reinforcement ratio is less than 1% and close to the minimum reinforcement (AFTES, 2005). Also the minimum segment wall thickness must satisfy the conditions imposed by the contact joints such as sufficient bearing surface area and sufficient space and clear distance for gaskets and caulking recesses. Minimum segment wall thickness must be compatible with the bearing surface area of TBM longitudinal thrust cylinders (AFTES, 2005). In addition to structural factors, lining thickness is also designed based on durability factors and DAUB (2013) recommends a minimum thickness of 300 mm for one-pass lining tunnels. Note that in CSO tunnels, if a sacrificial layer was considered for design life of the tunnel (100 or 120 years), the sacrificed layer thickness should be added to required structural thickness.

The outer diameter of tunnel is determined by adding the lining thickness to the inner dimension. The shield outer diameter is determined by adding the tail clearance and shield skin plate thickness, also known as overcut, to the tunnel outer diameter (RTRI, 2008). Shield outer diameter also limits the minimum curve radius of the alignment. A review of the more than 100 tunnel projects with different sizes (JSCE 2007) shows that when shield outer diameter is less than 6m, between 6–10 m, and more than 12 m, the minimum curve radius can be limited to 80 m, 160 m, and 300 m, respectively.

Length of the Ring

Depending on the diameter, the ring length can range between 0.75 and 2.50 m (DAUB, 2013). On one hand, it is desirable that the ring length to be narrow for transportation and erection simplicity, construction of curved sections, and to reduce length of the shield tail. On the hand, it is desirable for the ring length to be larger to reduce production cost, numbers of joints, total perimeter of segments and consequently gasket length and the number of bolt pockets where leakage can occur, as well as increasing the construction speed (JSCE 2007). Therefore, the ring length needs to be optimized for the efficiency

of tunnel works. Analysis of data from more than 60 projects presented in JSCE (2007) demonstrates that although in some cases increasing the diameter results in an increased ring length, there is no considerable relationship between the ring length and outer diameter of the segmental lining. This is mainly due to the fact that for smaller diameters, the available space for segment supply and handling defines the limitation of the ring length, whereas for larger diameters, segment weight and production are the limiting factors. The current practice is to use a ring length of 1.5 m for TBM tunnel diameters of 6–7 m, and a ring length of 1.8 m for tunnels of 7–9 m diameter. When a TBM larger than 9 m is used for excavation, most common ring length is 2 m. Recent projects with tunnel diameters greater than 9 m have taken advantage of 2.2 m long rings by optimizing the segmental lining thickness. Depending on the project conditions, in many cases, weight limitation given by transportation to the site on public roads is the main consideration.

Segmental Ring Systems

Parallel rings, parallel rings with corrective rings, right/left rings, and universal ring systems are among different systems used for tunnel segmental rings.

Parallel ring systems comprising of rings with parallel end faces and with circumferential faces perpendicular to the tunnel axis are not suitable for curved alignments. Practically all tunnel alignments have curves and for directional corrections even in case of curves with large radius, packers are the only solution to be adopted for in circumferential joints. Such segmental ring system cannot be properly sealed and are not suitable for curved alignments.

Parallel rings with corrective rings system is similar to parallel ring system, with corrective rings (either up, down, left or right ring) replacing packers for directional corrections. With this system, watertightness cannot be always guaranteed since the corrective drive cannot be precisely negotiated by only one type of corrective ring with only one end taper. This fact and requirement of different sets of formwork is among the main disadvantages.

Right/left systems are assembled from rings with one circumferential face perpendicular to the tunnel axis and the other one inclined to the tunnel axis. Difference between maximum and minimum ring length is called as taper. The sequence of right-tapered and left-tapered rings produces a straight drive or a tangent alignment, while a sequence of right/right ring results in a curve to the right and left/left rings produces a curve to the left with a minimum system radius. Upward and downward directional corrections are achieved through rotation of the tapered segment ring by 90° (ÖVBB, 2011). This ring system provides a proper sealing performance

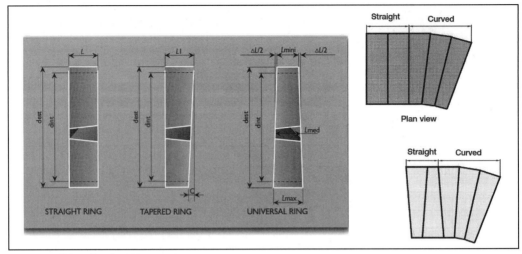

Source: Guglielmetti et al. 2007

Figure 2. Different ring systems and tapering and curve negotiation schematics

for an impermeable tunnel with only disadvantage of requiring different sets of formwork.

Nowadays, universal ring system is the most conventional system with circumferential faces of the ring inclined to the tunnel axis on one or both sides. Required ring taper is divided in both circumferential ends of the universal rings. As shown in Figure 2, all curves and directional corrections can be negotiated through the rotation of the segmental ring. The main advantage of this system is requirement of only one set of formwork (ÖVBB, 2011). The required ring taper (k) can be calculated with the following formula:

$$k = \frac{\phi_A b_m}{R} \qquad (1)$$

where ϕ_A is outer diameter of the segment ring, b_m is the average ring length, and R is the minimum curve radius.

Note that a correction curve drive that in case of deviation returns TBM back into designed tunnel alignment must be taken into consideration. The correction curve radius should be at least 20% less than the smallest desired curve radius horizontally and vertically (DAUB, 2013). In order to have a straight line using universal rings, it is necessary to turn each ring by 180° in reference to the previous one, alternatively having the key segment both on the top and the bottom. Using the right ring and the left ring, it is possible to always have the key segment on the top and, therefore, to be able to construct the ring from the bottom upwards. Nonetheless, in recent years and using advanced software for guiding TBM, universal rings can negotiate the straight drives with key segments always above springlines through adjusting

the drive error of less than a few millimeters in two or three rings.

RING SEGMENTATION AND SEGMENT GEOMETRY

Segmentation of tunnel rings during the design should be studied from different perspective: ring configuration in terms of number of segments, geometry of individual segments, and key segment geometry.

Ring Configuration

One of the main parameters for segmental lining design is the number of segments comprising a ring. Similar to the ring length, the shorter the length of each segment, the easier the transportation and erection process. From this perspective, itis good to divide a ring to several segments. However, longer segments and less number of joints results in much stiffer segmental ring and reduced production cost as well as less hardware for segment connection, less gasket length and less number of bolt pockets where leakage can occur. More importantly, the construction speed can increase significantly. Often the segment weight is a decisive factor in selection of maximum length of segments rather than the available space for handling segments inside TBM shield and in the backup gantries. The slenderness of the tunnel segment (λ), defined as the ratio between the breadth or curved length of segment along its centroid and its thickness, is a key parameter for segment length. Review of tunnel projects show that rings are divided into a number of segments that gives a segment slenderness of 8–13, with Fiber-Reinforce

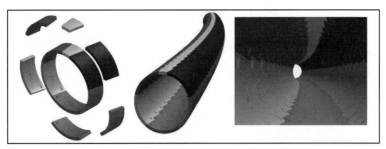

Figure 3. Most common ring configuration (5+1) used for mid-size tunnels under 6m diameter

Concrete (FRC) segments around the lower boundary of this range. However, using latest fiber technologies with double or triple hooked-end steel fibers, FRC segments with slenderness ratio of 10 are frequently adopted. Latest development shows a record of successful use of FRC segment with slenderness of more than 10 and up to 12–13 in four recent projects. In general, it is suggested to divide the ring into as many number of segments that a slenderness ratio of at least 10 can be obtained.

A review of many tunnel projects built in different sizes shows that for tunnels with a diameter of 6 m and below, a ring division into 6 segments is prevailed. In general, contractors and TBM manufactures prefer an even number of segments with odd number of ordinary segments and one key segment. This type of design is more compatible with configurations of TBM thrust jack forces pushing on the segments. The most common configuration is therefore, 5 ordinary segments and one smaller key segment, also known as 5+1 ring (Figure 3). Another common segmentation configuration for tunnels with a diameter below 6 m is 4+2 with 4 ordinary segments and 2 key size segments, one as key and the other as counter key segment. Rings are tapered up/down so there is always an opportunity to close the ring on top of springline with one of these two key size segment. Although in tunnels under 4m diameter, the ring can be easily divided into lesser number of segments, still a division of ring into 6 segments is preferred. Other than 5+1 and 4+2 configurations for a 6-segment ring, other configurations are sometimes adopted. 3+2+1 configuration with 3 ordinary segments (covering 72°), 2 counter key segments (covering 56.5°) and one key segment (covering 31°); or 6 configuration with all segments with the same size (covering 60°) and every other segment as the key segment in trapezoidal shape are among other configurations. Usually these configurations are used when the key segment, as a structural weak point in the ring, turns out to be too small in 5+1 configuration.

For tunnels of 8–11 m in diameter, a 7+1 configuration is the most common configuration. However, when tunnel diameter ranges between 6–8 m, 5+1 configuration may result in excessively long segments, and 7+1 configuration in too short segments. In some cases, size of key segment can be increased to reduce size of ordinary segments or a 6+1 ring can be adopted. Similar complexities are encountered when tunnel diameter is between 11–14 m as segmentation of a ring into 8+1 configuration is not preferred. Special solutions are required, such as dividing the ring into 8 segments (each covering 45°) with dividing also one of the ordinary segments into key and counter-key segments (covering 15°, and 30°). Using such configuration, excessively large key segments can be avoided, especially for such large diameter tunnel, while at same time the configuration is compatible with TBM thrust jacking pattern of an 8-segment ring. For tunnels larger than 14 m, a 9+1 configuration is the most common configuration.

Segment Geometry

Segment systems from the perspective of individual segment geometry can be divided into four main categories: hexagonal system, rectangular system, trapezoidal system, and rhomboidal system.

Hexagonal systems are assembled continuously from hexagonal elements, alternating bottom/top and left/right, with each element serving as a key segment. This is an old system that is not used anymore due to its watertightness deficiencies.

Rectangular systems are assembled in rings of rectangular or slightly tapered segments with a wedge-shaped key segment assembled from bottom to top, alternating between left and right (Figure 4a). This system can provide proper sealing performance and has some advantages such as simple longitudinal joint geometry. However, staggered longitudinal joints are not always guaranteed and star or crucified joints may be present that may cause leakage. The main disadvantage is that with this system, bolts are often used as the connection devices which is a time-consuming system compared to fast-connecting dowels. This system is still in use for large-diameter

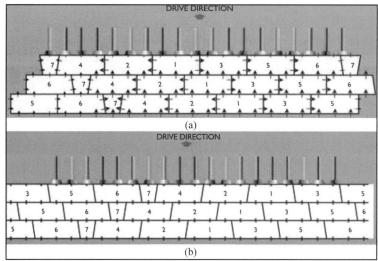

Source: Guglielmetti et al. 2007

Figure 4. (a) Rectangular systems, (b) Rhomboidal/parallelogrammic-trapezoidal systems

tunnels where shear capacity of the dowel system connection between the circumferential joints may not be sufficient. Typical details of bolt and dowel connection devices are shown in Figure 5.

Trapezoidal systems are assembled from an even number of trapezoidal segments in a ring often with the same length at centerline, with half the segments as counter key segments (wider on the side of the previously placed ring) and the other half are key segments (narrower on the side of the previously placed ring). After installation of all counter key segments as the first row like an open-tooth row, the second row is inserted in the gaps to form a complete ring. Advantages of this system are staggered longitudinal joints without encountering star joints and the fact that every other segment can be used as a key segment. The main disadvantage of this system is however, excessive TBM thrust jack forces on the first row segments (counter key segments) to keep a continuous penetration process while second row is installed (key segments). Also the ring is not built continuously which makes it difficult to place several key segments between the counter key segments.

Rhomboidal or parallelogrammic-trapezoidal systems are assembled from ordinary segments in the shape of parallelogram and a key and a counter key segment in the shape of trapezoid. As shown in Figure 4b, the tunnel is built continuously ring by ring, and often from bottom to top. The most common assembly procedure is to start with counter key trapezoidal element and placing parallelogrammic segments next to the counter key first and then next to previous parallelogrammic segments alternating

left and right. Ring assembly is completed with often smaller trapezoidal key segment. This system is now the most common system because of preventing crucified joints and improved sealing performance, continuous ring build from bottom to top, and compatibility with dowel connection system. Note that dowels are often used for rhomboidal and trapezoidal segments to avoid early crawling of the gaskets during the segments-approach phase of the ring assembly.

Key Segment Geometry

ITA WG2 (2000) and JSCE (2007) present two different key segment tapering geometries according to methods historically used for the ring assembly. One method that is not being used lately is to insert the segments from the inside of the tunnel, in which the longitudinal side faces of the key segment are tapered in the direction of the tunnel radius. ITA WG2 (2000) provides a geometry formulation which is specifically suitable for key segment insertion in radial direction. The other method, which is nowadays considered as the only common practice, is to insert the segments from the cut face side in the longitudinal direction of the tunnel in which the longitudinal side faces of the key segment are tapered in the longitudinal direction of the tunnel. Using this common method, the key segment tapering is defined by the angle of side faces with respect to joint centerline or longitudinal tunnel axis. Depending on designers and contractors' previous experiences, this tapering can be selected differently. A key segment taper angle of 8–12° is recommended based on a review of

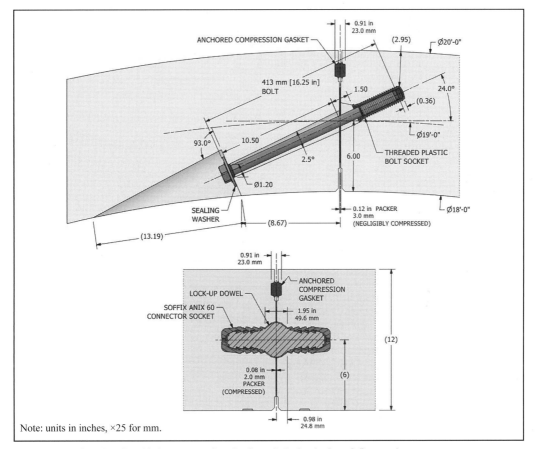

Figure 5. Typical details of joint connection devices: bolt (top), dowel (bottom)

recent projects. Note that adopting a rhomboidal or parallelogrammic-trapezoidal system, the same joint angle for key segment is suggested to be used for defining the geometry of other segments in the ring.

STRUCTURAL DESIGN PHILOSOPHY

Load and Resistance Factor Design

The design engineer should use load and resistance factor design (LRFD) method to design concrete precast tunnel segments. LRFD is a design philosophy that takes into account the variability in the prediction of loads and the variability in the properties of structural elements. LRFD employs specified limit states to achieve its objectives of constructability, safety, and serviceability.

Even though force effects may be often determined using elastic analyses, the resistance of elements using LRFD design methods is determined on the basis of inelastic behavior. Concrete precast tunnel segments should be designed using load factors and strength reduction factors specified in concrete

design codes such as ACI 318 (2014) or EN 1992-1-1 (2004). For load cases not covered in these codes, load factors and load combinations from other resources such as ACI 544.7R (2016) or AASHTO DCRT-1 (2010) can be used.

Governing Load Cases and Load Factors

The current practice in the tunnel industry is to design segmental tunnel lining for the load cases occurring during segment manufacturing, transportation, installation, and service conditions. Figure 6 shows segment stripping and storage as two early load cases. In the strength design procedure, the required strength (U) is expressed in terms of factored loads shown in Table 1 for presented governing load cases. In this document, aforementioned load cases are divided into three categories: production and transient loads, construction loads, and service loads. The resulting axial forces, bending moments, and shear forces are used to design concrete and reinforcement. For strength capacity of the

Figure 6. (a) Stripping segments from the forms in manufacturing plant, (b) Segments stacking for storage

fiber-reinforced concrete (FRC) sections, refer to ACI 544.7R (2016).

Structural capacity of the initially selected segmental ring is compared with the required strength in Table 1. If the strength is not sufficient, reinforcement ratio in Reinforced Concrete (RC) segments, residual tensile strength in FRC segments, and concrete compressive strength is adjusted to provide with additional required strength. If such measures are not the most economical ones, section thickness is adjusted and required strength is recalculated.

CONCLUSION AND BEST PRACTICE RECOMMENDATIONS

A crucial part of segmental lining design is initial consideration for segment and ring systems and their geometries. Nonetheless, these parameters and systems are further optimized during structural design steps by taking into account final details of the lining. Geometrical ring parameters include internal diameter (ID), thickness and length of the ring.

This paper summarized parameters that impact internal space required for the railroad, subway, roadway, water and wastewater (CSO) tunnels depending on each specific type of tunnel's intended function in service. Required construction tolerance in range of ±100 mm is recommended. For the initial selection of the lining thickness (t), a ratio of 1/23th of ID with upper and lower boundaries of 1/25th–1/18th ID are suggested, and 300 mm is recommended as minimum thickness of the lining for one-pass lining tunnels considering durability requirements. For tunnel diameters in the range of 6–7 m, 7–9 m and larger than 9 m, ring lengths of 1.5 m, 1.8 m and 2 m are recommended, respectively. In addition, 2.2 m is suggested as maximum ring length after optimizing segmental lining thickness for tunnels with diameters greater than 9 m.

Parallel rings, parallel rings with corrective rings, right/left rings, and universal ring systems

Table 1. Required strength (U) for governing load cases (ACI 544.7R, 2016)

Load Case	Required Strength (U)
Load case 1: stripping	$U = 1.4w$
Load case 2: storage	$U = 1.4(w \pm F)$
Load case 3: transportation	$U = 1.4(w \pm F)$
Load case 4: handling	$U = 1.4w$
Load case 5: thrust jack forces	$U = 1.2J$
Load case 6: tail skin grouting	$U = 1.25(w \pm G)$
Load case 7: secondary grouting	$U = 1.25(w \pm G)$
Load case 8: earth pressure and groundwater load	$U = 1.25(w + WA_p)$ $\pm 1.35(EH + EV)$ $\pm 1.5\ ES$
Load case 9: longitudinal joint bursting	$U = 1.25(w + WA_p)$ $\pm 1.35(EH + EV)$ $\pm 1.5\ ES$
Load case 10: additional distortion	$U = 1.4M_{distortion}$

w = self-weight
F = self-weight of segments positioned above
J = TBM jacking force
G = grout pressure
WA_p = groundwater pressure
EV = vertical ground pressure
EH = horizontal ground pressure
ES = surcharge load
$M_{distortion}$ = Additional distortion effect

have been presented among different systems used for tunnel segmental rings. Universal ring system, with circumferential faces of the ring inclined to the tunnel axis on both sides, is recommended as the best ring system with main advantages of requiring only one set of formwork, superior sealing performance and the ability to negotiate tight alignment curves.

Segmentation of tunnel rings from the perspective of ring configuration in terms of number of segments, geometry of individual segments, and key segment geometry have been discussed. Rings are commonly divided into a number of segments that gives a segment slenderness of 8–13, with slenderness ratio of 10 as the most frequent value. For tunnels with a diameter of 6m and below, a ring division into 6 segments is recommended with ring configuration of 5+1 (5 ordinary and 1 key segment) as the most common configuration. For tunnels of 8–11 m and larger than 14 m in diameter, 7+1 and 9+1 configurations are the most common configurations. However, when tunnel diameter ranges between 6–8 m or 11–14 m, some complexities are encountered. 6+1 and 8+1 configurations may be used which are not always preferred solutions since they may not be compatible with TBM thrust jack configurations pushing against segments. In such cases, special solutions need to be adopted.

Segment systems from the perspective of individual segment geometry have been discussed in categories of hexagonal, rectangular, trapezoidal and rhomboidal systems. The preferred system is the rhomboidal system assembled from ordinary parallelogram segments and one key and one counter key segment in the shape of trapezoid. Preventing crucified joints and improved sealing performance, continuous ring build from bottom to top, and compatibility with dowel connections are among the main advantage of the rhomboidal system. For the key segment the best practice is to insert key segment from cut face sides in the longitudinal direction with recommended key segment taper angle of 8–12°.

REFERENCES

AASHTO DCRT-1. 2010. *Technical Manual for Design and Construction of Road Tunnels— Civil Elements*. American Association of State Highway and Transportation Officials (AASHTO). Washington, DC.

ACI 318. 2014. *Building Code Requirements for Structural Concrete and Commentary*. American Concrete Institute (ACI). Farmington Hills, MI.

ACI 544.7R. 2016. *Report on Design and Construction of Fiber Reinforced Precast Concrete Tunnel Segments*. American Concrete Institute (ACI).

AFTES, 2005. *Recommendation for the design, sizing and construction of precast concrete segments installed at the rear of a tunnel boring machine (TBM)*. French Tunnelling and Underground Space Association (AFTES). Paris, France.

Bakhshi, M., and Nasri, V. 2017a. Design consideration. *Tunnels & Tunnelling North America Edition*. August-September 2017: 32–37.

Bakhshi, M., and Nasri, V. 2017b. Design of steel fiber-reinforced concrete segmental lining for the south hartford.

Bakhshi, M., and Nasri, V. 2016. ACI guideline on design and construction of precast concrete tunnel segmental lining. *ITA World Tunnel Congress (WTC) 2016*, San Francisco, USA, April 22–28, 2016.

Bakhshi, M., and Nasri, V. 2015. Design of segmental tunnel linings for serviceability limit state. *ITA World Tunnel Congress (WTC) 2015*, Dubrovnik, Croatia, May 22–28, 2015.

Bakhshi, M., and Nasri, V. 2014. Design considerations for precast tunnel segments according to international recommendations, guidelines and standards. *Vancouver TAC 2014: Tunnelling in a Resource Driven World*, Vancouver, Canada, October 26–28, 2014.

Bakhshi, M., and Nasri, V. 2013. Practical aspects of segmental tunnel lining design. Underground— The way to the future. *ITA World Tunnel Congress (WTC) 2013*. Geneva. May 31– June 7, 2013.

Blom, C.B.M. 2002. *Design Philosophy of Concrete Linings for Tunnels in Soft Soil*. PhD dissertation, Delft University of Technology, The Netherlands.

DAUB. 2013. *Lining Segment Design: Recommendations for the Design, Production, and Installation of Segmental Rings*. German Tunnelling Committee (DAUB). Cologne, Germany.

EN 1992-1-1. 2004. Eurocode 2: Design of concrete structures - Part 1-1 : General rules and rules for buildings. *European Standards (EN)*. Brussels, Belgium.

Groeneweg, T. 2007. *Shield Driven Tunnels in Ultra High Strength Concrete: Reduction of the Tunnel Lining Thickness*. MSc Thesis, Delft University of Technology, The Netherlands.

Guglielmetti, V., Grasso, P., Mahtab, A., and Xu, S. 2007. *Mechanized Tunnelling in Urban Areas: Design Methodology and Construction Control*. London: Taylor & Francis.

ITA WG2. 2000. Guidelines for the design of shield tunnel lining. *Tunn. Undergr. Sp. Tech.* 15(3): 303–331.

JSCE. 2007. *Standard Specifications for Tunneling: Shield Tunnels*. Japan Society of Civil Engineers (JSCE). Tokyo, Japan.

ÖVBB. 2011. *Guideline for Concrete Segmental Lining Systems*. Austrian Society for Concrete and Construction Technology (ÖVBB). Vienna, Austria.

RTRI. 2008. *Design Standards for Railway Structures and Commentary (Shield Tunnels)*. Japanese Railway Technical Research Institute (RTRI). Tokyo, Japan.

CSO tunnel. *Rapid Excavation & Tunneling Conference (RETC) 2017*. San Diego, CA, June 4–7, 2017.

Steel Fiber Reinforced Concrete (SFRC) for TBM Tunnel Segmental Lining—Case Histories in the Middle East

Guido Castrogiovanni and Gianpaolo Busacchi
COWI

Gianni Mariani
COWI (former)

ABSTRACT: The experience gained by COWI in some of the world's landmark mechanized tunnelling projects in the Middle East with the use of Steel Fibre Reinforced Concrete (SFRC) lining in sewer, storm water, and metro tunnels is reviewed in comparison with traditionally reinforced solutions.

The technical challenges offered by concrete mix design and its workability and mechanical properties testing, potential local damage during handling and installation of the tunnel lining, its corrosion and fire protection, and life service requirements in the encountered highly-concentrated chloride and sulphate environments are discussed together with considerations on the reduced projects' carbon footprint.

INTRODUCTION

COWI designed three of the most relevant mechanized precast segmental lined sewer, storm water, and metro tunnels projects in Middle East over the last ten years;

- Strategic Tunnel Enhancement Program, Abu Dhabi, UAE
- Abu Hamour Water Drainage Tunnel, Doha, Qatar
- Doha Metro, Red Line North Underground, Doha, Qatar

Important technical issues associated with the lining design of mechanized tunnels in the extremely aggressive ground and groundwater environment, with very high levels of chlorides and sulphates, were solved by COWI to provide the clients with the required 80–100–120 years' service life.

The most up to date service life design methods have been applied by COWI to determine the requirements for the concrete mixes, and the technical solution are discussed in the following paragraphs. The use of Steel Fibre Reinforced Concrete (SFRC) lining was successfully applied for the 3 aforementioned projects.

LIFE SERVICE REQUIREMENTS

Environmental conditions in the Middle East are particularly demanding in terms of deterioration mechanisms including chloride-induced corrosion, which require extensive experience with durability design of concrete structures.

To address the life service requirements in several tunnelling project, two different design strategies for concrete structures were developed and implemented;

- Strategy A: avoid the degradation mechanism threatening the structure due to the type and aggressivity of the environment during the service life, with the selection of non-reactive or inert materials.
- Strategy B: select an optimal material composition and structural detailing to resist, for a specified period of use, the degradation threatening the structure and allowing deterioration to a certain degree within the defined service life (also called a performance-based service life approach), generally requiring a mathematical modelling of the transport and deterioration mechanism.

The SFRC technical solution for mechanized precast segmental lined tunnels, corresponding to durability Strategy A (Avoidance of Deterioration), has successfully been implemented in the mentioned projects, avoiding the requirements of large concrete cover to carbon steel reinforcement and a maximum chloride migration coefficient, in possible combination with tailored external/internal coatings.

The groundwater, which governs the deterioration mechanisms, is often characterized by extreme values of dissolved salts and high water temperature as described in Table 1. In such high demanding conditions, the most obvious countermeasure to apply in order to assure the strict contractual requirements in terms of durability, would be to consider the application of protective coatings aiming at physically separating the traditionally carbon steel reinforced concrete segments from the external environment.

With the target to meet the contractual requirements while assuring the fulfilment of contractor's needs, COWI has developed alternative specific solutions that have presented clear technical advantages and have been a key factor for the successful delivery of the above-mentioned projects.

Table 2 and Table 3 summarize the typical contractual requirements in the Middle East, specifying the use of internal and external coatings in order to achieve an effective protection to the steel rebars against corrosions. It compares the potential and applied solutions, with "Standard" being the traditional solutions specified in employer's requirements and with "Innovative" the alternative solution proposed by COWI, and list the pros and cons of the internal/external coating removal.

Given the technical and operative complications related to the installation of protective coatings and the very high risk of not achieving the required performance during the whole life cycle, COWI has successfully focused any efforts in the definition of

.concrete mix designs able to replace the traditional carbon steel solution with the use of steel fibres.

SFRC—ADVANTAGES AND LIMITATIONS

Rebar Cage Removal and Cracks Issue

In general, standard steel reinforced segments show a greater margin of safety and are less sensitive to imperfections during the installation and assembly of the tunnel segments, whilst the SFRC technology requires a higher standardized and controlled set-up at the precast factory in terms of tolerances and acceptance criteria process.

When dealing with limited tensile stresses, the SFRC lining has undoubtedly advantages as it will develop a system of cracks that are evenly distributed and smaller in size compared to those in a traditionally carbon steel reinforced concrete segment. However, when the tensile stresses become significant, the SFRC will not work properly developing significant cracks that may require repairs works to potentially the replacement of the entire segment.

This issue becomes important from the structural and serviceability point of view if in combination with relevant steps between adjacent segments as it leads to the development of uncontrolled longitudinal cracks along the whole segment, triggered during the ring-assembling phase (pushing rams in action).

Even though the replacement of the carbon steel rebar cage with the use of steel fibres may show complications associated with the control of the tensile stresses in the tunnel lining, the reduction of resisting capacity can still be managed by adopting the following mitigation measures:

- A full-scale material testing investigation campaign and tight precast manufacturing construction tolerances specifications, part of the design process, where an effective collaboration between the designer and the contractor can be the key factor in the definition of successful solutions. It is the moment where the definition and the refinement of the

Table 1. Measured data and design values for chloride, sulphate, pH and temperature characterising the saline soil/groundwater around the structures

Parameter	Min.	Max.
Chloride, Cl^- (mg/l)	10,000	55,000
Sulphate, SO_4^{2-} (mg/l)	100	5,500
Temperature (°C)	28	32
pH	7	10.2

Table 2. Comparison between standard and innovative solutions for TBM tunnel linings

Solution	Type of Reinforcement	Concrete	Internal/External coatings
Standard	Carbon steel rebars	Ordinary Portland Cement	High-density polyethylene (internal)* Epoxy coating (external)
	Fibre reinforced concrete	Ordinary Portland Cement + Ground granulated blast-furnace slag + Microsilica	No coating is required
		Ordinary Portland Cement + Fly ash + Microsilica	

* The HDPE liner can be replaced by or be in combination with an internal sacrificial concrete lining. This would deteriorate over time to achieve the final defined service life.

Table 3 Innovative solution—pros and cons of coating removal

Innovative Solution	Pros	Cons
Internal HDPE coating and/or internal sacrificial lining removal	If correctly installed, it provides a very good protection against corrosion	A perfect installation of the coating is really hard to achieve. For various reasons (e.g patchwork of repairs) the HDPE lining does not provide the continuity of the liner and actually becomes critical for the concrete durability of the tunnel lining. The internal coating and/or concrete sacrificial lining is installed at the end of the tunnelling construction phase. It is always an activity on the critical path, with relevant construction time and costs. The HDPE liner, if not in combination with a concrete sacrificial layer, requires maintenance during the service life of the tunnel structure, with relevant cost on its life cycle.
External epoxy coating	Provides defence to segments against the aggressive environment	It can be damaged during storage, transportation, erection. Maintaining continuity of the epoxy coating is a big challenge. There is no way of repairing the damage once the segment is in place. The TBM brushes can damage the coating and there is no way to know about their presence or repair it.

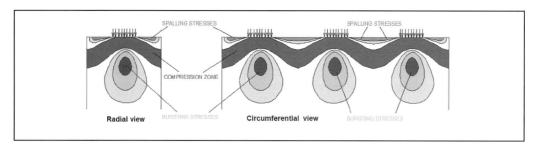

Figure 1. Stresses in tangential and radial direction under pushing ram

design choices can be addressed to improve the performances of the tunnel ring and reduce the production and construction costs.

- An extensive detailed calculations and verification of the segmental lining including the temporary phases, such as the TBM thrust from longitudinal reacting against the tunnel lining. In particular, the pushing rams assessment requires a 3D analysis to determine the tensile stresses that develop in the segmental lining, in combination with the typical construction imperfections associated with the assembling of the ring.
- The use of stainless steel reinforced segments as replacement or in combination with steel fibres where needed due to expected singularities, exceptional load cases and specific features (e.g., asymmetric loads), being the cross passageways or spur tunnels to access shafts one of these instances.

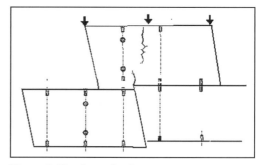

Figure 2. Crack mechanism

The use of SFRC can then become the typical solution to be applied to most of the tunnel contract length and be backed up with the stainless steel reinforced rings where peculiar circumstances exceed the capacity of the SFRC ones.

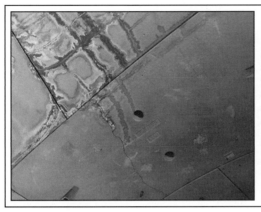 

Figure 3. Typical cracks

This strategy has been successfully implemented by COWI in the sewer, storm water, and metro tunnels projects in Middle East over the last ten years.

Corners Protections

The environmental conditions in the Middle East are particularly demanding in terms of protective cover to carbon steel, hence the use of rebar cages, with the relevant bars bending radii and the preassembly tolerances, causes large areas of the segments, particularly the corners, exposed to cracking and damaging, and potentially subject to brittle failure. In Figure 4, the area without any reinforcement is clearly visible.

According to the experience gained in the previous tunnelling projects where large covers were specified, unprotected precast segments corners are one of the main causes of splitting during rings handling and assembling, requiring a substantial amount of time allocated for the relevant repair works and, sometimes, becoming critical for the project delivery.

The use of steel fibres reduced considerably the corners damages during those phases, with the uniform distribution of fibres contributing to reducing the splitting and brittle failure risks.

In the abovementioned projects where COWI provided the detailed design of the SFRC lining following the strategy described in the previous paragraph, acceptance of segments at the precast factory was well above the 99% mark, with a considerable reduction of repair works following handling and assembling compared to similar project with carbon steel reinforcement.

Construction

In the recent sewer, storm water, and metro tunnels projects that COWI provided detail design service and support during construction activities, the

Figure 4. High risk of local failures in unreinforced area

contractor accepted the solution with sole SFRC lining for the following main advantages:

- Avoidance of a dedicated plant for steel cages production and specific QA/QC procedures for installation tolerances.
- Ease of mixing of the steel fibres with the designed concrete mix at the batching plant by means of a dedicated feeder.
- No need for stray current protection connection between all segments (required in metro tunnels design).
- Reduction of repair works to reinstate damaged segments corners with minimum cover to steel.

Although it shall be noted that SFRC segmental lining mandatorily require:

- A high concrete quality, and consequently, of a skilled contractor and concrete supplier.
- A significant number of material tests and quality control procedures during the refinement process, in a more expensive fashion

than when using the carbon steel reinforced concrete.

Design and Additional Risks

A significant risk during the design phase is related to the demanding verification and approval processes with clients and relevant stakeholders, generally reluctant to accept the use of SFRC lining because of:

- Not existing national or international SFRC standards, but only guidelines as a basis for future codes for concrete structures eg Model Code 2010.
- Limited examples of application in similar projects in the area.
- Risk of relevant cracking and potential repair works.
- Undefined behaviour in case of fire.

To address them, some of the mitigation measures adopted by COWI in the projects were:

- The detailed design of precast segment details (e.g., edges, corners, vacuum holes) to minimize the tensile stress levels.
- The detailed assessment of pushing rams phases and definition of the developed tensile stresses.
- Performance of specific tests for fire design.

SUSTAINABILITY

Amongst the segmental lining design requirements, the impact of sustainability through the different project phases led to make educated choices that delivered best value for money, managing costs throughout the life cycle of the tunnelling projects, with their own peculiarities as a sewer, storm water or metro.

The reduction of the size of the tunnel (smaller excavation, less material used, shorter construction period, less cost), the implementation of the single-lining solution (less cost, faster and smaller excavation, use of less material), the use of steel fibre reinforced concrete in the tunnel sections (less cost, use of less steel, lower embodied CO_2) were all technical solutions considered, studied and implemented by COWI in the aforementioned projects.

Concerning the specific experience of the metro projects, the design of the tunnel lining was developed to full compliance with the CEEQUAL and other sustainability objectives with its progressive approach. While some of the CEEQUAL certification credits are not directly applicable to tunnel construction, many credits are directly relevant including innovation in design, material re-use, use of recycled contents and construction waste management.

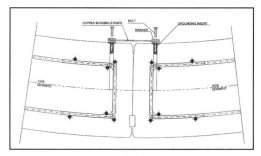

Figure 5. Longitudinal joint stray currents protection

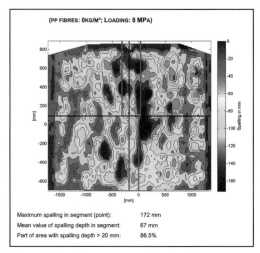

Maximum spalling in segment (point):	172 mm
Mean value of spalling depth in segment:	67 mm
Part of area with spalling depth > 20 mm:	86.5%

Figure 6. Spalling measured during fire test

A key element of the sustainability approach for the tunnel lining has been the reduction in the volume of cement used (leading to less CO_2 emissions), in combination with methods for reducing requirements for raw materials, aggregates and fresh water, the use of cement replacement materials, the reuse of concrete wash down water, the use of recycled aggregates, energy efficient machinery and equipment, and low emitting materials, all considered to be used during the construction of the tunnel sections.

The design of the tunnel lining based on the abovementioned sustainability criteria has the potential to achieve "Very Good" CEEQUAL awards.

CONCLUSIONS

In the Middle East, tunnelling projects see the use of the precast segmental lining reinforced with steel fibres being adopted as it is proven the best solution to solve the significant and peculiar durability issues.

The SFRC technology main advantages are evident and allow removing the reinforcement steel bars cages in the concrete segments, thus the need of a specific plant for the steel cages production and the relevant operational issues, reducing the production cost and increasing its rate. Contractors need to consider the greater margin of error related to the design of such technology, relying on high-level competence and solid technical support experience that will provide effective solutions to the fulfilment of the requirements and desired targets, to deliver best value for money and plan managing those costs throughout the life cycle of the project. Cooperation will be a key factor.

Steel fibres reinforced concrete segments are playing an important role in the tunnelling contractors decision process, also encouraged now by the clients seeing the advantages in important reduction of maintenance cost during the project lifetime and the its impact to the environment.

REFERENCES

[1] COWI; http://www.cowi.com/menu/project/bridgetunnelandmarinestructures/.

[2] The Abu Dhabi STEP Project; http://www.cowi.com/menu/project/bridgetunnelandmarinestructures/tunnels/abu-dhabi-step-project-includes-45-km-of-bored-tunnel.

[3] Tunnel Design Organization Awards to COWI for the STEP Project, 3rd Arabian Tunnelling Awards 2015; http://www.atcita.com/2015/awards.

[4] Abu Hamour Tunnel Named Global Best Project; http://www.cowi.com/menu/newsandmedia/news/bridgetunnelandmarinestructures/cowi-tunnel-named-global-best-project.

[5] Red Line North Underground—Doha Metro, Qatar; http://www.cowi.com/menu/project/railwaysroadsandairports/metros/red-line-north-underground-%E2%80%93-doha-metro.

[6] CEEQUAL; http://www.ceequal.com/awards/

TRACK 3: PLANNING

Session 1: Risk Management Challenges and Solutions

Steven Lotti and Rick Capka, Chairs

CEVP-RIAAT Process—Application of an Integrated Cost and Schedule Analysis

Philip Sander and Martin Entacher
RiskConsult

John Reilly
John Reilly International

ABSTRACT: Key processes necessary to identify and manage risks on complex tunneling projects have been developed over the last 20 years in order to implement risk-based approaches for better cost and schedule estimation. Cost and schedule, however, were mostly treated separately instead of integrating them in one model. This integration is highly relevant as schedule delays are very often the root cause for severe cost overruns. This paper presents a fully-integrated probabilistic cost and schedule model. The application is based on combination of two practice-proven approaches—the Cost Estimation and Validation Process CEVP® (Reilly et al. 2004/Washington State Department of Transportation) and the RIAAT (Risk Administration and Analysis Tool), creating a powerful tool for management of complex risk environments.

INTRODUCTION

Significant progress has been made over the last 20 years in the identification, characterization, mitigation and management of risk for complex projects. Risk guidelines have been developed (ITA 1992, 2004; ITIG 2006, 2012; Reilly 2001, 2003, 2008, 2013; Goodfellow & O'Carroll 2015) and are more routinely applied with increasing success, such that the general process and application of risk management principles are now generally clear. During this period, specific applications and detailed tools have been developed to assist with risk identification, characterization and mitigation, such as:

- Risk-based cost estimating, e.g., WSDOT's CEVP cost estimating/cost validation/risk management process (Reilly et al., 2004)
- Risk management processes and procedures (ITA 2004, ITIG 2006, Reilly 2008, Goodfellow & O'Carroll 2015)

Drawbacks in previous cost-risk processes were:

- A delay in obtaining results, since the model could only be run after completion of the cost-risk workshop, which frequently took several weeks
- A preliminary or approximate approach to the probable schedule element in terms of risk effects on the critical path and identification of the risk elements that contributed to those critical paths

RIAAT (RIAAT 2017. http://www.riaat.riskcon.at) solved both of these drawbacks since it is an integrated model which can be run in real time, during or at the conclusion of the workshop, to give results quickly. It allows efficient application of risk-based processes including risk characteristics (probabilities and consequences), correlations, interdependencies, linkage, risks occurring multiple times and schedule/critical path analysis (Sander et al., 2016).

Also added were full risk-based critical path schedule and cost integration in the risk-based cost and schedule estimating process and associated computer models. This is the subject of this paper.

DEALING WITH UNCERTAINTIES

Since empirical/historical data as input for risk analysis is often not available, risk probabilities and consequences can be difficult and complex to estimate. Normally, experts are involved in a workshop using, for example, Delphi technique. The risk-based method characterizes each risk with individual and specific distributions such as a large cost ranges for large uncertainties or a narrower cost ranges for smaller uncertainties. Using this approach, the uncertainty contributing to a particular cost estimate can be modelled more specifically and in greater detail than by use of a single-point deterministic estimate (Sander et al., 2009).

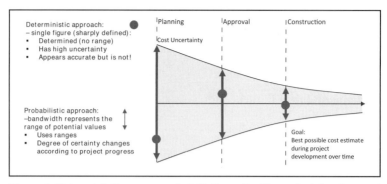

Figure 1. Deterministic versus probabilistic method in project development

Cost estimates, especially if only deterministic approaches are considered, can come with a high degree of uncertainty. This is especially true for early phases of projects when neither the exact quantities nor the exact costs or prices are yet known. Quantities will be determined for known project elements, but allowances must be made for unknown elements. Often a more detailed analysis is not yet available at this stage of the project due to a lack of precise information. With a deterministic approach, information about potential deviations (variability due to potential higher or lower values) for quantities and prices is not usually taken into consideration although this information is available or could easily be estimated (Figure 1).

At first sight, probability functions might seem more uncertain compared with what might seem to be a "totally defined deterministic value"—however, exactly the opposite is true (Rohr 2003) because the accuracy of a forecast is greater when the uncertainty component is included. The uncertainty is part of the answer and so not including it means that part of the answer is missing, therefore that answer is less accurate.

CEVP RIAAT PROCESS

Combining Both Approaches

CEVP is a Cost Validation and Risk Management process to address the concerns of:

- Why do project costs seem to always go up?
- Why can't the public and/or private owners be told exactly what a project will cost?
- Why can't projects be delivered at the cost you told us at the beginning?

CEVP opens the "black box" of estimating, ensures cost transparency and provides a profound decision making basis for senior management.

Figure 2. CEVP RIAAT process

The software RIAAT can fully implement the CEVP-Process. RIAAT combines a clear project structure and a convenient work flow with extensive modeling and simulation capacities. Key features include:

- Hierarchical project tree (WBS)
- Full integration of uncertainties for all cost components on all levels
- Live results and simulation updates within seconds
- Fully integrated cost and schedule model

Combining the CEVP Process with the simulation capacities of RIAAT (Figure 2) adds a powerful tool for the management of complex risk environments

CEVP—Cost Estimate Validation Process

In 2002, WSDOT, recognizing the need for a validated, integrated, cost-risk estimating process, developed the Cost Estimate Validation Process (CEVP®), to better estimate the range of cost and schedule for their complex megaprojects. The process has been described in detail (Reilly 2004, 2013)

In CEVP, estimates are comprised of two components: the base cost component and the risk component. Base cost is defined as the planned cost of the project if everything materializes as planned and assumed—the base cost does not include contingency but does include the normal variability of prices, quantities and like units. Once the base cost is established, a list of risks is identified and characterized, including both opportunities and threats, and listed in a Risk Register. This risk assessment

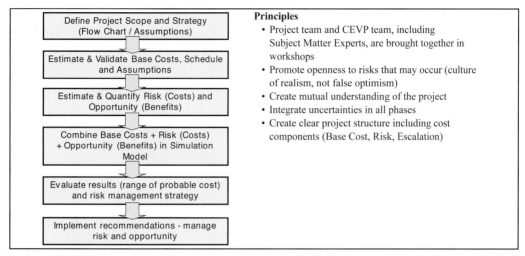

Figure 3. CEVP process

Principles

- Project team and CEVP team, including Subject Matter Experts, are brought together in workshops
- Promote openness to risks that may occur (culture of realism, not false optimism)
- Create mutual understanding of the project
- Integrate uncertainties in all phases
- Create clear project structure including cost components (Base Cost, Risk, Escalation)

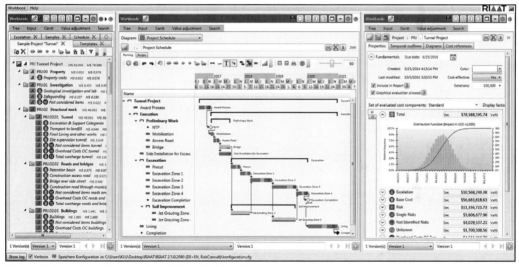

Figure 4. Sample main interface, RIAAT risk software

replaces general and vaguely defined contingency with explicitly defined risk events that include the associated probability of occurrence plus impact on project cost and/or schedule for each risk event. Risk is usually developed in a CEVP Cost Risk Workshop.

The validated base cost, base variability and the probable consequence of risk events are combined in a simulation model to produce an estimated range of cost and schedule, with probabilities of achieving a particular cost or schedule outcome (see Figure 3). The output is a rich data set of probable cost and schedule, potential impact of risk events, ranking of risks and risk impact diagrams.

RIAAT—Risk Administration and Analysis Tool

The Risk Administration and Analysis Tool RIAAT (RIAAT 2017) is a powerful software that enhances CEVP principles by integrating them into a continuous workflow. RIAAT uses probabilistic modeling. In spite of advanced mathematics in the background, results can be updated live during workshops and easily understood due to RIAAT's clear work breakdown structure. RIAAT supports full MS Excel Import/Export, advanced risk modeling and numerous options for visualization. Figure 4 shows the main interface of RIAAT—the following figures in this paper were generated using RIAAT.

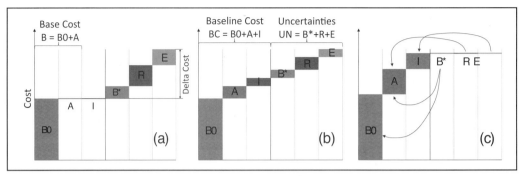

Figure 5. Waterfall diagram for cost component structure, planning phase (a), construction phase (b), project completion (c)

The RIAAT workflow is optimized, utilizing a normal estimating structure and well-understood elements such as base cost, risk and escalation which makes it a good fit to use in CEVP workshops. Cost components that need to be addressed in the cost estimate are:

- Base cost—the cost if "all goes according to plan" without contingencies
- Risk cost—the cost resulting from threats and opportunities that might occur
- Escalation cost—additional costs resulting from inflation

A best practice cost component structure for different project phases is shown in Figure 5. It consists of actual cost without uncertainties (left part of the waterfall diagram: B0—Base Value, A—Additional Cost, I—Indexation, B0 + A + I: *Baseline Cost*) and uncertain components (right part of the waterfall diagram: B*—Base Uncertainties, R—Risk Cost, E—Escalation, B* + R + E: *Uncertainties*). The sum of the uncertain cost components is also called *delta cost* and allows for inclusion of uncertainties in the project budget. While uncertainties are high in early project phases, they reduce to zero upon project completion. Escalation becomes indexation cost (contractual clause for compensation for inflation) and realized risks result in actual additional cost.

Construction schedules are fully integrated into RIAAT. Risks are assigned to tasks (elements) of the schedule from the project tree using drag&drop. Schedule results (e.g., completion dates, critical paths) are obtained using Monte Carlo simulation (since the performance of Monte Carlo simulation is no longer a problem, 100,000 iterations can be carried out in just a few seconds). Delays can be associated with time related cost which allows for an integrated cost and schedule analysis.

Process

The process used for the integrated cost and schedule model is shown in Figure 6. In the first step, Base Cost is estimated and validated, subjected to uncertainties, and integrated into the Work Breakdown Structure (WBS). Subsequently, identified risks and a markup for unknowns with cost and time impacts will be assessed and integrated into the WBS and the construction schedule. A probabilistic simulation of the construction schedule incorporates all risks with associated time impacts. The results include a construction completion date, delays with respect to specific milestones, critical paths and near-critical paths. The results of the construction schedule are linked back to the WBS, where the time impacts can be associated with time-related costs to evaluate the cost impact of program delays.

Figure 6 depicts the following steps:

1. Base cost estimate is reviewed, associated with uncertainties and integrated into the WBS.
2. Risks are assessed (probable cost & time impact) and integrated into the WBS.
3. Risks are assigned to tasks in the project's schedule. Subsequently, completion date, critical paths and delays from risks are simulated.
4. Cost impact from time delay is calculated with time-related cost and integrated into the WBS.
5. Project Cost including uncertainty is available an all WBS levels and for all cost components.

As shown in Figure 6 RIAAT is an ideal tool to fully utilize the added value of CEVP. The optimized use of cost components (e.g., base cost, risk), the ability

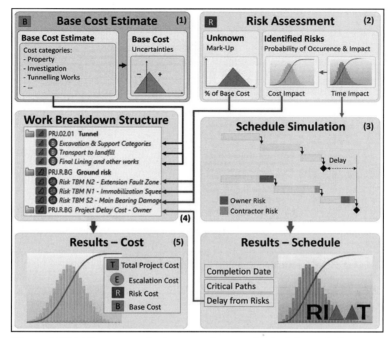

Figure 6. RIAAT—process for integrated cost and schedule analysis

for probabilistic simulation, the instant availability of results and the integrated cost and schedule analysis are very beneficial for the application of CEVP. The combination of CEVP and RIAAT is thus called the CEVP-RIAAT Process.

INTEGRATED COST AND SCHEDULE ANALYSIS—SAMPLE PROJECT

Project Description

A fictitious sample project is used in this paper to illustrate the process. It is based on experience from major European railway base tunnels. This 14-km twin-bore tunnel consists of several Tunnel Boring Machine (TBM) drives as well as Drill & Blast (D&B) drives in different geological formations, an access shaft, an emergency stop, various cross cuttings and (optional) inner linings. A linear project schedule is shown in Figure 7. In RIAAT, the base schedule is modeled as a Gantt diagram (Figure 8). The deterministic critical path is shown in red.

Base Cost and Risk Register

A deterministic base cost estimate is made by the design firm. It is reviewed, discussed and validated with the project team and a bandwidth is assigned to account for minor variability in the base cost estimate. Subsequently, risks are identified and assessed

in moderated workshops with the project team and subject matter experts. The process is structured using "risk fact sheets" to gather and systematize information such as risk description, qualitative and quantitative assessment, risk strategy and risk mitigation measures. The quantitative assessment typically consists of either probability of occurrence (0–100%) or expected rate of occurrence (e.g., 1, 2, 3, etc., modeled with a Poisson distribution) and cost/time impact using a three-point estimate (best, most likely and worst case). Complex risks (e.g., including correlations or dependencies) can be modeled using event or fault trees (ETA, FTA). The risk register is updated during the workshops to give the project team a clear picture of the ongoing process.

Table 1 shows the quantitative assessment of the top 10 risks. The risks are ranked according to their respective VaR95 (95% of not exceeding the depicted value in days delay), which is depicted in the range impact diagram in Figure 9. Range impact diagrams are used to compare risks with respect to their cost or time impact (in this case time). The width of each bar represents the bandwidth of a risk impact from the best case (left end of bar, VaR5) to the worst case (right end of bar, VaR95). Each bar represents a probability of 10%. The left end of Risk No. 1 (TBM Main Bearing Damage) represents VaR80. This is because the probability of occurrence

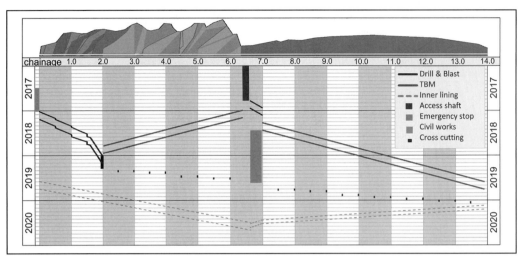

Figure 7. Linear base schedule—horizontal axis: station; vertical axis: time

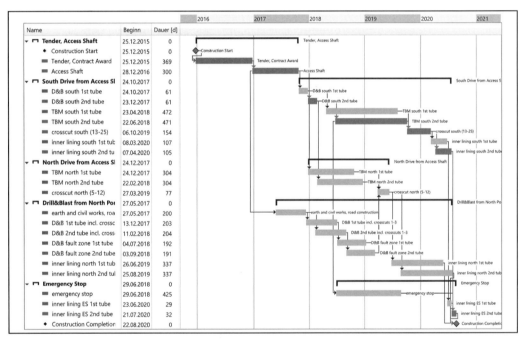

Figure 8. Base schedule in RIAAT, deterministic critical path without risks is shown in red

is as low as 20%. Hence, any probability lower than 80 equals to zero. Each of the top 10 risks has an associated % value. This value indicates the chance that the respective risk will be on the critical path according to the Monte Carlo simulation.

A single risk that is assigned to more than one task in the schedule will be dependent, i.e., the risk will impact both tasks if it occurs. The importance of the capability to model dependencies in schedules

was explained by Dorp & Duffey (1999). In the following example, independent risks such as "Main bearing damage" for four different TBMs are modeled separately as four different single risks to ensure independency. For clarity, similar independent risk events are not displayed in Table 1 and Figure 9.

After the risk register is complete, all risks with time impact are assigned to the base schedule (Figure 10). Colors can be used to indicate the type

Table 1. Sample quantitative assessment of top 10 identified risks

# Identified Risk	Probability of Occurrence	Rate of Occurrence	Cost Impact (USD × 1,000) Best	Most Likely	Worst	Time Impact (d) Best	Most Likely	Worst
1 TBM S2—Main Bearing Damage	20%	—	1000	2000	3000	90	180	400
2 TBM N1—Change in Exc.&Sup. Categ.	70%	—	500	3000	4500	20	120	180
3 TBM N1—Immobilization Squeezing	25%	—	1500	3000	5000	60	120	200
4 Contractor Appeal	50%	—	—	—	—	30	90	180
5 No Release of Design	30%	—	225	900	1350	30	120	180
6 TBM N—Delay installationon	25%	—	400	1200	2000	20	60	100
7 Extension Fault zone km 2.0	80%	—	0	840	1660	0	42	83
8 TBM S2—Extension of inner lining	—	3	150	200	250	5	10	20
9 Logistic Problems Crosscut S (13–25)	30%	—	150	375	600	20	50	80
10 CC N—Mountain water inflow >40l/s	—	3	222	886	1782	1	3	14

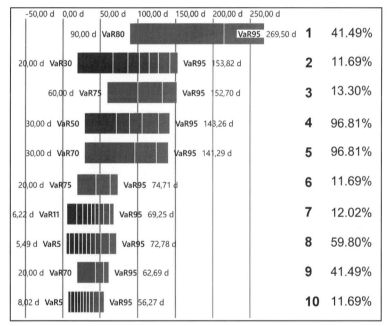

Figure 9. RIAAT range impact diagram for top 10 identified risks, bandwidth: VaR5–VaR95

of assigned risks, e.g., for owner risks, contractor risks and tender risks (pre-contract). The length of each task is not a deterministic number anymore, it contains uncertainties and is thus represented with a distribution function. Due to the assigned uncertainties, different critical paths become possible. The probabilities of occurrence for various critical paths are calculated using Monte Carlo Simulation.

RESULTS

Simulation results for the critical paths are shown in Figure 11. Different colors can be used to indicate alternative critical paths. A task with more than one color would be more than one critical path, e.g., in this example, the task "Tender, Contract Award" would be made up of all relevant colors and would be part of all possible critical paths. A graphical example for interpretation is given in Figure 12. In this example, there is a 60% chance that the completion date will be determined by the TBM south drive, but there is also a 30% chance that the TBM north drives will become critical. The D&B drive from the north portal only has a 12% chance of becoming critical.

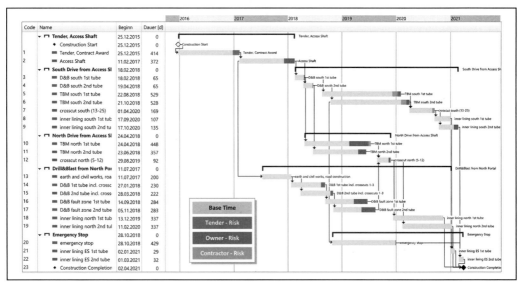

Figure 10. RIAAT schedule with assigned risks, colors can be used to indicate risk impact

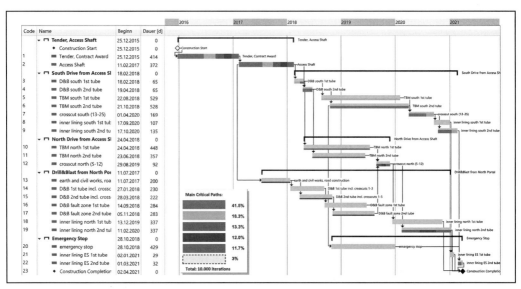

Figure 11. Results of critical path simulation in RIAAT, each color indicates one possible critical path

This will be the case when the fault zone turns out to be much longer than expected (risk 7).

The construction completion date and the deviation to the original construction completion milestone of the base schedule are shown in Figure 13. Direct time-related cost that is caused specifically by one risk event is calculated within the risk itself (see Table 1). In addition to that, a delay on the critical path causes additional time-related cost. This cost is now calculated using the overall project delay on

the critical path. In this case, this was done by taking into account only the portion of the critical path delay caused by the owner's risks (see Figure 14).

After including time-related cost, a probabilistic cost forecast for all cost components can be made. The results are shown in Figure 15. The vertical line represents the deterministic base cost without uncertainties. Taking into account uncertainties related to the base will result in the curve to the right of the vertical line. Adding risk cost results in the next curve.

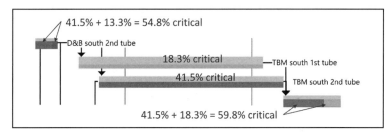

Figure 12. Interpretation of critical path simulation results

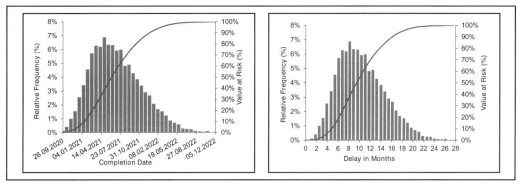

Figure 13. Construction completion date (left), and deviation to milestone (right)

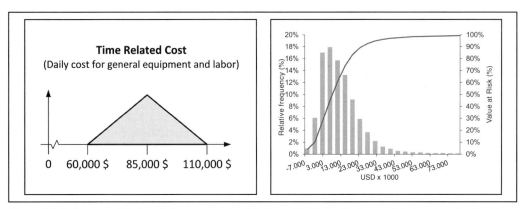

Figure 14. Delay on critical path from owner's risks is multiplied with time related cost and added to the overall risk cost of the project

Finally, escalation cost is added to obtain the total project cost (far right curve). Delta cost is obtained by comparing the total project cost with the deterministic base cost. In this case, a certainty level of VaR80 was chosen to determine the project's budget.

DISCUSSION AND CONCLUSIONS

This paper describes an integrated risk-based cost and schedule estimation, modeling and management process that has been used by a significant number of US and International projects and Agencies in the planning, design and construction phases of a significant number of complex projects. The process can be used to establish a transparent and realistic budget e.g., setting the budget for a program of projects at a probable outturn cost level e.g., a P80 level—an 80% chance that the projects will be delivered at or under this number (which also means that there is a 20% chance that they will be delivered over this number). The P-level will depend on the historical experience

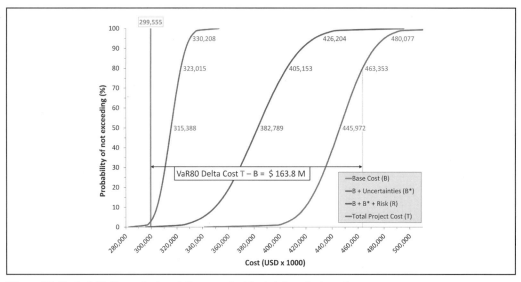

Figure 15. Probabilistic project cost (base cost, risk, total project cost)

of the Agency, its risk tolerance, and if the project is a large complex project—where perhaps P80 is appropriate, or a set of smaller more routine projects—where perhaps P60 is more appropriate.

Beyond the planning and design phases, the use of RIAAT can enable a risk-based approach to progress tracking and reporting and, management of construction change orders and cashflow. Integrated change order management can be applied, and probabilistic look-aheads can be used to update the project's budget certainty (or uncertainty).

These advances in risk-based cost and schedule estimation and project management are being implemented due to more widespread recognition of the need to apply risk-based methods, the advantages of using such processes and the publication of risk guidelines by international associations (ITA, ITIG, UCA), as well as U.S. Federal and State Agencies. In Summary we now have routine access to:

- Model results which can be used for budgeting in the planning and design phases.
- Budget control with integrated risk/change order management in construction.
- A fully integrated cost-schedule model that can analyze probabilistic risk impacts on programs (costs and construction schedules) and can enable an integrated probabilistic risk analysis and mitigation of potential delay and delay cost.
- Probabilistic schedule simulations can be used to determine major critical paths and their respective probabilities.

REFERENCES

[1] AACE International 2016. "Cost Estimate Classification System." AACE International Recommended Practice No. 18R–97.

[2] Dorp, J.R. & Duffey, M.R. 1999. "Statistical dependence in risk analysis for project networks using Monte Carlo methods." International Journal of Production Economics 58, pp. 17–29.

[3] Goodfellow, R. & Carrol, J. 2015 "Guidelines for Improved Risk Management Practice on Tunnel and Underground Construction Projects in the United States of America." UCA.

[4] ITA 1992. "Recommendations on the contractual sharing of risks" (2nd edition). International Tunnelling Association, published by the Norwegian Tunnelling Society.

[5] ITA 2004. "Guidelines for Tunnelling Risk Management." International Tunnelling Association, Working Group 2, Eskensen, S, TUST Vol. 19, No. 3, pp. 217–237.

[6] ITIG 2006 & 2012. "Code of Practice for Risk Management of Tunneling Works." January 2006 & May 2012 (2nd edition).

[7] Reilly, J.J. 2001. "Managing the Costs of Complex Underground and Infrastructure Projects." Tunnels e Perforazzoni, Ferrara, May.

[8] Reilly, J.J. 2003. "The Relationship of Risk Mitigation to Management and Probable Cost." Proc. International Tunneling Association, World Tunnelling Congress, Geldermalsen, Netherlands.

[9] Reilly, J.J. 2004 w. McBride, M., Sangrey, D., MacDonald, D. & Brown, J. "The development of CEVP® – WSDOT's Cost-Risk Estimating Process." Proc. BSCE Fall/Winter 2004.

[10] Reilly, J.J. 2008. "Chapter 4, Risk Management" in "Recommended Contract Practices for Underground Construction." Society for Mining, Metallurgy and Exploration, Inc. Denver.

[11] Reilly, J.J. 2013. Chapters on Risk, Cost and Schedule management in "Managing Gigaprojects." ASCE press, Edited by Galloway, Nielsen and Dignum.

[12] Reilly, J.J. 2016. "Megaprojects—50 Years, What Have We Learned?" Proc. ITA WTC 2016.

[13] RIAAT 2017. http://www.riaat.riskcon.at.

[14] Rohr, M.; Beckefeld, P. 2003, "Einführung eines Risikomanagementsystems als effektives Steuerungsinstrument im Bauunternehmen," Article, www.risknews.de, Issue 03.2003, pp 36–44.

[15] Sander, P., Reilly, J.J. & Moergeli, A. 2016. "Risk Management—Correlation and Dependencies for Planning, Design and Construction." ITA WTC 2016 Proc. April.

[16] Sander, P., Spiegl, M. & Schneider, E. 2009, "Probability and Risk Management," Tunnels & Tunnelling International.

Freezing of Glacial Soils for Cross Passage Construction in North America and Europe—A Comparison

Ulf Gwildis, Helmut Hass, Mike Schultz, and Mahmood Khwaja
CDM Smith

ABSTRACT: Ground freezing is a temporary ground improvement method used for underground excavations such as cross passages between subway tubes. The versatility of the method—freeze pipes can be installed at varying angles from the ground surface and from within the tunnels—allows its use under restricting conditions when other methods are not feasible. Design and execution approaches vary widely. This paper compares two recent cross passage construction projects in glacial geology—the Northgate Link light rail extension in Seattle and the Slowacki Tunnel in Gdansk—and draws comparisons from the application of ground freezing technology in North America and Europe.

INTRODUCTION TO GROUND FREEZING FOR CROSS PASSAGES IN GLACIAL GEOLOGY

During the Pleistocene glaciation the areas of present-day Seattle at the Puget Sound and Gdansk at the Baltic Sea were subject to repeated glacial advances from and retreats to the North. Each glacial cycle included abrasion and erosion of pre-existing soils and rock as well as material transport and deposition by ice, meltwater, and wind in environments ranging from terrestrial, fluvial, and lacustrine to marine. These geologic processes left sequences of different soil types—fine-grained lacustrine and marine sediments, till deposits that include cobbles and boulders, the latter often accumulated at the till boundaries, and coarse-grained meltwater (outwash) deposits to name a few—of a highly variable distribution. Deposits overridden by glaciers—the maximum ice thickness exceeded 1 km (3,000 ft) in both the Seattle and Gdansk regions—are typically very dense or hard, while the overlying meltwater sands (recessional outwash) of the retreating glacier as well as the even younger alluvial, lake, bay, and marsh deposits are less dense and less hard, in the case of the peat bogs at the Baltic shore, soft and highly compressible.

Glacially overconsolidated deposits generally have high shear strength as evident by the tall bluffs of the Puget Sound shoreline (Figure 1). A tunnel face in fine-grained glacio-lacustrine sediments may have some stand-up time under atmospheric condition; however, coarse-grained deposits should be expected to behave as a raveling or flowing ground under unsupported condition (Figure 2). The variability of glacial deposits is a factor to consider when developing an excavation plan for cross passages

between parallel TBM-driven tunnel tubes (Gwildis et al., 2014). Required at rail and road tunnels for evacuation, if other options for emergency exit to the ground surface are not feasible, construction of these short tunnel sections poses a challenge. Sequential Excavation Method (SEM) works in face conditions with some stand-up time and where water ingress can be controlled. Otherwise dewatering combined with ground improvement methods (spiling, grouting) is required. Ground freezing is another option.

The ground freezing method is based on withdrawing heat from the ground. Drilled freeze pipes are used to circulate a coolant for this purpose. With brine (e.g., $CaCl_2$) as coolant, the freeze pipes are part of a closed circulation system connected to a freeze plant. When using liquid nitrogen (LN_2), heat is withdrawn from the soil around the freeze pipes by vaporization of the cryogenic fluid, which evaporates into the atmosphere and needs to be continuously supplied. While the brine supply temperature generally ranges from $-20°C$ to $-40°C$ ($-4°F$ to $-40°F$), LN_2 starts to vaporize at $-196°C$ ($-321°F$), the latter resulting in faster freezing and lower soil temperatures at higher cost. A frozen cylindrical body expands outwardly from the freeze pipe. Spacing freeze pipes allows the build-up of a continuous freeze wall. Temperature monitoring pipes are used to verify the thermal design. The frozen soil is water tight and has increased strength that is dependent on soil type, temperature, and other factors. The frozen soil strength and its decrease over time due to creep behavior are input parameters for the structural design, which is required if the freeze wall serves as structural support. The versatility of this method—freeze pipes can be installed at various angles and lengths—combined with other factors (no or limited

Figure 1. Glacial deposits exposed at 100 feet tall near-vertical bluffs at the Puget Sound shoreline

Figure 2. Face collapse at Seattle SEM tunnel involving coarse grained glacial deposit with raveling behavior

surface impact, no dewatering, freeze wall thaws after completion of the work) makes it a competitive method for cross passage construction (Figure 3).

As for any complex underground construction technique, cross passage construction by ground freezing requires detailed subsoil investigation, design, planning, and construction monitoring to be successful. In glacial geology, the ground freeze design and planning needs to take into account the thermal and structural properties of the frozen and unfrozen soil types as well as the special geologic characteristics of the depositional sequence to avoid or reduce the risk of surprises and construction impacts. Considerations with specific relevance to the glacial geology setting are:

- Glacially overconsolidated deposits may be overlain by normally consolidated recessional outwash deposits and alluvium with high hydraulic conductivity. Evaluation of the hydraulic gradient, including the possibility of an increased gradient due to nearby construction dewatering activities, will determine if ground freezing is feasible or additional measures such as pre-treatment of the ground are required.
- Glacially overconsolidated deposits are hard to drill through and highly abrasive. Cobbles and boulders pose a risk that borings will need to be abandoned. Sudden and unanticipated changes in soil types when advancing a boring also pose challenges, especially when drilling from inside a tunnel beneath the groundwater table. The freeze pipe driller

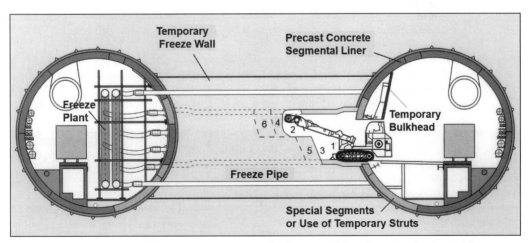

Figure 3. General concept of cross passage construction by ground freezing using horizontal freeze pipes

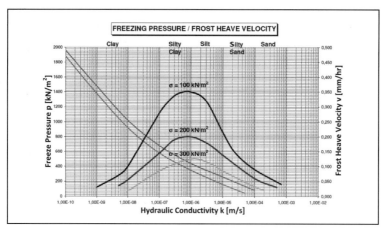

Figure 4. Graph of freeze pressure and heave rate as function of soil type

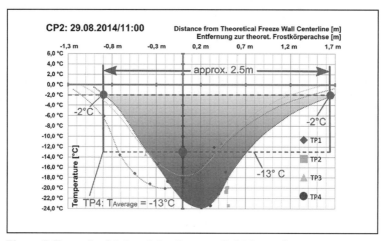

Figure 5. Example of determining freeze wall thickness from temperature monitoring data

should be experienced with these conditions. Accurate measuring of the freeze pipe as-built positions allows updating the initial thermal model and addresses the need for additional freeze pipes to fill any gaps between those already installed.

- Glacial deposits constitute a wide range of soil types including frost-susceptible soils. The amount of expansion of frost-susceptible soil types is a function of grain size distribution as well as confining pressure. Frost expansion can result in surface heave as well as freeze pressure on the tunnel liner in addition to earth pressure and hydrostatic head. Special laboratory testing quantifies freeze pressure and expansion of specific soil types

and provides the necessary design input (Figure 4).
- The variability of glacial geology constitutes an increased risk of the presence of unknown features, e.g., a coarse-grained erosion channel filling within a fine-grained lacustrine deposit providing the potential for high-inflow leaks and thermal erosion in case of insufficient freeze wall closure. This should be considered as part of a risk assessment when determining the start of excavation. Temperature monitoring pipes at varying angles and distances to freeze pipes allow updating the thermal model for determining at which time a continuous freeze wall of sufficient thickness has been created (Figure 5). Independent verification of freeze wall ring

closure by dewatering wells should be used, where possible.

NORTHGATE LINK TUNNEL CROSS PASSAGES, SEATTLE, WASHINGTON, USA

The Northgate Link construction contract extends Seattle's light rail system to the north adding around 5.5 km (3.4 mi.) of twin single-track tunnels excavated by two EPB TBMs. The tunnels have an internal diameter of 5.74 m (18-ft 10-in.) and are lined with 0.25 m (10 in.) thick gasketed precast concrete segmental liner. A total of 23 cross passages connects the twin tubes to meet the requirement per National Fire Protection Association (NFPA) 130 Standard for Fixed Guideway Transit and Passenger Rail System regarding a maximum spacing of 244 m (800 ft). The cross passages design required an excavated cross section 5.74 m (18-ft 10-in.) high and 5.23 m (17-ft 2-in.) wide, making them some of the largest cross passages in North America, relative to the diameters of the running tunnels (Pyakurel et al., 2017). Most of the cross passages were excavated entirely below the groundwater table in glacially overconsolidated glacial deposits, which for the purpose of underground excavation were contractually described in terms of Engineering Soil Units that group together soil types of similar characteristics and similar tunneling behavior. At the elevations of the cross passages often several Engineering Soil Units were present that included Till and Till-Like Deposits (TLD), Cohesionless Sand and Gravel (CSG), Cohesionless Silt and Fine Sand (CSF) and Cohesive Clays and Silts (CCS).

Based on geotechnical borehole logging as well as the results from pumping tests, the cross passages were classified using three SEM ground support categories, ranging from pre-support by spiling (grouted IBO self-drilling anchors) (Category 1) to pre-support by spiling in conjunction with dewatering and incremental excavation of a subdivided face (Category 2) to pre-excavation ground improvement by jet grouting or ground freezing (Category 3). Where the pumping test results indicated that high groundwater flow rates would make dewatering impractical, Category 3 was assigned. The tunnel contractor selected the ground freezing method for Category 3 cross passage construction, which was subsequently used for 11 cross passages (Figure 6).

Ground freezing design and execution was performed by two specialty contractors who both used a closed brine circulation system but two different approaches regarding freeze pipe layout, as schematically shown on Figure 6.

Method 1: Zone Freezing from the Ground Surface

The first five Category 3 cross passages were constructed by installing freeze pipes in roughly vertical borings drilled from the ground surface to depths up to 44 m (145 ft) in an evenly spaced grid (5 × 6) between the tunnel tubes (Figure 7). As-built surveys of the boreholes allowed identifying the need for additional freeze pipes to fill any gaps due to inevitable borehole deviations. The freeze pipes were connected to a freeze plant at the ground surface. Per this method the chilled brine is conveyed to the bottom of the freeze pipe via an inner tube, the so-called down pipe, while then flowing upwards in the annular space around the inner tube withdrawing heat from the ground via the steel surface of the freeze pipe. At higher elevations above the planned freeze wall, the ascending brine is not in contact with the outer freeze pipe surface anymore due to an air-filled, insulating annular gap. This so-called Zone Freezing confines the withdrawal of heat from the soil to the elevation range where the freeze wall is to be constructed, thereby reducing energy consumption and the risk of surface heave, as long as the insulating annular gap is wide enough to fulfill its function as intended. Additional borings with temperature monitoring

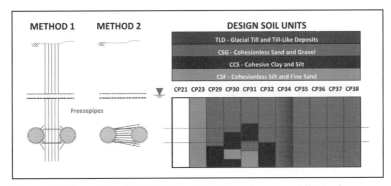

Figure 6. Schematics of Method 1 and 2 freeze pipe assemblies and distribution of design soil units

Figure 7. Freeze pipe head assembly at the ground surface (Photo credit: SoilFreeze, Inc.)

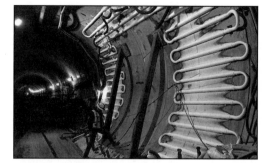

Figure 8. Active cooling of the tunnel liner intrados (Photo credit: SoilFreeze, Inc.)

devices at various horizontal distances to the freeze pipes allow to monitor the freeze progress and to update the thermal model that is used for determining when a continuous freeze wall of sufficient thickness and temperature has been created.

This zone freezing from the ground surface was supplemented by short angled freeze pipes drilled from within the tunnels and connected to small chillers within the tunnels. To compensate for the tunnel ventilation as a heat source and to further enhance the connection between the freeze wall and the tunnel liner to be watertight, active cooling of the inside of the tunnel liner at the break-through location of the cross passage excavation into the parallel tunnel tube was provided (Figure 8).

The Method 1 approach includes characteristics that deserve some thoughts. Because the freeze front expands laterally around each freeze pipe, the influence of freeze pressure on the tunnel liner should be considered more of a factor for vertical freeze pipes than for those drilled from within the tunnels. The approach relies mostly on temperature monitoring and thermal modeling to determine when freeze wall closure has been achieved and when opening the tunnel liner for starting the SEM excavation is safe to pursue. The pre-drilling of dewatering wells at the break-out location to verify freeze closure is obviously of limited use due to the freeze pipe orientations, i.e., physically verifying ring closure, as is common in horizontally drilled freeze pipe configurations, is not applicable. During cross passage excavation, the presence of the vertical freeze pipes within the excavation envelope requires special care to avoid freeze pipe damage and brine leakage. To allow sufficient working space, the freeze pipe sections within the excavation envelope have to be removed and both freeze pipe ends have then to be re-connected, which adds time to the process of cross passage construction. On the other hand, the Method 1 approach limits the amount of drilling operations from within the tunnels and also leaves a smaller foot print in the tunnel during the freeze process.

This makes Method 1 advantageous when cross passage construction is on the critical path and needs to be coordinated with the requirements of tunneling operations taking place at the same time.

Method 2: In-Tunnel Freezing

The next six cross passages were constructed by installing freeze pipes from one of the tunnel tubes in a ring-shape around the cross passage cross section. Given the geometric constraints due to the large cross passage cross section area relative to the tunnel diameters, in this specific case the freeze pipes had to be installed not just horizontally but at varying angles, generating a fan-shape in section view (Figure 6). Geometric constraints also meant that immediately outside the tunnel from which the freeze pipes were drilled, a short section of some of them would be within the excavation envelope, requiring special attention during excavation to avoid damage and leakage.

The drill locations had to be coordinated with the joints and bolt pockets of the segmental tunnel liner. The as-built positions of the borings were incorporated into a 3D CAD model. The actual freeze pipe positions and temperature monitoring data were as for Method 1 used to update the thermal model and to determine when ring closure had been achieved and freeze wall geometry and properties fulfilled the structural requirements. In this freeze pipe configuration horizontal dewatering wells allowed verification of ring closure prior to the break-out decision.

Drilling into glacial geology under high hydrostatic head from inside a tunnel using blowout preventers is generally considered a challenge that includes the risk of ground loss and the need for mitigation by grouting if that occurs. On the other hand, the obvious advantages of the Method 2 approach are that no surface installations are required that would cause third party impact and post-construction restoration needs. This comes at the price of spatial needs

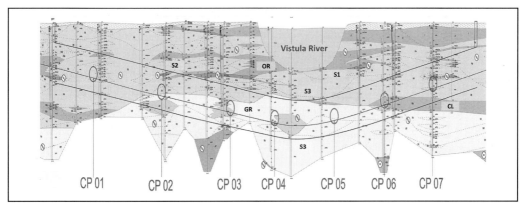

Figure 9. Geologic profile of the tunnel alignment

within the tunnels during installation works and during the ground freezing process.

For both ground freezing methods, the installation of shear bicone dowels between the running tunnel lining segments and vertical steel propping were used to transfer loads from the cross passages openings to the surrounding liner. Pyakurel et al. (2017) reported that freeze pressure induced deformations of the tunnel liner were observed, at one location reaching a maximum of 7 cm (2.75 in.), in which case additional horizontal struts were installed.

SLOWACKI TUNNEL CROSS PASSAGES, GDANSK, POLAND

The Slowacki Tunnel under the Vistula River was constructed as part of a 10 km (6 mi.) long highway project connecting the Baltic seaport of Gdansk with its international airport. The project included two parallel 1,072 m (3,517 ft) long Slurry TBM drives that constructed 11m (36.1 ft) internal diameter tunnel sections lined with 0.6 m (23.6 in.) thick gasketed precast concrete segments. Each of the tunnel tubes houses two traffic lanes. In order to provide emergency evacuation routes the tubes were to be connected by seven cross passages with a spacing of 175 m (574 ft) or less. The cross sections of the roughly 12 m (39 ft) long cross passages show an excavated height of 6.57 m (21.56 ft) at the interface with the tunnel liners to accommodate structural concrete collars and an excavated height of 4.93 m (16.17 ft) in between.

The cross passages were excavated entirely below the groundwater table at hydrostatic pressures up to 3 bar in complex interlayering and interfingering glacial meltwater and marine deposits and postglacial overburden deposits, the latter including organic silt and clay as well as peat (Figure 9). Most of the deposits at the tunnel elevation were medium dense to very dense, medium to coarse grained sands,

sandy gravels and gravels, providing for a high permeability environment with high-inflow rates in case of leakages.

Because of the varying conditions along the tunnel alignment (overburden, soil strata, hydrostatic head, surface loads) the structural requirements for the tunnel liner varied accordingly. Once the tunnel contractor had decided to use ground freezing, the design of the segmental liner was re-examined to determine whether additional stabilizing elements were required at the cross passages. For the segmental liner three types of steel reinforcement cages were developed, Type A for areas of shallow overburden and limited surface loads, Type B for areas with higher overburden loads, and Type C for the tunnel sections with cross passages. Type C segments had massively increased reinforcement with rebar configurations that also allowed drilling through the segments for the freezing works.

Hass et al. (2015) describe a complex 3D finite element (FE) model being used for evaluating the stresses in the tunnel liner for the various construction phases and loading cases, also considering the effect of the freeze wall around the cross passages openings, to determine the required reinforcement and temporary support. The modeling results defined the following structural requirements:

- Steel reinforcement cages with double layered tangential 16 mm (⅝ in.) reinforcement bars, resulting in a ratio of up to 209 kg steel/m³ concrete (352 lbs steel/yd³ concrete),
- Two types of temporary strutting frames, 'light' with 20 t (44 kips) of steel, 'heavy' with 108 t (238 kips) of steel (Figure 10), and
- Ground freezing support in conjunction with SEM excavation.

As the basis for the ground freezing design, a geological profile was developed for each individual cross

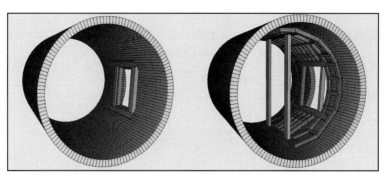

Figure 10. 3D FE modeling of temporary structural support (Strutting frame types 'light' and 'heavy')

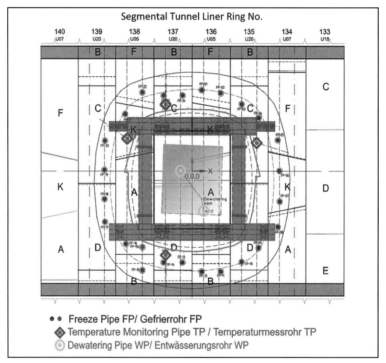

Figure 11. Assembly of freeze pipes, temperature monitors, and dewatering wells

passage. Based on the contractual requirements and the structural design of the freeze wall, a thickness of 1.8 m (5.9 ft) with an average temperature of −10°C (14°F) was determined to be sufficient to bear the loads during SEM excavation. The freeze wall thickness is generally defined as the thickness between the −2°C (28.4°F) contour lines to allow for some tolerance for the temperature devices.

To achieve the required freeze wall thickness, a layout of 24 horizontal freeze pipes and four horizontally inclined temperature monitoring pipes was developed. The layout considered the requirement that tensile reinforcement of the segmental liner was not to be damaged by coring, adequate distances of the core holes from the segment joints and pockets for bolted connections, and the position of the temporary strutting frames (Figures 11). The pipes were installed from the south tube following the completion of this first TBM drive by using drilling equipment with blowout preventers (Figure 12). During drilling of the freeze pipes and temperature monitors their as-built positions were determined and

Figure 12. Temporary strutting frame and freeze pipe heads (south tube)

incorporated into a 3D CAD model, which was then used for evaluating the need for additional freeze pipes where drilling deviations were unacceptable or borings had to be abandoned due to boulder obstructions. Thermal FE modeling based on as-built positions determined the initial freezing time to create a freeze wall as per structural design, the initial freezing time to reach the excavation line in the middle of the cross passage, the maintenance of temperature during construction, and the required freeze plant capacity for each cross passage during initial freezing and maintenance.

The boundary conditions for the thermal design included the initial soil temperature of 13°C (55.4°F), the temperature inside the tunnels of 20°C (68°F), and the shotcrete temperature during SEM excavation of up to 30°C (86°F) after two days. Additional FE modeling was conducted to evaluate the freeze wall connection to the tunnel liner of the north tube, where the freeze pipes end about 0.4 m (1.3 ft) away with the gap bridged by a lost steel drill bit. The calculations showed that a watertight connection could be achieved by using a 10 cm (4 in.) thick special insulation layer mounted on the tunnel liner intrados and that no additional cooling was required (Figure 13).

The freeze process was monitored by the temperature monitoring pipes drilled from the south tube. Additional temperature sensors were placed in small-diameter holes drilled into the liner segments of the north tube to allow verifying the freeze connection between the freeze wall and the tunnel liner

extrados (Figure 14). After the thermal model and the temperature monitoring data indicated that the required freeze wall thickness and temperature had been achieved, two dewatering wells drilled horizontally within the excavation lines of the cross passages were used for independent verification of freeze wall closure and water tightness before the break-out through the segmental tunnel liner of the south tube was allowed to start. Cross passage excavation was then performed as planned.

COMPARISON OF METHOD APPLICATION IN NORTH AMERICA AND EUROPE

In the case of both projects, the Northgate Link N125 contract and the Slowacki Tunnel contract, the tunnel contractor made the decision to use ground freezing for the construction of either some or all cross passages. Project approach, design, and implementation schedule of the ground freezing work was in both cases controlled by the overall tunnel construction schedule that took precedence.

While the ground freeze designers for the Seattle project were provided the subsoil characterization of the contract's Geotechnical Baseline Report, which grouped the highly variable glacial geology based on geotechnical characteristics and tunneling behavior into Engineering Soil Units, for the ground freeze design of the Gdansk project the designers developed geologic profiles for each cross passage based on the traditional approach of outlining the various glacial and overburden deposits per geologic as well as geotechnical criteria.

Both projects used the same types of design tools, from 3D CAD freeze pipe as-built presentation to FE modeling for ground freeze design. The main differences appear to lie in the level of detail used for stress analysis and thermal modeling. The Gdansk project used modeling to decide on the use of insulation vs. active cooling against the tunnel ventilation as heat source. This project differentiated between various types of tunnels segmental liner reinforcement and temporary supports, which is in part due to project specifics such as the higher variation of overburden thickness compared to the Seattle project.

For the ground freeze design of the Seattle project empirical values instead of project specific laboratory testing were used for the structural design. Innovative zone freezing by vertical freeze pipes helped to accommodate the critical path schedule. Temporary support of the tunnel liner at the cross passages had to be supplemented at one location due to freeze pressure induced deformations.

In Europe detailed modeling was used to achieve a high level of certainty before conducting the successful excavation. In North America the use of active cooling of the tunnel liner areas and deformation monitoring in conjunction with added

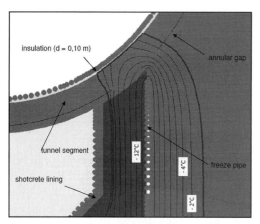

Figure 13. Modeling of freeze wall connection and tunnel liner intrados insulation

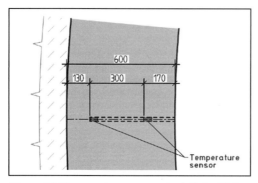

Figure 14. Temperature sensors installed within the segmental tunnel liner (north tube)

structural support as needed led to successful excavation. Sharing the lessons learned on either side of the Atlantic for use as a resource for future ground freezing projects is an objective of this paper.

CONCLUSIONS

Ground freezing is a proven and successful method for cross passage construction in unstable ground (Figure 15). The projects compared in this paper illustrate the versatility of this construction method and how considerations in the context of the tunnel construction contract determine specific project approaches. As shown by these examples, each ground freezing project has unique aspects that often

Figure 15. Road header excavation within freeze wall (Slowacki Tunnel)

lead to implementation of innovative design and construction solutions. The level of reliance on empirical data vs. project-specific lab testing, the level of modeling detail, the level of independent model verification, and the level of independent review and oversight layers all depend on the specific geotechnical conditions and the role the freeze wall plays in design and construction safety.

REFERENCES

Gwildis, U., Robinson, R., Sage, R., Sowers, D., 2014. Geotechnical Planning Advances for TBM Projects in Glacial Geology. *NAT Proceedings*. Los Angeles, California, June 2014.

Hass, H., Jagow-Klaff, R., Konarska, K., 2015. Application of Ground Freezing for Cross-Passage Construction of the Slowacki Tunnel. *Proceedings III Sympozjum, Tunel Drogowy Pod Martwą Wisłą,* Gdansk, Poland, June 2015.

Hass, H., Ostermeier, B., Fernández Remesal, J.M., 2015. Slowacki Tunnel Gdansk: Design and Construction of seven Cross-Passages with Ground Freezing Method and specific Tubbings. *Forschung + Praxis 46, U-Verkehr und unteriridisches Bauen, STUVA-Tagung 2015.*

Pyakurel, S., Klary, W., Gall, V., Long, N., Pooley, A., 2017. Risk Reduction, Management, and Mitigation from Experience-Based Learning During Construction of Cross Passages, Seattle, Washington. *RETC Proceedings.* San Diego, California, June 2017.

Unique Design and Construction Challenges for Near-Surface Large-Span Transit Station Caverns in Rock

Charles A. Stone and Eric C. Wang
HNTB Corporation

ABSTRACT: Continued growing demand for convenient access to prime locations in dense urban centers focuses the need for new transit station caverns within infrastructure and ground conditions presenting unique construction challenges. Paper highlights key principles and considerations for design of near-surface large-span mined caverns in rock, including entrances and shafts. Evaluation and mitigation strategies for the protection of adjacent and overlying infrastructure are shared. Practical examples from the authors' past experiences on Ohio River Bridges, Number 7 Subway Line Extension, Trans-Hudson Express Tunnel, East Side Access and Second Avenue Subway in New York are presented, and representative completed station caverns.

INTRODUCTION

Spurred by growing demand for convenient rapid transit access to locally preferred central business district locations, the need for new underground station caverns in congested urban centers often presents unique design and construction challenges. Design of station caverns, running tunnels, entrances, and ancillary structures must conform to project sponsor requirements, NFPA, FTA, and ADA guidelines. This paper identifies principles and considerations for design of near-surface large-span mined caverns in rock. Specifically, strategies for construction of large station caverns while protecting existing adjacent and overlying infrastructure near the cavern are described herein.

EXCAVATION METHODS FOR MINED STATION CAVERNS

Drill and Blast Excavation

One of the earliest mining methods, drill and blast excavation, has seen recent advancements in material and equipment while continuing to have widespread application. This method is labor intensive and time consuming, but quite flexible in accommodating different excavation geometries. Hence, it is often combined with other systems such as the enlargement of initial tunnel bores to form caverns. Specifically, the existing TBM excavation through the cavern footprint provides stable openings from which subsequent drill and blast slashes can be mined, extending to final cavern limits. In hard rock, sequentially excavated drifts using controlled periphery blasting and temporary support systems can safely develop final cavern geometry under shallow rock cover.

The introduction of automated jumbo drills has allowed remote operation, enhancing construction safety. Recent use of hydraulic multi-boom jumbo drill rigs has helped expedite the precision drilling of an excavation round. Newer computerized drilling jumbos, such as Robodrill, are capable of drilling individual blast holes within inches of their intended locations after initially surveying targets on the jumbo with a theodolite. Development of modern blasting plans specifying number, location, burden, spacing, depth, and diameter for various holes with associated timed delays, blasting caps, and powder factors provides improved fragmentation, perimeter control, and vibration management, addressing the primary concerns with urban construction. Drill and blast operation is designed in accordance with all applicable blasting laws and specific requirements of the project's contract documents.

Roadheader Excavation

Roadheader excavation of underground station caverns is applicable for weaker and less abrasive rock formations. This method is preferred at vibration-sensitive locations where continuous cavern excavation precluding explosives must be performed. Electro-hydraulically powered, roadheaders are well-suited for underground construction since they do not emit noxious fumes. Generally, excavation by roadheader is slower than drilling and blasting. Excavation rates reaching 39 cubic yards per hour have been reported in rock types with UCS up to 14,500 psi (Thuro and Plinninger, 1999). Cutting performance and tool wear are directly related to technical parameters such as installed power, cutter head and cutting tools, as well as geotechnical parameters such as intact rock properties, rock mass characteristics, weathering degree, hydrothermal alteration and geologic features.

However, by capitalizing on the considerable flexibility of roadheaders, multiple units may be used simultaneously to achieve satisfactory excavation rates. The ventilation requirements for roadheader excavation are typically less restrictive than for drill-and-blast excavation. Excavation of high caverns must be performed in a top heading and bench method to accommodate roadheader cutter boom limits. For East Side Access' GCT 3 East wye cavern under Madison Yard, where blasting vibration limits and blasting schedules were severely restricted due to the proximity of overlying sensitive structures, the roadheader method provided an effective alternative to drill and blast, expediting project completion (Wang et al., 2012).

Support Systems for Lining of Underground Station Caverns

Station cavern linings are constructed in two stages, the initial lining then final lining. The initial lining must be installed with a sequential excavation cycle, and must fully support all lining loads that develop during each individual cut in the overall excavation sequence until completion of the excavation and installation of the final lining. Initial support must be installed to continuously protect miners from rock falls during the excavation cycle. The final lining must be designed to retain all loadings that are anticipated during the service life of the project with acceptable safety margin under applicable building codes. Single-stage lining methods include the Norwegian Method of Tunneling (NMT), tunnel segmental linings, and occasionally, New Austrian Tunneling Method (NATM).

Station Cavern Initial Lining

Steel Sets

Current rock support design favors less costly "active" ground support systems over "passive" steel sets. Steel set support systems have traditionally been used where adverse ground conditions demand greater support capacity for excavation stability. Shear or fault zones in the rock mass typically require risk mitigation measures such as a robust ground support class with curved steel rib segments. This system of curved steel rib segments has historically provided practical protection under critical conditions where potential "heavy" or unstable ground conditions and the risk of rock fallout from shear zones or severely weathered or laminated strata necessitate flexible, portable, and instantaneous ground support. The system is often combined with geotechnical instrumentation to monitor performance, specifically measuring variations in steel rib stress and ground deformation with time.

Although costly and labor-intensive, this technique continues to be employed today for special conditions. Steel sets were used for shallow tunnel in faulted rock on Ohio River Bridges—East End Crossing (ORB-EEC) project in 2016, featuring minimum 16-foot thick limestone rock cover for a 58-foot wide excavation. The steel rib support system has also been employed when mining beneath critical sensitive structures, such as active subway lines or key utilities with minimal rock cover. For instance, steel ribs were used on the TBM-mined running tunnels under the existing Eighth Avenue Subway line tunnel with nominal cover for the East Side Access project in New York City (Wang et al., 2014).

Rock Reinforcement via Rock Dowels, Rock Bolts, and Cable Bolting

Ground support for underground mining operations typically involve steel rock reinforcement of rock mass surrounding excavation periphery. Rock reinforcement for mined large-span caverns is typically combined with surface protection to address localized stability between anchors. Available reinforcement products include:

- Rock Dowels—Solid threaded steel bars fully encapsulated with epoxy resin, "passive" reinforcement requiring displacement to fully engage the support element.
- Friction Rock Anchors—Manufactured products such as Swellex and Split Sets.
- Rock Bolts—Typically solid threaded steel bars with anchor heads, "active" support capable of locking in stress transferred into the rock mass.
- Cable Bolts—Multiple strand cables designed to be installed as a long grouted temporary or semi-permanent rock reinforcement, with or without pre-stressing prior to grouting.

Rock reinforcement holds the rock mass together such that ground loads are transferred into the rock mass adjacent to the cavern by arching action. The strength of the rock reinforcement elements must be greater than the load on the elements, and the quantity of support provided must be of higher capacity than the load demand due to the site specific geotechnical conditions. Rock reinforcement combined with shotcrete is excellent for immediate support of loose and potentially unstable ground, key blocks and laminar rock strata.

Sprayed Shotcrete Initial Lining

Use of sprayed shotcrete forms an integral part of the ground support system for station caverns constructed via the Sequential Excavation Method (SEM). The SEM construction method strives to retain the inherent strength of the surrounding rock mass by timely installation of an appropriately designed initial support system for individual cuts in the excavation sequence with compatible geotechnical instrumentation and monitoring program. The SEM lining system may consist of shotcrete, lattice girders, rock reinforcement, and spiling to control rock loads. Mining the cavern in smaller drifts facilitates rapid installation of compatibly-sized initial ground support to control ground movement by engaging the self-supporting capacity of the existing rock around the periphery of the opening with a shotcrete initial lining.

Excavation of the main cavern structure is performed from shafts, offline access structures, or as an enlargement of initial TBM driven running tunnels. Appropriate timely application of shotcrete typically utilizes an "observational" method, which relies on experienced field inspection and contracting personnel to accurately identify ground condition type and measure displacements induced by excavation. Accurate, real-time interpretation of geotechnical instrumentation data verifies support design and initial support requirements, and allows for adjustments based on changes in measured ground behavior.

The City Place Station for the Dallas Area Rapid Transit project provides an example of SEM station cavern construction. The excavation of the station chambers in Austin Chalk occurred 85 feet below the roadway deck of the new North Central Expressway in proximity to the adjacent foundations of the multistory City Place tower (Sauer et al., 1996). The two station platform tunnels are 79 feet apart and feature several cross adits of various sizes with relatively narrow pillars between.

Rock Reinforcement Combined with Shotcrete and/or Reinforced Ribs of Shotcrete

Rock reinforcement can be combined with steel fiber reinforced shotcrete support linings, providing a very flexible initial lining system capable of spanning a wide range of ground conditions. At the higher end of the support requirements the shotcrete lining incorporates reinforcing steel bars in the shotcrete ribs, accommodating the severe ground conditions (Chryssanthakis 2016). This initial lining methodology known as the Norwegian Method of Tunneling (NMT) can be used for final support.

The primary features of the NMT method are explained in several technical papers, culminating in a 2003 update (Grimstad et al., 2003). This popular system continues to be improved by the Norwegian Geotechnical Institute. The NMT method provides distinct advantage of an applicable rock support classification system for weak rocks in different stress regimes (Chryssanthakis 2016).

The ground support system for Second Avenue Subway's 72nd Street Station cavern in New York City included fully-grouted and tensioned rock bolts plus dowels, as well as multiple layers of steel fiber reinforced shotcrete (Fulcher et al., 2013). This design included detailed requirements for initial support, supplemented as required with additional support of the equivalent nature and materials.

Station Cavern Final Linings

The final lining must accommodate all the loading that it will be subjected to during its service life. Specific loads include rock, groundwater, building surcharge, live and superimposed dead loads, temperature, seismic, etc. Final lining construction depends on localized ground conditions, excavation diameter and geometry, specific site constraints, and functional requirements.

Cast in Place Station Cavern Final Concrete Linings

Traditional cast-in-place (CIP) concrete linings are the standard method for final lining construction in underground cavern structures with constant cross section and of sufficient length to enable repeated use of drainage fleece and concrete formwork systems. This final lining method requires expensive, complex, custom made formwork for drainage fleece installation and cast-in-place concrete placement. However, greater quality control of the concrete installation and more robust engagement of the final lining with the waterproofing membrane and surrounding rock mass is achievable with this method.

Completed in 2016, the 86th Street Station and 72nd Street Station caverns of New York's Second Avenue Subway are several recent examples of station caverns constructed following this method. Cavern and adit geometries were optimized featuring circular arcs with a limited number of radii to improve constructability, practicality of formwork and reinforcing installation, to reduce the overall volume of rock excavation and to optimize use of reinforced concrete (Voorwinde et al., 2015). Cavern and adit final linings for 86th Street Station were on the order of 1.5 feet to 2.5 feet thick, with drained invert slabs 2.0 feet thick.

Non-Structural Precast Concrete Underground Station Cavern Final Linings

Approximately 15 underground station caverns in the WMATA system in Washington, D. C. have been constructed of precast concrete lining segments between 1981 and 2001, reducing construction costs.

Precast concrete arched final lining is a technique whereby non-structural precast concrete panels are suspended from the rock arch which has been supported with rock reinforcement and shotcrete initial lining support system. Disadvantages associated with this architectural precast arch system include susceptibility to buildup of calcified material in the gap between the lining and rock mass, obstructed access to structural lining for future inspection, and repair of cracks and leakage. These station caverns must be over-excavated to accommodate the initial installation of the precast segments as well as future inspection of support elements and lining.

For non-structural precast concrete elements, a manufactured product acceptance plan, periodic inspection of the manufacturing process, verification testing and visual inspection are required to meet FHWA standards (FHWA 2012). This inspection should be based on the risks involved, findings from the qualification process and the quality of the manufactured elements.

Sprayed Shotcrete Station Cavern Final Lining

Sprayed Concrete Lining (SCL) has seen increased popularity in Europe over the past 15-years, especially for soft ground tunnels in the UK. Available options include: Double shell linings (DSL) combining a sacrificial primary lining carrying temporary loads and a secondary lining for permanent loads; Composite shell linings (CSL) featuring a primary lining for temporary as well as a portion of permanent loading developed through composite action with the secondary lining and; Single shell linings (SSL) with a single lining to carry both temporary and permanent loads.

Shotcrete final linings typically involve lattice girders as support for the steel reinforcement and to provide profile control for the necessary completed geometry of the cavern cross section. Shotcrete is typically applied in layers placed in distinct multiple passes to build up the necessary design shotcrete thickness of the final lining.

Key considerations prompting growing acceptance of sprayed concrete for permanent linings include (Pickett and Thomas 2013):

- Lining thickness can be reduced by installing spay-on membranes and engaging entire lining through composite action.
- Design assumptions can limit material savings available with CSLs.
- SSLs offer greatest potential savings.

Disadvantages of the SCL method include lack of precedence for fully composite shotcrete linings, the likely formation of cracking and water paths, and inferior water-tightness compared to CIP with a waterproof membrane.

Pneumatically Applied Concrete Final Lining

One recent innovative alternative known as Pneumatically Applied Concrete (PAC) has been employed on several underground structures including the wye caverns on the East Side Access project in New York (Gall et al., 2016). This lining method incorporates pneumatically applied wet mix shotcrete by hand as a final structural lining fully encasing the reinforcement bars without the requiring formwork. Specific applications favoring PAC are complex mined structures featuring non-uniform cross-sections such as wye caverns which capitalize on the opportunity to forego expensive, complicated formwork geometry. This method eliminates the custom formwork requirements of cast-in-place concrete construction.

DESIGN ISSUES FOR UNDERGROUND STATION CAVERNS

Drained vs. Undrained Station Cavern Design

Station caverns may be designed as either undrained or drained final lining systems. An undrained lining system has the capacity to develop the full hydrostatic pressure without sustaining damage or cracking the cavern final lining. Such systems require significant quantities of reinforcing steel and concrete thickness when the cavern span or depth of the cavern below water table is large. This design approach results in a large up-front cost which must be contemplated by the owner during the design phase. The alternative approach is a "drained" or "water pressure relieved" lining system. If the rock permeability is low, a drained system may be designed to minimize final lining thicknesses and cavern excavation volumes. Drained systems require a project sponsor commitment to long-term pumping, water handling, and maintenance costs, with the associated risk of structural problems should the continuity of pumping not be perpetually maintained. Certain types of rock, such as limestone, present a real risk of the drainage system clogging due to calcification. Drained design must be accompanied by a fully-drained, waterproof-lined structure, and compatible final concrete lining, ground water pressure relieved through an extensive system of a drainage composite behind and arch waterproofing membrane, piping, carrier piping, and under-slab crushed stone gravel filter layers to a sump and discharge facility in the invert, and piping designed for easy clean-out access.

The Army Corp of Engineers Manual (USACE 1997) offers guidance for the water pressure relieved hydrostatic pressure on projects in the United States.

Water pressure relief systems for shallow subway station caverns can be accomplished by installing drainage filter fleeces around the exterior periphery of the concrete final lining. Permanent maintenance of drainage fleeces and drain pipes is difficult, particularly in chalk, limestone, and shales. Therefore, drainage collection systems in the station cavern must be designed to ensure flushing and cleaning for the service life of the station cavern. Supplemental grouting may also be required, which must be performed such that drainage systems remain clear and unimpeded. Regarding the design of permanent concrete final linings, the Corp of Engineers criteria is as follows: "should be designed to withstand a proportion of the total external water pressure because the drains cannot reduce the pressures to zero, and there is always a chance that some drains will clog. With proper drainage, the design water pressure may be taken as the lesser of 25 percent of the full pressure and a pressure equivalent to a column of water three tunnel diameters high. For construction conditions, a lower design pressure can be chosen."

The authors' experience on underground construction projects indicates that while shafts and running tunnels are typically designed with an undrained methodology, shallow subway station caverns in rock are generally designed as less expensive structures with a drained final lining and water pressure relief system. A life-cycle cost analysis is highly recommended.

Generic Mined Cavern Station Configurations

Station configuration for large-span caverns in shallow rock cover must comply with project sponsor requirements, along with NFPA, FTA, ADA and other important guidelines. Cavern station configuration is factor that profoundly affects the overall cavern design. The overall construction cost of the cavern is directly tied to the excavation volume and the degree to which ground support must be provided during construction and for the service life of the project. The excavation volume is required to encompass the minimum architectural space for passenger circulation, platforms, operations, functional ventilation, power requirements, etc. These requirements must meet project sponsor criteria and be space proofed during early design phase. Furthermore, proximity to the top of rock surface has a profound impact on the amount of ground support required for the construction of the cavern. Normal operational circulation issues affect cavern configuration. Twin bore center platform layouts increase surge capacity and facilitate transfers, while single bore side platforms provide improved patron safety during periods of high platform occupancy.

Minimum and Maximum Cavern Rock Loads

Available methodologies to determine the ultimate cavern design rock loadings acting upon shallow station caverns include closed-form solutions, empirical methods, classification systems such as Rock Mass Rating and NMT Q system (Grimstad et al., 2003), and advanced numerical modelling methods considering ground-structure interaction. These methods result in loadings ranging from zero in excellent ground conditions to full overburden in exceptionally poor ground conditions, due to localized site conditions.

On the lower end of the rock loading spectrum, in some specific cases underground excavations can be carried out safely in rock without ground support, commonly known as driving a tunnel "bald headed." However, even in excellent geotechnical conditions, two to five equivalent feet of rock loading is often considered to account for periphery control in igneous rock and for strata control in sedimentary rocks. Sometimes, key geotechnical features such as a fifteen-foot thick shale zone above the cavern crown may dictate consideration of fifteen feet minimum required rock loading.

Development of full-overburden ground loads is unlikely except for very shallow caverns which derive geotechnical loads directly from shallow rock cover. Arching action will help shed load into rock adjacent to excavation. The extent of this arching action and its interaction with the structural lining can be analyzed with 2- and 3-dimensional discontinuous modelling programs such as UDEC and 3-DEC respectively. In exceptionally poor ground however, the underground cavern lining system is anticipated to encounter a significant portion of the full overburden rock load in localized areas.

Probabilistic-based design of rock cavern loading applies ground classification to capture the extreme variability of highly localized geotechnical conditions. Such ground classification typically varies with three to five increasing ground support classes encompassing "exceptionally good" to "exceptionally poor" ground conditions. Such classes are developed to cope with the anticipated ranges of ground conditions and boundary conditions at a cross section, (i.e., overburden thickness, tunnel intersections, building loading, etc.) This increased flexibility reduces the rock support costs for the entire cavern by addressing better ground conditions with more economical support systems. This optimization of initial/final cavern lining requires an initial investment of additional exploration at cavern locations.

A comprehensive, phased subsurface investigation program that meets or exceeds established guidelines is critical for defining ground conditions and determining potential rock loads for cavern design. Such a program includes review of existing data, test borings at appropriate spacings and depths, borehole and surface-based geophysics, observation wells with data loggers, and in-situ and laboratory testing. Data should be maintained electronically to facilitate incorporation into calculations, profiles, and design drawings.

Specifically, the authors suggest that test borings be performed along the cavern alignment, spaced 250 feet to 1000 feet apart. Tunneling data should be supplemented by four to fifteen additional holes per station cavern, depending upon the station length, to determine the site-specific data necessary for designing the underground station cavern. Additional targeted holes are required for the cavern entrances and shafts. The geotechnical program should include ATV logging in every station cavern borehole. Furthermore, inclined borings should be drilled after ATV logging, to determine joint set spacing of vertical joints.

Crown Pillar Stability Analysis

To capitalize on more competent ground conditions found at increasing depths and the associated benefit of reduced surface settlement at increasing depth should initially consider typical underground metro schemes, with 3.5% slopes entering and departing station limits. This scheme should be adopted during evaluation of local ground conditions for availability of adequate competent rock cover over the cavern. It is incumbent upon the designer to design for the ultimate stability of the crown pillar, the bridge of shallow rock cover over the cavern, to best manage geotechnical risk.

Rock loading data measurements from instrumented final lining and empirical geotechnical data of mined station cavern construction in Washington, D.C. and New York have indicated applicability of rock loading corresponding to a determined ground support class. Measured rock loads in the range of 3 feet to 50 feet of equivalent rock from these existing caverns have been correlated with geological data and cavern geometry using the Q value and scaled crown span (Desai et al., 2007). Rock load incorporates the impacts of all ground conditions and represents a convenient measure of ground behavior. Rock load can be estimated by conventional rules and verified by empirical, force-equilibrium, and discontinuum methods. In shallow excavations that require stiff support, the rock load can be expected to approach 80% to 100% of the in-situ vertical stress for severe ground conditions.

Methods of performing a crown pillar analysis in probabilistic terms have been developed during a period of over two decades by T. G. Carter (Carter 2014). However, economic and logistical pressures

continue to lead to collapses, daylighting to the surface for both mine workings and civil excavations.

Contribution of Initial and Final Lining to Rock Loads

Single-pass lining construction demand that the installed lining system meets all requirements of both initial and final linings. For a two-pass (stage) initial/final lining system, the designer must consider that the cavern ground loading is already 100% supported when the final lining construction commences.

For a two-pass final lining system, the designer must consider what percentage of the station cavern rock loading is to be applied to the concrete final lining during design. This issue is usually expressed by the project sponsor in its design guidelines. There are several possible approaches to this design issue:

- Allow zero contribution of initial lining support towards final lining loads.
- Allow only non-steel components of the initial lining system to contribute to final lining loads.
- Allow for creative analytical engineering analysis with 2-D and 3-D numerical modelling software.
- Allow a predetermined percentage contribution as stated in the project sponsor's design guidelines.

In the authors' experience, the project sponsor should be consulted in the development of the design philosophy. The project sponsor's requirements should be expressed clearly in the design criteria, and the designer should continue as planned to expedite the preferred design philosophy throughout the design phase.

Service Life of Final Lining

A key consideration for station cavern design and construction is service life and performance of the final lining. "There are no national codes specifically addressing tunnel and underground cavern design. For each project, a compilation of existing guidelines or standards is provided by following the applicable and widely recognized industry standards." (Zlatanic et al., 2010). Project performance criteria is typically addressed by the project sponsor and user objectives of a secure and dry underground environment with a specified service life ranging in the order of 100 to 120 years. The associated benefits include control of maintenance and operational expenditures. To achieve these objectives, the following systems should be incorporated and properly maintained for optimal performance of the final lining:

Fire Protection

Design of the cavern final lining should comply with Fire/Life Safety design criteria for protection during a fire incident. Project-specific fire loading curves should be used if available; alternatively, IFA Fire Guidelines may be used for design. Potential reduction of material properties after exposure to extreme temperatures should also be considered in the structural evaluation. Per Zlatanic et al. 2010, "For such extreme events, it is acceptable that the structure may require repair, but it is imperative that collapse or major structural failure be prevented. Explosive spalling of concrete and formation of toxic fumes are not permitted. All material should have a certified classification of non-combustibility."

Water-Tightness

Although, numerous waterproofing systems are commercially available, the selected waterproofing system should be compatible with proposed construction methods and consider relative movement across joints during the service life of the structure. The goal in meeting the water-tightness criteria is to provide an underground station cavern which is both safe and dry to achieve manageable maintenance and operations costs while preserving the integrity of structural lining and functional systems.

Muck Handling for Station Cavern Excavations

Planning for cavern excavation should incorporate a muck handling system that is compatible with the proposed excavation system. Shot rock generated from drill and blast excavation has noticeably different properties from the smaller rock fragments produced by roadheaders. Hence, a customized muck handling scheme is clearly needed. Typical options for conveyance of the larger shot rock include rail-mounted diesel muck cars loaded by Load-Haul-Dump (LHD) equipment and/or rubber-tired scoop-trams mucking directly from, or sometimes assisted by, backhoe excavators at the face. Conversely, the finer-sized muck created by the roadheader would be suitable for direct introduction onto a fixed-belt conveyance system. Obvious benefits with the fixed belt conveyance include continuous capacity except for routine maintenance, as well as the avoidance of congesting rail access which also serves as a means for transporting crews, material and equipment.

Schedule constraints can require adjustments to optimize the original muck conveyance system. For example, during station cavern excavation on the East Side Access project in New York, the original muck conveyance system was reconfigured by adding a jaw-type rock crusher plant to facilitate processing of over-sized muck to permit increased use of a fixed conveyor belt system (Wang et al., 2013).

Construction of underground station caverns in proximity to urban structures requires work restrictions and controls confining day-to-day operations, requiring extensive planning to meet construction schedules while satisfying environmental requirements.

Muck haulage through busy thoroughfares often presents construction challenges requiring practical and creative solutions to avoid tunneling delays associated with inability of muck conveyance to keep pace with excavation advance rates. The authors' experience includes use of robust long-distance muck conveyance systems featuring flexible and fixed conveyor belt components (which may be curved) extending from the headings through the tunnel and vertically up the shaft, efficiently transferring muck to a series of enclosed truss overland belt structures terminating at muck loadout structures or stockpiles. Considering the strict trucking restrictions imposed by local street ordinances, the successful replacement of truck haulage via the flexible and continuous conveyance alternative can be a public relations bonanza for projects such as East Side Access (Wang et al., 2013).

CONCLUSION

Unique design and construction considerations that are associated with large span station caverns in rock require a thorough understanding of available options, from conventional to more recent, to ensure safe and successful projects meeting owner and user objectives.

REFERENCES

Carter, T.G. 2014. Guidelines for use of the Scaled Span Method for Surface Crown Pillar Stability Assessment, Golder Associates.

Chryssanthakis, P. 2016. Performance of Reinforced ribs of Shotcrete (RRS) Under Different Stress Regimes, World Tunneling Conference.

Desai, D., Lagger, H., and Stone, C. 2007. New York Subway Stations and Crossover Caverns—Update on Initial Support Design, Rapid Excavation and Tunneling Conference.

Federal Highway Administration (FHWA). 2012. Acceptance of Non-Structural Precast Elements, Technical Brief, Publication No. FHWA-HIF-12-045.

Fulcher, B., Menge, S., and Grillo, J. 2013. New York City—Second Avenue Subway: MTA's 72nd Street Station and Tunnels Project Construction of a Large Span Station Cavern, Running Tunnels, Cross-Over, and Turn-Out Caverns, Shaft, and Entrances. Rapid Excavation and Tunneling Conference. pp. 22–45.

Gall, V., Thompson, A., Valdivia, A., Cao, W., Cicileo, C., and Schabib, J. 2016. Design and Construction Aspects of Pneumatically Applied Concrete Final Tunnel Linings Recent Experience at the East Side Access (ESA) Project in New York, World Tunneling Conference.

Grimstad, E., Bhashin, R., Hagen, A., Kaynia, A., and Kankes, K. 2003. Sprayed Lining, *Tunnels and Tunneling International*. p. 44.

Pickett, A., and Thomas, A., 2013. Where are we now with Sprayed Concrete Lining in Tunnels? *Tunneling Journal.* 44–53.

Sauer, G., Ugarte, E., and Gall, V. 1996. Instrumentation and its Implications—DART Section NC-1B, City Place Station, Dallas, TX. North American Tunneling Conference.

Thuro, K., and Plinninger, R.J. 1999. Roadheader Excavation Performance—Geological and Geotechnical Influences, 9th ISRM Congress, Paris, pp 1241–1244.

U.S. Army Corps of Engineers (USACE). 1997. Tunnels and Shafts in Rock, Engineer Manual 1110-2-2901.

Voorwinde, M., Dalton, L., Garavito-Bruhn, E., and Griffen, R. 2015 Final Lining at 86th Street: Second Avenue Subway. *Tunneling Journal.*

Wang, E.C., Velez, J., and Jordan, E. 2014. "Challenges of Fully Assembled TBM Back-up Through Completed Tunnel Bore. North American Tunneling Conference. pp. 1159–1168.

Wang, E.C., Velez, J., and Prestia, V. 2013. Long-Distance, Inter-boro Muck Conveyance Challenges. Rapid Excavation and Tunneling Conference. pp 765–772.

Wang, E.C., Velez, J., Shey, V., and Cao, W. 2012. Expediting TBM Re-launch using Existing Bores. North American Tunneling Conference. pp. 682–689.

Zlatanic, S., Chan, P., and Manuelyan, R. 2010. Methodology for Structural Analysis of Large-span Caverns in Rock. North American Tunneling Conference. pp. 441–458.

Bergen Point WWTP Outfall Tunnel Shafts—Risk-Mitigated Design for Excavation Support

Mahmood Khwaja and Michael S. Schultz
CDM Smith

ABSTRACT: Responsibility for shaft construction method selection generally lies with the contractor. Project specific logistics and site challenges for the Bergen Point WWTP tunnel shafts, and the need to mitigate risk to project cost and schedule, necessitated detailed evaluation during design for excavation support options for launching and receiving shafts. The contract documents were left somewhat flexible for the receiving shaft, but the contract documents are prescriptive for the launching shaft.

A brief description on the decision-making process leading to contract document preparation is provided along with the design considerations for both shafts is presented. A detailed review of hydrogeologic conditions and ground freezing design parameter requirements and tests is presented and the approach used in risk mitigation and finalizing the contract documents is discussed.

PROJECT BACKGROUND

Located in Suffolk County, NY, the Bergen Point Waste Water Treatment Plant (WWTP) Outfall Tunnel Project is being constructed to replace portions of an aging and failing 72-inch prestressed concrete cylinder pipe (PCCP). The existing PCCP section of the outfall starts at the WWTP effluent pump station and extends beneath the floor of the Great South Bay to the Barrier Island. The remaining portion of the outfall, 17,200 feet of concrete lined steel pipe then extends to the Atlantic Ocean. Replacement of the existing outfall also requires the construction of approximately 330 feet of new near surface 72-inch effluent pipeline and a 14,005-foot new outfall tunnel with a minimum 10-foot internal dimeter (ID) to replace the failing section of the existing outfall.

The near surface piping at the WWTP will convey the effluent from Effluent Ultraviolet Disinfection building to the tunnel launch shaft with a tie-in to the Final Effluent Pump Station. The pipeline alignment will be located between an existing 50-foot diameter water tank and the existing Final Effluent Pump Station in an open cut excavation.

The new tunnel starts at a shaft adjacent to the proposed effluent pump station and runs beneath the mudline of the Great South Bay to the connection point near the existing outfall sample chamber, located at Gilgo State Park on the barrier island just north of Ocean Parkway. Figure 1 shows the existing and proposed outfall alignments, easement limits, and the locations of the launching and receiving shafts.

The tunnel will be excavated in a southerly direction from the launching shaft located on Long Island to the receiving shaft on the Barrier Island. A slope of 0.1 percent has been selected to facilitate passive tunnel drainage gradient. The invert of the tunnel at the launching shaft is set at elevation (El.) –100.0 (all elevations are given in feet and correspond to National Geodetic Vertical Datum 1929). At this elevation and slope, based on the anticipated tunnel diameter, the minimum overburden between the tunnel springline and bottom of the lowest dredge depth of the three boating channels is 5.1 diameters; maximum hydrostatic pressure at invert is, approximately, 117 feet of head (3.5 bars).

SOIL CONDITIONS

In general, the subsurface soil profile from El. 0 to El. -20, consists of loose sand and silt, soft to very soft clay, and plastic silt. Extending to El. –65, below this loose to soft ground, the soil, predominantly, consists of medium dense to dense silty sand, clayey sand and sandy silts. Within this general matrix there are isolated pockets of clayey silts, poorly graded sands and well graded sands typical of outwash deposits. Underlying this material and extending to below the tunnel invert the soil continues to exhibit a distribution of an outwash deposit. However, the

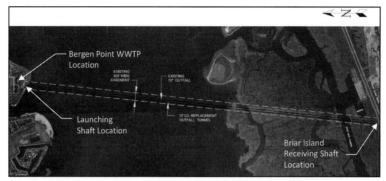

Figure 1. General project layout

composition of the soil grain size is finer and, generally, the consistency of the sandy soil is medium dense to very dense and the clayey soils are stiff to hard.

The project designated six unique soil groups or units, A through E and Fill. Table 1 provides a description of the units.

As shown on Figure 2 the face conditions will often be composed of two or more soil groups. At the shaft locations the geotechnical investigation program indicates the following:

LAUNCHING SHAFT

The soil conditions at the Launch Shaft at the WWTP site, Figure 2, generally consist of a thin layer of fill at the surface, El. +13. The fill thickness encountered was less than 10 feet. The fill is comprised of miscellaneous soils and organic material with a consistency that is generally loose or soft. This unit may contain boulders, wood, bricks or other construction and/or debris. Underlying the fill to varying depths is a layer of dense sand soils with varying amounts of silt, clay, and gravel components extending to about El. –60. This soil is generally medium dense to dense and is designated as Group B with intermittent layers of Group D. Underlying this Group B, and to El. –92 the subsurface profile is consistent with finer grained soils, predominately silt with varying amounts of clay or sand; soil is stiff to hard and is designated Group D. Underlying Group D, is soil consistent with soil designated under Group B.

RECEIVING SHAFT

At the Receiving Shaft site, Figure 3, the soil conditions show ongoing depositional process with little continuity between soils with common USCS designations. From the surface, El. +5, down to below the Receiving Shaft subgrade, approximately El. –105, there are pockets of low plasticity clay with organics (Group E), low plasticity silt (Group D), and well graded sand (Group B); smaller pockets of poorly to

Table 1. Typical subsurface soil units

Soil Unit	
Type	**Description**
A	Clean gravel and gravel sand, with little to no silt and clay
B	Clean sand, with little or no silt and clay
C	Gravel, with varying amounts of silt and clay
D	Sand, with varying amounts of silt and clay
E	Silts and clays
Fill	Miscellaneous materials

well graded gravel is present as well. The organic material noted in Soil Group B has been encountered in some of the land borings near the receiving shaft. Historic as well as project specific borings encountered these organics to about 20-foot depths.

GROUNDWATER

Groundwater levels are at or near the ground surface at the shafts and at sea level along the tunnel alignment. There are two groundwater aquifers generally recognized to exist in the project area; an upper glacial aquifer which includes the water table throughout most of Long Island and the upper portion of the Magothy aquifer. Groundwater levels along the tunnel alignment are expected to be a result of the upper glacial aquifer. However, some piezometers at the receiving shaft site encountered artesian conditions resulting from the Magothy aquifer.

At the launch shaft, the groundwater is shallow, tidally influenced and brackish. Groundwater levels at the shaft locations were measured in both monitoring wells and vibrating wire piezometers, ranging from El. 2.4 to 5.9 over a reading period that spanned several years. Several readings within 100 feet of the shoreline over a short-time period show that groundwater levels are influenced by, and coincidental with, tidal fluctuations of the Great South Bay; estimated lag time range from 4 to 6 hours.

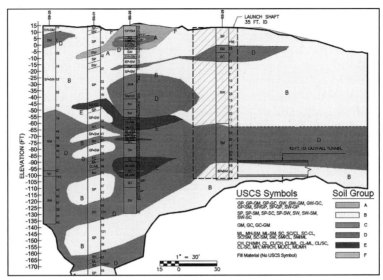

Figure 2. Subsurface profile—launching shaft

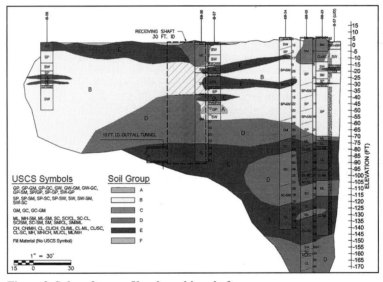

Figure 3. Subsurface profile—launching shaft

At the receiving shaft, groundwater levels range from El. 1.8 to 5.1, measured over several years; artesian groundwater conditions were observed in one of the monitoring wells.

PROJECT SITE CONSTRAINTS

Several factors associated with the tunnel site impose limitations on the proposed outfall replacement design and construction. A critical factor affecting the construction is that the overall plant operations

need to be maintained during construction. This requirement includes the ability of the County to process more than 30 million gallons of water per day. This directly affects the launch shaft location, which borders an existing 108-inch effluent pipe on the north, and the effluent pump station and existing outfall on the east. The conditions of the existing effluent pipe and the existing outfall are unknown and may be easily damaged due to shaft construction; the pump station and the pipelines must remain in operation during construction. To mitigate the risk

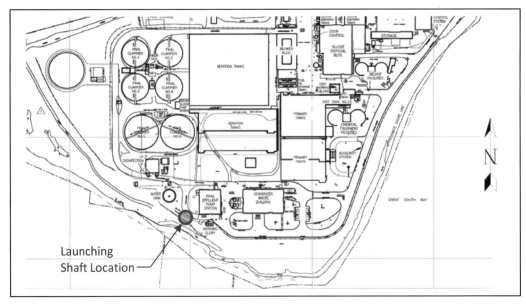

Figure 4. Launching shaft existing condition site plan

of damage, stockpiling of materials and passage of equipment will not be allowed above the pipelines.

In addition to the outfall tunnel replacement project, the Bergen Point WWTP is currently undergoing a plant expansion consisting of several separate construction projects. Coordination with the other contracts will be critical and are expected to have an impact on muck removal, site access, and available work space. The scale of the other construction project significantly restricts the available work area for the tunnel project to essentially a narrow section along the south side of the plant, adjacent to the shoreline of the Great South Bay. A power drop for the launch shaft site will be provided by the Owner near the north side of the waste water treatment plant. It will be the contractor's responsibility to determine the best location for routing the power to the launching shaft site; to coordinate installation of the power with the plant operations and at locations where it crosses the other contractor's work areas.

To the south of the tunnel site work area, the shoreline of the Great South Bay is lined with riprap and disturbance of the riprap is strictly prohibited during construction. Water levels at all shaft locations are at or very near the ground surface and are anticipated to vary in accordance with the water level in the Great South Bay. The existing condition site plan is depicted in Figure 4. The staging area allocated to the contractor is essentially parceled around the existing site.

Although the receiving shaft site is not as congested as the launching shaft site, it has its own

challenges. The receiving shaft is located on the Barrier Island near a protected wetland area. The existing outfall pipe is located to the east; stockpiling of material and the passage of construction equipment will be prohibited above the pipe. A power drop will not be available at the receiving shaft site and the contractor is expected to provide portable power units as required. The proximity of the protected wetland, limits the northern edge of the work area and wildlife and insects, including mosquitos, are anticipated to be some of the issues during the summer months.

EVALUATION OF SHAFT CONSTRUCTION METHODS

The section on "Project Site Constraints" highlights logistical challenges for construction of the outfall replacement tunnel. One of these challenges was a risk mitigated approach to shaft construction. For that purpose, white paper study was undertaken to identify potential shaft construction methods considered for the project; to evaluate the issues and potential risks related to each of the shaft construction methods. At the time, tunnel size had not been finalized, though it was estimated to be 10 feet ID. The study assumed that the excavated diameter would be in the range of twelve to fourteen feet. To facilitate launching and retrieval of similar sized TBMs, shaft size needed to be in the range of thirty to forty feet in diameter.

In general, the required work area for shaft construction and to support tunneling activities for this type of a project ranges between 1 and 3 acres, and is a function of the construction methods used. A 30-foot diameter shaft is estimated to be the preferred launch shaft size to accommodate tunnel construction. The exit shaft is primarily designed to remove the TBM, support lines, and to connect the existing outfall pipe, and may be designed smaller than the launch shaft. The final design shaft size is based on the existing ground conditions, the size of tunnel, method of shaft/tunnel excavation, and long-term usage.

The vertical alignment of the tunnel is described in the previous publications (Schultz, et al., 2017; Gilbert, et al., 2016) targeting this project. Based on the criteria presented, the proposed top of invert slab at the launch shaft is estimated at El. −102.5 (with tunnel invert set at El. −100), and the proposed top of invert slab at the receiving shaft is estimated at El. −90.5 (with tunnel invert set at El. −86.8). To address base stability and groundwater inflow issues during the excavation for a shaft in soil it is common to extend the lateral support walls below the shaft invert and into an impermeable soil layer. For these two shafts the clay stratum at each site could serve to mitigate upward flow if the excavation walls extend into the clay to form an open-ended cylinder that terminates in the low permeable soil.

Regardless of the approach selected for the shaft construction, there were common considerations that included: site access, staging area, environmental issues and overall site logistics. Environmental permitting, extensive mitigation measures and environmental monitoring would be necessary to protect existing vegetation and wildlife habitat during the shaft construction.

A broad spectrum of support of excavation systems were identified, most were eliminated from consideration due to the various requirements. It appeared that a stiff and watertight system was necessary; the following systems were considered: secant pile wall, slurry wall and ground freezing. Additionally, jet grouting was also evaluated for ground improvement to support shaft excavation. These shaft support methods are presented below:

Secant Pile Wall Method

The secant pile wall method consists of using overlapping, typically, concrete columns, to form structural or cutoff walls. The approach is to initially construct alternate columns (primary) and once these have attained a specified strength, fill in the space with secondary columns (Figure 5). Normally, the piles are the same diameter, typically in the 30-inch diameter range. The secondary piles overlap the primary piles by about 5 to 10 inches depending on the wall thickness required, pile diameter, verticality tolerances specified, and the depth of shaft excavation.

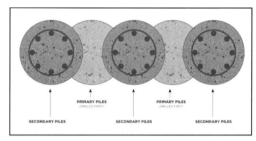

Figure 5. Secant pile installation

Only the secondary piles will have reinforcement. For maintaining watertight structure, verticality of the piles needs to be controlled well.

As shown in Figure 2, there is a low permeability stratum at about El. −100; if the piles are embedded into this stratum and the primary/secondary pile overlap is not compromised at higher elevations, risk of groundwater flow into the shaft excavation would be mitigated. At the receiving shaft (Figure 3), a significantly thicker, but similar stratum of finer grained soils is observed, but commencing at a higher El. −80.

Construction issues that were evaluated for use of secant pile approach included: staging area requirements, verticality control, base stability and the need for a grout plug or constant dewatering need to address buoyancy concerns at the invert. Furthermore, relying on the low permeability soil stratum at the two shafts can be risky as it is difficult to get comprehensive picture of the ground conditions with only three boreholes for each of the shafts.

Slurry Wall Method

The slurry wall (also known as diaphragm wall) method consists of linking, vertical primary and secondary, steel reinforced rectangular panels, to form a circular shaft. The panels are linked together with a non-circular or triangular joint at the intersections to form watertight joints. The sequence of panel installation is by first excavating for the primary panels (first set of alternate panels) and placing reinforcing cage and tremie-concrete into the excavation (Figure 6). The gaps are filled by excavating in between the primary panels and placing tremie-concrete into the excavation to form secondary columns. Typically, all panels have reinforcement; reinforcement in the primary panels needs to be positioned so as not to impact excavation for, and construction of the secondary panels.

Similar to the secant pile wall approach, the construction of a shaft by the slurry wall method requires an appreciable laydown area for construction staging, particularly for cage fabrication, lifting and installation. Typical risks associated with slurry wall construction are also similar in nature those associated with secant pile wall construction.

Jet Grouting Method

The jet grouting method consists of pumping grout, air, and drilling fluid from a rotating drill string equipped with side ports into the soil to create an in-situ soilcrete column mixture, generally ranging between 2 and 15 feet in diameter. Figure 7 is representative of typical jet grout columns. This method involves advancing a rotating drill string to the desired depth, at which point, high velocity air and

grout slurry is pumped into opposing nozzles and continuously rotated, thus mixing the soil and grout in situ. Compared to the secant pile method, the jet grout columns relatively quicker to install. An effective jet grouting result can be achieved when sample of the soil material can be tested by the manufacturer and the soil strata depths and thicknesses are known. Jet grouting is most effective in fine-grained to gravelly sand types of soil.

Use of jet grout can be very effective and efficient means to create modified soil to suit a specific project need, however, it is highly reliant on operator experience and ability. Without a good and effective

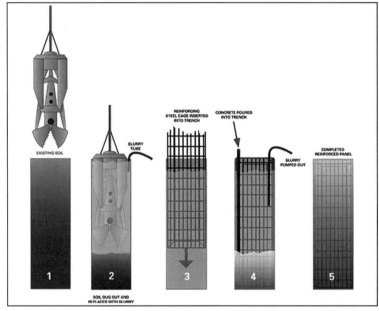

Figure 6. Schematic slurry wall installation

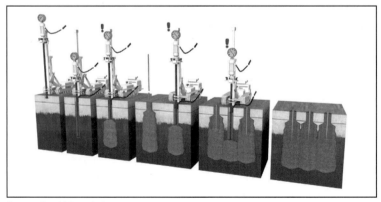

Source: Franki Foundation

Figure 7. Schematic jet grout column installation sequence

result, use of jet grouting can adversely affect the stability and water-tightness of the shaft. Additionally, jet grouting could be rendered ineffective if obstructions are encountered; sometimes this is hard to judge and predict in the field, and until the shaft excavation commences.

Ground Freezing Method

The ground freezing method involves drilling vertical freeze pipes to circulate a brine solution in a closed system to freeze the ground. The freeze pipes consist of two concentric pipes and an outer freeze pipe with an inner return line. The system circulates in a closed loop that includes a freeze plant. Brine is cooled to a temperature of about –20° C at the plant. The piping control system is designed to control the coolant flow and power requirements for both the freeze up phase and the maintenance phase of the operation. Once the freeze system attains closure, the system energy requirements can be reduced to maintain the freeze elements to stabilize the freeze zone. An instrumentation program is required to monitor and control growth of the freeze wall and control energy consumption.

Typically, shaft wall verticality is not of concern when using ground freezing as construction approach. Soil mass is frozen as a block and the shaft verticality is a function of excavation process control than of the freezing approach. This control allows for the concrete wall thickness to be reduced to the required thickness to accommodate the lateral loading without adding wall thickness to account for out of plumb wall elements such as a slurry wall panel, or secant piles. Because of the circular shape all the earth loading is acting radial to the center of the shaft thus all loads and resulting stresses on the wall are compressive or hoop stress.

Compared to other construction method, use of ground freezing limits construction related surface impacts; eliminates the need for dewatering and the associated permitting. The most significant advantage is that ground freezing forms an impermeable zone with almost no risk of ground water inflows; discrete elements systems, such as slurry walls and

secant piles, would require a very high degree of construction quality control to mitigate the risk of ground water inflows.

At the time that the design was being developed, the challenge with a prescriptive excavation support system utilizing ground freezing was that only a limited number of contractors in the United States had experience with such a technique for use in excavation support. The primary risk associated with the use ground freezing is encountering a soil stratum that has significantly different thermal properties such as a layer of shells, or a very high permeable layer of gravel that results in a high groundwater velocity. Additionally, encountering boulders or cobbles that cause a deviation in the planned location of any freeze element could potentially increase the distance between elements and thereby increase the freeze time substantially, introduce the risk of non-closure of a frozen ground.

There are methods of mitigating these risks by ensuring that continuous samples are taken during the geotechnical exploration phase; engaging a contractor who comprehensively understands the subtleties of ground freezing was of paramount concern. Bergen Point WWTP Outfall Tunnel Project design phase included a comprehensive ground investigation effort and the bidding process was specifically designed not to limit competition from international contractors.

GROUND FREEZING DESIGN PARAMETERS

To establish design parameters for ground freezing, CDM Smith undertook a comprehensive ground investigation and sample testing using chemical analysis. These design parameters, including internal friction angle (ϕ), cohesion, unconfined compressive strength (UCS), and modulus of elasticity (E), were baselined in the Geotechnical Baseline Report (GBR) and are included in Table 2 for both shafts. The elevation is provided in feet; parameters for cohesion, UCS and E are provided in metric units. Groundwater flow velocities are discussed in the GBR to help in mitigating the risk of retarding the

Table 2. Baseline design parameters for ground freezing

| Shaft | | | –10° C | | | | –20° C | | |
| | | | Cohesion | UCS | E | | Cohesion | UCS | E |
Name	El. (ft)	ϕ	MPa	MPa	MPa	ϕ	MPa	MPa	MPa
Launching	–10 to –70	26	0.62	1.98	150.2	25	1.74	5.47	334.7
	–70 to –109	21	0.22	0.64	33.2	21	1.01	2.92	159.7
	–109 to –113	8.5	0.94	2.18	131.2	16	1.88	4.98	221.2
Receiving	+6 to –25	21	0.22	0.64	33.2	21	1.01	2.92	159.7
	–25 to –70	26	0.62	1.98	150.2	25	1.74	5.47	334.7
	–70 to –95	8.5	0.94	2.18	131.2	16	1.88	4.98	221.2

rate of, or preventing, ground freezing. Also baselined in the GBR were the shaft stability requirements, as provided in Table 3; the requirements included: Structural stability of the break-in and break-out zones; Groundwater inflow; Shaft structural movement measured at the tunnel/shaft interface; Ground surface movement measured directly over the tunnel/shaft interface.

While the contractor is required to engage a ground freeze consultant for final design and construction engineering for the excavation support system, CDM Smith developed a representative model and performed numerical analysis to validate the

Table 3. Baseline design requirements for shafts

Design Condition	Baseline Requirement
Break-in/Break-out Stability	Minimum Factor of Safety of 2.5
Groundwater inflow	Maximum inflow at interface of < 25 gpm
Shaft deformation at tunnel/shaft interface	≤ 0.25 in
Ground surface deformation directly above tunnel/shaft interface	≤ 0.50 in, measured 3 feet to 5 feet from shaft wall centerline and 1 foot below ground surface

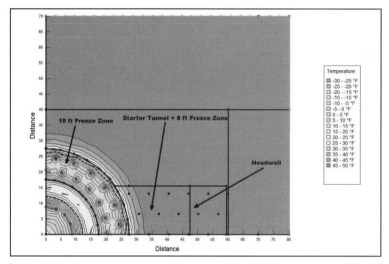

Figure 8. Ground freeze propagation (35th day)

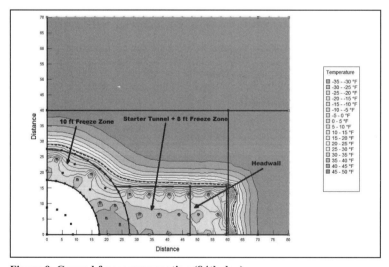

Figure 9. Ground freeze propagation (84th day)

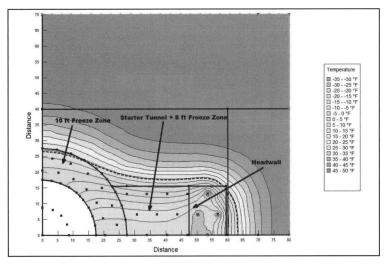

Figure 10. Ground freeze propagation (134th day)

ground freeze concept use for each of the shaft excavation support.

Although ground freeze numerical model was evaluated for several time periods, Figures 8 through 10 represents the ground freeze propagation for the launching shaft at the 35th day, the 84th day and the 134th day marks.

SHAFT CONSTRUCTION RECOMMENDATION

Construction of both shafts will encounter variable conditions as the soils are layered and soil excavations will intersect these layers. The range of soil categories at each shaft location are comprehensively captured in the contract documents. The size of the launch shaft was established based on minimum needs for installation of the vertical carrier pipes within the shaft and transition to the tunnel and to facilitate the construction of the connection to the WWTP as part of this project; the shaft size was, subsequently, checked against tunnel construction requirements for a typical TBM with excavated diameter in the range of 12 feet-6 inches to 14 feet-0 inches. The size of the receiving shaft has been established based on needs for installation of the carrier pipe and to facilitate removal of the TBM.

Based on the results of the ground investigation program, and in consideration of the various risk elements associated with environmental and logistical concerns, land restrictions, and construction access, it became apparent that the most risk mitigated approach for controlling project construction budget and schedule, would be to use ground freezing

to help support shaft excavation. Throughout the final design process, CDM Smith kept evaluating the veracity of the early-decision to use ground freezing as a prescriptive method of construction for the launching shaft and a suggested method for the receiving shaft. In addition, the contract documents provided contractor guidance on the use of ground freezing as a means for stabilizing ground excavation, controlling groundwater inflow, and for the tunnel break-in and break-out activities.

ACKNOWLEDGMENT

The authors acknowledge John Donovan and Suffolk County, New York, for the opportunity to present this paper. CDM Smith looks forward to supporting Suffolk County through the construction phase of this project.

REFERENCES

Schultz, M., Sanders, G., Taylor M.A., Donovan, J., *EPB or Slurry TBM? Suffolk County, Long Island NY Outfall Replacement Tunnel*, Rapid Excavation and Tunneling Conference, 2017, p76.

Gilbert, M., Schultz, M., Wright, Ben., Kelly, K., *Risk Issues Associated with New Outfall Tunnel under Great South Bay in Suffolk County, New York*, North America Tunneling Conference, 2016, p82.

Geotechnical Baseline Report, Suffolk County Outfall, Bergen Point Wastewater Treatment Plant, 2017.

BART Silicon Valley (BSV) Phase II, Tunneling Methodology— Comparative Analysis Independent Risk Assessment

Saqib A. Saki, James J. Brady, Robert J.F. Goodfellow, and Angel Del Amo
Aldea Services LLC

Alfred Moergeli
Moergeli Consulting, LLC

Krishna Davey
Santa Clara Valley Transportation Authority

ABSTRACT: A risk analysis approach can be used to compare the viability of two competing tunneling options even at different levels of design maturity. This paper describes the process used to provide the Santa Clara Valley Transportation Authority (VTA) with a comprehensive decision-making basis using comparative risk profiles for two tunneling alternatives; a single large diameter tunnel versus two smaller twin tunnels for extending BART service into downtown San Jose. Quantification of construction risk impacts were assessed in terms of cost and time for comparing the subsurface construction cost and duration of the two options. The analysis also compared the differences in O&M costs for the first 30 years of operation.

INTRODUCTION

The Santa Clara Valley Transportation Authority (VTA) is based in San Jose, California. It is an independent special district that provides multi-modal transit services. The Santa Clara Valley Transportation Authority is responsible for the design and implementation of highways and transit projects including the BART Silicon Valley ("BSV") Program. The BSV Phase II Extension project is a 6-mile extension which starts from the Phase I Berryessa Station as shown in Figure 1. It passes through Downtown San Jose to a new station in Santa Clara. This phase consists of 4.8 miles of running tunnels through San Jose. It includes four stations. The Alum Rock, Downtown San Jose and Diridon are underground stations, and Santa Clara is at grade. It has two intermediate ventilation structures and East and West tunnel portals.

VTA started the planning efforts for BSV Phase II in 2014 with an update to the project environmental studies. The continued ongoing community and public concerns about disruption during construction drew VTA's attention towards a single bore (SB) large diameter tunnel as an alternative to the twin bore (TB) option. The advances made by the tunneling industry with respect to developments in larger diameter, soft ground mechanized tunneling in urban settings encouraged VTA to initiate a feasibility study of a SB alternative. Project alignment, station configurations, emergency egress and

ventilation tasks were studied in the SB feasibility study which was completed in early 2016. The SB feasibility study concluded that a single bore option is technically feasible for the prevailing ground conditions and did not exhibit any fatal flaws (VTA BSV Phase II Tech Studies, 2017). It, and subsequent technical studies, further concluded that single bore might be a viable alternative to the twin bore configuration.

VTA selected Aldea Services, LLC (Aldea) to conduct an independent risk assessment to assist the decision-making process between the SB and TB tunneling options. The alternative configurations under consideration are a TB tunnel system and a deeper SB tunnel system. A risk assessment process was part of VTA's selection process to determine the preferred tunneling alternative. The assessment analyzed and described and compared the qualitative and quantitative risks associated with the two tunneling alternatives (RFP S16308, 2016). The assessment was carried out within a risk management framework that is intended to proceed throughout design and construction in accordance with the Guidelines for Improved Risk Management on Tunnel and Underground Construction Projects in the United States of America (O'Carroll and Goodfellow, 2015)

TUNNEL ALTERNATIVES

The two options are the TB option which constructs two single track 20-foot outer diameter subway

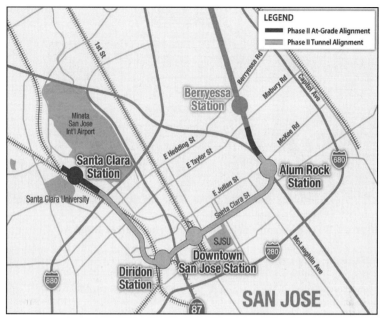

Source: VTA BSV Phase II Tech Studies, 2017

Figure 1. Phase II extension

tunnels, comparable with other tunnels in the BART (Bay Area Rapid Transit) system, and the SB option which constructs a single 45-foot external diameter subway tunnel that is designed to carry two tracks within the same tunnel, using a dividing wall between the trackways. Both alternatives are shown in Figure 2.

Twin Bore Alternative

The TB design consists of two circular tunnels constructed by two TBMs to interconnect the open-cut stations, mid-tunnel vent structures and portals. The tunnels will be connected to each other by cross passages at regular intervals along the alignment. The project had three proposed underground stations in the 65% Preliminary Engineering Phase; Alum Rock, Downtown San Jose station and Diridon/Arena station.

Single Bore Alternative

In addition to the feasibility study which found no fatal flaws, several follow-on SB technical studies further performed detailed evaluations of the SB tunnel option and indicated that a minimum internal diameter of 41 feet was desirable to meet the minimum clearances and vehicle envelopes stipulated in the BART Facilities Standards (BFS) through all of the necessary guideway configurations and

transitions along the project alignment (VTA BSV Phase II Tech Studies, 2017).

During early inter-agency coordination discussions, BART indicated a preference for side-by-side rather than stacked track configuration in the running track alignment. This arrangement required transitions from side by side running tunnel to the over/under configuration at the stations which controlled the diameter because the maximum open space was needed to facilitate the transitions. The SB Feasibility Study concluded a minimum depth of cover of 65 feet for the SB tunnel. Subsequent technical studies with more detailed evaluations indicated that a shallower minimum cover depth of 50 feet was constructible and appropriate for further evaluation of a SB tunnel as the design progressed (VTA BSV Phase II Tech Studies, 2017).

RISK ASSESSMENT PROCESS

The risk assessment process for an integrated project cost and schedule analysis seeks to identify all risks and uncertainties that might significantly affect the predicted project cost and schedule. It uses methods to quantify what each of those impacts might be by using estimates of minimum, most likely and maximum values of cost and schedule. A numerical simulation model is used to aggregate these impacts to obtain risk-based cost and schedule estimates for the project that are probabilistic distributions rather than

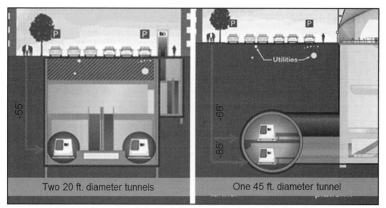

Source: VTA Board Presentation 09/22/17

Figure 2. Twin bore and single bore tunnel alternatives

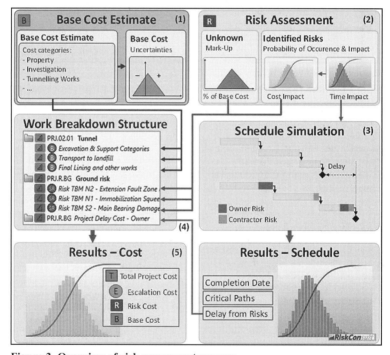

Figure 3. Overview of risk assessment process

single value estimates. This process is illustrated in Figure 3.

Qualitative Analysis

The comparative assessment process for project cost and schedule first worked to accurately derive comparative base costs for both alternatives and normalize the costs to a common date (December 31, 2016 in this case). Next, a workshop process was used, including stakeholders (such as BART, VTA,

the City of San Jose etc.) and nationally recognized subject matter experts to identify risks and uncertainties that might significantly affect the predicted project cost and schedule for both options. Risk assessment workshops were used to:

- Identify significant potential events and conditions (both risks/threats and opportunities) that could affect project cost and schedule.
- Assess risk impacts and likelihoods.

- Develop an integrated cost and schedule risk register as shown in Figure 4.
- Develop an integrated analytical cost & schedule risk model to quantify risks to cost and schedule with a probabilistic approach.
- Produce a distribution of probable cost and schedule outcomes for each option.

- Identify and discuss mitigation measures for significant risk components and estimate the potential risk reduction from each mitigation measure together with the residual risk after mitigation as shown in Figure 5.
- Identify, discuss and quantify potential opportunities and ways to exploit them.

Ref Item/ID	Categorization Main Category	Definition/Description Identified Hazard/ Identified Risk	Cause/Trigger	Effect	Phase & Owner Risk Materialization Phase	Risk Owner	Qualitative Assessment before Controls/Mitigation Likelihood / Time Impact / Cost Impact / Reputation Impact / H&S Impact / Environment Impact / Legal Impact / Functionality (Quality) Impact	Comments	Risk Rating	Risk Level
082	Compliance/Law/ Regulation	Tariffs/duties on non-US made goods	Federal requirements	Additional cost	Project execution	VTA	4 4	Nil	16	Threat High
007	Construction	Different ground conditions encountered from the ones assumed during Preliminary Engineering	Major unforeseen ground conditions	Additional cost & delays	Construction	VTA	3 4 3	Nil	12	Threat High
056	Procurement/Co mmercial/Scope	Overspecified requirements	VTA might follow a very strict prescriptive approach	Potential combination of: - No room for contractor's experience - Conflicts with contractor's planned and estimated performance	Construction	VTA	4 3	Nil	12	Threat High
078	Quality/Health & Safety/Environm ent (QHSE)	Disk cutters handling more time-consuming than originally expected	Cutters handled in TBM cutter head	Accidents	Construction	VTA	3 2 4	Nil	12	Threat High

Figure 4. Example risk register before controls (mitigation)

Risk Control/Mitigation Strategy (Combinations of) Measures/Controls: Apply a Safety System (e.g. by "STOP") to achieve System Safety STOP: S = Strategy / System / Substitution T = Technology O = Organization P = Personnel/Staff	System Safety	Realization	Residual Risk Likelihood / Time Impact / Cost Impact / Reputation Impact / H&S Impact / Environment Impact / Legal Impact / Functionality (Quality) Impact	Comments	Risk Rating	Risk Level	Controlling Action Owner	Milestone Target Date	Status
Mitigate Combination of 1. Minimize use of material subject to tariffs/duties on non-US goods 2. Ask for waiver welll ahead in time.	OP	Yes	3 3	By Design-Bid-Build (DBB) per VTA's & Aldea's decision	9	Threat Medium	PM	Part of Preliminary Engineering after system decision	Planning
Mitigate Carry out supplementary staged site investigations.	OP	Yes	2 3 2	By Design-Bid-Build (DBB) per VTA's & Aldea's decision	6	Threat Medium	PM	Part of Preliminary Engineering after system decision	Planning
Accept Educate VTA in time.	OP	Yes	3 2	By Design-Bid-Build (DBB) per VTA's & Aldea's decision	6	Threat Medium	PM	Part of Preliminary Engineering after system decision	Planning
Mitigate Combination of - Reaching out toTBM manufacturer(s) to work out this issue during design of TBM. - Investigate TBM manufacturers on use of real-time monitoring of main bearing. - Reduce required hyperbaric interventions to absolute minimum (emergencies only). - Plan for saturation diving interventions. - Plan for safe heavens for planned cutterheed maintenance/tool change. - Monitor cutters in real-time. - Plan for contingencies to handle combination of worn-out/blocked cutters. - Plan for contingencies to handle obstructions. - Back loading cutters. - Specific devices for cutter handling. - Proper training.	ST OP	Yes	2 1 3	By Design-Bid-Build (DBB) per VTA's & Aldea's decision	6	Threat Medium	PM	Part of Preliminary Engineering after system decision	Planning

Figure 5. Example risk register after controls (mitigation) implemented

		Impact (Opportunities)				Impact (Threats)					
	Exceptional	Major	Moderate	Minor	Insignificant	Insignificant	Minor	Moderate	Major	Exceptional	
Semi Quantitative	Over 12 months less project time	6 to 12 months less project time	3 to 6 months less project time	1 to 3 months less project time	< 1 month less project time	Additional project time exceeded by < 1 month	Additional project time exceeded by 1 to 3 months	Additional project time exceeded by 3 to 6 months	Additional project time exceeded by 6 to 12 months	Additional project time exceeded by >12 months	Time
	Cost Reduction/Rev enue Increase > 100,000,000	Cost Reduction/Rev enue Increase 50,000,000 - 100,000,000	Cost Reduction/Rev enue Increase 10,000,000 - 50,000,000	Cost Reduction/Rev enue Increase 1,000,000 - 10,000,000	Cost Reduction/Rev enue Increase < 1,000,000	Additional Cost < 1,000,000	Additional Cost 1,000,000 - 10,000,000	Additional Cost 10,000,000 - 50,000,000	Additional Cost 50,000,000 - 100,000,000	Additional Cost > 100,000,000	Cost

Risk Level = Likelihood x (highest) Impact		OPPORTUNITY					THREAT				
Very high High certainty of occurence. Probability: >75%	5	H -25	H -20	H -15	M -10	H -5	M 5	M 10	H 15	H 20	H 25
High Will occur. Probability: >50 - 75%	4	H -20	H -16	H -12	M -8	M -4	M 4	M 8	H 12	H 16	H 20
Medium May occur. Probability: >25 - 50%	3	H -15	H -12	M -9	M -6	L -3	L 3	M 6	M 9	H 12	H 15
Low May occur but not to be anticipated. Probability: 5% - 25%	2	M -10	M -8	M -6	M -4	L -2	L 2	M 4	M 6	M 8	M 10
Very Low Occurrence requires exceptional circumstances. Exceptionally unlikely, even in the long term future. Probability: < 5%	1	M -5	M -4	L -3	L -2	L -1	L 1	L 2	L 3	M 4	M 5
		-5	-4	-3	-2	-1	1	2	3	4	5

Risk Ranking Range		Risk Level
-11	-25	Opportunity High
-4	-10	Opportunity Medium
-1	-3	Opportunity Low
1	3	Threat Low
4	10	Threat Medium
11	25	Threat High

Figure 6. Risk matrix used for qualitative assessment

During the workshops, a numerical ranking method was used to quantify the range of each of those impacts using estimates of minimum, most likely and maximum values for each alternative for cost and schedule risks/opportunities as shown in Figure 6. 127 Total Risks, including 64 specific and 63 generic risks, were identified for the TB option, while 121 Total Risks, including 74 specific and 47 generic risks, were identified for the SB option. In the above usage a "generic" risk was a risk that came from Aldea's generic tunnel "seed" register that was determined to be applicable to the option. The "specific" risks were unique risks identified during risk workshops for the two options.

Quantitative Analysis

Probabilistic Risk Analysis Approach

The probabilistic approach was used to quantify risks. Conventional construction estimates are presented in terms of a single number. This form of estimating is termed "deterministic" cost estimating. A more reliable way of establishing budget costs is by use of probabilistic forms of estimating that can consider uncertainties and give a range of possible outcomes. These uncertainties can be in the form of pure quantity or material uncertainty or in the form of identified and unidentified risks. When these are

combined, a full probabilistic cost and schedule distribution can be developed. The advantage over standard deterministic methods is that it delivers more reliable contextual information because the result is a probabilistic distribution with a range for the risk potential (incl. best case and worst case). The analysis facilitates decision-making in line with the respective project stage. Since actual empirical data for risk analyses is often not available, the exact probability of occurrence can be difficult to estimate. However, use of probabilistic methods allows risks and costs to be depicted for each project phase with individual density distributions: larger distributions for larger uncertainties, narrower distributions for smaller uncertainties. Using this approach, reality can be modeled more accurately than with a single deterministic figure.

Probabilistic risk assessment allows the use of uncertain values and requires various inputs:

- The probability of occurrence depicted in Figure 7 describes the pure likelihood that a risk actually produces an impact. If the risk does not occur, the impact is always zero. If the risk does occur, a financial impact should be evaluated. For evaluation of the probability of occurrence there are two options available (pick only one per risk):

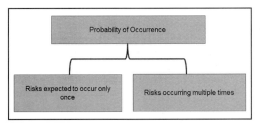

Figure 7. Probability of occurrence or average rate of occurrence

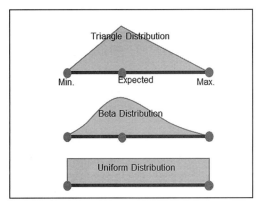

Figure 8. Cost impact distribution

– Risks expected to occur only once (in the project lifetime)
=> Probability of occurrence must be chosen as a percentage value.
– Risks potentially occurring multiple times
=> An average occurrence rate must be chosen. Through this value, a Poisson distribution is modeled which maps the potential

frequency of occurrence based upon the probability of no (0) occurrence.
• The cost impact is modeled by various distributions as shown in Figure 8. For simplification, three values can be used, e.g., in a triangular distribution: minimal impact, expected (most likely) impact and maximum impact. The area above the x-axis stands for the probability of occurrence of the specific hazard. Probability of occurrence of the two boundary values (min. and max.) is practically zero. The expected value is the most likely value.

However, the summation of risks cannot be calculated by simple mathematical additions. Combining risks defined by probability density functions requires statistical simulations (e.g., Monte Carlo Simulation, Latin Hypercube Sampling) to determine a probability density function for the combined risks and therefore depict the overall risk potential. If the information about a potential cost distribution is accounted for by including uncertainties (probability functions), the likely budget variability and impacts can be estimated.

Figure 9 shows an example for an overall cost distribution of a project. The distribution is developed by aggregating the effects of base costs, risks and escalation using the simulation method described previously. The figure shows that $5M would cover 70% of the project cost potential. However, even with such coverage, there is a 30% probability that the budget would be exceeded.

Introduction to Risk Modeling Process

Two numerical simulation models were developed using the RIAAT software (for more information, please see http://riaat.riskcon.at/) to aggregate these

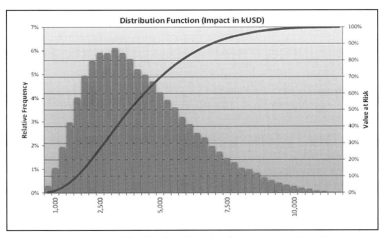

Figure 9. Example of a distribution function for aggregated project cost

impacts to obtain risk based cost and schedule estimates for each of the two project options. RIAAT performs numerical simulations to aggregate the contribution of each source of cost and schedule uncertainty to the overall project cost and schedule estimate. Cost impacts of schedule delays including potential changes to the critical path schedule are incorporated in the calculations. The result is an integrated cost and schedule model for the project that includes risk impacts together with the quantified uncertainties in these predictions.

The models aggregated the simulation results of Base + Uncertainty and Risk Costs. In the models, "Risk" includes both Identified Risk and Unidentified (or Unknown Known) Risk. The models also include cost elements to calculate the estimated cost impacts of schedule delays resulting from both Owner-caused and Contractor-caused delay risks. The results are presented in terms of probabilistic distribution ranges rather than single value estimates.

Escalation costs have not been included in the comparative analysis because their calculation is typically a financing calculation reserved for evaluating the time dependent cost of the entire project and the comparative analysis was not based on analyzing the entire project. Therefore, presenting the results of an escalation calculation would be premature and misleading at this time. Further, the schedule (time-related) outcomes of the comparative schedule analysis show critical differences in both the construction period and when subsequent revenue service will start. Any accurate escalation cost would have to be based upon both of these findings. In sum, the escalation calculation was determined to be premature at the comparative analysis phase. It is recommended that escalation calculations be performed for the entire selected project option as part of a budgetary/funding risk analysis and should incorporate the durations identified in the risk dependent schedule.

Quantitative Assessment Models Using RIAAT

The quantitative alternative comparison between the subsurface portions of the TB and SB options was performed using the RIAAT software to analyze 100,000 project cost simulations and 10,000 project schedule simulations for each option. The P_{80} level is the result found at the 80th percentile of outcomes, ranked from lowest to highest (i.e., in 100,000 simulations, P_{80} is the cost result of the 80,000 highest costing project simulation).

VTA chose the 80th percentile for logical reasons. Washington State DOT e.g., routinely looks at cost using probabilistic methods and their standard practice is to use the 60th percentile of cost for their projects so that they can be assured of maintaining their project budgets over half the time. Due to the one-off nature of this program, coupled with its size and complexity, it was considered appropriate to use a more conservative assessment. Based on that understanding, it was determined that using the 80th percentile of potential cost distributions would be appropriate for comparative purposes. The relative conservatism of comparing P_{80} outcomes had a beneficial effect of weeding out any tendency toward "optimism bias" during the process in that participants were never confused that the purpose of this task was not a VE exercise where proponents of the competing alternatives were there to "sell" their option by optimizing their way to a rosy outcome. The comparisons drawn are based on equally less than favorable outcomes and that has the benefit of examining overall Risk in the comparison phase. The conservatism also helped the process steer clear of being misconstrued as a budgeting exercise.

SUMMARY OF RESULTS

The simulations analyzed the comparative Base Costs which were then subject to variable uncertainty in future prices and quantities based upon the level of each option's design maturity/level of design completion (approximately 65% for TB versus 20% for SB). Risks that differentially affected the cost or duration of either option were rated to derive a probability of occurrence and range of possible consequences (should the risk be triggered) and loaded into the models. Risks used in the model underwent a "Basic Mitigation" assessment to filter out that portion of the original unmitigated risk that would be removed or reduced after acknowledging a basic level of oversight and diligence on the Owner's part. This was not the Aldea Team's usual practice, nor was it anticipated at the onset (typically unmitigated risks are used in order not to falsely claim mitigation benefits that have not yet occurred). However, as we progressed with this investigation it was realized that a basic level of mitigation was needed for comparative scenarios because otherwise all the risk uncertainty affected by design maturity level becomes effectively double-counted. In addition to specifically identified risks assessed during the workshops, the model also includes future Market Risk and Unidentified Risk which was based on assessments of project development factors; most notably design maturity. Finally, a Real Estate Savings Opportunity and a Business Interruption Risk based on assessments of the differences in local community impacts provided by each option were evaluated and modeled. Finally, there is Schedule Risk which is calculated by RIAAT based upon Owner-caused delays to achieving the project schedule; both Pre-Award and Post-Award of the Heavy Civil (Tunnel & Shafts) Contracts.

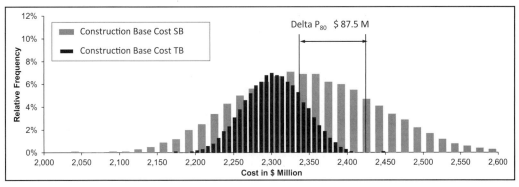

Figure 10. P₀ through P₁₀₀ comparison SB–TB (construction base + uncertainty cost)

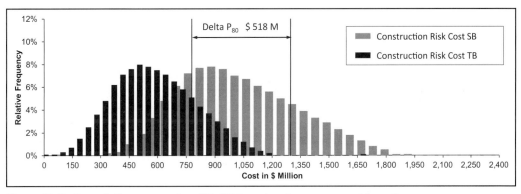

Figure 11. P₀ through P₁₀₀ comparison SB–TB (construction program risk cost)

Another difference in the evaluation between the two options is that TB was evaluated as a traditional Design-Bid-Build Contract due to the level of design progress (65% Design) while SB (20% Design) was evaluated as a Design-Build Contract to investigate the advantages in potential schedule savings that pursuing this type of contract delivery method might provide. While not "identical" this distinction accurately reflected the Owner's most advantageous approach to each option based on their current level of investment and development. Figures 10–12 present the full range of results (P₀ to P₁₀₀) for the simulated Base + Uncertainty Construction Cost (Figure 10), the Construction Program Risk Cost (Figure 11), and Heavy Civil Construction Completion Dates (Figure 12) for both options. P₀ is the lowest ranked result from the model's simulations defining the left end of each curve and P₁₀₀ is the highest ranked result defining the right end of each curve.

The P₈₀ level of comparison was selected by VTA as its organizational risk tolerance level which was to be used in the comparison of results. A summary of the P₈₀ results is presented in Tables 1 and 2.

RECOMMENDATIONS

The following recommendations were based on the results of this risk assessment and the Aldea Team's current understanding of the project based on information provided by VTA up to the date of this report.

- The integrated cost and schedule risk model should be updated at significant milestones during project development, execution and into commissioning (e.g., at pre-award, 25% completion, 50% completion and substantial completion for examples) to obtain improved project controls including risk-focused claim management and cash flow analysis.
- A strong focus on identifying and implementing risk mitigation measures and their resulting potential impacts on estimated total costs, and especially on schedule, should start and then continue with frequent involvement of VTA management and eventually include the contractor.
- Key Risk Indicators should be developed for significant risks and these should be monitored over time to indicate when risk

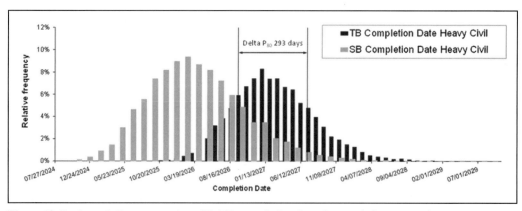

Figure 12. P$_0$ through P$_{100}$ comparison SB–TB completion dates heavy civil construction

Table 1. Comparison of twin bore and single bore options

Twin Bore vs. Single Bore Snapshot Comparison P80 Results Compared	Twin Bore	Single Bore	Delta	
COST				
Lower Base Cost		√	-3.5%	
Lower Base + Uncertainty Cost	√		-3.6%	
Lower Potential Risk Cost	√		-39.9%	
First to Revenue Service		√	-8.2%	
Lower O&M Cost (1st 30 Years - No Escalation)	√		-2.8%	
SCHEDULE			Delta	
First to Start of Construction		√	-540	calendar days
Shortest Heavy Civil Construction Duration (Tunnels & Shafts)	√		-247	calendar days
First to Heavy Civil Completion (Ready for Trackwork)		√	-293	calendar days

Table 2. P80 comparison summary spreadsheet

BSV Phase II Tunneling Alternatives - Comparative Analysis - Independent Assessment Results Summary DRAFT Final (Based on RIAAT_BSVII_Mitigated_V25_F01_R2, August 10, 2017)			Twin Bore (TB)	vs. TB 100%	Single Bore (SB)	vs. TB 100%	Delta SB -TB	
Base Cost	Based on Desigers' Estimates	Det.	$ 2,143,407,000	100%	$ 2,071,065,000	96.6%	$ (72,342,000)	
	Based on Desigers' Estimates + Uncertainties	P$_{80}$	$ 2,336,793,000	100%	$ 2,424,327,000	103.7%	$ 87,534,000	
Total Potential Risk Cost	Based on Risk Workshops & Aldea's Analysis incl. Unknown	P$_{80}$	$ 779,234,200	100%	$ 1,296,005,000	166.3%	$ 516,770,800	
O&M Costs (1st 30 Years)	($2016, i.e., No Escalation) Based on Aldea's Analysis	P$_{80}$	$ 1,758,099,000	100%	$ 1,807,957,000	102.8%	$ 49,858,000	
Heavy Civil Construction Completion Date		P$_{80}$	07-09-2027		09-19-2026		-293	d
Heavy Civil Construction Start Date		P$_{80}$	05-20-2021		11-27-2019		-540	d
Heavy Construction Duration (P$_{80}$ Heavy Construction End - P$_{80}$ Heavy Construction Start)		P$_{80}$	2241		2488		247	d

Note: Risk-based aggregated costs do not equal the sum of the sub-components because probabilistic metrics like P80's are not additive.

mitigation measures should be started for optimal mitigation.

- The contractor should be required to submit a qualitative Risk Register with his bid and then with his Base Schedule and update and review that Risk Register with VTA staff quarterly.

- Project Cost Controls that include risk components should be implemented to track cost development and to merge risk management and change order management. This will help VTA identify risks that did not materialize so that any associated contingency funds allocated for these risks can be released for other uses. It will also give VTA the opportunity to track risks that have materialized and what impact they have on the project and to improve its risk management practices for future projects.

CONCLUSION

VTA understood that any decision for selecting a tunneling method from competing options at two different levels of design maturity would need to be uncertainty-based in order to ensure that overall project risk was not overlooked due to lack of design completeness or inadequately evaluated due to lack of independent vetting. This analysis provided that context to the decision makers and can be set alongside non-cost factors such as the desire for geometric consistency within the BART system, or the desire to not disrupt downtown San Jose for extended periods during construction. One area that was impossible to resolve in this analysis was one of operation, where a fundamental disagreement on the means and criteria for safe and efficient operation of the system existed. Our analysis cannot resolve this area because the base criteria for quantifiable model input could not be agreed upon. Additional expert input has been solicited to discuss these operational issues, but further development of design will probably be necessary to bring resolution to these discussions.

In order to provide an analytical basis for a fair comparison between the two system geometries, we undertook the following:

- Both alternatives were carefully reviewed per their respective level of design,
- Base Cost and Base Schedules of both alternatives have been kept substantially unchanged,
- All issues that might affect both alternatives in an equivalent way have been transparently excluded,
- All cost estimates have been updated to the same basis (December 31, 2016),
- Schedules have been carefully reviewed, and respective activities have been loaded with identified risks,
- Intensive meetings with all stakeholders well ahead of the risk assessment assured a joint team and shared risk-based approach,
- A moderated qualitative risk assessment with workshops including all major stakeholders identified project risks across the full spectrum of project development; Planning to O&M,
- All identified project risks have been quantified,
- An integrated cost and schedule model was built and aggregated probabilistically,

- VTA, as well has both design teams, have been working together very professionally for the complete IRA/CA process,
- VTA's Project Management Team were exceptionally responsive to our requests for information and were essential to making this process work.
- An Extended Appendix included a full disclosure of all model inputs and results and concluded the IRA/CA.

All identified uncertainties were quantified wherever possible.

- Base Elements (for both cost & time) as well as Risks,
- Risk-loaded schedules with the resulting probabilistic impacts on cost,
- Owner's expectable "soft" cost (for own & external staff, business interruption, property value increase, etc.),
- Expected cost for Operations & Maintenance for the first 30 years.

All cost and schedule cost impacts have been probabilistically aggregated to Comparative Total Cost of Ownership at a Value at Risk 80% (P80) level. Therefore, supporting VTA for a risk-based, objective and unambiguous decision-making on the most advantageous solution. Contracted on March 9, the IRA/CA report was delivered on October 13, 2017 as per VTA's requirements and directions. This validated approach sets the stage for future risk-based project evaluations.

REFERENCES

BART Facilities Standards (BFS); Last Accessed Oct 31, 2017. https://webapps.bart.gov/BFS/bfs_spec.html.

BART Silicon Valley PHASE II, Tunneling Methodology Independent Risk Assessment, Pre-Proposal Conference Presentation, Nov 29, 2016.

O'Carroll, J., and Goodfellow, B., 2015. *Guidelines for Improved Risk Management on Tunnel and Underground Construction Projects in the United States of America.* UCA of SME, Denver, CO.

Some recent publications about Risk-Based Cost Estimation and Controlling like e.g., from http://www.moergeli.com/en/download => Risk Management (RM) III.

VTA's BART Silicon Valley—Phase II Single Bore Tunnel Technical Study, February 2, 2017.

VTA Board Workshop and General Public Final Review, Sept 22, 2017 (Live Recording—YouTube Video)

TRACK 3: PLANNING

Session 2: Project Delivery—Water and Wastewater

Michael Burnson and Nancy Nuttbrock, Chairs

DigIndy Tunnel System: Pleasant Run Deep Tunnel Optimization

Olivia Hawbaker
Citizens Energy Group

Maceo Lewis IV and Leo Gentile
Black & Veatch Corporation

ABSTRACT: The Pleasant Run Deep Tunnel is a 39,000-foot long component of the 28-mile long DigIndy deep rock tunnel system, the cornerstone of Citizens Energy Group's Long Term Control Plan (LTCP). Final planning optimized the conceptual design to achieve CSO capture objectives while eliminating expensive components and saving millions of dollars. After the elimination of several drop shafts originally planned, the final design included relocating drop shaft locations. The relocations were due to outreach with city stakeholders and included intermediate shaft relocation due to new major developments at the site. The relocations required the design team to revisit the optimization of the of the tunnel alignment to maintain minimized adit and consolidation sewer lengths.

PROJECT DESCRIPTION

Background

Each year there are 60–80 combined sewer overflow (CSO) events that contribute an average of six billion gallons of untreated waste into the waterways of Indianapolis, Indiana (City). In accordance with a Federal Consent Decree, Citizens Energy Group (Citizens), owner of the wastewater system, is implementing a Long Term Control Plan (LTCP) to address CSOs on behalf of the City (Indianapolis DPW, 2007). Citizens acquired water and sewer infrastructure from the City in August 2011 along with the responsibility of implementing the LTCP according to the terms of the Federal Consent Decree.

The LTCP includes using existing system capacity, expanding and upgrading the existing treatment facilities and a new storage and conveyance system. The storage and conveyance system, known as the DigIndy Tunnel System, is a 28-mile long network of six 18-foot diameter tunnel segments. The DigIndy Tunnel System begins on the south side of the city at the Southport Advanced Wastewater Treatment Plant (AWTP), extends along the city's waterways and ultimately will end on the north near the Indiana State Fairgrounds and on the east in the Irvington neighborhood of Indianapolis. The system will capture 97 percent of sewage overflows in the Fall Creek watershed and 95 percent of sewage overflows in the White River, Pleasant Run, Pogues Run, and Eagle Creek watersheds in a typical year, reducing overflow events to less than two and four per year, respectively. DigIndy will be the largest public works project in the city's history at a cost of approximately $2 Billion.

The DigIndy Tunnel System (Figure 1) will extend along White River, Fall Creek, Pogues Run, Pleasant Run, and Eagle Creek to create an underground storage and transport facility with the capacity to store over 250 million gallons of CSO during rainfall events. After a storm event, the tunnel system will be dewatered using a 90-million gallon per day deep tunnel pump station located at the Southport AWTP. Table 1 summarizes the tunnel system components.

Under the Federal Consent Decree, a timeline to achieve full operation (AFO) of CSO compliance was established. By 2017 the Deep Rock Tunnel Connector (DRTC) must be on-line, which includes the deep tunnel pump station, followed by the Eagle Creek Tunnel (ECT) in 2018. Prior to December 31, 2021 Lower Pogues Run Tunnel (LPgRT) and White River Tunnel (WRT) must be complete and functional in the system. Finally, by the end of 2025 Fall Creek Tunnel (FCT) and Pleasant Run Deep Tunnel (PRDT) must also be on-line in the system.

Advanced Facility Planning (AFP)

Pleasant Run is an urban stream with very low base flows that becomes dominated by CSO during wet weather events. The Pleasant Run Interceptor and connecting sewers direct combined flows to the Citizens Belmont AWTP. There are 50 CSO outfalls located along the stream.

Regulators, an existing manhole or structure with a weir wall or elevated outfall pipe, maintain

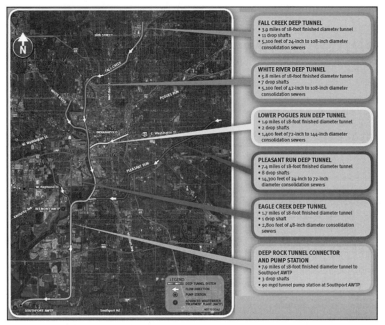

Figure 1. DigIndy tunnel system

Table 1. DigIndy system components

Component (AFO Year)	Tunnel Length (ft)	No. Drop Shafts	Adit Length (ft)	Total Volume (million gallons)
Deep Rock Tunnel Connector (2017)	39,300	3	800	75.2
Eagle Creek (2018)	9,200	1	0	17.5
White River (2021)	30,600	7	2,100	59.1
Lower Pogues Run (2021)	10,200	2	100	19.4
Fall Creek (2025)	20,300	11	5,700	40.8
Pleasant Run (2025)	39,000	8	1,500	75.0
Pleasant Run Extension (2025)	2,300	0	0	4.4

flow in the collection system and allow overflows to discharge into the CSO outfall pipes. Some regulators are directly connected to interceptor sewers, whereas other regulator structures have a small capture sewer that conveys the normal dry-weather flow to the interceptor sewer.

The initial concepts for Pleasant Run Deep Tunnel (PRDT) were described in the City's 2007 Raw Sewage Overflow LTCP and Water Quality Improvement Report. The 2011 AFP (Indianapolis DPW, 2011) refined the concept to capture CSOs along Pleasant Run. The major components of the system included:

- 42,600 feet of main tunnel.
- 4,000-foot long Bean Creek shallow tunnel.
- 10 drop shafts.
- 20,271 feet of consolidation sewer.

Through the AFP evaluation, the number of CSOs required to be directed to PRDT was reduced from 50 to 30. This was achieved by adjusting regulator weir heights in the existing Pleasant Run Interceptor and associated capture sewer diameter increases. Value Engineering (VE) during AFP also recommended alternatives to be investigated during final design. These included:

- Increasing tunnel slope to 0.2 percent to improve scour and reduce depth of drop shafts.
- Evaluating the need for several drop shafts by adjusting alignment and increasing consolidation sewer lengths.
- Partial tunnel lining.
- Eliminating a length of parallel sewer at two CSOs.

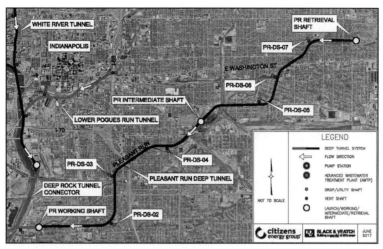

Figure 2. PRDT 75-percent design alignment

Design Optimization

Design concepts developed during AFP and its associated VE were reviewed and the design optimized to meet the following goals: help improve constructability, minimize disruption to the community, help increase level of control (LOC), and reduce cost. The following paragraphs provide the highlights of the salient design features adopted by Citizens to achieve these goals (Citizens, 2016).

PRDT is now in final design. The tunnel will extend approximately 7.5 miles east-northeast, beginning at the Pleasant Run Extension from DRTC near the intersection of Bluff Street and Pleasant Run Parkway. PRDT ends at Pleasant Run Golf Course near the intersection of Pleasant Run Parkway and Arlington Avenue. (Figure 2).

Alignment. Final design efforts included several changes to the initial horizontal and vertical AFP alignment, which generally parallel the Pleasant Run waterway and the associated CSO outfalls. The changes helped improve constructability, reduce length of consolidation sewers, and eliminate shallow tunnel segments. The selected alignment is shown on Figure 2.

The first change to horizontal alignment was to realign the tunnel from the working shaft eastward beneath W. Southern Avenue instead of beneath Pleasant Run. This change helps improve constructability, allowing the contractor to launch the tunnel boring machine (TBM) on an initial 6,000 foot straight run. The realignment also helps eliminate the proposed Bean Creek shallow tunnel by relocating drop shaft PR-DS-02 discussed further below.

The second modification to the AFP alignment includes a straight run beneath English Avenue upstream of the proposed Intermediate Shaft. This alignment reduced the number of subsurface easements, assisted with the elimination of the drop shaft and provides a more constructible segment for excavation and muck handling.

In addition, the alignment changes reduced the number of subsurface easements and impacted properties from 143 to 46. The alignment optimization lessens the impact to the community to a degree and helps Citizens avoid administrative cost of preparing and securing rights-of-entry and easements.

Intermediate Shaft. The PRDT Project includes four different types of tunnel shafts for construction and operation. These include the TBM working shaft, the intermediate shaft, the TBM retrieval shaft, and drop shafts. The TBM working shaft for PRDT will be situated near the east end of the Pleasant Run stub that was constructed as part of the DRTC project. The site is located at the southeast corner of the Bluff Road and West Pleasant Run Parkway North Drive.

For several reasons, an intermediate shaft is included. The intermediate shaft for PRDT is situated approximately at the midpoint of the alignment and planned to be at the former Citizens Coke & Gas Plant site. The alignment optimization resulted in several turns that may affect muck handling, as was the case on DRTC. The intermediate shaft will allow a secondary access point to set up or maintain the TBM, transport materials, personnel, and equipment into the tunnel, and remove the tunnel spoil and waste material during excavation. The intermediate shaft will allow a central location for spoils removal along the tunnel which will more efficiently convey tunnel spoils.

The second reason is to allow the contractor the option of backing up the TBM after the full alignment is excavated. This technique is being

Figure 3. PRIS alignment alternatives

successfully employed on other segments of the DigIndy Tunnel System. For example, the 9,000-foot long Eagle Creek Tunnel was excavated as part of the DRTC contract. The TBM completed tunnel excavation, the utilities stripped from the tunnel and the TBM reversed to the main DRTC tunnel to complete its mining. While there is real cost to this effort, the cost is expected to be offset by eliminating a large-diameter baffle drop shaft.

The originally proposed PRDT Intermediate Shaft was located adjacent to Pleasant Run Creek in the central portion of Citizens Energy Group's Prospect Coke & Gas Plant property. During design, the redevelopment at the Prospect Coke & Gas Plant was introduced, requiring shaft relocation and tunnel realignment to accommodate future site development. The design team developed alternative shaft locations and corresponding tunnel alignments in order minimize impacts to future site development at the Prospect Coke & Gas Plant property. (Figure 3)

The five alternative locations were selected on the basis of minimizing impacts to future site development, avoiding a railroad easement, limiting the amount of affected property owners, and maintaining proper tunnel geometry for the tunnel boring machine (TBM) constructability.

Of the five alternatives, Alternative 4 is the only alternative that does not need railroad easements at the Prospect Street intersection, provided that the 40-foot tunnel easement is varied at the intersection to avoid current railroad easements. Alternatives 1, 2, and 3 traverse through both railroad easements at

the Prospect Street intersection. Since three of the alternatives require railroad easements anyway, a fifth alternative was added to locate the Intermediate Shaft in the Salvage Yard section of the Prospect Coke & Gas Plant property. Advantages and disadvantages are summarized on Table 2.

After selecting Alternative 5 for the alignment, the shaft location was moved to the Salvage Yard portion of the Coke and Gas Plant. The team developed three alternatives for an exact location of the intermediate shaft (Figure 4) with modified alignments (Figure 5). One alternative attempted to tuck away the shaft and would require a good amount of clearing for the geotechnical investigation and construction. Another alternative is located out of the 100-year floodplain and is located on the western portion of the site, maintaining much of the site for future construction while mitigating permitting needs and increasing constructability. This option, shown in yellow, was selected as the final alternative. Construction site limits are outside the floodway and estimated to be approximately 6 acres. The shaft is outside of the 100 and 500 year flood plain and is also away from "unknown contamination" documented on site.

Consolidation Sewers. Citizens' internal hydraulic modeling group performed a detailed evaluation of the combined sewer collection system in the Pleasant Run basin. The team characterized the Pleasant Run CSO system as very interconnected, lending it to optimization by raising regulator weir elevations so the capacity of existing infrastructure

Table 2. PRDT intermediate shaft alternatives advantages and disadvantages

Intermediate Shaft Location	Tunnel Segment Length* (feet)	Advantages	Disadvantages
Original	5,897		
Alternative 1	6,689	• Shaft is farthest away from creek improvements project.	• Tunnel alignment requires railroad easement. • Longest tunnel segment length alternative.
Alternative 2	6,458	• Balanced tunnel alignment geometry and shaft location.	• Tunnel alignment requires railroad easement.
Alternative 3	6,175	• Tunnel alignment affects the fewest property owners.	• Tunnel alignment requires railroad easement. • Shaft is closest to creek improvements project.
Alternative 4	6,130	• Tunnel alignment does not require railroad easement.	• Tunnel alignment contains a reverse curve with no straight segment in between. • Tunnel alignment affects the most property owners.
Alternative 5	6,125	• Shaft is located on separate parcel away from future site development. • Shortest tunnel segment length alternative.	• Tunnel alignment requires railroad easement.

* Tunnel segment length refers to distance of the tunnel alignment from St. Paul Street to the railroad easement at the intersection of English Avenue.

Figure 4. PRIS salvage yard alternative locations

can be maximized. Expanded modeling allowed the design team to better utilize existing infrastructure and reduce the number of drop shafts and length of collection consolidation sewers (CCS) and associated diversion structures. Table 3 summarizes the optimization strategies by drop shaft carried forward into the design.

The system hydraulic modeling helped develop weir height and regulator capture pipe diameter adjustments to eliminate 5,000 feet of CCS and two drop shafts.

- 27 of 50 CSO regulator weirs are raised from 1 to 4 inches.
- 17 of 50 regulator capture pipes will be increased by 9 to 30 inches in diameter.

Figure 5. PRIS salvage yard alternative alignments

Drop Shafts. Up to 10 drop shafts were envisioned during the AFP phase of the project. Some shafts were recommended for elimination during AFP Value Engineering. The hydraulic analysis conducted by Citizens during final design helped confirm that two shafts could be eliminated for final design through optimizing flows in the Pleasant Run Interceptor system. Further, all drop shafts will be tangential vortex-type (Table 3.). Not only is this Citizens' preference, the reconfiguration of the existing interceptor system reduces flows such that direct-drop or baffle-type planned during the AFP are no longer required. The following drop shafts were reevaluated by Citizens during final design:

- PR-DS-02 was relocated to the southeastern side of Garfield Park, a historic and heavily used recreation area in the City. This alternative eliminated the extensive near surface consolidation sewer through Garfield Park and minimizes disruption to the park and the costly shallow ground tunnel previously referred to as the Bean Creek Branch Collection Sewer.
- The proposed site for drop shaft located near the intersection of East Pleasant Run Parkway South Drive and Napoleon Street, was eliminated. By raising the weir heights at CSO 022 and CSO 149, hydraulic modeling showed that drop shaft PR-DS-03 can be eliminated, while conveying these flows to drop shaft PR-DS-04. To accommodate these

flows at drop shaft PR-DS-04, the weir height at CSO 151 will need to be raised and the size of the consolidation sewer leading to PR-DS-04 increased. Additionally, a new interceptor connection will be needed upstream of PR-DS-04, a capture pipe size will need to be increased, and a new regulator to the existing interceptor will be constructed.

- The proposed drop shaft located near the intersections of English Avenue and East Pleasant Run Parkway North Drive, was also eliminated. Flow will need to be diverted to drop shaft PR-DS-07, upstream of CSO 076, by constructing a regulator. By increasing the weir heights at CSO 075 and CSO 076 and upsizing their respective capture pipe diameters, the model shows that drop shaft PR-DS-06 can be eliminated. These flows will be conveyed downstream by the interceptor.

The capture pipe sizes will need to be increased and weir heights raised at CSO 074, CSO 077, and upstream of CSO 075 near the intersection of East Pleasant Run Parkway North Drive and Southeastern Avenue.

Level of Control. A Storm Water Management Model (SWMM) was been developed of the CSO outfalls, diversion structures, relief structures, consolidation sewers, drop shafts, adits, and the main PRDT tunnel to estimate design flows and hydraulic behavior to support the PRDT design as well as to verify the compliance with the defined levels

Table 3. CSO design optimization strategies

Drop Shaft	CSO	Design Optimization
DS-01	019, 120	Raise regulator weir, eliminate 1,200 feet of 36-inch CCS, increase capture pipe to 24 inches.
DS-02	015, 016	Relocate drop shaft. Eliminate Bean Creek shallow tunnel.
DS-03	022, 149	Redirect flows to DS-04. Eliminate drop shaft.
DS-04	022, 025, 023, 149, 151, 119, 108, 127	Collect flows from CSOs, increase capture pipe size, remove diversion structures.
DS-05	028, 029, 072, 073	Raise regulator weirs to eliminate CSOs, increase capture pipe size, remove diversion structures.
DS-06	075, 076	Raise regulator weirs up to 5 feet, increase capture pipes up to 42 inches. Eliminate drop shaft.
DS-07	077, 078	New weir, regulator structure.
DS-08	080, 081, 224	Increase capture pipe up to 36 inches, weir height 4 feet.
DS-09	083, 084, 154	Additional consolidation sewer, raise weir up to 2.5 feet.
DS-10	088, 089, 090, 091, 092, 229	Increase capture pipe up to 18 inches, raise weir up to 1.5 feet.

Table 4. PRDT drop shaft nomenclature

AFP	Type	Preliminary Design	Final Design
DS-1	Direct Drop	PR-DS-01	PR-DS-01
DS-2	Vortex	PR-DS-02	PR-DS-02
DS-3	Vortex	PR-DS-03	—
DS-4	Vortex	PR-DS-04	PR-DS-03
DS-5	Vortex	PR-DS-05	PR-DS-04
DS-6	Vortex	PR-DS-06	—
DS-7	Vortex	PR-DS-07	PR-DS-05
DS-8	Vortex	PR-DS-08	PR-DS-06
DS-9	Vortex	PR-DS-09	PR-DS-07
DS-10	Baffle	PR-DS-10	PR-DS-08

of control (LOC) in the Long Term Control Plan (LTCP).

The modeling included the hydraulic analysis of the tunnel system but not hydrology. Some aspects of the PRDT design depend on other projects such as the Deep Rock Tunnel Connector (DRTC), Fall Creek Tunnel System (FCTS), and White River/ Lower Pogues Run Tunnel (WRLPgRT) systems. Dewatering information for the tunnel system was provided by Citizens. All scenarios evaluated assumed that the Deep Tunnel Pump Station (DTPS) with a rated capacity of 90 million gallons per day (mgd) has 11.7 mgd capacity available to PRDT, and that the Southport AWTP has capacity to accept all the pumped flows. Additionally, all the scenarios considered that the entire volume of the PRDT was available for storage of the PRDT CSOs.

The PRDT SWMM model predicts 14 untreated overflow events along Pleasant Run and 98.1 percent capture during the 5-year design period. Both results comply with the defined LOC.

Construction Cost. The design optimization strategies described in this paper are expected to help reduce the overall project cost by millions of dollars. The strategies include:

- Eliminating two drop shafts.
- Replacing a 30-foot diameter baffle drop structure with a vortex drop structure at the Retrieval Shaft.
- Reducing the length of connecting sewers by 5,000 feet.
- Eliminating the shallow ground tunnel beneath Bean Creek through Garfield Park.

REFERENCES

Citizens Energy Group. 2016. Pleasant Run Deep Tunnel. Basis of Design Report. Citizens Project Number 92TU0534.

Indianapolis Department of Public Works. 2007. Raw Sewage Overflow Long Term Control Plan and Water Quality Improvement Report.

Indianapolis Department of Public Works. 2011. Advanced Facilities Plan for Pleasant Run Deep Tunnel. DPW Project Number CS-32-004A.

Planning, Design, and Construction of CSO Pumping Station Structures

Geoffrey A. Hughes and Rafael C. Castro
JCK Underground Inc.

Carlton M. Ray, Moussa Wone, and Ronald E. Bizzarri
DC Water

John F. Cassidy
Greeley and Hansen, LLC

ABSTRACT: During the early stages of many Combined Sewer Overflow (CSO) tunnel programs, Owners are required to make critical decisions related to the configuration of underground structures to house dewatering pump systems. Key considerations may include siting, sequence of work, schedule, operational capacity, design criteria, constructability and contract packaging. Several of these items were addressed when building a 250 million gallons per day (MGD) pumping station within a 132-foot diameter by 175-foot deep shaft for DC Water's DC Clean Rivers Project. This paper discusses the planning of underground structures for pumping stations in general terms and provides examples from the DCCR Project to demonstrate how active, flexible program management was used to overcome challenges and achieve a successful project execution.

INTRODUCTION

The abatement of Combined Sewer Overflows (CSOs) is commonly achieved by constructing tunnel systems to intercept, convey and temporarily store combined sewage. Following wet weather events, dewatering pumps lift the sewage from the deep tunnel system to the surface for treatment and discharge. When undertaking programs to implement such systems, key decisions must be made during the early phases of planning to establish the configuration of pumping facilities and to house them within the underground structures. Apart from physical sizing, these decisions are influenced by factors such as program schedule, sequencing, design criteria, contract packaging, constructability, and spatial limitations. Furthermore, where responsibilities for future stages of design and construction are assigned to separate parties, it is essential that interface points be clearly defined so that integrated components are developed with mutual compatibility.

As a core element of DC Water's DC Clean Rivers (DCCR) Project, the 250 million gallons per day (MGD) Blue Plains Tunnel Dewatering Pumping Station is housed within a 132-foot diameter by 175-foot deep shaft and provides a case study example of the early key decisions made, types of challenges faced, and solutions found (see Figure 1). Successful execution of the project not only resulted

from the management of factors during the planning stage, but also the flexibility to overcome challenges during execution not least of which was the refinement of interface loading criteria between shaft and pumping station after the shaft lining was constructed.

This paper presents general considerations common to all projects with pumping station coordination challenges, details and lessons learned during the DCCR Project and conclusions and recommendation for future work.

GENERAL CONSIDERATIONS

Organization and Planning

The first decision an Owner faces when confronted with the need to abate CSOs, is finding the right people to get the job done. In general, establishing a general program organization will involve a blend of existing staff, new direct hires and contracted expertise. The ad-hoc nature of the work commonly lends itself to the latter option, but retaining in-house staff is also essential for strategic leadership, internally validating technical decision, providing direction and maintaining program momentum. Furthermore, since larger programs can take decades to complete it may be quite attractive for an Owner to self-perform and establish a special division within its organization focused on program implementation.

Alternately, depending on the Owner's regulatory remit, the option may exist for the Owner to assign virtually all responsibility for implementation to others. The make-up of the organization will also be influenced by the Owner's existing organizational structure.

Whether the approach is in-house, integrated or contracted, the types of expertise employed must vary to suit each stage of program advancement. At the earliest phases, a program management organization (PMO) will concentrate its effort on producing a Long Term Control Plan (LTCP) and obtaining regulatory approval to proceed. Closely tied to this task is the need to identify and secure property to house the facilities and to perform Environmental Impact (EI) assessments to the satisfaction of regulators. To achieve these tasks, the PMO's focus is largely on the evaluation of hydraulics, alternatives, and economic and environmental impacts. It is essential, however, that the PMO have a fundamental appreciation for detailed design and constructability, such that this stage of the work does not overly constrain later phases, including statements regarding aquifers which may constrain construction dewatering and identifying work site areas that may limit the ability to construct structures on parallel timelines.

The PMO's role will continue through the work to monitor technical development and will likely also involve establishing and monitoring the program's schedule, budget, and verifying the quality of installed work.

Design and Delivery Method

Where an LTCP has selected the option of tunnel storage and pumping, suitably qualified design experts must be retained. Assignment of responsibility will be a factor in selecting the choice of delivery method and the apportionment of scope to each delivery package. Where the owner prefers to maintain close control over the final configuration of facilities, it may be that design-bid-build is preferred, whereas design-build can allow more latitude and potential to innovate. There are numerous other delivery options such as: assigning the program manager the responsibility for final design tasks; adding construction to the single point responsibility with Construction Management At Risk (CMAR); or using cost reimbursement fixed fee construction contracts as used for the Portland, OR CSO program [Edgerton 2008].

For any of these procurement options, at each division of responsibility it is necessary to ensure the points of coordination between the responsible parties are definitive. The Owner and its PMO must work hard to identify these interfaces and risk allocations—a structured risk management program can be useful to assist this process. In general, the greater the number of contract divisions, the greater

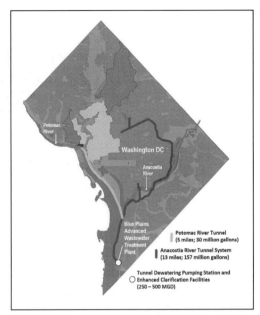

Figure 1. DC Clean Rivers Project location of tunnels and pumping station

will be the risk for omitting or creating conflicts at interfaces or for misaligned goals. Additionally, consideration should be given to the option of using different delivery models for the various components of the program which complicates the ability to manage the interfaces among designs being completed by a number of different design teams under different contractual vehicles.

The benefits of closely aligning interdependent engineering disciplines naturally favors the formation of two groups: a geo-structural and system hydraulics group focused on tunnel, shaft and diversion structures; and a group focused on pumping station function and fit-out (mechanical, electrical, plumbing, controls, etc.). With a corresponding division of design packages, the parallel development by the two groups requires close and continuing coordination—particularly for items related to size, shape and structural interfaces.

Following the identification of pumping station operating mode and capacity, preliminary designs will be developed along with the design criteria that will govern later work. Where final design is continued by the same entity, the design criteria are commonly memorialized in a design report. Where design will be by others, the design criteria may become part of the scope for the final design contract or mandatory requirements (MRs) in tender documents for design-build contracts [Hughes 2015].

When managing the paths of design for the two groups, it is preferable to synchronize the progress of each to facilitate the process of fixing coordination points at the progressive stages of completion. Careful consideration is necessary when using different delivery models such that early planning is synchronized to set the design criteria and mandatory geometrics at interface points. Inter-group reviews and meetings should also be conducted at an appropriate frequency to encourage mutual corroboration.

This division of expertise noted for the two design groups also extends to the construction phase where some contractors are better qualified and experienced than others to manage the pumping station fit-out. Construction contract packaging may also be influenced by the need to limit total package cost, encourage competitive bidding and promote the participation of local contractors.

Schedule and Spatial Considerations

With completion schedules often fixed by decree, program milestones will sometimes limit the feasibility of using delivery options with longer lead times. In many instances, design-bid-build may not be practical and options such as design-build will be favored instead, provided, of course, that the agency has the authorization to procure with such methods.

Seldom will spatial constraints not impact the selection of shaft sites and tunnel alignments, and since the combined sewage needs to be treated, the tunnel dewatering pumping stations are necessarily sited within or adjacent to existing and operational wastewater treatment facilities. If schedule permits, the pumping station fit-out can follow construction of the storage tunnel and pumping station structures. However, where compliance milestones are driving a particularly aggressive schedule, simultaneous installation under separate contracts can be accomplished by specifying an interim milestone for early completion of the pumping station structure and vacating a portion of the work area. The Narragansett Bay Commission CSO Abatement Program included such provisions, whereby the tunnel contractor completed installation of the pumping station cavern and shafts and turned over a portion of its main staging area to a follow-on tunnel dewatering pumping station contractor [Hughes 2008]. On smaller sites with only enough room to perform work tasks in series, the schedule must be lengthened accordingly.

Where adjoining construction activities are proposed, managing interfaces will involve clear distinction of boundaries for physical connection of adjacent structures. Suitable milestones must be identified to sequence the connections and define the means of making them. The method of connection may need prescriptive direction regarding structures and materials to be left-in-place such as waterproofing, rebar and pipe flanges. For underground connections, details regarding the type and configuration of

ground support should be addressed. Temporary fill or bulkheads may also be needed to prevent flooding of temporary works and specific commissioning sequencing (involving other contractors or the Owner) where live connections are anticipated.

Deeper systems employing a pumping station shaft typically involve three primary structural elements: the support of excavation (SOE) systems, such as slurry walls, to support the ground during excavation; an inner final lining within the SOE to support both the permanent or long-term earth loads (walls and base); and the interior floors, vertical walls and other pumping station structural elements. Key decisions must be made regarding which element is better assigned to the designer/builder responsible for the shaft (or cavern) as opposed to the follow-on fit-out—particularly with respect to the inner final lining.

To better coordinate components such as key ways, dowels and structural interconnections, responsibility for the final inner shaft lining would be grouped with the interior fit-out. On the other hand, if the designer/builder for the shaft structure SOE were responsible for the final inner lining, there is a reduced potential for later complications associated with construction performance, e.g., groundwater inflow, SOE verticality/quality. Furthermore, there may be efficiencies gained from combining the service of the temporary and long-term shaft wall elements. Often these latter benefits lead to the division of work with a single designer and builder designing and building both the temporary and permanent structural shaft elements. This requires close coordination with the follow-on pumping station design and construction.

Fitting out pumping station and screening facilities within previously constructed shaft structures requires the prior identification of layout dimensions for intermediate floor slabs, beam pockets, columns, internal walls, sumps, elevator pits, etc. At each connecting point the design of each "side" must also consider the effects of interactive loading both locally (e.g., stress concentrations, moments) and globally (e.g., total bearing, buoyancy of the entire structure). Other types of coordination details may involve the provision of wall penetrations for piping and utilities, identifying temporary variations in the elevation of the ground surface or groundwater, identifying elements of temporary construction that may be left in place (structural collars at tunnel connections, soil support capping beams, guide walls, temporary utilities, construction facilities, etc.).

Simultaneous use of sites by contractors will also need controls regarding occupation and temporary site conditions including boundaries, shared access roads and sequenced occupation of laydown areas. The site logistics of security, site safety, environmental controls (street sweeping) and temporary site utilities should be allocated or shared. In some circumstances, it may make sense for the Owner to retain responsibility for these types of task (or assign to a Construction Manager), for example when there are many contractors, the site is located within the Owner's existing facility or the Owner chooses to have a primary role regarding site safety.

Configuration

While some factors have far greater influence on design than others, each project will have its own unique set of circumstances driving the type and configuration of pumping station selected. Factors will include:

- *Internal Facilities*
  - *Pump Type, Number and Size*—operational capacity, flow variation and redundancy, manifold, wet well, orientation of motor units, etc.
  - *Piping and Control Valve Layout*—control flow to minimize hydraulic inefficiency, turbulence, vortices, cavitation; maximize operability of the system, protect facilities with fail-safe design, prevent damage from transient conditions.
  - *Operating Equipment*—control panels, instrumentation, HVAC ducts, plumbing, overhead cranes/hoists, screening and grit handling.
  - *Occupied Space*—access and clear space for equipment operation, maneuver room for component removal and replacement, compliance with building codes (interpretation by local permitting authorities), provision of safe zones, stairwells, elevators and redundant means of emergency egress.
  - *Internal Supporting Structures*—floors, columns, walls, roofs, sumps.
- *Structural "Shell"*
  - *Geology & Ground Support*—circular structures favored in weak potentially unstable ground, larger spans require proportionately greater support, vertical/horizontal interfaces lead to irregular stress concentration, greater flexibility to vary shape in competent ground, groundwater pressure and allowable infiltration.
  - *In-Situ Stress*—determining local stress magnitude and direction; predicting impact on temporary structures, rate of recovery of over-consolidated soils and long-term loading.

Two general layouts have been adopted: single circular/elliptical shafts (e.g., Atlanta West, Washington DC, Portland, North Dorchester, Hartford); and caverns with connecting shafts (e.g., Chicago, Milwaukee, Providence, Easterly-Cleveland, St Louis). Where both options appear feasible, comparative cost-benefit analyses are prudent. Caverns are typically more suited to deeper facilities within relatively competent rock and sequential excavation can also make this option viable where rock quality is less favorable. The single larger diameter shaft offers the benefits of a simpler configuration and lining method.

Until the early-2000s, perceptions of risk regarding the support of large diameter (100-foot+) shafts at depths greater than 100 feet through soil, markedly influenced decision-making in favor of smaller diameter access shafts and caverns. Demonstrated advances in slurry wall technology (principally regarding verticality and increased panel thickness) in the past 15 years have, however, made larger diameter shafts more attractive. Dual-cell shaft configurations have also opened options to build more elongated footprints (e.g., Brightwater).

Design Criteria

A designer's personal judgment and interpretation will depend on the individual and may vary from project to project. Judgment is also influenced by whether the final design is developed before or after the price for construction is fixed; the tendency for conservatism being tempered in the latter. Since the engineer stamping the drawings is the one who makes the ultimate decision, a baseline for criteria and loading cases must be established to guide interpretation. The list of considerations for selecting design criteria includes but is not limited to:

- *Dimensions & Configuration*—Drawings are the better means of representing minimum/maximum dimensions and tolerances. Wherever practical, duplicative text should be avoided in complementary specifications or reports.
- *Physical Conditions and Parameters*—Soil properties, interior and exterior fluid levels, flooding conditions, transient or long-term loading, etc. should be specified. Include guidance on the application of Earth Pressure Coefficient K, the magnitude of which may vary for temporary conditions and by depth. Where geotechnical data and baselines are contained elsewhere within the documents, ensure potential conflicts or ambiguities are avoided.
- *Design Codes*—Provide general guidance of which codes, practices and methods to observe. For ACI strength design, distinction is needed to identify whether to observe ACI

318 or ACI 350 and the degree of discretion the designer may exercise for certain parts of these codes, e.g., ACI 350 environmental durability factors, maximum allowable strain in temporary vs permanent conditions, etc.

- *Loads and Load Combinations*—Identify which to use and supplement with additional cases as appropriate. Specific direction may also be prudent to clarify load cases for certain circumstances and the means by which loads should be calculated, assuming that an interface with frozen ground will contribute no skin friction to counteract uplift.

- *Seismic Requirements*—In addition to standard performance parameters such as MDE, ODE, IBC site class, and max allowable strain, it is advisable to include a preferred methodology, e.g., Hashash et al, 2001.

- *Durability and Minimum Design Life*—Specify certain methods of demonstrating design life, e.g., ACI-365. Specify level of sulfate exposure, include specific restrictions such as minimum concrete cover, state which means of mitigation can or can't be employed to achieve the design life. The selection of design life must be realistic and will vary according to the type of component material and environment.

- *SOE Systems*—Continued advances in quality control for Support of Excavation (SOE) support systems such as slurry walls, tremie slabs and rock dowels, have led to their more frequent use as components of permanent design. Direction should be provided regarding the degree to which SOE systems can be incorporated, allowable/prohibited methods, etc. Where support elements are to act compositely or share loads, clarify how elements may or may not interact. Where SOE systems are not incorporated into the finished work, provide limitations to prevent detrimental impacts on permanent structures, e.g., minimum "sacrificial" thickness of concrete adjacent to frozen soil.

- *Watertightness*—Allowable rate of infiltration and guidance on allowable/prohibited methods.

CASE STUDY—BLUE PLAINS TUNNEL DEWATERING PUMPING STATION

Project Background, Organization and Procurement

DC Water is a public agency which provides drinking water distribution and wastewater collection and treatment to more than 640,000 residents, 17.8 million annual visitors, and the 700,000 people employed in DC. It also collects and treats wastewater for more than 1.6 million suburban customers from Maryland and Virginia. DC Water operates the Blue Plains Advanced Waste Water Treatment Plant (BPAWWTP) located at the southern end of the city with a dry-weather flow capacity of 370 MGD and a peak pumping capacity of 1.08 billion gallons per day. DC Water's collection system contains both combined and separate sewers and, like many communities with CSOs, DC Water has been required by a Federal Consent Decree (CD) to develop a LTCP to reduce the CSOs that discharge into adjacent waterways during wet weather.

DC Water's LTCP is referred to as the DC Clean Rivers Project, which includes an 18-mile long tunnel system designed to capture, temporarily retain, and convey CSOs to the BPAWWTP. At the BPAWWTP, a new 250 MGD (expandable to 500 MGD) Blue Plains Tunnel Dewatering Pumping Station (BPTDPS) will lift the temporarily stored flows to the surface for treatment and discharge.

To manage the DCCR Project, DC Water formed a new DCCR management group and retained a Program Consultants Organization (PCO) to assist with program management, preliminary design and procurement. Within DC Water, a separate management group—the Department of Engineering and Technical Services (DETS)—was overseeing expansion of the BPAWWTP and was assigned the responsibility for planning and oversight of the BPTDPS, for which it retained a separate Pumping Station Program Manager (PSPM) to provide consulting assistance.

To achieve the CD milestones, DCCR elected to deliver the most critical primary elements of program—the tunnels, shafts and pumping station—using design-build delivery. As well as schedule, the principal determinants for contract packaging included work site availability, completion of the geotechnical investigation, sizing contracts to encourage bid competition, and design and construction discipline compatibility. Identified as the most time sensitive elements of the program, the 4.6-mile long Blue Plains Tunnel and pumping station structures (SOEs and final inner linings) were packaged as Division A. To meet the CD milestone A, a separate package designated by DCCR as Division Y, would be constructed concurrent to Division A and would include the Pumping Station facility fit-out (pump selection and layout, internal structural framing, mechanical, electrical, plumbing, etc.) (Figure 2). An interim milestone was established in Division A requiring the completion of the Blue Plains Tunnel Dewatering Shaft (BPT-DS) structure no later than 1,280 days after Notice To Proceed such that the shaft and the adjacent surface area could be turned over to the

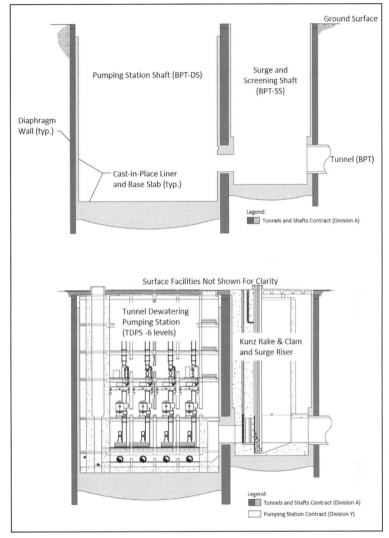

Figure 2. Contract packaging included a tunnel and shafts contract (Division A) and a Tunnel Dewatering Pumping Station contract (Division Y). Division A elements shown above; Division Y shown in lower section within the Division A shaft.

BPTDPS design-builder to begin construction of the pumping station internals (see Figure 3).

Within the first 1,280 days, the Division A design builder used the entire 4.2-acre site to support construction of the 132-foot diameter pumping station, 75-foot diameter screening shaft and launch the TBM to construct the Blue Plains Tunnel. After the 1,280-day milestone, the Division A site was reduced to approximately 3 acres and used to stage and complete the Blue Plains Tunnel. The approximately 1.2 acres allotted to the Division Y design builder was used to begin construction of the structural

internal elements. At completion of Division A, the Division Y design-builder expanded to occupy the entire 4.2 acres to construct the remaining elements of the pumping station and adjacent supporting high rate treatment facility (see Figure 4).

DC Water's design build procurement method involved the development of a 30% preliminary design and its incorporation into Request for Proposal (RFP) tender documents. The RFP detailed the scope of work and established the boundaries for final design by defining certain design criteria as MRs. The design-builder (DB) retained a designer

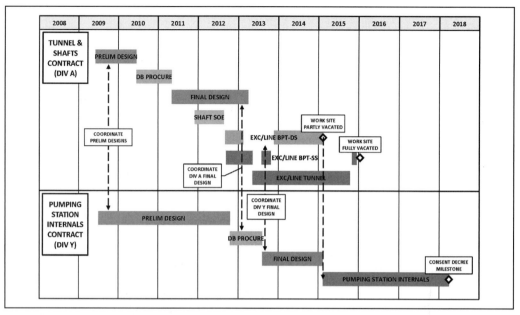

Figure 3. Divisions A and Y summary schedules indicating coordination of design and schedule required by the two teams

to prepare a tender design, act as the Engineer of Record (EOR) and assume the responsibility for developing the final design.

Both Division A and Y packages contained significant MRs, such that the latitude for modification during final design would be significantly constrained. Since the design and construction of the tunnel elements would necessarily take much longer than the pumping station, design development under Division A progressed at a considerably more accelerated rate than Division Y. In practice this meant that conceptual locations of internal elements and rough-order-of-magnitude loads had to be nominally fixed to establish a basis of bid. Accordingly, the Division A contract contained MRs that were based on Division Y's less than 5% design progression, highlighting points of coordination related to those items noted in the "Design Criteria" and "Configuration" lists above. At the time the Division A RFP documents were published for procurement, the Division Y conceptual design report had just been finalized. This meant key decisions regarding pumping station configuration which affected the design of the pumping station shafts were required to be made and committed to during the pumping station conceptual design phase. These key early decisions and commitments included:

- Finished inside diameter and depth of the pumping station and screening shafts.

- Minimum shaft wall thickness to support pumping station super structure and internal floor system.
- Discharge piping (multiple 48-inch lines) shaft wall penetration plan locations and elevations.
- Pumping station column dimensions, locations and maximum loading on the shaft base slab.
- Pumping station floor slab and beam pocket locations (5 floors internally connected to the shaft walls), dimensions and loading onto the shaft wall.

Final Design

Based on the 30% design, the RFP documents proposed a 60-foot inside diameter (ID), 160-foot deep screening shaft (BPT-SS) and a 132-foot ID 175-foot deep dewatering [pumping station] shaft (BPT-DS), separated by a 30-foot long connecting tunnel. During the collaborative bid process, the successful DB obtained approval to eliminate the connecting tunnel by re-locating the screening shaft and forming a figure-eight configuration with a section of shared initial soil support wall. The DB preferred this configuration to increase the space available to stage trailing gear during TBM mobilization and launch. It also elected to increase the size of the BPT-SS to an inside diameter of 76-feet, to provide more space to support its mining operation once the pumping station shaft was turned over to the pumping station fit-out DB.

At the BPT-DS, the DB elected to use 5-foot thick slurry walls and a waterproof membrane together with the mandatory 3-foot minimum thick cast-in-place interior lining. For the permanent condition these elements were designed to share ground loads in a composite manner, with the inner wall designed for external hydrostatic loads equivalent to a 500-year flood. At the base, a dome shaped invert slab varied in thickness from 16- to 26-feet and was composed of five layered slabs. The excavation of soil was to be performed in-the-dry with dewatering wells employed to depressurize the formation and maintain basal stability for the lowest 50 feet of excavation.

Since the design and construction of the BPT-DS and BPT-SS were essential to advance the tunnel work, priority was given to their design shortly after award of the contract. Design of the slurry walls was split from the final lining and invert slabs to allow "just-in-time" completion. For design of the final lining and invert slab, updated BPTDPS details of the internal structures were obtained from the Division Y PSPM and found to contain several modifications from the prior stage of design that had been included in the Division A RFP documents, such as: the addition of new structural elements (e.g., elevator shafts, stairwells, sumps, screening shaft dividing walls, etc.), increases to the estimated loads, the identification of new areas where exposure to wastewater would occur (requiring ACI 350), wall penetrations for discharge pipes, utilities, etc. But, the most notable coordination challenge proved to be the lateral connections of the wall to the internal mid-level floors and floor beam.

The BPT-DS EOR predicted that during excavation and after placement of the internal lining, the wall support system would sustain approximately 2- to 3-inches of diametral convergence due to ground loading. It proposed that this movement be accommodated with laterally flexible connections to the internal structures filled with compressible material. However the PSPM opposed this option based on the concern that allowing flexibility would cause increased vibration during pump operation. It was therefore, directed that the interfaces at these structures be rigid fixed connections. Impacts to Division A design included the need for additional reinforcing steel in the walls at the rigid connections, the bottom reinforcement, shear key and ring beam. The additional reinforcement and loading on the shaft walls were not originally contemplated and required a change to the Division A contract.

The BPT-DS EOR predicted that during excavation and after placement of the internal lining, the wall support system would sustain approximately

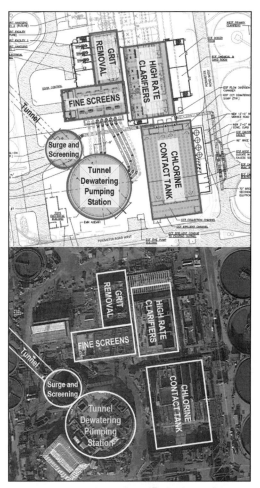

Figure 4. Layout of facilities and aerial image during Division Y construction

2- to 3-inches of diametral convergence due to ground loading. It proposed that this movement be accommodated with laterally flexible connections to the internal structures filled with compressible material. However, the PSPM opposed this option based on the concern that allowing flexibility would cause increased vibration during pump operation. It was therefore, directed that the interfaces at these structures be rigid fixed connections. Impacts to Division A design included the need for additional reinforcing steel in the walls at the rigid connections, the bottom reinforcement, shear key and ring beam. The additional reinforcement and loading on the shaft walls were not originally contemplated and required a change to the Division A contract.

During the final design of the pumping station interior structures, the Division Y DB's EOR raised concerns regarding potential transfer of lateral loading from the shaft walls to the pumping station floor slabs and beams. In response, the Division A EOR conducted a 3D structural numerical analysis to determine the interactive loads imparted at the connections between the Division A and Division Y structures. This analysis necessitated that the DCCR PCO estimate expected earth pressure recovery over the life of the structure and was, in part, based upon data gathered from strain gauges and earth pressure cells fixed within the slurry walls. The ground loading data was evaluated to estimate the "current state" of relaxed ground resulting from slurry panel installation and subsequent shaft excavation. The difference from the "current state" and the original design criteria earth loads of K_o and normal groundwater (not depressurized) were used in a 3D structural numerical analysis. The results of this analysis were used to provide the Division Y DB with loading information for its EOR to finalize the design of the BPTDPS floor slabs.

When planning the Division A DB procurement, it was recognized that design changes such as those noted above were likely to arise during completion of final design. Accordingly, a contingency allowance was set in the contract to reimburse this anticipated additional work.

CONCLUSIONS

The Blue Plains Tunnel was completed on time and on budget. For the most part, the tunnel and shafts were constructed with little issue. However, the coordination between Division A and Division Y presented challenges, the resolution of which required significant effort from DC Water and its program managers, construction managers and design-builders.

Where approaches are adopted that split out responsibilities for design and construction, coordination points must be managed and monitored until completion of the work. The greater the number of coordination points, the greater the burden of risk associated with their management and the increased potential for opposing interpretations requiring reconciliation.

Out-of-phase design progress is sometimes critical and essential to facilitate schedule, but bears with it the disadvantage that assumptions must be fixed when based upon incomplete evaluations and analysis. Balancing early decisions which fix methods and design with the development of procurement contracts that allow innovation is difficult, but can be accomplished.

Design build delivery of final design intimately ties together the critical paths of design and construction. Coordinating changes during execution of the work carries a relatively high risk of adversely impacting the final quality of the facility and delaying its construction. Incorporating measures and provisions in the contracts to facilitate coordination between connected design-build works can minimize the consequences but not eliminate this risk.

Regardless of contract delivery method, the selection of design criteria for pumping station design requires careful planning and execution. Some situations simply can't be addressed by adding "by-others" notes to the scope of work or contract drawings.

LIST OF PARTICIPANTS

- Owner: DC Water—DCCR and DETS Staff
- Tunnel and Shaft Structures, Division A:
  - Program Consultants Organization (PCO)—Greeley and Hansen and McMillen Jacobs Associates
  - Design Build Contractor (DB)—a joint venture of Traylor Brothers, Skanska and Jay-Dee Contractors (TSJD)
  - Engineer of Record (EOR)—Halcrow (CH2MHill)
  - Construction Manager—EPC Consultants, Inc.
- Pumping Station Internal Facilities, Division Y:
  - Pumping Station Program Manager (PSPM)—AECOM
  - Design Build Contractor (DB)—a joint venture of PC Construction Company and CDM Constructors
  - Engineer of Record (EOR)—CDM Smith, Inc.
  - Construction Manager—Arcadis

REFERENCES

Edgerton W.W. ed. 2008. Recommendation Contract Practices for Underground Construction. *SME.*

Hashash Y.M.A., Hook, J.J, Schmidt, B., Yao, J.I-C. 2001. Seismic design and analysis of underground structures. *Tunneling and Underground Space Technology 16, pp 247–293, Elsevier Science Ltd.*

Hughes G., Kroncke M., Wone M. 2015. Practical Aspects of Final Design Development using Design-Build Procurement. *RETC Proceedings.*

Hughes G., Kaplin J., Halim I., Albert P. 2008. Design and Construction of the Fields Point Tunnel Pump Station for the Narragansett Bay Commission CSO Abatement Program, Providence, Rhode Island. *NAT Proceedings.*

Emerging Demand for Subsea Tunnels in Chile

Victor Figueroa and Nicolás Zegpi
SKAVA Consulting S.A.

ABSTRACT: Chile has an economy that strongly depends on natural resources, mainly mining and agriculture. Water is a strategic asset for the latter. Considering the water strategic economic value and that due to drought, aquifer and reservoir depletion, water supply reliability has become a permanent problem in the country's most productive regions. The mining sector alone has planned for at least 19 projects using seawater in their processes. Considering the geological and tectonic context of Chilean coastline, these projects face significant difficulties. That is why subsea tunnels for intakes and outfalls are part of the solution to this challenge. This paper describes Chilean water context and design and constructability approaches for three different cases.

INTRODUCTION

Clean water is a fundamental resource for human society and for some productive sectors is a strategic resource. In Chile, the sustained economic and social development, in previous decades, has generated constant increases in water demands, primarily due to the economic dependence to natural resources, mainly form mining and agricultural areas. In those areas, almost every product depends on water availability.

Chile is a long narrow country, hence the geography and climate varies from the most arid desert in the world in the extreme north of the country to one of the rainiest places of the world in the south. Also, there is a clear difference in the temporal water availability. Thus, each region has different water management policies. During the last years droughts have taken place in some regions, although these droughts are mostly seasonal, there are indications that show a more long-term trend.

Even though there is not clear projections for water consumption in the future, the limited availability of this resource has led to develop mechanisms that enlarge its availability. This mechanism in most cases involves the use of sea water in the productive process. In some cases, the salty water is used directly in the productive process but in others, it is desalinated.

The largest lack of water in the country is located in the northern regions. Although the northern regions are located where water is scarcer, the mining industry is pushing an important economic and social development, resulting in a constantly larger demand for this resource. This is the main reason why mining companies are planning to start using seawater in their processes. Chile's far north is characterized by a very abrupt topography; the

coastline is usually defined by high and rough cliffs. Hence, taking off seawater is usually not simple. Additionally, environmental constraints in intervening the coastline explains the increasing use of subsea tunnels.

All of those tunnels are being built in seismic zones, most of them, with active faults near the project.

Water Resources in Chile

Even though around 75% of the planet's surface is water, only 2.5% is fresh water, but it is mostly located in the big ice caps, glaciers, lakes or atmosphere. That results in only around 0.75% of the water available for human consumption in rivers, lakes, groundwater and wetlands. According to the Food and Agricultural Organization of the United Nations (FAO) Chile is the 15th country with larger per capita precipitation. However, in the previous years, it has been changing and becoming more unstable, generating uncertainty in water availability. Even worse, this trend is expected to continue. As can be seen in Figure 1, in Chile there is in an important difference between water availability and usage in different regions, making water management more difficult.

To manage the water resources Chilean legislation incorporated the concept of ownership of water rights with free market transactions among private owners. Thus through legislations the market has assigned an economic value to water. The main objectives of the Chilean water code are: 1) creating solid water exploitation rights, 2) creating markets of water resources, and 3) reduction of the role of the state in the matter.

The water rights are granted to anyone who can demonstrate usage, until the hydrological

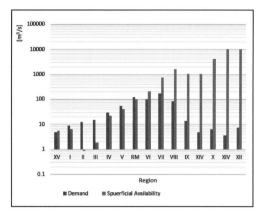

Figure 1. Water availability and demand from North to South (Data from Ref 2)

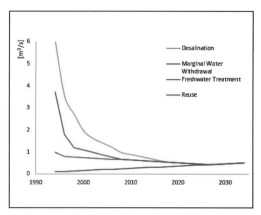

Figure 2. Water production cost evolution (Data from Ref 4)

availability in any given basin is reached with a certain safety factor. Then, the users have to trade them economically.

The water usage in Chile is largely dominated by the forestry and farming sector with 82% of their usage. Then, the industrial sector takes around 7% of water exploitation rights, but this sector groups many different productive services, so the tendency of the water usage has not been studied properly, but it is expected that it will increase within the following years. Currently, the mining sector represents around 3% of the national demand, it is expected that it will increase to around 200% in its water usage in the following 25 years. On the other hand, the sanitary sector accounts for 8% for human consumption and transporting and treating waste water.

Seawater Desalination

Clean water availability constitutes a global problem and it has become one of the most important issues due to population growth in the past century. According to predictions, two-thirds of the global population will be located in hydric scarce regions in 2025 (Drioli & Macedonio, 2012). The unique methodologies to increase the water supply provided by water cycle are desalination and reutilization.

Desalination is the physical process that allows salts and water to be separated from a watery solution and, consequently, to use salty water as clean water. The process to separate salts and water through evaporation and condensation is well known since ancient history but it has not been permissive in large scale due to the high-energy consumption level.

Desalination technologies were developed at the mid-twentieth century. The most important technologies developed were Multi Effect Distillation (MED) and Flash Multi Stage Distillation. Those are still today the most used desalination technologies.

Over time, continuous improvements in the desalination process reduce its energy consumption and cost. As can be seen in Figure 2, desalination is becoming a viable option to produce clean water.

World Desalinated Water Usage and Trend

Currently, more than 12.500 desalination plants exist in the world, as shown in Figure 3; most of them are located in the Middle East, Africa and the dry region of USA.

As desalination costs have been decreasing, the installed and commissioned capacity has increased exponentially in the last decades. The growth in desalinated water capacity can be seen in Figure 4.

Desalination capacity has grown due to technological and cost consequences, and it has grown as a consequence of variability of rain patterns. Making desalination a stable source of water during all seasons, this characteristic gives desalination an enormous potential, considering the expected changes in climate. Also in some countries, for example Saudi Arabia, long and complex pipelines have been constructed to supply water to inland cities, demonstrating that desalinated water can be a solution not only for coastal populations. Additionally, there are still big areas where the technology could be applicable and it's still not being used. Missions (2017) believes that each zone with a climate characterized with less than infrequent rain, see Figure 4 could be supplied with desalinated water. Therefore, the desalinated water capacity in the world is still expected to keep growing rapidly in the following decades.

Desalinated Water in Chile, Usage and Trend

Although there are indications that desalination has been used long ago in the Chilean mining industry, desalination is a relatively new technology in Chile. Causes of this development are diverse, the main

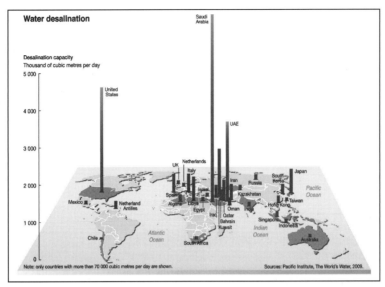

Figure 3. Desalination plants distribution (Taken from Ref 5)

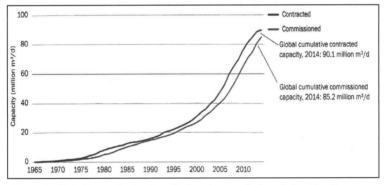

Figure 4. Global cumulative installed and commissioned desalination capacity 1965–2014 (Taken from Ref 6)

factor is the growing drought in the center and north of the country that has led to hydric resource exhaustion for the industrial sector and even in some cases restricting it to the population consumption. Added to this, another factor is the increase in social conscience in environmental and natural resources protection from the communities and regulatory codes. This has led to the different economic sectors to find solutions to develop business looking for alternative water supply sources. In this context there are three main sectors which operates desalination or seawater impulsion plants, mining industry, sanitation and industrial. Currently Chile is the Latin-American country with the largest installed capacity for water desalination with 17 plants reaching a capacity of approximately 700 [Mm³/day].

Mining Sector

Currently the mining sector is the most important sector in desalinated water usage, with 11 existing operations which use sea water in their process. Those operations are located in the far north of Chile, with a total desalinated water installed capacity of 2.038 [l/s] and a total of 3.657 [l/s] of direct seawater usage in the process. Presently, copper mining is the principal one followed by iron and some nonmetallic mining operations.

Sanitation Sector

The scarcity of water problem in the country far north has led to finding technological alternatives. Given that, in this region, traditional water sources are becoming more expensive with time or even are

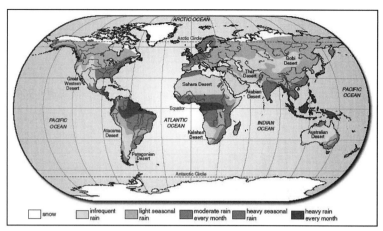

Figure 5. Global rain patterns (Taken from Ref 5)

depleted, and desalination cost has been constantly reducing, desalination has become a viable option to solve population water consumption problem. Added to that the authorities are studying the option to give tributary incentives to the companies which apply to this alternative.

As an example in Antofagasta city, the cost to generate clean water from the traditional sources is almost at the same price as the desalination. This is mainly because the Loa River is declared exhausted by the authorities more than 15 years ago and to produce water from the Andes is too expensive. That's why the biggest desalination plant of the country is located in this city.

Today the sanitation sector operates 4 desalination plants with a total capacity of 1.360 [l/s] and a market participation of 39.3%.

Industrial Sector

At present, the industrial sector is the third most important but largely surpassed by the first two, with only 2 operations and an installed capacity of 64 [l/s].

Summary

These three sectors are currently the most important, and there is an expectation of a big increase of desalinated water demand for each one of them. A summary of the operations and installed capacity in Chile is shown in Table 1.

Mining is not only the most demanding in amount of desalinated water, but also is largely the most important relative to investment, because the cost in transportation of water to the mining operations which are located at an average distance of 170Km and 2100 meters height. In addition, the mining industry has an important investment in seawater usage which has a participation of 51.4% against

Table 1. Chilean current installed capacity summary

Sector	Operations	Installed Capacity (l/s)	Participation (%)
Mining	11	2038	58.9
Sanitation	4	1360	39.3
Industrial	2	64	1.8
Total	17	3462	100

the 48.6% of the desalinated water. Generating other important investment source for subsea tunnels.

Seawater and Desalinated Water Market Projections

At least 12 operations with seawater or desalinated water usage are projected with a new desalinated water installed capacity of approximately 8100 [l/s] and a total seawater usage of approximately 4500 [l/s]. This means growth superior than 200% of the current desalinated water capacity, more projects which involve water desalination and seawater usage are expected to be presented. Based on Chilean Copper Committee the sea water usage in copper operations during 2016 was 2.446 [l/s] and is expected to grow to 20.089 [l/s] in 2017. As can be seen in Figure 5, as well as seawater consumption increases, continental water also increases, but in a much more moderate ratio because the hydric resources are nearly exhausted.

As said above continental water resources in some northern regions are almost exhausted and droughts are becoming more common and are increasing in intensity making continental hydric resources less reliable. Hence, an increase in population is directly related to an increase in desalinated water demand and this solution is even more necessary to increase the reliability of water production for human consumption. Due to droughts, desalination water market is expected to grow in zones where it wasn't expected to need it. Currently the government has already announced 7 desalination projects with a total installed capacity of 1.021 [l/s] with the most important located in Arica and Copiapó. In addition, to the government projects, private markets are increasingly considering desalinated water to be a bigger business opportunit. This is giving the definitive impulse to the sanitary sector to consider desalination a real option and invest on it.

The industrial sector is becoming aware of a potentially attractive business in desalinated water, based on the necessity of clean water in industrial process, human consumption and mining process. Lots of new projects will not be profitable if there isn't new sources of water supply. Also, some mining projects, from small to medium, can't afford a seawater desalination plant and transportation system, therefore the desalination project to supply medium and small mining projects are being developed. Also, the energy sector has been increasing the usage of seawater in their process and, in some cases like thermal plants, has commenced to adapt to desalinate water.

Summary

Currently, the mining sector has the predominant usage demand for desalination water capacity and will maintain its predominance, projecting a demand for 58% of the total capacity against a 28%demand for the sanitation sector and a 14% demand for the industrial sector.

But more important than that is the expected new installed capacity for desalinated water in all these sectors totals 14.092 [l/s]. The projected operations and their capacity are shown in Table 2.

As can be seen in Table 3 and Table 4, in terms of both capacity and number of operations the current state will be more than doubled.

Chilean Geology and Topography

As said previously, Chile is a long and narrow country located in the south-west of South America. Therefore, Chile is a very diverse country in climate, topography and geology. In addition, most of Chilean territory was generated through the uprising caused by the subduction of the Nazca plate under the South American plate, giving Chile very particular characteristics. One of the most important characteristics is the presence of the Andes Mountains all along the country. Due to subduction, Chile is one of the most tectonically active places in the world with large amounts of active volcanos and frequent earthquakes. During the last century, Chile was struck by 28 earthquakes with a magnitude higher than 6.9 in Richter scale, including the strongest seismic event

in the world. Also, the combination of large scale seismic events and long coast lines makes Chile very prone to tsunamis. As can be seen in Figure 7, Chile is subdivided in 5 main regions, Far or Big North, Near or Small North, Central Chile, South and Far South.

Given that the most arid regions, and therefore the most important related to desalination and water pumping station, are located in the Far North and the Near North, only these regions will be described.

In northern Chile most of the coastline is rugged and is characterized by high cliffs with heights from 250 m to 800 m, also the topography under the sea is very steep. Beyond the high cliffs is the Atacama Desert, one of the driest places on earth, with specific zones that do not receive rain precipitation at all. After the desert are the Andes Mountains and the Altiplano in the northern region. The Altiplano, due to altitude, receive an important amount of precipitation and in some places, this water manages to cross the desert even reaching the sea, creating very fertile and rich valleys.

Another important characteristic in northern Chile is the seismicity due to interplate fault

Table 2. Desalination projects summary

Sector	Operations	Installed Capacity (l/s)	Participation (%)
Mining	12	8.122	57.6
Sanitarium	7	3.997	28.4
Industrial	3	1.973	14.0
Total	22	14.092	100

Table 3. Current operation and projected operations

Sector	Current Operations	Projected Operations	Total Expected Operations
Mining	11	12	23
Sanitarium	4	7	11
Industrial	2	3	5
Total	17	22	39

Table 4. Current and projected desalinated water installed capacity

Sector	Current Installed Capacity [l/s]	Projected Installed Capacity [l/s]	Total Expected Installed Capacity [l/s]
Mining	2.038	8.122	10.160
Sanitarium	1.360	3.997	5357
Industrial	64	1.973	2037
Total	3.462	14.092	17554

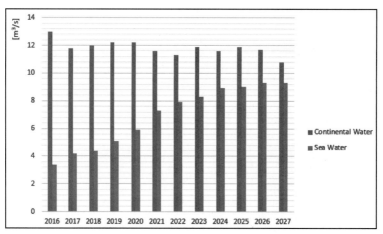

Figure 6. Continental water versus seawater projections

displacements. This region has an important number of active faults, with potential to generate superficial seismic activity with great accelerations.

Seawater catchment can be done by two different ways, pumping wells or open catchment. Although pumping wells are considerably better related to water quality and temperature variations, for large scale desalination or pumping plants the pumping wells are not recommended because experience shows that it's difficult to procure the volume of water required using pumping wells. However, open catchment presents the following disadvantages:

- Water pretreatment is needed.
- Subsea emissary needs to be long enough to be at least 8 m under the sea level in low tide.
- Stand on a zone without seaweed which can settle the emissary.
- Suction velocity should be slow enough to avoid interfering with sea life.

The verticalness and geology of the shore, along with the seismicity and susceptibility to tsunamis generates great difficulties in connecting the desert plains with the sea. These challenges make open catchment trough tunneling almost the only option to make this connection in a permanent and secure way.

CASE STUDY

In this paper three cases study will be presented. Due to confidentially agreements, project names and locations will not be presented.

The First case is an electric project, this project takes advantage of the important solar radiation in the desert during the day and the height of the coastal cliff, using part of the energy generated during the day through photovoltaic solar panels to

Figure 7. Main geographical Chilean regions

pump seawater to a reservoir located above the cliff. The objective of pumping water to a higher reservoir is to use this water to produce energy during nights as a hydric power plant. That is the reason why, a reversible flow subsea tunnel is needed. Because of the nature of this project, the verticality of the coast line is an advantage, reducing the length of the tunnels needed to reach the altitude needed for production. This project is expected to generate 300 MW using hydraulic energy. The scale of this project makes it unviable to use pumping wells to catch the

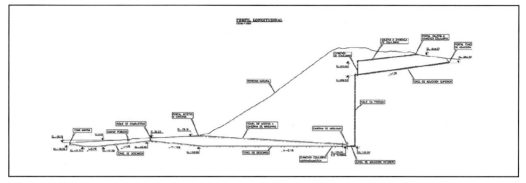

Figure 8. Electric project longitudinal profile

water and the seismicity makes it unviable to generate the catchment through a superficial emissary. Thus, subsea tunneling was the selected option to generate the catchment. Given the large amounts of water pumped and used to generate electricity, tunnel dimensions were big enough to excavate it through the Norwegian Tunneling Method and Drill & Blast methodology. The tunnel projected for this use is 2.5 km long, with a section rounding 6 × 5 m. with around 250 m located under the sea level. This tunnel reaches an altitude of approximately 700 m. To solve the problems involved in subsea tunneling through the drill and blast methodology, a series of considerations were projected, including minimum depth, injections and lake tap at the end of the tunnel. A simplified longitudinal profile is shown in Figure 8.

As can be seen in Figure 8 there are five main tunnels and one cavern in this project. The Adduction tunnel, at left side, is projected to discharge water from the sea during day and catch water from the reservoir to produce energy during night, this will be constructed in dry condition but tunnel is designed to be under the pumped water level, so especial water condition should be taken account in the design of this tunnel. One vertical pressure shaft, this shaft is around 500 m long and is designed to gain pressure for energy production during night time and conduct sea water to the reservoir during day. The subsea discharge tunnel, this is the longest tunnel in this project with around 1.7 km. This tunnel is designed for seawater catchment during day and release back to the sea the water used for production during night. As can be seen, this tunnel is completely under the sea level, and around 250 m are under the tide line. The other two main tunnels in this project are the cavern access tunnel and the Surge Tank.

Considering that the longest tunnel is completely designed under the sea level and even that the rock mass is expected to be a good to fair rock mass special drilling measures and pre-grouting are considered to limit the water inflow.

The second case presented in this paper is a desalination plant to supply water to the operation of a copper mine project. In this case the water needed for the mining process will be taken under the sea level and pumped to the mining project located over 1600 m s n m. As well as the first case, in this case an open catchment through a subsea tunnel was the selected option. Given that the volume of water to supply a desalination plant is considerably lower than the volume of water required in the first case, the selected methodology was to construct through a Micro TBM and Pipe Jacking. This project consider two tunnels under the sea level, one for catchment and one for discharge. Both tunnels with an inner diameter around 950 mm and approximate lengths of 350 and 550 m. For both tunnels, the maximum cover will be 20 to 25 m. The main advantages of the methodology proposed are:

- Circular excavation with limited over excavation.
- Limited disturbance in rounding rock mass (allowing to maintain a low permeability)
- Higher production (around 6 m/day expected)
- Reduce environmental impact (because of non-explosive usage)
- Higher safety (given the immediately concrete support)

On the other hand the main disadvantages are:

- The methodology present a very limited adaptability to geological changes, therefore the site study must be very rigorous and accurate (even more considering Chilean geology)
- Lack of local experience
- Equipment needs to be imported or adapted
- Complex maintenances
- Needs high skilled personnel

Given the construction methodology have a very limited adapt and that Chilean geology is very unpredictable, even in short lengths, in this case the main factor to consider prior the construction is the geotechnical and geological exploration. Prefeasibility study shows that 75% of risk lies on insufficient preliminary study, poor construction execution and unappropriated design.

The third case presented is for a big desalination plant for human consumption, projected to reach a final capacity of 1200 [l/s] constructed in three stages, the first two consisting of 450 [l/s] and the last one with 300 [l/s]. Given that this project is expected to produce water through inverse osmosis, the main residuum will be brine generating the necessity of discharge tunnels. To construct this project two options are evaluated, excavation through Drill and Blast and Mechanized excavation. Drill and Blast methodology provides more flexibility and adaptability but disturbs the rock mass increasing permeability, generating infiltrations in the tunnel increasing excavation time and costs. For this project a series of subsea tunnel are being evaluated for adduction and discharge. Feasibility study was evaluated with four tunnels with lengths between 200 and 600 m and diameters from 900 mm to 1500 mm depending on the stage evaluated and construction methodology.

CONCLUSIONS

Technological changes have generated a rapid increase in desalination and seawater pumping techniques, making this technology a real option to solve supply problems. The reduction in cost and the increase in water necessity for population consumption, mining and industrial operations are generating an increasing necessity for new water supply sources.

Desalinated water and seawater usage are becoming more common and it is expected to continue growing all around the world, but it is expected to grow even faster in Chile. The main factor pushing desalination to grow in market share are the increasing need in mining process and the lack of reliability in the traditional sources, mainly due to the droughts, in the sanitation sector.

Geology and topography in Chile make the use of subsea tunneling the best option for water catchment for bigger desalinated projects and pumping stations. This added to above is generating an important increase in subsea tunnel construction. Although it has not been addressed in this document, there are projections to build road projects in which it is being evaluated to build some subsea tunnels.

Many different technologies and methodologies can be used in subsea tunnel construction, depending on the characteristic of the project, like number of tunnels, required diameters, type of soils or rock where the project is located, access, schedules and many different factors. The option selected is crucial and needs to be studied carefully, given that in large degree, it determines time and cost of the construction. So, prospection and exploration are even more important than in common tunneling.

REFERENCES

[1] FAO, AQUASTAT water information system.
[2] Minería Chilena 2017/18. Catastro de plantas desalinizadoras y sistemas de impulsión de agua de mar.
[3] DGA Chile, 2016. Atlas del Agua de Chile.
[4] Drioli & Macedonio, 2012. "Membrane Engineering for Water Engineering."
[5] GWI., Desalination Markets 2007. A Global Industry Forecast (CD ROM), Global Water Intelligence, Media Analytics Ltd., Oxford, UK, 2007, www.globalwater intel.com.
[6] Mission, 2017. Desalination and Water Recycle. http://12.000.scripts.mit.edu/mission2017/desalination-and-water-recycling/.
[7] IDA, 2015. "Global Clean Water Desalination Alliance "H20 minus CO2" Concept Paper. https://www.diplomatie.gouv.fr/IMG/pdf/global_water_desalination_alliance_1dec2015_cle8d61cb.pdf.

Implementing Major CSO Solutions via Deep Rock Tunneling: Louisville Ohio River Tunnel (ORT)

Jonathan Steflik, Adam Westermann, Donnie Ginn, and Todd Wanless
Black & Veatch

Jacob Mathis and Greg Powell
Louisville and Jefferson County Metropolitan Sewer District

ABSTRACT: Louisville and Jefferson County Metropolitan Sewer District (MSD) will complete their Long Term Control Plan, part of an $850 million 20-year Integrated Overflow Abatement Plan (IOAP) by December 2020 to reduce combined sewer overflows (CSOs). Originally scoped as three separate CSO basin projects, the Ohio River Tunnel (ORT) was developed in response to challenges encountered throughout design and now involves construction of a deep rock tunnel with a capacity of approximately 37 million gallons.

This paper discusses the challenges involved in completing the design of the ORT Tunnel and Shafts contract package on an accelerated 8-month schedule. The design phase for the basins was originally planned for approximately two years, but was compressed due to evolving design requirements for three CSO storage basins. The schedule will allow adequate construction time to meet MSD's consent decree deadline. Key project elements include a 2.5-mile long, 20-foot finished diameter deep rock tunnel, four drop structures to convey consolidation flows to the tunnel, downstream pump station shaft, and adits connecting drop shafts to the tunnel in downtown Louisville.

INTRODUCTION

In August 2005 MSD entered a Consent Decree with the United States Environmental Protection Agency (EPA) and the Kentucky Environmental and Public Protection Cabinet. The Consent Decree was developed in response to an enforcement action related to the Federal Clean Water Act. Objectives set forth in the Consent Decree are to eliminate unauthorized discharges from MSD separate sewer systems, combined sewer systems, and water quality treatment centers.

On April 15, 2009, an Integrated Overflow Abatement Plan (IOAP) was approved by EPA and MSD and was entered into federal court. The solutions identified in the IOAP include storage, treatment, conveyance/transport, and sewer separation projects. The IOAP states that the storage and conveyance system be designed such that, in a typical year, each existing CSO regulator structure does not overflow more than a specified number of times. A total of thirteen CSO storage basins were recommended as gray solutions. Preliminary construction costs and construction completion deadlines for each of the basins were developed as part of the IOAP. In order for compliance to be met, MSD must place each of the basins in service by the deadline specified in the IOAP.

PROJECT BACKGROUND

The IOAP originally defined the project as three separate CSO storage basins: 13th Street and Rowan Street CSO Basin; Story and Main CSO Basin; and Lexington and Payne CSO Basin. The following sections describe the basin projects, issues and challenges encountered during the design phases, and the decisions leading up to combining the three basins into the Ohio River Tunnel.

13th Street and Rowan Street CSO Basin

The 13th Street and Rowan Street CSO Basin (Rowan Basin) project was originally proposed as an off-line CSO conveyance and storage system devised as part of the IOAP to limit the number of CSO events for eleven regulator structures in a typical year. The Rowan Basin was to be located west of downtown Louisville's Central Business District, north of Rowan Street between 10th Street and 13th Street. The IOAP mandated that the Rowan Basin capture, convey, and store 4.4 million gallons (MG) of CSO volume. This requirement was increased to 11.2 MG after hydraulic modeling was performed by MSD's modeling consultant. The IOAP deadline for the Rowan Basin to be operationally complete is December 31, 2020.

Risk Assessment and Evaluation of Alternatives

On December 15, 2014, MSD retained a consulting engineer to design the Rowan Basin. As part of the preliminary design, MSD requested the engineer develop alternative conveyance routes and storage methods for evaluation of, including constructability, cost, and a variety of risks, including impacts to the community, including impacts to traffic flow, businesses, residents and tourism. Six alternatives were evaluated which included a CSO basin and open cut conveyance pipeline along Main Street, a CSO basin and microtunnel conveyance pipeline under Main Street, and four alternatives involving a deep rock storage and conveyance tunnel.

The locations of critical structures are of paramount interest to those who live and work in the affected neighborhoods. Political and civic leaders and local area activists played an important role in the execution of the project. Direct communication with key stakeholders and the public was necessary throughout design to avoid or mitigate these issues.

Significant potential social and community ramifications associated with the Rowan Basin project were addressed early and throughout the project. Consensus-building was critical to MSD's success in outreach efforts. Construction activities, duration, and impacts were described to the community through public meetings and engagements with key stakeholders and Louisville Metro Council members. MSD received input in these meetings regarding desired mitigation measures.

The surface of West Main Street would have been disrupted at five intersections if a conveyance pipeline were installed using near surface microtunneling construction methods. Open-cut jacking and receiving pits required for microtunnel installation would have significantly impacted vehicular and pedestrian traffic at each intersection and certain crosswalks.

If a conveyance pipeline were installed in West Main Street using open-cut construction, the public would have experienced severe disruption. Permit requirements would likely have stipulated

the open-cut installation be performed in segments. This would have limited disruption to a defined work zone, but could have increased construction duration and cost.

Rowan CSO Basin and Microtunnel Under Main Street Alternative

After a Key Decision Workshop, MSD chose to proceed with development of the CSO basin and microtunnel conveyance pipeline under Main Street alternative, as shown on Figure 1. A preliminary geotechnical investigation was performed, which included drilling three vertical soil borings. In addition, a Preliminary Environmental Assessment (PEA) was conducted to identify whether additional environmental assessments should be completed. The preliminary Opinion of Probable Construction Cost (OPCC) was updated for this alternative, and a preliminary project schedule was developed. Ten technical memoranda were prepared regarding the basin and microtunnel design.

Short Deep Rock CSO Tunnel Alternative

After preliminary technical memoranda for the Rowan CSO Basin were developed, MSD met with Louisville Metro Government and key stakeholders in downtown Louisville to discuss potential construction impacts and schedule for the proposed basin and microtunnel installation in Main Street. Louisville Metro expressed concern over the disruptions to businesses and access along Main Street that would occur and asked MSD to explore less disruptive alternatives.

In response to Louisville Metro's feedback, MSD requested the engineer update the OPCC for a deep rock CSO tunnel alternative developed during the conceptual design phase. The conceptual tunnel was approximately 2,600 feet long, 160 feet deep with a finished diameter of 28 feet. The OPCC was updated for both the Rowan CSO Basin and the deep rock CSO tunnel alternatives using newly obtained rock survey and historical information for existing utility and structure locations. The discovery

Figure 1. CSO Basin and microtunnel alternative

of previously unknown utilities and building foundations in the Main Street right-of-way required the microtunnel to be lowered by approximately 12 feet from the conceptual design. Lowering the microtunnel resulted in lowering the Rowan CSO Basin, which resulted in significantly increased rock excavation costs.

The updated OPCC for the deep rock CSO tunnel alternative was approximately 10 percent higher than the updated OPCC for the Rowan CSO Basin and microtunnel conveyance pipeline. Because the updated OPCCs were comparable and in order to minimize disruptive community impacts, MSD decided to change the project to a short, deep rock CSO tunnel, as shown on Figure 2. MSD directed the engineer to proceed with detailed design of the short CSO tunnel. Eighteen Technical Memoranda were

prepared along with geotechnical investigations for the short CSO tunnel.

Story and Main CSO Basin

The Story Avenue and Main Street CSO Basin project was originally proposed to be an off-line CSO conveyance and storage system developed as part of the IOAP. The Story and Main CSO Basin was to be located in the Butchertown neighborhood of Louisville, at approximately Franklin Street and Buchanan Avenue. The project site was located on the unprotected side of the floodwall, as shown on Figure 3. The IOAP requires the Story and Main CSO Basin to capture, convey, and store 8.3 MG of CSO volume. The IOAP deadline for the Story and Main CSO Basin to be operationally complete was December 31, 2020.

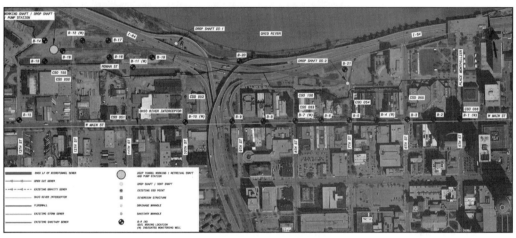

Figure 2. Short CSO tunnel

Figure 3. Drop shaft DS03 site location—previously Story and Main CSO Basin project site

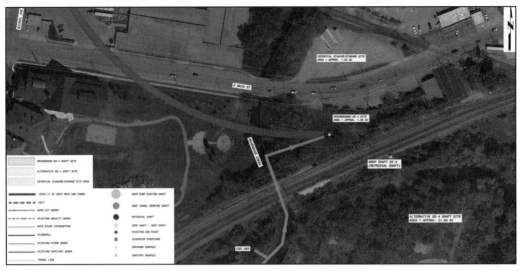

Figure 4. Drop shaft DS04/retrieval shaft location—previously Lexington and Payne CSO Basin project site

Added Scope from Stakeholder Feedback

The Story and Main CSO Basin was to be an above grade circular basin. The need to protect facilities on the river side of the floodwall, in addition to multiple scope items added in response to key stakeholder feedback, increased the OPCC by over 40 percent from the original budget cost. The proposed basin site is in an area that is experiencing significant redevelopment. Several of the influential stakeholders expressed concern over the appearance of an above grade basin. Due to the increased construction costs and operational concerns, MSD decided to investigate combining the Story and Main CSO Basin with the Rowan CSO Basin as a deep rock tunnel.

Lexington and Payne CSO Basin

The Lexington Road and Payne Street CSO Basin project was originally proposed to be an off-line CSO conveyance and storage system developed as part of the IOAP. The Lexington Road and Payne Street CSO Basin was to be located in the Irish Hills neighborhood of Louisville, at the site of a former metal recycling facility, as shown on Figure 4. The IOAP requires the Lexington and Payne CSO Basin to capture, convey, and store 13.7 MG of CSO volume. The IOAP deadline for the Lexington and Payne CSO Basin to be operationally complete was December 31, 2020.

Risk Assessment and Hazardous Materials on Proposed Site

The site selected for the Lexington and Payne CSO Basin was on a parcel that had been used as a metals recycling facility. A Phase I Environmental Assessment revealed widespread contamination of soils and groundwater at the site. In addition, initial negotiations with the property owner proved difficult to reach a selling price near fair market value. MSD and the engineer investigated using a site east of Beargrass Creek, north of the CSX railroad tracks and south of Main Street, for a tunnel boring machine (TBM) retrieval shaft and drop shaft for the Lexington and Payne CSO Interceptor, which previously would have conveyed CSO to the Lexington and Payne CSO Basin. The alternative site would mitigate risks associated with the original basin site.

OHIO RIVER TUNNEL

On September 30, 2016, MSD decided to extend the short deep rock tunnel to combine the three CSO basins into the Ohio River Tunnel (ORT). The ORT mitigates issues encountered during design on each of the basins. The tunnel significantly reduced the amount of construction in downtown Louisville, especially along Main Street. Technical issues associated with building an above grade circular basin on the unprotected side of the floodwall as well as concerns from influential stakeholders in the neighborhood were eliminated at the Story and Main Basin Site. The retrieval shaft location was moved from the Lexington and Payne Basin site, reducing risks associated with impacted soils.

In order to meet the IOAP deadline and allow adequate time for construction, the design completion deadline was established as the beginning of June 2017, leaving approximately eight months to complete design and permitting. Construction would need to begin no later than October 2017 to allow the tunnel and pump station to be operationally complete by December 31, 2020. Adherence to the design and construction schedules was critical.

Project Description

The ORT Project is an off-line CSO conveyance and storage system located in downtown Louisville. The project consists of a 37 MG capacity deep tunnel and an integrated deep pump station located at 13th Street and Rowan Street, designed to convey captured CSOs to the existing Ohio River Interceptor (ORI) sewer. The project area generally extends from 13th Street and Rowan Street to approximately 1200 East Main Street.

The key components in the ORT Tunnel and Shafts Package for construction are listed below:

- One Pump Station Shaft, 40-foot finished internal diameter through the overburden soil and rock.
- One Working Shaft, 40-foot finished internal diameter through the overburden soil and rock.
- Approximately 13,360 feet of 20-foot internal diameter main tunnel.
- Three drop shafts (DS01, DS02, and DS03 with integral vortex flow insert and vent piping.
- Approximately 225 feet of 8-foot internal diameter adit and deaeration chamber for drop shaft DS01 to the main tunnel in rock.
- Approximately 1,160 feet of 20-foot internal diameter tunnel bifurcation from the main tunnel to Drop Shaft DS02.
- Approximately 62 feet of 8-foot internal diameter adit and deaeration chamber for drop shaft DS03 to the main tunnel in rock.
- One Retrieval Shaft, 25-foot minimum internal diameter through the overburden soil and rock.
- One cascade type drop Shaft (DS04) constructed within the Retrieval Shaft following construction of the main tunnel.

The Rowan Pump Station, with submersible pumps for dewatering the ORT, will be constructed under a separate contract. Near surface consolidation sewers in the Butchertown neighborhood, the Irish Hills neighborhood, and downtown Louisville will be constructed under two additional contracts.

Project Geologic Setting

A geotechnical investigation was conducted by drilling a series of soil and rock borings along the proposed ORT alignment. Additional soil and rock information from the recently completed Ohio River Bridges project was also evaluated. The unconsolidated deposits and bedrock to a depth of approximately 300 feet below ground surface (bgs) are the most relevant to this project.

Soil

The overburden along the project alignment consists of man-made fill, alluvium, and glacial outwash deposits. Generally, the alluvium is 10 to 30 feet thick in the Ohio River Valley and commonly overlies glacial outwash deposits. Underlying glacial outwash deposits are approximately 50 to 130 feet thick, depending upon the location.

Bedrock

Devonian, Silurian, and Ordovician-age carbonate bedrock underlies most of Jefferson County, Kentucky. In descending order, these bedrock formations include the Devonian New Albany Shale; the Sellersburg and Jeffersonville Limestones; the Louisville Limestone; the Waldron Shale; the Laurel Dolomite; the Osgood and Brassfield Formations; and the Saluda Dolomite and Bardstown Member of the Drakes Formation. The Pump Station Shaft, Working Shaft, Retrieval Shaft, and drop shafts will encounter nearly all of these bedrock units. The ORT will be constructed within the Laurel Dolomite, the Osgood Formation, the Brassfield Formation, and the Drakes Formation.

Geologic Structure

Erosional processes stemming from the transport of glacially derived sediments have scoured and eroded the bedrock surface along the ORT alignment. As a result, relief in the bedrock surface widely ranges in elevation, as can be seen on Figure 5. Moving east to east along the project alignment, bedrock slopes downward from its highest point of El. 409 near the Pump Station Shaft, to its lowest point of El. 326 west of Slugger Field.

Previous Tunnel Construction Experience in Louisville

Two bedrock tunnels have been constructed in the eastern portion of the Louisville metropolitan area, near Prospect, Kentucky. These projects are the East End Crossing Bridge and Tunnel, and the Louisville Water Company (LWC) Phase II Riverbank Filtration (RBF) Tunnel. The locations of these tunnels are shown on Figure 6, along with the ORT Project alignment. These tunnel projects provided valuable institutional experience and lessons learned from rock tunneling construction in the area.

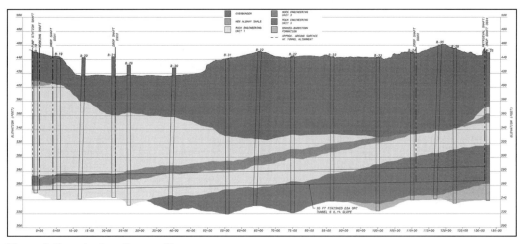

Figure 5. Tunnel subsurface profile

Figure 6. Previous tunnel construction experience in Louisville

East End Crossing Tunnel

The East End Crossing project was constructed to provide a highway connection between Clark County, Indiana, and Jefferson County, Kentucky. As part of the project, a 1,800-foot twin heading transportation tunnel was constructed. Mining began in early 2014, and the tunnel opened in December 2016.

The tunnels were constructed utilizing the New Austrian Tunneling Method (NATM). The overall excavation dimensions of each tunnel are approximately 70 feet in width and 50 feet in height. The tunnel is situated approximately 90 feet below grade

and was blasted through Silurian-age Louisville Limestone and Waldron Shale. The Waldron Shale, which was encountered along a section of the tunnel crown, was weakened during blasting. In September 2014 loosened bedrock slabs began to fall from the tunnel crown. The project was delayed for approximately two months while additional tunnel ground support systems were implemented. A series of 20-foot long rock anchors were drilled and epoxied into the tunnel crown. A membrane was installed along the entire length of both tunnels before a 16-inch thick cast-in-place concrete liner was installed.

Louisville Water Company RBF Tunnel

The LWC Phase II RBF Tunnel combined a deep bedrock tunnel with shallower horizontal collector wells capable of delivering up to 60 mgd of groundwater to the water treatment plant. The project included four collector wells and multiple well screen laterals to deliver water to the tunnel through drop shaft connections. The 7,750-foot long tunnel was mined through bedrock by a TBM. The average tunnel depth is 150 feet between the ground surface and the tunnel crown.

Natural gas was detected in one geotechnical boring completed as part of a study investigating the feasibility of extending the Phase II RBF tunnel. For the tunnel itself, natural gas was not encountered in any of the 43 geotechnical borings completed along the alignment.

The tunnel was excavated with a 12-foot diameter Robbins open gripper, main-beam rock TBM. Locomotives and muck cars were typically operated in the tunnel. A maximum single shift production of 84 feet and a maximum single day production of 144 feet were recorded.

Post-excavation grouting was required to mitigate the inflow of approximately 300 gallons per minute (gpm) from an unanticipated fault zone. In other areas, bolt holes intercepted the fault or fault splays and produced groundwater inflows of up to 35 gpm. Delays in mining occurred as it was necessary to grout these features. Some crown instability of thin-bedded limestone also occurred during mining, which required additional rock bolts and the installation of welded wire fabric.

Stakeholder Involvement

MSD and the engineer met with Louisville Metro Department of Public Works (Metro) to determine requirements for right-of-way encroachment, road closures, maintenance of traffic plans, pavement repair plans, working hours, and noise limits. The engineer incorporated feedback from Metro into the contract documents to ensure building and encroachment permit applications would be reviewed and approved in a timely manner.

MSD held public outreach meetings originally scheduled for the three CSO basins to present ORT design concepts, including tunnel alignment and structure locations. MSD and the engineer met with the Louisville Downtown Partnership (LDP), a nonprofit economic development agency, to identify and coordinate with key stakeholders who will be impacted by the construction of the ORT. LDP facilitated meetings with the Kentucky Arts Center, the Muhammad Ali Center, the Kentucky Science Center, and other key stakeholders in downtown Louisville.

Waterfront Development Corporation has identified the Working Shaft and Pump Station Shaft site for the Louisville Waterfront Park Phase IV. MSD and the engineer met with representatives from Metro and Waterfront Development Corporation to confirm locations and layouts for the shafts, pump station, vent piping, and final grading. MSD coordinated with Waterfront Development to create an acceptable building design that will be incorporated into Waterfront Park.

Permitting Requirements

Each permit and approval requires unique information to mitigate impacts. This section addresses federal, state, and local permits acquired during design, including regulatory agency and scheduling considerations. Additional permits obtained by the contractor were required during construction.

United States Army Corps of Engineers

MSD and the engineer met with the United States Army Corps of Engineers (USACE) to confirm which permits would be required and specifically what information would need to be included in each application. USACE facilitated a meeting with numerous potential divisions that could require permits. In addition, representatives from the USACE real estate group attended the meeting to discuss easement procedures for the portion of the tunnel alignment that crosses the McAlpine Locks and Dams parcel under USACE jurisdiction. It was determined that the ORT would need USACE approval for NEPA procedures, a Section 408 permit, and a Section 10 permit.

A Section 408 Levee Modification Permit was required to cross the floodwall with the proposed deep rock tunnel alignment. A monitoring plan and monitoring points on the floodwall at the Working Shaft site and the Drop Shaft DS03 site were required for approval. Additionally, because the ORT will connect the unprotected side of the floodwall to the protected side of the floodwall, a watertight, concrete flood barrier was integrated into the design of the Retrieval Shaft/Drop Shaft DS04. The watertight barrier will protect the dry side of the flood wall in case the Ohio River rises to the top of floodwall elevation and gates at the drop shafts fail.

The National Environmental Policy Act (NEPA) requires federal agencies, including the USACE, to assess the environmental effects of their proposed actions prior to making decisions. Using the NEPA process, the USACE evaluates the environmental and related social and economic effects of their proposed actions. Due to the tunnel alignment under the Ohio River, NEPA approval from USACE was required.

Section 10 of the Rivers and Harbors Act of 1899 requires USACE approval for any construction activities of structures in or under a navigable body of water. Because the ORT tunnel alignment crosses under the Ohio River for approximately 5,000 feet, a Section 10 permit was required. MSD and the engineer met with USACE early to determine what information would be required with the permit application. Approval of the Section 10 permit was contingent on NEPA approval and Section 408 approval.

Kentucky Transportation Cabinet and Federal Highway Administration

MSD and the engineer met numerous times with the Kentucky Transportation Cabinet (KYTC) and the Federal Highway Administration (FHWA) to discuss an Air Rights Agreement for alignment crossings of the ORT under elevated portions of Interstate 64, the Clark Memorial Bridge, and Interstate 65.

At the initial meeting, MSD and the engineer presented the horizontal alignment under the majority of the elevated portion of Interstate 64 adjacent to the Ohio River as shown on Figure 7. The intent was to keep the tunnel alignment in public right-of-way to minimize issues acquiring subsurface easements. KYTC expressed concerns about constructing and maintaining the infrastructure in state right-of-way. Even though the TBM construction would not influence existing infrastructure or impact potential future infrastructure, KYTC did not wish to establish a precedent by allowing the parallel installation of utilities in the right-of-way for significant distances. KYTC requested the alignment be modified to limit the length of tunnel under KYTC infrastructure. In addition to limiting construction in the right-of-way,

KYTC mandated that no drill and blast construction take place under KYTC infrastructure.

The engineer modified the horizontal tunnel alignment such that five TBM crossings under KYTC rights-of-way were required. The first KYTC crossing is under the elevated section of Interstate 64 near the 9th Street exit, north of the Working Shaft and Pump Station Shaft site. The second KYTC crossing is under the elevated section of Interstate 64 near 7th Street and River Road. The design of this crossing was modified from the previous drill and blast adit to a TBM tunnel bifurcation from the main tunnel to Drop Shaft DS02. The third KYTC crossing is under the U.S. Highway 31 Clark Memorial Bridge.

The fourth KYTC crossing is under the elevated section of Interstate 64 near Bingham Way and River Road. The fifth KYTC crossing is under the newly constructed Interstate 65 overpass at Witherspoon Street. KYTC and FHWA met with MSD and the engineer to provide review comments and request final modifications to the horizontal alignment, as shown on Figure 8.

After the horizontal alignment was established, KYTC expressed concern about the tunnel vertical alignment. The Waldron Shale formation was anticipated to be excavated in the crown of the tunnel for approximately the first 1,000 feet of mining. Due to previous tunneling experience in the Waldron Shale on the East End Bridge Tunnel, KYTC was not comfortable granting air rights below KYTC right-of-way without Type IV initial support (steel rib beams with lagging). MSD and the engineer responded to the feedback by lowering the invert of the Pump Station Shaft and the invert of the main tunnel by 20 feet. The revised tunnel subsurface profile is shown on Figure 9. Lowering the tunnel eliminated

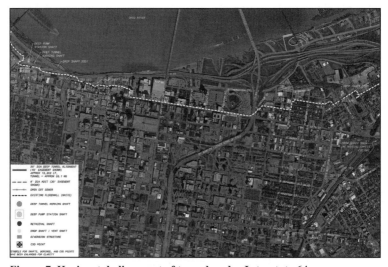

Figure 7. Horizontal alignment of tunnel under Interstate 64

Figure 8. Final horizontal alignment of tunnel

mining in the Waldron Shale, and actually provided construction float by decreasing initial support installation time.

Once the horizontal and vertical alignments were acceptable to KYTC and FHWA, the engineer created boundary descriptions for the Air Rights Agreement. The boundary descriptions do not limit KYTC to construct, develop, or otherwise use the rights-of-way above the tunnel; the boundaries are 25 feet above and below the tunnel springline. Similarly, the horizontal offsets for the descriptions are 25 feet on either side of the horizontal tunnel alignment centerline.

The engineer met with KYTC structural engineers in the field to determine monitoring instrument types and locations on and around KYTC structures along the tunnel alignment. The monitoring plan was submitted along with cross sections showing the tunnel horizontal and vertical alignment in relation to KYTC structures.

Kentucky Division of Water

MSD determined that groundwater from the tunnel during construction could not be discharged to the existing sanitary sewer system. A Kentucky Pollutant Discharge Elimination System (KPDES) permit will be required; however, the contractor will be the owner and operator of the water treatment and discharge system. MSD notified the Kentucky Division of Water (KDOW) that the contractor who would be submitting the permit application would need an expedited review process in order to avoid delays during construction.

In addition to the communication regarding the KPDES permit, MSD obtained a Notice of Intent (NOI) Sanitary Sewer General Construction Permit and a Floodway Construction permit from KDOW.

United States Fish and Wildlife Service

As part of the permitting process, MSD obtained an Endangered Species determination letter from the United States Fish and Wildlife Service. Due to the redominantly urban and industrial settings of the construction sites, review and approval of the permit application was straightforward.

DESIGN EXECUTION

Detailed design of the Ohio River Tunnel was completed in approximately eight months. The engineer allocated global resources in multiple office locations to allow more work to be accomplished on the project each day than would typically performed in an eight-hour work day and 40-hour work week.

Design execution required frequent communication both on a consistent and as-needed basis with the internal design team, the MSD design and operations team, key stakeholders, permit reviewers, and regulators. A rigorous quality control program that was established at the beginning of the design phase was followed and updated as design progressed.

The engineer developed a Preliminary Construction Schedule as part of the Basis of Design Report. The Ohio River Tunnel system was projected to require approximately 31 months to construct from Notice to Proceed until commissioning. The engineer recommended including at least six months of float in the construction schedule.

Near the end of the design phase, MSD and the engineer held a Value Management Workshop. The Value Management Workshop team recommended the design team attempt to find ways to add float to the construction schedule. The finished diameter for the Rowan Pump Station was reduced from 50 feet to 40 feet by the Rowan Pump Station engineer. The change in shaft sizes presented an opportunity to

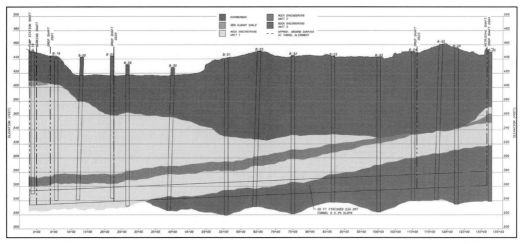

Figure 9. Lowered tunnel subsurface profile

increase float in the project construction schedule. Adding float to the construction schedule was accomplished by lowering the invert elevation of the Pump Station Shaft, the Working Shaft, and the downstream end of the tunnel by 20 feet. Lowering the Pump Station Shaft and downstream end of the main tunnel recovered 23 calendar days of float in the construction schedule and allowed the tunnel and shafts contractor to turn over the Pump Station Shaft and Working Shaft earlier than anticipated. Lowering the tunnel recovered float by eliminating excavation in the Waldron Shale, which significantly reduces the amount of Type IV initial support required and increases anticipated TBM advance rates. Lowering the Pump Station Shaft and downstream end of the main tunnel increases the tunnel slope, which eliminates the need for a cunette, and further reduces construction duration and cost. As stated previously, the lower tunnel design allowed MSD to obtain the KYTC Air Rights Agreement required to construct and operate the tunnel in the interstate rights-of-way in a timely manner.

CONCLUSION

The Ohio River Tunnel (ORT) project combines the IOAP storage volume requirements from three previously identified CSO storage basins: the Rowan CSO Basin; the Story and Main CSO Basin; and the Lexington and Payne CSO Basin. Detailed design of the project was completed in approximately eight months to accommodate construction completion before the IOAP deadline of December 31, 2020.

Due to the compressed design schedule, the design team recognized the necessity of early and consistent communication with key stakeholders. Approvals were obtained by closely coordinating

and quickly responding to feedback, allowing the engineer to manage scope changes an accelerated design schedule throughout the design process.

Allocating essential global resources in multiple office locations allowed the design team to consolidate efforts required to complete the design deliverables.

A rigorous quality control program was essential to delivering bid-ready contract documents on time. The heightened sense of urgency due to the schedule likely contributed to timely and constructive feedback from technical experts, both internal to the design team and at regulatory agencies. Previous tunneling experience and familiarity with the regional and local geology contributed to the design team's ability to confidently make design recommendations.

REFERENCES

Black & Veatch, Rowan CSO Basin Planning Evaluation Conceptual Engineering Report—Final Draft. August 2015.

Black & Veatch, Ohio River Tunnel Basis of Design Report—Revised Draft. January 2017.

Black & Veatch, Ohio River Tunnel, Tunnel and Shafts Package Geotechnical Baseline Report. May 2017.

Hazen and Sawyer, Lexington and Payne CSO Basin and Interceptor Draft 60% Design Report. August 1, 2016.

HDR Engineering, Inc., Story and Main CSO Storage Basin 60% Draft Technical Memo. August 1, 2016.

Louisville and Jefferson County Metropolitan Sewer District, Integrated Overflow Abatement Plan, September 30, 2009. 2012 Modification.

Narragansett Bay Commission Phase III CSO Abatement Program—Pawtucket Tunnel

Todd Moline
MWH Constructors

Christopher Feeney
Stantec

ABSTRACT: The Narragansett Bay Commission (NBC) operates collection and treatment facilities in two service areas in Rhode Island. In 1992, NBC began a three-phase Combined Sewer Overflow (CSO) Abatement Program to reduce untreated CSO discharges to Narragansett Bay. Phases I and II of the Program included construction of a 16,500-ft, 26-ft diameter rock storage tunnel and cavern pump station. Phase III will include construction of a second deep, rock storage tunnel (13,000-ft, 28-ft diameter), drop shafts, and a cavern pump station. This paper presents planning and optimization of Phase III construction utilizing data and case studies from Phases I and II construction.

INTRODUCTION AND BACKGROUND

The Narragansett Bay Commission (NBC) embarked on a three-phase Combined Sewer Overflow (CSO) control program in 1992, aimed at lowering annual CSO volumes and reducing annual shellfish bed closures in Narragansett Bay in accordance with a 1992 Consent Agreement with the Rhode Island Department of Environmental Management (RIDEM). NBC established the CSO Control program with the goal of reducing annual CSO volumes by 98 percent, and achieving an 80 percent reduction in shellfish bed closures. Phases I and II of the program, which focused on the Fields Point Service Area (FPSA) in Providence, Rhode Island were completed in 2008 and 2015, respectively. The program to date has succeeded in lowering annual CSO volumes and reducing annual shellfish bed closures to levels complying with the terms of the Consent Agreement.

The largest facility in Phase I was a deep rock storage tunnel in Providence designed to store CSO volumes during wet weather events for subsequent pump out and treatment at the Field's Point Wastewater Treatment Facility (FPWWTF). Phase I was completed in 2008 at a cost of $360M. Phase II, which began construction in 2011, consisted of interceptors to connect additional outfalls to the Providence Tunnel plus several sewer separation projects. The final component of Phase II was completed in 2015 at a cost of $197M.

The third and final phase of the program (i.e., Phase III), which commenced in 2016, is focused primarily on the Bucklin Point Service Area (BPSA) in the communities of Pawtucket and Central Falls (see Figure 1). Because of the projected cost of Phase III and its impact on sewer rates, NBC has elected to reevaluate the original Phase III plan to optimize the proposed facilities and evaluate the affordability of the program. The proposed Phase III implementation strategy, while prioritizing water quality benefits, also limits the financial impact on NBC rate payers. Stantec, on behalf of NBC, incorporated an Integrated Planning Framework (IPF) and affordability analysis to evaluate the financial impact of the program. The communities of Providence, Pawtucket and Central Fall have median household incomes that are below the state median. Phase III also incorporates construction experience gained from the first two phases of the CSO program and input from a stakeholders group engaged during the reevaluation process.

The reevaluation process effectively extended the implementation schedule, dividing Phase III into four (4) sequential sub-phases (i.e., Phase IIIA, IIIB, IIIC, and IIID). The extended schedule for Phase III reduces the overall financial burden of the program on rate payers and provides flexibility to address

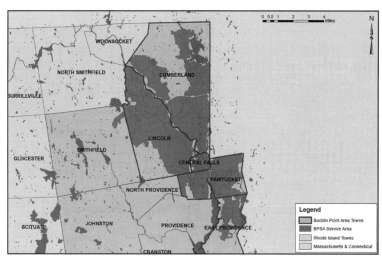

Figure 1. Bucklin Point service area

Table 1. Capital cost and annual volume reduction for amended recommended plan by phase (typical year simulation)

Sub-Phase	Refined Alternative 5 ($Millions - 2018)	Annual Overflow Volume Reduction (Typical Year) (%)
IIIA	$476.51 M	82%
IIIB	$31.17 M	7%
IIIC	$164.45 M	4%
IIID	$83.43 M	N/A—FPSA Outfalls
Total	**$755.56 M**	**93%**

changes in economic conditions and prioritize water quality improvements. The final recommended plan was optimized based upon hydraulic modeling, constructability analysis, and operating efficiencies. The overflow capture efficiency was optimized; resulting in the elimination of CSO interceptors, reduction in facility dimensions, and a drop shaft. The net result was a cost reduction to the Program of $60M.

Table 1 provides the conceptual opinion of construction cost for each sub-phase and estimated annual reduction in CSO volumes. The most significant water quality benefit is achieved following implementation of Phase IIIA (i.e., 82% reduction of total CSO volume).

PREVIOUS PHASES—PHASES I AND II

Phase I of the NBC CSO Control Program was constructed from 2000 to 2008. Several papers were published on the construction of the Providence Tunnel (Castro et al. 2007; Bradshaw et al. 2007; Hughes et al. 2008; and Kaplin et al., 2009). The project was awarded the 2009 Underground Constructors of

SME Project of the Year. The primary components were the construction of a 16,500-ft, 26-ft diameter storage tunnel in bedrock (Providence Tunnel), and an adjacent underground tunnel pump station also in bedrock. In addition to the storage tunnel and tunnel pump station, seven drop shafts (each with an accompanying vent shaft) were constructed along the tunnel alignment to direct combined sewage overflows to the tunnel for storage and subsequent treatment. Near surface facilities at each drop and vent shaft location consisted of one to two diversion structures to divert flows from existing CSOs, consolidation conduits, a gate and screening structure, an approach channel, and a vortex generator at the top of each drop shaft. The alignment of the Phase I tunnel and related facilities is shown on Figure 2. Construction methodologies utilized in the previous phases included large diameter tunnel construction, drill and blast tunneling, ground freezing, jet grouting, secant piles, raise bore shaft construction, and microtunneling.

Phase II was constructed from 2011 to 2015. It consisted of an additional drop shaft and vent shaft; a 1,800-ft, 8-ft diameter conveyance tunnel; additional near surface facilities; and over 22,000-ft of interceptor sewers. The objective of Phase II was to divert overflows from outfalls located along the Woonasquatucket and Seekonk Rivers in Providence, RI. Overflows were diverted from the Woonasquatucket River to the Phase I storage tunnel via a new interceptor and a new drop shaft constructed during Phase II. Overflows were diverted from outfalls along the Seekonk River to the storage tunnel via an additional new interceptor and an existing drop shaft constructed during Phase I.

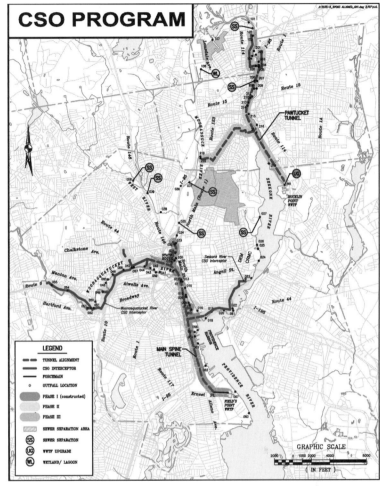

Figure 2. NBC CSO Program (Source: NBC Conceptual Design Report 1998)

The majority of the interceptors were installed by microtunneling.

Providence Tunnel

The tunnel was mined through a bedrock using a 30-ft diameter, single-shield tunnel boring machine (TBM). The tunnel contractor elected to use a two-pass liner system. The tunnel was lined immediately with precast concrete segments after mining. Upon the completion of mining the entire tunnel, a final cast-in-place concrete liner was installed using a slip form. The inner, finished diameter of the tunnel after the installation of the cast-in-place liner was 26-ft. After initial startup of the TBM, production rates had a mean value of about 36 ft per day and a median value of about 42 ft per day. The highest recorded production rate was 76 ft per day. The cavern, which is about 150-ft long, by 60-ft wide by 60-ft tall, was

constructed using drill and blast methods with permanent support provided by rock bolts and shotcrete.

Shafts were constructed for the storage tunnel, the tunnel pump station cavern, and the conveyance tunnel. The storage tunnel shafts consisted of: 1) a 50-ft diameter TBM launch shaft, which upon completion of the tunnel was converted to 26-ft diameter access shaft containing a screening facility and sump for the pump station; and 2) a 50-ft diameter TBM retrieval shaft, which became a 26-ft diameter access and ventilation shaft. A 37-ft diameter work shaft was constructed for the tunnel pump station cavern, which became a 32-ft diameter access shaft to the pump station for an elevator, stairs, HVAC, piping, utilities, and access. A second 11-ft diameter equipment access shaft was constructed to facilitate removal of large equipment and materials in the future.

The tunnel and cavern shafts ranged in depths from 240-ft to 290-ft from the ground surface. The shafts were excavated through approximately 165-ft of overburden soil and 80-ft to 120-ft of bedrock. Ground freezing was used as the means of temporary ground support during excavation of the overburden soils in all of the shafts. After excavating overburden, reinforced concrete shaft liners were installed from the bottom to the surface using a continuous slip form. Overburden soils consisted of urban fill underlain by alluvial and glacial deposits. Each shaft was advanced to its final depth through bedrock using drill and blast methods. Rock bolts and shotcrete provided support to the shaft walls in rock. Permanent reinforced concrete shaft liners were installed later for the final shaft dimensions.

An 8-ft diameter adit tunnel and a deaeration chambers was constructed from the storage tunnel to each drop and vent shaft using drill and blast methods. The section of adit that connected to the drop and vent shafts had an enlarged diameter to serve as a deaeration chamber. The lengths of the adits ranged from 112 ft to 1,800 ft. The mining of each adit and deaeration chamber began from within the storage tunnel and advanced to the bottom of the drop and vent shaft.

The seven drop shafts of Phase I ranged in finished diameter from 5 ft to 9 ft. The design utilized a vortex generator at the top of each drop shaft to convey flow from the surface to the tunnel. The vent shafts were 2-ft diameter. In all but one case, the drop and vent shafts were constructed under separate contracts by the contractors constructing near surface facilities and consolidation conduits. The drop and vent shafts were installed from the ground surface using techniques typically for installation of large-diameter foundation shafts. A steel casing was advanced through the overburden as temporary support while the soil was excavated within using a clamshell. The type of rig used to install the steel casing varied. When the top of bedrock was encountered, the casing was seated into the rock and the rock was mechanically broken, and excavated from the ground surface. The contract documents specified the use of a cast-in-place reinforced concrete shaft liner. However, all contractors requested and were allowed to substitute precast concrete cylinder pipe or high density polyethylene (HDPE) pipe as the permanent shaft liner. The liner was installed within the temporary steel casing and opening in the bedrock, and the annulus between the liner and the casing or bedrock filled with grout.

The drop shaft and vent shaft closest to the TBM launch shaft were installed by the tunnel contractor, using raise bore techniques. Upon the completion of the conveyance tunnel, a drop shaft and vent shaft made from precast reinforced concrete pipe was installed in the work shaft and the annulus backfilled with controlled low-strength material.

PREVIOUS PHASE SUCCESS AND CHALLENGES

The Phase III recommended solution consists of the same type of underground solution (i.e., deep rock tunnel, tunnel pump station, drop/vent shafts, adit tunnels, near surface structures, and consolidation conduit) in the same geologic formations. The current design team has a significant benefit of previous subsurface investigations, designs, construction data, ground conditions, production rates, and published papers on the Providence tunnel and facilities. The previous phases' success and challenges will be incorporated into the planning and design of Phase III. The geotechnical field investigation program has been designed to confirm and supplement subsurface data collected during previous construction phases and form historical regional geological data.

Published papers on construction of the Phase I main tunnel and shafts, and the Fields Point Tunnel Pump Station have been reviewed for information on ground conditions and ground behavior in response to excavation (Castro et al. 2007; Bradshaw et al. 2007; Hughes et al. 2008; and Kaplin et al., 2009).

The following ground behaviors excavated during Phases I and II are of particular interest for design of the Pawtucket tunnel:

- A high permeability seam of cobble- and boulder-size material in the glacial till deposit near the top of bedrock was a source of higher-than-anticipated groundwater velocity that delayed closure of a freeze wall for a shaft support system.
- Ground water previously contaminated by petroleum products delayed closure of a freeze wall for a shaft support system.
- Top of rock included numerous sand-filled fractures that produced heavy inflows of groundwater, which made it difficult to effectively seat the shaft support systems into rock.
- Blasting rock in the cavern resulted in considerable over breakage.
- Artificial fill mapped at several drop shafts and was noted to be highly variable in composition.
- Elevated levels of naturally occurring arsenic in tunnel muck impacted disposal costs.
- Glaciolacustrine deposits composed of fine sand and silt were encountered at or below the bottom of deep excavations (up to approximately 35 feet deep) during the installation of

near surface structures. These deposits exhibited zero- to low-strength behavior when subjected to high pore pressures induced by high groundwater exit gradients at the base of excavations, or to external vibrations.

PHASE III—PAWTUCKET TUNNEL

The Pawtucket Tunnel is the largest and most complex component of the overall Phase III plan. Consequently, Phase IIIA is expected to require three years for the design phase and five years for the construction phase. All of the other sub-phases are expected to require two years for design and three years for construction.

Phase IIIA provides the most significant water quality benefit in terms of CSO volume reduction (see Table 1). It is also the sub-phase with the highest capital cost and most significant impact to NBC household affordability. Phase IIIA includes the design and construction of the Pawtucket Tunnel, tunnel pump station, drop shafts, gate and screen structures (GSS), consolidation conduits, regulator modifications, trash racks at defined outfalls, and green stormwater infrastructure projects. The Pawtucket Tunnel will provide a storage volume of 58.6 MG to eliminate overflows from the 3-month design storm for all BPSA overflows. The Pawtucket Tunnel and the other Phase III components are included in Figure 3 which shows their relative locations to Phase I and II. During the optimization, the tunnel alignment was shifted to the eastern alignment to avoid a river crossing at the wider reaches of the Seekonk River. In addition, the design team is evaluating a shaft style tunnel pump station as an alternative to the previously constructed cavern style pump station. Comparisons of construction risk, cost, impacts to cavern design due to orientation of principle rock stresses, hydraulic efficiencies, and life cycle cost are being performed.

The Pawtucket Tunnel consists of the following:

- Deep rock tunnel: TBM driven tunnel 150-ft to 200-ft below grade extending north of the Bucklin Point Wastewater Treatment Facility (BPWWTF) in East Providence to Front Street in Pawtucket, RI. Figure 3 shows the alignment orientated in a north-south direction along the eastern banks of the Seekonk and Blackstone Rivers.
- Tunnel dimensions: Approximately 13,000-ft, 28-ft internal diameter for a total volume of 58.6 MG.
- Launch and Receiving work shafts: 50-ft diameter shafts at each end of the tunnel to support mining. The purpose of the two work shafts is to stage tunnel construction to launch and remove the tunnel boring machine

(TBM). Following construction, the launch shaft will be utilized as a tunnel screenings/sump location to protect the tunnel pump station. The receiving shaft will be utilized for ventilation, tunnel odor control, and drop shaft for an upstream outfall.

- Three drop shafts: 6-ft to 8-ft ID, 150-t to 175-ft deep. Drop shafts will convey flow from outfall locations to the tunnel for storage.
- Tunnel pumping station:
  - Utility Shaft—32-ft diameter, 260-ft deep
  - Access Shaft—32-ft diameter, 260-ft deep
  - Pump Cavern—60-ft wide by 60-ft high by 120-ft long
- Tunnel dewatering pumps: Two-stage pumping operation with four (4) 10 to 12 MGD pump pairs. Firm capacity is defined as three (3) pump pairs in operation with one pump pair on standby. Pump capacity will be evaluated further in design to determine the appropriate balance between dewatering time and available secondary capacity at treatment plant. An on-going evaluation is being conducted to compare cost, construction risk, and life-cycle cost of a shaft style pump station versus a cavern pump station.
- Consolidation conduits: Several consolidation conduits (48-inch to 72-inch ID, approximate length 4,700-ft) to convey flow from nearby outfalls to consolidated drop shaft locations.
- Regulator modifications: Regulator modifications for several outfalls along the alignment.

Phases IIIB and IIIC address the remaining outfalls in the BPSA; Phase IIID includes sewer separation and CSO interceptor to address the remaining outfalls in the FPSA in Providence. The largest remaining project element is included in Phase IIIC, which is a 7,775-ft, 10-ft diameter conveyance tunnel connecting a large outfall (i.e., OF-220) to the Pawtucket tunnel. The location of the conveyance tunnel is also shown on Figure 3. The conceptual design for the stub tunnel includes a deep rock tunnel, 70-ft to 200-ft below grade and a 6-ft to 8-ft diameter drop shaft.

Area Geology

As previously noted, the geologic formation is consistent with the conditions encountered in previous phases. The depth to bedrock is reduced as the alignment progresses to the north, where rock ranges from 12 feet to 15 feet deep to visible at the surface. Along the tunnel alignment, bedrock is overlain by varied landforms and their associated surficial deposits: artificial fill, river terraces, kame terraces, outwash plains, and ground moraines (see Figure 4). Existing

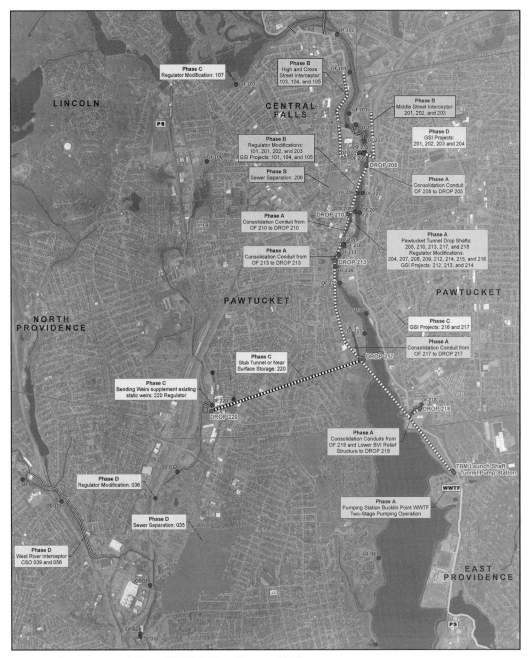

Figure 3. Recommended Phase III CSO Plan (Stantec 2017)

information regarding the surficial stratigraphy of the Providence area are briefly described below from most recent to earliest (Stantec/Pare 2016):

- Artificial Fill: Areas filled for construction and waste disposal. Artificial fill is present at

several drop shaft locations along the alignment composed of varying material and placement methods.

- River Terraces: This unit is mapped as a medium to coarse sand, with thin beds of gravel occasionally found. It could be

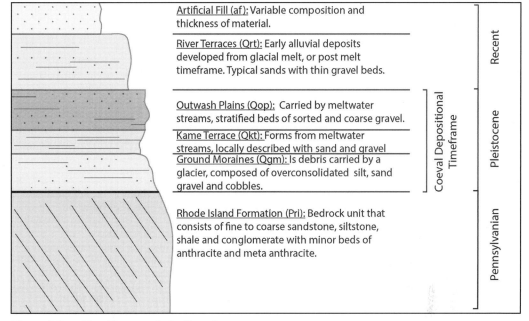

Figure 4. Local landforms and stratigraphy

considered an early alluvial unit, deposited during melting or post glacial timeframe.

- Kame Terraces: Kame deposit forms are described as being composed of irregular shaped mounds of sand and gravel.
- Outwash Plains: Locally, this deposit is described as a sorted sand with locally coarse gravel beds and can also be called glaciofluvial.
- Ground Moraines: Formed from the accumulation of debris carried by a glacier, and is typically composed of overconsolidated silt, sand, gravel and occasional cobbles and boulders. This widespread deposit is generally no more than 15 ft thick, but can be more. Locally, this unit is described as being a relatively thin layer of till with lenses of sorted material, and typically is resting on bedrock.

The bedrock along the alignment consists of gray to black, fine- to coarse-grained sandstone, siltstone, shale, and conglomerate, with minor beds of anthracite and meta-anthracite. The formation consists of five primary lithologies (Haley & Aldrich, 1999):

- Conglomerate: Gray, hard to very hard with a primary gravel size of fine gravel or larger. Clasts lithologies include quartzite, granite and other miscellaneous rock types. The

matrix appears to be a quartz and clay cement (sericite and chlorite).

- Sandstone: Gray, moderately hard to very hard with coarse to fine sand clasts. The matrix is cemented by quartz, calcite and clay (sericite and chlorite).
- Siltstone: Gray to black, moderately hard to hard, with silt grains dominating. This rock type constituted almost 15% of the core retrieved for the Providence tunnel alignment.
- Shale: Dark gray to black, soft to hard, with clayey grains dominating.
- Anthracite Coal: Black, generally medium hard to hard with very soft zones. This material is generally composed of carbon thinly laminated with ash. The coal contains too much ash (i.e., noncombustible material) to technically qualify as coal according to standard definition.

All rock types encountered were found with interspersed graphite (i.e., graphitic shale). Additionally, quartz veins up to 3-ft in thickness were observed in the core samples obtained along the Phase I Main Spine tunnel alignment, three miles southwest of the current site (Haley & Aldrich, 1999). Bedrock at the base of the Providence Tunnel, three miles southwest of the current site, is predominantly sandstone, shale, and siltstone with lesser amounts of conglomerate and coal.

Rock outcrops are visible in the river channel as it narrows to the limits of tidally influenced Seekonk River. Rock outcroppings have been mapped along the alignment. Areas of rock outcrops are visible at the upper reaches of the Seekonk River at the confluence of the Blackstone River. Rock is visible at the historic Slater Mill Dam to the Main Street Dam. Limited subsurface data is available and aquifer elevations, weathering depth and degree of weathering in the rock, rock fracturing characterization are unknown at this time. Geotechnical investigations for Phase IIIA are on-going and include rock cores, geotechnical laboratory testing, borehole wireline geophysics, and seismic refraction.

Project Delivery

A major objective of Phase III is to increase the focus on risk and value management. The goal is to reduce unanticipated and unnecessary financial impacts to the Program. The underlying program objectives and guiding design criteria have remained unchanged. NBC is currently evaluating the proposed project delivery method for the tunnel and pump station cavern construction, comparing traditional design-bid-build (DBB) to fixed-price design build (DB). An objective of the planning process is to identify the impacts to cost, schedule, and risk allocation for each project delivery method. Stantec has utilized the following approaches to determine the potential impacts of using DB:

1. Developed an updated program schedule, opinion of probable construction costs (OPCC), and projected cash flow assuming the use of the design-bid-build (DBB) method.
2. Reviewed and revised as necessary: the DBB program schedule, OPCC, and projected cash flow assuming the use of DB.
3. Studied effects of land acquisition, permit, and procurement requirements on the program schedule when using alternative project delivery methods.
4. Performed conceptual risk assessments of each project delivery method.
5. Performed a conceptual value assessments of the proposed alternative project delivery methods.

Two conceptual risk assessments of the Pawtucket Tunnel were performed, assuming the use of DBB and DB. Selected items of risk pertaining to the tunnel were identified, and the risk levels associated with each item's potential impact on cost and/or schedule were evaluated. Most of the identified risks are typically associated with an award to a contractor based solely on price without contractor engagement during design.

Potential significant risks of DB are bids exceeding OPCCs due to over-allocation of risks to the design-builder, impacts of delays related to the permitting process, and differing site conditions. The risks of DB are primarily associated with procurement and preconstruction activities. Typically, less risk is associated with construction because the DB has direct responsibility for design.

The following are benefits attributed to the DB process:

- Potential for compression of schedule because of the ability of the design-builder to phase the design and construction.
- Selection of the design-builder is based on both cost and qualifications.
- Design-builder has sole responsibility for design and construction of both temporary and permanent facilities.
- Cost of design and construction is defined at contract award.
- Cost is determined through a competitive process.
- Schedule is fixed at contract award.
- Public acceptance tends to be positive with a lump-sum contract award.

The following are potential negatives attributed to the DB process:

- Potential reduced lack of interest by DB firms due to high proposal preparation costs.
- Procurement process takes substantially more time compared to DBB (although typically offset by overall time savings of the DB process).
- Reduced Owner involvement in final design.
- Owner still liable for ground conditions that exceed thresholds defined in GBR.
- Potential conflict with standard approach to regulatory permitting.

CONCLUSION

The Pawtucket Tunnel is the largest, most complex element of Phase III. The deep rock tunnel has a storage capacity of 58.6 MG to reduce overflows and protect water quality. The Phase III Program has been reevaluated and optimized to meet the regulatory framework outlined in the Consent Agreement with RIDEM. An IPF methodology was utilized to extend the implementation schedule to reduce the overall economic impact on rate payers. The ground conditions and geologic formations encountered during previous underground construction experience has been used to focus the geotechnical investigation work plan to provide a comprehensive record of ground conditions to designers and constructors. Design implementation incorporates an increased focus on value and risk management. Alterative project delivery systems are being evaluated to mitigate risk and increase price certainty. The optimization process successfully reduced the cost of the Program by approximately $60M and reduced the economic impact to rate payers with an extended implementation schedule.

The current program schedule shows the Pawtucket Tunnel project undergoing design through spring of 2021, with contractor procurement starting immediately thereafter. If the project is procured using the design-build project delivery method, requests for qualifications are anticipated to be issued in late 2018 to early 2019.

ACKNOWLEDGMENT

The authors would like to thank the leadership and support of the Narragansett Bay Commission and Stantec for permission to publish this paper. Special appreciation is due to Raymond Marshall, Richard Bernier, and Thomas Brueckner of NBC; and Mathew Travers, Melissa Carter, Sean Searles, and Anthony Accardi of Stantec. The authors also thank Kathryn Kelly of Narragansett Bay Commission for her contribution to this paper.

REFERENCES

Castro, R.C., G. Hughes, F. Vincent and P.H. Albert. 2007. "Drop Shafts for Narragansett Bay Commission CSO Abatement Program, Providence, Rhode Island." Rapid Excavation and Tunneling Conference, Proceedings. pp. 1047–1057. Littleton, CO. SME.

Bradshaw, A.S., H.J Miller, and C.D.P Baxter. 2007. "A Case Study of Construction-Related Ground Movements in Providence Silt." Proceedings of the FMGM 2007 Conference, ASCE, Boston, Massachusetts.

Hughes, G., J. Kaplin, I. Halim, and P.H. Albert. 2008. "Design and Construction of the Fields Point Tunnel Pump Station for the Narragansett Bay Commission CSO Abatement Program, Providence, Rhode Island." North American Tunneling Conference, Proceedings. pp. 824–833. Littleton, CO. SME.

Kaplin, J.L., J.P. Peterson, and P.H. Albert. 2009. "Underground Construction for a Combined Sewer Overflow System in Providence, Rhode Island." Rapid Excavation and Tunneling Conference, Proceedings. pp. 22–34. Littleton, CO. SME.

Haley & Aldrich. 1999. Geotechnical Design Memorandum; Main Spine Tunnel, Connecting Adits and Shafts. Vol. I of III of the Geotechnical Preliminary Design Report; Phase I—Preliminary Design. CSO Control Facilities Program, Providence, Rhode Island.

Stantec/Pare. 2016. Preliminary Geologic and Geotechnical Conditions for the Pawtucket Tunnel. Technical Memorandum prepared for Narragansett Bay Commission Phase III CSO Program.

Ship Canal Water Quality Project: Meeting Federal Requirements

Shannon M. Goff and Gregg W. Davidson
McMillen Jacobs Associates

Dylan Menes
Seattle Public Utilities

ABSTRACT: The Ship Canal Water Quality Project will control combined sewer overflows (CSOs) from seven basins in Seattle, WA. Both the City of Seattle and King County have federal EPA Consent Decrees with which this project must comply by 2025. Following an extensive options evaluation, the preferred option selected jointly by both agencies was to build a tunnel to convey and store excess flows from their associated basins. Design of the project's tunnel and pump station elements commenced in 2015. This paper presents the development of the Storage Tunnel design to date, some of the challenges encountered during the design phase, an overview of the coordination between the various design-bid-build contracts and the current project status and schedule.

INTRODUCTION

The Ship Canal Water Quality Project (SCWQP) will provide offline storage of combined stormwater and wastewater in a Storage Tunnel. The tunnel is to be constructed between the Ballard and Wallingford neighborhood combined sewer overflow (CSO) areas on the north side of the Lake Washington Ship Canal (Ship Canal). Both the City of Seattle (City) and King County (County) have federal EPA Consent Decrees. The SCWQP will comply with their decrees by reducing CSO frequencies to no more than one per year over a 20-year moving average from seven basins. The SCWQP was selected as the preferred option for a joint project between the two agencies through a triple-bottom-line analysis that considered economic, social, environmental, and operational impacts of the joint project. Utilizing a shared Storage Tunnel and pump station was found to be a cost-effective way to meet the project's objectives and would result in fewer impacts to communities in multiple neighborhoods and have more operational flexibility than other options. The project is being delivered via a Joint Project Agreement (JPA) between the two agencies.

PROJECT OVERVIEW

The main components of the SCWQP include the Storage Tunnel and appurtenances, conveyance facilities to convey City and County CSO flows into the tunnel, and a tunnel effluent pump station and force main to drain flows from the tunnel. Figure 1 provides an overview of the SCWQP location and facilities.

After the SCWQP is constructed and operating, CSOs will occur only during extreme storm events (no more than once per year per outfall) when the capacity of the tunnel is exceeded. During less extreme events, stored flows will drain from the tunnel to the County's West Point Treatment Plant for treatment after the rainfall event ends and/or conveyance capacity is available.

The SCWQP program consists of five construction contracts that will be packaged and advertised in stages between 2018 and 2023 to meet the Consent Decrees. The major contracts, in order of construction, include the Ballard Early Works Package, an effluent pipe from the pump station that will be part of a Seattle Department of Transportation (SDOT) bike trail project, the Storage Tunnel, the Tunnel Effluent Pump Station (TEPS), and the Wallingford and Ballard Conveyance packages. The SCWQP includes the following elements of work:

- 29 million-gallon Storage Tunnel (14,000-foot-long, 18-foot 10-inch ID), excavated by pressurized-face tunnel boring machine
- Microtunnel beneath the Ship Canal (600-foot-long, 96-inch ID, curved alignment)
- Five diversion structures for diverting combined sewage away from existing CSO outfalls and into the Storage Tunnel
- Five drop structures to convey combined sewage from the surface into the Storage Tunnel

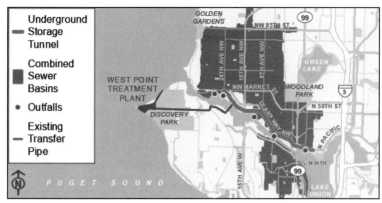

Source: https://www.seattle.gov/Util/EnvironmentConservation/Projects/
ShipCanalWaterQuality/index.htm

Figure 1. Project overview

- Ancillary facilities including mechanical/electrical vaults, odor control rooms, and flow meter vaults
- 48 MGD Tunnel Effluent Pump Station (TEPS) located at the west end of the Storage Tunnel
- Conveyance facilities including:
  - 2,400 linear feet of gravity sewer lines ranging from 24 to 72 inches in diameter
  - 1,900 linear feet of 48-inch-diameter force main
- Site remediation of contaminated tunnel shaft sites
- Improvements to an existing pedestrian pier for use in barging of tunnel spoils
- Rehabilitation of an outfall discharge pipe in Ballard
- Utilities relocation and improvements throughout the project area
- Bike improvements in coordination with the SDOT Burke-Gilman Missing Link project

The tunnel horizontal alignment has been located within the street right-of-way to the largest extent practicable. A Tunnel Effluent Pump Station (TEPS) will be located at the tunnel boring machine (TBM) launch shaft (west end of the tunnel, or West Shaft). Five shafts in total will be constructed along the alignment for connecting to diversion and drop structures, and a microtunnel will be bored beneath the Ship Canal between the Fremont Shaft on the north side of the Ship Canal and the 3rd Avenue W Shaft on the south side of the canal.

REGULATORY REQUIREMENTS

The federal Clean Water Act (CWA) requires express authorization for the discharge of any pollutant from a point source into navigable waters of the United States. To meet this discharge authorization requirement, the CWA established the National Pollutant Discharge Elimination System (NPDES) program. The Washington State Department of Ecology's regulations govern individual NPDES permits to meet water quality criteria, and state CSO control law requires local governments to develop reasonable plans and compliance schedules to achieve the "greatest reasonable reduction" of CSOs at the earliest possible date. State regulations indicate that "the greatest reasonable reduction" means control of each permitted CSO outfall in such a way that no more than one untreated discharge may occur per year using a 20-year moving average to assess compliance. The project must complete construction and have operational control by the end of 2025.

PROJECT DEVELOPMENT

The SCWQP was selected as the recommended option by the City and County for controlling CSO basins in the City's Ballard CSO area (Outfalls 150, 151, and 152), Fremont area (Outfall 174), and Wallingford area (Outfall 147). It also will control the County's overflow locations at 3rd Avenue West (DSN008) and 11th Avenue NW (DSN004). In the options analysis phase, each agency developed independent and joint solutions to overflow control. The City's Long-Term Control Plan (LTCP) and the County's CSO Control Plan Amendment evaluated conceptual CSO control options for costs, technical feasibility, and community impacts using a triple-bottom-line analysis that rated the options. The joint City and County SCWQP was found to be a cost-effective way to meet the project's objectives, would result in fewer impacts to communities in multiple neighborhoods, and would have more operational flexibility than other options.

The City and King County are delivering the SCWQP via a Joint Project Agreement (JPA). SPU will be the lead agency for constructing, owning, and operating the project. This joint approach offers the capability of greater flexibility to control the outfalls in the project area because the larger storage volume can be used to optimize storage for each basin depending on variability of rainfall and flows in each basin. The SCWQP may be operated based on system flows, levels, and predictive rainfall forecasts that are simulated in a hydraulic and hydrologic model to optimize gate and other flow settings.

Following approval of the Facility Plan for the project by the Washington State Department of Ecology, procurement of the consultant for final design services of the Storage Tunnel and TEPS was undertaken in mid-2015. The selected design team, led by McMillen Jacobs Associates, mobilized in November 2015. A series of transition documents were produced in late 2015 and early 2016 to transition between the preliminary design (led by Jacobs Engineering Group Inc., formerly CH2M Hill) and the final design teams. Final design commenced in January 2016. A value management process was completed after 30% design development. Key changes resulting from the design transition phase included a shallower tunnel alignment, reconfigured and shallower West Shaft and TEPS structures, and greater separation of the Storage Tunnel and TEPS design contract phases. Another key decision from the value management process resulted in the optimization of the TEPS pump configuration and arrangement.

The Storage Tunnel package design will be completed in June 2018, with the design of the TEPS package continuing into 2019. Procurement of the Storage Tunnel package contractor is expected to take place from 4Q 2018 to 1Q 2019, with construction notice to proceed now anticipated in late-1Q 2019. Procurement of a Construction Management/ Program Support Services team, to be led by Jacobs Engineering Group Inc., was completed in early 2018.

EXISTING ENVIRONMENT

Surface Conditions

The 2.7-mile tunnel alignment will have an average depth of approximately 80 feet, and be located primarily under street right-of-way spanning the neighborhoods of Ballard, East Ballard, Fremont, Wallingford, and Queen Anne. Sites for the tunnel

shafts, CSO diversion structures, and conveyance alignments were identified as part of the SPU Final LTCP and have been refined through the final design process.

West Shaft Site

The West Shaft site, located near the intersection of 24th Avenue NW and Shilshole Avenue NW, will serve as the western terminus of the Storage Tunnel. The West Shaft, which will receive flows from SPU's CSO basins in Ballard (Outfalls 150, 151, and 152), will be used to launch the TBM and will house the TEPS, which will be used to empty the tunnel. The West Shaft will have an inside diameter of approximately 80 feet and a depth of 120 feet. It is designed as a slurry wall shaft with a cast-in-place concrete permanent lining. Currently, the West Shaft site is predominantly occupied by a parking lot. The site was used as an industrial facility and pier from the late 1880s up to 1970. Past uses of the property include a steel and brass foundry, a shipyard, a machine shop, a wood pipe and tank factory (including possible wood treatment), and a boat building facility. The Ballard Early Works Project, the first construction contract to commence, will remove contaminated soils across the upper 15 feet of the site replacing them with compacted, clean fill in preparation for the Storage Tunnel contractor. Additionally, the Ballard Early Works Project will complete improvements to an existing pedestrian pier at 24th Avenue NW so that spoils removed from the West Shaft site can be transported via barge. When the SCWQP is completed, the reconstructed pier will be converted back to a public amenity.

11th Avenue Shaft Site

The 11th Avenue NW Shaft site is in the East Ballard neighborhood near the intersection of NW 45th Street and 11th Avenue NW. This site will house the drop shaft that will receive flows from the County's overflow structure in 11th Avenue NW (DSN004), a 10-foot-long adit connecting the shaft to the tunnel, and related ancillary facilities. The 11th Avenue site is within the right-of-way of NW 45th Street. The drop shaft is located adjacent to the tunnel and is designed as a large-diameter, drilled shaft. The drilled casing for the shaft will be 10 feet in diameter and installed to a depth of 60 feet. The final lining will be provided by grouted-in-place pipes: a larger diameter pipe for conveyance and a smaller diameter pipe for ventilation. A secondary smaller diameter drilled shaft with a single grouted-in-place pipe is present to provide additional air release during tunnel filling. It is located over the tunnel crown approximately 30 feet upstream of the drop shaft. Historical maps indicate that prior to the construction of the

Ship Canal, a shallow arm of Salmon Bay extended to the northeast from the main body of the bay to near the location of the shaft. The property just to the south was built up using slag fill and previously housed a rail terminal.

Fremont Shaft Site

The Fremont Shaft site is in the Fremont neighborhood near the intersection of 2nd Avenue NW and NW 36th Street. This site will house the shaft, which will receive flows from the City's CSO Outfall 174, a 12-foot-long adit connecting the shaft to the tunnel, and related ancillary facilities. The Fremont Shaft site is within the right-of-way of NW 36th Street. The Fremont Shaft will have an inside diameter of 32 feet, a depth of 83 feet, and will connect to the Storage Tunnel with an adit. The Fremont Shaft will also launch the microtunnel boring machine (MTBM) south for the connection under the Ship Canal to the 3rd Avenue West Shaft in Queen Anne. The shaft is designed as a secant pile shaft with cast-in-place concrete permanent lining. Historical records indicate that the street existed in 1904, but there were no previous buildings or structures on the site.

3rd Avenue West Shaft Site

The 3rd Avenue West (W) Shaft site is located on the south side of the Ship Canal in Queen Anne near the intersection of 3rd Avenue W and W Ewing Street. This site will house the 3rd Avenue W Shaft and facilities connecting the shaft to the existing sewer system in the area. The shaft will have an inside diameter of 20 feet, a depth of 66 feet, and will be used to receive the MTBM. It is designed as a secant pile shaft with cast-in-place concrete permanent lining. The 3rd Avenue site is within the W Ewing Street right-of-way, in an area that is currently used as a parking lot for the West Ewing Mini Park. Prior to the construction of the Lake Washington Ship Canal, a stream meandered through this area, passing just north of the 3rd Avenue W Shaft footprint.

East Shaft Site

The East Shaft site is in the Wallingford neighborhood near the intersection of Interlake Avenue N and N 35th Street. This site will be the location of the East Shaft and ancillary facilities including an odor control room and electrical building. The East Shaft will serve as the eastern terminus of the Storage Tunnel and will be used to receive the tunnel boring machine. The East Shaft will have an inside diameter of 32 feet and a depth of 62 feet. It is designed as a secant pile shaft with cast-in-place concrete permanent lining. The East Shaft site is currently predominantly occupied by a parking lot. A portion of the parking lot is built on the foundation of a structure

that existed on the site until 2015. The property has been owned by the City since 1947, and has been used as a temporary fire station, a print shop, and a maintenance facility. Prior to the city's ownership, the property was used as a plant nursery.

Subsurface Conditions

The subsurface exploration program included 73 borings drilled and sampled along the alignment, completed in five phases. Phases 1 through 3 were completed during the preliminary design development in 2014 and 2015. Phase 4 was completed during the design transition phase in early 2016, and Phase 5 was carried out after a Value Engineering study in 2016 to better define conditions at the various shaft sites. The subsurface exploration program is summarized in the project Geotechnical Data Report (GDR). The Contract Documents for the Storage Tunnel will also include a Geotechnical Baseline Report (GBR) that will seek to establish a contractual basis for the allocation of geotechnical risk during the performance of the work.

The tunnel will be constructed in a mixture of highly abrasive, very dense or hard glacially overconsolidated glacial till (gravel, sand, and silt), outwash (sand and gravel), and interglacial fluvial (sand and gravel) and lacustrine deposits (silt and clay). The geologic profile for the storage tunnel is presented in Figure 2. Groundwater pressures along the tunnel invert will be between 3.5 to 5 bar. Pressurized-face tunneling methods, along with a gasketed segmental lining, will be required to resist groundwater and soil pressures. The glacial overconsolidation of these soils has resulted in high soil density and strength as well as high horizontal stresses. Additional considerations for tunneling include the presence of cobbles and boulders and highly abrasive conditions.

The access shafts will be constructed through similar soils, but will also encounter looser and softer soils near the ground surface. The potential for liquefaction and lateral spreading exists at three of the shaft locations. Confined aquifers have been identified near the base slabs at two of the shaft locations. The shaft excavations will likely require relatively tight shoring with dewatering, excavation in the wet with tremie slabs, or ground improvement to provide a stable excavation base. The excavations for the near-surface structures (diversion structures and ancillary vaults) will also likely require tight shoring or localized dewatering.

PROJECT COMPLEXITIES

Permitting and Land Acquisition

Permitting for the project is complex because of the scope of the work and the various facilities located on agency-owned and private parcels, within the

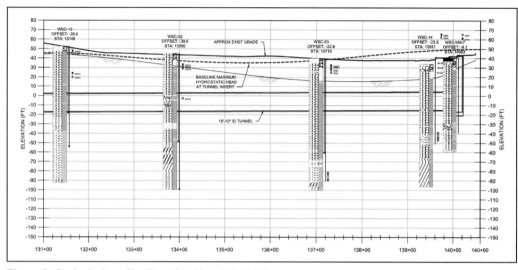

Figure 2. Geological profile (Sta. 131+00 to 140.48.66)

right-of-way, and within the shoreline. Additionally, the project area is within federally adjudicated fishing areas of the Muckleshoot Indian Tribe and the Suquamish Tribe. Permitting agencies include SDOT for Street Improvement permits; Seattle Department of Construction and Inspections (SDCI) for Shoreline, Land Use, and Construction permits; and the Army Corp of Engineers for the Ship Canal. The permitting process has been an integral part of the design phase as many of the permits are on critical path with regards to schedule milestones. Mitigation measures undertaken for schedule delays include presubmittal conferences with permitting agencies and agreements for interagency reviews. Temporary and permanent land acquisition has required extensive effort and coordination with a target of the land acquisitions being acquired by completion of the storage tunnel design.

Hydraulic Design

The storage and control volumes for the storage tunnel are based on an integrated model between the City and County systems developed using proprietary Mike Urban water modeling software. The model has been run using several scenarios based on historic rainfall data and include a factor for climate change. The control volumes represent the amount of flow to the storage tunnel from each outfall location for a given event. The storage volume represents the volume required based on the allowable capacity of King County's North Interceptor, to which the storage tunnel will drain. This modeling and the resulting control and storage volumes have been instrumental in the selection of the optimum tunnel diameter. During the design transition phase, and the early part

of the final design, a range of tunnel diameters from 14 feet to 18 feet was considered. Following the completion of the storage tunnel 60% design and the updated integrated model, the City selected an internal tunnel diameter of 18 feet 10 inches to provide additional storage capacity for larger storm events and account for uncertainties in the modeling so as to insulate against future unanticipated increases in flows.

In addition to the integrated model, several of the critical facilities are being modeled using computational fluid dynamics (CFD) as well as physical models. These models are being completed concurrently with the design of these facilities to validate the design. The critical facilities include the 11th Ave drop structure due to the high flows (170 MGD) anticipated at this location as well as the 3rd Avenue drop structure. Additionally, CFD and physical modeling will be used to validate the TEPS design because of its unique configuration.

Tunnel Effluent Pump Station

The preliminary design for the SCWQP included a dual circular cell (a figure-8 or "Snowman" arrangement) for the West Shaft to facilitate TBM mining as well as to provide a TEPS wet well with an adjacent dry well. During the design transition phase, this configuration was changed to the shape of a "donut"—one larger diameter shaft (the wet well) that entirely encapsulates the dry well within it. There are several benefits to this revised configuration—the primary benefit to the storage tunnel contract is the smaller footprint of the shaft, which results in a larger surface working area around the shaft during tunnel construction. Additional benefits include the dry well shaft being in compression from the hydrostatic loads and therefore less susceptible to leaking through the concrete walls.

Microtunnel

The preliminary design for the SCWQP included a two-drive microtunnel alignment beneath the Ship Canal with three shafts (one intermediate) needed to keep the microtunnel away from an existing deep utilidor. To avoid the need for the intermediate shaft, a curved alignment (650-foot-radius) was adopted during the design transition phase. The curved alignment still allows the microtunnel to go around the existing utilidor, but eliminates the need and expense of an intermediate shaft.

Vibration and Settlement During Construction

Based on currently available data, building damage from vibration during tunnel excavation is not anticipated due to the depth of the tunnel. As is typical of tunnel projects, the SCWQP will require excavation that could result in minor ground settlement in localized areas. Minor settlement at the surface is anticipated to be less than 0.1 to 0.3 inch over the tunnel alignment. Protective measures such as settlement monitoring and pre- and postconstruction surveys will be used to minimize impacts to critical structures along the alignment.

Ground settlement could occur in areas where soils are excavated and dewatering occurs. Activities such as pile driving, sheet-pile installation, and other activities could cause vibration and subsequent ground settlement. Excessive settlement could impact or apply loads to nearby roadways, rail lines,

utilities, and structures. More detailed analysis is being conducted to determine areas where soils could settle. Any settlement from constructing the shafts or conveyance elements is expected to be minor and would be repaired either during or after construction.

As the tunnel is primarily located within the right-of-way, very few structures are within the anticipated zone of influence. Critical structures along the alignment with regards to settlement include the County's Ballard Siphon wet weather barrel, the Ballard Bridge, and Aurora Avenue Bridge.

The Ballard Siphon wet weather barrel is a deep, 7-foot-diameter sewer pipe under the Ship Canal between Ballard and Interbay neighborhoods. It serves to supplement two existing, shallow, 3-foot-diameter wood stave pipes built in the 1930s. It is housed within a 9.3-foot internal diameter tunnel constructed in 2013 as part of the Ballard Siphon Replacement Project (BSRP), which also included relining of the two wood stave pipes with 30-inch-diameter high density polyethylene (HDPE). Normal flows are through the older pipes, with the wet weather barrel providing additional capacity when needed. The wet weather barrel will be utilized by the SCWQP when sending stored flows to the West Point Treatment Plant. As a result of a change to the vertical alignment of the Storage Tunnel, it now passes over rather than under the wet weather barrel, thereby eliminating any potential settlement effects on it. However, the alignment crosses only 5 feet above the exterior of the tunnel housing the wet weather barrel, so tunneling will require careful and stringent monitoring in this area.

The Ballard Bridge is a double-leaf bascule bridge across the Ship Canal. The bridge was built in 1917 with wooden approach structures. Subsequent improvements to the bridge include the 1939 replacement of wooden approaches with concrete and steel structures, the widening of the north ramp to provide access to and from Leary Way NW in 1958, and seismic retrofits to various portions of the bridge and approaches completed between 1994 and 2014. The bridge was placed on the U.S. National Register of Historic Places in 1982. The bridge is owned and operated by SDOT. The Storage Tunnel alignment is directly beneath Pier 4 of the northern bridge approach. Piers 3 and 5 are located approximately 70 feet on either side of the alignment. The superstructure is a continuous span between the abutment and Pier 5, with rocker bearings located offset from the substructure supports and at the end of the span at Pier 5. Under the current tunnel alignment there is approximately one tunnel diameter between the bottom of the Pier 4 piles and the crown of the tunnel.

The Aurora Avenue Bridge (SR 99 George Washington Memorial Bridge) is a cantilever and truss bridge across the Ship Canal at Aurora Avenue

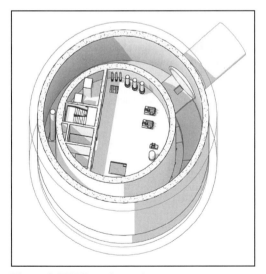

Figure 3. TEPS configuration

North with concrete approach structures from both the north and south. It was originally built in 1931, with subsequent improvements to the bridge including repair of bearings, drainage and floor beams in 1994, seismic retrofit work from 2011 to 2012, and cleaning and painting works commencing in 2015. The bridge was placed on the U.S. National Register of Historic Places in 1982. The bridge is owned and operated by the Washington State Department of Transportation. The Storage Tunnel alignment under the bridge is approximately three tunnel diameters beneath span N-7 of the northern bridge approach. The tunnel alignment is approximately 12 feet south of the centerline of N 35th St.

Critical structures with regards to vibration include several facilities that are within only feet of the required excavations. These include several intermediate- and high-pressure gas lines, a 36-inch-diameter brick sewer from the early 1900s, and a railroad adjacent and sometimes through the construction sites.

Site Constraints

Because of the urban environment, the project sites are limited in area and are subject to neighborhood constraints such as after-hours work, noise and vibration impacts, and difficult truck routing. The West Shaft site, which is the TBM excavation site, is particularly restricted at approximately 1.8 acres. Off-site storage options are being explored to augment this available working area, and a temporary pier area is being permitted to allow additional work area for the contractor.

Based on current plans, approximately 90,000 cubic yards of spoils would be generated from site demolition, shaft and other facility excavation, foundation installation, and ground improvement activities. An estimated 290,000 cubic yards will be excavated during tunnel construction. Spoils that are unsuitable for reuse by the SCWQP will need to be disposed of at an appropriate facility. The primary anticipated spoil removal method will be by barging via the reconstructed 24th Avenue pier.

Risk Mitigation

Risk registers were developed separately for the tunnel and non-tunnel elements of the SCWQP during the preliminary design phase. These registers have been expanded and refined during the final design phase. In addition, risk mitigation measures have been identified for those design, construction, and operations risks that were considered to have significant potential impacts on either cost, schedule, quality, or operations. A quantitative risk analysis was also conducted to inform the owner's risk reserves. Robust risk management practices will continue

through the design, construction, and commissioning phases.

Contract Coordination

The design and construction of the various contracts associated with and adjacent to the SCWQP requires significant coordination. During the final design phase, significant effort was made to avoid schedule overlap and coordinate on contract interface and handover points. This has led to the initially separate Fremont Conveyance construction contract being integrated into the Storage Tunnel contract. The Fremont Conveyance design was accelerated to align with the Storage Tunnel design at 90% and 100% design milestones and will be bid as a single construction contract.

Other coordination examples on the construction side are at the West Shaft site where the Ballard Early Works Package will remediate the site and prepare the pier for barging operations during tunnel construction. At a similar time, SDOT's Burke Gilman Missing Link Project will be working near that site with the installation of the new bike trail and some associated modifications to the nearby streets. Another example is at the Fremont Shaft site, where a preceding City drinking water project will be relining a 24-inch-diameter water main in an existing utilidor beneath the Ship Canal.

PROJECT STATUS

The SCWQP is currently in detailed design. The Storage Tunnel 90% design will be completed in February 2018. As part of the design review process, the City intends to undertake an industry review of selected 90% design documentation. Following this review, the 100% design will be delivered in June 2018. Development of the Storage Tunnel contract procurement documents will take place during the second half of 2018, with an Issue for Tender package expected to be issued in 4Q 2018. The completion of the TEPS 100% design will be scheduled to allow the TEPS contractor to mobilize to the West Shaft site after the tunnel contractor has demobilized.

CONCLUSION

The Ship Canal Water Quality Project, after several years of intense effort and hard work, is now well on its way to becoming a reality and meeting both the City of Seattle and King County federally required EPA Consent Decree dates. The first construction start milestone will be achieved in 2018 thereby setting the stage for design completion and construction commencement of the remaining contracts. This project has represented a significant commitment for both the City of Seattle and King County—the tunnel component of the project is the major part of that commitment and both all parties are confidently looking forward to a successful start, and completion, of construction in the next few years.

ACKNOWLEDGMENTS

The authors would like to acknowledge the many contributions of the McMillen Jacobs Associates design team, their subconsultants, and City of Seattle's Seattle Public Utilities and King County's Department of Natural Resources staff in the design development of the Ship Canal Water Quality Project.

REFERENCE

Seattle Public Utilities. 2017. Ship Canal Water Quality. https://www.seattle.gov/Util/Environment Conservation/Projects/ShipCanalWaterQuality/index.htm.

The Coxwell Bypass Tunnel, Cleaning up Toronto's Waterfront

Daniel Cressman, Olive Cantina, and David Day
Black & Veatch

Samantha Fraser, Robert Mayberry, and Caroline Kaars Sijpesteijn
City of Toronto

ABSTRACT: The City of Toronto is currently undertaking the detailed design of the Coxwell Bypass Tunnel (CBT); tendering for construction was initiated in 2017. The CBT represents the first stage of the Don River and Central Waterfront (DR&CW) Project, a long-term project designed to reduce combined sewer overflows (CSOs) and clean up the Inner Harbour of Lake Ontario, the Don River, and Taylor-Massey Creek. This paper outlines the scope of the CBT and provides context to its purpose in the larger $2 billion DR&CW Project. The CBT involves the excavation of 10.5 kilometers of 7.3 meter diameter tunnel through shale rock of the Georgian Bay Formation, five storage shafts, and 11 connection shafts along the alignment of the CBT. An overview of the CBT design and procurement strategy implemented is provided, as well as the risk-based analysis and decision-making process used to specify a shielded rock tunnel boring machine (TBM) with a precast tunnel lining (PCTL) system.

INTRODUCTION

Toronto is a growing metropolitan area with a population of approximately 2.8 million within the city limits and 6 million residing in the Greater Toronto Area. Toronto is the largest city in Canada and the fourth largest in North America. The city is located on the northwestern shore of Lake Ontario and utilizes the lake as a source of drinking water and for recreational activities, such as boating, swimming, running, biking, and a place of tourist attractions and beaches. The City of Toronto has invested in the waterfront by constructing new transit lines and pedestrian and cycling paths along the Lake Ontario shoreline with the goal of bringing people to the waterfront. Refer to Figure 1 for a visual of the Lake Ontario shoreline.

Figure 1. City of Toronto—Lake Ontario shoreline

To improve water quality in Lake Ontario as well as the Don River and Taylor-Massey Creek, the City of Toronto is embarking on an extensive wet weather flow (WWF) management project. The Don River and Central Waterfront (DR&CW) Project will provide approximately 600,000 cubic meters (m³) of WWF storage to obtain 1-CSO level of WWF control. 1-CSO level of control is a Ministry of the Environment and Climate Change (MOECC) standard that refers to the allowance of, on average, one overflow per year per outfall. To satisfy these design criteria, the scope of the DR&CW Project has been developed to include construction of a 22-kilometer (km) tunnel system, 12 tunnel and WWF storage shafts, 28 WWF connection structures, seven offline storage tanks, a new pump station, and a new high rate treatment plant. The 22 km of tunnel will range in finished diameter from 4.4 meters to 6.3 meters, be excavated through both rock and soft ground soil conditions, and convey flow through gravity operation.

The ground conditions along the tunnel alignments consist of shale bedrock of the Georgian Bay Formation overlain by Quaternary (surficial) deposits described as Glacial-Fluvial and Glacial-Lacustrine formations. Procurement of the 22 km tunnel system, throughout these mixed conditions, will be split into three tunneling contracts. The first tunneling contract, the Coxwell Bypass Tunnel (CBT), will consist of 10.5 km of tunnel to be constructed adjacent to

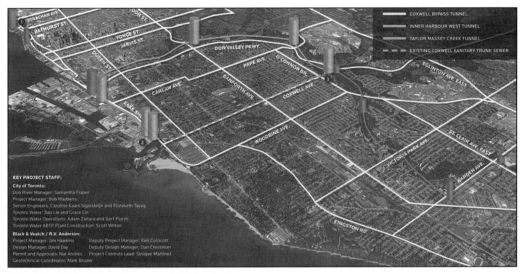

Figure 2. DR&CW project tunnel alignment

the East Inner Harbour of the Lake Ontario shoreline and the Don River at approximately 50 meters deep. In addition to forming part of the DR&CW WWF system, the CBT will be required to provide bypass capacity to the critical Coxwell Sanitary Trunk Sewer. The bypass will allow for maintenance activities and redundancy of sewage flow conveyance to the Ashbridges Bay Treatment Plant (ABTP). The tunnel will be excavated through shale bedrock of the Georgian Bay Formation. This section of tunnel is proposed to be constructed with a single shield rock tunnel boring machine (TBM) and lined with precast concrete segments to a 6.3 meter finished diameter. There are five shafts on the CBT ranging between 50 meters and 56 meters deep and between 20 meters and 22 meters inside diameter. Four baffle drops are to be constructed to accommodate the future connection of WWF outfalls (to the tunnel system) and to allow flow from the existing Coxwell Sanitary Trunk Sewer to be cascaded down 40 meters into the new bypass tunnel.

The second tunneling contract, the Taylor-Massey Tunnel (TMT), will consist of 6.0 km of tunnel extending east from the CBT termination shaft adjacent to Taylor-Massey Creek. The tunnel will be excavated through Quaternary (surficial) deposits and will require use of a pressurized face TBM, either earth pressure balance (EPB) or slurry. The tunnel will be lined with precast concrete segments to a 4.4-meter inside diameter. There are six shafts on the TMT ranging between 13 meters and 60 meters deep and between 5 meters and 12 meters inside diameter. Five baffle drops are to be constructed to accommodate connection of WWF outfalls (to the tunnel

system) and to allow flow to be cascaded down an approximate 40-meter drop in the TMT.

The third tunneling contract, the Inner Harbour West (IHW) Tunnel, will consist of 5.8 km of tunnel extending west from the CBT along the Inner Harbour of Lake Ontario. Like the CBT, the IHW Tunnel will be constructed through shale bedrock of the Georgian Bay Formation and is proposed to be constructed with a single shield rock TBM and lined with precast concrete segments to a 6.3-meter finished diameter. There are three shafts on the IHW Tunnel ranging between 40 meters and 50 meters deep and between 15 meters and 30 meters inside diameter. Connection to the existing Western Beaches Tunnel (WBT) is proposed at the terminus shaft.

The first stage of tunnel construction, the CBT, was scheduled to begin prequalification of general contractors in spring 2017 with the project tendering in fall 2017. The contract documents include a Geotechnical Baseline Report, a Disputes Review Board (DRB), and a requirement for bid preparation documents to be held in escrow.

THE DR&CW PROJECT

DR&CW Tunnels

The alignment of tunnels to be constructed as part of the DR&CW Project is provided on Figure 2. The proposed CBT starts at ABTP, located just west of the intersection of Coxwell Avenue and Lake Shore Boulevard East. The CBT will be connected to a WWF pump group within the new integrated pumping station (IPS) at ABTP. From the IHES-2(B) launch shaft, at ABTP, the CBT will follow Lake

Table 1. DR&CW tunnel summary

Tunnel Description	Length (meter)	Diameter (meter)	Geology
Coxwell Bypass (CBT)	10.6	6.3	Rock
Taylor-Massey (TMT)	6.0	4.4	Soft Ground
Inner Harbour West (IHW Tunnel)	5.6	6.3	Rock

Table 2. DR&CW shaft summary

Shaft ID	Tunnel Contract	Depth (meter)	Diameter (meter)	Purpose
IHES-2(B)	CBT	56	20.0	TBM Launch
LDS-3(B)	CBT	50	20.0	Storage
BB-1	CBT	50	20.0	Storage
NTTPT-1	CBT	52	20.0	Storage
CX-1(A)	CBT	54	22.0	TBM Retrieval
CX-1(B)	TMT	16	12.0	Launch
ST-4	TMT	22	10.5	Storage
MCS-2	TMT	61	12.0	TBM Retrieval
MCS-1	TMT	13	12.0	TBM Launch
LDS-3(A)	IHW Tunnel	49	20.0	TBM Launch
IHWS-2(D)	IHW Tunnel	43	15.0	Storage
WBS-1	IHW Tunnel	38	30.0	TBM Retrieval

Shore Boulevard 2.7 km west to the mouth of the Don River. The alignment turns north at the Don River following Bayview Avenue, adjacent to the Don River, for 6 km to the North Toronto Treatment Plant, Shaft NTTPT-1. From NTTPT-1, the alignment turns east for 1.9 km to CX-1(A), the tunnel terminus shaft located within Taylor Creek Park. The tunnel slope is 0.15 percent to maximize sediment transport characteristics through the WWF tunnel. The established slope places the tunnel at a relatively consistent and significant depth, ranging from 50 meters to 56 meters. A baffle drop structure is provided at the upstream end of the CBT to drop flow approximately 40 meters from the existing Coxwell Sanitary Trunk Sewer into the CBT at the CX-1(A) shaft. The land use changes significantly along the alignment from an industrial area currently undergoing rapid development to a high density residential area in the Lower Don River and transitioning to parkland in the upstream portions of the alignment along the Don River. This has created unique challenges related to coordinating and obtaining permits and approvals for the project.

The TMT starts at Coxwell Ravine Park, where a connection to the CX-1(B) shaft on the CBT is provided. The tunnel follows the Taylor-Massey Creek east for 4.6 km, predominantly through parkland. At Warden Avenue, the MCS-2 shaft is provided, and the tunnel alignment turns north following Warden Avenue for 1.4 km, adjacent to Taylor-Massey Creek, to the tunnel terminus shaft within a Toronto Transit Commission parking lot at St. Clair Avenue East and Warden Avenue. As with the CBT, the tunnel slope

is 0.15 percent to maximize sediment transport. The depth along the alignment varies between 13 meters and 60 meters.

The third tunnel, the IHW Tunnel, starts at the mouth of the Don River, where connection to the LDS-3(B) shaft on the CBT is provided. The tunnel follows the Inner Harbour of Lake Ontario west for 5.6 km to its terminus shaft in Inukshuk Park south of Strachan Avenue and Lake Shore Boulevard. The tunnel slope is 0.15 percent to maximize sediment transport characteristics through the WWF tunnel. The depth along the alignment is relatively consistent at approximately 50 meters deep. The alignment of the IHW Tunnel travels through the congested downtown core underneath Queens Quay, a popular tourist destination. Table 1 provides a summary of the tunnel lengths and dimensions.

DR&CW Shafts and Connections

The project includes several shafts that are required to mitigate the risk of geysering and ensure safe hydraulic performance of the system. The shafts have been sized to ensure this safe hydraulic performance is maintained. Although sized for hydraulic purposes, the shafts will be used to facilitate construction and long-term operation and maintenance. It is proposed to construct (i) the CBT with a single TBM launched west from the IHES-2(B) shaft; (ii) the TMT with a single TBM launched south from the MCS-1 shaft, retrieved at the MCS-2 shaft, and relaunched east from the CX-1(B) shaft; and (iii) the IHW Tunnel with a single TBM launched west from

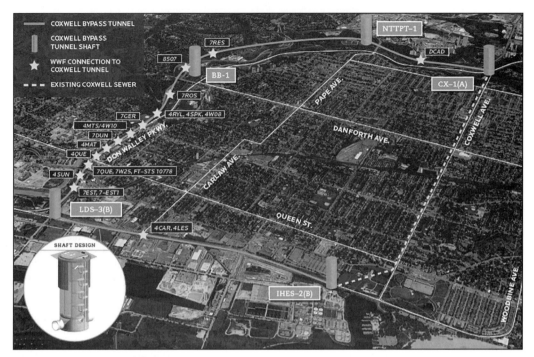

Figure 3. Coxwell tunnel alignment

the LDS-3(A) shaft. Intermediate shafts within the drive are identified as storage shafts because their primary purpose is WWF storage as well as facilitating long-term operation and maintenance. Table 2 provides the general dimensions of the tunnel shafts.

In addition to the tunnels and shafts, a total of 46 combined sewer overflow (CSO) and stormwater outfalls, which currently discharge into the Inner Harbour of Lake Ontario, the Don River, and Taylor-Massey Creek, will be connected to the tunnel system. To connect the WWF outfalls to the tunnel, consolidating the outfalls to 25 drop shaft locations is proposed. The drop shafts are proposed to be either a vortex with deaeration chamber or a baffle drop structure. The choice between the two is based on site-specific constraints at each drop shaft location.

THE COXWELL BYPASS TUNNEL

The CBT is the first tunnel section of the DR&CW Project to be constructed. The tunnel alignment, the location of shafts, and future connections to the tunnel are provided on Figure 3.

Topography

The topographical elevation along the CBT alignment varies from an elevation of 78 meters at ABTP or the IHES-2(B) shaft to an elevation of 125 meters at select locations outside of the Don River Valley.

Along the shoreline of Lake Ontario, the topographical elevation is relatively consistent at approximately 78 meters, or a few meters above the Lake Ontario elevation of 74.5 meters. Moving north up the Don River, the topographical elevation climbs consistently, with a high point where the tunnel moves out of the Don River Valley, to an elevation of 92.2 meters at the CX-1(A) terminus shaft.

Geological Setting

Surficial soils in the project area, deposited during the Quaternary period, were laid down by glaciers and associated glacial rivers and lakes. The surficial soils consist predominantly of glacial till, glaciolacustrine and glaciofluvial sand, silt and clay deposits, and beach sands and gravels. Recent deposits of alluvium have been found in river and stream valleys and their flood plains. Fluctuation in the glacial front resulted in a complex distribution of glacial till layers separated by interstadial deposits of sand, silts, and clays. The Quaternary soil deposits overlie Ordovician-aged bedrock of the Georgian Bay Formation, which consists predominantly of shale with interbeds of limestone and siltstone (Golder, 2016a).

The Georgian Bay Formation typically consists of highly weathered to fresh, grey, very fine to fine grained fissile shale, weak to medium strong shale with widely spaced jointing and subhorizontal

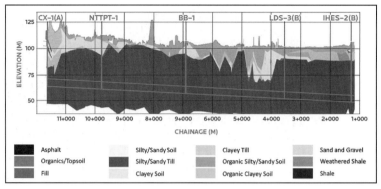

Figure 4. Coxwell tunnel stratigraphic profile

bedding planes, interbedded with slightly weathered to fresh grey, fine grained strong to extremely strong calcareous siltstone and limestone layers (hard layers) (Cao et al., 2014). The highly weathered zone typically occurs at the soil rock interface transitioning to fresh as the depth of bedrock cover increases. The Georgian Bay Formation is characterized by its high horizontal stresses and has been observed to exhibit time-dependent deformation (TDD). During the preliminary design, geotechnical investigation measurement of the Major Horizontal Stress (P) provided a range of 1.6 to 5.6 megapascals (MPa) and a Minor Horizontal Principle Stress (Q) range of 0.1 to 3.8 MPa. The term "Time-Dependent Deformation" has been used to describe the concurrent behavior of stress relief and the swelling mechanism. The time-dependent swelling mechanism is a result of the dilution of pore water salt concentration that causes the space between the clay particles to expand (Lo and Micic, 2010).

Tunnel Geology

The vertical stratigraphic profile of the CBT is provided on Figure 4. The CBT is expected to be excavated entirely through shale bedrock of the Georgian Bay Formation, at approximately 50 meters deep. The majority of the tunnel alignment runs parallel to a buried bedrock valley located adjacent to Bayview Avenue. The CBT crosses this valley in two locations, in the southern portion of the alignment, adjacent to the Don River, and at the northern portion of the alignment approaching the CX-1 shaft. In these two locations, the bedrock cover is expected to be less than 6 meters.

ENVIRONMENTAL/PERMITTING CONSTRAINTS

While tunneling has the potential to greatly reduce the complexity of permits and approvals as compared to open cut construction of a sewer within a highly

urbanized area, the overall expanse of the project and the nature of tunneling itself opens a variety of challenges. As such, it is important on a project of this magnitude to have input from stakeholders during the design phase in order to work collaboratively to the end result.

With more than 20 engaged regulatory agencies and property owners affected by the works, the CBT Project's effects are widespread. It has therefore been viewed as critical to engage these stakeholders, follow up on past communications from the Environmental Assessment, and ensure that comments and concerns are addressed accordingly.

Among the main concerns of the MOECC and the local conservation authority (TRCA) are the potential effects of active dewatering from construction shafts. The design team has made efforts through the hydrogeological assessment to quantify dewatering rates with the goal of avoiding impacts to surface waters and settlement of structures. The selection of shaft and tunnel support methods has considered these issues and minimizes the extent of dewatering that would otherwise be required.

The design team has also optimized the tunnel alignment to minimize the total area required outside of the City of Toronto's right-of-way.

STRUCTURES

Tunnel Lining

For the CBT, the use of a one-pass, precast concrete tunnel lining (PCTL) was evaluated against the use of a two-pass lining system. The typical approach in the Greater Toronto Area has been to excavate the shale of the Georgian Bay Formation with an open-faced main-beam type rock TBM and install temporary support directly behind the main shield. The final lining, cast-in-place concrete, is typically installed a minimum of 100 days after excavation. This two-pass system allows for the full elastic stress relief and the majority of the TDD of the rock mass

Table 3. Tunnel lining considerations

Tunnel Lining	Advantages	Risks
Two-Pass	• Cost-Effective • Typical Approach in Georgian Bay Formation	• Ground water seepage (risk increased in reduced rock cover) • Overbreak in tunnel crown • Slaking in tunnel invert • Buried valleys • Gas (methane) • Degradation of exposed shale
One-Pass (PCTL)	• Schedule Efficient • Cleaner Tunnel	• Support pressure required from lining is substantial • Time-dependent deformation and elastic stress relief places stress on lining • Squatting of PCTL rings • Grouting TBM in place • Washout of annular grout

to take place prior to the installation of the final lining. As an alternative, the design team looked at the use of a single shield rock TBM with a PCTL to mitigate certain risks associated with the two-pass system. As previously discussed in the section titled Tunnel Geology, the CBT alignment crosses a buried bedrock valley with reduced rock cover in two locations, and in addition to these two locations, the tunnel alignment runs parallel to the buried valley on Bayview Avenue. The risk of encountering weathered bedrock, significant ground water inflows or surficial soil deposits through these buried valleys was thought to be a significant risk. In an attempt to mitigate this risk as well as others, the design team analyzed the use of a PCTL, as shown in Table 3.

Selection of a PCTL was considered to mitigate some risks and take certain risks associated with a two-pass lining system off the table. However, use of a one-pass system introduces risks of its own. The biggest risk identified was TDD. This deformation decreases over time (logarithmic relationship) and historically has been mitigated by delaying the installation of the final liner until after a set amount of time or until convergence monitoring demonstrates that movement has decreased below an allowable rate. The use of a single-pass system does not allow for this historical approach.

To evaluate this risk, a detailed numerical analysis to estimate the loads that would be exerted on the PCTL was conducted. The modeling was done using a custom module programmed by Golder for FLAC 2D, a finite difference program. TDD has been found to only occur in zones where the in situ stress has dropped below a threshold value and increases with greater levels of stress relief. The custom module calculates the TDD at each integration point according to the amount of stress relief that has occurred.

The results of the analysis indicated that the displacements that will occur after lining installation will range between 0 percent and 10 percent of the total displacements, corresponding to up to approximately 2 millimeters (mm) that will exert a static load on the tunnel lining (Golder, 2016b). Through selection of an appropriate annular grout and segment design, the displacements, after lining installation, can be accommodated in the PCTL design. The one-pass PCTL was selected as the preferred tunnel lining method because the inherent risks associated with the use of PCTL can be mitigated through incorporation of the appropriate design measures and the uncertainty associated with ground condition risks, apparent in the one-pass system, are avoided. The performance design of the precast lining is provided on Figure 5.

In consideration of the work done (analyzing TDD and the associated ground loads on the PCTL), the PCTL design provides minimum performance criteria that must be provided in light of these loading scenarios:

- 300 mm thick steel fiber reinforced concrete with fc = 40 MPa and fiber dosage of 40 kilograms per cubic meter (kg/m^3).
- 200 mm annular space to be grouted with a two component solution from the tail shield, 3 to 15 MPa strength at 28 days.
- Cast-in gasket designed and tested to sustain 12 bar pressure.

The final design and detailing of the PCTL will be the responsibility of the Contractor after evaluation of its segment handling and TBM thrust loads. This design approach ensures that the tunnel lining is suitable for its end use and that the expected ground loads will give the Contractor flexibility in the design

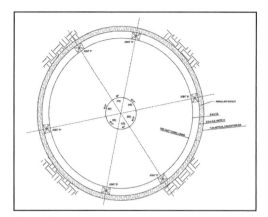

Figure 5. Typical tunnel cross section

to consider its means and methods of manufacturing, transportation, and installation within the tunnel.

Shafts

The proposed shaft excavation and support methods consist of watertight shoring through the surficial soils to mitigate the use and impact of dewatering; refer to the Environmental/Permitting Constraints section. The proposed watertight shoring consists of either secant piles or slurry walls, depending on the depth of shaft. The shoring systems will be keyed into sound rock to cut off the ground water. Below the shoring, the shafts will be vertically cut in the sound rock, with face support provided by a combination of rock bolts, welded wire mesh, and/or shotcrete.

The final lining of the shafts will be done using concrete cast directly against the rock and shoring. The base of these shafts is subjected to large vertical uplift pressures from ground water. To counteract these pressures, the base slab is currently being designed to work as an inverted dome to resist the majority of the water pressure in direct compression (similar to a pressure vessel design). The roof of most shafts will be buried below finished grade. These shafts will be constructed using custom precast concrete girders and slabs to save on costs associated with forming and pouring concrete at great heights above the base of the shafts. Figure 6 illustrates the features of a typical shaft.

Drop Structures

The DR&CW tunnel system will intercept flow from 46 CSO and storm water outfalls. A variety of different possible methods for safely dropping WWF into the tunnel were analyzed and evaluated.

The use of either a vortex drop structure with a deaeration chamber or a baffle drop structure was selected as the preferred method to drop flow

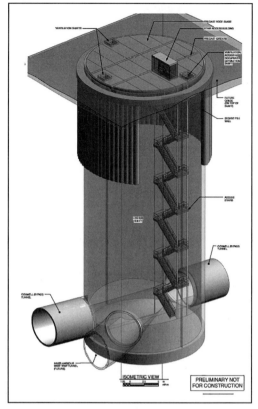

Figure 6. Typical shaft design

from the surface to the depth of the CBT. These two types of drops were selected because they provide the proper hydraulic performance, sufficient energy dissipation, reduced air entrainment, and a proven design for the flows being conveyed to the tunnel (typically >5 cubic meters per second (m^3/s) and <45 m^3/s). The configurations of a typical vortex drop structure and baffle drop structure are provided on Figures 7 and 8, respectively.

The rationale for choosing between the vortex drop structure with a deaeration chamber and the baffle drop structure was dependent on the location of the tunnel connection as follows:

- Connection to Tunnel Shaft—The baffle drop structure is preferred in situations where flow is consolidated to a tunnel shaft. The incremental cost of installing baffles within the tunnel shaft was less than the vortex drop.
- Connection to Tunnel—The vortex drop with a deaeration chamber is preferred for a connection directly to the CBT. This approach is cost-effective, in comparison to the baffle drop, and it fits well within the site-specific

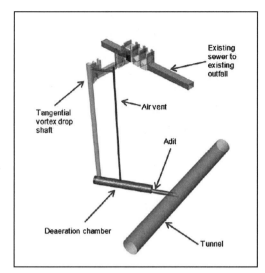

Figure 7. Typical vortex drop with deaeration chamber

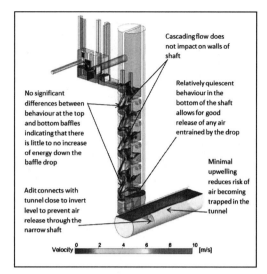

Figure 8. Typical baffle drop structure

constraints associated with the downtown core and valley construction sites on the CBT. The surface diversion chamber and vortex drop fit perpendicular and within the street right-of-way. Elsewhere in the DR&CW system, the TMT specifically, the use of a baffle drop structure has been identified as preferred for connections to the soft ground tunnel. A baffle drop connection to the soft ground tunnel is cost-effective, mitigates risk associated with constructing the deaeration and adit tunnels in soft ground, and fits within the site constraints (open parkland) on the TMT.

Although it is currently not being considered to complete connections of the CSO and storm water outfalls to the CBT in Stage 1, four baffle drop structures on the CBT are to be constructed within the shafts to facilitate connection of WWF outfalls in the future. The design of these baffle drop structures has incorporated results of physical modeling completed by Northwest Hydraulic Consultants in Canada and Hydrotec in England.

Similar to the construction approach taken with the baffle drop structures, currently, construction of the vortex diversion chambers is not included in Stage 1 of the CBT. However, the 11 WWF connections along the alignment of the CBT, including the drop shafts, ventilation shafts, deaeration tunnels, and adit tunnels are to be constructed within the CBT contract to allow for connection to the existing WWF outfalls to the CBT in the future.

CONSTRUCTION SEQUENCE AND PROCUREMENT

The CBT completed prequalification in summer 2017 and tendered in fall 2017. At the time this paper was prepared, the CBT was out for tender with an estimated construction cost of approximately $450 million. The CBT contract includes the requirement for the successful bidder to procure a TBM and the PCTL from prequalified contractors. In an effort to advance the construction schedule, a separate site preparation contract has been released to prepare the tunnel launch shaft, site IHES-2(B), for shaft excavation. The site preparation work will include removal of trees, construction of the site access roads, installation of site hoarding, removal of a berm across the site, and grading of the site. The CBT is proposed to be constructed with a single TBM drive excavating from IHES-2(B) to CX-1. The City of Toronto has included the use of a DRB and the requirement for bid preparation documents to be held in escrow.

The advanced site preparation contract has been initiated to ensure the CBT project is properly coordinated with the new IPS project and Outfall project, both of which are scheduled to tender in 2018 and are located in close proximity to the CBT tunnel launch shaft. Refer to Figure 9 for a depiction of the various site locations.

The other DR&CW stages of tunnel construction and projects to provide WWF storage are scheduled to be procured in four additional stages providing a total of three separate stages of tunnel construction, one stage for WWF connection structures, and one stage to construct offline tanks and prevent WWF surcharging north of the tunnel alignment:

Figure 9. Concurrent projects

- Stage 1—CBT
- Stage 2—TMT and Connections
- Stage 3—Offline Storage Tanks
- Stage 4—IHW Tunnel
- Stage 5—Tunnel Connections on IHW Tunnel and CBT

Future stages of tunnel construction will be scheduled as funding is confirmed by Toronto Water.

CONCLUSION

The DR&CW Project is an ambitious undertaking by the City of Toronto, designed to improve water quality in Lake Ontario and bring people to the waterfront. The first stage of the project to be designed, tendered, and constructed is the CBT, with construction starting in summer 2018. This paper presents the design approach taken in preparation of the tender documents and details the risk-based decision-making utilized to select a precast tunnel lining system with a shielded rock TBM for excavation of the tunnels.

DISCLAIMER

At the time this paper was prepared, tendering of the CBT was in progress, and therefore, this paper must not be considered as containing the official information. Final design and documents for tendering will be available through the City of Toronto.

ACKNOWLEDGMENTS

The writers thank the City of Toronto for allowing them to present the DR&CW and CBT Projects and would like to acknowledge the contribution of a number of individuals to the paper and efforts of the entire project team.

REFERENCES

Cao, L.F., Peaker, S.M., Ahmad, S., and Sirati, A. 2014. *Engineering Properties of Georgian Bay Formation in Toronto*. GeoRegina 2014, Saskatchewan, Regina, Canada.

Golder Associates Ltd. 2016a. *Coxwell Bypass Tunnel—Preliminary Geotechnical Design Report—DRAFT*.

Golder Associates Ltd. 2016b. *Tunnel & Shaft Lining Evaluation Study—DRAFT ver. 2*.

Lo, K.Y., Micic, S. 2010. *Evaluation of Swelling Properties of Shales for the Design of Underground Structures*. ITA-AITES 2010 World Tunnel Congress, Vancouver.

Session 3: Geotechnical, Environmental, and Sustainability

Brent Duncan and Zuzana Skovajsova, Chairs

Assessing Resilience Impacts from Integrated Above- and Below-Ground Urban Infrastructure

Priscilla P. Nelson
Colorado School of Mines

ABSTRACT: This paper discusses an appropriate framework and metrics for infrastructure analysis that can include complex systems representations for all sectors—physical, social and environmental. In order to make better decisions concerning the use of underground space, particularly in urban environments, the functions and operations of the human and physical infrastructure systems must be understood in an integrated framework with common and meaningful metrics and representations. Considering the importance of economics, sustainability and vulnerability to extreme events, decision makers need an understanding of the valuation for underground space as a resource in order to consider life-cycle engineering and trade-offs and pros and cons of above- and below-ground infrastructure investments.

INTRODUCTION

Physical infrastructure systems provide water, sewer, transportation, energy, and communications—the critical services that are the essence of our increasingly urban society. These distribution and transmission systems deliver the services we rely on and expect—they contribute public good, even though they are often managed by private entities. Since being initially designed and installed as simple, linear and uncoupled systems, they have been added to, repaired, and connected in new ways so that the decomposable systems of the past have become the tightly coupled, nonlinear and intractable complex systems of the present. They develop emergent behaviors that defy control in an absolute sense, particularly when these systems are asked to perform under conditions of crisis and disasters.

Our national infrastructure may be valued at between $50 and $80 Trillion, perhaps more. This is equivalent to $200k to $300k for each U.S. citizen as his/her birthright, and this suggests that we are warranted to consider that the nation's infrastructure is a pre-investment upon which the economic engine runs, the quality of life is assured, and career developments of each individual are leveraged.

However, the US public and private infrastructure is aging. State-of-practice design and operation from the past has led to robust-enough systems for which we have sufficient experience to permit simplifying assumptions that enabled operation with minimal monitoring. There were sufficient reserves for acceptable service under known stress. But as we interconnect aging systems into larger networks, and observe decreasing performance levels, reductions in excess capacity and new stresses (e.g., poorly understood interdependencies, attack), we learn that our systems have lost robustness. As our system complexity has increased, many of the design simplifications are no longer acceptable, and new concepts of design and control provide an opportunity for new approaches to system management.

Sustainable urban underground development must meet current human needs while conserving spatial resources and the natural and built environments for future generations to meet their needs. This requires a systems perspective for integrated above and below ground resource use and management, and must include consideration of cost effectiveness, longevity, functionality, safety, aesthetics and quality of life, upgradeability and adaptability, and minimization of negative impacts while maximizing environmental benefits, resilience, and reliability (Bobylev 2009).

Over the past century, we have experienced dramatic changes in demographics, and we can look forward to a world in which existing sociotechnical systems will be operated with increasing sophistication of control through dedicated information and communications technology (ICT) links, making them effectively cyberphysical systems. The interconnection of aging physical infrastructure systems into larger networks, and the loss of redundancy associated with high efficiency operations has led to reduced reliability and poorly understood interdependencies. To complicate matters, our cyberphysical infrastructure has not been maintained, causing unexpected vulnerabilities and cascading failures (ASCE, 2017; AWWA, 2001). To reduce the vulnerability and increase our understanding requires a commitment to research that can only be accomplished through a multi-disciplinary approach focused on the normal, day-to-day operations as well as the response of these systems to the impact of extreme events of natural, technological, or human-initiated origin. The problem is multi-faceted and is embodied within the expertise of such disciplines as engineering, mathematics, natural sciences, information and computer science, decision and risk management, economics and other social and behavioral sciences.

As extreme events frequency and the magnitude of resulting disasters have increased, unexpected performance response, and lack of resilience have been noted (Sanford Bernhardt and McNeil, 2008). While there has been success in modelling complex response and predicting behaviors of our urban sociotechnical networks under stress, the models have grown so complex that data is not available to validate the model predictions (NRC, 2009).

However, as a nation and a world, we will become increasingly urban, with competition for surface space becoming ever more intense. The underground must add to the desirability and vitality of the urban experience—and our underground designs must uplift and inspire those who use and visit the urban environment. Therefore, it is important to make underground infrastructure systems reliable and resilient. We also need to understand why the public may view underground space as undesirable, so we need to understand:

- How is underground space planning best integrated with surface space planning, which depends on the subsurface geology, geographic constraints, past usage, society and culture?
- How can we effectively assess life-cycle performance, which requires that we are able to attach a value to underground space as a resource (separate from mineral rights and material resource development)?
- How can we model the performance of our complex systems, combinations of old and new with increasing interdependencies, under normal and stressed operations?
- Can we understand how natural and technological hazards evolve into disasters, and how to make our communities and above and below-ground infrastructure systems more resilient to extreme events?

It is clear that we need to understand our sociotechnical system dynamics and resilience at a fundamental level or we will learn the wrong lessons from the past. Resilience is a significant concept to many fields including psychology, economics, ecology, or even governance systems. According to the Resilience Alliance (http://www.resalliance.org/576

.php) and as applied to ecosystems, metrics for resilience have three defining characteristics: the amount of change the system can undergo and still retain the same controls on function and structure; the degree to which the system is capable of self-organization; and the ability to build and increase the capacity for learning and adaptation.

Here, resilience is defined as the ability (sufficient capacity and/or flexibility) of a system to experience unexpected shocks or perturbations, and to respond and recover functionality at some acceptable level of performance or action. We have an urgent need for improved understanding of the genesis and evolution of resilience, in particular in urban and coastal regions. We need to build and enhance social and ecological capital and community resilience, as well as to increase system adaptive capacity (including self-organization) and improve the cost-effectiveness of investments in sociotechnical (human, cyber and physical) infrastructure systems.

RESILIENCE FRAMEWORK

The focus for resiliency is on functionality and ability to adapt and restore functionality, including planned and spontaneous responses. To understand the evolution of a resilient response in an urban environment, an interdisciplinary approach is needed that captures attributes of the complex environmental, human and physical systems in a region. In addition, the concept of what is an appropriate responding region itself needs to be investigated through development of layered and registered data resources. These models require the assembly of varied and deep information reflecting current and future conditions, response and usage so that we can expand our knowledge and validate our discoveries and predictions for system performance response. With these assembled information and modelling resources, we can develop a framework of variables and relationships that will support a cross-disciplinary exploration of resilience on a common, integrated, cross-sector basis, and build knowledge as we develop and test theory and models about the resilience of complex sociotechnical systems. Only then can we answer important questions including:

- What observations (evidence) can we make (identify) to indicate qualitatively whether a specific system or network will demonstrate resiliency?
- What metrics can be used to evaluate the capacity of a system or network for resilient response?
- How does resiliency develop or evolve in response, and what factors control or influence the development? Is it a process with

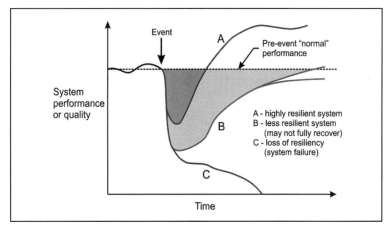

**Figure 1. Conceptual definition of performance response functions
(A) highly resilient system, (B) less resilient system (may not fully recover)
(C) loss of resiliency—system failure (after Nelson and Sterling, 2012)**

thresholds, tipping points, state changes, or is it a continuous function?

- What can we understand about when investment or adaptive management is warranted to improve resiliency of a system of systems, networks, and interdependent systems?

A number of sector-specific models (e.g., telecommunications, electric power) of system performance are currently available (Lee et al., 2007) that are more comprehensive in terms of both their geographical scope and the level of detail they capture, and more sophisticated in terms of how effectively they represent, to operational personnel, the information they can produce. But not all sectors have been addressed, and the models are relatively primitive (e.g., employing disparate, ad-hoc implementations, and not interoperable across geographical service areas) for other sectors. A close analogy between physical infrastructure systems (the focus here) and social systems (e.g., disease) and virtual systems (e.g., Internet) is notable, as is the observation that many of the solution methods in complex systems science are common to all applications. Lacking is a consistent methodology for modelling cross-sector behaviors of critical infrastructures under stress.

As a foundation for integrated study of complex system resilience, it will be important to develop Performance Response Functions (PRFs) that serve as the backbone curves for system response. Performance response concepts have been introduced before (e.g., Bruneau et al., 2003; Silverman, 2004; Fwa, 2005; Rose, 2009; Reed et al., 2010), but PRFs have a greater potential for breakthrough insights in evaluating what performance means, establishing cross-sector performance metrics and variables (the

resilience framework), and understanding how system performance response functions (PRFs) record or reflect important aspects of system behavior at different temporal and spatial scales.

As an example, consider Figure 1, which shows PRFs for socio-technical system performance. The green area represents the loss in performance of system A with respect to a specific event (e.g., storm, earthquake, terrorist act), measured as quality degradation from pre-event "normal" performance over time. The vertical scale is some metric for system performance, which could be based on service delivered, an econometric measure, etc. The response depends on system capacity relative to event magnitude and scale, how well the system has been maintained, how intense the event is, the pre-preparation of the community for such an event, and the geography and social structure of the community and region. In the case of system A, the impact was minimized in intensity and duration, and recovery was rapid reflective of a high level of resilience. In the case of system C, the system failed and recovery was not possible.

It is important that resilience be appreciated and characterized as a system response representing the return to service or functionality, and the restoration of trust and well-being. As such, we need fundamental investigations into PRF metrics that include contributions from the social environment (human/organizational capital and capacity), the physical environment (infrastructural systems), and the natural environment (eco-systems). Any measurement of resilience will depend on the definition of the region being considered to include spatial and temporal scales, boundaries, and the model level of detail or granularity.

Conceptually, resilience is a very useful concept but the data required for its application is not often obtained, and the assessment of integrated urban resilience is not yet widely recognized nor utilized by practitioners. Neither has the linkage between currently defined outcomes/metrics been made with standards or policy incentives an important aspect of implementation.

PRF ANALYSIS AS A METAMODEL RESILIENCE FRAMEWORK

PRFs can also be viewed as a kind of metamodel outcome that reflects the functionality of the system(s) but without the sensitivity toward privacy and security that exists for some descriptive data sets. Convolution of performance response functions for different but interdependent above- and below-ground systems informs the development of a new science of resiliency that can effectively be applied across sectors and systems. This work involves establishing metrics for characterizing infrastructure robustness and fragility, but the metrics must also be applicable to sociotechnical organizations to begin to capture and model community resilience as an integrative and vital concept for our increasingly urban world. This work will lead to enhanced interpretations of the behavior of interdependent infrastructures, thus contributing to systems theory of integrated sociotechnical system behavior, particularly under conditions of increasing density and underground development. In this way, the importance of investment in the urban underground can be demonstrated as a key element of sustainable and life-cycle approaches to better planning and construction in our urban environments.

While we can construct PRFs for specific components of the infrastructure (Croope and McNeil 2011) and we can observe PRFs for regions or communities using data censored by time (Hallegatte 2008), constructing PRF's to assist decision makers and allocate resources requires us to understand scale, aggregation, interactions and interdependencies. PRFs can also serve as a framework to consider use of new technologies, evaluate strategic investments, or introduce stresses to systems. PRF concepts may be applied to individual systems, or all systems in a region. With coupled models, PRFs can be used to explore how the performance/behavior changes as a function of degree and type of interconnectivity of systems in a sector and across sectors, and across temporal and spatial scales.

Models and metrics for resilience have been the subject of much recent work (Gilbert, 2010), and examples of metrics for resiliency and/or PRFs include:

- Services—infrastructure function delivery (e.g., pressure, volume, rate, quality, reliability).
- Human activity (e.g., trips taken, tickets bought, calls made, population density, other demographics).
- Economics (e.g., income statistics, sales tax paid, targeted purchases).

Given a record of spatially distributed pertinent information over many time intervals, the geography and variation of a metric may yield understanding about how the region responds, where resources come from that aid in recovery—ultimately laying the bases for a prediction of comparative resiliency among different communities and societies. When integrated over space and time, the resulting character of the PRFs may indicate typical shape functions. If so, then a new science of complex system analysis of PRFs and resilience may be explored in which a fundamental understanding of how metric functions vary as a function of spatial, temporal and intensity effects and regional boundaries (and perhaps characteristics of boundaries) can be achieved. This would include building an understanding about how PRFs differ (or are the same) across scales, sectors, and systems. The prospect of a science of resilience and PRFs may actually have its own algebra for representing multi-system performance responses.

Demands for infrastructure service vary over time but have typically been assumed constant (Tierney and Trainor, 2004). This assumption must be carefully examined, e.g., the supply side of transportation systems can be measured by metrics such as accessibility, travel time and capacity using risk assessment methods, but the supply-demand relationship has not been captured. PRFs can be used to explore individual system vulnerability, develop summary indicators of net resilience of all systems providing satisfaction to this demand, and to assess topological properties of a networked infrastructure as well as interactions between network structures when subjected to disturbance. These properties include the shortest path distance, clustering coefficients, network density, vertex degree, node and link "betweenness" centralities, network connectivity loss and efficiency.

By keeping track of the history of the system through a memory kernel as dictated by data and event modelling, it is possible to develop new quantitative performance models of infrastructure systems (Lee et al., 2007). However, the challenge is to define more specific measures which will integrate across resilience computations in economics or social sciences.

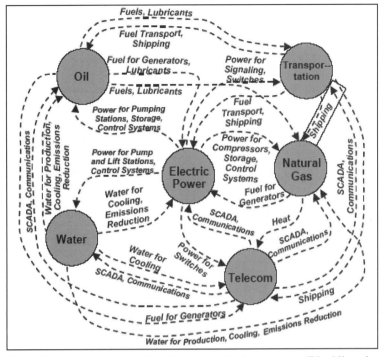

Figure 2. Interdependencies for six sectors of infrastructure (Rinaldi et al., 2001)

INTERDEPENDENCIES AND RESILIENCE

The major elements of sustainability are increasingly a fundamental requirement to the successful undertaking of large capital construction programs—effectively constituting a social license for a "balanced solution." We need to create a methodology and the resources that will establish the value of subsurface space as a resource in urban environments—one that enhances sustainability and urban system resiliency (for example, see Allouche et al., 2008 for the impact of Hurricane Katrina on buried utility services]. Our sociotechnical systems are interdependent, so that disruption of one infrastructure system can impact the operation of other systems. Sources of interdependencies are varied and include technological, cyber, geographic and spatial, economics and business, social/human, political/policy/legal, organizational, resources and supply chains, and security.

Figure 2 illustrates interdependencies that may be developed among six physical infrastructure sectors, and illustrates the complexity of performance and behavior of these systems. In some cases, geospatial mining tools for data-rich sectors have been used to explore the interdependencies, e.g., among transportation and energy systems by merging geospatial and nonspatial data (Shih et al., 2009). Such compound system resiliency analysis will likely lead to new understandings of compound system performance and sources of vulnerabilities.

The resilience of our urban communities depends on many factors that extend beyond the physical system complexities and interdependencies. Therefore, social network research is needed to provide linked and registered metrics for event impacts, yielding change trajectories over time (Anex, et al., 2006, Bobylev, 2009, 2016). Social networks in use now represent a new resource that allows researchers access to social data quickly, and software designed to analyze word function and content on social media (such as blogs) and on-line information sources can be used to capture and analyze the public's acute reactions to extreme events (Sherrieb et al., 2010), providing an exciting window into social resilience.

CONCLUSIONS

Are human creations durable enough to stand up to time? Can we measure their strength and adaptive capability? What is the value of underground space and how can it be best used to improve urban infrastructure service and resilience. It is important to tackle these as principal questions that has perplexed scholars in the natural, social, and engineering sciences as well as the humanities—what are the elements of survival of social and built systems? Such questions have engaged scholars since antiquity, and emerge in both classical descriptions and more modern studies of the growth trajectory, and decline of great civilizations.

The complexities of technical and architectural development on an increasingly urbanized planet is elevated in importance as people and resources are increasingly concentrated. Critics such as Perrow (2007) argue that these characteristics of modernity are just the advance guard of more and greater catastrophes. Recent cataclysms support his views: the collapse of the Twin Towers of the World Trade Center fatally damaged even very large buildings in their environs, while more recently, the Japan Tohoku earthquake and tsunami wrecked habitation, industry, and commerce on a broad scale and forced unprecedented multiple simultaneous nuclear emergencies. These events emerged as surprises that, though envisioned (Mitchell, 1996), surpassed the scale of extant planning and demanded solutions that were as expedient as they were inventive. The burgeoning growth of resilience studies offers an advanced and scientifically valid approach to understanding capacities for system survival. Wildavsky (1988) suggested that in a dilemma between anticipation and resilience, resilience was to be preferred in most cases because many threats cannot be anticipated. Others (e.g., Kendra and Wachtendorf, 2007) argued that the distinction between anticipation and resilience is not so absolute, because resilience owes much to anticipating needs as well as to creativity and improvisation. What to plan for, how durable our creations need to be, what we must look forward to, and how we can improvise in crisis comprise the basic knowledge of resilience—survival—that we will study as we build a science of resilience—one that includes both above- and below-ground resources as an integrated environment for the urban quality of life.

ACKNOWLEDGMENTS

Appreciation is given to UTRC2 (University Transportation Research Center, CCNY) for funding support under the project "Metrics and Performance Response Functions for Assessment of Resilience of Urban Infrastructure Systems."

REFERENCES

Allouche, E.N., Sterling, R. and Hall, D. 2006. Assessment of Damage to Urban Buried Infrastructure in the aftermath of Hurricane Katrina. Proceedings, Pipeline 2006, ASCE, Chicago, IL.

American Society of Civil Engineers. 2017. Report Card for America's Infrastructure. http://www.infrastructurereportcard.org/.

Anex, R.P., Realff, M.J., and Wallace, W.A. 2006. Resilient and Sustainable Infrastructure Networks (RESIN). NSF Workshop Report, http://www3.abe.iastate.edu/biobased/RESIN.htm.

AWWA. 2001. Reinvesting in Drinking Water Infrastructure, AWWA Water Industry Technical Action Fund. http://www.mcwane.com/upl/downloads/resources/americas-water-infrastructure-challenge/reinvesting-in-drinking-water-infrastructure.pdf.

Bobylev, N. 2009. Urban underground infrastructure and climate change: opportunities and threats. Proceedings of the 5th Urban Research Symposium, Marseille, France. http://www.urs2009.net/docs/papers/Bobylev.pdf.

Bobylev, N. 2016 "Underground Space as an urban indicator: Measuring use of subsurface, TUST, Vol 55, DOI: 10.1016/j.tust.2015.10.024.

Bruneau, M. et al. 2003. A framework to quantitatively assess and enhance the seismic resiliency of communities. Earthquake Spectra, 19(4): 733–352.

Croope, S. and McNeil, S. 2011. Improving Resilience of Critical Infrastructure Systems Post-Disaster for Recovery and Mitigation. Transportation Research Record. DOI: 10.3141/2234-01.

Fwa, T.F. 2005. The Handbook of Highway Engineering, CRC Press.

Gilbert, S.W. 2010. Disaster Resilience: A Guide to the Literature. National Institute of Standards and Technology Special Publication 1117: 113.

Hallegatte, S. 2008. An Adaptive Regional Input/Output Model and its Application to the Assessment of the Economic Cost of Katrina, Risk Analysis, 28(3): 779–799.

Kendra, J.M. and Wachtendorf, T. 2007. Improvisation, Creativity, and the Art of Emergency Management, Understanding and Responding to Terrorism, NATO Security through Science Series E: Human and Societal Dynamics, Volume 19. pp. 324–335.

Lee, E.E., Mitchell, J.E., and Wallace, W.A., 2007. Restoration of Services in Interdependent Infrastructure Systems: A Network Flows Approach. IEEE Transactions on Systems, Man, and Cybernetics, 37(6): 1303–1317.

Mitchell, J.K. 1996. Improving Community Responses to Industrial Disasters. In The Long Road to Recovery: Community Responses to Industrial Disasters, United Nations University Press: pp. 10–40.

National Research Council. 2009. Sustainable Critical Infrastructure Systems— A Framework for Meeting 21st Century Imperatives. National Academy Press.

Nelson, P.P. and Sterling, R. 2012. Sustainability and Resilience of Underground Urban Infrastructure: New Approaches to Metrics and Formalism. Proceedings, ASCE Geo-Congress, 3199–3208.

Perrow, C., 2007. The Next Catastrophe: Reducing Our Vulnerabilities to Natural, Industrial, and Terrorist Disasters. Princeton University Press, p. 388.

Reed, D.A., Powell, M.D. and Westerman, J.M. 2010. Energy Supply System Performance for Hurricane Katrina, J. Energy Engineering, ASCE. 136(4): 95–102.

Rinaldi, S.M., Peerenboom, J.P. and Kelly, T.K. 2001. Identifying, Understanding and Analyzing Critical Infrastructure Interdependencies. IEEE Control Systems Magazine. 21(6): 11–25.

Rose, A. 2009. A Framework for Analyzing the Total Economic Impacts of Terrorist Attacks and Natural Disasters. J. of Homeland Security and Emergency Management. 6(1), Article 9.

Sanford Bernhardt, K. and McNeil, S. 2008. Agent-Based Modeling: An Approach for Improving Infrastructure Management. Journal of Infrastructure Systems. 14(3): 253–261.

Sherrieb, K., Norris, F.H., and Galea, S. 2010. Measuring Capacities for Community Resilience. Social Indicators Research. 99(2): 227–247.

Silverman, B.G. 2004. Toward Realism in Human Performance Simulation, ESE Dept. Paper 1-1-2004, U. Penn, http://works.bepress.com/barry_silverman/7.

Tierney, K. and Trainor, J. 2004. Networks and Resilience in the World Trade Center Disaster. In Research Progress and Accomplishments, Multidisciplinary Center for Earthquake Engineering Research, State University of New York at Buffalo, pp.158–172.

Wildavsky, A. 1988. Searching for Safety, Transaction Publishers, 35.

Design, Construction, and Risk Management Strategies for Shallow Tunnels in Urban Settings

Kumar Bhattarai
SA Healy

ABSTRACT: Shallow tunnels possess serious design and construction challenges especially in urban areas. These challenges are further amplified when crossing under live railroad tracks, major road or highways, foundations of existing high-rise buildings, critical utilities, major commercial districts, and so on. The construction risks include excessive settlements of structures and utilities or even sink-hole formations leading to potential disputes and litigations, a significant increase in construction cost and schedule delays. This paper presents risk management strategies during contracting, design, and construction phases for shallow tunnels excavated using various tunneling methods.

INTRODUCTION

Shallow tunnels in urban settings involve a number of serious challenges involving design, construction, social, legal, commercial, technical, right-of-way, public interfaces and relations, traffic management, and public safety. Other challenges include utility relocations and conflicts and environmental compliance. These challenges are further magnified when crossing under live railroad tracks, roads or highways, foundations of existing high-rise buildings, critical utilities, major commercial districts, and so on. The construction risks include excessive settlements of structures and utilities or even sink-hole formations leading to potential disputes and litigations and a significant increase in construction cost and schedule delays.

In the United States, recently a number of shallow tunnels have successfully been completed as roads, transit and railways, water, combined sewerage overflow, and so on. The major projects are Westside CSO Tunnel Project in Portland Oregon; Alaskan Way Viaduct Tunnel in Seattle, Washington; Anacostia River CSO Tunnel in Washington DC; and Dulles International Airport People Mover Tunnels in Dulles, Virginia.

The common method of project delivery for the tunnels has been traditional design-bid-build; however, over the last decade, alternative delivery methods such as Design-Build and P-3 models have become a more frequent choice of project delivery, especially for large projects. The increasing popularity of the alternative delivery methods came with a growing interest by project owners to seek ways to minimize dispute, transfer design and construction risks to the contractors, and deliver the project on cost and schedule.

SHALLOW TUNNEL RISKS

Depending upon the location of a project, tunneling risks would vary; however, some of the risks and impacts can generally be listed as below:

- Inadequate ground investigations leading to inadequate structural design
- Soil and rock type and their properties such as strength, elasticity, swelling, in-situ stress
- Fractured/weathered rock and loose soil with a shallow cover
- Excavation method—SEM, bored one pass/ two pass, roadheader, blasting
- Adequacy of instrumentation and monitoring
- Blasting impacts
- Temporary support failure in tunnels
- Settlement, deformation or collapse of buildings, bridges, and utilities close to tunnel and excavations
- Ground different from that is anticipated leading to differing site conditions claims
- Choice of a TBM with inadequate features, not compatible to the ground encountered during excavation, leading to soil failures or sinkholes
- Bearing failure leading to a TBM being frozen in place at the heading
- Uncontrolled groundwater drawing leading to settlement of properties close to the excavation—a source of protest, delays, and lawsuits
- Tunnel failures due to inadequate support or delay in support installation and formation of sink holes

CASE STUDIES

Canada Line Rapid Transit Project

The Canada Line Rapid Transit Project is a 19.5-kilometer long CN $2 billion rapid transit P-3 project that connects Downtown Vancouver with Vancouver International Airport (YVR) and Richmond City Centre.

In July 2005, the P-3 project was awarded to InTransitBC and was completed on 11 August 2009, a remarkable 110 days ahead of the schedule, in time for the 2010 Winter Olympics. SNC Lavalin Inc. (SLI) was responsible for all design, procurement, and construction.

The underground section included:

- A 2.5-km long twin bored tunnels;
- A 6.5-km long cut-and-cover tunnel; and
- Eight underground stations.

The 2.5-km long 6.0-m diameter bored tunnel section runs from 2nd Avenue Olympic Village Station to Waterfront Station in downtown Vancouver. The tunnel was excavated by a mixed-face EPB machine and supported by the 250-mm-thick precast concrete segments. The bored tunnel section was designed and constructed by SLCP-SELI (JV).

Alignment Within the Till with Shallow Cover

After exiting 2nd Avenue Olympic Village Station, the tunnel alignment enters the Till layer under the False Creek water body, 25 m below high water level. The tunnel alignment then moves along Davie Street at depths ranging from 12 m to 25 m with high-rise towers and deep basements on each side of the alignment. Within the last 150 m of the Till section, the tunnel enters a combined vertical and horizontal curve (212 m radius) before a shallow transition zone, 7 m below the foundations of a residential tower of over 30-stories high.

Risk Identification and Assessment

The contract required that the JV to prepare a risk assessment for the bored tunnel alignment. Weekly risk workshops were held to discuss existing and new risk items and mitigation progress among the staff of JV, SLI, and the Owner Canada Line Rapid Transit Inc. The JV developed mitigation measures and the Risk and Safety Plan stating all the risks and mitigation measures for each section of alignment.

The JV identified five areas of the highest risk as they were close to the raft foundations of the buildings in the zone of unfavorable subsurface conditions. The first two buildings were adjacent to False Creek. The sections, described above, through the Till were assessed as having the highest risk as

the tunnels passed under the twin residential towers over 30-stories high with five levels of underground parking. The risk of impacting the towers was mitigated by lowering the alignment from the foundations by 7 m.

Several mitigation measures were suggested in the risk workshops to reduce the risks in the unfavorable geological conditions underneath the dense urban environment. They were:

- Realignment and increase the tunnel separation by one tunnel diameter underneath the towers
- Use of ripper teeth and cutters to increase the cutter life
- Installation of a piezometer constantly monitor water pressure
- Twenty-four-hour monitoring (surface, buildings, and extensometers) during the most critical sections such as tunneling underneath the towers
- Specialized crews for hyperbaric interventions stationed on site
- Operations and boring parameter control: phases and values to be checked daily with PLC automatic data logger at all sections especially at the 300-m section with a full face of sand and erodible silts under high water pressure

The project employed a systematic approach to risk management and fostered effective communication in sharing the knowledge and experience between the engineers and TBM operators. This strategic implementation of risk management led to the completion of an urban tunnel built in the very difficult ground and groundwater conditions that are outside of the normal operating conditions of the EPB TBM, with minimal ground loss from 0.06% to 0.25% in the 750 m-long soft ground section. In addition, the settlements of the two towers were 0.6 and 0.8 mm in lieu of the predicted settlement of 8 mm. Overall settlements were in the order of 3 mm—albeit, 10 mm maximum at one stretch. The risks were properly assessed and mitigated. The residual risks were low and effectively managed (Henry and Moccichino 2008).

Anacostia River Tunnel Project

The 12,500-foot long Anacostia River Tunnel was designed with an inside diameter of 23 feet and it was constructed using an Earth Pressure Balance (EPB) TBM, manufactured by Herrenknecht AG. The tunnel was driven south from the CSO19 site near RFK Stadium to the Poplar Point Pumping Station. The groundcover ranges between 2 and 4 tunnel diameters in soft ground, the shallowest being about two diameters when crossing underneath the Anacostia River. The tunnel also passes underneath the CSX railroad tracks, new Interstate 695, and the WMATA Green Line. The project also includes six shafts and four adits. The tunnel was designed and constructed by Impregilo Healy Parsons Joint Venture (JV).

The soil materials encountered in the project area comprise, from top downward: Fill, Alluvium, Patapsco/Arundel Formation (undivided) of the Potomac Group (P/A) and Patuxent Formation of the Potomac Group (PTX).

The tunnel was excavated in ground conditions varying from heavy and sticky clay to saturated sands, in the P/A and PTX formations. To achieve good production and a proper control of the process, soil conditioning with additives was injected into the material at the face. Proper backfill grouting of the liner was also fundamental in controlling settlements, especially during mining under sensitive structures or under the Anacostia River.

The JV employed a dedicated Risk Facilitator to develop the Project Risk Register (PRR) and facilitate the risk management process. Every potential risk was identified and scored in accordance with its probability and severity. The PRR was used as a living working document and updated during the various phases of the tunnel design and construction. Risk identification and mitigation were performed throughout the tunnel design and construction phases involving management and technical staff of the JV, Owner, consultants, and subcontractors.

The major tunnel risks were: (1) continuous recharge of water head underneath the river; (2) shaft break-out/break-in; (3) settlements of structures and utilities; (4) boulder quantities; (5) machine failure/break-downs; and (5) tunnel flooding. The other project risks were: right-of-way (ROW); public relations; traffic management and public safety; utility relocations and conflicts; environmental compliance; archaeological discoveries; labor harmony; and interfaces with the adjacent construction work.

The JV produced a series of Construction Impact Assessment Reports and implemented a series of mitigations during crossing beneath sensitive structures. Compensation grouting associated with real-time monitoring was implemented when crossing the 48" and 108" force mains; real-time monitoring was also implemented during the excavation under the CSX tracks, Green Line and DDOT structures.

All instrumentation and monitoring was managed and displayed on a GIMS system in the TBM control cabin and at the site offices, and DC Water's facilities. Instrumentation included extensometers, ground monitoring arrays, utility monitoring points, piezometers, tiltmeters, and seismographs. In general, ground settlements were minimal and less than the predicted values.

The risks were properly assessed and managed. The tunneling was completed successfully and the project is expected to be completed by the end of 2017.

RISK MANAGEMENT—CURRENT PRACTICES

An Owner's vision of risk retaining/transfer is generally stated early in the procurement process through Request for Qualification (RFQ) and Request for Proposal (RFP) documents. During the selection process, a preferred Proponent is selected based on: (1) its demonstration of an ability to finance and withstand financial shocks during the project execution; (2) qualification, experience, and commitment; and (3) ability to manage those risk items. After the preferred Proponent is selected, risks items are then negotiated and documented in the final agreement. For D-B-B and D-B projects, risks that are mainly transferred or retained are the design and construction risks. For P-3 projects, maintenance risk, operations risk, financial risk, ownership risk, market risk, etc. are also allocated or transferred.

Industry Practice Overview

The Owner uses the following clauses in the Contract to allocate risks:

- Escrow documents
- Responsibility and obligation of the Owner and the Contractor
- Payment schedule and limitation of payment
- Bond, insurance, warranty, damage, and destruction
- Dispute resolution clauses
- Differing site condition clause
- Liquidated damages for delay
- Changes by the Owner
- Geotechnical Baseline Report
- Liquidated damages for delays

The GBR has been a part of the contract documents for more than two decades to resolve claims and disputes in the underground construction industry. Despite the wide use of the GBR, the industry is still experiencing disputes, cost increase, delays and failures, especially in large underground projects.

In the last three decades, there have been significant efforts in the industry to promote risk management as a Best Management Practice (BMP) vehicle for tunneling works. Most notable publications are:

Risk score matrix

Probability of occurrence	Impact score 1	2	3	4	5		
5	5	10	15	20	25	Expected	> 75%
4	4	8	12	16	20	Likely	51% - 75%
3	3	6	9	12	15	Occasional	26% - 50%
2	2	4	6	8	10	Unlikely	6% - 25%
1	1	2	3	4	5	Remote	< 5%
Rating	1	2	3	4	5		

	NEGLIGIBLE	MINOR	MODERATE	MAJOR	CATASTROFIC
Cost	$< 0.2M	0.2M<$< 1M	1M<$<5M	5M<$<20M	$>20M
Time	T<1w	1w<T<4w	1m<T<3m	3m<T<6m	T>6m

Figure 1. A typical risk scoring matrix

- Risk Analysis Methodologies and Procedures—Federal Transit Administration (2004)
- Project Management Oversight Procedure 40 Risk Assessment and Mitigation Review—
- Federal Transit Administration (2008)
- A Code of Practice for Risk Management of Tunnel Works—International Tunneling Insurance Group, 2nd Edition (2012)
- Design-Build Subsurface Projects, Second Edition—Brierley, Corkum and Hatem (SME 2010)
- Subsurface Conditions—David J. Hatem (1998)
- Guidelines for Improved Risk Management (GIRM) on Tunnel and Underground Construction Projects in the United States of America—O'Caroll and Goodfellow (SME, 2015)
- Geotechnical Baseline Reports for Construction: Suggested Guidelines—Randal J. Essex (ASCE 2007)

Use of a risk register is being increasingly adopted as an approach to identify, mitigate, and manage risk. In D-B-B contracts, use of a risk register has been adopted but has mostly been limited to the design phase, albeit occasionally continued to the construction phase (Wone et al., 2015). However, its application as a systematic approach has been expanded to construction only in large D-B and P-3 projects.

A risk register is developed during the design phase and updated as the project progresses. The risk is allocated, controlled, mitigated, and managed for each of the risk items, and generally transferred to the party best qualified to manage that risk. A qualitative or quantitative analysis is performed to quantify, mainly, the impact of a risk in terms of cost and schedule.

One common risk management practice is to estimate risk scores—likelihood (probability) × severity (impact)—before and after mitigation in the construction submittals (Figure 1). These submittals would show how risks in terms of cost and schedule are reduced or mitigated by applying certain innovative design and construction techniques. A mitigated risk will show a low to very low risk score.

For a project with a significant underground or geotechnical component, both qualitative and quantitative analyses are performed to estimate the cost and time impact.

Qualitative analysis assesses the probability and severity of identified risks and prioritize the risks using a risk evaluation matrix. The project team with various stakeholders and experts collaborates to set these probabilities and impact levels. Quantitative analysis is to evaluate and quantify the impact of the identified risks on the project cost and schedule. Each risk is reviewed and assigned the cost and schedule impact corresponding to the level of probability provided. A contingency value is then calculated to mitigate these risks (Figure 2 and Figure 3).

In the last two decades, a formal risk management approach was gradually adopted in different forms in the tunnel projects in the US, including some in shallow settings. Some of these case histories are discussed below.

In the case of Westside CSO Tunnel Project, the Owner opted for a Cost Reimbursable Fixed Fee Contract (CRFF), with contingencies for unmitigated risks. The collaborative risk management approach of the contractor's team (Impregilo and Healy JV) contributed to the project being completed within the original contract price without the use of any of the project contingencies.

In the case of Anacostia River Tunnel (ART) Project, the Owner required the D-B contractor to propose their own risk management methods and prepare an interim risk register during the bid, submit an initial risk register within sixty days after the notice to proceed (NTP), and update the risk register regularly. The risk register was required to identify

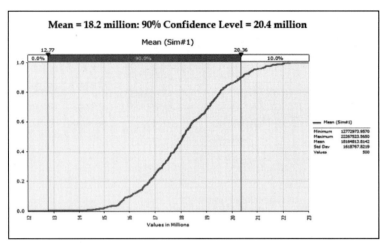

Figure 2. A typical quantitative risk analysis—cost impact probability distribution

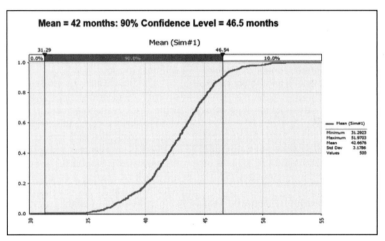

Figure 3. A typical quantitative risk analysis—schedule impact probability distribution

major geotechnical, construction, third party risks, and provide the information on risk identification, affected project areas, risk response plans, and the status of the identified risks. The collaborative risk management system is currently being implemented and the project is expected to be completed within time and budget with the limited use of the project allowances.

In the case of Lake Mead Intake 3 Project, the procurement documents required the D-B Contractor to implement a risk management program that identifies hazards, assesses risks, develops mitigation measures for identified risks, and allows the ability to rank risks to assess the effectiveness of each mitigation measure. The contract also required the

prequalified Proponents to demonstrate their understanding of the project risks and provide technical solutions that mitigate the risks in the proposal.

In the case of MTA—Purple Line Project, the contract procurement documents required P-3 Proponents to propose a risk management plan to identify the significant risk items, potential severity and probability of risks, conduct a risk sensitivity analysis, and propose a risk management strategy to eliminate or reduce major risks.

In the case of North East Boundary Tunnel Project, during the procurement, the Owner required prequalified contractors to propose a Risk Facilitator and a risk management plan in the proposal and prepare a risk register and conduct regular risk

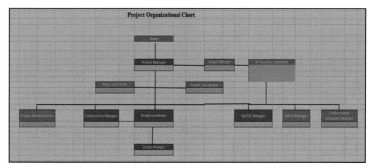

Figure 4. A typical project risk management organization chart

management meetings after the NTP, per the GIRM guidelines. The proposed risk management plan was required to identify the methods to track risks and state how a contingency/mitigation plan would be developed and implemented.

RISK BASELINE REPORT AS A CONTRACT DOCUMENT

Risk management process—risk register—has so far been used as a Best Management Practice (BMP) vehicle to manage risks systematically, however, not as a contract document. As Owners are now increasingly comfortable using a risk register in construction, the risks can be included in a defined Risk Baseline Report (RBR) as a part of contract documents alongside with the GBR. Both documents should be made legally and technically consistent, and can be used alongside when managing risks and resolving differing site conditions claims. It is especially important for shallow tunnels with the issues of tunnel face and crown instability, surface settlements, and impacts to structural and aesthetic integrity of the existing structures and utilities within the zone of influence of tunneling.

This section discusses and recommends the three-stage process that can be adopted by Owners and Proponents to prepare and include the RBR as a contract document.

The three-stage process involves: (1) the Owner prepares Risk Baseline Report—Owner (RBR-O); (2) the Proponents prepare Risk Baseline Report—Proponent (RBR-P) based on the Owner's RFP requirements; and (3) the Owner and the selected Proponent jointly agree on the final Risk Baseline Report for Construction (RBR-C).

Risk Baseline Report—Owner

The Owner's contract requirements for the RBR-P and the Proponents risk management plan should typically include the following items:

- The Proponent's overall risk management process
- Risk tracking process throughout the project
- Risk contingencies—cost and schedule—in the Bill of Quantities
- A plan to develop and implement contingency and/or mitigation
- Adequate information for the Owner to understand how the risks are managed
- An RBR that includes risk items, risk matrix, risk register, risk management process, and risk contingencies—including for the Owner provided list of risks
- A Risk Facilitator and state the role and responsibilities
- A risk management organization chart (Figure 4)

Owners should consider requiring Proponents to provide a method(s) to estimate contingencies, such as the Monte Carlo Method, including the type of distribution to be used for the analysis. Typically, three-point estimates (minimum, most likely, and maximum)—free of contingencies—are used as inputs for the distribution and quantify the cost and schedule contingencies needed to mitigate the risks. The Latin Hypercube Sampling is typically used to ensure the full range of each input distribution is sampled to compute the output distribution.

The topics (headings) in the risk registers should include: (1) design; (2) partnership; (3) stakeholders; (4) permitting; (5) commercial; (6) construction; (7) utilities; (8) environmental; (9) health and safety; (10) quality; and (11) public image. Further breakdown can be listed to clarify the risks associated with the risk headings (Table 1).

Risk Baseline Report—Proponent

The Proponents should prepare the RBR-P based on the Owner's contract requirements.

The RBR-P should include a risk management plan, a risk ranking, and a matrix based on probability

Table 1. Suggested outline of the Owner provided risk headings of the risk register in the RBR-O

Risk Heading	Risk/Hazard Description
Shaft stability—break in/out	Material inflow, instability
Potential damage to existing building/structures	Damages, major failures
Excessive need for ground treatment (quantity overrun/ method problems)	Additional time and material
Freezing	Heave or settlement of existing structures
Comply with the 100-year durability requirement	Reconstruction, repair
Slower than anticipated TBM advance rates	Schedule delays, additional costs
Adverse geomechanical conditions	More frequent interventions leading to construction program delay. Extra costs to complete the works
Excessive water ingress / flooding of tunnel or receiving areas	Damages at existing structures, utilities or SOE
	Tunneling stops, delays
Subsidence, sink-holes, excessive settlement, or collapse at surface	TBM may get stuck or buried
	Injury or death of site personnel
Excessive groundwater drawdown	
Settlement due to ground (volume) loss	Settlement trough at surface leading to damage to surface buildings and structures

and severity in terms of cost and schedule, and the PRR (Figure 5) with a risk impact and mitigation plan. The RBR-P should include contingencies for each risk item.

Proponents should add additional risk items that are not in the RBR-O. The Proponent's risk register should identify major geotechnical, construction, third party risks, and so on, providing the information on risk identification, affected areas, risk response plans, and cost and benefits.

The Proponents should also include a risk management plan and contingencies—both schedule and cost—for each of the added risk items. The contingencies can be based on the 90% confidence level developed using the Monte Carlo Method quantitative analysis (Figure 2 and Figure 3). The Proponent may also suggest an alternative probabilistic distribution for its quantitative analysis than the one stated in the RFP.

The Proponent's risk management plan in the RBR-P may include discussion of:

- Risk Facilitator's qualification and experience
- Risk Facilitator's responsibilities and reporting structure
- Initial analysis of the project risks
- Project Risk Register
- Schedule and frequency of risk workshops; a risk partnering approach
- Proactive identification, evaluation, control, monitoring, management, reporting of risks in all phases and at project levels
- An approach that anticipates and mitigation of risks rather than the correction of unforeseen matters

- A proactive communication approach among all the involved parties to facilitate problem solving
- Risk evaluation process—avoid, mitigate, actively accept, passively accept
- Risk monitoring and reporting plan including risk action logs and reports

Risk Baseline Report—Construction

The Owner will evaluate the RBR-P including the contingencies and rank the RBR-P along with other proposal documents. The owner will select the preferred Proponent based on the qualifications and the price proposal. The Owner will then negotiate the RBR-C—that includes the PRR and risk management plan—and the contingencies that will be applied during the construction. Additional risks that the Owner identifies should also be included in the PRR. The RBR-C should clearly state how each risk has been allocated and the ownership of the risk is specified. Also, it should state how each risk is monitored and the trigger levels for each risk is defined. Also, it should state how it is consistent with the GBR, risk insurance clauses—bonds, insurance, indemnification, warrantees, and guarantees—and other contract terms and conditions, specifications, and drawings.

Conforming with the agreed Partnership process, the Contractor and the Owner should jointly participate in the administration of the RBR-C and the risk management process during the life of the contract. Risks may be revised, added, or eliminated as work progresses, as well as probabilities, severity, impacts, and mitigation plans.

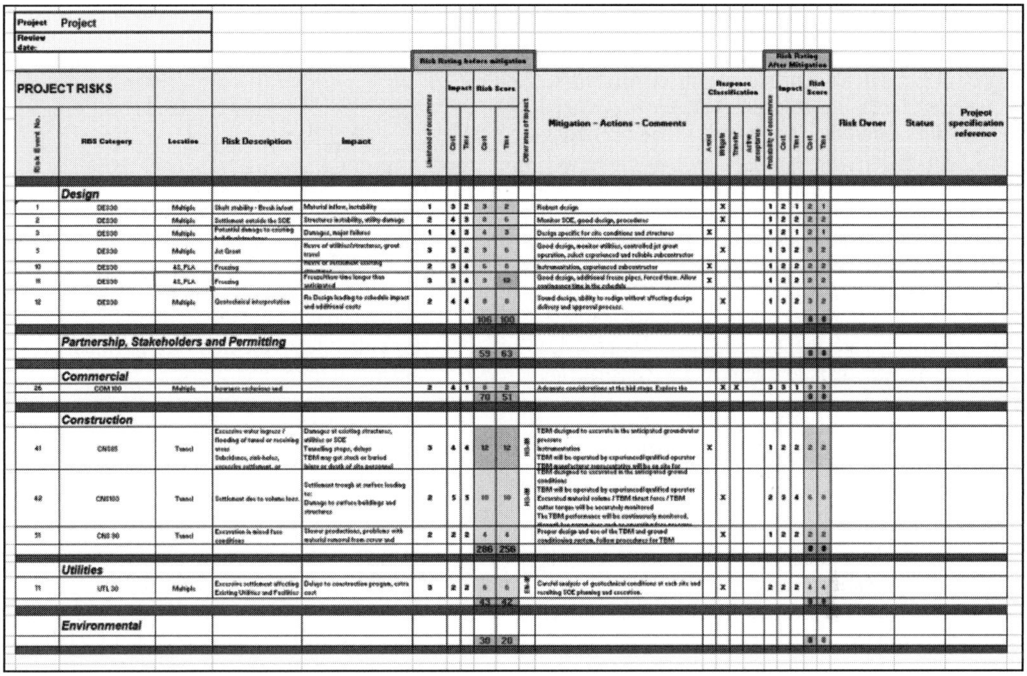

Figure 5. A typical project risk register of RBR-P

Methods to Administer Risk Baseline Report and Payment Mechanism

The RBR should establish and clarify contractual understanding of the risk (risk baseline) and how the risks will be allocated in the contract. The baseline should be defined in terms of a type of risk, ownership, risk scores before and after mitigation, life of the risk items, the milestone of their completion, and the contingencies associated with them. The payment should be based on the ability of the Contractor to manage and mitigate the risk within the cost and schedule—free of contingencies associated with that risk.

The risk register in the RBR should identify the risk owner, who will be responsible for risk management and mitigation, and risk scores before and after mitigation, and the contingencies—cost and schedule—allocated to an individual risk item. The risk contingency should be presented and agreed as a separate line item in the Bill of Quantities. The contingency fund should be used to pay the risks those appear during construction. As the construction progresses and moves towards the completion, the contingency fund required to mitigate the risks decreases.

The contingency fund should be controlled by the Owner and defined as an allowance in the Contract. The release of the contingencies, upon completion of the risk item, should be done in the following ways:

- Divide contingencies for each major risk item in an equal proportion, or any other proportions defined under the contract, between the Owner and the Contractor, if the risk is completed within the baseline cost and schedule, without the use of contingency fund
- Distribute the residual contingency based on the agreement if the fund/schedule contingency was used in the completion of the risk item

The Design-Build Contractor should sign an agreement with their designer/subcontractor to distribute the contingency for a risk item associated with the relevant design/construction element. This will foster partnership and create a win-win situation at all levels of the construction organization.

The contingencies—cost and schedule—should be tracked regularly and discussed in the project risk workshops. The Owner's Risk Manager and the Contractor's Risk Facilitator should jointly track the risks: their completion and mitigation status via the PRR over the defined life of the risks.

OTHER CONTRACT DOCUMENTS RELATED TO RBR

The RBR should be consistent with other contract documents, including relevant clauses of General and Special Conditions, such as: (1) GBR; (2) liquidated damage for delays; (3) performance and payments bonds; (4) insurance program; (5) escrow bid documents; (6) indemnification clauses; (7) partnering; (8) differing site condition clause; (9) changes by the Owner; (10) changes by the Contractor; and (11) drawings and specifications. The RBR should include the risk items that refer to these contract provisions and state the ownership of these risks, such as who will manage them as the construction progresses.

The listing and discussion of risk items should be consistent in both the GBR and RBR; some of which are:

- Geologic setting and subsurface conditions
- Engineering characterization of rock, soil, groundwater, subsurface gas, faults, etc.
- Man-made features of construction significance
- Tunneling excavation conditions, ground behavior and support system—bored, SEM, drill and blast, trenchless construction
- Shaft, station, and adit excavation conditions; ground behavior; and support system
- Building and utility protection measures

During construction, these risks and the associated contingencies should be monitored regularly and the corresponding contract clauses interpreted and checked for consistency when invoking the differing site condition and payment mechanism clauses.

CONCLUSIONS

Execution of successful shallow tunnel projects show that the success depends upon understanding risks, formulating effective risk management strategies, and fostering win-win contractual agreements based on sharing both risk and rewards. Some lessons learned include:

Pre-Procurement Stage—Public Agency

- Employ an experienced and competent engineering firm (Engineer) and a Risk Manager in the Owner's organization
- Perform workshops to identify major risk items, their ownership, rankings, and mitigation methods
- Require the Engineer and the Owner's Risk Manager to prepare a detailed project risk register, perform qualitative and quantitative risk analysis, and prepare a detailed risk management plan
- Prepare the GBR and the RBR-O for procurement

Procurement Stage—Public Agency

- Provide both risk sharing options for the Proponents: (1) share geotechnical risks based on the GBR, and (2) sharing major risk items using the RBR-O
- Provide an option for value engineering submission
- Provide effective criteria for payment provisions and time extensions
- Provide effective contractor selection criteria that requires the Proponents to demonstrate their managerial and technical abilities to manage geotechnical and other risks
- Employ a best-value hybrid selection process—involving both qualification and price based selection process
- Provide Risk Baseline Report as a contract document
- Quantify risk contingencies and establish a contingency fund
- Include a line item for risk contingencies in the Bill of Quantities and Prices Form
- Require the Proponents to demonstrate their risk management plan in their proposal
- Require the Proponents to submit RBR–P along with the proposal

Procurement Stage—Proponents

- Prepare Proponent's own risk register and the RBR-P
- Make provisions based on the RBR for mitigating/sharing risks (and rewards) and insurance requirements—between various project contractual parties: Owner/Proponent, Proponent/Subcontractor, Proponent/Design Consultants
- Involve insurance companies when preparing the RBR-P, as appropriate
- Assess and quantify geotechnical and other risks share/transfer provisions from the RFP
- Propose an efficient and effective organization chart
- Provide value engineering proposals, as appropriate

Procurement Stage—The Preferred Proponent and Public Owner

- Negotiate a best-value contract, as appropriate
- Negotiate a win-win RBR-C

Construction

- Update the PRR regularly
- Manage and mitigate geotechnical and other risks effectively
- Foster partnering and involve all parties including the Owner in the risk workshops and risk management process
- Involve risk insurance companies, as appropriate
- Complete the project on time and within the cost

REFERENCES

Bhattarai, K., Nicaise, S., Min, S., Thibault, K., and Trapani, R. 2016. Geotechnical Risk Management Strategies in Public Private Partnership Method for the Delivery of Tunneling Projects, a North American Perspective. *Proceedings of Tunneling Association of Canada Conference.* Ottawa, Canada.

District of Columbia Water and Sewer Authority. 2013. Requests for Proposals—Anacostia River Tunnel Project.

District of Columbia Water and Sewer Authority. 2016. Requests for Proposals—Northeast Boundary Tunnel Project.

Feroz, M., Moonin, E., and Grayson, J. 2010. Lake Mead Intake No. 3, Las Vegas, NV: A Transparent Risk Management Approach Adopted by the Owner and the Design-Build Contractor and Accepted by the Insurer. *Proceedings of North American Tunneling Conference.* pp 559–65. Portland, Oregon.

Goodfellow, B., O'Carroll, J., and Konstansis, S. 2014. Risk Registers and Their Use as a Contract Document. *Proceedings of North American Tunneling Conference.* pp 670–80. Los Angeles.

Henry, B., and Moccichino, M. 2008. Risk Management of the Canada Line Transit Tunnels. *Proceedings of North American Tunneling Conference*, pp 322–333, San Francisco.

International Tunneling Insurance Group. 2012. A Code of Practice for Risk Management of Tunnel Works. 2nd ed.

Maryland Department of Transportation (MDOT) and Maryland Transit Authority (MTA). 2014–2015. Request for Proposals to Design, Build, Finance, Operate, and Maintain the Purple Line Project through a Public-Private Partnership Agreement.

O'Carroll, J., and Goodfellow, B. 2015. Guidelines for Improved Risk Management (GIRM) on Tunnel and Underground Construction Projects in the United States of America. Underground Construction Association of SME.

Reilly, J. 2007. Alternative Contracting Methods. *Proceedings of Rapid Excavation Tunneling Conference*, pp 418–29, Toronto, Canada.

Smith, K. 2017. Anacostia: Keeping It Clean. *North American Tunneling Journal*, Dec/Jan 2017, pp 14–17.

Southern Nevada Water Authority. 2007. Requests for Proposals: Lake Mead Intake Number 3—Shafts and Tunnel.

Wone, M., Ray C.M., Corkum, D., Teetes, G., and Lisse, S. 2008. Risk Management in Capital Improvement Projects—The DC Clean Rivers Project Approach. *Proceedings of Rapid Excavation Tunneling Conference*, pp 1010–21, Washington DC.

Small-Diameter TBM Tunneling: Risk Management Approach to Face Geological Uncertainties

Giuseppe M. Gaspari and Andrea Lavagno
Geodata Engineering S.p.A.

ABSTRACT: The increasing urbanization and consequent demand for underground services is booming the world trenchless industry, from developing countries to the most advanced economies, and North America isn't the exception. However, the particular variability of the geology must face the hazard of uncertain ground conditions as micro-tunnels, pipe jacking and small diameter TBMs drill through the alignment.

It is the case of alternate shale rock formations and glacial tills layered under the major cities of Southern Canada Provinces and Northern United States. Recent case histories and experiences involved Geodata in developing a Risk Management approach specifically tailored for small diameter tunneling.

The choice of mechanized tunneling technology is itself already a mitigation measure, as TBMs can effectively face hydrogeological-geotechnical unexpected conditions, but not enough: a number of mitigation measures have been tailored in order to improve the excavation process reliability.

INTRODUCTION

The world's cities today are closed networks of transportation systems, utilities, and residential and industrial buildings. Millions of people live and work in such major cities, often in restricted and congested spaces. Confirmed trends of growing urban population will constantly demand a proper allocation and re-distribution of the limited urban space to the various urban functions, both existing and new.

As already demonstrated by the development worldwide in the last century, the resolution of the constant conflict between the demand (for infrastructures and services) and the supply (limited urban space) has often led the planners, politicians, architects, and engineers to consider tapping a seemingly invisible resource: the underground space.

Nowadays, the underground space is created for storage, security, commerce, underground electric stations, and various other purposes. However, a remarkable difference shall be highlighted between the transportation infrastructures (such as rail, road, and metro) and the water systems (in particular new water-mains and CSO, Combined Sewer Overflow). In fact, while for the first ones the tunneling market has been registering a continuous increase in dimensions of tunnels, excavated by either large TBMs (Tunneling Boring Machines) or conventional SEM (Sequential Excavation Method), for the second ones, the recent development of the trenchless technology is improving the efficiency and reliability of smaller TBMs that are more and more capable to face different geological contexts in both urban and sub-urban areas.

There are other important urban tunnels that perform multiple functions, such as the SMART system in Kuala Lumpur, where a 13 km long tunnel, excavated using two 13.3 m diameter TBMs, serves both as a road tunnel for traffic deviation and a storm-water diversion duct to mitigate the high risk to flooding in the center of the city. In addition, recent water projects are currently being built with larger internal diameters than ever, even overpassing the 6 m, in order to face the increasing risks related to climate change in our cities: it is the case of the Anacostia River Tunnel Project and the Northeast Boundary Tunnel in Washington D.C. or the Coxwell Bypass Tunnel in Toronto (ON).

However, it should be pointed out that, apart from those major projects, widely discussed on general media and technical papers, the demand for mechanized excavation is also increasing for installation of smaller pipes in urban area, not just water-related but also for gas-supply and waste-disposal.

In summary, there is an ever-increasing potential for application of small-diameter mechanized tunneling in urban areas because, in theory, any linear infrastructure that can be developed on surface can also be readily developed underground, perhaps also with reduced life-cycle costs. Nevertheless, the experience is still limited, the case histories are not often discussed in technical conferences and the consensus is not universal on the specific methodologies and mitigations applied in different geological and social environments, thus leading the industry to face in the close future several unexplored challenges.

MAIN CHALLENGES AND RISKS OF SMALL DIAMETER TBM TUNNELING

Definitions and Characteristic of Small TBM Tunnels

The internal diameter for smaller tunnels analyzed in this paper is usually limited below 4.5m. However, the expansion of our cities and the increased demand of service is extending more and more the length of those water/gas supply systems, thus imposing small TBMs to tunnel long stretches underground facing geological variabilities and uncertainties that in the past have not been that common for smaller machines.

Additional challenges come from the typical urban context, imposing shallow covers under pre-existing structures, often characterized by strategic public or private significance. Furthermore, the under water table condition is a common risk for those tunnels, that consequently suggests trenchless technology adoption, including pipe-jacking and micro or mini TBMs. All of those methods aim to control the tunnel face pressure, an important aspects in both rock formation and soil deposits under water.

Given the difficulty in identifying suitable available land for the construction shafts, the clear trend in the last decade is to increase the distance between shafts, thanks to the new robotic inspection technologies that allow to space more and more manholes respect to the old standards. It is the Authors general rule of the thumb that alignment characteristics such as the restraint not to have launching and receiving shafts closer than 2–2.5 km would impose the use of small TBMs instead of pipe-jacking. The result is the registered high demand of small TBMs in the tunneling market, in particular in North-America.

In the past years, the risk management approach in large-scale projects has been widely analyzed and applied, thanks to the publication of ITA (International Tunneling Association) 2004 *Guidelines for Tunneling Risk Management* and to The International Tunneling Insurance Group 2012 *A Code of practice for Risk Management of Tunnel Works*. Both documents, however, do not specifically describe risk management approaches for small TBMs. Respect to standard or large machines, in fact, there are limitations and hazards to be considered unambiguously since the early design stages of a small-TBM project, in order to approach every risk assessment and its mitigation in tailored manner.

Adoption of the Risk Management Plan for Small TBM Tunneling

In general terms, tunnel engineering focuses nowadays on the Risk Management approach, especially when dealing with mechanized tunneling. As the design and construction of tunnels is generally associated with a high level of risks due to a whole series of uncertainties involved, risk should not be ignored, but managed through the implementation of a specific RMP, Risk Management Plan (Guglielmetti, Grasso, Mahtab, Xu, 2008, *Mechanized Tunneling in Urban Areas*)

Considering the uncertainties of any underground work, and in particular those imposing the use of small TBMs, it is important for a RMP-based project to consider all potential risks connected with hydrogeological-geotechnical conditions and provide proper mitigation measures in order to reduce either the impact or the probability of such risks. Those methods should be planned since the earliest design steps, in order for their efficient and technically sound implementation at the time of construction, thus reducing impacts in terms of production costs and time, including those connected with workers safety, environment preservation and social influence.

In the RMP approach, the choice of a mechanized tunneling technology for underground works is itself already a mitigation measure, as small Tunnel Boring Machines (TBMs) can effectively face hydrogeological-geotechnical unexpected conditions. Furthermore, the isolated and at times completely sealed work-area offered to workers guarantees safer operations as they do not come in direct touch with the surrounding ground, apart from face maintenance operations.

However, even if the most important advantages of mechanized tunneling respect to conventional tunneling are concentrated in the quality/controllability of the process and of the final product results, it has to be considered the trade-off in terms of flexibility of the temporary support typology and installation method, that in small diameters can face relevant issues and potential delays due to limited working space available. Another disadvantage of small TBMs is represented by the numerous constraints in operating pre-consolidation, as well as pre-supporting and investigations ahead of the face, during the construction process, due to the proximity of the TBM cutter-head to the tunnel face.

Due to such limitations, mechanized tunnel projects always require a good hydro-geological/geotechnical characterization to be interpreted from extensive geo-investigations in advance of tunneling and to be constantly updated and refined during the different design phases.

Small TBM Tunneling Major Geotechnical Hazards and Countermeasures

A small TBM can only be effective as an excavation method if the excavated ground can be foreseen with sufficient level of estimate and if the machine type is chosen, designed and operated to face the specific listed hazards of the project. Should a sound interpretation of the soil/rock behavior be missing, it will result in misleading decisions during construction, thus resulting in losses in terms of safety and productivity.

In order to achieve proper efficiency in mechanized tunneling excavation, a number of hazards should be identified and correspondent risks have to be quantified at the time of designing the TBM.

Those risks are usually analyzed in light of their effect on the excavation and they include the following:

- Presence of hard rocks characterized by high resistance, that can only be faced by increasing the thrust and torque and adjusting the rotation speed of the machine;
- Presence of weak soils or very weathered rocks without self-supporting capability, that will require the machine to adjust the advancing parameters in order to reduce volume losses and instabilities ahead of the face, in order to avoid sink-holes and excessive surface settlements;
- Presence of rock with highly abrasive minerals, that can only be excavated planning adequately consumption rates for the excavating tools and disc cutters utilization/substitution;
- Presence of mixed-face conditions, that will require a proper calibration of the thrusting jacks to maintain the alignment precision and the cutter-head efficiency;
- Presence of explosive and/or dangerous gases spread out in the rock mass and/or localized in pockets (such as methane or sulfuric gas), that will require special equipment for both TBM and workers;
- Under water-table excavations, with consequent risk of localized water-pockets identification in rock fractures or risk of dewatering and consolidation effects in soils, thus inducing settlements and instabilities, to be counteracted with sealed excavation;
- Presence of artesian conditions in the groundwater, that can flood and damage the TBM if not properly counteracted at the face with the pressure in the chamber and closed excavation mode;
- Presence of cavities, either empty or filled with loose materials and/or water, such karst conditions, that can cause flooding and deviations from the design alignment.

The first half of above listed risks is mostly connected with the geotechnical properties of the excavated ground and can usually be counteracted with an appropriate interpretation of the geotechnical investigation data results from site and lab tests. It is in fact possible to design appropriate thrust, torque, rotation speed, face pressures (in case of closed mode

TBMs) and pre-cast segmental lining or temporary supports in order to counteract those risks with sufficient safety factor.

However, for the second half of the list, it is quite difficult to identify deterministically the amount and location of dangerous gases, groundwater, artesian conditions and cavities, thus they can create real troubles to the excavation of the tunnel. In addition, it is important to point out a scale problem, as quite often the small dimension of the tunnel might easy be similar to the "pocket" in which such geological an hydro-geological singularities can be identified. Finally, efficient countermeasures to be adopted a priori in every case are not possible, but they will need to be tailored on the specific project and its very peculiar characteristics.

Some of the possible countermeasures could respectively be listed as below:

- Designing adequate (but very expensive) anti-explosion systems for the TBM and carefully defining the excavation procedures and the TBM parameters for tunneling advance;
- Choosing closed-mode TBMs such as slurry shields or earth pressure balance (EPB) in order to keep sealed excavation from waterinflows and reduce the risk of instabilities and dewatering;
- In order to counteract the artesian water risks, different advantages and disadvantages are listed in literature for the different type of mechanized tunneling methods, so that it is always required to verify the specific project constraints and decide the best countermeasure on the base of an overall analyses of the risk register and the program of advancement for the tunnel;
- Identification of cavities is nowadays still an open-to-discussion issue, as different methodologies are available but exact identification of karst conditions is only possible with systematic and diffused probe-drillings in advance of tunneling; however, this is a method that proved often not suitable for TBMs due to the machine equipment constraints and the time-consuming methodology.

As a matter of fact, for those specific hazards, decisions made at design phase always require to be constantly updated during construction, accordingly with the new inputs coming from the site on the performance of the designed solutions and applied supports. The consequent validation of the designed TBM parameters, of the cutter consumptions and of advance rates shall eventually induce design optimizations accordingly to the so called "observational method," commonly applied for conventional tunnels.

Table 1. Example risk matrix and tolerance criteria, following International Tunneling Association (ITA) 2004 *Guidelines for Systematic Risk Management*

	RISK (PxI)				
5	5	10	15	20	25
4	4	8	12	16	20
3	3	6	9	12	15
2	2	4	6	8	10
1	1	2	3	4	5

PROBABILITY (P) low <----------> high

low <------------------IMPACT (I)-------------->high

Risk score	Risk type		Action to be taken
15-25	Unacceptable		The risk shall be reduced at least to "Unwanted" type, regardless the necessary costs
5-12	Unwanted		Risk mitigation measures shall be identified and implemented (application of ALARP principle)
3-4	Acceptable		The hazard shall be managed throuhout the project.
1-2	Negligible		No further consideration of the hazard is needed

The main objective of the RMP-based design and construction method is that of reducing the risk to a residual value that is considered acceptable by both the Owner and the Contractor. Countermeasures shall then be identified and implemented following the ALARP principle (As Low As Reasonably Possible), allowing for less impact to be caused by geological uncertainties and for a proper mitigation of those hazards still having an "unwanted" risk type, as per the classification reported in Table 1.

Given the general context above, mechanized tunneling is more and more considering the implementation of probe drillings or other indirect exploration methodologies application in order to analyze the ground expected behavior ahead of the TBM face. This investigation technique constitutes an input for the design updates that follow step by step the construction advancement, resulting in a much more efficient tunnel excavation if planned in principle since the initial phases of the project.

The next sections will focus specifically on probe drilling and other investigation methods to mitigate the geological and hydro-geological risks in small TBM tunneling.

RISK MANAGEMENT APPROACH INTRODUCED INNOVATIONS

Risk Management Framework and Risk Register for Small TBM Tunneling

Construction of tunnels of small diameter through TBMs is generally associated with a high level of risk due to a number of uncertainties in design and construction as per previous section. The risk should be managed through the implementation of a project Risk Management Plan ("RMP"). The RMP includes the methodology required for Risk Analysis and the

requirements of the Risk Register. The objective is to ensure that all risks are identified and reduced to acceptable levels, and subsequently managed in an effective way.

The Risk Analysis for small TBM tunnels derives from the general one, including the following components:

- Identifying and describing the hazard factors that can potentially impact the construction process, thus affecting the expected project construction schedule and budget.
- Assessing the initial project risk, i.e., the risk to which the project is exposed in absence of any mitigation measure, quantifying the risks and their costs (assigning an occurrence probability and an index of severity of the consequences)
- Identifying pro-active planned actions (mitigation measures) to reduce the probability and/or intensity of its occurrence to acceptable levels (eliminate or reduce risk)
- Evaluating the residual risk, i.e., the risk to which the project is exposed after the mitigation measures have been adopted.
- Defining guidelines to effectively manage and assign the project residual risk to the contractual parts.

The Risk Analysis process leads to the development of the Risk Register. The Risk Register is first derived in the preliminary stages of the project, and continuously revised throughout the project design phases. As new risks are identified during the design or the perceived nature of identified risks change over time, risks are added or modified in the Risk Register.

In case of small TBMs, the Risk Register should clearly identify the geological variabilities along the tunnel alignment and might consider to mitigate risks of mixed face conditions introducing intermediate receiving and relaunching shafts. Another important factor is the introduction of a scale factor to all the geo-risks in order to properly quantify the vulnerability of smaller TBMs to hazards such as cutting tools wear, cobbles, or boulders in soils and buried valleys or fracture zones in rock. The impact of those hazards is definitely higher for smaller TBMs than "standard" ones and this should be clearly considered because risk analyses are often driven by experience and literature case histories, that not always report the same tunnel diameter.

Adapting the Risk Matrix Approach to Small TBM Tunneling

The specific risks associated with small TBM tunneling can be identified either through a qualitative or a quantitative method. The qualitative method is based on the definition of Risk Matrixes that can visually describe the product between impact and probability to fall into a pre-defined risk tolerance (Table 1).

Different categories of hazards are commonly identified for urban tunneling: (a) general hazards; (b) environmental hazards; (c) construction hazards; (d) health & safety hazards; (e) lifecycle hazards. Three main evaluation criteria are adopted: (i) Engineering judgment; (ii) Probabilistic analyses; (iii) Deterministic analyses.

Probability of Occurrence

Frequency of occurrence is assessed following standardized intervals, according to ITA's *Guidelines for Tunneling Risk Management* (2004). The frequency can be referred to the anticipated number of times the particular hazard would be encountered during the construction of the project. Table 2 shows the typical

probability definition of a small TBM tunnel related to the construction hazard category.

While the definition of the probability appears a quite similar procedure for different hazard categories, however for those hydro-geological, geotechnical and environmental hazards, falling into different categories, a particular care is required in projects involving small TBM tunneling. In fact, the probability of occurrence is suggested by the Authors not to be applied to the whole project, but to separate segments of the alignment with anticipated similar ground/environmental conditions. This procedure allows the designer and the contractor to effectively put in place those mitigation measures that most suitably permit a reduction of the specific risk to an acceptable level.

Assessment of Impact

ITA's *Guidelines for Tunneling Risk Management* (2004) suggest that consequences be classified into five factors which address a breadth of impacts from encountering these hazards (Figure 1). For most of the projects with small TBM tunneling analyzed by the Authors, the following impact factors would be referenced:

- **Economic loss:** economic losses to third parties and/or to the Owner.
- **Delays:** delays in construction and in operational efficiency of the project.
- **Health & Safety:** personal injuries and loss of life as well as permanent health conditions.
- **Environment:** environment pollution, tailing/mucking/disposal, damage to flora and wildlife.
- **Social:** residents' complaints and political influence on the project schedule and budget.

A rating criteria has to be applied to each factor in order to quantify/qualify the intensity of the impact.

Table 2. Hazard probability (frequency) of occurrence

	Probability (P)—Hazard Frequency of Occurrence (during the whole construction period)				
Frequency Class	Descriptive Frequency Class	Interval	Central Value	Description	
5	Very likely	>0.3		Likely to occur repeatedly during construction of the underground project (expect it to happen)	
4	Likely (probable)	0.03 to 0.3	0.1	Likely to occur several times during construction of the underground project (more likely to happen than not)	
3	Occasional	0.003 to 0.03	0.01	Likely to occur at least once during construction of the underground project (about 1 in 10)	
2	Unlikely	0.0003 to 0.003	0.001	Unlikely to occur during construction of the underground project (about 1 in 100)	
1	Very unlikely (remote)	<0.0003	0.0001	Extremely unlikely to occur during construction of the underground project (about 1 in 1,000)	

ECONOMIC RISK (PxI)

PROBABILITY (P) low<----->high					
	5	10	15	20	25
	4	8	12	16	20
	3	6	9	12	15
	2	4	6	8	10
	1	2	3	4	5
	low<------------------IMPACT (I)------------->high				

DELAY RISK (PxI)

PROBABILITY (P) low<----->high					
	5	10	15	20	25
	4	8	12	16	20
	3	6	9	12	15
	2	4	6	8	10
	1	2	3	4	5
	low<------------------IMPACT (I)------------->high				

HEALTH & SAFETY RISK (PxI)

PROBABILITY (P) low<----->high					
	5	10	15	20	25
	4	8	12	16	20
	3	6	9	12	15
	2	4	6	8	10
	1	2	3	4	5
	low<------------------IMPACT (I)------------->high				

ENVIRONMENTAL RISK (PxI)

PROBABILITY (P) low<----->high					
	5	10	15	20	25
	4	8	12	16	20
	3	6	9	12	15
	2	4	6	8	10
	1	2	3	4	5
	low<------------------IMPACT (I)------------->high				

SOCIAL RISK (PxI)

PROBABILITY (P) low<----->high					
	5	10	15	20	25
	4	8	12	16	20
	3	6	9	12	15
	2	4	6	8	10
	1	2	3	4	5
	low<------------------IMPACT (I)------------->high				

Figure 1. Typical risk matrixes as proposed for the innovative approach adopted in small TBMs

Figure 1 shows the typical proposed matrixes for small TBM tunneling projects as modified respect to the known literature. In fact, Health and Safety and Environmental risks are given relatively low risk tolerance in comparison to the other risk factors, suggesting a higher sensitivity in requirements for risk mitigation measures. It is suggested that tolerance criteria be discussed in dedicated risk management workshops, collecting the experience and the local risk perception directly from the Owner.

In particular, when applying the impact criteria reported in Table 3, few more innovation have been introduced by the Authors in small TBMs tunneling project to account for the specific reduced tolerances:

- For the Health & Safety/Environmental hazards, no hazard causing at least one single fatality (i.e., Impact is "3" or more) is considered as an acceptable one, no matter how limited the probability is;
- For the Health & Safety/Environmental hazards, no single (or limited) minor injuries is tolerated when their occurrence probability level rises to the point that they are likely to occur repeatedly during the construction (expect it to happen, i.e., Level "5").
- In addition, Environmental hazards tolerance is lower than Economic or Delays risk tolerances, which means that the unacceptable (i.e., "red) area is greater for the former than for the latter.
- Finally, for Social hazards the tolerance (i.e., "red area") is the same as the Economic or Delays ones, but an acceptable risk is considered at the lowest probability whatever the impact level and at the lowest impact whatever its probability (i.e., daily complaints by few local residents/retailers).

MITIGATION METHODS FOR SMALL TBM TUNNELING

If unacceptable risks are identified for any given hazard, risk mitigation measures need to be implemented in order to reduce the initial risk to an acceptable level of "residual risk."

Several mitigation measure categories can be identified in the design phase, as part of the Initial Response Strategy. For small TBM tunneling, they would fall into one of the following categories:

- **Avoid risk:** these measures can be performed through different actions, i.e.: terminate the project element, design the risk out, or change the project element leading to the risk.
- **Reduce risk:** this can be done through the actions of sharing, controlling and monitoring risks
- **Transfer risk:** these actions include to insure, allocate and/or contract

Table 3. Hazard impact criteria

		Impact (I)—Hazard Consequence				
Consequence (severity) Class	Descriptive Consequence Class	Economic Loss to Third Parties and/or Owner [$CDN Million] A	Delays [days/ hazard] B	Health & Safety (injury to third-parties or workers) C	Damages to Natural Environment D	Social Impacts (political, cultural, consensus & satisfaction) E
5	Disastrous (very high)	>10	>300	Several injuries and/or fatalities	Permanent, severe damage	National/city negative impact
4	Severe (high)	1–10	30–300	Multiple severe injuries, single fatality	Permanent, minor damage	Major city impact or minor national impact
3	Serious (medium)	0.1–1	5–30	Single serious injury, single fatality and/or several minor injuries	Long-term effects	Serious city impact
2	Considerable (low)	0.01–0.1	1–5	Single serious injury, several minor injuries	Temporary severe damage	City impact
1	"Insignificant" (very low)	<0.01	<1	Single (or limited) minor injury	Temporary minor damage	District impact, insignificant impact

- **Retain risk:** when risks are retained, the owner typically assigns contingency in the project planning.

It should be noted that design steps taken throughout the Preliminary and 50% Final Design stages of a tunneling project with small TBMs should have ready avoided or reduced a number of perceived risks associated with the conceptual design, thanks to an extended geological-geotechnical investigation campaign. Additional mitigation measures are usually defined in subsequent design phases, substantiated with increased knowledge from the ongoing hydro-geological and environmental investigations.

Such additional mitigation measures, particularly in case of very variable geological conditions and site/budget constraints not allowing for intermediate shafts, might often include the possibility of extended geo-investigations during construction through means of probe drillings ahead of the tunnel face.

Direct Investigation Methods in Small TBM Tunneling: The Use of Probe Drillings

Probe drillings are just one of the available technologies in the market to investigate in advance the ground ahead of the tunnel face. In order to state their advantages and disadvantages respect to other techniques, application ranges and applicable mechanized tunneling methods will be discussed.

However, the most important requirement is to plan in advance those inspections as well as to foresee the exact areas of their application and their daily impact on the production.

As a preliminary observation, it must be considered that the type, scope, number and density of probe drilling should in every case be determined by a geological, geotechnical, hydrogeological characterization, focused on the specific project, based on previous investigations made during the design stage, and/or on the result of previous probe drills if any.

Probe-drilling investigation methods in mechanized tunneling can primarily be divided into two families: Direct investigations and Indirect investigations. As shown in Figure 2, many different options are available, with different applicability fields, providing different geological/geotechnical information, thus a complete outlook cannot take into account a single-oriented classification. All of them are also often adopted for small TBMs.

In case of Double Shield TBMs in general, the applicability of the methods of prospecting the geology ahead of the TBM face during its advancing is intensively impeded by the following issues:

- The cutting head is characterized by limited open sections due to the necessary stiffness required by the head itself, due to the excavation of high hardness rocks.

Drillholes from the TBM tunnel itself:
- Drilling exectuion: horizontal->through the TBM cutting head, or inclined->behind the cutterhead in Open-TBM, through the shield in Shielded TBMS, or radial through the lining
- Information obtained: lithological nature of the ground, presence of water, presence of voids and/or decompressed zones
- Reliability of the obtained information: variable, depending on equipment used (smaller drill equipment and rods can guarantee lower precision in directioning, and information obtained. Core-drilling is generally less reliable than in surface or other tunnel's drilling, due to space constraints and smaller equipement to be used.
- Cost/benefit: very variable, depends on information that needs to be collected

Drilling from other tunnel
- Drilling execution: only possible if a pre-existing tunnel or ADIT is available, and acces to it is feasable
- Information obtained: variable and multiple information on either the geological or hydrological or geotechnical conditions, depending on density and what is the scope and distance from tunnel
- Reliability of the obtained information: can be very high and detailed if accessibility is good
- Cost/benefit: depends on density and what is the scope, distance from tunnel, and of course availability of other tunnel

Vertical Drilling from surface
- Drilling execution: possibile to use a wide range of drilling equipment, with the highest performance rates and reliability if space on surface is not constrained by underground or aerial utilities
- Information obtained: wide range of infos depending on equipment (core recovery, water pressure and inflow, etc.), but limited to the drilling points (interpolation/interpretation needed)
- Reliability of the obtained information: can be very high
- Cost/benefit: depends on distance from surface, density of the drillholes

Sub-horizontal Drillling from surface
- Drilling execution: much more difficult to perform and guarantee alignment, less frequent equipment availability
- Information obtained: if alignment reliability is provided by accurate guidance system installed in the drill bits, the sresults of the drill can provide continuous information about ground close to the alignment, without interfering with excavation, and with limited need of interpretation and/or interpolation
- Reliability of the obtained information: can be very high
- Cost/benefit: depends on position of the alignment vs surface, usually it can be very costly and equipment is not easily available

Geophisical methods from surface
- Execution: no interference with excavation
- Information obtained: depends on parameters to be obtained, distance from tunnel crown
- Reliability of the obtained information: depends on parameters to be obtained, distance from tunnel crown
- Cost/benefit: depends on position of the alignment vs surface, but cost is generally lower

Figure 2. Advantages/disadvantages of direct and indirect investigation methods in small TBMs

- The head is furthermore mostly occupied by excavation tools (cutters), thus making it difficult to access to the front with the drilling equipment and rod;
- The presence of one or two rays of thrust cylinders, the exis.ence of cylinders for transversal grippers, the space occupied by the grippers themselves, etc. restrict the space available for the positioning of the probe-drilling system (driller body).
- The installation system of the segments (the erector) impede most of operations from the shield.

In general, Double Shielded TBMs can only allow for probe-drilling with larger diameters: the Authors have not identified a single case history with opposite conditions. However, smaller diameter TBMs can definitely allow for probe-drillings, but the bigger the diameter the more suitable inclined drills are and the more efficient the operations can occur. For small diameter TBMs, inclined probe-rills tend to become very difficult or even impossible, for geometric reasons in positioning/orientating the drills, due to the minimum length of drill-rods, the equipment dimensions and to the limited space available in the shield. Even the option of geo-structural mapping of the face and/or of the sidewalls makes it difficult or impossible way when the operator has to work in the limited space and with the limited visibility offered by a small diameter machine. All of the above makes in fact only one option feasible for small diameters' TBMs, that is the option of horizontal perforations, without recovery, namely the method of lower efficiency and greater impact on the tunneling process schedule.

As a consequence, it is commonly accepted that for the excavations in medium to hard rocks, by means of mechanized methods with shield (TBMs), the direct investigation systems have an average applicability which is a function of the excavation diameter: the larger the diameter, the higher the rank of viable modes (drill-holes with core recovery—boreholes; drill-holes without core recovery; geo-structural mapping of the face and/or of the sidewalls) and the increased the efficiency and productivity of the method. On the contrary, the smaller the diameter, the lower the choice of applicable technologies (until it is reduced to the sole option of horizontal drilling without core recovery) and the worse the efficiency.

Advantages and Disadvantages of Alternative Investigations Methods in Small TBMs

Considering the issues risen in the previous section for small TBMs, the same inputs that can be obtained from probe-drilling can in all cases be achieved through the application of indirect techniques, with effective results based on the efficiency of a single method for the determination of the required information.

The following describes the advantages and disadvantages, in general terms, and information obtained by each method with a description of its own reliability. It is in fact important that the perforation of the tunnel is compared with other methods of prospecting, to be carried out not by the tunnel but from different positions, and therefore constitute an alternative to drilling from the TBM, particularly if this is characterized by a small diameter. A number of commonly adopted options is analyzed in Figure 2 as listed here below:

- Horizontal drilling from another tunnel, or geophysics from another tunnel or existing adits;
- Sub-vertical perforations from the ground surface;
- Sub-horizontal perforations from ground surface;
- Geophysical investigations from the ground surface.

In general terms, one can observe that it is difficult to univocally determine if and when the method of probe-drilling from the tunnel itself ensures greater reliability (and a lower cost/benefit ratio) respect to the information obtained with indirect investigation techniques. The influencing factors on the advantage of alternative technologies to probe-drillings are heavily influenced by:

- Alignment configuration and correspondent ground surface conditions;
- Presence of pre-existing underground accesses and adits to facilitate the exploration;
- Underground or suspended aerial utilities restraining the equipment operations or requiring relocations;
- Type of information that are required to be collected: presence of water, identification of any faults and their extension and filling, properties of soils, risk of gas: not all methods are equally effective.

Furthermore, each of these techniques has a highly variable efficiency based on the quality of the data interpretation and management, particularly in case of a poor back-analysis set-up that should include also the TBM parameters, including but not limited to: thrust and advancement parameters, torque, rotation speed, tool wear, drivers record of observations, etc.

For all that, it is difficult to establish, a priori, the best investigation method, without studying a specific case and needs: a more careful assessment must be carried out according to the specific purpose of the investigation in the particular project, considering the contractual obligation, the geotechnical information, the health & safety risks and the social/environmental constraints.

However, it is possible to highlight that the execution of perforations during the advancement in mini-TBMs (diameter: 2.0–4.0 m) is rather uncommon in the industry given the great difficulties that this activity implicates. In fact, the projects experienced by the Authors confirm that the applications of this technique are limited and not often reliable. Examples of systematic probe-drilling application in mechanized tunnels of small diameter are especially related to open-type TBMs with diameter included between 3.5 m and 4.0 m: in those cases the absence of the shield and the availability of small spaces allowed the positioning of a drilling machine capable of performing holes both along the tunnel axis and inclined.

Regarding the shielded micro-TBM, the difficulties increase considerably especially if the use of drills was not planned since the initial design phases of the project, allowing the machine to be specifically designed and built providing as a minimum the following items:

- Holes in the mini-TBM shield.
- Collars for the directional support of the drill rods.
- Dedicated supports for the drilling machine.
- Systems of connection/disconnection of the drilling's hydraulic systems, in order to reduce downtime per drilled meter of tunnel.

Almost all cases of shielded mini-TBM adopting probe drillings, registered problems are mainly related to the reduced space mostly occupied by: engines, thrust cylinders, mucking belt, hyperbaric chamber if necessary. In these conditions it is also very difficult to locate a drilling machine that operates parallel to the axis of the tunnel alignment; however, for the same reasons, in many cases it is impossible to place the drill in an inclined position, with the aim of directing the holes in a diverging direction from the excavation section.

CONCLUSION

In light of the projects on small TBMs in North America where the Authors have been involved, there is a wide range of available methods to reduce risks related to geological and geotechnical hazards, most of which tends to the execution of additional geotechnical, geological and hydrogeological investigations during the tunnel advancement, either direct or indirect. However, often those mitigation measures are applied just to accomplish a qualitative reduction of risks by expecting deterministic information on particularly hazardous alignment stretches. It is instead recommended by the Authors that such additional investigation play active role in an overall Risk Management Plan (RMP) dedicated to the project. This article has the purpose of suggesting specific applications of the known Guidelines on the RMP for small TBM tunneling.

In general terms, it is difficult to define establish a priori the effectiveness of the different methods of investigation applied to mechanized tunneling of small diameters in order to guarantee greater reliability and higher cost/benefit ratio. In particular, the choice between probe-drilling and other techniques, direct or indirect, depends on several factors, including but not limited to: the alignment configuration, the ground surface conditions, the presence of pre-existing underground tunnels/adits, the interference with existing utilities. Most of all, the efficiency of the chosen method is highly affected by the type of information that is required to be collected and the type of data-processing, interpretation, back-analyses and predictions required by the specific project, with particular reference to the contractual obligations, the baselined geological and geotechnical information, the health & safety risks and the social/environmental constraints.

Under normal conditions, the fundamental goal for the design and subsequent construction of a tunnel is to assure that the work is realized within the budget and constraints of time and cost, is stable and durable over a long time, and corresponds to the technical specifications and requirements of the Client. These objectives are really very important, but they are not comprehensive enough for small TBM tunneling, particularly in urban environment.

Indeed, in this case it is also necessary to take into account a set of completely distinct elements or factors that frequently influence the choice of the design and construction. The presence of these elements requires that particular attention be paid to the rules like:

- Disturb as little as possible the integrity of the ground surface and the built-up environment above.
- Take into account all existing structures and all underground services, such as the sewage system and superstructures.
- Respect the limits specified for surface settlements, which is a function of the ground type and the pre-existing conditions as well as of the TBM type (slurry or EPB usually) and exact diameter.
- Avoid absolutely tunnel face collapse, which can cause property and/or personnel damage.

In fact, any accident in a densely populated urban area can have a very serious impact on public opinion and, in the extreme case, it may cause damage to properties leading to a complete stop of the project for months or even years. Clearly, the risks related to such hazards need to be minimized, when it cannot be possible to avoid them totally, choosing alternative solutions.

REFERENCES

Boone S.J., Westland J., Busbridge J.R. and Garrod B., 1998, *Prediction of Boulder Obstructions, Tunnels and Metropolises*, Nego Jr and Ferreira (eds), Balkema, Rooterdam.

Guglielmetti, Grasso, Mahtab, Xu, 2008, *Mechanized Tunneling in Urban Areas*, Taylor&Francis, London.

ITA (International Tunneling Association) Working Group 2, Research & Development, eds. 2004, *Guidelines for Tunneling Risk Management*.

The International Tunneling Insurance Group, 2nd Edition, May 2012, *A Code of practice for Risk Management of Tunnel Works*.

Rehabilitation of Tunnels: An Owner's Perspective

David Tsztoo, Anthony Yu, and Teena Redhorse
San Francisco Public Utilities Commission

ABSTRACT: In today's tunneling industry, an Owner must consider program objectives and capital resources in their evaluation of tunnel rehabilitation versus new tunnel alternatives. Despite recent innovations to improve cost and efficiency for new tunnel construction, there are other considerations in the Owner's decision making process. This paper will explore why a tunnel rehabilitation project can be the favorable option resulting in a lower budget, shorter design and construction timeline, reduced pre-construction efforts, less environmental and community impacts, and increased construction/operational flexibilities—now and for the future.

INTRODUCTION

Since ancient times, storing and conveying water via underground water systems and tunnels has been touted for its strategic advantages. Underground cisterns and tunnels assured protection of essential fresh water supplies from man made contamination, and especially from enemies who would gain an advantage by disrupting an adversary's water supply in a siege. Modern day engineers do not usually specify tunnels for reasons of avoiding deliberate contamination or siege. However, they are just as enamored with the idea of building tunnels for water conveyances.

Such was the case for the San Francisco Public Utilities Commission (SFPUC). The SFPUC provides drinking water to 2.6 million residential, commercial, and industrial customers in the San Francisco Bay Area. The SFPUC's water system includes 160 miles of transmission pipelines and tunnels from the SFPUC's largest reservoir, the Hetch Hetchy Reservoir inside Yosemite National Park, to San Francisco. About 80 miles of the system consist of tunnels. A 19 mile section of the tunnel network, called the Mountain Tunnel, was thought to be at risk of a catastrophic failure in its concrete lined section of 11 miles. For much of the project planning process, a new replacement tunnel was presumed to be required according to the tunnel planners.

Originally constructed between 1917 and 1925, the Mountain Tunnel has since been in continuous service since 1925. The 19 mile tunnel is located downstream of the Hetch Hetchy Reservoir water source in Yosemite National Park, between the Early Intake and Priest Reservoir (Figure 1). Approximately 11 miles of the tunnel is lined with unreinforced concrete. Though the condition of the tunnel was being monitored through water sampling and periodic inspections, there was growing concern

that the tunnel lining was at risk of a partial collapse that could interrupt water delivery for up to nine months. The SFPUC was faced with a decision to construct a new bypass tunnel to replace the lined section of the Mountain Tunnel or to rehabilitate the existing tunnel.

In order to decide between constructing a new tunnel or rehabilitating the existing one, the SFPUC embarked on a detailed alternatives analysis study, developing a set of performance standards for the tunnel, and identifying different alternatives. The study included four alternatives for in-depth review (McMillen Jacobs 2017):

- Rehabilitate the existing tunnel, focusing on repair and contact grouting of the 11miles of concrete lined section.
- Relining with smaller diameter steel pipe in the 11 miles of concrete lined section.
- Construct a new bypass tunnel within the tunnel right of way to replace the 11 miles of concrete lined section.
- Construct a new bypass tunnel outside the tunnel right of way to replace the 11 miles of concrete lined section.

A 60-day shutdown of the tunnel was scheduled in early 2017 to perform a visual inspection, gather detailed information regarding the location and type of defects in the lining, and establish the current structural condition of the tunnel (Figure 2). The inspection indicated the many defects in the existing tunnel lining and the feasibility of repair (McMillen Jacobs 2017).

The study also performed geotechnical investigations to ascertain site conditions for the new tunnel alignments, a condition assessment of the existing tunnel, and a detailed hydraulic analysis to

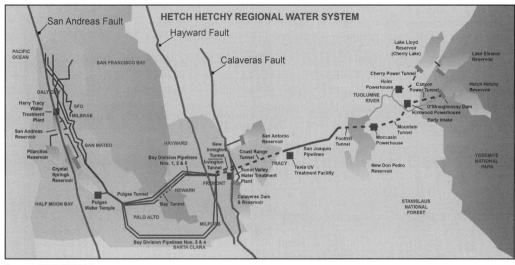

Source: SFPUC 2017.

Figure 1. Mountain Tunnel is downstream of Hetch Hetchy Reservoir

assess the flow capacity of the existing tunnel and evaluate effects of various improvements. The overall assessment was that existing tunnel was found to be not beyond repair and many sections to be in relatively good condition (McMillen Jacobs 2017). The hydraulic analysis indicated that relining the tunnel would significantly reduce hydraulic capacity (McMillen Jacobs and Black & Veatch 2017). So, the selection of the Preferred Project came down to comparing the Rehabilitation and the New Bypass Tunnels against the performance standards and the other considerations of construction, environmental, permitting, cost and schedule described in this paper. After thoroughly reviewing all this information and data at hand, the SFPUC ultimately decided the best course of action was to rehabilitate the Mountain Tunnel.

WHY REHAB?

The purpose of this paper is to explain why tunnel rehabilitation can sometimes meet the needs of water agency owners better, or just as well, as new tunnels or bypass tunnel projects. Many owners are resource limited. They have competing program priorities, limited capital resources, and limited staff expertise to undertake and properly manage mega size new tunnel projects.

Water agency capital budgets typically include many improvement projects competing for the same capital. The project list can include improvements to treatment facilities, transmission pipelines, pump stations, and other facilities besides tunnel projects. Not every project gets immediately funded and some are postponed for later funding in a ten year plan.

Source: Robin Scheswohl/SFPUC 2017.

Figure 2. Cataloging every lining defect during the 2017 tunnel inspection

Owners want to stretch every dollar to cover as many priority projects as possibly.

Water agency Owners typically have to sell bonds to fund their capital projects. The sale of bonds is a long term financial commitment that many water districts are more hesitant to undertake in these fiscally conservative times. Smaller water districts, in particular, are not willing to mortgage the future operations of the district to significant long term debt.

Some districts have to also contend with stakeholder and customer opposition to rising of water rates to service the debt payback. After years of increased water rates to pay for improvements on other parts of the SFPUC water delivery system, ratepayers would be understandably wary of additional increases to their water bill to pay for a new Mountain Tunnel.

IT'S ABOUT LESS MONEY AND TIME

When choosing to upgrade major transmission facilities, most water districts would look seriously at cheaper alternatives rather than undertake a total replacement facility. Most Owners would want to avoid a situation where the entire program budget is dominated by mega size projects, with outsized impacts on annual cash flows to service bond debt and unpopular customer rate increases to cover the bond repayments—as may be the case of large capital improvement programs, including new, expensive tunnel projects.

Given a choice between constructing a new tunnel, or bypass tunnel, or simply a rehabilitation of the existing tunnel, Owners would naturally first look at the rehab option. Rehab projects are significantly cheaper than building from new. It is estimated that the upfront capital outlay of a rehab can be 30% to 50% of the cost of a comparable size and length of a new or bypass tunnel. Cheaper upfront capital investment in initial construction can mean increased financial capacity to invest in periodic maintenance of a rehabilitated asset for the long term.

Because the scope of work for a rehab is significantly less, the rehab may be completed in approximately half the time or significantly less time for comparable size facilities. This also translates to significantly less expense for project management and other soft costs. A relatively simple rehab project will not require large consultant services contracts for highly skilled and expensive design consultants. Nor would just-as-expensive construction managers and inspectors need to be hired on large, multi-year contracts to do contract administration and in-tunnel inspections.

There is significantly less cost and schedule associated with property acquisition for a rehab project. No permanent tunnel easements or rights of ways are needed for new tunnel alignment. Permitting for temporary construction surface access and staging areas for non-agency owned property take less effort because either the rehab project does not require them or require less of them. In the case of the new tunnel alignment proposed for Mountain Tunnel outside the tunnel right of way, there was the need to acquire tunnel easements and temporary surface rights for staging and access through private property and national forest lands. The property issues included concerns about non-cooperative private owners, working with the United States Department of the Interior, and possibly with United States

Congress for the final approval—a potentially long and daunting process.

Rehab projects can be done without the need for subsurface exploration. A new tunnel by comparison must conduct an extensive geotechnical investigation along any new alignment, at locations where tunnel portals, adit tunnels, and vertical shafts are contemplated for the design. All this is dependent on gaining temporary access usually through private or other agency owned property. The longer the new alignment, the more complicated and time-consuming will be the effort. All such considerations are essentially cost and time savings for Owners who elect to go with the rehab option.

CONSTRUCTION CONSIDERATIONS

With any new tunnel, there is also an increased risk of cost overruns due to technical challenges, unforeseen conditions, large upfront costs for tunnel boring machines and other support equipment, constraints on the electrical grid for construction power, and environmental unknowns. Each new day of new tunnel excavation can yield a new set of unknowns and the potential differing site condition change order.

By contrast, tunnel rehab projects are somewhat repetitive and have fewer unknowns that could promote a lot of cost overruns. The tunnel repairs are typically classified into about four types ranging from patching of small holes to routing, cleaning, and backfilling of much larger cavities with welded wire mesh and shotcrete (Figure 3). The design details and the construction for these repairs are fairly similar from rehab to rehab project. It is fairly easy for contractor labor crews to get proficient after the first few repairs of each type of repair, gain efficiencies from lessons learned, and possibly make remarkable reductions in the unit cost and unit time of the repairs. This would make subsequent repairs of the tunnel more efficient and less likely to overrun.

The scope of a rehab project also lends itself to flexibility during construction. Tunnel rehab construction focuses primarily on repair of the concrete lining. Such repairs are discrete as opposed to continuous construction along the tunnel alignment that must be completed in blocks or sections before any return to service can be contemplated. Discrete repairs can be done in prioritized batches, or as much as the Owner's shutdown window constraints and budget allows.

POLITICS OF NEW VS. OLD

Sometimes, project leadership has been known to advocate for new replacement projects for reasons other than the practicality of cost and time, or even beyond the performance standards. New projects promise improvements that can be considered as

Source: SFPUC 2017.

Figure 3. Cleaning the lining defect to exposed rock and reinforcing with wire mesh prior to shotcrete

panacea for problems ranging from the old facility as allegedly beyond repair, to local hiring for massive number of construction jobs, to having the opportunity to associate one's name with a "legacy" project.

By contrast, there is less glamor and notoriety associated with a repair or rehab project. Tunnel rehab projects tend to be very similar and familiar. The main work is the patching or repair of various defects ranging from small holes to larger sections big enough for an inspector to crawl into. The repair is usually performed by shotcreting followed by systematic contact grouting through injection holes drilled through the repaired sections and into the native rock surrounding the tunnel in order to fill the annular spaces between the rock and the lining. In case of the current SFPUC Mountain Tunnel, the repair details are almost a carbon copy of similar repair details developed 15 years ago for the East Bay Municipal Utility District's Claremont Tunnel.

ENVIRONMENTAL CONSIDERATIONS

New tunnel projects generally must undergo an extensive environmental review process before the project can be approved for construction. For projects in California, this involves the development, internal reviews, and public review process of an Environmental Impact Report (EIR) over several years. If a federal agency is involved as a project participant or for the approval of property for the tunnel easements and temporary construction staging areas, then an Environmental Impact Statement (EIS) is also required. The EIS can be done concurrently with the EIR, but is typically completed with lag of an additional 6 months to over one year after the EIR certification to allow the findings and related information in the EIR to be re-used in the EIS.

The environmental impacts of new tunnels are generally unavoidable and beyond simple mitigation offsets that are easily accepted by the communities that must host the construction of the tunnel facilities and staging areas. Neighborhoods must endure the visibility, dust and noise, and share the streets and highways with the hundreds of truck traffic on a daily or weekly basis over many years of construction, and may want more mitigation for their endurance.

New tunnel construction projects of large diameter and long alignment have the additional issue of dealing with exceedingly large volumes of tunnel spoils. Such spoils must be transported often at long distances to commercial disposal fill sites, or shorter distances to agency owned fill sites. The longer haul to commercial sites often involves disclosure of the volume of regulated diesel emissions that have health effects on the public. The shorter haul to agency owned sites may involve less diesel emissions but requirements on the back end of the project for site restoration and native plant re-establishment.

Rehab projects generally have little to no significant environmental impacts that must be described, mitigated, and publicly vetted before the project can be approved for implementation. Repairs, contact grouting, and other work to rehab or improve an existing tunnel are completed underground and with relatively little surface visualization. The standard construction impacts of noise, traffic, and environmental pollution can be avoided or mitigated by design to the point of insignificance. Rehab projects typically require smaller surface areas for temporary staging of construction trailers, equipment parking and storage of materials.

The environmental review process for a rehab project typically involves less documentation and less public review. Because the environmental impacts are so much less or can be avoided, it is often possible to publish a Mitigated Negative Declaration (MND) for the rehab project. If federal agencies are involved, the comparable federal document is the Environmental Assessment (EA). Both documents are easier and less time consuming to develop and may save about one year compared to the EIR/EIS process for the new tunnel (AECOM 2017).

ENVIRONMENTAL PERMITS

Separate from the environmental documentation process, is the issue of obtaining the environmental permits to allow construction to proceed. For example, in California, a Biological Opinion must be evaluated and obtained, and then California Department of Fish & Wildlife permits must be obtained. If the project involves federal participation or obtaining of easements, then United States Fish & Wildlife Service permits must be obtained. Such permits typically require special expertise either in-house or hired as consultants to interact with the permitting officials in order to work out conditions of approvals and details of the final project descriptions before the permit can be approved. It is not uncommon for such permits to consume over a year of critical schedule after the environmental review process is completed.

If the project requires new tunnel easements and temporary surface easements for staging areas in national park land or forest preserves, as was the case for the new Mountain Tunnel options, then the permitting process is more complicated. The federal agency may pre-condition the granting of property easements on the approvals of the environmental review process, and in the process add more conditions of approval to the environmental permits.

MEETING PERFORMANCE STANDARDS

It is generally assumed that new tunnel projects can be designed to satisfy any set of performance standards for the project. While this is basically true, it over shadows the fact the most performance standards are derived from the design and operations of the existing facility, with perhaps a few upgrades.

In the case of SFPUC Mountain Tunnel Improvement Project, some of the Performance Standards were based upon the performance of the existing tunnel back when the tunnel was relatively new. A few standards were derived from the current operations of the tunnel. There were eight performance standards used as criteria for the selection of the Preferred Project during the project alternatives analysis (McMillen Jacobs 2017):

- Service Life: This standard requires the typical tunnel design for 100 years of service life. Although the best way to meet this standard is to construct a new tunnel, the rehab option can also achieve a 100 year service life. The Mountain Tunnel design consultant's solution was to fix all the defects in the concrete lining with welded wire reinforcement and shotcrete, and perform contact grouting to fill all the annular spaces between the lining and the surrounding rock (Figure 4). When completed the lining should be as structural sound as a new lining and the rehab tunnel should last another 100 years with normal, periodic maintenance.

- Water Quality: This standard limits the overall turbidity from Mountain Tunnel to occurrences of over 1 NTU to no more than twice per year, and occurrences of over 100 NTU to no more than once every 5 years. During normal operations, groundwater intrusion is the main culprit for degrading water quality. For both new and existing tunnels, a way has to be found to limit this intrusion. For concrete tunnels, the best way to cut off the intrusion seepage pathways is to do an adequate

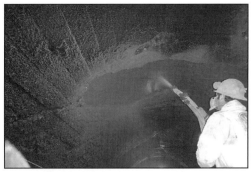

Source: SFPUC 2017.

Figure 4. Repairing the lining defect with shotcrete

job of grout injection of the native ground surrounding the tunnel, or contact grouting. For the rehab project, the entire 11 miles of concrete lined section will be aggressively contact grouted, essentially sealing the rehab tunnel from seepage.

As an improvement, Mountain Tunnel will also install new large control valves at the downstream portal to keep the tunnel full of water when the tunnel is not running. With the tunnel full and pressurized, there would be little to no hydraulic gradient for the initiation of groundwater intrusion. In addition, a short section of very leaky tunnel, upstream of the South Fork Siphon crossing underneath the Tuolumne River, will be replaced by a new 450 foot long bypass tunnel section. This will eliminate the one worst section where groundwater intrusion occur the most.

- Water Conveyance Capacity: This standard requires a hydraulic capacity of 740 cfs or 478 MGD. Advantage goes here to a new tunnel, in that a new tunnel can be sized to accommodate any flow capacity. However, in the case of the Mountain Tunnel rehab, flow capacity will be enhanced by the complete repair of wall defects, and invert paving and possibly smoothing, to improve hydraulic efficiency. The rehab project should be able to recover 706 cfs of initial capacity, or over 95% of this performance standard. The SFPUC found this very sufficient and efficient when the consideration of budget cost and schedule savings over the new tunnel are factored in. Also, the addition of downstream control valves to keep the tunnel flow at full volume will eliminate the erosive effects of the current tunnel operations, with intermittent surges and turbulent transitions between full flow and open channel flow inside the tunnel on a daily basis.

- Minimum Flow: This standard requires a minimum flow rate of 300 MGD be available at all times outside of planned and unplanned outages. This is actually a fairly easy criterion to satisfy for a new tunnel and a rehab tunnel. For an existing tunnel, the key is to do the repairs of lining defects competently so that lining fallout does not occur and block flow capacity. This is accomplished by routing the defects back to structurally sound concrete, and backfilling the cavity with welded steel reinforcement, and high strength shotcrete. The resulting repair would be as structurally sound as new lining.

- Operational Flexibility: This standard includes four key operations. Mountain Tunnel must accommodate reductions in demand such that the tunnel may operate in open channel flow for extended periods. The tunnel needs to operate at full portion to meet water supply needs. The tunnel needs to accommodate power generation and local recreational needs, such that the tunnel may operate with substantial fluctuations in daily and hourly flows to the extent possible. The tunnel needs to accommodate full dewatering every *five years* for 100-day shutdowns for needed inspections of the Hetch Hetchy Aqueduct. The new tunnel and rehab tunnel can both be designed to handle all of these operational needs. In the case of the Mountain Tunnel rehab, downstream control valves will be added to maintain full volume flows so the erosive effects between full flow and open channel flows can be significantly avoided. With the downstream control valves, keeping the tunnel full of water, flows can be ramped up and down relatively quickly without developing the vacuum or surge pressures that promote erosion of the lining.

A related operational consideration during construction is the owner's requirement for emergency return to service. The implementation of tunnel repairs can be done in finite prioritized batches. Such repairs are discrete as opposed to continuous construction along the tunnel alignment that must be completed in large units before any return to service can be contemplated. This is a very important consideration for any owner that encounters an event that requires an emergency return to service. Such emergencies usually require the curtailing of construction and return to water services over a few days. The 2017 Mountain

Tunnel inspection and interim repair contract had a 3-day requirement for the contractor to return the tunnel back to the owner for emergency return to service.

- Planned Outages: This standard requires the reliable operation of the tunnel with an inspection frequency of 20 years with outage durations limited to 30 days, and major repairs at no more than once every 20 years with outage durations limited to 100 days. This is fairly easy for a new tunnel or a rehab tunnel to satisfy. After completion of both types of projects, the key is to not ignore the periodic maintenance required to keep the tunnel lining in good physical condition, and eliminate the need for major maintenance that often results from neglect. In the case of the rehab tunnel, the repairs need to done competently so that lining fallout does not occur and require major maintenance.

For Mountain Tunnel, simple planned inspections would only require outage durations of less than 10 days. A 30 day outage would allow time for some patchwork repairs of the lining. These short duration outages should be conducted concurrently with the 5 year periodic outages for the Hetch Hetchy Aqueduct inspection interval under the Operational Flexibility Performance Standard. Periodic inspections every 20 years with outage durations of 100 days for the Mountain Tunnel should only be planned if major repairs are needed. Again, the goal of the inspection outages should be to catch the incipit defects when such defects are still small in size and fairly easy to repair.

- Unplanned Outages: This standard limits the interruption in water delivery from a catastrophic event to no more than 90 days. Although there is uncertainty with any catastrophic event, the new tunnel and rehab tunnel can both be designed to make the lining as robust as possible to withstand shakeout from the forces of remote earthquake faults, or inadvertent damage from man-made events. Such is the case with the Mountain Tunnel rehab. The 2017 inspection found the existing tunnel lining to be an average of 14 inches thick. The rehab will structurally repair all the defects and the entire 11 miles of lined section will be contact grouted to make sure the lining is in intimate contact with the surrounding granitic rock. By doing so, any need for repairs after a catastrophic event will be mitigated and the forecast interruption for repairs should be less than 90 days.

- Seismic Reliability: This standard requires the reliable delivery of the Minimum Flow without interruption following a near tunnel seismic event. This is the easiest of the Performance Standards to satisfy in that the tunnel does not cross any active earthquake faults and the Sierra foothills location of the tunnel is in a region of low seismic activity.

RECOMMENDED INSPECTIONS AND MAINTENANCE

Comprehensive and competent repairs and contact grouting during the rehab construction should produce a tunnel whose lining is free from defects and with a renewed service life that compares with a new tunnel lining. It is important for owners to support the renewed tunnel with proper, periodic inspection monitoring, water quality testing, and maintenance. The inspection should be conducted at reoccurring intervals of between 5 to 20 years, as required by tunnel condition. The inspections may have to be conducted at shorter intervals if it is noted during the initial inspection that the erosion is occurring more aggressively than anticipated. The key is to catch any new defects in the lining while they are still incipiently developing. Such defects should be small in scope and more easily addressed in subsequently scheduled tunnel shutdowns that are well planned and budgeted in advance. If the repairs can be scheduled periodically at intervals of no more than 20 years or concurrent with the inspections, then the scope of repairs will be less significant, cost less per shutdown, and be able to be accomplished in fewer shutdowns of shorter duration, with better control of the scheduling and costs of the work—all good considerations for budget and operation minded owners.

CONCLUSION

Done right, the rehab tunnel project can result in a renewed tunnel that can match the 100 year service life and other performance standards of a new tunnel but at a fraction of the cost and schedule. The new tunnel can be a more glamorous project to design and construct, but the renewed tunnel will typically require a less complicated and time consuming environmental review process and fewer environmental permit conditions for completion. In the case of Mountain Tunnel, the rehab project can almost match the required flow capacity of the new tunnel with the same accommodations for operational flexibility, planned and unplanned outages, and seismic reliability. As with any new or renewed tunnel, the key to facility longevity, without major headaches, is the attention and commitment to performing periodic inspection and maintenance. A well designed and executed periodic repair program with the newest engineering and construction methods can yield a similar results in a fiscally responsible way. The rehab option will successfully preserve the SFPUC's Mountain Tunnel and meet its needs for many years to come.

REFERENCES

McMillen Jacobs Associates, 2017. *Mountain Tunnel Improvements Project Alternatives Analysis Report*, prepared for the San Francisco Public Utilities Commission, July, Draft Report.

McMillen Jacobs Associates, 2017. *Mountain Tunnel Improvements Project Inspection Report*, prepared for the San Francisco Public Utilities Commission, June, Final Report.

McMillen Jacobs Associates, 2017. *Mountain Tunnel Improvements Project Condition Assessment Report*, prepared for the San Francisco Public Utilities Commission, June, Draft Report.

McMillen Jacobs Associates and Black & Veatch, 2017. *Mountain Tunnel Improvements Project Hydraulic Analysis for Conceptual Improvement Alternatives*, prepared for the San Francisco Public Utilities Commission, June, Final Report.

AECOM, 2017. Opportunities and Constraints *Report for the Mountain Tunnel Improvements Project*, prepared for the San Francisco Public Utilities Commission, July, Environmental Report.

Sewer Tunnel Beneath Meramec River to Fulfill Regional St. Louis Treatment Plan and Environmental Vision

Everett L. Litton and Mark J. Stephani
WSP USA Inc.

Jerry L. Jung
Metropolitan St. Louis Sewer District

ABSTRACT: Phase II of the Lower Meramec River System Improvements Project, also known as the Lower Meramec Tunnel (LMT), consists of a 6.8-mile-long, 12-foot excavated diameter, 78 to 286-foot-deep sanitary sewer tunnel. The LMT is part of Metropolitan St. Louis Sewer District's (MSD) Project Clear; a program planned to span 23 years to improve water quality throughout MSD's service area. The tunnel's main objective is to intercept flows and to take offline the interim Fenton Wastewater Treatment Facility (WWTF). This paper presents project details, geologic conditions, hydrogeologic challenges, initial support, final lining and the project's bidding timeline. Other topics include: incorporating lessons from Phase I (Baumgartner Tunnel), risk mitigation, unique chert characterization and considerations to address flooding within the project area.

BACKGROUND

As a result of East-West Gateway Council of Governments' studies performed in 1970 and 1972, MSD's service area was extended to include pollution abatement within the Lower Meramec River Basin in south St. Louis County. The first step in improving water quality within the Lower Meramec River Basin included development of the 201 Facility Plan, which was originally prepared in 1979. Previous investigations concluded that a regional concept involving a major interceptor sewer conveying flow to a single regional treatment facility that would discharge its effluent to the Mississippi River, was the most cost-effective wastewater scheme within the Lower Meramec River Basin.

The Meramec River, one of the longest free-flowing waterways in Missouri, is widely used for recreational boating, fishing, canoeing and floating and is the central natural element for a series of parks located along the river's banks. At one time, the Meramec River was considered one of the most polluted rivers in Missouri. Through efforts made by a variety of local organizations, MSD, the Missouri Department of Natural Resources (MDNR) and other state and federal agencies, significant strides have been made in cleaning the river. In 1979, the Meramec River received wastewater discharges from more than 40 treatment facilities which served over 90 percent of the development within the area. MSD has since removed most of these facilities and now utilizes three treatment facilities: the Lower Meramec, Fenton and Grand Glaize WWTFs. Of these, the Fenton and Grand Glaize WWTFs were intended to be interim treatment facilities and functioning until construction of the interceptor sewer could take them offline.

In 1985, a major update to the 201 Facility Plan was prepared. The purpose of this update was to include revisions to better facilitate the long-range goal of pollution abatement and to include the results of ongoing activities within the planning area. The update revised the interceptor sewer alignment and separated the concept into three phases, Phase I (Baumgartner Tunnel), Phase II (LMT) and Phase III (Grand Glaize Tunnel), to better integrate with the existing and proposed interim treatment facilities.

In 1998, an alternative analysis was performed which initiated design of the Baumgartner Tunnel. The Phase I tunnel was designed by Horner & Shifrin, Inc. with final design completed in 2003. The 12.5-foot excavated diameter, 20,200 feet long tunnel (Nickerson et al, 2005) was constructed by the Baumgartner Tunnel Joint Venture, a joint venture of Frontier-Kemper Constructors and Gunther Nash, from 2004 to 2007.

Between 2005 and 2012, HDR Engineering, Inc. evaluated alternatives for expanding the Lower Meramec WWTF, including an alignment study for Phases II and III. In 2014, MSD awarded a design services contract to HDR Engineering, Inc. and, as a sub-consultant, WSP USA, Inc. (formerly Parsons Brinckerhoff) who are serving as the lead tunnel design firm.

Source: The State of Our Missouri Waters, Meramec River Watershed (MDNR, 2015)

Figure 1. Meramec River Watershed

PROJECT DESCRIPTION

The LMT Project, formally known as the Lower Meramec River System Improvements—Baumgartner to Fenton WWTF Tunnel Project, consists of an approximately 35,932 feet (6.8 miles) extension of the Baumgartner Tunnel at a slope of 0.1%. The project is located in south St. Louis County and traverses portions of unincorporated St. Louis County, City of Sunset Hills and the City of Fenton. The depth of cover ranges from 78 to 286 feet, largely controlled by surface topography. The excavated tunnel diameter is anticipated to be approximately 12 feet with an 8-foot internal diameter final lining. The tunnel will be constructed entirely in rock with an invert elevation of 255.1 feet NAVD88 where it connects with the Baumgartner Tunnel and at elevation 291.1 feet where it terminates near the Fenton WWTF.

The proposed tunnel is anticipated to operate under both open channel and submerged flow conditions. Dry weather flow velocities of the Phase I and Phase II portions of the tunnel are anticipated to range from 2.5 ft./sec. to 4.9 ft./sec., which provides sufficient velocity between the minimum (2 ft./sec.) and maximum (10 ft./sec.) flow velocities as identified in the Recommended Standards for Wastewater Facilities (10 State Standards, 2014) for grit and solids suspension.

Drop structures have been spaced along the alignment to accommodate as much gravity flow to the tunnel as practical and economical. Selection of the drop structure type has considered technical and functional criteria that influence the capital and life-cycle costs associated with constructing and operating the drop structures for the duration of their service life. The following criteria were applied in the process of selecting a recommended drop structure type:

- General construction and cost considerations;
- Hydraulic performance;
- Air management and odor control;
- Durability; and
- Operations and maintenance.

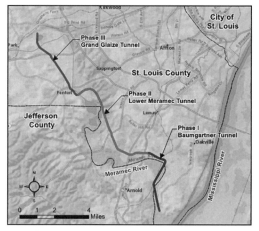

Figure 2. Program vicinity map

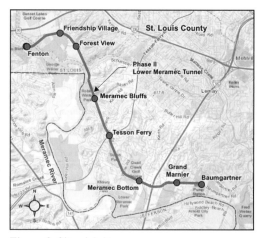

Figure 3. Site vicinity map

Table 1. Drop structure design flows

Drop Structure	Tributary Area, acres	10-Year Design Flow*, cfs
Grand Marnier	500	3.2
Meramec Bottom	2,530	12.5
Meramec Bluffs	170	1.8
Forest View	630	6.0
Friendship Village	700	2.2
Fenton	5,950	35.1

* Design flow based on larger value of simulated cloudburst and synoptic storm events.

Table 2. Construction shaft dimensions

Shaft Name	Purpose	Approx. Project Station, feet	Min. Inside Diameter, feet
Baumgartner Construction Shaft	TBM Launch Shaft and Primary Staging Area	202+99	35* (soil) 24* (rock)
Optional Tesson Ferry Construction Shaft	Intermediate Construction Shaft	362+58	20 (soil) 18 (rock)
Fenton Construction Shaft	TBM Retrieval Shaft	562+31	32 (soil) 28 (rock)

* Dimensions are existing based upon previous construction. Existing shaft will be utilized during Phase II construction.

Four types of drop structures were considered: tangential vortex, vortex with helical ramps, plunge inlet and baffle type drop structures. Both plunge and tangential vortex drop structures were considered suitable. For locations where the design flow rate was expected to exceed 10 cfs, tangential vortex drop structures (Jain and Kennedy, 1983) were recommended. Smaller plunge style drop structures were utilized elsewhere.

During construction of the Baumgartner Tunnel forethought was considered and the terminal tunnel boring machine (TBM) receiving shaft, the Baumgartner Shaft, was built to facilitate construction of LMT.

The existing Baumgartner Shaft is located on MSD property at a formerly decommissioned sludge lagoon facility. The shaft is approximately 190 feet deep with the upper 130 feet constructed as a slurry wall which was keyed 4 feet into bedrock where supplemental grouting was performed. The lower 60 feet of the shaft was constructed through limestone bedrock. The shaft is currently unlined and will serve as LMT's TBM launch shaft during construction.

Up to 27 acres are available at the Baumgartner Shaft site to utilize as the contractor's primary work/staging area. The site is amply suitable for positioning a crane, materials storage, lay down area, contractor office trailers, sub-station, workshop and parking. Additionally, a decommissioned lagoon located on the site is intended to be used for muck disposal thus minimizing haul traffic and disposal requirements.

Three shafts are proposed near the Fenton WWTF, with the following key functions:

- Accommodate tunnel construction for TBM retrieval and the ability to launch a TBM for the Grand Glaize Tunnel (Phase III), if required;
- Intercept the Fenton WWTF flows via drop structure, deaeration chamber and connection adit to the LMT; and
- Provide adequate venting at the upstream terminus of the LMT to mitigate potential surge and transient flow conditions.

The Fenton Construction Shaft is located on property owned by MSD near the Fenton WWTF. The shaft is in a low-lying area within the Meramec River floodplain adjacent to Opps Lane and near the bank of the Meramec River. The size of the site is approximately 8 acres which provides sufficient laydown space for construction activities. The shaft is anticipated to be approximately 136 feet deep with a depth to bedrock of approximately 63 feet. Slurry or secant pile wall construction will be specified for initial support in overburden with supplemental grouting at the bedrock-soil interface to provide a relatively watertight construction operation.

The Fenton Construction Shaft site is located within the 100-year floodplain and significant flooding has been experienced, including recent flood events in December 2015 and May 2017. A portion of the site will be raised for flood protection and resiliency purposes. To maintain relatively consistent nearby grade and to limit the amount of floodplain fill, the final configuration of the shaft cover will be set above grade and the 500-year flood elevation to protect against flooding. Fill material will be placed to raise the grade around the shaft structures above the 100-year flood elevation. Additionally, vent shafts for the entirety of the project will extend above the 500-year flood elevation.

GEOLOGIC SETTING

Geologically LMT traverses upland and lowland areas of the Meramec River valley. The lowland valleys are primarily characterized by alluvial deposits ranging from silts to lean clays and from sands to gravels at greater depths. The lowland valleys are generally present at the beginning and end of the alignment near Baumgartner and Fenton, respectively. The upland area, which is higher in elevation and within the middle portion of the LMT alignment, generally consists of relatively minimal loess overburden except within channels flowing to the Meramec River. Bedrock conditions within the vicinity of the project area consist of the Salem Formation, followed by the Warsaw Formation, which is underlain by the Burlington-Keokuk Limestone.

The Salem Formation consists of gray to brown limestone and dolomite which is calcarenitic, fine to coarse-grained, occasionally cross bedded with thin to massive bedding and oolitic in places. The Salem Formation also contains scattered light gray chert. The Salem Formation is slightly weathered to unweathered with areas of moderately to highly weathered rock expected near the bedrock-soil interface. The Salem Formation overlies the Warsaw Formation. The contact with the underlying Warsaw Formation is gradational.

The Warsaw Formation consists of an upper and a lower portion. The upper portion of the Warsaw Formation consists of dark gray, fissile shale. The lower portion of the Warsaw Formation consists of calcareous shale and gray limestone which is finely to coarsely-crystalline, occasionally dolomitic and contains geodes and scattered light gray to white chert nodules and stringers. The Warsaw Formation is slightly weathered to unweathered with areas of moderately weathered rock expected near the bedrock-soil interface. The Warsaw Formation conformably overlies the Burlington-Keokuk Limestone and is commonly considered to be the basal Meramecian Series Formation in the region. The lower contact is gradational into the underlying Burlington-Keokuk Limestone.

The Burlington-Keokuk Limestone consists of the Burlington Formation and the overlying Keokuk Formation. Locally they are so similar in appearance and properties that they are considered a single unit. The Burlington-Keokuk Limestone consists of white to blueish gray limestone, which is medium to coarsely-crystalline, medium to thick bedded, fossiliferous and chert bearing. The Burlington-Keokuk Limestone is slightly weathered to unweathered with areas of moderately weathered rock to be expected near the bedrock-soil interface and at depth. The chert appears in the form of light gray nodules and layers. The amount of chert present in the Burlington-Keokuk Limestone has been reported to comprise more than 50 percent of the rock in some layers (Thompson, 1986).

SUBSURFACE INVESTIGATION

The subsurface investigation program has implemented a phased investigation approach. Geotechnical data was originally collected during the alignment study in 2012 with subsequent data collected in 2014, 2015, 2016 and 2017. The investigation program was further developed and refined to fill data gaps and for risk mitigation in subsequent phases.

The investigation program has consisted of 63 borings along the proposed tunnel alignment ranging from 160 to 312 feet below ground surface. Twenty-two of the borings were inclined at 15 degrees from vertical to improve the likelihood of encountering near-vertical joint sets. The remainder of the borings were drilled vertically. Borings were drilled through the overburden using either a casing advancer system or hollow-stem augers. Rock was cored using HQ sized tooling. Sampling of the soil overburden was generally not conducted except at locations under consideration for shaft construction.

Packer testing was performed to evaluate the hydraulic conductivity of joints and bedding planes in the bedrock. Tests were typically accomplished at 10-foot intervals using a straddle-packer system starting from the bottom of the boring and continuing sequentially up at least two tunnel diameters above the anticipated crown and more extensively at potential shaft locations. Tests were performed in general accordance with the Engineering Geology Field Manual (USBR, 2001).

Rock was logged using either an optical or acoustical televiewer to provide data on features and discontinuities observable along the sidewalls of the borings. Open standpipe piezometers were installed at several locations along the alignment. Piezometer screens were generally set within the bedrock at depths where rock conditions, packer results and downhole geophysical data indicated zones of higher hydraulic conductivity.

Laboratory testing was performed on samples of soil and rock obtained from the borings. Most of the testing was accomplished on rock samples within one tunnel diameter above or below the tunnel horizon. Rock tests included porosity, density, slake durability, free swell, point load (diametral), point load (axial), uniaxial and triaxial compression, Brazilian tensile, direct shear, chert content, Mohs hardness, Cerchar abrasivity, punch penetration, TBM Suite (abrasion, brittleness and drillability) and thin-section petrographic analysis. Soil testing was limited to borings at potential shaft locations and included moisture content, Atterberg Limits, grain size, specific gravity and triaxial compression testing.

ANTICIPATED TUNNELING CONDITIONS

Bedrock conditions anticipated to be encountered in the tunnel consist of the Warsaw Formation and the Burlington-Keokuk Limestone. The Warsaw Formation is composed of limestone and shale with minimal amounts of chert whereas the

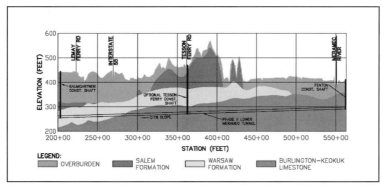

Figure 4. Geologic profile

Burlington-Keokuk Limestone is typically limestone and chert which, in places, composes up to 40% to 60% of the rock mass. Figure 4 summarizes the anticipated extent of each geologic formation based upon an interpretation of the geotechnical investigation.

Rock core samples and acoustical/optical tele-viewer data shows bedding planes as the primary structural feature. The bedding planes are mostly horizontal to sub-horizontal with cross-bedding joints existing randomly. The vast majority of the joints observed are near horizontal with a dip angle less than 15° and dip directions ranging from 0° to 360°. Rock Quality Designation (RQD) values average 87% with approximately 60% of data greater than 90%. The rock core is generally characterized as fresh (unweathered) to slightly weathered.

Rock Mass Rating (RMR) classifications were performed based on observation of rock core samples obtained from the borings and subsequent laboratory tests. RMR values range from 40 to 69 with an average of 55. Median unconfined compressive strength within the Warsaw Formation and the Burlington-Keokuk Limestone is 7,546 psi. and 7,289 psi., respectively. Table 3 presents the proposed initial ground support types along the tunnel alignment.

An open-face, main-beam TBM is anticipated to be utilized for construction of the tunnel with drill and blast excavation anticipated for construction of the deaeration chambers, adit connections and construction shafts. The TBM will be required to be capable of probing and grouting ahead of the face and mandatory probing and grouting will be required along approximately 15% of the alignment. Additional probing and grouting will not be required unless certain specified inflow criteria are encountered.

BAUMGARTNER TUNNEL CONSTRUCTION

Based upon previously published papers (Nickerson et al., 2005 and Abkemeier and Groves, 2011) and project documentation provided by MSD, the following generally describes the Baumgartner Tunnel construction and some of the challenges encountered. The Baumgartner Tunnel is approximately 20,200 feet long, was excavated to 12.5-foot diameter and lined with 96-inch diameter, reinforced concrete pipe with a polyvinyl chloride (PVC) liner. The project is located near the confluence of the Meramec and Mississippi Rivers and included two large diameter shafts at the Lower Meramec WWTF and three permanent access shafts. Five drop structures were constructed as part of the project.

The tunnel commences in St. Louis County, crosses to Jefferson County, then terminates in south St. Louis County after crossing the Meramec River twice. The tunnel was bored at depths of about 160 to 180 feet below ground surface and passed through the Warsaw Formation and the Burlington-Keokuk Limestone. A rebuilt 3.8m (12.5 ft.) TBM was utilized during construction with a refurbished conveyor.

On June 24, 2004, during drill and blast excavation of the screen shaft at the Lower Meramec WWTF, groundwater inflow was observed through several of the vertical production holes drilled in the shaft bottom. The inflow into the screen shaft was estimated between 200 to 250 gpm through approximately eight drill holes scattered across the bottom

Table 3. TBM initial support

Support Type	Support Description	Percentage of LMT Length, %
Type I	2 Pattern rock dowels*	40
Type II	4 Pattern rock dowels with steel channel*	59
Type III	Steel ribs with partial timber lagging	1

* Spot rock dowels, welded wire fabric and mine straps will be included as supplemental support and will be installed as required.

of the shaft. A subsequent grouting program was initiated from within the shaft entirely around the perimeter to mitigate the risk of a larger groundwater inflow. Thirteen primary grout holes were drilled in the sidewall of the shaft with subsequent secondary, tertiary and quaternary grout holes also drilled.

The feature encountered within the screen shaft was theorized to be a horizontal water-bearing feature thought to be associated with the transitional contact between the Warsaw Formation and the Burlington-Keokuk Limestone. Additional exploration was performed during construction, which demonstrated a zone within the transitional contact between the Warsaw Formation and the Burlington-Keokuk Limestone had the ability to transmit a large volume of water. Hydrogen sulfide and methane gas were found to be in solution within the groundwater. A successful grouting program was performed at the lift station shaft and along the tunnel from Station 37+50 to 70+00 with grouting under the Meramec River accomplished using angled holes from the river bank.

ANTICIPATED CHALLENGES

Given the fact that the LMT is an extension of the Baumgartner Tunnel, design has taken an approach to focus on lessons learned from previous tunnel construction in similar geologic conditions. From the onset of the design, risk management has been implemented to mitigate known challenges as previously encountered. The subsurface investigation program was developed to mitigate the following anticipated challenges, which include:

- Water-bearing zones within the rock formations;
- The presence of hydrogen sulfide and methane gas; and
- Abrasive chert.

Groundwater Control

Due to the large, unanticipated groundwater inflows observed during construction of the Baumgartner Tunnel, the subsurface investigation program for LMT was tailored to include aquifer testing (pump tests) to assist in mitigation of this risk. A focus was placed on the transitional contact between the Warsaw Formation and the Burlington-Keokuk Limestone. Additionally, supplementary geotechnical data was collected at the location of lowest rock cover along the alignment, which occurs at the tunnel's crossing of the Meramec River near the Fenton WWTF.

To quantify the amount of anticipated groundwater into the excavated tunnel and shafts, packer testing was performed. 493 packer tests were performed along the tunnel alignment. Data obtained from these tests were used to estimate the rock's equivalent hydraulic conductivity.

Approximately 70 percent of the packer tests displayed a "very low" flow rate with corresponding hydraulic conductivities less than 1×10^{-5} cm/sec. Extremely high flows were observed within a few of the packer tests, however generally these flows occurred more than two tunnel diameters above the tunnel crown. Several moderate to high hydraulic conductivities were observed within two diameters of the tunnel with most of them occurring near two locations. One near the transitional contact zone of the Warsaw Formation and the Burlington-Keokuk Limestone and the other near the crossing of the Meramec River.

With an objective to better quantify the hydraulic conductivity of the observed "moderate to high" zones and ultimately improve the reliability of the groundwater inflow estimate, aquifer tests (pump tests) were performed at the two areas of interest. At each aquifer test site one pumping well and four piezometers were utilized. Prior to the collection of background water level measurements, the pumping wells and piezometers were developed. Baseline measurements were established, which included water level measurements in each piezometer and pumping well every four hours for one month. The water level readings were then compared to barometric pressure, precipitation and the stage of the Meramec River over the same period.

A step-drawdown pumping test was performed at each aquifer test site to establish the pumping rate to be used in the subsequent constant discharge test. Stressing the aquifer by pumping water from it at several flow rates, or steps, helped identify the pumping flow rates the aquifer could sustain. The pumping well was allowed to fully recover after the step-drawdown test and before the constant discharge test began.

After the pumping well and piezometers were fully recovered from the step-drawdown tests, constant discharge tests were performed. Water quality measurements, including temperature, pH, specific conductance and turbidity were measured before and during the constant discharge test. On February 7, 2017, the constant discharge test was performed near the crossing of the Meramec River and the pumping process lasted approximately 24 hours. Additionally, the constant discharge test was performed near the transitional boundary at the Meramec Bottom site on February 9, 2017 and lasted approximately 49 hours. For each of the two aquifer tests, the pumping rates were adjusted slightly throughout the test to maintain the discharge rate to within about 10 percent of the target flow rate.

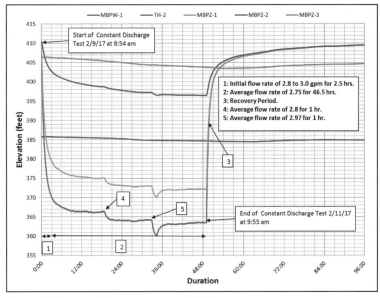

Figure 5. Meramec bottom constant discharge test

Results of the long-term aquifer testing were considered to provide a more comprehensive evaluation of the rock mass hydraulic conditions that more closely simulate groundwater drawdown conditions during tunnel construction. During construction, the groundwater will drain from the surrounding rock mass; a condition that is simulated by the long-term aquifer test. Packer test results, on the other hand, evaluate the rate at which water inflow can penetrate the rock mass under a range of pressures, yielding a hydraulic conductivity value. While packer testing is common within the tunnel industry to estimate hydraulic conductivity, its relatively small zone of influence and the difference between pressure injection and gravity drainage are shortcomings of the procedure.

Presence of Chert

Chert is composed primarily of microcrystalline silicon dioxide (SiO_2) characterized as being hard (Mohs hardness of 7) and highly abrasive. Chert deposits are typically found as either irregular nodules or as layers of rock. Because seawater is seldom saturated with silica, most chert beds are believed to have originated from the remains of tiny organisms such as diatoms and radiolarians. The hardness and abrasiveness of chert typically causes a reduction in TBM penetration rates and more frequent cutter changes that may lead to a longer TBM mining schedule.

During the geotechnical investigation, chert was typically present within the rock core in layers

or nodules and varied in thickness from about 0.5 to 14 inches. Chert concentration was estimated visually as a percentage of the total core run length and ranged from 0 to 65 percent. Chert concentrations were present, but considered minimal in the Warsaw Formation, but significant chert concentrations were observed within the Burlington-Keokuk Limestone.

Unique laboratory testing, including x-ray diffraction and acid immersion were performed on select rock cores to verify the accuracy of the visual chert estimates. Core samples were selected close to the tunnel horizon and approximately 5-feet long core sections that contained both chert and limestone were tested. A total of 22 rock core specimens were tested in the laboratory to confirm the accuracy of the visual classification.

The samples were dried, weighed and subsequently immersed in acid containing 15% solution (by weight) of hydrogen chloride (HCL). The purpose of the testing was to allow the limestone material to dissolve in the acid solution while the chert and quartz materials remained intact. The remaining portion of the material included insoluble minerals and remnant soluble minerals that were not dissolved during the acid treatment. Samples were immersed until no reaction between the HCL solution and the specimen was observed. Verification testing was performed to ensure the acid solution was not neutralized. Once complete, the solid particles were again dried and weighed. The particles were pulverized to a powder and X-ray diffraction was performed to determine the percentage of quartz versus chert.

Figure 6. Specimen photo before and after acid immersion

Assessments were made on cutter consumption rate (Bruland, 1998 and Nilsen & Ozdemir, 1993) for the two rock types the TBM will penetrate: The Warsaw Formation and the Burlington-Keokuk Limestone. Intact rock properties were input to the models to calculate the cutter consumption rate. A cutter consumption rate of 235 to 380 yd³/cutter was estimated for the Warsaw Formation and 225 to 243 yd³/cutter for the Burlington-Keokuk Limestone. These values generally agree with construction records from the Baumgartner Tunnel and indicate the chert is expected to have an impact on TBM performance.

Anticipated Gases

Hydrogen sulfide and methane gas are known to exist in solution within the groundwater and was observed during construction of the Baumgartner Tunnel (Abkemeier and Groves, 2011). These gasses may have formed in soft organic clay deposits present under the sediment in the river valleys or they may also be naturally occurring within shale deposits. Although the formations anticipated for construction are generally made up of limestone and dolomite, the Warsaw Formation has a shale component.

According to the Center for Disease Control and Prevention (CDC) NIOSH Pocket Guide to Chemical Hazards (2007), the Recommended Exposure Limit (REL) for hydrogen sulfide is C 10 ppm, where the "C" indicates a ceiling value that should not be exceeded. The Occupational Safety and Health Administration (OSHA) Permissible Exposure Limit (PEL) is C 20 ppm. Methane is generally not considered toxic; however, it is flammable. Methane is also an asphyxiate and may displace oxygen in an enclosed space. The explosive limits for methane percent by volume in air is 5 to 15%. OSHA requires an action level of 10% of the lower explosive limit (LEL) for safety plans to ensure a safe noncombustible atmosphere.

During implementation of the subsurface investigation program, gas meters were used to sample the air at the top of the borehole during drilling. Readings were taken at the beginning of each day to measure the vapors that accumulated overnight. Positive readings were recorded for both methane and hydrogen sulfide, with the highest readings obtained of 0.5% by volume for methane and 0.5 parts per million (ppm) for hydrogen sulfide. Daily gas monitoring at the surface is not considered a reliable method for evaluating the potential for encountering gas at the tunnel horizon, however due to these initial readings as well as gas encountered during Baumgartner Tunnel construction, additional sampling and further characterization of the gasses was recommended.

A detailed groundwater sampling and testing program was developed. A total of 19 open standpipe piezometers were available along the tunnel alignment and were sampled in December 2016. Based on the results of the initial sampling event, eight wells were recommended for subsequent sampling occurring once a month from March through July, 2017.

Water samples were extracted from the wells via a bladder pump at low flow rates (approximately 0.4 to 0.6 L/min). The pump was lowered to a predetermined depth in each well with a drop tube connected to a screen below the pump to be situated at a depth approximately five feet above the bottom of the well to prevent pumping silt that may have accumulated. The bladder pump discharged water into a flow cell via ¼-inch diameter polyethylene tubing; the flow cell was connected to a YSI 556 MPS meter that measured temperature, specific conductance, dissolved oxygen, pH and redox potential (ORP). The water was pumped into the flow cell using disposable polyethylene tubing. Once the temperature, specific conductance, dissolved oxygen, pH and ORP measurements (measured at three minute intervals) stabilized for three consecutive readings, the groundwater was determined to be formation water.

Groundwater samples collected were submitted to a laboratory under appropriate chain-of-custody for testing. Dissolved methane in groundwater was

Figure 7. Configuration and setup of groundwater sampling

detected at a concentration exceeding method detection limits in several of the wells during at least one sampling event. Where detected, methane ranged from 15.2 µg/L to 3,590 µg/L. Hydrogen sulfide was present within four monitoring wells in at least one sampling event ranging from 66.4 µg/L to 3,230 µg/L.

The presence of dissolved methane and hydrogen sulfide in groundwater samples from the groundwater near the tunnel horizon indicate that these constituents may be present in the breathable air space within the tunnel excavation. Actual concentrations of these gases in the tunnel air are difficult to estimate due to assumptions regarding temperature and pressure, variations in concentrations, variations in groundwater inflow rates and actual ventilation conditions. However, based on the results of this testing, the conditions during construction will be baselined as potentially gassy. As such, tunnel ventilation will be required to supply sufficient air to the tunnel to provide a safe and breathable environment. Design of the tunnel ventilation will need to consider mitigation of potential impacts from methane and hydrogen sulfide concentrations that may enter the tunnel by off-gassing from groundwater flowing into the excavation.

DELIVERY APPROACH AND PROJECT SCHEDULE

LMT will be delivered utilizing a traditional design-bid-build approach with contractor prequalification requirements similar to the procurement of other MSD tunnel projects. The contract documents will allow the contractor to submit alternate designs within the parameters of the contract documents and significant portions of the work will be performance based. A comprehensive Geotechnical Data Report (GDR) has been prepared for the project and a Geotechnical Baseline Report (GBR) will be provided to baseline the anticipated conditions during construction.

The project schedule is anticipated as follows:

- Final Design Complete—Summer 2018
- Permitting and Easement Acquisition—Fall 2018 to Spring 2020
- Contractor Prequalification and Bidding—Spring 2020
- Notice to Proceed—Summer 2020

ACKNOWLEDGMENTS

The authors would like to thank the Metropolitan St. Louis Sewer District and HDR Engineering, Inc. for their continued support during the project. Special thanks to Shannon & Wilson (Bill Kremer, Scott Garbs and Tom Abkemeier) for their assistance throughout the investigation program and to Larissa Maynard and Kyle Williams of WSP for their assistance in preparing portions of this paper.

REFERENCES

[1] Abkemeier, T. and Groves C. 2011. Construction Grouting of the Baumgartner Tunnel. Rapid Excavation and Tunneling Conference: 2011 Proceedings. Edited by S. Redmond and V. Romero. Littleton, CO: SME.

[2] Bruland A. 1998. Hard Rock Tunnel Boring—Advance Rate and Cutter Wear. Department of Building and Construction Engineering, Norwegian University of Science and Technology. N7034. Trondheim, Norway.

[3] Center for Disease Control and Prevention, NIOSH Pocket Guide to Chemical Hazards, September 2007.

[4] Engineering Geology Field Manual. Reprinted 2001. U.S. Department of the Interior, Bureau of Reclamation, 2nd Edition.

[5] Jain, S.C. and Kennedy, J.F. July 1983. Vortex-Flow Drop Structures for the Milwaukee Metropolitan Sewerage District Inline Storage System. Iowa Institute of Hydraulic Research Report No. 264.

[6] Nickerson, J., Abkemeier, T., McCleish J. and Willig N. 2005. Baumgartner Shafts and Tunnel. Rapid Excavation and Tunneling Conference: 2005 Proceedings. Edited by J.D. Hutton and W.D. Rogstad. Littleton, CO: SME.

[7] Nilsen B. and Ozdemir L. 1993. Hard Rock Tunnel Boring Prediction and Field Performance. Rapid Excavation and Tunneling Conference: 1993 Proceedings. Littleton, CO: SME.

[8] Recommended Standards for Wastewater Facilities. 2014. Policies for the Design, Review, and Approval of Plans and Specifications for Wastewater Collection and Treatment Facilities by the Great Lakes—Upper Mississippi River Board of State and Provincial Public Health and Environmental Managers. Albany, New York: Health Research, Inc.

[9] The State of Our Missouri Waters—Meramec River Watershed. June 2015. Missouri Department of Natural Resources, HUC-8: 07140102.

[10] Thompson, Thomas L., Paleozoic Succession in Missouri: Part 4 Mississippian System, Missouri Department of Natural Resources, Division of Geology and Land Survey, Rolla, MO, 1986.

Sustainable Infrastructure Tunneling: Construction Materials Considerations from the Early Project Stage

Fabio Pellegrini, Brendan Daly, Andreea Enescu, and Nicolas Swetchine
LafargeHolcim

ABSTRACT: While tunnels offer substantial environmental and social benefits during their use phase their construction can have a heavy impact on the environment.

Beyond geotechnical, hydrogeological, tunnel-boring mechanical and other considerations, smart handling of enormous material streams is a major concern. Proper materials management—inbound and outbound—is key to optimizing resource efficiency and sustainability. Planning for best-handling practices and optimal reuse of excavation materials reduces the amount of waste to be dumped, improves environmental performance and reduces costs and risks during tunneling and landfilling.

A prerequisite to implement such a sustainable materials-management plan is to integrate construction material science throughout the initial project stages. Evaluation of viable solution strategies must be part of the prefeasibility studies and each design phase.

SUSTAINABLE CONSTRUCTION—A KEY TREND FOR BUILDING MATERIALS

The environmental and economic impacts of a construction project happen in two distinct phases. The first is during the construction stage, and the second is during the constructed structure's use phase. Life-cycle cost is the sum of all recurring and one-time costs (explicit and implicit), which are amortized over the design life of the project. Construction costs, as well as operating and maintenance costs, are considered and weighed against other design alternatives. The more durable and longer lasting the project, the lower the life cycle cost.

The construction sector of tomorrow will be innovative, climate-neutral and circular in its use of resources. It will be respectful of water and nature. It will be inclusive—enhancing quality of life for all (LafargeHolcim Sustainable Strategy—The 2030 Plan," 2016).

This aspiration needs to be seamlessly applied to the entire life cycle of construction products and services: from the sourcing of raw materials to the end-of-life of the products. Implementing such a sustainable development roadmap with construction materials goes well beyond the performance of product manufacturing operations. It encompasses the entire construction value chain and the life-cycle of construction.

Beyond a superb material and construction performance over the life cycle, the optimization of resources in the tunnel construction phase is essential to its overall sustainability performance.

IMPORTANCE OF SUSTAINABILITY IN UNDERGROUND PROJECTS

The first phase of the project (construction phase) is the focus of this paper. The second phase (use phase), however, is equally important. For the Gotthard Base Tunnel project reviewed below, the development and use of a long-lasting concrete solution in the tunnel was of paramount importance to ensure long-term success. The sustainable reuse of aggregates mined from the tunnel for producing that long-lasting concrete formulation provided the long-term durability requirements of the project owner.

A large amount of excavation material accumulates during tunnel construction. Maximizing the reuse of good-quality excavated material as aggregates to produce concrete on-site, as sub-base for roads, tracks or other infrastructure applications and as landscaping material around the project site are of crucial importance for sustainable life-cycle management.

Because underground construction generates large volumes of spoil materials and consumes large volumes of natural resources, many infrastructure projects are negatively perceived by the public. To address these perception issues, tunnel owners and developers need to evaluate the various options for beneficially reusing excavated material to reduce local impacts and advance environmental stewardship through sustainable construction materials solutions. These sustainable construction solutions not only save money and the environment, but also promote a positive community perception for local development.

Considering resource-efficient and effective construction solutions in the planning of tunnel projects is especially important as excavation materials are too often considered as waste to be dumped in landfills with no value versus their beneficial treatment and reuse in situ. A key point is to assess the cost of double handling the material and logistics associated with material supply for concrete production.

FROM FIRST IDEA TOWARDS EXECUTION: WHEN TO START THINKING ABOUT CONSTRUCTION MATERIAL SOLUTIONS AND NOT MERELY MATERIAL SUPPLY

Excavation material management plays a very important role in determining a project's logistics requirements and needs to be addressed at a very early stage of the tunnel planning process. Determining

Source: © AlpTransit Gotthard AG

Figure 1. Sustainable construction of the world's longest railway tunnel in Switzerland included the beneficial reuse of excavation material in situ

the quantity and especially the quality (material acceptable for beneficial reuse) of the raw material from tunnel excavation operations is a key first step. Geology and construction methods (tunnel boring machine versus drill and blast) must be considered thoroughly to properly design the optimum solution and maximize the economy of the project.

Excavation material management is a unique project within the overall tunnel-construction planning process and needs to be managed by specialists combining geotechnical expertise, excavation methods know-how and production competences in aggregates and concrete. Innovative solutions are required to achieve optimal results.

LESSONS LEARNED FROM THE GOTTHARD BASE TUNNEL PROJECT

The Gotthard Base Tunnel in the eco-sensitive Swiss Alps is the world's longest railway tunnel and a key component of the New Transalpine Railway, one of the largest environmental protection projects in Europe.

All project partners had to develop integrated solutions—from material processing to the extremely high level of durability required for achieving 100-year life span goals.

Construction of the tunnel's two 57-km long, 9.2-m diameter parallel tubes produced 28.2 million tons of excavation material (Hitz and Kruse, 2016). To put this into perspective, if this amount of material was loaded onto wagons, it would result in a wagon train stretching from Switzerland to Chicago in the USA.

The Swiss authorities and the owner of the project defined at a very early stage the following overall

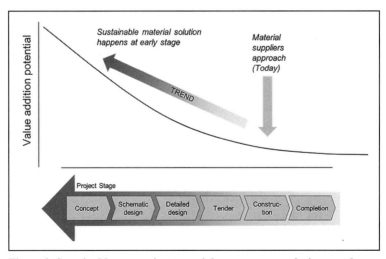

Figure 2. Sustainable excavation material management solutions need to be considered during the very early stages of the project

Source: © AlpTransit Gotthard AG

Figure 3. Environmental concerns were a high priority and addressed right from the start of the Gotthard Base Tunnel design process

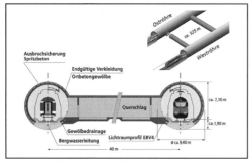

Source: © AlpTransit Gotthard AG

Figure 4. Construction of the two parallel single-track tubes of the 57-kilometer Gotthard Base Tunnel produced 28.2 million tons of excavated material

objectives for the tunnel's excavated material treatment in situ:

- Maximization of excavation material reuse
- Minimum environmental burden
- Economically advantageous management
- No negative impact on excavation methods and performance/schedule of execution

Based on these project objectives and specifications, the excavation material from the Gotthard Base Tunnel was classified into the following categories:

- A-Material, appropriate for reuse as aggregates for concrete production: 9.6 million tons, which was 33.3% of the total quantity of excavation material.
- B-Material, appropriate for embankments: 6.2 million tons, which was 21.7% of the total quantity of excavation material.
- Uncontaminated soil/material for landfill and land restoration: 12.7 million tons, which

Source: © AlpTransit Gotthard AG

Figure 5. All the aggregates for producing the project's concrete came from recycling of the tunnel's excavated material, thus limiting the amount of primary aggregate resources

was 44.3% of the total quantity of excavation material.
- Waste disposal/dumping in special sites: 200,000 tons, which was 0.7% of the total quantity of excavation material.

Testing Reuse of Material and Concrete Mix Design Qualification System

Under the lead of the project owner AlpTransit Gotthard Ltd. (AlpTransit Gotthard website) and three years before entering in the tendering phase, a global leader in construction materials (LafargeHolcim) was supporting the optimization of concrete mix designs based on an analysis of excavation materials from the project region from other minor tunneling projects (LafargeHolcim solutions for the world's longest railway tunnel, 2015).

Concrete production for shotcrete and in situ casting was tested under ideal conditions in underground laboratory facilities (Hagerbach Test Gallery), taking into consideration the different geology characteristics of the tunnel to be excavated (granite and gneiss), the temperatures (rock 45 °C; air 30 °C) and relative humidity (70%–90%). Concrete mix-designs surpassing three-year durability testing for a life expectation of 100 years were admitted in the final tunnel design specifications (Schmid and Kradolfer, 2016).

CONCLUSION

Why waste beneficial excavation materials from tunneling operations when they can be reused as valuable raw material resources to improve a project's sustainability and economic performance? The optimum management of excavation materials and their beneficial reuse for a myriad of construction material solutions provide owners and developers significant

Source: © AlpTransit Gotthard AG

Figure 6. More than one-third (10 million tons) of the tunnel's excavated material was recycled into aggregates for manufacturing concrete on site in a tailor-made underground ready-mix plant

Source: © AlpTransit Gotthard AG

Figure 7. Excavated material not used in concrete production was reused for different purposes, including landscaping of tunnel access routes and for sunbathing and relaxing on these islands

opportunities for achieving sustainability goals in state-of-the-art tunnel development projects.

As demonstrated with the Gotthard Base Tunnel, the key to success is addressing at the early stages of project design the construction materials science and innovative sustainable construction solutions. This will allow owners and developers to shape the right combination of assets and products to maximize the reuse of excavation materials, minimize environmental burden, and provide economically advantageous materials management, with no negative impact on excavation methods and performances/schedules. Capitalizing on such remarkable experiences will lead to more sustainable development—now and long into the future.

REFERENCES

AlpTransit Gotthard Ltd., *www.alptransit.ch*.

Hagerbach Test Gallery, The Versuchsstollen Hagerbach VSH, http://hagerbach.ch.

Hitz, A.; Kruse, M., 2016. "The Mountain from the Mountain—Management of the Material Excavated from the GBT," Chapter II, Article 9, "Tunneling the Gotthard," Edited by the Swiss Tunneling Society.

"LafargeHolcim Sustainable Strategy—The 2030 Plan," 2016, *http://www.lafargeholcim.com/2030-plan*.

Schmid, H.C.; Kradolfer, R.W.W., 2016. "Test System for Concrete Mixes," Chapter III, Article 9, "Tunneling the Gotthard," Edited by the Swiss Tunneling Society.

"The Gotthard Base Tunnel: LafargeHolcim Solutions for the World's Longest Railway Tunnel," *http://www.lafargeholcim.com/gotthard-base-tunnel-world-record*.

Project Plans for the California WaterFix Tunnels

John Bednarski, Jay Arabshahi, and Sergio Valles
Metropolitan Water District of Southern California

Shanmugam Pirabarooban
State of California, Department of Water Resources

ABSTRACT: This paper describes the proposed California WaterFix tunnels project in the Northern California Delta region from the context of the recently completed environmental (EIR/EIS) process for the project. The conceptual engineering effort led to development of a project that will include more than 70 miles of large diameter tunnels. The new system ensures water supply reliability while minimizing environmental impacts to the surrounding Delta area. The project's main twin-bore 40-foot ID tunnels will utilize a single-pass precast concrete segmental liner. Preliminary design of the project components is planned to commence in 2018. Current implementation plans will be discussed in the paper. The overall budget for the tunnels, pump plant, river intakes and appurtenant facilities is $14.9 billion in 2014 dollars.

INTRODUCTION

Water drawn from the Sacramento-San Joaquin Delta provides water supply to 66 percent of California's population (25 million people) and supports approximately 3 million acres of the State's agriculture industry. The existing through-Delta water system is outdated and unreliable with environmental risk to some fish and wildlife species. The California WaterFix (WaterFix) has been established to retrofit and modernize California's water delivery system through the Delta in an environmentally sensitive manner. The planned facilities construct new diversion points in the north Delta, and provide a means to transport water supplies under the Delta, rather than through sensitive natural channels.

Under the WaterFix program planning process, several alternatives were developed to convey water from the Sacramento River in the north to the existing pumping facilities in the south Delta through an isolated conveyance system. The new conveyance system would become an integral part of the State Water Project (SWP) and the federal Central Valley Project (CVP) by transporting water to the export pumping plants for each of these projects. Under current plans, the WaterFix design and construction efforts will be managed by a new special purpose joint powers authority (JPA).

The initial conceptual study efforts on the overall program commenced in 2007 and examined various options for the proposed conveyance system. Several alternatives were analyzed between 2007 and 2015. The Conceptual Engineering Report published in July, 2015 (and included as part of the EIR/EIS process), identified the Modified Pipeline-Tunnel Option (MPTO) as the preferred conveyance alternative. MPTO includes three river intakes along the Sacramento River, various sizes of pipelines and tunnels, junction structures, a combined pumping facility and two forebays that are capable of delivering up to 9,000 cubic feet per second (cfs) from the Sacramento River to the SWP and CVP. The river intakes are located near Hood in Sacramento County approximately 40 miles from Clifton Court Forebay (CCF) in Contra Costa County. The MPTO tunnel system is entirely gravity-fed over the approximate 35 mile alignment, from the intakes to the combined pump plant. Figure 1 depicts the current configuration of the proposed WaterFix facilities.

GEOTECHNICAL CONSIDERATIONS

The Delta is an arm of the San Francisco Bay estuary that extends into the Central Valley. The geology of the Delta has been shaped by the landward spread of tidal environments resulting from sea level rise after the last glacial period. Since the last glacial age, flood-borne deposits, supplied by the major river systems in the Delta, have overlaid the region with sediment deposits and biomass accumulations. Taken together, the region, prior to the advent of agricultural interests in the late-1800s, was largely a tidal wetland and alluvial floodplain consisting of consolidated silts, sands and clays overlain with peat and peat muds.

During the development of the planning documents for WaterFix, approximately 240 boring and cone penetrometer tests were conducted at the

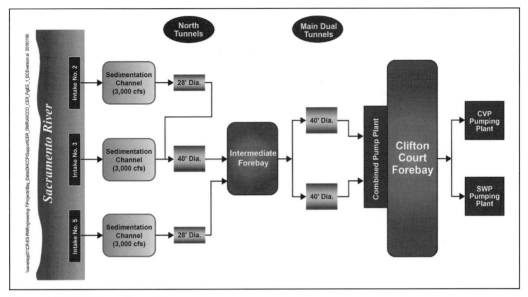

Figure 1. WaterFix configuration

intakes, forebays and along the various conveyance alignments. Most of the investigations were conducted at depths between 100 and 200 feet, well within the foundational depth of planned facilities, including the tunnels and pump plants. Based on these investigations, and the use of existing historical information on the Delta, a preliminary geologic understanding of the Delta in the vicinity of WaterFix facilities was developed.

At tunnel depths ranging from 100 to 150 feet below the ground surface, dense layers of silts, sands and clays are anticipated. This material would be suitable for the planned tunneling activities. At the ground surface, widely varying depths of peat and other organic material are expected. Data indicates that depths of peat in the Delta along the alignment vary from non-existent to about 40 feet deep, with the deepest deposits located in the center of the Delta near Bouldin, Venice and Mandeville islands. Construction in peat conditions would require specialized design approaches because of the unstable nature of the material.

In some locations along the alignment, there are geotechnical data gaps of several miles, due to the inability to gain access to private property during the planning phase of the project for geotechnical investigations. To mitigate these data gaps and other known uncertainties related to geology along the alignment, the project would rely on existing information, along with the implementation of a new two-phase geotechnical investigation program. Under this multi-phased investigation plan, up to 2,000 additional investigations would be conducted,

consisting of borings, cone penetrometer and other physical data collection methods. The initial phase of the effort would focus on determining if variations exist in what otherwise appear to be relatively consistent subsurface conditions. Based on the findings from the first phase of work, additional investigations are planned to fine-tune information, and to gather sufficient information so that accurate estimates of subsurface construction methods and costs can be determined. Additionally, this information would be used to finalize design approaches and construction methods for ground conditions that are overlain with peat and contain high groundwater levels.

OPTIMIZATION OF PROGRAM FEATURES

The current WaterFix configuration reflects an optimization process that took place in 2015. The primary goals of the optimization effort included 1) reducing environmental impacts of the proposed facilities along the Sacramento River, and 2) identifying a project configuration that would place the concrete segmental tunnel liner systems into compression during system operation. Prior design concepts had caused the liner system to be in tension during operations. The resulting revisions to the MPTO WaterFix facilities, referred to as the Clifton Court Option (CCO), were structured to address both of these issues. The CCO alternative retains the project's major design criterion of: 1) maximum velocity of 0.2 feet-per-second at each intake fish screen, and 2) maximum total system flow of 9,000 cubic feet-per-second (cfs) from the Sacramento River (3,000 cfs per each intake). Figure 2 depicts the anticipated differences

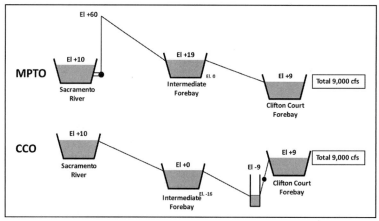

Source: Bednarski, 2015

Figure 2. Simplified hydraulic configuration comparison MPTO vs. CCO (• = pump station location)

in the hydraulic profiles between the original MPTO and the CCO alternative in a simplified side-by-side comparison. Project designers found that the MPTO hydraulic configuration led to the tunnel lining systems being subjected to internal tension due to high hydraulic grade lines. Conversely, the CCO configuration, with system pumps immediately upstream of the Clifton Court Forebay, places the tunnel lining systems in compression due to the lower anticipated hydraulic grade lines.

The significant components for the MPTO that are revised under CCO modifications include the following:

- Combining and relocating the three individual pump stations from the Sacramento River to the terminus of the project at Clifton Court Forebay.
- Modifying the Intermediate Forebay (IF) to work in conjunction with the new pump configuration
- Revising piping, gates and controls at each of the three river intakes
- Revising tunnel diameters for the three North tunnels
- Modifying tunnel segmental lining systems to take advantage of reduced hydraulic grade conditions

CCO Tunnels

For the purposes of designing the segmental liner for the tunnels, the overall tunnel system can be divided into two regions, namely the North Tunnels section and the Main Tunnels section. North Tunnels deliver water from the three river intakes to the IF, and the Main Tunnels convey water from IF to the Clifton Court Forebay (CCF).

North Tunnels. The CCO alternative relies on gravity flow from the Sacramento River to the IF, and then down to the CCF pumps station. As such, hydraulic losses into the North tunnels were reduced from those that are experienced in the MPTO alternative in order to successfully implement the CCO alternative. Consequently, North tunnel sizes under the CCO were increased from the MPTO as shown in Figure 3. The diameters are approximate and should be further refined in preliminary design.

Main Tunnels. Under the CCO alternative the size of the twin main tunnels remains unchanged from the 40-foot ID that is utilized in the MPTO.

Tunnel Segmental Liner Criteria

Early in the planning process for the overall tunnel system, it was determined that a single-pass tunnel liner system could be utilized as a cost effective lining system. The tunnel liner system consists of precast concrete segmental liner with bolted-gasketed joints, and there is no steel second-pass liner in the tunnels. For the Main Tunnels, it is anticipated that a 9-piece ring configuration would be used with segment thickness of approximately 20 inches. The segments (up to 7,000 psi strength) will be cast and steam-cured in concrete segment plants under strict quality control measures. Reinforcement will consist of traditional steel reinforcement and steel fiber as required to increase durability and provide crack control.

Under the single-pass liner design, typical joint between segments will include a gasket to seal against water seepage and alignment bolts for tunnels subject to compression load only. If the segment ring is

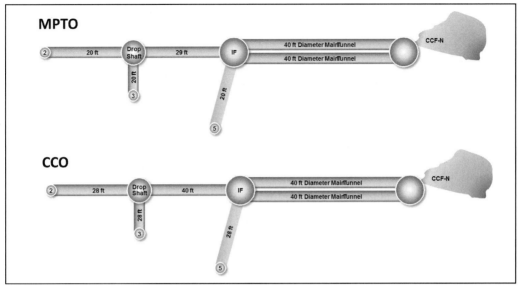

Source: Bednarski, 2015

Figure 3. Tunnel diameters: MPTO vs. CCO

subjected to internal tension load, as was anticipated under the MPTO arrangement, special positive connection across the joint and tension reinforcement are necessary to transfer the tensile force throughout the segments. Historically, it is uncommon that a bolted-gasketed tunnel liner system is subject to net tension in soft ground conditions. However, under the MPTO, this was the case. Therefore, substantial research and analysis were conducted during the study phase to ensure feasibility and constructability.

In addition to strength requirements, leakage control through the liner is essential to ensure liner performance. Excessive leakage through the liner would lead to potential soil erosion, hydraulic fracturing and loss of liner support. In the long run, deterioration of the tunnel liner could occur. In addition, water leakage from the tunnel to the surrounding soil translates to economic loss.

The performance criteria for the tunnel liner system dictated that the liner be designed for all the following load cases to ensure reliable performance during the minimum 100-year design life of the system:

- Full external ground load and external ground water pressure
- Net internal pressure (difference between internal hydraulic pressure and external ground water pressure)
- Ground strain associated with seismic design
- Segment handling loads such as lifting, hosting, TBM pushing

- Crack and leakage control performance criteria

Advantages of the CCO Tunnels

Under the CCO scenario, the net internal hoop tension on the segmental liner can be substantially reduced or eliminated. This will significantly reduce overall tunnel costs, and reduce leakage risks.

Advantages of CCO for tunnel design can be summarized as follows:

- 40-foot Main Tunnels (30 miles × 2 = 60 miles) are subject to compression-only loading for the majority of the tunnel alignment between IF and CCF. The elimination of tension on the liner implies that special high-strength tension bolts are not required at the joint and additional hoop reinforcement is not necessary in the segment. Additionally, the T-lock liner inside the tunnels will not be required. Under this situation, liner construction utilizes conventional proven tunneling methods for better production and lower costs than previously planned under the MPTO.
- Leakage from the tunnel under the CCO configuration to the surrounding soil is virtually eliminated with the tunnel lining system always under compression. An assessment completed in February 2017 of the potential leakage rates from the tunnels concluded that there would be negligible leakage from the

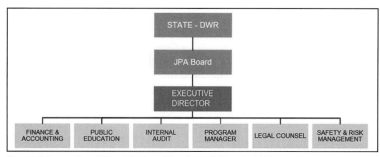

Source: Metropolitan Water District of Southern California 2017

Figure 4. Anticipated organization chart

tunnels or inflow to the tunnels. In fact, when taken as a complete system, it is estimated that there would be a net inflow of 3 cfs to the tunnel over the roughly 73.5 miles of project tunnels, or an inflow rate of 18 gallons per minute per mile of tunnel. Inflow to the tunnels and leakage from the tunnels calculated based on anticipated conditions for filling, dewatering and operation are projected to be minimal and well within typical ranges for tunnels of the size and length proposed for WaterFix.

- For the North Tunnels (between river and IF), net tension will likely remain given the variable river elevations. However, the tensile force magnitude is substantially reduced for CCO because the HGL was reduced. Maximum probable high water HGL is 10 feet, which is only 20% of the net internal pressure of MPTO (50 feet). Hoop stress is also reduced as some of the North Tunnel diameters are smaller than the Main Tunnels. Because the tension force is reduced, joint bolting and hoop reinforcement will be reduced. In addition, other tension-resisting devices (e.g., shear cones) may become viable because the tensile load is decreased. The T-lock liner will most probably not be needed on the North tunnels for leakage control.

- Eliminating net tension along the majority (or all) of the Main Tunnels and decreasing tension in the North Tunnels will benefit the entire program. The CCO alternative optimizes liner design, reduces construction costs, increases tunneling production rates, shortens construction schedule and eliminates some of the long term potential risks associated with tension design of large-diameter high-pressure segmental liner.

DESIGN AND CONSTRUCTION OF THE WATERFIX

The design and construction of WaterFix would be managed under contract with DWR through a proposed Delta Conveyance Design and Construction Joint Powers Authority designated the Design and Construction Authority, or "DCA." This approach was successfully used in the mid-1990s when DWR contracted with the Central Coast Water Authority to design and construct a portion of the Coastal Branch of the California Aqueduct. The Central Coast Water Authority was established as a public entity organized under a joint exercise of powers agreement and constructed water treatment and conveyance facilities to bring State Water Project supplies to Santa Barbara and San Luis Obispo counties.

In coordination with DWR, the DCA would design, construct and deliver completed WaterFix facilities to DWR upon completion of system commissioning. The DCA would be a public agency, organized as a special purpose public agency pursuant to the Joint Exercise of Powers Act, consisting of certain public water agency members. A detailed agreement between DWR and the DCA would govern the roles and responsibilities of the parties to carry out the design and construction of WaterFix. The overall goal of the DCA would be to safely design, construct and deliver the project on time, on budget and in accordance with approved specifications, while managing risk prudently.

Recognizing DWR staff resources are stretched to an extreme level due to the necessary commitment to complete significant repairs to the Oroville Reservoir spillways as a result of damage during heavy runoff in 2017, there is a need to employ different but proven approaches to pool resources for the design and construction of WaterFix. Staff resources are needed for a period of about 13 to 17 years and would ultimately be reduced at the end of construction. Pooling experienced expertise in a manner that

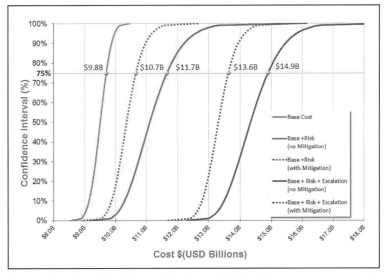

Source: Metropolitan Water District of Southern California 2017

Figure 5. Confidence curves showing 75% confidence interval

avoids the need to hire significant additional new staff at DWR is prudent. In any major infrastructure process, there is a period of acquiring needed additional staff, and then once the project approaches completion, there is a period of downsizing. Utilizing the DCA to pool experienced resources to manage activities and contractors is preferable and can avoid the expansion and contraction of staff at DWR. The DCA would sunset as WaterFix is completed.

Organization Structure

The anticipated organizational structure is shown in Figure 4 and the responsibilities of the offices within the structure are described below.

Project Governance

The DCA would be responsible for delivering the project in accordance with baseline specifications for the project, including design specification, budget, schedule and mitigation obligations. As design work progresses, changes to the baseline specifications would be requested by the DCA at its discretion for approval by DWR. In addition, certain "material changes" on the project would require DWR approval. These include:

- **Cost:** Any actions that cumulatively could cause more than a 5% increase in budgeted costs for each major design feature or management item

- **Schedule:** Any actions that could cumulatively add 6 months to the approved project schedule

- **Operation:** Any actions that could impact the water delivery capability, reduce project life, or significantly increase operations and maintenance costs of the project; and

- **Permits:** Any actions that could be inconsistent with, or would require an amendment of, a major permit for the project

PROJECT CONFIDENCE AND RISK ASSESSMENT

As a component of the risk assessment process, and to assist with creating the budget contingency, the WaterFix project team evaluated the risks associated with the project budget to establish a baseline confidence level that the project would be completed within the estimated budget. This is a common practice with large construction projects, with the resulting confidence curves being used as one of the factors in determining overall project risk.

For WaterFix, Aldea Services developed confidence curves for a variety of different cost scenarios, ranging from base cost, which does not consider mitigation costs or risk, to a total cost that includes the base cost, risk, mitigation and inflation. The resulting confidence curves, which were based in part on the risk assessment workshops and probabilistic analyses are presented in Figure 5. The results of these analyses indicate a 75 percent confidence level that

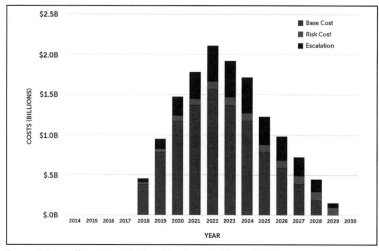

Source: Metropolitan Water District of Southern California 2017

Figure 6. Annual construction expenditures for base, risk and escalation

Table 1. Cost comparison: risk adjusted cost at 75% confidence level vs. initial cost estimates

Item	(1) 5RMK Estimate (Billions)*†	(2) Jacobs Eng Estimate (Billions)*†	(3) Risk Adjusted Estimate with Mitigation at 75% Confidence Interval (Billions)*‡
Construction	$9.50	$8.86	$10.66
Contingency	$3.38	$3.15	
Construction Subtotal	$12.88	$12.01	$10.66
PM/CM/Eng	$1.91	$1.91	$1.91
Land acquisition	$0.15	$0.15	$0.15
Grand Total	$14.94	$14.07	$12.72

Source: Metropolitan Water District of Southern California 2017
* Program estimates in 2014 dollars
†~36% Contingency on construction for 5RMK and Jacob Engineering estimates
‡ Based on risks known at time of assessment

the project would be completed within the budget estimate, based on information available at this stage of the project. A typical confidence level for projects of similar scope and size is 60 percent; however, because of the size and complexity of the program; a more conservative confidence interval of 75 percent was targeted.

At a 75 percent confidence level, the chart in Figure 6 shows how the base costs (blue) along with risk costs (red) and inflation costs (purple) are distributed over the estimated construction period on a year-by-year basis. The risk (red) costs are a direct calculation from the risk analysis and inflation is based on the average inflation rate over 20 years prior to the analysis and applied to the scheduled construction period. The chart is consistent with the

risk adjusted cost estimate and schedule included in the 2015 conceptual engineering report. As funding is available, additional information would be gathered, and the program would be refined during design and the risk management process would be adjusted to the charted confidence curves.

Table 1 presents three cost estimates that have been prepared for the WaterFix program over the last several years. All costs shown in this table are in 2014 dollars. Column (1) is the original bottoms-up Class 3 program estimate that was prepared in 2015 by 5RMK. Column (2) shows the results of Class 3 bottoms-up construction estimate prepared by Jacobs Engineering as a check estimate on the original 5RMK estimate. Excellent continuity between the two independent estimates was achieved. The

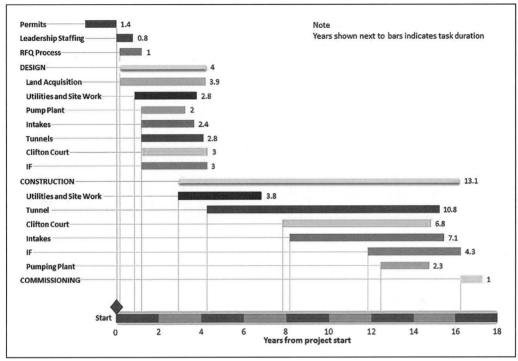

Source: Metropolitan Water District of Southern California 2017

Figure 7. WaterFix summary schedule

5RMK and Jacobs Engineering estimates include a contingency of approximately 36 percent. Program Management (PM), Construction Management (CM), and Engineering (ENG) costs are held constant at $1.91 billion and land acquisition costs at $150 million. This table used three separate estimates to show the program can be completed within the proposed budget of $14.94 billion. Finally, Column (3) shows the comparison between the risk-adjusted cost estimate for the program with at a 75 percent confidence level in comparison to the 5RMK and Jacobs Engineering cost estimates.

PROJECT SCHEDULE

The current high-level program summary schedule is presented in Figure 7. The schedule is primarily based on the information in the 2015 Conceptual Engineering Report as well as other available data for similar large-scale construction projects. The schedule includes estimates of 12 to 15 months to fully staff the DCA, up to four years to complete the design phase and about 13 years to complete construction. Once the DCA is established and the design is advanced, the design and construction teams would look for opportunities to reduce the overall schedule.

Upon project authorization, detailed schedules would be prepared for various project activities, based on the detailed Work Breakdown Structure (WBS) and applicable project documents. These detailed schedules would identify major milestones, time-sensitive areas and critical path activities. Any issues that have a real or potential impact on the schedule would be highlighted and would include the source of the issue and any mitigation measures taken to minimize or eliminate the impact. Schedule reports would be issued on a regular basis (at least monthly), as determined during program start-up.

CONCLUSIONS

The reliable and continued supply of high quality water through the Delta faces many risks, including fishery declines, earthquakes, floods, and rising sea levels. Despite previous actions and efforts by local, state and federal entities to address these issues, as well as other challenges in the Delta, the region's ecosystem has continued to decline. WaterFix addresses these long-standing issues by providing a pathway to reliable water supplies with infrastructure that is designed to withstand earthquakes and adapt to flood and rising sea levels, while protecting habitat, species and the Delta ecosystem.

Under the CCO alternative, water will flow by gravity from the river to the pump station, from which point it is lifted into Clifton Court by two identically sized pump facilities. By utilizing gravity flow through the North and Main tunnels, operating pressures in the tunnels are reduced, thereby simplifying the design of the tunnel's segmental liner system. Relocation of the main pumping plants from the river intakes reduces the amount of construction required in an environmentally sensitive area, and eliminates the need for permanent high voltage transmission lines, and long term operational activities in these areas.

The project has undergone an unprecedented level of public review, comment and scientific input. Extensive analyses and risk assessments have been conducted to better understand and mitigate risks commonly associated with infrastructure projects of this size. For WaterFix, the key risk areas have been identified, and tools to mitigate these risks have been incorporated into the project's risk management process. The physical project meets the attributes of a potentially successful project based on analysis and comparison to mega project. The proposed program management has evolved in a way to increase confidence in the ability to minimize and manage risks to enable the project to be delivered within current budget and schedule estimates.

REFERENCES

Bednarski J., Arabshahi J., Lum H., Valles S., Enas G., 2015, Evolution of a Mega Project: Update on the Bay Delta Tunnels Project, *Rapid Excavation and Tunneling Conference 2015.*

Metropolitan Water District of Southern California 2017, Modernizing the System: California WaterFix Infrastructure, *Joint Meeting of the Special Committee on Bay-Delta and Water Planning and Stewardship Committee—July 10, 2017.*

TRACK 3: PLANNING

Session 4: Project Delivery—Transit and Highways

Claudio Cimiotti and Emily Chavez, Chairs

The Future of Transit Tunneling in Washington, D.C.

Brian H. Zelenko, Harald Cordes, and William H. Hansmire
WSP USA

ABSTRACT: In post-World War II Washington, DC, political forces came together to establish the Washington Metro System. Tunneling and underground construction methods, which were greatly influenced by local geology, have significantly improved over the nearly 50 years since Metro started construction in 1969. Half of the original 103-mile system is underground. To meet current transit demands that have exceeded expectations and future projections, new transit lines, redundancy in the downtown core, and connectivity to high-speed rail into the growing city along Amtrak's Northeast Corridor are being considered. Current tunneling methods (including DC Water's tunnel program), emerging technologies, and alternative delivery methods (including Los Angeles Metro experience with design-build project delivery) are discussed relative to the future of transit tunneling in Washington, DC.

BACKGROUND AND HISTORY

By the time the latest expansion of the Washington Metropolitan Area Transit Authority (WMATA) system, known locally as Metro and which opened in 1976, is completed in 2020, the system will serve 97 stations and operate on 117 miles of track on six interconnecting lines. Metro provides a critical transportation link to a population of approximately 6 million people within a 1,500-square-mile Washington Metropolitan area and has allowed job growth to expand to all corners of the region. In the 1950s and 1960s, when the system was first conceived and construction started, most jobs were centered in downtown Washington, and most of the workforce commuted by bus or car. Today, transit-oriented development has increased residential, commercial, and government facilities near most of the existing 91 stations increasing the importance of the Metro system as a critical transportation link for the region. According to recent American Public Transportation Association (APTA) data, Metro is the second busiest transit system in the United States (after New York City's transit system).

During the nearly 50 years of construction on the Washington Metro system, technology improvements and lessons learned from the global tunneling industry were implemented. Portions of this Background and History have been adapted from other work by the authors (Roach et al., 2017) and expanded for use herein. For soil tunnels, these improvements included the change from a "two-pass" tunnel lining with a typical initial lining of steel ribs with wood lagging or segmental precast lining erected within the tunnel shield followed by a final lining of cast-in-place concrete to a "one-pass" lining system of pre-cast concrete gasketed segmental lining. Soil tunnels were originally excavated using open-face tunnel shields, but changed over time. By the late 1980s, tunneling using closed, pressurized-face tunnel boring machines (TBMs) using a "one-pass" tunnel lining consisting of segmental precast concrete permitted tunneling in a wider range of soil conditions with much less risk of damaging overlying utilities or structures, greater safety for tunnel construction personnel, and without the need to dewater the soils. However, due to concerns about conditioning clays with Earth Pressure Balance (EPB) TBMs, open face machines were still favored by contractors for their better advance rates, lower cost, and ability to deal with boulders which were occasionally encountered in the Coastal Plain Terrace Deposits.

Ground improvement techniques that made tunneling possible in weak or very wet soils at the start of Metro construction were largely limited to cement and chemical grouting and, to a lesser extent, ground freezing. Many early (pre-1978) WMATA tunneling contractors used chemical grout for use in fine-grained soils. This changed over time as technologies evolved for jet grouting (replacement of soil by grout), compaction grouting, and compensation grouting where grouting is undertaken as tunneling takes place.

For tunnels in rock, the work started when tunneling technologies were transitioning from the traditional method of drill-and-blast excavation with rock support using structural steel (steel sets) to more modern methods. The first rock tunnel running north from Dupont Circle Station was a single double-track tunnel excavated by drill-and-blast methods. Later, tunnel contractors used hard-rock TBMs between and through the stations. WMATA personnel had concerns about the delays associated with gripper-type TBMs getting stuck for extended periods in weak rock formations (there was no reverse gear) and with replacing worn disc cutters. Consequently, they preferred conventional rock excavation techniques. Drill-and-blast excavation continued out of necessity to excavate station caverns in rock, as well as many smaller-size excavations such as for cross-passages between tunnels.

Design and construction of rock tunnel linings evolved to follow principles of rock reinforcement by using rock bolts and shotcrete, eliminating the use of steel sets. These practices progressed further where rock bolts were replaced by un-tensioned rock dowels fully encapsulated with cement or resin and shotcrete for initial support and, in some cases, shotcrete as the permanent lining. Groundwater leaks were a major problem during early construction. Dry tunnels without leaks became possible when PVC membranes were placed between the initial and final linings of rock tunnels (Sauer et al., 1987). It was part of the early introduction and adaptation of European tunneling methods and initial use in the U.S. (Heflin 1985) of the New Austrian Tunneling Method (NATM), also known as the sequential excavation method (SEM). This was the first time in the United States where the owner fully accepted the method and used it for later projects. This approach to tunneling integrated the several techniques that had evolved globally over the years and was used to successfully excavate tunnels in rock and, for the first time in the U.S., in soil (Heflin et al., 1991).

The contract for the Fort Totten Station and tunnels in 1988 was the only contract out of three offered by WMATA where the contractor chose to utilize NATM. It was the first use of NATM in soft ground (sands and clays) for a transit tunnel in the United States. The project included 958 linear feet of twin tunnels, in addition to a portion of the station excavation located west of the Fort Totten Station. Excavation and ground behavior were monitored during construction by an extensive geotechnical instrumentation program. Specified initial support included shotcrete applied in three stages, with welded wire fabric and lattice girders. Soil anchors were also required for initial support. Additional support was provided by forepoling bars, forepoling sheets, and face shotcreting.

The tunnel research work that began on the first projects of the Washington Metro set the stage for geotechnical instrumentation and monitoring for tunneling that is undertaken throughout the world today.

Metro engaged the University of Illinois at Urbana-Champaign Department of Civil Engineering to conduct field research on soil tunneling, rock tunneling, and cut-and-cover excavations. The lessons learned for instrumentation and procedures for tunneling were published as *Methods for Geotechnical Observations and Instrumentation in Tunneling* (Cording et al., 1975).

Washington Metro engaged a board of experts to peer review and advise on all aspects of design and construction. The Washington Metro Tunnel Board of Consultants (including Ralph B. Peck, Don U. Deere, and A. A. Mathews) became the forerunner for what is today a common practice within the underground industry (Roach et al., 2017).

Approximately half of the original 103-mile system was constructed underground, including approximately 19 miles of rock tunnels, 12 miles of mined soft-ground tunnels, and cut-and-cover construction. A total of 48 stations are underground, 11 of which were mined in rock, one was mined in soil, and the rest were constructed by cut-and-cover methods. The 19 miles of rock tunnels were constructed for the Red Line extending from Dupont Circle Station north to the Rockville Station, approaches to and under the Potomac River between Foggy Bottom and Rosslyn Stations, and the north end of the Red Line between Wheaton and Glenmont Stations.

The local geology straddles the boundary between the Coastal Plain and the Piedmont Physiographic Provinces. From the boundary, also known as the Fall Line, which extends through the metropolitan area from SW to NE, there is an increasingly thickening wedge of Coastal Plain deposits to the southeast underlain by older Piedmont rocks. Coastal Plain soils include interbedded alluvial soils, ranging from clays, silts and sands to cobbles and boulders, deposited by rivers and tributary systems within channels, bars, floodplains, terraces and alluvial fans. The alluvial deposits are underlain by older Potomac Group soils, which were deposited in a marine environment, and are generally denser and stiffer than the alluvial deposits. Northwest of the Fall Line, Piedmont residual soils, weathered and hard intact bedrock are evident near ground surface. These geologic conditions had a significant influence on tunneling and underground construction methods. Also influencing construction were the presence of the Potomac and Anacostia Rivers, Rock Creek, and other tributary streams and creeks.

EXPANSION OF THE SYSTEM

In recent years, there have been several expansions of the original Metro system that have included tunneling and underground construction. These projects include the Dulles Corridor Metrorail, also known as the Silver Line, the Rosslyn Station Expansion and

Source: Vojtech Gall/GZ Consultants

Figure 1. Steel pipe arch canopy support for WMATA Silver Line extension in Tysons Corner

Access project, the Medical Center Station Metro Crossing project, and the Purple Line, a new light rail transit system overseen by the Maryland Transit Administration (MTA) that will provide a circumferential connection between three suburban Metro Stations (Bethesda, Silver Spring, and Greenbelt) in Maryland.

The Dulles Corridor Metrorail Project, estimated at over $5B, extends Metro service from West Falls Church Station through Tysons Corner to Wiehle Avenue Station (Phase I) and to Dulles International Airport, terminating at a station beyond the airport in Loudoun County, Virginia (Phase II). SEM tunneling was employed for a 1,700 LF section through Tysons Corner in Phase I of the project. SEM was used, including a single- and double-grouted steel pipe arch canopy pre-support due to shallow cover and soft ground (Figure 1). This design-build project, overseen by the Metropolitan Washington Airports Authority (MWAA), is funded by various public and private entities, including the Dulles Toll Road, demonstrates how third parties can work together with WMATA to address regional transportation needs and expand the Metro system.

Rosslyn Station is an important underground hub station in Arlington, Virginia, that serves the Orange, Blue, and Silver Lines. The station was constructed in the early 1970s and was opened in 1977. As part of a new residential and commercial development, an access improvement project was undertaken in 2009 and completed in 2013 to accommodate expanded station capacity associated with the new development. The work included a new station entrance and a new track-level mezzanine that expands the passenger capacity of the station. Construction included a new vertical elevator and stair shaft entrance from the ground surface, connected by a 35-foot-wide by 40-foot-high mezzanine constructed using SEM techniques leading toward the existing station. Passengers walk

through a 19-foot-wide by 15-foot-high passageway between the new concourse and the existing escalator entranceway. The SEM approach to design and construction made the complex underground structures possible.

The Rosslyn Station expansion is an excellent example of transit-oriented development leading to privately-financed expansions of the system. In Maryland, the Purple Line will include a new underground entrance to Metro's Bethesda Station to facilitate connectivity between Metro and the Purple Line.

OTHER RECENT AND FUTURE TUNNELING IN WASHINGTON, D.C.

The DC Water Clean Rivers Project is a $2.6B program to reduce the occurrence of combined sewer overflows into Washington's local waterways. The project includes several large diameter conveyance and storage tunnels in soft ground of the Coastal Plain deep under the city. Some of these new tunnels pass below existing WMATA metro lines.

The first major tunnel construction started in 2013 and included a 5-mile-long, 23-ft inside diameter (ID) bored tunnel along the Potomac and Anacostia Rivers from DC Water's Blue Plains Advanced Wastewater Treatment Plant to a pump station in the Navy Yard. Subsequent contracts included a 2.4-mile long, 23-ft ID tunnel from RFK Stadium under the WMATA Green line and the Anacostia River to Poplar Point, and a half mile long 20-ft ID tunnel in the Bloomingdale neighborhood along First Street (Figure 2). All these tunnels have been successfully excavated in the Potomac Formation soils with state-of-the-art EPB TBMs. The most recent tunnel contract of the Clean Rivers Project was awarded

Source: Brian Zelenko/WSP at TBM Naming Ceremony on April 14, 2015

Figure 2. DC Water's First Street Tunnel EPB TBM "Lucy" as introduced by Mayor Muriel Bowser

in 2017 and will include a 5-mile long tunnel connecting the previously constructed Anacostia River and First Street Tunnels. This tunnel will also utilize an EPB TBM with pre-cast concrete gasketed segmental lining. The depth of this new tunnel system is between 70 to 160 ft below the ground surface and includes multiple deep shafts and connection tunnels excavated by SEM. Unique to this project is the fact that DC Water used Design-Build procurement for all the tunnel contracts. To date DC Water's Clean Rivers Project has been very successful in terms of work quality and schedule adherence. Those projects with soft ground tunneling and Design-Build procurement are solid precedents for any future transit tunneling in Washington, D.C.

A final DC Water tunnel scheduled to be procured in 2020 along the Potomac River will provide additional precedent for tunnel excavation by TBM in rock, mixed-face and soil. As with the other tunnels in DC Water's Clean Rivers Project, the Potomac River tunnel is expected to be procured as a Design Build contract.

CURRENT TRANSIT TUNNELING IN OTHER MAJOR U.S. CITIES

As urban areas grow in population, several major cities in the U.S. have been expanding their transit systems with tunneling and underground construction. These projects, which minimize impacts to surface streets and structures, are instructive in helping to envision the future of transit tunneling in Washington, D.C. As with the recent DC Water Clean Rivers tunneling program, owners in recent years have been favoring alternative delivery methods, such as design-build.

New York City

Similar to the Rosslyn Station expansion for WMATA, the New York City Metropolitan Transit Authority (NYC MTA) has added station access and increased the size of stations for a number of locations throughout their system, including Columbus Circle, Hudson Yards, Lexington Avenue/63rd Street Station, and South Ferry, just to name a few. The MTA's 2015–2019 Capital Program includes $7.1B to expand the network through major investments. The MTA receives approximately 35% of its income from a regional sales tax, regional tax on mortgage receipts, and a tax on businesses that refine or sell petroleum-based fuel. In recent years underground transit improvements have included:

- NYC MTA 2nd Avenue Subway (Phase 1)—$4.4B project includes a 2-mile long segment from 96th Street to 63rd Street and includes new stations at 96th, 86th and 72nd Streets. The 2nd Avenue Subway (Phase 2), which will extend the system north to 125th Street, has started preliminary design.
- NYC MTA No. 7 Subway Extension and Station—$2.1B project extended the subway line from Times Square Station to the Jacob K. Javits Convention Center at Tenth Avenue and West 34th Street.
- Long Island Railroad (LIRR) East Side Access Project—$10.8B project, still under construction, extends LIRR service from Queens and Long Island to Grand Central Terminal.

Los Angeles

The Los Angeles County Metropolitan Transportation Authority (LA Metro) has embarked on a major expansion through Measure R, a voter-approved ballot measure that resulted in a half-cent sales tax for Los Angeles County to finance new transportation projects and programs which has been in place since 2009.

Four segments that include tunneling and underground construction are underway as design-build contracts:

- Crenshaw to Los Angeles International Airport (LAX)—$1.3B, 8.5 miles long project includes 2.1 miles of tunnels and three underground stations;
- Purple Line Extension Segment 1—$1.6B, 3.9 mile-long tunnel project includes three underground stations and extends to Beverly Hills;
- Purple Line Extension Segment 2—$2.4B, 2.6 mile-long tunnel project includes two underground stations and extends the Purple line to Century City; and,
- Regional Connector—$1.8 B, 1.9-mile-long project through downtown L.A. will connect two existing transit lines and will include three new underground stations.

In 2016, Los Angeles County voters approved Measure M, which will provide an additional $120B in transportation funding over the next 40 years. With the 2028 Olympics set to be in Los Angeles, further extension of the Purple Line (Segment 3) to UCLA, the location of several sports venues are anticipated.

San Francisco

San Francisco is served by both the Bay Area Rapid Transit (BART) system and the San Francisco Municipal Transportation Agency (SFMTA). Several expansion projects are underway or have been proposed.

- SFMTA Central Subway Project—$1.6B, 1.7-mile-long, tunnel project will extend the Muni Metro T-Third Street Light Rail Line from SOMA to Union Square and Chinatown. The project includes tunneling with two EPBMs and three underground stations constructed by SEM and cut-and-cover methods.
- BART 2nd Transbay Tube—this proposed project would build a second tube between Oakland and San Francisco south of and parallel to the existing immersed tube tunnel. It would connect to the new Transbay Transit Center and provide connections to Caltrain and the new California High Speed Rail.

Seattle

Over the past 20 years, Sound Transit has expanded Seattle's transit system with a series of tunnel and underground projects. Since its creation in 1993, the agency has passed three major funding ballot measures. The most recent measure in 2016, Sound Transit 3, provides $53.8B in funding to support expansion of the transit system. The measure is partially funded by increases to sales, vehicle excise, property taxes as well as federal grants. Recent work has included:

- Northgate Link Extension—$1.9B project includes twin tunnels that will extend 3.6 miles north from University of Washington Station at Husky Stadium.
- East Link Extension—$3.7B project, which includes a 2,000 LF tunnel constructed by SEM, will extend light rail 14 miles from downtown Seattle to downtown Bellevue and Redmond.

FUTURE TRANSIT TUNNELING IN WASHINGTON, D.C.

The future of Metro's transit tunneling in Washington, D.C. is tied to future capital investments and improvements to the Metro system. This was discussed in WMATA's Momentum Strategic Plan (2013–2025) and subsequently in planning and analyses conducted soon after the Momentum Plan for a 2040 Regional Transit System Plan. Although the 2040 Strategic Plan has not yet been finalized and adopted by Metro's Board, some of the planning associated with it has been posted on Metro's Planning Blog website, PlanItMetro. As highlighted by a 2013 Washington Post article which included an interview with Shyam Kannan, Metro's chief planner, Metro identified a need to expand capacity in the system's core to meet future transit ridership demands. This could be accomplished by adding a new core loop that includes approximately 10 new stations and adding capacity and connections at four existing stations.

This investment faces several hurdles. One of the big hurdles that Metro faces is the lack of a local dedicated funding source, unlike many of the other cities and transit agencies noted above. Without a dedicated funding source, it will be difficult to achieve the goals of the Momentum Plan. Today, four years since the issuance of the Momentum Plan, a local dedicated funding source has still not been established. Secondly, future investment will be tied to ridership and recently transit ridership numbers have been declining across the country, especially for Metro. According to APTA, national transit ridership was down approximately 1.9% over the past two years. Metro's ridership has been down by 14% during the same period. For Metro, this is partially due to recent repair work, such as the year-long SafeTrack program, and ongoing reliability challenges. It is also due to the increasing popularity of Alternative Works Schedules (AWS), particularly by federal workers, reductions in late-night and weekend service, as well as the popularity of Uber and other ride share companies. However, in the larger, longer view, there will be a demand for safe, efficient, and reliable service that can only be achieved by new tunnels.

Additionally, the future may also include an extension of planned High Speed Rail by Amtrak in the Northeast Corridor south to a re-developed Union Station, possibly extending further south to Virginia and the Southeast Corridor. Recent news reports have noted that entrepreneur Elon Musk, CEO of SpaceX and Tesla, is planning to privately finance and develop a new high-speed underground transit system, Hyperloop, between Baltimore, MD and Washington, D.C. Maryland has issued a conditional utility permit to his company, The Boring Company, allowing construction of a 10.3-mile section of tunnel along the proposed route. These future projects are under consideration because long tunnels and complex underground works are more feasible and can be built with less risk than in the past.

Metro

In the last two years, Metro has focused its efforts on restoring its system to try to correct years of deferred maintenance on an aging system and improving safety and reliability for the travelling public. Programs such as Back2Good, to get the system in good repair, and SafeTrack, an accelerated track work plan to address safety recommendations by the Federal Transit Administration (FTA) and the National Transportation Safety Board (NTSB), have been implemented. Through SafeTrack, Metro completed approximately three years' worth of

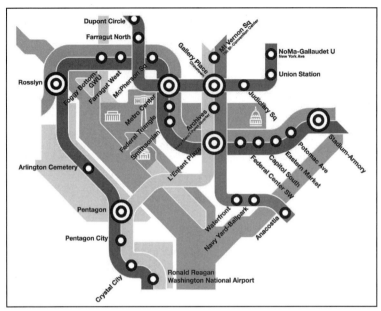

Source: WMATA, *Momentum, The Next Generation of Metro, Strategic Plan 2013–2025*

Figure 3. Metro's core and core stations

maintenance and repair work in approximately one year. Much of the systems maintenance work is related to leaking tunnels constructed in the 1970s and 1980s. Today's tunneling industry has the ability and technology to construct nearly watertight tunnels, representing a huge improvement over past construction methods.

The Momentum Plan identified several key issues, which are common to other major cities in the U.S. which have invested in transit improvements:

- Improve regional mobility and connect communities
- Enhance access
- Add operational redundancy
- Build for the future
- Add new sources of predictable funding

As noted in Momentum, the region is forecasted to increase 30% in population and 39% in employment over the next 30 years. And Washington, D.C. is one of the few metropolitan areas where growth is occurring in the city core, inner suburbs, and outer suburbs. The Momentum plan identified Core Station Improvements, including pedestrian underground connections, that Metro stated should be completed by 2025 to have maximum impact, increase system and core capacity, and improve the effectiveness of the rail network. These improvements are envisioned at stations such as Metro Center, Gallery Place, Union Station, L'Enfant Plaza, Farragut West, and

Farragut North (Figure 3). In 2012 dollars, they were estimated to have an order of magnitude cost of $1B.

Other conceptual improvements, included in the Momentum Plan, were further evaluated by Metro during the planning effort for a 2040 Regional Transit System Plan. These improvements would be required, in addition to the 2025 Plan Core Station Improvements, to serve the regions transit needs with the projected growth. It noted that as transit patronage reaches full capacity on lines converging at Rosslyn and L'Enfant Plaza Stations, new east-west and north-south transit tunnels through Arlington and Washington, D.C. would be required to accommodate trips and improve capacity through the system core. Some of those improvements, with associated order of magnitude cost in 2012 dollars, related to future transit tunneling include:

- New connection of existing lines at Pentagon with a new Pentagon Station ($600M)
- Silver/Orange Line Express Line from West Falls Church to a 2nd Rosslyn Station to a relocated Blue Line ($2.3B)—this would allow passengers on the Silver and Orange Lines better access to the eastern side of downtown Washington. A second Pentagon Station would allow passengers on the Orange and Silver Lines to reach the Pentagon without having to switch to the Blue Line. The underground portion of this new line would begin in the Piedmont bedrock at the Rosslyn

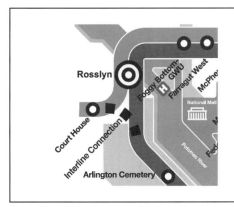 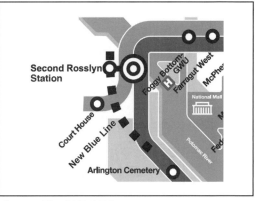

Source: WMATA, *Momentum, The Next Generation of Metro, Strategic Plan 2013–2025*

Figure 4. Conceptual new Metrorail Blue Line connections at Rosslyn Station

Station and then transition to Coastal Plain soil deposits at the second Pentagon Station. Tunneling could be completed by an EPB TBM or a Hybrid TBM (Figure 4).

- Relocated Yellow Line ($2.7B)—this would improve north-south capacity and require a new tunnel south from the Pentagon under 10th Street SW and NW and then west to Thomas Circle allowing the Green and Yellow Lines to operate in separate tunnels. As this alignment is generally in the Coastal Plain Deposits, tunneling could be accomplished using either a Slurry or EPB TBM.
- Relocated Blue Line ($3.3B)—this would improve east-west capacity by creating a new Blue Line alignment through Rosslyn to Georgetown and then along M Street NW to Thomas Circle. As this alignment is primarily in the Piedmont sections of Washington, D.C., tunneling could be accomplished using a hard rock TBM. Station construction could be completed fully by SEM or cut-and-cover methods. Construction of a station in Georgetown could be completed by SEM construction with significant attention made to protect existing structures through a congested part of the city with the use of ground improvement.

As indicated on the PlanItMetro website, one possible way to increase capacity in the downtown core would be the conceptual development of a Core Loop (Figures 5 and 6) that extends north from a second Rosslyn Station in a deep tunnel to Georgetown, extends east toward Union Station to a second Metro station at Union Station possibly constructed below the existing station while providing connections to the existing Red and Green Lines along the way. As shown in the conceptual figure, the Core

Loop could then extend south and west to address the needs from the Relocated Yellow Line noted in Momentum. This work would involve tunneling and underground construction for the stations. These are conceptual drawings and future planning could lead to alternatives such as a continuation of the new line from Union Station to the east, perhaps two miles to RFK Stadium.

Union Station Redevelopment

The Union Station Redevelopment Corporation (USRC), in coordination with Amtrak, is actively planning for a $7B phased expansion and modernization of Washington Union Station. The historic station, which opened in 1907, is proposed to handle triple the number of passengers and double the train service. Major improvements are being planned to add a 3-million SF air-rights development, named Burnham Place after Union Station architect Daniel Burnham, over the train shed. Additional improvements include increasing capacity along the Northeast Corridor to New York and Boston with the addition of new underground passenger concourses below the existing train shed, and plan for possible High Speed Rail, extension of Maryland Regional Rail Trains (MARC) to Virginia, as well as possible extension of Virginia Railway Express (VRE) trains to Maryland.

Constructed in the 1970s, Metro's busiest station is at Union Station on the Red Line. The near-term station access improvements planned by Metro will be included in the Union Station re-development. According to the U.S. DOT Federal Railroad Administration (FRA) 2017 Concept Screening Report, a second Metro Line servicing Union Station is being considered in long-term planning (beyond 2040) in order to meet future travel demand. The proposed conceptual alignment is parallel and south

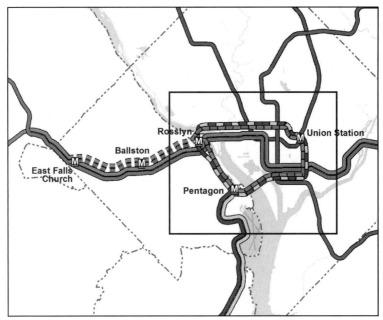

Source: WMATA Planning Website and Blog. https://planitmetro.com/
Figure 5. Proposed 2040 Metrorail network—Core Loop

of Massachusetts Avenue passing along the southern edge of Columbus Plaza, directly south of Union Station (Figures 7 and 8). A second Metro Station would need to be constructed below the existing station at Union Station. This station would extend roughly between North Capitol Street and Louisiana Avenue, below existing surface parking lots just south of Massachusetts Avenue. Tunneling using a Slurry or EPB TBM would likely be required, combined with SEM for station construction within the Coastal Plain deposits. Alternatively, the second Metro Line could be located on the north side of Union Station to tie-in with the planned air-rights development. This would allow Union Station to be served by two Metro Lines; however, the new Metro station would not be a transfer station due to the distance from the existing Metro Station on the Red Line.

High Speed Rail

High speed rail connecting to the Northeast Corridor, which is only being considered in the fourth phase of Union Station Re-Development, would likely enter the station at a new deep underground level below the existing station and platforms. High speed rail tunnels entering Union Station from the north would likely need to begin somewhere around the Anacostia River crossing, approximately five miles north of the station. As the tunnels enter the station, they would need to be aligned in coordination with the existing

large footings which are supporting the existing historic station structure on Coastal Plain Terrace Deposits. For HSR to continue south to Virginia, it is possible that twin-bored tunnels would follow an arc route out of the station and below the National Mall before crossing the Potomac River into Virginia. The FRA, in cooperation with the Virginia Department of Rail and Public Transportation, is preparing a Tier II Environmental Impact Statement (EIS) for the 123-mile portion of the Southeast High Speed Rail Corridor from Washington, D.C. to Richmond, VA. The study begins at the southern terminus of the Long Bridge across the Potomac River in Arlington, VA.

Due to the shallow connection at Union Station, it is anticipated that twin-bore tunnels would be a preferred approach, rather than a single large-bored tunnel with two tracks which would require more complex underpinning. Due to high groundwater levels and Coastal Plain deposits, a pressurized-face TBM, either slurry or EPB, would likely be used to build the HSR alignment south of Union Station.

CONCLUSION

Washington, D.C. expects an increase in population, employment, and transit oriented development over the next 30 years that will put a strain on the existing Metro transit system. This led WMATA to develop the Momentum Strategic Plan 2013–2025 that identified core station improvements, new east-west and

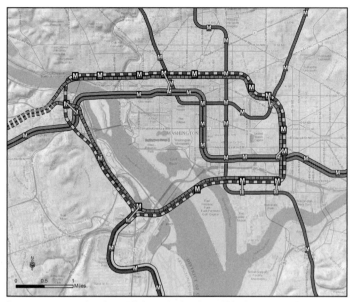

Source: WMATA Planning Website and Blog. https://planitmetro.com/

Figure 6. Proposed 2040 Metrorail network—Core Loop detail

Source: U.S. DOT Federal Railroad Administration (FRA), 2017, *Washington Union Station Expansion Project, Concept Screening Report*

Figure 7. Conceptual proposed WMATA Metrorail line south of Union Station

north-south transit lines that would improve access and effectiveness of the network in the system core. These proposed improvements, amounting to over $10B (in 2012 $), would require tunneling and underground structures. In addition, Washington Union Station is about to begin a $7B phased expansion and modernization program that will include significant underground construction at the station and

may include tunneling for High Speed Rail from the Northeast Corridor, through Union Station and south to Virginia across the Potomac River. These future projects are under consideration because long tunnels and complex underground works are more feasible and can be built with less risk than in the past.

A reliable funding source continues to be an impediment for the Washington Metro to meet

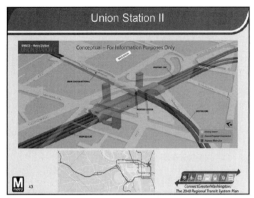

Source: U.S. DOT Federal Railroad Administration
(FRA), 2017, *Washington Union Station Expansion
Project, Concept Screening Report*

**Figure 8. Conceptual future WMATA Metrorail
line and station south of Union Station**

its future needs. Other major cities such as New
York, Los Angeles, Seattle, and San Francisco
have addressed the funding hurdle for transit capital programs, together with federal support, and are
underway with major tunneling and underground
construction works. A recent decline in transit ridership, particularly in Washington, D.C., would need
to demonstrate signs of recovery before decision
makers would be willing to make the substantial
investments required to improve the transit system.
However, planning and conceptual design work
should continue given the clearly identified future
demand in the system core. State of the art tunneling and underground construction methods that have
been used in other major cities for recent transit
tunneling projects have been successfully used in
Washington, D.C. for recent Metro expansion as well
as the DC Water Clean Rivers Program. Therefore,
Washington, D.C. is poised for significant transit
tunneling work in the future.

ACKNOWLEDGMENTS

The authors would like to acknowledge the invaluable discussions and thoughtful reviews provided by
Allison Davis and James Darmody of WMATA.

REFERENCES

American Public Transportation Association
(APTA) Quarterly Ridership Report 2017
2nd Quarter http://www.apta.com/resources/
statistics/Pages/ridershipreport.aspx.

Cording, E.J., Hendron Jr., A.J., MacPherson,
H.H., Hansmire, W.H., et al. 1975. In
*Methods for Geotechnical Observations and
Instrumentation in Tunneling.* Report to the
National Science Foundation by the Department
of Civil Engineering, University of Illinois at
Urbana-Champaign. No. UILU-ENG-75-2022,
December.

Davis AICP, Allison, WMATA Director, Strategic
Planning, Office of Planning, in-person discussion on October 19, 2017.

Heflin, L.H. 1985. WMATA Use of the new Austrian
tunneling method (NATM) for lining and support. In *Proceedings 1985 Rapid Excavation
and Tunneling Conference*, Vol. 1. Littleton,
CO: SME. pp. 381–391.

Heflin, L.H., Wagner, H., and Donde, P. 1991. U.S.
approach to soft ground NATM. In *Proceedings
1991 Rapid Excavation and Tunneling
Conference*. Littleton, CO: SME. pp. 141–155.

Los Angeles Metro Capital Projects. https://www
.metro.net/projects/capital-projects/.

New York City MTA Capital Program Summary.
http://web.mta.info/capital/.

O'Connell, Jonathan, December 18, 2013, "Metro
Planners consider building 'inner loop' of new
stations to alleviate congestion," *Washington
Post*.

Roach, Michael F., Lawrence, Colin A., Klug, David
R., Fulcher, W. Brian, eds., 2017. *The History
of Tunneling in the United States*, Society for
Mining, Metallurgy & Exploration (SME).

San Francisco MTA Central Subway Project. https://
www.sfmta.com/projects-planning/projects/
central-subway-project.

Sauer, G., and Garrett Jr., V.K. 1987. Achieving a dry
tunnel. In *Proceedings 1987 Rapid Excavation
and Tunneling Conference*, Vol. 1. Littleton,
CO: SME. pp. 461–478.

Seattle Sound Transit Projects. https://www
.soundtransit.org/.

WMATA, *Momentum, The Next Generation of
Metro, Strategic Plan* 2013–2025 https://www
.wmata.com/initiatives/strategic-plans/upload/
momentum-full.pdf.

WMATA Planning Website and Blog. https://
planitmetro.com/.

U.S. DOT Federal Railroad Administration (FRA)
Southeast High-Speed Rail DC to Richmond
Environmental Impact Statement (EIS) https://
www.fra.dot.gov/Page/P0729.

U.S. DOT Federal Railroad Administration (FRA),
2017, *Washington Union Station Expansion
Project, Concept Screening Report*, https://
www.fra.dot.gov/eLib/Details/L18793.

Subsurface Investigations, Design, and Construction Considerations for the Montreal Transportation Agency Côte-Vertu Underground Storage and Maintenance Garage

Giovanni Osellame and Jean Habimana
Hatch

ABSTRACT: The Montreal Transportation Agency (Société de Transport de Montreal—STM) is currently constructing the Cote-Vertu underground storage and maintenance garage that will facilitate daily operations and offer parking spaces for the new Azur trains. The project comprises three parallel tunnels of 1200 ft length each, 25 ft width with a pillar with of 16.5 ft; a 2,000 ft long connecting tunnel to the existing tunnel which will have caverns that have spans as large as 62 ft with very shallow rock cover. The contract requires the use of a roadheader to excavate the garage to ensure the stability of the pillars. The project crosses near known major faults.

The paper addresses the geotechnical investigation programs, the design considerations and the selection of construction methods as well as provide updates on ongoing construction activities that started in May 2017.

PROJECT DESCRIPTION

Current planning by the STM considers extending the Blue Line to the East of Montreal by adding five stations. This extension increases ridership on the Orange Line and in order to satisfy the additional demand, new third generation Azur Metro trains are being acquired by STM. As a result of the acquisition, the STM identified the need for an underground storage and maintenance facility located 1.5 km to the west of the Orange line Côte-Vertu station, which is discussed in this paper.

The project advanced from the planning and feasibility stage in 2014–15 to detailed design and preparation of bid documents during 2016 and 2017. Presently, the surface excavation works have commenced.

The location of the facility was selected so as to be able to accommodate the needs of the future extension of the Orange Line toward the north. Planning and land availability led to the selection of an underground layout with limited surface footprint.

Figure 1 shows the location and layout of the facility and its connection via a tunnel to the existing Orange line tunnel. Key components of the project include:

- An area serving as the junction between the connecting tunnel and the existing Orange line tail track;
- An 8.5 to 10 m wide double-track 800 m long slightly rounded D- shape tunnel connecting the facility to the existing Orange line tail track
- Three parallel 8.5 m wide double-track slightly rounded D-shape tunnels, of which tunnel 1 is 150 m in length and serves as the maintenance area and tunnels 2 and 3, each 300 m in length, serve for the storage of the metro trains
- An open-cut area linking the three tunnels to the connecting tunnel. The open cut was designed to accommodate two other tunnels (tunnels 4 and 5) that are planned for the future extension.
- An open-cut workshop and service facility
- One ventilation shaft with two fans

GEOLOGICAL INVESTIGATIONS

The first investigation phase was planned during the preliminary design to help guide the selection of the location of the facility and to characterize the overall ground conditions at the selected location. This phase included boreholes at accessible areas and geophysical seismic profiles. Of particular concern at this phase, was the proximity of the facility to the known major Outremont fault and the incidence of any secondary faults. This phase included 19 boreholes with an average depth of 40 m of which 7 were vertical and 12 inclined. In situ testing consisted of rock permeability Lugeon testing, rock dilatometer tests to determine the rock mass deformation, joint and fractures characteristics and orientations using both the acoustic and optical survey methods.

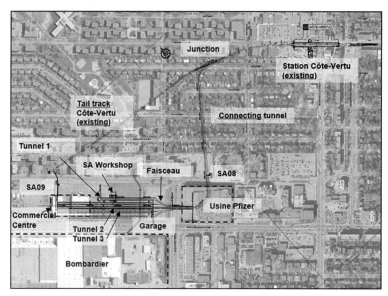

Figure 1. Plan view of the underground garage storage and maintenance facility

Laboratory testing consisted of the usual suites of rock mechanics tests as well as rock abrasivity tests (Cerchar Abrasivity Index (CAI)). The limitations of the first phase included primarily land accessibility issues and only one borehole along the connecting tunnel. During this phase, workshops were also held to assess the geotechnical risk associated with the presence of potential faults within the facility footprint.

The second phase was carried out at the beginning of the detailed design phase and consisted of 9 inclined and 2 vertical boreholes at the junction area, along the connecting tunnel and within the garage area. The goal of the second phase was to provide a degree of confidence that the main garage tunnels where within a stable rock mass exempt of major faults, to provide the key parameters required for the design and to provide sufficient information so as to be able to establish excavation and ground support categories. This phase included mostly 45 to 60 degree inclined boreholes and with similar in situ and laboratory testing as in phase 1.

Investigating Fault Zones

The «Outremont fault zone» is located immediately to the east of the Cote-Vertu station and secondary associated systems were encountered and mapped during the excavation of the station and the tail track tunnel sections. Investigation procedures included mostly inclined boreholes with televiewer surveys and detailed stratigraphical description and correlations between the distinct formations; the relative dip of the formations which are normally almost horizontal was also used to predict any potential problem areas.

GEOLOGICAL SETTING

Regional Geology of Montreal Island

The geology of Montreal Island has been presented in detail by T.H.Clark and Grice (1972) and later by Durand (1978). Briefly, the geological setting of the Montreal Island consists of a variety of Pleistocene and Recent deposits that overlie early Paleozoic sedimentary rocks and Precambrian rocks, both of which are cut by Mesozoic intrusions. The principal events affecting the near surface rock assemblage were faulting, gentle folding, and minor metamorphism in the Mesozoic era and multiple glaciations and isostatic movements during the Pleistocene epoch (Grice 1972).

Bedrock geology in the region consists of Ordovician dolomite, limestone, and shales that are either exposed at the surface or underlie the Pleistocene and Recent deposits, except where they have been cut through by small bodies of dikes and sills associated with Monteregian Hills Mesozoic intrusions, one of which created Mont Royal. During the Mesozoic, the sedimentary rocks were faulted and gently folded and display a predominant bedding structure that dips a few degrees to the northeast. Except at Mont Royal, which is an 83 m high remnant of an intrusive plug immediately to the west of the downtown area, the bedrock surface has been reduced to low relief by glacial erosion.

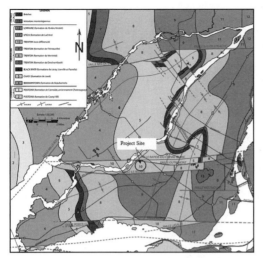

Figure 2. Geology of the Montreal area (after Clark, 1972)

There are five (5) major rock formations in the Montreal area: (1) pure crystalline limestones or dolomites with negligible shale, (2) fossiliferous and micritic limestones with shale interbeds, (3) shales and (4) a massive intrusive rock which forms Mont Royal, and (5), dikes and sills which are present to a lesser extent. All these formations can be expected to be weathered and fractured close to the surface and some are locally cut by dikes and sills.

There are several East-west faults such as the Rapide-du-Cheval-Blanc, Outremont, Sainte-Anne-de-Bellevue and l'Île Bizard Faults that are considered to be related to the Monteregian intrusive event during the Cretaceous and/or to the Taconic orogeny.

Altered and fractured zones do not ordinarily exceed a meter in width, however, adjacent to the principal faults, they have been observed to be wider and deeper. For instance, the construction of Laurier Metro Station crossed the White Horse Rapids fault, which was 180 m wide (Durand 1978).

Solution cavities are rarely encountered in the carbonate rocks; however, pronounced irregularity of the rock profile has been affected by glacial activity through the displacement of the limestone beds particularly along the shaley interbeds. The principal lithological weakness is low strength along the shale beds within limestone units and along structural joints and fractures zones.

Local Geology

Figure 2 indicates that the facility is located within the limestones of the Trenton group identified as «Non-differentiated Formation», by Clark 1972.

The identification is interpreted as signifying the large stratigraphic variation between the faults of the Rapide-du-Cheval Blanc (Cheval Blanc and d'Outremont fault) and the l'île Bizard fault. It is surmised that this variation is due to a series of particular geological structures controlled by a tension zone, a number of secondary faults and « horst » and « graben » features.

The bedrock of the «Non-differentiated Formation» is composed of the Tétreauville and Montréal limestone formations.

The Tétreauville limestones are described as moderately 10 to 40 cm thick greyish regular beds composed entirely of very finely textured micrite, with the absence of any fossils, and interbedded with 1 to 10 cm thin shale beds. At the base of the formation, the shale component increases up to 50 percent with thicker 30 to 40 cm beds.

The limestones of the Montréal formation are subdivided into two members: the Rosemont at the top and Saint-Michel at the base. The Rosemont member is composed of thin beds of grey alternating fossiliferous and crystalline limestone with thin shaley interbeds. The Saint-Michel member is distinctive by its alternating beds of crystalline limestone and shale and the absence of fossils.

Dykes and sills associated with the Monteregian intrusives and with thicknesses varying from a few centimeters to meters are distributed within the rock mass at all depths. Rock mass discontinuities consist of a series of vertical to sub-vertical joints and horizontal joints related to the bedding planes

The majority of the tunnels are located within the limestones of the Tétreauville formation with the exception of certain reaches of the connecting tunnel which lie within the Montreal formation.

Stratigraphic Correlations and Faulting

The «Outremont fault» is located some 400 m to the east of the projected facility. The analysis of all the information has led to the geological interpretation presented in Figure 3. This shows for instance, the presence of two formations at the same depth where stratigraphically a vertical distance of 50 m should separate them and as well the pronounced dip of up to 24 degrees of the normally horizontal strata. This condition suggests that immediately to the east of the garage facility, the presence of «trans-tension type of flower geological structure » described as a series of normal faults produced by tensile stresses within a large shear zone. While the tunnels of the garage facility are expected to lie outside of this zone, it is anticipated that the connecting tunnel will cross at least three of these features.

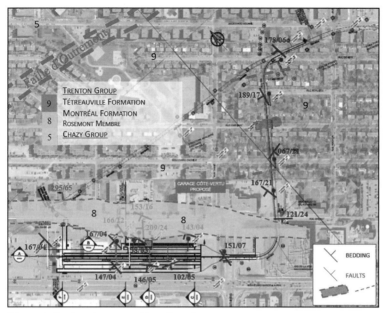

Figure 3. Project layout and local geology

GROUND CHARACTERISTICS AND BEHAVIOUR

Overburden and Ground Water

The overburden is composed of compact to dense glacial deposits consisting of a well graded mixture of gravel, sand, silt and a trace of clay. Cobbles and boulders are anticipated and recent excavations have encountered boulders up to 1 m in size.

The ground water table varies seasonally from a 1.5 to 2.5 m depth.

Intact Rock and Rock Mass Characteristics

Intact rock parameters have been established by laboratory testing and vary per rock type. A summary is derived from the GBR.

Tétreauville formation, the mechanical properties varied as follows: Unconfined compression (UCS) indicated an average value of 91 MPa, Brazilian BTS, an average of 7.6 MPa, Young's modulus 28 GPa and Poisson's coefficient 0.13. Cerchar rock abrasivity and quartz content were 1.1 and 6 percent respectively.

The shale component of the rock mass exhibited lower properties in the range of 44 MPa for the UCS, 4.7 MPa for the BTS, 17 GPa for Young's modulus and 0.12 for Poisson' coefficient. Cerchar rock abrasivity for the shale was 0.5.

The rock mass is dissected by a conjugate set of vertical joints, planar joints or discontinuities corresponding to the bedding planes and discontinuous

surface joints, spacing of the vertical joints varies from 1 to 3 m. Joint surface characteristics are generally described as closed with no alteration and undulating to planar.

Ground water inflows were evaluated for the tunnels part of the project using the methods indicated in Heuer (2005) and resulted in a construction period infiltration rate of 7 L/s (111 Gal/min) and flush flows of the order of 16 to 20 L/s (254 to 317 Gal/min) for the connecting tunnel and for the three garage tunnels.

DESIGN CONSIDERATIONS

Subsequent to the completion of the feasibility study in 2016, the STM established a project office and proceeded to the detailed design phase, the preparation of contract documents, including a Geotechnical Baseline Report and the awarding of the first construction contract in the spring of 2017. Currently, the surface overburden excavations have been completed and open-cut rock excavations are proceeding.

Final Alignment Selection

The location of the project was selected so as to provide an adequate access and ease of launching of future trains into the extended network. During the design stage, the storage and maintenance facility tunnels layout was shifted some 15 m to the west from the feasibility location due to the location of a poor quality rock mass or fault zone along the eastern limit.

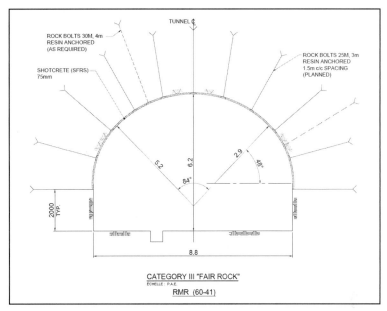

Figure 4. Typical double-track tunnel section

The choice of the vertical profile geometry was mainly based on geological conditions namely the continuity and rock quality of the top of rock, and the rock cover criteria. It was also based on the requirement for connectivity to the existing and future network.

Based on previous experience in Montreal area, and due to the close proximity to potential fault zones, a tunnel rock cover criterion of 10 m was established and applied for the tunnels. The criterion assumed a 2 m of fractured rock and an 8 m cover of good to excellent quality rock which generally respects the ratio of one to one between the span of the excavation and the sound rock cover as most of the excavations are in the range of 8 to 9 meters.

For spans that are greater than 10 m and large openings of up to 18 m such as at the junction, a sequential excavation is required and the ratio is respected by a three sequenced excavation consisting of an initial heading face which respects the width to sound rock cover criterion of 1 and is followed by the rock support. Subsequent sequenced headings or slashes complete the opening.

Figure 4 shows the typical tunnel cross section for a double track consisting of a slightly rounded D- shape that is approximately 8.8 m wide and 6.2 m high. This shape has been optimized from the previous standard D-shape to improve the performance of the concrete lining. The junction to the existing tunnel at Cote-Vertu Station has a maximum span of up to approximately 18 m wide and as high as 9 m and a shallow rock cover that is 3 to 3.5 m.

Figure 5 shows a profile of tunnels 1 and 2, the rock pillars and surface excavation geometry. It should be noted that pillar width was limited by land availability to accommodate the three parallel tunnels and projected expansion for two additional tunnels. Due to this geometry and to reduce rock mass disturbance, mechanical excavation using a heavy 135 T class roadheader was specified and is being implemented.

Characterization of Expected Ground Conditions and Design of Initial Support System

The ground conditions in the area of the storage and maintenance facility tunnels are expected to be generally of good to average quality rock mass and to lie outside of the suspected fault zones. A 10 m wide fault zone was considered as a contingency. Along the connecting tunnel, rock mass quality is very variable from good to poor and three fault zones are expected to be crossed.

Rock mass classification was carried out using the RMR system and class II to V were identified.

Table 1 provides a summary of the four classes of excavation that were considered and design of the matching initial support systems for each class. Design of the initial support requirements and sequence of excavation was done by a combination of empirical, analytical and numerical modeling

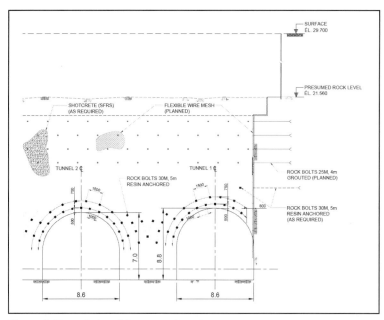

Figure 5. Tunnels 1 and 2 at portal section

Table 1. Summary of expected ground conditions and initial support systems

Class of Excavation	II	III	IV	V
RQD	Good	Fair	Poor	Very Poor
RMR	61 to 80	41 to 60	21 to 40	<21
Percentage of the length				
Garage facility tunnels 1, 2 and 3	52 percent	44 percent	3 percent	1 percent
Connecting tunnel	51 percent	31 percent	13 percent	5 percent
Support System	Pattern bolting at 1.8 m spacing with resin anchored 3 m long bolts (25 mm) and SFRS shotcrete (75 mm)	Pattern bolting at 1.5 m spacing with resin anchored 3 m long bolts (25 mm) and SFRS shotcrete (75 mm)	Pattern bolting at 1.2 m spacing with resin anchored 3 m long bolts (25 mm) and SFRS shotcrete (100 mm)	Main support with W 200 × 52 and W250 × 45 steel sets at 1 m spacing, SFRS shotcrete (200 mm). Pre-excavation support with spiles as required
Excavation sequence/ advance	Garage facility: 5 m mechanical Connecting tunnel: 2.5 m drill and blast	Garage facility: 5 m mechanical Connecting tunnel: 2 m drill and blast	Garage facility: 2.5 mechanical Connecting tunnel: 1.5 m drill and blast	All tunnels: 1.0 m mechanical

methods using FLAC 2D and 3D such as the modelling of the sequential excavation and support required for the large spans at the junction area.

Advance rates for mechanical and drill and blast were designed based on the previous Laval metro extension experience with a typical 2.5 m advance round length for drill and blast operation in very good to fair ground and only 1.5 m in poor ground conditions.

Particular Design Areas

Areas which required more detailed analysis were the large ventilation tunnel and the large spans at the junction. Analysis were carried out using FLAC2D for the ventilation tunnel and FLAC3D for the junction tunnel sections. This led to a multi-sequenced mechanical excavation and support consisting of lattice girders and steel fiber reinforced shotcrete (SFRS) and higher capacity rock bolts and SFRS.

CONSTRUCTION CONSIDERATIONS

For the underground excavations of the three tunnels of the garage Cote-Vertu, mechanical excavation using a 135 T class roadheader was specified. The objective was to maintain the integrity of a tight pillar width required to remain within the available site right of way and reduce the footprint of the garage.

For the connecting tunnel and due to tight schedule requirements, drill and blast or mechanical excavation method was specified. The Contractor has selected to excavate this tunnel by mechanical means using the roadheader.

Present equipment mobilization and usage is one roadheader model MT720 which is currently excavating the access ramp and part of the open cut excavations so as to start work on the connecting tunnel as soon as possible. A second roadheader model MT720 is expected to mobilize in January 2018.

During the preparation of the Contract documents and the GBR, the performance of the roadheader was based on the experience gained on the Laval metro extension project in 2003–2005. Of particular interest and usefulness is the data published by Restner U and Reumueller B. (2004) on the performance of the ATM105 roadheader used on that project. The analysis which was undertaken compared the ratio of UCS to BTS as a measure for the evaluation of rock toughness evaluated as low to average while the fracture energy was used as an additional factor for the evaluation of rock cutting behavior. It must be noted that the test results do not often account for the positive effect of the softer shaley beds on the excavation prediction rate.

For the garage Cote-Vertu, the predicted net cutting/excavation rate (NCR m³/h) which is a measure of the solid volume of rock cut per hour the roadheader is actually in contact with the face varied between 30 and 36 m³/h for a rock mass with a UCS of 90 and 79 MPa respectively. For the rock mass with a higher UCS and possible igneous intrusions, predictions were 20 m³/h. Currently the MT720 roadheader is excavating open-cut sections and the NCR varies from 25 to 30 m³/h.

Drill and blast method is currently used for all surface open-cut excavations and the main challenge being the need to limit vibration through the highly populated and urban areas and to comply with the City of Montreal maximum peak particle velocity (PPV) vibration criteria of 25 mm/sec. The techniques employed consist of closely spaced large diameter line drilling holes using a ratio of diameter to spacing of 0.56 combined with cushion blasting to achieve the final walls. Mass excavation uses 5-m benches, a 0.6 kg/m³ powder factor, electronic delays with a limit of one hole per delay. To date, the vibration criteria has been met.

CONCLUSIONS

The geological investigation program was implemented to optimize the selection of location of the garage facility with respect to the ground conditions. The ground characterizations resulted in five classes of excavation whereby 95% of the tunnel length can be excavated by a regular full face excavation with advances varying from 1 to 2.5 m and rock support consisting of 3 m long resin anchored bolts and SFRS shotcrete. Heavier initial support with the use of steel sets and SFRS is planned for the approximately 5% of the length. The large span tunnel sections with openings of up to 18 m require a sequenced excavation respecting a cover to span criterion of one and support consisting of a combination of lattice girders and SFRS and a higher capacity resin anchored rock bolts and SFRS shotcrete.

The results of the evaluations indicate that mechanical excavation methods using a MT720 class roadheader are suitable for this type of rock mass. Drill and blast construction methods are used for the open-cut excavations and adequate measures are being taken to meet the City of Montreal vibrations limits requirements.

REFERENCES

Clark, T.H. (1972), Région de Montréal, (Publication no. 152). Québec, Ministère des Richesses Naturelles.

Durand M. (1978). Classification des phénomènes et cartographie géotechnique des roches rencontrées dans les grands travaux urbains à Montréal, Canada. International Congress of Engineering Geology, Madrid, Spain, Vol. 1, 45–55.

Grice, R.H. (1972). Engineering Geology of Montreal, International Geological Congress, Canada, B-18, 15pages.

Restner, U., Reumueller, B, « Métro Montreal— Successful operation of a state-of-the-art roadheader—ATM 105-ICUTROC—competing with drill and blast operation in urban tunnelling » EUROCK 2004 and 53rd Geomechanics Colloquim Schubert et al (ed.) 2004.

Heuer, R.E. (2005), Estimating rock tunnel water inflow-II, Proceedings of Rapid Excavation and Tunnelling Conference, p. 394–407.

Windsor–Detroit Tunnel: Application of State-of-the-Practice Maintenance and Safety Standards

Eric C. Wang and Ruben Manuelyan
HNTB Corporation

Paul Mourad
City of Windsor

ABSTRACT: To improve prioritizing of facility cyclic maintenance and operations costs, the Windsor-Detroit Tunnel Corporation tasked HNTB Corporation with developing recommendations for implementation of best practices for tunnel systems condition assessment and subsequent planning for maintenance activities (monitoring, repair and replacement) for the Detroit-Windsor Tunnel facility. This bi-national facility circa 1930s features two-lane roadway tunnel alignment including: approach, shield-driven tunnel, and immersed tube sections, respectively; with associated portal and ventilation structures. The August 13, 2015 ruling, that made National Tunnel Inspection Standards (NTIS) part of the Code of Federal Regulations, provided opportunity to update facility's existing inspection, maintenance and operation manual.

INTRODUCTION

Tunnels are vital elements of the transportation network. Most of the tunnels currently in service in the US and Canada have come to age, often being subjected to extreme operational and environmental conditions, and new ones are being constructed at an accelerated pace. To establish confidence in highway tunnel safety and performance, a comprehensive and systematic program for inspection, maintenance and operation is needed.

DEVELOPMENT OF STANDARDS FOR HIGHWAY TUNNEL INSPECTION PROGRAM

The 1967 collapse of the Silver Bridge on the Ohio River in West Virginia spurred the FHWA to establish in 1971 the National Bridge Inspection Standards (NBIS) mandating biennial inspection of bridges. The NBIS provides guidance for uniform inspection procedures and techniques, personnel qualification, inspection frequency, and the reporting of findings.

Tunnels, which constitute a unique type of structures that are encountered in both highway and railway infrastructure, were not addressed in the NBIS, and several decades passed until the Federal Highway Administration (FHWA) and the Federal Transit Administration (FTA) initiated in 2001 a joint effort to develop a system for preparing a similar inventory, inspection and maintenance program for these structures. This first endeavor resulted in the publication of:

- The *Highway and Rail Transit Tunnel Inspection Manual (2005)* addressing inspection procedures for civil, structural and functional systems.
- The *Highway and Rail Transit Tunnel Maintenance and Rehabilitation Manual (2003)* providing information on maintenance and rehabilitation practices for highway and transit tunnels.

The continued usage of many highway and rail tunnels past their reasonably expected design life, as well as failures occurring in newly completed ones, such as the 2006 fatal ceiling collapse of the Boston Central Artery's I-90 Connector Tunnel, raised the issue of regulatory requirements for inspection of tunnels and gained the attention of the Federal Authorities.

To address these issues, and as predicated by the *Moving ahead for Progress in the 21st Century Act* (MAP-21) declaring that "it is in the vital interest of the country to inventory, inspect and improve the condition of the Nation's highway tunnels," the FHWA developed the National Tunnel Inspection Standards (NTIS), which became effective on August 2015. These standards included:

- The *Tunnel Operation, Maintenance, Inspection and Education Manual* (TOMIE).
- The *Specifications for the National Tunnel Inventory* (SNTI).

Table 1. Sample tunnel inventory coding system [SNTI, 2015]

Identification

Item ID	Inventory Name	Code
I.1	Tunnel Number	0224700903568
I.2	Tunnel Name	Arch Cape Tunnel
I.3	State Code	41
I.4	County Code	124
I.5	Place Code	43000
I.6	Highway Agency District	05
I.7	Route Number	00101
I.8	Route Direction	0
I.9	Route Type	3
I.10	Facility Carried	US101
I.11	LRS Route ID	000900100S00
I.12	LRS Mile Point	89
I.13	Tunnel Portal's Latitude	45.475886
I.14	Tunnel Portal's Longitude	12.3575887
I.15	Border Tunnel State or Country Code	(blank)
I.16	Border Tunnel Financial Responsibility	(blank)
I.17	Border Tunnel Number	(blank)
I.18	Border Tunnel Inspection Responsibility	(blank)

Age and Service

Item ID	Inventory Name	Code
A.1	Year Built	1937
A.2	Year Rehabilitated	1998
A.3	Total Number of Lanes	2
A.4	Average Daily Traffic	5000
A.5	Average Daily Truck Traffic	500
A.6	Year of Average Daily Traffic	2010
A.7	Detour Length	28
A.8	Service in Tunnel	3

Table 2. Tunnel element coding system (SNTI, 2015)

Element Number	Element Name	Tunnel Description
10001	Cast-in-Place Concrete Tunnel Liner	The tunnel ends have a cast-in-place concrete liner
10003	Shotcrete Tunnel Liner	The tunnel interior has a fiber reinforced shotcrete lining
10051	Concrete Portal	The tunnel has a cast-in-place concrete portal at each end
10111	Concrete Slab-on-Grade	The tunnel has a cast-in-place concrete slab on grade
10600	Tunnel Lighting Systems	The tunnel has a lighting system
10601	Tunnel Lighting Features	The tunnel has light fixtures
10850	Traffic Sign	The tunnel has 2 traffic signs at each end

defined in TOMIE as "In order to track the conditions of tunnels throughout the United States and to ensure compliance with the NTIS, the FHWA established a NTI database to contain all of the initial tunnel inventory and inspection data. The inventory and inspection data will be available in the annual report to Congress. This data will also allow patterns of tunnel deficiencies to be identified and tracked, which will help to ensure public safety. The NTI database provides information for a data-driven, risk-based approach to asset management that can be used for informed investment decisions."

CURRENT REGULATORY STANDARDS

The NTIS published in the Office of the Federal Register on July 14, 2015 as part of the Code of Federal Regulations—23 CFR Part 650 Subpart E, became thirty days later the official document regulating the inspection programs for all US highway tunnels. This new highway regulation applies to all tunnels located on public roads, both on and off Federal-aid highways; and establishes the standards for uniform and consistent inventory and inspection procedures. The NTIS establishes requirements for reporting and correction of critical findings, and national tunnel inspection training and certification of tunnel inspectors.

The National Tunnel Inventory database established by FHWA contains all the initial tunnel inventory and inspection data. This information is made available in an annual report to Congress. The primary objective of ensuring public safety is addressed by this database which helps identify and monitor trends in tunnel deficiencies. Ultimately, it provides

These manuals provide standardized methods of data gathering on the condition and operation of tunnels, which are necessary for tunnel owners to make informed decisions as part of asset management programs for maintenance and repair of their tunnels.

The TOMIE provides uniform and consistent procedures for operation, maintenance, inspection and evaluation of tunnels.

The SNTI establishes a recording and coding system for elements which comprise the National Tunnel Inventory (NTI) database to be submitted to the FHWA in accordance with the NTIS. Refer to Tables 1 and 2 for examples of coding systems for tunnel inventory and tunnel elements, respectively.

In conjunction with the SNTI, the NTI database was established by FHWA, the purpose of which is information for a data-driven, risk-based approach to asset management which can be used for informed investment (repair/ replacement) decisions.

Per initial tunnel inventory jointly performed by FHWA and FTA more than 350 highway tunnels have been identified in the United States. Almost 40% of these tunnels are older than 50 years, and about 5% were built more than a century ago. Key factors justifying the need for inspection programs include the number, relative age and associated traffic impact of these tunnels; to ensure safe and operational conditions are maintained. [TOMIE, 2015]

Inspection of highway tunnels encompasses structural, civil and functional systems. Unique functional systems for highway tunnels include lighting, ventilation, drainage, fire detection and alarms, fire suppression, communication, traffic control and toll collection.

Responses by approximately three dozen highway tunnel owners to an informal survey prepared by FHWA and FTA reveal frequency of tunnel inspections by owners ranging from daily to 10-year intervals with an average of once per 24-months, as indicated in TOMIE, 2015.

Prior to development of FHWA's TOMIE Manual in July 2015, several key organizations had issued technical advisories, best practices, rehabilitation guidelines. These intermittent "as-needed" notices eventually led to an August 2009 "Domestic Scan" survey of several well-known US Tunnels to determine best practices for roadway tunnel design, construction, maintenance, inspection, and operations. The resulting eight recommendations identified in TOMIE are:

1. Develop standards, guidance, and best practices for roadway tunnels.
2. Develop an emergency response system plan unique to each tunnel facility which considers human behavior, facility ventilation, and fire mitigation.
3. Develop and share inspection practices among tunnel owners.
4. Consider inspection and maintenance operations during the design stage.
5. Develop site-specific plans for the safe and efficient operation of roadway tunnels.
6. A tunnel includes a long-term commitment to provide funding for preventive maintenance, upgrading of systems, and training and retention of operators.
7. Share existing technical tunnel design knowledge within the industry.
8. Provide education and training in tunnel design and construction.

A prior survey of European tunnels had evaluated "several innovative design and emergency management plans."

APPLICATION TO THE WINDSOR–DETROIT TUNNEL

As stated in Section 1, Maintenance and Repair Section 4(2) of the International Bridges and Tunnels Regulations, issued by the Canadian Ministry of Justice, "An owner of an international tunnel shall ensure that a detailed visual inspection or an inspection of the electrical, communication, mechanical and plumbing systems of the tunnel is conducted in the manner set out in the Highway and Rail Transit Tunnel Inspection Manual." In 2015, this manual was superseded by the TOMIE, hence an assessment of the inspection and maintenance program currently in place for the Windsor-Detroit Tunnel was conducted based on compliance with the TOMIE. The currently implemented procedures with respect to best practices for tunnel systems condition assessment, monitoring, repair and replacement in accordance with the latest industry accepted codes, standards, manuals and guidelines were examined, and recommendations were provided for updating these procedures.

Project Background

Construction History

Completed in 1930 as the first vehicular tunnel constructed between two nations. The bi-national subaqueous 5,160 linear foot two-lane roadway tunnel structure was built using three methods, as shown on the plan and profile in Figure 1.

- Two approach tunnel sections: Rectangular reinforced concrete cut-&-cover box tunnel construction extending from each portal to the mined tunnel sections below each ventilation structure. Specifically, the Windsor and Detroit approach sections were 602-ft and 627-ft long, respectively. Refer to Figure 2.
- Two Shield-driven tunnel sections: Circular structural steel and reinforced concrete filled segmental liner sections connected from the approaches to the immersed tube section under the Detroit River. The shield-driven sections on Windsor and Detroit sides measured 1243 and 466 linear feet, respectively. Refer to Figure 3.
- Single immersed tube tunnel section: One 2,197 linear foot long subaqueous octagonal-shaped section (Figure 4), constructed in backfilled trench covered by armor stone "rip-rap" protection.

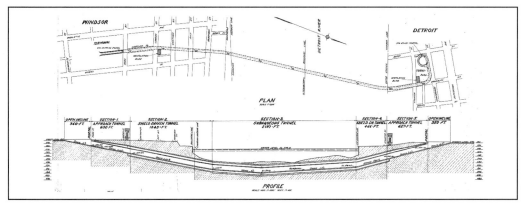

Figure 1. Windsor–Detroit Tunnel plan and profile (Detroit & Canada Tunnel Company, 1929)

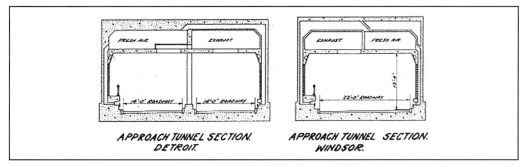

Figure 2. Cut-and-cover approach structures (Detroit & Canada Tunnel Company, 1929)

Existing Tunnel Facility Description

The existing facility comprises a tunnel, ventilation buildings and several ancillary structures described below.

Tunnels. An existing two-lane roadway tunnel that consists of relatively short cut-and-cover approach sections extending from each portal to circular segmental steel-lined shield-driven tunnel sections below respective ventilation buildings. These circular tunnel sections connect at both ends to the octagonal-shaped immersed tube section under the river completing the running tunnel structure.

Ventilation system. Ventilation of portal and approach tunnel sections is achieved with overhead ventilation ducts above the roadway ceilings. The ceiling cavity has a longitudinal center bulkhead, separating the exhaust duct section from the fresh air duct section. Ceiling mounted extraction openings exhaust air from both roadways. Sidewall ducts from the overhead fresh air duct extend down through the outer walls to the fresh air openings near the roadway surface.

Ventilation buildings. The Windsor ventilation building is located on University Street at the

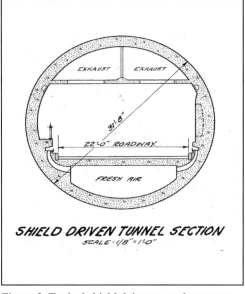

Figure 3. Typical shield driven tunnel structure (Detroit & Canada Tunnel Company, 1929)

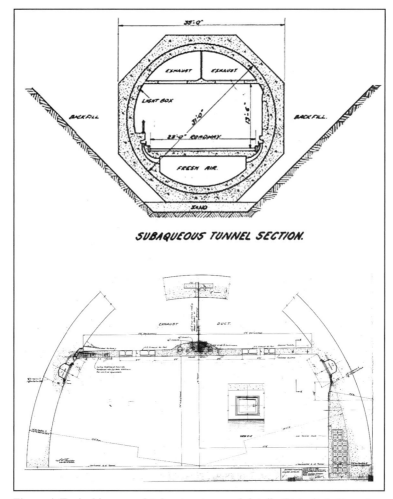

Figure 4. Typical immersed tube structure and details (Detroit & Canada Tunnel Company, 1929)

landside end of the shield-driven tunnel segment. This four-story structure is configured to extract roadway exhaust or smoke, and supply fresh air in two separate tunnel directions. The landside ventilation direction is through a series of approach tunnel sections connecting through a series of tunnel sections that lead toward the middle of the tunnel length under the river. The mid-tunnel ventilation connection starts with the shield-driven tunnel sections with the ductwork running through all the sections and ultimately connecting to the subaqueous tunnel sections at the middle of the tunnel. The ventilation building supply and exhaust fans are located above the approach tunnel closest to the shield-driven tunnel section. The ventilation building includes a basement level which accesses the upper and lower air-plenum ducts of the shield-driven tunnel section.

The Detroit ventilation building is located at the corner of Atwater Street and Randolph Street and consists of four stories and a basement level. The basement level provides access to the upper and lower ducts of the tunnel. This vent structure is configured to extract roadway exhaust and smoke and supply fresh air in two separate tunnel directions. One ventilation direction is through a series of approach tunnel sections connecting through ventilation ducting toward the portal. The other ventilation direction is connected through a series of tunnel sections that lead toward the middle of the tunnel length.

The tunnel connections from the basement of the ventilation buildings serve the upper ventilation duct above the tunnel ceiling and supplies fresh air to the lower fresh air duct.

Scope

HNTB facilitated the development of Joint Operating Agreement (JOA) governing the bi-national facility to include implementation of best practices for tunnel systems condition assessment and subsequent processes and activities (monitoring, repair, replacement) in accordance with the latest industry accepted codes, standards, manuals and guidelines. Specific tasks include:

Task 1.1—Review of existing available documents for compliance with the best practices and provide suggestions for required updates including codes and standards. The deliverable was Summary Report of Findings as Input to JOA.

Task 1.2—Review of existing JOA section pertaining to Maintenance, Repair, Purchase and Replacement of Equipment and Property.

Task 1.3—Review of existing JOA section related to Maintenance and Repair of Tube and Ventilation Buildings.

Task 1.4—Review of existing JOA section regarding Maintenance and Repair of Plazas, Garage, Storage Buildings, and Supervisor's Offices.

All related findings were summarized in Summary Report of Findings as Input to JOA deliverable.

Task 1.5—Recommended maintenance procedures. Based on the results of Tasks 1.1 through 1.4, we prepared a suggested maintenance plan in accordance with latest industry standards and practices for the main elements of the tunnel, ventilation buildings, plazas, and associated equipment, the applicable requirements from FHWA National Tunnel Inspection Program and the Canadian International Bridges and Tunnels Regulations (2012). The recommendations were submitted as part of Recommended Maintenance Procedures deliverable. As part of the Recommended Maintenance Procedures manual, practical checklists were provided in an easy-to-use inspection guidelines format as shown in Table 3.

Following initial review of documents and prior to commencing the assessment, a site meeting was conducted with operations and maintenance department managers to further clarify existing conditions and concerns at the facility. During February 2017 site visit, the following existing conditions were observed:

- Grouting repair in the exhaust ventilation ducts above the roadway tunnel ceiling, see Figure 5.
- Invert drainage in the lower (supply) ventilation ducts beneath tunnel roadway slab (Figure 6).

Evaluation of the Current Rehabilitation Program

Prior to August 2015, inspection had been performed in accordance with guidelines provided in the Highway and Rail Transit Inspection Manual per the International Bridges and Tunnels Regulations. This manual provided guidelines for tunnel inspection procedures and documentation which were superseded by the formal National Tunnel Inspection Standards (NTIS) effective August 2015 including the TOMIE and SNTI. Issuance of this recent tunnel industry standard featuring specialized training of tunnel inspectors has provided an opportunity to evaluate the current rehabilitation plan which had followed the previous manual guidelines. Specifically, prior to NTIS, facility upgrade and repairs had typically been performed in an ad-hoc or as-needed manner. While the updated recommendations manual will be used for future inspection, evaluation, maintenance and operations; the current tunnel assessment was performed following earlier inspection activities in preparation for the following phased tunnel rehabilitation:

Phase I—Concrete repairs. Phase I work included the repairs (chipping, patching and crack sealing) of the concrete tunnel liner above the ceiling slab in the River section exhaust duct. It also included the replacement of the weep drain control system. This project was started in April 2017 and completed June 2017.

Phase II—Electrical upgrades. Phase 2 work included the removal and installation of the electrical power and communication systems above the ceiling slab in the River section exhaust duct. This project was performed from May to September 2017.

Phase III—Ceiling replacement. In August 2015, HNTB performed a peer review of the ceiling slab replacement conceptual study to assess the proposed concepts for the ceiling rehabilitation and replacement presented in Final Draft of the WSP formerly Parsons Brinckerhoff's (PB) report entitled *Detroit-Windsor Tunnel Ceiling Slab Replacement Conceptual Study Report*, dated June 22, 2015, supplemented by additional and back-up information and to provide comments, suggestions, and potentially value engineering ideas.

In addition, the review provided an independent check of the construction cost estimate and schedule of the recommended alternative. The purpose of the review was to validate the recommendations included the subject report from technical, constructability, construction staging, cost and schedule aspects, and for compliance with National Fire Protection Association Standard NFPA 502 (Standard for Road Tunnels, Bridges, and Other Limited Access Highways, 2014 edition) while considering

Table 3. Inspection application summary table (WDT Task 1.5, 2017)

Item Number	Facility Elements	Civil/Structural	Architectural	Facility Systems
	Toll Plazas			
	Tollbooths			
	Overhead Canopies	(A)	(A)	
	Overhead Sign Structures	(A)		(J)
	Pavement and Sidewalks	(B)		
1	Doors and Windows		(C)	
	Walland Floor Finishes		(C)	
	Weighing Scales			(J)
	Lighting	(A)	(C)	(J)
	Signage	(A)		
	Visual Messaging	(A)		(J)
	Portals			
	Retaining Walls	(B), (D), (1)		
	Pavement and Sidewalks	(B)		
2	Railing	(A)		
	Signage	(A)		
	Wall Finishes		(C)	
	Lighting	(A)	(C)	(J)
	Ventilation Buildings			
	Slabs, Walls and Columns	(B), (G)		
	Stairs	(B)		
	Railing	(A)		
	Elevators			(J)
3	Monorails	(A)		(J)
	Hangers	(A)		
	Masonry Walls		(B)	
	Mechanical Equipment			(J)
	Electrical Equipment			(J)
	Communication/SCADA			(J)
	Tunnels			
	Roadway Slabs and Walkways	(B), (G)		
	Ceiling Slabs	(A), (B), (G)		
	Walls	(B), (G)	(C)	
4	Benches	(B), (G)		
	Liner	(B), (G). (H)	(C,	
	Finishes		(C,	
	Pavement	(B)		
	Signage	(A)		
	Lighting	(A)	(C)	(J)

(A) Visual and hands-on (steel) (E) In-situ Non-destructive (concrete)
(B) Visual and hands-on (concrete) (F) Scanning
(C) Visual and hands-on (architectural) (G) Laboratory testing
(D) In-situ sampling (concrete) (H) Functional testing

the project constrains. The review was done as a desk top review only with no field verifications. In August 2017, bids were received for ceiling replacement which includes the removal and replacement of the concrete ceiling slab and steel hanger system, removal and replacement of the collector drain system, and associated electrical work in the River section exhaust. Project commenced October 20, 2017 and is scheduled to be completed by July 2018.

RECOMMENDED INSPECTION AND MAINTENANCE PROCEDURE

As recommended in the Task 1.5 deliverable previously described, the Final Recommended Maintenance Procedure Manual provides procedures for periodic inspections, reporting, repair and maintenance in accordance with TOMIE. Under these guidelines, specific intervals are given for several

Figure 5. Exhaust duct—leakage repair grouting

Figure 6. Invert drainage into supply duct beneath roadway

types of inspections defined as Routine, Damage, In-depth and Special. The development of not only inspection frequencies but also their detailed scope will help facilitate coordination of condition surveys and prioritize rehabilitation and upgrade activities.

CONCLUSION—BENEFITS

Recent enactment of NTIS provided a timely opportunity to review existing documents addressing operations and maintenance responsibilities for an 85-year old bi-national tunnel facility. The primary benefit of the tunnel assessment task was to help establish an updated tunnel maintenance and inspection manual providing procedural guidelines in accordance with current requirements and best industry practices. This revised manual also provided tabulation of recommended inspection procedures and applicable techniques as well as applicable types of inspection for individual elements of the tunnel facility for clear user-friendly reference by client staff and their inspection sub-consultants. Moving forward, key rehabilitation and upgrade decisions will be necessary as this critical infrastructure connection serving 12,000 vehicles daily between Detroit and Windsor continues to age. This manual would inform those investment decisions for the tunnel facility rehabilitation to maintain safe and efficient operations in a cost-effective manner.

REFERENCES

Canadian Minister of Justice, International Bridges and Tunnels Regulations dated March 14, 2012 (Current October 13, 2017).

Detroit & Canada Tunnel Corporation and The Detroit and Windsor Subway Company Ltd. and The Corporation of the City of Windsor and Windsor Tunnel Commission—Joint Operating Agreement dated August 28, 1998.

Detroit & Canada Tunnel Company, Record Drawings, 1929.

Detroit-Windsor Tunnel Ceiling Slab Replacement Conceptual Study Report prepared by WSP (formerly PB) dated June 22, 2015.

Detroit-Windsor Tunnel, Peer Review of Ceiling Slab Replacement Conceptual Study prepared by HNTB Corporation dated August 31, 2015.

US Department of Transportation, FHWA and FTA—The Highway and Rail Transit Tunnel Inspection Manual, 2005 (superseded by TOMIE 2015).

US Department of Transportation, FHWA and FTA—The Highway and Rail Transit Tunnel Maintenance and Rehabilitation Manual (2003).

US Department of Transportation, FHWA Publication No. FHWA-HIF-15-005 Tunnel Operations, Maintenance, Inspection, and Evaluation (TOMIE) Manual, July 2015.

US Department of Transportation, FHWA Publication No. FHWA-HIF-15-006 Specifications for the National Tunnel Inventory, July 2015.

Windsor-Detroit Tunnel—Best Practices for Tunnel Condition Assessment, Monitoring and Maintenance. Task 1.5—Recommended Maintenance Procedures prepared by HNTB Corporation dated June 28, 2017.

Consideration of Single Bore as a Construction Option for VTA's BART Silicon Valley Phase II Extension Project

S. Zlatanic
HNTB

K. Davey
Valley Transportation Authority

ABSTRACT: In 2009, Phase II Extension of Valley Transportation Authority's (VTA) total 16-mile BART Silicon Valley Extension Program (BSV Phase II) had planned for nearly 5-mile-long twin-bore tunnel through downtown San Jose to the City of Santa Clara with 33 cross-passages, three cut-and-cover underground stations and one at-grade station. In 2014, as the project was being revived through VTA planning efforts and environmental reviews, to better address growing community concerns, VTA had undertaken evaluation of single bore construction option with emphasis on tunnel alignment, diameter, depth, station configuration, and tunnel and station ventilation and fire life safety aspects including emergency egress. If finally approved, the single bore concept will be used for the first time in the US to house both guideways and stacked station platforms within the tunnel.

INTRODUCTION

The BART Silicon Valley Phase II Extension project is a 6-mile extension from the current Phase I Berryessa Station terminus through Downtown San Jose to a new station and terminus in Santa Clara. This extension includes 4.8 miles of running tunnels through San Jose; four stations, of which three are underground (Alum Rock, Downtown San Jose, and Diridon) and one is at grade. (Santa Clara); two intermediate ventilation structures, and East and West tunnel portals (see Figure 1).

In 2014, VTA restarted the planning efforts for Phase II with an update to the project environmental studies. As part of this planning effort, VTA initiated the single bore feasibility and technical studies, inspired primarily by continued ongoing community and public concerns about potential disruptions during construction, as well as by technological advances and developments in larger-diameter, soft ground mechanized tunneling in dense urban settings. In October 2016, VTA initiated the Single Bore Tunnel Technical Studies (the Studies) to evaluate single bore tunnel diameter, track alignment, depth of stations, passenger vertical circulation, operations, tunnel and station ventilation, and emergency egress. Findings of the Studies considered BART's comments, especially in terms of operations, maintenance, and safety, while closely following BART Facility Standards (BFS) and provisions of NFPA 130.

SUBSURFACE CONDITIONS

Geology and Hydrology

The geologic setting of the Santa Clara Valley consists of a sedimentary basin bounded by the Santa Cruz Mountains on the west and the Diablo Mountains on the east. Basement rocks present across the basin consist of Cretaceous Franciscan formation in the central and western portion of the alignment and Miocene/Pliocene sedimentary rocks at the eastern end of the alignment. The boundary between these basement rocks appears to be delineated at the North Silver Creek Fault. The valley has filled with Quaternary and Holocene alluvial and bay margin deposits consisting of clays, silts, sands, and gravels, and a variety of combinations thereof. These deposits were placed in coalescing alluvial fan deposits created during different depositional periods, with relatively calm deposition producing fine grained clay and silt deposits and periods with higher energy deposition producing sand and gravel deposits. (Wentworth et al, 1999)

As is typical with alluvial deposition, lateral continuity of a given strata is not extensive, as the depositing streams migrate back and forth over the alluvial fans. Some broad trends can be discerned, however on a large scale within the basin. The deeper unconsolidated sediments within the valley are typically sands and gravels and comprise the primary groundwater storage and production zones in the valley. This primary aquifer is overlain throughout

Figure 1. VTA's BART Silicon Valley Phase II Extension Project

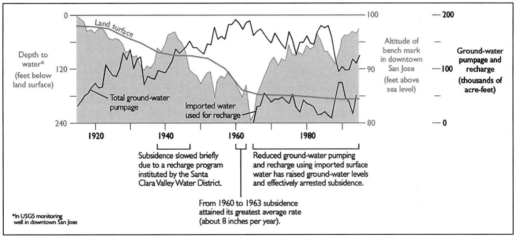

Source: Ingebritsen et al. 1999

Figure 2. Groundwater levels and subsidence in Santa Clara Valley

most of the valley by a thick regional aquitard that was deposited during a calmer depositional period. Overlying the regional aquitard is another sequence of coarse-grained sediments commonly called the Upper Aquifer within the Valley. The Upper Aquifer is in turn overlain by generally finer grained deposits up to the surface, where human activity has modified the near surface area now classified as fill.

Groundwater in the Santa Clara Valley is a major source of domestic, industrial, and agricultural water supply. The aquifers and active recharge basins developed by the Santa Clara Valley Water District (SCVWD) to manage and maintain sufficient water supply serve as large-scale conveyance systems for much of the water in the basin. Over the nearly

100 years that the SCVWD has been maintaining the groundwater basin, they have observed significant subsidence (up to 13 feet in downtown San Jose and developed alternative water supplies to offset the reliance on groundwater supply and arrest the settlement within the valley (see Figure 2).

The SCVWD continues to monitor water levels throughout the valley and monitor settlement through individual instruments called multiple point borehole extensometers, as well as through remote satellite monitoring using interferometric synthetic aperture radar (InSAR) technology. The valley remains somewhat vulnerable to future settlement if water imports decrease and the reliance on locally produced groundwater increases. Also, artesian water

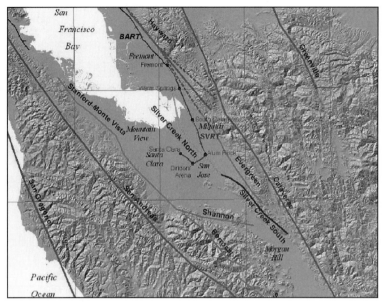

Source: Silicon Valley Rapid Transit Program, Central Area Stations 65% Geotechnical
and Seismic Design Criteria Report for Underground Stations, January 2009

Figure 3. Faults present in the project vicinity

pressures have been observed in places (Poland and Ireland 1988). This condition exists when aquifers below the ground surface are hydraulically connected to aquifers along the edges of the valley at higher elevations. In this case the water pressures in the valley are increased due to the connection with aquifers above the valley resulting in water pressure that causes the column of water to rise above the ground surface. This condition is well known in the north-central part of the Santa Clara Valley, but has also been observed at locations along or near the tunnel alignment. Upward hydraulic gradients caused by artesian conditions need to be properly managed; to evaluate this condition, water levels could be measured at required frequency in project wells and piezometers to document seasonal fluctuations of the groundwater and provide the data to develop appropriate design provisions (tunnel and station construction need to accommodate potential for water pressures to create "soft" or "running "ground conditions during the excavation).

Faulting and Seismicity

The Santa Clara Valley is bounded on east and west by the Hayward/Calaveras Fault system and the San Andreas Fault zone, respectively. These fault zones are the most significant seismic sources near the Phase II alignment, but there are additional smaller faults to carefully consider (Figure 3). Other smaller faults within the valley have been mapped, including the Stanford Monte Vista fault on the west side of the Valley and the Silver Creek Fault in the eastern portion of the valley (Wentworth et al., 2010). The proposed alignment crosses the Silver Creek Fault very near the location where the alignment crosses Coyote Creek. While the major seismic sources are expected to be from the major faults, the Silver Creek Fault may impose additional risk factors that will need to be addressed, including potential creep and basin effects for the seismic response of the tunnels and structures.

Holocene activity on the Silver Creek is not well documented, however Quaternary displacement has been observed in seismic profiles that cross the fault. Offset seismic reflectors have been correlated with stratigraphic zones dated using Carbon-14 methods to help constrain the timing of offset. The resolution of reflectors in younger sediments is not adequate to confirm or deny Holocene offset. Further evaluation of the potential creep and seismic potential along the Silver Creek Fault needs to be performed. A relatively recent 2010 US Geological Survey evaluation of the offset reflectors implies that Silver Creek Fault may be capable of a creep rate of up to 2 millimeters per year (vertical or horizontal) (Wentworth et al. 2010.

The Silver Creek Fault also delineates the west side of the Evergreen Basin in the eastern portion of the Valley. The geology of the Evergreen Basin is distinctly different than in the Valley to the west, with alluvial and sedimentary deposits that are much

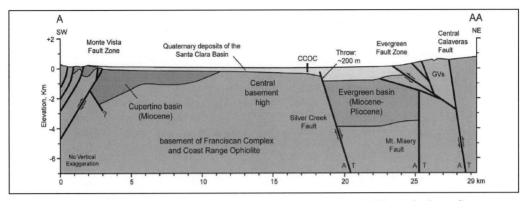

Figure 4. Geology of Evergreen Basin, Section A–AA from Wentworth, Williams, Jachens, Graymer and Stephenson (2010, USGS Open File Report 2010-1010)

thicker (Figure 4). Such different thicknesses and materials have been known to locally amplify the seismic response to structures that overlie a basin, imposing "basin effects" on the seismic response. The potential importance of basin effects along the alignment will also need to be evaluated during the final design of the project.

In addition to the site-specific issues related to seismicity, the USGS has recently further developed a regional seismic prediction model to include the effects of interaction between interrelated fault zones (Field, E.H., and 2014 Working Group on California Earthquake Probabilities, 2015). The result of this new model is that probabilities of significant seismic events have increased, the potential impacts of which would need to be included in the seismic design for the project to the most recent standards.

Water Wells

The original tunnel alignment traverses some of the oldest parts of San Jose where the original water supply consisted of shallow wells completed in the Upper Aquifer. Up until the 1970s it was not a requirement to register or obtain a permit to have a personal well on one's property. The SCVWD maintains a database of known water wells in the district. Because registration of wells was not required before the 1970s, many property owners along the alignment may not have notified the SCVWD, or may be unaware that they have wells on their property at all. In other cases, they may rely on groundwater wells for onsite irrigation or other supplies. A formal well survey is warranted for a comprehensive assessment of presence of water wells along the tunnel alignment.

Impacts on Single Bore Alignment

The BSV Phase II alignment right of way was extensively studied in earlier investigation stages for the twin bore configuration. During the previous phases of investigation, many deep borings were completed at the proposed station sites due to anticipated cut and cover construction and dewatering needs. Some of the borings extend to over 150 feet depth and are valid to provide data for the single bore alternative. The other previous borings for the tunnel sections, however, were focused on the twin bore configuration, and only extend to about 20 feet below the planned twin bore tunnel bottom. Therefore, as the single bore alternative would extend to greater depths than the twin bore alignment, the existing borings are limited in providing the overall characteristics of the subsurface for tunneling construction, and additional investigation in certain areas or around key structures may be necessary to supplement the investigation and mitigate risks associated with locations that are not fully characterized. In addition, since the previous twin bore design was completed, a few important changes in criteria and seismic considerations have occurred; also, important advancement in the mechanized tunnel technology including tunnel boring machine (TBM) systems have taken place.

At the eastern portal of the proposed single bore tunnel alignment, the ground consists primarily of fine-grained Holocene and Pleistocene sediments. These ground conditions dominate the tunnel profile from the portal through the Alum Rock Station. As the alignment proceeds from Alum Rock Station toward downtown, the tunnel horizon skirts along the top of the Upper Aquifer zone, described above, where, at the low point of this segment, the alignment is expected to encounter the sands and gravels of the Upper Aquifer in 20–30% of the tunnel face at the invert. This condition continues to the Downtown San Jose Station and Crossover where coarse-grained sands and gravel occur much higher in the tunnel profile. From the Downtown San Jose Station to the west portal of the tunnel, the ground conditions will be highly variable consisting of

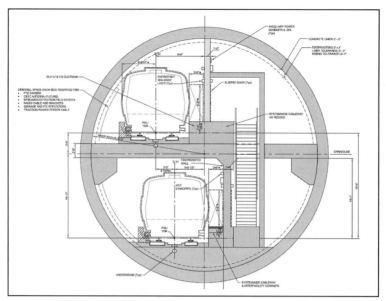

Source: VTA's BSV Phase II, Single Bore Technical Studies Summary Report, Final, April 2017

Figure 5. Typical tunnel section

intermittent sands and gravel deposits surrounded by finer grained sediments.

The groundwater conditions will impose higher groundwater pressures on the tunneling operations due to the increased depth of the single bore profile. Overall, the magnitude of increase will on the order of one bar of pressure or less, and a single bore TBM and its concrete lining will need to be designed to accommodate this additional pressure.

The above discussions on the existing boring depths, groundwater conditions, and the relevance to the lower profile of the single bore alternative are also applicable to the station access shafts associated with the single bore which would be placed off the tunnel alignment (the shafts are tangential to the single bore, and located outside of the public right-of-way). Excavation shoring, bottom stability, dewatering, and protection of existing buildings/facilities in the congested downtown environment are significant construction considerations.

SINGLE BORE DIAMETER, ALIGNMENT AND DEPTH

Single Bore Diameter

The Studies arrived at a preliminary tunnel internal diameter of 41 feet to satisfy vehicle envelopes and required clearances stipulated in the BFS through all necessary guideway configurations and transitions along the project alignment. For operator's (BART)

preferred track alignment, which maximizes 'side-by-side' rather than 'stacked' track configuration (Figure 5) within the large diameter tunnel. The tunnel section at transitions between the two track configurations essentially governs the minimum tunnel diameter.

The 41-foot minimum tunnel diameter meets BFS requirements for the standard vehicle running clearance envelope over the length of each transition (four total) between stacked and side-by-side track positions. Additional space allowances comprise a radial tunnel construction tolerance of 6 inches to account for normal variations in the precast concrete segmental liner construction and TBM steering tolerances. The critical section meets the BFS safety and maintenance provisions, accommodates walkways and provisions for cross passages, and is adequate in terms of BART systems space planning within the running tunnel environment. Also, a 41-foot tunnel inside diameter provides for prudent design allowances for future design development of elements that are of 'advanced design nature' including structure fire resistance; potential stand-off distances to accommodate threat assessment/vulnerability criteria as it may be adopted by VTA and/or BART in the future; additional clearances and seismic design details as may be required where the tunnel crosses the Silver Creek Fault Zone, and provisions for special track slab that may be required at certain locations along the guideways to minimize impacts of

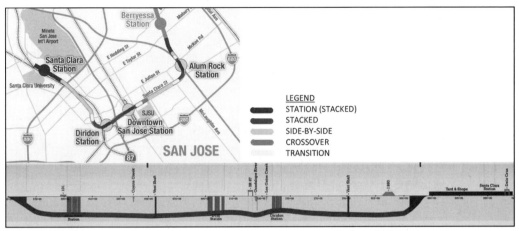

Source: VTA's BSV Phase II, Single Bore Technical Studies Summary Report, Final, April 2017

Figure 6. BSV Phase II alignment (plan and profile)

noise and vibration to overlying sensitive structures and facilities.

Single Bore Alignment

The horizontal geometry of the northbound and southbound tracks for the single bore tunnel is accommodated primarily within the original 2009 twin bore design's pre-defined right of way corridor. Through early discussions with BART operations staff, it was suggested that a side-by-side, same level track configuration is preferable to a stacked configuration, as it more closely resembles existing configurations throughout the BART system. As such, the extent of side-by-side track configuration has been maximized—at the East Portal, between Alum Rock and Downtown San Jose Stations, and west of Diridon Station to the West Portal. The tracks are maintained in a vertical stacked configuration at all stations to accommodate stacked platforms, and between Downtown San Jose Stations and east of Diridon Station due to insufficient distance between stations to accomplish horizontal or vertical transitions (Figure 6). The transition from horizontal side-by-side to a vertical stacked configuration is accomplished considering the vertical and horizontal geometric constraints while satisfying BFS requirements. Direct fixation track utilized within the bored tunnel currently assumes the use of resilient ties, consistent with BFS standards.

An initial run time analysis to assess travel times through the single bore indicates that the single bore alignment provides shorter travel times for all runs, primarily due to the shorter overall length (approximately 600 feet) of the single bore alignment. East of Downtown San Jose Station, the change from a No. 10 scissor crossover to a No. 15 (with potential

for crossover No. 20) presently allows a higher crossover speed, which helps to offset its location farther east from the station (2,000 feet). For trains utilizing the crossover, the location is anticipated to limit the maximum speeds between Alum Rock and Downtown San Jose Stations to 50 mph, rather than the 70 mph possible on the previous design alignment. This results in a three second penalty in travel time to westbound trains, and a one to two second benefit to eastbound trains. Trains bypassing the crossover, as would be typical during normal operations, would realize no travel time impact, and would see an overall improvement in run time ranging from 10 to 15 seconds resulting from the shorter alignment.

Single Bore Depth

The alignment developed for purposes of the Studies was optimized to avoid known underground obstructions under Coyote Creek and near SR 87/Guadalupe River Crossing. A parametric analysis was performed to assess the influence of ground loss and geologic conditions on tunnel depth. The overall geologic profile developed for the original twin bore design was (Figure 7). Where the potential single bore tunnel alignment deviates from the twin bore alignment, geologic conditions represented in the twin bore profile appear to be reasonably consistent with potential alignment adjustments. Depth of the single bore tunnel is expected to be greater than that of the twin bore tunnel. With greater depth, the single bore tunnel will encounter more granular ground in the eastern portion of the alignment, and these granular materials typically behave differently than fine-grained, cohesive ground with respect to potential settlement.

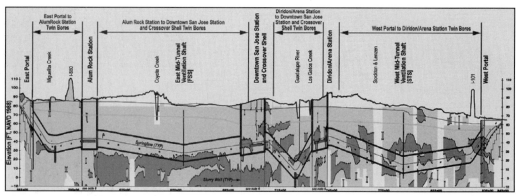

Source: Silicon Valley Rapid Transit Program, Central Area Guideway, 65% Submittal: Combined Geotechnical Baseline Report, December 31, 2008 Issued for Record

Figure 7. BSV Phase II geological profile developed for twin bore design

The Studies focused on assessing the impact of the deeper geologic conditions on the potential settlement estimates already published for the twin bore option. The criteria used to evaluate potential settlement issues are the same as those used for the twin bore design, namely maximum settlement of 1 inch at the ground surface, angular distortion less than 0.002 and horizontal strain less than 0.00075.

Tunnel-related settlement is typically evaluated assuming the shape of the settlement trough, perpendicular to the tunnel, follows a Gaussian distribution, according to the method developed by Schmidt (1974) and O'Reilly and New (1983). One of the key parameters in this approach relates to the "volume loss," which corresponds to the volume of extra material excavated as the ground surrounding the tunnel excavation responds to the excavation activity.

Previous single bore feasibility studies (by others) assumed a volume loss parameter of 0.5%, which is based on a long history of tunnel excavation using smaller diameter TBMs with a broad range of operating parameters. The volume loss value is typically quantified as a percentage of the overall tunnel face area. Modern TBM technology has advanced in recent years to the point that with good operations, volume losses of 0.2% to 0.3% are more typical today, especially for larger diameter tunnels, and excavation operations can be controlled to reduce the volume loss even further under certain circumstances.

To evaluate a potential range of tunnel-related settlement, a parametric analysis was performed by varying the volume loss, k-values, tunnel depth and geologic conditions at each depth. One of the key geologic factors for each section is the relative percentage of granular ground conditions versus cohesive ground conditions. Granular ground will typically be more prone to running conditions than

cohesive ground, and the shape of the estimated settlement trough will be slightly different, and is described by a "trough width" factor. The calculated results are compared to the settlement criteria established in the twin bore design and used in previous feasibility studies (by others). Most critical sections were selected for the parametric analyses; in general, the sections analyzed coincided with the section having the highest percentage of granular material in and above the tunnel horizon, thus providing a "maximum" estimated settlement estimate based on the least favorable stratigraphy. In addition, potential building impacts were evaluated.

In summary, it was determined that a minimum depth of 50 feet (between ground level and top of the tunnel) is required for a single 45-foot outside diameter bore constructed using the latest state-of-the-art modern pressurized face TBM and tunneling practices to control ground losses and minimize settlements.

The following factors were considered:

- Geologic and hydrogeological conditions at multiple critical locations along the alignment
- Results of settlement analyses performed for the estimated prevailing soil types within and immediately above the tunnel horizon
- Allowances for additional settlements due to station (shafts and connecting adits) excavation
- Estimates of potential building impacts

The settlement profiles analyzed indicate that a 50-foot cover depth, combined with the tunnel boring machine (TBM) volume loss performance of 0.35% or less, should generally meet accepted tolerable building settlement criteria. In, addition this minimum ground cover would meet required safety factor for TBM blow out condition potentially arriving from maintaining the appropriate machine face pressures.

Based on the settlement analyses at the section locations evaluated during this study, the geologic and tunneling conditions in the western portion of the alignment appear to be slightly more susceptible to tunnel related settlement than the eastern portion of the alignment. The settlement profiles analyzed from these sections indicate that at all depths, a volume loss performance of 0.35% or less should meet the settlement criteria for overall settlement, angular distortion, and horizontal strain. The results of the parametric analysis west of Diridon Station indicate that the ground in this area is more granular than in the eastern part of the alignment, and thus the cover requirement would govern. It should be noted that the area west of Diridon is also the location of horizontal curve that may be more challenging to tunnel through and maintain tight control of volume loss. Although the selected minimum tunnel depth appears practical, additional detailed studies will be performed to estimate total impacts due to shaft and adit construction at each station location, and those compared to settlement, angular distortion, and horizontal strain criteria.

CONSTRUCTABILITY ISSUES

Ground conditions have a significant influence in the specification of a TBM. A detailed evaluation of the new profile stratigraphy will be needed to confirm if the site geology and hydrology indicates a strong indication for an Earth Pressure Balance (EPB) TBM, a slurry TBM, or hybrid. The proportion of sand and gravel deposits would be expected to go up if the profile intersects more of the Upper Aquifer unit, therefore one may make a case for a slurry TBM as an optimum choice for tunnel excavation. Detailed design will need to evaluate choices for any of the tunnel construction alternative mechanized methodologies. Although there is a large amount of existing investigation data already available, if a deeper profile is selected, additional deeper borings at half a dozen critical locations will be necessary to adequately characterize subsurface conditions for design and contract geotechnical baseline purposes.

Since ground conditions have a significant influence in the specification of a TBM, a more detailed evaluation of the new profile stratigraphy will be needed to determine if the resulting changes in the anticipated fractions of coarse and fine alluvium that would be excavated along a deeper tunnel profile would favorize suitability of certain TBM type. The groundwater will impose higher groundwater pressures on the tunneling operations due to the increased depth of the single bore profile. Overall, the magnitude of increase will be on the order of one bar of pressure or less, and a single bore TBM and its concrete lining can be readily designed to accommodate this relatively minor additional pressure. Also, the presence of groundwater along the tunnel profile is not an issue for modern pressurized face tunnel boring using watertight, gasketed, precast tunnel linings. Also, the tunneling process itself generally does not impact the quality or level of the encountered groundwater. However, at stations, shafts, and any mined excavations outside of a bored tunnel, the presence of groundwater is a significant construction consideration. Where excavation dewatering cannot be used due to high permeability, subsidence risk, water supply impact, or other restrictions, it will be necessary to employ some combination of watertight diaphragm shoring systems, cement-soil mixing, jet grouting, chemical grouting and/or ground freezing.

STATION CONFIGURATION

Station depth is the most critical element of station configuration analysis, especially in terms of passenger experience, fire and life safety, and maintenance requirements. Egress during station normal and emergency situations, as well as the configuration and operation of ventilation shafts, are of outmost concern to BART. From a passenger experience point of view, station depth influences vertical circulation, travel distances, and wayfinding; from BART perspective, it would also affect equipment maintenance access, related procedures, and long term operating costs.

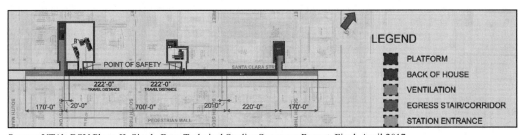

Source: VTA's BSV Phase II, Single Bore Technical Studies Summary Report, Final, April 2017

Figure 8. Downtown San Jose Station plan with 700-foot long platform section

Downtown San Jose (DTSJ) Station configuration is evaluated to accommodate the required depth of the single bore while maintaining passenger safety and convenience. DTSJ station design concept distributes the station and platform access points between two entrances (Figure 8). One of VTA's existing properties within the Mitchell Block, a property on Santa Clara Street with a building currently used as a Chase Bank and VTA offices, is used as main western entrance.

The proposed configuration of the DTSJ Station west entrance consists of an open entrance shaft housing the vertical circulation elements. It provides direct sightlines from the platform access adit to the exits at street level. Passengers would be able to observe all areas of the station entrance shaft and orient themselves within the greater urban context, making wayfinding exceedingly intuitive (Figure 9). To the east, two smaller sized parcels on Santa Clara Street, between 1st and 2nd Streets, are potentially designated for the secondary entrance (Figure 10). This entrance is centrally located between the two light rail transit (LRT) stations and allows for convenient transfer between the two modes of transportation. These two entrances would evenly distribute passengers and comply with BFS egress and ventilation, and NFPA 130 requirements. For the vertical bi-level configuration at each station, the platforms are stacked and serve one trainway each.

The 41-foot tunnel internal diameter, determined for the critical section at transitions, provides an allowance of 15-foot 6-inch wide by 8-foot (unobstructed) upper and lower (stacked) platforms, located 66.5 feet and 85 feet below the street, respectively. Also, there are provisions for lighting, select signage, and additional finish surfaces within the station section, which accommodates vehicle clearance envelope and guideway elements. Figure 11 depicts the typical configuration of the tracks, track equipment, train dynamic envelope, platforms, signage, utility chases, and seating over the 700-foot long platform. The platforms and trainways are separated by the 2-foot 6-inch thick structural slab. BFS requires a minimum platform width of 8 feet at non-continuous obstructions and 10 feet at continuous obstructions.

Accessed via level ground from the entrance shafts, single bore station platforms are unencumbered by vertical circulation elements (VCE) and column obstructions. Amenities such as seating, map cases, and trash receptacles are located outside the actual 15.5-foot wide platform. Over 8 feet of vertical clearance (the minimum specified by BFS), is available to passengers. This translates to a Level of Service (LOS) A, or 29 square feet of platform area per passenger. BFS requires a minimum of LOS C, or 7 square feet of floor space per passenger. Egress

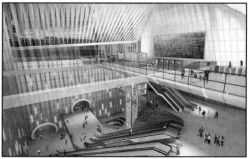

Source: VTA's BSV Phase II, Single Bore Technical Studies Summary Report, Final, April 2017

Figure 9. DTSJ Station west entrance

Source: VTA's BSV Phase II, Single Bore Technical Studies Summary Report, Final, April 2017

Figure 10. DTSJ Station east entrance

and access will be streamlined by eliminating VCE queueing on the platform areas; rather, the VCEs will be housed within the vertical access shaft, where they would be configured to minimize queueing and cross-flow of passengers.

In the trainway, all BFS requirements including spaces for the vehicle clearance envelopes, required tolerances, emergency refuge spaces, and third rail working envelopes are all provided. In the event of a train fire or other emergency condition, the two trainways are separated by the intermediate fireproofed slab. The non-incident trainway and station platform would be isolated from any smoke generated by fire incident.

In addition to BART Facility Standards (BFS) criteria, the station needs to comply with provisions of CBC and NFPA 130. The following requirements were found to be the most restrictive and governed the station access and egress planning:

- Evacuation of the station occupant load from the platforms in four minutes or less

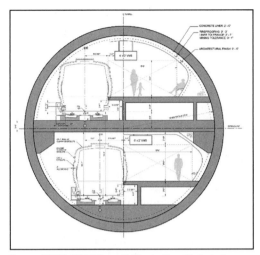

Source: VTA's BSV Phase II, Single Bore Technical Studies Summary Report, Final, April 2017

Figure 11. Typical single bore section at stacked station platforms

- Evacuation from the most remote point on the platforms to a Point of Safety (POS) in six minutes or less
- No point on the platforms shall be more than 300 feet from POS
- There shall be a minimum of one exit within 20 feet from each end of the platforms

Ventilation requirements for the station and platforms are also a determining factor for the station layout. Ventilation equipment and exhaust shaft layouts were coordinated during the Studies to ensure proper allocation for these components underground and at grade.

STATION AND TUNNEL EMERGENCY VENTILATION AND EGRESS

Evacuation and emergency egress systems and facilities for underground transit systems must be designed to facilitate patron self-rescue activities and responsible agency response procedures. BFS and applicable state and national codes and standards, including NFPA 130, provide the governing prescriptive and performance-based criteria that were applied to develop and assess the emergency egress measures and systems for a single bore dual guideway tunnel and station configuration. During a tunnel fire incident, the tunnel ventilation system must provide smoke and fire hazard control to establish a tenable environment indefinitely at the defined Points of Safety—along the identified emergency egress/incident response paths in the tunnel and within specific passageways and circulation areas in each underground station. Proper functioning and maintainability of the tunnel ventilation system and its capability for smoke and fire hazard control must be demonstrated by detailed engineering analysis.

For transit stations throughout the United States, especially for heavy rail ones, it is impractical to evacuate deep stations to the street in six minutes. Key sections of NFPA 130 (2014) employed in evaluating the single bore configuration define performance standards to establish Points of Safety, rather than prescriptive solutions:

- Section 5.3.35, the Point of Safety definitions include an enclosed exit that leads to a public way or safe location outside of the station, trainway, or vehicle.
- Section 5.3.1.2 requires the design of a means of egress to be based on an emergency condition requiring evacuation to a Point of Safety.
- Section 5.3.4 permits the concourse in an enclosed station with an emergency ventilation system to be defined as a Point of Safety when designed to provide protection for the concourse from exposure to the effects of a train fire at the platform, as confirmed by an engineering analysis.

To analyze station emergency egress scenarios for a single bore with stacked guideways and platforms connected to an adjacent off-street station, first a prototype station then a site-specific Downtown San Jose station model were developed. It was determined that queuing at the bottom of escalators and stairs within shafts that house the vertical circulation elements (VCEs) must be evaluated to confirm the time duration required to clear both upper and lower platforms during an emergency. Ventilation system operation and makeup air flow from the surface through the shaft are key elements for establishing a tenable environment from the lateral passageways to each station platform, all the way to the street level, and thereby allowing the passageways and shaft concourse areas to be defined the Points of Safety under NFPA 130. The ventilation system modeling for both the station environment and the system wide network flow conditions was completed utilizing computational fluid dynamics (CFD) and subway environmental simulation (SES) applications, respectively, to identify Point of Safety locations within the station areas including platforms and the running tunnel guideways.

Maximizing Passenger Safety

The principal objective of the fire/life safety systems of a transit station, which include the layout, protection and function of the emergency egresses and ventilation systems, is maximizing passenger safety during the station design fire and/or smoke emergency conditions. One key fire/life safety element used within the latest BART Facilities Standards and the referenced industry standards, including the National Fire Protection Association (NFPA 130) *"Standard for Fixed Guideway Transit and Passenger Rail Systems"* (2014), and the California Building Code (CBC, 2016), is Point of Safety. The generally accepted, functional definition of a Point of Safety as applied to transit stations is:

- A location at the end of an egress path outside of the structure and within a public way where patrons are safe from exposure to fire hazards (relative level of safety)
- A location within a protected enclosure where patrons are able evacuate to an exit discharge at a surface public way location

POS is only one element of the industry accepted standard analysis approach, for which the BFS provides criteria, for demonstrating passenger safety performance of a station and determining acceptable conditions (a tenable environment) are met during station design fire and/or smoke emergency conditions. The comprehensive, industry-accepted approach to assessing the safety performance of a station involves comparing the calculated Available Safe Exit Time (ASET) to the Required Safe Exit Time (RSET). Passenger safety is achieved when the ASET exceeds the RSET during the station fire and/or smoke emergency condition adopted by project design criteria as a credible design scenario. Key design inputs affecting the RSET include the given passenger load and the number and configuration of the egress routes. The key design inputs that drive the ASET include the design fire size, rate of fire growth, and the ventilation system design. A properly designed station and its emergency ventilation and fire life safety systems will provide an ASET that exceeds the RSET and thereby ensure passenger safety for the station-adopted emergency condition.

BART Criteria for Passenger Load and Design Fire Growth

The current station passenger load being used for egress evaluations assumes a crush load of 2,500 patrons on the arriving train at the incident platform, combined with added passenger loads from: a) those waiting on the platform in each direction derived from the 2035 peak period forecast and one six-minute missed headway, and b) a train with 625 passengers (one quarter of train crush load) on the opposite platform. These are the fundamental parameters governing the configuration of the station egress elements, including:

- The width, the number and the distribution of vertical circulation elements along the center platform for a twin bore cut and cover station
- The width and the number of lateral passageways off each stacked platform, leading to station entrance structures, for a single-bore station housed within the large diameter tunnel

These latest station passenger load numbers, which apply for any station fire life safety egress analyses for either a twin bore or single bore option, represent a significant increase in the assumed station patronage compared to the passenger loads in effect during the previous design phase (2008). This requirement, when considered in conjunction with BFS requirements for faster design fire heat release rate (instantaneous fire growth rate), represents more challenging conditions for which new ventilation design and fire life safety egress analyses will be needed for either twin or single bore option.

It is noted that the 2008 ventilation and fire life safety designs and emergency egress analyses for both tunnels and stations had a) lesser patronage requirements (2000-passenger train crush load, 2030 ridership projections) and b) included medium fire growth scenario.

Downtown San Jose Station Egress

[o]Emergency egress analyses from the Downtown San Jose Station (West) were prepared including a simulation of an evacuation using Legion Spaceworks software. The analyses were conducted pursuant to criteria in the NFPA 130, which is referenced in the BFS. Evacuation analysis and simulation for the Downtown San Jose Station confirmed that the station layout would comply with NFPA 130 requirements to clear the platforms in four minutes or less, and to reach a Point of Safety in the corridors leading to the emergency stairs or the station entrances in six minutes or less. The design also meets BFS by having exits near both ends of each platform and multiple exit routes from each platform. In summary, both spreadsheet analysis and Legion simulations confirmed that the conceptual site-specific plan for the DTSJ Station provides sufficient capacity to meet BFS and NFPA 130 requirements (Figure 12).

Tunnel Egress

The egress walkway width, cross-passage location, and geometry (especially the cross-sectional area of an enclosed guideway) are the key elements that impact the pedestrian flow capacity for self-evacuation and rescue operations within the tunnel. The single bore tunnel houses both track guideways and has three types of track

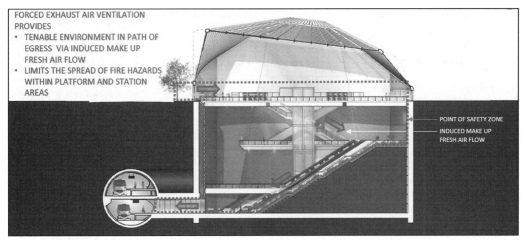

Source: VTA's BSV Phase II, Single Bore Technical Studies Summary Report, Final, April 2017

Figure 12. Tenable environment in the station in path of egress via induced make-up fresh air flow

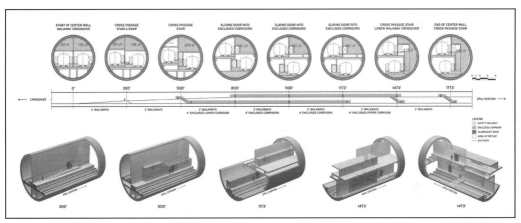

Source: VTA's BSV Phase II, Single Bore Technical Studies Summary Report, Final, April 2017

Figure 13. Single bore cross-passage configuration at transitions

configurations—side-by-side, stacked, and transition sections—each with a unique cross-passage layout. Cross-passage configuration for the side-by-side guideway resembles cross-passages in a cut- and-cover box tunnel configuration, with a center wall, as shown in the BFS. Stacked and transition guideway sections, however, introduce cross-passages with stairwells, which are not addressed in the BFS. These configurations were evaluated for expected performance in comparison to side-by-side guideways.

The entire single bore alignment has a total of 76 cross-passages: 51 configured side-by-side and 25 that include a stairwell. Evacuation flow modeling developed for incident response train operations in the running tunnels confirmed that pedestrian flow conditions are established by guideway emergency walkway flow conditions. Widened stair widths

included in the model were effective at maintaining flow from incident guideway to non-incident guideway where rescue train would be positioned. Based on pedestrian flow calculations and pedestrian flow simulations, enhanced stair widths of 48 inches ensure good evacuation performance at the relatively short sections of stacked tunnel configuration. The widened stairs perform similarly to cross passages of the conventional, side-by-side, center wall design.

In summary, the tunnel egress analyses for the various cross passage and guideway configurations required for the single bore concept indicate that structures and systems could be designed to provide evacuation flow conditions equivalent to those prescribed by BFS; at transitions, these would include the longitudinal fire/smoke protected corridors (Figure 13).

CONCLUSION

VTA's Single Bore Tunnel Technical Studies provided verification of major project elements related to this construction option including tunnel diameter, track alignment, depth of stations, passenger vertical circulation, operations, tunnel and station ventilation, and emergency egress. The findings are of pre-conceptual nature and with a primary objective of making sure:

- The single bore methodology for tunneling and use at the Downtown San Jose Station is feasible and practical,
- BART Facilities Standards are met,
- BART staff comments are recorded and either addressed in the Studies or tabled as a subject of future studies.

BART inputs in terms of operations, maintenance, and safety have been very valuable and as of January 2018, are being evaluated and implemented. It is likely that this construction option will undergo further modifications which are minor in nature. If finally implemented, this will be first application of the single bore methodology for transit stations and guideways in the U.S. Final decision, which will be jointly made between VTA and BART, is expected in March 2018, when preferred alternative will be selected for advancement into FTA's (Federal Transit Administration) Entry in Engineering phase.

REFERENCES

Hanson, R.T., Li, Zhen, and Faunt, C.C., 2004, Documentation of the Santa Clara Valley regional ground-water/surface water flow model, Santa Clara County, California: U.S. Geological Survey Scientific Investigations Report 2004–5231, 75 p.

Poland, J. F. and Ireland, R. L., 1988. Land Subsidence in the Santa Clara Valley, California, as of 1982. U.S. Geological Survey Professional Paper 491-F, 65 p.

Wentworth, C.M., Williams, R.A., Jachens, R.C., Graymer, R.W., and Stephenson, W.J., 2010, The Quaternary Silver Creek Fault beneath the Santa Clara Valley, California: U.S. Geological Survey Open-File Report, 2010-1010, 50 p. [http://pubs.usgs.gov/of/2010/1010/].

Wentworth, C.M., Blake, M.C., McLaughlin, R.J., and Graymer, R.W. (1999), "Preliminary Geologic Map of the San Jose 30 × 60-minute Quadrangle, California: a digital database," United States Geological Survey, Open-File Report OF-98-795.

Ingebritsen, S.E., and Jones, D.R., 1999, Santa Clara Valley, California—A case of arrested subsidence, in Galloway, D., Jones, D.R., and Ingebritsen, S.E., editors, Land subsidence in the United States: U.S. Geological Survey Circular 1182.

Field, E.H., and 2014 Working Group on California Earthquake Probabilities, 2015, UCERF3: A new earthquake forecast for California's complex fault system: U.S. Geological Survey 2015–3009, 6 p., https://dx.doi.org/10.3133/fs20153009.

Schmidt, B. "Prediction of Settlement Due to Tunneling in Soil: Three Case Histories," RETC Proceedings, Vol. 2 (1974).

O'Reilly, M.P. and New, B.M., "Settlements Above Tunnels in the United Kingdom—Their Magnitude and Prediction. Institution of Mining and Metallurgy, (1083).

(VTA's Central Area Guideway, 65% Submittal: Combined Geotechnical Baseline Report, December 31, 2008 Issued for Record)

The Scarborough Subway Extension (SSE)—Large Single-Bore Transit Tunneling in Toronto

Matthew Geary and Tomas Gregor
Hatch

Edward Poon
Toronto Transit Commission

ABSTRACT: Hatch is performing preliminary and detailed design for the Toronto Transit Commission's (TTC) Scarborough Subway Extension (SSE), a 6 km long extension of the existing Line 2. SSE will be delivered by the Design-Bid-Build procurement method and is currently projected for tendering in late 2019. The running structure of the extension consists of a large single bore 10.7 m internal diameter tunnel housing twin subway tracks, mined by earth pressure balance (EPB) or slurry TBM. The large single bore tunnel solution represents a significant departure from the traditional TTC arrangement of twin tunnels for subway. This paper discusses the technical background, design and construction considerations, operational constraints and assessments undertaken that led to the selection of the single bore concept for Toronto's next major subway expansion.

INTRODUCTION

The planned Scarborough Subway Extension (SSE) is a 6 km long extension of the existing Line 2 (Bloor-Danforth) from the existing line terminus at Kennedy Station to a new terminus, the Scarborough Centre Station, located west of McCowan Rd and just south of Highway 401. The new extension consists of an approximately 200 m long cut and cover box structure between the existing Kennedy Station and the end of a single bore tunnel which also houses a new fan plant and requires demolition of the existing Kennedy tail track. A TBM extraction shaft at the end of the new cut and cover section is the start of the 10.7 m internal diameter (ID) tunnel, connecting the extraction shaft to the cut and cover Scarborough Centre Station. The project includes several facilities supporting the subway operation, namely two standalone Traction Power Substations (TPSS), eight Emergency Exit Buildings (EEB-1 to EEB-8), and two ventilation fan plants. All planned work terminates south of Highway 401. The project includes a cut and cover tail track within the TBM Launch Shaft, and Scarborough Centre Station with integration of a new bus terminal at the station. The project is jointly funded by all three levels of government: the City of Toronto, the Province of Ontario and the Canadian federal government. The Toronto Transit Commission (TTC), is managing the project on behalf of the three funding partners with the help of Scarborough Link JV (Hatch, WSP, Parsons)

program management team. Hatch has been retained to perform preliminary and detailed design of the tunnel for the design-bid-build (DBB) project procurement. Figure 1 presents the current SSE design alignment and related facilities.

PROJECT BACKGROUND AND DEVELOPMENT

Tunnel Alignment—Geology and Features

The soil deposits for the SSE project are a result of a complex glacial depositional system that took place during the Wisconsin glacial period. This fluctuating glacial advance and retreat produced a distribution of heavily over-consolidated hard plastic glacial till layers, separated by interstadial stratified deposits of very stiff to hard glacio-lacustrine clays and very dense, non-plastic silt and sand. A mix of soft ground tunneling conditions will be encountered at tunnel horizon, with typical ground cover above the tunnel crown ranging from approximately 10 m to 30 m. The entire alignment is located below the water table.

Some notable project features encountered along the alignment include:

- Secant pile protection walls and potential jet grouting within the vicinity of the extraction shaft for existing building protection and to accommodate future transit construction plans

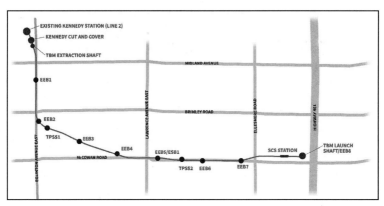

Figure 1. SSE tunnel alignment

- Crossing under buildings with potential for pre-construction mitigations pending results of detailed design
- Crossing a 180 m wide dedicated electrical power transmission right-of-way (hydro corridor) within close proximity to high voltage transmission line towers
- Multiple creek crossings and a woodlot crossing
- Crossing the existing Scarborough Rapid Transit elevated transit structure, within close proximity of the structural piers. The elevated Scarborough Rapid Transit line will remain fully operational during SSE construction. A Level 2 impact assessment and pre-construction mitigations will be established during detailed design (and may include a compensation grouting program)

The TBM will mine from the launch shaft and proceed through pre-installed headwalls for Scarborough Centre Station. Support of excavation for the EEBs is anticipated to include both secant piling and slurry walls (dependent upon location and structure depth). The EEB support of excavation is expected to be installed prior to tunnel passage, allowing regularly spaced stoppage locations along the alignment where cutterhead interventions, TBM maintenance and repairs can be performed. The EEBs can be treated as either unexcavated mine-through locations (with excavation and precast concrete tunnel lining removal occurring after completion of tunneling), or pre-excavated open-air stoppages—this will be left to the contractor's means and methods. An enlarged EEB approximately halfway along the alignment has currently been designated as a pre-excavated open-air stoppage.

The Initial Twin Tunnels Concept

The Transit Project Assessment Process (TPAP) environmental process started in 2013 for the SSE project. During 2014 a number of alignments were being considered as part of the TPAP, with all underground options assuming a construction methodology using twin bored tunnels and cut and cover for stations, track cross-overs and storage tracks. The twin tunnel methodology has been implemented successfully in Toronto for both of the previous TTC subway expansions, the Sheppard Subway and the Toronto-York Spadina Subway Extension (TYSSE).

While there were several SSE alignments under consideration, the preferred alignment at that time included at least three stations and approximately 7.5 km of running tunnels.

Large Single Bore Concept

The concept of a Large Single Bore (LSB) tunnel accommodating both the running train structure spatial requirements as well as station platforms had previously been considered in Toronto for TYSSE as well as for the Eglinton Crosstown Light Rail Transit project. In both cases, the idea was rejected for various non-technical and technical reasons. In May of 2014, Hatch produced an internal study addressing some of the concerns and demonstrating the opportunities of the single bore solution for the SSE. Hatch was retained through an open and competitive bidding process by the TTC for design of the SSE tunnels and related facilities in December 2014.

LARGE SINGLE BORE TUNNEL AND INTEGRATION WITH STATIONS

In early 2015 the LSB tunnel housing both subway tracks and station platforms was proposed as a new innovative concept for TTC's consideration. This solution was considered technically feasible as a result of tunneling technology developments and advancements over the past 10–15 years. Over that period, several tunneling projects were successfully delivered world-wide using bored tunnels with up to 15 m TBM cut diameter. The LSB solution appeared to have very attractive attributes warranting more detailed investigation. The traditional twin, single track tunnels solution, as implemented on Sheppard Subway and TYSSE, was carried along as a benchmark for comparison to the LSB. The main attractions of the LSB configuration included:

- A significant reduction in public and third parties inconvenience due to reduction of open on-street excavations for stations and track cross-overs
- A lower construction cost. Tunneling using TBMs has historically proven to carry a lower construction cost compared to open-cut construction. While the cost of a large diameter running tunnel between stations is higher compared to twin tunnels, the savings in the avoidance of cut and cover construction (stations and track cross-overs) more than offsets the larger tunnel cost in the running structure. The estimated project cost saving at this very preliminary stage was approximately 25% of the capital cost when comparing several alignment options, excluding elements common to both twin tunnels and LSB such as subway vehicles, traction power, track work, and surface facilities.

Early Concepts for Station Arrangements

Stacked Platform Station

For station integration with the LSB, one of the early concepts developed employed vertically separated platforms (referred to as stacked) housed within a 12.2 m ID tunnel. The intention of the stacked arrangement was to maximize the space within the LSB for station and platform activities thereby minimizing additional open-cut construction required for the station. See Figure 2 for a conceptual rendering of the stacked station cross section. This approach was successfully employed on the Line 9 Barcelona Metro expansion (Roig et al., 2010). With the SSE alignment mainly travelling within the existing road right-of-way, open-cut station works could be placed off-street accessing the LSB horizontally through large scale engineered openings in the precast concrete tunnel lining (PCTL).

Upon further review by TTC, the stacked station arrangement presented serious operational constraints as track cross-overs must be located in

Figure 2. Stacked Platform Configuration at Stations

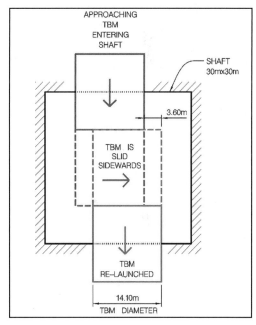

Figure 3. Plan schematic diagram of TBM operation at staggered station shaft

the immediate vicinity of stations they service. To achieve a comparable operational flexibility at a stacked station, a complicated train ramp system would need to be implemented, increasing both the cost and operational complexity. For stations where no track cross-over was required, the stacked arrangement remained viable.

Staggered Platform Station

In response to the stacked station issues, a new solution was developed to provide a station arrangement

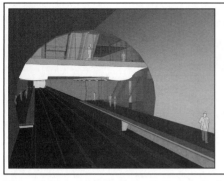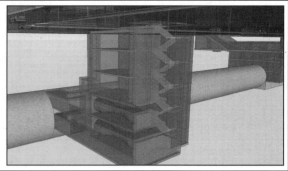

Figure 4. Staggered platform station renderings

that would accommodate an adjacent track cross-over. The solution employs a staggered side platform whereby the bored tunnel does not have to house any more than two tracks plus a single side platform within its envelope. By introducing a shaft of approximately 30 m length at the central portion of the station the bored tunnel can be horizontally staggered, and station facilities and passenger access can be accommodated. The TBM must break into this pre-excavated shaft, be physically shifted sideways by 3.6 m (on a prepared sliding platform), and then re-launched on the adjusted alignment. See Figure 3 for a schematic of the process. Translation of TBMs within shafts is performed regularly; TBMs are often skidded through partially completed stations and re-launched. TBMs are also commonly slid sideways during twin TBM removal from extraction shafts where a shaft opening window is only on one side of the shaft.

The central open-cut shaft is then used to house the main station entrance structure (see Figure 4).

The staggered platform concept required a minimum tunnel ID of 12.8 m, only marginally larger than that required for the stacked concept. The staggered station layout offered the following advantages:

- A short open-cut section provides a TBM maintenance zone at no additional cost
- The track alignment allows placement of minimal cost track cross-overs within the bored tunnel envelope at either or both ends of the station directly adjacent to the platform ends
- The layout provides the passengers with a clear view of both platforms before descending the last flight of stairs/escalators for easy orientation
- The central shaft provides an opportunity to include two entrances to the station on opposite sides of the street

- More open space above trainways which is desirable for fire ventilation system purposes

After space proofing design and inputs from TTC systems, a tunnel ID of 12.9 m was established as capable of accommodating both the stacked and staggered station arrangements.

Station Depth Considerations

Figure 5 illustrates the minimum depths for a cut and cover station with twin tunnels and for the single bore station. For the LSB tunnel, the minimum cover has been established as $0.75 \times$ the excavated diameter. The diameter to cover ratio established for the LSB has been achieved on several previous projects.

Alignment depth for the twin tunnels arrangement is generally driven by the cut and cover station depths, and not by minimum tunnel cover requirements; the opposite is true for the LSB tunnel. The difference in top of rail elevation for the two solutions, when placed as high as practicable, is approximately 5 m.

Some initial concerns were raised regarding an increase in station depth and resulting passenger impacts in the form of ridership and passenger flow. No documented evidence could be produced to support a relationship between station depth and system usage—upon examination of the 194 stations on the Moscow Metro system (Goncharov, 2012), no correlation could be established between ridership and station depth. For passenger flow, the introduction of 5 additional meters depth translates to an additional 20 seconds of travel on an escalator; this additional time was not expected to influence travel decisions. Travel routes and station locations in plan are expected to have a much greater influence on station usage than station depth.

EXPRESS SUBWAY

While discussions among designers and TTC with respect to the station layout efficiencies, construction

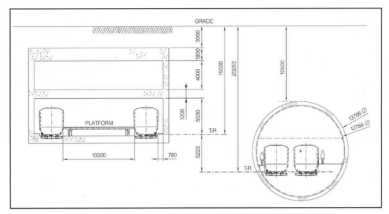

Figure 5. Depth of stations for twin tunnels and single bore tunnel

sequencing, operational constraints and fire ventilation systems were ongoing, in January 2016 the City of Toronto recommended the reduction of the SSE scope to only one station, at Scarborough Centre Station, and the truncation of the alignment length to approximately 6 km. The principal attraction of the LSB solution was the reduction of capital cost and surface disruption at the station locations. With a relatively long tunnel (6 km), and only one station, it quickly became apparent that the potential advantages of the LSB solution may not applicable to this truncated SSE configuration. The terminal Scarborough Centre Station works include the complex integration of an underground bus terminal over the station necessitating cut and cover construction methodology. Open-cut construction of the station eliminated the potential for station integration with the LSB, thereby diminishing the cost-effectiveness of the LSB solution.

Selection of 10.7-m Single Bore

Following a Peer Review Panel assessment, the LSB tunnel diameter was revisited and a new evaluation was performed on the LSB sizing to accommodate only the running tunnel requirements, without the needs of stacked or staggered stations arrangements. The 10.7 m ID tunnel was confirmed as an acceptable dimension for the new express subway alignment. After reassessment and comparison of cost and feasibility of the 10.7 m LSB versus twin tunnels using numerous permutations of truncated SSE alignments proposed by TTC, it became clear that the cost of the 10.7 m LSB options were lower when compared to twin tunnels. One of the major cost differentiators was track cross-over construction—for the LSB, trackwork and associated signals/systems could be housed with the bored tunnel, whereas a twin tunnel arrangement would require significant cut and cover operations within major arterial roads. The project

design is currently proceeding with a 10.7 m diameter single bore tunnel and a cut and cover methodology for Scarborough Centre Station. A conceptual rendering of the 10.7-m ID running tunnel alongside the TYSSE twin tunnels is provided in Figure 6.

DESIGN CONSIDERATIONS

Several issues were raised by the project team during evaluation of the LSB concept for SSE that were investigated and addressed to ensure technical feasibility of the approach. Some of the key design considerations are presented below.

Large Single Bore Tunnel in North America

The development in TBM technology enabling construction of tunnels of this size is relatively new, with most of the projects taking place in the past 15 years. Very few of these large machines have been used in North America in mixed and soft ground, although the 12.9 m diameter machine for the Port of Miami Tunnel has successfully completed its two drives, the 9.8 m Evergreen Line tunnel drives in Vancouver are complete, and the 17.5 m diameter machine for the Alaskan Way Tunnel has completed mining. These recent projects have demonstrated a precedence for the proposed SSE LSB concept viability.

Tunneling-Induced Settlements

Concerns were raised regarding the settlement magnitudes resulting from the LSB tunneling. For tunneling induced settlements, a zone of influence (ZOI) is assumed above the tunnel and is defined by the width of the settlement trough; this trough width is defined by the ground conditions at the face of the tunnel (O'Reilly et al., 1982). Generally, the settlement trough width is wide in clays with smaller vertical movements, while in sands the trough width is narrower but with larger vertical movements.

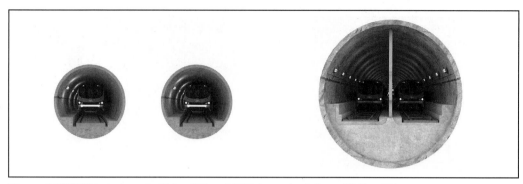

Figure 6. TYSSE twin tunnels (5.4 m ID) and SSE large single bore (10.7 m ID)

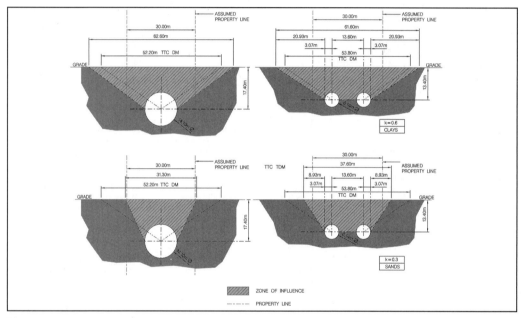

Figure 7. ZOI for single and twin tunnels

For calculating the ZOI, a trough width parameter K is used, with K generally varying between 0.3 for sands to 0.6 for clays. Figure 7 illustrates the ZOI for these two K values for both the LSB and the twin tunnels located at the proposed minimum depths below the surface.

With the tunnel faces in clay deposits, the width of the ZOIs for the two tunneling options are essentially equal. If the tunnel faces are in sands the width of the ZOI is smaller for the LSB than for the twin tunnel option. It should be noted that these ZOIs are conservative and terminate at zero settlement at the trough limits. Generally, the LSB application for SSE will result in a reduction of tunneling ZOI when compared to twin tunnels.

The prediction of surface settlement for tunneling is based on an assessment of the amount of control provided at the excavation face. This amount is quantified in terms of the volume of ground loss that occurs at the face, and is defined by its percentage of the excavated volume. Historically, face loss percentages have ranged up to 3 or 4%, however, since the introduction of pressurized face TBMs these values have now been reduced to well below 1%. For the Sheppard Subway, constructed in Toronto almost 20 years ago using EPB TBMs, the face losses ranged up to 0.8% (Garrod et al., 1998). The pressurized face technology successfully limited inward movement of the ground towards the tunnel face. The remaining areas where reductions in settlements can be achieved are in the control of ground movements towards the tunnel along the length of the TBM and along the tunnel lining. Significant developments have been achieved in annulus grouting techniques, with grouting carried out concurrent with mining advance using tailshield grout lines and a two-component grout mix. This contrasts with the Sheppard Subway where grouting was conducted through the lining resulting in up to one ring length of ground being unsupported as the machine moved forward. These annulus grouting improvements have been incorporated in the recently completed tunnel drives on the Eglinton Crosstown Light Rail Transit project. This resulted in face losses for the first tunnel drive below 0.4%, and as the tunnel progressed and the operators became more experienced, these reduced to less than 0.2% for the second half of the drive (Solecki et al., 2016). More recently, attention has focused on the small annular space around the TBM shield. From experience gained tunneling beneath the Schulich Building on the TYSSE project, it has been demonstrated that the filling of this overcut annular space with bentonite as the TBM advances brings further reductions in ground loss. This improvement was manifested in the very small vertical movements (1 or 2 mm) that were measured for the Schulich Building as the TBMs passed beneath it (Kramer et al., 2015).

Another factor for consideration in the settlement performance of larger TBMs is the relative size of these annular spaces compared with the face area of the tunnel. This is illustrated in Table 1.

The annulus space areas are approximately one half the ratio to excavation face area for the LSB compared to a single TBM on the twin tunneled TYSSE project. It can be concluded that LSB tunnels can achieve lower face losses, in terms of face area percentage, when compared to metro sized twin tunnels.

The benefit of this effect has been demonstrated on the Alaskan Way Viaduct Replacement Tunnel.

Table 1. Annular space dimensions

	Single TYSSE TBM	SSE LSB TBM
Area of Excavation (m^2)	29.55	111.2
Area of Annulus around TBM Shield (m^2)	0.37	0.89
Area of Annulus around TBM Shield (as % of face area)	1.20%	0.60%
Area of Annulus around Tunnel Lining (m^2)	2.49	7.09
Area of Annulus around Tunnel Lining (as % of face area)	8.40%	4.50%

This tunnel was excavated using a 17.5 m diameter EPB TBM, with settlement performance ranging from 0.04% to 0.11% face loss (Cording et al., 2017).

TBM Features and Operation

A pressurized face TBM will be specified for the tunneling works on the SSE project. Two main types of pressurized face TBMs are available: Earth Pressure Balance (EPB) TBMs and Slurry TBMs. Either technology will be considered acceptable if all design specification requirements are met. Only closed mode operation of the SSE TBM will be permitted due to the sensitivity and the number of buildings, structures, and utilities that overlie the bored tunnel.

The following is a preliminary listing of TBM features under consideration to successfully excavate the SSE tunnel and control ground movements:

- Excavated diameter of approximately 11.9 m to construct the 10.7 m ID tunnel
- Airlock and compressed air systems providing capability for hyperbaric interventions
- Overcut annulus injection system through TBM shield, with pressure cell measurement capability in the shield for monitoring purposes
- Tailshield grouting system for two-component grout to fill the annular void around the tunnel lining
- Shield and articulation capability for a minimum turning radius of 275 m (compatible with minimum alignment radius of 300 m allowing for curve correction capability)
- Thrust rams that can controlled individually or in groups to steer the TBM within a 150 mm radius bullseye
- Bentonite or slurry injection capability into excavation chamber during stoppages/shutdowns (manual or automatic) with PLC trigger in response to falling pressure sensor readings
- Real-time data logging systems with remote web access
- Fire/Life and Safety systems

Boulders

The significance of boulders on a tunnel drive is generally a function of the boulder to TBM size ratio, and this ratio is reduced with the introduction of the LSB. Project experience in the Toronto area with more typical transit sized TBMs (6 m-6.5 m) has demonstrated no significant impact on TBM performance or major stoppages due to boulders, though boulder fragments have been frequently observed within the excavated material. The higher power and size of the LSB will allow it to deal with boulders more efficiently than standard twin tunnel sized TBMs.

Precast Concrete Tunnel Lining

The PCTL detailed design will be included in the contract documents and the successful contractor will be responsible for procuring a TBM compatible with the provided lining design. The tunnel has a 10.7 m ID and a minimum lining thickness of 425 mm. The PCTL is composed of reinforced concrete cylindrical rings, supporting the ground along the alignment for all expected ground conditions, hydrostatic pressures and handling/installation forces. Each PCTL ring is 2.0 m in length and composed of eight segments.

Safety

The LSB design allows for safety and service walkways fully compliant with TTC standards and NFPA 130. EEBs are spaced along the alignment in compliance with NFPA 130. In addition, a fire rated central wall is planned as separation between the tracks, and access doors will be provided at regular spacing within the central wall for access to each side of the tunnel in areas between EEBs, track cross-overs and Scarborough Centre Station. Ventilation design will satisfy all regulatory requirements.

PLANNED PROJECT IMPLEMENTATION

Project Delivery Strategy

It is currently assumed that the SSE project will be procured as a single DBB contract, including tunnels and related facilities, Scarborough Centre Station, systems procurement and installation. For the last two TTC subway expansions, an owner-procured approach for TBMs and PCTL has been employed. In a departure from this practice, the TBM and PCTL will be procured by the contractor for SSE. The driving factor behind the owner procured approach in the past was to shorten the project delivery schedule. In the case of TYSSE, this included advanced works contracts tendered for the TBM launch shafts. For SSE there is a significant amount of work required to establish and construct the launch shaft and site, and this work can be carried out by the contractor in parallel with the TBM and PCTL manufacture without introducing delays to the project schedule. The contractor will be responsible for TBM selection and procurement, while satisfying the minimum features and requirements defined within the contract specifications. Strict requirements for the TBM manufacturer's qualification to fabricate the TBM will also be specified. A fully detailed PCTL design will be included in the contract. A Geotechnical Baseline Report (GBR) will form part of the contract documents.

CONCLUSION

Preliminary design of the SSE tunnel is currently being finalized and completion of detailed design is anticipated at the end of 2018. the SSE is currently projected for tendering in late 2019 using the DBB procurement method. Since the early introduction of the LSB concept, a range of technical and operational concerns have been addressed, comparisons have been estimated, and a variety of alignment configurations have been analyzed and assessed. Moving forward, the SSE project offers a unique opportunity to introduce the recent and proven advances in large diameter tunneling to the Toronto area for the next major TTC transit expansion.

ACKNOWLEDGMENTS

The authors would like to thank the TTC for permission to publish this paper, and all project team members for their efforts to date on this challenging project.

REFERENCES

Cording, E.J., Nakagawa, J.T., Painter, C.Z., McCain, J.T., Vasquez J., Sowers, D., and Stirbys, A.F. 2017. Managing Ground Control with Earth Pressure Balance Tunneling on the Alaskan Way Viaduct Replacement Project. In Rapid Excavation and Tunneling Conference Proceedings 2017.

Garrod, B.L., and Delmar, R. 1998. Earth Pressure Balance TBM Performance—A Case Study. In 15th Annual Canadian Tunneling Conference Proceedings, Vancouver, BC.

Goncharov, A., The Scheme of Tunnels Depth and Stations Structure of the Moscow Metro. http://www.alexeygoncharov.com, 2012.

Kramer, G.J.E., Bidhendi, H., Cording, E.J., Walters, D., and Poon, E. 2013. TYSSE Tunneling Test Section and Excavation Beneath Schulich Building. In Rapid Excavation and Tunneling Conference Proceedings 2015.

O'Reilly, M.P. and New, B.M. 1982. Settlements above tunnels in the United Kingdom—their magnitude and prediction. In Proceedings of Tunneling '82 Symposium, London, p.p. 173–181.

Roig, J., Fernandez, E., Ruiz, M., and Sanz, A. 2010. In Tunnel Station Concept on Deep Subway Tunnel Projects in Urban Areas: Barcelona Line 9. In World Tunnel Congress Proceedings 2010.

Solecki A., Taghavi A., and Hassan I., 2016. Redefining Settlement Control Industry Standards with Modern Mechanized EPB Tunneling. In World Tunneling Congress Proceedings 2016.

The Next Big Tunnel Under Downtown Seattle

Gordon Clark
HNTB Corporation

Joseph Gildner
Sound Transit

ABSTRACT: The Alternative Analysis and early engineering studies are underway to allow Sound Transit to define a new light rail line. The new system will reach from West Seattle to Ballard and is expected to be operational between 2030 and 2035. The 11.8-mile segment will include 14 stations. It is expected to have 3.4 miles of either twin bore or large single bore tunnel and include six underground stations, 7.5 miles of elevated guideway with six elevated stations, and 0.6 miles of at-grade track with two at-grade stations. Two major water crossings will also be required over the Duwamish River in the south and over or under Salmon Bay in the north.

INTRODUCTION

On the eve of completing the SR99 road tunnel, one of the largest diameter tunnels in the world, a team of planners and engineers led by HNTB are already performing feasibility studies for the next big tunnel through the heart of the city (See Figure 1). The transit tunnel will be part of a new 11.8-mile Sound Transit light rail line between West Seattle in the south and Ballard in the north. The complete project is estimated in 2014 USD at $5.5B. The tunnel is expected to extend 3.4 miles from the International District on the south edge of downtown, to the north side of Seattle Center, passing directly under the densest part of the central business district at depths up to 150 feet below the street.

BACKGROUND

In November 2016 Seattle area voters approved a $54 billion Sound Transit bond measure to fund additional light rail, commuter rail, park-and-ride spaces, and bus-rapid transit. The West Seattle to Ballard Link Light Rail Extension will be one of the first projects funded under the new plan. It was developed as three separate projects or segments under an earlier system planning effort called *ST3* that identified representative alignments from (1) West Seattle to Downtown segment; (2) Downtown segment; and (3) Ballard to Downtown segment. The representative alignments were developed to provide voters an indication of the mode of transportation (light rail), corridors or areas to be served, number and general location of stations, and a conceptual cost estimate for each of the segments. The plans include the key attributes as shown in Table 1.

Sound Transit combined the three segments in Table 1 under the project name of "West Seattle to Ballard Link Extensions" (See Figure 1). In July of 2017, Sound Transit awarded the HNTB-led team a 3-phase planning and design contract and issued NTP for the first phase in October of the same year. The three phases are as follows:

- Phase 1 Alternatives Analysis and selection of Locally Preferred Alternative
- Phase 2 Draft Environmental Impact Statement & Conceptual Design
- Phase 3 Final Environmental Impact Statement & Preliminary Design

Phase 1 will include completing an Alternatives Analysis study in parallel with advanced engineering studies. The engineering studies are to determine the technical feasibility of several high-risk and key components. The engineering studies will focus on five areas, numbered to correspond to the locations shown in Figure 2.

1. High Level Bridge Crossing over the Duwamish River
2. Salmon Bay Crossing and location and configuration of the Ballard Station
3. Alignment of transition trackwork in SODO (south of downtown) District
4. Tunnel and station configuration in downtown Seattle
5. Location and configuration of the West Seattle Station

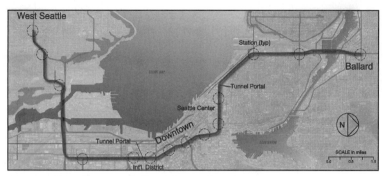

Figure 1. West Seattle to Ballard light rail project

Table 1. ST3 representative alignment

Segment	Capital* Cost	Length	Tunnel	At-Grade	Aerial
1. West Seattle to Downtown	$1.4–$1.5B	4.4 miles	0.2 miles	0.2 miles 1 station	4.0 miles 4 stations
2. Downtown	$1.6–$1.8B	1.8 miles	1.8 miles 4 stations		
3. Ballard to Downtown	$2.3–$2.5B	5.6 miles	1.4 miles 2 stations		4.2 miles 3 stations
	$5.3–$5.8B	**11.8 miles** **14 stations**	**3.4 miles** **6 stations**	**0.2 miles** **2 stations**	**8.2 miles** **6 stations**

* 2014 USD

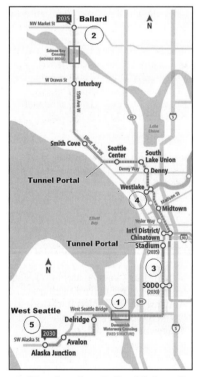

Figure 2. West Seattle to Ballard light rail alignment

Phase 1 is expected to be completed by April 2019. Phase 2 will follow for a planned duration of 18 months. Phase 3 is also planned for an 18-month duration and is scheduled to be completed by April of 2022. Final Design and Construction are planned to be completed to allow the portion of the system from West Seattle to Downtown to open by 2030 and the portion from Downtown to Ballard to open by 2035.

UNDERGROUND STRUCTURES

With 3.4 miles of tunnel alignment and six mined or cut-and-cover stations, the underground portion of the project represents less than 30 percent of the alignment, but it is expected to require over 60 percent of the budget at roughly $1 billion USD per mile. There are many challenges facing the designers of the next big tunnel under downtown Seattle.

Design Challenges

The tunnel will pass through the densest portion of the central business district of Seattle. The two downtown streets under consideration for the alignment (5th Ave and 6th Ave) are both fairly narrow with a 60-foot-wide right-of-way. The alignment is flanked on each side by buildings that reach 40 to 70+ stories above the street and that have parking garages that extend 50 to 100 feet below the surface (See Figure 3). During the construction of the foundations and parking garages of these skyscrapers, the contractors made extensive use of temporary tie-backs to support the excavations.

While these high-strength steel cables were de-tensioned before commissioning of the buildings, they remain in the street to depths of 120 feet or more below the surface. The spacing of the tie-backs varies but on average is on a 10-foot horizontal and vertical grid which means hundreds of them are within the boring horizon.

Whether the tunnel configuration is a single large bore, twin side-by-side tunnels, or twin tunnels stacked vertically, the tiebacks will interfere with the construction unless the alignment is much deeper than currently planned. The tie-backs will need to be removed prior to a Tunnel Boring Machine (TBM) passing through the ground on the existing alignment to avoid the machine being damage or disabled.

Construction of an underground transit station in the middle of the street will be disruptive to traffic on the busy streets. The central business district of Seattle employs a work force of over 200,000 workers, and most of these commute daily by bicycle, car, and bus on the surface streets. North-South streets such as 5th and 6th Avenues are especially crowded because the downtown area is in the center of an hourglass shaped urban area bounded by Puget Sound on the west and Lake Washington on

the east. Trucking out tunnel spoil or muck from a mined or cut and cover station will be very challenging. Finding locations for entrances to the stations will be more challenging and may require demolition of existing structures. There will be opportunities for transit oriented development. Locating ventilation structures will be especially challenging considering the emphasis on air quality and the increasing number of high rise apartment buildings in the central business district. Not only are the streets filled with cars but beneath the surface of the streets is a dense array of utilities including water, sewer, gas, electric, steam, and fiber optic communication lines.

Existing Light Rail Tunnel

The Seattle Bus Tunnel, built over 25 years ago, consisted of approximately one mile of twin bore tunnels and four cut-and-cover stations which began at International District/Chinatown and passed under 3rd Ave and under Pine Street to Westlake Station. The tunnels were later converted to handle both buses and light rail vehicles and additional light rail only tunnels were constructed that extended north to Capitol Hill, University of Washington, Roosevelt, and Northgate. Future Sound Transit plans call for the tunnel facility to be used for light rail only.

The new Sound Transit tunnel (shown in red in Figure 4 and Figure 5) will parallel the existing bus/light rail tunnel (shown in blue on Figure 4 and in orange on Figure 5). The new tunnel will pass under the existing tunnel at Pine Street in an expanded Westlake Station. The existing bored tunnel under Pine Street has approximately 30 feet of cover with the top of rail some 45 feet below the street surface. The new tunnel will pass beneath the existing tunnel. This will place the top of rail for the new tunnel in a range of 80 to 120 feet below the surface.

Geology

Seattle is in the central portion of the Puget Lowland, an elongated topographic and structural depression

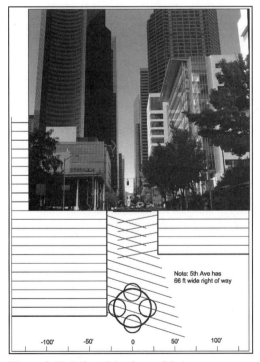

Note: 5th Ave has 66 ft wide right of way

Figure 3. Building tiebacks on 5th Ave

Figure 4. Light rail alignments Downtown Seattle

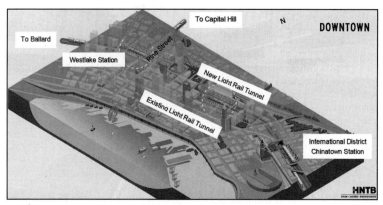

Figure 5. Downtown Seattle light rail tunnel alignment

bordered by the Cascade Mountains on the east and the Olympic Mountains on the west[1]. The land is characterized by a series of north-south trending ridges separated by deep ravines and broad valleys which are the result of glacial scouring and sub-glacial erosion. Geologists believe that the Puget Sound area has been subjected to six or more major glaciations during the past three million years. The last glacial period (Vashon), approximately 13,000 years ago, resulted in a drift sequence of lacustrine deposits, advance outwash, till, and recessional out-wash and these materials are generally found in the surface and shallow subsurface geology. Below these Vashon sediments are glacial and non-glacial materials (Pre-Vashon) which also include generally similar deposits. Along the proposed tunnel alignment, as shown in Figure 6, the tunnels, stations, and shafts will encounter medium to dense sands, stiff to hard silts and clays and glacially overridden dense sands, and till-like soils. In many instances, these materials are mixed and interbedded and the contacts are not easily defined. Boulders on the order of one to six feet in diameter are known to exist in some of these layers. Methane may also be present, particularly in some of the silt layers. The older clay can have numerous slickensides, and the dense sand, once unconfined, will likely have a tendency to flow. The varying permeability of the various layers increases the likelihood of encountering perched water and saturated coarse (running) sands and fine (difficult to grout) silts. Some of the older silts may also present difficult conditions for tunneling and the mining of cross passages.

The subsurface geology along the proposed alignment includes glacial and nonglacial soils and overlying post glacial deposits. The thickness of the post glacial deposits range between 30 and 60 feet and consists of fill and estuarine deposits.

The Puget Sound area is considered to be a region of moderate to high earthquake potential.

The largest recorded earthquakes to affect the area were deep subcrustal events centered in Olympia in 1949 (magnitude 7.1) and the Nisqually (magnitude 6.8) centered east of Olympia and south of Tacoma in 2001. The offshore Cascadia Subduction Zone is capable of producing a Richter 9.0 event[2], and some believe it is overdue for a significant movement. Recent geologic evidence indicates that the newly discovered Seattle Fault, which crosses Puget Sound just south of the downtown, may be active and capable of producing earthquakes with magnitudes on the order of 7.5. While tunnels are typically less impacted that surface structures due to seismic activity, the tunnel and underground stations must be designed to withstand earthquakes generated from these two primary sources.

Tunnel Configurations

One of the first studies to be undertaken by the HNTB led team of engineers and planners working on the Alternative Analysis phase of the project will be to recommend a configuration for the new tunnel. Several options will be considered. Because of the deep building foundations and narrow right-of-way along both 5th and 6th Avenues, there may not be sufficient space to construct traditional side-by-side twin bored 20-foot diameter tunnels without impacting the adjacent building foundations. One option would be to vertically stack twin bored 20-foot diameter tunnels (See Figure 7).

One of the challenges with stacking the twin tunnels is configuring the cross passages. Rather than being a short horizontal passage way between the two tunnels for evacuation purposes, the stacked tunnels would require a vertical passage involving either a stairway or long sloping ramp. Construction of cross passages is often a high risk operation, and creating a vertical cross passage would possibly further complicate the construction and increase the

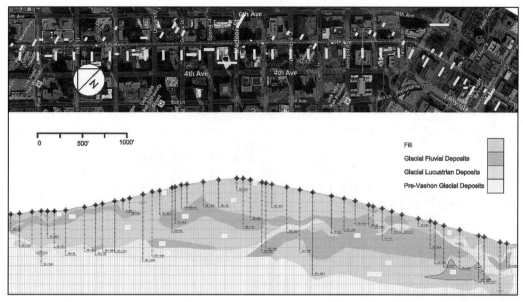

Figure 6. Profile of glacial geology under 5th Ave in Downtown Seattle

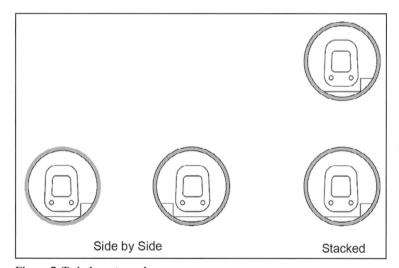

Side by Side   Stacked

Figure 7. Twin bore tunnels

risk. If the tunnels remained stacked, the stations would have to have stacked side platforms at the new Madison Station and possibly at the new a station adjacent to the existing Westlake Station.

Another option that will be considered is a large single bore tunnel that would include two side-by-side tracks separated by a center wall. This would eliminate the need to mine cross passages, as they would be contained within the tunnel. Track configuration could be either stacked as shown in Figure 8 or side by side. Several station configurations are possible with a single large tunnel. These include a tunnel large enough to contain both trackwork and side platforms. A smaller single tunnel could contain the trackwork with the station platforms adjacent to the tunnel in separate structures that could be constructed using either cut-and-cover or mined operations.

Sections of the tunnel alignment under downtown may be over 120 feet deep, and a mined station with elevator-only access via shafts will be considered in conjunction with twin bored tunnels.

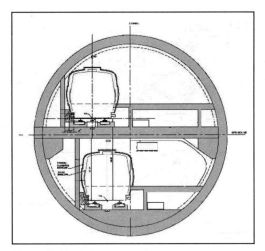

Figure 8. Single bore tunnels

If a single bore tunnel is used then station platforms could be integral to the tunnel, entrances via shafts and only the connections between the shafts and the platforms would need to be mined. The tunnel alignment north of the Westlake Station would likely not be as deep and more conventional cut-and-cover stations construction is possible.

The location of the tunnel portal south of downtown will likely be just south of the International District but will need to be coordinated with existing facilities and the ultimate alignment through SODO. The tunnel portal north of downtown will be near the intersection of West Republican Street and Elliott Ave West (See Figure 2).

CONCLUSIONS

This project will result in the next big tunnel under downtown Seattle that will connect a new light rail alignment from West Seattle to Ballard. The 11-mile alignment will include seven aerial, one at-grade, and six underground stations. The new line will provide transfer points to the Downtown Seattle Transit Tunnel at the existing International District/ Chinatown Station and at the existing Westlake

Station. In addition to over 3 miles of new transit tunnel, the new line is planned to include a rail-only high-rise bridge structure over the Duwamish Waterway and a new movable bridge over the Ship Canal in Ballard.

The goal of current phase of the process is to select a preferred alternative. HNTB is leading a team of consultants performing the development and screening of various alignment and station location options and performing advance engineering to determine the feasibility of critical elements. This phase will be followed by the preparation of environmental documents which will be accompanied by conceptual engineering leading to Preliminary Engineering in the final phase of the HNTB contract.

ACKNOWLEDGMENTS

The author gratefully acknowledges the cooperation of Sound Transit in allowing the presentation and publication of this information. Several key personnel are involved in developing this challenging light rail project. The author appreciates the support of Sound Transit's Project Director, Mr. Cathal Ridge, the Design, Engineering and Construction Management Director, Mr. Ron Endlich, the Corridor Design Manager, Mr. Dirk Bakker, the Operations Director, Mr. Wesley King, and Sound Transit's Light Rail Development Manager, Ms Sandra Fann. The author also acknowledges the assistance of Jim Parsons, Project Manager for the HNTB led team of consultants. The author acknowledges the input from our geotechnical consultants, Golder Associates and Shannon and Wilson, Inc., particularly Joseph Hachey and Robert "Red" Robinson.

REFERENCES

[1] Booth, D.B., R.A. Haugerud, and K.G. Troost. 2003. "The Geology of Puget Lowland Rivers," in D. Montgomery, S. Bolton, and L. Wall, eds. Restoration of Puget Sound Rivers, pp. 14–45. University of Washington Press, Seattle, USA.

[2] Pacific Northwest Seismic Network; Cascadia Subduction Zone; 2017 https://pnsn.org/ outreach/earthquakesources/csz.

Crossing the Chesapeake Bay 21st Century Style: The Parallel Thimble Shoal Bored Tunnel

Enrique Fernandez, Alejandro Sanz, Juan Luis Magro, and Enrique Alcanda
Dragados

ABSTRACT: The Chesapeake Bay Bridge Tunnel (CBBT) was constructed in the late 1950s and early 1960s, as a new fixed link at the entrance of the Chesapeake Bay. The bridge-tunnel complex incorporates two immersed tube tunnels under the Thimble Shoal and Chesapeake Channels to facilitate marine navigation. Originally conceived as an immersed tube, the Parallel Thimble Shoal Tunnel will be constructed as a bored tunnel to reduce traffic congestion on the existing facility. The paper presents the advantages of this sustainable approach through the use of pressurized face tunneling.

SHORT HISTORY OF THE THIMBLE SHOAL TUNNEL

The initial permanent link between Virginia's Eastern Shore and the Norfolk/Virginia Beach area started on the 1930s with ferry operations that were discontinued following construction of the Chesapeake Bay bridge Tunnel, which opened to the traffic in April, 1964.

This new link included: 12 miles of low-level trestle, two 1-mile tunnels, two bridges, almost 2 miles of causeway, four manmade islands and 5-½ miles of approach roads. At 23 miles in length, it remains the longest bridge tunnel complex in operation in the world.

According to CBBT's historical records, "the construction was accomplished under the severe conditions imposed by hurricanes, northeasters, and the unpredictable Atlantic Ocean." In spite of these adversities, construction was completed in 42 months. On completion the new link was recognized as one of the seven engineering wonders of the modern world and as an "Outstanding Civil Engineering Achievement" by the American Society of Civil Engineers (ASCE).

The two original tunnels were constructed as Immersed Tube Tunnels (ITTs), which was considered the most viable construction methodology at that time.

By 1995, the increased demand on traffic across the complex led CBBT to develop

Figure 1. CBBT construction

capacity improvements in the form of a parallel bridge. Closure of the facility would result in a 400 mile detour around the Bay. Construction of the parallel bridge was completed in April 1999. The new parallel bridge highly improved mobility and safety, except at the two tunnels which were not duplicated and became the bottleneck of the scheme.

In May 2013, the CBBT Commission approved a resolution for the construction of the Parallel Thimble Shoal Tunnel, utilizing a Design-Build contract as a way to maximize cost savings and to shorten the project delivery schedule.

Initially, CBBT's consultant envisaged an immersed tube tunnel, similar to the original scheme.

Figure 2. CBBT scheme

However, during development of the procurement documents, a bored tunnel alternative was proposed by tunneling industry experts due to the significant development of closed face tunneling machines since the original ITT construction. It should be noted that the first Earth Pressure Balance (EPB) TBM was developed in Japan in 1974 and had a diameter of just 12.2 feet. The continued development of EPB means that TBMs larger than 43.3 feet in diameter (the requested diameter for Thimble Shoal Tunnel) have proven their performance in non-cohesive soils under the water table.

Initially, four international bidders were prequalified for the Parallel Thimble Shoal Tunnel Project, two bidders for bored and immersed tube tunnel options, one bidder for the immersed tube tunnel option and the Chesapeake Tunnel JV (CTJV—joint venture of DRAGADOS USA and Schiavone Construction) for the bored tunnel option.

Figure 3. Immersed tunnel portion towed to the final location

WHAT ARE THE CLIENT NEEDS?

CBBT needs to increase the bridge/tunnel complex's capacity to avoid traffic congestion and reduce travel time, primarily during the peak summer vacation period.

Immersed tunnel technology requires significant dredging activity to prepare the foundation layer for the precast concrete elements forming the tunnel. These are typically manufactured in a dry dock and floated to their final location where they are sunk and connected, prior to being buried under a layer or rock armor that protects it against ship impacts and ocean currents. Additionally, construction of an immersed tunnel can reduce channel depth with negative impacts to navigation. To avoid this, a deep dredge line is often required to maintain or increase existing navigation capacity. As a result, the limits of the dredged area increase dramatically requiring a greater separation between the new and existing tunnels. See Figure 4.

The only way to address this issues was to significantly enlarge the existing islands to the size to construct the new tunnel portals and this represented significant marine works, approximately 5,25 acres per island. Marine works, tunnel bed dredging and island extension create tremendous environmental impact and navigation channel disruption for both civil and military vessels. A bored TBM tunnel solution is technologically feasible and the required size of the tunnel, 43.3 feet is well within the limits of proven TBM technology. One of the major advantages to a bored tunnel solution was the elimination of island expansion resulting in project savings to CBBT of approximately $300 Million.

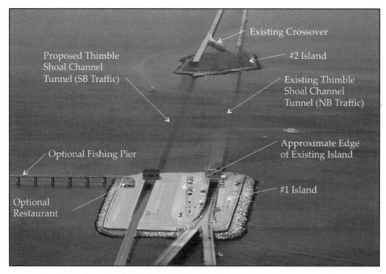

Figure 4. Proposed immersed tunnel scheme

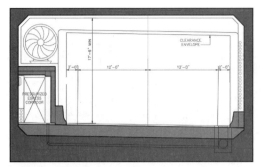

Figure 5. Immersed tunnel cross section

Figure 6. Portal Island No. 1

Figure 7. Bored tunnel cross section

The Portal Islands Nos. 1 and 2 are sufficiently large that the new tunnel portals including space required for TBM assemble, launch and disassemble can be accommodated within the existing island footprints. Maintaining the project within the existing islands footprint simplified the Route 13 highway geometry and allowed for the existing crossovers and trestle approach to be maintained for the most part.

An additional secondary benefit to the close proximity of the new and existing roadways is that the existing roadway geometry is maintained which will force approaching vehicles to reduce the speed at the entrance and exit of the tunnel, increasing traffic safety. With an immersed tube tunnel the roadway maintain a parallel alignment and therefor vehicles will be traveling at higher speeds and the risk of accidents will increase.

The bored tunnel cross section includes: two travel lanes, two limited width shoulders and a dedicated emergency egress corridor.

THE DRAGADOS TEAM PROPOSAL

The Dragados Team's proposed concept consisted of excavating the tunnel using a TBM without expansion of the existing islands, and minimal traffic disruption in both the existing tunnel and the Thimble Shoal navigation channel.

Extensive analysis was undertaken during development of the tender design to confirm that the bored tunnel solution was constructible. The primary risks that we identified during the tender and have been managed through design development were:

- The existing artificial islands were constructed using a perimeter rock berm comprising individual layers ranging from a granular core to large armor stones. These layers provide scour protection and maintain the island geometry during construction. The interior was then filled with hydraulically placed fill composed of clean sands with fines with a gradation of 6 percent. This results in the core of the islands being made of soft and permeable material with low bearing capacity while the island perimeters are made of stones, varying from gravels to armor stone with a minimum weight of 10 tons.
- The armor stones on the islands perimeter may have been displaced during islands construction resulting in "man-made" obstructions for the new project. These could potentially impact the bored tunnel construction or the portals works.
- The geotechnical conditions under the islands and the bay comprise various sub-horizontal soil layers ranging from soft loose to stiff dense. Under Portal Island No. 2 there is an organic material (peat) layer with high water content that will required ground treatment to facilitate tunneling operations.

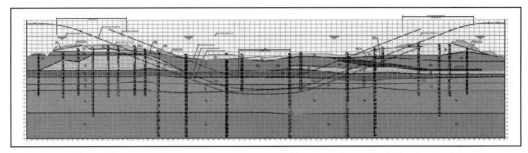

Figure 8. Tunnel geological profile showing natural ground

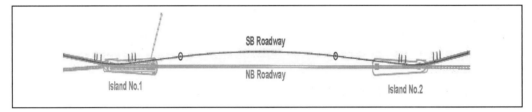

Figure 9. New tunnel alignment in red

- The available surface available on the existing islands for building new connecting structures is limited, with approximately 65 feet between the existing tunnel approach structure and the islands' perimeter splash wall. The new approach structure on Portal Island No. 1 will accommodate the TBM assembly and launch. Conversely, the new approach structure on Portal Island No. 2 will accommodate the TBM reception and disassembly.
- The existing approach structures and tunnel are founded on soft material and were not designed to accommodate differential settlements. The design and construction sequencing of the proposed approach structures had to be developed to minimize impact to the existing structures, in conjunction with a comprehensive instrumentation and monitoring program to verify the performance of the new and existing structures during construction.

Excavation with a TBM within the islands and under the bay will require a closed face TBM. Considering the ground properties and boundary conditions an Earth Pressure Balance (EPB) TBM was selected. The bored tunnel internal diameter was set as 39 feet to accommodate the roadway and egress corridor. In consideration of a precast segmental tunnel lining and necessary excavation/equipment clearances a TBM diameter of 43.3 feet was selected for the tunnel excavation.

In consideration of the severe watertightness criteria prescribed by CBBT's technical requirements, single pass segmental lining was developed to minimize the TBM diameter, control excavation and minimize volume loss in the small ground pillar left between the new tunnel and existing structures. This single pass lining approach combined with demanding watertightness criteria required a careful design of the precast concrete segmental lining.

Once the TBM size was fixed, the proposed tunnel vertical and horizontal alignment and new approach structure geometries were defined. The new approach structures, used to launch and receive the TBM, were located between the existing portals and the islands perimeter. The proposed approach structure follow the existing islands' geometry until the ends of the islands where additional works will be required to allow the TBM to pass through the existing armor stone protection. The close proximity of the existing and new tunnels within the islands required different solutions for each section of the new tunnel.

The crossover area, where the new tunnel connects with the existing structure is in the narrowest part of each island. This will require sequential excavation and complex construction process to avoid impacting the adjacent live traffic operations.

After the crossover, the approach structure vertical alignment slopes down to 6% which is 2% steeper than the existing approach structures. This steeper alignment allows the tunnel to gain sufficient ground cover as quickly as possible to facilitate TBM mining. This, however, generates asymmetric

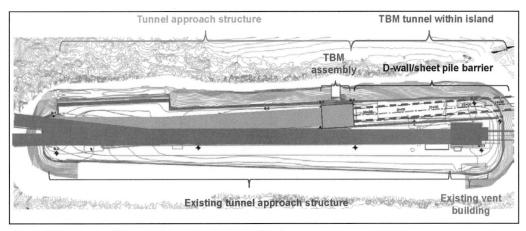

Figure 10. Plan view of Portal Island No. 1 with main elements

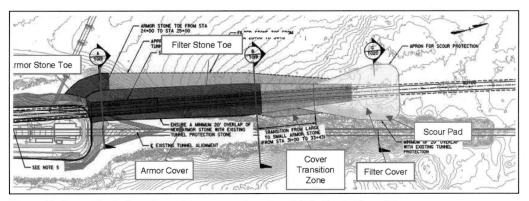

Figure 11. Berm No. 1 plan view, with cross sections A and B indicated in red

ground pillar between the new and existing approach structure and requires strutted slurry walls with a thick jet grouted plug below the approach structure base slab.

The TBM tunnel starts excavation within the island and runs parallel to the island perimeter for about one third of the island length. In this section, the TBM tunnel alignment gradually separates from the existing one. To further mitigate the potential impacts of the existing approach structures a settlement cut-off wall comprising either additional slurry wall panels or sheet piles will be implemented. The aim of this protective barrier is to create a stiff curtain to cut off possible settlements induced by the volume loss generated by the tunnel excavation adjacent to the existing infrastructure. This also provides a suitable tunneling environment to allow pressurized face mode TBM operations in the low cover areas within the islands. The settlement cut-off wall will be extended some distance into the bay to provide protection to the existing immersed tunnel in the areas where the two tunnels are close to each other.

Then, due to insufficient ground cover and to allow the TBM excavation out of the artificial islands, engineered berms will be created. Construction of the new berms will also allow for the removal of potential obstacles (rock armor and scour protection) along the TBM alignment. This material located in the islands perimeters could impact TBM operations during boring. Construction sequencing of the engineered fill berms to maintain and protect the existing islands from scour and settlement at all stages of construction is a key part of the project. In Portal Island No. 2, the new engineered fill berm will be extended vertically above the sea level to allow for execution of a jet grouting of the organic layer under the island. This will limit the potential ongoing settlement of this layer and further impact to the existing infrastructure and prevent differential settlements in the future.

Once all the preliminary works are done, the TBM can be assembled in Portal Island No. 1 and excavate the tunnel in closed mode to Portal Island No.2 where it will be dismantled. Following

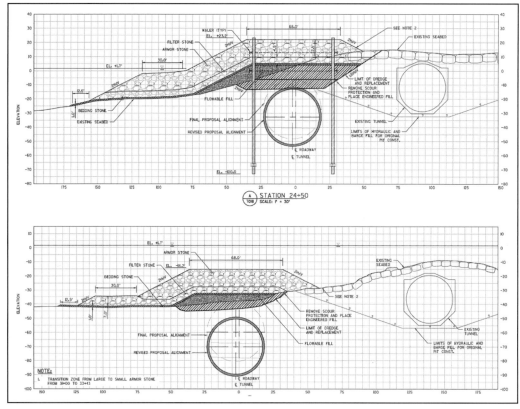

Figure 12. Berm No. 1 cross sections A (up) and B (down) indicated in Figure 11, showing existing immersed tunnel (right) and new TBM tunnel (left)

excavation, the tunnel invert will be backfilled with concrete to prevent buoyancy and the internal structures and tunnel systems will be installed.

REQUIRED INLAND WORKS AN PORTAL FACILITIES

Prior to TBM excavation there are multiple island works to be undertaken. Considering the limited footprint available, requirement to maintain traffic operation in both the highway and navigation channel, an optimized construction sequence has been developed:

- Instrumentation and Monitoring (I&M) of the existing infrastructure is the first activity to be implemented. This will provide the baseline for measuring settlements of the sensitive existing assets during construction.
- Supplemental ground investigation to confirm ground properties and identify potential obstructions has to be carried out.

- Tunnel approach structure support elements and settlement cut-off walls between the existing and new tunnels (slurry walls and sheet piles) can now be built. This is followed by the jet grout base plugs and then the portals can be excavated, installing permanent and temporary bracing as excavation progresses.
- The artificial engineered fill berms will be constructed concurrently with the approach structures. Existing scour protection (armor stone, rip rap, etc.) will be removed from within the proposed footprint of the engineered fill berms prior to fill placement. This will allow for potential obstructions along the TBM alignment to be removed. This is a sensitive operation that will occur in multiple stages to minimize impact to the existing immersed tube tunnel. Limited lengths will be exposed to allow protection of the existing structure in the event of a storm. The berms will also incorporate temporary sheet piles

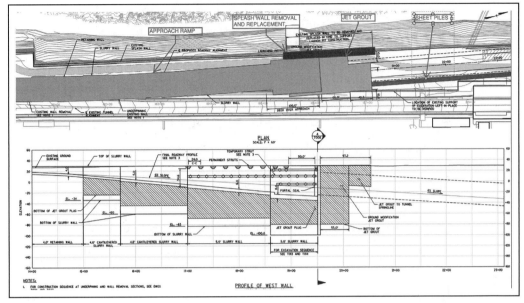

Figure 13. Plan view and longitudinal profile of portal works in Portal Island No. 1

in some areas to facilitate placement of the engineered fill.

- The portal works will require isolated removal of the existing island splash wall, and underpinning or modifications of the existing tunnel approach structure Support Of Excavation (SOE).
- In addition to this, some temporary foundations and mooring facilities will be required to allow TBM assembly and operations.

Once the major civil works have been completed, it will be possible to install the temporary facilities to support TBM tunneling. Given the limited size of the island and considering that the first section of the tunnel will be excavated in a sensitive low cover area, the TBM will be fully assembled prior to launch to avoid long stoppages in this area.

One key aspect is the ability to guarantee accessibility in all stages of the project, also during TBM assembly and disassembly. TBM assembly is normally a delicate operation with large equipment, but in this case it is a significant challenge due to the low bearing capacity of the artificial island soils, the limited capacity of the highway trestles that cannot accommodate the heaviest TBM elements, and the limited access to the launch portal. This TBM assembly operation was integrated into the design by incorporating temporary foundations required for the heavy lifts into the permanent design, thereby allowing the transport of the TBM elements by water.

Another important element to be considered when dealing with large TBMs is the muck handling. These machines generate large amount of spoils that needs to be removed from the site at the same rate due to the limit storage area. For this purpose, a temporary muck pit accommodating 9,500 cubic yards capacity, placed on the east side of Portal Island No. 1 was required. This has a footprint equivalent to the new approach structure, and will be connected to the TBM by means of a series of overland conveyors. This muck pit provides storage capacity for the TBM to excavate when muck removal from the site is not possible.

Tunnel logistics are critical in such a confined site, especially tunnel lining segments and consumables handling. The low bearing capacity of the ground combined with the heavy precast concrete segmental rings will limit segments storage capacity on site to the area used (and reinforced) for the TBM assembly. The segment storage area has a limited capacity of 16 rings.

Other TBM support facilities such as: ventilation, water supply and tanks, backfill grout plant, workshops, etc. will be placed around the islands, minimizing impact to island accessibility.

TBM SELECTION, TECHNICAL REQUIREMENTS, AND TBM CAPABILITIES

Tunnel boring with a face pressure TBM is always challenging. This is particularly the case for pressurized mining mode (or closed mode) which is usually applied when the tunnel boring occurs in low

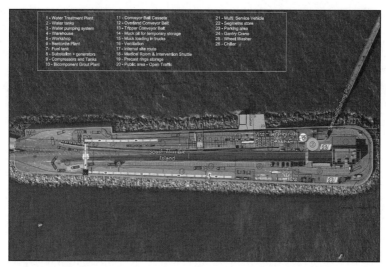

Figure 14. Portal Island No. 1 site set up

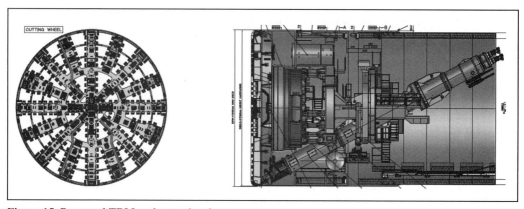

Figure 15. Proposed TBM and cutterhead

cohesion soils, to counterbalance earth and water pressure with the TBM itself.

As an example, this mode of operation allows tunnel construction in highly congested urban areas where the tunnel excavation could generate otherwise surface settlements and consequently, damage in buildings and other infrastructures and utilities. It is also the most recommended method to excavate tunnels under the water table to avoid water inflows into the tunnel that usually becomes an issue in tunnels bored in open mode.

The Parallel Thimble Shoal Tunnel is a perfect match for a pressurized TBM and is also the proper technical solution to avoid impacting the existing immersed tunnel during construction of the parallel tunnel. A pressurized face TBM will manage the ground conditions while controlling the water pressure, which in this case is that of the Atlantic Ocean.

As mentioned before, the EPB excavation mode was selected from the two face pressure TBM options, EPB or slurry shield, in accordance with the anticipated geology which is mostly fine materials. One additional advantage of the EPB technology is the wide range of grounds where this method can be applied. Evolution of the soil conditioning technology has further increased the range of subsurface conditions that can be handled, including very fine to coarse grain size soils. The use of slurry mode TBMs is much more limited by the type of ground to be encountered.

TBMs in the range from 40 to 50 ft. diameter are increasingly used to build road tunnels. More than 30 TMB tunnels with diameters in excess 45 feet have been completed. This experience and the lessons learned from our previous projects and other projects, publicized in tunneling conferences and technical publications, has allowed this solution to be selected. Without this technology, all the parties would have had to accept high risk by selecting the original immersed tunnel solution. Technology development and innovation are always of great benefit to challenging projects like this one. The specific TBM requirements for the Parallel Thimble Shoal Tunnel are shown in Table 1.

A shaft screw conveyor has been selected, as opposed to a ribbon type adopted for the SR-99

Table 1. TBM technical data

TBM Type	EPB
Excavation diameter	43.29 feet
Operating pressure	5.5 bar
Opening ratio	45%
Installed power	10,000 kVA
Variable speed	0–2.3 rpm
Nominal torque	41,565 kNm
Maximum torque	52,164 kNm
Breakout torque	55,863 kNm
Max stroke	10 ft
Thrust cylinders	38 No
Thrust force at 380 bar	181,458 kN
Screw conveyor diameter	4.25 ft
Manlock	2 No
Foam injection points (CH, chamber, screw)	16+12+4
Wire brushes	4 rows

Alaskan Way Viaduct Replacement project as boulders larger than one foot are not expected along the alignment of the Parallel Thimble Shoal tunnel. Conventional shaft screws should be able to digest boulders this size considering the large diameter of the cutterhead itself. The maximum boulder size that the screw conveyor can digest is 20 × 15 inches.

Following this rationale and considering also that the ground is expected to be less abrasive than that encountered in Seattle, the innovative Atmospheric Cutter Changing Devices (a number tools replaced from within hollow arms of the cutterhead at atmospheric pressure) featured by Bertha in Seattle TBM has not been considered for this project. Atmospheric Cutting Tools in accessible cutterheads have proven their efficiency throughout the years by facilitating the cutting tool replacement in atmospheric mode and minimizing the amount of hyperbaric interventions required along the tunnel alignment. This therefore reduces the risk to tunneling works and impact to adjacent structures. They allow to replace the cutting tools without entering physically in the chamber, simplifying and making safer this critical activity.

The annular gap left by the TBM excavation will be backfilled with a two-component grout which was first used in Japan on the seventies and currently applied by TBMs worldwide. This technology allows to control the volume and pressure of grouting, so it can be compared with the theoretical amounts to confirm that annulus is completely backfilled. This

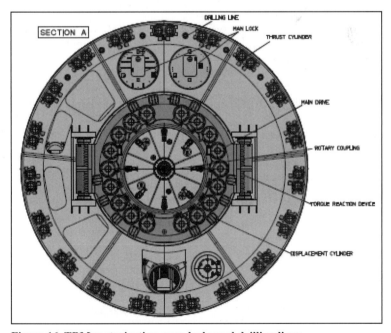

Figure 16. TBM motorization, man locks and drilling lines

allows for compensation for any potential ground loss left behind during excavation by the cutterhead, as well as minimizing water infiltration into the tunnel. Two-component mixtures not only gain gel consistency quickly but eventually achieve the required strength to maintain the tunnel liner and prevent its deformation.

These technical features will allow us to operate the machine in true Earth Pressure Balance mode, which is a perfect fit for the anticipated ground conditions. The following criteria will be used to ensure efficient TBM mining from the start to the end of the tunnel drive, regardless of the surrounding subsurface conditions:

- **Face pressure control:** maintaining the chamber permanently full of muck, with the right density. Excess water, liquids and air as a sub product of the soil conditioners (degradation of air bubbles of the foam) that accumulates at the top of the excavation chamber will be removed by means of an automatic purge system activated from the TBM operators cabin.
- **Shield pressure control**: continuous injection of a pressurized thick bentonite polymer mixture to preserve the annulus created between the excavation cutterhead diameter and the shield. This will minimize ground loss and therefore deformations.
- **Reconciliation of excavation quantities** by means of real-time weight and volume readings, comparison screens and charts in the operators cabin and muck sampling to confirm density used in the theoretical calculation. Weight and volume reconciliation will be performed continuously while mining with a reconciliation at the end of each push. This will ensure that the excavation muck extracted through the screw conveyor matches the theoretical one, so no volume loss is produced.
- **Tail backfilling** of the ring annulus with bi-component grout, controlled by both, volume and pressure, to ensure perfect backfilled of the gap and compensate for any potential ground loss created while mining.
- **Daily and per shift hands-on meetings**, to brief the crews, who ultimately will make this tunnel happen, in addition to a continuous training program by both, TBM manufacturer and CTJV Management.

TUNNEL LINING

Design of a 42' diameter tunnel lining using segments with fiber reinforcement exclusively has been

Table 2. Precast concrete segmental lining design basic data

Parameter	Value
Lining Thickness	18-inches
Internal diameter	39-feet
Design life	100 years
Ring length	78-inches
Ring configuration	9+1 (9 standard segments plus 1 key)
Ring geometry	Universal
Segment geometry	Trapezoidal segments
Min. lining radius	819 feet

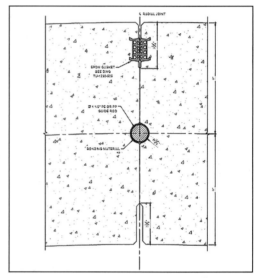

Figure 17. Radial joint details

very challenging, particularly with a hydrostatic head of 110 feet below the Chesapeake Bay This will be the world's largest diameter tunnel with steel fiber reinforcement segment lining. To achieve this ambitious goal, it has been necessary to study and analyze all the load cases that could push the mechanical properties of the steel fiber reinforced segment up to its maximum performance.

Segmental Lining Configuration

The precast concrete segmental lining design basic data are shown in the Table 2.

The segmental lining proposed is an universal ring with trapezoidal segments which mitigates the risk of encountering cruciform joints thus providing the ring with a high level of watertightness.

The ring consists of 9 segments and a key and was designed considering both handling and stacking

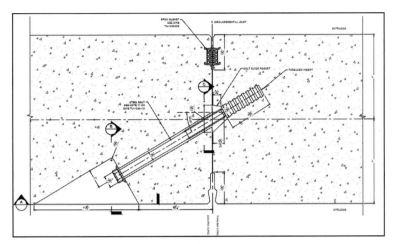

Figure 18. Circumferential joint detail

Table 3. SFRC resistance requirements

Property	Value (psi)	Value (Mpa)
Young's modulus short term	4931×10^3	34×10^3
Young's modulus long term	2901×10^3	20×10^3
Cylinder compressive strength	7500	52
Splitting tensile strength	600	4.2
Residual flexural tensile strength L/600	700	4.8
Residual flexural tensile strength L/150	500	3.4

loads and ring erection. The design is a compromise between a ring formed of larger segments which are more difficult to handle and stack but with a reduced ring assembly time, and one with smaller segments which easier to handle but slow down the ring assembly process.

The lining has been designed to accommodate a tunnel alignment radius of 819 feet which ensures that both the vertical and horizontal curves are feasible.

Steel Fiber Reinforced Concrete Resistance Requirements

SFRC resistance requirements considered in the design were defined based on a global design strategy and on previous experiences. They are listed in Table 3.

Design Criteria

The most significant design concepts are described below:

Durability

The tunnel lining is designed for 100 years of durability and an allowance for steel fiber corrosion of the

1 ⅛ inches of the segments intrados due to carbonation and the extrados by ⅗ inches due to chlorides.

Watertightness

The segment lining was designed to provide a watertight lining. Each segment features a single EPDM (Ethylene Propylene Diene Monomer) gasket, 4 connection bolts on circumferential joints and 2 more bolts on the radial joints. The bolts guarantee a suitable compression range for the high performance of the gasket. The majority of the standard steel bolts will be removed once the segmental lining has been grouted in place; any bolts remaining permanently will be stainless steel.

TBM Thrust

The large the tunnel diameter, the higher the thrust it transfers to the tunnel segments. As a result, in most of the tunnels with a diameter greater than 30 feet, hybrid reinforced segment lining solution is adopted (steel fibers and rebar). It was therefore necessary to develop some of the following strategies, to design a segment lining using only steel fibers as reinforcement:

- Optimizing the maximum thrust jacks pads contact surface. It was necessary to analyze the ring segmentation and clearance from the pads to the segment crossing radial joints so that the pad area could be maximized, thus reducing the tension transferred by the TBM Jacks.
- Minimizing the offset of the thrust jacks. Both manufacturing tolerance of the TBM and variations in the position of the segment ring inside the tailskin are the major factors for jack offsets. 2 connection bicones, 4 bolts in circumferential joints, guide rods and bolts in radial joints were introduced to reduce lips and offset during segment assembly.
- Analysis of post-cracking behavior. Different post-cracking models were carried out to understand the behavior of the fibers during crack propagation.
- TBM Thrust Analyses during tunnel excavation. Detailed studies were carried out to anticipate the different TBM thrusts that could be reached during tunnel mining. A database of different Dragados projects was analyzed to determinate thrust peaks during tunnel boring.

Buoyancy

The tunnel has to resist flotation along the entire length for both long-term condition and short-term (construction) conditions. The construction of engineering fill berms will address the buoyancy problem and protects the tunnel from the impact of vessels and extreme storm events.

Ship Collision

Although the likelihood of a vessel impact occurring at any particular location along the tunnel alignment is low, the consequences of such an impact on the structural behavior of the segment lining have been analyzed. A codified probabilistic approach was adopted for both the required tunnel capacity and to establish the maximum likely impact occurrence along the tunnel alignment.

A two-dimensional model was run to ensure that the ovalization created by the maximum ship impact loading were within acceptance limits which confirmed the performance of the segments gaskets without compromising tunnel watertightness.

CONCLUSIONS

Current TBM technology allows construction of large road tunnels in the range of 40 to 50 feet diameter and above. A significant number of tunnels have been constructed in difficult ground of comparable sizes. Most of these have been constructed in urban settings or under the water table.

The proposed TBM, working in EPB mode, will be appropriate for the challenging ground conditions expected to be encountered for the Parallel Thimble Shoal Tunnel. This method will generate minimal environmental impact, including sea life as well as marine traffic through the channel as the amount of marine works required to construct project has been minimized.

Both construction logistics and the permanent tunnels portals can be accommodated within the existing islands without disrupting the daily traffic operations and without enlarging the existing Portal Islands.

Construction works can be done 24/7, if required, as opposed to the immersed solution where limited working time is envisaged due to the weather restrictions, channel navigations, limited activities during the night, etc.

The new tunnel will be deeper that the existing one, reducing impact in future dredging works to increase the channel capacity.

Consequently, the bored tunnel should be the way of the new century to go to build underground infrastructures in difficult ground, minimizing disruption to the surrounding areas and contributing to the sustainable growth of the region.

TRACK 3: PLANNING

Session 5: Planning for Success: Risk Management and Contracting Strategies

Andrew Bursey and Nick Karlin, Chairs

The Risks Associated with TBM Procurement and the Next Steps Towards Industry Change

Lok Home
The Robbins Company

Gary Brierley
Doctor Mole Incorporated

ABSTRACT: Risk management in the world of TBM tunneling is, in itself, a risky business. The underground often presents obstacles and complex projects spanning miles of tunnel multiply those risks. However, there are ways to manage and reduce risk in our industry; i.e., by ensuring that thorough geotechnical studies are done and that contingency plans are in place. The TBM itself can be designed with risk reduction in mind, using tools that expand visualization of the ground around the machine and arm the contractor with ways to get through challenging ground conditions with minimal delays. This paper will explore risk in TBM tunneling from the viewpoints of the consultant, the contractor and the equipment manufacturer. It will also seek to make recommendations as to how risk can be better managed in today's tunneling industry.

INTRODUCTION

The majority of tunnels for civil engineering applications are now being constructed using some form of mechanical excavation. Beginning in the 1960s with rock TBMs, the tunneling industry has introduced both Earth Pressure Balance (EPB) and Slurry Pressure Balance (SPB) soil machines; Mixed Soil and Rock machines and a huge variety of different mechanical devices for the construction of small diameter tunnels. Over time, these machines have become more powerful and more adaptable to a wider variety of ground conditions; so much so that tunnels are now being constructed in ground conditions and in the vicinity of third party impacts that would have been considered beyond the state of the art just ten years ago (see Image 1 for an example—a mixed ground Crossover TBM used successfully at Mexico City's Túnel Emisor Poniente II).

All of the above is highly advantageous for the tunneling industry but it has also placed a much higher level of risk on the performance characteristics of the tunneling machines, on the contractors operating those TBMs, and on the manufacturers of those machines. Most of the risks for a tunneling project are associated with creating the space inside of which the finished facility will be constructed. In order to create that space the tunnel contractor must make many decisions about the best way to *excavate the ground*, the best way to *control the ground* at the face of excavation, and the preferred method for *supporting the ground* around the tunnel in a manner

Figure 1. Modern TBMs like this one used at the TEP II in Mexico City are capable of successfully excavating in soft ground and hard rock

that is safe for the workers and stable with respect to all of the overlying and adjacent existing structures. If it is proposed to use some form of TBM in order to build the tunnel, then the TBM becomes central to all three of the above activities and becomes an integral part of managing the risks associated with those activities.

The primary objective of this paper is to discuss how the TBM manufacturer can and should work together with a tunneling contractor in order to minimize and then to manage (i.e., control) many of the risks associated with a tunneling project. In

795

Figure 2. The McNally roof support system utilizes steel slats extruded from pockets in a TBM's roof shield

general, and as all members of the tunneling fraternity are well aware, there are lots of risks associated with every tunneling project which need to be identified, allocated, and managed as a result of the various contracts between the Project Owner and the Designer, between the Project Owner and the Prime Contractor, and between the Prime Contractor and various subcontractors and equipment suppliers; including the TBM manufacturer. As with any contract, the responsibilities of the various parties need to be clearly stated and the basic framework of the contract should create a fair and equitable working relationship between the parties. This becomes a most interesting challenge for the TBM manufacturer since he is providing an extremely complicated and expensive piece of equipment that is central to project success; not only as a result of its mechanical performance and durability but also as a result of how it is *operated* by the Contractor. Hence, many things must go *right* in order for the TBM to contribute in a positive manner to the successful outcome of a tunneling project. In order to address the topic of risk as related to the TBM manufacturer this paper is divided into the following sections:

- TBM Performance Characteristics
- The TBM Contract Document
- Managing TBM Risks During Construction
- Summary and Conclusion

Following the above, this paper will also discuss some of the next steps associated with industry change foreseen by the authors.

TBM PERFORMANCE CHARACTERISTICS

As stated above the TBM contributes to project success in two very important ways:

1. Its mechanical performance and durability, and
2. Its ability to help control potentially adverse ground reactions.

For instance, for a rock tunnel the TBM must be able to dependably excavate the rock and to allow for all aspects of equipment maintenance in a predictable manner. Prior to bid, the TBM manufacturer provides the bidding contractor with operating criteria and expected TBM technical capabilities based upon geological information provided by the owner's consultants and the project TBM specification; this is then incorporated by the contractor into his bid. In general, rock tunnels do not have negative impacts on adjacent existing structures and in most situations, it is relatively easy to install adequate support in the tunnel utilizing equipment provided as part of the TBM unless the tunnel is in highly faulted ground, stressed ground, or mixed-face conditions (see Figure 2—an example of an easy-to-install ground support system for open-type TBMs known as the McNally roof support system).

However, operating criteria for EPBs, SPBs and Mixed Soil and Rock TBMs are far more complicated than rock TBMs because of the enormous variations in different types of soil. Soil behaviors can vary from firm ground needing little face control and/or support to flowing ground and high water pressure, which can create huge problems both for the machine itself and for adjacent structures. In addition, soil TBMs have a more difficult interface between the machine's inherent performance characteristics and how that machine is *operated* by the Contractor, particularly in highly variable subsurface deposits. Hence, a perfectly good TBM can be operated in a manner that causes problems for the equipment, problems for tunnel production, and problems for third parties.

The bottom line for all of the above is the preparation of a listing of required TBM capabilities. These capabilities should be mutually agreed upon both by the tunnel contractor and by the TBM supplier, and must also meet the consultant's criteria. The specifications must be completed prior to commencement of TBM manufacturing. Listed below is a general example of TBM requirements:

- Proper Size with Sufficient Drive Power
- Cutterhead Design and Excavating Tools
- TBM Shield and Working Chamber
- Ground Conditioning at the Face for Either Rock or Soil or Both
- Thrust Capacity and Steering Control
- Spoil Removal within the TBM and along the Tunnel
- Spoil Weight/Volume Verification
- Bearing Seals and Tail Seals
- Shield Gap and Annulus Injection System
- Facilities for Ground Support Installation
- Guidance System and Alignment Control
- Data Loggers and TBM Performance Monitoring

All of the above TBM technical capabilities are incorporated into a technical proposal prepared by The TBM supplier with extensive input from the tunnel contractor. In essence, this single document represents one of the most important parts of the planning effort for a successful tunneling project built with a TBM. When the TBM disappears through the shaft wall or portal face the assumption is that it is equipped with all of the technical capabilities needed to make it to the exit end of the tunnel. If that is not the case, then significant project delays are in the offing, either as a result of reduced rates of advance or because of TBM modifications needed while in the tunnel. A TBM can be modified while underground using a suite of options known as Difficult Ground Solutions (DGS), to be discussed later in the paper. However, these features are much more effective at reducing risk if they are included on the TBM before it is launched.

THE TBM CONTRACT DOCUMENT

In order to accomplish the performance capabilities listed above the TBM supplier must design and manufacture a TBM for each specific application. With respect to the TBM's mechanical performance and durability the TBM is expected to operate "effectively" under very harsh conditions and for the duration of construction and it goes without saying that the different parties associated with a project will have radically different concepts about the meaning and the expectations associated with the word "effectively." One of the most common causes of claims, disputes, and lawsuits is the occurrence of "unfulfilled expectations" by one or more of the parties in a contractual relationship. Hence, and as a result, one of the most important goals of contract preparation is to forthrightly and unambiguously control project *expectations* in the contract wording. It is also important as a part of contract preparation to establish the fair and equitable distribution of project risks among the contracting parties.

The two most important sources of the risks associated with TBM performance are how well the TBM interacts with "anticipated" ground conditions both with respect to tunneling productivity and with respect to possible negative impacts on overlying and adjacent existing structures. Hence, the contract document for the TBM supplier should have well-developed descriptions for both anticipated ground conditions and for major third party interactions as provided in the project-specific Geotechnical Data and Baseline Reports. In addition to the geotechnical and third party considerations there are numerous other items that should be established in the TBM contract document and given below is an annotated listing of some of those items:

- **Warranty**—Clearly the TBM should be expected to perform reliably and at progress rates provided by the TBM supplier, and a warranty paragraph to that effect should be included.
- **Limitation of Liability**—However, a warranty only applies to the TBM itself and not to liquidated or consequential damages, force majeure, duty to defend, or project delay. Hence, the TBM supply contract should contain a valid Limitation of Liability paragraph addressing those topics.
- **Differing Site Conditions**—The TBM contract should also provide access for the TBM supplier to the legitimate application of the DSC clause. If the ground is found to be materially different as indicated by the prime agreement, then the TBM may need to be modified after the drive has begun.

- **Dispute Review Board**—The TBM supplier should also have access to some form of dispute resolution as part of its contract.
- **Safety**—The TBM Supplier is not responsible for on-site safety unless the TBM itself contributes to a problem. Hence, whenever TBM supplier personnel are on-site they are there as "guests" under the Prime Contractor's safety plan as explained in OSHA regulations.
- **Flowdown Requirements**—The TBM supplier must be extremely careful about flowdown requirements from the prime agreement which may or may not be applicable to the TBM supply contract. In general, the TBM supplier should not accept a blanket statement that all obligations contained in the Prime Contract apply to the TBM supplier. Some examples of problematic flowdown requirements are Indemnification, Duty to Defend, Liquidated Damages, Hazardous Materials, Default and/or Termination Provisions, and Waiver of Rights.
- **Standard of Care**—The TBM supply function also involves a large component of engineering services and the TBM supplier should only be deemed to be liable for those services if they were performed "negligently." This "Standard of Care" is also closely related to the TBM suppliers' proposed Scope of Services as explained below.

Probably the most important part of the TBM supplier's agreement is a detailed description of his Scope of Services. Almost no matter what is written in the body of the contract the TBM supplier can control its potential liabilities by explaining in detail the services it intends to provide and, equally importantly, those services and/or project activities for which it is not responsible. For instance, the TBM as supplied will have certain performance capabilities but that does not mean that the TBM will be operated and/or maintained in a proper manner in the field. Improper TBM operation and maintenance can be a significant risk for a tunneling project and the TBM supplier must limit its liability for inappropriate TBM operation. The TBM manufacturer cannot be held responsible for the damage caused by an unqualified TBM operator or by unqualified modifications to the TBM (see Figure 3—an example of a modification installed by a contractor that may have contributed to significant equipment downtime).

MANAGING TBM RISKS DURING CONSTRUCTION

The risk profile for a tunneling project can be divided into four steps:

1. Risk Identification
2. Risk Avoidance and/or Minimization
3. Risk Allocation
4. Risk Management.

All activities associated with Risk Identification, Risk Avoidance, and Risk Mitigation take place during the planning and design stages of a project wherein the Owner and his Design Consultants attempt to formulate a risk profile that is described in the risk literature As Low as Reasonably Practical (see SME Guidelines for Risk Management, 2015). This is an extremely important responsibility on the part of the Owner and its Designers as it represents a sincere desire by those parties to provide a contract document for bidding where the risks for *all* parties to the contract have been minimized as much as possible. At that point, the Owner's remaining responsibility is to *allocate* any remaining risks between itself and the Prime Contractor in a fair and equitable manner in the contract document for construction. After award, this process continues as the Prime Contractor continues to allocate its risks to various subcontractors and equipment suppliers. Hence, and for this paper, the question remains how much tunneling risk can be fairly and equitably allocated by the Prime Contractor to the TBM supplier.

For instance, and as discussed above, the TBM cannot be expected to perform in a ground condition that is known to be materially different as indicated

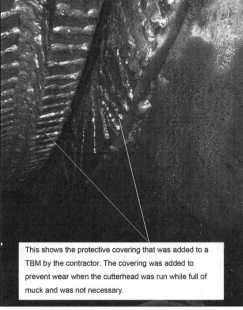

This shows the protective covering that was added to a TBM by the contractor. The covering was added to prevent wear when the cutterhead was run while full of muck and was not necessary.

Figure 3. An example of contractor-added machine modifications that were unnecessary

by the contract document. Other examples of dramatic differences between TBM performance characteristics and operational requirements would be as follows:

- **Gassy Ground**—The TBM can be equipped with gas monitors but the Prime Contractor is still responsible for ventilation issues and evacuation procedures.
- **Over-Excavation**—The TBM can be equipped with monitors that show how much spoil is being removed from the tunnel but that doesn't necessarily stop the TBM operator from over-excavating. Presently, there is no single monitoring system available that can accurately measure the volume and density of material being removed from the tunnel. Therefore several monitoring systems should be utilized on each project (Robinson et al, 2012).
- **Guidance**—The TBM will be equipped with a laser guidance system but survey errors may still cause the machine to go off of alignment.
- **TBM Maintenance**—Poor TBM maintenance by the Prime Contractor may cause TBM utilization to suffer or premature failure of components to occur through no fault of the TBM supplier.
- **Operator Training**—The TBM supplier can offer training but the Operator qualifications and capabilities are the responsibility of the Contractor. Improper operation of equipment is one of the leading causes of tunneling delays.

The complete list of TBM performance capabilities versus TBM operational responsibilities is long and, as described above, can result in "unfulfilled expectations" for a tunneling project. The main issue raised by all of the above is: How can we write a good contract that clearly defines the design of the machine and the TBM supplier's responsibility, as well as the contractor's responsibility and scope of machine operation?

NEXT STEPS TOWARDS INDUSTRY CHANGE

Looking at TBM Procurement Differently

There must be a more objective way for owners and contractors to view risk, other than looking for the lowest equipment price and highest willingness to accept risk from a TBM supplier. In fact, a correctly designed TBM is the key to a project's success, and correct machine design, even with increased initial cost, is part of that formula to success. Field results have shown, time and again, that a TBM built with "risk insurance"-type features (such as probe drills,

shield lubrication, etc.) can have a huge impact on a project's success in terms of schedule, cost, and safety. It is far better to build features into the machine from the start as part of a comprehensive risk management strategy, than to add them in the tunnel after an unforeseen event has occurred or the machine has become stuck.

Even when risks are considered low, it is still better to equip the machine from the outset with the tools needed to get through unforeseen conditions. These tools have been tested in the field and can mean the difference between project success and failure. Robbins currently is equipping several shielded hard rock TBMs with Difficult Ground Solutions (DGS)—a suite of options that can prevent a machine from becoming stuck and can enhance visualization of the ground around the TBM (Harding, 2017). For example, two-speed gearboxes allow a rock machine to shift into a high torque, low RPM mode to get through fault zones and collapsing ground without becoming stuck (see Figure 4—an example two-speed gearbox torque-speed curve).

Shield enhancements such as external shield lubrication can further keep a machine from becoming stuck. Radial ports in the machine shield can be used to pump Bentonite between the machine shield and tunnel walls to reduce friction (see Figure 5).

Emergency thrust systems are another addition that can be deployed when ground convergence occurs. Additional thrust jacks between the normal thrust cylinders can supply added thrust in a short stroke to break loose a stuck shield (see Figure 6).

Remedying Contract Structures to Reduce Risk and Cost

As mentioned throughout this paper, a contract structure that clearly defines the responsibility of the supplier and the responsibility of the contractor while allocating risk fairly is what is needed. Contractors must take responsibility to allocate the appropriate amount of risk given the limited capabilities of a given machine.

Part of more accurate risk estimation lies in the industry's ability to find and utilize consultants who are up-to-speed on the latest in TBM technology and mixed ground capabilities, and can therefore accurately specify the technical capabilities required of a given machine.

Another aspect of inexperience and improper risk allocation is the extreme specifications that are being created for many current projects. These specifications vastly overestimate the given risks of a project (e.g., if test results show 200 MPa rock, they will want to have a solution capable of excavating 300 MPa. If tests show 100 l/sec water inflows they will want a solution capable of handling 200 l/sec). These types of specifications increases

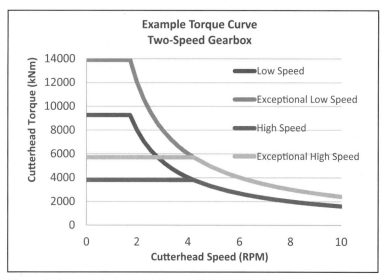

Figure 4. Example torque-speed curve for a TBM with two-speed gearboxes

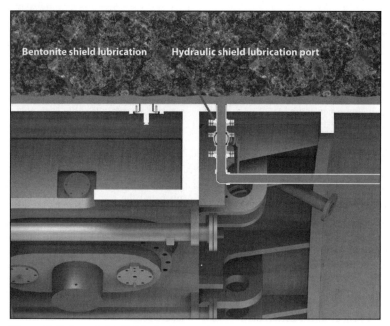

Figure 5. External shield lubrication system

the complexity of a TBM needlessly, and thereby increase the cost of the end product to the owner.

Risk-based cost and schedule estimation is being used on more projects, and will be an important part of the process moving forward (Sander et al, 2017). But even with these tools and the industry guidelines available—such as those produced by the UCA of SME (O'Carroll & Goodfellow, 2017)—an increase in industry knowledge of those tools is needed. If these tools are not used, the unequal allocation of risk will continue.

Operating the TBM Differently

When an adequate GBR is lacking and/or when risks can't be properly quantified, a push for continuous

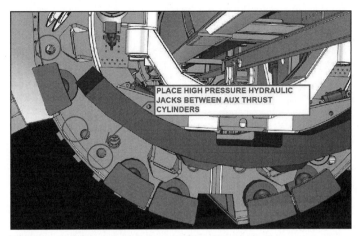

Figure 6. Possible locations for additional thrust jacks

probe drilling should be made by all parties involved. Writing continuous probe drilling into the contract can and has effectively reduced risk—but we need more buy-in from the industry. Through continuous probing, crews can generate an in-tunnel GBR concurrent with advance. This GBR could be used to analyze trends and predict upcoming transition zones. The requirement for an in-tunnel GBR would effectively force contractors to take the time to analyze what is ahead of them—a small price to pay when a big feature is detected in time to save the tunneling operation.

In addition, the TBM supplier should have more ways to address improper operation of the TBM by the contractor. The supplier should have access to the Dispute Review Board as readily as the contractor so that justification of a lawsuit, or lack thereof, can be determined by all parties involved.

SUMMARY AND CONCLUSION

The tunneling industry has seen enormous advancements in the performance capabilities of all forms of Tunnel Boring Machines (soil, rock, mixed rock/soil and small diameter) to the point that tunnel success has become intimately related to those capabilities. As an additional result of those advancements tunnel designers and tunnel contractors are continuously pushing the envelope for the size, length, depth, and alignments of tunnels in difficult ground and in the immediate vicinity of sensitive, existing third party structures. Hence, tunnel designers and contractors are becoming highly reliant on the knowledge and experience of TBM suppliers to rise to the challenge of those increasingly challenging projects; however, there are limits. The TBM supplier's financial opportunity for providing the equipment cannot be allowed to outstrip its responsibilities for project risks. Machine capabilities are still limited and

cannot be expected to serve as the primary excuse for unrealized project expectations. In the final analysis all parties involved with the successful completion of a tunneling project including project Owners, Designers, Prime Contractors, Subcontractors, Suppliers and Insurance Companies must accept their fair share of risk commensurate with the benefits associated with their contribution to the finished facility. From that perspective, the TBM supplier is not high on the list of project beneficiaries and, therefore, cannot be expected to assume unreasonable project liabilities relative to their role in the project. Hence, unreasonable attempts to transfer project risks to the TBM supplier must be controlled in no small measure so as to actually protect the integrity of the tunneling industry.

REFERENCES

Goodfellow, B. & O'Carroll J. 2017. Guidelines for Improved Risk Management on Tunnel and Underground Construction Projects in the United States of America. *Underground Construction Association of SME: smenet.org.*

Harding D. 2017. Difficult Ground Solutions (DGS): New TBM Solutions carve a Path to Success. *Proceedings of the ITA-AITES World Tunnel Congress, Bergen, Norway.*

Robinson R., Sage R., Clark R., Cording, E., Raleigh, P. & Wiggins C. 2012. Conveyor Belt Weigh Scale Measurements, Face Pressures, and Related Ground Losses in EPBM Tunneling. *Proceedings of the North American Tunneling Conference.*

Sander P., Entacher M., Reilly J., Brady J. 2017. Risk-Based Integrated Cost and Schedule Analysis for Infrastructure Projects. *Tunnel Business Magazine, August 2017.*

Alternative Delivery Drives Alternative Risk Allocation Methods

John Reilly
John Reilly International

Randall Essex
Mott MacDonald

David J. Hatem
Donovan Hatem LLP

ABSTRACT: Project principals seek project outcomes which meet required development plans and budgets. Underground construction demands a focus on commercial and political requirements when unanticipated subsurface conditions impact project tschedules, budgets, public perception, and public acceptance. Design-build (DB) and Public/Private Partnership (P3) delivery increase related complexity. This paper explores answers to the following questions:

- Why select DB or P3s for subsurface projects?
- Why is risk allocation important?
- What is our risk allocation experience for DB/P3?
- What are applicable international practices?
- What "lessons learned" apply to DB and P3s?
- Can a Geotechnical Baseline Report be adapted for DB and P3?
- How can risk allocation be incorporated into contract documents?
- What risk allocation guidelines can improve DB and P3?
- The way forward?

WHY ARE OWNERS SELECTING DB AND P3S FOR SUBSURFACE PROJECTS?

There are many objectives that motivate the continued growth in Owner selection of Design-Build (DB) and Public-Private Partnership (P3) approaches, most of which are beyond the scope of this paper and are addressed elsewhere (Hatem and Gary, 2017; Reilly 2009b). Primary objectives include containment of Owner risk exposure for cost overruns, and the opportunity to transfer substantial design and construction risk to the design-builder and/or the concessionaire (in P3s) (Reilly and Hatem, 2017).

Given those objectives, how should risk allocation for subsurface conditions be addressed in DB and P3s? Some Owners contend that transfer of all or most subsurface conditions risk to others is entirely consistent with those objectives; while other Owners are receptive to more significant degrees of risk sharing for subsurface conditions.

Presently, there are no generally-accepted or standard approaches to subsurface conditions risk allocation in DB and P3. Owner practices vary significantly, with the tendency to shift significantly more risk than in traditional Design-Bid-Build (DBB) delivery (Reilly and Hatem, 2017; Hatem and Gary, 2017; Briglia and Loulakis, 2017).

For example:

- Owners may furnish more or less subsurface information.
- Owners may or may not disclaim (in whole or in part) the design-builder's right to rely upon Owner-furnished subsurface information.
- The contract documents may or may not include a geotechnical data report or a Geotechnical Baseline Report (GBR) prepared by the Owner's consulting engineer.
- The contract documents may or may not include provisions that allow for sharing of risk due to materially differing subsurface conditions identified during design or encountered during construction (Gransberg and Loulakis 2012; 2016; Hatem and Corkum 2014; Loulakis, et al. 2015; Hatem 2014b; 2016b 2017; Essex 2016).

These varied approaches are attributable to project-specific factors that may deter reasonable risk allocations. These factors may defy, if not contradict, standardized or generalized guidelines on this subject—for example some project Owners desire near-absolute risk transfer in a quest to completely insulate themselves from cost overrun risk.

WHY IS RISK ALLOCATION IMPORTANT FOR SUBSURFACE PROJECTS?

On major underground projects, the most significant cost and time disputes involve disagreements as to whether the encountered subsurface conditions are materially different from those anticipated at time of bid. Therefore, contractual terms pertaining to subsurface conditions risk allocation are essential as a mechanism to fairly allocate subsurface risk exposures—such risk allocation should be accomplished with contractual clarity in a fair and balanced manner.

Achieving such objectives on underground projects should account for (a) the interrelationships and interdependencies (the "I and Is") among the compatibility and suitability of (i) permanent work design, (ii) construction means and methods design and implementation, and (iii) the character and evaluation of anticipated subsurface conditions; and (b) the respective roles and responsibilities ("R and Rs") of the Owner, engineer and constructor relating to those interrelationships and interdependencies. Risk allocation for subsurface conditions in DB and P3s should account for roles and responsibilities of the primary project participants relative to interrelationships and interdependencies.

In DBB, roles and responsibilities relating to subsurface conditions investigation, preparation and furnishing of permanent works design typically are undertaken by the Owner and its engineering team. The interrelationships and interdependencies inextricably involve all of those roles and responsibilities. Many, if not most, subsurface conditions disputes in DBB arise because those roles and responsibilities are assigned to different project participants who perform those activities at sequentially distinct points of time, isolated and independent of each other (Reilly and Hatem 2017).

The integration of design and construction roles and responsibilities in DB and P3s logically should present an opportunity for significant improvement in risk allocation and minimization of disputes pertaining to subsurface conditions—especially in circumstances in which the DB Team have primary roles and responsibilities relating to the interrelationships and interdependencies.

WHAT IS OUR RISK ALLOCATION EXPERIENCE FOR DB/P3 SUBSURFACE PROJECTS?

Development of Risk Management in U.S. Underground Construction

In the late 1990s, the American Underground Construction Association, the predecessor of UCA, was concerned with managing the cost of complex underground projects. Risk was discussed as a major cost factor in conferences with Owners, the International Tunneling Association (ITA) and others. This led to work on the development of risk management guidelines over the next several years, including WSDOT's Cost Estimating Validation Process (CEVP) 2002, BTS 2003, ITA 2004, ITIG 2006, Reilly 2008 (UCA Better Practices), Goodfellow and O'Carroll 2015 (UCA Guideline).

Application to DB and P3 Projects

These risk management guidelines have been implemented in some DB and P3 contracts to the benefit of those project Owners. Examples include Lake Mead Tunnel 2008 (DB); Seattle Alaskan Way Tunnel 2011 (DB), Seattle SR520 Floating Bridge 2010 (DB), the Ottawa Light Rail Project (P3), and the Port of Miami Tunnel (P3).

The initial application of risk management generally started with a priority of the Owners to control cost (Reilly 1999b, 2001c, 2003c) which led to the development of better cost estimating and risk identification, by validating base costs and combining base costs with risk costs (Reilly 2004c). A result was the explicit identification and characterization of risk leading to better risk management and response. Combined with the increased use of DB and P3 contracts, this required Owners to address risk management in the planning and design phases and to link or transfer some of those risks to the contractor (Reilly and Hatem 2017), leading also to a need to define risk "Owners" and address the treatment of interface risks.

Risk Transfer/Ownership Considerations

While it is clear in the planning phase that all risks are the responsibility of the Owner, how to manage the transition of responsibility for risks during the bidding/tendering and construction phases is less clear. Generally Owners are reluctant to make the full risk register a contract-referenced document and so the practice to date has been, generally, to release a sub-set of the full risk register without quantifying those risks, although there is one case where the full risk register was given to the contractor (Goodfellow 2009).

The success of transfer of risk responsibility is mixed for the projects referenced above—running from a) full open and collaborative risk management by Owner and contractor to b) poor communication of risks, including interface risks, in construction. P3 projects would seem to require better and clearer risk allocation since the consequences of poor risk management are potentially more severe (e.g., the capitalization of increased interest costs and revenue losses associated with project delays).

Operational risks, though peripheral to this paper's focus, are also of concern for P3 projects. The Sydney Cross-City Tunnel P3 failed because the cost of the project was underestimated (initial capital cost estimate $AU330m, delivered capital cost $AU680) and the revenue from tolls was also underestimated. Several other Australian P3 projects similarly failed to meet cost and revenue objectives (Tunnel Talk 2013).

Europe is quite inhomogeneous regarding DB and P3 projects—both contract types are quite rare in the German-speaking countries although Germany has some experience with P3 highway projects. There are projects in continental Europe related to highways, including "brown field" and/or upgrading projects which use a shared risk allocation approach. In these the contractor assumes design and construction risks, except for some risks e.g., archaeological risks, which are normally shared, at least with respect to schedule extension but where no additional monies are awarded to the contractor.

Risk and Alliancing

Better examples of meeting cost goals through risk management would seem to come from the early experience of Allianced contracts (Reilly 2009b) where reductions in cost of 25% have been reported. Alliancing requires the Owner, engineer and contractor to meet cost and schedule goals by working together under pain-gain requirements, using joint risk identification and management procedures. Examples of such projects include the Sydney Northside Storage Tunnel (Henderson 1999, Evans and Peck 2004) and the Channel Tunnel Rail Link (Reilly 2009b).

HOW DO INTERNATIONAL AND NORTH AMERICAN RISK PRACTICES COMPARE?

This is a broad topic and varies greatly from country to country. Risk management (Reilly 2008a) includes risk identification, risk analysis, risk response (mitigate, avoid, transfer or accept) and the use of risk-based cost estimating procedures (Reilly 2017). Data was available to the authors regarding these topics in the US, Canada, UK, Germany, and Australia. Codes, Guidelines and Codes of Practice

for risk management have existed for some time in general (e.g., ISO 31000) and specifically for underground construction (e.g., ITIG 2006, 2012) in recent decades.

Examples—U.S. Practice

Risk management in construction has advanced from the varied "ad-hoc" policies of Owners and contractors to structured risk processes such as CEVP and codified Owner policies for risk management (WSDOT, FHWA, CA etc.). Owners have varied requirements for risk management in construction—some include specific risk requirements or reference codes such as ITIG (ITIG 2006, 2012), while others have been silent on risk requirements. Linkage or transfer of Owner-defined risks in the planning and design stages to the contractor has been mixed but there are some positive examples—the Southern Nevada Water Authority's (SNWA) Lake Mead Tunnel and Intake No. 3 project included risk management requirements in the contract documents that applied elements of the ITIG code. The contractor, Salini-Impregilo, entered into a partnership relationship with the Owner, fully sharing and collaborating on risk planning and the construction risk register. In other projects the contractors have been reluctant to share risk information and coordinate risk identification and response with the Owner, even where the contract documents have called for this requirement.

U.K.

The implementation of Alliancing and Framework contracting (e.g., target cost contracts with shared pain-gain incentives) for underground construction has contributed to improved risk planning and management as a normal business practice. NEC and FIDIC based contracts recognize risk management and risk requirements and have been accepted for some time, gaining increased urgency after the Heathrow Tunnel collapse in 1994 with a loss of $141 million (Wallis 1999).

Germany and Austria

One author (Reilly) worked with a German company (Hochtief) in the late 1980s and 1990s where the application of basic risk management in tunneling was discussed and that company's experience applied, to a limited extent, in the North American market. Hochtief was also active in Australia and is reported to have adopted some of the Australian risk approaches related to early Alliance contracting.

This author's focus on risk was intensified in the late 1990s and a series of papers on risk in underground construction was produced (Reilly 1999b et. seq.). A presentation was made to the Geotechnik Colloquium in Salzburg in 2000 summarizing advanced project management, delivery and improved approaches to cost and risk management. A colleague (Sander) reports that these ideas have been discussed but the application of risk management in Germany and Austria is not universally implemented—results vary by country and the policies/understanding of risk by the involved Agencies.

Australia

Australia has used risk management in underground construction, in particular as part of the Alliancing process since the mid-1990s. In Alliancing (Ross 2005, Henderson 1999) the Owner, engineer and contractor together commit to deliver the project to a specific cost and schedule with pain-gain provisions (e.g., the parties share costs incurred above the target and rewards of cost underruns) related to those and other goals. That incentive has driven a need for advanced cost and risk management, similar to the WSDOT CEVP process noted earlier.

South America

This author is currently advising on, and implementing, advanced cost and risk management for an airport expansion project in Peru. This includes implementation of CEVP risk-based cost-estimating and advanced risk management using the integrated cost-schedule Risk Identification and Administrative Tool (RIAAT) process (Sander et al. 2018) This requires full training of the Owner's staff in the application of these processes.

General

Development, application and implementation of risk identification, risk management, risk-based probable cost estimating and communication of these processes to Owners, agencies and contractors worldwide is mixed but improvements to risk management for DB and P3 projects is advancing with progress in specific applications, Codes of Practice, and a broader awareness of the need for risk management. It is expected that this trend will continue as risk management is better and more broadly understood and accepted by Owners and contractors in terms of the need for more reliable delivery, management to achieve cost goals and revenue requirements, particularly for P3 contracts.

UNDERGROUND DBB "LESSONS LEARNED" APPLICABLE TO DB AND P3

Why is underground construction termed "risky business"? Tunnels tend to be long, horizontal facilities. Given that geologic conditions are inherently variable, no reasonable amount of exploration

will capture the true nature of the conditions to be encountered. An unanticipated change in ground conditions that slows or stops excavation advance can have an extraordinary impact on the overall cost and schedule of the work. This aspect is what sets underground construction apart from other heavy civil works projects. History has demonstrated that neither contractors nor their insurers are in the business of carrying unlimited risk. Owners can attempt to allocate these risks to the contractor, but this rarely works to the Owner's advantage.

Years ago, Owners provided bidders with the data from their site exploration programs, and made the bidders responsible for making their own interpretations of the data. The competitive bid environment demanded optimistic interpretations, which led to legal battles when those optimistic conditions were not realized. In an effort to reverse the litigation trend, the industry developed a number of improved contracting practices including GBRs, Escrow Bid Documents (EBD), and Dispute Review Boards (DRB). See USNCTT (1974, 1984), UTRC (1989, 1991), Matyas, et al (1995), Essex (1997, 2007) and DRB Foundation (2007).

Lessons were learned as improved contracting practices evolved through DBB contracting. The following lessons are equally applicable to DB and P3s:

1. The variability of the subsurface, and the contracting practices that balance the attendant risks, transcends the form of project delivery. Despite this, some Owners and advisors believe that through DB or P3, ground risks can (and should) be fully allocated to the contractor or concessionaire. The reality is that ground risks will always exist. If the contractor sustains a compensable loss, two dynamics may occur: (1) a contractor who is undercapitalized due to unpaid claims may seek other ways to save costs to the detriment of the project and the Owner; and (2) the contracting party will inevitably find ways to recover the additional costs.

2. Owners need to appreciate the above realities, and be prepared to risk-share—i.e., allocate risks to the contracting party for anticipated subsurface conditions, and be prepared to shoulder the costs if more adverse conditions are encountered.

3. The Owner should plan to execute a substantially complete site exploration program as a basis for bid development (Robinson, et al, 2001). Where this was not followed, such as the Deep Tunnel Sewerage Scheme in Singapore (Wallis, 2006) and a number of Design-Build-Operate-Maintain (DBOM) projects in Australia, the claim-related costs

and delays dwarfed the "savings" from reduced exploration.

4. Owners should not disclaim the accuracy or applicability of available subsurface information. To do so is contrary to the objective of fair risk allocation for *known* or *anticipated* subsurface conditions. Obtaining sufficient subsurface information is an essential starting point for this process. Again, there are a number of projects in Australia where insufficient data was provided at time of bid, and bidders were instructed to carry out their own investigations post contract-award. Unfortunately, their fixed-price contracts did not account for the costs of conditions disclosed by the subsequent explorations. Notably, few of those contractors exist today.

5. There are many ways to compensate the contractor for the work performed. Contract documents can create meaningful linkages between the baselines in the GBR and the payment provisions. The more discrete the bid item descriptions, the more readily the contract can "flex" with the variable conditions. The use of target cost compensation models, with share-pain and share-gain provisions, can incentivize the parties to work toward common goals that will benefit relationships and reduce the impact of unanticipated conditions—to the benefit of both parties.

6. The Owner should establish a contingency fund for unanticipated conditions. How those amounts are defined, and how they are paid out, will vary depending on the nature of the project and operational constraints of the Owner. These can take the form of provisional bid items and quantities, and time and material reimbursement. A "ladder" method was implemented on the Port of Miami Tunnel (Chen, 2009) where the financial responsibilities for extra work were shared according to pre-established rules.

7. How unspent contingency funds are allocated back to the parties can create positive incentives for the parties to work together to minimize claims. For example, unspent contingency funds might be returned on a 60/40 contractor/Owner basis. This could incentivize the contractor to minimize claims in favor of a bonus.

8. The absence of a Differing Site Condition (DSC) clause in the contract does not eliminate the potential for claims, despite the efforts of some Owners. Contractors have found other legal means of recovering unanticipated costs, such as the theory of implied warranty—i.e., that the contractor had the right to assume that the boring logs were accurate.

HOW CAN/SHOULD A GBR BE ADAPTED TO DB AND PPP DELIVERY?

GBRs were initially developed during the DBB era. Under DBB, the Owner and its engineer prepare a 100% complete design and fully integrated GBR, with a clear picture of the anticipated construction methods inherent in the design approach. As such, the Owner's team is able to prepare the physical baselines consistent with the known subsurface strata and conditions, as well as the behavioral baselines consistent with the anticipated means and methods.

However, with DB and P3, the engineer of record joins with the contractor not the Owner. Unless the DB or P3 reference design and specs are so prescriptive as to direct design details and construction methods (in effect, a DBB scheme), the DB team should either: (1) be given the opportunity to create the behavioral baselines consistent with their design and construction approach; or, (2) be given the opportunity to comment on, and offer revisions to, draft baselines prepared by the Owner (Essex, 2018, Pending). The key is to obtain the contractor's and its designer's views of the ground within the context of their proposed design and construction approaches. Without this exchange, the Owner's behavioral baselines will likely be disconnected from what the contractor plans to do. This breeds ambiguity, not clarity. This recently occurred on a sewer project in Washington DC—the Owner's engineer prepared a "final" GBR that was neither complete nor amendable during the bid phase—an unfortunate combination. On the Niagara Tunnel project in Ontario (Delmar et al., 2006) and a transit project in the DC area, bidders were given a GBR-A for bidding that solicited feedback from the bidders on specific items (GBR-B), following which a ratified GBR was finalized as the official GBR to the contract. This three-step process is discussed in Essex (2007).

The best way forward is to have Owners properly informed about the benefits of collaboration and clear identification/mitigation of construction-related risk. A "one-way" GBR created under a DB or P3 framework will generate ambiguity, increased risk for the contractor, and will trigger a relatively higher bid price. A collaborative process that allows the contracting party to provide input to the GBR, through one means or another, is more likely to promote common understandings and expectations, reduced bid contingencies, a better construction risk register, and a lower risk of disputes.

HOW SHOULD RISK ALLOCATION BE INCORPORATED INTO CONTRACT DOCUMENTS?

This section addresses risk allocation related primarily to the risks of subsurface conditions to be encountered during construction. There are a number of means to incorporate risk allocation and risk sharing into the contract documents:

- Through the GBR
- Through the payment provisions
- Through the nature of the design
- Through one or more construction-focused risk registers

GBR

A well-written GBR will help clarify how subsurface risks are to be allocated between the parties. Clearly presented physical and behavioral baselines will help to clarify who carries what risks under the contract. In some cases this will be clear-cut but in others more complex.

A clear-cut instance is where A and B ground types are anticipated with baseline percentages for each. The costs of excavation and support are lower for the A ground. If the B ground percentage exceeds the baseline, then the discrepancy can be addressed through the B unit price or a change order.

Encountering contaminated ground might warrant a more complex approach. One approach would be to have the contractor responsible for maintaining a full-time health and safety officer on site, with a health and safety program ready to be implemented if required. Baseline quantities could either be finite or zero, with unit prices or provisional sums set to compensate for the costs of testing, handling, treating and disposing of the contaminated quantities. In either instance, the contractor is compensated for the work actually performed.

Payment Provisions

The payment provisions should clearly explain the work that is to be accomplished, and should synchronize with the GBR's baselines. Different forms of payment can be taken:

- Unit bid items with planned or baselined quantities
- A mix of unit prices and lump sums with an upset price
- A single lump sum with a negotiated, cost-loaded CPM for progress payment purposes

An Owner could choose a number of options to account for contingency items:

- Provisional or optional bid items and quantities
- Provisional sums that are linked to specific categories of work, such as the contamination example explained above, or other work items such as probe drilling, grouting, or the handling of groundwater inflow rates in excess of the baseline
- An undisclosed allocation of funds held by the Owner separate from the bid items

An example of a provisional bid item is where high groundwater inflows ("X") are anticipated in a rock TBM tunnel. The baseline flow rate is "X" but is not to be paid for separately. A provisional table is included that solicits bidders' prices per shift or per hour if the actual quantities exceed X, 1.5X, or 2X. In this way, the Owner can forward-price the cost impacts of exceeding the baseline. This approach may or may not be possible initially for certain public Owners but the concept should be explored.

Design Conservatism

An Owner may elect to constrain the use of certain means and methods if those approaches infer excessive risk for the Owner. In this manner, the Owner is willing to pay a higher price to mitigate a risk it may not be willing to endure, despite this risk technically being the contractor's responsibility. An example is the use of a pressurized face TBM when a closed face, non-pressurized TBM might work. Another example is the required use of slurry walls, secant pile walls, or ground freezing to excavate shafts "in the dry," rather than risk potentially adverse effects associated with a leaky support system or a dewatering program.

Risk Registers

Current practice calls for developing and maintaining a joint risk register during project execution in order to help both parties plan and implement the work in a risk-mitigating manner as was done on the Lake Mead Tunnel and Intake project (Grayson et al. 2015). A concept worth considering is to maintain two construction risk registers—one that addresses risks associated with conditions within the baselines and therefore the contractor's responsibility, and the other related to risks associated with conditions that fall outside the baselines that are therefore the Owner's financial responsibility. The more realistic the GBR baselines, the more likely that change orders will be required. The second risk register will lead to better preparation and better cost control by reducing the "surprise" factor.

WHAT GUIDELINES SHOULD INFLUENCE RISK ALLOCATION DECISIONS ON DB AND P3 SUBSURFACE PROJECTS?

Roles, Responsibilities and Risk Allocation

On major subsurface projects, subsurface conditions risk allocation should be correlated with the assignment of roles and responsibilities for (a) subsurface investigation and evaluation, and (b) design adequacy. There are critical interdependencies and interrelationships among the scope and character of subsurface investigations. These include: how subsurface information or data is characterized, evaluated and reported; the influence of such evaluations on the final design approach; and the compatibility, suitability and constructability of the final design relative to the anticipated conditions (Loulakis, 2013; Hatem, 2017; Reilly and Hatem, 2017). In DB and P3s, the crux of the matter falls to the design-builder's roles and responsibilities in addressing the above elements (Reilly and Hatem, 2017).

For these reasons, contractual decisions in DB and P3s concerning subsurface conditions risk allocation are made in recognition of the design-builder's substantial roles and responsibilities in terms of evaluation of the available subsurface information and design development. These decisions are (and should be) made on a project-specific basis.

Risk Allocation Approaches

Subsurface conditions risk allocation in DB and P3s should be achieved through a variety of project-specific approaches that take into account the respective roles, responsibilities and risks of the Owner and design-builder. (Hatem, 2017). There are a number of workable means to identity and allocate subsurface risks between the parties; the key is that the ultimate allocation of risks be fair and equitable, considering the uncertainty of the (un)known subsurface conditions at time of bid (Hatem, 2017a).

Owner Practices That May Negate Risk Transfer

The principle underlying the opportunity for substantial risk transfer to the design-builder for design adequacy and subsurface conditions arises from the design-builder's control of the design development process—as compared to the relatively minimal roles and responsibilities of the Owner. However, that principle may be undermined and subverted by Owner practices in one or more of the following respects:

- The Owner design requirements are based on a relatively high (30% to 60% or more) degree of design development and/or are overly-prescriptive and detailed.

- The design-builder's ability to exercise judgment, discretion, and innovation in the design development and finalization is unreasonably restricted or constrained. The Owner mandates that subsurface data, reports or information that it furnishes be utilized by the design-builder as the exclusive or primary basis for the design-builder's final design.

- The Owner affords the design-builder limited site access or opportunities to conduct its own subsurface investigations.

- The Owner exercises an overly-broad review and right of rejection over the design-builder's design submittals.

- The Owner (or its consulting engineer) imposes judgments upon the design-builder relating to the characterization and evaluation of subsurface conditions and/or the appropriateness of the design in the specific context of anticipated subsurface conditions.

Notwithstanding the contractual provisions, the cumulative effect of these practices can put the Owner in a position of substantial control of the design process and other important variables that impact design adequacy and the level of subsurface conditions risk (Gransberg and Loulakis, 2012; Loulakis, 2013; Hatem, 2016a; Lien and Rose, 2017; Briglia and Loulakis, 2017; Reilly and Hatem, 2017).

These practices produce a misalignment of the control and risk equilibrium reasonably expected in DB and P3s that, in turn, produce unfair and imbalanced risk allocation. Those practices are likely to result in dispute boards, arbitration panels, and other dispute resolvers determining that the imbalance has unfairly constrained the design-builder, and that the cost overrun exposure, even though *contractually* allocated to the design-builder, may *ultimately* be re-allocated (in whole or in part) to the Owner (Hatem, 2014a; Loulakis, et al., 2015; Reilly and Hatem, 2017).

Scope Validation and Progressive DB

In time-limited bid periods, the design-builder may not have sufficient time to validate the scope of the design given the inability to assess the suitability of Owner-furnished subsurface information. Some Owners in such instances may allow for equitable adjustments to the design-builder. (Briglia and Loulakis, 2017, pp. 118–119; Hatem and Gary 2017).

On other projects, the progressive DB approach may be utilized which allows the Owner and the preferred design-builder to collaborate in: (a) defining the scope of subsurface investigation; (b) evaluating the products of that investigation; (c) developing a design that is compatible with the anticipated conditions; and, (d) developing contractual risk allocation approaches specific to those anticipated subsurface conditions and design approaches.

Both the scope validation and the progressive DB approaches are intended to improve the process, and clarify at what time final contractual decisions as to subsurface conditions risk allocation may be informed and determined. The improvement is accomplished by providing a meaningful and interactive approach where the Owner and design-builder can more effectively synchronize their respective roles and responsibilities. In the context of the scope validation and progressive DB approaches, the particularization, clarity and documentation of contractual risk allocation decisions may be enhanced through the Owner and design-builder's collaboration in the preparation of a GBR.

Risk Allocation Guidelines

Industry guidelines are needed to achieve fairness and balance in design adequacy and subsurface conditions risk allocation in DB and P3s (Hatem 2014a; Reilly and Hatem, 2017; Gransberg 2017 pending). Those guidelines should address the performance practices in DB and P3s that are consonant with the participants' contractually-defined roles and responsibilities, and principles of fairness and balance in risk allocation.

WHERE DO WE GO FROM HERE?

It is clear that both good and poor practices are being implemented. The industry's charge is to gather the right forums with experienced participants, and share the good and bad lessons learned so that planners of new projects are adequately informed. Getting Owners to attend may be the ultimate challenge. Another path is to create short, succinct guidance documents, perhaps published through the Construction Institute of the ASCE, written by committees similar to those who created the cornerstone documents of the 1980s and 1990s. The messages need to be clear and to the point, and put into publications that should be on every Owner's shelf. Another possibility it to provide free guidance documents available online, similar to the DRB Foundation's products. Education is the key—the industry just needs to get to it.

REFERENCES

[1] Briglia, S. and Loulakis, M., 2017. Geotechnical Risk Allocation on Design-Build Construction Projects: The Apple Doesn't Fall Far From the Tree, *Journal of the ACCL*, 11(2).

[2] Chen, W-P., 2009. Port of Miami Tunnel Update—A View from Design Builder's Engineer. In: Rapid Excavation and Tunneling Conference, Las Vegas. SME, pages 687–699.

[3] Delmar, R., Charalambu, H., Gschnitzer, E., Everdell, R. 2006, The Niagara Tunnel Project—An Overview. Proc, Tunnelling Association of Canada.

[4] DRB 2007 "Dispute Review Board Practice and Procedures Manual" Dispute Resolution Board Foundation www.drb.org/concept/manual/.

[5] Essex, R. 2016 Risk Management in Underground Construction. In: Geotechnical Baseline Reports in Underground Construction.

[6] Essex, R. 1997 Technical Committee on Contracting Practices of the Underground Technology Research Council, American Society of Civil Engineers Geotechnical Baseline Reports for Underground Construction..

[7] Essex, R. 2007, Technical Committee on Contracting Practices, Geotechnical Baseline Reports for Construction Underground Technology Research Council, American Society of Civil Engineers.

[8] Essex, R. 2018 Pending, Geotechnical Baseline Reports as a Risk Management Tool. Chapter in International Construction Contract Law, 2nd Edition, by Klee, L., Wiley Blackwell.

[9] Evans and Peck, 2004, "Northside Storage Tunnel Project, Post Implementation Cost Review, Final Report" 15 November.

[10] Goodfellow, R. J.F. 2009 "Transfer of a Project Risk Register from Design into Construction, Lessons Learned from the WSSC Bi-County Water Tunnel Project, Proc. RETC 2009.

[11] Gransberg, D. and Loulakis, M. 2016. Answering the $64,000 Question: Geotechnical Risk in Design-Build Projects. GEOSTRATA, ASCE Geo-Institute, July/August.

[12] Gransberg, D. 2017 Guidelines for Managing Geotechnical Risks in Design-Build Projects. In: NCHRP 24-44. [Pending]. The Transportation Research Board anticipates publication of those guidelines in 2018.

[13] Grayson, J., Nickerson, J. and Moonin, E. 2015 "Partnering through Risk Management: Lake Mead Intake No. 3. Risk Management Approach," RETC June.

[14] Hatem, D. 2017a Design-Build: A Realistic Solution for Owner Cost Overrun Risk? Tunnel Business Magazine (October 2017).

[15] Hatem, D. and P. Gary, ed., 2017 Public-Private Partnerships and Design Build: Opportunities and Risks for Consulting Engineers, 2nd ed. Washington: American Council of Engineering Companies, pp. 343–562.

[16] Hatem, D. 2014a PPP and DB: Who is Responsible for Risk: A Call for Guidelines. North American Tunneling Journal, October.

[17] Hatem, D. 2014b Design-Build and Public-Private Partnerships: Risk Allocation of Subsurface Conditionsw. GEOSTRATA, ASCE Geo-Institute, August.

[18] Hatem, D. 2016a Diverse and Bifurcated Design Roles: Distinguishing Design Responsibility and Design Risk Allocation. Donovan Hatem LLP Design and Construction Manage Professional Reporter, Dec.

[19] Hatem, D. 2016b Risk Allocation for Subsurface Conditions in Design-Build and P3. In: Risk Management in Underground Construction. Miami.

[20] Hatem, D. 2017 Risk Allocation and Professional Liability Issues for Consulting Engineers on P3 and Design-Build Projects. In: D. Hatem and P. Gary, ed., Public-Private Partnerships and Design-Build: Opportunities and Risks for Consulting Engineers, 2nd ed. Washington: American Council of Engineering Companies, pp. 343–562.

[21] Hatem, J. and Corkum, D. 2014 Purpose and Preparation of Geotechnical Baseline Reports in Design-Build and Public-Private Partnership Subsurface Projects. In: Geo-Congress 2014, American Society of Civil Engineers. Atlanta.

[22] Henderson, A 1999 'Northside Storage Project', Proc 10th Australian Tunneling Conference, Melbourne March.

[23] ITA 1992. "Recommendations on the contractual sharing of risks" (2nd edition). International Tunnelling Association, published by the Norwegian Tunnelling Society.

[24] ITA 2004. "Guidelines for Tunnelling Risk Management." International Tunnelling Association, Working Group 2, Eskensen, S, TUST Vol. 19, No. 3, pp. 217–237.

[25] ITIG 2006 and 2012. "Code of Practice for Risk Management of Tunneling Works." January 2006 and May 2012 (2nd edition).

[26] Lien and Rose, Design-Build Performance Specifications, and Spearin: How Modern Trends in Project Delivery Have Impacted a Contractor's Defenses, The Construction Lawyer, Journal of the ABA Forum on Construction Law, Vol. 37, No. 3, Summer 2017.

[27] Loulakis, M. 2013 Legal Aspects of Performance Based Specifications for Highway Construction and Maintenance Contracts. Legal Research Digest 61, National Cooperative Highway Research Program, pp. 45–50.

[28] Loulakis, M., Smith, N., Brady, D., Rayl, R. and Gransberg, D. 2015 Liability of Design-Builders for Design, Construction and Acquisition Claims. Legal Research Digest,

[29] Matyas, R. M., Mathews, A.A., Smith, R.J., Sperry, P. E., 1995 Construction Dispute Review Board Manual McGraw Hill Construction Series.

[30] NAS 1984 Geotechnical Site Investigations for Underground Projects. National Academy of Sciences, Volumes I and II.

[31] Reilly, J.J. 1999a "Policy, Innovation, Management and Risk Mitigation for Complex, Urban Underground Infrastructure Projects" ASCE New York, Metropolitan Section, Spring geotechnical Seminar, May.

[32] Reilly, J.J. 1999b w. Isaksson, T and Anderson, J 1999a "Tunnel Procurement-Management Issues and Risk Mitigation" Proc. 10th Australian Tunneling Conference, Melbourne, March.

[33] Reilly, J.J. 2000c 'Tying it together—Policy, Management, Contracting and Risk—an Overview, Proc. North American Tunneling Conference 2000, Boston, June, pp 159–168.

[34] Reilly, J.J. 2001c 'Managing the Costs of Complex, Underground and Infrastructure Projects', American Underground Construction Conference, Regional Conference, Seattle, March.

[35] Reilly, J.J. 2003c "The Relationship of Risk Mitigation to Management and Probable Cost," Proc International Tunnelling Association, World Tunnelling Congress, Geldermalsen, Netherlands, April.

[36] Reilly, J.J. 2004a "Management and Control of Cost and Risk for Tunneling and Complex Infrastructure Projects," Proc. North American Tunneling Conference, American Underground Construction Association, Atlanta, April.

[37] Reilly, J.J. 2004c w. McBride, M, Sangrey, D, MacDonald, D and Brown, J. "The development OF CEVP®—WSDOT's Cost-Risk Estimating Process" Proc. Boston Society of Civil Engineers, Fall/Winter.

[38] Reilly, J.J. 2005a "Cost Estimating and Risk Management for Underground Projects," Proc. International Tunneling Conference in Istanbul, May, Balkema.

[39] Reilly, J.J. 2008a, "Chapter 4, Risk Management" in "Recommended Contract Practices for Underground Construction," Society for Mining, Metallurgy, and Exploration, Inc. Denver.

[40] Reilly, J.J. 2009a 'Probable Cost Estimating and Risk Management Part 2' Proc. International Tunneling Association, World Tunnel Conference, Budapest May.

[41] Reilly, J.J. 2009b, 'Alternative Contracting and Delivery methods—Update', Proc. International Tunneling Association, World Tunnel Conference, Budapest May.

[42] Reilly, J.J. 2013b Author of the Foreword, co-author of the Chapters on Risk, Cost and Schedule management in "Managing Gigaprojects," ASCE press 2013, Edited by Galloway, Nielsen and Dignum.

[43] Reilly, J.J. 2016b "Megaprojects—50 years, What Have We Learned?" ITA WTC2016 Proc. April.

[44] Reilly, J.J., Laird, L., Sangrey, D. and Gabel, M. 2011a "Use of Probabilistic Cost Estimating CEVP® in the management of Complex Projects to Defined Budgets" ITA Conference, Helsinki, May.

[45] Reilly, J.J., 2017 "A Short History of Risk," introduction and summary, Risk Management in Underground Construction Conference, Washington DC, November sponsored by TBM Magazine, http://undergroundriskmanagement. com/presentations/.

[46] Reilly, J. and Hatem, D. 2017 Design Build and Public-Private Partnerships: Managing Cost Overrun Risk for Project Owners, Design and Construction Management Professional Reporter (Donovan Hatem LLP December 2017).

[47] Robinson, R.A., Kucker, M.S., and Gildner, J.P., 2001. Levels of Geotechnical Input for DB Contracts for Tunnel Construction, Rapid Excavation and Tunneling Conference, June, San Diego.

[48] Ross, J. 2005, "Project Alliancing Practice Guide," Project Control International Pty. Ltd.

[49] Sander, P., Reilly, J., Entacher, M. 2018 "CEVP-RIAAT Process—Application of an Integrated Cost and Schedule Analysis" Proc NAT.

[50] Tunnel Talk 2010 "Adding the Insurance Payout Consequence" https://www.tunneltalk.com/ Discussion-Forum-Aug10-Insurance-payouts -for-failures.php.

[51] Tunnel Talk 2013 " Australia suffers toll concession failures" Article, 16 July, https://www. tunneltalk.com/Discussion-Forum-16Jul13 -Australia-PPP-toll-tunnel-crisis.php.

[52] UCA 2015 / Goodfellow, R. and Carrol, J. "Guidelines for Improved Risk Management Practice on Tunnel and Underground Construction Projects in the United States of America." UCA web publication www.smenet .org/uca.

[53] UTRC 1989 Underground Technology Research Council, American Society of Civil Engineers Technical Committee on Contracting Practices, Avoiding and Resolving Disputes in Underground Construction..

[54] USNCTT 1974 Better Contracting for Underground Construction, U.S. National Committee on Tunneling Technology, National Academy of Sciences.

[55] Wallis, S.1999 "Heathrow failures highlight NATM (abuse?) misunderstandings" Tunnel 3/99 pp 66–72.

[56] Wallis, S. 2006, The Good, Bad, and Mixed on the DTSS. TunnelTalk, Direct by Design, April.

[57] WSDOT 2002 "Cost Estimating Validation Process (CEVP®) Initiation Report" and "Cost Estimating Validation Process (CEVP®) Evaluation Report."

Construction Cost Estimating Using Risk-Based Approach

Michael S. Schultz and Greg Sanders
CDM Smith

ABSTRACT: Owners, seeking to gain a better understanding of underground risks and the possible scheduling, and budget implications of those risks on the allocation of capital funds on projects are asking for a deeper understanding of the likelihood of major risk events happening. Advanced approaches to risk-based cost and schedule analysis provide more cost certainty in designer's cost estimates and a means to quantify the risks associated with underground construction. Risk Registers are being used as a tool to help identify and estimate the probability of occurrence of certain risk events and to cost out the significant risks associated with tunneling work that carry over into construction. Taken one step further these potential "high impact" risk events are looked at in more detail to make more informed decisions on budgets and schedules. These tools can provide a more realistic understanding for the purposes of budgeting, reserving contingency funds, and making cash flow projections. This paper reviews the approaches being used and will provide project examples where risk-based cost estimates were prepared and how they were used. Applications to alternative delivery approaches (design-build, progressive design-build, etc.) are also briefly discussed.

WHAT'S WRONG WITH CURRENT COST ESTIMATING?

The accuracy of construction cost estimates and the possibility of large cost overruns on tunnel projects, primarily due to Differing Site Conditions encountered during construction, and the negative aspects that accompany them, are a concern for all stakeholders but especially owners, contractors and engineers. Currently engineers estimates are often initial "baseline" estimates only and may or may not reflect the likely-hood of impacts to budgets and schedules due to an "event." If baseline estimates do reflect these uncertainties they are often hidden and represent only the estimator's opinion. An estimating process that can more "accurately" identify and quantify the range and probability in potential consequences of significant risk events has many benefits (Figure 1).

The current standard of practice is for owners to carry a contingency in their project budgets. The contingency is often a factor or percentage of the cost estimate that is simply added to the project budget based on a deterministic project cost estimate. This is consistent with an ASTM E2516 Class 1 Estimate, or the AACE Recommended Practice No. 56R-08 Class 1 estimating standards. Owners are often required to have the funding available to carry the full cost of construction plus contingency in their budgets. Funds are therefore encumbered until project completion or when portions of the contingency fees can be "released" based on when a certain level of project completion or milestone in the construction is reached. Contingencies of 10 percent to even

20 percent of the total construction cost (as bid) are typical. For large projects maintaining a contingency of many millions of dollars though the life of the project could result in delaying other projects or prevent the allocation of the funds to other pressing needs.

ESTIMATING RISK AND UNCERTAINTY

The first step to understanding the risk-budget relationship acknowledging that there are different levels of certainty associated with each construction activity line item in the estimate and that different techniques are often used to price each line item. For example, an estimator may use empirical calculations and the baseline rock strength value when estimating TBM excavation rates and average production values from multiple past projects for the installation of a cast-in-place tunnel liner. Clearly these two-line items are based on different levels of uncertainty applied to the current project. To account for the combined uncertainties of all the line items, the designer prepares a baseline cost estimate to estimate within that level of accuracy and confidence level. For a Class 1 Estimate, in accordance with ASTM E2516, the expected accuracy range is a value that represents a variation after application of contingency at a 50 percent level of confidence for a given scope. Figure 1 shows the possible range in costs compared to the baseline.

When the baseline estimate is prepared the designer generally is asked to provide a recommended budget for the owner to provide the necessary

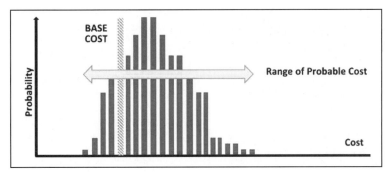

Figure 1. Comparison of baseline vs. range of probable cost

approvals and funding. The baseline budget should represent an understanding of the risk allocations that have been decided on by the Owner and possibly with input from the Contractor. The contingency that the Owner then carries into Construction should be representative of that risk allocation. Contingency funds are not always sufficient to cover budget short falls due to unforeseen change conditions claims and continue to be an issue for construction projects despite the use of this approach.

A comprehensive approach to estimating future unknown conditions is to incorporate the residual risks carried into construction from the risk register into the estimating process by using risk-based cost and schedule estimating. Risk-based cost and schedule estimating uses the risk analysis to quantify the variable impact for specific risk items. The process consists of identifying individual construction events that are associated with a specific identified risk. Once these events are identified, they can be evaluated using probability distributions to account for the probability of occurrence and associated cost and schedule implications for each. The interrelated cost items are then aggregated though the use of Monte Carlo simulation software that produces a project cost distribution that represents the likely range of cost. Using the TBM advance rate example, the estimator could use the entire range of rock strengths test data to develop a range of cost for the tunnel excavation and schedule impacts instead of just the baseline value.

Recent papers have suggested that the reason for the risk-based estimating approach not being widely adopted and used is because contractors may be reluctant to adopt methods that may increase their bids by including cost associated with risk (Reilly et al., 2015). While this may be true, this reason does not apply to the preparation of the engineer's estimate because the purpose of the engineer's estimate is to estimate the most probable construction cost and not represent the lowest bid price.

RISK-BASED COST ESTIMATING METHOD APPROACH

The authors have been using risk-based estimates for tunnel engineer's estimates of probable construction costs for the past five years or so. Over this time the process continues to be refined using experiences gained from projects. The current methodology includes: the initial project scope and definition followed by developing a baseline cost estimate, developing the risk register, performing an initial statistical analysis, introducing measures within the specifications and Geotechnical Baseline Report (GBR) to mitigate and further quantify risk, refining the risk distribution within the estimate based on the risk mitigation measures, and presenting the final estimated cost. These steps are shown in Figure 2 and discussed in detail below:

- Step 1: Define project scope.
- Step 2: Prepare the base estimate and develop the cost data as you would in a deterministic estimate. Individual unit cost is based on design quantities, historical data and quotes from sub-contractors and other vendors at a line-item level.
- Step 3: Prepare the risk register by listing the identified risks for the project along with other information about the risks, such cost magnitude, schedule impacts, and likelihood of occurrence. Although the risk register is created during the risk identification process at the risk workshop conducted with the client, it is periodically updated throughout the project. Items that have a cost impact are then associated with the affected line-item in the base estimate at each project milestone.
- Step 4: Select the proper distribution method for each identified risk line item combination. CDM Smith typically uses Normal, Pert and Binomial distributions. Normal distributions are traditional bell curves with a parameter mean and standard deviation. The normal

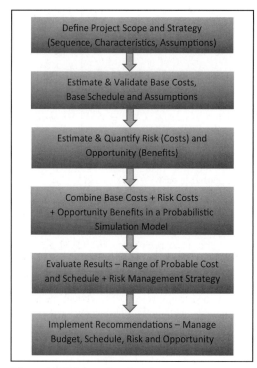

Figure 2. Risk-based estimating process

distribution is used to represent the uncertainty of a model's input whenever the input is the result of many random processes acting together in an additive manner. For example, the variation in the wage rate for a tunnel laborer. The Pert distribution is defined using a specified minimum, most likely, and maximum values. Pert distributions provide a similar approximation as a normal distribution but can be biased from the mean. The Pert distribution is typically used when a line item has a finite lowest likely value, but the maximum value could be orders of magnitude higher than the mean. This is typically used for line items such as tunnel advancement rate. Binomial distributions are used to model the probability of an event occurring.

- Step 5: Once the cost model has been populated with the risk distributions, the Monte Carlo Simulation is performed. The estimated value for each cost is calculated by summing the estimated cost of each line item 10,000 times with each trial using a randomly generated cost based on the probability distributions for each cost element. Each trial then generates an estimated project cost. These costs are recorded and a probability density function is created from which a mean value

is extracted. The mean, then, is the average cost of the project if it was to be performed 10,000 independent times. That cost is called the Most Likely Estimated Cost. In addition to the Most Likely Estimated Cost, the probability density function generated by the statistical analysis also allows the engineer to determine the cost at any confidence level. In a typical project, a confidence level will be chosen to calculate the likely lowest and highest estimated project cost. These values are then used to provide the required contingency recommendation. The choice of confidence levels reflects the client's risk posture. For example, if the 90 percent level is chosen, the client is accepting a 10 percent risk that the project will overrun the budget.

- Step 6: Risk reduction measures are incorporated into the GBR or specifications to reduce the cost of specific line items. The distributions with the estimate are adjusted to meet the revised risk probability. Additionally, the risk registers are reviewed at specific project milestones to regularly identify new risks not previously identified and to consider risks beyond cost, schedule and scope. The Monte Carlo simulations should then be rerun multiple times to refine the process. As shown in Figure 3, this can be used as an iterative process to help further identify and mitigate low probability but high impact events further during design and residual risks carried into construction.

PROJECT EXAMPLE

The following section provides a brief example of a line item from a risk-based engineering estimate. The project for which the estimate was prepared consisted of an approximately 4921 feet (1,500 meter) long outfall tunnel extending from a waste water treatment plant to a marine diffuser location. One of the risks identified within the risk register during the risk workshop was the likelihood of encountering a boulder or obstruction during tunnel excavation. The geotechnical investigation included 20 conventional borings and 26 CPT test locations. The borings encountered a marine deposit overlying interlayers of clayey silt to silty clay, sand and silt, and silty sand except at the offshore locations, where a gravel deposit was encountered overlying the interlayered deposits and underlying the marine deposit. These deposits are inferred to be of glacial origin based on available geological information. The SPT blow count value obtained within the granular soils was more than 50 blows/0.3 m. It was also noted in the boring logs that although cobbles or boulders were not encountered during drilling within

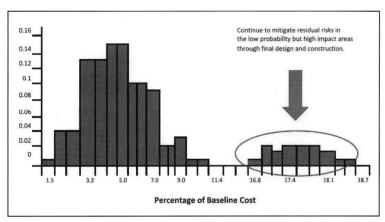

Figure 3. Comparison of baseline vs. range of probable cost

Table 1. Example risk register entry

Hazard	Cause of Hazard	Potential Consequence	Risk Likelihood	Financial	Project Schedule	Regulatory/Legal	Health and Safety	Operating & Maintenance	Risk Score	Control Measures To Be Implemented
TBM breakdown in the tunnel or stoppage	Ground conditions, boulder timber or other obstruction present along the alignment	Schedule delay, possible claim	1	3	1	0	0	0	3	Anticipated, ground conditions are addressed in the GBR, obstructions are baselined at 1 occurrence

the granular deposit, the deposit may contain cobbles and boulders.

During the risk workshop the likelihood of encountering a boulder was assigned an "unlikely to happen" rating with minor schedule impact and significant financial impacts as shown in Table 1. In the estimate line item for boulders, a binomial distribution with a five percent probability was assigned to the likelihood of encountering one boulder during the tunnel excavation. A tunnel crew was created for the removal of the boulder and associated standby time for the TBM and related support equipment assigned to the line item for boulder removal. A Pert distribution was created for the boulder removal line item that assumed a typical boulder would most likely take most of a 10-hour shift to breakup and remove. The Pert distribution was limited with a minimum value of 0 hours assuming the TBM would be able to break up the boulder without requiring intervention by the crew; a maximum value of 50 hours was assigned as reflected in Figure 4.

As can be seen from this example a greater level of effort is required for the preparation of a

risk-based estimate. However, unlike a deterministic estimate, risk and uncertainty in the risk based estimate are explicitly and quantitatively identified and modeled. In this example there is still a risk that more boulders might be encountered, but that risk had been discussed with the client during the risk workshop and was determined to be an acceptable shared risk. The shared risk was then communicated to the contractor in the GBR where they were told to expect to encounter one boulder that would require intervention for removal.

Alternately, if a higher likelihood would have been assigned to the boulder risk during the risk workshop the line item for the boulder occurrence could have been configured with a different distribution. For example, a Pert distribution could have been used to include a most likely risk of encountering three boulders with a minimum of 0 and a maximum of 15. This would most likely result in a higher estimated project cost and estimated contingency to account for the more severe risk environment. The project manager would then be aware of the risk of boulders and know that a set contingency cost was in

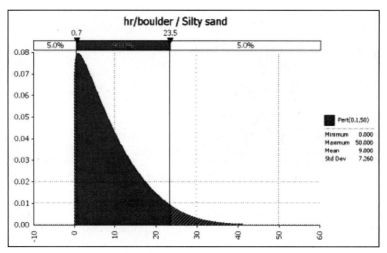

Figure 4. Example pert distribution

place to account for the boulder risk. If the portion of the alignment where boulders were anticipated was excavated without incident the Project Manager would then be able to reallocate that portion of the contingency amount to other needs.

COST ESTIMATE COMPARISON

To provide an example of how risk-based estimating has performed the following section includes portions of the bid results for three different projects where risk-based estimating was used. Each of these projects were selected not based on their results but based on of the variety of construction techniques and the availability of the bid data.

- Project A: Project A is an 850-foot-long, 48-inch diameter microtunnel with three 35-foot-deep shafts located in the Midwestern United States. The tunnel crosses below a river which floods several times a year. Ground conditions consist of silty sand with a possible mixed face composed of a limestone outcrop within the last 50 feet of the tunnel. This project was one of the first projects where CDM Smith used risk-based estimating. At the time the risk factors were only applied to a limited number of line items.
- Project B: The project is in the southeastern United States and consists of a raw water pump station and 1,100-foot-long, 7-foot-wide horseshoe-shaped intake tunnel. The ground conditions are limestone and claystone with high anticipated groundwater inflow. This project was the first project where risk-based estimating was used

Table 2. Bid results

Project	A	B	C
Number of bidders	2	6	5
High bid	$2,964,000	$16,297,000	$131,688,000
Median	$2,899,000	$14,037,000	$123,643,000
Low bid	$2,835,000	$12,579,000	$102,197,000
Estimated max cost	$3,056,000	$16,061,000	$129,726,000
Estimated most probable cost	$2,792,000	$14,814,000	$122,018,000
Estimated low cost	$2,641,000	$14,055,000	$115,200,000

extensively throughout the estimate by CDM Smith.
- Project C: The project is in the northeastern United States and includes a tunnel launch shaft and a 14,500-foot-long, 10-foot-diameter outfall tunnel. The ground conditions consist of sandy silts and clays of glacial origin.

All the projects shown in Table 2 are still in construction so the final cost is not available and the contingency amounts are not provided. Figure 5 provides a comparison of the combined bid results of all three projects. Based on the deviation between the median bid and the estimated probable cost, it can be concluded that the probable cost provides a reasonable approximation of the median bid price. Likewise, assuming the contractors have accounted for the different-levels of assumed risk in their bids, the risk based estimates also provide an accurate range of the anticipated cost due to the range

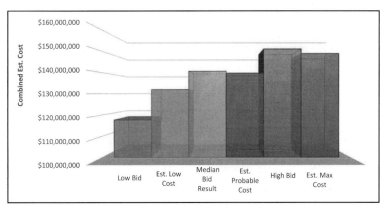

Figure 5. Combined bid results for projects A, B and C

of assumed risk. It is our belief that the estimated contingency likewise provides a reasonable and justifiable method for determining the required contingency amount to account for the anticipated risk. However, as can be seen by comparing the estimated low cost and the low bid the method has not been shown to provide a reasonable approximation of the winning bid price.

APPLICATIONS

Contractors, owners and engineers can all benefit from using risk-based cost and schedule estimating and analysis. The more uncertainty and complexity involved in the underground construction the more beneficial is the use. As shown by this paper, risk-based estimating can be an effective and useful tool for the preparation of an engineer's estimate. The argument has been made that in a traditional design-bid-build environment, risk-based estimating is unlikely to produce the low bid. Reilly stated in his 2015 paper, the more complex a project is, the more information is needed, and this information is critical to contractor's success. Perhaps, if the contractor prepares both a deterministic estimate and a risk-based estimate it would be very helpful for contractors when preparing a bid. By carrying out the additional step of performing a risk-based estimate a contractor will obtain a better understanding of who "owns" each potential risk. This information will allow the contractor to make a more informed decision when developing their bid strategy possibly leading to a lower contingency price.

For alternative delivery projects involving design-build, progressive design-build, public private partnerships and other types of alternative contract delivery, the risk-based cost and schedule analysis could be used as an effective means of the project team being able to communicate project cost

and schedule risks between the parties. This would be true between the design engineer and the contractor and could be extended to include the Owner and/or the Construction Manager as well. In addition, this approach can be extended to identify opportunities for cost and schedule savings if conditions encountered are better than the most probable conditions.

In an ideal case, if the risk-based cost and schedule approach is taken and is agreed to by all parties as the basis of payment for construction cost, similar to the manner in which Geotechnical Baseline Reports (GBRs) are now incorporated as Contract Documents to identify DSC claims, this could lead to better agreements on risk allocation and perhaps fairer payment for the actual cost of construction and reduce the disputes and conflicts that often arise from them.

CONCLUSION

The use of risk-based cost estimating methods provides a means of more accurately estimating the probable construction cost of a project during design and leads to a better understanding of the risks and risk impacts to budget and schedule during construction. Also, risk-based estimating can also help the engineer and owner to create more informed strategies to manage disputes and claims during construction. Traditional deterministic cost-estimating methods are useful and probably are still needed for the preparation of bids, but the use of risk-based estimating can also provide contractors with more relevant understanding of who "owns" each potential risk during the formation of their bid strategies. The next step in the evolution of risk-based estimates is to use them for design-build and other alternative delivery projects to evaluate how the technique can be used as a useful tool for better contracting and dispute resolution.

REFERENCES

ASTM, Designation E2516-11, Standard Classification for Cost Estimate Classification System.

Nolder G "Improving Cost Estimation with Quantitative Risk Analysis."

Reilly, J.J. 2001, "Managing the Costs of Complex, Underground and Infrastructure Projects," American.

Underground Construction Conference, Regional Conference, Seattle, March.

Reilly, J.J., 2008, "Alternative Contracting Methods—Part II," Proc. North American Tunneling Conference 08, San Francisco, June.

Reilly, J.J., Sander P., Moergeli A., 2015, "Construction—You Need Risk-Based Cost Estimating," Proc. North American Tunneling Conference 15, New Orleans, May.

Sander, P., Spiegl, M. & Schneider, E. 2009, "Probability and Risk Management," Tunnels & Tunneling.

International.

Shane J. 2015 "Risk-Based Engineers Estimate," Minnesota Department of Transportation, March St. Paul, Minnesota.

WSDOT 2009, "Cost Estimating Manual for WSDOT Projects," Guideline Document.

Procurement of and Contracting for Underground Construction Projects in North America

Gary Brierley
Doctor Mole Incorporated

David Corkum
McMillan Jacobs Associates

ABSTRACT: Beginning around 1990 the construction industry began contemplating using the Design/Build method of project procurement for infrastructure projects. In general, this transition from Design/Bid/Build to Design/Build was driven initially by a desire to help reduce project schedule and claims and then by an effort to facilitate project financing via PPPs. However, major underground projects can be complicated and risky for Design/Build as a result of the significant interface that takes place between construction activities and ground conditions which literally impacts 100% of the project. This paper is intended to discuss many of the important planning, design, and construction variables that need to be evaluated when deciding to use either the DBB or DB methods of project procurement for a tunneling project.

INTRODUCTION

Almost all construction prior to the beginning of the twentieth century was performed by what could be described as Design/Build. At that time, if someone wanted to procure a heavy-civil project they would hire a "master-builder" who would incorporate all aspects of design and construction into a single contract for delivering the Owner's finished facility. These so-called master-builder firms were large multidiscipline organizations composed of planners, designers, constructors and often financial advisors capable of complete project delivery.

Beginning around 1900, however, construction projects started to become increasingly complex and project Owners wanted more control over both the performance characteristics of the finished facility and the cost for constructing those facilities. Concerns were also expressed by project Owners that the designs had become more elaborate than necessary and that overdesign inured to the financial benefit of the master-builder. This was especially true for public Owners who were procuring projects intended to be used by the general public for very long periods of utilization while being good stewards of the public money. Also fresh in the minds of public Owners were the scandals associated with the building of railroads where single entities controlled the design, construction, material supply, and financing of the undertaking to the disadvantage of the public. The above concerns led to the formation and proliferation of a number of large engineering firms capable of working directly with public Owners to plan and design projects in an environment that was devoid of Contractor influence. Thus, came into being the Design/Bid/Build (DBB) method of project procurement, which involved the preparation of detailed project designs followed by the low bid Contractor providing only construction services.

Beginning around 1990, and driven again by a desire for "best value," the public sector began to develop a renewed interest in the possibility of using the Design/Build (DB) approach for providing infrastructure projects. One important subset of infrastructure, however, consists of the design and construction of tunneling projects which are radically different than almost any other form of construction. Tunnels are constructed entirely *within* the ground and are, as a result, forced to deal with ground characteristics and ground behavioral impacts that can be difficult to anticipate in advance. In addition, most of the costs and most of the risks associated with a tunneling project are related to the creation of the underground space and the temporary support of that space that is needed in order to construct the finished facility. Hence, one of the biggest issues with respect to the design and construction of a tunneling project is the preparation of a contract package that accurately characterizes all of the work required both to create the required underground space and then to install the finished structure inside that space.

In actuality, there are aspects of both the DBB and the DB approaches to project procurement that can be used effectively to produce a tunneling project and the primary objective of this paper is to point out what a project Owner needs to know and needs to do in order to use either approach for the successful completion of a tunneling project. By successful, the authors of this paper are referring to what is the best method for providing the least costly method for constructing a tunnel in the least amount of time and in the ground conditions that actually *exist* along the proposed tunnel alignment. In order to accomplish that objective this paper to divided into the following five sections:

- The Planning Effort
- Design/Bid/Build Project Procurement
- Design/Build Project Procurement
- Construction-Phase Services
- Summary and Conclusion

THE PLANNING EFFORT

Without a doubt, one of the most difficult aspects of producing a successful tunneling project is to adequately determine and provide all of the characteristics of the finished facility that the Owner needs to accomplish in order to fulfill the long-term function of that facility. In general, this phase of the work is referred to as the *Planning Effort* and most project Owners need to retain the services of one or more consulting engineering firms to help them accomplish that phase of the work no matter how the project is procured. For DBB, the results of the project planning effort would heavily influence the final design and be incorporated directly into the contract package, thereby providing the Owner with a high level of control over that aspect of the project. For a DB project, a considerable amount of thought must be expended by the Owner deciding which aspects of the design must be *prescribed* by the Owner as compared to those aspects that can be accomplished by the Design/Builder through *performance* criteria set within general design parameters.

In essence, the planning effort for a DBB tunnel is virtually identical to that for a DB project. In each case the project Owner must work together with a group of experienced designers and consultants to decide exactly what constitutes all of the important factors of the finished facility. In the final analysis, the project Owner wants the finished facility to serve its intended function and to operate in a proper manner for many decades, which is a primary objective of the Planning Effort.

It is also during the Planning Effort when the Owner begins to develop a risk profile for the project, which is a function primarily of ground conditions and third-party impacts. In order to accomplish the above the Owner must complete a comprehensive subsurface investigation and develop a detailed assessment of existing structures, utilities, and rights-of-way that could be impacted by tunnel construction. It is also during this period of time when the Owner must evaluate possible environmental issues and develop a list of community concerns.

It is beyond the scope of this paper to describe all of the work required to accomplish the Planning Effort, but suffice it to say that a good Planning Effort sets the stage for everything that follows for either a DBB or DB project. Below is a listing of items that need to be understood, incorporated or produced as a result of the Planning Effort:

- Applicable federal, state and local regulations, codes, permits, and design standards, real estate and right of way acquisitions
- The results of subsurface and environmental investigations as summarized in Geotechnical and Environmental Data Reports and, possibly, a Geotechnical Baseline Report for Design
- Required utility relocations and other structures that might be impacted during construction
- A listing of important third-party impacts including community impact mitigation requirements
- A comprehensive listing of risk-related issues
- Contract packaging and interfaces
- A detailed listing of the prescriptive requirements related to the finished facility and to the performance requirements necessary for the temporary structures

As stated above, the Planning Effort is identical no matter how the Owner intends to ultimately "procure" the project. Indeed, at this stage of the program it may very well be that the Owner has no preconceived notion of its procurement approach and, a significant part of the Planning Effort may be related to developing the approach best suited to project procurement. The culmination of the Planning Effort would then result in a recommendation based on the pros and cons of the various procurement approaches with the only difference relating to what the output for that effort would be called: i.e., for a DBB procurement it is called the Basis of Design Report (BODR) and for a DB approach it would be called the Tender Documents. (See Figure 1.)

DESIGN/BID/BUILD PROJECT PROCUREMENT

If a project Owner elects to procure a tunnel project utilizing DBB, then he starts with the BODR and proceeds to retain a group of well-qualified tunnel designers to finalize design and to produce the Final Contract Document for bidding. Indeed, for some Owners, the ability to select a team of designers that

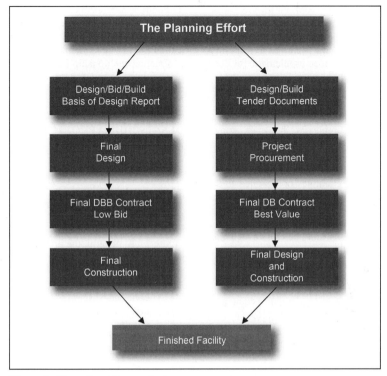

Figure 1. Flow chart for project procurement

will be the Owner's advocate and counselor throughout the project is a strong reason for deciding to use the DBB method of procurement. Clearly, if the tunnel's primary purpose is water or wastewater conveyance, then competent hydraulic engineers will be a key component of the design team. For example, the cost-effective design of a water tunnel subjected to significant *internal* pressure requires an expertise that is best not learned on the job. Similarly, highway engineering skills will be required for traffic management and life safety systems if the tunnel is for a highway, and railroad engineering experience is required if the project is for light or heavy rail. Indeed, the skill sets associated with the technical design elements of the finished facility could easily be perceived by the Owner as dwarfing the tunneling requirements.

According to conventional wisdom, the primary advantages of the DBB method of project procurement are as follows:

1. **Owner Control**—The Owner has complete control over design and can respond more fully to financial, political, and community concerns.
2. **Detailed Contract Document**—The Owner provides a detailed description of all of the

work that needs to be performed by the low-bid Contractor.
3. **Quality Control**—The Owner retains more control during construction, especially with respect to the required characteristics of the finished facility.

The generally accepted operative theory of DBB procurement holds that a good design and a solid set of contract documents result in a transparent, level playing field for qualified Contractors to evaluate the magnitude of its responsibilities and to accurately price the work and its associated risks. Thus, contractors bidding projects awarded on the basis of lowest cost are rewarded for being efficient and maximizing productivity. For simple, straight-forward projects this theory often rings true but as a project becomes more complex this level field gets bumpier.

It is also important to realize that a significant portion of the design effort required to execute a DBB tunnel project has been, and will continue to be, delegated to the Contractor. Even if a construction methodology has been called for in the Contract Document (the support of excavation, for example) the details of that method are usually left to the Contractor and its specialty engineers to determine. For example, in keeping with the aforementioned

SOE example; it may be appropriate for the Owner to specify that a diaphragm wall be *designed* and *constructed* by the Contractor. Owners generally favor the ability to call for a preferred SOE system or to exclude one that has been problematic for them in the past. However, by delegating this design to the Contractor, they are still receiving efficiencies that the Owner's design engineer would not be able to ensure.

With increased complexity, however, comes greater risks. Known or identified risks can be retained by the Owner and addressed with various payment schemes (unit price items or time-and-material items for example) or assigned to the Contractor to be absorbed in its price, but neither one of these risk allocation techniques is without peril for the Owner. Unit price pay items can be gamed by a Contractor by artificially pricing certain items higher if he expects to encounter them more often and/or by pricing certain items lower if he expects to encounter fewer of those items. One of the biggest criticisms of DBB is that, because the design takes place before the Contractor has been selected, the design may not reflect that Contractor's favored approach. One way to mute this criticism is for the Owner to specify alternate methods of construction from which the Contractor can choose in order to work with his preferred alternative.

Another approach in DBB that allows for Contractor input is to allow for so-called "Value Engineering" proposals. Under a typical Value Engineering scenario, the Contractor bids and is then prepared to construct the project as designed by the Owner's engineer, but, at some point after award, the Contractor submits a Value Engineering proposal seeking to change some aspect of the design. Presumably the VE proposal represents an approach that is faster, less expensive, and/or less risky to construct, which, depending upon the terms of the contract's VE provisions, those savings in time and/or money may be shared with the Owner. For tunneling projects, it is also important to make a design distinction as to whether the VE proposal affects only temporary structures or the finished facility.

And, finally, DBB projects, where the contractor selection is based purely on lowest price, have a natural tendency to develop into adversarial contests between the Owner and the Contractor. For low bid contracts, the bidding contractors are pressured to interpret the contract documents as favorably as possible and to try to shed as much risk as possible by alleging differing site conditions or by finding ambiguities in the contract document. This tendency by the Contractor results in a competing tendency by the Owner to protect its pocketbook from what it sees as unreasonable claims, and for the Designer to close ranks with the Owner around criticisms of its contract document. Hence, this adversarial relationship makes a fair and objective contract interpretation and the related decision-making process difficult. This potentially downward spiral of the contractual relationship based on the unsatisfactory resolution of claims has driven the tunneling industry to explore other procurement approaches such as Design/Build as is described below.

DESIGN/BUILD PROJECT PROCUREMENT

If a project Owner decides to procure a tunnel project utilizing DB, then he starts with a set of Tender Documents and proceeds with selecting a short-listed group of well-qualified Design-Builders (Designer and Contractor teams) that he believes will provide him with all of the design and construction services necessary to produce his finished facility (See Figure 1). In general, the primary advantages of DB are as follows:

1. **Single Point Responsibility**—The Owner contracts with only one entity which designs *and* constructs the proposed facility.
2. **Contractor Control**—Most often the Contractor is the lead partner for the DB team and is, therefore, able to focus the design on innovative concepts tailored to their means and methods that reduce both cost and schedule.
3. **Increased Coordination**—DB offers more opportunity for cooperation both between the Designer and the Contractor and between the DB team and the Owner.

However, and this is a big however, the vast majority of the DB literature is related to vertical (building) construction for private Owners as compared to infrastructure construction (especially tunnels) for public Owners. For instance, private Owners own the land on which their project is being constructed and are also free to make decisions rather quickly about various project components whereas Public Owners are constrained to do what is best for the entire community; whatever that might entail. And finally, as stated above, 100% of the tunneling project comes into contact with the ground and for a long tunnel in an urban area, with literally hundreds of third party impacts to overlying and adjacent structures, existing utilities, and rights-of-way for highways and railroads.

Hence, rather than prepare a final Contract Document for Bidding, the Owner plans to work closely together with a group of DB teams in order to provide a Final Contract Document for Best Value. In actuality, it is more difficult to produce a DB contract document as compared to one for bidding because the Owner is inundated with and must critically evaluate an enormous amount of information from each of the DB teams about the "best" methods for design and construction but also for cost and schedule issues and for a myriad of items associated with risk reduction, logistical considerations, third party impacts, and finally, the scope of required subsurface and environmental investigations. In addition, and most importantly, the Owner must also implement a procurement procedure that maintains the strictest of confidence for all discussions with each proposer.

The final outcome of all of the above is that the project Owner is able to choose a DB team that it believes will provide Best Value with respect to accomplishing all of the issues identified as a result of the Planning Effort. In addition, the project Owner also has the opportunity to work together with each team to prepare a Final Contract Document for "Best Value" that considers the specific needs of each DB team relative to project success. All in all, this is a significant accomplishment and should, as claimed by the proponents of DB, lead to a more cooperative and productive contractual relationship. In order to produce the contract package between the project Owner and the DB entity, the project owner can also make reference to several existing sources of information. For instance, the Engineers Joint Contract Documents Committee, the Design Build Institute of America, and the Associated General Contractors of America all have sample documents for this agreement that can be made applicable to a tunneling project. In general, each of these documents states that the Design/Builder has the right to rely on information accumulated by the project Owner during the Planning Effort such as the following:

- Property descriptions, easements and rights-of-way surveys
- Topographic and utility surveys and a detailed listing of potential private property impacts
- Zoning, deed, and land use restrictions
- A complete listing of governmental authorities having jurisdiction over the project, the approvals required by those authorities, and the entity responsibility for obtaining those approvals
- All environmental studies and assessments and any impacts defined by those studies
- All subsurface information accumulated to date for the subject project

After choosing one of the available standard formats as listed above, the project Owner will want to tailor that document to its particular situation and that process will involve a large number of studies and evaluations; the foremost of which involves what to do about the subsurface information. Without a doubt, the project Owner *must* provide the Design/Builder

with a comprehensive and accurate description of all of the subsurface *data* acquired at the subject site. For any substantial tunneling project in an urban area this will amount to many hundreds of pages of boring logs, laboratory test results, geophysical investigations, and information about previous construction activities; noting that nothing about the *data* should be withheld and/or disclaimed. A separate decision must also be made about the level of subsurface *interpretation* that the project Owner wishes to make available to the Design/Builder. There are no hard and fast rules associated with this decision, but the American Society of Civil Engineer's document entitled Geotechnical Baselines for Construction provides useful guidance.

Another important aspect of contract preparation for a Design/Build contract is the underlying attitude of the project Owner as relating to working with a Contractor. In general, the project Owner must believe that the *primary* goal of a DB contract document is to establish a fair and cooperative teaming arrangement that will actually *help* the DB Entity be successful in delivering a satisfactory finished project to the Owner. Some project Owners view the Contractor as an adversary that must be controlled and who is intent on filing a continuous series of claims for extra compensation. Although such a situation is possible, a primary focus of a DB contract document should be to provide the Contractor with all of the information that is necessary for him to be successful. Nothing is more important for the success of a DB project than the development of a close and cooperative working relationship between the project Owner and the Design/Builder.

CONSTRUCTION-PHASE ISSUES

One of the most important aspects of a successful DBB or DB tunneling project (second only to the subsurface investigation) is the performance of construction-phase services. During construction it is imperative that all activities in the field be observed and documented, both to make certain that the work is being performed in accordance with the contract document and to provide the factual project records that are necessary to help resolve claims and disputes. In general, and similarly to the Planning Effort, the construction-phase services for both DBB and DB projects are largely the same.

No matter how a tunneling project is procured, both the Owner and the Constructor need to have teams of experienced professionals in the field recording construction activities, monitoring ground behavior, and preparing thorough records about what is taking place in the tunnel on a daily basis. The most important items that need to be recorded are the details of the tasks being performed and the costs and schedules associated with those tasks. For a tunneling project the cost of a particular activity is largely related to the time required to accomplish that activity. Hence, schedule delays become costly and one of the most important roles of the Construction Management (CM) staff is to provide a continuous commentary about the cause and effect relationships associated with what is happening in the tunnel. It is also important for the CM staff to document and to help control environmental and community impacts especially for existing utilities and structures.

As has been discussed above, all tunneling projects can be divided into the work required to produce the underground space and the work required to construct the finished facility. The work required to produce the underground space is largely related to the Contractor's means and methods of construction which, for a both a DBB and a DB project, will be assigned primarily by the Contract Document to the Contractor and to the specialty designers and subcontractors working for the Contractor. The Owner does, however, have a right to make certain that all of the prescriptive requirements for the finished facility are being implemented in the field as indicated by the contract document. It is exceedingly difficult and expensive to repair errors in a finished facility that is located underground and, as a result, the details of the finished facility must be constructed as specified.

One of the biggest differences, however, between DBB and DB is a realization by the project Owner that the Design/Builder is largely in "control" of the design for both the temporary and final facilities and for the project Owner to think hard about when the Owner's involvement with construction transfers from observing what is happening in the field to directing and/or to interfering with the Design/Builder's responsibilities as defined by the contract. Overly aggressive involvement in design and construction activities by the project Owner is one of the leading causes of project delay and project disputes for a DB contract.

Despite all best efforts, however, it is inevitable that some claims and disputes will develop during construction and it is the firm belief of the authors of this paper that *every* tunneling project, both DBB and DB, should retain the services of a highly experienced and active Dispute Review Board (DRB). The DRB must be retained from the beginning of construction and must be allowed to visit the site and hold meetings at regular intervals throughout construction. One of the most expensive and potentially destructive events that can occur during tunnel construction is a significant project delay and

both the project Owner and the Constructor must do everything in their power to try to avoid that from happening. Hence, the DRB's ability to facilitate cooperation and decision-making in a timely manner is an extremely important component of project success for either a DBB or a DB tunneling project.

SUMMARY AND CONCLUSION

What has become increasingly clear over the past decade is that a project Owner need not commit to DBB or DB exclusively for either an entire infrastructure program or even for a single project. In the final analysis it is feasible to fit a tunneling project essentially anywhere into the spectrum between pure DBB and pure DB. Permanent facilities, structures, and systems are often most seamlessly procured if the project Owner offers the constructor a complete 100% design. However, those elements of the project that are highly dependent on the methodology or sequence of construction can be better planned, procured and constructed if the Contractor controls the design. For those projects where Owner control for the finished facility is paramount and where risks are relatively simple, straightforward, and quantifiable, then Design-Bid-Build would be a good choice. For a project with relatively straightforward finished facilities but highly complicated and/or difficult temporary facilities, then DB procurement could be the preferred alternative. Hence, with good planning and effective risk control, the tunneling industry now has many options for a project Owner to set the stage for successful project completion.

REFERENCES

Brierley, G.S, Corkum, D.H., and Hatem, D.J. 2010. Design-Build Subsurface Projects, 2nd Ed. Littleton, CO: Society for Mining, Metallurgy, and Exploration, Inc. (SME).

Loukakis, M.C., Smith, N.C., Gransberg, D.D. 2016. What Does the Case Law Say? www.geoinstitute.org.

Hatem, D.J. 2014. Design-Build and Public Private Partnerships: Risk Allocation of Subsurface Conditions. www.asce.org/geo.

Brierley, G.S. 2004. Design/Build Procurement Practices for Underground Construction. Washington, D.C., Underground Infrastructure Advanced Technology Conference.

Essex, R.J., 2007. Geotechnical Baseline Reports for Construction—Suggested Guidelines. American Society of Civil Engineers (ASCE).

Current Trends in Procurement Delivery of Major Tunnel Projects

Steven R. Kramer
COWI North America

ABSTRACT: This paper explores the current trends in tunneling procurement and delivery being used across the United States. This includes a comparison between conventional and alternative delivery methods as they have been applied to tunnel projects in transportation and water/wastewater. The paper will illustrate methods such as design-build, progressive/modified design build, construction management at risk and public-private partnerships and how alternative delivery methods have been successfully used and challenges faced as they are applied to a wide range of tunnel projects across the U.S. The author will share personal experiences and the experiences of other individuals and firms using actual projects constructed.

INTRODUCTION

The selection and use of alternative delivery methods continues to expand across the underground and tunnel industry within the U.S. More owners are evaluating the full range of alternative delivery methods in comparison to traditional delivery options. This growth is beyond a comparison between design-bid-build and design-build. Owners are seeking the method that will best allow them to deliver their projects on-schedule, on-budget while also meeting many institutional needs. This has created an environment where owners, engineers, contractors and financiers need to be able to objectively evaluate the advantages and disadvantages of a specific approach for the desired deliverable. The selection needs to consider the obvious impacts of cost and schedule but frequently also needs to evaluate softer factors such as the political climate, community/neighborhood issues and environmental issues. This may require a team of individuals with varying skills beyond design and construction professionals to assess these issues.

This paper will describe the continuum of delivery methods that can be applied to underground and tunnels projects, where success has occurred and the complexity of issues that should be addressed in choosing a method.

TYPES OF PROJECT DELIVERY APPLICABLE TO UNDERGROUND CONSTRUCTION

Figure 1 shows the major types of delivery methods and where public-private partnerships can be used with these methods. In the U.S., the full range of delivery methods has been applied for tunneling and trenchless projects. There is a movement towards alternative delivery for large tunnel projects, typically larger than $250 million in construction costs. For transportation tunnel projects in highways and for transit systems, various forms of alternative delivery are now the method of choice. This is a recent change as many transit tunnel projects especially in the metro New York region were or are being delivered with conventional design bid build methods (e.g., Second Avenue Subway, Seven Line and East Side Access). In comparison, conventional delivery is frequently the preferred method for water and wastewater tunnels especially at public agencies. Alternative delivery options are not the primary methods for nor as well-proven in water and wastewater tunnel projects in the U.S. and therefore their acceptance is slower. This may be changing with recent successfully delivery using design-build of tunnel projects at DC Water in Washington, DC and other water/wastewater agencies.

As shown in Figure 1, Public-Private Partnerships (P3s) can be incorporated into a variety of alternative delivery models. So far, we have seen the incorporation of availability payments and concessions used with design build for tunnel projects.

Today's tunnel market is steady throughout the U.S. but not experiencing explosive growth. For owner's, there is an adequate market supply where competitive bids are usually received from multiple contractors. Large international contractors has successfully entered the U.S. tunnel markets. The international players are frequently teaming with U.S. contractors and winning projects. Several of the international contractors have financing arms and low-cost access to capital that can supply funds from the same company who is conducting the construction. U.S. tunnel contractors will more likely seek a financial or banking partner when a P3 option is

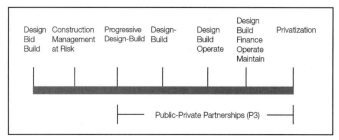

Figure 1. Major types of delivery methods for public infrastructure projects

being requested by the owner. The low-cost to capital due to current low interest rates may influence the use of P3s and allow entry by new contracting partners.

APPLICATION OF PROCUREMENT AND DELIVERY OPTIONS TO TUNNEL PROJECTS

Let's take a look at the various conventional and alternative delivery methods and where some of the methods have been recently used.

Design Bid Build (DBB)

In the past, conventional delivery was the preferred method for delivery of tunnel projects in the U.S. The agency/public owner and a design engineer would develop the design to 100% and then issue the project for bidding through a competitive procurement process. The lowest bidder, possibly prequalified was selected to construct the tunnel. The owner and engineer typically serve as a team with design-bid-build.

Recent Tunnel Examples

- *New York City Transit / East Side Access*: 11 miles of 22 ft. diam. transit tunnel at a cost of $10.8B.
- *New York City Transit / Second Avenue Subway*: 2 miles of 22 ft. diam. transit tunnel at a cost of $4.5B.
- *Northeast OH Regional Sewer District, Cleveland, OH/Euclid Creek Tunnel*: 18,000 ft. of 24 ft. diam. wastewater tunnel at a cost of $199M
- *Indianapolis, IN, Deep Rock Tunnel Connector*, Phase I: 42,000 ft. of 20 ft. diam. wastewater tunnel at a cost of $179M

The above examples illustrate the price variability based on location, geology and risk of the project. New York public agencies have typically used design-bid-build as their preferred method of delivery. Some studies indicate that tunnel costs in New York can be three to four times higher than other major cities (Rosenthal 2017 and City and State Magazine New York 2017). Recently, New York agencies have begun to explore alternative delivery for several major tunnel projects such as the Gateway Program for improving rail and transit infrastructure in New York City (The Gateway Program Development Corporation and Davis 2017) and the next phase of New York City Transit / Second Avenue Subway. It remains to be seen if alternative delivery can lower the cost and decrease the time of tunnel construction in New York City.

In the water and wastewater industry, these agencies have also preferred to use design bid build for delivery of their tunnel projects. Agencies in many Midwestern cities such as Cleveland, Columbus, Indianapolis and Chicago have a long and successful history of delivering wastewater tunnels to reduce combined sewer overflows (CSO's) using design-bid-build. Owners cite the following benefits using conventional delivery:

- More ability to influence/control the design
- Greater understanding of costs prior to advertisement
- Larger ability to transfer risk to contractor based on completed designs
- More familiarity with contractual terms and conditions
- Desire to specify/use new technology

As alternative delivery methods are further utilized, the relationship between owners and consulting engineers changes. In the U.S., consulting engineers were traditionally working as owner's representatives and assisting them with minimizing design liability and financial risk. With alternative delivery methods, some engineers may be working for constructors as designers and other engineers may be serving as an owner's representative. This can create some possible conflicts for a consulting engineering firm, where one group of employees provides services to the owner on one project and another group

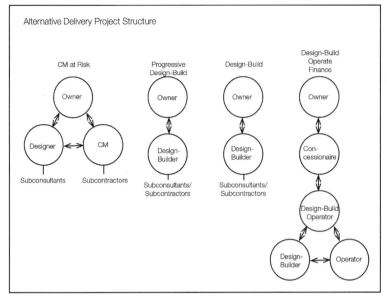

Figure 2. Alternative delivery project structures

of employees is working for a contractor an another project for the same owner. The contractor maintains his same relationship working for the owner/agency in design bid build.

Figure 2 illustrates the relationship between the owner and the various players for different delivery methods. For the construction manager at risk method, the owner may have the designer and the construction manager report directly to them. There are also variations where the construction manager and designer are part of the same team. In design-build, the design-build team reports to the owner. Typically the designer is a subconsultant to the design-builder. In some cases, the designer has elected to be a joint venture partner with the constructor. For progressive design-build and design build, the design-builder directly reports to the owner. When using public private partnerships (P3s), the concessionaire and/or the design-build operator reports to the owner. A concession is a P3 project delivery structure involving a lease or to be constructed public asset to a private concessionaire for a specified period of time (U.S. Department of Transportation 2010).

Owners have moved towards alternative delivery methods to achieve the following benefits:

- Single entity responsible for design, construction, technology integration and overall project delivery performance
- Shorter schedule and less schedule risk (reduced interfaces)
- Early cost certainty (less potential cost growth)

- Efficient administration for Owner
- Fewer potential disputes and change orders

It is important to recognize that in each of the structures shown in Figure 2 the relationships between the players is changing when compared to conventional delivery. The owner and the designer may be on competing sides if problems arise during project delivery. However, the relationships between all parties—owner, designer and constructor can be strengthened by using various forms of partnering.

Figure 3 shows the potential for saving time using alternative delivery. Some owners have achieved a time savings of 10% to 30% using alternative delivery. The time savings can be generated in two areas:

- Accelerated project schedule by not having to wait for completion of the design before hiring the design build team
- Ability to start construction sooner by determining early lead items and not waiting for the entire design to be completed.

CM at Risk (CMAR)

Constructor is selected based on qualifications shortly after the designer selection. The team provides pre-construction services including constructability, estimating, value engineering, and scheduling services during design. Typically between 50% to 60% stage of design development, CMAR provides

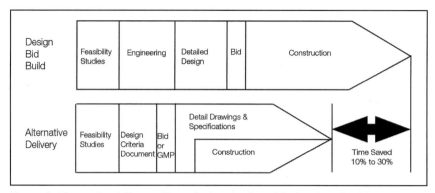

Figure 3. Alternative delivery allows the integration of design, long lead item procurement and construction activities

a price and schedule proposal based on an Owner defined scope.

CMAR has been used for many infrastructure projects ranging from transportation to wastewater treatment to public facilities. The application to CMAR has been more limited in underground construction. The method has a following in some parts of the U.S. more frequently for vertical construction.

Recent Tunnel Example

City of Atlanta / Water Supply-Quarry/Chattohoochee Tunnel. The project will convert an existing quarry into a 2.4 billion gallon raw water storage facility. A 24,000 ft. long, 13 ft. diameter tunnel will connect to the storage facility and thereby provide a back-up water supply for the City. The five-year $330M project was awarded in May 2014 (Klepal 2016). It is one of the few tunnel projects being delivered using CMAR. The CMAR takes the burden off the owner in managing and coordinating the project. The CMAR serves as an advocate for the City and at the best interest of the owner.

Progressive Design-Build

Owner involves the design-build team during the earliest stages of the owner's project development, ensuring they are part of the project team developing design solutions. Promotes collaboration between owner, designer and contractor. Typically the progressive design-build team will provide a lump sum price to the owner at say 50% to 60% of the design or the point when the design is enough understood where a lump sum price can be provided to the owner. The owner can elect to accept this price or issued the design to open procurement for bidding.

Recent Tunnel Example

City of Atlanta / Plane Train Extension at Hartsfield Airport. The project will extend the Plane Train

system approximately 700 ft. from the existing tunnel at the baggage claim area to the past the SKYTrain station. The tunnel extension provides additional flexibility to Plane Train system. Owner selected progressive design-build due to the complexity of the project, the desire to stay involved in the design development and influence the design solutions. Also, the preferred method of constructing the tunnel was not clear. A single qualification based procurement process was used and a contract was awarded to a design/construction team in 2017. The winning team was led by Clark Construction Group (Yamanouchi 2017).

The selected team will prepare the design to 50% to 60% completion and then negotiate a guaranteed maximum price (GMP). Project is estimated to cost $100M.

Design-Build (DB) Lump Sum

The risk and responsibility for providing both design and construction are with one entity under one contract with the owner.

Recent Tunnel Examples

DC Water/Clean Rivers Program. The program includes 15 miles of 23 ft. diameter sewers at an approximate cost of $2.6B (DC Water). The program is designed to reduce combined sewer overflows (CSOs) into the Anacostia and Potomac Rivers and Rock Creek in Washington, DC. DC Water used a program manager to develop the design to a selected stage and then issued multiple design-build contracts. The owner wanted contracts large enough to attract well-qualified and experienced contractors but also small enough to not be overly restrictive (e.g., bonding, insurance and management). The tunnel contracts were limited to about $300M in capital cost. As shown in Table 1, some of the tunnel contracts were completed and others are in construction.

Table 1. Delivery method for selected U.S. tunnel projects recently constructed or underway

Owner/Project	Contractor	Delivery Method	Tunnel	Location	Approx. Capital Cost/Completion Year*
Chesapeake Bay Bridge Tunnel District / Parallel Thimble Shoal Tunnel	Dragados & Schiavonne	DB	5,700 ft. long, 42 ft. Diam. 2 lane highway tunnel	Cape Charles, VA	$756M / 2022
Citizens Energy / Deep Rock Tunnel Connector	Shea-Kiewit	Conventional[†]	CSO, Phase 1: 42,000 ft., 20 ft. Diam.	Indianapolis, IN	$179M / 2017
City of Columbus/ OARS (2 contracts)	1. Kenny / Obayashi 2. Trumbull	Conventional[†]	CSO: 23,300 ft., 20 ft. Diam.	Columbus, OH	1. $264M 2. $90M 2017
DC Water / Blue Plains Tunnel	Traylor, Skanska, Jay Dee	DB	CSO: 24,300 ft., 23 ft. Diam.	Washington, DC	$330M / 2016
DC Water / Anacostia River Tunnel	Salinini Impregilo, SA Healy & Parsons	DB	CSO: 12,500 ft., 23 ft. Diam.	Washington, DC	$254M / 2018
DC Water / First St. Tunnel	Skanska, Jay Dee	DB	CSO: 2,700 ft., 23 ft. Diam.	Washington, DC	$158M / 2016
FDOT / Port of Miami Tunnel	Bouygues	P3 (Availability Payment)	Twin 4,200 ft. Highway, 39 ft. Diam.	Miami, FL	$1.062B / 2014
LA MTA / LA Metro Regional Connector	Skanska—Traylor	DB	Subway: 1.9 miles, 22 ft. Diam.	Los Angeles, CA	$1B / 2021
LAMTA / LA Metro Purple Line	Skanska—Traylor-Shea	DB	Subway: 3.9 miles, 22 ft. Diam.	Los Angeles, CA	$1.6B /2024
NEORSD / Euclid Creek Tunnel	McNally/Kiewit	Conventional[†]	CSO: 18,000 ft., 24 ft. Diam.	Cleveland, OH	$198.6M 2015
NEORSD / Dugway Storage Tunnel	Salini Impregilo	Conventional[†]	CSO: 15,000 ft., 24 ft. Diam.	Cleveland, OH	$153M / 2019
NYCT / 2nd Ave. Subway	Shiavonne, Shea, Skanska	Conventional[†]	Subway: 2 miles, 22 ft. Diam.	NY, NY	$4.5B / 2016
NYCT / East Side Access	Dragados, Judlau	Conventional[†]	Subway: 11 miles, 22 ft. Diam.	NY, NY	$10.8B / 2023
Southern Nevada Water Authority/ Lake Meade Intake & Tunnel	Salini Impregilo & SA Healy	DB	15,000 ft., 23 ft. Diam.	Las Vegas, NV	$447M / 2015
SFMTA / Central Subway	Barnard, Salini Impregilo, SA Healy	DB	Subway: 8,230 ft., 20 ft. Diam.	San Francisco, CA	$233.5M / 2016
VDOT / Midtown Tunnel	Skanska, Kiewit, Weeks	P3 (DBFOM)	3,700 ft. Highway, 42 ft. wide	Portsmouth, VA	$2.1B / 2017
WSDOT / Alaskan Way Viaduct Replacement Tunnel	Dragados	DB	2 miles, 57.5 ft. Diam.	Seattle, WA	$4.25B / 2017

* Operational date may differ from tunnel completion date.
† Conventional refers to the use of the design-bid-build delivery method.

Chesapeake Bay Bridge Tunnel (CBBT) District/ Parallel Thimble Shoal Tunnel. Owner selected design-build to allow innovation in design through an alternative technical concept (ATC) process and increase price competition. Using the ATC process, design-build teams can meet with the owner on a confidential basis to explore alternative design and construction methods to improve delivery, cost and/ or operation of the asset. The Parallel Thimble Shoal Tunnel Project will construct a new two-lane tunnel under Thimble Shoal Channel. When complete, the new tunnel will carry two lanes of traffic southbound and the existing tunnel will carry two lanes of traffic northbound. The new tunnel will be 5,700 ft. long with an external diameter of 42 ft. The winning bid was $756M for this 5-year project and construction began in October 2017 (CBBT).

On the west coast, there are currently several major tunnel projects being constructed using design- build. This includes the expansion of the Los Angeles and San Francisco metro systems and the replacement of the Alaskan Way Viaduct with a new highway tunnel. These complex tunnel projects are demonstrating and furthering the use of alternative delivery.

Another major tunnel project called California Water Fix is evaluating delivery methods. This program will require several TBM's and multiple projects. The agencies involved are determining if alternative delivery methods will help them control costs, maintain schedule and better manage its risks.

Design-Build-Operate-Finance-Maintain/Public Private Partnerships/Concession

This is a contractual arrangement between a public agency and a private sector entity. Through this agreement, the skills and assets of each sector are shared in delivering a service or facility for the use of the general public. In this form of project financing, a private entity receives a concession from the public sector to finance, design, construct, own, and operate a facility

Recent Tunnel Examples

Virginia Department of Transportation (VDOT) VDOT / Elizabeth River Crossing (Parallel Midtown Tunnel): Project comprised of a new two-lane 3,700 ft. of immersed tube tunnel under the Elizabeth River and adjacent to the existing Midtown Tunnel (Rush 2016). The new tunnel provides two travel lanes in the westbound direction and thereby allows the existing tunnel to be converted to two travel lanes in the eastbound direction. The tunnel project is one part of five components involving three facilities in the Hampton Roads region of Virginia. Total project price was $2.1B which included a design, build,

finance, operate and maintain concession. The project is part of a 58-year public private partnership with VDOT. The tunnel project was completed and opened to traffic in 2016.

Florida Department of Transportation (FDOT)/ Port of Miami Tunnel: Project included twin 4,200 ft. long 39 ft. diam. tunnels. The owner selected a P3 with an availability payment. Availability payments are reimbursements made by a public entity to a private concessionaire for its responsibility to design, construct, operate and/or or maintain a facility for a set period of time (Istrate and Puentes 2011). FDOT did not have current funds available to complete the project. Using the P3 approach, FDOT transferred the responsibility to design, build, finance, operate and maintain the project to the private sector. The equity for the project was divided among several banks (90%) and the design-build contractor. The owner also selected a unique tiered risk ladder to help manage unforeseen geological conditions. The geologic risk was shared between the owner and design-build contractor.

Construction of the Port of Miami tunnel was completed in 2014 at a cost of $1.062 billion (Kramer 2016). The portion of costs for design and construction of the tunnel was $667 million. Upon opening of the tunnel, FDOT began availability payments to the concessionaire for a 30-year period until 2044. These payments will be contingent upon actual serviceability and quality of the tunnel. If the tunnel is closed or the road is in bad condition, part or all of the payment for that period may be withheld. A unique element of this project was no toll will be collected for use of this roadway. Most transportation P3s utilize some type of toll collection to repay the capital costs.

Privatization

The owner elects to sell the public asset to a private entity for construction, future operation and maintenance. As occurred with the Chicago Skyway or the Indiana Toll Road, the outright sale of tunnel assets has not yet occurred.

Table 1 summarizes several of the currently in construction and recently completed tunnel projects across the US. It illustrates that owners are using the full array of procurement and delivery methods to design and construct tunnels. As discussed earlier, the data in Table 1 tends to indicate that larger projects, especially transportation tunnels are moving towards using alternative delivery methods while water/wastewater tunnels are still primarily being procured using conventional delivery.

Table 1 also indicates that owners are using all types of delivery methods for constructing tunnels and P3s are just starting to be used. It is noteworthy that the contracting community has adapted to the changes in the marketplace and contractors are applying both conventional and alternative delivery for tunnel projects. The data shows that the international tunnel contractors are playing a significant role in the construction of U.S. tunnels.

The concept of Early Contractor Involvement (ECI) has been applied to many of the alternative delivery methods. With ECI, the contractor "buys-in" to the design solution to obtain the best quality and value with single-source accountability. The project achieves a shorter project schedule than with design-bid-build with more reliability and predictability. A collaborative atmosphere is created that promotes equitable risk sharing and management. The owner needs to be willing to relinquish some control over value, quality, function and other project objectives.

There are new variations being evaluated for the next round of major tunnel projects to be procured. Prior to some of the recent tunnel projects using forms of P3, tunnel projects were viewed as too complex and possibly too risky to use this delivery method. With the recent technical successes of the Port of Miami Tunnel and the Elizabeth River Crossing, concession formulas are now being evaluated for other major tunnel projects. The access to new capital would be an approach to allow some tunnel projects to move forward into construction during this time of limited funding at public agencies.

At the Virginia Department of Transportation (VDOT), they considered using a procurement approach for the Hampton Roads Bridge-Tunnel that would allow bidders to evaluate both design-build and design-build with some of concession. After evaluation of various delivery models, it was determined that a concession type of procurement using toll revenue would not be able to finance the project. This $3.3 billion project includes the expansion of the current bridge-tunnel and its approaches from four lanes to six lanes in both directions from the I-664 interchange to the I-564 interchange, with a new, three or four lane bridge tunnel built to carry traffic eastbound (Hampton to Norfolk, VA). VDOT recently released a RFQ (Request for Qualifications) under its Public-Private Transportation Act to procure the project using the design-build delivery method (VDOT 2017).

The Gateway Tunnel program in metro New York is also looking at using a concession method for constructing the tunnels. This $30 billion program includes constructing a new tunnel beneath the Hudson River, replacing the Portal Bridge and rehabilitating the existing tunnel beneath the Hudson River. This project is often cited as one of the top transportation projects in the U.S. The Gateway program is a critical link to provide train and commuter rail service in the northeast. The project is a joint effort between Amtrak, Port Authority of New York New Jersey, New Jersey Transit and the federal government. The costs have continued to increase as preliminary engineering is performed. Some form of P3 may allow the project to move forward without funds from the federal government or only partial funding from the federal government.

The earlier described projects and Table 1 are indicative of the state of delivery options for tunnel projects in the U.S. It is clear that all types of procurement and delivery options are now being considered by public owners. There remains a preference to utilize conventional delivery in some parts of the country and for water/wastewater tunnels. The movement to alternative delivery for transportation tunnels has occurred, especially for the larger and more complex projects. At the same time, some transportation agencies are evaluating how to include P3 options as part of their programs as way to maximize financial resources especially when toll options are available. The tunnel industry has now followed other parts of the infrastructure business where design-build first took hold and then progressed to the full range of alternative delivery options.

CONCLUDING THOUGHTS

To summarize, here are the key trends in U.S. tunnel procurement and delivery:

- Owners are now evaluating the full range of delivery options from conventional to all types of alternative delivery including the use of private financing
- With alternative delivery options, the risks of delivery and use are shifting to the designers and contractors. The costs of these risks are now being added into bid prices.
- In the past 10 years, the use of alternative delivery has substantially increased especially for transportation options. Many if not most of the highway and transit projects are now being bid using one of the alternative delivery options.
- For water and wastewater tunnels at public agencies, there is a mix of conventional and alternative delivery options.
- Designers and contractors have adapted their approaches to the new delivery options.

REFERENCES

Chesapeake Bay Bridge Tunnel District (CBBT) website, www.cbbt.com/parallelthimbleshoaltunnel.

City & State New York (NY) magazine and website, Oct. 2017.

Davis, Jeff, "What is the Gateway Program," ENO Transportation Weekly, June 30, 2017.

DC Water, Clean Rivers Project website, www.dcwater.com/clean-rivers-project.

Klepal, Dan, "Blasting Starts for Atlanta's Five-Mile Tunnel Project to Protect Water," The Atlanta-Journal Constitution, March 29, 2016.

Kramer, Steven R. and Nicholas, Paul, "Procurement and Delivery Strategies to Increase Competitiveness on Tunnel Projects," Proceedings of RETC 2017, San Diego, CA, June 4–7, 2017.

Rosenthal, Brian M., "The Most Expensive Mile of Subway Track on Earth," The New York Times, December 28, 2017.

Rush, Jim, Publisher/Editor, "Tunnel Updates," Tunnel Business Magazine, Feb. 2016, pg. 44.

Rush, Jim, Publisher/Editor, "Second Lane of Elizabeth River Opens to Traffic," Tunnel Business Magazine, Oct. 2016, pg. 9.

The Gateway Program Development Corporation website, www.gatewayprogram.org.

Virginia Department of Transportation, Request for Qualifications Relating to the I-64 Hampton Roads Bridge-Tunnel Expansion Project, Dec. 15, 2017.

Yamanouchi, Kelly, "Hartsfield-Jackson Begin Work on Plane-Train Tunnel Expansion," The Atlanta-Journal Constitution, July 26, 2017.

Fort Wayne Utilities Three Rivers Protection and Overflow Reduction Tunnel—Project Bidding Successes and Lessons Learned

T.J. Short and Mark Gensic
Fort Wayne Utilities

Leo Gentile and David Day
Black & Veatch Corporation

ABSTRACT: The $200 million Three Rivers Protection and Overflow Reduction Tunnel (3RPORT) is the primary element in the City of Fort Wayne's Consent Decree program known as Fort Wayne Tunnel Works. 3RPORT is the largest project undertaken in city history. The 3RPORT project was successfully bid and awarded in 2017 and employed key strategies to complete the design and bid the project during a window between other major tunnel projects in the USA. Key strategies included early engagement, favorable bidding of contractors through information sharing, outreach, and one-on-one discussions. Other strategies included prequalification, accelerated design, closely monitoring tunnel construction activity, project lifecycle risk management; and actively coordinating design, program management, legal and specialty consultant teams.

HISTORY

The City of Fort Wayne began the implementation of its combined sewer overflow (CSO) reduction program in 2008 as part of a Consent Decree signed with the United States Environmental Protection Agency, the Indiana Department of Environmental Management, and the U.S. Department of Justice. The Three Rivers Protection and Overflow Reduction Tunnel (3RPORT) is the primary improvement project in the Consent Decree known as Fort Wayne Tunnel Works and is the largest project that the City has undertaken in its history. Fort Wayne Utilities (FWU) began preliminary planning for the project in 2012 and initiated design on the project in 2014. Per the deadlines listed in the Consent Decree, the project is required to be complete and in operation by December 31, 2022.

PROJECT OVERVIEW

The City of Fort Wayne is located along the St. Joseph, St. Marys and Maumee Rivers and covers 110 square miles with over 25 percent being served by combined sanitary and storm sewer system. FWU CSO solutions included a three-prong strategy:

1. Reduce Through Separation
2. Collect More
3. Treat More/Store More

Within the combined sewer system area, there are a total of 41 outfalls that contribute CSO discharge into the three rivers during rain events. The goal of the Long-Term Control Plan is reduce the number of overflow events per year on all of Fort Wayne's three rivers by 2025. The CSO discharging into the St. Joseph River will be reduced to one overflow during a rain event per typical year and four overflow events per typical year into the St. Mary's and Maumee Rivers. Overall, there will be a 90-percent reduction of overflow events within a typical year. Once constructed, 3RPORT will be a 25,000-foot long 16-foot finished diameter tunnel constructed in rock about 175 to 200 feet below grade.

The key to FWU's CSO program was to convey combined sewage to its Water Pollution Control Plant (WPCP) located east of downtown (Figure 1). The City had constructed three large impoundments, or wet weather ponds (WWP), adjacent to the WPCP in the early 1920s to collect and retain CSO flows for treatment. The three ponds provide about 350 million gallons of storage. The City also expanded capacity of its WPCP in 2016 to 90 MGD in anticipation of higher wet weather flows. 3RPORT will convey flows from CSOs to the WWP for treatment by the expanded WPCP.

CAPTURING THE BIDDING WINDOW

FWU and its engineers recognized that in order to maximize potential contractor interest and competition in 3RPORT while still completing the project by the Consent Decree deadline, it was advantageous to steer the bid period away from other major tunnel

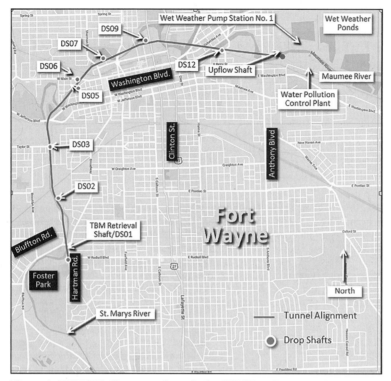

Figure 1. 3RPORT alignment, drop shafts, WPCP and wet weather ponds

projects in the Midwest and eastern USA. And while not a small CSO program, 3RPORT is a single project. Other CSO programs included multiple large projects and longer construction durations creating more attractive construction packages. Late 2016 and early 2017 appeared to be a window for targeting 3RPORT bids as shown Figure 2.

Once the decision was made to target this date range, the focus became completing the design. A decision was made to focus on the tunnel and shafts package and defer the deep dewatering pump station design.

Final planning and design was accelerated by several months to advertise bids in a window between other Midwest tunnel projects in order to increase interest and maximize qualified contractor competition.

GARNERING CONTRACTOR INTEREST

Once the design approach was clarified to focus on a tunnel and shafts package, FWU focused on attracting the contracting community to take a look at 3RPORT. FWU developed a public awareness and contractor outreach program, Tunnel Works, well in advance of bid advertisement. In mid-2015, FWU began publishing a newsletter available online. A contractor open house was held in January 2016

inviting national tunneling contractors, local contractors and other service providers to hear details about the project, view preliminary drawings, and view the alignment and shaft locations accompanied by FWU and its engineering team. A key benefit from holding an open house was the opportunity for national and international tunnel contractors to meet local contractors and vendors early on. The City of Fort Wayne is keenly interested in maximizing local participation by contractors and vendors to benefit the local economy. Eleven tunnel contractors and dozens of local firms attended the open house.

PREQUALIFICATION

Following the contractor open house, FWU prepared a request for qualifications to engage contractors whose experience most closely matched the 3RPORT project. This would provide a transparent process to focus the bidding pool and encourage shortlisted bidders to thoroughly review, understand and ultimately bid the project. FWU believed that obtaining about five qualified bids would constitute a shortlist of qualified contractors and encourage competition.

A contractor prequalification process was used to prequalify prospective tunnel contractors to ensure only qualified contractors would bid, information could be gathered and shared efficiently

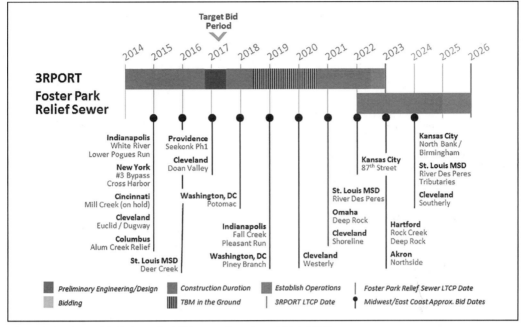

Figure 2. Estimated bid dates for USA tunnel projects 2014–2025

during final design, contractors would prioritize their bidding of FWU's project, and that the project could be awarded more efficiently after bids were received.

The 2016 Pre-Qualification Schedule was as follows:

- March 22 Request for Qualifications Issued
- May 18 Statement of Qualifications (SOQ) Due Date
- June 17 Shortlist of Contractors List Announced

To qualify for the shortlist, contractors were required to submit a Statement of Qualifications, meet pass/fail qualifications and demonstrate a minimum level of knowledge, skill, and experience with key personnel, similar successful projects to 3RPORT and timeliness. Pass/fail criteria included bonding capacity and minimum experience with similar size rock tunnels. Scored evaluations included firm's past project experience and performance; team experience and performance; legal status and responses to five open ended questions.

Staff Experience

Specific Criteria for personnel were established with preferred length of experience specified. Positions to be provided and preferred minimum experience (Table 1).

In order to further evaluate experience, additional information was requested for key supervisory

Table 1. Position preferred minimum experience

Position	Years' Experience
Project Manager	10
Project Superintendent	15
Shift Superintendent (Walker)	5
TBM Operator	5
Quality Manager	5
Safety Manager	5
Field Engineer	5

staff: project manager, project superintendent and shift superintendent. At least three project summaries demonstrating the individual's experience in the role for which they were listed were requested.

Similar Project Experience

Prospective contractors were required to provide references and descriptions for at least three projects ongoing or completed in the last 15 years and having the characteristics listed below. If a project was completed as a joint venture (JV), the firm must have been the joint venture lead to qualify.

To achieve the maximum points possible, reference projects must exceed the minimum criteria and clearly demonstrate the project's keys to success, safety program and overall contractor performance. Reference projects submitted were required to include the following minimum characteristics:

- Rock tunnel construction
- Over 10,000 feet in length (at least 2 of 3 projects)
- Moderate to heavy groundwater inflow encountered
- Lining system (all types)
- Finished diameter of 14-ft or greater

The descriptions were also required to include: project name, award date, completion date contract value at award and at completion, components of the work self-performed, project manager, superintendent and safety manager, owner, and resident engineer. A detailed project description was to also include project characteristics (number of shafts, depth, lining type and type of TBM), the contractor's approach to the project and keys to the project success, Safety program description and performance. FWU also requested which major project elements were self-performed and which were subcontracted. Lining type, indicate cast-in-place, pre-cast or not applicable. Other evaluation criteria included the following:

- List of Rock Tunnel Equipment Owned
- Summary of Project Bonding Capacity
- Litigation and Incomplete Project History

Open Ended Questions

Six open ended narrative questions were posed under the prequalification and later used as a basis for one-on-one discussions with shortlisted contractors.

Question 1: Describe the approach you would take to construct an 18-ft bored diameter tunnel given the geological profile provided and the potential for high groundwater inflow rates.

Question 2: Describe steps City Utilities can take to best partner with your firm for a successful project. Give examples from past projects of effective partnering tools, and tools that fell short of expectations.

Question 3: Describe your approach to engaging local contractors and the local workforce in the tunnel Construction. Give examples from past project of approaches and outcomes.

Question 4: Explain your management approaches to a project the size of 3RPORT. Include an organizational chart that identifies the reporting structure and lines of communication and give examples from past project on the relationship between the management approach and the project result.

Question 5: Provide information based on past projects where risk assessments and method statements have been utilized to manage and/or mitigate risks during construction. Give examples of successful risk management approaches from past projects. Contractor may include example(s) of contractor's risk registers during construction and monthly updates of same.

Question 6: Provide information on past projects which have compromised baseline schedule as a result of construction problems, i.e., shaft and/or tunneling failures. How did the situation occur? How was the situation corrected and did it affect the project completion milestone?

Evaluation

Evaluation of contractor responses consisted of three areas: mandatory, scored and non-scored. Each area is described below.

Mandatory Evaluation

The Mandatory Evaluation is a pass/fail assessment and failure to meet any of these mandatory criteria resulted in a scoring of nonresponsive. These criteria included the following:

- Submitted the Statement of Qualifications in the prescribed format.
- Completed the Contractor Qualification Statement in full including the open-ended questions.
- Completed a minimum of 30,000 feet of tunneling in rock with diameters of 14-ft or greater over the past 15 years as primary contractor or Joint Venture lead.
- Has had no surety complete work on its behalf in the last 15 years.

Scored Evaluation

The Scored Evaluation was based on the quality of the answers provided in the Statement of Qualification and interviews conducted with provided references. Applications were scored by a committee made up of some design team members using the following maximum possible points:

- Contractor Past Project Experience and Performance—30 pts

- Supervisory Team Experience and Performance—35 pts
- Open Question Responses—35 pts

All references listed were contacted and information contained in the Application confirmed. Each reference was questioned with regard to the Contractor's overall performance, a discussion of problems with performing the work, organization, adequacy of equipment, timeliness, change orders, and quality of personnel.

Non-Scored Evaluation

Non-Scored Evaluation uses the remaining information to better understand the tunnel contracting environment prior to bidding the project and inform the contractors of FWU's project delivery. Non-scored items included the following:

- Contractor ownership information
- Supporting team experience
- List of rock tunneling equipment
- Bonding capacity
- Litigation and project completion history
- Any additional information provided in appendices

Contractor Submittals

FWU received eleven statements of qualifications (SOQs) in May 2016. Two of the eleven SOQs were submitted as JVs with each of the two firms exchanging the lead. A committee lead by FWU reviewed and evaluated the SOQs. Scoring was applied according to the criteria above and a shortlist based on results of the SOQ review and evaluation. Eight contractors/JV teams were shortlisted. Eventually, two of the shortlisted contractors declined to submit bids. The two contractors who had submitted as JV then decided which was going to be the lead, eliminating another shortlisted team. FWU was then left with five shortlisted contractors.

DISCUSSIONS WITH SHORTLISTED CONTRACTORS

The 3RPORT project requires construction within very challenging geotechnical conditions. Bedrock consists of limestone and dolomite with rock mass ratings ranging from 43 to 73, or fair to good rock. However, high groundwater pressures up to 6.5 bar and tunnel heading flush flows from 3,000 to 10,000 gallons per minute were baselined. These ground conditions require that complex alternatives be considered to mitigate the risks relative to the project cost and schedule. For this project, the owner, the design professional and the program manager agreed that pre-bid consultations with the prequalified contractors would have significant benefits as design alternatives were evaluated. In addition, it was believed that consultations with the prequalified contractors would provide the bidders with some insight to the methodology and design considerations prior to receiving the bid documents.

During final design, one-on-one interviews with shortlisted contractors were conducted to open a dialog on constructability, risk management, tunnel boring machine selection, shaft and liner construction techniques and schedule. These discussions were held separately with representatives of each contractor/JV, FWU, and the design engineer. FWU described these discussions as invaluable means to talk openly about the project, anticipated challenges, constructability issues, and schedule. Information shared during the discussions was held in confidence and was not shared with other contractors. Specific topics included the following.

- Managing significant groundwater inflows under high hydraulic head conditions
- Potential lining methods—one pass vs. two pass
- Grouting for groundwater control and back-filling of lining
- Slurry TBM experience
- Shaft construction—support of excavation and final lining alternatives
- Measurement and payment for grouting and other items including potential baselines
- Anticipated electrical power needs
- Muck management

The overall results of these discussions were incorporated into the contract documents where appropriate.

OTHER TOOLS

FWU employed other tools to further understand the project risks and develop mitigation measures.

Pre-Excavation Grouting Consultant

The design engineer discovered during the geotechnical program that the geology of the project area consists of porous limestone and dolomite that contains a prolific aquifer. Grouting for shafts and the tunnel and adits will be critical to the constructability. The draft contract documents were reviewed by and comments received from an independent grouting consultant. The comments were discussed and incorporated into the technical specifications and measurement and payment sections.

Functional baselines were developed for addressing uncertainties in the efforts that would be required to complete pre-excavation grouting for shafts and adits. Baselined quantities were required to be included in the general construction lump sum as well as additional bid items with specified quantities. The additional quantities were provided to compensate the contractor for efforts that are required beyond the prescribed base effort.

TBM Selection

FWU and its design team reviewed options for TBM in the porous bedrock under high head conditions. Initially both non-pressurized shielded TBM and slurry TBM were considered as options for mining the tunnel. Following discussions with contractors, the design engineer and program manager, FWU decided to specify a slurry TBM as a risk mitigation measure due to anticipated high groundwater quantities.

Based on previous projects that utilized slurry TBMs in bedrock with significant groundwater inflows, it was understood that interventions for cutter head inspection or maintenance activities may be challenging and difficult to estimate. The effort to complete the interventions was baselined by requiring that bidders include a baselined effort in the lump sum and an additional quantity of intervention effort was included as contingency item. This strategy was similar to the strategy regarding pre-excavation grouting and was intended to provide some certainty to the bidders and a fair and equitable method to compensate the contractor during construction for its efforts.

Construction Risk Register

Risk registers were developed and used during design and procurement to help prioritize attention to risks to the project. The construction risk register has been developed by the contractor and will be reviewed during quarterly meetings. The intent of the meetings and register review is to identify and discuss construction risks to help ensure that each risk has been mitigated or minimized to the extent possible by FWU and the contractor.

Total Probable Maximum Loss Study

FWU retained legal and technical experts to assist with risk, insurance, and contract review. The consultant reviewed the major risks to construct the 3RPORT project with the objective of developing a list of scenarios associated with key insurable risks. The probable maximum loss (PML) was estimated based on these scenarios. The PML estimates are associated with a 90% confidence level (P90), which means that if a given event occurs, there is a 10% probability that the estimate loss could be exceeded.

The costs provided in the amendments based on 50% (most likely) and 70% confidence levels to show the range of costs depending on the level of confidence required. Different risk scenarios were considered for a range of areas of the projects, as the hazards are different for the shafts, connecting adits, starter and tail tunnels, and bored tunnel. The risk scenarios examined include pump station shaft collapse due to support of excavation failure, issues associated with difficult ground during rock tunneling, TBM fire and internal inundation from the tunnel face and external flooding. The PML was used as a basis for the contractor's insurance.

Disputes Review Board

A formal Disputes Review Board (DRB) has been assembled to review construction progress and evaluate potential claims. As is typical for projects of this nature, the DRB consists of members each selected by FWU, the contractor and one independently. The DRB will meet quarterly and have representatives from FWU, design engineer and construction contract manager in attendance.

LESSONS LEARNED

Some of the lessons learned include starting the contractor information and outreach process early, benefits of prequalification, closely monitoring the tunnel construction industry, actively managing risks, attending conferences to establish contacts and raise visibility, value of one-on-one discussions with contractors and actively coordinating design, program management and specialty consultant teams.

RESULTS

The 3RPORT project was successfully bid and awarded in 2017 and construction is ongoing towards meeting the Consent Decree deadline. Through its outreach and visibility program, FWU received a high level of interest from the national and international tunnel contracting community. The prequalification process had been used before by the FWU staff, but due to the size and complexity of the project, the process was more challenging than any prequalification process that was done previously. However, at the end of the process, it was felt that the contractors were better prepared for bidding. Cost competitive bids were received from multiple highly qualified contractors and FWU was able to move forward through the award and construction agreement approval stages efficiently. The project is under construction and an update on progress will be included in the presentation associated with this paper.

Interlake Tunnel—A Future Design-Build Project

Ronald D. Drake
EPC Consultants, Inc.

ABSTRACT: The Interlake Tunnel is a planned water conveyance tunnel between two reservoirs in Monterey County, California to provide flood control and additional water supply for the "salad bowl" agriculture industry. The tunnel is under development by the Monterey County Water Resources Agency and will be two miles long with an inside finished diameter of 10-feet. The project will be constructed through the highly fractured Monterey shale formation. This paper describes the technical design and construction details of the tunnel and adjacent spillway modification project, and the project delivery plan and schedule.

INTRODUCTION

Project Overview

The Interlake Tunnel is a proposed design-build, soft ground, 2-mile-long tunnel located in Monterey and San Luis Obispo Counties, approximately 27 miles northwest of Paso Robles, California. The concrete-lined tunnel will have a finished diameter of 10 feet, and will be constructed to transfer wet-year flows from the Nacimiento Reservoir to the adjacent San Antonio Reservoir providing added flood control and increasing the net total storage available in the reservoirs. In addition, a separate project involves raising the maximum height of the spillway at the San Antonio reservoir to provide approximately 45,000 acre-feet of additional storage. The Interlake Tunnel Project and San Antonio Spillway Modification Project are two separate and distinct projects but are being developed concurrently under a single financing plan for construction, operation and maintenance.

The Interlake Tunnel has been mandated by Monterey County Water Resources Agency (MCWRA) to be procured using design-build contracting, in compliance with California Assembly Bill 155 (AB155), for the final design and construction of the project.

The San Antonio Spillway Modification Project will be procured as a design-bid-build contract. The spillway modification project is dependent on the construction of the Interlake Tunnel Project, without which the spillway modification is unnecessary.

History

The Interlake Tunnel has been under consideration since 1978 to manage flood-control releases from the Nacimiento Reservoir and to utilize available storage capacity in the adjacent San Antonio Reservoir.

The project has continued to be a top regional priority and was identified in the Monterey County Water Resource Agency Capital Facilities Plan prepared by Boyle Engineering in July 1991[1]. No definitive action was taken to advance the project until May 2014 when a group of farmers in the Salinas Valley, known as the "Salad Bowl of the World," revitalized the urgency for the tunnel project due to the heightened awareness of water needs resulting from the multi-year drought.

Project Team and Financing

The projects are now under urgent development by MCWRA, requiring aggressive, multi-faceted efforts to prove feasibility, obtain environmental clearances and permits, procure interim and permanent financing, procure design and construction services, and efficiently navigate the regulatory and political challenges. Consultants supporting MCWRA in the development of the projects are EPC Consultants, Inc. (Program Manager), Horizon Environmental (Environmental and Permitting), McMillen Jacobs Associates (Tunnel Preliminary Engineering and Spillway Modification Design), GEI Consultants (Geotechnical Investigations) and Amec Foster Wheeler (Hydrologic Modeling).

Interim funding for development of the projects consists of $3 million from Monterey County and a $10 million grant from the California Department of Water Resources. This interim funding is advancing the projects through preliminary engineering, environmental clearance and permitting. Permanent financing will be secured via Proposition 218 property tax assessments on lands receiving benefits from the added water storage and flood control provided by the projects.

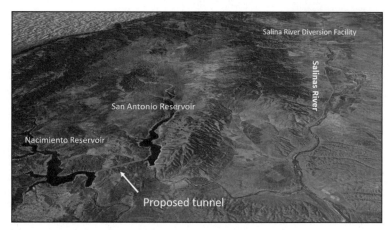

Figure 1. Location of reservoirs and proposed tunnel

PROJECT OPERATING FUNCTION

The tunnel project will augment the existing surface water supply for the Salinas Valley by enhancing the operating performance of the two upstream reservoirs: Nacimiento, built in 1957; and San Antonio, built in 1966. The current average annual controlled releases from these reservoirs to the Salinas River are approximately 200,000 acre-feet (see Figure 1).

The tunnel project as originally conceived, will provide additional flood control and increased conservation releases by conveying excess inflows into the Nacimiento Reservoir through the tunnel to the San Antonio Reservoir. The Nacimiento Reservoir has a more productive watershed and fills about three times faster than the San Antonio Reservoir, often leaving the San Antonio Reservoir partially filled when the Nacimiento Reservoir reaches capacity and spills. During wet years, the tunnel connection will divert a portion of the water from the Nacimiento Reservoir to the San Antonio Reservoir thereby increasing the net storage of both reservoirs and reducing the number of spill events and flow volumes that cause flooding downstream.

A reservoir simulation hydrograph prepared by ECORP Consultants[2] (Figure 2) demonstrates the effects of the tunnel to avoid flood releases from the Nacimiento Reservoir and increase the net storage of the reservoirs. The hydrograph, based on historical data from a typical wet year, shows the baseline fluctuation in storage in both reservoirs from inflows and releases in acre-feet (shown as dashed lines for both reservoirs). The Nacimiento reservoir fills to capacity and spills while the San Antonio reservoir has available storage capacity. By diverting potential flood water in the Nacimiento Reservoir through the tunnel, flood spills are avoided and the San Antonio Reservoir storage is increased dramatically.

Hydrologic reservoir simulation modeling was performed over the 47-year operating history of the reservoirs to evaluate how wet year flows through the tunnel would affect the reservoir operations based on historical data. The preliminary analysis indicates that on a yearly average, the tunnel would transfer approximately 50,000 acre-feet of water to San Antonio Reservoir and reduce the number of flood events from Lake Nacimiento by 60 percent.

With the increase in net storage, the reservoirs can be operated to provide increased conservation releases for downstream use and to recharge groundwater aquifers enhancing water supply sustainability in the Salinas River basin. The increased downstream flows also have the potential to provide an added ecological benefit to the migratory steelhead trout in the Salinas River.

The operating objectives of the Projects are summarized as follows:

- Minimize flood releases from the Nacimiento Reservoir and reduce associated downstream flood damages
- Increase the overall surface water supply available from Nacimiento and San Antonio reservoirs by maximizing the opportunity for water to be collectively stored in the reservoirs
- Improve the hydrologic balance of the groundwater basin in the Salinas Valley and reduce seawater intrusion
- Continue to meet environmental flow requirements
- Minimize impact on existing hydroelectric production
- Preserve recreational opportunities in the reservoirs
- Protect agricultural viability and prime agricultural land

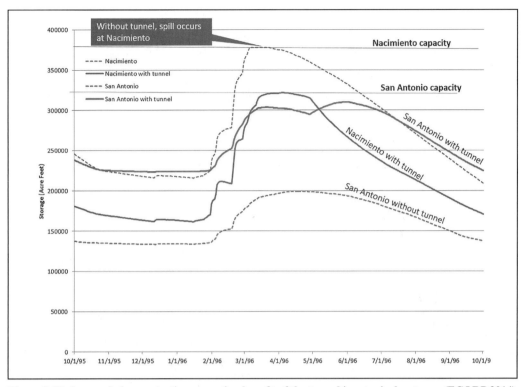

Figure 2. Hydrograph demonstrating operating benefit of the tunnel in a typical wet year (ECORP 2014)

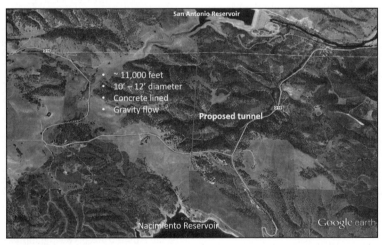

Figure 3. Interlake Tunnel site plan

PROJECT DESCRIPTION

Tunnel Description

The Interlake Tunnel will connect the two reservoirs as shown in Figure 3 with a gravity-flow 10-foot diameter tunnel, approximately 11,000 feet long at a –0.4 percent gradient from Lake Nacimiento.

The conceptual design of the tunnel is to operate in pressure flow mode based on head relationships between inflow and outflow with a maximum flow of 1,700 cubic feet per second.

The tunnel's Nacimiento intake will be designed to maximize the ability to move wet-year water to avoid flooding while maintaining a reasonable lake

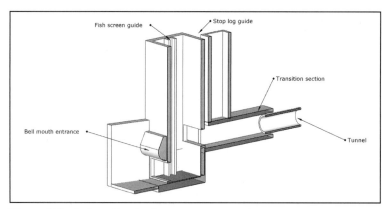

Figure 4. Nacimiento Intake Structure cross section

level for recreational use. Conceptual design has established a minimum lake surface elevation, before tunnel operation, of 760 feet MSL (above mean sea level), a minimum tunnel operating head of 15 feet, and a tunnel intake invert elevation of 745 feet MSL. When the San Antonio reservoir storage approaches the spillway elevation, transfer of water through the tunnel will be stopped.

The Nacimiento Intake structure at the tunnel's south portal shown in Figures 4 and 5 will include an approach channel with debris control; stop log channels; a wet well to accommodate vertical or sloping fish screens; fish screen operating and maintenance equipment; and appurtenant facilities including power supply, support building, instrumentation and SCADA systems, access road, and security fencing.

The outlet structure located at the north tunnel portal in Lake San Antonio will be equipped with an energy dissipater facility. The current design anticipates a downstream tunnel control facility above the San Antonio outlet, equipped with a control valve assembly with hydraulic actuators. The north portal outlet will have appurtenant facilities including a service building, SCADA/telemetry instrumentation and controls, power supply, access roads and utilities, as shown in Figure 6.

The hydraulic structures of the tunnel project include the Nacimiento Intake Facility, San Antonio Outlet Valve Facility, and San Antonio Energy Dissipater. The layout is shown in Figure 7.

San Antonio Spillway Description

With the operation of the tunnel to provide more inflow to the San Antonio Reservoir, there is a possible opportunity to increase the storage capacity in the San Antonio reservoir by modifying the existing spillway with the addition of a crest control device. This concept has the effect of "raising the dam" to increase storage. It is estimated that a 7.5-foot rise of the spillway could increase the storage capacity

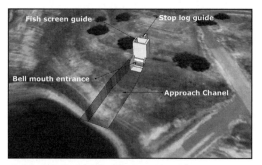

Figure 5. Nacimiento Intake site plan

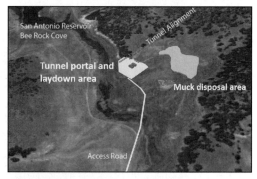

Figure 6. San Antonio portal site plan

of San Antonio Reservoir by approximately 45,000 acre-feet. This relatively inexpensive added storage increases the benefits of the tunnel by providing additional surface storage for flood control and conservation releases.

The spillway modification is anticipated to be steel radial arm crest gates installed near the current spillway crest, located as shown on Figure 8. Three gates are anticipated to span approximately 80 feet and will likely be equipped with mechanical lifting

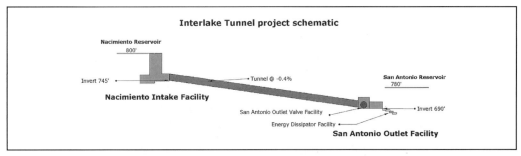

Figure 7. Interlake Tunnel hydraulic structures

Figure 8. San Antonio spillway modification concept

devices that are manually operated due to the relative infrequency of use. Repair and modification of the existing concrete spillway will also be required to comply with the safety requirements from the Division of Safety of Dams (DSOD).

The San Antonio Spillway Modification Project is dependent on the Interlake Tunnel to fill the San Antonio Reservoir beyond its current capacity.

The reservoir operations schematic provided in Figure 9 summarizes the details of storage pools, elevations of spillways and reservoir facilities, elevations of the proposed tunnel invert, and the added storage opportunity in the San Antonio Reservoir. Adequate hydraulic head for flow through the tunnel will be dependent on the surface water elevations in both reservoirs which will change as water is diverted from the Nacimiento Reservoir to the San Antonio Reservoir.

GEOLOGIC SETTING

Regional Geology

The projects are located in the south-central Coast Ranges of California which consist of relatively rugged north-northwest trending mountain ranges and intervening valleys, in a tectonically active region. The northwest orientation of the valleys is influenced by the nearby San Andreas Fault system which defines the boundary between the Pacific and North American tectonic plates.

The dominant bedrock formation is the Miocene Monterey Formation, which consists of strongly lithified deep marine sediments that predominantly include diatomite, siliceous shale, mudstone and chert with lesser amounts of carbonaceous rocks, and sandstones. The rocks are typically thinly bedded, highly folded and closely fractured. The Monterey Formation has naturally occurring hydrocarbons and is the most prolific oil-producing formation in California.

Dibblee (2006)[3] mapped several Quaternary landslides overlying the Monterey Formation rocks in the project area, including several near the Lake Nacimiento tunnel portal and one near the Lake San Antonio tunnel portal, as shown in Figure 10. The Rinconada Fault is near the north tunnel portal and is considered potentially active by the USGS and capable of a magnitude 7.5 earthquake. The Rinconada Fault zone crosses through the west abutment of the San Antonio Dam near the spillway.

Ground Water

Ground water occurs within the fractured shales and claystones of the Monterey Formation. Local to the

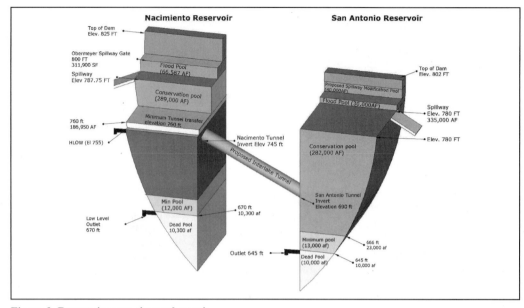

Figure 9. Reservoir operations schematic

tunnel environment, there are ground water aquifers which supply private wells for residential and agricultural use. Approximately 20 wells have been identified within a 3,000-foot zone on either side of the tunnel alignment which are being evaluated to help define the groundwater regime. Protection of the ground water supply for the private well owners is a significant objective for the design and construction of the tunnel project.

Geotechnical Exploration Program

A geotechnical exploration program is underway to determine the bedrock structure, faults and shear zones, rock types and strengths, bedding planes and joint patterns, permeability, groundwater levels and landslides potential. Geologic field mapping, exploratory boring, lab testing, and seismic refraction surveys will be used to characterize subsurface conditions. Five borings are planned for the tunnel alignment: one at the north portal, two at the south portal, and 2 along the tunnel alignment (Figure 11).

A Geotechnical Data Report (GDR), Geotechnical Interpretive Report (GIR), and Geotechnical Baseline Report (GBR) will be prepared by the Project's engineering consultants to report the findings of the geotechnical investigations and provide interpretation of the anticipated ground conditions for design and construction of the projects.

DESIGN AND PRELIMINARY ENGINEERING

McMillen Jacobs Associates has been retained by MCWRA as engineering consultants to perform

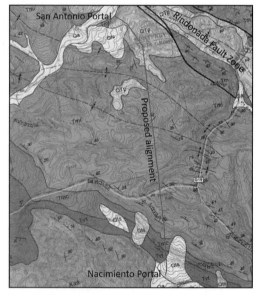

Figure 10. Geologic map indicating landslides and fault locations (Dibblee, 2006)

preliminary engineering for the tunnel project and final design of the spillway modification project. Subconsultant GEI Consultants is conducting the geotechnical exploration program and is leading the effort to coordinate with the California Division of Safety of Dams regarding the design and approval of the San Antonio spillway modification project.

Figure 11. Interlake Tunnel boring locations (GEI Consultants, June 2017)

Tunnel Preliminary Engineering

The engineering consultants are performing preliminary engineering and design to support the environmental clearance effort and to design the projects to the level needed to determine project costs. Design-Build documents will be prepared for use by MCWRA to solicit Design-Build Statements of Qualification and Requests for Proposals for the tunnel project. A significant engineering challenge is to design the projects within the capital cost constraints deemed viable for project financing. Permanent financing for the project will be based on capital and operations and maintenance cost estimates requiring the engineering consultants to complete sufficient design detail. MCWRA has required 95% confidence of probable costs certified by the engineering consultants to achieve the desired level of cost certainty for the permanent project financing. The level of design and permitting effort is therefore expected to be approximately 75 percent to accomplish this objective.

The design-build documents developed during preliminary engineering will include the following:

- Plans
- Specifications
- Geotechnical Data Report (GDR)

- Geotechnical Interpretive Report (GIR)
- Geotechnical Baseline Report (GBR)
- Final design criteria memorandum
- Preliminary engineering report
- Hydraulic analysis and reservoir modeling reports
- Information for Bidders and Design-Build contact form

Spillway Modification Final Design

McMillen Jacobs Associates is preparing 100% plans and specifications for the spillway modification project to be bid as a design-bid-build contract under California's Public Contract Code requirements. The 100% design approach removes the regulatory permitting risks for the contractor by obtaining Division of Safety of Dams (DSOD) approval before the contract is bid. Coordination with DSOD and compliance with updated dam safety regulations is a critical component of the design effort. A physical model study of the spillway performance is planned to be part of the final design. The spillway modification project is anticipated to be bid simultaneously with the tunnel project. The tunnel design-builder is not precluded from bidding on the spillway modification project.

Engineer's Report

An Engineer's Report detailing the facts describing the projects supported by scientific data, model findings, and cost-benefits analysis will be prepared by the engineering consultants to support the permanent financing initiative. The cost estimates will include detailed capital costs, environmental mitigation costs, and operating and maintenance costs. The Engineer's Report will recommend tax assessment methodology defining appropriate boundaries for zones of benefit and assessment formulas for acreages based on land use and how the acreage is benefited by the projects. Under the provisions of California's Proposition 218, an election of property owners within the assessment zones will provide property tax assessments used to secure bond financing for the projects.

Design-Builder Design Scope of Work

The design scope of work in the Design-Build contract is expected to include the following requirements:

- Perform additional geotechnical exploration and laboratory testing
- Complete the final design of the tunnel and appurtenant facilities to include 100-percent plans and specifications for review and approval by MCWRA, and for submittal to permitting agencies for approval.

- Prepare permit applications for the various permits required by the regulatory and permitting agencies.

TECHNICAL DESIGN AND CONSTRUCTION CONSIDERATIONS AND CHALLENGES

Hydrologic Modeling

A fundamental element of the projects planning and design is reservoir operation simulation modeling to confirm the hydraulic design and operation of the tunnel and to forecast the changes in flood control and downstream releases. Assessments of environmental impacts from the projects will be incorporated into the Environmental Impact Report's Project Description. A basin-wide surface and ground water hydrologic modeling program has been developed for Monterey County by the US Geological Survey (USGS). Multiple simulation scenarios will be run using this model to optimize the proposed reservoir operations plan with the tunnel and spillway modification. The modeling will also predict the surface/groundwater interaction as a result of the projects and contribute to an overall water management plan for the Salinas River basin. The modeling work is integrated with the engineering design and environmental clearance processes. Hydrologic modeling is on the critical path for the development of project design and environmental clearance documents.

Protection of Wells and Ground Water Supply

The tunnel project will cross underneath private properties with existing residential and agricultural wells supplied by underground aquifers. There are approximately 20 wells identified within 3000 feet of the alignment whose owners have raised significant concerns about potential loss of their wells because of the tunnel project. In response to these valid concerns, design and construction of the tunnel project will have requirements to protect the ground water supplies and to provide replacement water should any wells be adversely affected by the projects. MCWRA has passed a resolution which identifies mitigation measures for possible impacts to private wells caused by the Interlake Tunnel and Spillway Modification Project which include:

- Implementation of construction techniques to avoid impacts to ground water during construction and operation of the tunnel
- Design of a tunnel lining system that prevents water from entering the tunnel at no more than 1 GPM per 1000 feet of tunnel or less.
- Pre-excavation and post-excavation grouting and waterproofing performance specifications.

- Preconstruction surveys of properties and monitoring of wells
- Groundwater modeling to evaluate potential ground water inflows into the tunnel and probable effects on wells.
- Placement of supplemental water storage tanks on properties where wells are impacted by construction.
- A notification system for property owners to report any changes to well conditions
- A contingency plan to provide supplemental water as may be required for affected properties.

Fish Expulsion Systems

The Nacimiento Reservoir has a large population of white bass which are an aggressive predator and are therefore prohibited by government code from being removed alive from the reservoir. This provision applies to the Interlake Tunnel project requiring the incorporation of fish expulsion systems for white bass at the tunnel intake structure in the Nacimiento Reservoir. MCWRA has been in negotiations with the California Department of Fish and Wildlife (CDFW) beginning in 2015 to seek relief from this requirement because of the high added capital and O&M costs for fish screen facilities. The anticipated results of negotiations with CDFW will add fish screen and appurtenant facilities to the Nacimiento intake structure and identify additional funding sources.

Regulatory Compliance and Environmental Mitigation

When the tunnel project was re-initiated in 2014 by the farmers in the Salinas Valley, it was proposed as a simple project that could be permitted under a Negative Declaration of Environmental Impacts under the California Environmental Quality Act (CEQA). Public interest in the project forced MCWRA to initiate preparation of an Environmental Impact Report (EIR) under CEQA which has added time and cost to the project's development plan. The project has since been determined to have a Federal nexus with the U.S. Army Corps of Engineers (USACE) as the lead federal agency under Section 404 of the Clean Water Act. A determination has not been made whether the project will be authorized under a nationwide permit. As the federal lead agency, USACE will need to demonstrate the project's compliance with the federal Endangered Species Act (Section 7), the National Historic Preservation Act (Section 106), and NEPA. The most significant element of the environmental clearance process is compliance with the Endangered Species Act (ESA). The primary issue is the potential impacts to steelhead trout in the Salinas

River due to increased downstream flows generated by the projects.

MCWRA has implemented a coordinated environmental clearance and permitting process to obtain all necessary regulatory approvals and permits before the Proposition 218 election. Regulatory approval is required from several state and federal agencies including USACE, U.S. Fish and Wildlife Service, National Marine Fisheries Service, State Office of Historic Preservation, State Department of Water Resources, State Water Resources Control Board, Central Coast Regional Water Quality Control Board, and California Department of Fish and Wildlife. The mitigation measures associated with the regulatory approvals and permits are not known and could add significant costs which would jeopardize the financial feasibility of the project.

Cost Control to Maintain Viability for Financing

A critical aspect of the projects is cost control to achieve viable permanent project financing. The projects will be paid for by their beneficiaries through property tax assessments with no tolerance for cost growth. This mandates cost control as a fundamental design and construction criteria.

Under California law Proposition 218, the benefitting property owners will vote to approve property tax assessments to fund the bonds for permanent financing for construction, operations and maintenance of the projects. Both the tunnel and spillway modification projects are expected to be included within the Proposition 218 initiative. Approximately 280,000 equivalent acres have been identified in the Salinas River Valley that will be assessed property taxes based on formulas that proportionally weight active and passive land use factors and special benefits determinations. Most of the benefitting properties are agriculture-producing acres whose owners have a keen awareness of the cost of water and flood prevention. In the end, the property owners will determine if the additional water and flood control benefits of the projects are worth the cost. Without a successful Proposition 218 election, the projects are not financially viable.

Construction Considerations

Fluctuating lake levels can impact the construction of the tunnel project. During a winter storm, the inflow into the Nacimiento Reservoir can increase the water level by over 30 feet in one day. The invert elevation of the tunnel is established at 745 ft. MSL and 690 ft. MSL in the Nacimiento and San Antonio Reservoirs respectively which are below the spillway elevations in each reservoir. It is anticipated that the tunnel will be constructed during dry periods when the reservoir elevations are below tunnel invert elevations. Timing of construction to avoid high lake levels impacting tunnel construction is an important cost and schedule consideration of the tunnel project.

Disposal of tunnel muck is anticipated to be on-site near the north portal adjacent to Lake San Antonio assuming that the tunnel will be excavated completely, up-grade, from the north portal, as shown in Figure 6. If the design-build contractor elects to excavate the tunnel or portion of the tunnel from the south portal at Lake Nacimiento, alternate muck disposal sites will be required involving truck transport and identification of acceptable muck disposal sites.

The design-build contractor will be required to obtain necessary construction permits for erosion control, storm water control, grading, and building permits from appropriate jurisdictional agencies in Monterey and San Luis Obispo County and the California State Water Resources Control Board. Details of the permitting requirements will be included with the Information for Bidders (IFB) documents.

Electrical power is available in the tunnel project vicinity but not at the tunnel portal sites. Access to the portal locations has been identified and any restrictions will be defined in the environmental documents.

CONSTRUCTION PROCUREMENT APPROACH

Design-Build Mandated by SB 155 for the Interlake Tunnel

Assembly Bill (AB) 155 was passed in the California Legislature in 2014 to provide MCWRA with authorization to use a design-build project delivery system for the tunnel project. The design-build delivery method was justified in the legislation "by shifting the liability and risk for cost containment and project completion to the design-build entity and provides for the more timely and efficient project delivery in light of the current drought conditions and the Governor's emergency declaration."[4] SB 155 did not mandate that the San Antonio Spillway Modification project be procured by design-build.

California Design-Build Procurement Requirements

Senate Bill 785 (2014) defines the design-build project delivery using a best value procurement methodology considering cost, schedule, relevant experience, and other factors. The Interlake Tunnel will follow the requirements summarized as follows:

MCWRA will prepare and issue:

- Set of documents setting forth the scope and estimated price of the project
- RFQ to prequalify or short-list design-build entities.

Table 1. Project budget

Description	Millions	Percent
Administration and Management	$10.6	13%
Engineering, Design Environmental Clearance	$6.4	8%
Tunnel and appurtenances	$48.0	60%
Spillway Modification	$15.0	19%
Total	**$80.0**	

Table 2. Project schedule milestone summary

CEQA Approval	September 2018
Permits Issued	April 2019
San Antonio Spillway Design	October 2019
Tunnel Preliminary Engineering	November 2018
Engineer's Report	January 2019
Proposition 218 Financing election	February 2019
Issue Design-Build RFP	October 2018
Tunnel D/B Notice to Proceed	March 2019
Construction Complete	September 2020

- RFP that invites prequalified entities to submit competitive sealed proposals in the manner prescribed.

The RFP will have the following elements:

- Identification of the scope of work of the project, the estimated project cost, and the methodology and selection procedures that will be used by MCWRA to evaluate proposals on a best value approach.
- The relative importance assigned to each of the rating factors for selection identified in the request for proposals.
- Procedures for negotiations with responsive proposers

MCWRA will award a design-build contract for the tunnel project to the responsible design-build entity whose proposal is determined by MCWRA to have offered the best value to the public on the condition that permanent project financing has been secured via the Proposition 218 process. MCWRA does not plan to pay a stipend fee to unsuccessful bidders.

Project Labor Agreement

The AB 155 legislation requires the performance of the design-build contract in conjunction with a project labor agreement (PLA) in accordance with California's public contract code. MCWRA negotiated a PLA in May 2015 with the Monterey/Santa Cruz Counties Building and Construction Trades Council and the Tri-Counties Building and Construction Trades Council and their affiliated local Unions. The design-build contractor is required to be a signatory to the PLA.

PROJECT BUDGET

The current project budget, including contingencies is shown in Table 1.

A key consideration of the project budget is the annual debt service on the permanent financing which can be measured as the cost per acre-foot of new water for the agricultural properties. MCWRA is working diligently to minimize added costs associated with regulatory requirements or environmental

mitigation to keep the projects economically viable for the agricultural property owners.

PROJECT SCHEDULE

The current project schedule completion milestones are reflected in Table 2.

CONCLUSION

The Interlake Tunnel and Spillway Modification Projects are a contracting opportunity for design-build and design-bid-build contracts respectively. The project development team is working to advance the projects through preliminary engineering, design, environmental clearance, and permitting requirements. With a successful election to fund property tax assessment bonds, the projects will be soliciting Requests for Proposals for design-build and design-bid-build contracts beginning in the 4th Quarter, 2018. Notice to Proceed for the tunnel project is forecast for the 2nd Quarter of 2019.

Due to the lack of local, state, or federal funding, the Interlake Tunnel project is an example of infrastructure development initiated and financed by the project's beneficiaries. There are numerous other opportunities where the value of a project's benefits motivates its beneficiaries to initiate the development effort. Important components for successful project development are a politically astute sponsor and an appropriately experienced program management team to plan and manage the project through environmental clearance, permitting, design, construction and financing.

ENDNOTES

[1] Monterey County Water Resources Agency, Water Capital Facilities Plan, July 1991, prepared by Boyle Engineering Corporation.

[2] ECORP Consulting, Inc., Interlake Tunnel Project Simulation Modeling, July 2014.

[3] Geologic map of the Paso Robles quadrangle, San Luis Obispo County, California / by Thomas W. Dibblee, Jr. edited by John A. Minch.

[4] State of California Governor's Declaration of Drought State of Emergency, January 17, 2014.

TRACK 4: CASE HISTORIES

Session 1: Sewer/Water 1

Ben DiFiore and Brian Harris, Chairs

A Case Study of Risk Mitigation Measures on the West End Trunk Line Microtunnel

Alex Prieto, Rory Ball, and Jason Marie
Mott MacDonald

Gerald DeBalko
Pennsylvania American Water

ABSTRACT: Risk mitigation measures adopted in the design and construction of a 1,050-linear foot microtunnel segment (59-inch casing pipe, 30-inch carrier pipe) of the overall 16,200 linear foot West End Trunk Line Project proved successful. As subsurface conditions created challenges during the microtunnel boring machine's (MTBM) mining process, the design and construction management approaches were implemented and validated. The paper discusses how the challenges were overcome, including the MTBM becoming wedged within the excavation, very slow periods of excavation advance rate and unexpectedly high thrust loads as compared to the calculated values.

PROJECT OVERVIEW

The West End Trunk Line (WETL) Microtunnel is a 1,050-linear foot segment of gravity sewer within a utility easement adjacent to State Route 372 in Coatesville, Pennsylvania. The WETL Microtunnel is the second phase of a larger program (WETL Upgrade Project) undertaken by Pennsylvania American Water that aims at increasing the collection system's conveyance capacity to the Coatesville Wastewater Treatment Plant. The microtunnel portion of the program was divided into a separate construction contract due to the relatively significant depth for the gravity sewer. This portion of the overall 16,200 linear foot alignment traverses an easement located on ArcelorMittal's Steel Production Facility and was too deep for conventional open cut excavation to be cost effective. The microtunneling segment employed a two-pass system consisting of a 59-inch outer casing pipe that would be jacked from the launch shaft and an internal 30-inch diameter carrier pipe supported on spacers that would be inserted after completion of the casing pipe installation.

SUBSURFACE CONDITIONS

Geologic Setting

The Geotechnical Data Report for the project indicated that the project site resides within the Lowland Section of the Piedmont Physiographic Province and sits above the Conestoga Formation. The topography in this area consists of broad, moderately dissecting valleys separated by broad hills. The Piedmont Province extends from North New Jersey down to Central Alabama, largely composed of limestone and dolomite rock with the potential to contain karstic features. Separately, the Conestoga Formation is mostly within Pennsylvania and characterized as medium-gray, impure limestone with black graphitic shale partings and conglomeratic at the base. Unlike the Piedmont Province, the Conestoga Formation is moderately resistant to weathering, creating irregular and widely spaced joint patterns.

Subsurface Investigation

The subsurface investigation included four borings, field testing and laboratory testing. The four borings were spaced roughly every 330-feet along the alignment and were taken to depths approximately 10-feet below the tunnel invert. Each boring provided the surface elevation, boring depth, sample type, blow count, N-value, total recovery, solid core recovery, rock quality designation (RQD) and a Unified Soil Classification System (USCS) description. Bedrock depth was based upon refusal and was found to underlie a sandy silt or silt layer as well as a thin layer of surficial fill. The tunnel at its shallowest point had cover of at least 13-feet and was well within the bedrock. The results from these borings corresponded well to several historical borings taken in 2011.

In conjunction with the borings, field tests performed included infiltration testing, packer testing and groundwater monitoring. The results from these tests provided the anticipated groundwater elevation

Table 1. Laboratory test results (samples used fall within the tunnel horizon)

Parameter	Average Value	
Unconfined Compressive Strength	4,129 psi	
Modulus of Elasticity	6.4×10^6 psi	
Triaxial Shear Strength	$\phi = 31.3°$	C = 2,314 psi
	$\phi = 37.9°$	C = 3,140 psi
Splitting Tensile Strength	858 psi	
Point Load Strength (UCS)	250.6 psi	(6,441.9 psi)
Punch Penetration Index (Peak Slope Index)	20.3 kip/in	
Cerchar Abrasivity (CAI)	1.33	
Slake Durability	1st Cycle 99.15%	2nd Cycle 98.725%

(~330.5-feet) and the hydraulic conductivity of the strata. The hydraulic conductivity was used to help determine the condition of rock mass discontinuities. The rock mass discontinuities in the tunnel horizon were classified as very tight near the launch shaft and open joints closer to the reception shaft.

Rock samples from the exploratory borings were tested in the laboratory for unconfined compressive strength (UCS), modulus of elasticity, triaxial shear strength, splitting tensile strength, point load strength, punch penetration index, cerchar abrasivity and slake durability. The UCS for this limestone classifies as medium to strong rock (Hoek 1994) and slightly abrasive to medium abrasive (Thuro 2007). See Table 1 for averaged laboratory test results. The results from the borings, field tests and laboratory tests confirmed that the tunnel drive would be through karstic limestone.

DESIGN AND CONSTRUCTION CONSIDERATIONS

MTBM Selection and Details

The Contractor selected to use a Herrenknecht M-906M AVN1200TC-1505 with an overcut of 60.67-inch. The MTBM was a slurry driven system comprised of a cutterhead and five cans, extending over 45-feet in total length:

- The selected cutterhead (Figure 1a) was refurbished using a closed shield with five single disc rollers and five double disc rollers. The cutterhead had rear access to facilitate changing of the cutters.
- Can 1: steering can and contained the laser guidance target. This can also had an articulation joint for controlling the line and grade.
- Can 2: storage can where fresh disc cutters were kept when disc cutter changes became necessary. This would save significant time

for workers compared to carrying them into the tunnel during shutdowns.
- Can 3: power pack for the cutterhead, slurry system and all necessary components located in the MTBM.
- Can 4: another storage can.
- Can 5: 'gripper' can. This can had a hydraulic, telescopic jacking station and a hydraulic gripper. The telescopic jacking station was the equivalent to a half-extended Intermediate Jacking Station (IJS). This would allow the MTBM operator to pull the cutterhead away from the rock face if the cutterhead was unable to produce enough torque to free itself. The gripper was there to help prevent the tunnel from moving while retracting the cutterhead from the rockface by extending its grippers outward and creating reaction into the rock.

After the initial launch, the MTBM did not complete the drive in this configuration. Several modifications were necessary to faciliate advancement. These modifications are illustrated in Figure 1b and will be discussed later in this paper.

The casing pipe selected for this drive was a 59-inch outer diameter Permalok steel casing. Based on this material selection, as well as the ground conditions, the tunnel length and the anticipated slurry mix, the maximum jacking resistance calculated was 392-tons. This was well within the selected jacking frame (AVN 1200/1500) capacity of 573-tons. The jacking frame capacity, along with the additional 573-ton thrust capacity from the telescopic gripper can, was the reason the Contractor chose to eliminate the need for an IJS.

ANTICIPATED RISKS AND CHALLENGES ENCOUNTERED

After Mott MacDonald's initial review of the means and methods, the following tunneling risks were identified for monitoring by an onsite, full-time tunneling inspector during MTBM setup, MTBM launch, MTBM excavation, casing installation, MTBM retrieval and annular grouting:

- Specified equipment, means and methods employed
- Potential voids in the karstic limestone
- Soil seams in the ground
- Fractured rock zones
- Cutterhead tool wearing and associated impacts to steering and jacking forces
- Inadvertent fluid returns to an adjacent water storage lagoon
- Unexpected change in ground conditions

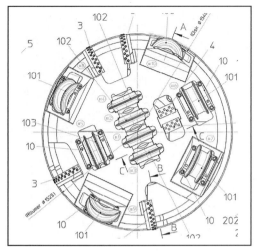

Figure 1a. MTBM cutterhead #1

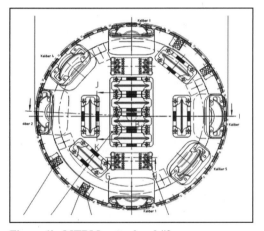

Figure 1b. MTBM cutterhead #2

During the tunneling operation three major challenges were faced:

1. MTBM wedging within the first 30-feet of tunneling,
2. Significantly slower than expected advance rates,
3. Unexpectedly high thrust loads.

Each of these challenges was successfully overcome while providing important lessons learned that could help reduce risks on similar future projects.

Challenge #1—MTBM Rescue and Re-Launch

The first major challenge began almost immediately after the MTBM launch with the MTBM getting

lodged against the surrounding rock. On November 10, 2015 MTBM Can One and Two were launched without issue. As Can Three began contacting rock along its circumference, the thrust loads began to climb at an unusually high rate. Excavation continued until the jacking forces became excessively high and advancing was no longer an option. The Contractor decided that the MTBM would be returned to the surface for inspection and determine any necessary modifications. Removal of the MTBM was very difficult and took nearly a week. Can Four and Five were successfully removed quickly from the jacking frame, while Can One, Two and Three remained trapped within the ground. After Can Three (Figure 2a) was finally freed, removal of Can One and Two occurred with relative ease. With the MTBM fully removed, it provided an opportunity to examine the ground conditions because the tunnel was unlined at this point (Figure 2b). The 30-feet of exposed rock confirmed that the ground was karstic limestone as expected. Up on the surface, the MTBM received the following modifications:

- Replacement of cutterhead with a refurbished cutterhead with 14 single disc rollers.
- Removal of one of the empty storage cans to reduce the overall length of the MTBM.
- Removal of a ½-inch skin from Can Three.

These modifications helped to alleviate the high thrust loads. Regarding the cutterhead, a cutterhead with only single disc rollers chips differently than double disk rollers and is typically used for harder rock. Regarding the storage can, removing the empty storage can reduces the length of rigid sections of the tunneling machine that pass through the overcut annulus, thereby allowing more space for steering adjustments. This precautionary measure was implemented because the advantage of having two empty cans was outweighed by the potential for higher thrust loads. Thirdly, the most important modification to the machine was the removal of a ½-inch skin from Can Three. The importance of this is two-fold: the overcut around the can was larger afterwards, creating less side friction and the dimensions of the can has more room for out-of-round tolerance. This can be particularly beneficial to the the powerpack, which is a heavy can and therefore subject to potential squatting of the shield.

In addition to these MTBM modifications it was known that the annulus had not been supercharged with lubrication during launch as suggested in ASCE, *Standard Design and Construction Guidelines for Microtunneling (ASCE 2015)*. Supercharging the annulus would have helped limit the potential for rock fragments to flow back underneath the MTBM and increase thrust loads.

Figure 2a. Can #3 stuck against rock

Figure 2b. Exposed ground conditions

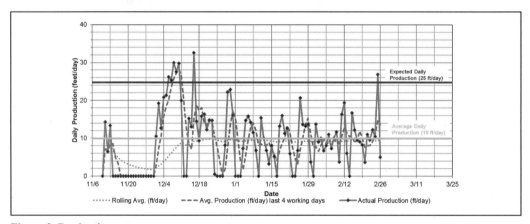

Figure 3. Production rate

With the completion of the modifications, the MTBM was relaunched approximately one week after being rescued on December 1, 2015. The advance rates steadily increased for the first few days and eventually reached the contractor's baseline daily production of 25-feet per day (Figure 3). By the 11th day of excavation the advance rate started to slow down marking the start of the Challenge #2: slow advance rates.

Challenge #2—Production Rates

During tunneling, the Contractor utilized one 12-hour shift per day. Advance rates hovered around 12-feet to 16-feet per shift when no delays were encountered. Unfortunately, there was significant downtime attributed to equipment malfunctions or replacements that hampered progress. Downtime reduced the average daily production to just 10-feet per shift. By January 1, 2016, the Contractor utilized two eight-hour shifts per day to help with production. However, this increase in working hours did not help the production rate. Lower than expected production

rates were a concern for all parties involved and needed to be further evaluated. Using the data collected by the tunneling inspector, a breakdown of events (Figure 4) was produced to see if time was being lost due to ground conditions or non-production activities. The following were determined to have the most significant impact on production:

- MTBM rescue and re-launch.
- Checking and changing the disc rollers.
- General equipment maintenance.
- High jacking loads and associated IJS delays.

The MTBM rescue and re-launch was previously described above in detail. This section will focus specifically on the disc cutters and tunneling equipment. The high jacking loads are the subject of Challenge #3 and will be discussed in the next section.

The MTBM was designed with an access door into the cutterhead plenum chamber. From this location, personnel would be able to remove the back-loaded disc cutters and install fresh ones. The process of changing or checking cutters required a significant amount of time due to the work involved. This process includes draining slurry from the cutterhead plenum, getting personnel into the chamber, inspecting the cutters, removing damaged cutters, installing fresh cutters, and restarting the MTBM system. A typical cutter check would take a minimum of four hours and frequently took longer. Typically, the reported damage to cutters was on only one or two-disc cutters at a time, which led to more frequent stoppages to make cutter changes. The type and quantity of damage to the disc cutters also provided insight into some possible scenarios that were occurring during excavation. The cutter wear observed was placed into three categories: (1) Typical wear (Figure 5a); (2) Flattened Faces (Figure 5b); and, (3) Broken Bolts and Cutter Disc Housings (Figure 5c). Other equipment issues were typically resolved within a few hours and were straightforward fixes.

The Typical Wear category was the theoretical wear that should occur if excavation is going as planned. The disc wear would be well rounded and evenly distributed around the circumference. This wear was expected to occur on all discs, but should have occurred faster to the gauge discs due to the

increased distance traveled. The Typical Wear condition was generally seen as expected on most cutters.

When the cutters developed a Flattened Face, it suggests that the disc is not rolling, but being dragged along its path. This could severely limit the chipping potential of the MTBM. A flattened edge could be the result of a broken disc bearing from a frontal impact load, such as too much thrust applied to the rock face. The flattened cutter shown in Figure 5b was located on the outside of the face, and could have created more work for the gauge cutters. This cutter was changed on January 13, 2016.

The last failure noted was a Broken Disc Cutter Housing. The housing connects the disc cutter to the cutterhead through bolts. A disc cutter with a broken housing may not be performing any chipping. This could have significantly slowed down the advance rate as other cutters would need to compensate this lost work. The broken housing in Figure 5c indicated an impact load must have occurred. Based on the direction of fracture, an impact to the side of disc seems a likely culprit. This cutter was changed on January 9, 2016 because the cutterhead was observed to be locking up and the bolt was subsequently never found. One hypothesis for the broken bearing is the

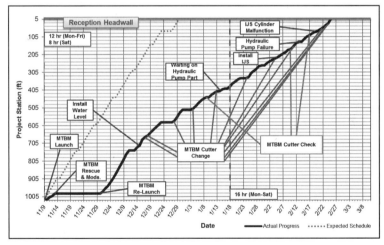

Figure 4. Tunneling schedule and issues encountered

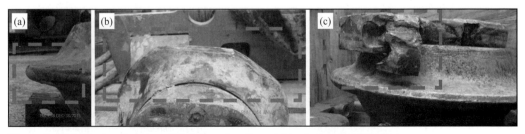

Figure 5. (a) typical cutter wear, (b) flattened cutters, (c) broken cutter housing

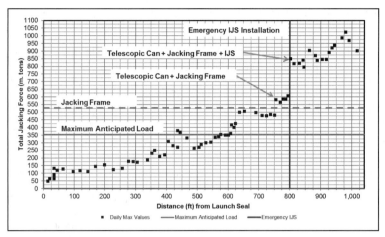

Figure 6. Jacking forces over the tunnel drive

outside face cutter had been flattened for a while, creating rock ledges along the circumference of the machine. Gauge cutters were being impacted from the front by the rock face. This would create strain leading to shearing failure along the plane shown in Figure 5c. When personnel went to inspect the cutters on January 9, 2016, the fractured housing was easily spotted and removed, whereas the flattened side of the outside face cutter was not visible and not removed until January 13, 2016 when another cutter check was performed. The flattened discs and broken cutter housings are likely related to high thrust forces in rock. The reasons for the high thrust forces were investigated and several potential causes are discussed in Challenge #3.

Challenge #3—High Jacking Forces

After the three re-launch MTBM modifications, the overcut annulus was lubricated with bentonite, polymer and water mix injected from Can Two. This was an improvement from no lubrication, but still raised concerns about the expected skin friction and quality of the supercharged annulus. When the MTBM was pulled out, a significant seam was noted running parallel to the tunnel at eleven-o'clock (Figure 2b). It is possible this seam continued and more seams were encountered; increasing the chance for lubrication loss. Pumping from more than one can could have significantly improved the effectiveness of the lubrication. The calculations had assumed a friction coefficient of steel against lubrication and only pumping from one can made it hard to realize that assumption.

As the thrust loads climbed at a higher than expected rate, the contractor implemented additional mitigation measures. Additional ports were attached to the lubrication system. By the end of December 2015, the jacking forces had surpassed the maximum

expected thrust of 392-tons, only 430-feet into the drive. Over the next month, the thrust forces would continue to increase and eventually the telescopic can was required to further advance the excavation. It was realized that an IJS would be necessary to efficiently complete the drive. On February 2, 2016, with another 200-feet remaining, the emergency IJS was installed to supplement the jacking frame and telescopic can. The final maximum thrust was about 854-tons (after subtracting the 276-tons required to close the IJS). This was over double the expected thrust. Although the thrust force was very high, it was relatively consistent. Figure 6 shows a relatively linear relationship between the distance excavated and the total jacking force. This suggests the ground conditions did not vary significantly.

One option the operator was unable to exercise was mining with a higher MTBM torque which can tend to improve the advance rate if rock chipping becomes more effective. However, when the torque was increased by the MTBM operator during the high thrust loads, the cutterhead would lock up. This was a concern as it could lead to MTBM roll, damaged cutters and an increase in downtime.

LESSONS LEARNED

Primarily, the lessons learned from this project focus on equipment and system issues that contractors can consider during project planning.

The emergency IJS was needed when the thrust forces surpassed the jacking frames capacity. When the realization occurred that an IJS was needed several challenges arose. First the Contractor needed to begin fabricating an appropriate jacking station in the shop. If the Contractor does start this fabrication early enough, time can be spent waiting for the IJS to arrive on site. This did occur and caused a work delay. The second issue is related to the thrust loads already isolated to the pipe string prior to the installation of the IJS. When the IJS was installed, the majority of the thrust used to push the pipe string was now ahead of the IJS and not between the IJS and the jacking frame. The lesson here is that even when the expected loads are manageable, installing an IJS at certain distances behind the MTBM can help manage unanticipated thrust loads better.

The second lesson learned revolved around the disc cutters and their arrangement on the cutterhead. The disc cutter arrangement was left up to the Contractors means and methods. Exploring different cutter arrangements prior to launch may have indicated that another type of cutterhead would produce better results and faster advance rates. Potential alternatives would be the number of disc cutters, single disc versus double disc and their spacing. These parameters could affect the size of the rock fragments and the ease in which the rock fractures. Additional cutters around the gauge could also reduce the wear on the gauge cutters and keep the overcut annulus at the specified dimensions for proper tunneling.

Lastly, each segment of the MTBM (referred to as cans) should always be checked for the roundness and desired diameter both during fabrication and before launch in case any geometric changes occur. It is possible that on this project the dimensions of one of the cans led to the MTBM becoming trapped shortly after launch.

CONCLUSION

The advantages of a full time tunneling inspector on site permitted the project team to capture these issues in detail and develop a data-based hypothesis for why they occurred. In addition, the daily monitoring and daily reports allow for real time tracking of conditions so that the issues can be addressed in a proactive manner. The following list is an example of some of the risk mitigation measures that can be recorded and the benefit it provides:

- Photographs—A visual record of events and conditions that can be refered to in the future.
- Notes—Any information such as quotes, descriptions of events or first hand data can help support any other risk mitigation measure.
- Rock Samples—Collecting samples and noting the locations when spoils started to

come off the separation plant can be useful in determining if ground conditions impacted the work.
- Slurry Measurements—Recording parameters such as slurry weight, viscoscity, sand content and pH daily; and using these parameters to check how MTBM operations are proceeding according to plan and not impacting the work.
- MTBM Can measurements—Taking measurements of the MTBM before launching can help check the correct equipment is being used and eliminate issues such as roundness or tolerances before launching the machine.
- Lubrication Records—Records of the quantity of lubrication pumped into the tunnel annulus provides a better understanding of whether the annulus was supercharged to reduce blowback of rock cuttings around the shield.
- Schedule Tracking—Collecting information on how time is spent throughout the day helps keep track of the utilization rate and if issues unrelated to tunneling are slowing down the work.
- Real Time MTBM Data Analysis—Keeping an updated log of data from the MTBM can provide insight into reasons for concern moving forward as well as explaining issues that may have occurred in the past.

Several of these construction management techniques were employed on the West End Trunk Line Microtunnel and helped to address issues. In the end, the combination of the contractor's hard work and determination along with a collaborative approach with the construction management led to a successful project for Pennsylvania American Water.

REFERENCES

ASCE, 2015. Microtunneling Operation. In *Standard Design and Construction Guidelines for Microtunneling. Edited by Julie McCullough and Xavier Callahan.*

Hoek, E., 1994. Rock mass properties. In *ISRM News Journal, 2(2), 4–16.*

Kasling, H. and Thuro, K., 2007. Determining rock abrasivity in the laboratory. In *Engineering Geology.*

High-Capacity Hoisting at Rondout West Branch Tunnel Project

Donald Brennan and Derek Brennan
Kiewit Infrastructure Co.

ABSTRACT: New York City Department of Environmental Protection (NYCDEP) is currently completing the repair of an existing water tunnel feed to the City of New York. The repair involves the construction of a by-pass to the existing tunnel. The depth of the new work and the type of equipment and materials required that the construction team revisit the traditional approach to shaft servicing. This paper will discuss the hoisting systems used at Shaft 5B Rondout West Branch Bypass Tunnel project near the City of Newburgh, New York. The shaft hoisting equipment was designed to handle the initial drill and shoot and the tunnel boring machine muck removal as well as the supply of all construction equipment including concrete segments and 40 foot sections of steel interliner pipe. Personnel were also transported from the surface to the foot of shaft located approximately 900 feet below. The hoisting equipment consisted of a total of 14 separate hoists and winches with a total of over 3,800 horsepower. The main muck and supply hoist consists of a double drum, variable frequency drive unit running at 1,100 feet per minute with a line pull of 62,210 pounds.

INTRODUCTION

The existing Rondout West Branch Tunnel (RWBT) is part of the Delaware Aqueduct which has been in service since 1944 and accounts for more than 50% of New York City's water supply. In early 2015, the New York City Department of Environmental Protection contracted Kiewit-Shea Constructors, AJV (KSC) to construct the Rondout West Branch Bypass Tunnel project (Rondout) in order to mitigate the leakage of 35 million gallons of water a day from two areas of the existing tunnel; the Roseton and Wawarsing areas.

The scope of the Bypass Tunnel 2 project is separated into two distinct phases. Phase 1 consists of completing the shaft sinking to a depth of approximately 900 feet at shaft 5B, 700 feet at shaft 6B and the excavation of approximately 12,500 linear feet of tunnel at a diameter of 21 feet 7 inches. This phase also includes the installation of 9,200 linear feet of 16 foot diameter steel interliner pipe through the new bypass tunnel with cast-in–place concrete liner for a finished diameter of 14 feet. Access chambers at the top of shafts 5B and 6B will be constructed for access and housing the mechanical and electrical equipment for supporting pumps and valves.

Included in phase 2 is the additional excavation from shafts 5B and 6B to the RWBT, and the drainage tunnel to remove groundwater infiltration from RWBT. This work will be completed during a scheduled shutdown of the RWBT and will include approximately 100 feet of excavation of the two connection tunnels. Additionally, construction of the permanent plugs within the RWBT will be undertaken. Figure 2 shows the layout of the work.

Access to the work through the deep shaft at site 5B was critical to the success of the project. The hoisting system needed to be multi-functional to handle the different types of construction from drill and shoot of the shaft, starter and tail tunnel, the assembly of the TBM, the excavation of the TBM tunnel including the supply of the precast segmental liner, the grouting operations, the installation of the steel interliner pipe and the cast-in-place concrete as well as ingress and egress of personnel. The production of the TBM operations was critical to the schedule of the project. This required a quick cycle for muck haulage and supply of the precast segmental liner in order to keep pace with the excavation time of the TBM. For TBM assembly, components were limited to a maximum weight of a 115 tons for a single lift. Various cranes were analyzed for hoisting, however none would meet all of the citeria. Therefore, the project team selected a specialized hoisting system designed in house in conjunction with Timberland Equipment Limited that consisted of a total of 14 separate hoists and winches with a total of over 3,800 horsepower.

STRUCTURAL AND MECHANICAL DESIGN

The structural portion of the high capacity hoisting system at the Rondout project consisted of eight separate designed and fabricated major structural components weighing over 625 US tons. These components were designed and supplied to access the thirty foot diameter shaft to accommodate all of the

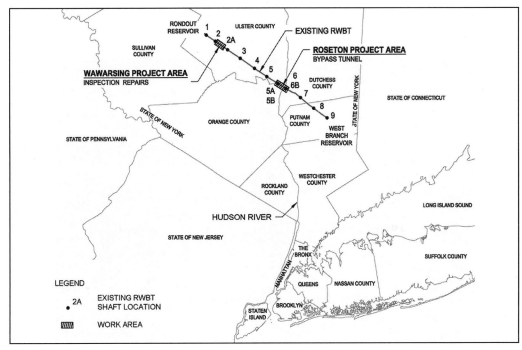

Figure 1. Project location map

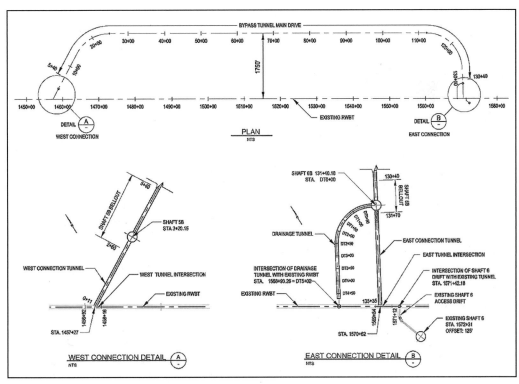

Figure 2. Bypass tunnel alignment plan

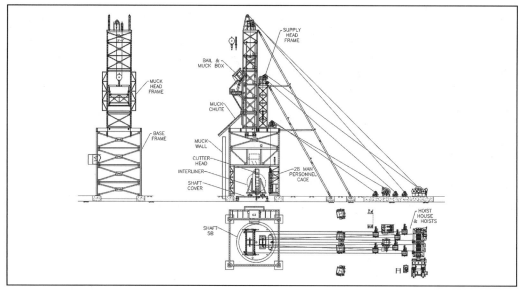

Figure 3. General layout of Rondout Shaft 5B head frame assembly

previously mentioned operations throughout phase 1 of the project.

The main components consisted of:

1. A movable 140- ton heavy-duty shaft cover
2. Shaft runner beams with Hilman rollers
3. The collar mounted base frame structure
4. Muck head frame, dump scroll, chute and back wall
5. Supply head frame
6. Muck and supply back legs with erection frame
7. Twenty-eight person personnel cage and access platform
8. Muck car bails and muck cars

Movable 140-Ton Heavy-Duty Shaft Cover

The movable shaft cover measures approximately 63 by 26 feet and weighs approximately 68 US tons. The cover rolls on Hilman rollers in an east west direction powered by a ten-horse power, electric, Stanspec dual directional car puller. The dual directional puller utilizes ⅞-inch diameter wire rope with a 40,000-pound line pull at approximately 15 feet per minute.

The overall length of the shaft cover is necessary because there are two 36-inch gage rail tracks installed on the cover to accommodate the roll-on of various TBM components that, along with carrier supports, will weigh approximately 280,000 pounds. The cover is made up of three separate compartments, each with a very specific purpose.

The most eastern compartment is fixed and measures thirty feet in length and only serves the purpose of supporting the dual rail tracks. The northwestern section measures 33 by 18 feet and consists of five removable panels that will be removed during a mucking operation so that this section of the shaft will be opened for the raising and lowering of the muck bail and boxes.

The remaining area of the cover is designated as the supply door and measures 33 by 8 feet in size. This section of the cover is supported by a series of four offset hinges so that the supply door may be rotated and moved to the north in such a way to completely open the supply area for the transport of tunnel rail, tunnel supplies and precast segmental liner. An electric Thern winch and an overhead sheave deck mounted on the north side of the shaft cover and base frame is used to open and close the hinged supply door as required.

An eight-foot high safety fence is installed between the muck deck and the supply door so that hoisting of these two functions may take place simultaneously.

Shaft Runners Beams with Hilman Rollers

The shaft runner beams consist of two W21×132 beams. The beams are spliced together to form a continuous section, 100 feet in length. Each 100-foot long section is supported on four concrete footings founded on base rock foundations.

There is a total of ten Hilman rollers, each complete with Accu-Roll guides and a capacity of 50 metric tons, that run on top of the two runner

Figure 4. Base frame assembly

beams and, in turn, support the movable shaft cover as previously described. The 50 metric ton capacity of each Hilman roller is required to support the shaft cover and the point loading from the various TBM components during installation.

The Collar-Mounted Base Frame Structure

The base frame structure is a large fabrication that measures 42 feet wide by 39 feet deep with a height of approximately 54 feet. The total base frame weight is just short of 76 US tons. The entire base frame is supported on four wide flange legs resting on concrete foundations on a solid rock base surrounding the shaft collar.

The required function of the base frame is to support the main muck and supply head frames, the personnel access platform, the muck dump scrolls and chute as well as provide an east and west opening that will allow the entrance of all tunnel supplies, all TBM components and the entrance of truck loaded interliner pipe that will measure 16 feet in diameter by 40 feet in length.

The east and west side openings in the base frame measures 30 feet wide by 24 feet high. The western portion of this 24-foot high opening allows a rail mounted carrier to enter the interior of the base frame while supporting a 22-foot diameter TBM cutter head. The same opening is used to allow the interliner pipe to enter as well. After the truck loaded interliner pipe has entered the interior of the base frame, the hoist blocks are attached to the east end of the interliner pipe and the pipe is rotated and hoisted to a vertical position so that it may be lowered down the shaft to the invert and its' final destination within the driven tunnel.

The second floor of the base frame structure supports the hoist house control room from where all hoisting functions and shaft cover movement is controlled. A stair tower and walkway allows access to the control room and to the muck and supply head frame.

Muck Head Frame, Dump Scroll, Chute and Back Wall

The muck head frame is a pre-fabricated structure that measures 12 by 21 feet with an over-all height of 94 feet. The entire muck frame complex which includes the dump scroll, chute and sheave deck is supported on heavy beams that are a part of the upper portion of the previously described base frame.

To allow the rotation and the dumping sequence of the 20 cubic yard muck boxes, a portion of the north face of the muck head frame is completely open and designated as the dump slot. Because of the complete lack of bracing in this portion of the frame, an external frame work truss is attached to the two legs of the muck head frame to reinforce the legs of the frame in this area.

The dump scroll consists of two heavy plates attached to the north-east and north-west legs of the muck head frame. This heavy plate structure supports two tracks that allow the bail supported muck cars to rotate and automatically dump 20 cubic yards of tunnel muck onto the chute which is also supported from the northern legs of the muck head frame. This dumping method is a mining technique, well known, as a "Kimberly" dump.

The dumped material from the 20-cubic yard box is slowly displaced and falls onto the dump chute which measures approximately 27 feet in length on an angle of 42 degrees to the horizontal. A portion of the dump chute is supported on the top deck of the base frame in addition to the muck head frame support. The upper surface of the muck chute and the impact area is lined with two layers of steel plate and a layer of anti-abrasive material designed to protect the chute, provide a low coefficient of sliding friction and lower the sound of material being dispersed.

Figure 5. Head frame assembly; supply frame, upper stair tower, upper muck frame

The top section of the muck head frame consists of a series of heavy beams that support the sheave deck and three wire rope sheaves. The main sheave that is attached to the muck bail has a pitch diameter of 46.5 inches and supports a 1-½-inch diameter Briden Endurance Dyform 34LR Max wire rope with a breaking strength of 166 US tons. This wire rope is two parted around a 57-inch diameter Crosby block and returns to the sheave deck where it is anchored in a fixed support frame.

The two remaining sheaves are designated as guide rope sheaves and have a pitch diameter of 28-⅛ inches and support a 1-⅛-inch diameter Briden Dyform 6 wire rope with a breaking strength of 82.5 US tons. The two guide ropes are used to guide the muck bail while running in the shaft and are anchored at the foot of shaft by way of a hydraulically operated cylinder designed to measure the required tension in the rope guides.

Supply Head Frame

The supply head frame is a pre-fabricated structure that measures 9 by 22 feet with an over-all height of 46 feet. The supply frame, like the muck head frame, is supported on heavy beams that are a part of the upper portion of the previously described base frame. The supply sheave deck is of similar design and configuration as the muck sheave deck with identical sheaves and wire ropes for ease of design and functionality with the double drum hoist for synchronized loads.

The supply head frame is used to support the required services of the underground works for the transport of tunnel rail, tunnel supplies and precast segmental liner. The two guide ropes are used to guide the precast segment handling device and other supplies while running in the shaft which are also anchored at the foot of shaft.

Muck and Supply Back Leg with Erection Frame

Both the muck and supply head frames have inclined back legs designed to support the loading from the two 1-½-inch diameter main ropes as well as the four 1-⅛-inch diameter guide ropes. The loading from these wire ropes can impose a force of up to 102,000 pounds of compression into the back-leg structure. The lengths of the muck and supply back legs are 159 feet and 105 feet respectively and they are horizontally supported laterally at mid and one-third points along their lengths.

It is virtually impossible to erect these long back legs in one piece so a back leg erection frame is designed and fabricated to allow reasonable sections of the back legs to be erected as units including the attached lateral support struts. The erection frame is outfitted with varying length wire rope slings so that dual sections of the back legs may be raised from a horizontal position on the ground to a predetermined and accurate inclined position in the air to accommodate connections to the two head frames and the foundation anchors on the ground. A total of 149,000 pounds of back leg structure was successfully erected using this unique erection technique.

Twenty-Eight Person Personnel Cage and Access Platform

Personnel ingress and egress to and from the shaft invert is undertaken with the use of a specially designed personnel cage capable of transporting a total of 28 people. The personnel cage is a two-level cage with 14 persons per 41 square foot level. The lower section of each level is enclosed with solid

Figure 6. Back leg installation

plate whereas the upper half of each level has a fine mesh enclosure. The entrance to the cage is by way of three sliding doors, two on the lower level and one sliding door on the upper level.

The personnel cage is raised and lowered with a single wire rope that is supported from a personnel sheave located on the top deck of the base frame structure. The base frame structure supports a sheave deck and three wire rope sheaves. The main sheave that is attached to the personnel cage has a pitch diameter of 38.75 inches and supports a 1-¼-inch diameter Briden Endurance Dyform 34LR wire rope with a breaking strength of 122 US tons. The two remaining sheaves are designated as guide rope sheaves and have a pitch diameter of 28-⅛ inches and support a 1-⅛-inch diameter Briden Dyform 6 wire rope with a breaking strength of 82.5 US tons. The two guide ropes are used to guide the personnel cage while running in the shaft and are anchored at the foot of the shaft.

Access to either level of the cage at the collar level is by way of a personnel access platform with an internal stairway that is prefabricated and located inside the south end of the base frame structure. This access platform also allows for a method of traversing from the west side to the east side of the shaft area. Access at the foot of shaft is by way of a 30-foot high stair tower.

The personnel cage is equipped with a broken rope safety device installed on each end of the upper framework of the cage. The safety device utilizes a linkage system and a heavy coil spring to activate a pair of high strength serrated wedges that, when activated, grip the 1-⅛-inch diameter rope guides. The cage is guided throughout the shaft area with a total of four bronze thimbles located on the four corners of the cage.

Muck Car Bails and Muck Cars

The excavation of the tail and starter tunnels is with the use of the drill and shoot method. The excavation

of the main tunnel drive is with the use of a 22-foot diameter Robbins TBM. The TBM deposits muck into a series of 20-cubic yard lift off box muck cars from a trailing conveyor belt system. The cars measure approximately 15 by 6 by 7-feet high and run on a chassis with two bogies on a 36-inch gage track. The six or seven car train is transported with a 35-ton diesel powered locomotive.

When the train reaches the area at the foot of the shaft, each successive car enters into a bail structure that is designed to raise the lift off box only to the surface dump area. The bail system consists of an inner and outer bail. The inner bail is pinned at one corner to the outer bail so that it can rotate and dump the contents of the 20-cubic yard box while traversing through the dump scroll on the north face of the muck head frame. The combination of bail, box and tunnel muck weighs approximately 106,000 pounds and travels to the surface at a maximum velocity of 550 feet per minute.

HOIST REQUIREMENTS AND SPECIFICATION

The main mucking and material hoist is a double drum electrical variable frequency drive (VFD) hoist that has two drums capable of running simultaneously locked together or independently. The hoist consists of three separate modules; the main drive module, the drum hoist module and the HPU module. The main drive module has two 1300 HP electric drive motors directly coupled to a two speed gear reducer with two high speed input shafts and one low speed output shaft. The primary feed for these drives is 4160V. The drum module has a 480V 500 HP drive motor that transmits torque through a gear reducer with a final bullgear reduction acting directly on the supply drum. Depending on the mode of operation, the two 1300 HP drives can drive one or both drums through the single shaft at various speeds and capacities. Similarly depending on the mode of operation, the 500 HP drive can drive one or both of

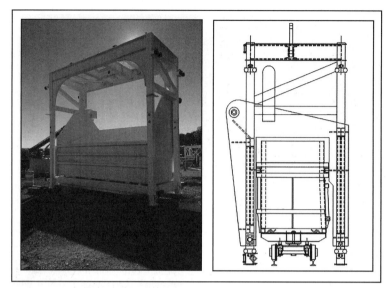

Figure 7. Muck bails for drill & shoot and TBM operations

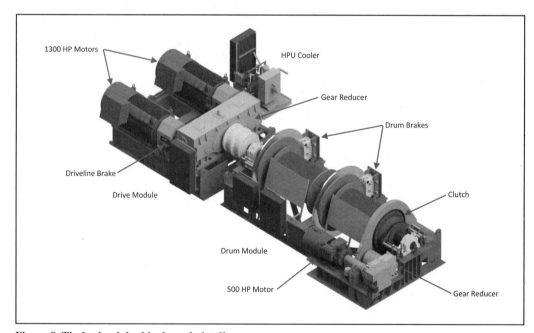

Figure 8. Timberland double drum hoist diagram

the drums through the bullgear with slower speeds and less capacity. With that said when the hoist is configured correctly, the muck drum can be driven at high speeds from the two 1300 HP motors at the same time the supply drum can be driven by the 500 HP motor.

The HPU module provides hydraulic pressure for the brake system, the clutch system and the

shifting for the two speed gear reducer as well as cooling the gear oil. Each drum and driveline has an independent brake system spring applied, hydraulically release that is capable of stopping and holding 150% of the linepull. The clutches are engaged and disengaged from each drum independently using hydraulic cylinders allowing the hoist to be operated in various modes.

Figure 9. Timberland E-house building and hoist layout prior to hoist house installation

The electrical setup for the main hoist is extensive. The incoming feed to the electrical hoist pad is 13,200 volts which then steps down to 4160 volts for the two 1300 HP or medium voltage drives and steps down again to 480 volts for the 500 HP or low voltage drive. Each of the three motors is driven with a VFD. The two 1300 HP drives are regenerative drives that reduce the harmonics to less than 5% distortion so that the power generated by the drives can be put back into the grid as reusable power. The low voltage 500 HP drive dissipates the energy produced through a separate brake resistor. A requirement of the hoist manufacturer was to provide an electrical switchgear building to house all of the drives, control equipment and communication systems with plug and play connections for the main hoist feeds. This allowed the manufacturer to permanently install all of the major electrical components, test the equipment and ship the building as a single unit to avoid tear down, packaging and resetting up of the equipment. This was a significant time and cost savings on the job and simplified the site commissioning.

The double drum hoist has six modes of operation;

- Mode 1: The muck / supply mode is set up to have the two drums run independently. The muck drum primarily for high speed mucking cycles while being able to hoist materials on the supply drum. The muck drum has a line speed of 1,100 fpm with a linepull of 62,210 pounds and the supply drum has a line speed of 500 fpm with a linepull up to 32,650 pounds.
- Mode 2: The double drum heavy lift mode operates with both drums clutched in together utilizing the two 1300 HP drives providing a total lifting capacity of 124.4 tons at 50 fpm.
- Mode 3: Also utilizing the two 1300 HP drives, the double drum high speed mode operates with both drums clutched in together

Figure 10. Hoisting equipment inside hoist house

providing a total lifting capacity of 62.2 tons at 250 fpm.
- Mode 4: The muck drum heavy lift mode runs the muck drum in low speed at 50 fpm and a capacity of 62.2 tons utilizing the two 1300 HP drives.
- Mode 5: The supply drum heavy lift mode runs the supply drum in low speed at 50 fpm and a capacity of 62.2 tons utilizing the two 1300 HP drives.
- Mode 6: Emergency mode runs both drums from the 500 HP drives if the two 1300 HP drives or gearbox were to fail.

Personnel ingress and egress is achieved with Timberland's model PH250 and a 28 man two level cage. The hoist will lower personnel the 900 feet at a rate of about 240 fpm. The hoist is a single drum, electric over hydraulic hoist system which incorporates a hydrostatic drive and two independent braking systems similar to the brakes on the main hoist. The system is powered utilizing two independent 250 HP electric motors direct coupled to the hydraulic pumps providing a spare pump drive all the time. The control for this hoist is fed back to the operator's

Figure 11. Timberland double drum installation

cab in the headframe using a single joystick and HMI screens. For emergency purposes, the 480 volt power for the personnel hoist system is feed from the essential services switchgear on site which includes backup generator power. For this reason the PLC for the personnel hoist is independent from the remainder of the hoist system and is located at the HPU with a local operator's station for setup and operator control during a power outage.

The hoist system includes six guide rope winches, two for each of the three main drums; muck, supply and personnel. These winches were primarily used during the sinking of the shaft as the depth continued to increase. Once the bottom was established the guide ropes are fixed and the drums locked. Each winch is a single drum electrically powered with a 20 HP motor and a gear reducer between the motor and drum with a linepull of 7000 lbs. The winch has a single spring brake within the motor and a locking dog that is engaged once the guides are set.

Four stage hoists were utilized in the system as well to raise and lower a full diameter shaft work deck for the installation of all the utilities and have a linepull capacity of 36,980 lbs. each. Each hoist is controlled with a single 20 HP electric motor through a gear reducer and a set of open gears. The brake system consists of a fast acting motor brake and an air controlled band brake. The control for these four hoists are also in the operator's cab within the headframe.

OPERATIONAL CONTROLS

The controls for all of the hoists were setup in an 8 foot by 8 foot cab to be operated with one operator utilizing three joysticks for the main hoists, two HMI display touchscreens and a series of control panels. Traditionally the operator's station is positioned in the hoist house overlooking the hoists, however it was decided to place the cab inside the headframe structure over the shaft cover. This allowed the operator an unobstructed view of the cover and rigging

Figure 12. Timberland personnel hoist and 28-person cage

Figure 13. Operator's cab and lowering of the TBM cutterhead

operations as well as providing the safest viewpoint for personnel. The main PLC cabinet for the hoist system was located in the cab along with the load cell control cabinet and CCTV equipment. Additionally there were controls designed and installed by KSC that were tied into the hoist control system for the anti-two block systems, the load cell system, personnel cage proximity switch and door locks and muck dumping proximity switch.

SETUP AND ERECTION

The assembly of the hoist and head frame consisted of three phases of work; foundations, pre-assembly and erection over the shaft. The foundations were poured for the runner beams, base frame, back legs and the hoists which included 77,000 pounds of rebar and 1,200 cubic yards of concrete. The most challenging piece of the foundations was assuring the embedded bolts for the double drum hoist were accurately laid out as the two skids weighed a 100 tons combined and needed to be placed so that the drive output shaft and the drum shaft were perfectly aligned. Pre-assembly of the structure included the shaft cover, the muck and supply towers, the sheave decks, the base frame and the back legs all of which took place on site yard while the site utility, shaft utility and foundation work were being performed. Once shaft utilities were completed, the assembly over the shaft began with the base frame. The base frame stair tower was installed prior to installing the supply tower and lower muck tower for easy access from the deck to all of the tower mounting base plates. Similarly, the upper stair tower was installed prior to the upper muck tower eliminating the need to work out of a 185 foot man lift. Lastly the back legs were pinned to the towers and secured to the foundation slab. The structural pre-assembly took about 8 weeks followed by approximately 7 weeks of assembly over the shaft including the muck wall structure.

Once all of the structural steel was in place, rope up of all the hoists were completed along with

the electrical and mechanical installations throughout the hoist and head frame assembly. The technical hook ups and commissioning took an additional few weeks to complete. The crews used throughout the setup were made up of ironworkers, laborers, mechanics and electricians along with assistance from the hoist manufacturer.

COMMISSIONING AND TESTING

There were three phases to the commissioning of the hoists; at the factory, on site and finally the tuning of the medium voltage drives. Timberland performed a factory acceptance test in their plant prior to the equipment being shipped to site. All systems and functions were tested to ensure minimal troubleshooting on site. Similarly, thorough site acceptance tests were completed once all of the components were installed and functional which included KSC's control system. The final step was the testing and tuning of the drives for each mode of operation. Tests were performed at different load capacities and speeds as well as floating the load, changing the load simulating loading a muck box, and emergency stopping. By code, the hoist needed to be load tested to a minimum of 110% which we accomplished using a water bag system consisting of two water bags per load block. The water bags were also used to calibrate the load cells as well.

Our largest challenge was the time required for the tuning of the drives which was completed over several weekends. There were dozens of trial runs for each mode while fine tuning the parameters for each drive to adjust for the drift during floating of the load, the inertia settings, and meshing of the low speed and high speed. Initially we experienced extreme oscillation in the drives during testing while in low speed due to the sensitivity of these parameters.

OPERATION AND CHALLENGES

The job faced many operational challenges that a multiple hoist system had to accommodate. The

Figure 14. Completed headframe and hoist house

initial challenge was coming up with a system designed to accommodate all of the operations for the various aspect of the project from high speed to high capacity to size constraints with equipment and 40 foot lengths of steel pipe lowered within a 30 foot diameter shaft. The logistics of keeping up with supplies to match the performance of the TBM excavation as well as quickly changing over to a grouting operation had to be well planned to be well executed. There was also the difficulty of having a fixed hoist system along with a smaller shaft that needed to accommodate many utilities with the conveyance of the thirteen various hoists. The most time consuming challenge of the fixed hoist was getting the material and equipment under the hooks and the rigging required to properly place these items down the shaft.

CONCLUSION

The Rondout West Branch Bypass Tunnel is a complicated and challenging project and the hoisting at

shaft 5B was no exception. But with a bit of ingenuity and the right technical expertise the challenges were overcome. At the time this paper was written, the project had successfully completed the drill and shoot operation mucking an 18 CY skip at the rate of seven minutes per cycle with room to speed up, hoisted various pieces of equipment in and out of the shaft and lowered the heaviest components of the TBM.

REFERENCES

JA Underground: Professional Corporation, dba Jacobs Associates. 2014. Rondout-West Branch Bypass Tunnel Construction and Wawarsing Repairs Project; Geotechnical Baseline Report.

Timberland Equipment Limited. 2015. Timberland Proposal No. 02-0905-302 Rev C. Owner's Manual for Model 620X2-2-2X1300EVFD, PH250-1-2X250EH and GP70-1-20E.

Boring Hard, Abrasive Gneiss with a Main Beam TBM at the Atlanta Water Supply Program

Tom Fuerst
Robbins

Don Del Nero
Stantec

ABSTRACT: Atlanta, Georgia's water supply program is a priority project involving a 5.0-mile long tunnel connecting up with the Chattahoochee River, which will establish an emergency water supply for the city. A 12.5 ft diameter Main Beam TBM is boring the area's deepest tunnel through hard, abrasive Gneiss rock at rates of up to 100 ft per day. This paper will examine the project specifics and design, as well as the performance of the TBM. It will then draw conclusions as to the optimal TBM design for excavation in the area's exceedingly hard geology based on this project and past projects in the area.

INTRODUCTION

The average North American public utility has only a three-day back-up supply of clean drinking water. The City of Atlanta, Georgia, USA, was, until recently, no exception to that rule. In fact, just three cast-iron water mains built in 1893, 1908, and 1924 conveyed raw water to treatment facilities for ultimate use by 1.2 million customers in the city and surrounding areas. The overtaxed system, paired with the increasing risk of drought, prompted the city's Department of Watershed Management into action. In 2006, the department took steps to purchase the Bellwood Quarry from Vulcan Materials Co.—a 300 ft deep, vertical-sided quarry where granitic gneiss was mined for a century to become structural blocks for Atlanta's buildings as well as crushed stone aggregate for roads.

The USD $300 million project would turn the inactive quarry into a 2.4 billion gallon raw water storage facility connected up with the Chattahoochee River and various water treatment facilities, bolstering the city's emergency water supply to 30 days at full use and to 90 days with emergency conservation measures. The price is a small one to pay by many estimates: if the city were to lose its water supply, the estimated economic impact could be at least USD $100 million for just one day.

To make the program a reality would require excavation of Georgia's deepest tunnel (400+ ft), starting at the quarry and running under two treatment facilities for 5.0 miles to an intake at the Chattahoochee River. It would also require construction of two pump stations at the Quarry and Hemphill Reservoir, five blind-bored pump station shafts at the Hemphill site up to 420 ft deep, as well as two more pump station shafts, one riser shaft, and one drop shaft. The quarry would ultimately store raw water before it is withdrawn for treatment at the Hemphill and/or Chattahoochee water treatment plants, connecting the quarry to the Hemphill Water Treatment Plant (HWTP), the Chattahoochee Water Treatment Plant (CWTP) and the Chattahoochee River. After construction, the area around the quarry would then be turned into Atlanta's largest park totaling 300 acres complete with hiking and biking trails, baseball fields, and an amphitheater (see Figure 1).

The project schedule, primarily driven by the condition of the City's existing water infrastructure, compelled the city to consider Alternative Project Delivery (APD) instead of traditional design-bid-build. The project schedule required a start date for construction of January 2016 and a substantial completion date of September 2018. The method selected was Construction Manager at Risk (CM@R), where the contractor acts as a consultant to the owner during the development and design phase and as a general contractor during the construction phase. The setup resulted in a unique process to start TBM manufacturing, in particular, before the tunneling subcontractor was mobilized at the site. The decision to use a new TBM by the City of Atlanta was primarily risk-based.

The PC Construction/HJ Russell (PCR) JV was selected as the CMAR for the project, who then purchased a 12.5-ft diameter Robbins Main Beam TBM for the tunnel. The designer for the construction works including tunnel and shafts, JP2—consisting of Stantec, PRAD Group, Inc., and River 2 Tap—specified the hard rock TBM. The project arrangement is unique: in fact the CMAR purchasing the machine and the Engineer-of-Record specifying the machine may represent the longest such tunnel to be delivered under this setup in North America.

Detailed Project Description

Phase 1 involves design and construction of one TBM tunnel, two pump station shafts, one drop shaft, one riser shaft, five blind bored (top-down) pump stations shafts, and a quarry highwall rockfall protection system to provide long-term protection of the tunnel inlet. Specific details include the following (Del Nero et al., 2016):

- The TBM tunnel is approximately 5,500 ft long and partially concrete lined with a finished diameter of 10 ft;
- The primary pump station shaft at the quarry is approximately 250 ft deep with a finished diameter of 35 ft. The low level pump station shaft has a finished diameter of 20 ft and is approximately 340 ft deep. The primary and low level pump station shafts are connected to the tunnel and quarry via adits;
- The drop shaft at the quarry is about 320 ft deep with a finished diameter of 25 ft above

El. 805 ft and 4.5 ft below El. 805 ft. The drop shaft is connected to the Quarry low-level pump station shaft and riser shaft through adits. The drop shaft provides a flow capacity of 90 million gallons per day;
- The riser shaft at the quarry is about 320 ft deep with a finished diameter of 25 ft above El. 805 ft and 12 ft below El. 805 ft.
- The quarry riser shaft is connected to the quarry drop shaft through an adit;
- The five pump station shafts at the HWTP are about 420 ft deep and 9.5 ft in bored diameter;
- Each pump station shaft will have a 76-in. diameter grouted steel casing to house the pump; and
- The five pump station shafts are connected to the main tunnel by five, 7.9 ft adits with lengths ranging from 20 ft to 30 ft.

Major elements of the Phase 1 Extension project include one TBM tunnel, an extension of the Phase I tunnel, and one combined drop (baffle type) and construction shaft. The construction shaft is 250 ft deep with a finished diameter of 30 ft. The TBM tunnel is approximately 18,470 ft long and partially concrete lined with a finished diameter of 10 ft (see Figure 2).

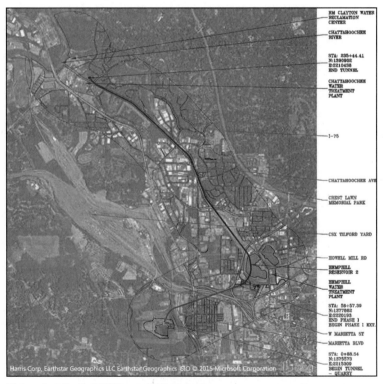

Figure 1. Atlanta WSP tunnel alignment

GEOLOGY

The project is located in the Piedmont Physiographic Province. The geology of the Piedmont in the greater Atlanta area generally consists of medium-grade metamorphic rocks with granitic intrusions. These crystalline rocks are some of the oldest rocks in the Southeastern United States.

Detailed geotechnical and hydrogeological field investigations were made, comprising 25 deep borings along the tunnel alignment and 30 shallow borings in the areas for shaft construction. The borings revealed granitic gneiss rock averaging 25,000 psi UCS along with five transition zones between soil and bedrock where groundwater was expected. These zones had the potential for high-yield fracture flows, as local wells in the area had been reported to yield up to 500 gallons per minute.

Lithologies Along the Tunnel Alignment

The majority of the proposed tunnel alignment is located in the in the Clairmont Melange (Bedell et al., 2017). The following text is from the Geologic Report as provided by PetroLogic Solutions as part of the preliminary geotechnical investigation. The order of the geologic unit descriptions proceed from the quarry to HWTP and then through to the Chattahoochee Water Treatment Plant (CWTP).

The majority of the Contorted Unit consists of a sphene-epidote-muscovite-biotite-quartz-feldspar gneiss, medium-grained, schistose in part; inter-layered with sphene-epidote-muscovite-quartz-feldspar-biotite schist, medium- to coarse-grained;

garnets may be present, but are small and scarce. Hornblende gneiss/amphibolite lenses and layers (commonly boudinaged) are common. Contains, in many places, lenses and discontinuous layers of unfoliated granite on a scale of feet and tens of feet. Concordant and discordant quartz veins are common. Pegmatitic layers and coarse pegmatites up to 60 in. thick are abundant and characteristic; shear foliation in the gneiss/schist wraps around the coarse pegmatites and small bodies of granite, which are generally not sheared.

This rock mass is extremely contorted; foliations are quite variable over short distances, and are generally low-angle and undulatory. Random fractures are abundant; through-going joint sets are scarce and not well-developed.

The zoned feldspar gneiss consists of an epidote-muscovite-biotite-quartz-feldspar gneiss, fine- to medium-grained, with disseminated very coarse zoned feldspar crystals; very feldspathic overall; deep weathering is characteristic.

The Brevard Zone black mylonite is generally composed of biotite, quartz, and feldspar. This unit is typically extremely fine-grained and weakly foliated. Where the foliation is better developed, the rock is shown to be very contorted. In most outcrops, the black mylonite is dark gray to black and locally contains thin light colored layers of white mylonite (see rock unit 2B description). Weathering of this unit generally yields a reddish brown to red, uniform fine clayey residuum.

The Brevard Zone white mylonite is interpreted to be sheared granite. This mylonitized granite is

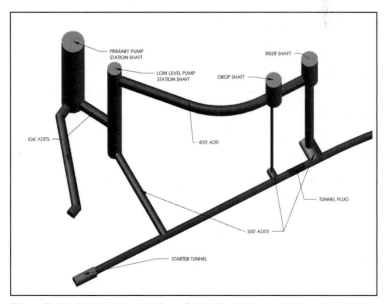

Figure 2. Shaft, adit and tunnel configuration

composed of muscovite, quartz, and feldspar; much of the feldspar is pink and coarse-grained. Shearing was pervasive and produced a well-developed shear foliation. Reduction in grain size was not as extreme as in Rock Unit 2A. Weathering of this unit generally yields a white to tan, uniform fine clayey residuum.

Rocks encountered through shaft sinking and TBM tunneling at the quarry all reflect the information as provided in the preliminary geologic report. Foliation is quite contorted over the scale of the excavation, and degrees of schistosity vary across the excavation.

TBM PROCUREMENT AND ASSEMBLY

The open-face, main beam TBM procurement played an essential part in meeting an aggressive project schedule. Considering the CMAR would be contracted before the tunnel subcontractor, an evaluation was conducted to determine if the city, CMAR or tunnel subcontractor should supply the TBM. Additionally, the evaluation included considering whether a used or new TBM should be supplied. Ultimately, the evaluation determined the best risk-reward scenario was for the purchase of a new TBM by the CMAR under an on-site first time assembly (OFTA) model.

The basis for this approach included, but was not limited to, the following:

- The CMAR entity would be more familiar than the city with the purchase of major equipment;
- A new TBM could be manufactured before the tunnel contractor was contracted or even mobilized, saving as little as two months and as much as six months in project schedule;
- A used TBM, supplied by the tunnel subcontractor, would need to be refurbished, delaying the start of TBM mining at the portal located at the base of the quarry (i.e., there was no shaft to be excavated, which would otherwise delay the start of tunnel mining);
- An OFTA scenario saves 4 to 5 months as compared to factory assembly and testing;
- An open-face, main beam TBM is a proven machine in the ground conditions in the Atlanta area; and
- An open-face main beam TBM is less sophisticated than an Earth Pressure Balance or Slurry TBM, so an engineer derived design specification did not involve excessive risk.

The new TBM for the project was manufactured by The Robbins Company. TBM components were shipped in truckloads to the jobsite in summer 2016 to be assembled using Onsite First Time Assembly (OFTA)—a method developed by the

Figure 3. Overview of the quarry with tunnel portal in view

TBM manufacturer to build machines on location rather than in a manufacturing facility. The method has the potential to shave months off of the delivery schedule and millions in USD.

The PCR JV sub-contracted the Atkinson/Technique JV (ATJV) to work on the assembly and launch of the machine, and to operate the TBM during tunnel construction. This was done in concert with a small contingent of TBM manufacturer assembly staff as required in the TBM specifications. Components were moved via 70 truckloads down steep roads to the bottom of the massive quarry—an approximately 11-hour journey from the facility in Solon, Ohio (see Figure 3).

Crews including TBM manufacturer personnel worked in the blazing heat of summer at the bottom of the quarry, which was below sea level, on days where highs hit 110 degrees Fahrenheit at 100 percent humidity. The challenges of working under these difficult site conditions, paired with the new contract relationship, caused some delays over the originally planned schedule. However, the TBM's start of boring still commenced months ahead of what would have been possible with full shop assembly followed by reassembly at the quarry portal.

The machine was launched in the second week of October 2016 following a large ceremony in which Atlanta's Mayor Kasim Reed and local and national media were in attendance (see Figure 4).

MACHINE DESIGN

Specifications for the robust hard rock machine were made following detailed geotechnical and hydrogeological field investigations comprising 25 deep borings along the tunnel alignment and 30 shallow borings in the areas for shaft construction. The borings revealed granitic gneiss rock averaging 25,000 psi UCS along with five transition zones between soil and bedrock where groundwater was expected. These zones had the potential for high-yield, fracture

Figure 4. The completed Main Beam TBM during the launch ceremony in October 2016

flows, as local wells in the area had been reported to yield up to 500 gallons per minute.

It was decided that the TBM cutterhead would be designed for 19-in. cutters to excavate the hard rock geology with greater penetration while offering longer cutter life and greater wear volume—40% more over 17-in. disc cutters based on the TBM manufacturer's research and empirical observations. The key TBM parameters are shown below:

- Bore diameter—3.8 m
- Disc cutter diameter—19 in.
- TBM weight—220 tons
- Total number of cutters—26
- Number of face cutters—16
- Number of center cutters—4
- Number of gage cutters—6
- Maximum individual cutter load— 70,000 lbs.
- Average instantaneous penetration rate of 5.8 ft/hr to 6.7 ft/hr
- Operational cutterhead thrust— 1.82 million lbs
- Operational cutterhead RPM—11.5
- Operational torque—727,516 ft-lbs
- Average ring life—32 ft
- Primary voltage—13,200 volts
- Installed power—1,770 hp
- Variable frequency drives—4
- TBM boring stroke—6.6 ft
- TBM dust control water flow rate—18.5 gpm
- Turn radius—1,200 ft

Tunnel support along the alignment was classified into three types: A, B, and C. The TBM would erect ground support consisting of two rows of double corrosion protection dowels as both excavation support and permanent support in Type A ground. Type B ground support would consist of four friction dowels with welded wire mesh, and Type C support would consist of steel ribs with welded wire mesh

as lagging. Both Type B and Type C ground would be concrete lined after excavation. Tunnel lining would have a 100-year design life, making the finished internal diameter 10 ft. Ultimately the tunnel is expected to be lined along approximately 40% to 50% of its length.

SITE PREPARATION AND TUNNEL EXCAVATION

A substantial scaling program of the quarry walls was undertaken to provide safe egress and ingress to the quarry bottom and TBM location throughout assembly and machine launch. Scaling around the quarry rim took place from April through August 2016. To secure the approximately 300 ft tall rock face above the tunnel portal at the base of the quarry, a stabilization system was designed. The system covers the full depth of the quarry and a width of approximately 400 ft centered over the portal. 3 mm mesh and rock dowels have been installed in the locations identified to have potential rock wedge failures. Canopies were installed as additional protection to workers at the portal. At the tunnel "eye," where the tunnel breaks into the ground from the portal, 20 ft long spiles were installed along the tunnel crown to stabilize the transition area. While scaling of the quarry could last indefinitely in such a large drill and blast excavation, following initial inspection, ATJV implemented a scaling protocol that requires visits quarterly to inspect the rock mass and quarry rim.

Since the October launch, the machine has advanced at rates up to 15 ft per hour and 50 ft per day in two eight-hour mining shifts with a daily maintenance period. As of late December 2017 the TBM had advanced 12,000 ft.

The machine is generally in favorable geology, but like other Atlanta area tunnels, is in very hard and very abrasive rock. Cutter wear has been good; well within estimates. The crew typically performs three to four cutter changes per day. Muck removal is being accomplished with five-car muck trains. One train covers 6 linear ft of tunnel. While much of the excavation has been uneventful the crew did come across an old abandoned well directly in the tunnel alignment, which was promptly sealed off (see Figure 5).

CONCURRENT CONSTRUCTION AND UPCOMING CHALLENGES

Construction of the shafts and pump station at the Hemphill site, where an existing reservoir is located, is being done concurrent with tunneling. The construction of the five blind-bore shafts was found to pose a significant risk to the unlined reservoir. As such, a shaft pre-excavation grouting program was designed for the soil to rock transition zone and rock

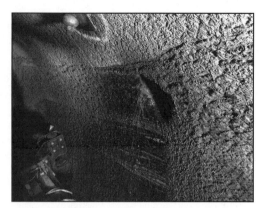

Figure 5. Abandoned well directly along the tunnel alignment

Figure 6. Completed Peachtree shaft

zone to greatly reduce the chance of communication between the reservoir and the five pump shafts. Grouting from the surface was planned with special care taken to additionally avoid any risk to the tunnel structure. By October 2017 the Peachtree shaft at the Hemphill site had been completed (see Figure 6).

The shafts are being constructed using blind bore techniques since surface blasting is prohibited at HWTP due to adjacent reservoirs. Upon the completion of the five, 11-ft diameter steel casings in overburden, the 9.5-ft diameter 400-ft deep blind bore shafts are then being drilled from the bedrock surface to tunnel depth through the steel casings in overburden. Upon completion of the blind bore drilling, a 76" ID steel pipe casing with 1-in. wall thickness will be lowered into the shaft and grouted in place. Ultimately the shafts will be directly connected to the tunnel via five, 8-ft diameter adits with lengths ranging from 20-ft to 30-ft. As of December 2017, all of the deep pump station shafts have been completed, and four of the five blind-bored shafts, to a depth of 425 ft, have been completed.

As for tunnel construction, several zones requiring systematic probe drilling still lie ahead at various locations along the alignment including under the Hemphill Reservoir. However ground conditions are expected to remain favorable overall. Project completion is scheduled for September 2019.

CONCLUSIONS

The Atlanta WSP is an important example of excavation in abrasive hard rock conditions. It also is an example of a unique and successful project structure with the CMAR purchasing a TBM and subcontracting tunneling work. Many of the challenges for the tunneling operation are upcoming and will be reported on in detail at the NAT conference.

REFERENCES

Bedell A., Jiang T., Warburton W., Horton K., Del Nero D., Jones B. 2017. Atlanta's Latest Mega-Tunnel. Proceedings of the 68th Highway Geology Symposium.

Del Nero D., Jiang T. & Bedell A. 2016. Atlanta's Mega Water Project. Proceedings of the Tunnelling Association of Canada (TAC) Conference.

PetroLogic Solutions, Detailed Geologic Mapping Along The Proposed.

City Of Atlanta Raw Water Tunnel Alignment, Fulton County, Georgia, Atlanta, Georgia, 2014.

World's Largest Tunnel Gates and Reservoir Connection Go Online as Part of Chicago's Tunnel and Reservoir Plan (TARP)

Miguel Sanchez and Faruk Oksuz
Black & Veatch

Dave Schiemann
U.S. Army Corps of Engineers, Chicago District

Patrick Jensen
Metropolitan Water Reclamation District of Greater Chicago

ABSTRACT: Chicago's Tunnel and Reservoir Plan (TARP) is a nearly $4.0 billion and over 30-year long program and arguably the largest and longest combined sewer tunnel and reservoir system in the world. The McCook Main Tunnel construction and gates installation is now complete for connection of the McCook Reservoir to the existing TARP Mainstream tunnel system. This paper describes the overall project and the final stages of commissioning including installation and operation of world's largest underground roller (or wheel) gates for flow control within the bifurcated tunnel sections that operate under 92 meters (300 feet) of water pressure.

INTRODUCTION

The McCook Main Tunnel connects Chicago TARP's Mainstream Tunnel to the McCook Reservoir. The tunnel system consists of a 10 m (33 ft) finished diameter and 490 m (1,600 ft) long hard rock tunnel constructed from a 27.5 m (90 ft) diameter and 92 m (300 ft) deep main gate shaft. The gate shaft houses six high head 4.4 m by 9 m (14.5 ft by 29.5 ft) wheel gates installed in the bifurcated and steel lined section of the tunnel. The tunnel also includes portal and energy dissipation structures as it daylights into the reservoir.

Construction of the McCook Main Tunnel in live flow conditions has been a challenging task for the McCook Main Tunnel Project participants, the project owner U.S. Army Corps of Engineers (USACE), local sponsor Metropolitan Water Reclamation District (MWRD) of Greater Chicago, designer Black & Veatch, and contractor Kiewit Infrastructure Company (Kiewit). Once completed in two stages in 2017 and 2029, the McCook Reservoir will hold 38 billion-liters (10 billion-gallons) of combined sewer overflows (CSO) and flood waters from the city of Chicago and 36 surrounding communities in Cook County, Illinois.

The construction of the tunnel system was divided into two contracts. The first contract for the gate and construction access shaft was completed in August 2011. The second contract included tunnel excavation and concrete and steel lining that was completed in September 2014; installation of gates and hydraulic cylinders that was completed in June 2017; and the last major construction activity is the removal of a temporary concrete plug (bulkhead) and lining that tunnel section to bring the overall tunnel and gates system online before December 31, 2017.

This paper focuses on the installation, testing and commissioning of the high head wheel gates. Live tunnel connection details to TARP Mainstream Tunnel were addressed in a previous paper submitted at the World Tunneling Conference in 2016.

CHICAGO TARP SYSTEM

The MWRD has been addressing CSOs and flooding in Chicagoland since the late 1960s and formally adopted the Tunnel and Reservoir Plan in 1972 to protect the region's most precious drinking water supply, Lake Michigan. Phase I of TARP, which included construction of 175 km (109 miles) of deep storage and conveyance tunnels with diameters up to 10 m (33 ft), was completed in 2006. In addition

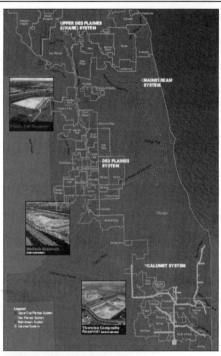

CHICAGO'S TUNNEL AND RESERVOIR PLAN (TARP)

> Adopted by MWRDGC in 1972

> Captures Combined Sewer Overflow (CSO) and Floodwater

> Phase I Tunnels - 110 miles of tunnels completed:
>   • Mainstream
>   • Upper Des Plaines (O'Hare)
>   • Des Plaines
>   • Calumet

> Phase II Reservoirs – 18 BG of storage:
>   • McCook (under construction)
>   • Majewski Reservoir (complete)
>   • Thornton Composite (complete)

• The largest water infrastructure undertaking in Chicago

Figure 1. Chicago tunnel and reservoir plan (TARP) and McCook Reservoir aerial view

to the protection of Lake Michigan from CSO discharges, Phase I resulted in substantial improvements in surface water quality as well as the quality of life for lake and riverfront communities in Chicago. Water quality improvements and flooding mitigation will be further enhanced as Phase II reservoirs are placed in service, including the three large reservoir systems, McCook, Thornton, and Majewski, as shown in Figure 1.

The McCook Reservoir is the largest reservoir in the TARP system. Once completed, this $1.016 billion reservoir facility will receive 38 billion liters (10 billion gallons) of CSO and floodwater via the McCook Main Tunnel which connects the TARP Mainstream Tunnel to the McCook Reservoir and from Distribution and Des Plaines Inflow tunnels which will bring flow from the Des Plaines Tunnel of TARP.

MCCOOK MAIN TUNNEL LAYOUT

McCook Main Tunnel daylights into the McCook Reservoir at the northeast edge and extends east towards the existing Mainstream Tunnel (Figure 2). The tunnel was excavated using sequential drill-blast and lined with concrete and steel in sections for long term stability and to minimize infiltration and exfiltration.

The McCook Main Tunnel was excavated in its entirety in bedrock, consisting of massive, relatively homogenous Silurian and late Ordovician dolomites.

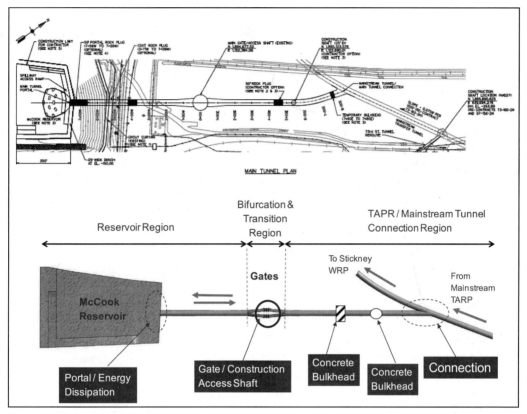

Figure 2. McCook Main Tunnel layout and components

These rocks form a relatively uniform 100+ meter (300+ feet) thick sequence across the site and incorporate the Racine Formation, Sugar Run Formation, Joliet Formation, Kankakee Formation, Elwood Formation and Wilhelmi Formation of Silurian age. The tunnel is located in the Kankakee and Elwood Formations.

The McCook Main Tunnel system has the following key components:

1. Main Tunnel Section: Approximately 490 m (1,600 ft) long, 10-m (33-ft) inside diameter hard rock tunnel, bifurcated into two tunnels for 88 m (290 ft) through the gate shaft section.
2. Main Gate/Access Shaft: 27-m (88-ft) diameter, 90 m (295 ft) deep circular shaft located near midpoint of the Main Tunnel and houses the bifurcated tunnel section. This shaft was used for construction of the tunnel and houses the high head wheel gates for controlling flow between TARP Mainstream Tunnel and McCook Reservoir.

3. Construction Shaft (contractor option): A 7.6 m (25 ft) diameter and 87 m (285 ft) deep construction shaft was located at 91 m (300 ft) downstream or west of the Mainstream tunnel connection. Kiewit elected to build this shaft to facilitate the live connection work. As the tunnel and gate shaft excavation and lining were completed, a temporary concrete bulkhead was installed to isolate the live connection section from the rest of gate and reservoir works to the east.
4. Gates: A total of six wheel gates operating under 100 m (300 ft) of water pressure head were installed in the Main Gate/Access Shaft. Each gate measures 4.4 m by 9 m (14.5 ft by 29.5 ft) with associated hydraulic cylinders, power units, and associated gate controls. Each bifurcated section of the Main Tunnel contains one main gate and two guard gates—one upstream and one downstream of the main gate. The gates, hydraulic cylinders and controls were manufactured under a separate contract and were provided to the contractor as government furnished items.

Figure 3. Bifurcated and steel-lined gate section of McCook Main Tunnel

The gates were designed by Black & Veatch and fabricated by Oregon Iron Works (now Vigor Works LLC).

5. Main Tunnel/Mainstream Tunnel Connection— This is the connection section of Main Tunnel to the existing, live 10-m (33-ft) diameter Mainstream Tunnel that remained in service throughout construction. Removal of the temporary concrete bulkhead and lining of that tunnel section will complete the connection.

6. Main Tunnel/McCook Reservoir Connection— the Main Tunnel connection to the McCook Reservoir included portal excavation and stabilization work at the quarry highwall face and an energy dissipation structure. The portal was excavated from the reservoir side and supported with rock bolts, wire mesh and shotcrete.

7. Control Building—a surface facility to house gate operating controls, hydraulic power units and provide limited storage.

The McCook Main Tunnel system design, construction, commissioning and operation were coordinated with the overall McCook Reservoir water control plan as well as the reservoir excavation, quarry highwalls stabilization, groundwater protection system construction, Distribution and Des Plaines Inflow Tunnel connections, and other reservoir features. Hydraulic structures were designed to withstand erosion or cavitation during reservoir filling and emptying cycles and to handle flows up to 850 cubic meter per second (30,000 cubic feet per second) and velocities approaching to 12 meter per second (40 feet per second).

WORKING WITH LIVE TUNNEL FLOW CONDITIONS

The connection to the Mainstream Tunnel was challenging due to limited amount of time available to access the connection area as the Mainstream Tunnel had to remain live or in service at all times. The connection is located near downstream terminus of Mainstream Tunnel which drains over a 65 km (40.5 miles) of tunnel network virtually encompassing the highly developed city of Chicago. MWRD operates the Mainstream tunnel system and the Mainstream Pump Station to collect and pump out the CSOs and subsequently treats the flows through the Stickney Water Reclamation Plant. When it rains, the tunnel fills up rapidly, and there is also a constant base flow in the tunnel of about 113 to 150 million liters (30 to 40 million gallons) per day. Kiewit designed a base flow bypass system consisting of upstream and downstream steel check dams (partial bulkheads) with a 914mm (36 inch) nominal diameter HDPE bypass pipe across the connection section of the Mainstream Tunnel (Figure 4). The connection was lined with reinforced concrete starting with the invert, then walls and finally crown. The connection works in-progress were exposed to inundation multiple times where all personnel and equipment were evacuated out of the area upon notice from MWRD operators or the weather service dispatcher. All installations were successfully completed despite multiple and complete flooding of work areas and the tunnel and subsequent reinforcement and lining work (Figure 5) were successfully completed by the end of 2016 as detailed in the referenced 2016 WTC paper.

Several lessons learned were valuable to the team as the Mainstream Tunnel was opened up

Figure 4. Upstream check dam (left) and bypass pipe and downstream check dam

Figure 5. Bull nose at main tunnel (left) and mainstream tunnel connection and concreting of invert toward main tunnel

for the first time after 35 years of service. Tunnel liner was in near perfect circular shape without any sign of damage, major cracking or water seepage. Despite being downstream, there was no sediment or grit accumulation observed, primarily due to high velocity tunnel flows created with large dewatering pumps. It is noted however, there was significant grit accumulation in the Main Tunnel once the construction work was suspended over the spring and summer of 2015. The tunnel liner concrete exhibited very high unconfined compressive strength, on the order of 83 MPA (12,000 psi) compared to initial placement specification of 28MPA (4,000 psi), also known as the MWRD's RA mix. Visual inspection, field and laboratory testing of the concrete liner and verification of its existing condition allowed the designer to shorten the limits of excavation for connection by approximately 12 m (40 ft) at the downstream end.

HIGH HEAD WHEEL GATES

Six high-head, vertical-lift wheel gates were installed in the Main Tunnel to control the flow of CSOs and floodwater between the reservoir and Mainstream Tunnel. The gates are housed in the Main Gate/ Access Shaft (MGAS) in the bifurcated section of the Main Tunnel (Figure 6). Each bifurcation has one main gate and two guard gates—one upstream and one downstream of the main gate. All gates have an opening of 4.4 m by 9 m (14.5 ft by 29.5 ft). The main gates are the primary feature to control water flow and are designed to seal in both flow directions. The guard gates are designed to seal against hydrostatic head on one side only. The upstream guard gate holds back water on the Mainstream Tunnel side and the downstream guard gate holds back water on the reservoir side. The guard gates provide redundancy to the main gates and can also be used to isolate a main gate when needed for maintenance.

Figure 6. Gate leaf tunnel suspended from a crane (left) and gates in bifurcated section of main tunnel (right)

The gates are operated by a hydraulic operating system. Each gate is raised and lowered by a hydraulic cylinder that is mounted above the gate in a vertical position. The cylinders are actuated by two hydraulic power units (HPUs) at ground level in the control building near the MGAS.

GATE DETAILS

Each gate consists of three sections (leaves) that are pinned together to form the gate. There are four pin connections between the lower and middle leaves and four pin connections between the middle and upper leaves. The upper leaf is connected to the hydraulic cylinder by means of a single pin. Each leaf has four wheels, two on each side (Figure 6). These wheels are the bearing portions of the gate when it is subjected to load.

The gates use neoprene seals to control the water. The seals along the sides and top of the gate are center bulb seals. The seals have a fluorocarbon (PTFE) coating on the bulb to reduce friction during gate operation. This coating will eventually wear away but does not compromise the sealing function of the seal. The seals on the main gates have a 0.95-cm (⅜-inch) preset deflection for flow towards the reservoir, and a 0.32-cm (⅛-inch) preset for flow in the reverse direction.

The seals on the guard gates have a 0.95-cm (⅜-inch) preset. A pressure groove behind the seal is provided so hydrostatic pressure will force the bulb against the sealing surface. Holes in the seal bars allow pressure to build up in this groove. This groove also assists in lessening wear on the seals during gate movement when no differential pressure is present. A wedge seal is mounted at the bottom edge of the gate. The joint between the gate sections is sealed with flat natural rubber seals. All seals are split at the joint between gate sections to allow installation and

removal of the gate in sections. The seals are detailed to allow a 0.16-cm (¹⁄₁₆-inch) preset compression between the gate sections in order to minimize leakage at these joints.

Each leaf is a steel structure made up of welded horizontal plate girders that span from wheel to wheel. It is composed mostly of A572 grade 50 steel, with some portions that are 304L stainless steel. The gate is metallized with a zinc-aluminum coating to help protect it from corrosion. The metallized coating is sealed with a vinyl sealer. The gates are open on their loaded side (Mainstream Tunnel side for the main gates). This keeps the gate from being buoyant when submerged. Drain holes in the girder webs and in the gate bottom keep water from ponding on the girders as the water level goes down.

The wheels for each gate are 0.9-m (3-feet) in diameter and made form ASTM A705, UNS 13800, Condition H1025, 380 BHN stainless steel. They rotate about a fixed axle that is of the same material, but tempered to a harder condition (430 BHN) so that wear will more likely occur in the wheel. The wheels are also softer than the wheel track plates so that wear will more likely occur in the wheel. The wheels are crowned with a 15-m (50-feet) radius so that they bear continuously on the wheel track plates even when the gate deflects under load. The wheels have a force fit bronze bushing that rolls on the greased interface between the bushing and the fixed axle.

The gate guides position each gate correctly and provide a seating and sealing surface for the gate. They consist of upper and lower guides. The upper guides are only used during gate installation and removal. They help to control gate motion in the upper part of the shaft. The upper guides are bolted to the concrete of the gate well walls. They consist

Figure 7. Hydraulic cylinder on a trailer (left) and a hydraulic power unit in control building (right)

of ASTM A36 angles that have been galvanized to provide a durable coating.

The lower guides are the primary traveling surfaces of the gate. The wheels of the gate run along wheel tracks. There is a front track and a back track. The distance between the tracks is slightly larger than the wheel, to keep the wheel from binding but also to reduce play in the gate itself. The wheel tracks run from the tunnel invert all the way to the cylinder support level. These wheel tracks are ASTM A693, UNS 13800, Condition H950, 430 BHN stainless steel. This is a harder material than the gate wheels so that damage occurs in the wheels rather than in the track. On the side of each guide is a roller track plate on which the guide roller contacts. These track plates are bolted to an ASTM A304L grade stainless steel that forms the guide slot. This also is the sealing surface of the slot. When seated, the gate presses up against the wheel track and compresses the seals against the sealing surface. The guides are anchored to the shaft concrete with embeds and rebar anchors cast into a 34 MPA (5,000 psi) second placement concrete.

The MGAS concrete consists of a 34 MPA (5,000) psi reinforced concrete. This concrete anchors the steel liner and forms the sides of each of the six gate wells. It was designed to resist water pressure around the exterior of the shaft. Also, the walls between the gate wells were designed to resist the full height differential water pressure on them (i.e., one gate well full and the adjacent one empty). The concrete was placed in sections. Between each vertical construction joint is a PVC waterstop system to resist leakage through the joint. The horizontal construction joints were prepared to allow for bonding between the concrete lifts and do not have a waterstop.

Each of the guard gate wells have large access spots, enough to access the tunnel invert using a 4-person crane basket. The main gate wells have access spots as well, but much smaller.

HYDRAULIC OPERATING SYSTEM DETAILS

The cylinders raise and lower the gates through the use of hydraulic pressure generated by the HPUs. Fluid pressures in the cylinder are monitored by two pressure transducers, one on the bore side of the piston near the top of the cylinder and one on the rod side of the piston near the bottom of the cylinder. The position of the cylinder rod is monitored by a position sensor. Each cylinder is made up of a shell, a piston head, a rod, a clevis, supply and return piping and a manifold with check valves. The bore diameter of the cylinder is approximately 1 m (3 feet), the rod diameter is approximately 0.3 m (1 foot) and the total stroke in the cylinder is approximately 9.5 m (31 feet) (Figure 7). Each cylinder weights approximately 65 kips and is capable of producing a jacking load of 2,100 kips. The cylinder rod is coated with a protective anti-corrosion metallic coating. The cylinder shell is coated with epoxy paint. The hydraulic fluid is carried by piping between the HPUs and cylinders.

The cylinders are actuated by two HPUs in the control building near the MGAS (Figure 7). Each HPU typically controls three gates on each side of the bifurcated tunnel. Each HPU consists primarily of an internal reservoir, two pumps, a manifold, a PLC cabinet with screen, and various internal instruments for the function of the HPU. There is a manual crossover between the HPUs to allow for redundancy in hydraulic operation.

GATE TESTING AND OPERATION

Several tests were conducted to check installation and operation of the gates, as follows:

- Gate Leaf Dry Run Test—While suspended from a crane, a gate leaf was slowly lowered and raised between the ground surface and tunnel invert to verify the proper alignment of the upper and lower gate guides and sill plate at the invert.
- Functional Performance Test—The fully-installed gate system was tested to verify the operation of the HPUs, hydraulic cylinders, gates, and hydraulic piping under operating conditions.
- Intermediate (Flat) Seals Air Pressure Test—The flat seals between the gate leafs were subjected to air pressure up to 1.2 MPA (175 psi) to verify performance of those seals prior to filling tunnel with water.
- Wet Test—A 79-m (259-ft) column of water obtained from a canal was pumped into the gate shaft and tunnel on one side of a pair of guard gates, and then one side of the main gates to test sealing of gates at full hydrostatic head conditions.

The normal condition for all gates is the raised position, allowing flow to move unimpeded between the reservoir and Mainstream Tunnel. The gate speed is approximately 0.3 m (1 foot) per minute, so it takes approximately 30 minutes for the gates to move from the open position to the closed position. When it is desired to isolate the reservoir from Mainstream Tunnel, both of the main gates will be closed. In the event that the main gates do not close (or only one closes), the respective guard gates will be closed (upstream guard gates if Mainstream Tunnel water level will be higher, downstream guard gates if the reservoir water level will be higher). The gates are designed to hold back large differentials of water pressure; however, the gates will typically be moved or operated when there is relatively equal water levels on both sides of the gate.

The motion of the gates is controlled by the HPUs in the control building at the ground surface. Generally, the gates are operated in pairs (i.e., both main gates move together, both upstream guard gates move together, both downstream guard gates move together). The gates can be operated locally from the control building or remotely from the Stickney Water Reclamation Plant via SCADA.

CONCLUSION

Removal of the temporary concrete bulkhead and final stages of commissioning will take place from late 2017 through the first quarter of 2018. Once the bulkhead is removed and construction of the remaining reservoir features is completed, including an inflow/outflow connection and aeration facilities, the reservoir will be available to take water before the end of 2017.

The McCook Main Tunnel and Gates are one of kind underground structures that are used for management of high volume and high pressure flows in tunnels. Design, fabrication, storage, delivery, installation, and commissioning of tons of concrete and steel structures at depths up to 300-feet were truly an engineering and construction feat for the many that worked on this project, and for those that will be benefiting from the water quality improvements and flooding mitigation for years to come.

ACKNOWLEDGMENTS

The authors acknowledge the efforts and contributions of all project participants. In addition to authors listed for this paper, we specifically acknowledge Gordon Kelly (USACE), Kevin Fitzpatrick (MWRD), Brent Bridges and Mark Petermann (Kiewit), Charles Strauss and Clay Haynes (Black & Veatch), Carmen Scalise (Metropolitan Water Reclamation District of Greater Chicago), Matt Trotter (Kiewit Infrastructure Company), and Mike Padilla (U.S. Army Corps of Engineers, Chicago District) for their contributions to this paper.

REFERENCES

Oksuz, F., and Trotter, M., 2015. *McCook Main Tunnel Construction,* New York: George A. Fox Tunneling Conference.

Oksuz, F. and Trotter, M. 2016. *McCook Main Tunnel Connection*, San Francisco, World Tunneling Conference.

Establishing Access into Chicago's Main Stream Tunnel Under Live CSO Flow

Lukasz Dubaj
Kiewit Infrastructure Co.

ABSTRACT: The McCook Reservoir Main Tunnel System Project includes a 1600' long, 33' diameter tunnel connecting the McCook Reservoir to the existing Mainstream Tunnel (MST) which is part of Chicago Metropolitan Water Reclamation District's (MWRD) Tunnel and Reservoir Plan (TARP). This paper focuses on establishing access between the two tunnels during live CSO flows in order to complete construction.

Establishing access to the live tie in included the installation of a temporary access platform, installation of temporary bulkheads at the upstream and downstream ends of the MST along with the assembly and finally installation of a 36" HDPE pipe that would allow up to 42 million gallons of CSO water to flow to the nearby water treatment plant while allowing to safety perform the tunnel tie-in.

PROJECT INTRODUCTION AND OVERVIEW

Kiewit Infrastructure Co. (KIC) was awarded the McCook Reservoir Main Tunnel System (McCook) project in 2012, the project scope included the construction of a new 1600' long, 33' diameter tunnel connecting the McCook Reservoir to the existing Mainstream Tunnel (MST) Chicago Metropolitan Water Reclamation District (MWRD). The scope of the tunnel tie-in connection included assessing the existing MST tunnel conditions, demolishing the existing liner, installing new ground support and placing a new liner connecting the newly excavated tunnel with the existing MST. Without knowing the conditions that existed within the MST, establishing safe and reliable access was critical to the success of the work. When working in the tie-in connection, crews worked three eight hour shifts per day seven days per week. The typical crew comprised of a foreman, four miners and two operators and the equipment used underground included a CAT 953 loader and two CAT 314 excavators.

After evaluating numerous concepts, the project team proceeded with a flow through pipe and bulkhead system to allow access and safe work in the MST while allowing a maximum of 42 million gallons of CSO water per day (MGD) to flow through the pipe and further downstream in the MWRD system. A steel and timber deck provided access above the MST flow while the flow through pipe and bulkheads were being assembled and installed. Detailed planning and diligent coordination between KIC and MWRD was required as the water level in the MST needed to be below elevation –255 feet at all times

to allow entry into the MST and during the installation of the bulkheads the water level was required to be below elevation –258 feet. The MST invert elevation is –263.94. All personnel were required to wear personal flotation devices (PDF) when on the access deck during the installation. Any time personnel were within 20' of the access deck, installing the flow through pipe or bulkheads they were required to be tied off and wear PFD.

Access in the MST was established in two primary phases, the first was the installation of an access deck that allowed MST flows to remain uninterrupted while completing work on the tunnel arch and sides. The second phase included the assembly of a 36" HDPE flow through pipe, installation of upstream and downstream steel bulkheads and installation of the flow through pipe, removal of the sewer material between the bulkheads and finally the removal of the access deck which allowed safe access to the MST, specifically the invert, while allowing sewer water to continue to flow downstream. The installation of the access deck and installation of the bulkheads and flow through pipe are broken into further detail in the following sections: access deck mobilization, installation of access deck steel beams, installation of access deck mats, installation of bulkhead framework, assembly of flow through pipe, installation of flow through pipe and supports, installation of steel bulkhead plates and access deck demobilization.

To summarize the work before the installation of the access deck, controlled blasting was used to excavate five rounds in limestone rock. One full face round along with four concentrated blasts to create a pilot tunnel that would connect the McCook Tunnel to the MST. With the pilot tunnel excavated

Figure 1. Panoramic view within the MST from left to right is the upstream end, McCook tunnel connection and downstream end

Figure 2. Steel beams mobilized to site

and supported, continuous access from the McCook tunnel invert to the top of the bench and to the pilot tunnel was needed, a ramp was built with excavated spoils and material brought in from the surface using a CAT 953 loader.

After the access deck was installed and just prior to the start of the installation of the bulkhead and flow through pipe the existing MST concrete liner was demoed, new contract required ground support was installed in the MST. Ten controlled blast rounds were excavated and supported which provided additional access to the installed deck. After all of the blasting was completed, the top heading of the McCook tunnel was connected to the MST with an 81'-6" opening providing clear access to the 160' long steel and timber deck.

Access Deck Mobilization

Twenty-two loads were needed to mobilize the steel beams and deck mats to the job. All materials were unloaded at a nearby site and transported to the main construction shaft site as needed using a CAT 930 loader. The deck primarily consisted of 31'-3" long W14×109 beams with web stiffeners and 8" thick timber mats. Each steel beam weighted about 3,500 lbs. The timber mats varied slightly in size as per the design but in general were 8'-0" wide by 8'-0" long and each weighing about 2,200 lbs. The steel beams could be picked using a center plate

(knife plate), by two knife plates on each side of the beam or on the ends (beam support plates). Using a Liebherr 1130 crane, the beams were lifted from one end and lowered down vertically because the shaft diameter was 20' at its narrowest point. Timber deck mats were also staged off site and moved to the site as needed. The mats had two 1" coil rods running through them with about 3" sticking out on each end. A plate was attached to either side of the coil rod and could be adjusted between an "up" or "down" position. The plates were used for rigging in the "up" position and were stored away in the "down" position. Typically, the timber deck mats were set down the shaft in a bundle of four. For both the steel beams and timber deck mats a CAT 314 excavator was used to move the material from the shaft bottom to the tie-in location and another CAT 314 excavator was used to set them into place.

Installation of Access Deck Steel Beams

Prior to the installation of the steel beams, a small wall form was erected, poured and stripped as part of the existing MST liner was blasted off. To start, control points were surveyed in on the side of the MST where it meets the McCook tunnel. Once control was established, the first three beam locations were

Figure 3. Steel beam mobilized from the McCook tunnel to the MST

Figure 4. Final steel beam set into place in the MST

Figure 5. Installing access deck timber mats

surveyed in. Next the three beams were set, leveled and secured to the left rib only using Hilti anchors. The MST tunnel had a slight curve to it and therefore the spacing of the beams was 7'-11" on the left rib, 8'-0" at tunnel centerline and 8'-1" on the right rib (tunnel orientation is always based on looking downstream in MST). The steel beams were then leveled again and secured on the right rib. The concrete that was poured back prior to the installation of the first beam did not conform to the original curvature of the MST tunnel liner, and therefore, the beam support plates did not have full bearing on the concrete surface. Also, the Hilti anchors had a gap between the steel and the concrete that varied between 0" and 3½." After discussing the infield concerns with the designer, it was determined that the support plates did not require full bearing on the concrete surface and installation of the beams was able to proceed. The installation went much smoother after fourth beam was set because the beams could be secured to the existing MST liner. With one concern resolved, another came up. The existing 33'-0" diameter MST liner had ½" to 1" full circle indentations (presumably from a lap angle or dutchmen used in a curve) which forced the crews to set the beams lowered than originally planned. In the end this wasn't a huge concern but the deck did not have a smooth transition between timber mats as planned. The orientation of the beam was also important because the beams had four couplers welded to it that secured the deck mats from floating away. OSHA scaffold planks were used as a temporary means of access before timber deck mats were installed.

Installation of Access Deck Timber Mats

After the first three steel beams were secured, timber deck mats were installed. The mats were made of 8"x8" No. 1 Southern Yellow Pine timber anchored together with two coil ties. There were three different widths of timber mats due to the cure of the MST. Mat A, was 7'-0" wide and was installed closest to the left rib, next Mat B and Mat C were 8'-0" wide

Figure 6. In the MST, looking downstream at the installed access deck and handrail

and were in the center and directly adjacent to Mat A, and finally Mat D was 6' wide and was installed on the right rib. All of the timber mats were 7'-9" long. When one complete bay (mats A, B, C and D) was installed, this created a 1'-0" opening in the deck between Mat C and Mat D so that anchors for the 36" HDPE pipe could be installed. The deck mats were secured with a bolt and ½" thick plate connected to a coupler that was welded to the steel beam. At every location where the mats were secured to the deck, a ¾" thick plywood sheet with an opening in the middle was installed in an effort to protect the nut, plate and bolt from equipment. After 12 deck mats (3 bays) were installed, the CAT 314 excavator could move out onto the newly constructed access deck and continue installing steel beams and followed by deck mats. This process was continued completing the deck bay by bay. After all of the deck mats were installed, the top of the deck was located just over 12'-0" above the invert of the concrete in the MST. Once installation was completed, the 1'-0" opening in the deck was covered up and secured with plywood and handrails were installed on either end of the deck. Inspection were completed throughout the

Figure 7. Survey layout for beam "A" and "B" on the upstream end

Figure 9. W14×35 piles being installed at the upstream bulkhead

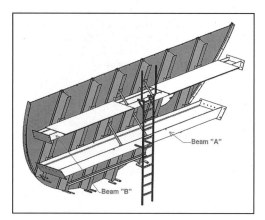

Figure 8. Upstream bulkhead

installation of the steel beams and timber mats and once installed inspections of the access deck were conducted twice a day by the shift superintendents.

MST Bulkhead Framework Installation

Prior to the installation of the bulkhead framework, additional survey control points were installed in the crown and above the bulkhead frame. String line and plumb bobs were tied to installed spads to extend down to the elevation of the two main beams for each bulkhead. Survey was a critical component of the operation as there was a tight tolerance for the bulkhead beams (±3" in elevation) and the entire bulkhead was sloped at a 20 degree angle. After all of the bulkhead materials were mobilized into the tunnel, the CAT 314 excavator was used to move, carefully set, and level the lower beam "A" into place near the upstream end of the access deck. Once braced against the existing concrete liner, a ladder was used to access beam A from the access deck. The beam was again checked against the known string line

lengths from the crown and spring line. Next, steel shims were wedged between the beam end plates and concrete so that the end of the beams had full bearing on the MST liner. Finally, Hilti anchors were installed to secure the beam in place. With beam "A" finished, the excavator was rigged up to beam "B" and just as the previous beam, set and leveled and checked against the known string line lengths. Again, shims were added if needed however the beam sat relatively flush against the concrete surface and minimal shims were needed. Four Hilti anchors were installed at each end to secure beam "B" into place. Both downstream and upstream bulkhead beams "A" and "B" were identical. Each beam "A" was a W36×135, 26'-5" long and weighed 3,904 lbs. while each beam "B" was a W21×62, 32'-2" long and weighed 2,000 lbs. Beams "A" and "B" on the upstream and downstream sides (4 in all) were installed in one 8 hour shift. Before the installation of the bulkhead framework continued, the installation of beams "A" and "B" were verified and signed off.

After beams "A" and "B" were installed at both ends of the access deck, the next step was to install six W14×35 piles at each bulkhead so the piles brace against the beams. A water hose was used to spray the invert bottom to loosen up any silt and sediment so that a tight seal between the steel pile and MST concrete invert bottom could be created. The upstream piles were longer than the downstream piles as the upstream bulkhead was 2'-2" taller than the downstream. The piles were installed, aligned and temporarily secured to the beams with clamps. Although it was difficult to check the middle two piles because of the continuous flow of CSO water, all were verified to be bearing against the invert concrete surface. Once verified, a six inch long fillet weld was placed connecting the pile to Beam "A" and an identical weld connected the same pile to Beam "B." When welding the connections between Beam "A" and the piles access was quite difficult due to the lack of space. The installation of all 12 piles for the upstream and downstream bulkheads took one day. Next, four steel angle knee braces were set and welded on each bulkhead. The knee braces created a connection between the two middle piles and each beam (A & B). A weld was placed to join the two sections of the bulkhead.

Each bulkhead had two rolled rib braces on each side. The brace consisted of two plates with a 1" gap in between them welded to a 33' diameter rolled plate. The rib braces were installed at a 20 degree angle mimicking the piles and providing support for the bulkhead plates. Each brace was secured using six Hilti anchors. As previously mentioned, about every 50 feet in the MST, the existing concrete had 1" indentations into the existing concrete. On the upstream end of the tunnel, the bulkhead lined up on a concrete indentation. This indentation created a gap between the rib brace and the concrete. The void was filled with steel shims and wedges and dry packed with hydraulic cement. Once the piles and rib braces were installed and verified, the first bulkhead plate was installed with the excavator at the upstream bulkhead. The bottom of the plate was cut to fit the curvature of the 33' diameter MST liner and the top had a 36" diameter half circle cut out to allow the flow through pipe to set inside. An identical plate was installed at the downstream prior to launching the flow through pipe. Finally, a ¼" thick gasket was stretched out around the steel plate with the semicircle cut into it in preparation for the flow through pipe installation.

Assembly of Flow Through Pipe

Support anchors were surveyed and installed in the crown of the MST. The anchors were laid out to follow the slight tunnel curve to provide access for future work when setting forms and pouring the new final liner. A total of 11 anchors spaced at 16'-0" were installed using two Genie Z34/22 manlifts. The anchors were verified after the installation to ensure correct placement and were also pull tested to ensure the installation met the quality requirement to support the pipe when in service. Additional hardware was installed at each anchor support, this included four thread-on eye nuts, coil rods, shackles and a turnbuckle. The turnbuckles were key in the installation of the flow through pipe as they provided any fine adjustments necessary in the pipe support height.

The procurement and assembly (fusion) of the flow through pipe was subcontracted to ISCO Industries. The 36" diameter high-density polyethylene (HDPE) pipe was used as a conduit to transport tunnel flow through the work area after the bulkheads were installed. Four, 50 foot long, sections of pipe were delivered to the project and lowered down the construction shaft. Due to the shaft only being 20' in diameter, the pipes needed to be lowered down vertically which for safety and quality concerns required a custom built lifting frame. A McElroy TracStar 900 Fusion Welder was also lowered into the tunnel and mobilized approximately 40' from the edge of the downstream end of the access deck. The welding machine was used to create three butt fusion welds in the tunnel to create the full 200 foot long pipe. All pipe welding was completed within one shift. With the laborers assistance, the operator used a CAT 314 excavator to maneuver two sections of pipe onto rollers located at each end of the fusion welder. The pipes ends could then be moved closer to the welding machine, cut flush, heated and ultimately fused together. Once the first weld was completed the now 100'-0" long pipe was carefully moved upstream to allow for the next joint to be welded. The excavator then picked up the next piece of 50'-0" pipe and placed it on the welder roller located at the downstream side of the machine and the welding process would start over. Additional rollers, separate from those attached to the fusion welder were used to assist in the maneuvering of the welded pipe

Figure 10. Installed flow through pipe anchors

Figure 11. HDPE pipe being mobilized to MST just before starting to fuse pipe sections

sections. The welding process continued one more time until all of the welds were completed and the welded flow through pipe stretched past each end of the access deck.

After all flow through pipe welding was completed and just prior to the launching, pipe support clamps were installed at 16 foot centers. Clamp type supports featured a ½" thick, 36" diameter pipe clamp with supports allowing the pipe to hang, or be supported from the bottom and side. Although the pipe was initially hung from the anchors to allow work on the tunnel invert and sides, the additional bottom and side supports were needed to allow the arch concrete to be completed.

Installation of Flow Through Pipe and Supports

Before the flow through pipe was launched, inflatable plugs and lifting plates were installed. At each end rubber plugs were installed and inflated inside of the 200 foot long pipe. This prevented water from coming into the pipe and kept the end of the pipe above water and assisted in maneuvering the pipe during install. Next, a lifting tugger plate was installed on downstream end of the pipe. Finally, the upstream end of the flow through pipe was carefully pushed off of the upstream end of the access deck using a CAT 314 excavator allowing the end of the pipe to set into the water. The pipe was launched in between the first and second pile from the right rib (looking downstream), the same location where it would eventually be installed. The excavator slowly moved back and forth to push about 20-feet of pipe into the water at one time. At no point did the pipe that was being pushed upstream get hung up which created a smooth process and taking minimal time. When only about 20-feet of pipe was left to launch, a long wire rope cable (tail) was attached to the lifting eye. This tail allowed for a shackle to be easily attached and hooked to an air winch.

Once the flow through pipe was completely launched, an air winch was bolted to the timber access deck. The winch was installed about 50 feet from the downstream edge of the timber access deck. The winch pulled the flow through pipe under the deck and into position. Also, a fairlead (6" bronze bushed sheave) was secured to the downstream end of the access deck. A CAT 314 excavator was moved into place about 5-feet from the edge of the downstream end the access deck and a snatch block with a shackle was connected to the excavator boom. The wire rope from the winch was fed through the fairlead, through the snatch block connected to the excavator, ran under the 160' long access deck and was finally connected with a shackle to the long tail that was previously secured to the end of the pipe. As the pipe began to inch closer and closer to its final resting top, a second CAT 314 excavator was used at the

upstream end to assist the pipe getting through the lower bulkhead sheet that was previously installed. Still secured to the winch with the assistance of the excavator, the pipe laid completed under the access deck with about 4-feet extended past the upstream bulkhead and another 16-feet extended past the downstream bulkhead. A wire rope was attached to the lifting plate on the upstream end of the pipe and anchored to the side of the tunnel. With all of the pipe support hardware already installed, the second excavator carefully lifted a portion of the pipe with a nylon sling and a labor secured the pipe clamp to the shackle that was connected to the supports hanging from the tunnel arch. With almost all of the clamps secured to the supports, the winch cable was disconnected from the downstream end of the pipe and allowed the excavator to move out of the way so the rest of the pipe supports could be installed. Although the pipe supports were surveyed in and the clamps on the pipe were carefully measured, small adjustments in the alignment were made to the clamps so the clamps were directly above the supports. Also, adjustments in the height of the pipe were made using the previously installed turnbuckles. The correct alignment of the clamps was critical to the success of the flow through pipe so that once section would not be overloaded.

Installation of Steel Bulkhead Plates

With the bulkhead framework, bulkhead plates below the flow through pipe, flow through pipe and flow through pipe supports installed, the remainder of the bulkhead plates were installed. All of the 1" thick bulkhead plates had a small hole at the top of the plate which allowed for a shackle to be rigged

Figure 12. HDPE pipe launched upstream

and the plate hoisted with the CAT 314 excavator. The upstream bulkhead installation was started first. The plate installation was completed from the outside ends towards the middle. With the three center plates left to install in the upstream bulkhead, the attention was then shifted to the downstream bulkhead installation. The downstream bulkhead was installed in the same manner as the upstream one. The three outer plates were installed at the downstream bulkhead, the plates were again verified and the crew went back to the upstream bulkhead to install the final three plates. After a final inspection was completed of both bulkheads, the final plates were installed and steel wedges were used to secure the plates into place. One of the main reasons the bulkhead plates slide into place as easily as they did was because of the vaseline. However, a few of the plates had a harder time going in, when this happened, the crew would first lift the plate back up and slide it back into place a few times. After the installation was completed, trash pumps were used to dewater the area in between the two bulkheads. During the dewatering process, a few of the timber deck mats were removed from the access deck with the excavator to provide additional/easier access down to the concrete invert. Although the trash pumps worked well at first, the sludge became too thick to pump. The decision was then made to remove the sludge with an excavator and muck it out to the surface and send the sludge off site to be properly disposed of. After the area was completely dewatered and a considerable amount of hand mucking of sewer sludge completed to expose the concrete invert, a steel angle was welded across the bottom of the four center piles on both bulkheads. Then, the toe of each pile was backed by a 1" thick "L" lug and secured with two Hilti anchors to the concrete liner. Next, one foot tall mini-bulkheads were installed about 10 feet behind each bulkhead and. A small submersible pump was

Figure 13. Upstream steel bulkhead plates being installed

Figure 14. Removing CSO sediment from MST

set in between the bulkhead and min-bulkhead. This allowed for any water that would leak past the main bulkhead to be trapped in between the two bulkheads and pumped outside of the working area. Lastly, a ladder was installed to allow access to the top level of each bulkhead. With the installation of the bulkheads complete, a few leaks on both the upstream and downstream were noticed. The two main places for leaks were where the bulkhead was bearing against concrete and around the flow through pipe. A combination of oakum, lead wool, and hydraulic cement was used to chink off the leaks. With time, many of the leaks sealed themselves off with the assistance of sediment that came through the MST.

Access Deck Removal and Demobilization

Just prior to starting the bench drill and shoot operation, the access deck was removed. Removal of the decks started at the upstream end and continued downstream until about 8 bays remained at which point the 4 bays closest to the downstream bulkhead were removed. Typically, two bays of timber mats were removed followed by two steel beams. This was limited by the reach of the CAT 314 excavator. Timber deck mats and steel beams were transported one at a time with a separate CAT 314 excavator to the shaft bottom. It was important that timber mats and steel beams were thoroughly cleaned and checked before they were flown out of the shaft so that no loose material flew down the shaft and the surface was not contaminated with CSO sediment.

Figure 15. Completed installation of the upstream bulkhead with the access deck removed

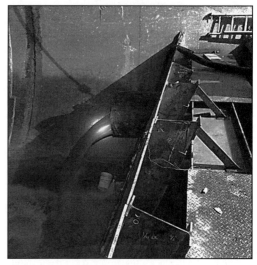

Figure 16. CSO water flowing through HDPE pipe at the downstream end

KEYS TO SUCCESS AND LESSONS LEARNED

There were a number of keys to the successful completion of establishing access in the main stream tunnel along with a few lessons learned. Some of the keys to success included methodical surveying throughout all of the operations, constant communication with the designers and fabricators, utilizing turnbuckles to support the follow though pipe and finally skilled craft specifically operators. The most notable example of these keys to success was explaining to the designer of the steel beams that part of the existing liner had blasting off during the initial tie-in and the end plate would not have full bearing on the main stream tunnel liner. Understanding the field conditions, the designers were able to quickly provide a solution by working with the construction team to clarify the requirements for the minimum bearing of the steel against the tunnel liner resulting in no downtime. All of these examples contributed to the success, however, the biggest single factor to successfully establishing access was the detailed planning and discipline execution in the field. Individuals were dedicated to the planning efforts about one year before work had begun.

With regards to the lessons learned, skilled welders were key to the steel bulkhead installation. A number of welds were redone for various reasons including not meeting the specified length or thickness. Also, welds were undercut which likely was a result of the travel speed being too fast. The availability of miners that could weld to the standards that were required was a challenge and slowed down production on the steel bulkheads. Another challenge came after the entire installation was completed, the flow through pipe extended about 15 feet past the downstream bulkhead. The project team decided to cut a 10 foot section of pipe off so only about 5 feet extend past the bulkhead. After the pipe was cut, the water near the bulkhead became more turbulent and increased the difficulty to plug leaks. An additional pipe support could have been designed and installed so the flow through pipe does not deflect and the water discharging from the follow through pipe caused less problems with leaks.

THANK YOU TO ALL THOSE INVOLVED IN THE SUCCESS OF THE PROJECT

Finally, a very special thank you to all of the staff and craft that worked on the McCook Main Tunnel System project and those that supported the project off site. The work was easier to execute, changing conditions were easier to implement and problems were easier solved because of everyone's dedication to the safety and success of the project. Thank you for bringing a positive attitude to work every day.

TRACK 4: CASE STUDIES

Session 2: Transportation 1

Clint Wilson and Elmar Feigl, Chairs

After the Tunnel Drive, Finishing the Highway to Replace SR 99 Below Seattle

Gregory Hauser
Dragados USA

Susan Everett
Washington State Department of Transportation

Joseph Clare
Mott MacDonald

ABSTRACT: The Alaskan Way Viaduct Replacement, Bored Tunnel Project included a 9,270 LF tunnel and construction of a four lane, double decked highway to carry traffic below the City of Seattle Washington. Despite a major two year delay, tunnel mining with the world's largest earth pressure balance TBM was completed on April 4, 2017 with virtually no impact to surface structures and utilities.

This paper will describe major events of tunnel mining leading up to the successful hole through and a description of the work sequence to complete the tunnel interior structures including mechanical, electrical, plumbing, fire and life safety, the communication and support systems to allow service in 2019.

INTRODUCTION

Washington State Route 99 (SR-99) parallels the waterfront of Seattle on the Alaskan Way Viaduct, (AWV) an elevated double deck reinforced concrete structure built in the 1950s. In 2001, the magnitude 6.8 Nisqually earthquake caused substantial damage to the AWV. The Washington State Department of Transportation, (WSDOT) along with the City of Seattle and King County agreed on a bored tunnel replacement for the AWV and the state requested design-build proposals in 2010 as part of an overall program to replace the damaged viaduct and maintain traffic volumes through the city.

In December 2010 the proposal submitted by Seattle Tunnel Partners, (STP) was selected by WSDOT to provide a replacement highway with two lanes in each direction within a tunnel that would be the largest TBM mined tunnel in the world. At 57.5 feet (17.5m) in mined diameter, the tunnel for the Alaskan Way Replacement Project was a tremendous undertaking and posed settlement risks for the downtown buildings that the tunnel would be driven under. As a result, WSDOT required strict controls for settlement and monitoring of the tunnel progress and impact to downtown Seattle and the city's buildings, roads, utilities, and citizens.

The proposal from STP, a Joint Venture of Dragados USA and Tudor Perini Corp., included a 52-foot ID tunnel with a cast-in-place double deck

highway constructed as the tunnel was being driven and with an original completion date of December 2015. The tunnel boring machine (TBM) was manufactured by Hitachi Zosen Corp. from Osaka, Japan and delivered to the site in April 2013. The TBM was launched on July 30, 2013. In early December 2013 the TBM began overheating. By December 7, TBM production was halted for further investigations and by February 2014 it was determined that major repairs would be required. The cause of the damage is under litigation and is not a subject of this paper.

To repair the TBM, STP constructed an 80-foot diameter, 120-foot deep shaft to access the TBM, removed the cutter head and main drive unit, replaced the seals and main bearing, and made other repairs to the TBM. The TBM mined out of the north wall of the access shaft on January 4, 2016 and continued north to a planned Safe Haven #3, crossed under the existing viaduct, and beneath downtown to the north portal site. During the two-year period that the TBM was down for repairs, and throughout the remaining tunnel drive, activity on other parts of the project continued which helped reduce the overall impact to the project and minimized the overall project delay. While mining was delayed, STP continued with construction of the north and south operations buildings, the cut and cover roadways that transition to the tunnel, and the interior highway structure within the tunnel.

PLACING THE INTERIOR CONCRETE HIGHWAY STRUCTURE

With the halt of tunneling operations, STP continued to proceed forward with mobilizing the concrete formwork on the site and started the concrete placement of the internal highway structure. At this stage, the tunnel had been driven over 1,000 feet and it was determined that STP could safely remove the jacking launch frame, clear the launch shaft for form work placement, and the form travelers for the interior structure.

The launch jacking frame and 13 false sets, (precast tunnel rings) erected in the launch shaft were removed to clear the tunnel opening to allow the form travelers to be erected. STP built a ramp from the floor of the launch shaft to the invert of the tunnel to facilitate segment delivery to the heading. Even though tunnel precast segment deliveries were halted at this time, a means of segment delivery through the constructed interior structure was needed for tunnel segment delivery once tunneling resumed.

In addition to the ramp from the shaft invert to the tunnel invert, a flat concrete invert was placed along the tunnel to allow for concurrent segment delivery and concrete placement for the interior structure. The flat invert allowed tunnel vehicles to pass concrete ready mix trucks, concrete pumps, rebar delivery flatbeds, cranes, other equipment and personnel who continued placing rebar and concrete for the interior roadway structure during tunnel segments delivery. The flat invert was wide enough for passing vehicles, maintained the segment delivery vehicle stability, eliminating the need to weave around obstacles and preventing rollovers.

Invert concrete pours occurred usually every Saturday after the tunnel crews left for the day, and the pours extended up to the back of the trailing gear. The actual sequence for the interior structure is as follows:

Figure 1. Placing the flat invert in the tunnel

Figure 2. Pre-cast concrete segments being delivered to the TBM heading

1. Remove the jack frame and false sets
2. Install the forms for the cast in place concrete
3. Pour the flat invert in sections up to the stopped TBM and trailing gear
4. Once the initial pours were completed, install the ramp from the shaft invert to the flat tunnel invert
5. Anchor dowel holes in the pre-cast segments for the corbel were drilled off the trailing gear as the TBM advanced, enabling hook dowels placement along with the reinforcement.

The mobilization and startup of the interior structure work commenced while the TBM repairs were underway. This allowed the building trades time to familiarize themselves with working in a tunnel and the interior structure construction.

STP started receiving formwork from Peri GmbH in June 2014 and, once on site, started installing the forms and the travelers into the tunnel and prepared for the initial concrete pours. The first of the corbel pours, in 54 linear feet increment pours, was placed on Sept. 2, 2014. All pours were limited by design to 54 linear feet maximum and no continuous pours were allowed until the concrete had achieved initial strength for form removal. The last corbel was poured at the North Portal on Sept. 22, 2017.

Figure 3. Corbel form in place and showing re-bar and utilities placed in the pour, and blue concrete bonding agent

Figure 4. Corbel reinforcement

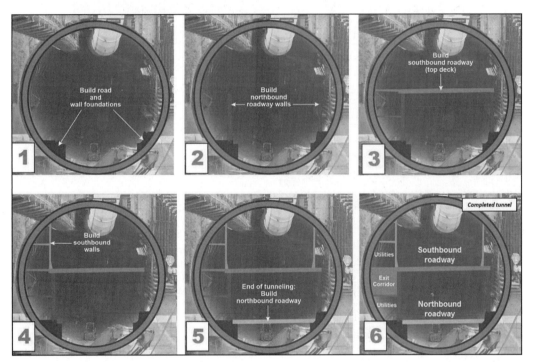

Figure 5. Interior structures construction sequence

Construction of the interior structure progressed in much the same way as a high-rise building. As the tunnel mining was proceeding, work installing the interior structure followed, with the mechanical, electrical, plumbing, fire and life safety systems picking up the rear. The sequence of the interior concrete placement is depicted in Figure 5. Figure 6 depicts the lower Northbound roadway formwork for walls. Figure 7 depicts shotcrete application of the upper Southbound roadway walls to the tunnel segmental lining as a closure.

Construction of the lower northbound road deck consisted of placing a pre-cast road deck onto the lower corbels. Construction of this effort was started in November 2017 and is forecasted to be complete in March 2018. The pre-cast deck is installed in 8-foot-wide panels. Post tensioning ties 25 panels (200 feet) together.

Figure 6. Lower, Northbound, wall forms and traveler

Figure 7. Shotcreting the closure pour between the CIP upper walls, (Southbound), and the tunnel segments

MECHANICAL, ELECTRICAL AND INFRASTRUCTURE INSTALLATION

As the interior structure of the roadway proceeded north, crews worked immediately behind them installing mechanical, electrical, and plumbing (MEP), fire and life safety systems. From south to north, vertical alignment of the tunnel has a downward gradient and then a rising gradient resulting in

a low point one-third of the way north. As a result, drainage systems, including four pump stations, are in the invert below the roadway deck to process road and fire sprinkler runoff. The drainage system has a pumping capacity of 1,090 gallons per minute to handle storm flow. In addition, there is up to 480,000 gallons of emergency storage below the roadway. To mitigate against fires, WSDOT prohibits petroleum tankers and all vehicles carrying hazardous materials, from using the tunnel.

The MEP installation work consisted of installing hangers, mesh and fire proofing, strut support channels, lighting, digital message signs, cameras, jet fans, antennas, and the fire detection and suppression system. Tunnel ventilation systems installation includes jet fans in the roadway space and exhaust dampers in the exhaust ventilation duct at approximately 100 ft intervals along the length of the east side of the tunnel. The ventilation duct along the east side of the tunnel, Figure 8, connects to the operations buildings located at the south and north ends of the tunnel where extraction fans exhaust gases from the tunnel during emergencies. The tunnel control system automatically opens the exhaust dampers, located in the side of the exhaust duct along the tunnel, at the location of a fire keeping fire smoke from propagating along the length of the tunnel.

Electrical conduits placement, on the tunnel liner segment, occurred ahead of the north bound deck pre-cast panel installation.

Electrical rooms are spaced along the tunnel length on the west side of the tunnel, adjacent to the southbound roadway. These rooms contain and route

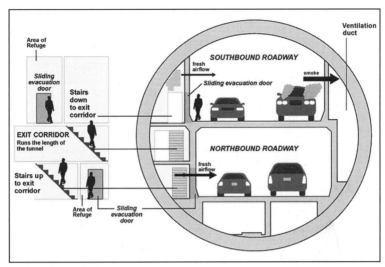

Figure 8. Roadway, ventilation, and evacuation cross section

electrical systems for powering the tunnel lighting, jet fans, cameras, digital message signs, communication systems, emergency power, fire control systems, air quality and operational systems. Electrical room installation consists of placing concrete for the ceiling and walls, installation of conduit and equipment and connecting the rooms to each other and to the operations buildings. Also along the west side of the tunnel are the evacuation doors and egress corridor shown in Figure 8.

COORDINATION WITH CITY OF SEATTLE AND THIRD PARTIES

WSDOT began third party coordination during project alternative development. After the Nisqually earthquake that severely damaged the aged viaduct, the city of Seattle, King County and the Port of Seattle collaborated with WSDOT on the development and selection of an alternative to replace the capacity of the viaduct. The long journey to a preferred alternative included a public vote that rejected all alternatives, extensive public outreach, and legal challenges. During the selection process, tunneling technology advanced making the single bored tunnel a realistic alternative. In early 2009, the parties jointly recommended the bored tunnel alternative and FHWA, the City, and WSDOT co-signed the environmental impacts statement record of decision in August 2011.

WSDOT entered into a Memorandum of Agreement (MOA) with the city of Seattle that outlined responsibilities for plan review, construction, and method of plan approval. According to this MOA, WSDOT reimburses the City for their costs of review, inspection, and coordination of city infrastructure built by WSDOT. The MOA was included in the design/build project as a reference document. The plan review process was included as a contract requirement.

The city of Seattle Fire Department (SFD) and the city of Seattle Department of Transportation (SDOT) participated in all aspects of alternative selection. WSDOT performed fire heat and ventilation studies with SFD to reach agreement on the fire size used for tunnel ventilation and fire control design. The Seattle Fire Department has participated in twice a month fire and life safety meetings since 2007. Though the meeting's name has changed (fire and life safety to mechanical/electrical/plumbing to commissioning/start-up), the emphasis remained—safe tunnel, safely operated. At these meetings, raised issues are addressed and subgroups are formed for issues resolution.

When WSDOT needed a deviation or clarification of city of Seattle Fire Code or NFPA 502 standards, WSDOT reached concurrence with the SFD and documented the concurrence through signed letters. WSDOT and SFD reached agreement on the following issues before the Request for Proposal (RFP):

- Design Fire Properties
- Fixed Fire Fighting System (FFFS) discharge density
- Area of FFFS simultaneous discharge
- Locations requiring FFFS and areas not requiring FFFS
- Locations requiring Fire Resistant Wiring and acceptable types of wiring

- Width of 44 inch for egress passages, egress doors 42 inch wide
- Fire rated enclosure locations and two and three-hour rating locations
- Allowance of piping and conduit in egress area for tunnel operation and the requirements for this allowance.
- Emergency exit spacing requirements
- Egress doors allowed on slopes greater than 5 percent.

WSDOT, working closely with the SFD, developed 17 emergency response scenarios for tunnel operation, along with emergency responder communication plans. The SFD and the Seattle Police Department (SPD) endorsed the plans.

SDOT operates the road network that connects to the tunnel. WSDOT developed, working closely with SDOT, a corridor operation plan that specifies Digital Message Sign (DMS) use, communication and responsibilities of tunnel and roadway operation during emergencies and normal operations. A maintenance and operation agreement finalizes all decisions. SDOT engineers inspect city infrastructure daily. SDOT engineers meet with WSDOT engineers weekly to review city infrastructure design, and quarterly on operation and emergency plans. The WSDOT/SDOT operation meeting frequency is increasing to twice a month during the year before tunnel opening.

WSDOT developed detailed base maps included in the RFP. These base maps included extensive research of available data and recent pothole data. The Design/Build contractor was responsible for utility verification during final design. WSDOT and the Seattle Tunnel Partners, the Design/Build contractor, obtained plan approval on utility relocations and repairs through Seattle Public Utilities (SPU)

and Seattle City Light (SCL), King County and private utility providers. During design, WSDOT, SPU and SCL met at least once a week to review utility designs. In addition to keeping essential utilities operable, WSDOT and STP strived for a utility design that was compatible with future projects that were also in design. A strict deadline for review comments was set on all parties. The organization of the review timelines, comment deadlines and comment resolution was critical to the project schedule. Utility design is one of the more challenging aspects of this project's design. Washington State law dictated cost responsibility for utility relocation.

CONCLUSION

The initial plan by STP and WSDOT was to have the tunnel open to traffic by December 2015. Events have prevented meeting that original schedule, but even though there were problems, these problems were addressed and the tunnel was completed with virtually no impact to any of the existing structures, buildings, roads, or utilities. Concrete work for the interior structure should be completed by April of 2018 and then MEP will continue through commissioning of all systems and open to traffic by early 2019.

Through all of this, all parties to the Project continued to work together in the best interest of the Project to deliver a safe, reliable, structurally sound alternative route through Seattle and allow the elimination of an unsightly and seismically vulnerable behemoth.

Seattle will gain a new highway out of sight to most of the city and the removal of the existing AWV will open the city to the waterfront and Elliott Bay, a tremendous improvement and enhancement that everyone can be proud of.

Cross Passage Freezing from the Ground Surface

Aaron K. McCain, Brenton Cook, and Larry Applegate
SoilFreeze, Inc.

Nate Long
Jay Dee Contractors, Inc.

ABSTRACT: Ground freezing techniques were used to successfully excavate cross-passages that were part of the Sound Transit Northgate Link Extension in Seattle, Washington. In order to keep the tunnels unobstructed during mining operations and minimize in-tunnel freezing equipment, the decision was made to freeze the cross passages from the ground surface. Freezing from the ground surface to support excavation for tunnel cross passages had not occurred in the United States prior to this project. This paper presents the approach taken to navigate the project challenges overcome by freezing from the ground surface to provide stable and watertight shoring for the project.

INTRODUCTION

The Northgate Link Extension project is the third phase of the Sound Transit master plan to extend light rail throughout the greater Seattle metropolitan area. When completed, the Northgate Link Extension will extend the existing light rail system north by about 4.3 miles; from the University of Washington to the Northgate neighborhood. A general overview of the project alignment is provided as Figure 1. The project was funded by a combination of federal grants and allocations from local taxes. The general contractor on the project was JCM Northlink LLC (JCM), a joint venture between Jay Dee Contractors, Frank Coluccio Construction Company, and Michels Corporation. SoilFreeze, Inc. was retained as a subcontractor to provide ground freezing design and construction services.

The project involved the construction of two parallel 18-foot, 10-inch (inside) diameter tunnels using EPB tunnel boring machines and precast segmental concrete tunnel liners. The two tunnels maintained roughly the same elevation along the alignment and were spaced roughly 40 feet apart (centerline to centerline). A total of 23 cross-passages connecting the two tunnels were included at approximate 800-foot intervals. Cross passage excavations were elliptical in shape with dimensions of approximately 18.5 feet in height and 17 feet in width. The cross-passages were 19 to 20 feet long as measured between the exterior of the tunnels at springline elevation. Cross passages were intentionally located directly beneath public right-of-ways thus avoiding buildings. Many of the cross passages were located beneath narrow roads in residential neighborhoods with mature

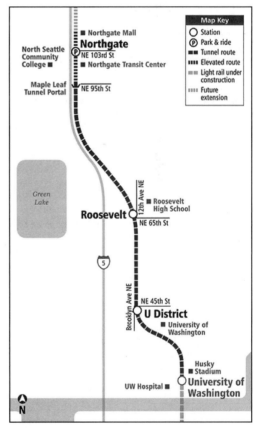

Figure 1. General overview of the Northgate Link Extension project (image credit: SoundTransit)

899

vegetation, existing utility structures and piping, and limited parking areas. Excavation and support for each cross-passage was completed using the Sequential Excavation Method (SEM).

The subsurface soil and groundwater conditions at the cross-passage locations were described in the Northgate Link Geotechnical Baseline Report (GBR) prepared by Jacobs Associates and the Northgate Link Geotechnical Data Report (GDR) prepared by Shannon and Wilson, Inc. Soils along the project alignment were glacially overridden and characterized as very dense or hard. Subsurface soils consisted of Cohesionless Sand and Gravel (CSG), Cohesionless Silt and Fine Sand (CSF), Cohesive Clay and Silt (CCS), and Till and Till-Like Deposits (TLD).

The contract documents identified three categories of ground support systems for the cross-passage excavations based on the soil conditions identified at each location. Category 3 ground support systems required ground improvement before excavation could begin. Category 3 cross-passages were typically the most difficult soil conditions located near a transition between very permeable sandy CSG and the less permeable CCS or CSF. The boundary between the two dissimilar soils would have rendered dewatering efforts ineffective, as sufficient groundwater drawdown could not be achieved. Initially five cross passages were identified as Category 3, but freezing was used to stabilize the soils for an additional five after soil explorations identified difficult soil conditions at those locations.

The five cross-passages initially requiring Category 3 ground support and frozen from the ground surface are as follows:

- Cross Passage No. 21 (CP 21): The center line of the excavation was approximately 113 feet below the ground surface (bgs).
- Cross Passage No. 29 (CP 29): The center line of the excavation was approximately 125 feet bgs.
- Cross Passage No. 30 (CP 30): The center line of the excavation was approximately 123 feet bgs.
- Cross Passage No. 31 (CP 31): The center line of the excavation was approximately 114.5 feet bgs. In addition, CP31 was located at the low point of the tunnel alignment and included a sump pit which extended about 16 feet below the centerline of the cross passage.
- Cross Passage No. 32 (CP 32): The center line of the excavation was approximately 76 feet bgs.

GROUND FREEZING FOR CATEGORY 3 CROSS PASSAGES

Ground improvements for Category 3 ground support systems were originally specified to be jet grouting or ground freezing. JCM was concerned with community and environmental impacts of the jet grouting in the densely populated urban/residential environment. In addition, freezing presented a more flexible schedule than jet grouting as the work could be done before or after the TBM had passed the location. As a result, JCM opted to pursue ground freezing as the preferred method to improve soils conditions.

Ground freezing involves circulating calcium chloride brine, chilled to –15 °F (–26 °C) or colder, through an array of steel pipes installed within the subsurface. Heat is extracted from soils immediately around each chilled steel pipe, freezing groundwater within the soil matrix. As soils freeze radially outward from individual chilled pipes, the frozen soils from adjacent pipes eventually overlap to form an impermeable barrier with increased strength and stiffness. Ground freezing only needs to change the temperature of in-situ soils in contrast to more intrusive methods such as injection, permeation, or replacement. Advantages of ground freezing include: the ability to pinpoint the location of the freeze pipes and thus the extent of the ground freezing; the ability to verify the extent of ground freezing through ground temperature monitoring; and minimal disturbance and spoils management at the ground.

At the request of JCM, SoilFreeze provided an approach that installed the majority of freeze pipes from the ground surface so that larger freeze equipment would be located out of the tunnels. Smaller freeze systems installed within the tunnels would supplement the primary freeze system elements from the ground surface. JCM felt that freezing from the ground surface would achieve two principal goals: 1) limit the work and equipment that needed to be located within the tunnels and 2) remove the bulk of the freeze system installation off the critical path of the project by installing the freeze pipes independent of the completion of the tunnels.

DESIGN AND INSTALLATION OF THE GROUND FREEZE SYSTEM

The choice to freeze from the ground surface contained technical and outreach challenges that were successfully addressed and mitigated through careful, innovative design and progressive, forward-thinking during the construction process. The primary challenge of freezing from the ground surface was to maximize the benefit of freezing for successful excavation of the cross passages while minimizing impacts to existing near surface infrastructure and

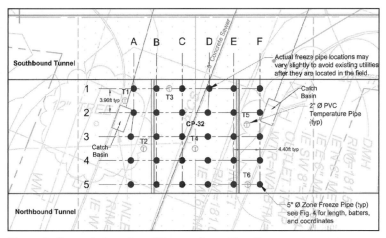

Figure 2. General plan view of freeze pipe grid pattern

the newly installed tunnels. Specific challenges can be categorized in these two areas: 1) within the freeze around the tunnels and 2) at the ground surface.

Specific challenges within the freeze zone included:

- Development of a frozen soil zone within time frames required by the construction schedule.
- Coordinating freeze pipe installations to minimize the number of pipes intersecting the path of the cross-passage excavations.
- Freezing the areas in the "shadow" of the tunnel, beneath the tunnel haunches where freeze pipes from the surface cannot access.
- Ensuring that frozen soils sufficiently adjoin to the extrados of the warmer tunnel liner segments.
- Ensuring that sufficient freeze continues during excavation and subsequent construction of the cross passage.
- Methods to manage and/or mitigate pressures that develop on the tunnel liner segments.

The specific challenges at the ground surface included:

- Optimize the freeze system to minimize electrical demands due to power supply restrictions.
- Coordination with utility entities to protect, maintain, and not interrupt, existing infrastructure crossing through work zones.
- Mitigating potential impacts to nearby residences resulting from equipment noise, limited vehicular access, and damage to landscaping.

- Methods to manage and/or mitigate ground movement associated with freezing activities.

Design of the Frozen Soil Shoring

In general, the design of a frozen soil shoring system requires two types of analysis. The first is a thermal analysis where the freeze pipe spacing, extent of frozen soil with time, and frozen soil temperatures within the frozen soil shoring are determined. This information is then used to estimate frozen soil geometry and strengths to be implemented in a constitutive geotechnical analysis. The constitutive geotechnical analysis assesses deformations, stresses, and strength-based factors of safety that could be anticipated within a given frozen soil shoring system.

Highly specialized two-dimensional (2D) engineering software was used for the thermal analysis. The thermal software utilizes finite element methodology to accurately evaluate thermal variations in the ground. Specifically, the thermal software calculates transient (time dependent) frost growth around chilled freeze pipes within a 2D plane.

Three-dimensional (3D) constitutive geotechnical analyses were calculated using finite element software. This computer software is capable of analyzing stress and strain behavior for complex 3D soil geometries under static or dynamic loading conditions. Soil-structure interaction between unfrozen soils, frozen soils, groundwater, and structural elements can be effectively analyzed.

Iterative thermal analyses were used to conclude that a grid of 30 freeze pipes was sufficient to freeze the soils between the two tunnels within a six to eight week time frame. This time frame was governed primarily by the cohesive soils that were present at each location. Cohesive soils typically have a higher moisture content, and therefore, a higher latent heat. Freeze pipes were arranged in a grid pattern consisting of six columns of five pipes as shown on Figure 2. The profile of the approach is provided as Figure 3. The spacing between each column of freeze pipes was approximately 4½ feet while the spacing between rows of freeze pipes was about 4 feet.

The outermost columns of freeze pipes, in plan, were positioned outside the excavation limits of the cross-passages. This left approximately 20 pipes that would be encountered during excavation for each cross passage. A grid pattern with less pipes would

not adequately freeze the soils above and below the cross passage within the project schedule.

The circular shape of each tunnel restricted access from the ground surface to soils below the springline elevation of the tunnels directly adjacent to the extrados of either tunnel liner, thus forming a 'freeze shadow'. Short haunch freeze pipes were installed through the tunnel liners below the springline of both tunnels to freeze just beyond the limits of the excavation. Although a "shadow" did not exist above the tunnels, it was decided that haunch freeze pipes would also extend above the springline of the tunnels instead of drilling additional freeze pipes from the ground surface.

Typical locations of haunch freeze pipes are shown in Figure 4. The number and location of the haunch freeze pipes varied at each location based primarily on the potential for groundwater gradients following the tunnel alignments. Installation of the haunch freeze pipes did require a chiller to be installed in each tunnel, however the size of the chiller was significantly smaller than one that would be required to freeze the entire freeze zone. The in-tunnel chillers were placed on elevated platforms installed along the interior the tunnel liners, near springline elevation. The in-tunnel chillers did not impede traffic or work within either tunnel.

A final design hurdle involving freeze pipe positioning was to develop a system that would ensure that frozen soils would remain frozen directly adjacent to the tunnel liner extrados. This interface represented the weakest point of the frozen soil shoring.

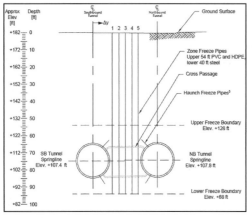

Figure 3. General profile view of freezing approach

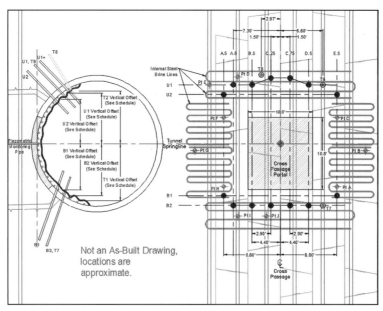

Figure 4. In-tunnel haunch and interface pipe layout

Figure 5. In-tunnel freeze system in operation

Ventilation systems actively circulating air through the tunnels warmed the interior face of the segmental concrete tunnel liners. This air circulation acted to create a warm internal boundary that could generate enough of a thermal gradient through the concrete liner to thaw a thin layer of soil and induce a seepage path into the excavation. To combat the warmth imparted by the ventilation system, steel pipes circulating chilled brine were affixed to the interior face of the tunnel liners surrounding the proposed cross passage portals, as shown in Figures 4 and 5. Chilled brine was supplied to these pipes using the same chiller systems that were installed for the haunch freeze pipes. Additionally, insulation blankets were affixed to the interior surfaces of the tunnel liner to limit warm air exposure.

The system of vertical freeze pipes installed from the ground surface in combination with the in-tunnel freeze pipe systems, were analyzed using both the thermal and constitutive geotechnical models previously discussed. Thermal models predicted a zone of ground improvement significantly smaller than the specified ground modification extents presented in the project drawings. Despite the frozen soil zone being smaller than specified, constitutive models indicated that maximum frozen soil deformations during excavation would be less than ½-inch at the crown of the cross passage, with little to no deformation at the invert of the cross passage. In addition, the calculated strength based factor of safety for the frozen soil shoring generally exceeded 5.0 at each cross-passage location. Reviewers eventually allowed for a reduction of the modified soil zone extents for three reasons: 1) the location of each freeze pipe would be known; 2) the radius of frozen soil around each pipe is relatively uniform and predictable using thermal modeling; and 3) the extent of the frozen soil could be verified and monitored using thermocouple sensors placed at key locations.

SoilFreeze employed zone freeze pipes as the primary freeze elements that would be installed from

the ground surface. Zone freeze pipes are a patented system designed to isolate freezing within a targeted zone at depth. Zone freeze pipes work by circulating chilled brine through a 4-inch steel vessel with high heat transfer characteristics that forms a 'freezing length' at depth. The portion of the zone freeze pipe extending from the top of the steel vessel to ground surface is comprised of high density polyethylene (HDPE) brine delivery lines confined within a gasketed, water tight, polyvinyl chloride (PVC) sleeve. The HDPE lines are buffered from the PVC sleeve and the surrounding soils using air to form an 'insulated length' with intrinsically low heat transfer characteristics. The insulated length of the zone freeze pipe served to minimize freezing of soils extending above the targeted freeze zone to the ground surface, thus reducing the required chiller capacity and maximizing system efficiency. This resulted in lower energy demand and also minimized the freeze system footprint within the narrow above-ground streets. It was estimated that the zone freeze pipe technology reduced the required chilling capacity and power requirements by at least half. Assembly of the insulated PVC portion of a zone freeze pipe installation is shown in Figure 6.

Figure 6. Installation of the PVC portion of a zone freeze pipe

Installation of the Freeze System

The installation process for each freeze system posed its own unique set of challenges. When the freeze pipe grid was superimposed on existing conditions at each site, conflicts were apparent requiring adjustments. The following is a summary of the conflicts at the ground surface for each cross-passage location:

- CP 21—The location of the cross-passage was partially located beneath a low traffic, two-lane roadway at the southeast corner of the University of Washington campus. The entire eastern half of the cross passage was located beneath a vegetated slope that could not be disturbed, requiring that all work had to be completed within an area that was less than half the size of the designed freeze pipe grid at depth. In addition, an 8-inch ductile iron water line and manhole structure were located within the work area.
- CP 29—This cross passage was located beneath a narrow neighborhood side street. There were no buried utilities at this site, however overhead power was present on the north side of the road and local service lines crossed over the road and above the freeze pipe installations.
- CP 30—This cross passage was located beneath a narrow side street. Large heritage trees were located on the north side of the road, with additional trees just to the south. The "heritage" trees had protected root zones that extended out into the work zone. Utilities consisted of overhead power on the south side of the road, two buried fiber ducts, a 2-inch gas line, a 12-inch cast iron water line, and a 20-inch cast iron water line. Neither of the water lines were located within the work area.
- CP 31—This cross passage was located below the northbound lane of a residential avenue. Accordingly, the work area was located within the northbound lane and extended slightly beyond the east curb line. The buried utilities extending through the work area consisted of a 42-inch diameter steel riveted water line, an 8-inch cast iron water line, and a 2-inch gas line. A fire hydrant was removed prior to mobilizing to the site.
- CP 32—The shallowest cross passage was located directly below a small residential road located near the intersection with a busier thoroughfare. Overhead power lines were present to the south of the work area and buried utilities crossing the work area consisted of storm drain laterals, an 8-inch concrete sewer line, a 4-inch gas line, and a small 2-inch water line lateral.

Where conflicts were identified, the location of the freeze pipes at ground surface were adjusted. In a few cases the adjustments were minor and the freeze pipes could still be installed vertically. However, in most cases the adjusted freeze pipe locations were significant enough that the pipes needed to be installed at an angle (or batter) to maintain the required spacing at depth. Freeze pipes that were installed at angles such that the freeze vessels passed through the original design grid point in plan at the mid-level elevation of the cross-passage excavation. Due to the depth of the cross-passages, installation angles were minor, typically ranging between 2 and 10 degrees from vertical. The battered freeze pipes typically resulted in slightly closer freeze pipe spacings at the top of the frozen soil column and a greater spacing at the bottom of the frozen soil column at depth. Thermal models were adjusted to include spatial variations of the freeze pipes with depth at each cross passage. Ultimately it was determined that the extent of the frozen ground was not significantly impacted by these spatial variations.

After layout of the freeze pipe grid was completed and conflicts were identified at each site, JCM returned to the site and installed 12-inch diameter PVC sleeves to mark freeze pipe installs that were located directly adjacent to utilities. In general, this process was safely completed with a vactor truck. The PVC sleeves extended to depths just below the utilities and served as a guide during the installation of the freeze pipes.

The second advantage provided by the PVC sleeves installed by JCM was to provide redundant insulation around the freeze pipes to reduce freezing around utilities. Freezing does not adversely affect most utilities including gas, fiber-optic, sewer, power, and running water. However, during negotiations with the various utility agencies, there was an expressed concern about the impact of the freezing temperatures. Therefore, the additional insulation provided around the utilities was an added benefit of the PVC sleeves.

Freeze pipe installations from the ground surface were completed by Cascade Drilling. A steel casing was advanced using sonic drilling techniques to design depths. For each installation, a steel freeze vessel with HDPE brine supply/return lines attached was incrementally lowered into the cased hole while 20-foot sections of PVC sleeve were stacked and joints waterproofed. After the freeze pipe was lowered to depth, the steel casing was retracted and the annular space around the freeze pipe was backfilled with grout to the ground surface.

The work zones required to install the freeze pipes blocked the narrow roadways at CP 21, 29, 30, and 32, however a single lane was maintained past the drilling zone at CP 31. Drilling was completed during hours dictated by the city and site cleanliness was maintained throughout the installation process. No complaints were received from the general public during the installations. After freeze installations were completed, roads remained closed at CP 29, 30, and 32. Pipes installed at CP 21 were recessed below the street level and covered with a steel sheet which allowed the road to remain open during freeze operations.

Tracking the locations of each freeze pipe at depth was critical, therefore the steel drill casing was surveyed prior to the installation of the freeze pipe. A downhole gyroscopic survey tool was used to determine the spatial locations of each installation between the tunnels for varying depths. If the space between two adjacent freeze pipes was excessive as determined by an updated thermal model utilizing as-built data, an additional freeze pipe was installed in the gap to maintain the necessary freeze coverage. A total of fourteen additional freeze pipes were added for all five cross passages due to drilling inaccuracies. The downhole survey tool was also implemented mid-drilling process for pipes located immediately adjacent to the tunnels to verify that the hole alignment had not drifted towards the tunnel. All drill holes verified in this manner were found to be on target and were installed to depth.

Within the tunnels, JCM was responsible for installation of the haunch freeze pipes to better control schedule and space constraint impacts within the tunnels. JCM installed the short haunch pipes by pre-drilling through a pneumatic packer, removing the drill tooling, then driving the steel freeze pipes to depth. The location and alignment of the short haunch freeze pipes were surveyed by JCM using conventional optical surveying methods.

FREEZEDOWN AND EXCAVATION

Freezedown

Start of freezedown and freezedown durations varied for each cross-passage location. Chilling units servicing the freeze pipes at ground surface were initially powered by generators. At some cross-passage locations, chillers at the surface were eventually switched over to the local power grid. Chiller units at the ground surface had footprints of 8 feet by 16 feet. One chiller unit was sufficient to freeze each cross-passage, except at CP 31 where additional freezing capacity was needed to support the excavation for the deeper sump pit.

Chillers and generators operated full-time during freezedown, excavation, and final lining construction

Figure 7. Freezedown from the ground surface

phases of each cross-passage. Accordingly, mitigation of mechanical noise for each freeze system was a requirement by the city. This was accomplished by using specially designed low noise fan blades on each chiller unit; working with the power company to get a power drop at each site thereby eliminating the need for generators; and building a sound dampening structure around the generators and chillers. The sound dampening structures were constructed by JCM and consisted of plywood walls internally lined with sound dampening blankets and insulation. Each sound dampening structure was large enough to provide ample air circulation to the chillers, while remaining small enough to fit within the limited footprint at each site. These measures successfully dampened the mechanical noise to less than 63 decibels, the operational threshold allowed by the city.

In general, the frozen soil shoring was formed within the time frame calculated during the design process. Brine and ground temperatures were closely monitored during the freezedown process, using thermocouples installed at various depths around each site. Recorded ground temperature data was used to calibrate as-built thermal models. Measured brine temperatures were used as input to the thermal models and the soil and freeze pipe parameters were slightly adjusted so that the calculated temperatures closely matched the observed temperature trends. The calibrated thermal models were instrumental in estimating spatial extents of the ground freezing at varying depths, as well as extrapolating future development of the frozen soils.

Ground Movement

One of the challenges on this project was monitoring and controlling the ground deformation associated with freezing the ground for extended periods of time. The expansion that occurs when pore water undergoes the phase change to ice does not typically cause significant volumetric expansion of a

Figure 8. Cutting and capping the steel vessel within the excavation

soil unit. However, when freezing is maintained for a long period of time, cryogenic suction will draw water towards the freezing front and can form lenses of pure ice. If conditions permit, such ice lenses can result in segregation heave and excessive deformations at the ground surface.

The zone freeze pipes were never designed to completely prevent freezing of soils extending above the freeze zone at depth. As a result, near surface soils eventually froze during prolonged freezing operations. As the near surface soils froze, some minor heaving at the ground surface was captured by optical monitoring points and extensometers. Observed ground movements were minor and slightly varied at each site. The observed movement was highly dependent on surface water runoff, available groundwater, site soils, and instrumentation accuracies. Ground deformation was limited and controlled using a number of techniques which included:

- Managing surface water to limit surface runoff from entering the work area and collecting.
- Adjusting/raising the temperature of the brine. This was a delicate balance between being cold enough to maintain freezing at depth while simultaneously warming near surface soils to limit the development of ice lenses.
- Use of heat trace tape and blowing warm air into the upper portion of the zone freeze pipes.

Where there were no utilities located within the freeze grid, ground movement was not controlled and any deformation was repaired as part of the site restoration. Where utilities were present, particularly the 42-inch riveted steel water main at CP 31, movement was successfully maintained below thresholds defined in the project specifications. No damage to any utility was observed.

Deformations were also observed along the tunnel liners at depth. Some deformation of the tunnel was acceptable and remediation actions were limited to varying supplied brine temperatures. After taking such measures, observed deformations stopped and rebounded slightly after excavations began.

Excavation

Prior to excavating each cross passage, short probe holes were drilled through the tunnel liner to verify that groundwater had been completely cut off, and the tunnel liner segments could be removed without groundwater intrusion. Excavation was completed using SEM construction methods and a remotely operated roadheader. Four inches of an initial shotcrete lining was placed over frozen soils exposed during SEM work. Conservatively, two of the four inches of initial shotcrete lining was intended to be sacrificial and provided an insulating barrier prior to application of the final shotcrete lining. This initial layer, paired with welded wire fabric, facilitated successful placement and strength gain of the dry shotcrete mix.

As the excavation progressed, rows of freeze pipes were deactivated and purged of brine ahead of the excavation work. This allowed the freeze pipes in the path of the excavation work to be cut and removed without releasing the calcium chloride brine into the excavation which could thaw frozen soils, potentially impacting the integrity of the frozen soil shoring. Once the freeze vessels were cut and deactivated, soils along the base of the cross-passage excavations would no longer be actively frozen for the remainder of the project. This was anticipated during the design phase and the outer rows of freeze pipes and the haunch pipes were strategically positioned to remain active and compensate for the additional heat introduced during the excavation and subsequent construction of the cross-passage.

During the excavation, the majority of the pipes installed from the surface were cut, re-capped, pressure tested, and then buried in the shotcrete. Two of these cut freeze pipes were brought back online immediately over the portals at each tunnel to help maintain the freeze along the crown of the excavation, directly adjacent to the concrete tunnel liners. In addition, several of the interior freeze pipes were replumbed with heavy duty rubber hoses that connected freeze vessel segments at the crown of the cross passages with freeze vessel segments extending beneath the excavation. The hoses were buried into the walls of the excavation and covered with shotcrete. This innovative approach was sufficient to keep the soil mass below the excavation actively frozen for the duration of the project.

After the excavation was completed, the freeze system remained in operation for approximately eight weeks during which time the waterproofing system was installed, reinforcing steel was placed, and the permanent final concrete lining was cast in-place. During this time, the frozen soil system remained operational to maintain the robust and water tight shoring.

CONCLUSION

Serious project challenges were successfully overcome with freezing from the ground surface providing safe, stable, and water tight shoring for five Category 3 cross-passage excavations along tunnel alignments for the Sound Transit Northgate Link Extension project. Each of the above-ground freeze systems were successfully installed and operated within crowded and difficult urban environments with minimal impacts to the population. Zone freeze pipes were successfully employed to limit freezing impacts to near surface infrastructure, optimize chilling capacities, and focus freeze efforts at depth to develop a robust frozen soil shoring system.

Figure 9. A completed cross passage after waterproofing

Downhole survey techniques were established that can be implemented on future projects requiring drilling of freeze elements in close proximity to sensitive structures at significant depths. Methodologies to mitigate propagation of frozen soils and limit frozen soil induced heave deformations around existing infrastructure were successfully employed.

Lessons Learned in Dry Ground Excavation Using an EPBM

David C. Girard and Ran Chen
J.F. Shea Construction, Inc.

ABSTRACT: Earth pressure balance (EPB) sensors are mounted in the excavation chamber of an EPB tunnel boring machine (EPBM) to monitor face support pressure. From this location, these sensors do not accurately portray the actual earth pressure in front of the cutterhead when exposed to relatively dry ground conditions. Mining and ground settlement monitoring data show that contact force can be used as an alternative parameter for maintaining sufficient face support pressure in granular soils above the water table. Avoiding unnecessarily high EPB pressures can improve the advance rate and minimize cutterhead wear, improving operational efficiency. Some lessons learned from the Crenshaw project and some suggestions for future projects are presented.

INTRODUCTION

The Crenshaw/LAX Corridor project (Crenshaw project) is a part of Los Angeles County Metropolitan Transportation Authority's (Metro) major capital expansion program, aimed at improving public transit service and mobility in Los Angeles County. This design-built project was awarded to Walsh/Shea Corridor Constructions (WSCC) for $1.27-billion teamed with HNTB/Arup as the Designers. The Crenshaw project runs through South LA and will service Los Angeles, Inglewood, and El Segundo along with portions of unincorporated Los Angeles County. The project's guideway connects the existing Metro Exposition Line at Expo/Crenshaw Station to the existing Metro Green Line at Aviation/LAX Station. Overall, the project has eight stations, two of which were bid options that have been exercised. The guideway includes at grade trainway, aerial structures, cut and cover box structures, and bored tunnel, as shown in Figure 1.

The scope of this paper is to summarize the main challenges encountered, and lessons learned during the twin bore EPBM excavations, with the intent of generating some interesting topics with our peers for purposes of improving future tunnel construction and machine design.

GEOLOGICAL INFORMATION

The Crenshaw project is located in the northern part of the Los Angeles Basin, which is directly underlain by unconsolidated Quaternary-age sandy sediments [1]. These generally are subdivided into unconsolidated Holocene-age sediments (Young Alluvium) and late-Pleistocene materials (Old Alluvium). Young Alluvium typically consists of surficial sediments including clay, sand, and gravel. Old

Figure 1. LA Metro Crenshaw/LAX Corridor Project alignment

Alluvium typically consists of sediments including pebble-gravel, sand, silt, and clay. The northern part of the project alignment along Crenshaw Boulevard is directly underlain by Young Alluvium over Old Alluvium. Cohesionless soil with blow counts consistently below 30 blows per foot and cohesive soil

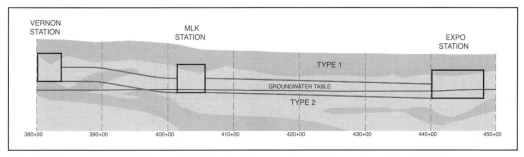

Figure 2. Geological profile for underground stations and tunnel (reproduced from GBR)

Table 1. Estimated volume percentages of Fill, Type 1, and Type 2 Soil

Soil Type	Exposition Sta.	Bored Tunnel	MLK Sta.	Bored Tunnel	Vernon Sta.
Fill	1%	0%	1%	0%	1%
Type 1	42%	0%	30%	15%	66%
Type 2	57%	100%	69%	85%	33%

Table 2. Soil engineering properties (from GDR)

Soil Type	Approximate Blow Count Range (blows/ft) (blows/30 cm)	Total Unit Weight (pcf) (×10³ kg/m³)	Effective Stress Strength Parameters c' (psf) (Mpa)	φ' (deg)	Undrained Shear Strength (psf) (Mpa)
Type 1 Soil	2–42	125	250	20	500
Young Alluvium		2.0	5		10
Type 2 Soil	5–29	125	250	30	
Young Alluvium		2.0	5		
Type 1 Soil	12–100	125	600	28	3600
Old Alluvium		2.0	13		75
Type 2 Soil	13–100	125	0	35	
Old Alluvium		2.0			

with blow counts consistently below 15 blows per foot are considered Young Alluvium. Sub-surface soil in the excavation zone of the underground station and tunnel is classified for excavation purposes as Fill, Type 1 Soil, and Type 2 Soil based on the engineering parameters derived from laboratory testing. Type 1 and 2 Soils are within both Young Alluvium and Old Alluvium. A generalized subsurface profile for the underground excavation is shown in Figure 2.

The estimated volume percentages of the Fill, Type 1 Soil, and Type 2 Soil encountered during excavating the underground stations and bored tunnel are summarized in Table 1. Type 1 Soil consists predominantly of fine-grained silt, clay, and organic soil. Type 2 consists predominantly of a mixture of fine- to coarse-grained sands and gravels and includes some cobbles and boulders. The engineering properties of these two types of soils are shown in Table 2.

GDR grain size distribution plots were overlain by EPB and Slurry applicability ranges which showed that either tunneling method was suitable for the ground to be encountered. All pre-bid costs considered, EPB was chosen as the preferred tunneling method.

It is important to mention that the groundwater table along both tunnel alignments was relatively low during construction, which was approximately at tunnel spring line at Exposition Station, 4 to 5 feet above the tunnel invert at MLK Station and below the invert at Vernon Station, as shown in Figure 2. The effects of this groundwater condition on EPBM advance will be described in this paper.

HIGH CUTTERHEAD TORQUE AND REASONS

Prior to, and after tunnel excavation began on the first drive, engineers and operators recognized a challenge: high cutterhead torque and low advance rate, as shown in the top plot in Figure 3. To understand and meet the challenge, these factors were

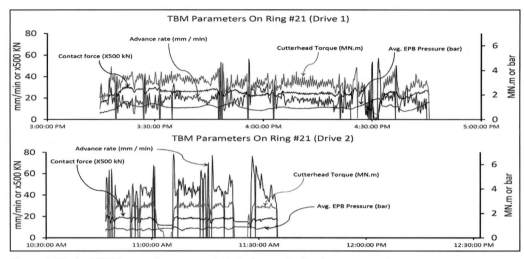

Figure 3. Typical TBM operating parameters (before and after improvement)

considered: face support pressure, soil conditioning, and in-situ soil condition.

For this project, DIN 4085 [3] was used to calculate high, low and target face support pressures. This method applies a reduction factor μ to the major principal stress term of the Rankine's earth pressure, as shown in Eq. (1). Here the reduction factor μ depends on the ratio of the depth to the size of the open face, as shown in Eq. (2).

$$p = \mu K_a \sigma_V - 2\sqrt{K_a}c \qquad (1)$$

$$\mu = -0.005\alpha^3 + 0.0155\alpha^2 - 0.178\alpha$$
$$+ 0.9947, \text{ if } \alpha_i < 10;$$

$$\text{or } \mu = 0.25, \text{ if } \alpha \geq 10; \qquad (2)$$

$$\alpha = d/D \,;$$

d is the depth of the point of interest; and
D is the size of the opening which needs to be supported.

The initial face support pressure was calculated using D equal to the tunnel bore diameter. Further evaluation of face support pressure was done using logged data for total thrust. Accounting for shield friction, the remaining thrust force was projected on the 60% closed cutterhead which resulted in support pressure safety factors in excess of 4.0, instead of the desired value of 1.5. This excessive face support force was caused by two potential reasons. Frist, pressure transducers in the mixing chamber were not able to accurately represent the face support pressure

occurring in front of the cutterhead, but were showing values of material which was already breasted by the rotating steel surface which was 60% closed. Also contributing to the inaccuracy was the relatively dry and dense ground, and that the material was not fully mixed at the tool gap to turn into a uniform mixture, which resulted in a significant pressure gradient between the face and chamber bulkhead where pressure transducers were mounted. In the typical case where groundwater is present in the entire face; soil conditioner has a longer life when mixing with wetter material, the mixture will be more uniform at the tool gap, and this gradient will be small. Also, hydrostatic pressure will be more readily detected by transducers. Second, the initial face support pressure target might be overestimated by using Eq. (1) and (2) above based on the assumption of opening D as the tunnel bore diameter. Since the ground is partially supported by a 60% closed cutterhead, the more reasonable opening size for D in Eq.(2) should be the maximum bucket opening size of the cutterhead. From Eq. (1) and (2), a smaller D value yields a smaller μ value, which results in a lower required face support pressure.

Because of the unnecessarily large face support force, friction between soil and steel at the TBM face was greater than would normally be expected for a machine of this diameter, resulting in correspondingly high cutterhead torque and steel wear, which agrees with R. Godinez etc.'s analysis [4], "*the torque due to cutterhead front and back face resistance accounts for 62% of total torque....*"

Also, conservative face support pressures restricted material from entering into the mixing chamber. Given a greater percentage of cutterhead open area, this issue would be minimized proportionally as truer face support readings would be reflected and opening restrictions reduced.

Surface and subsurface settlement data was monitored as tunneling progressed. Particular attention was given to Multi-point-borehole-extensometers (MPBXs) where settlements could be detected just 5 feet above the crown of each tunnel drive. As these instruments were approached and passed, it became clear that calculated face support pressures were more than adequate for preventing settlement from over-excavation or insufficient support pressure in the mixing chamber (plenum) of the EPBM.

The need for proper soil conditioning was another contributing factor in the fight against high torque. Foam injection is a common method for transforming granular material into a plastically flowable mixture to enhance the fluidity of the spoil in the plenum, reduce abrasion, and reduce cutterhead torque. Laboratory tests [2] showed that the moisture content (*w*) needed to be raised to the minimum level of 13.25% for Type 2 soil in order to achieve a satisfactory slump with the most efficient use of soil conditioner, as shown in the right photo in Figure 4.

Figure 5 shows phase diagrams for an average cubic foot of Type 2 soil which was assumed to have a moisture content of 10%. This average is based on mixing a face of material where the water table is at tunnel springline, with saturated soil below springline having *w*=16%, and *w*=3% above springline to account for some capillarity. To produce plastic, flowable EPB muck, *w* in this case must be raised over 3% before introduction of foam to prevent bubble destruction resulting from using consumed foam to provide the additional moisture, illustrating the phrase "you have to get your hair wet before shampooing." The final phase shows the increase in entrained air from foaming which reduces internal friction by providing greater void space in the soil matrix, leading to increased slump, reduced shear strength, and a material that can reliably transmit pressure to EPB soil pressures transducers.

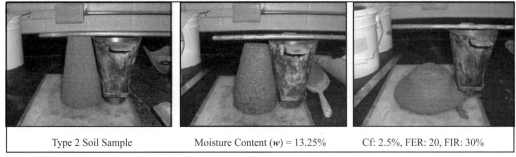

| Type 2 Soil Sample | Moisture Content (*w*) = 13.25% | Cf: 2.5%, FER: 20, FIR: 30% |

Figure 4. Type 2 soil testing (photos courtesy of BASF)

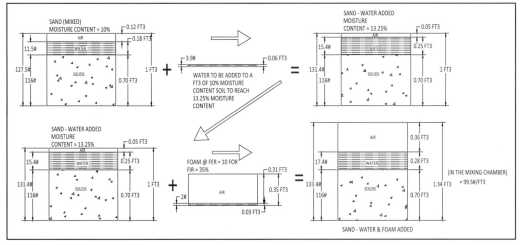

Figure 5. Phase diagrams of transition to EPB muck

Figures 2 and 5 also illustrate the increasing need for added water as the water table lowered with respect to tunnel bores as both tunnels advanced toward MLK and Vernon stations. Prior to mining, methods for wetting the ground in front of the machine were considered, resulting in no practicable solutions. Similar ground entirely below the water table would typically have w of approximately 16%, and the added benefit of hydrostatic pressure to assist movement of fluidized muck toward the screw auger inlet due to lower effective stress. Tunnel drives above the water table, south of MLK station had appreciable amounts of clay which contained w in excess of 20%, alleviating many of the problems associated with the dryer sands encountered between Exposition and MLK stations. For the reaches between Exposition and MLK stations, the lack of apparent cohesion was evident by the almost immediate realization of surface settlement as the EPBM shield passed beneath surface monitoring points. Here, shield gap bentonite proved to be most useful for holding the disturbed, cohesionless ground until the final lining emerged from the tail shield and received annulus grout, keeping settlement within contractual limits.

Compounding the dry ground problems, both tunnel drives involved excavating through jet grouted blocks of ground where cross passages were to be excavated. These blocks provided good safe-havens for interventions, but also resulted in reduced muck flow within the cutterhead as soil-cement re-hydrated and plugged the center of the cutterhead within the spokes, and restricted the pathway to the screw auger inlet. Because full face support pressure had to be maintained within the grout blocks, advance rates slowed while mining these rock-like conditions, adding to the plugging effect of the soil-cement. Plugging reduced the available volume within the plenum, leading to rapid drops in EPB pressure for a given rotation of the screw auger. This led the EPBM operator to reduce screw auger output to stabilize EPB readings, resulting in reduced advance rates, exacerbating the high torque issues. A summary of problems, causes and possible solutions is presented in Figure 6.

SOLUTIONS AND ACTIONS

Based on the cause-effect analysis above, potential solutions to the root cause and indirect causes of high cutterhead torque are summarized, some of which

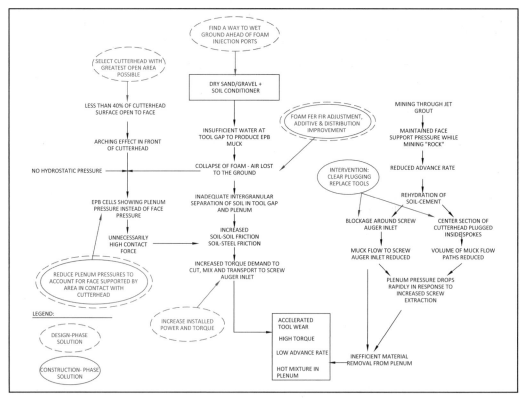

Figure 6. Cause-effect chart

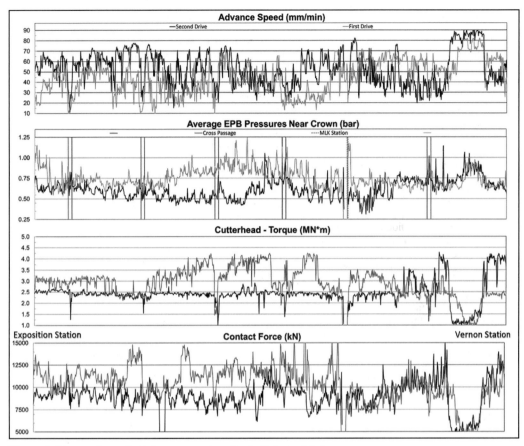

Figure 7. Comparison of mining parameters for both tunnel drives

were actions taken to improve excavation performance after tunneling began.

Given no reasonable means of pre-wetting the ground in front of the cutterhead prior to excavation, additional water had to be supplied by the ground conditioning system. Because water alone is not an effective means of friction reduction in granular soils, bentonite or lime slurries can be used to impart flowability into the mixture. An additional bentonite line was considered for installation to assist the existing lines which could inject bentonite into the mixing chamber, but was not constructed, nor was bentonite used for mixing with the muck due to existing agreements with land fill sites which received the material. Bentonite slurry is not a contaminant, but it does greatly reduce the ability for sites to receive large quantities of material in a short period of time. The slurried muck must be spread and dried until the water content nears optimum before it can be properly compacted. Because muck from the Crenshaw project was predominantly sand, the foamed muck

broke down in a matter of hours and required little compactive effort to densify the resulting fill.

Because some torque reduction was realized at the perimeter of the cutterhead from shield gap bentonite injection, it was decided to shift one of the six foam injection ports closer to the center of the cutterhead. Foam production was improved by increasing air flow to and through the foam generators via larger filters, and by providing a dedicated water supply. Foam generators were rotated to vertical positions with various bead sizes tested inside them to determine the best results for the revised throughput. Different types of surfactant and polymer were tested in the improved foamers on the second drive, with minimal to no perceptible improvement realized. The importance of increasing intergranular space with simple surfactant foam, once the material reached a moisture content that would sustain foam, was of greatest operational significance with respect to soil conditioning.

Of greater significance to overall production than soil conditioning improvements was a 30%

reduction in contact force on the second tunnel drive which produced a corresponding 30% reduction in cutterhead torque. The comparison between tunnel drives can be seen in Figure 7, with emphasis on the reaches from Exposition to MLK station. This reduction also resulted in a 0.2 bar average decrease in face support pressure as measured by bulkhead transducers. Reduction was done only after observing geotechnical instrumentation, and arriving at the still-conservative face support factor of safety of 1.6 in front of the cutterhead by using contact force as the verification. This reduction in plenum pressure made it easier for material to receive soil conditioner and pass through cutterhead openings into the mixing chamber, while maintaining positive pressure at the crown to indicate that the chamber was still filled. Additionally, scraper tools were modified to help break up the face to reduce further packing of the already-dense Old Alluvium.

Also on the second drive, every jet grout zone was used as a stopping point for inspecting the cutterhead. Interventions were conducted once the majority of the zone had been mined through to prevent subsequent soil-cement plugging after intervention. Blockages within the mixing chamber were removed and cutting tools replaced. Muck flow passages to the screw auger inlet were restored to full volume giving the operator greater flexibility with screw extraction in the form of stabilized plenum pressures.

SUMMARY

In dry sandy ground, reducing cutterhead torque by minimizing contact force while maintaining sufficient face support pressure is very critical. Tests need to be implemented to find optimal foam usage to produce a well-conditioned mixture which will reliably transmit pressure. Engineers and operators need to utilize contact force and settlement monitoring information in addition to face support pressure readings from bulkhead pressure transducers to avoid unnecessary confinement and friction at the cutterhead face. Routine mixing chamber material buildup removal and cutting tool maintenance is necessary to stabilize plenum pressures at peak extraction rates. The EPB machine is a dynamically balanced system; it is challenging but certainly possible to operate this system in the dry ground condition.

ACKNOWLEDGMENTS

The authors acknowledge Los Angeles County Metropolitan Transportation Authority for guiding and supporting this project. We are also thankful to our colleagues who provided expertise that greatly assisted the project.

REFERENCES

[1] Hatch Mott MacDonald, Geotechnical Baseline Report of Crenshaw/LAX Transit Corridor Project, 2012.
[2] BASF, Soil conditioning Tests (Type 1 & 2 Soil) for Crenshaw/LAX Project, Los Angeles, CA, 2011.
[3] DIN 4085—Calculation of earth-pressure, 2011.
[4] R. Godinez, H. Yu, M. Mooney, E. Gharahbagh, G. Frank, Earth Pressure Balance Machine Cutterhead Torque Modeling: Learning from Machine Data, *2015 RETC Proceedings.*

Phase 1—Second Avenue Subway Project, New York: Light at the End of the Tunnel: Delivering a Mega Project on Schedule While Maintaining Budget

Michael Trabold and Anil Parikh
AECOM

Richard Giffen
ARUP

ABSTRACT: Phase 1 of the Second Avenue Subway is a $4.5 billion project which opened on schedule for revenue service on January 1, 2017. To complete the project on schedule while maintaining the budget was only achieved by utilizing different strategies with all parties working together to make it happen—owner, designer, construction manager, and contractors.

This paper will discuss the design, planning, and then execution of completing such a complex mega project on schedule while maintaining budget. It will cover strategies that were used by the owner, designer and contractors to improve coordination amongst team members, which streamlined review processes, and removed obstacles to complete the project.

INTRODUCTION

The Second Avenue Subway (SAS) is a new subway line on the east side of Manhattan for the New York Metropolitan Transit Authority (MTA). The project was first conceived in the late 1920s to replace an existing elevated line. After a number of starts and stops, parts of the line were constructed in the early 1970s but due to the New York fiscal crisis in 1975, the project was halted. The current project consists of 8.5 miles of track, 16 new stations, one renovated station, and connections to existing subway lines. Due to its length and complexity, the Second Avenue Subway was subdivided into four phases. The first phase which has recently opened consists of the renovation of an existing station at 63rd Street, new mined cavern stations at 72nd and 86th Streets, and a new cut and cover station at 96th Street. This phase had an estimated cost of $4.5 billion and as of May 2017 had approximately 176,000 daily riders.

Phase 1 was further divided into 10 different contracts to allow more contractors to participate thus increasing contractor competition. The split also allowed for smaller contracts to be awarded in a staggered fashion. The first contract to be awarded in 2007 was a tunnel boring contract to spearhead the construction. The other contracts were awarded in the following years starting with 96th Street and ending with 86th Street. Each station was typically built using two contracts, one for heavy civil structural

and the other for architectural finishes, mechanical, electrical, and plumbing (MEP) work.

DESIGN

The design contract was awarded in 2001 to the AECOM-Arup Joint Venture (AAJV) with an initial task to develop a preliminary engineering level design for the complete alignment before commencing final design of Phase 1 in 2006.

AAJV's approach was to form a fully integrated multi-disciplined team to work closely with the client to challenge the status quo by suggesting new ideas and submitting proposals for discussion. Some of these ideas, such as platform edge doors and large scale system upgrades, were ultimately not adopted but many others were.

An early task was to review the clients design standards and procedures against those from similar systems all over the world. From this, AAJV was able to develop a project specific Design Criteria Manual that took only the best and latest international design approaches that could most easily work with the client's operational requirements.

In particular, the column free station layout has received widespread praise from the client, government agencies and the public. Existing New York City subway stations are renowned for steel columns at close centers which restrict visibility, require significant maintenance and create a claustrophobic atmosphere. By contrast, the new stations with their

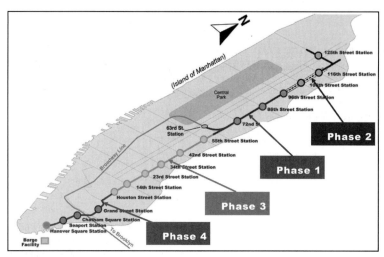

Figure 1. Project map

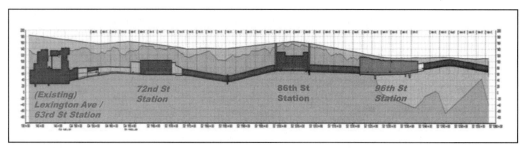

Figure 2. Phase 1 elevation and geological profile

abundant station specific artwork have been compared to art galleries.

In addition, AAJV was able to incorporate other enhancements from international best practices such as drained liners to reduce water pressure loads, fiber reinforced concrete, acoustic panels in the station roof arch to deaden ambient noise, improved ventilation, lighting and way finding.

A particular constraint for the design team was the geology. The Phase 1 alignment varies from high rock at the southern end of the alignment to soft ground at the northern end, (Figure 2). The high rock is generally overlain by fill material approximately 8 to 10 feet deep. The rock gives way sharply to soft ground around 92nd Street. The soft ground is predominantly varved silt and clay overlain by organics and approximately 15 feet of fill material. This ground is very sensitive to vibration, movement and changes in ground water levels. This geological profile drove the decision to design the northern station at 96th Street as a cut-and-cover station and the other stations at 72nd Street and 86th Street as mined.

Another challenge was the adjacent properties and densely populated environment. Buildings are either high rise construction on piles or masonry row house construction on shallow foundations. The masonry buildings are typically five to six stories high with one basement, and most are over 100 years old (Figure 3). Almost all buildings are residential with commercial tenants on the ground floor. The buildings are generally built right to the property boundaries along the right of way. Extensive utility networks under the roadway and sidewalks had to be accommodated, relocated and protected. This required a complex sequencing and extensive coordination with the utility agencies.

The station layout and locations dictated that entrances and ancillary structures were located adjacent to or within existing buildings requiring extensive investigations, monitoring, protection and strengthening. AAJV developed an approach of providing a suggested temporary works scheme with the contract bid documents to minimize risk. The contractors were able to adopt and modify or revise to suit their preferred approach. This was successful as

Figure 3. Phase 1 typical existing buildings

all underpinning and load transfer operations were completed without incident.

Phase 1 was awarded using the design-bid-build method of procurement with regular design and constructability reviews undertaken with the client, construction manager and third parties at the 30%, 60%, 90% and 100% design stages. AAJV ultimately produced over 20,000 documents across the 10 contracts.

CONSTRUCTION SUPPORT—DESIGN

In the early stages of the heavy civil structural construction contracts, the design team conducted the typical submittal reviews of the contractor's shop drawings, mainly the temporary roadway decking submittals. Once construction started in earnest it became clear that the contractor needed a quicker turnaround of their submittals and face to face meetings with the design team to answer their questions and clear up any open items. The face to face meetings helped expedite the contractor's shop drawings submissions for slurry walls, secant piles, and roadway decking design. The 96th Street station was a cut and cover station and the slurry walls were used for both temporary support of excavation and for the permanent walls, which made the rebar detailing very complex. Couplers were used for the reinforcing steel bars in the wall that had to connect with the rebar in the invert, roof slabs, and mezzanine beams. The mezzanine framing consisted of T-beams 2 feet wide and 4 feet in height; each had both top and bottom rebars that had to tie into the slurry wall. This made for a congested rebar cage and required that each row of couplers to have a specific elevation to insure it could be connected after excavation was complete and the slabs were placed. This required the designer and contractor to work closely together on the layout of the rebar to insure the design requirements were met, and to make sure

the rebar cage was constructible. The concrete for the slurry wall, which was placed by the tremie method, would have to flow between the rebar and not cause any voids in the slurry walls, which made the rebar detailing important. During the development of the shop drawings for the rebar cages, the designer had regular meetings with the contractor to go over the detailing of the cages and make adjustments as needed. These meetings continued with the 72nd Street and 86th Street Station mining contracts. The owner welcomed these meetings because the critical issues affecting the project schedule were resolved much faster than the conventional submission review process.

The running tunnels were constructed with a tunnel boring machine (TBM) and the final liner was cast in place concrete, not precast segments. The caverns for 72nd St and 86th St were built by drill and blast method; with top headings and drifts.

The design team had field representatives on site during the construction of the TBM tunnels and

Figure 4. Project team at weekly workshop meeting

Figure 5. Unfinished fare array

mining of the caverns at 72nd and 86th St Stations., The design team members worked hand and hand with the contractors in implementing the initial support—rock bolts and ribs—prior to the installation of the initial liner of shotcrete. Similarly in areas of fractured rock or unexpected conditions, the design team worked with the contractor on site to evaluate the existing condition, develop any modifications to the initial support to suit the field condition, and allow the work to continue in a safe manner.

Finishes and MEP Contracts

One of the challenges during the shop drawing approval process of the architectural finish and MEP contracts was that multiple disciplines and New York City Transit's (NYCT) various user groups were required to review the same submittal. The designer played a major role in resolving conflicts between the reviewers and coordinating comments from all reviewers. The contractor received one coordinated response from the designer which helped expedite the shop drawing approval process.

The team developed an in house electronic document management system to monitor the work flow between the contractors, construction manager's office, NYCT review personnel, and the design team. This system housed all submittals, requests for information (RFIs), contract documents, review and response forms, and was accessible to the entire project team.

To facilitate and stream line the submittal and review process, the design team held weekly workshops with the contractor and their sub-contractors to discuss recently submitted submittals, upcoming submittals, and go over RFIs. Face to face meetings between the designer and contractor made the process a team approach, put faces to names, and

opened the lines of communication. All parties could add items to the agenda, which brought the critical submittals or RFIs to the forefront and provided quicker responses.

As construction advanced, those meetings were moved from the designer's office to the contractor's office on site. This allowed for a quick site visit to discuss and resolve any challenging issues. It also brought the field superintendents into the conversation on how best to achieve the design intent and still have a buildable design. The workshops resulted in less RFIs, faster turnaround of submittals and better communication between the designer and contractor.

One of the main concerns of the MTA Capital Construction (MTACC) which is a department within the MTA that manages the capital projects for the authority, towards the end of 2015 was the coordination of the trades in the various rooms: mechanical equipment, HVAC ducts, lights, electrical panels, and fire alarm systems to name just a few. With the completion deadline of December 2016 looming, MTACC requested that the design team create a task force, visit every room at the three new stations to ascertain the status of the construction of each room, review the design drawings to see if all equipment would still fit, and determine what work still needed to be completed. The multi discipline task force featured architectural, mechanical, electrical, and communication personal; it was provided with iPads that contained the amplified design drawings as well as the approved shop drawings. Using a color coding system—with green for completed/coordinated room, yellow for potential coordination issues with unbuilt work, and red for coordination issues—the design team was able to alert the construction manager's (CM's) office to potential conflicts in the field, as well as identify which shop drawings needed to be submitted sooner so the coordination issues could be worked out. Over the course of eight weeks the team visited over 700 rooms and came up with 200 findings in the yellow/red categories.

The design team also assembled an independent review team to review the amplified drawings to check for any design issues and code compliance issues that could arise. The team assembled had not worked on the design, brought a fresh perspective to the project, and identified areas of concern with both the design and code compliance. Their findings were reviewed by the design team, and if appropriate, then addressed by issuing design sketches to the contractor.

Another issue that could delay the opening of the new line was observations made by NYCT's Code Compliance Department and the various user groups. With a project of this size, the earlier these observations or code compliance items were identified, the quicker they could be addressed and closed out.

Each observation made by NYCT was routed through the CM's office. The CM reviewed and then assigned the observation to either the contractor or the designer for responses and action. The design team assembled a team dedicated to review and respond to each item that was deemed in the designer's court by the CM's office. The review team was led by an architect familiar with the design and applicable codes, and was joined by senior engineers in mechanical; electrical; plumbing; and fire alarm who were able to turn around most issues in 72 hours. For more difficult issues that required resolution with NYCT personnel, a weekly Friday morning meeting was held to resolve the issues. Weekly meetings meant that the issues would not sit for a long period and would be addressed in a timely manner.

One key area that exemplified the team work needed to complete the project on time was the station's fare array areas. Each new station has two new fare array areas. These areas at the station entrance mezzanines include not only the turnstiles and entrance gates but also way-finding maps; static signs; digital customer information signs; On the Go kiosks (interactive way-finding maps); Metrocard vending machines; Fire Department of New York (FDNY) Lock Boxes; closed circuit security cameras; and a station service center (SSC), formerly known as a token booth. They also had architectural ceilings, original art work, fire alarm devices and custom made light fixtures. These areas required coordination between both NYCT's Stations and Automated Fare Array (AFC) departments, multiple members of the design team, and both the station and systems contractors.

The fare arrays were a complex area to arrange and critical to the operations of the station. It took multiple meetings on site with numerous NYCT personnel, the contractors, and the design team

Figure 6. Completed fare array at 96th St. Station

(which included architectural, lighting, mechanical, electrical and communication trades) to fit all the pieces together. The sequence of construction was important, as certain schedule critical items had to be resolved in the field before the remaining work could be completed. An example of a schedule critical item is the turnstiles. They had to be precisely located in the field before the security cameras could be installed, which meant the architectural ceiling, the conduits and back boxes for the cameras could not be installed until the turnstiles were laid out and installed to insure the cameras were in the correct location to satisfy the security requirements of NYCT.

OWNER'S APPROACH

The planning for Second Avenue Subway Project started in 1995 with the environmental work. The approach MTA took was that it would be very difficult to build the entire 8.5 miles of subway with 16 stations at one time mainly due to funding constraints. The goal was to build the project in phases such that MTA (the owner) could put each phase into operation while constructing the next phases.

The design contract was awarded to the joint venture team of AECOM-Arup in late 2001. The goal was to complete the conceptual design and preliminary engineering for the full length of Second Avenue subway in three years. The challenges during the planning and design phase was to get NYCT's user groups and maintenance divisions acceptance on design decisions such as location of station entrances, space requirements, platform widths, etc. MTA implemented the following steps to manage the completion of conceptual design, preliminary engineering design, and final design:

1. The Owner's Program Management staff and key technical staff were co-located with the design team.
2. Establish a decision making process by forming up Working Group Committees, Technical Advisory Committees and Executive Committees. The project team wrote Technical Advisory Committee papers to memorialize the decisions made at these meetings which were the basis for going forward with the design. Technical papers included the options evaluated, cost, schedule, environmental, real estate, operations and maintenance impacts, and the recommendation of the project team. The benefit for these committees was that the decisions were made much faster and were able to maintain the design schedule.

Figure 7. Information flow chart

3. Prepared contract packaging options at the early stages of the design and got the approval from NYCT before proceeding with the final design. Contract packaging options included high level cost, schedule, cash flow projections, and pros and cons of each option. The owner also considered bonding requirements, competition and market condition at the time of starting construction.

4. Performed in-depth constructability reviews with construction professionals during the preliminary and final design phase to identify risk and mitigation measures for tunnel and stations construction. These peer review teams were both AAJV personnel and recognized industry experts.

5. Established risk assessment process from the early stages of the design. A risk register was prepared and maintained throughout the life of the project.

Preliminary engineering design was completed in December 2004. The project did not go into final design until April of 2006 due to delayed funding authorization. During this time the owner requested the design team to continue refining the design to reduce the project cost. This was done as part of the extended preliminary engineering design. The first construction contract to construct tunnels from 92nd Street and 2nd Avenue to 63rd Street/Lexington Avenue was awarded in March 2007.

In early 2008, MTA decided to increase the number of contract packages in order to increase the competition and attract more bidders. This decision created additional challenges with respect to interface coordination between the stations finishes contracts and systems contracts.

MTA scheduled bi-weekly senior manager meetings with the contractor in order to manage the timely resolution of project issues. These meetings were attended by the senior staff of the contractor, AAJV, and MTACC.

Community Outreach

The MTA Capital Construction implemented a comprehensive outreach program during the construction. SAS community outreach liaisons were assigned to each station and reported to an outreach director who reported directly to the president of MTACC. The SAS outreach program included periodic briefings to the local community boards, monthly meetings with the community leaders, quarterly public workshops, newsletters on project progress, and a 24 hour hot line. In addition, MTACC opened a community information center to update the visitors on project progress, SAS exhibits, and lectures related to project's construction activities.

MTA implemented good neighbor initiatives program. The implementation of this program greatly reduced construction impacts and helped generate a greater public tolerance for impacts that couldn't be as easily mitigated.

MTA conducted monthly underground tours of the work site beginning in December 2011. The project stakeholders (residents, business owners and elected officials) were able to see the complexity and immensity of the work and progress underway.

MTACC in July of 2014 opened the Community Information Center (CIC) to serve as a one-stop shop for Second Avenue Subway information. In addition; the CIC hosted interactive exhibitions, daytime school presentations, contests and evening lectures about the subway's history and construction.

Project Completion to Revenue Service

The project was scheduled to be completed in December 2016. Fifteen months prior to the project completion, the MTACC chairman along with the MTACC president, met with the contractors and made an agreement to accelerate the project and maintain the project completion date of December 2016. This was a major challenge at the time based on the remaining work to be completed along with the systems integration and testing.

In order to resolve the project issues in a timely manner and ensure timely completion of the project, MTACC implemented the following actions.

• MTACC assigned Program Managers at each project site to oversee the project progress, assist the CMs to resolve issues that arose between the stations and systems contractors, and to resolve issues between the contractors and NYCT's maintenance and operations staff.

• MTACC required the stations contractors and systems contractor to submit the project completion schedule, which was monitored by MTACC's chief project scheduler and his

Figure 8. Good neighbor initiative

Figure 9. Community information center

staff on a daily basis to track the progress of the work. Issues driving the critical path work activities were raised to upper management to ensure that it was resolved immediately.

- There were significance challenges in coordinating the interface work between the station's contractors and system's contractor. Weekly meetings were held with each station's contractor and the system's contractor to resolve interface and access related issues. The meetings were attended by the executives of the contractors, major subcontractors, CMs, AAJV, NYCT staff, and MTACC's president and his staff.
- NYCT's operations and maintenance staff was required to perform inspections and

witness all tests prior to acceptance. NYCT provided dedicated staff to provide observation items as work was progressing and to witness the tests performed by the contractor.

- NYCT's Code Compliance staff was assigned on a full time basis nine months prior to project completion to inspect over 1,400 spaces within the four stations.
- MTACC's Project Executives monitored on a daily basis, the completion of the project's critical work. This included completion of the required tests for systems acceptance and commissioning, completion of the safety-critical items for the code compliance acceptance.
- MTACC's testing and commissioning group was responsible for monitoring the completion of all required level of testing prior to acceptance. They prepared the number and type of tests to be performed for each system, and coordinated with NYCT staff so they could witness all tests.
- Governor Cuomo also monitored the project closely, holding daily teleconferences with MTA and NYCT personnel; and making weekly station visits during the last four months of the project. His personal involvement energized the work crews and cut through red tape—for example, by eliciting faster responses from suppliers to ensure delivery dates.

SUMMARY

The Governor of New York inaugurated Phase 1 of the Second Avenue Subway on New Year's Eve 2016 with a symbolic ride and a speech that reminded the officials present that "We need to show people that government works and we can still do big things." The opening of the new subway line on time was important to show the public that large, complex projects can be completed on schedule and while maintaining the project budget. This can only occur when all parties involved work together to accomplish that goal. For Second Avenue Subway that was accomplished by improving communication between the designers and contractors, and by a forward thinking owner who was willing to try different approaches to complete the project on schedule.

North Hollywood Station West Entrance—A Successful Connection of Metro Red Line Subway and Orange Bus Line

Tung Vu
VN Tunnel and Underground, Inc.

Milind Joshi
Los Angeles Metro

Alex Gonzalez
Skanska Civil West

Richard Silos
AECOM

ABSTRACT: The North Hollywood Station is currently the final stop of the Los Angeles Metro's Red Line Heavy Rail subway. Metro's Orange Line Bus Rapid Transit (BRT) Terminal is located on west side of Lankershim Boulevard across from the North Hollywood Station. Typically, there is a very high volume of patrons transferring from Orange Line Bus to Red Line Train going to Union Station in the mornings and vice versa in the afternoon. The heavy pedestrian traffic crossing Lankershim Boulevard to connect these two lines created a safety hazard for the patrons as well as traffic congestion for the vehicles in the north-south directions. The North Hollywood Station West Entrance was designed to address the above safety and congestion issues by providing a direct underground connection from the station concourse level to the ground surface via a stairway, single escalator, and two elevators. This new 150 feet long, 50 feet wide, and 40 feet deep underground entrance, constructed using cut-and-cover method, also houses ancillary equipment rooms to serve its operation. This paper will discuss in detail the design and construction challenges of this design-build project including: design optimization, deep excavation in a narrow footprint, tight budget and schedule, and connection to the existing structure and system.

INTRODUCTION

The North Hollywood Station is currently the final stop of the Los Angeles Metro's Red Line Heavy Rail subway. All patrons who wish to travel further north must cross Lankershim Boulevard to connect with the Orange Line Bus Transit Terminal. This resulted in about 50 to 100 people attempting to cross Lankershim Boulevard every 10 minutes or so, which posed an unsafe situation and contributed to traffic congestion. The connection time was also long and inconvenient due to the traffic signal timing at this busy intersection. To improve the pedestrian safety and vehicular traffic in this area, the Los Angeles County Metropolitan Transportation Authority (LA Metro) advertised the MRL/MOL North Hollywood Station West Entrance design-build project in 2013. The new entrance will also reduce connection time between the lines; hence improving capacity to accommodate the growing ridership. The contract was awarded to the Skanska-AECOM team in early

2014 as the lowest bidder. The design-builder team has overcome several design and construction challenges to complete the project in time. The new entrance was opened to the public on August 15, 2016. The project was well received by the community as it has greatly improved the pedestrian safety and traffic in the area.

DESIGN CHALLENGES

The main challenge of this project is to fit many components within a small footprint. These include the passageway, stairs, elevators, escalator, mechanical room, electrical room, elevator machine room, fresh air intake and exhaust ducts, and oil-water separator pit. The passageway connects to the existing North Hollywood station at the concourse level, which is located at approximately 40 feet below grade. The ground conditions at the site consist of interlayers of loose to medium dense silty sand and medium dense, fine to coarse, poorly graded sand. The underground

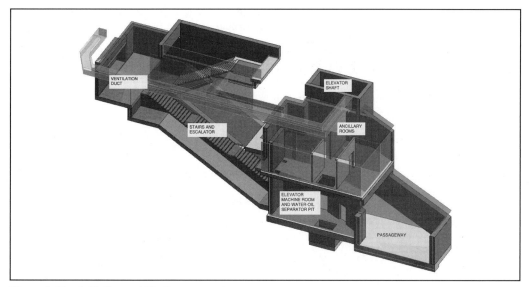

Figure 1. North Hollywood Station west entrance

structure is located above groundwater level. A 3D view of the underground entrance structure is shown in Figure 1.

In the request-for-proposal documents, the ancillary rooms were located underneath the stairs and escalator and at the same level as the passageway. This room arrangement would require the entire site to be excavated to 40 feet below grade to build these rooms and backfilled to the ground surface once they are constructed. During the bid proposal, major rearrangement and optimization of the room layout was done by the design team to make use of the available space and provide the contractor with competitive bidding costs to win the project.

First, the ancillary rooms were relocated to above the roof of the passageway with exception of the elevator room, which was kept at the passageway level as required by Metro's vertical transportation group. This allows the contractor save both excavation and backfill volumes. However, locating the ancillary rooms on the second floors required service stairs, which were difficult to accommodate within the very confined footprint. Fortunately, the existing emergency stairs No. 8, which were used for the North Hollywood station, were no longer needed after completion of the new entrance. These stairs were located within the site footprint and were to be demolished prior to the entrance construction. The design team reutilized a portion of this existing stairs that connects to the North Hollywood station to provide the access to the ancillary rooms on the second floor of the new entrance structure.

Second, the entrance structure was designed during preliminary engineering to be several feet away from the existing North Hollywood station structure to connect to the exit point at the surface. The design team reconfigured the entrance structure to build it against the existing station wall. This allowed the contractor not only to save costs by reutilizing the existing support of excavation system of the subway station, but also gaining additional space for the elevator machine room and oil-water separator pit.

Third, during the final design, a concern was raised by the community about a need for the increase of capacity when the Orange Bus Line is converted to a light rail line. As a result, knock-out-panels were introduced on the entrance walls to allow to expand the entrance structure to accommodate an additional escalator or stairs if the need for increased capacity exists. In addition, the design team also overcame other challenges such as fitting the location of the exhaust shaft to be 40 feet away from intake shaft and other entrance points, resolving conflicts of new traffic signal pole foundation with the existing station structures, etc. Details of structural analysis are discussed in the section below.

The main structures of the new entrance consist of the underground reinforced concrete structure and the elevator and escalator steel canopies above ground. The structures were designed in conformance with the Metro Rail Design Criteria (MRDC) and California Building Code. The steel canopies were analyzed using RISA 3D. Due to its variable shape, the underground structure was designed with SAP2000 3D using cell elements. The new underground entrance structure is connected to the existing station through the existing embedded couplers

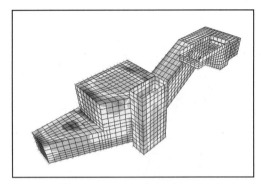

Figure 2. Results from SAP2000 3D model

and rebar dowels. The connection was modeled using link elements. The properties of these link elements were specified based on the structural capacity of the existing dowels and thickness of concrete elements at the connections. The ground support was modeled using compression-only soil springs at the base slab. Figure 2 shows an example of results from the SAP2000 3D model.

The results from the SAP2000 3D model were used to design the structural elements of the entrance structure for the static loads. For the seismic design, some typical sections of the entrance structure were analyzed for the seismic racking that is typically required per MRDC for cut and cover structure. The racking deformations that were previously developed for the design of the existing North Hollywood station for the maximum design earthquake and operating design earthquake were used for this new entrance structure. Even though the entrance structure is located above the permanent groundwater level, a hydrocarbon resistant (HCR) membrane was also designed to cover the entire underground

structure to prevent water and gas leaks into the structure.

As discussed above, two knock-out-panels were designed on the walls to allow the future expansion of the entrance to include an additional escalator or staircase. A knock-out-panel (KOP) is typically a predefined opening on the walls of the structure where the boundary elements around the opening are designed to allow safe removal of the KOP without impact on the remaining structure. Rebar arrangement and construction joints are designed to facilitate removal of KOPs in the future. In addition, the anchored water barrier strips are embedded in the concrete wall around the KOP to allow the future removal of the HCR membrane at the KOP without compromising the remaining water/gas proofing system. The KOPs have been designed for several existing subway stations in Los Angeles where an increase in future ridership is anticipated. New entrances will be built and connected to the existing stations through these KOPs to accommodate the growing demand.

CONSTRUCTION CHALLENGES

The construction experienced several major challenges. The project site is very small and located between busy streets. Figure 3 shows the approximate the worksite footprint and the laydown area. Due to the confined worksite and the fact that the project was being constructed between two active subway and bus lines, the entire project was built from only one access point. Material was transported by forklift to and from the laydown yard crossing Chandler Blvd. The construction team had to carefully choreograph all construction activities to overcome this logistical constraint while maintaining production efficiency in order to stay on schedule.

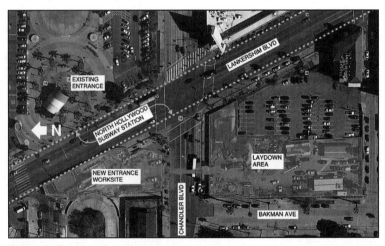

Figure 3. Construction site map

Figure 4. Extended existing soldier piles for reutilization

Figure 5. Small size excavator for low headroom excavation

This condition existed the entire time of construction, from the very beginning of the soldier pile installation to the final placement of the street level architectural colored concrete at the plaza.

The entrance was designed to allow the use of existing soldier piles of the North Hollywood station. This provided a big time and cost savings to the overall project. In all, there were twenty (20) soldier piles that were reused. However, the challenge was that these piles were cut off to a depth of 6 feet below grade when the North Hollywood station was completed. In order to fully utilize these existing piles, they needed to be extend to the existing street surface level before start of excavation. A great amount of effort was spent on locating these piles, trenching, and welding to extend them to the street level. Figure 4 shows an example of some soldier piles after they were extended. The reutilization of the existing soldier piles proved to be faster and cheaper than installing new twenty-four inch by fifty plus foot piles.

The next challenge with the support-of-excavation was installing a temporary roadway deck at the very north end of the excavation. The original excavation layout called for a 100% open cut excavation; however, because the turning radius at the north end of the excavation and the adjacent roadway was not wide enough to accommodate large trucks, the City of Los Angeles Department of Transportation (LADOT) required a temporary steel deck to support the roadway allowing for a 26-foot radius turn versus 14-foot. The addition of the steel deck resulted in an excavation that was now only 90% open cut. To make up for the production impact caused by the steel deck, the construction team utilized smaller excavation equipment: a mini-excavator and a skid-steer as shown in Figure 5. Their smaller pieces of equipment were able to work below the steel deck and maintain our production and thus schedule time.

Figure 6. Cutting 4-foot thick reinforced concrete KOP

One of the biggest challenges was demolition and removing the existing KOP in order to make the connection of the new entrance structure with the existing North Hollywood Station. To reduce the risk of potentially causing any damage to the existing station structure and utilities, the construction team decided to use a diamond wire-saw versus the conventional method of using a 2600 lbs hydraulic jackhammer ram at the end of an excavator arm. The diamond wire-saw method was carefully planned to precisely cut the station wall to avoid damages to the existing structure as well as to avoid additional concrete to be removed with handheld tools. The end result was a big success. The demolition did not cause damage to the existing structure and the exact size of the KOP was removed. Figure 6 shows how the 4-foot thick KOP was cut and removed.

In addition, due to incompleteness of the as-built documents of the existing station, the construction team could not locate the existing tie-in electrical

Figure 7. Installing escalators

conduits and the existing rebar couplers (or formsavers) to make connection with the new entrance structure. The team spent 20 days locating the electrical conduits with the use of specialized camera equipment. It was discovered that the existing conduits were located outside of the tie-in point. This resulted in beginning the construction work from the opposite end and finishing at the tie-in point. And in order to maintain schedule and reduce the 20 days impact, the construction team built the exterior concrete walls simultaneously. On the 30-degree sloped area, the exterior walls of the stairs and escalator were braced against each other, which meant that both walls were placed simultaneously enabling the formwork to support itself from both sides.

The next challenge was figuring out how to install the new escalator in the very tight space while coordinating with the other concrete work. The new entrance consisted of two new escalators, one 60-foot long and the other 40-foot long. The longer 60-foot escalator was at the bottom of the entrance and needed to be installed prior to the construction of the new entrance roof. This was a significant challenge for several reasons. First, timing of the delivery of the escalator as to not impede the schedule was critical. Second, lowering the escalator into place and avoid hitting the support-of-excavation struts and roof rebar was difficult. Last, it was critical to ensure the onsite crane had the capacity to pick and set the escalator into place. To overcome these challenges, the escalator arrived on site two months early and the roof rebar was shortened by installing 90-degree rebar with couplers at the ends in order to add on the required rebar splice length later. Last, the escalator was fabricated in four sections each within the cranes lifting capacity. Figure 7 shows how the escalators were installed.

In conclusion, the construction of the North Hollywood Station West Entrance was challenging due to various constraints and a very tight footprint. The linear construction activities made up 90% of the time due to these constraints. However, due to proper planning and good project team work, the construction was successfully completed on time and budget. Figure 8 shows the entrance at its completion.

Figure 8. North Hollywood Station west entrance completed

Figure 9. Patrons crossing Lankershim Boulevard prior to North Hollywood west entrance open

Figure 10. No patrons crossing Lankershim Boulevard after entrance open

SUCCESS OF THE PROJECT

The MRL/MOL North Hollywood Station West Entrance project was designed and constructed to ensure safety for the patrons commuting between Metro Orange Line and Metro Red Line. The new underground entrance was engineered to utilize a narrow strip of Metro owned property and connect to the North Hollywood station through the existing knock out panel. The design-builder has overcome numerous challenges and accomplished the objective of the project. The new entrance has not only created safe transfers between two different modes of transportation for the patrons, but it has also reduced their transferring time. It has helped reduce the traffic back-up for the vehicular traffic. The project provides the patrons with conveniences with the operation of new escalators, elevators, and ticket vending machines. The design complies with ADA requirements for the patrons.

The construction was completed on time without causing any significant inconveniences to the patrons. It was opened to public on August 15, 2016 and has been well accepted by the community. Figures 9 and 10 illustrate the pedestrian traffic improvements due to this new entrance structure. Due to its success and contribution to the community, the project was awarded the Outstanding Active Transportation Project of the Year by the Metropolitan Los Angeles Branch of the American Society of Civil Engineers in July 2017.

ACKNOWLEDGMENT

The project team would like to thank LA Metro, City of Los Angeles Bureau of Engineering, and LADOT for their great support during the design and construction of the project.

Risk Management in Early Subaqueous Transit Tunneling

Vincent Tirolo Jr.
STV Inc.

ABSTRACT: The management of risk on tunnel projects today is a formal matrix driven process initiated by the insurance industry We assume today when developing a risk matrix most of the risks can be identified. We have this confidence because of the high level of communications within the tunnel industry. However, even today, we may be unaware of some potential risks. Consider subaqueous tunneling in middle to late 19th and early 20th Century. Few subaqueous tunnels had been built, experience was very limited, and there were no tunneling organizations or conferences. How did tunnel engineers of that era assess risk with little or no case history information? How did they entice government and private interests into funding such high risk enterprises? We will explore three areas where these early practitioners were able to build confidence of both business and government in these high risks enterprises. The experience and management of risk on subaqueous tunnels that began in the 1860s resulted in a boom for the industry that continues today.

INTRODUCTION

Management of risk in tunneling today is a formal process involving the code of practice such as the 2012 ITIG Code of Practice for Risk Management of Tunnel Works, 2nd Edition, which was based in part on the Association of British Insurers and the British Tunnelling Society 2003 document, The Joint Code of Practice for Risk Management of Tunnel Works in the UK. These codes of practice generally recommend a formal approach to risk management of a project that centers on the use of a formal risk register. Both of these documents were developed because insurers where threatening to leave the tunnel market because of the high frequency of major tunnel failures where repair costs could exceed the original cost of construction. The insurers believed owners, engineers and contractor were using insurance as a risk management tool to lower construction cost. Most of these tunnel projects were government sponsored infrastructure projects.

The initial risk register is developed during the design process when uncertainties are the greatest and continues through construction. As the design and construction process develops, certain risks are mitigated or retired as tasks are completed but new risks may be added. It is only when the project is completed that all uncertainties are retired from the risk register. This is similar to the evolution of a client's perception of project cost described by Wood (2004). A client starts a project with an upper and lower bound on cost, e.g. greatest uncertainty. At tender the project is better defined so uncertainties are less and these bounds are reduced significantly.

However, it is only at project completion that the true cost of a project is known. The purpose of risk management is to reduce those bounds and thus the uncertainty of tunnel project cost as early as possible in the trajectory from inception to completion of a project. Goodfellow, O'Carroll and Konstantis (2014) recommend how risk registers could be applied to Design-Build projects in North America. Risk registers can be divided into procurement, permits, design, environmental construction, and functional/operational categories. Often the most difficult aspect of risk assessment is assigning the primary ownership of the risk.

When developing a risk matrix hundreds of uncertainties are identified, often at workshops, and "scored" based on likelihood, cost and or schedule impact. The resultant risks are numerically ranked from negligible to intolerable. Mitigation measures are used to reduce "intolerable" scores into a tolerable range. The assumption in developing a risk matrix is that most of the risks can be identified. In the early 21st century with our annual or biennial international forums such as the World Tunnelling Conference (WTC), The Rapid Excavation and Tunneling Conference (RETC), and the North American Tunneling (NAT; information on hundreds of tunnel projects is freely disseminated to engineers who specialize in tunnel design and construction. WTC 2016, NAT 2014 and RETC 2017 all had sessions specifically devoted to risk management. However, even in this modern environment, Duddeck (1987) emphasized that the worst case is being unaware of potential risks. Therefore it is important to develop a strategy for mapping uncertainty, evaluating risks and preparing for immediate action to prevent damaging or even dangerous unanticipated situations.

Now consider the situation in the middle to late 19th Century and the early 20th Century in subaqueous tunneling. The Thames Tunnel was the first successful subaqueous tunnel and began in 1825 and was completed by 1843 after four major flooding events and seven miner fatalities. How do you assess risk with little or no case history information? How can you interest government and private interests in funding such high risk enterprises? If we used a modern risk matrix, we would have determined that most of the risks these early pioneers encountered would be rated intolerable. Yet we know to the great benefit of large urban centers such as New York City and London, tunnel works did occur in spite of all these difficulties and uncertainties. Not only did they occur but dozens of subaqueous tunnels were mined with great success. How did that happen? What limited risk management tools did these earlier pioneers in the subaqueous tunnel industry use to achieve this unlikely success? We will identify some of these tools in this paper.

GROUND CONDITION: LOOK FOR GOOD GROUND

The major obstacle for mining below rivers was water flooding into the tunnels. Although Lord Cochrane invented and patented (No. 6018) the use of compressed air for underground works in 1830 and used in numerous bridge pier and shaft applications, Glossop (1976) states it was not used in subaqueous tunnel construction until 1879 at the Port of Antwerp by H. Hersent and under the Hudson River in New York by De Will C. Haskin. Neither application used a shield. Prior to 1879 attempts to mine a tunnel under a river involved placing the tunnel into soil or rock judged to be impermeable.

Commonly we are all taught that Karl Terzaghi is" the father of modern soil mechanics." The operative word in that expression is modern. Certainly soil and rock mechanics existed before Terzaghi. Rankine (1863) in his Manual of Civil Engineering has a chapter devoted to Earthwork. Foundations are discussed in the chapter on Masonry. Rankine also has a chapter titled "Of Various Underground and Submerged Structures." In Section I, "Of Tunnels" Rankine stated:

"The most favorable material for tunneling is rock that is sound and durable without being hard. Great hardness of the material increases the time and cost of tunneling, but gives rise to no special difficulty. A worse class of materials are those which decay and soften by the action of air and moisture, as some clays do; and the worse of those which are constantly soft and saturated with water, such as quicksand and mud."

Quicksand is what the Tunnelman's Ground Classification for Soils would now call flowing ground.

When mining tunnels the key to success was mining in the best ground possible and when confronted with "constantly soft and saturated with water, such as quicksand and mud" mining with poling boards in multiple timber supported drifts and maintaining drainage of the face. Rankine did mention Brunel's shield for the Thames Tunnel. Finding good ground was always the key to success in these early attempts on subaqueous tunneling. After reviewing soil samples from the site, Brunel, located his tunnel profile into a stratum of stiff blue clay sandwiched between the upper loose stratum water bearing gravels and a deep stratum "quicksand." This proved not to be the case as shown in Figure 1. Often less than 1 meter of clay protected the tunnel from inflows from the Thames.

The tunnel only had less than 4 m of cover below the river's mudline. The brick lined rectangular tunnel was mined using an 11.4 m wide by 6.8 m high × 2.74 m long shield composed of twelve individual cast iron frames, each subdivided into three working chambers through these difficult ground conditions. The tunnel was 459 m long. Mining a rectangular tunnel of this size through similar soil conditions would be a challenge, even today. Completing this first subaqueous shield driven tunnel was an amazing feat of engineering and construction by Marc and his son Isambard Kingdom and the miners they employed. The tunnel is now part of the London Underground. Both Brunel's are honored with plaques, Figure 2, near Rotherhithe Station.

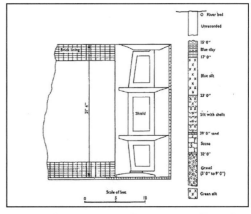

Figure 1. Subsurface profile at Thames Tunnel (from Glossop 1976)

Figure 2. Plaques in London commemorating the work of Marc and Isambard Kingdom Brunel

Marc Brunel had the first patent of a shield, No. 4204, in 1818. The patent was for a circular shield. In Henry Law's (1857) account the construction, He stated that Marc Brunel had considered the use of a circular shield for the tunnel, but because of the shallow cover over the tunnel Brunel abandon decided to use a rectangular shield rather than a circular shield.

The lessons learned from the Thames Tunnel where incorporated in the next subaqueous crossing of the Thames, the Tower Tunnel by James H. Greathead. This was a much smaller tunnel than the Thames Tunnel. Its external diameter was only 2180 mm. The tunnel was mined entirely within London clay with a minimum of 6.7 m of cover below the mudline. The tunnel was 411 m long and was successfully completed in one year. Moreover, as pointed out by Copperthwaite (1906), the Tower Tunnel has a number of notable features:

1. First use of a circular shield for subaqueous tunneling.
2. First use of a cast iron segmental liner.
3. Screw jacks were used to propel the shield (Greathead choose screw jacks over hydraulic jacks for simplicity)
4. Cement grout was injected into the annulus between the cast iron liner and the shield to minimize ground subsidence.

Kirby, et al (1990) describe the earliest subaqueous tunnel in the USA as a water tunnel mined under Lake Michigan from 1864 to 1867 in dense blue clay.

Until the successful introduction of compressed air and a tunnel shield by Greathead on the City and South London Railway in 1886, control of ground risk required locating the tunnel profile of subaqueous tunnel in impermeable rock or soil. It was possible in some circumstances to add fill over tunnels to mitigate of the risk associated with shallow or permeable cover. In London, early subway tunnels were mined deep and often required passengers to access stations using deep stairs or elevators.

Geologic conditions and geologic uncertainties were then and are now, generally the most significant risks in tunneling. The tools we have to evaluate these risks are much more sophisticated than the tools available in the late 19th and early 20th centuries. The tools at that time were limited to wash borings, probes and test pits. We also have the advantage of ground improvement to mitigate risks. In the late 19th and early 20th centuries the method of excavate/dredge and replace was sometimes an option for ground improvement. However, more frequently especially for subaqueous tunnel work, placement of clay fill (e.g., clay blankets over the existing mudline) was the most common mitigation measure. The success of the Thames and Tower Tunnel and other projects located in "good ground" reassured investors and governments that subaqueous tunnel were feasible enterprises that could ease transportation congestion in major cities.

SCHEDULE RISK

Funding for subaqueous tunnels in the middle to late 19th Century and the early 20th Century was primarily by private investors. Many major commercial business centers, including New York and London, where separated from residential areas by major waterways. Toll bridges and railroad tunnels across waterways. At the time bridges required government permits. For example, bridges across waterways in the United States required a Corps of Engineers issued federal permit which limited commercial rights. Tunnels did not require federal permits. Efficient transportation to commercial business centers also opened new areas of these cities to residential development. Thus tunnels offered an opportunities for investment. Monies were available but what was critical then, as it is now, is return on investment. Return on investment is directly related to the construction schedule. The Thames Tunnel, in part, took 18 years to complete because of disruptions in funding. When Greathead (1895) remarked:

"Though a great engineering triumph with the appliances available at it's time of construction, and a lasting testimony to the genius and spirit of Brunel, the mode of construction of the Thames Tunnel has not been attempted elsewhere; and there can be no doubt that for nearly half a century that work served as a warning to engineers and capitalists not to embark in any undertaking of a similar character, and no other subaqueous tunnel was constructed."

This schedule risk caused difficulty in funding the small pedestrian tunnel across the Thames, the Tower Subway. Some innovations of the Tower Tunnel have been discussed earlier, but its major impact was to bring credibility back to subaqueous tunnel scheduling. The Tower Tunnel and shafts were completed within one year and the tunnel was mined with a single shield. The maximum progress was 2743 mm (six 457 mm rings) per shift; three shifts per day; 24 hours a day.

Greathead's next tunnel project was the City and South London Subway. Construction started in October 1886 and was completed in November 1890,

ahead of schedule. This project crossed the Thames and the two single track tunnels were each over 5000 m long. There were two sets of cast iron segmental liners. One external diameter was 3320 mm and the other external diameter was 3430 mm. Mining was 9 to 21 m below the ground surface. The profile was primarily in London Clay except for approximately 229 m that were in water-bearing sands and gravels that required the use of compressed air. The first successful use of compressed air in a shield driven tunnel. Eighteen shields and multiple headings were used to maintain the project schedule. Greathead is honored in London with a statue and a plaque, Figure 3, for his contributions to the tunneling.

In the late 19th Century and the early 20th Century the use of multiple shields, and when necessary compressed air, greatly reduced (but did not completely eliminate) schedule risk. The Pennsylvania Railroad Company (PRR) tunnel construction program in New York City is an example of schedule certainties. All six PRR tunnels were 7000 mm external diameter tunnels lined with cast iron segments and mined with shields and compressed air. Two tunnels totaling 4023 m were mined under the Hudson River using four shields. Four tunnels totaling 4755 m were mined under the East River using a total of eight shields. The geologic conditions were much more difficult under the East River. The East River tunnels were completed 1908 two years behind schedule.

Figure 3. Statue and plaque commemorating James Greathead

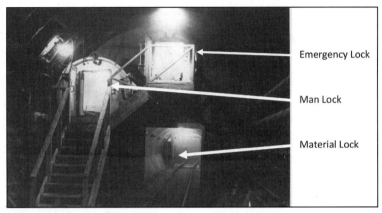

Emergency Lock

Man Lock

Material Lock

Figure 4. Compressed air lock system

Tunnels will always be enterprises with risk but what Greathead and others proved is that the risk was manageable. The use of shields and compressed air reinforced the belief that risk was primarily a function of ground conditions not tunneling engineering and technology.

IMPROVING THE SAFETY OF MINERS

Those of us working today know that safety is considered a priority on all of our projects. Any injury is investigated to understand its cause and address conditions to avoid a recurrence. Death on the job results in project stand-down safety workshops and more. It is hard to believe that this has not always been the case. In the late 19th and early 20th Century tunnel fatalities were considered an eventual part of the business. Even in the middle of the 20th Century, it was common said that there would be one fatality per mile (1600 m) of constructed tunnel.

In the middle to late 19th Century and the early 20th Century deaths far exceeded one fatality per mile. Deaths on the job were commonplace. Tunneling was a high risk operation and the miners that carry out the work were always in the most danger. Seven miners died on the Thames Tunnel Project.

The introduction of compressed air into subaqueous tunneling in 1879 by DeWitt Clinton Haskin Hudson River tunnels (H & M tunnels) and H. Hersent at the Port of Antwerp (Glossop1976) greatly reduced both ground risk and schedule risk. But this reduction in risk came at a price, the safety of the miners and engineers who worked in the tunnels. Today we would call a life-safety risk "Intolerable." In middle to late 19th Century and the early 20th Century this risk would have been classified using today's risk classes as "Undesirable." In 1879, the fatality rate was one miner per month primarily from

compressed air sickness (Ernest Moir 1896). Thirty miners died on the PRR East River Tunnels.

We live in a different world. In the world of middle to late 19th Century and the early 20th Century, human life, especially those of those in the poor working/immigrant class, were held in less regard than those in the middle and upper class. Social Darwinism was the working premise of the day. Fortunately for the tunneling industry, tunneling has always been a hands on operation requiring the participation of both miners and engineer's "in the hole" for projects to be completed successfully. The great tunnel engineers of the day such as Marc and Isambard Kingdom Brunel, James Greathead, Charles M. Jacobs, and others; are products of the middle and upper classes that took a special interest in tunnel safety because they worked in the tunnels. Many of these early tunnel engineers' own lives were shortened, e.g. Greathead and both Brunels, by exposure to the tunneling conditions of the time.

By the early 1880s, the situation reached the point where miners were refusing to work in the tunnels. Ernest William Moir had been an assistant to James Greathead and had first been exposed to compressed air tunneling on City and South London Subway. In 1891 Moir was working on the re-start of the Hudson River tunnels. Moir introduced the medical lock to that project and the rate of fatalities dropped from one miner per month to two miners lost out of a work force of 120 over 15 month period. A similar situation developed in 1904 on the PRR East River Tunnels. At that time Moir was employed by the tunneling contractor, S. Pearson & Son. He returned to New York and became a strong advocate for tunnel safety. Eventually the death rate reduced to less than 0.5% for tunnels of that era. This is still higher than in the modern era. For example on the Third Tube of the New York –New Jersey Lincoln Tunnel, there were no fatalities due to caisson disease (decompression illness) and the lost time incident rate was 0.0318%.

This early work done by Moir and others was the beginning of the scientific study of the use of high pressure compressed air and other breathing gases in tunneling. Today these high pressure techniques are used primarily for hyperbaric interventions. Best practices for high pressure interventions are contained in the British Tunnelling Society and International Tunnelling and underground Space Association 2015 Guidelines.

CONCLUSIONS

The history of subaqueous tunneling in in middle to late 19th Century and the early 20th Century is a great success story. In less than 100 years, the technology of subaqueous tunneling evolved from a high risk experiment to a well understood technology. In New York City alone from 1903 to 1933 thirteen twin, three, and four tube subaqueous tunnels were mined using the shield and compressed air technology. These tunnels all involved risk. The early pioneers in the subaqueous tunnel industry understood the importance of risk mitigation on its most fundamental basis. Their tools were limited. First they focused on convincing private industry and governmental authorities that subaqueous tunnels were feasible in good ground. Next they proved that not only were subaqueous tunnels feasible but that they could be constructed economically and on schedule. Finally, the life/safety risks of shield tunneling with compressed air were addressed by Ernest Moir and other pioneers in tunnel safety. A great success story of risk management that we tunnel engineers of today should recognize and appreciate.

REFERENCES

Association of British Insurers and the British Tunneling Society (2003), *The Joint Code of Practice for Risk Management of Tunnel Works in the UK.*

Copperthwaite, William C. (1906), *Tunnel Shields and the Use of Compressed Air in Subaqueous Works*, D. Van Nostrand, New York.

Duddeck, H. (1987), "Risk Assessment and Risk Sharing in Tunnelling," *Tunnelling and Underground Space Technology,* Vol. 2, No. 2, pp. 315–317.

Glossop, R (1976), "The Invention and early use of compressed air to exclude water from shafts and tunnels during construction," *Géotechnique, Vol. 26, No. 2*, pp 253–280.

Goodfellow, R., O'Carroll, J. and Konstantis, S. (2014), "Risk Registers and Their Use as a Contract Document," *Proceedings, North American Tunneling 2014, SME*, pp 670–680.

Greathead, James H. (1895), The City and South London Railway; with some Remarks upon Subaqueous Tunnelling by Shield and Compressed Air, *Proceedings, The Institution of Civil Engineers, Vol.123, Session 1895–96, Part 1*, pp. 39–72.

Hewett, B.H.M. and Johannesson, S. (1922), *Shield and Compressed Air Tunneling*, McGraw-Hill, New York.

International Tunnelling Association and British Tunnelling Society (2105), Guidelines For Good Working Practice in High Pressure Compressed Air, ITA Working Group No. 5, Health & Safety in Works.

International Tunnelling Insurance Group (ITIG) (2012), *A Code of Practice for Risk Management of Tunnel Works.*

Kirby, R.S., Withington, S., Darling, A.B. and Kilgour, F.G. (1990), *Engineering in History*, Dover Publications, New York, p. 488.

Law, H. (1857), *A memoir of the several operations and the construction of the Thames Tunnel.* By Sir Isambart Brunel.

Moir, Ernest w. (1896), Tunnelling by Compressed Air, *Journal of the Society of Arts*, pp. 567–585

Rankine, W. J. M. (1863), *A Manual of Civil Engineering, 2nd Edition*, Charles Griffin and Company, London, pp.

Wood, A. Muir (2004), "Ahead of the Face," *The 2004 Harding Lecture*, the British Tunnelling Society.

Shallow SEM Tunneling Under Major Roadways and Through Active Landslides in Edmonton Glacial Tills

John Kuyt, May ElKhattab, and Eden Almog
Arup

Ian Cisyk
EllisDon

ABSTRACT: The Edmonton Valley Line LRT is a $1.8bn(CAD) 13km light rail project connecting downtown Edmonton to Mill Woods. The scheme includes a 450m section of twin bore 7m diameter mined tunnels between the downtown surface stations and the signature bridge crossing over the North Saskatchewan River and valley. This paper addresses the technical project challenges, including shallow excavation underneath Jasper Ave with 3m cover, management of face stability and groundwater inflows from intra-till sand lenses, and accommodating active landslide movements on Grierson Hill. Challenges were addressed through use of advanced soil structure interaction numerical modeling, spiles and canopy pipes for pre-support in glacial tills, and instrumentation and monitoring response plans to address changing ground conditions with contingency support measures.

INTRODUCTION AND PROJECT OVERVIEW

The Edmonton Valley Line LRT project is a new light rail transit line for the City of Edmonton, connecting downtown Edmonton to the Mill Woods community in the southeast of the city. The approximately 13km long alignment includes 11 stops along the route, with a key elevated station incorporating a transit center and Park & Ride in the southeast and a transfer point to existing LRT in the downtown at Sir Winston Churchill Square. The project is being delivered through a public-private partnership (P3) between the City of Edmonton and TransEd Partners, consisting of Bechtel, EllisDon, Bombardier, and Fengate Capital Management Ltd. Engineering design support is provided by Arup, Associated Engineering, and IBI Group. Pre-construction work began in 2016, with expected start of operations in 2020 along the line.

The focus of this paper will be on the Quarters Tunnel portion of the project as shown in Figure 1, including the tunnel approach along 102nd Ave. into a cut and cover tunnel, followed by the mined tunnel section crossing underneath Jasper Ave. and continuing to descend underneath 95th St for a total length of approximately 450m, and concluding with the portal and bridge approach structure located on Grierson Hill within an active landslide.

The mined tunnel itself consists of two SEM excavated tunnels incorporating a steel fiber reinforced shotcrete initial lining for support during construction activities, followed by installation of a smoothing shotcrete layer, waterproofing membrane, and a cast-in-place steel fiber reinforced concrete final lining. Following completion of the tunnel lining, subsequent works will include the installation of the rails and track structures, overhead catenary cabling and structures, tunnel systems, cabling conduits, and emergency water and egress facilities. The tunnel arrangement is shown in Figure 2.

The main technical challenges in the design and construction of the Quarters Tunnel include shallow urban tunneling under existing buildings and infrastructure, particularly a shallow crossing underneath Jasper Avenue which was required to remain open to traffic with less than half a tunnel diameter. Other challenges included ground conditions which consisted of highly variable glacial tills with intra-till sand lenses and an active landslide at the river portal.

GEOLOGICAL CONDITIONS

In the City of Edmonton area, Mesozoic (Lower Cretaceous) mudstones consisting primarily of

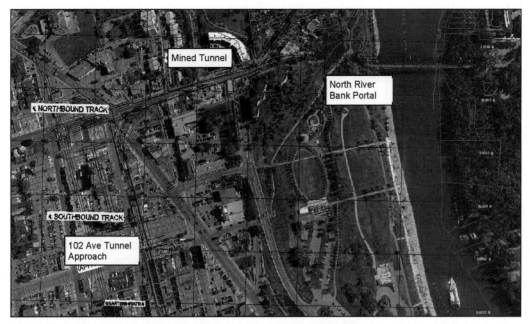

Figure 1. Edmonton Valley Line LRT, Quarters Tunnel site and vicinity

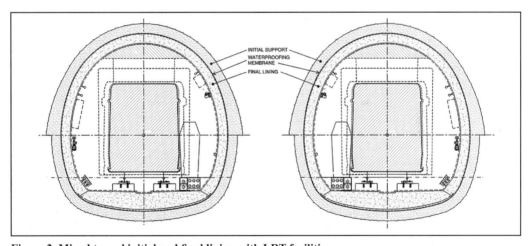

Figure 2. Mined tunnel initial and final lining with LRT facilities

interbedded limestone and sandstone make up the upper reaches of the bedrock. The bedrock is overlain by cohesive till deposits that vary from basally-deposited lodgment tills to basally-deposited ablation till (high plasticity tills), each with isolated intra-till sand pockets that can be water charged. The tills are overlain by ice contact-deposited glacio-lacustrine silts and clays. Figure 3 presents the soil stratigraphy along the tunnel alignment, determined from both historical and project geotechnical investigations.

The typical soil unit characteristics at the project site are as follows:

1. **Fill**: The thickness of the fill along the tunnel alignment was generally less than 2m, but extreme values as thick as 3.3m were encountered. The fill soils vary between organic topsoil, clay, sand, and gravel along the tunnel alignment.

2. **Glacio-Lacustrine Clay (Lake Edmonton Clay)**: Glacio-lacustrine clay extended to

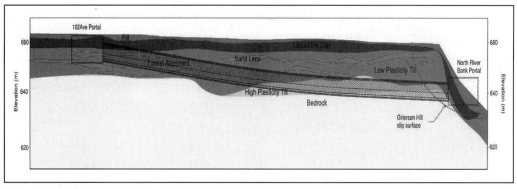

Figure 3. Soil stratigraphy along tunnel alignment

depths varying between 2.4m and 6m below the existing ground surface with a maximum thickness of 5.4m. The glacio-lacustrine clay contains silt laminations and partings, and is generally described as firm to very stiff high plasticity silty clay with some random zones of medium plasticity.

3. **Glacial Till**: The glacial till along the alignment consisted of clay, silt, and sand size particles with some gravel sizes and trace cobbles and boulders, generally in slightly unsaturated and overconsolidated conditions. A distinction was made between the till with low to medium plasticity in the upper horizons and high plasticity in the lower horizons.

4. **Intra-Till Sand**: Dense to very dense intra-till sand pockets were encountered within the glacial till. These intra-till sand lenses were identified on the previous Metro Line Light Rail project in Edmonton (Elwood et. al, 2011), and have been addressed as a significant risk for the tunneling excavation works. For this project, the intra-till sand typically occurred as pockets that vary in size from approximately 0.3m to 2.5m in thickness and extent from small localized features, eg. 0.5m, to large interconnected pockets several hundred meters in length. Previous experience demonstrated that the intra-till sand pockets could be sub-artesian and relatively free draining, despite fines contents greater than 50% at times. Therefore, the pockets had the potential for rapid release of groundwater upon exposure, although the duration and volume of inflows were typically expected to be brief due to the limited storativity of the sand pockets based on their volume. During design, it was anticipated that up to 10 to 15% of the total volume of the twin tunnel excavations within the glacial till would be comprised of the intra-till sand.

5. **Bedrock**: The bedrock throughout the tunnel alignment consisted of clay shale described as an extremely fine-grained argillaceous claystone with occasional interbedded bentonite, sandstone, siltstone and coal. The bedrock quality was highly variable and relatively soft, in particular when adjacent to bentonite seams. The sandstone layers were extremely weak to very weak and largely intact, containing fine to coarse-grained sand sized particles. Highly fractured hard coal layers were encountered with thicknesses from 0.2m to 1.2m, with an average of 0.6m.

Geotechnical Design Parameters

Geotechnical parameters have been derived based on field and laboratory testing, including: index properties, direct shear, Consolidated Isotropic Undrained Compression (CIUC) triaxial, Uniaxial Compressive Strength (UCS), pressuremeter, and 1D Oedometer consolidation tests.

Finite element analysis for soil-structure interaction utilized a Mohr-Coulomb constitutive model at varying strain levels to account for soil softening via shear modulus degradation. All soils were modeled such that the stiffness and strength are defined in terms of effective properties.

The range of shear stiffness is important with respect to tunnel design and an estimation of the range of stiffness expected was required in order to adequately implement the results from the pressuremeter tests. Based on the strength ratio for the geotechnical materials present along the alignment, the tunnel radial strain levels required to maintain stable conditions were anticipated to be between 0.7% and 2%, neglecting the tunnel support stiffness. Considering the effects of tunnel pre-support and initial liner in 3D finite element models, strains were anticipated to reduce significantly to below 0.5% generally, with localized strains up to 2% near the

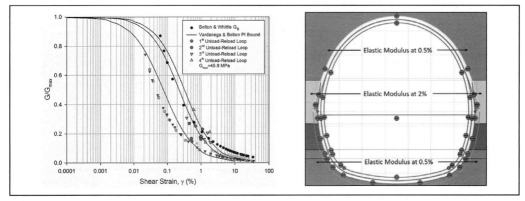

Figure 4. Strain-dependent stiffness behavior, (left) shear modulus degradation curve for Lacustrine Clays, (right) lower and upper bound stiffness applied to tunnel model example

Table 1. Summary of geotechnical parameters

Geological Unit	Drained Conditions		Stiffness at 0.5% Strain E_{upper} [MPa]	Stiffness at 2% Strain E_{lower} [MPa]	Hydraulic Conductivity k [m/sec]
	Cohesion c' [kPa]	Peak Friction Angle φ_p [deg]			
Fill	5	28	22	10	N/A
Glacio-Lacustrine Clay	5	28	22	10	1.0×10^{-8}
Glacial Till (low to medium plasticity)	10	35	100	30	1.0×10^{-7}
Glacial Till (high plasticity)	10	31	37	15	1.0×10^{-8}
Intra-till sand	0	31	75	25	1.0×10^{-6}
Clay Shale Bedrock	50	22	200	N/A	1.5×10^{-8}

axis level. On this basis, the strain-dependent stiffness behavior of the ground was approximated in the model using lower and upper bound elastic moduli. Figure 4 shows the strain-dependent stiffness of the Lacustrine Clays encountered at the Jasper Ave crossing and an example of how the strain-dependent stiffness was applied in the finite element model.

A summary of the geotechnical parameters for the ground is presented in Table 1.

Key Geotechnical Risks

As with any tunneling project, a risk assessment was performed and a number of key geotechnical tunneling risks were identified. Each risk was then addressed by implementation of systematic support, localized support, and/or contingency measures based on site observations.

1. **Jasper Avenue abandoned sewer**: A historically abandoned 675mm diameter sanitary sewer and manhole, backfilled with cementitious grout, was located directly within the shallow tunnel excavation beneath Jasper

Avenue. Due to access restrictions on the roadway above, the sewer and associated structures had to be demolished within the tunnel excavation. Consequently, several impacts to tunneling were considered including potential instability of the structures once partially demolished, incomplete backfilling within the pipe, and the potential for face instability or groundwater inflows resulting from loose granular material or voids around the original installation works.

2. **Glacial Till Instability, Intra-till Sands**: Based on historical projects in the City of Edmonton, glacial till instability was identified as a significant risk. Instability was expected to occur due to the cohesionless intra-till sands or block/wedge failure within the cohesive till mass. Intra-till sands tended to rapidly release stored groundwater through seepage at the tunnel face when the ground was left exposed for a prolonged period. This could cause fines erosion in the soil mass, increasing soil permeability and potentially resulting in a loss of ground stability.

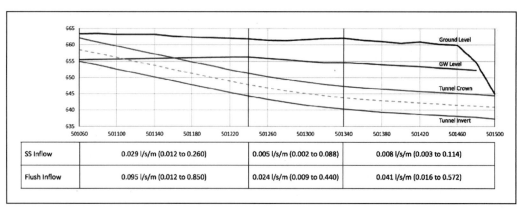

	501060	501100	501140	501180	501220	501260	501300	501340	501380	501420	501460	501500
SS Inflow	0.029 l/s/m (0.012 to 0.260)					0.005 l/s/m (0.002 to 0.088)			0.008 l/s/m (0.003 to 0.114)			
Flush Inflow	0.095 l/s/m (0.012 to 0.850)					0.024 l/s/m (0.009 to 0.440)			0.041 l/s/m (0.016 to 0.572)			

Figure 5. Mined tunnel predicted inflow rates

3. **Glacial Till Instability, Blocky Fissures**: The second form of potential instability may result from translation of glacial till blocks along pre-existing fissures within the soil mass. The most common source of block instability was anticipated when intra-till sand lenses are present immediately above the tunnel crown. The weight of glacial till blocks underneath the intra-till sand lens exceeds the cohesion of the sand, releasing the soil mass to collapse into the tunnel.

4. **Rock Bench Instability**: Rock instability at the bench of the tunnel was a concern near to Grierson Hill due to lack of sufficient confinement of the bedrock. Although substantial instability was unlikely, there were risks associated with localized over-break requiring backfill with shotcrete and/or a loss of bench sequencing. It was essential to maintain a level of confinement on the bedrock to avoid elastic rebounding of the material leading to dilation at the joints and fractures. Dilation of the joints caused the clay shale to approach the residual shear strength and behave similar to a mass of cobbles that separate along the pre-existing discontinuities. Resistance to dilation and raveling of the rock mass was provided by the addition of face bolts designed such that the cone of influence interacts with the adjacent bolt.

5. **Groundwater Inflows**: Groundwater inflow through exposed intra-till sands was identified as a potential major risk. Based on the estimated hydraulic conductivities and measured groundwater levels within the intra-till sands, representative tunnel inflow values per tunnel have been predicted for three sections along the alignment, see Figure 5. The predicted inflows consider initial contact (flush) flow rates and steady-state seepage conditions.

These seepage rates do not consider the effect of piping that can occur within uniformly graded fine sands and silts. Although seepage rates may appear to be low initially, finer particles may be washed out of the soil matrix under sustained hydraulic head and seepage forces, resulting in higher permeability, greater inflow rates, and potentially ground instability with formation of large voids. As a contingency measure, a supply of dry-mix shotcrete was maintained on site for rapid application to the face during adverse flow conditions.

6. **Methane Gas Emissions**: Despite no methane being detected during the geotechnical investigation, it was identified that there was a potential for methane gas to be released when the coal seams were encountered during excavation. The risk of methane gas emissions has been assessed to be low since the coal seams daylight in the Grierson Hill slope, allowing dissipation of any gas. The design of the ventilation system combined with ongoing air quality monitoring was used to provide safe working conditions.

7. **Grierson Hill Slip Surface**: Seams of highly plastic bentonite within the clay shale bedrock were identified below the design elevation of the tunnel invert, creating localized planes of weakness that may result in overall slope instability on and near Grierson Hill. There was a risk that a slope failure slip surface may pass through the tunnel, therefore a safe zone was adopted within 30m of the North River Bank portal where slope stability measures were required prior to tunnel excavation. Furthermore, a more robust system of excavation support was used in the approach to the portal, as shown in Figure 6. For the permanent lining, the connection between the

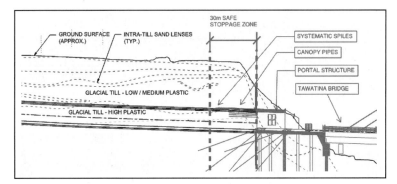

Figure 6. Mitigation measures due to Grierson Hill landslide slip surfaces

tunnel and portal structures was designed to accommodate residual lateral movements up to 80mm in any direction after stabilization measures were installed.

TUNNEL EXCAVATION AND SUPPORT METHODS

The mined tunnel was constructed using sequential excavation method (SEM) techniques with a double-shell support system consisting of an initial lining and a final lining separated by a waterproofing membrane. During the excavation stage, immediate ground support is provided through the pre-support measures and a shotcrete lining to provide a passive structural shell that resists ground and seepage loadings. The support scheme used followed a minimum requirement for each area of the alignment, with additional support elements that could be installed based on observed ground conditions and monitoring data; further details of this process are presented in the following sections. After completion of excavation activities, installation of a shotcrete smoothing layer, waterproofing, and a cast-in-place final lining were used to meet the tunnel design life operational requirements and support ground, groundwater, and service loads.

Before each round of advance, pre-confinement of the ground core ahead of the tunnel face was used to limit ground movements. Measures included the use of steel overhead pre-support elements, either canopy pipes or spiles, and fiber-reinforced polymer face bolts to enhance stiffness and shear strength for possible wedges forming ahead of the face and reduce face extrusion deformations.

Following pre-support, excavation and placement of a steel fiber reinforced shotcrete initial lining with steel lattice girder was completed using a 'staggered full-face' approach with a top heading and bench sequence. Rapid completion of the tunnel lining ring behind the face was required to limit ground movements, necessitating a full-face excavation and

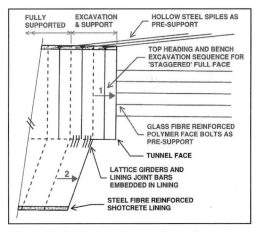

Figure 7. Mined tunnel excavation and support method

lining approach. However, to enhance control of the ground ahead of the face and improve constructability for the lattice girder and lining, the excavation was split into top heading and bench segments with an offset of one advance, with an elephant's foot used to support the top heading lining during bench excavation.

Design Approach

Design of the excavation and initial support system was carried out using established empirical and analytical design methods along with numerical modeling for detailed analysis and verification. A number of critical design sections along the alignment were used to consider varying depths and combinations of ground in mixed-face conditions. Each method was used to define the loadings acting on the support system, assess tunnel stability, and develop a detailed understanding of the ground-structure interaction, allowing optimization of the support design to strike

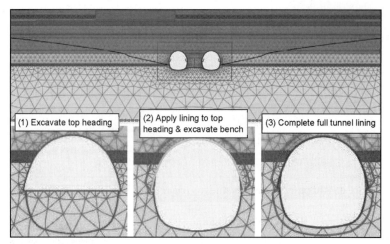

Figure 8. Typical mixed face design section (top), with excavation and lining sequence (bottom)

a balance between allowable ground movements and required support pressure. An extract of the design models at one of the design sections, along with the modelled construction sequence, is shown in Figure 8. Additionally, a number of different support categories and contingency measures were developed to allow an adaptable construction sequence appropriate to varied and changing geologic conditions along the alignment.

Construction Sequence and Support Measures

The SEM construction sequence, shown generally in Figure 9, was implemented on site using the support measures prescribed in the design with adjustments to suit the ground conditions encountered as assessed through the geologic mapping and performance of the works. Each support measure can be used for one or several purposes in maintaining the integrity of the tunnel within the glacial tills, including use as a contingency measure to mitigate observed movements.

- A steel fiber reinforced shotcrete lining was installed as the primary support for the tunnel, acting as a passive support that picks up load after placement and curing and maintains support until the waterproofing and cast-in-place final lining are completed. Sequencing of the shotcrete installation was adapted to suit installation of the various other support measures in the tunnel. As a contingency measure in case of excessive lining deformation, wire mesh was available to provide additional strength and stiffness to the lining in critical sections.
- As a part of the shotcrete lining, a sealing layer was utilized to secure working areas

and prevent loose or raveling material from falling into the excavation while support is installed. Increased coverage of the sealing layer and/or increased thicknesses were applied on site when required to suit the ground conditions.

- In case of loose materials in the tunnel crown and face, pocket excavations were available as a contingency measure, where excavation and application of a sealing layer were done in sequence for localized 'pockets' in the face. This was particularly useful in managing identified features in the face such as the abandoned sewer, as well as when intra-till sand lenses were encountered in the crown.
- An elephants' foot overcut with increased lining thickness and lining joint bars were used to account for the split between top heading and bench sections of the lining, with the elephants' foot temporarily transferring hoop loads outward into the ground while the bench is installed and the bars joining the two lining sections together across the structural joint.
- Steel spiles were installed as pre-support measures, both as a systematic support scheme for areas with low cover above the tunnel and as an ad-hoc support scheme or contingency measure where loose material, including intra-till sand lenses, are present in the tunnel face or crown. Variations in the density of spiles were used to adapt to different conditions.
- Canopy pipes are large diameter hollow steel grouted pipes, which were installed as a systematic support to reduce loading and

Figure 9. SEM tunneling: (left) excavation, (center) spile installation, (right) shotcrete application

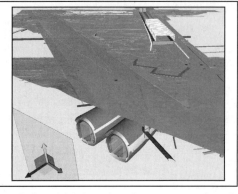

Figure 10. Jasper Ave crossing 3D views: (left) underground view, (right) section view at center of road

maintain stability of the tunnel in areas of very low cover and break-in/break-out activities at the tunnel portals.

- Glass fiber reinforced hollow face bolts were installed as pre-support elements in a configured support scheme during extended stoppages to maintain face support as drained conditions developed ahead of the tunnel face, particularly where intra-till sands were present to act as a drainage path. They were also available as a spot-installation measure to stabilize locally loose or blocky materials in the tunnel face, and as a contingency measure for systematic support to increase confinement of the face if needed.
- A domed face excavation applied with an increased thickness of sealing layer shotcrete was also used to provide support during shorter stoppages, where a lesser degree of support was needed prior to resuming tunneling a short while later. The domed shape was optimized in the design to increase the level of support provided to the face.
- Steel lattice girders were used throughout the tunnel top heading for control of the excavation profile and shotcrete thickness and as an end support for the installed spiles.
- Probe holes were installed systematically ahead of the face and downward below the invert to detect changes in geological strata, identify any intra-till sand lenses, and drain any saturated ground ahead of the face. In case of large volumes of water inflows discovered, perforated drainage tubes were also available to dewater the ground, or permeation grouting could be applied through the face bolts.
- All of the above measures were combined in the construction following the designed plans, with decisions on which support measures and categories to implement being carried out by the engineers and contractors based on the ground conditions and monitoring results.

Jasper Ave Crossing

The crossing under Jasper Ave was one of the most technically and logistically challenging portions of the project, as the tunnel crossed under the most sensitive surface infrastructure in the project with very

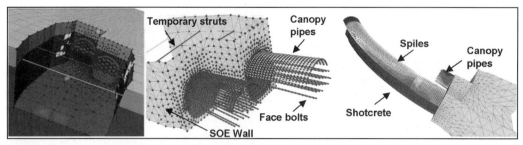

Figure 11. Jasper Ave FLAC3D model setup: (left) overview, (center) portal breakout, (right) tunnel excavation

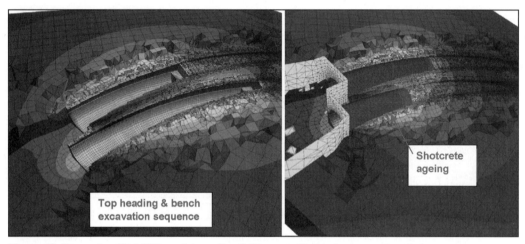

Figure 12. Jasper Ave FLAC3D model results: (left) ground behavior in construction sequence, (right) shotcrete ageing effects

shallow depth immediately after launch as shown in Figure 10. Initially, pre-support from canopy pipes installed through the portal headwall provided tunnel stability as the portal breakout was completed. Due to the constraints of the tunnel alignment to immediately progress around a sharp 85m radius curve at a 6% descending grade, subsequent rows of canopy pipes could not be used as the geometry of installation and support was incompatible with the curve. Instead, an innovative solution using a spile umbrella consisting of 4–5 overlapping layers of spiles with tight spacing along the perimeter. These were used to carry the loads from the overlying soil and road, as the tunnel was too shallow to allow arching mechanisms within the ground, and to provide a rigid support to minimize movements in the roadway and active shallow utilities.

Due to the sensitive nature of the shallow crossing at Jasper Ave, a detailed FLAC3D numerical finite difference soil structural interaction model was used to verify the initial assumptions regarding ground relaxation effects and ground behavior under

the planned support regime. The model incorporated portal and tunnel excavations and support features, including fiber reinforced shotcrete, canopy pipes, spiles, and face bolts. Expected production rates in construction, including excavation round length and advance rates, were incorporated into the model to assess time-dependent shotcrete hardening behavior and changing porewater pressure conditions within the glacial tills. Figure 11 shows the model with the launch portal, excavation stages, pre-support measures, and lining placement within the tunnel, and Figure 12 shows model results with ground movements resulting from the tunnel construction sequence.

In addition to the requirements for supporting surface infrastructure, special mitigation measures were required to address the abandoned sewer and manhole within the tunnel, as shown in Figure 14. To verify the location of the pipe and manhole and drain any free water, probe and drainage holes were installed from several meters away. Once the location was known, face bolts were installed along

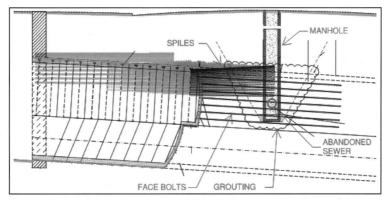

Figure 13. Special support measures for demolition of abandoned sewer

aligned trajectories through the manholes and pipe to pin them in place during excavation and removal. Additionally, grouting through the hollow face bolts was used to backfill any voids and improve the strength and stability of any granular or loose materials. The advance of the spile canopy was used to pin the features in place at the perimeter, preventing instabilities from extending outside the tunnel. As the excavation advanced to the pipe and manhole, full top heading excavations were substituted with pocket excavations to allow gradual exposure of the features, as shown in Figure 13. As each section of pipe or manhole was exposed, it was carefully cut away from adjoining sections and removed from the tunnel. In this way several advances progressed until removal was complete, with the face and perimeter of the tunnel maintaining stability throughout.

INSTRUMENTATION AND MONITORING

Instrumentation and monitoring was used to monitor the behavior of the tunnel during construction. Monitoring was also used to check the compatibility of construction activities with the design, and detect potential risks. This section presents the methodology adopted to devise the instrumentation and monitoring plan, including list of instruments used, trigger levels, and monitoring response plan.

Instrumentation and Monitoring Plan

The instrumentation and monitoring plan set out instrument types and monitoring frequencies devised based on project requirements and the tunnel design. Due to the strict project requirements for building and utility movements, the design of the tunnel was governed by limiting ground movements. A building and utility damage assessment was performed to determine the allowable movements. Tunnel movements were estimated using New & Bowers (1994) which assumes a ribbon-shaped zone of ground loss

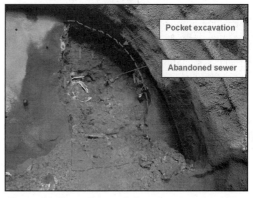

Figure 14. Demolition of abandoned sewer with pocket excavations for stability

and has been proven to work well for mined tunnels in over-consolidated cohesive deposits. 2D and 3D numerical modeling of the tunnel construction sequence indicated that a volume loss of 0.5% and a trough width parameter of 0.55 could be reasonably adopted for the ground movement assessment.

A combination of surface and in-tunnel monitoring instruments were used on the project to fully monitor the progress of tunnel excavation. Table 2 presents a summary of the instrumentation used and their frequency.

Trigger levels include an Action Level and a Maximum Level. The Action Level is defined as level of settlement or distortion at which corrective measures are required, including procedural modifications that permit construction to continue without reaching the Maximum Level. The Maximum Level is defined as level of settlement or distortion at which no further settlement or distortion is permitted. Maximum Levels for SDMPs, PMPs, Inclinometers, extensometers, and in-tunnel convergence were derived from ground movements associated with

Table 2. Summary of instrumentation installation frequencies

Instrumentation Type	Instrumentation Frequency
Building Monitoring (Optical Survey Prism and biaxial tilt meters)	At least one optical survey prism and one biaxial tilt meter on each building within Zone of Influence
Pavement Monitoring Points (PMPs) and Soil Deformation Points (SDMPs)	Array of at least 5 PMPs and SDMPs evenly spaced at a maximum spacing of 15 m along the ground surface.
Multiple Point Extensometer (MPBXs)	Maximum spacing of 100 m centers
Inclinometers (SAAs)	A minimum of 2 inclinometers at maximum spacing of 50 m
In-tunnel convergence	Array of 7 convergence points at maximum spacing of 15 m

the tunnel design; whereas, trigger levels for building monitoring instruments were based on project requirements. The Action Level was considered to be 80% of the Maximum Level, allowing some margin to implement mitigation measures.

Based on the defined action and maximum levels, a monitoring response plan was created to define the response process for exceedance of trigger values. Figure 15 presents the initial response process initiated when instrumentation results are received, as well as the further action and investigation process for responding to maximum levels.

Changes to the excavation methods were implemented following the plan when trigger levels were exceeded, using a series of contingency support items to improve the ground behavior such as localized use of shotcrete for added support, pocket excavations and face doming, use of additional support incl. spiles and face bolts, and the use of permeation grouting or drainage to adjust the ground conditions.

Surface Ground Movement Back Analysis

Ground movements were estimated based on a volume loss of 0.5% and a trough width parameter of 0.55. However, during construction, surface movements exceeded those initially estimated. Figure 16 presents surface vertical movements recorded at two arrays of PMPs along the tunnel alignment compared to the estimated and back analyzed settlement troughs, along with the geological profile at the array location. Observed movements indicate that an actual volume loss of 0.6% and trough width parameter of 0.65 are appropriate for the tunnel. Typical volume losses in Edmonton tunnels have historically ranged between 0.2 to 0.8%, so the observed volume loss is still within the expected range. The trough width parameter is greater than anticipated because the glacial till in this area is behaving similarly to a medium to high plasticity clay, rather than low to medium plasticity.

Larger surface movements were observed at specific sections of the tunnel alignment. The calculated volume loss at one PMP array was 0.85%, which is higher than observed volume losses at other tunnel locations. This effect can be caused by construction activities resulting localized loosening of non-cohesive materials above the tunnel crown coupled with short term consolidation of the medium plasticity till. The intra-till sand lenses above the crown were drained through probe holes during construction, which could have resulted in consolidation settlement of both the lacustrine clay and the till. The recorded vertical movements and settlement troughs, including predicted movements with and without consolidation and back-analyzed movements, are shown in Figure 17 along with the geological profile for this PMP array.

Although higher volume losses than anticipated were achieved during the tunnel construction, these were still considered to be generally low for SEM tunnel construction (<1%). At higher levels of volume loss, generally the ground movements increase but the tunnel lining loads decrease due to increased stress relaxation ahead of the advancing face. The behavior of the tunnel lining confirmed this expectation, as movements were well within limits and no cracking was observed. Ground settlements were slightly greater than expected, however the actual movements of buildings and utilities did not exceed the allowable limits; generally, the buildings did not settle as much as the free ground due to the benefits from structural rigidity.

CONCLUSIONS

This paper presented the design and construction of a 450m long twin bore mined tunnel forming part of the Edmonton Valley Line LRT project. The paper focused on the design and construction of the tunnel excavation and installation of initial support in a densely populated urban environment, under shallow cover in challenging geological conditions. The design of the tunnel excavation sequence and support was governed by challenging geotechnical risks from natural materials and historical infrastructure and strict requirements for limiting ground surface movements with minimal cover. Extensive monitoring of the tunnel advance has informed the selection and adaptation of the excavation and support system based on ground conditions and surface movements. The resulting ground movements were largely in line with design assumptions, however several areas with slightly higher movements than expected were

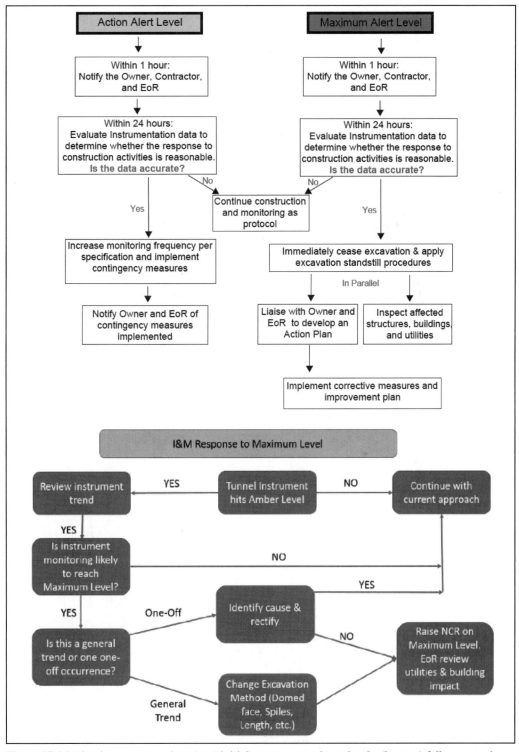

Figure 15. Monitoring response plan: (top) initial response to trigger levels, (bottom) follow-on action plan flowchart

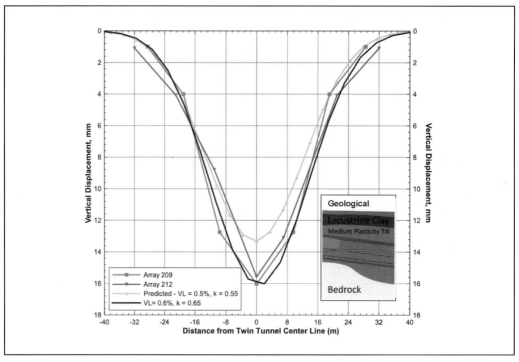

Figure 16. Surface vertical movements at array 209 and 212

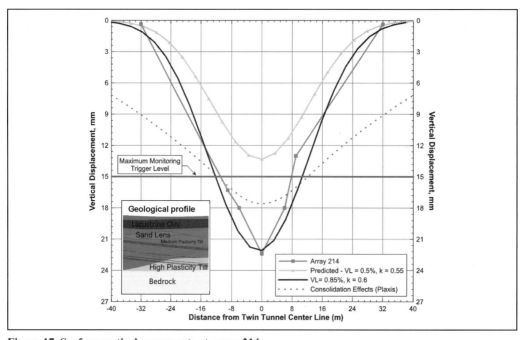

Figure 17. Surface vertical movements at array 214

encountered, which has been attributed to loose materials present within the Edmonton Glacial Tills and associated short term soil consolidation. However, the maximum ground movements observed were considered to be very limited (<25mm generally) and buildings and utilities did not exceed the allowable movements.

ACKNOWLEDGMENTS

Special thanks to the City of Edmonton as the owner and sponsor of the project. Thanks as well to Dr. David Elwood for his contributions to the design team with his wealth of knowledge and experience about ground conditions and tunneling in Edmonton Glacial Tills. The authors also thank Ken Sullivan (Bechtel) for his contribution to this paper.

REFERENCES

Elwood, D., Martin, C.D., and Dafni, J., 2011. Ground characteristics for the design of the Edmonton North LRT Twin Tunnels. Proc. 14th Pan-American Conference on Soil Mechanics and Geotechnical Engineering, Toronto, Ont.

New B.M. and Bowers K.H. (1994). "Ground movement model validation at the Heathrow Express trial tunnel." Proc. Tunnelling 1994. IMM, London, pp 301–327.

TRACK 4: CASE STUDIES

Session 3: Sewer/Water 2

Joel Kantola and Julian Prada, Chairs

Indianapolis Deep Rock Tunnel Connector and Pump Station Start-up

John Morgan
Citizens Energy Group

Alexander Varas
AECOM Technical Services, Inc.

ABSTRACT: The intent of this paper/presentation is to provide a review of the design, construction, startup challenges faced with constructing the Deep Rock Tunnel Connector (DRTC) and the DRTC Pump Station (DRTC PS). The 15.5 km (9.6 mi) TBM excavated DRTC is a 5.5 m (18') diameter concrete lined tunnel with an overall CSO storage volume of 340,000 cubic meters (90 MG). The DRTC PS, constructed 85 m (260 ft) below ground surface in a rock cavern 30.5 m (100 ft) long by 18.3 m (60 ft) wide and 24.4 m (80 ft) high, has a firm pumping capacity of 340 MLD (90 mgd). Both projects were constructed under separate contracts and are now in service.

An important aspect of the paper is this was the first rock tunnel for Indianapolis. As such, the tunnel industry was not familiar with Indianapolis. Nor was Indianapolis accustomed to risk sharing on projects as is done in the tunnel industry. Many of the factors noted in this case history are not new to the tunneling industry. However, as an Owner not familiar with the industry as a whole we learned how to better partner. The end result is a successful project.

INTRODUCTION

The DRTC and DRTC PS projects are the results of an amendment to the City of Indianapolis Long Term Control Plan (LTCP) entered into with the United States Environmental Protection Agency (USEPA) and the United States Department of Justice (USDOJ) in 2006. The amendment provided for the elimination of the "Interplant Connection Project" a soft ground tunnel 11.2 km (7 mi) long with an inside diameter of 3.7 m (12 ft). The amendment was proposed and formally approved by USEPA in 2008/2009. The Interplant Connection Project would have conveyed combined sewage from the Belmont Advanced Wastewater Treatment Plant (BAWT) to the Southport Advanced Wastewater Treatment Plant (SAWT) in a tunnel without having any storage capacity. The DRTC would allow for storage of the combined sewage and serves as a storage vessel for SAWT. While a storage tunnel had been planned and included in the original LTCP, altering the "Interplant Connection Project" into a deep rock storage tunnel allowed for a longer tunnel with a finished tunnel diameter of 5.5 m (18 ft). The re-configured deep rock tunnel allowed for: more potential bidders, an increase in tunnel boring machine availability, and a more cost effective solution to the combined sewer overflow issues.

Within the Indianapolis Collection System the BAWT handles the majority of combined sewage with the SAWT handling the majority of separated sewage. BAWT is somewhat limited in ability to expand the treatment facility due to the waterways that surround three of the four boundaries of the plant. SAWT is located in a flood plain and is surrounded by an earthen levee. The DRTC project included levee work to expand the SAWT facility and make available more area for future plant expansion.

The DRTC project is a major component of the LTCP to address combined sewer overflows (CSOs) from outfalls along the White River. The DRTC is the southernmost portion of a deep large diameter conveyance and storage tunnel system to provide overflow relief during wet weather events.

The DRTC project includes approximately 15.5 km (9.6 miles) of 6.1 m (20 ft) excavated diameter deep rock tunnel, three tangential vortex drop shafts and associated vent shafts, three utility shafts, and a launch shaft at the south end at the SAWT facility, and a retrieval shaft at the north end. Also, constructed under a separate contract and included in the DRTC project, is the CSO tunnel dewatering DRTC PS at the downstream terminus of the CSO tunnel system near the existing SAWT facility (Nasri 2011 and 2012). The DRTC PS, constructed 85 m (260 ft) below ground surface in a rock cavern 30.5 m (100 ft)

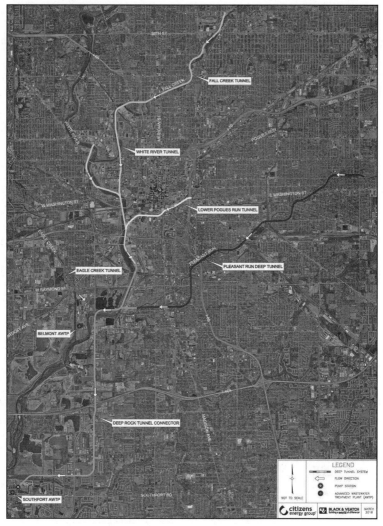

Figure 1. DigIndy tunnel system map

long by 18.3 m (60 ft) wide and 24.4 m (80 ft) high, has a firm pumping capacity of 340 MLD (90 mgd). AECOM was retained by the City of Indianapolis, Department of Public Works to provide detailed design of the project. A joint venture of JF Shea and Kiewit Infrastructure (S-K, JV) built the tunnel and Oscar Renda Contracting and Southland Tunneling (RS, JV) built the pump station.

Indianapolis' deep tunnel system is known as DigIndy. The DigIndy system is comprised of 6 separate segments. They are Deep Rock Tunnel Connector, Eagle Creek Tunnel (Line AA), White River Tunnel, Lower Pogues Run Tunnel, Fall Creek Tunnel and Pleasant Run Tunnel (Figure 1). The entire system, with a storage capacity of nearly

1.0 million CM (250 MG) will capture CSO over-flows from nearly 40 CSO locations.

Construction for the work began with Deep Rock Tunnel Connector and Eagle Creek Tunnel, 2011–2017. White River Tunnel and Lower Pogues Run Tunnel followed in 2016 and both tunnels are scheduled to be in service by the end of 2021. The Fall Creek Tunnel, with an NTP in 2021, will follow the White River Tunnel alignment, Fall Creek Tunnel is next with a Notice to Proceed (NTP) in 2021 and has a a completion date of December 2025. At that time the TBM will be brought to the surface and moved to begin the construction of Pleasant Run Tunnel. Pleasant Run Tunnel will have a late 2021 Notice to Procced (NTP) and will be completed by the end of 2025.

THE DEEP ROCK CONNECTOR TUNNEL PUMP STATION (DRTC PS)

The DRTC CSO dewatering pumps are located approximately 85 m (260 ft) below grade in a deep pump room cavern excavated in rock. CSO flow captured and stored in the tunnel will be pumped from the pump station to the SAWT facility. The DRTC PS consists of a below ground pump room (cavern), a pump station building, a grit and screening removal equipment, a 10.7 m (35 ft) diameter access/discharge shaft, and a 4.9 m (16) diameter equipment shaft (Figure 2).

The span of the cavern is about 18.3 m (60 ft), its height is around 24.4 m (80 ft) and its length is approximately 30.5 m (100 ft). The crown arch consists of three circular arc segments. The crown arch geometry, due to its design configuration, created a balance between the rock excavation volume and the opening stability. Also, the rock cover above the crown consists of sound limestone with a thickness of at least one third of the cavern span.

The components of the DRTC PS include: screen and grit removal system constructed in the DRTC Launch/Screen and Grit Shaft, a 1.83 m (6 ft) diameter connector tunnel between the Screen and Grit Shaft and the below ground Pump Cavern, a subterranean Pump Cavern containing four main tunnel pumps, two emergency dewatering pumps, one header dewatering pump, two sump pumps, a Main Access Shaft, an Equipment Shaft, an at-grade Pump station Building, an at-grade Main Access Shaft Building, and an at-grade Discharge Chamber. Also, an overhead gantry crane is provided inside the pump cavern for lifting the pumps and other equipment, and transporting them to a drop zone within the cavern from which they are lifted through the Equipment Shaft up to the surface.

THE DEEP ROCK TUNNEL CONNECTOR (DRTC)

DRTC was excavated by a single main beam rock TBM and supported primarily with rock dowels. The shaft excavation was supported in the overburden soil with a slurry wall and within the rock by shotcrete and rock dowels. The tunnel launch and retrieval shafts were excavated through 33 m (108 ft) and 23 m (75 ft) overburden, respectively, by conventional soil excavation methods and approximately 37 (121 ft) and 43 m (141 ft), respectively, to the tunnel invert elevation of rock by drill and blast method. Within the overburden, the slurry wall acts as the permanent wall for the shaft. Beneath the overburden, a 0.6 m (2 ft) cast in place concrete lining was installed as a permanent support for the shaft. The pump room was excavated, by drill and shoot method, and supported by permanent rock bolts and shotcrete. Construction of the drop shafts was performed by advancing an over-sized steel casing through the overburden and rock to the required depths.

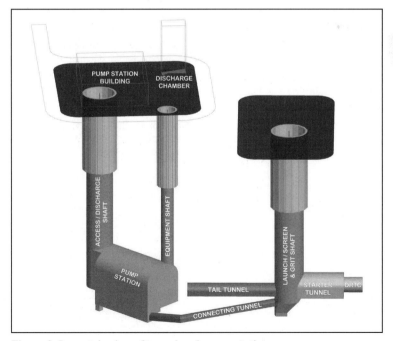

Figure 2. Isometric view of tunnel and pump station

The DRTC project included approximately 15.5 km (9.6 miles) of 5.5 m (18 feet) finished tunnel and eleven (11) shafts. The shafts include collection of three (3) combined sewer overflows via tangential vortex drop structure and three (3) corresponding vent shafts, one combined sewer overflow plunge shaft, three utility shafts, and a tunnel launch shaft and a tunnel retrieval shaft. The three tangential drop shafts are designed to deliver nearly 2,000 MLD (550 mgd) of CSO flow to the tunnel (CSO 008—1,136 MLD (300 mgd); CSO 117—379 MLD (100 mgd); CSO 118—568 MLD (150 mgd). The tangential drop structure vent shafts allow the captured air to escape prior to the flow entering the tunnel.

The design intent of the utility shafts was to assist the Contractor in allowing for construction operations (air handling/concrete delivery for tunnel liner construction/potential power drop). For this project, the Contractor chose not to use the structures for TBM power delivery. The Contractor elected choice was to install booster stations throughout the tunnel. As this was part of a "means and methods" decision the choice was entirely up to the Contractor.

The utility shafts were meant to provide tunnel access for inspection and any potential maintenance issues that may arise once the system is put into operation. The need came as a result of the southern downstream segment of the DRTC having nearly 9.6 km (6 miles) of tunnel with no overflows to intercept resulting in no shafts allowing access. However, due to more advanced surge modeling, not available during the design of the project, the utility shafts were identified as a surge point and were subsequently plugged and abandoned prior to the tunnel going into service.

The Launch and Retrieval, will serve the obvious purposes for construction and mucking operations. The Retrieval Shaft will serve as the Launch Shaft for the TBM on the White River Tunnel construction project phase. The Launch Shaft will serve as the DRTC PS Screen and Grit Shaft.

From the beginning, the Contractor's plan was to construct all shafts early in the project. Once all shafts were constructed, the focus shifted to the TBM (assembly, launching, and tunneling). The overall plan was to eliminate any problems related to launching the TBM and to help focus on the TBM assembly and launching once it arrived.

PROJECT ALIGNMENT CONSIDERATIONS

The overall project alignment is driven by the location of the combined sewer overflow (CSO) locations that is needed to be intersected. The project team during design also kept land acquisition challenges in mind. With that criteria established, the intent was to keep the project alignment within the public right

of way whenever possible. As many have learned, land acquisition and tight schedules do not work well together. If a land owner is resistant to an easement, either temporary or permanent, the options available are limited. The time impact of the judicial system does not favor a fast track project. As such, our goal was to limit our land acquisition needs when selecting the alignment. By doing so we were able to keep the project on schedule to meet the completion dates within the LTCP for this project.

SUBSURFACE INVESTIGATION

The original plan envisioned a geotechnical investigation every 2,000 feet along the tunnel alignment and at each CSO structure location. However, based on a Value Engineering recommendation for the DRTC Advanced Facility Plan (AFP) geotechnical samples were taken every 1,000 feet of tunnel alignment to help gather additional geotechnical information.

This proved very beneficial in helping bidders understand the formation(s) that would be tunneled through during the course of the project. The majority of interested bidders conducted detailed investigations of the rock cores during their examination of the project site.

The underlying carbonate formations encountered in geotechnical boreholes at depths of up to 90 m below grade consist of, in descending order, North Vernon limestone, Vernon Fork limestone, Geneva dolomite and Wabash dolomite. The DRTC tunnel horizon is located within the Vernon Fork and Geneva Formations.

This provided more defined geotechnical information regarding the tunnel's vertical profile to potential bidders and would provide additional information to what exists between bore holes. This enhanced geotechnical played an important role in receiving nine (9) DRTC project bid proposals in August 2011. The other factor in receiving the bids was our part in industry outreach that made many in the tunneling community aware that this project was the first of many tunnel projects for Indianapolis.

PROJECT COORDINATION

Beginning with Notice to Proceed for the DRTC (December 2011) and for the DRTC PS (April 2014) weekly progress meetings have taken place. Besides allowing for the normal flow of contract communications, these weekly meetings between Contractor, Designer, Construction Inspection (CI) Team and Owner have allowed for detailed discussions at key contract milestones. Also, given the fact that the both Contractors had never worked in Indianapolis prior to this project the Owner found that helping the Contractor with points of contact to be a critical role in the success of the project. As both projects progressed to completion dedicated project coordination meeting were held with both contractors and other interested parties. These meetings proved particularly helpful when work areas common to both projects were shared, as was the DRTC Launch Shaft which was transformed into the DRTC PS Screen and Grit Shaft. Another important coordination activity occurred when the DRTC PS required water from the tunnel project for the pre-testing of the CSO main tunnel pumps. Weekly meetings and daily communications helped in attaining the successful testing of the pumps.

PROJECT COMPONENTS

Working with the regulatory agencies during the amendment to the Consent Decree, it was determined that a CSO tunnel dewatering rate of 340 MLD (90 MGD) would be acceptable. This is done via 3 primary, and 1 spare, 113.5 MLD (30 MGD) main tunnel dewatering pumps. Power usage for each of the 3 pumps is 1,500 kwatts. The pumps are vertical centrifugal operating at 1,900 horsepower each.

The overflows intercepted by the DRTC include CSO 008, 032, 117 and 118. The predecessor project was only going to intercept CSO 117. Adding the other overflows to this project resulted in CSO being collected 4 years earlier than originally indicated in the LTCP. This aspect helped in obtaining the modification to the LTCP.

PROJECT STATUS

Both DRTC and DRTC PS are nearly completed and the DRTC PS is going through final demonstration testing. Both projects will be operational by the end of 2017.

CONCLUSIONS

As with most projects, many challenges have been encountered along the way. The Contractor, construction inspection team, design team, Owner's project management team and Owners operations staff have work closely together throughout the project. This project helped develop a lessons learned document that has proved beneficial for the remaining tunnel projects that were designed.

Early in 2018 the DRTC PS will proceed through demonstration testing with the understanding that there will be minor operational modifications.

REFERENCES

Nasri, V., and Morgan, J. (2011), Indianapolis Deep Rock Conveyance and Storage Tunnel. Rapid Excavation Tunneling Conference 2011, San Francisco, CA, June 19–22, 2011, pp. 402–413.

Nasri, V., Morgan, J., and Varas, A. (2012), Indianapolis Deep Rock Pump Station Cavern. North American Tunneling Conference 2012, Indianapolis, IN, June 20–23, 2012, pp. 435–443.

Nasri, V., Morgan, J., and Varas, A. (2012), Indianapolis Deep Rock Conveyance and Storage Tunnel and Pump Station Cavern. International Tunneling Association World Tunneling Congress 2012, Bangkok, Thailand, May 18–23, 2012.

Morgan, J. and Shutters, T. (2014), Deep Rock Tunnel Connector Project Construction Status Update. North American Tunneling Conference 2014, Los Angeles, CA, June 24–26, 2014, pp. 624–632.

Nasri, V., Varas, A., Miller, M. and Castillo, J. (2017), Rapid Excavation Tunneling Conference 2017, San Diego, CA, June 5–7, 2017, pp. 326–335.

South Coast Water District's Wastewater Tunnel Rehabilitation Project, Complexity Coupled with Environmental Sensitivity

Shimi Tzobery, Kevin Kilby, and Trimbak Vohra
Parsons

Richard McDonald
Drill Tech Drilling and Shoring

Rick Shintaku
South Coast Water District

ABSTRACT: Located beneath multimillion-dollar homes along the Laguna Beach cliff, the construction of the 10,4740-ft long sewer tunnel rehabilitation project poses unique geotechnical, environmental, and public relations challenges. The 60-year-old sewer tunnel has severely deteriorated over the years and could result damages to the active sewer pipeline. The new shaft and access tunnel are located on an extremely constrained site in a high-traffic tourist area, thus adding complexity to construction. This paper presents the construction approaches implemented in the first phase of the project and how the team addresses the unique community-sensitive requirements.

INTRODUCTION

South Coast Water District (SCWD or District) operates the 2-mile-long Beach Interceptor Sewer Tunnel and sanitary pipeline located between Three Arch Bay and Aliso Beach in the City of Laguna Beach, California. Based on the District, the tunnel contains a 24-inch gravity sewer pipe with an average daily flow of 1.25 million gallons supporting residences and commercial businesses located along the Pacific bluff. Completed in 1954, the tunnel was mostly hand-dug, with some areas blasted, primarily through bedrock, to an opening size of approximately 6 feet wide by 6 feet high. The tunnel alignment is generally parallel to the cliff and includes a series of portals and adits constructed for tunnel access along the beach stretch. The tunnel is mostly unlined and unsupported, except in areas of low cover and difficult ground where heavy timbers and wooden lagging were placed. The original sewer pipe consisted of a 21-inch vitrified clay pipe (VCP) with mortar joints. However, the mortar joints were reported to have undergone chemical change and decomposition due to hydrogen sulfide gas (H_2S), which resulted in cracked pipes and joints. Therefore, the VCP pipeline was replaced with the currently existing 24-inch fiberglass reinforced plastic mortar pipe (RPMP), commonly known as "techite pipe."

Over the past 60 years, the tunnel has deteriorated mostly due to rock mass weathering, groundwater seepage, and rotting of the timber and lagging support. Rock falls and failures in ground support increased the risk of damage to the sewer pipeline and potential effects of a catastrophic spill in the adjacent Pacific Ocean beaches. These failures also present a potential of safety issues to the District's maintenance workers. SCWD has inspected and evaluated the existing tunnel and sewer pipe conditions, potential geologic hazards, and feasible alternatives to repair the tunnel and pipeline. The evaluation identified areas of concern in relation to the continuous deteriorating conditions of the tunnel and the RPMP pipeline. Previous studies and designs identified two elements as critical to this facility. First, the existing sewer line must be protected from falling rock. Second, the stability of the sewer line must be ensured to allow for safe inspections and lateral repairs. With these guidelines, SCWD staff members have performed annual tunnel inspections and urgent repairs for maintaining worker safety and ensuring the integrity of the Beach Interceptor sewer pipeline. Typically, the repairs include localized shotcrete lining and maintenance to timber struts and lagging. A special emergency sewer tunnel stabilization and RPMP protection work were completed in 2007 for a limited 750-foot-long section located between Adits 14 and 16A. The efforts were performed using beach access while strictly following community limitations and environmental restrictions.

Figure 1. SCWD's wastewater tunnel rehabilitaion project—4 phases alignment

**Figure 2. Typical timber supported
section of existing sewer tunnel**

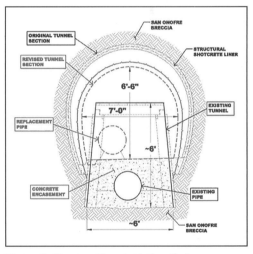

**Figure 3. Cross section depicting timber
supported tunnel showing original design
vs. revised design**

As part of SCWD's long-term capital program, the rehabilitation of the entire Beach Interceptor Sewer Tunnel was identified for prioritized execution and completion. It was recognized that, until the entire stabilization of the tunnel and protection of the pipe are completed, limited access and extra precautions are required for maintenance and repair works on the pipeline or sewer laterals. In 2016, SCWD made efforts to plan, design, and permit the activities required for the tunnel project using the existing tunnel access points at Portal 2 and 4, as well as a new access shaft. The original modified horseshoe

tunnel section was designed based on requirements for the 2007 emergency repair work. The shape and minimum size of the modified tunnel was reviewed as part of a value engineering study. The revised tunnel section was reduced to a final size of 6.5 feet wide by 7 feet high. The section retains the modified horseshoe shape with enough space for a new pipe and personnel walkway for maintenance. The original rehabilitation concept of protecting the existing 24-inch RPMP by concrete encasement, stabilizing and enlarging the tunnel, and installing a second pipe above the encasement floor was maintained. The

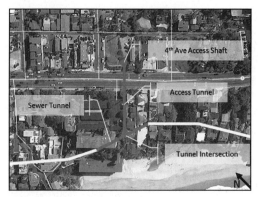

Figure 4. Phase I overview

entire project was set to be completed in four phases and within 5 years.

Phase I includes the construction of a 104-foot-deep new access shaft within the 4th Avenue SCWD property, a 335-foot-long new access tunnel, an enlarged tunnel intersection, and a 300-foot-long section of RPMP protection and existing sewer tunnel rehabilitation. In 2016, bids were solicited from prequalified contractors. The general contractor, Drill Tech Drilling and Shoring (DTDS), was awarded a notice to proceed (NTP) for the construction of Phase I. The SCWD, construction manager (Parsons), and engineer (Mott MacDonald) have collaborated with DTDS in successfully implementing the Phase I of this critical environmentally challenging tunneling project.

GEOLOGY

Geology for Phase I of the SCWD tunnel rehabilitation project involves two main components: Quaternary terrace deposits and San Onofre breccia. As described in the Geotechnical Baseline Report (GBR), the entirety of the access tunnel, the sewer tunnel, and the lower half of the access shaft are situated in the San Onofre breccia; the top half of the access shaft is situated in the Quaternary terrace deposits, as briefly described below:

- Quaternary terrace deposits consist of sands, silts, and clays in varying amounts, which form thick to massive beds and alternate between granular soils and cohesive soils. The bottom 7 feet of the unit consists of a clean sand beach deposit that directly overlays the wave cut platform of the San Onofre breccia. This sand has inclusions of boulders eroded by wave action. The overlying beds of sand and silt have been deposited by slumping and debris flows.

- San Onofre breccia consists of massive to thickly bedded siltstone breccia with very thin beds of shale. The angular clasts range from coarse sand to boulders and are derived from the Catalina schist source basement rock offshore to the west. The bedding is defined by mean clast size within the unit as a whole. The bedding within the breccia is defined by alternating larger and smaller clast sizes as the cobbles and boulders were confined within layers separated by layers with finer grained clast size. The beds are locally undulating with an overall dip to the southwest.

The existing sewer tunnel shows evidence of seepage in the southern tunnel section; this section is typically supported by wooden sets and lagging. The seepage within the limits of Phase I is presumably introduced through joints in the rock mass from a water table above. The influence of the groundwater on the breccia has been detrimental to the competency of the rock, causing a weathering rind of varying thickness to form on exposed surfaces. The seepage also suggests that a perched water table is located above the contact between the San Onofre breccia and the Quaternary terrace deposits. However, no significant groundwater was encountered during the excavation of the access shaft and access tunnel.

ENVIRONMENTAL SENSITIVITIES AND CONSTRAINTS

The City of Laguna Beach is a known as a tourist locale: approximately 6 million tourists and locals visit the City and beaches annually. The 4th Avenue shaft site located in the area is confined and is situated in an affluent dense residential area and a commercial business area. As a lead agency under the California Environmental Quality Act (CEQA), SCWD prepared and completed on October 2010 an updated environmental impact report (EIR) for the existing tunnel stabilization and sewer pipeline replacement. The EIR identified potential environmental concerns, evaluated their impacts to the project, and developed project-specific mitigation measures as part of the execution. SCWD has committed to the implementation of key environmental mitigations to be executed by the project team for addressing the unique community-sensitive project requirements.

To reduce potential construction noise impacts and to minimize impacts to city traffic and local transportation by the project's heavy truck demands, the construction working times and yearly duration were strictly limited. In accordance with the EIR limits, construction hours are limited to between

Figure 5. 4th Ave shaft site—ventilation system with dust collector

7 a.m. and 6 p.m. No construction is allowed on weekends and federal holidays unless an emergency arises. In addition, no work is allowed during the weeks of Memorial Day, Independence Day, Labor Day, and Thanksgiving holidays, and a 2-week break for the Christmas holiday. These off hours and durations are incorporated into the construction schedule and observed by the project site staff.

Noise and visual impacts to the city and neighboring community during construction were identified as a major concern. As part of the project staging, a high-quality 12-foot-tall sound wall had to be built to surround the 4th Avenue shaft site to mitigate the expected construction noise and to reduce the visual impact of the work. The wall was constructed to blend in with the surrounding neighborhood using a carefully chosen wall color and landscaping. Per the request of the neighbors, a remotely controlled sliding sound wall gate was installed at the shaft site (in lieu of a swinging gate) to ease the closing of the gate, save the swinging space area, and ensure adequate noise attenuation. The gate presented its own challenges as it was built on a steep street slope.

In addition, all construction vehicles and equipment must be equipped and maintained with properly operating sound attenuating devices such as mufflers and ambient adjusting backup alarms. To further reduce noise, SCWD invested in a new electric feeder station, including three new transformers, to eliminate the continuous operation of diesel generators on site. This electrical power allows the integration of an electric crane in lieu of a diesel crane to hoist spoils out from the shaft. The electric crane reduces both the noise and emissions produced on site.

From an air quality perspective, the South Coast Air Quality Management District's (SCAQMD's) rules and regulations are followed in order to reduce short-term fugitive dust impacts on nearby residents. Restrictions include the control, prevention, and collection dust from all active portions of the construction site, such as installing ventilation dust collector systems, limiting on-site vehicle speed, as well as

Figure 6. Sound wall sliding gate at access shaft site

Figure 7. Installation of project electrical feeder

housekeeping and dust watering requirements. If dust is visibly generated and travels beyond the site boundaries, dust-generating activities must cease until winds have abated or activities have been modified. To reduce vehicle emissions and congestions in the neighboring streets, on-road ve3hicles are not allowed to idle for more than 5 minutes in private roads and residential areas. In addition, queuing of construction trucks accessing the sites is prohibited. The contractor also provides a vanpool to the site and encourages ridesharing for the construction crew.

In addition to having two independent firms provide continuous monitoring for strict environmental compliance, three noise surveys have been conducted to establish a baseline before starting the construction work and during construction. Survey results indicate that the construction noise mitigation measures are successful in reducing the construction noise to levels that are below or close to the daytime noise standards set by the City as acceptable for residential and commercial uses.

Figure 8. Inclinometer installation at access shaft site

Figure 9. Continuous vibration monitoring using site seismographs

GEOTECHNICAL AND GROUND VIBRATION MONITORING

To verify ground stability and ensure that no settlement impact is attributed to the project, a broad variety of geotechnical monitoring instrumentation have been installed and monitored at the site; soil deformation monitoring points, pavement monitoring points, and geotechnical inclinometers were installed adjacent to the assess shaft to monitor ground movement as excavation progressed. Baseline readings were established and the geotechnical monitoring instrumentations are read based on a specified frequency. Three seismographs are continuously monitoring ground vibrations generated from construction activities at the excavation area. One monitor is stationary at the shaft site; the other two monitors "leap frog" to keep ahead of the tunnel excavation. The monitors have a real-time telephonic alert system that sends text messages to identified recipients when trigged values are reached. Instrumentation for the access tunnel includes convergence monitoring points for detect ground movement and settlement underground. Convergence monitoring points were installed and monitored in areas of potentially high stress such as tunnel intersections. All geotechnical and vibration monitoring readings have been monitored and evaluated by both the contractor and the construction manager to identify the source of vibration and mitigate vibration caused by construction activities.

NEW ACCESS SHAFT CONSTRUCTION

The 4th Avenue shaft has a 20-foot internal diameter (ID) and 104 feet of depth from the working surface grade to top of the shaft slab. The new shaft was constructed to provide access to the existing sewer tunnel and support the construction of approximately 2 miles of sewer tunnel rehabilitation work for subsequent phases of the project. Prior to shaft sinking operations, a permanent reinforced concrete shaft collar was constructed. The shaft collar functioned both as support for anticipated equipment loading adjacent to the shaft and as a physical barrier for fall protection and flood control. The reinforced concrete collar is 36 inches thick and 10 feet high, and extends approximately 3.5 feet (minimum) above grade. The collar was constructed using internal and external prefabricated steel forms. Concrete embeds were placed atop the collar to achieve a 6-foot tall barrier above ground surface as required by CalOSHA for shafts exceeding 100 feet in depth.

Shaft excavation was primarily performed using a Hyundai 3.5-ton mini excavator staged at the bottom of the excavation. The soil overburden encountered in the upper 60 feet of the shaft was excavated using a conventional excavator bucket. The underlying bedrock was broken using a breaker attachment alternating with the bucket for loading. Excavated material was hoisted from the bottom of the shaft using a Sany 40-ton rough-terrain crane staged at the shaft collar and a self-dumping 3.5-yd^3 muck box. During excavation operations, a mini drill was on standby to assist in excavating the bedrock by drilling a series of relief holes, as needed. This excavation method was only used for approximately 6 feet of the shaft depth; the breaker attachment on the excavator was generally capable of excavating the bedrock materials encountered.

Figure 10. Shaft collar structure

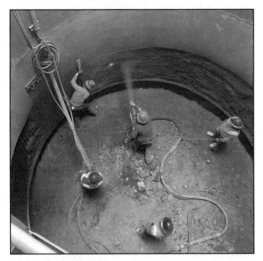

Figure 11. Placement of final liner shotcrete in the access shaft

The original design required the installation of initial ground support using grouted steel liner plates in the soil overburden and a 2-inch-thick flash coat of shotcrete in the bedrock or breccia material, with a reinforced cast-in-place liner as final support for the entire shaft depth to achieve the specified 20-foot ID. To reduce the construction duration of the shaft, the contractor proposed and was allowed to install an alternate complete shaft support system where the complete and final liner was placed as shaft excavation progressed over the entire shaft depth. This was accomplished by installing a 12-inch-thick shotcrete liner, reinforced by welded wire fabric (WWF) and lattice girders for shape control in sequence with the shaft excavation. This sequential excavation method was used and repeated in typical 5-foot vertical lifts based on actual ground conditions until the bottom of excavation was reached, thus enabling the contractor

to furnish a finished shaft product nearly concurrent with the completion of excavation.

Pre-excavation grouting was planned to control a potential presence of running sands and water at the soil/bedrock transition zone at approximately 60 feet below the surface. A localized probe hole drilling investigation was conducted to evaluate the actual ground and groundwater conditions at the transition area and to determine the best suited mitigation efforts. Four potential conditions and assigned mitigation plans were put in place in advance of reaching the soil/bedrock transition area:

- Condition No. 1—No Water Present and No Caving or Running Soil Present: Continuously implement the shaft excavation 5-foot lifts with special care. If any evidence of raveling occurs, the mitigation efforts established for Condition No. 2 will be used.
- Condition No. 2—No Water or Perched Water is Present, Moderate Caving/Running Soils: Reduce the excavation lift heights, excavate in segments and immediately install ribbed expanded metal stay-in-place forms and dry mix shotcrete for initial support. Grout and void fill behind the forms as necessary. After the ground is stabilized, excavate and install the approved final support (lattice girders, WWF, 12-inch shotcrete).
- Condition No. 3—No Water or Perched Water Present, Extreme Caving/Running Soils: Install pre-excavation groutable rebar spiling prior to reaching the transition zone. Special care is required during excavation. Reduce the excavation lift heights, as necessary. If any evidence of raveling occurs, use the mitigation efforts established for Condition No. 2.
- Condition No. 4—Flowing Ground Conditions, Flowing Water-Laden Soil: Install pre-excavation overlapping jet grout columns from within the shaft along the internal perimeter before reaching the transition zone. Special care is required during excavation. Reduce the excavation lift heights, as necessary, and place final support as excavation progresses.

Approximately 15 feet above the soil/rock transition, an exploratory 24-inch-diameter probe hole was drilled using a small-size Tescar CF2.5 drill rig. As witnessed by the contractor and inspection team, the probe hole showed no evidence of water, but it did show some slight caving closer to the soil-rock interface. As agreed, Condition No. 2 was present and appurtenant mitigation efforts had to be employed. After the soil rock transition was exposed in the shaft excavation, excavation lift heights were controlled

Figure 12. AM-50 Alpine roadheader

and material was initially supported by use of Stay-Form and dry-mix shotcrete layers applied using a Lova reed dry pot. This initial support was installed until the ground was secured to allow installation of the permanent reinforced shotcrete liner in sequence. This process was implemented for approximately 10 feet of shaft depth above the transition zone. After the rock formation was reached, no caving was experienced and the typical ground support sequence in 5-foot lifts progressed to the bottom of the shaft. A 3.5-ft-thick reinforced concrete slab was placed at the bottom of the shaft as designed. The slab extended to the 22-foot outside diameter (OD) of the shaft excavation and was "locked-in" by the 12-inch-thick shotcrete liner placed on the wall above.

NEW ACCESS TUNNEL CONSTRUCTION

Construction of the access tunnel started with the excavation of a 12-foot starter tunnel and a 10-foot tail tunnel. Due to limited space in the shaft, the excavation method involved a mini excavator with hydraulic breaker and milling head attachments. The bulk of the material was removed with the breaker, and the milling head was used to trim the contours of the ribs and back. This configuration allowed for minimal over excavation and tight tolerance for the liner. The remaining excavation of the access tunnel was completed using a AM-50 Alpine Roadheader. A Genie all-terrain forklift was used for muck removal and an electric crane (Arva 30-ton) was used to hoist the spoils from the shaft.

Prior to the excavation of the access tunnel, the contractor proposed to increase the size of the tunnel to allow a variety of muck handling options, including load haul dumps and rail-mounted muck cars. Given the benefits and potential cost saving for subsequent project phases, SCWD approved this change and negotiated the cost. The chosen option increased the height of the tunnel by 2.5 feet to a total of 11 feet tall, as well as increasing the width by 8 inches for a

total of 10 feet 8 inches. In addition, the grade was adjusted to approximately 1.5 percent throughout the access tunnel by lowering the access shaft by an additional 4 feet to a total of 104 feet.

For the shaft-tunnel intersection ("brow"), grouted No. 8 rebar spiles were installed for both the starter tunnel and the tail tunnel to create a pre-support canopy over the tunnels. The ground support for the 10-feet-long tail tunnel consisted of W4×13 horseshoe-shaped 3-foot-centered steel sets with synthetic-fiber-reinforced shotcrete infill. For the 12-foot-long starter tunnel, the ground support includes horseshoe-shaped DSI Type 50 lattice girders and synthetic-fiber-reinforced shotcrete. The lattice girders were installed on 3-foot centers and were encapsulated in a minimum of 6-inch-thick shotcrete as the final liner. The remainder of the access tunnel was transitioned to a modified horseshoe shape. The ground support consists of 6-inch-thick shotcrete as a final liner. The shotcrete was installed daily within 2 feet of the tunnel heading as the tunnel excavation progressed. All shotcrete was placed using a hand nozzle and Schwing shotcrete pump.

The alignment and the grade of the tunnel were maintained by the use of two Leica pipe lasers. Each laser was mounted in two block outs built into the shaft walls. They were aligned by the surveyor installed survey points and were set to shoot at 3 inches inside the springline on each rib. These

Figure 13. Initial excavation of the tail tunnel

points were used to measure offsets and to confirm the internal boundaries and thickness of the final liner.

TUNNEL BREAKTHROUGH AND SEWER TUNNEL REHABILITATION

When the access tunnel heading was approximately 20 feet from the sewer tunnel intersection, the access tunnel excavation was temporarily ceased. The location of the sewer tunnel was confirmed by drilling a series of probe holes from within the access tunnel. Temporary protection around the sewer RPMP in the existing tunnel was erected and installed. The temporary pipe protection consisted of W4×13 steel sets placed over the pipe at 4-foot centers and wood lagging in between to form a reinforced "teepee" around the pipe. That temporary protection extended approximately 20 feet on either side of the intersection point. Temporary pipe protection materials were transported by hand using Adit 10 beach access. With this protection in place, the excavation of the access tunnel was carefully resumed (mostly by hand) until tunnel breakthrough was reached and the sewer tunnel was intercepted.

Immediately after the tunnel breakthrough, an in-place emergency sewer bypass system was deployed to mitigate a potential catastrophic spill if the operational sewer pipe were to break. This involved a strategic and logistic placement and installation of bulkheads and pumps in the existing sewer tunnel. To contain sewage spills within the limits of the tunnel if pipe break were to occur, bulkheads composed of concrete blocks and mortar were located at each adit entrance, upstream and downstream of the sewer pipe access flanges. Sump pumps capable of handling solids were staged in a

Figure 14. Completed access tunnel

compartment formed around the upstream access flange to convey sewer water to the downstream access flange for emergency repair work, if needed.

Prior to tunnel excavation activities, the existing sewer pipe had to be encased in reinforced concrete for permanent protection. This encasement also establishes the base for the modified sewer tunnel invert. The original concrete pipe protection design called for an exposure of a 6-inch-wide continuous strip down to bedrock along each side of the pipe. A systematic investigation indicated the pipe bedding was consisted of approximately 17 inches of sub-rounded river rock gravel underlying approximately 10 inches of crushed VCP. To avoid delays on Phase I, all pipe bedding within the limits of the existing tunnel was removed, and a complete concrete encapsulation of the existing sewer pipe was implemented. The digging and removal of the bedding was executed by hand, and the spoil was transported for off-haul via portable conveyor belts. Encapsulating concrete was then placed in two lifts for locking the pipe into place and prevent potential floating and movement during placement. A rebar mat was placed and secured using rebar standees preaffixed to the first concrete lift. A coverage of a minimum 8 inches at the crown was achieved, as designed. The concrete pipe protection was placed in 20-foot sections starting at the intersection, working toward the far northbound and southbound limits of the tunnel. This process was repeated until the concrete protection was installed complete throughout the tunnel limits of Phase I.

With the sewer pipe completely protected and encased in concrete, sewer tunnel enlargement and stabilization commenced, starting with the intersection area between the access tunnel and the sewer tunnel. The tunnel intersection was designed as an expanded room extending as wide as 21'-6" and as tall as 11'-10". This design allowed heavy equipment to make both northbound and southbound turns to support approximately 2 miles of tunneling work in subsequent project phases. The intersection area was excavated in full by the roadheader and was temporarily supported to allow complete erection and assembly of the intersection support structure as one piece. The initial support was achieved by a 3-inch flashcoat layer of synthetic-fiber reinforced shotcrete. The sewer tunnel intersection was then permanently supported by a series of W5×19 and W8×31 interconnected arched steel sets. After the steel sets were erected, the final support was completed using a final layer of synthetic-fiber-reinforced shotcrete in two separate applications.

Rehabilitation of the sewer tunnel has progressed both northbound and southbound up to the limits of Phase I (i.e., 150 feet of tunnel in each direction from the point of tunnel intersection). The

Figure 15. Sewer tunnel intersection

sewer tunnel rehabilitation followed the existing pipe as a bench mark to achieve the desired tunnel alignment and clearance orientations. To accommodate the equipment and ventilation system, the cross section of the sewer tunnel rehabilitation was expanded at the contractor's convenience to an 8'-8" wide by 9'-0" tall straight leg horseshoe, in lieu of the minimum 7'-0" wide by 6'-10" tall tunnel required. The revised sewer tunnel cross section was supported by W5×19 steel sets at 5 feet on center, or W5×19 steel sets at 3 feet on center as dictated by the ground conditions encountered. All sewer tunnel steel sets supports were encapsulated in 8 inches of synthetic-fiber-reinforced shotcrete applied in two separate applications. The sewer tunnel intersection and tunnel rehabilitation work was excavated using the AM50 roadheader and low-profile genie forklift with muck box attached for spoils handling.

SUMMARY AND CONCLUSIONS

The construction of the SCWD's 2-mile-long Beach Interceptor Sewer Tunnel and sanitary pipeline raised several project concerns and considerations, including tunneling safety challenges, logistical issues, and strict regulatory environmental requirements. For the first phase of the project, the following challenges and issues were successfully handled and addressed by the project team:

- The Access Tunnel crosses approximately 70 feet below the high-traffic Pacific Coast Highway. A special double permit was obtained for crossing Caltrans' right-of-way, which required a strict monitoring regime. Special care was taken during the crossing to ensure that ground support installation and continuous settlement monitoring efforts were adequately coordinated and implemented.

- Permit restrictions require 6 weeks per year of construction shutdowns surrounding holidays. Risk mitigation meetings were held with the contractor and the construction manager to discuss and assign mitigation measures and activities required prior to and during the time off. These mitigations steps include completing the ground support installation, securing the tunnel, applying face support, as needed, supplying on-site inventory of ground support elements, conducting environmental and geotechnical monitoring, and performing condition inspections during the shutdown weeks.

- The access shaft site location within highly congested neighborhoods of Laguna Beach was subject to heavy traffic delays to the concrete supply. To mitigate the potential impact to the ground support shotcrete quality, the contractor tested and obtained approval for wet shotcrete mixes from three ready-mix suppliers located in different areas around the project site. In addition, an approved dry shotcrete mix was held on site in reserve for small shotcrete quantities, as needed, and for emergency use.

- The team was deeply aware of the impact that this project could have on the neighboring Laguna Beach community. The proximity of this work to homes and the recreational beaches required a commitment by the project team for community awareness and a highly-coordinated public outreach program to minimize negative impacts to the community. The numerous project-specific environmental requirements and continuous monitoring activities were addressed and compiled with the EIR requirements. In addition, the project team has collaborated to assist the SCWD with public outreach efforts such as holding open-house presentations at public meetings, responding to questions/complaints, drafting and distributing notices, and providing ongoing information, as needed.

The project team recognized that a key factor in the successful and effective project execution was the teamwork and cooperation among all parties as exhibited throughout the construction of the first phase of the Beach Interceptor Sewer Tunnel rehabilitation project. In addition to the challenging execution of the access shaft and tunnel work, it was important to maintain positive community relations in unison with actual construction nuisance prevention measures to build trust and to help eliminate potential problems before they become major issues.

ACKNOWLEDGMENTS

The authors wish to thank the Owner, South Coast Water District; the Design Engineer, Mott MacDonald; and the Construction Manager, Parsons, with Denton Mudry Environmental and LGC Geotechnical, for their vital support during this challenging Phase I of the tunnel rehabilitation project. In addition, the authors would like to thank the Contractor, Drill Tech Drilling & Shoring, and its tunnel workers for their contributions to the project.

REFERENCES

SCWD, 2010, Final Environmental Impact Report (EIR) and Response to Comments, Tunnel Stabilization & Sewer Pipeline Replacement Project (SCH# 2008031094).

Mott MacDonald (2016a), Geotechnical Data Report, Phase I, Tunnel Stabilization and Sewer Pipeline Replacement Project, South Coast Water District.

Mott MacDonald (2016b), Geotechnical Baseline Report, Phase I, Tunnel Stabilization and Sewer Pipeline Replacement Project, South Coast Water District.

Norris Cut Force Main Replacement Tunnel, Miami, Florida

Robin Dill and Roger Williams
AECOM

Lin Li, PE
MDWASD

Eloy Ramos
Nicholson Construction

ABSTRACT: The Miami-Dade Water and Sewer Department utilized a design-build delivery approach for a replacement sewage force main under Norris Cut through very challenging coralline limestone geology. This case study will discuss alternatives considered for the replacement, how a preferred method was selected, and how various risks were addressed in the design and construction. Ultimately the mile-long tunnel was mined using a custom-built, 10-foot diameter hybrid (slurry and EPB) tunnel boring machine. The tunnel was lined with pre-cast concrete segments for initial support and 60-inch fiberglass-reinforced mortar carrier pipe was grouted inside the tunnel for the force main replacement.

INTRODUCTION AND BACKGROUND

In 2012 Miami-Dade Water and Sewer Department (MDWASD) engaged AECOM to evaluate viable and cost-effective vertical and horizontal alternatives for replacing an existing damaged 54-inch Pre-Stressed Concrete Cylindrical Pipe (PCCP) sanitary sewerage force main (FM) that extended between Fisher Island (FI) and the Central District Wastewater Treatment Plant (CDWWTP) on Virginia Key (VK). The FM which was the sole means of transmitting approximately 24 MG average daily flow to the CDWWTP, was subject to a condition assessment which identified structural deficiencies in the 9,000 linear foot segment of pipeline that connects the two land masses. The deficiencies in the PCCP, manifested in the form of breaks in the pre-stressed bands, were identified in a total of twelve pipe segments, eleven of which were under the Norris Cut Channel, a pristine protected waterway, lush with benthic resources. Figure 1 shows the location of the existing FM.

The pipe was determined to be in a state of incipient failure with wire breakage at or very near the point that could produce yielding of the pipe under seasonal high operating pressure. Failure of the FM would prove catastrophic to the environmentally-sensitive community by discharging millions of gallons of raw sewage which would adversely affect safety, human health, and the pristine coastal environment, which included tourist-filled beaches, benthic resources and protected habitats. The economic

magnitude of a failure would be immense, impacting local businesses, and resulting in beach and hotel closures with significant consequences to Miami Dade's vibrant tourism industry.

Given the risks associated with a lack of system redundancy and urgency of schedule, the MDWASD prioritized the implementation of a new replacement FM using a design-build (D-B) approach. The project had to be completed immediately, and commissioned without disruption to the service of the existing FM. As such, MDWASD tasked AECOM with the evaluation of replacement alternatives, and development of robust design criteria to facilitate the selection of a Design-Builder through a best value selection process.

This paper provides an overall summary of the activities during the planning, design development, geotechnical investigation, procurement, and construction of the FM replacement.

GROUND CONDITIONS AND SUBSURFACE RISKS

A previous study had been performed in the project area that provided relevant information. This investigation was for a microtunnel constructed across Government Cut (Dill 2012). The Government Cut project involved installation of a 60-inch replacement FM inside a 72-inch diameter steel casing which was installed using a pipe jacking and microtunneling approach. This replacement was for an upstream

Figure 1. Existing force main

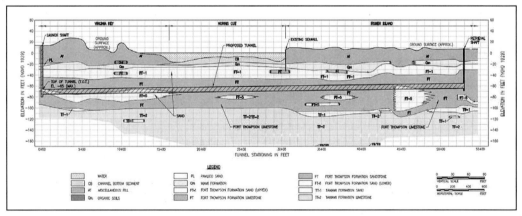

Figure 2. Geologic profile along proposed Norris Cut FM alignment

segment of the same sewer FM that conveyed flow from Miami Beach under Government Cut eventually reaching the CDWWTP. The replacement FM was approximately 1,175 feet in length and 90 feet below sea level. The microtunnel was completed successfully in 2013 and provided valuable insight into subsurface conditions and ground behavior. The previous study had included several borings on Fisher Island and Virginia Key.

A geotechnical investigation was undertaken for definition of subsurface conditions along the Norris Cut FM alignment. The goal of the program was to provide enough data to provide an adequate basis for the Design-Build Proposers to reasonably

price the proposed construction. Supplemented by the previous borings conducted for the Government Cut project, additional borings were performed along the alignment including three water borings across Norris Cut. In total, 16 test borings were used to define the ground conditions along the alignment corridor shown on Figure 2. In addition to the borings, a geophysical survey consisting of bathymetry, seismic refraction and seismic reflection was completed as well as laboratory testing.

The subsurface conditions along the FM alignment were found to be generally consistent with those for the Government Cut project, and tremendously challenging from a tunneling perspective.

967

Figure 3. Fort Thompson coralline limestone exhibits karstic features and permeabilities as high as 5 cm/sec

The alignment was found to pass through the Fort Thompson Formation, including beds of very porous coralline limestone with inclusions and layers of loose to medium sand. Up to six distinct layers were found with significant variability in the vertical direction. The limestone bedrock layers of this formation are characterized as karstic, exhibiting vugs and solution cavities of various sizes.

The Fort Thompson Formation bedrock layers exhibit extremely variable intact and rock mass properties. The intact core samples can have unconfined compressive strengths ranging from as little as 50 pounds per square inch (psi) to as much as 12,000 psi, although typical strengths are below 4,000 psi. The Fort Thompson Formation is part of the Biscayne aquifer, a surficial aquifer system in southeastern Florida. It is further classified as one of the most permeable aquifers in the world (Parker 1955). A combination of solution pipes, bedding planes, vugs, as well as matrix porosity contribute to the high permeability of the Fort Thompson Formation. In the project area, the maximum in-situ permeability of the Fort Thompson bedrock was found to be on the order of 5 centimeters per second (cm/s), while the maximum in-situ permeability for sand layers is approximately 5×10^{-2} cm/s.

The extremely pervious nature of the Fort Thompson Formation results in significant challenges and risk in constructing deep excavations and tunnels in the Miami area. It is essentially impossible to dewater the formation, and allowing groundwater entry into the excavation can be disastrous. As an example, during the Government Cut project, despite the use of a secant pile wall system and implementation of a tremie-concrete slab for the launch shaft, the Contractor experienced leakage into the shaft upon launching the MTBM, even after constructing a break-in seal. This required the shaft to be temporarily flooded while remedial grouting was performed, causing a significant delay.

Another tunneling risk involves the existence of pockets and continuous layers of loose sand within the coralline limestone. These loose sand zones interspersed within the bedrock cause uneven face pressures with an advancing TBM and can cause significant steering challenges as well as sudden loss of fluid and face pressure, especially if conditions change unexpectedly. Previous project experience and the extremely pervious ground conditions drive the need for a sealed shaft construction approach and require the use of a closed-face, pressure-balanced tunnel boring machine.

In order to manage risk on the project and in accordance with industry practice, a Geotechnical Baseline Report (GBR) was written for the contract. Since the contract was to be D-B delivery procurements, it was intended that this document be provided to all proposers before pricing was submitted. The GBR presented a subsurface profile based on the geotechnical investigation data (see Figure 2), and included properties for the various geologic formations and subunits anticipated to be encountered. The GBR also presented design and construction considerations including baselines related to the challenging ground conditions as discussed above. Minimum design requirements were presented in the design criteria drawings and specifications.

ALTERNATIVES EVALUATION

With AECOM's assistance, MDWASD evaluated a total of thirteen replacement alternatives. One of the critical challenges was finding space on FI for construction staging along the corridor of the existing FM. This exclusive residential community is fully built-out and limited land is available for construction purposes. To facilitate development of alignment alternatives, feasible shaft and staging areas were first identified, then various construction methodologies were considered based on the available locations. The proposed alternatives largely focused on trenchless approaches including TBM tunneling with a segmental lining, pipe jacking/microtunneling, and horizontal directional drilling (HDD) using dual bores. Multiple alignments for each alternative were identified based on feasible shaft and staging areas. One open cut alignment was also considered.

The options were evaluated and ranked based on reliability, environmental impact/permitting, stakeholder acceptance, constructability, hydraulics, and schedule impacts. Based on the ranking results, the preferred alternative was a two-pass, TBM-mined tunnel approach. The selected alignment consisted of approximately 5,300 linear feet of tunnel. The selected tunnel launch shaft site was at the north side of the CDWWTP as shown on Figure 4. The proposed

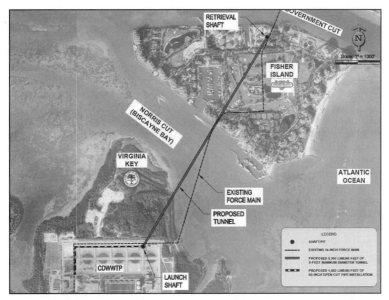

Figure 4. Preferred alternative for force main replacement—two pass TBM-mined tunnel

tunnel extended under Norris Cut to a retrieval shaft on Fisher Island.

The preferred alternative received a very favorable hydraulic rating relative to other alternatives. It also allowed construction of shafts in the uplands of VK and FI, allowing permits to be acquired in 3 to 6 months; situated shafts where construction activities are isolated; and did not have a major impact on the daily activities of the sensitive residents in the island community. More importantly, it proposed to locate the reception shaft at a site adjacent to a private dock, allowing year-round construction, and enabling construction in less than two years.

DESIGN CRITERIA DOCUMENTS

For the design criteria documents, it was decided to leave the actual tunnel size up to the Contractor, but specify that it be a minimum 9-foot minimum inside diameter, using precast concrete segmental liners for initial support. The bidders were allowed to propose a new or refurbished TBM, and were required to use a hybrid-type machine capable of operating in either Earth Pressure Balance (EPB) or slurry mode. Grouting capabilities out in front of the TBM were also required.

Based on an assessment of the ground conditions revealed by the subsurface investigations, it was decided to place the alignment within the Fort Thompson formation as shown on Figure 2. The design criteria documents required that the top of the tunnel elevation be no higher than El –65. This requirement assured that the tunnel would be mined

in bedrock at a safe distance below existing seawalls and building foundations with adequate rock cover.

The design criteria documents also required the Contractor to use sealed shaft methods of construction, consisting of secant pile or slurry wall systems, although other equivalent systems would be considered. The specified carrier pipe was a 60-inch diameter fiberglass reinforced mortar pipe to be grouted inside the segmentally-lined tunnel using low strength concrete or lightweight cellular concrete.

PROCUREMENT

Considering the aggressive schedule that had to be met, coupled with the high degree of risk of certain elements of the project, the D-B procurement was structured to be a two-step "best value" approach that selected the Design-Builder on the basis of both technical solution and price. Instead of price being the primary factor in selection, the more important considerations were meeting schedule and mitigating risks through innovative design and construction features.

The Request for Qualifications (RFQ), as the first step, was prepared to describe the required and preferred qualifications from prospective Design-Build teams. The Owner's objective of the RFQ was to attract and shortlist enough qualified D-B teams to allow for a sufficient competition in providing innovative technical proposals at reasonable pricing during the next step. The selection criteria were also established to emphasize what elements of the project were most important to the Owner. The technical proposal scoring was based on a 100 point scale.

The shortlisted teams were asked to submit technical and price proposals. Before opening pricing, technical proposals were reviewed, and the teams were interviewed and individually scored by each selection committee member. The total technical score for each team was calculated by adding the scores from each committee member. The price proposals were then opened, and bid prices were adjusted by dividing by the total technical score.

The two-step "best value" process which is provided under Florida law had been used successfully for the first time on the Government Cut project, and was favored by MDWASD.

Procurement Results

Overall, five teams submitted qualification packages and three teams were shortlisted. The three teams all submitted technical and price proposals.

The selected team based on adjusted bid price consisted of Nicholson Construction with Bessac proposed for the tunneling component of the project. Nicholson was responsible for overall project management and shaft construction.

CONSTRUCTION

Launch Shaft

The launch shaft was constructed using the secant pile technique. As the shaft was designed to work in hoop compression, the secant pile shaft walls were generally unreinforced. Out of a total of 68 piles constructed at the launch shaft, 10 piles had steel reinforcement and three had GFRP (glass fiber reinforced polymer) cages. The GFRP cages were located

at the TBM drive zone to maintain stiffness of the shaft walls during tunnel break-in without damaging the TBM. Supplemental structures included a reinforced thrust wall, hoop reinforcing beam and pile guide wall. Figures 5 and 6 detail the launch shaft configuration.

The secant piles were constructed using conventional bored and cast in-situ techniques using a fixed-mast Bauer BG39 drill rig. The piles were excavated inside the temporary casing using a combination of augers, coring barrels, and digging buckets as required based upon the ground conditions to the design tip elevations. The secant piles were fully cased to prevent any sloughing of any material into the excavation. The sectional heavy wall casing served two purposes: maintaining borehole stability and limiting deviation due to the stiff drill string. Figure 7 shows a detailed sketch of the sequence of installation.

The water tightness of the system was critical as the bottom of the shaft was 85-ft below the water table in very pervious ground conditions. The water tightness relied completely on the quality of the interlock between the piles. The verticality of the piles was probably the most important aspect of the construction. Pile verticality was monitored using two independent quality control devices and the targeted 1:200 or 0.5% tolerance was obtained for all 68 piles.

Launch Shaft Excavation

To aid with in-the-wet excavation efficiency, the overburden and limestone layer located within the footprint of the shaft was pre-drilled using the Bauer BG39 converted with a continuous flight auger attachment. The predrilling was performed without

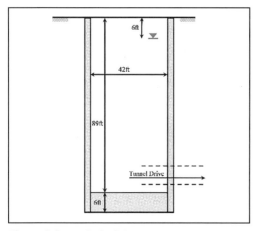

Figure 5. Launch shaft layout

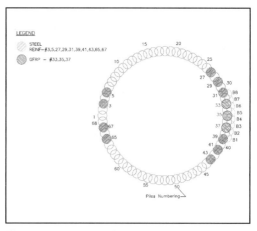

Figure 6. Launch shaft layout

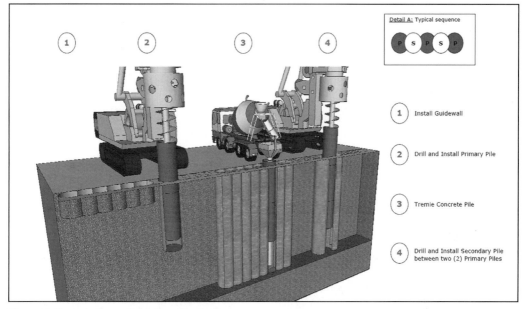

Figure 7. Secant pile methodology (typical)

removing material in order to maintain the stability of the work platform.

Line drilling was also implemented to remove any concrete bulges in the sand layer or concrete that had permeated into the limestone layer. The equipment used was a Casagrande C-12 rig tooled with a down the hole hammer. After the completion of the pre-drilling operation, an excavator was used to remove as much material as possible from the top of the shaft based on its reach capacity. The sand layer was excavated with a crane equipped with a mechanical clamshell and the harder limestone was excavated with a combination of vibro-chiseling and excavation with the clamshell. The excavation proceeded in-the-wet until the bottom of the shaft was reached.

Launch Shaft Tremie-Slab

At the completion of excavation, a 5-foot thick reinforced tremie-slab was poured. Prior to concreting, steel dowels were drilled and epoxied into the wall by divers to transfer the uplift force from the slab to the piles. A re-injectable flexible grout tube with ports was secured to the secant piles near the top of the slab, to act as a secondary measure of water tightness.

Ground Freezing

Localized ground freezing was used to control groundwater inflow during TBM break-out. A block

Figure 8. Excavation of the launch shaft using mechanical clamshell

Figure 9. Launch shaft excavated after tremie-slab poured

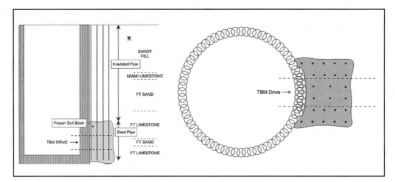

Figure 10. Ground freezing break-in zone

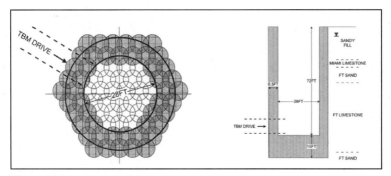

Figure 11. Retrieval shaft layout

of ground approximately 30 cubic feet was frozen at the TBM break-in as shown in Figure 10.

By freezing the water, limestone, sand and gravel in this area, a strong solid block was created. The frozen block adhered to the launch shaft, creating a watertight seal between the block and shaft walls. The frozen block was installed before the launch of the TBM and maintained frozen until after the break-in. The ground freezing allowed for the inspection of the cutter head and cutting teeth after the break-in without the need of a hyperbaric intervention.

The soil freezing technology converts water into ice by the continuous circulation of a brine fluid within a system of small diameter, closed-end pipes installed in a pattern consistent with the shape of the area to be treated. The highly permeable ground and possible moving groundwater were the major concerns regarding the feasibility of sustained ice block formation. To minimize this risk, a permeation grouting program was implemented. Nicholson drilled and grouted 8-inch diameter holes using low mobility grout to reduce the flow of groundwater into the break-in block using a rotary-hydraulic drill rig tooled with a duplex drilling system.

Temperature nodes were installed in temperature pipes at critical locations throughout the break-in block area and in other areas of the freeze system

and equipment to monitor the freeze down process. The electric powered chillers were then energized and the refrigerant pumped through the system to remove the heat from the ground. As the brine is chilled to –15°F to –20°F, the soil around each pipe freezes radially outward until the entire block of soil surrounding the array of freeze pipes was solidly frozen. When temperatures at the sensor locations were down to the design temperature (+5°F to 10°F or colder), the entire block was frozen solid and ready for tunneling.

Retrieval Shaft

The Retrieval Shaft was constructed using the Deep Soil Mixing (DSM) technique. DSM is a soil improvement technique that is used to construct in-situ soil cement by blending the soil (or rock fragments) with cementitious materials. The construction process involved the insertion of a large-diameter auger into the ground while grout is injected through ports in the tooling. The injected grout is mixed using paddles on the auger that mix the ground and grout as it penetrates down to the design depth to form a column. The DSM was intentionally designed to be constructed using a Bauer BG39 drill rig, the same rig used to construct the secant piles at the launch shaft.

Figure 12. DSM core sample from the trial program

Figure 13. Retrieval shaft excavation

The DSM technique was selected in order to overcome the logistical challenges associated with the location of the site, its size and surrounding environment. The retrieval shaft was located on FI which is only accessible by boat. As such, it was preferable to use a construction technique that was not dependent on ready-mix concrete and that allowed the excavation of the site in-the-dry, without the need for dewatering.

A trial program consisting of drilling six test columns adjacent to the site was implemented. The trial was focused on the quality, strength, and performance of the soil cement material which had to be formed through a layer of hard limestone (USC breaks of over 4,000 psi) located at depths between 70 and 87 feet. There was no established precedent for mixing material this hard at this depth. The results

Figure 14. Precast segments stored on site

determined a strength of 200 psi was consistently attainable for the production columns. A photograph of a DSM core is shown in Figure 12 indicating a high quality, well-distributed mixture of cement and limestone.

Verticality of the grout columns was critical to the integrity of the shaft structure and was closely monitored during construction. The DSM methodology allowed both the support of excavation walls and the bottom plug slab to be constructed from the surface simultaneously prior to any shaft excavation. Once the DSM walls and bottom plug were completed, the excavation of the shaft occurred in-the-dry. Prior to the start of excavation of the shaft, guide walls were constructed at the top of the shaft in order to guide the advancement of the excavation within the required tolerance. The excavation of the shaft was performed by means of a mini-excavator equipped with a bucket and a grinder. The excavated material was loaded out of the shaft on a skid pan hooked to a crane.

Tunnel Segmental Lining

The tunnel was lined with reinforced concrete segments assembled in the tail skin of the TBM. The segment thickness was 7.48 inches. The segmental lining was manufactured by Nicholson and Bessac in a precast plant located in Medley, approximately 22 miles away from the jobsite. 8100 concrete segments were produced to line the Norris Cut Tunnel.

The lining consisted of six segments to form one ring. A total of 1350 precast concrete rings were installed. Each ring weighed approximately 5.3 tons using 7,000 psi concrete. Grade 60 steel reinforcement was used for the segments, which also included the use of EPDM tailor-made watertight gaskets, anti-return pins, a radial-bolting system, guiding rods and bituminous packers.

Segments were batched by ring and stored using wooden wedges in order to avoid damages on the extrados. They were stacked following the loading sequence on the service train as shown in Figure 14.

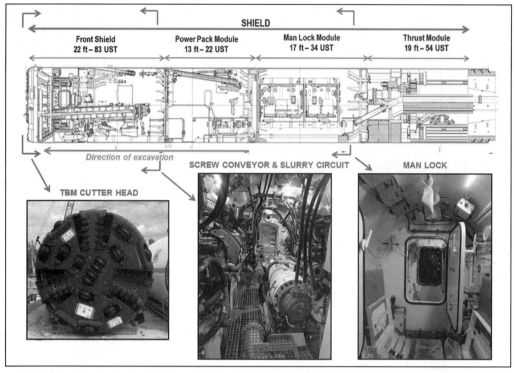

Figure 15. TBM shield was divided into four modules to accommodate dual mode operation

Tunnel Boring Machine

A new 10.2-foot diameter Herrenknecht/Bessac TBM was manufactured to bore the tunnel. As required by contract, the machine was a dual mode hybrid TBM capable of converting from slurry mode to EPB mode within the tunnel in a short time. This type of machine is quite unique in this diameter range. Only a few of this size have operated worldwide. In general, the dual mode capability consisted of an EPB machine with a slurry suction box connected to the screw conveyor discharge gate.

It is more common to see dual mode TBMs of much larger diameter with space that allows for an easier installation of all the equipment necessary for conversion between operating modes. Since the TBM used for the Norris Cut Tunnel project had an outside diameter of only 10.2 feet, Herrenknecht and Bessac had to innovate in the design of the machine to fit all the equipment (pumps, transformers, erector and other components) in such a tight space. As such, the overall length of the shield had to be increased up to 71 feet.

The shield is divided in four modules, and designed for groundwater pressure up to 45 PSI (3 bars):

- **The first module** is the articulated shield, the hollow main drive and the cutter head. The cutting wheel is a closed-face type, equipped with 15 double 14-inch diameter disc cutters, 20 scrapers and 6 buckets. Cutting discs included hard-faced tungsten carbide inserts to reduce the wear and enhance the lifetime. The face access for cutter inspection/change was possible through a door with an opening size of 18 inches by 18 inches.

  In slurry mode, the TBM excavates like a standard microtunneling machine with a cone crusher on the cutting head spokes, feeding a slurry network. No air cushion could be provided within the chamber due to the tight space. A reverse water flushing system was designed to clean the suction port of the cutting chamber.

  In EPB mode, the excavated material goes through the screw conveyor and is discharged into the hydraulic mucking system as aided by a suction box.

- **The second module** holds the power pack, the probe drilling rig and the diver-pit to allow wet cutter change.

- **The third module** integrates a double chamber lock, allowing manned access for four

Figure 16. Diver pit was provided inside tunnel shield for cutter change in-the-wet

Figure 17. Lowering the TBM into the launch shaft

hyperbaric workers. The control command of the TBM was also located in this shield section.

- **The fourth module** is the thrust/segment erection system. The erector was a remote-controlled, center-free rotary-arm-type system, mounted on the tunnel axis with a vacuum segment gripping system.

The shield was built by Herrenknecht, whereas the back-up gantries were manufactured by Bessac. Both companies worked closely together to custom-build the TBM and back-up systems to be compatible and work efficiently.

Cutting Tool Replacement

Porous ground formations typically aren't capable of withstanding compressed air confinement. This renders manned hyperbaric interventions into the TBM cutter head chamber very difficult and risky. Replacing cutting tools usually requires replacing the slurry in the cutter head chamber by compressed air to perform a dry intervention. The air pressure counterbalances the ground water pressure and thus prevents water ingress into the chamber. The Fort Thompson Formation was considered too permeable to allow a standard in-the-dry procedure to carry out cutter head maintenance. Therefore, Nicholson and Bessac built a diver pit within the second module of

Figure 18. Launching of the TBM using a pipe jacking approach

Figure 19. Jacking pipe was used for the first 170 feet of tunnel to facilitate TBM gantry assembly

the TBM (see Figure 16), allowing wet interventions within the cutter head chamber.

Fortunately, there was never the need to use the diver pit. One brief intervention within the cutter head chamber was performed, but was accomplished in a dry mode. Prior to the intervention, a low density high viscosity slurry containing vermiculate was poured into the cavities where it filled the fissures and treated the Fort Thomson Formation. The treatment produced an air-tight seal that provided two hours of safe intervention time to inspect the cutting tools. The hard-faced carbide tungsten disc cutters performed admirably during the entire TBM drive, as none required changing or maintenance.

TBM Launching Sequence

Nicholson and Bessac designed and implemented a highly innovative launching sequence which improved safety, saved significant time, drastically reduced the size of the circular launch shaft, and prevented complicated additional underground structures. The sequence involved the jacking of pipes

for the first 170 feet of tunnel, during the launching step of the TBM. This technique avoided lowering the backup unit gantries one by one. Instead, the gantries could be completely assembled within the jacked pipe section of tunnel such that the TBM mining process could begin efficiently with the complete back-up support system functioning.

The jacked pipe was connected to the TBM thru precast liner rings pre-installed in the TBM thrust module. Once the pipe jacking was completed with the backup gantries assembled, the TBM segmental lining continued by pushing from the pre-installed segment. The transition between the two tunneling methodologies was successfully accomplished without incident.

Alignment Challenges

Probably the most difficult challenge experienced during the excavation was to maintain the design alignment while boring through the mixed face ground conditions. There were three main areas along the drive (see Figure 2) where the ground was

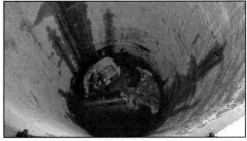

Figure 20. TBM Breakout and retrieval shaft re-excavation for TBM removal

composed of loose sand below the springline and hard limestone at the crown. The lack of bearing capacity of the sand to react against in order to push the TBM up combined with the relative hardness of the limestone caused the TBM to dive down. This made it extremely difficult to maintain the required 2% slope (upward) along these portions of the drive. This challenge was mostly overcome by using the maximum steering capacity of the machine together with reducing the speed of the excavation to a third of the speed used for the areas where single face ground conditions existed. However, a "bump" in the alignment did occur. This required the backup gantry system to be modified with flexible connections to safely pass through the irregular tunnel geometry in the "bump" area.

Another significant challenge was to maintain the alignment tolerance when transitioning from the mixed face condition to the single face condition.

TBM Break-Out

The break-out into the retrieval shaft from the very pervious Fort Thompson Formation was challenging and would typically require a separate break-out seal be constructed to improve the ground adjacent to the retrieval shaft wall to obtain the proper water tightness. However, another approach was taken. In order to guarantee a successful and safe breakout, the wall thickness of the deep soil mixing was increased to 10 feet at this location. Also, once the retrieval shaft was completed, the shaft was excavated to depth; inspected and confirmed watertight; backfilled with 50 feet of the same excavated material; and then flooded.

The TBM broke though into a backfilled shaft balanced with equal water pressure. Three stages of grouting (tail void grouting, cement grouting, and polyurethane injection) were performed in the overcut areas around the TBM modules prior to re-excavating the shaft. Upon re-excavation, the water ingress turned out to be less than 20 gallons per minute.

TBM Dismantling

Typically a TBM is removed entirely from the ground after break out into the retrieval shaft. On the Norris Cut project, only the front part of the shield was retrieved, leaving behind the shield skins of the rear modules, which served as initial tunnel support in place of the segmental lining until the carrier pipe was placed and grouted. All the equipment inside the rear modules was dismantled and removed before pipe installation.

CONCLUSIONS

The Norris Cut FM Tunnel project was completed successfully under extremely challenging subsurface conditions, in some of the most pervious ground in the world. Using a D-B approach, MDWASD, with the assistance of AECOM, assessed the project risks and developed a conceptual design to address them while still leaving flexibility in the design criteria documents. Through a best value selection process which promoted innovation, a D-B team was procured to execute the project. The selected team, consisting of Nicholson/Bessac implemented cutting-edge solutions to overcome the difficult ground conditions and set precedents for new viable construction techniques for deep heavy civil construction in the greater Miami area.

Several conclusions and lessons learned from this project include the following:

- The design-build best value selection process worked well given the challenging ground conditions and the Owner need to retain a highly experienced team with innovative ideas to address the project risks. This best value procurement approach has become a model used by MDWASD for all their design-build projects.
- The custom-built dual mode TBM used by Nicholson/Bessac was well suited for the challenging ground conditions and was designed to address project-specific

underground risks. For future projects in the Miami area it is essential that a TBM be custom made and tailored accordingly to the project requirements and prevailing ground conditions.

- The use of pipe jacking to launch a TBM is a viable method to use on small diameter tunnels, allows for a reduced shaft diameter, and eliminates the need to install initial segment rings manually, thus enhancing safety. However, this capability must be custom-built into the TBM.
- Deep soil mixing (DSM) is a viable technique for deep shaft construction in the Miami area. The DSM method can be used through hard limestones with strengths over 4000 psi, and allows for an in-the-dry excavation approach since the DSM technique can be used to construct a bottom plug slab in advance of excavation.

- Controlling the tunnel alignment while a TBM advances through mixed ground (sand below the springline and hard limestone at the crown) can be extremely challenging and requires well-trained and experienced TBM operators.

REFERENCES

Dill, R, Watson, K., and Vega, E.A. 2012. MDWASD Design-Build Procurement Approach and Results for the Government Cut Pipeline Replacement Project. *North American Tunnel Conference Proceedings*, SME.

Parker, G., Ferguson, G., Love, S. 1955. Water Resources of Southeastern Florida with Special Reference to the Geology and Ground Water of the Miami Area, *U.S. Geological Survey Water-Supply Paper 1255* (http://sofia .usgs.gov/publications/papers/wsp1255/PDF/ wrsf_1255.htm).

Operations and Maintenance of Waller Creek Flood Control Tunnel

John Beachy and Ramesh Swaminathan
City of Austin, Watershed Protection Department

ABSTRACT: This paper will discuss the operations and maintenance of the Waller Creek Tunnel, the City of Austin's largest flood control project. Located in downtown Austin, the mile-long, concrete lined 26.5' (avg. dia), 70-foot deep inverted siphon began accepting flood waters in May 2015. The next twelve months were the wettest in Austin's history, with the city receiving 59 inches of precipitation. Limited access to the tunnel is confined to the Outlet located on Lady Bird Lake, attraction to visitors of Austin. In February 2017, the tunnel was dewatered for sediment and debris removal. Key elements of this project included evacuating millions of gallons of water, removing over 3200 tons of sediment from a 40-ft deep shaft using combination vacuum/ compressor trucks, and inspecting the condition.

WALLER CREEK TUNNEL BACKGROUND

The Waller Creek Tunnel (WCT) program represents the single largest capital improvement program in the history of the City of Austin's (COA) Watershed Protection Department (WPD). The project will alleviate flooding, enhance water quality, and create the landscape canvas for a world-class chain of parks in a presently impaired part of Downtown Austin. This paper will provide general background of the WCT program, different modes of operation and finally the details of the first dewatering and removal of sediment accumulated within the tunnel after an approximate 22 months of operation.

The City of Austin lies within what is popularly known as "Flash Flood Alley," and is one of the most flood-prone regions in the continent. High rainfall intensities are common due to an almost unlimited source of moist air from the Gulf of Mexico. This warm moist air collides with the cooler air from the north and moisture from the Pacific. This is further aided by Balcones Escarpment's hilly terrain that acts as a ramp for the moisture. The net result of this phenomenon called the Orographic Effect, is the frequent yet sudden torrential downpours in the spring and fall months in the Central Texas area.

The Waller Creek watershed is the most developed tributary watershed of the Colorado River within the limits of the COA. The lower reach of Waller Creek traverses the City's downtown corridor, where flood events routinely inundate significant areas along the banks of the creek. Encompassing over six square miles of urban landscape, the Waller Creek watershed includes over 3,700 acres of parks, single family, commercial, and institutional land uses. The WCT project will capture and divert floodwaters to a tunnel beneath the city and discharge them to the Colorado River at Lady Bird Lake.

Figure 2 illustrates a general layout of the Waller Creek floodplain in both its pre- and post-tunnel construction conditions. The WCT program is expected to vacate approximately 28-acres of land from the existing 100-year floodplain which is currently the basis for a dramatic ecological and economic redevelopment of the lower reaches of Waller Creek that is anticipated to dramatically transform the architectural landscape of Downtown Austin.

Tunnel Features

Tunnel

The below-grade concrete-lined tunnel (see Figure 3) extends from Waterloo Park at 12th Street to the Colorado River at Lady Bird Lake. It consists of an inverted siphon approximately 5,600 linear feet long, with diameters increasing from 20.5

Figure 1. Waller Creek at 6th Street in aftermath of 1915 Flood. Photo Courtesy of Austin History Center.

feet at the Waterloo Inlet Facility, to 22.5 feet at the 8th Street Creekside Inlet, and finally to 26.5 feet at the 4th Street Creekside Inlet till it ends at the Outlet Structure. The tunnel has a reverse horse-shoe cross section with a flat invert.

Waterloo Inlet Facility

Primary diversion occurs at the Waterloo Park Inlet Facility, located at the northern most reach of the tunnel. The facility is designed to screen large debris from floodwaters entering the tunnel. In addition, the facility houses the required pumping infrastructure

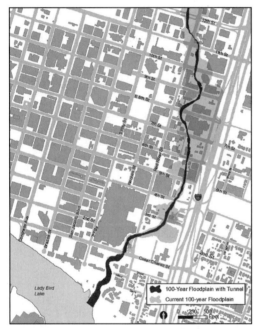

Figure 2. Waller Creek flooplain in downtown Austin, current and post current tunnel conditions

for dry weather and post-flood recirculation of water from within the tunnel.

Creekside Inlets

Additional flood waters that drain to the creek downstream of the Inlet Facility can enter the tunnel via two creek side inlets (CSI) (see Figure 4). These facilities are located at 8th and 4th streets and contain debris screening nets to prohibit trash and debris from entering the tunnel.

Outlet

Flood waters are finally discharged to Lady Bird Lake at the outlet structure via a submerged 40-foot diameter shaft and then over an Ogee Weir. The outlet contains two 42-inch recirculation pipes that hydraulically connect the tunnel with Lady Bird Lake to allow for dry weather recirculation. Primary maintenance access into the tunnel is through the outlet shaft.

Interim Operation

The construction sequence of the WCT led to a condition of interim operations beginning on May 15, 2015, when the tunnel was first filled with rainwater. This occurred prior to several design facilities, components, and systems being fully constructed and operational. Components of the WCT were utilized as they became complete. Contractors and City staff employed creative strategies to ensure both the operational needs of WCT facilities were being met and construction progress could continue. Examples of coordination efforts include intermittent recirculation pumping, utilization of portable generators to power facility components, and installation of temporary meters and level gages. This interim operation condition was in place during the Sediment Debris and Removal Project. Ongoing construction resulted in limited tunnel access and staging areas, particularly at the shafts for the 4th and

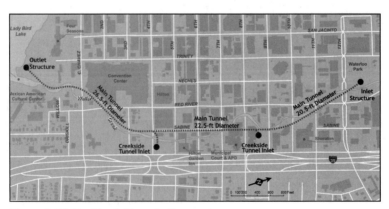

Figure 3. Waller Creek Tunnel alignment

8th Street creek side inlets. Constant communication was critical as the work required for dewatering and sediment removal overlapped with the construction of the 4th and 8th Street Creekside Inlets. Critical activities included the removal of steel falsework inside the 4th Street shaft that was directly above the work being performed inside the tunnel. It is to be noted that this critical work was possible only when the tunnel was dewatered.

Modes of Tunnel Operation

Dry Weather Recirculation

During dry weather conditions, recirculation of the approximately 22 million gallons of water held

Figure 4. Waller Creek Tunnel 8th Street creek side inlet during initial construction

within the tunnel is required to ensure that anoxic conditions do not occur (see Figure 5). Water is drawn from Lady Bird Lake into the tunnel through the two 42-inch pipes at the outlet facility. Four recirculation pumps located at a lower wet well of the inlet facility have capacity to convey up to 28 cubic feet per second (CFS) through the tunnel. The pumps convey the water through a dissolved oxygen (DO) curtain prior to discharging back into Waller Creek.

Flood Operations

During wet weather conditions, flows from the upper reaches of Waller Creek cause a rise in the water surface at the inlet pond, causing water to crest the WCT Morning Glory Spillway. This event triggers the Flood Operations Mode (see Figure 6), a sequence that closes the discharge gates at the pump houses, ceases recirculation operations, and initiates debris management to ensure the influent mechanical bar screens do not get clogged. Water is conveyed through the tunnel and safely discharges into Lady Bird Lake over an Ogee weir located at the outlet facility. This condition continues until levels in the creek lower and the water surface falls below the Morning Glory Spillway. Once water levels recede, recirculation operations (post-flood mode) for the tunnel are initiated, and nutrient-rich creek water is rapidly evacuated and replaced with stable lake water.

Figure 5. Dry weather flow recirculation

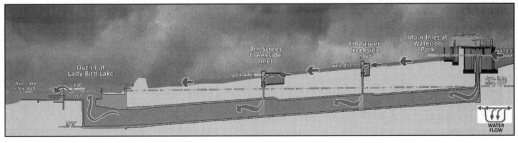

Figure 6. Flood operations

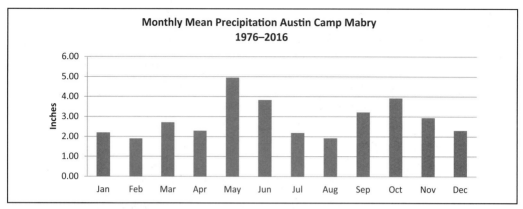

Figure 7. Precipitation distribution in Austin metropolitan area

Tunnel Dewatering

Tunnel dewatering, a significant maintenance operation is planned to occur periodically, and is the only time that the tunnel will not be filled with water. This operational mode requires major coordination to other activities in the Downtown district, including bypass of upstream flows through the Waterloo inlet facility into the creek to the greatest extent possible. The tunnel is evacuated using external portable pumps.

WHY DEWATER THE TUNNEL?

There are multiple factors that helped inform the decision to initiate a largescale maintenance project on the WCT while various tunnel components themselves were being constructed. The primary reason is to maintain the flood diversion and conveyance function. Due to the intrinsic inverted siphon design of the tunnel and the outsized nature of the inlet pond and outlet lagoon, suspended solids from the creek water will settle at inverts of the various tunnel components. Over time, this sediment accumulation can negatively impact the hydraulic capacity within the tunnel and will need to be removed to maintain flood diversion function. As indicated earlier, the tunnel was flooded in 2015 and the initial intent had always been to dewater to inspect and remove sediment after approximately one year of tunnel operation. This combined with the historic rainfall that the region experience was a compelling reason to take on this effort.

A secondary reason was to perform a condition assessment of the tunnel's internal features which can be critical to future maintenance activities within the tunnel. Staff identified several key data needs including photo and Close Circuit Television (CCTV) video documentation of as-built conditions of underwater tunnel components, Light Detection and Ranging (LiDAR) imaging to establish accurate internal dimensions, and sediment deposition profiles after the initial period of operation. This information helped create a composite baseline condition that will help inform future maintenance activities and to serve as bench marks in the future.

Finally, a critical aspect of the tunnel program is to vacate FEMA floodplains for the currently impacted areas in the lower Waller Creek watershed. As-built survey data obtained by a registered professional surveyor is required as part of the Letter of Map Revision (LOMR) application to FEMA for floodplain vacations. Given the significant effort required to access the WCT, performing the survey at a time coinciding with the sediment and debris removal work was considered sensible.

SEDIMENT AND DEBRIS REMOVAL PROJECT PLANNING

The maintenance process for the WCT presented multiple unique challenges, including the need for specialized equipment; time critical windows to coincide with historic climate patterns in the Austin area; and safety constraints due to below-grade confined space working environments. Due to limited resources within WPD's Field Operations Division, WPD partnered with an external contractor, National Power Rodding (NPR), to perform the Sediment and Debris removal project.

Working in a Live Tunnel

One of the biggest challenges encountered during this project was working within a tunnel that was susceptible at both ends to inundation during rain events (Waller Creek at the inlet and Lady Bird Lake at discharge). Working in this environment required constant monitoring and preparedness. This is particularly critical given the nature of the relatively small Waller Creek watershed, whose urban characteristics lead to rapid runoff into the creek with little

Figure 8. Waller Creek Tunnel outlet facility

to no lead time. WPD worked with NPR to establish trigger levels for a Significant Rain Event (SRE) that would require the removal of all personnel and equipment from areas of potential inundation. To reduce exposure to this concern WPD choose to perform the work during February through March, a typically drier period in Austin (see Figure 7).

Unknown Conditions

As mentioned previously, the inverted siphon below-grade design of the tunnel requires that it be filled with water at all times. Due to this, WPD staff was unable to obtain an accurate estimate of sedimentation levels in the tunnel prior to the commencement of the Sediment and Debris Removal Project. In February 2016, WPD staff performed a dive into the bottom of the outlet shaft to inspect sediment accumulation. Due to very limited visibility caused by the silts and clay nature of sediment which tend to get suspended, the dive team made a very approximate estimate of about six inches of sediment accumulation. WPD extrapolated this level of sedimentation throughout the tunnel to provide a best-known estimate of 700 cubic yards of material for the purpose of project deliverables.

Logistical Challenges

The WCT design makes for several logistic challenges that required careful planning and coordination. Primary maintenance access to the tunnel is through the outlet shaft and limited supplementary access is available through three surface access points. During the Sediment and Debris Removal Project, the three surface access points were highly limited due to ongoing active construction projects. Additionally, the outlet shaft access point is located along the shores of Lady Bird Lake adjacent to several major attractions including the Butler Hike and Bike Trail, the Austin Rowing Club, and the Four Seasons Hotel. These prime amenities attract thousands of residents and out of town visitors. Furthermore, the project time window coincided with South by South West (SXSW), one of Austin's flagship music, film and interactive festival that attracts hundreds of thousands of visitors. Several of the events associated with this festival occurred in venues directly adjacent to the WCT outlet, resulting in additional challenges to the project such as addressing noise concerns, increased vehicular and pedestrian traffic, and road closures.

Staging for heavy equipment is highly limited at the WCT outlet and is always at risk of potential inundation (see Figure 8). Therefore, significant planning and coordination were required to ensure that only the necessary vehicles and equipment were brought onsite.

Safety

The safety of personnel working within the tunnel was of paramount importance. NPR utilized Baer Engineering to develop a Health and Safety Plan for the Sediment and Debris Removal Project. The following are key strategies employed to ensure the safety of personnel and equipment working within the tunnel during the course of the project:

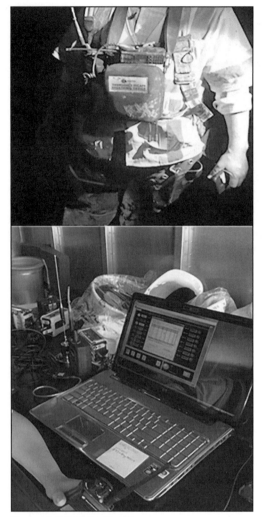

Figure 9. WPD Staff with emergency escape breathing device (top) and air monitoring screen (bottom)

Drowning and Engulfment

The SRE Response Plan was developed to address the potential for rising water within tunnel. The initial stages of the project put personnel in work areas near areas of the tunnel facilities filled with water; therefore, life jackets were required when this hazard was present.

Ventilation

NPR provided a 30,000 (approximately) cubic feet per minute ventilation fan in order to supply fresh air to the tunnel if required. This fan was to be installed at the WCT inlet facility. Once the tunnel was dewatered it was found that sufficient air exchange

occurred without the need for the installation of the ventilation fan.

Tunnel Access

All personnel entering the tunnel were required to attend pre-entry safety training and wear appropriate Personal Protective Equipment (PPE) including harness, hard hat, radio, and an emergency escape breathing device pack as shown in Figure 9. Prior to tunnel entry, a daily safety meeting was conducted during the entire length of the project, the air quality of the tunnel was recorded, and a confined space permit was completed. Personnel and equipment were lowered and hoisted into the tunnel using a crane, rigging, and suspended personnel basket. Communication within the tunnel and to the surface was conducted through two-way radios.

Air Monitoring

NPR, along with their safety sub-consultants, developed a robust monitoring plan that utilized state of the art monitoring equipment to measure and relay below-grade conditions to an above ground station manned by trained personnel. The monitors were daisy chained the entire length of the tunnel and programmed to cascade any point measurement of increased levels of the volatile and other harmful compounds to a monitoring truck located at the top of the Outlet structure. If any of the multiple monitoring devices detected any volatile substances at level higher than OSHA established limits, it activated hazardous condition outlined in the Health and Safety Plan and all personnel were required to evacuate the tunnel until conditions were determined to be safe. Fortunately, no hazardous atmospheric conditions were encountered during the project.

Water Quality

To ensure compliance with federal, state, and local water quality parameters, NPR utilized a series of settling and filtration devices to prevent the discharge of sediment-laden water into Lady Bird Lake. The contractor elected to provide three open top weir tanks and four bag filters and installed them in a series. Limited space at the outlet facility again proved to be a challenge and placing the weir tanks required them to be hoisted by the crane, rotated in the air, and then placed into the correct location on the jobsite.

SEDIMENT AND DEBRIS REMOVAL PROJECT EXECUTION

Initial Setup, Mobilization, and Pumping

The official start to the 75-day project was February 8, 2017. The tunnel was isolated from Lady Bird Lake by closing the slide gates on the lake

Figure 10. National Power Rodding vacuum truck

Waller Creek Tunnel Sediment and Debris Removal Project—Fact Sheet	
Tunnel Characteristics	
Total Length	5,600 lf (±)
Total Volume	22 MG (±)
Diameter	20.5 ft to 26.5 ft
Design Conveyance—100-year flood reoccurrence interval	8,200 cfs
Project Results	
Total Volume Water Pumped	Over 60 MG
Tons of Sediment Removed	3200 (±) tons
Total Truck Trip	Over 300
Rain Events	7

Figure 11. Waller Creek Tunnel sediment and debris removal project fact sheet

side of the 42-inch tunnel recirculation pipes and establishing a perimeter fence around the facility. This was followed by mobilization of equipment to evacuate 1.6 million gallons (MG) of water from the outlet lagoon to the top of the outlet shaft 414 (Mean Sea Level). This allowed for the following activities to occur: clearing the sediment from the lagoon floor; installing a secondary fall protection fence around the outlet shaft; placing secondary isolation of Lady Bird Lake on the tunnel side of the 42-inch recirculation pipes including inflatable plugs and blind flanges; and mobilizing water quality equipment, a crane, and larger pumps to evacuate the larger volume of water within the tunnel itself. Tunnel contents were pumped directly into the lake until approximately ten percent of the water remained. This sediment-laden portion of the water was pumped through a filtration system to prevent sediment discharge into the lake. A recurring theme during these activities and for the duration project was the relatively minor rain events that caused the tunnel to be filled and then re-evacuated multiple times.

Sediment Removal

Once the water was removed from the tunnel, the sediment removal process began. Given the thick nature of the sediment-laden water, specialized vacuum trucks were used (see Figure 10). These trucks came equipped with compressors to provide additional suction lift, which allowed the mud-laden water to be removed from the bottom of the outlet shaft approximately 41 feet below the floor of the outlet lagoon. An almost immediate revelation was the underestimation of the amount of sediment accumulated within the tunnel. The outlet shaft alone had more than three to four feet of accumulation, making it unsafe to lower any kind of earth moving equipment, such as skid steer loaders, or personnel. This initial obstacle needed to be overcome prior to determining the amount of sediment present in the rest of the tunnel. Determining the amount of sediment was crucial to executing the rest of the project. The team

decided to utilize remotely operated closed-circuit television (CCTV) cameras at the three upstream access points to determine the extent of the sediment levels. Access through the upstream end of the tunnel at the Waterloo Inlet allowed NPR to conclude that the upper third of the tunnel did not have substantial sediment deposition.

Additional trucks were brought in to increase the rate of sediment removal from the outlet shaft. Crews worked 12-hour days utilizing additional pumping hoses to pull the sediment out of the tunnel. In total, it took the team nine days to remove the mud from the outlet shaft alone. Skid steers fitted with blades were then lowered into the shaft and used to push the sediment from within the tunnel to the outlet shaft (see Figure 12). Efficiencies increased rapidly once the outlet shaft was cleared, and during peak production modes, over eight vacuum trucks were being used to transport over 30 truckloads per day. Overall 323 truck trips were made and approximately 3,200 tons of sediment and debris was removed from the tunnel (see Figure 11). In all, sediment removal took more than 16 days excluding days lost due to rain events and re-evacuation of the water.

The majority of the tunnel sediment was hauled to and deposited at a local compost company. Some of the initial loads that contained floatable trash were taken directly to the landfill.

As sediment was removed from the lower end of the tunnel, staff was able to visibly observe that the sediment was being deposited in layers as stratification was evident in the cross sections of the sediment. It also became clear that the sediment had shifted during the numerous rain events that the tunnel received in a dewatered condition. WPD staff utilized the debris stains on the walls of the tunnel to measure the extent of the tunnel and found that deposition was occurring throughout the tunnel, with the greatest extent occurring nearing the 4th Street CSI connection and the tunnel outlet.

Figure 12. Skid steers pushing sediment in the tunnel

With the relatively dry weather window rapidly closing, a decision was made by the project team to remove as much mud as possible fully knowing that there would be a certain minimal amount of sediment left behind. The focus was shifted to ensure all the required inspection elements of the project were completed.

Project Inspections

As mentioned earlier, a key goal for this project was to perform a range of inspections within the tunnel. This included initial TV inspection of the tunnel utilizing a CCTV camera mounted on an XUV vehicle with modified lighting; a team of outside surveyors and specialist brought in to establish an as-built hydraulic condition of the tunnel required for the final Letter of Map Revision (LOMR) submittal to FEMA; a firm specialized in Light Detection and Ranging (LiDAR) to gather specific as-built conditions of the tunnel's interior diameter (see Figure 13); and finally an engineering firm hired to perform forensics on the tunnel's concrete liner as part of the one-year tunnel warranty. These efforts required significant coordination to enable all the different aspects to be completed. Weather proved to be the biggest challenge during the inspection phase. The tunnel needed to be free of water and majority of the sediment to accommodate the inspections. It took up to four days after a rain event to get the tunnel into a condition suitable for inspection. Given some

of our specialized forensic inspection team members were traveling from out of town this created some challenges to find a weather window to ensure that they would be onsite when they could perform their work. Originally the inspections were to be completed sequentially to minimize interference among the different groups and limit the number of personnel in the tunnel at any one time. To meet favorable weather windows, the schedule for all the inspection activities was greatly compressed to reduce the risk for a rain event interruption. Space constraints at the top of the Outlet structure, limited the ability to locate equipment and as a result, all tunnel access for equipment and personnel was provided by a single crane. Putting all the inspectors into the tunnel in the morning required close coordination amongst all parties to allow for the proper sequence of staff and equipment to enter. An initial attempt to provide access to the multiple inspection teams resulted in major coordination issues including ability to lower specialized equipment and movement of teams in and out of the tunnel. Due to this, the team developed a more strategic entry schedule for various groups to enter the tunnel that accounted for staging time, use of specific equipment inside tunnel such as lifts and number of personnel inside the tunnel at any particular time (maximum of 12). Ultimately all of the data collection goals were met.

RESULTS AND DISCUSSIONS

Rain Events

Unpredictable rain events proved to be the most significant challenge for the project. In total, there were seven rain events that required complete work stoppage during the project. Even though this was generally consistent with historic rain pattern, very small events with less than a quarter-inch of precipitation caused multiple days of impact to the project schedule. Although the project was scheduled during a historically drier time of year for the Austin area, and the amount of rainfall did not exceed the expected amount, the frequencies of the rain events still proved to be problematic. The project team immediately realized that the best strategy would be determining a milestone that could be reached by the next wet weather period.

Sediment Levels

WPD's initial estimation based on the diver's data proved to be well short of the amount of material found within the tunnel. One of the contributing factors to the sediment levels within the tunnel could be the historic rainy period that occurred from May 5, 2015 to the commencement of the project in 2017. This timeframe included the wettest 12 periods on

Figure 13. LiDAR Image of WCT with sediment (Courtesy AG&E Structural Engenuity Consulting Structural Engineers)

record for the Austin area with the city receiving over 59 inches of precipitation.

Special Events

Elements outside the control of the team did not allow for special events beyond the SXSW festival calendar to be included in the project planning phase. In the future, WPD will look to minimize potential conflicts with adjacent stakeholders both to be a better neighbor and increase project efficiencies.

CONCLUSIONS

Based on lessons learned from this initial Waller Creek Sediment and Debris Removal Project, WPD discovered that the two greatest challenges with the project consisted of the rain events and adjusting to the magnitude of sediment accumulation material found within the tunnel. In the future WPD will be implementing strategies to further minimize the risks encountered in debris removal efforts. This includes scheduling the project in late summer, as this time period coincides with the historically driest period of the year for the Austin area a period drier than early spring and in addition does not conflict with any of the major events held in Austin. Additionally, WPD will look to utilize different technology such

as surface sonars and submersible drones, to better understand sediment levels in the tunnel prior to future sediment and debris removal projects. This will allow for better resource allocation plans to reduce sediment removal times. Furthermore, WPD is in the initial phase of exploring alternatives to prevent minor storm events from entering the tunnel and thereby reducing the amount of sediment accumulation. In addition, WPD is exploring systems in the Outlet shaft that will help resuspend settled solids that can enhance the self-cleaning ability of the tunnel during major storm events. With a few key exceptions, project went very well and will be replicated in future projects including the safety plan, the use of modified vacuum trucks, and the Significant Rain Event plan.

ACKNOWLEDGMENTS

The City of Austin—Watershed Protection Department would like to thank all parties and entities that contributed to the success of the Waller Creek Tunnel Sediment Debris Project including: National Power Rodding Inc., Baer Engineering, Oscar Renda Contracting Inc., S.J. Louis Construction, Inc., City of Austin—Public Work Department, and City of Austin—Parks and Recreation Department.

Prefabricated Tunnel Pipe Liners—A Modern-Day Approach to Efficient Installation

Jesse Schneider and Brian Kelley
Kelley Engineered Equipment, LLC

ABSTRACT: Many modern tunnels require the installation of prefabricated tunnel pipe liners after the excavation and initial ground support are complete to yield the long design life that clients require. Since excavation cost increases significantly with tunnel diameter, owners and designers naturally want to minimize the tunnel diameter for the required final liner size. This often leads to limited annular space between the excavation diameter and the final liner, which requires the use of specialized equipment to carry and position the liner pipe in its final location. Outlined in this paper are various methods and examples of equipment that can be utilized for installation of prefabricated tunnel liner pipe.

INTRODUCTION

Prefabricated tunnel pipe liners serve many purposes, usually providing a smooth inner lining as well as increasing design life for the tunnel. Although some tunnels do not require a second pass lining, this paper will focus on the process of installing the final liner pipe after the tunnel is driven and the initial or permanent ground support has been completed.

One common objective for every party with stake in a tunneling project is to minimize the excavation diameter for a given final liner inside diameter. By evaluating the current technology available for installing prefabricated tunnel pipe liners, the designer, owner, and contractor must determine the required excavation diameter based on a multitude of project constraints.

The focus of this paper is primarily for hard rock and segmentally lined tunnels after a tunnel boring machine is completed with its drive. Some discussion about other conventional tunneling methods and their corresponding pipe installations will be discussed as well. It is assumed that primary ground support is already installed in the tunnel, and the tunnel pipe liner is not acting as a construction ground support element.

TYPES OF TUNNEL PIPE LINERS

The scope of this paper is to evaluate pipe installation methods, not necessarily covering liner types in detail. This section is merely intended to highlight the different known liner types to the authors, not to select liner types for a given project.

Steel Pipe

One of the most common final lining materials is steel, mainly due to its superior mechanical properties. Some advantages and disadvantages of the pipe liner types are as follows.

Pros:
- Excavated diameter is minimized since final liner wall thickness is very small
- High tensile, compressive strength, and high ductility

Cons:
- Welding of joints can be costly and time consuming
- Potential corrosion, or other required measures to prevent corrosion
- Manufacturing cost is high for larger sizes

Precast Concrete Pipe

Precast concrete pipe is another common material used for final lining. With its own set of challenges and benefits, listed here are a few.

Pros:
- Low material cost
- Long design life

Cons:
- Liner weight is generally higher
- Excavation size requirement is larger since wall thickness is larger.

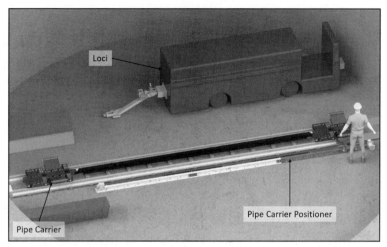

Figure 1. Bottom of shaft pipe carrier positioner

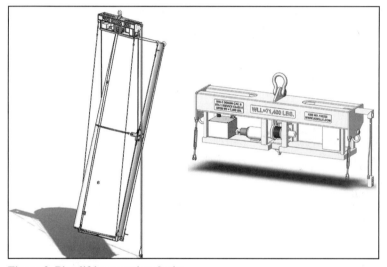

Figure 2. Pipe lifting rotation device

Centrifugally Cast Glassfiber Reinforced Pipe (HOBAS®)

HOBAS® pipe is a trademarked brand of HOBAS Pipe USA. It is a unique product compared to steel and concrete. Listed below are some of the traits for its usage in tunnels.

Pros:
- Lightweight
- Corrosion resistant

Cons:
- Maximum diameter 126"
- Maximum length 20 feet

PIPE INSTALLATION SYSTEM COMPONENTS

Pipe Carrier Positioners

After pipe is delivered to the site, the next step to install a second pass tunnel liner is to get the pipe delivered to and positioned on the pipe carrier. For some projects, this is easier than others. Often a shaft is used to access a tunnel. As with tunnels, shaft excavation diameter is generally minimized which can lead to inadequate space for the prime mover (loci, loader, etc.). In the absence of a tail tunnel, a pipe carrier positioner can be utilized to move the pipe carrier independent of the prime mover. See Figure 1 for an example of a chain driven system that moves the pipe carrier into the tunnel. The prime mover must be shifted to the side with a car passer, or hoisted out of the shaft.

Pipe Lifting Systems

A common challenge with shaft accessed tunnels is the delivery of the pipe down the shaft. With competing objectives of pipe length and shaft size for economical purposes, it is common that the pipe length that is desired for a project is too long to be transported down the shaft in the final horizontal position.

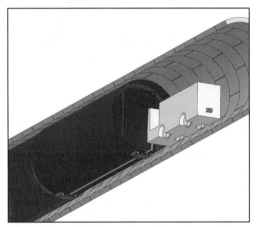

Figure 3. Rail dolly pipe carrier

For this situation, a lifting system capable of tipping the pipe nearly 90 degrees can be utilized as shown in Figure 2.

These devices can be equipped with integral hoists for rotating the pipe. However, if a crane with multiple hoist lines is available, the second hoist line can be used to rotate the pipe above and below ground, eliminating the need for the integral hoist shown in Figure 2.

Rail Dolly Pipe Carriers

The Rail Dolly Pipe Carrier without lifting and side shift capabilities, is the simplest form of transportation of pipe in tunnels. These carriers, as shown in Figure 3, are simple and efficient for transporting the pipe into the tunnel.

Either one or multiple pipes can be carried in the tunnel at a time, depending on loci capacity, tunnel grade, and placement methodology. These can be used in conjunction with a needle beam style (or similar) pipe placer to deliver pipe to its final general location before having final alignment adjusted with the pipe placer. Figure 4 shows a Rail Dolly Pipe Carrier that can deploy wheels to run on rock or concrete invert. This allows rail to be removed as the pipe is being placed.

Another type of rail dolly carrier engages on the inside of the pipe and uses a counterweight to offset the pipe weight as shown in Figure 5. This style can also allow for removal of rail, since the rail wheels are not below the pipe being carried.

Rubber Tired Pipe Carriers

Most TBM tunnels utilize rail for haulage which makes rail transportation of pipe the best choice. In road header, drill/blast, and New Austrian Tunneling

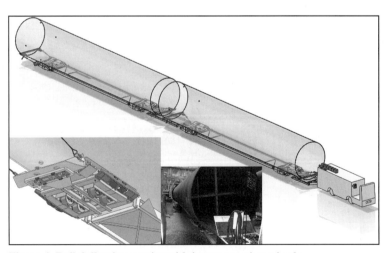

Figure 4. Rail dolly pipe carrier with invert running wheels

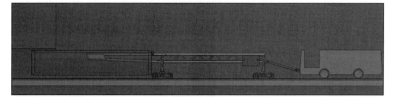

Figure 5. Rail-mounted pipe carrier with inside attachment

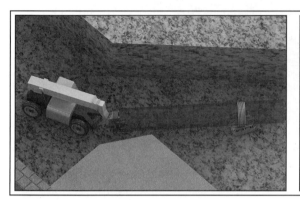

Figure 6. Rubber tired carrier/placer

Figure 7. Needle beam style pipe placer

Method (NATM) style tunnels many times rail is not used. In this case, a rubber tired option is preferred to avoid the temporary installation of rail through the entire tunnel. Rubber tired pipe carriers become more complicated and therefore more costly. The primary cost driver here is steering capability, which is required without rails for guidance. Steering generally is achieved with radio remote or umbilical controlled hydraulic systems. Also, rubber tires are much larger in diameter and width than a rail wheel with equivalent capacity. The wheel size therefore increases the overall pipe carrier size.

Needle Beam Pipe Placers

When pipe transport duration to the heading becomes critical, a separate pipe carrier and pipe placer can be used to accelerate the cycle time. This needle beam style pipe placer in Figure 7 is an example of a pipe placer that stays in the pipe at the placement zone.

Generally, this style of pipe placer will "walk" inside the pipe from one pipe to the next. A needle beam can be configured in a manner that allows the legs to pass over an internal obstruction on the pipe, such as a rebar cage. This style is shown in Figure 8.

A pipe rounder can be incorporated in these placers if required by the roundness specification of the project or for fit up needs.

Combination Carrier/Placers

When transport cycle time is not as critical, a Combination Carrier/Placer can be utilized. The example shown in Figure 9 is a low profile, rail mounted carrier that uses a loci to push the pipe in the tunnel. Hydraulic arms provide raise, lower, and side shift capabilities. Polyurethane wheels that contact the pipe are hydraulically driven to allow rotation of pipe in the tunnel for backing bar alignment, and/or grout port alignment.

The pass-through style shown in Figure 10 is beneficial if internal picking is possible, with extraction of the front dolly underneath the pipe as part of the installation process. Annular space for the rail is the only space constraint with this type of carrier, which is its primary advantage.

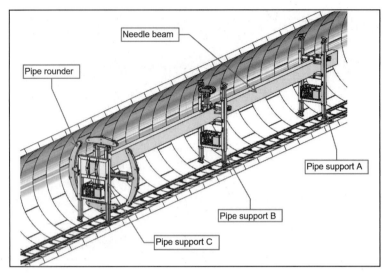

Figure 8. Needle beam style pipe placer

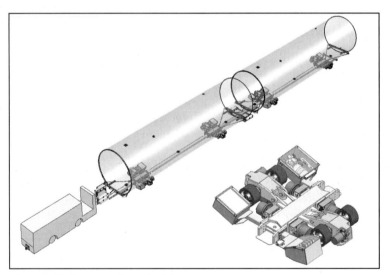

Figure 9. Rail-mounted carrier/placer

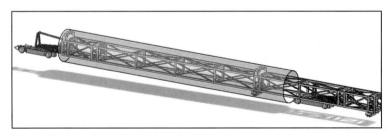

Figure 10. Pass-through style carrier/placer

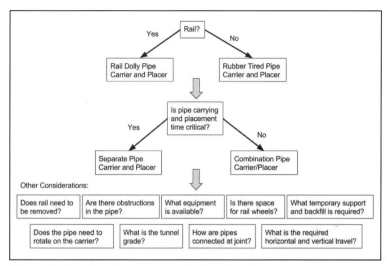

Figure 11. Pipe installation system decision flow chart

Table 1. Tunnel diameter to pipe diameter comparison

Contractor	Project Name	Tunnel Inside Diameter (inch)	Pipe Outside Diameter (inch)	Diametral Difference (inch)	% Pipe Diameter / Tunnel Inside Diameter	Type of Pipe	Carrier/Placer Type
Frontier-Kemper	Seymour-Capilano	149.6	120.8	28.8	80.74%	Steel	Rail mounted carrier/placer
Michels/Jaydee/Colluccio	Bay Tunnel	146	110	36	75.34%	Steel	Rail mounted carrier, Needle beam pipe placer
SAK Constructors	Lemay Tunnel	135	112	23	82.96%	Concrete	Rail mounted carrier/placer
Kiewit/Shea JV	Rondout	218	192	26	88.07%	Steel (w/rebar)	Rail dolly pipe carrier, Needle beam pipe placer
Traylor/Shea JV	San Vicente	126	104.5	21.5	82.94%	Steel	Rail dolly pipe carrier w/ invert running wheels, Needle beam pipe placer
Aecon/McNally	Port Mann	102	84	18	82.37%	Steel	Rail mounted carrier/placer
	Average =	146.1	120.6	25.6	82.07%		

HOW TO SPECIFY A PIPE INSTALLATION SYSTEM

The process of choosing and specifying a pipe installation system can be complex and multi-faceted. The following decision tree is meant to be a tool that will determine the type or style of pipe carrier and placer needed.

MINIMIZATION OF EXCAVATION DIAMETER

To determine the industry standard final liner diameter to excavated tunnel diameter ratio, sample case histories have been tabulated in Table 1.

The average Diametral Difference between the Tunnel Inside Diameter and the Pipe Outside Diameter for these samples is 25.6 inches. This is equivalent to an 82% ratio of the Pipe Diameter to the Tunnel Inside Diameter. While this sample set is small, it is indicative of approximate industry standard sizing when designing an internal liner.

The technology exists to carry and place pipes with low profile rail wheels. As the wheels become smaller, higher alloy wheels and higher capacity bearings are required, which result in higher cost. Smaller wheels are also more likely to de-rail. However, for overall project economic purposes, the lower profile carrier can lead to increased tunnel area utilization which is better overall.

CONCLUSIONS

The efficient installation of prefabricated tunnel pipe liners is possible with the use of modern equipment and should be utilized to minimize the tunnel excavation diameter for a given final liner pipe inside diameter. If excavation size can be reduced, the total cost of tunnels can be reduced. Tunnel designers and owners can and should consult with tunneling contractors and equipment designers to choose tunnel sizes for the required final lining size needed.

ACKNOWLEDGMENTS

The authors would like to acknowledge Matt Swinton of Kiewit Underground for his cooperation and information provided related to Kiewit projects with prefabricated tunnel liners.

Pre-Excavation Grouting at the Hemphill Site–Atlanta WSP Tunnel, Atlanta, Georgia

Adam L. Bedell, Konner Horton, and Don Del Nero
Stantec Consulting

Brian Jones
City of Atlanta, Department of Watershed Management

ABSTRACT: The City of Atlanta's Water Supply Program Tunnel Project, comprises a 24,000-ft long, 13-ft diameter, hard rock tunnel. A complex aspect of the project involves connecting five blind bore shafts to the tunnel in a location close to current drinking water reservoirs. During the supplemental geotechnical investigation following initial contract award, results from additional borehole geophysics were reviewed. Unfavorably oriented fracture sets forced a change in the original pre-excavation grouting program designed for that site. Real-time grout monitoring and geophysical data were compared to provide assurance that the program as designed correlates with the in-situ ground conditions. Following grouting, results were modeled, and all grouting data was reviewed to determine if grouting was complete.

PROJECT BACKGROUND

The current water supply program operated by the City of Atlanta's Department of Watershed Management (DWM) consists of four aged raw water pipelines, one of which dates to 1893. Based on previous assessments completed by the DWM, the entire water system is at, or will soon reach, its recommended useful life. As such, the City acquired the Bellwood Quarry in 2006 with the intention to create a water storage facility with a volume of approximately 2.4 billion gallons to serve approximately 1.2 million people.

The project location is shown in Figure 1, which is generally in the Northwest part of downtown Atlanta, Georgia. The overall project has been divided into two phases. The Phase 1 project connects the Quarry and the Hemphill Water Treatment Plant (HWTP), and the Phase 1 Extension project connects the HWTP to the Chattahoochee Water Treatment Plant (CWTP) and the Chattahoochee River. A 24,000-ft tunnel with a finished diameter of 10 feet connects all three elements. The HWTP location is also where the City's two most proximate drinking water reservoirs are located.

Procurement Method

The Construction Manager at Risk (CMAR) model was used as the overall project contracting method. Specific to this procurement method involves producing a pricing set of design drawings, specifications, and data and baseline reports, which represent a partial design (typically 60 to 70 percent is used) to allow the CMAR to start pricing the work as design progresses towards final design. As the design evolves and changes are made, assumptions in bid pricing from various subs to the CMAR are reflected in various bid stages.

As changes to the evolving design are made and submitted to the CMAR and Owner, specific details are delineated and given to all parties describing changes in the design. This allows for various sub-contractors bidding on certain, niche parcels of work (as released by the CMAR) to either adjust their price based on the revised design or keep their submitted prices based on their perceived risks, design detail revisions, and overall effect of design package revisions.

Hemphill Site

The Hemphill site is the point on the project where the conveyance and storage parts of the project needs meet. During the initial phase of the subsurface investigation, a pump station shaft was intended to provide transmission of raw water to and from the tunnel to the HWTP as well as a construction shaft to provide access during construction. During this period, it was communicated from the City that any disturbance from the excavation processes to the existing unlined reservoir was an unacceptable consequence. A pre-excavation grouting program was designed as a risk mitigation tool to reduce the potential for any communication between the reservoir during either of the excavation shafts.

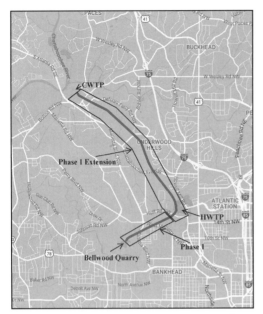

Figure 1. General project location map showing the location of the HWTP in relation to the rest of the project

Schedule and blasting restrictions steered the design away from traditional shaft sinking techniques towards using blind boring methods. Blind bores allowed shaft excavation at the Hemphill site to be decoupled from tunnel excavation, subsequently removing it from the critical path of the project. A result of this change in connecting the surface components and the tunnel was that the location of the pumps now needed to be a lot closer to the tunnel. The size and breadth of the pre-excavation grouting program was reduced in accordance with the change in shaft size selection, construction methodology and arrangement.

Geologic Conditions

The Atlanta Water Supply Program is location within the Piedmont physiographic province. Many underground components for the project are within a single geologic unit, the Clairmont Melange. Characteristic of the mélange are interbedded biotite-quartz-feldspar schists and gneisses, with minor granitic lenses. Foliation is very well developed and highly contorted wrapping around the granitic lenses while often displaying a sheared texture. Strike and dip of foliation commonly varies by 35 degrees and the mélange is locally described to as "consistently inconsistent."

During the initial subsurface investigation, boring RWB-15 was drilled based on accessibility while along the tunnel alignment at the Hemphill site. Two

other borings, RWB-25 and RWB-26, were drilled based on the original proposed pump shaft and construction shaft locations. Boring RWB-15 showed extremely poor rockmass conditions, as the hole was reamed 6 times due to stability issues. Packer testing was not performed nor were down hole borehole geophysics due to concerns of lost tooling. While the other borings, RWB-25 and RWB-26, showed some signs of similar geologic conditions (increased weathering along fracture planes, decreased RQD within discrete intervals, and increased permeability values), nothing observed nor tested was as pervasive or severe as RWB-15, which was proximate to the tunnel while the others were more distal.

As the design evolved from traditionally excavated shafts (drilling and blasting) to blind bores, it was determined that the rockmass, close to RWB-15 (as suggested poor ground conditions) warranted additional borings. Additional drilling occurred, proximate to RWB-15, and along the tunnel alignment close to the blind bore shaft locations. Borehole stability proved not to be an issue with the additional borings and packer testing and downhole geophysics were performed. Rock cores collected exhibited characteristics of the lineament hit by RWB-15, but fracturing was less penetrative while packer testing results indicated similar to slightly less permeabilities as shown in RWB-25 and RWB-26.

DESIGN AND GROUTING CONSIDERATIONS

There were two underlying premises behind the pre-excavation grouting program. The first was the decree from the Owner that under no circumstances shall the existing City water reservoirs be affected by blind drilling processes. As the #2 reservoir was constructed prior to the 1930s and less than 100 feet away from the shafts, it was unlined and while all information from the subsurface investigation suggested that there was not a connection to the local water table, a risk mitigation measure was required. Second, as common with blind bore shaft sinking techniques, the area around the shaft is traditionally grouted to lower the potential for catastrophic fluid loss. During excavation, the shaft is filled with water to maintain hydrostatic balance and provide for a stable excavation as ground support is not installed. As the large diameter reaming process proceeds to the target elevation, water is maintained in the drilled shaft to keep the excavation open.

During supplemental drilling, the pricing set of documents needed for soliciting bids for the work by the CMAR were then issued and the tunnel contractor, Atkinson-Technique JV solicited bids for the pre-excavation grouting work package. The initial pre-excavation grouting layout for the blind bore shafts were used. Geophysical results from RWB-25

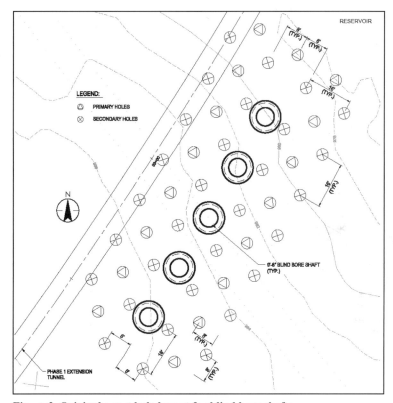

Figure 2. Original grout hole layout for blind bore shafts

and RWB-26 suggested moderately open foliation joints (common within this unit) and that inclined grout holes oriented towards the TBM tunnel would be sufficient to intersect open, variable (relatively flat lying) foliation joints.

Once the data analysis was complete, it was determined that a change to pre-excavation grouting program was needed. Geophysical information from supplemental borings HDB-2 and HDB-3 indicated two open, high angle joint sets with apertures ranging from two to four inches, in conjunction with open foliation joints along the "inconsistent" foliation. As designed, the existing pre-excavation grouting program had a high likelihood of missing the newly identified joint sets.

Grout holes were to be drilled at a bearing of 260 degrees at 10 degrees from vertical. This orientation provided the highest probability of hitting both newly identified high angle joint sets (Set 2 and Set 3) while also targeting the known foliation joint set, as shown on Figure 3.

Primary grout holes were spaced at 16 foot centers with secondary holes split spaced in between them. This pattern results in 8-foot spacing between primary and secondary grout holes. Grouting started with primary grout holes and once complete, drilling

and grouting of secondary grout holes occurred. Each row had targeted grout elevations from which stages below would be grouted under pressure. This creates a block of treated ground that surrounded the future blind bore drilled shafts. One row near the tunnel alignment contained vertical holes and was drilled and grouted last to help seal off the grouted block as shown on Figure 4 and Figure 5.

Grout Details

Pre-excavation grouting work would be paid for as unit rates per bid quantity estimates. Estimated drill footages for both overburden and rock, estimated grout pump times, cement bag estimates, and grout stages were provided. Grout mixes provided were by volume starting at 2:1 and progressing to a maximum 0.5:1 water to cement ratio. Refusal criteria was 0.25 gpm or less for 5 minutes at the full grouting pressures, which were 0.8 psi per foot from the point of injection. For the program, Type III cement was required. Bentonite was not used in any of the grout mixes.

The steps for thickening the grout mix were straightforward per grouting industry standards. Pressure and flow rate were tracked to determine if a

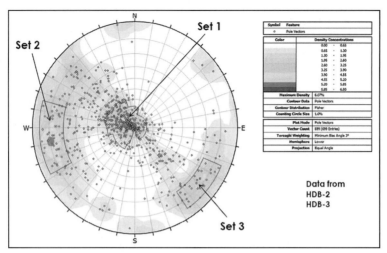

Figure 3. Revised stereoplot from additional analysis from HDB-2 and HDB-3. This was produced after the pricing set of drawings had been issued for bid. Joint sets 2 and 3 were the newly identified joint sets.

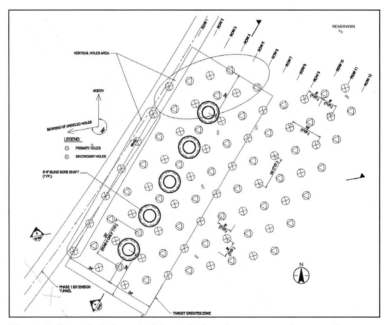

Figure 4. Grout hole layout of the redesigned pre-excavation grouting program

mix change was warranted. If pressure increased while flow rate decreased, then the grout mix stayed the same as the grouting system was functioning properly. When the pressure reached the target injection pressure and the flow rate was below the refusal injection rate, refusal on the stage was called and the packer assembly was moved. If flow rate was constant and pressure did not increase, the grout mix would be stepped down and thickened and injection would continue at the specified grout mix until changes in pressure or flow rate were observed. Typically, mix changes were made after 100–200 gallons of grout were injected or if it was immediately apparent that the interval was open and a thicker grout would be needed.

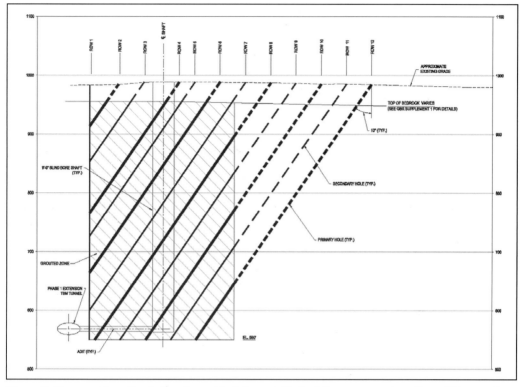

Figure 5. Cross-section of redesigned grout hole layout. Dashed lines represent portions of the grout holes that would not be grouted under pressure.

CONSTRUCTION

Hayward Baker mobilized to the Hemphill Site in April 2016 with drilling scheduled to commence in May. Overburden casing was advanced 5 feet into rock to ensure the pvc casing was properly socketed into competent rock. Hayward Baker proceeded to install all surface casing for primary grout holes. Grouting was staggered between primary rows as to lower the potential for compromising grout holes that had not yet been grouted through communication from an active grout hole. Production drilling began in July 2016. Grouting of primary grout hole rows started in July 2016 and ended in August 2016. Drilling production secondary grout holes started in August 2016 and grouting all secondary holes and Row 1 (which was designated to be the last row drilled and grouted) concluded in October 2016. In total, 84 grout holes were drilled between primary and secondary portions of the program.

Hayward Baker utilized their proprietary grout monitoring equipment during all grouting. Cement was stored in bulk and then batched and mixed onsite. Grout was delivered to grout carts, small highly mobile tracked equipment that allowed for quick grout injection at various specified grout stages.

The main challenge associated with the program was associated with the ground. It was known that the ground was highly fractured around RWB-15, but conditions around HDB-2 and HDB-3 were better and they were off the main lineament. With packers traveling along the hole, small pieces of rock would bridge the grout hole. This would prevent the packer from any further traverse towards the targeted depth. This was overcome through either re-drilling the hole and removing the obstruction or by grouting the interval and then re-drilling the interval.

Grout hole drilling was the critical path item of the project. Drilling hours were extended to allow for 2nd shift drilling to reduce the potential for grouting downtime, i.e. the grout carts have nowhere to go. During production grouting, this only occurred once.

IS GROUTING COMPLETE?

Towards the completion of secondary grouting, the natural question asked was whether grouting was considered complete. Schedule impacts concerning other sub-contractors was at risk as there were two other sub-contractors who were scheduled to complete work prior to North American Drilling arriving onsite to begin blind bore shaft drilling. At this point, direct evidence of grouting efforts was not available. The decision was based on numerical grouting results and analysis. The following breaks down what grouting factors were analyzed.

Grout Takes

The overall grout take for the program was about 53,000 gallons. Primary grout holes took approximately 35,000 gallons of grout and the secondary grout holes took about 18,000 gallons of grout.

The secondary holes took about half of the grout that the primaries took. This reduction in grouting quantities follows the trend of what one would want to see-a progressive reduction in grouting quantities as you traditionally progress past primary grout holes through subsequent grouting.

Grout Curtain

The grout curtain, as identified by 2 rows of grout holes around the perimeter of the drilled grout holes, took 68% of the total grout injected. This curtain is comprised of both secondary and primary grout holes. This was judged as reasonable as the holes are on the edge of the area and not within the middle. Grout injected along the edge is not confined and will travel outside the treatment area as far as injection pressure, fracture aperture, and cement particle size will allow.

Location of Future Blind Bores

The cross-section of the treated area from around the future blind bores was looked at in relation to grout injected. Grout takes on 10 feet of either side of the blind bores along the inclined grout holes, including the area within the blind bore (~30 ft. total) were assessed. Grouting data from grout stages that fit within this window around the blind bores was analyzed to see grout takes in the immediate vicinity of the bores, not the area overall. Of the total grout injected over the treatment area, the area immediate to the blind bores took only 29% of the injected volume. Of the grout injected proximate to the blind bores, 68% was 2:1 by volume, 17% was 1.5:1 by volume, 8% was 1:1 by volume, and 7% was 0.75:1 by volume. The percentages of mixes used over the entire treated area are within 1–2 percentage points of the values just listed for the 30 ft. area around the blind bore shafts.

Packer Testing Data

The packer testing data from HDB-2 and HDB-3 reflect most of the permeabilities in the range of 1×10^{-5} cm/sec to 1×10^{-7} cm/sec, with the remaining zones around 1×10^{-4} cm/sec. Typically, permeabilities lower than 1×10^{-5} cm/sec are considered

not groutable. It was judged that the data reflects this as just under 70% of all the grout injected was 2:1 by volume. The thicker mixes pumped reflects the open fractures observed from the borehole geophysical results. Typically, when 0.75:1 by volume was injected, refusal occurred quickly, as one would expect.

Grout Model

All grouting data was compiled into Civil3D, and modeled, as presented in Figure 6. The first graphic is only primaries, the 2nd graphic only secondaries, and the 3rd graphic is the composite. The blind bore shafts are shaded all the way down. What was observed is that the models correspond the grouting data. Following just primaries, large grout takes intersecting with the blind bore shafts was not observed. All the larger grout takes are within the treatment zone, but distal to the blind bore shafts. Following just the secondaries, there are few large to moderate grout takes, but as with the primaries, these are within the treatment zone, but distal to the blind bores. The composite log is judged to demonstrate good overall coverage of the grouting program.

Correlation

In some instances on the grout logs, zones were observed where takes were slightly larger (~180 ft, ~250 ft, and ~ 340 ft below ground surface). These were interpreted as the foliation joints identified in the field investigation. When looking at individual hole grout takes, there were sometime depths that would correspond between holes, but never really across more than 2 holes or so. Also, if a zone at 400 ft on row 4 took 1,500 gallons, the holes around it (both primary and secondary) were reviewed to see if there were any corresponding elevated values (even if not as large). Similarities in grout take volumes was not readily observed within the area proximate to the blind bores. This condition is also not apparent from the model.

iGrout Logs

Lastly, when looking at the Hayward Baker's iGrout logs, a large portion of the time is spent achieving refusal. At thinner mixes, this is when pressure filtration takes over and the water is squeezed out of the grout and the finer fractures are filled. All the logs from the area proximate to the blind bore shafts reflect proper refusal without pre-mature thickening of the grout mix or poor injection trends (the trends reflected are what they should be—high initial rate of grout injection with flow rates slowly dropping while pressure remains constant until the refusal criteria is met).

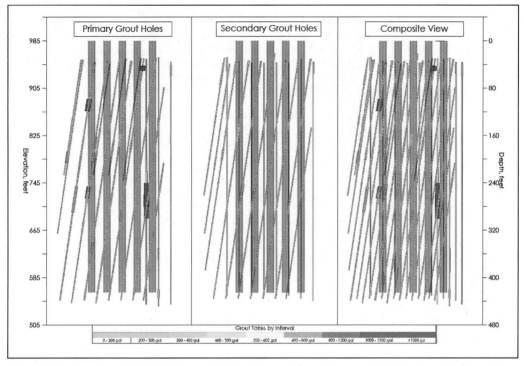

Figure 6. Simplified model depicting grout takes observed for primary grout holes, secondary grout holes, and a composite view

Is Grouting Complete?

The rational for initiating a tertiary grouting program was not observed. One could inject more grout into the ground (you can always inject more), but the program would be at a point of diminishing returns. Also, based on the overall grout takes, and the grout takes around the blind bores, there was not data available that suggested that there could have been increased take around the blind bores. It was judged that the data set reflects a grouting program implemented as designed. Grout takes decreased between primary and secondaries, there were a few large takes (all outside the blind bore area), but overall the takes were not large, and the amount of thinner mix used reflects the permeabilities of the rockmass from the borings. The data suggests good encapsulation around the treatment area with grout penetration and many of the grout holes communicated to one another. It was judged that impacts to the blind bore construction schedule and the additional cost for a tertiary grouting program are greater than a small benefit from additional grouting that may be gained.

FIELD VERIFICATION

Hayward Baker demobilized offsite in October 2016. North American Drilling began reaming blind bore shafts number 1 and number 4 in February 2017. Large diameter reaming to the final diameter began in May 2017 and was completed in July 2017. During the large diameter reaming process, drill cuttings are recirculated to the ground surface. Once on the surface, cuttings are transported to sedimentation ponds via the drill return water where the large and small particulates settle out of suspension and water is then pulled from the ponds for further use.

During the large diameter reaming process, drill cuttings were inspected. Numerous, irregular shaped pieces of grout have been picked from the drill cuttings pile. In addition, fluid levels in the shafts and at the reservoir were monitored during drilling. Both the reservoir water level and the water level in the blind bore shafts were constant during shaft reaming. The presence of grout pieces in the blind bore drill cuttings and stable water levels in both shafts and the reservoir further demonstrate that grout penetration was sufficient during the injection process.

CONCLUSION

An evolving design coupled with interpretation of supplemental geotechnical information required a change in the design pre-excavation grouting program as originally submitted to the CMAR for bidding purposes. Grouting results were quantitatively analyzed as well as modeled. Grout logs indicated proper grout injection and grout-rockmass interaction. Prior to blind boring operations, the grouting program was considered complete. During shaft reaming, grout chips and pieces have been recovered from the drill cuttings pile and observed stable water levels in both the shafts and the reservoir substantiate grout penetration into the rockmass and provide assurance that the grouting program was implemented as designed.

Currently at time of press, the tunnel has been excavated by the 5 blind bore shafts and 4 of the 5 blind bores have been excavated. During mining, some of the grout holes were observed in the crown of the tunnel, but grout was not observed along fractures. Following 4–6 weeks after excavation, calcium was observed along fractures within the tunnel. Calcium leaching from the grout injected from the surface has been observed in the tunnel by the location of the blind bore adits, providing further substantiation to the effectiveness of the program.

Hard Rock Tunnel Design Improvements over 42 Kilometers Spanning Multiple Projects

Alston M. Noronha and Mark H. Bradford
Black & Veatch

ABSTRACT: Over the past eight years, the Black & Veatch tunnel design team has designed five hard rock tunnels for three cities within Indiana and Kentucky. This paper discusses a few design enhancements over the course of these 42 kilometers of tunnels in Indianapolis, Indiana; Fort Wayne, Indiana; and Louisville, Kentucky. Design improvements include optimizing tunnel slope relative to its length and depth, reducing drop shaft surface footprint by combining drop and vent shafts into one shaft, incorporating drop shafts within larger Tunnel Boring Machine (TBM) shafts, and refining the design of the bifurcation junctions of two full size tunnels excavated by a TBM.

OVERVIEW OF THE SIX PROJECTS

The projects discussed in this paper pertain to three different cities, namely Indianapolis, Indiana; Fort Wayne, Indiana; and Louisville, Kentucky. Each of these cities have designed and started construction on a tunnel system to help alleviate their combined sewer overflow (CSO) issue, clean up their water bodies, and meet consent decree with the United States Environmental Protection Agency (EPA) and their respective departments of environmental management.

The tunnel projects in chronological order are:

1. Deep Rock Tunnel Connector (DRTC)—Indianapolis, IN
2. White River Lower Pogues Run Tunnel (WRLPgRT)—Indianapolis, IN
3. Three Rivers Protection and Overflow Reduction Tunnel (3RPORT)—Fort Wayne, IN
4. Ohio River Tunnel (ORT)—Louisville, KY
5. Fall Creek Tunnel (FCT)—Indianapolis, IN
6. Pleasant Run Deep Tunnel (PRDT)—Indianapolis IN

Except for DRTC, all other projects listed above have been designed by Black & Veatch (B&V) and will be discussed in the following sections.

Indianapolis, Indiana

The city of Indianapolis, like a lot of developing cities across America, has older sewer infrastructure in the central part of the city, which is unable to keep up with the growing population and its needs. Both sanitary and storm water are carried to two main treatment plants through the same sewer infrastructure. During precipitation events, these pipes get overwhelmed and CSO water discharges into the river. In an attempt to clean up the city's water ways, a Long-Term Control Plan (LTCP) has been put in place, which includes construction of approximately 45 km (28 mi) of 5.5 m (18 ft) diameter tunnel in rock, up to 88 m (290 ft) below ground. The upstream end of the tunnel system is just north of the Indiana State Fairgrounds and it terminates at the Southport Advanced Wastewater Treatment Plant (downstream end). There are a few tunnels that branch to the east and west, as shown on Figure 1.

The DRTC is the southernmost part of this LTCP. Its main components are:

- Two large shafts varying in diameter from 13.4 m (44 ft) at the surface to 10.7 m (35 ft) in rock.
- 15 km (9.5 mi) of 5.5 m (18 ft) finished diameter tunnel (6.1 m (20 ft) excavated diameter).
- Nine small shafts, 1.1 to 2.4 m (3.5 to 8 ft) diameter, to serve as utility shafts, drop shafts, or vent shafts.
- Two tunnel bifurcation chambers.

The WRLPgRT is the upstream tunnel section from DRTC. Its main components are:

- One large shaft varying in diameter from 10.7 m (35 ft) at the surface to 9.1 m (30 ft) in rock
- 12.4 km (7.7 mi) of 5.5 m (18 ft) finished diameter tunnel (6.1 m (20 ft) excavated diameter)

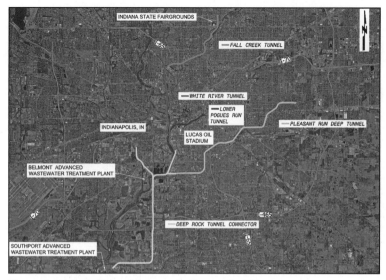

Figure 1. DigIndy Tunnel System, Indianapolis, IN

- Fifteen small shafts, 1.2 to 2.4 m (4 to 8 ft) diameter, to serve as drop or vent shafts.
- Two tunnel bifurcation chambers.

The FCT is the upstream most tunnel section on the north side and connects to WRT. Its main components are:

- One large shaft varying in diameter from 10.7 m (35 ft) at the surface to 9.1 m (30 ft) in rock
- 6.2 km (3.8 mi) of 5.5 m (18 ft) finished diameter tunnel (6.1 m (20 ft) excavated diameter)
- Twenty small shafts, 1.2 to 2.7 m (4 to 9 ft) diameter, to serve as drop or vent shafts.

The PRDT is the upstream most tunnel section on the east side of the DigIndy Tunnel System, and connects to DRTC. Its main components are:

- Two large shafts varying in diameter from 10.7 m (35 ft) at the surface to 9.1 m (30 ft) in rock
- One large shaft varying in diameter from 10 m (33 ft) at the surface to 8.8 m (29 ft) in rock
- 11.9 km (7.4 mi) of 5.5 m (18 ft) finished diameter tunnel (6.1 m (20 ft) excavated diameter)
- Twelve small shafts, 0.9 to 2.4 m (3 to 8 ft) diameter, to serve as drop or vent shafts.

All four Indianapolis tunnels mentioned above have associated near surface diversion structures, and deep rock adits connecting drop shafts to the main tunnel.

Fort Wayne, Indiana

The City of Fort Wayne has entered into a Consent Decree with the EPA, U.S. Department of Justice (DOJ), and the Indiana Department of Environmental Management (IDEM) to implement a CSO LTCP to reduce the volume of combined sanitary sewage that is discharged into the waterways within the City of Fort Wayne. Part of this solution is the 3RPORT project. The tunnel is anticipated to be fully operational by the end of 2022. The proposed tunnel will receive flows from existing combined sewer outfalls to reduce combined sewer overflows to the St. Marys and Maumee Rivers to four overflow events within a typical year. The tunnel will then convey the flow to the Wet Weather Pump Station No. 1 for transfer to and storage in the Wet Weather Ponds or directly to the Water Pollution Control Plant for treatment. The upstream end of the tunnel system is at Foster Park and it ends at the Water Pollution Control Plant (downstream end), as shown on Figure 2. The main components of the 3RPORT are:

- A pump station shaft varying in diameter from 20.7 m (68 ft) at the surface to 19.5 m (64 ft) in rock
- A Tunnel Boring Machine (TBM) working shaft varying in diameter from 10 m (33 ft) at the surface to 8.8 m (29 ft) in rock
- A retrieval shaft varying in diameter from 8.8 m (29 ft) at the surface to 7.6 m (25 ft) in rock
- 7.5 km (4.7 mi) of 4.9 m (16 ft) finished diameter tunnel, between 53 and 70 m (175 and 230 ft) below ground.

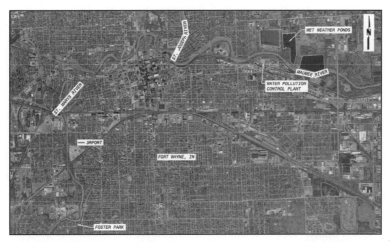

Figure 2. 3RPORT System, Fort Wayne, IN

- Eleven small shafts, 0.6 to 2.4 m (2 to 8 ft) diameter, to serve as drop or vent shafts.
- Associated near surface diversion structures, and deep rock adits connecting drop shafts to the main tunnel.

Louisville, Kentucky

Louisville and Jefferson County Metropolitan Sewer District (MSD) entered into a Consent Decree with the EPA and the Kentucky Environmental and Public Protection Cabinet. The Consent Decree was developed in response to an enforcement action related to the Federal Clean Water Act (CWA). In an effort to prevent and control sewer overflows an Integrated Overflow Abatement Plan (IOAP) was approved by EPA. Part of this plan includes the Ohio River Tunnel (ORT) project, which is an off-line CSO conveyance and storage system located in downtown Louisville, Kentucky. Integrated into this project, is a deep pump station located at 13th Street and Rowan Street designed to convey captured CSOs to the existing sewer system and the Ohio River Interceptor (ORI). The project area generally extends from 13th Street and Rowan Street (downstream end) to approximately 1200 East Main Street (upstream end), as shown on Figure 3. The main components of the ORT are:

- A pump station shaft and TBM working shaft varying in diameter from 14.6 m (48 ft) at the surface to 13.4 m (44 ft) in rock
- A retrieval shaft varying in diameter from 10 m (33 ft) at the surface to 8.8 m (29 ft) in rock
- 4.4 km (2.7 mi) of 6.1 m (20 ft) finished diameter tunnel, between 53 and 67 m (175 and 220 ft) below ground.

- Three small shafts, 2.1 to 2.7 m (7 to 9 ft) diameter, to serve as drop and vent shafts.
- One tunnel bifurcation chamber.
- Associated near surface diversion structures, and deep rock adits connecting drop shafts to the main tunnel.

TUNNEL DEPTH AND SLOPE

The depth at which a tunnel should be constructed in bedrock is influenced by subsurface conditions such as rock mass characteristics, hydraulic conductivity, and the presence of groundwater and naturally occurring gasses. Rock mass characteristics or properties, such as boreability, abrasiveness, strength, massiveness, and durability, influence the selection of tunnel depth. The hydraulic conductivity of the rock mass generally has the most influence on the selected depth of the tunnel due to the risk associated with tunnel construction within a high hydraulic conductivity rock mass. In addition, several hydraulic considerations can affect the vertical alignment design geometry of the tunnel system, such as adit-to-tunnel connections, tunnel slope, and constructability issues. Tunnel depth at the downstream end of a tunnel system, generally at the working shaft and at a deep pump station, has an influence on the hydraulic requirements of the pump station. The depth and presence of existing underground infrastructure also influences the vertical alignment of the tunnel.

The slope of the tunnel is closely controlled by tunnel constructability, rock stratigraphy, and system hydraulics. Slope also influences sediment transport during initial tunnel filling. Tunneling uphill from the working shaft is recommended to induce a gradient for groundwater to flow away from the tunnel face, and to mobilize sediment and grit that

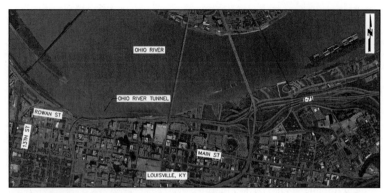

Figure 3. Ohio River Tunnel, Louisville, KY

might otherwise tend to settle along the tunnel. This slope away from the tunnel face lowers the risk of the TBM being inundated by water at the tunneling face if it should hit a major water-producing feature in the tunnel, assuming the tunnel dewatering pumps and sumps operate as required. A minimum slope of 0.1 percent is recommended for constructability.

The vertical alignment of the tunnel should also be maintained within the most favorable rock conditions for tunneling efficiency as well as to limit groundwater infiltration during and after construction. Rock near the soil-rock interface is commonly weathered and fractured, increasing its hydraulic conductivity and decreasing its quality. The slope of the tunnel should be chosen to maintain sufficient rock thickness between the tunnel crown and the soil-rock interface along the alignment in order to avoid this soil-rock weathered zone.

The main factor driving vertical tunnel alignment is geology, and in most cases not differential construction costs for various tunnel elevations associated with long term operations and maintenance (O&M). This is because the increase in construction and O&M costs due to a deeper tunnel in better quality rock, with lower hydraulic conductivity, is lower than the cost increase associated with constructing a shallower tunnel in poorer rock conditions or in zones of higher tunnel water inflow. The shallower tunnel in this case would also increase project risk and could result in higher claims during construction. Identifying a tunnel zone with good rock quality and minimal water inflow is key to finalizing an elevation for the tunnel. If the geotechnical data show that a deeper tunnel reduces overall construction risk, then a deeper tunnel should be recommended.

Indianapolis, Indiana

The Indianapolis tunnel system was designed at a depth to tunnel invert between 56 to 88 m (185 and 290 ft) below ground surface. This is between 28 and 55 m (92 and 180 ft) below the top of rock.

This depth was selected primarily based on the overall rock formation with least hydraulic conductivity, based on packer testing from the geotechnical investigation. The slope of all the tunnels are 0.1 percent. A tunnel slope greater than 0.1 percent could give the contractor more options for managing groundwater tunnel inflow. However, because the depth of overburden soil increases from the pump station to the upstream end of the tunnel system, increasing the slope would result in the upstream end of the tunnel being too close to the soil-rock interface, introducing more risk in the process, or requiring the downstream end of the tunnel system to be deeper. The deeper the tunnel is, the more the cost associated with deeper shafts and the higher the operating costs associated with deeper pumping. The length of the tunnel system in Indianapolis was the key factor that governed the tunnel slope. Since the longest tunnel segment is approximately 13 kilometers (8 miles) a small increase in tunnel slope would mean a relatively larger elevation increase from the downstream to the upstream end. At 0.1% slope, over the longest tunnel segment, the elevation increases by 13 m (42 ft).

Fort Wayne, Indiana

The 3RPORT, at 7.5 km (4.6 m), is shorter compared to the Indianapolis tunnel system. The tunnel invert elevation is between 53 and 69 m (175 and 225 ft) below ground surface, which is between 33 and 46 m (110 and 150 ft) below top of rock. Given its shorter length, the slope of this tunnel was designed at 0.15%. This increased slope aids in sediment transport in the tunnel and helps save on shaft construction costs in the upstream section of the tunnel.

Louisville, Kentucky

The ORT is the shortest of all tunnels discussed in this paper at 4.4 km (2.7 m). The tunnel depth was mainly governed by the Waldron shale layer. The Waldron shale has relatively lower rock quality as compared with the other rock formations in the project area. With potential slaking issues and the friable nature of this shale, the tunnel was designed at an elevation below the Waldron Shale layer. The depth of the tunnel ranges between 55 and 67 m (180 and 220 ft) below ground, which is between 23 and 53 m (75 and 175 ft) feet below top of rock.

Hydraulic modeling of the tunnel system at 0.1% slope indicated less than desired velocities in sections of the tunnel leading to deposition of sediment along the tunnel invert. Sediment transport in the tunnel is dependent on the rate of inflow to the system during a storm event, and where the majority of that flow is entering the system. With a minimum tunnel diameter of 6.1 m (20 ft) and a minimum pumping capacity of 50 cubic feet per second (cfs) (32 million gallons per day (mgd)) at the deep pump station, dewatering velocities will be less than 0.06 m/s (0.2 feet per second (fps)).

One option to mitigate sediment transport issues was to incorporate a cunette in the tunnel invert for these sections. However, further hydraulic modeling and design showed that increasing the tunnel slope to 0.2% mitigated these low velocities in the tunnel and would do away with the need for a cunette. In addition to sediment transport advantages, the higher tunnel slope decreases depths of shafts towards the upstream section of the tunnel, thus saving cost.

The tunnel slope increase has hydraulic implications with the tunnel flow going supercritical above certain slope. Each tunnel was modeled to make sure the hydraulics and transients work for that particular tunnel system.

DROP SHAFT CONFIGURATION ENHANCEMENTS

Several factors go into selecting a drop shaft site and the configuration of the drop shaft. Some of these criteria are area requirements, property ownership, existing utilities and infrastructure at the site, contamination considerations, natural and environmental impacts to the site, construction access, community impacts during construction and long-term impacts, construction considerations, long-term shaft site utilization, and geotechnical considerations. Keeping these in mind, the drop shaft configuration used for most of the shafts on the projects mentioned in this paper is a combination of vortex drop shaft with a vent shaft.

Since most of the CSOs are in or around urban areas, the number of sites available to construct a drop shaft are generally limited. In spite of having a detailed plan for site selection, we can run into several issues on the few available sites to construct the shafts. Given that most of the shaft designs incorporate a two-shaft system with flow dropping down to the tunnel through the drop shaft and air being

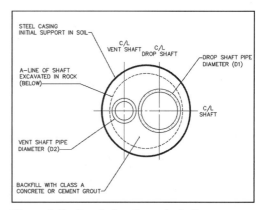

Figure 4. Sectional plan through shaft

vented from the tunnel elevation through the vent shafts, Black & Veatch designed a single shaft that incorporates both drop and vent vertical conduits, as shown in Figure 4. At the tunnel elevation, the drop and vent shaft locations are as per design, which is typically tens of feet away from each other. However, the vent conduit is bent back and brought into the same shaft as the drop, as shown in Figure 5. These two conduits then run from the tunnel elevation to the ground surface. This approach reduces the shaft surface footprint and reduces the number of surface penetrations from two down to one. This configuration is beneficial for shaft sites that have limited area for construction. This configuration only works for flows up to a certain range. As flows get larger, so do the drop and vents sizes. Constructing a large enough diameter single shaft has limits beyond which it becomes too expensive to go with this configuration. The 3RPORT and ORT projects have incorporated this drop shaft configuration into the design.

DROP SHAFT WITHIN TBM SHAFT

The design of the horizontal alignment for tunnels on CSO projects is governed by many factors, but one of the most critical is routing the tunnel near the proposed drop shaft sites. This practice helps to minimize cost associated with drill and blast construction of adits to connect the drop shafts to the main tunnel. Sometimes the beginning and end of the tunnel alignment where the TBM is launched and retrieved, are located in areas that meet the needs of working and retrieval shafts, and are also located where flow needs to be dropped to the tunnel. By eliminating the need to construct a dedicated drop shaft to drop flow to the tunnel, it further helps to minimize cost. This arrangement occurred a total of five times at TBM shafts in three tunnels that the team designed in the past eight years. The TBM shafts where this occurred include:

- 3RPORT Retrieval Shaft
- Ohio River Tunnel Retrieval Shaft
- Fall Creek Tunnel Retrieval Shaft
- Pleasant Run Deep Tunnel Working Shaft
- Pleasant Run Deep Tunnel Retrieval Shaft

The design flow rate at these five shafts ranges from 55 to 620 cfs (36 to 400 mgd). The venting of tunnel air was a critical component in evaluating the final design of the drop shaft configuration within the retrieval shafts where flow was being dropped. Venting of entrained air must be considered since the flow with entrained air is being directly conveyed to the main tunnel where the entrained air could potentially be trapped within the tunnel and take up storage volume or could become pressurized causing a geyser risk. Large diameter TBM shafts can be configured as Plunge Drop, Vortex Drop, or Baffle Drop shafts depending on the design drop flow rate,

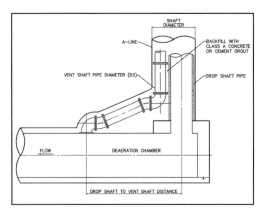

Figure 5. Vertical sectional

amount of air entrainment, and tunnel air ventilation needs. Table 1 summarizes these previously mentioned considerations as well as the finished inside diameter of these TBM shafts.

The tunnel venting needs control the final section needed in the retrieval shaft after the TBM is removed from the tunnel. A large portion of the shaft area is needed to accommodate the air that is being forced out of the DigIndy and 3RPORT systems. These systems have a high number of drop shafts filling the tunnel system at a high rate for the capacity of the system. The upstream end is the primary air release point for these systems and these tunnel systems have significant transients which in turn increases the air venting needs. Nearly 90% of the shaft area is needed to vent the tunnel air to the atmosphere and maintain air velocities below 9.1 m/s (30 fps) for these systems. A baffle drop structure requires more area in the shaft as compared with a vortex drop. Therefore, the option of a baffle drop would require upsizing the retrieval shaft and significantly add cost to these projects. However, the ORT has only four drop shafts delivering flow to a tunnel that has a significant storage volume. Transient modeling during design development showed that the ORT filled "like a bathtub" with no transient issues influencing the air venting needs of the system. The air venting needs of the ORT only required approximately 20% of the retrieval shaft area to maintain reasonable air venting velocities. This allowed the option of utilizing a baffle drop structure for the ORT project, where only a vortex or plunge drop could be utilized at the Fall Creek and 3RPORT retrieval shafts. The baffle drop structure within the retrieval shaft only reduced the area available for venting by 55%, leaving 45% of the shaft area open for venting

Table 1. TBM drop shaft considerations

Tunnel	TBM Shaft	Finished Inside Diameter	Flow	Location	Expected Air Entrainment	Tunnel Venting Needs	Shaft Options
3RPORT	Retrieval	6.4 m (21 ft)	620 cfs (400 mgd)	Upstream End	High—Large Flow	Very High—Significant surge and tunnel transients	Vortex
ORT	Retrieval	7.6 m (25 ft)	275cfs (178 mgd)	Upstream End	Minimal—Moderate flow into large diameter tunnel	Low—Minimal flow entering large tunnel volume	Vortex and Baffle
FCT	Retrieval	7.3 m (24 ft)	56 cfs (36 mgd)	Upstream End	Minimal—Small flow into large diameter tunnel	Very High—venting air from 22 km (14 mile) of tunnel	Plunge and Vortex
PRDT	Working	10.7 m (35 ft)	100 cfs (65 mgd)	Along Tunnel	Minimal—Small flow into large diameter tunnel	Minimal—large shaft over tunnel alignment	Vortex and Plunge
PRDT	Retrieval	7.3 m (24 ft)	275 cfs (178 mgd)	Along Tunnel	Minimal—Moderate flow into large diameter tunnel	Moderate—venting air from 11 km (7 mile) of tunnel	Vortex and Plunge

tunnel air. The baffle drop was preferred since it does not have an approach channel nor a transition channel, both of which are associated with a vortex drop, and a baffle drop entrains less air in the flow as compared with a plunge drop.

The option of a plunge drop at the 3RPORT retrieval shaft was quickly removed from the design options due to amount of air entrainment that the large flow would have introduced to the tunnel, which already had surge and transient issues. The large air entrainment would have only worsened the surge issues increasing the geyser risk at the 3RPORT retrieval shaft, which is at the entrance to a popular park in Fort Wayne. The design of the vortex shaft with the large shaft, as shown on Figure 6 and Figure 7, is integral to the overall shaft lining to minimize the reduced cross section of the shaft available to vent air from the tunnel during filling.

The PR working and retrieval shafts are located along the PR leg of the DigIndy System. More than 60% of the CSO's in the DigIndy System are located upstream of the PR junction with the main spine of the tunnel system. A baffle wall is located in the top half of the PR tunnel where it connects to the main spine of the system. This baffle wall was installed to reduce the chance of a trapped pressurized air pocket being forced up the PR leg which would increase the air venting needs of the PR working and retrieval shafts. Once the water level in the tunnel reaches the bottom of the baffle wall, only the remaining air in the crown of the PR leg needs to be vented from the tunnel at the working shaft or Retrieval shaft. Dropping the 100 cfs (65 mgd) flow directly to the tunnel inside the PR working shaft as a plunge drop was evaluated and it was determined that the amount of entrained air that would come out of solution downstream of the shaft would be minimal. Even if a small pocket of trapped air formed and traveled back up the tunnel to the working shaft, the 10.7 m (35 ft) diameter shaft allows for the safe expansion and venting of that air without the risk of a geyser forming. A similar evaluation was done for the PR retrieval shaft flow and the analysis returned a similar result. More air comes out of solution from the relatively larger flow, but the amount is a fraction of the tunnel air venting requirement due to the potential filling rate of the overall tunnel system. The potential filling rate of the tunnel system necessitates a large venting capacity at the upstream ends of all the tunnel branches, including the PR leg. Baffle edge sizing parameters developed during baffle drop modeling that was performed for the Fall Creek White River tunnel system, indicate that 4.1 m (13.5 ft) is the required baffle edge length to pass the flow. This edge length plus 0.5 to 0.6 m (1.5 to 2 ft) for the wet and dry side divider wall would reduce the cross section of the shaft available to exhaust air by approximately 68% and result in

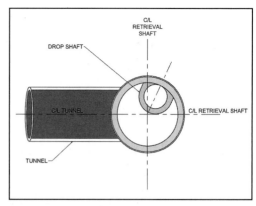

Figure 6. Sectional plan through rock

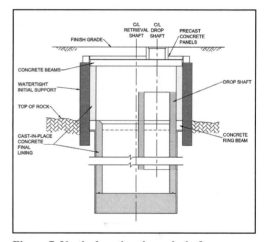

Figure 7. Vertical section through shaft

air venting velocities in excess of 9.1 m/s (30 fps). The 1.7 m (5.5 ft) diameter vortex drop shaft or 2 m (6.5 ft) diameter plunge drop shaft would impact the venting area by less than 17% and would keep venting velocities below 9.1 m/s (30 fps).

MAIN TUNNEL BIFURCATIONS

The need to collect flow from CSOs along feeder streams such as Eagle Creek, Pleasant Run, Lower Pogues Run, and White River, and to increase tunnel storage capacity to meet Federal Consent Decree requirements made it necessary for full diameter branch tunnels to be included in the DigIndy system and ORT. The DigIndy system has four branches and ORT has only one. These branches require a wye or bifurcation in the tunnel, which must be lined after mining. The original design of the DRTC had only one such bifurcation for the future PR tunnel, but Line AA was added to the construction contract to collect CSO flow along Eagle Creek.

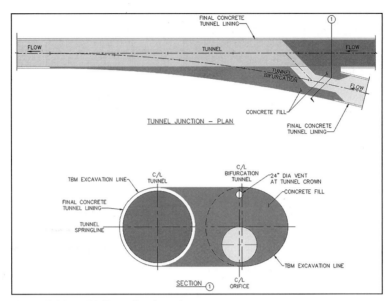

Figure 8. Tunnel bifurcation junction

The original AECOM design for the bifurcation lining required an arched chamber with a flat floor to be excavated and lined with a cast-in-place, reinforced concrete structure transitioning the two full diameter tunnels to a single tunnel. This design required blasting out the rock in the floor and crown to the required shape. The amount of blasting needed to excavate the chamber was not substantial, but required a significant amount of time and effort due to the detailed shape. The Contractor, SK-JV, proposed an alternate design consisting of a similarly shaped chamber with a flat floor, but the arch was lined with reinforced shotcrete. The contractor also included rock bolts in the arch, walls, and floor to minimize the thickness of the chamber lining to approximately 300 mm (12 in). This minimized the excavation needed for the chamber, but the extra labor and cost for excavation, forming, shotcrete application, and rock bolts was still there. The baffle in the crown of the branch tunnels was added later to address the air venting mentioned previously. The bifurcations for Lower Pogues Run and White River were designed utilizing this same approach at the request of the owner (Citizens) to maintain uniformity across the tunnel system.

In an effort to make construction of the tunnel wye or tunnel bifurcation more efficient, Black & Veatch modified the design for the ORT project as shown on Figure 8. The ORT design is similar to that of an adit connection to the main tunnel where the invert elevations match and the branch tunnel flow enters the main tunnel through a smaller diameter opening at an angle of approximately 60 to 70 degrees. This design requires minimal rock removal at the upstream edge of the bifurcation and at the invert, where the smaller diameter opening crosses the invert to connect to the main tunnel. The large opening is filled with the same structural concrete as the tunnel lining concrete to eliminate the need for additional rock bolts for rock support and only minimal reinforcing steel is needed because the lining is circular. The smaller diameter opening that matches the main tunnel invert elevation still allows for small equipment to access the upstream parts of the tunnel branch for inspection and cleaning, and minimizes the risk of large volumes of air being directed from the main tunnel, up the branch tunnel section.

CONCLUSIONS

Continuous improvement should be the mindset of any designer. Looking for ways to make construction more efficient and less costly, is one of the key drivers to tunnel design advancements over the past several decades. All necessary checks, like hydraulic and transient modeling should be in place to validate design improvements. This coupled with knowledge of similar projects in the area help to refine design parameters. Designers should maintain a lessons-learnt document on every tunnel construction project and incorporate them into successive designs, constantly refining and improving tunnel designs. Maintaining good communication, with the owner, stakeholders, manufacturers, and contractors is also key to a successful tunnel design.

TRACK 4: CASE STUDIES

Session 4: Transportation 2

Michael Torsiello and Patrick Finn, Chairs

Risk Mitigation Through the Use of In-Tunnel Ground Freezing for Seattle Light Rail Tunnel Cross Passage Construction

Rick Capka
Sound Transit

Nate Long
Jay Dee Contractors, Inc.

ABSTRACT: Effectively controlling groundwater is an essential pre-requisite to the excavation of soft ground tunnels using the Sequential Excavation Method (SEM). This paper evaluates the use of in-tunnel ground freezing, in lieu of more traditional dewatering methods, as the primary means of controlling challenging groundwater conditions for the construction of several cross passages along the 4.3 mile Northgate Link Extension light rail project in Seattle, WA for the Central Puget Sound Regional Transit Authority (Sound Transit). The paper will include a review of the risk-based decision making process used during construction to select in-tunnel ground freezing over dewatering to mitigate a differing site condition, and it will analyze the outcome of that decision by reviewing the successes and challenges associated with the tunnel construction.

INTRODUCTION AND PROJECT OVERVIEW

The Northgate Link Extension project is a 4.3 mile long light rail extension in Seattle, WA. The project includes approximately 3.4 miles of twin-bore tunnel, one tunnel portal, two underground stations, an elevated guideway and station, and a 450-stall parking garage (refer to Figure 1). The project, which begins on the campus of the University of Washington (UW), is a northern extension to an existing 20 mile long light rail system that connects the downtown Seattle corridor with the City of Sea-Tac, including the Seattle-Tacoma International Airport.

The 3.4 mile tunnel alignment consists of two, segmentally lined, 18-ft 10-in diameter bored tunnels; one each for the future northbound and southbound trains. Based on the chosen alignment and existing geographical conditions, tunnel depth ranges from approximately 20-ft at the northern portal to over 130-ft under UW campus. Along the alignment, 23 cross passages (numbered sequentially in the contract from 21 through 43) were located perpendicularly between the twin tunnels at a maximum spacing of 800-ft in order to meet National Fire Protection Agency (NFPA) emergency egress requirements. With the exception of several cross passage locations on UW campus, these 23 cross passages were located within public right-of-way.

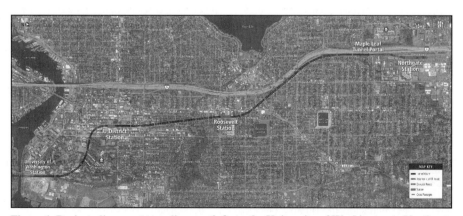

Figure 1. Project alignment traveling north from the University of Washington to Northgate

---

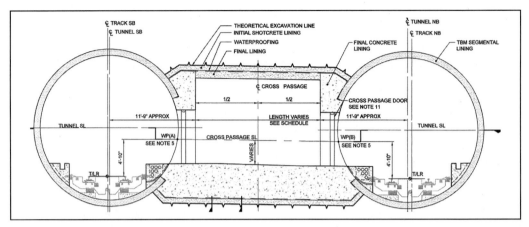

Figure 2. Cross section of a typical cross passage between twin TBM tunnels

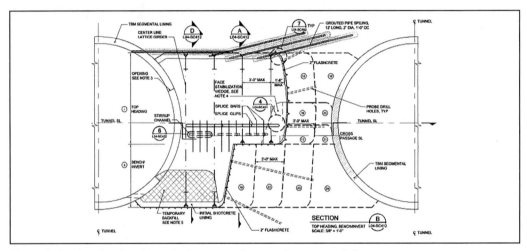

Figure 3. Section showing the typical support of excavation approach for a cross passage

The geology along the tunnel alignment is highly variable, consisting primarily of glacial and interglacial soils above a depth of bedrock estimated to be between 1,800 and 2,500-ft deep. For construction purposes, the Geotechnical Baseline Report (GBR) organized these soils into seven engineering soil units (ESUs), of which the following four were primarily expected to be encountered during tunnel construction: Till and Till-Like Deposits (TLD), Cohesionless Sand and Gravels (CSG), Cohesionless Silt and Fine Sand (CSF), and Cohesive Clay and Silt (CCS). Hydrogeologic conditions are equally variable, with recharging aquifers (granular water-bearing soils) and aquitards (finer-grained sediments) dispersed throughout the alignment and a groundwater head at tunnel invert defined to be up to 2.9 bar.

JCM Northlink LLC, a joint venture between contractors Jay Dee Contractors Inc., Frank Coluccio

Construction Company, and Michels Corporation, was contracted by Sound Transit through the traditional competitive bidding procurement process to build the twin-bore tunnels, cross passages, portal structure, two underground station excavations, and associated tunnel mechanical, electrical and architectural finishes.

CROSS PASSAGE DESIGN AND BASELINE CONDITIONS

The 23 cross passages were designed to provide efficient emergency egress between the twin-bore tunnels while also providing enough space to house required electrical and mechanical systems for the light rail tunnels. The cross passages have a typical excavated cross-section that is approximately 16 feet wide and 18 feet tall, and they range in length from

Table 1. Cross passage ground support category definition

Ground Support Category	Category Description	# of Cross Passages
1	Favorable ground and groundwater conditions. Basic SEM support measures including probe drilling, grouted pipe spiling, face stabilization wedge, and shotcrete/lattice girder	1
2	Moderate ground and groundwater conditions. Dewatering required, either from the surface or within the tunnel, prior to excavation. SEM support measures including probe drilling, grouted pipe spiling, face stabilization wedge, and shotcrete/lattice girder	17
3	Difficult ground and groundwater conditions. Water bearing fine grained soils that are difficult to dewater and highly unstable. Ground improvement, either jet grouting or ground freezing, required prior to excavation. SEM support measures including probe drilling, grouted pipe spiling, face stabilization wedge, and shotcrete/lattice girder	5

20 to 28 feet, depending on the spacing between the twin tunnels (refer to Figure 2).

Based on these dimensions and the anticipated ground conditions, the cross passages were designed to be constructed using the sequential excavation method (SEM). SEM for these cross passages generally consisted of various pre-support measures, such as spiling or dewatering, followed by a sequenced top heading and bench excavation, with the subsequent installation of steel lattice girders and shotcrete for temporary support (refer to Figure 3). Following excavation and temporary support, each cross passage was to be lined with a PVC waterproofing membrane and a cast-in-place reinforced concrete final lining.

Although the ultimate SEM sequence and support requirements varied at each cross passage based on the local subsurface conditions, the designer established three different ground support categories of cross passage in an effort to define the scope of work. These three categories were subsequently defined as Unit Price Items under the contract, and each cross passage was baselined to be one of the three categories (see Table 1).

HYDRAULIC CONDUCTIVITY DIFFERING SITE CONDITION (DSC)

Cross passage Nos. (CP) 34 through 39, located immediately to the north of the underground Roosevelt Station site, were part of the 17 Ground Support Category 2 cross passages. All six of these cross passages were located under the water table and within the zone of a large, recharging aquifer of cohesionless sand and gravels (CSG). The Contractor's approach to dewatering each of these cross passages consisted of a deep well system (two to three wells per cross passage) installed from the surface and designed to lower the groundwater table below tunnel invert. The designs of these deep well systems were based upon the CSG aquifer and the baselined hydraulic conductivity, or the soil's resistance to water flow, range for the CSG of 2×10^{-4}

Table 2. Design vs. required dewatering flow rates

Cross Passage No.	Design Dewatering Flow Rate (gpm)—based on GBR	Required Dewatering Flow Rate (gpm)—based on pumping tests
34	630	1,400
35	600	1,700
36	650	1,800
37	500	N/A (no pumping tests)
38	440	2,300
39	200	400

to 2×10^{-2} feet per minute (ft/min). The designed productivity, or flow rate, of each deep well system required to effectively lower the groundwater table can be found in Figure 5.

The Contractor proceeded to install the deep well systems at each of these five cross passages, starting at CP 38. In an effort to confirm the performance of the system once installed, the Contractor elected to perform a 24-hour pumping test on the wells. The results of this pumping test indicated that the hydraulic conductivity for the aquifer was significantly higher than the upper bound of the range baselined in the contract.

In order to verify these results, Sound Transit directed a series of extended pumping tests at five of the six cross passages potentially impacted by these results. The conclusion reached after these series of tests was that the hydraulic conductivity in the CSG aquifer was substantially higher than the range baselined in the GBR. Consequently, a much higher volume of water would have to be removed from the aquifer in order to effectively dewater each cross passage location (refer to Table 2).

With the discovery of this DSC well into the execution of the construction contract, and well into the construction of these cross passage dewatering systems, the project team was faced with a complex technical challenge that had the potential to

Table 3. DSC mitigation options key benefits and risks

Option	Key Benefits	Key Risks
Enhanced Dewatering	• Utilizes existing system infrastructure, therefore minimizing directed change • Potentially less schedule impact, assuming it works	• Revised system won't work as designed • System failure will lead to rapid recharge of groundwater and destabilized ground • Insufficient discharge capacity within city infrastructure, or high volume discharge results in overflow event for city • Decreasing well efficiency over time results in rise of groundwater and ground instability
Jet Grouting	• Works well in CSG soils • Once in place, does not rely on electrical or mechanical systems for it to work • Provides long term ground water barrier beyond excavation	• Environmental challenges with containing and disposing of jet grout spoils • Jet grouting operation damages bored tunnel linings • Shadow effect from jet grouted columns results in groundwater pathways resulting in destabilized ground • Surface obstructions, including underground utilities and overhead power lines delay progress • Jet grouting operation damages city infrastructure • Schedule
Ground Freezing	• Works well in CSG aquifer • Provides stable, safe conditions for excavating cross passages • Although requires constant power, if power systems fail ground remains stable for sufficient period of time to address failure	• Ground water velocity too high, resulting in ineffective freeze down • Surface obstructions (for surface ground freezing) delays progress • Damage to bored tunnel lining due to freeze, subsequently leading to long term groundwater issues • Surface heave from freeze damages infrastructure • Schedule

significantly impact the direction and outcome of the project.

RISK-BASED DECISION-MAKING PROCESS TO MITIGATE DSC IMPACTS

The project team, which included the design consultant, construction management consultant, contractor and owner, proceeded in earnest to pursue a technical solution to the DSC that would facilitate the safe excavation of the cross passages while minimizing impacts to the overarching project goals of quality, cost and schedule. The following three options were ultimately vetted by the project team in the months immediately following the discovery of the DSC.

1. **Enhanced Dewatering System:**
   Given the reality that significantly higher volumes of water would have to be removed at these cross passages in order to effectively dewater, the first technical consideration was to supplement the existing dewatering systems with additional wells and/or deeper wells in order to provide the necessary capacity. Due to the nature of the aquifer and the coarse grained soils, dewatering with this additional capacity appeared technically feasible. However, discharging these high volumes of water within the local city infrastructure posed a significant challenge due to the limited capacity of the combined sewer system. The project team pursued the possibility of constructing temporary discharge infrastructure and upgrading a combined sewer pump station in an effort to work around this constraint.

2. **Jet Grouting:**
   Ground improvement using jet grouting was defined as an option for Category 3 cross passages under the contract. Consequently, it was evaluated by the project team as a potential alternative to dewatering to mitigate the DSC.

3. **Ground Freezing:**
   Like jet grouting, ground freezing was defined as another ground improvement option for Category 3 cross passages under the contract. For the five cross passages originally defined under the contract as Category 3, the Contractor had already selected ground freezing as the means of ground improvement.

Table 3 summarizes the project team's evaluation of the key benefits and risks associated with each of the three options considered for mitigating the DSC.

After analyzing the benefits and risks associated with these options, Sound Transit elected to conduct a three-day workshop that included the support of external technical experts in tunneling, dewatering, and ground freezing. The options outlined above were debated in detail during the workshop, and the technical experts assisted the project team with deciding to pursue ground freezing as the means to improve the ground conditions at five of the six cross passages (CP 34–38) impacted by the DSC.

IN-TUNNEL VS. SURFACE GROUND FREEZING

After it was determined that the best course of action would be to ground freeze cross passages 34–38, the project team had to make another important decision; whether to install the ground freeze pipes vertically from the surface or to install them horizontally from inside of the tunnel. Successful surface freeze pipe installations had already been performed for cross passages 21 and 29–32, however, these installations were fraught with a variety of difficulties, including:

- Existing overhead power lines, trees and buried utilities significantly increased the complexity of the drilling pattern.
- Despite the efforts of the experienced and qualified drilling subcontractor, deep drilling depths induced considerable deviations at the tunnel level. A modified, slower, downhole survey system had to be put in place once the extent of the deviations was realized. Deviations also led to having to install additional freeze pipes to cover the missed areas.
- Getting temporary power from the local public utility took longer than anyone anticipated. In some cases, it was over six months from the applications to when the power was usable. This led to having to use generators to begin the freeze down and maintain schedule. These generators had to run 24 hours a day, 7 days a week until line power was available. At one location, line power was impossible and the generator had to run during the entire operation of the freeze.
- Roads were closed for months while the freeze was installed and operated. High impact drilling operations were carried out in quiet urban neighborhoods, less than 10 feet from property lines (refer to Figure 4).

In addition to these experiences, cross passages 35–38 are located directly under major arterial streets and the freeze drilling and installation would impact thousands of cars per day. Based on these complications, in-tunnel freeze pipe installation was evaluated as an alternative approach. This, however, had its own unique set of challenges, including the following:

Figure 4. Freeze pipe drilling on the surface at CP30

- Drilling within the tunnel excludes access for any other major work activities in that section of tunnel and puts the freeze pipe drilling on the critical path of the cross passage excavations. This increases schedule risk and requires a substantial schedule adjustment to the finish works of the tunnel.
- The southbound TBM tunnel was being excavated, and would still be under excavation during the freeze pipe drilling, so additional safety precautions had to be taken to ensure that the drilling operations did not contact the moving conveyor belt.
- The cross passages are relatively large compared to the running tunnel. This leaves very little headroom to install the freeze pipes above the cross passage and amplifies the equipment constraints.
- Drilling from the tunnel into high groundwater head has a chance to cause a catastrophic tunnel failure.
- Once the freeze pipes have been installed, freeze plants must be installed in the tunnel. These freeze plants, while not as impactful as the drilling operations, take up space and cover up areas of the tunnel that still require finish work to be completed.

These potential challenges were evaluated against the known difficulties of installing the freeze system from the surface. After careful consideration of these risks, along with their associated project cost and schedule implications, the project team elected to pursue the in-tunnel freeze system option.

GROUND FREEZING DESIGN AND CONSTRUCTION

Once it was decided to freeze from inside the tunnel, a robust design needed to be assembled. JCM

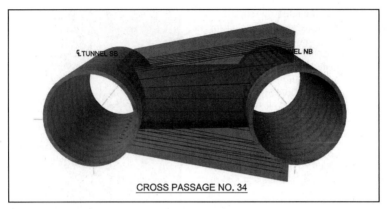

Figure 5. Conceptual design of the in-tunnel ground freeze design

selected specialty subcontractor Moretrench to perform this work. Moretrench reviewed the subsurface baseline conditions, as well as the additional information that was gathered during the dewatering well drilling and subsequent pump tests. Using all the available data, they were able to produce a design that would be suitable to the complex ground conditions (refer to Figure 5).

This design consisted of primary freeze pipes, installed from the southbound tunnel, with a row secondary freeze pipes installed in the lower quadrant of the northbound tunnel. The primary freeze pipes ranged in length from 19-ft to 41-ft, with -10 to 15 degree drilling angles. The secondary pipes and a freeze loop installed on the surface of northbound tunnel ensured a complete watertight seal. All the freeze pipes would be drilled with a lost casing, and afterwards a welded steel pipe was inserted into the casing and pressure tested. Temperature pipes would be installed in strategic locations to monitor the formation of the frozen soil mass. Freeze down time was evaluated based upon the spacing of freeze pipes and the in-situ soil properties. The anticipated worst case was 8 weeks to achieve sufficient structural strength to begin excavation. The watertight seal to the tunnels would be confirmed by comparing a piezometer installed within the center of the cross passage to one installed within the aquifer but well outside the frozen zone's influence.

Drilling below the high water table posed considerable challenges and precautions had to be taken to control groundwater and soil inflows. To achieve this level of control, stuffing boxes with rubber wipers were installed at every drilling location. Each of these stuffing boxes was custom made to match the drilling angle of the holes. Multiuse ports were installed in each of the stuffing boxes for drainage and grouting. The original plan was to use positive displacement to complete all the drilling, however, drilling at the first cross passage (34) was initially

much more difficult than anticipated. During this first round of drilling it was determined that material would have to be removed during drilling. This material would have to be accurately tracked to ensure that no ground loss would occur, which could have led to settlement at the surface. Drilling holes above the cross passage was also very difficult due to the geometry of the tunnels and freeze pipes. A large platform had to be installed in the tunnel to provide the drill access to complete these holes (refer to Figure 6). Once all the freeze pipes were installed and tested, they were connected to multiple manifolds of the brine distribution system that circulates chilled brine from the freeze plants. Specially designed tunnel chillers were installed in the tunnel on the opposite side of the cross passage from the excavation and shotcrete arrangement. At this point the system was activated and freeze development began.

CROSS PASSAGE CONSTRUCTION RESULTS

Freeze down durations remained consistent with the thermal models at all locations and cross passages were ready as excavation resources became available. Cross passages were excavated in sequential order from CP34 to CP38, except for CP37 which was started before CP36 due to access constraints. Refer to Table 4 for freeze install and excavation dates.

Cross passage excavation was performed using a standard SEM sequence with separate top heading and bench/invert excavations, and lattice girders spaced at four feet. The bulk of the excavation was performed with a Brokk demolition robot using a variety of attachments, including a hydraulic breaker and a TEI drill (refer to Figure 7).

Excavation near the extrados of the cross passage was difficult, due to the high strength of the frozen soil and the proximity of the freeze pipes. Care was taken to stay clear of the freeze pipes, but in two

Figure 6. Drilling upper row of freeze pipes at CP37

Table 4. Cross passage freeze installation and excavation dates

Cross Passage No.	Freeze Drilling Start	Freeze Down Start	Excavation Start	Excavation Finish
34	05/16/16	07/25/16	08/24/16	11/28/16
35	07/12/16	08/23/16	12/01/16	12/23/16
36	08/03/16	09/28/16	01/05/17	01/27/17
37	08/29/16	10/11/16	12/02/16	01/10/17
38	09/19/16	10/25/16	01/12/17	02/03/17

Figure 7. Excavating back to the segments at CP34

instances they were struck by the hydraulic breaker. Fortunately, these instances did not result in a brine leak and excavations were able resume quickly. Support of excavation was provided by dry mix steel fiber reinforced shotcrete, supplemented with welded wire fabric to provide reinforcement during early strength gain. An additional 2" of flashcrete was applied to the frozen soil to act as a sacrificial insulating layer and protect the 10" shotcrete initial lining.

One of the major advantages of horizontal in-tunnel freeze pipes realized during construction was the lack of freeze pipes inside the excavation envelope. During the excavations of CP21 and CP29-CP32 (vertical freeze pipe locations), a significant amount of time was required to cut and cap, or reroute, the freeze pipes that were installed within the excavation limits. The removal of these pipes also reduced the effectiveness of the freeze system below the excavation. Although there were no stability issues as a result of this reduced capacity, it is not impossible to imagine a scenario where delays compound into a possible failure of the frozen soil beneath the excavation. Although these particular productivity and quality benefits were not clearly understood when the decision was made to go with in-tunnel freezing, the project ultimately benefited from the simplified excavation sequence.

One unexpected challenge with the in-tunnel freeze was the amount of incomplete freezing of the center of the latter cross passages. Beginning with the breakout of CP37, the project team discovered unfrozen soil at the center of the excavations. This soil was unstable and raveled much quicker than anticipated when uncovered. Cross passages 36 and 38 had drain holes installed in the center of the excavations before the segments were removed. These drains were kept open and the slowly draining water reduced the pressure over a few days, allowing excavation to proceed.

Ultimately, cross passages 34–38 were successfully excavated without additional support measures. No ground loss occurred during any of the excavations and these cross passages were some of the fastest excavated on the project. The exceptional results at these cross passages validates the decision to move to ground freezing for the excavation support.

CONCLUSION

In summary, when faced with a significant differing site condition that posed a threat to the safe and timely excavation of several cross passages, the project team collaboratively implemented a deliberate risk-based analysis process to select ground freezing as the most appropriate method of ground improvement to mitigate the differing site condition. Subsequently, a horizontal in-tunnel freeze system was determined to be the most effective means to achieve this ground improvement. Once installed, the in-tunnel freeze system performed as designed and allowed the cross passages to be excavated safely and efficiently without impact to the overall project schedule.

TBM Passing Under Existing Subway Tunnels in Los Angeles, California

Jason Choi
WSP USA

Matthew Crow and Gary Baker
LA Metro

Ron Drake
EPC Consultants, Inc.

Patrick Jolly
Arcadis

Christophe Bragard
Traylor Bros., Inc.

ABSTRACT: The development of a comprehensive underground subway system in Los Angeles is proceeding after the construction of the first tunnels by LA County Metro in the 1990s. Expansion of the subway system has posed challenges with respect to interference with existing subway tunnels and transit operations. Most recently, Metro's Regional Connector Transit Corridor (RCTC) twin bored tunnels required passing under Metro's existing Red/Purple Line tunnels with about 5-ft of clearance. The structural integrity, safety and revenue operation of the existing subway line had to be maintained during construction of the undercrossing. This paper summarizes site conditions, monitoring program, contingency plan, execution and outcomes of the successful crossing under Metro's Red/Purple Line with an EPBM.

INTRODUCTION

In Los Angeles County, California, six rail transit lines are currently in operation by the Los Angeles County Metropolitan Transportation Authority (Metro): Blue, Red, Green, Gold, Purple, and Expo Lines. These lines mostly connect regions of the Los Angeles metropolitan area to downtown Los Angeles. The Regional Connector Transit Corridor (RCTC) project is currently being constructed in downtown Los Angeles to improve mobility for transit riders traveling throughout Los Angeles County. Upon completion, RCTC will provide direct connections between Metro Gold, Blue, and Expo Lines with a new 1.8-mi long underground rail connection with three new underground stations at 1st Street/ Central Avenue, 2nd/Broadway, and 2nd/Hope Street. The west end of RCTC will connect to the existing Metro Blue/Expo Line tail tracks at the 7th/ Metro Center Station. The east end of RCTC will split into north and east legs via an underground wye structure near the intersection of 1st and Alameda Streets connecting on the surface with the Metro Gold Line to Pasadena and Azusa to the north and to the Metro Gold Line to East Los Angeles on the east leg (Figure 1).

Along its alignment, the RCTC twin tunnels pass beneath Metro's existing Red/Purple Line tunnels at the intersection of 2nd and Hill Streets (Figure 1). Figure 2 shows the cross section of the under-crossing plan which had approximately 5- to 7-ft of vertical clearance between the RCTC tunnels and the Red/Purple Line tunnels. A 60-ft zone of influence on either side of the existing tunnels was used for the crossing study, excavation, and ground control plans (Figure 2).

The RCTC twin tunnels are excavated by 21.6-ft diameter Earth Pressure Balance Machine (EPBM) installing 18.8-ft ID and 20.6-ft OD bolted precast concrete segmental lining with double gaskets. The bored tunnels connect stations and cut-and-cover

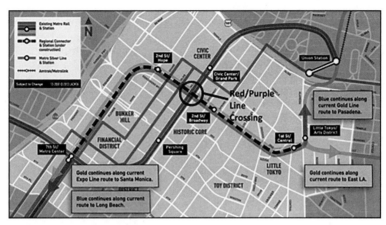

Figure 1. Project alignment

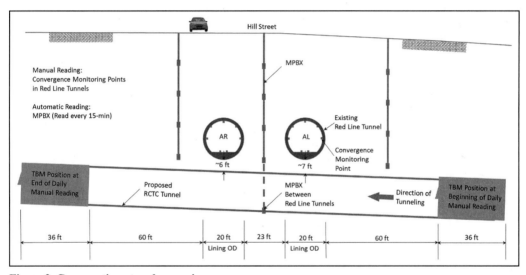

Figure 2. Cross section at undercrossing

sections of the alignment for a total distance of 4,940-ft. The depth of ground cover ranges between approximately 25- to 130-ft. Two tunnels, L-track and R-track tunnels, are approximately 27.5- to 34.3-ft apart center to center and the vertical grade varies from 0.2- to 5.0-%. The Red/Purple Line crossing is located within a vertical curve of the RCTC tunnels transitioning from 0.2- to 4.6-% grade over 480-ft reach. The excavation of the L-track tunnel began on February 2, 2017, and completed on July 21, 2017 and the R-track tunnel began September 25, 2017 and is scheduled to complete in January 2018. The L-track tunneling crossed beneath the Red/Purple Line tunnel in early May 2017 and the R-track tunnel crossed beneath the Red/Purple Line tunnels in mid-November of 2017. Figure 3 shows the typical section of the RCTC tunnels.

Located in downtown Los Angeles, the existing Red/Purple Line is a heavy rail subway line constructed beginning in 1990. The 19.8-ft finished diameter twin tunnels were excavated using a 21.1-ft OD open face circular shield tunneling machine "digger shield" equipped with a mechanical digger on a rotating arm. The ground was initially supported by W6×20 steel ribs at 4-ft spacing and continuous wood lagging in the flanges. The steel circular ribs were erected in the shield and pushed from the tail as the machine advanced. The circular ribs were expanded against the ground using spacers (dutchman). An HDPE water/gas proofing membrane was installed followed by a 12-in thick cast-in-place reinforced concrete final lining and invert. Figure 4 shows the typical cross section of the existing Red/Purple Line tunnels.

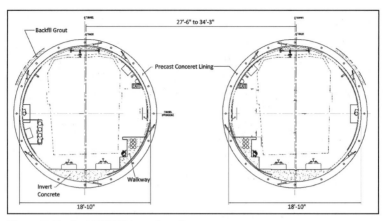

Figure 3. Typical RCTC tunnel section

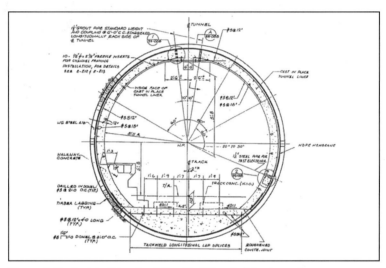

Figure 4. Typical section of existing Red/Purple Line tunnels

The key elements used in planning and coordinating the undercrossing without impact to the Red/Purple Line tunnel structures or train operations were:

- Understanding geologic conditions
- Design evaluation
- Instrumentation and monitoring program
- Contingency plan
- Coordination among stakeholders
- Tunnel workmanship

GEOLOGICAL CONDITONS

The ground near the crossing site consists of artificial fill, Pleistocene- and Holocene-age alluviums, and Pliocene-age sedimentary strata of the Fernando Formation. Artificial fill consists of mixtures of gravel, sand, silt, and clay, with construction debris. Beneath the artificial fill, the alluvium layer overlies an erosional contact with the Fernando Formation at variable depths below the surface. The alluvial deposits are subdivided into two units: finer grained and coarser grained. Both are dominantly comprised of sandy soils, while the coarser grained contains more poorly graded and well graded sand, more gravely soils, and much less silty and clayey soils. Fernando Formation consists of a poorly bedded to massive clayey siltstone to silty claystone that is poorly cemented and extremely weak to very weak (UCS 35 to 700 psi) per ISRM, 1978. The formation includes some bedding, sandy layers, concretions, and nodules that range from moderately to strongly cemented and range in strength from weak to strong per ISRM, 1978. The Fernando Formation commonly exhibits a

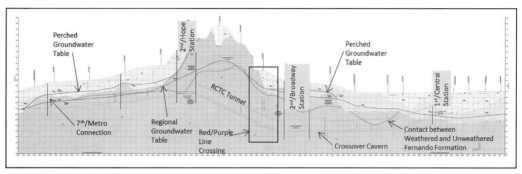

Figure 5. Geologic profile

weathered profile, described as varying from "moderately weathered" to "highly weathered," in comparison to the underlying "slightly weathered" to "fresh" bedrock. The crossing is entirely within the slightly weathered to fresh Fernando Formation with the ground cover ranging from 73- to 79-ft above the RCTC tunnel crown. Figure 5 shows the geologic profile along the entire alignment. As a tunneling ground, Fernando Formation around the crossing was described as a firm to slow raveling ground with local wedge failure according to the project Geotechnical Baseline Report. Shown on Figure 5, perched groundwater exists within lower portions of the alluvial deposits, due to the relatively low permeability of underlying Fernando Formation, and a regional water level is present within the Fernando Formation. These groundwater elevations can vary due to seasonal rainfall. The groundwater pressure at the springline of RCTC tunnels at the crossing was estimated about 1.8-bar.

DESIGN ANALYSIS

Prior to tunneling under the existing Red/Purple Line tunnels, numerical analyses were performed by the RCTC final designer, Mott McDonald, to assess the impact on the existing Metro Red/Purple Line tunnels due to RCTC tunneling. To predict the ground response due to tunneling, two stages of numerical modeling were conducted using FLAC3D (Figure 6). First, volume loss due to RCTC tunneling was correlated to EPB face pressure assuming the design ground parameters. In this stage, it was concluded that a volume loss of 0.3-% would be anticipated under the EPB face pressure of 10-psi, also predicted to a maximum vertical green field settlement of 0.35-in at the elevation of the Red/Purple Line tunnel invert. Second, sensitivity of the existing tunnels' response to a range of anticipated volume loss was evaluated. The numerical analysis concluded:

- Structural response of the Red/Purple Line tunnel linings meets Metro's design criteria.

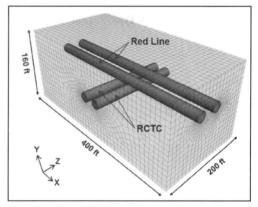

Figure 6. FLAC3D model of RCTC and Red/ Purple Line at crossing by Mott McDonald (2015)

- Impact mode on the Red/Purple Line would be a dip in the tunnel alignment approximately 400-ft long to a maximum dipping of 0.2-in.
- Pre-tunneling mitigation measures such as compensation grouting is not required.
- Existing circumferential crack widths are estimated to potentially increase by 0.02-in with no new cracks forming. Should existing cracks widen or new cracks develop wider than the predicted levels, remedial actions such as crack injection may be required.
- The magnitude and distribution of the estimated invert movements and the slope changes pose little risk to train operations.
- Recommended monitoring program to include:
  - Pre- and post-construction condition surveys that include crack measurements and leak detection.
  - Monitoring of transverse and longitudinal Red/Purple Line tunnel movements during RCTC mining.

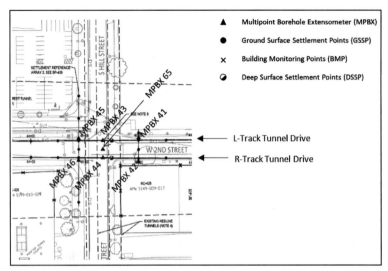

Figure 7. Location of MPBXs

– Monitoring of ground displacements in the immediate vicinity of the Red/Purple Line tunnels using surface monitoring point and multipoint borehole extensometers (MPBX).

MONITORING PROGRAM

Prior to the Red/Purple Line crossing, the RCTC contractor was required to demonstrate proof of tunneling performance. This was achieved by geotechnical instrumentation and monitoring of the TBM performance and ground/structure movement in the tunnel reach before the undercrossing. The RCTC project installed a comprehensive instrumentation system that was rigorously monitored during the 3,200-ft of tunneling leading up to the Red/Purple Line tunnels. The geotechnical instrumentation included Ground Surface Settlement Points, Deep Surface Settlement Points, Building Monitoring Points, Utility Monitoring Points, Liquid Level Monitoring, and a horizontal inclinometer Shape Array (SAA). Also, the RCTC EPBM was equipped with extensive instrumentation to measure and monitor earth support pressures, excavated volumes, and backfill grout to ensure there was virtually no ground loss or void occurrence during tunneling.

Horizontal inclinometers (Shape Arrays) and MPBXs were used in the relatively difficult alluvium deposits at the start of the L-track and R-track tunnels to measure TBM performance relative to settlement. In this difficult reach, about 0.5-in of settlement at 3-ft above the tunnel crown was observed and interpreted to be approximately 0.18% of ground loss by the tunneling (Hansmire et. al., 2017). Within the Fernando Formation prior to tunneling beneath

the Red/Purple Line tunnels, the ground movement at 3-ft above the RCTC tunnel crown was measured up to 0.1-in settlement by MPBXs. This data demonstrated that the crossing under the Red/Purple Line tunnels would be performed without risk of settlement to the existing tunnels.

At the crossing, there was an array of MBPXs installed 50-ft before, in between, and 10-ft after the Red/Purple line tunnels. These MPBXs were scheduled to take readings every 15 minutes while the TBM shield was within the influence zone (Figure 2). Beyond this zone, the MPBX reading frequency was reduced gradually to daily and weekly. A total of seven MPBXs were installed and monitored within the influence zone. Figure 7 shows locations of the MPBXs relative to the crossing.

To monitor the Red/Purple Line tunnels at the locations of the crossing, arrays of Convergence Monitoring Points (CMPs) were installed in 80-ft sections within each existing tunnel (Figure 8). The installation of CMPs started early March 2017 and required seven nights (four nights for installation and three nights for baseline readings) of Red/Purple Line track allocation to access the existing revenue tracks without interference to revenue operations. Manual survey readings of CMPs were performed every night during EPBM tunneling within the influence zone under the Red/Purple Line tunnels. The CMP targets were measured a week after and a month after passing the influence zone to verify no settlement had occurred. Figure 8 shows the CMP arrangement and distribution.

The Red/Purple Line tunnels were examined prior to the crossing with pre-condition surveys, photo documentation, and installation of crack gauge

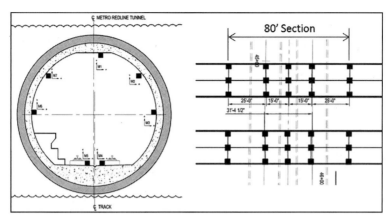

Figure 8. CMP arrangement and distribution

instruments to measure and document any movement in the tunnel lining.

Allowable limits were established for movements measured by MPBXs and CMPs consisting of Action Levels which required corrective measures be taken including procedural modifications that permit tunneling to continue without exceeding the specified Maximum Level. Maximum Levels were defined at which no further movement will be acceptable and an orderly cessation of tunneling and immediate implementation of remedial measures is required. Action and Maximum Levels for the MPBXs and CMPs are shown on Table 1.

COORDINATION MEETINGS

Weekly coordination meetings began approximately one month before the crossing among stakeholders including RCTC Project Team, RCTC Contractor, Metro Engineering, Metro Operations, Safety, and Rail Fleet Services. Initial coordination discussions focused on the construction activities and procedures associated with tunneling beneath the Red/Purple Line structures utilizing the EPBM. These discussions were extremely helpful for Metro Operations and Rail Fleet Services to understand the possible impacts of tunneling underneath the Red/Purple Line tunnels which enabled the team to better assess the actual risks to rail operations. From these discussions, Metro Operations recommended a speed restriction from 55- to 25-MPH for added precaution when any part of the TBM shield is directly under the Red/Purple Line tunnels.

Coordination meetings continued through the crossing process which included updates on instrumentation monitoring and track allocation requests to access the existing tunnels to verify that the tunnels had not moved.

The coordination meetings developed allowable limits for monitoring points (Tables 2 and 3)

Table 1. Allowable limits

	Action Level	Maximum Level
Longitudinal angular distortion	1/1000	1/600
Diametric convergence	0.25-in	0.5-in
Vertical in-tunnel displacement	0.2-in	0.35-in
Movement at 3-ft above TBM	0.25-in	0.75-in

and notification procedures to key Metro Operations, Safety and Rail Fleet Services personnel for any exceedance of allowable limits. In addition to immediate notification of the designated personnel, exceedance of allowable limits required the stoppage of all tunneling activities, and the initiation of a meeting to discuss mitigation actions. Furthermore, should an extraordinary event be notified, as defined by abnormal TBM performance indicators, a streamlined notification protocol would be implemented to immediately stop the TBM operation; notify Metro Train Control to stop the Red/Purple Line operations; notify the key personnel; and hold a tele-conference within an hour.

CONTIGENCY PLAN

Based on the RCTC design-build contract requirements and the discussions during coordination meetings, a contingency plan for Red/Purple Line crossing was prepared and documented. The contingency plan included Daily Review (DR) meetings with representatives from the RCTC Contractor and Metro Project Team to jointly review the instrumentation data and other records to ensure that the best control of ground movement was being achieved and proper construction progress was being maintained. The daily reviews included the overall tunneling operation, effectiveness of adjustments made to the construction procedures, TBM main parameters records, reconciliation of muck weights, instrumentation

Table 2. Movement of lowest sensor of MPBX during RCTC undercrossing

3' Above New Tunnel Crown	Heave	Settlement	Action Limits	Maximum Limits
MPBX 41	0.08"	0.26"		
MPBX 42	0.02"	None		
MPBX 43	0.15"	None		
MPBX 44	0.03"	0.05"	0.25"	0.75"
MPBX 45	0.03"	0.02"		
MPBX 46	0.02"	0.09"		
MPBX 65	0.07"	None		

Table 3. Movement of CMP within Red/Purple Line tunnels during RCTC undercrossing

Maximum Reading: May 3 to December 19, 2017	Red/Purple Line AL Tunnel	Red/Purple Line AR Tunnel	Action Limits	Maximum Limits
Angular distortion between arrays	0.0003	0.0003	0.001	0.0017
Diametric angular distortion within array	0.0003	0.0003	0.001	0.0017
Diametric convergence	−0.10" / 0.22"	-0.14" / 0.08"	±0.25"	±0.5"
Vertical displacement	−0.20" / 0.10"	-0.19" / 0.19"	±0.20"	±0.35"

data, shift and progress reports, and quality control reports. A Management Action Team (MAT) was established consisting of authorized representatives from RCTC Project Team, RCTC Contractor, Metro Engineering, Metro Operations, Safety, and Rail Fleet Services, and RCTC Designer with authority to issue a stop train notice and initiate shutdown of the TBM operation if a Maximum Level was breached. A MAT meeting would be called when a Maximum Level was breached or when a Shutdown Initiation was issued by DR meetings. MAT meeting attendees would review all monitoring data, updated geological models, relevant information and mitigation measures or contingencies to determine the immediate and following measures to be implemented.

Upon reaching Action or Maximum Level, an automatic alarm system was established to notify the RCTC Contractor and the RCTC Project Team within five minutes of the alarm. The RCTC Contractor was responsible to confirm the validity of the alarms within 12 hours for MPBX alarms or 24 hours for CMP alarms due to the time required for manual survey verification within track allocation constraints. Upon confirmation, the causes for the alarms were to be investigated with corrective actions implemented. Any alarm was considered as valid until proven otherwise.

The following corrective measures were identified to ensure that the excavation and construction can resume safely without inflicting damage and ensure safe operation of the Red/Purple Line following an Action/Maximum level alarm. Shutdown Initiation would be discussed and amended as required by the DR meeting for Action Level breach and by the MAT

meeting for Maximum Level breach. The possible identified corrective measures included:

- Instrumentation and Inspection
  - Increased frequency of monitoring
  - Installation of additional instrumentation
  - Implementation of additional geotechnical investigation
  - Tunnel structure inspection
- Engineering Evaluation
  - Perform additional calculations and design reviews
  - Assessment of the structure.
- Tunnel Construction Procedures Modifications
  - Review face pressure, annulus bentonite process and/or grout pressure requirements
  - Where lost ground and incomplete grouting result in voids around the tunnel excavation, perform approved measures to fill voids.
  - Perform training and/or re-training of personnel as necessary
  - Proceed to face inspect of tools and geology (either a manned intervention or by camera only)
  - Grouting and/or ground improvement from the tunnel and/or from the surface

To minimize the risks associated with the crossing, good tunnel workmanship and a diligent and vigilant implementation of a TBM performance monitoring program was essential. The TBM used on this project had state-of-the-practice systems that were specifically designed to support the ground at all times. These systems along with the procedures and protocols planned for the project were expected to minimize any ground loss that could lead to settlement. Multiple checks and balances in the established procedures and protocols ensured a safe and successful construction of the tunnels underneath the Red/Purple Line. The RCTC Project Team was required to be diligent and vigilant in the performance of this critical work using the proven tools to monitor TBM performance which had been demonstrated on the initial reach of the tunnel drive before the Red/Purple Line under crossing.

EXECUTION AND OUTCOMES

Tunneling Progress

Since beginning of the L-track tunnel excavation in February 2017 at the launching wall in the 1st/Central station excavation, the RCTC TBM excavated approximately 3,000-ft before approaching the Red/Purple Line tunnels. In early May 2017, the TBM arrived at the Red/Purple Line crossing, and successfully excavated through the Red/Purple Line tunnel influence zone (Figure 2). The R-track tunnel launched at the same launching area in September 2017 and crossed the Red/Purple Line tunnels in mid-November 2017. The TBM was operated 24 hours per day in a continuous mining mode through the influence zone of the crossings.

The L-Track tunnel crossing through the 220-ft influence zone took 51 hours from May 3, 2017 1:00 PM to May 5, 2017 4:00 PM. The TBM shield was directly under the Red/Purple Line tunnels a period of 20 hours (May 4, 2017 6:00 AM to May 5, 2017 2:00 AM) for the 105-ft reach under the Red/Purple Line tunnels.

The R-Track tunnel crossing through the 220-ft influence zone took 50 hours from November 16, 2017 3:30 PM to November 18, 2017 5:30 PM. The TBM shield was directly under the Red/Purple Line tunnels a period of 16.5 hours (November 17, 2017 1:30 AM to November 17, 2017 6:00 PM) for the 105-ft reach under the Red/Purple Line tunnels.

During these under crossing periods, the Red/Purple Line train operation was in normal schedule without any interruption; no abnormality or incident was reported; no crack movement was detected by the crack gauges; no new cracks were found during post-condition survey; no check grouting was required based on the performance of backfill grout; and no tunnel or ground movement exceeded limits. During tunneling within the influence zone for both tracks, the average face pressure was measured 2.0- to 3.3-bar; the average backfill grout pressure varied 1.0- to 4.4-bar; and the total backfill grout

volume was measured 153- to 162-ft^3 which was 95- to 106-% of the theoretical backfill grout volume. Tunneling under the Red/Purple Line tunnels was prompt without any need for implementing mitigation measures.

The tunneling progress for the Red/Purple Line crossing is summarized as follows:

L-Track Tunnel

- May 3, 2017 1:00 PM—TBM head arrived at 60-ft from Red/Purple Line tunnel.
- May 4, 2017 6:00 AM—TBM head arrived at and was directly under Red/Purple Line tunnel.
- May 5, 2017 2:00 AM—TBM tail just passed Red/Purple Line tunnel.
- May 5, 2017 4:00 PM—TBM tail passed 60-ft from Red/Purple Line tunnel.

R-Track Tunnel

- November 16, 2017 3:30 PM—TBM head arrived at 60-ft from Red/Purple Line tunnel.
- November 17, 2017 1:30 AM—TBM head arrived at and was directly under Red/Purple Line tunnel.
- November 17, 2017 6:00 PM—TBM tail just passed Red/Purple Line tunnel.
- November 18, 2017 5:30 PM—TBM tail passed 60-ft from Red/Purple Line tunnel.

Monitoring Results

As described above, seven MPBXs (No. 41, 42, 43, 44, 45, 46, and 65) and five CMP arrays were the primary instruments monitored closely during the undercrossing. The MPBXs were automatically actuated and read by dataloggers every 15 minutes and the CMPs were manually surveyed every night during the two periods of 51-hour L-Track tunneling and the 50-hour R-Track tunneling. After the crossings were excavated, the CMPs were surveyed manually at nights of May 9, May 12, June 8, November 27, and December 19, 2017 to verify that no movement had occurred. The resulting data from the lowest sensor of MPBXs installed about 3-ft above the new RCTC tunnel crowns showed 0.02- to 0.15-in of heave and none to 0.26-in settlement during the crossings. As shown in Table 2, the lowest sensor of MPBX 41 indicated the settlement exceeding Action Limit of 0.25-in prior to the TBM arrival at the sensor, but it was accepted due to manual survey accuracy for the MPBX head, and the action decision was to continue tunneling and monitoring. The CMP arrays indicated movements of the Red/Purple Line tunnel well within the allowable limits (Table 3).

CONCLUSION

Metro's RCTC twin bored tunnels were successfully constructed under Metro's existing Red/Purple Line tunnels built in the 1990s with a clearance between the old and new tunnels as low as 5-ft. The structural integrity, safety and revenue operation of the existing subway line was maintained during tunnel construction which was performed without any induced movement or new or larger cracks in the lining of the existing tunnels. No voids were developed under the existing tunnels and there was no interference to transit revenue operations. The tunnel crossings were a challenging task for engineering and construction management and the successful crossings are attributable to the following key elements: 1) thorough engineering evaluations performed using geotechnical investigations, previous experience, and design analysis; 2) planning and implementation of a practical, diligent, and vigilant instrumentation and monitoring program; 3) active stakeholder engagement and coordination; and 4) good tunnel workmanship employed by the RCTC Contractor in the performance of the work.

ACKNOWLEDGMENTS

Since Metro issued the Notice to Proceed on July 7, 2014, revenue operations are anticipated in 2021. Gary Baker is Metro's Project Manager. The design/build contractor is Regional Connector Constructors (RCC), a joint venture of Skanska USA Civil and Traylor Bros., Inc. Geotechnical instrumentation is by Geocomp. Final Design is by Mott MacDonald (MM). Preliminary Engineering was undertaken by CPJV, a joint venture of AECOM and WSP USA. During construction, CPJV is providing engineering support services and Arcadis provides Construction Management Support Services (CMSS) to Metro. EPC Consultants, Inc is part of the CMSS team. Bill Hansmire serves as Metro's Design Manager for the overall Project. Dick McLane is responsible for all TBM tunneling. Christophe Bragard is responsible for construction engineering for underground work including tunnels, shafts, and geotechnical instrumentation.

REFERENCES

Duntton, M. 2015. *Tunneling Impact Assessment—Metro Red Line Tunnels.* 294646_DU03_07.1_RE_0028. Contract No. C0980. Mott MacDonald

Hansmire, W.H., McLane, R.A, Frank, G., Bragard, C.Y., 2017. *Challenges for Tunneling in Downtown Los Angeles for the Regional Connector Project.* Proceedings of the World Tunnel Congress—Surface challenges—Underground solutions. Bergen, Norway.

What Comes Down Must Go Up—Dewatering-Induced Ground Movement on the SR-99 Bored Tunnel Project, Seattle, WA, USA

Joseph Clare
Mott MacDonald

ABSTRACT: The Alaskan Way Viaduct Replacement Program was a $3.1 billion program to replace an aging viaduct in Seattle, Washington, USA. The bored tunnel project was a $1.3 billion design-build project that included a 2.8 km long, 17.4 m diameter road tunnel. Construction of a 24.3 m diameter, 36.5 m deep access shaft for TBM repairs relied on dewatering to depressurize a confined aquifer below the shaft base. Early design predictions estimated that settlement would be limited to the immediate vicinity of the shaft. Actual depressurization lead to unexpected widespread settlement which included numerous historic buildings and aging city infrastructure. Extensive monitoring relied on traditional and remote instrumentation of private and public facilities which proved useful to record ground movement.

PROJECT INTRODUCTION

The Alaskan Way Viaduct section of State Route 99 (SR-99) has been a fixture of the downtown Seattle waterfront for over five decades. The two-level viaduct carries about 110,000 vehicles a day and provides a convenient route to and through downtown Seattle. Among its transportation functions, the viaduct provides an important north-south route for neighborhoods west of Interstate 5 (I-5). The SR-99 corridor plays an important role in freight mobility, providing a major truck route through downtown, to the Ballard-Interbay and greater Duwamish manufacturing and industrial centers. However, the viaduct's days are numbered. In 2001, the Nisqually earthquake caused damage to the viaduct and wear and tear from daily traffic have also taken their toll on the facility (WSDOT 2009).

In January 2009, WSDOT chose a bored tunnel alternative to replace the viaduct using a design build approach. At the time, this tunnel would be the world's largest bored tunnel at 17.48 m (57.3 feet) bored diameter and 15.8 m (52 feet) interior diameter (Sowers 2014). Large tunnels were not new to WSDOT as the nearby Mt. Baker Ridge I-90 tunnel remains the world's largest soft ground tunnel at 19.8 m (65 feet) interior diameter (Robinson 2002). Figure 1 depicts the tunnel interior layout of the new SR-99 tunnel and Figure 2 shows the alignment through the city of Seattle.

GEOLOGY AND HYDROGEOLOGY

The start of the tunnel was in the Pioneer Square neighborhood located immediately south of the central business district of Seattle. This is an historic part of Seattle with less than favourable Holocene and fill soils. Fill soils are prevalent with underlying alluvium, beach deposits and reworked glacial deposits. The fill often contains wood debris associated with past lumber mill activities that occurred in Seattle's early settlement (Williams 2015). Over consolidated glacial soils (glacial till and sand and gravel) are the primary soils underlying the fill and Holocene soils. A thick layer (greater than 30 feet thick) of glacial clay and silt underlie the tills. For the first 300 m of tunnel, the TBM mined through these fill and Holocene deposits at a 4.5 percent down gradient to eventually be fully surrounded within the glacial soils for the remainder of the tunnel route.

The groundwater regime of the Seattle area is complex. Several glaciations have influenced the

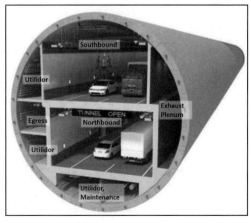

Figure 1. Tunnel structural section

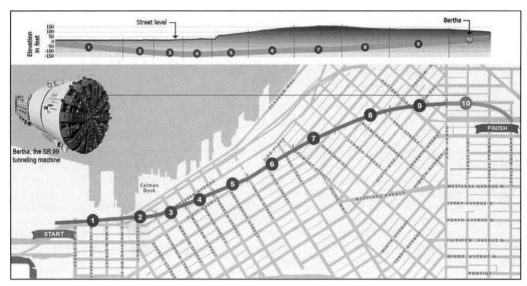

Figure 2. SR-99 route through city of Seattle

geology to create sand and gravel layers forming aquifers; and silts, clays, and tills, that act as aquitards. At least three aquifers, locally separated by aquitards, may be present within the upper 70 m at the at the south end of the tunnel alignment (S&W 2015). An unconfined upper aquifer lies within the near surface Holocene fills and granular deposits. A middle aquifer is confined within the glacially overridden sand and gravel materials. A lower confined aquifer lies beneath the locally thick clay and silt aquitard. The thickness of this lower aquifer is not known and generally extends below 45 m below the surface.

TBM STOPPAGE AND REPAIR

WSDOT selected a design-build contract approach for the SR-99 tunnel to deliver the tunnel. In December 2010, Seattle Tunnel Partners (STP), was competitively selected by WSDOT. Prior to tunnelling, the contractor implemented the south end settlement mitigation plan (SESMP) which consisted of closely spaced concrete drilled shafts along the east and west sides of the tunnel for over 300m of the tunnel alignment. These concrete shafts were intended to mitigate damage to the adjacent viaduct from any ground loss occurring during tunnelling.

The contractor, STP, launched the earth pressure balance tunnel boring machine (TBM) in the summer of 2013. In early December 2013, after mining approximately 300 m, the TBM struck an abandoned steel well casing left over from an earlier investigation phase of the project. The TBM stopped to conduct inspections and removal of the obstruction. Mining resumed for a short length in early

2014 and then stopped once again to investigate high temperature readings and a slow advance rate. This stoppage revealed that the seal system on the main bearing was damaged and replacement would be required. Because of the damage, the STP elected to construct an Access Shaft to facilitate additional inspection and repairs to the TBM.

Access Shaft Design

The shaft location was adjacent to the existing viaduct structure, just north of the TBM location, and within the previously constructed SESMP piles. The SESMP piles were constructed prior to tunnelling and form a row of 500 mm diameter concrete piles along the east and west sides of the tunnel alignment.

The Access Shaft design would comprise a circular secant pile shaft along with the creation of a box around the rear of the TBM shield. A 2000-ton mobile lift tower would be constructed to hoist the TBM cutterhead and drive assembly to enable repairs. Only the cutterhead and forward shield section would extend into the shaft. The remaining mid and tail shield would remain within the shaft wall and just outside the shaft and within this box.

Construction of the Access Shaft to repair a then world's largest TBM was itself a major undertaking. Although the TBM was still relatively shallow—one diameter below the surface—but at nearly 17.5 meters, the shaft would be required to be 37 m deep with an inside diameter of about 25 m. Shaft shoring system comprised unreinforced concrete secant piles varying in diameter from 1 to 3 meters to form a compression ring without the

Figure 3. Access Shaft schematic depicting the SESMP piles and dewatering wells

need for internal bracing. Installation of 3-m-diameter unreinforced secant piles comprised the larger compression ring of the shaft. Smaller 1m diameter secant piles were used to fill in gaps and to incorporate the existing SESMP piles into the shaft structure. Installation of the 3m secant piles was ultimately successful but was challenging due to ground instability and taking precautions to not cause impact to the adjacent viaduct structure.

To create the box behind the TBM shield, jet grout columns were installed between the existing SESMP piles and along the top of the segmental tunnel lining behind the TBM shield. Chemical grout was utilized to seal below and behind the shield. The base of the shaft was not designed to resist groundwater uplift and no shaft seal was planned for the TBM shaft entry. To mitigate against ground loss during TBM entry and to compliment the structural design of the shoring system, a dewatering system was necessary. The Access Shaft design is depicted in Figure 3.

Initially, the plan was to limit dewatering to the interior of the shaft and to only dewater the upper and middle aquifers (confined within the shaft) with the base of the shaft terminated in clay. However, during installation of the dewatering wells it was determined that the clay beneath the shaft contained interbeds of silts. Out of concern for piping through the clay base plug, the plan was modified to install four 67-m deep wells outside of the shaft to depressurize the lower aquifer below the clay base. Dewatering of this aquifer began in late 2014 along with excavation of the shaft interior.

The Access Shaft dewatering scheme consisted of:

- Access Shaft Wells: Four dewatering wells within the shaft to facilitate shaft excavation. These wells extended approximately 6 m below the bottom of the secant pile shoring. Prior to excavation, these wells cumulatively pumped about 54 to 109 m^3/day reducing to 27 to 54 m^3/day as excavation progressed. After excavation was complete, discharge was less than 11 m^3/day.
- Deep Wells: Four deep dewatering wells were installed outside the shaft perimeter and extended into the lower aquifer. These four wells were used to depressurize the lower aquifer below the silt/clay base of the shaft to address shaft bottom stability concerns and pumped up to 4088 m^3/day.
- SESMP Box Wells: Four wells were installed within the confines of the box constructed around the TBM shield on the south side of the access shaft. These wells discharged up to 490 m^3/day prior to shaft excavation and increased up to 654 m^3/day during shaft excavation.
- Silt wells: To address piping and shaft bottom stability concerns, three wells were installed within the silt/clay unit within and below the shaft base. These three wells discharged up to 43 m^3/day however one of the wells accounted for 80 percent of the discharge.

Instrumentation

The Pioneer Square area of Seattle has had a long history of settlement with uneven and cracked sidewalks, paved areas, and cracks and settlement of the older masonry buildings. This historical settlement has occurred prior to and continues to this day independent of the project. Historical settlement is largely due to placement of fills and development on top of historical tidal lagoon sediments. Past logging mill operations and waterfront development has resulted in large deposits of wood debris and loose fills that are undergoing continued compression and decomposition over time. The ongoing settlement has typically been localized and near the ground surface resulting in clearly observable cracks and gaps in surfaces and buildings.

The project team recognized the occurrence of historical settlement early on and for this reason installed deep benchmarks extending through the Holocene soils. Six deep benchmarks were installed spaced outside the tunnel alignment, each comprised of a 25-mm diameter fiberglass rod grouted into a 100-mm diameter borehole within the glacially over consolidated soils. A 76-mm diameter casing with viscous grease surrounds the rod above the grouted zone in the Holocene soils to allow free movement of the rod inside the casing. The intent of these deep benchmarks was to provide stable project control that was not affected by settlement of the Holocene soils.

WSDOT's technical requirements formed the basis for an instrumentation program for the tunnel alignment. Soldata, now Sixense, was competitively selected to propose and deliver an instrumentation program specifically for the tunnelling aspect of the project. Instrumentation was developed to monitor surface, near surface, and deep subsurface behaviour to provide early detection of ground deformation such that the operations of the TBM could be adjusted. Many of the surface, building, and utility mounted instrumentation were monitored for ongoing risk assessment and to mitigate the potential for third-party claims. Additionally, measurements of ground deformation provided a means of calibrating the extent and magnitude of predicted settlement for use on the project and to advance the industry.

Instrumentation used to monitor ground deformation included:

- Automatic Motorized Total Stations (AMTSs) consisting of remote controlled automatic total station devices that conduct repeatable measurement of reflector targets on structures and surfaces to monitor movement.
- Automatic Reflectorless Settlement Points (ARSPs), similar to the above AMTS but relying on natural reflection of surfaces to measure settlement.
- Manual Survey Monitoring Points (MSMPs) consisting of targets or rods that are measured using manual surveying techniques

to estimate movement of utilities (USPs), structures (SMPs), surface and near surface ground (NSSPs).
- Multiple Point Borehole Extensometers (MPBXs) consisting of borehole installed sleeved rods, anchors, and displacement sensors to measure vertical ground movement at different depths.
- Vibrating Wire Piezometers (VWPs) consisting of instruments installed in boreholes at various depths to measure piezometric pressure in the ground.
- Observation Wells (OWs) consisting of an open borehole with a data logger installed to measure groundwater levels.
- Viaduct Monitoring—A multitude of instrumentation was installed on the existing viaduct to monitor tilt, settlement, and changes in crack widths.

In addition to the above, Sixense, STP, and WSDOT collaborated to utilize satellite radar interferometry to evaluate settlement caused by tunnel excavation. Interferometry synthetic aperture radar (InSAR) techniques were used to measure settlement over a broad area of the tunnel alignment. Satellite images were taken approximately every 10 days and evaluated utilizing a Sixense proprietary software system, Atlas, for monitoring settlements. In an urban area, the typical density of natural reflectors is about 10,000 points per square kilometer. Settlement is measured when images are compared to previous or to a baseline image. This data and analysis was beneficial to provide a high density of displacement over a wide area overlapping and extending beyond the extent of all project installed instrumentation.

WSDOT geometric office was tasked with providing control survey for the project. In this role, WSDOT brought in control survey from outside the project area to establish control elevation on project installed deep benchmarks. Three primary deep benchmarks were established by drilling through the Holocene deposits and founded in the underlying over-consolidated glacial sediments. WSDOT was responsible for periodic monitoring of the Viaduct bents (columns) and surface points throughout the project area. WSDOT also conducted ground based LiDAR scanning block by block of the tunnel alignment to evaluate movement of infrastructure in addition to the point specific monitoring points.

DATA ANALYSIS

Construction of the Access Shaft to repair the TBM began in spring of 2014 and excavation commenced in October of the same year. The dewatering system was activated in the fall of 2014 and the depressurization of the lower aquifer began in early November

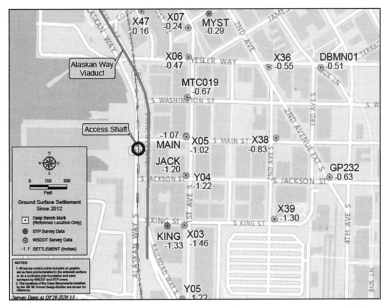

Figure 4. Measured settlement in inches of WSDOT points between 2012 and June 2015

2014. As a result of the dewatering, piezometric elevations of the lower aquifer were lowered by as much as 43 m near the Access Shaft. Drawdown of the middle aquifer was over 30 m.

In mid November 2014, STP survey crews measured differences between secondary control points when checked against the primary (deep benchmarks) control points. Subsequent measurements confirmed the difference and WSDOT survey teams were dispatched to evaluate the entire project control. Digital level surveys performed by WSDOT confirmed that the deep benchmarks had settled. Settlement of the benchmarks measured to 18 mm in January 2015.

Following confirmation that the benchmarks had settled, WSDOT and STP began a rigorous process of re-establishing project control by bringing in control from well outside the project area and evaluating all the project instrumentation and critical infrastructure including the existing viaduct structure.

In addition to re-establishing project control, WSDOT survey teams manually surveyed surface points throughout the project area. The results indicated that widespread settlement of the area had occurred. Settlement was confirmed up to 500 m east and 1000 m north and south of the Access Shaft location.

This settlement was also confirmed in the vast amount of the project instrumentation. Remote imaging from InSAR revealed the striking extent and distinct boundaries of the settlement. The data

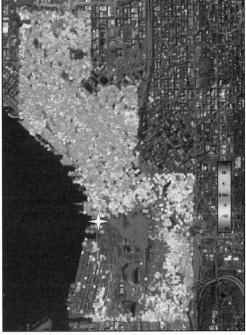

Figure 5. InSAR measured deformation in vicinity of Access Shaft (denoted with star) spanning the period of dewatering 7/10/2012 to 11/24/2015. Red (darker) area shows settlement, green (lighter) shows no change (Hochard 2015).

set in InSAR image in Figure 5 shows the settlement from just a few weeks following the start of deep dewatering (Hochard 2015). Cumulative settlement was approximately 30 mm in the vicinity and to the east of the Access Shaft. Dewatering continued for another year to allow the repair of the TBM to proceed.

With the TBM repair complete and resumption of mining in early 2016, STP began to undertake remedial actions to backfill the Access Shaft and halt the dewatering efforts. The four deep dewatering wells were sequentially turned off in January 2016 followed by the four wells in the box on the south side of the shaft in February 2016. Prior to turning off the wells, the team prepared a plan to turn off the wells sequentially and monitor instrumentation twice daily to measure any elastic rebound. WSDOT conducted twice daily measurements of the Viaduct structure and conducted level loops throughout the Pioneer Square and downtown parts of the city. Communication proved to be key in this activity with stakeholders including the historic Pioneer Square neighborhood and the city of Seattle. Daily reports were distributed to stakeholders summarizing the status of activities and providing selected instrumentation results.

Figure 6 provides the results of monitoring the groundwater recovery with the stepped shutdown of the four deep depressurization wells, DDW1 through DDW4. The figure shows the initial small rise in piezometric elevation with the first well turned off, followed by increasing elevation rise as additional wells were turned off. A 30-m peizometric elevation rise was measured in the first 5 days of turning off the wells.

Along with monitoring the groundwater recovery, monitoring of the ground throughout the area was performed to monitor any rebound. Figure 7 depicts the measured deformation of viaduct bent 101 west located just north of the Access Shaft. The steep decline corresponds with the start of dewatering in November 2014 followed by a continued slight settlement as dewatering continued. Figure 7 also shows the steep rise in ground rebound measured on the viaduct column 101 west just north of the Access Shaft as the wells in the lower aquifer were turned off. The majority of measured rebound occurred in two months following ceasing dewatering. Within a year of ceasing dewatering, deformation of this point returned to pre-dewatering levels.

As the groundwater recovery and ground rebound tapered off, the frequency of monitoring was also reduced. The amount of rebound that occurred can also be seen in the InSAR image Figure 8. The instrumentation shows that the amount of rebound has nearly equaled the initial settlement deformation.

SUMMARY

The depressurization of the lower aquifer for the Access Shaft construction appears to have caused

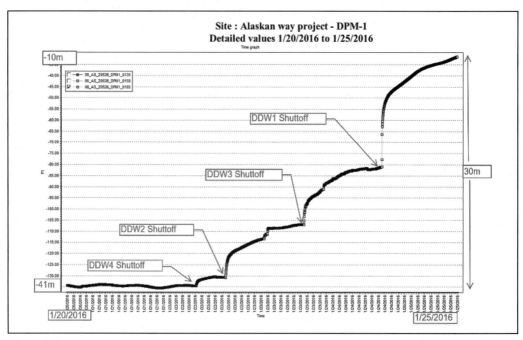

Figure 6. Stepped recovery of piezometric level of lower aquifer during dewater well shutoff sequence

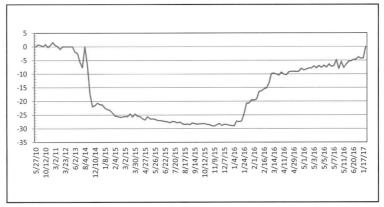

Figure 7. Measured deformation (mm) of viaduct column (bent) 101W in vicinity of Washington Street just north of the Access Shaft from 2010 to 2017

Figure 8. InSAR measured deformation in vicinity of Access Shaft (denoted with star) spanning the period of dewatering 1/08/2015 to 9/07/2016. Blue (darker) area shows rebound, green (lighter) shows no change (Hochard 2016).

compression of the deep glacially over consolidated soils resulting in settlement over a relatively large area. The initial ground deformation resulting from depressurization of the lower aquifer was not anticipated by the project team. This was largely due to the presence of glacially over consolidated sediments and a local history of not having deformation of these deposits in the past. Because this deformation originated from deep sediments, the impact to the surface was spread over a wide area and did not result in localized differential settlement.

Following cessation of dewatering, monitoring indicated the recovery of the lower aquifer and a corresponding ground rebound. Instrumentation monitoring after dewatering was useful to identify the extent and magnitude for comparison with initial measurements. The widespread installation and use of instrumentation for the project provided useful background information of long-term and continued settlement of the Holocene fill sediments in the Pioneer Square area. Much of this was limited to areas with known compressible and decomposing organic fill deposits and not project related.

In conclusion, the use of a multi-tiered large format instrumentation program has proved useful to the project team to not only manage the intended tunnelling aspect but also useful to monitor for unanticipated construction activities. The use of remote InSAR monitoring proved useful to measure wide areas outside of the immediate project limits and for comparison over long time periods.

REFERENCES

Hochard, Guillaume. December 22, 2015. *Automatic Settlement Monitoring Using Satellite Interferometry, Alaskan Way Tunnel Atlas—Report 15.* Nanterre, France. Soldata.

Hochard, Guillaume. October 17, 2016. *Automatic Settlement Monitoring Using Satellite Interferometry, Alaskan Way Tunnel Atlas—Report 18.* Nanterre, France. Soldata.

Robinson, R.A., Cox, E., Dirks, M. 2002. Tunneling in Seattle—A History of Innovation. *Proceedings of the North American Tunneling Conference.* Edited by L. Ozdemire. A.A. Balkema.

Shannon & Wilson, May 14, 2015. *Interim Settlement Evaluation Report, Alaskan Way Viaduct Replacement Program.* Washington State Department of Transportation. Seattle, Washington.

Sowers, D., Galisson, L., Audige', E. 2014. Current and best practice for monitoring tunnelling projects in urban areas—Alaskan Way Bored Tunnel, Seattle, WA. *Proceedings of the Australalasian Tunnelling Conference* September 2014.

Williams, David B. 2015. *Too High and Too Steep: Reshaping Seattle's Topography.* University of Washington Press. p.31.

WSDOT, September 2009. *Alaskan Way Viaduct Replacement Project History Report.* Washington State Department of Transportation.

Widening of a Road Tunnel Without Interruption in Service

Marco Invernizzi
Alpi Engineering LLC in Association with Rocksoil Spa

Pooyan Asadollahi
Parsons Corporation

ABSTRACT: Considering the growth in urban population, widening transportation tunnels is essential to ease traffic flow in congested routes. While changing the route or halting traffic may not be feasible, the only practical option would be to maintain the tunnels in service during the widening construction activities. This paper describes a practical example of a tunnel widening technique adopted in Nazzano Tunnel, a 2-lane tunnel in Rome, Italy, to which an additional lane was added. The planning and design of this project required particular sensitivity to the contractual obligations related to traffic and work deadlines to minimized risk exposure. The construction technique including the precut steel shell, ground improvement resins and segments as well as the different construction stages and productivity are presented.

INTRODUCTION

This paper presents a practical example of a tunnel widening technique through which transportation tunnels can be maintained in service during construction. The technique was adopted in Nazzano Tunnel near Rome, Italy, from Station 522+000 to Station 523+200 of "A1 Milan-Naples Motorway: Orte-Fiano Romano Section." This is the first true experimentation of so-called 'Nazzano Method' carried out between 2004 and 2007 (Lunardi et al., 2007 and 2014; Tonon, 2010).

To ease the traffic flow through the tunnel and improve the quality of service offered, the tunnel was planned to be widened from two to three lanes. After works was completed, each direction has: one fast lane with 11.5 feet width; two slow and normal lanes each with 12.3 feet width; and one 10-foot-wide hard shoulder. The original 2-lane tunnel had radius of 20 feet, cross-section area of 1,022.6 square feet and road width of 31.2 feet while the widened 3-lane tunnel has radius of 30.7 feet, cross-section area of 2,583.3 square feet and road width of 48.8 feet.

Figure 1 depicts initial and final states of the road before and after widening operation. Figure 2 illustrates plan view of the motorway and the tunnel as well as the cross-section of original and widened tunnel.

The three following options were considered for the tunnel widening:

1. Changing the route: which had environmental issues
2. Tunnel widening with traffic halted: which had traffic problem (1,800 vehicle per hour on average)
3. Tunnel widening under traffic: which was the only feasible solution

The main construction activities of the project were as follows:

- Underground work: twin tunnel (North and South carriageways) with length of 1,106 feet
- Work on surface:
  - Prefabricated artificial tunnels and anchored sheet micro-piles walls
  - Culverts and retaining walls
  - Cutting and embankments

Ground along the tunnel belongs to the Plio-Pleistocenic marine series constituted over three sedimentary events and has friction angle of 22 to 26 degrees with cohesion of 0.4 to 1 ksf and unit weight of 127 pounds per cubic feet.

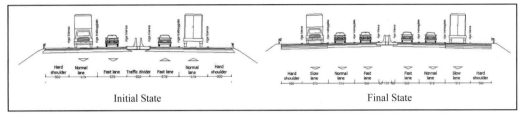

Figure 1. Cross-sections of each direction before and after widening

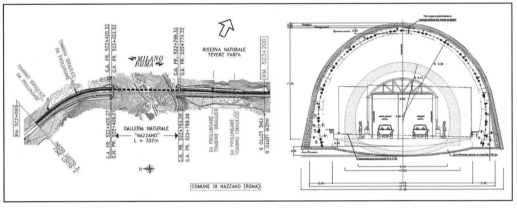

Figure 2. (left) Plan view of the motorway and Nazzano Tunnel; (right) cross-section of original and widened tunnel

CONSTRUCTION STAGES

Planning of construction stages was done by respecting the contractual obligations as far as traffic concerns. Three main stages were as follows (see Figure 3):

- Stage 1:
  - Northbound Tunnel:
    - Tunnel widening
    - Absence of tunnel invert
    - Traffic on two lanes 10.7 feet
    - Traffic protection by means of a steel shell
  - Southbound Tunnel:
    - No work
- Stage 2:
  - Northbound Tunnel:
    - No work
    - Absence of tunnel invert
    - Traffic on four lanes 11.5 feet wide (three lanes plus one south lane)
  - Southbound Tunnel:
    - Tunnel widening in the presence of traffic
    - Construction of tunnel invert (in the absence of traffic—15% of total time of tunnel construction)

- Level of service like the final one (three plus three lanes available over 85% of construction time) able to manage the peak of traffic volume
- Stage 3:
  - Northbound Tunnel:
    - No work—when the execution of the invert in the southbound tunnel was completed, the same work was done for the northbound tunnel reversing traffic flow
    - No invert
    - Traffic on four lanes 11.5 feet wide (three lanes in direction south plus one lane in direction north)
  - Southbound Tunnel:
    - Execution of the invert in two halves
    - Traffic on two lanes 10.7 feet wide

METHODOLOGY OF THE TUNNEL WIDENING

Technologies that were necessary to widen a tunnel without interrupting traffic includes mechanical precutting, steel traffic protection shell and active arch construction with a multi-purpose equipment as schematically depicted in Figure 4. Steel traffic protection shell had a length of 197 feet; six modules

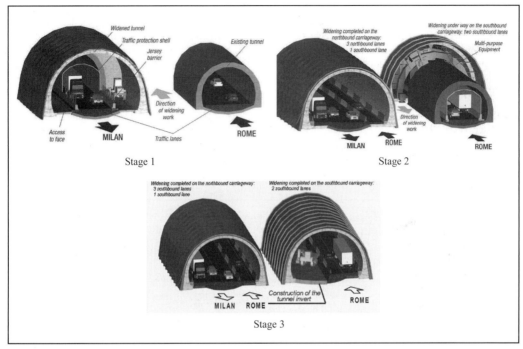

Figure 3. Construction stages

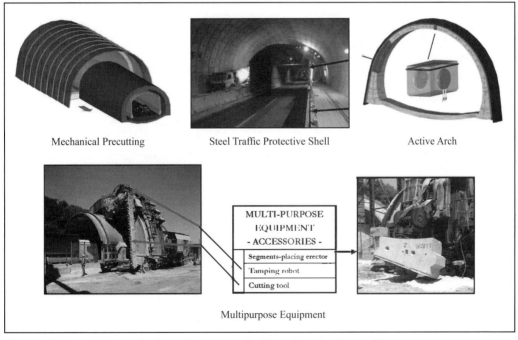

Figure 4. Technologies required to widen a tunnel without interrupting traffic

with length of 32.8 feet each, two of which are reinforced at the tunnel face. The transportation system of the protection shell is rail mounted on New Jersey type reinforced concrete support elements. The active arch construction includes 19 reinforced concrete prefabricated segments with compressive strength of 5,800 psi.

CONSTRUCTION ASPECTS OF THE TUNNEL WIDENING

Construction phases include (see Figure 5 for layout of the yard during each operation phase):

- Mechanical precutting execution:
  - Execution of precutting shell
  - Mucking by mechanical shovel
- Excavation of the face and demolition of the existing tunnel:
  - Excavation and demolition of the roof
  - Excavation and demolition of the sidewalls
- Placing of the active arch segments:
  - Placing of the sidewall segments
  - Hooking the segment of the erector of the multi-purpose equipment
  - Installing the segments

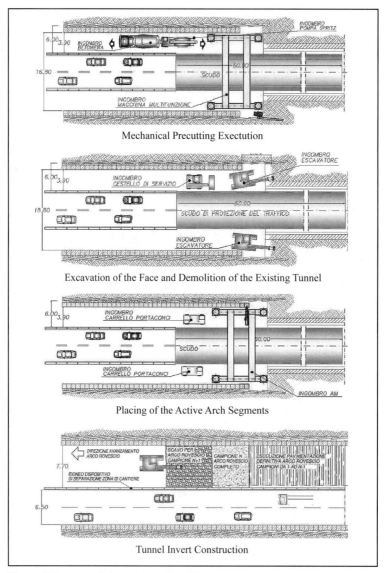

Mechanical Precutting Exectution

Excavation of the Face and Demolition of the Existing Tunnel

Placing of the Active Arch Segments

Tunnel Invert Construction

Figure 5. Layout of the yard during different operation phases

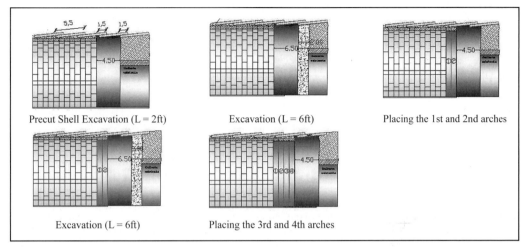

Precut Shell Excavation (L = 2ft)   Excavation (L = 6ft)   Placing the 1st and 2nd arches

Excavation (L = 6ft)   Placing the 3rd and 4th arches

Figure 6. Advance cycle

- Construction of the tunnel invert in two stages:
  - Excavation
  - Assembling of steel cages
  - Concrete casing

The stages and the sequence of them, as carried out to widen the Nazzano tunnel, are summarized in Figure 6. Each cycle is comprised of the execution of the precutting stage and four subsequent alternating stages of excavation and erecting the final lining of pre-fabricated concrete segments. The advancement rate was three feet per day.

ACKNOWLEDGMENTS

The Owner of Nazzo Tunnel is Autostrade of Italy. The widening operation was designed by Rocksoil Spa and constructed by Cossi Construction Spa. The authors thank Cristiano Bonomi for his contribution to this paper.

REFERENCES

Lunardi, P., Lunardi, G., Cassani, G. 2007. Widening the Nazzano motorway tunnel from two to three lanes + an emergency lane without interrupting traffic. Proceedings of the International Congress on "Tunnels, drivers of change," Madrid, 5th–7th November 2007.

Lunardi, P., Selleri, A., Belfiore, A., Trapasso, R. 2014. Widening the "Montedomini" tunnel in the presence of traffic: the evolution of the "Nazzano" method. AFTES International Congress on "Tunnels and Underground space: risks and opportunities"—Lyon, 13th–15th October 2014.

Tonon, F, 2010 "Methods for Enlarging Transportation Tunnels while Keeping Tunnels Fully Operational" ASCE Journal of Practice Periodical on Structural Design and Construction, Vol 15, Issue 4, Nov 2010.

The Crenshaw Corridor—Cross Passages: Design vs. Construction

Luis Piek, Harnaik Mann, and John Kuyt
Arup

Patrick Finn
JF Shea

ABSTRACT: The Crenshaw/LAX Light Rail Transit Corridor Project included twin bore tunnels constructed by EPBM with cross passages between them for emergency egress and for locating MEP and communications equipment. Prior to commencement of the tunnel bores jet grouting was used to treat the Late-Pleistocene Alluvial Soils at the five cross passage locations. The initially specified design requirements for ground treatment versus the final jet grouting results indicated the treated ground was significantly better than specified, meaning a dramatic reduction in the structural support of the tunnel opening and cross passage shotcrete support was viable. After redesigning the cross passage support in close collaboration, the design and construction teams utilized additional subsurface investigations and an observational approach to allow construction to proceed safely, leading to further efficiencies in executing the works. The initial design approach is discussed, as well as the steps leading to the finished support. Additional details regarding the finishing works is also covered.

INTRODUCTION AND PROJECT OVERVIEW

As most tunnel projects near the final stages of completion, it seems that some of the best stories get buried in the work, forgotten to the world spare a select few who can still manage to recall all the details. This paper is an effort to record the 3-year design-build effort of the managers, designers, constructors, owners' representatives, and laborers whose dedication delivered the project from concept to its final form.

The Crenshaw/LAX Light Rail Transit Corridor Project is the first in a series of significant transportation projects undertaken by the Los Angeles County Metropolitan Transportation Agency (Metro) as part of the Measure R half cent sales tax passed in 2008, and the second Measure M bond measure passed in 2016, which will fund a variety of transportation programs in Los Angeles through the coming 40 years. Procured as a Design-Build contract, the $1.27 billion 8.5 mile-long light rail project was awarded to the Walsh/Shea Corridor Constructors (WSCC) team in September 2013 with HNTB and Arup as designers. Underground works include three subsurface stations, a total of approximately 1 mile of twin-bore 18'-10" inner diameter tunnels with five connecting cross passages, and approximately two miles of underground cut and cover guideways. The stations are two-story, 44-foot-high structures and are between 400 and 800 feet long (with crossover), buried 60 feet below grade. The project also includes

at-grade guide ways, at-grade stations, and bridges, as described in Ong, K. and Crow, M. NAT 2014 and Chan et al. RETC 2017. Figure 1 outlines the scope of the project, with the underground sections highlighted in blue.

The tunnels were bored using a single 21.5 ft Herrenknecht EPBM. The sequence of works included launching the EPBM from Expo station and boring south, breaking into Martin Luther King Jr. Station and continuing to Leimert Park Station. The EPBM was then recovered, the gantries pulled back to Exposition Station, and relaunched to complete the second bore. This meant that the ground conditions in the region of the cross passages could be twice reviewed before their start of excavation. The bored tunnel sections are shown in the lower right, with Exposition, Martin Luther King Jr., and Leimert Park Stations in order from right to left.

TUNNEL AND CROSS PASSAGE CONFIGURATION

When laying out the bored tunnels, cross passage locations are often dictated by NFPA 130 requirements, requiring no more than 800 ft between egress locations. Metro Fire & Life Safety Design criteria requirements are very similar, requiring a nominal spacing of 750 ft, but no more than 800 ft between egress points. Based on these requirements, between Exposition Station and Martin Luther King Jr. Station 4 cross passages are required. There is more of a difference between the NFPA code and the Metro

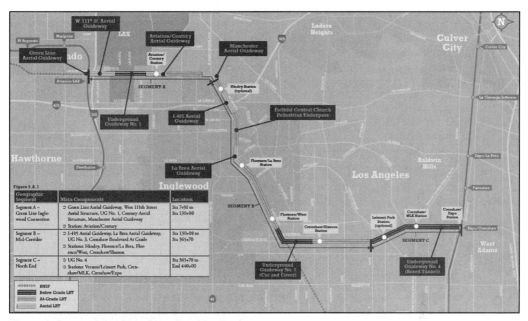

Figure 1. Scope of the Crenshaw Corridor Project

fire & life safety criteria requirements with respect to the maximum distance between surface exits from the tunnels—2500 ft from NFPA 130 and 1250 ft for, Metro criteria. This meant that between the Martin Luther King Jr. Station and Leimert Park Station, a fifth cross passage was required.

Fortunately, the entire length of the twin bores was within the public right of way for this project, extending beneath a large boulevard with a painted center divider, which provided an opportunity to build the cross passages without impacting third party surface structures and reducing the overall construction risks.

When creating underground metro structures, it is quite impressive how quickly the available space is occupied, first by the train's dynamic envelope, then with a variety of equipment, utilities, walkways, and maintenance equipment—all the while leaving enough clearance for riders to safely exit in the event of an emergency. Cross passages therefore tend to be a perfect space to locate the various ancillary equipment needed to support the metro operational systems. However, as more equipment is placed into the cross passages, they become larger in size and result in complications to their design and construction. For this project, a combination of BIM tools (e.g. Revit) and rapid prototyping processes using 3D printing allowed optimization to locate these services, which include: three radio cabinets, a control power panel, radio power cabinet, communications interface cabinet, transformer, two blue light stations, two CCTV cameras, methane detector, H2S sensor and panel,

fan and damper, telephone, smoke detector units, and a 4-way wet standpipe with fire-hose connectors and valve assembly. All of these services were fit into a 9'-6"W × 16'-0"L × 7'-0" to 9'-3"H room, all while maintaining the required 6'-6" wide clear walking path along the cross passage length. In particular, the physical 3D print of the cross passage turned out to be one of the best communication tools available to the team, especially in explaining the layout and assisting the numerous "over the shoulder" reviews by other discipline leads and Metro Fire Life Safety. It is strongly suggested other designers take note— BIM is often now required on most projects, and it doesn't take much effort to turn them into 3D prints. As the expression went on site, "If a picture is worth a thousand words, a 3D print is worth a hundred meetings."

GROUND CONDITIONS AND GROUND TREATMENT DESIGN

The Crenshaw/LAX Transit Corridor is located in the northern part of the Los Angeles Basin, and below fill layers, the project works encounter two primary soils varieties: Type 1 alluvium—baselined as 15% of the tunnel length—consisting of fine grained silts, clays and organic soil; and Type 2 alluvium—baselined as 85% of the tunnel length—consisting of a mixture of fine- to coarse-grained sands and gravels, including some cobbles and boulders which are interlaced in beds that pinch and pull along the alignment. The groundwater levels were baselined at maximum historic levels, providing a static head on the tunnels in the range of 20–30 psi, however actual conditions encountered during construction varied between 10–15 ft below the baseline. Seismic effects were considered for two earthquake levels: the Operating Design Earthquake (ODE), defined as a 50% chance of exceedance in a 100-year time span (equivalent to a return period of approximately 150 years); and the Maximum Design Earthquake (MDE), defined as 4% chance of exceedance in a 100-year time span (equivalent to a return period of approximately 2,500 years).

While the Type 1 soil types indicated sufficient standup time to safely excavate the cross passages, the Type 2 course grain soil types lacked sufficient cohesion to safely excavate the ground without some type of ground treatment. Despite the relatively small 18'H × 16.5'W horseshoe-shaped cross section, it was determined at the design stage that the ground treatment should target a UCS of 300 psi to safely be able to allow excavation of the cross passages. Ground treatment specialists selected jet grouting using triple fluid with cement grout, based on a fines content of 15% or higher within the Type 2 ground. A sodium silicate jet grout solution was considered, but ruled out as it could not sufficiently meet the performance requirements. Analysis showed that the entirety of the bore did not need to be treated, only the region stretching from crown to crown of the twin bores, as the majority of traction forces generated during cross passage excavation occur only within this zone. A minimum cover of 8 ft above the crown and below the invert of the cross passage excavation envelope was also specified as shown in Figure 3. The final layout of the jet grout columns utilized 8 ft columns spaced on 6 ft centers.

INITIAL CROSS PASSAGE AND STEEL SUPPORT DESIGN

To support the segmental lining in the twin bore tunnels during cross passage excavation a design was developed for two rolled structural steel beams placed either side of the ring opening, connected with a number of cross beams to transfer load from the cut rings into the steel frame and adjacent rings. The rolled steel beams used plate connectors and bolts to leave the invert area clear, allowing uninterrupted use of the temporary construction rails. Strand7, a 3D finite element software, was used to model the steel support structures, bolt connections and individual segments and connectors between the rings with soil springs to capture the soil-structure interactions and the transfer of load from the segments onto the steel work during cross passage construction. One of the major benefits of the Strand7 software was the ability to capture the construction stages in detail, allowing the performance of the connectors to be critically

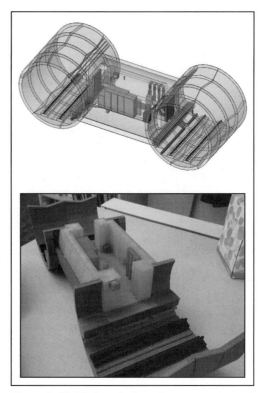

Figure 2. 3D BIM rendering of cross passage and resulting 3D Print

evaluated and confirm their capacity to carry the loads with sufficient redundancy. Figure 4 illustrates the layout of the support as initially designed.

Some particular features of the initial support of the cross passage were the top arch is designed solely in steel fiber reinforced shotcrete (SFRS), with a flat invert and light reinforcement cage to prevent upward heave. Also, to minimize the cross passage cross section, the upper corners of the opening in the segmental lining were chamfered, leaving enough body on the back side of the segments for the placement of the waterproofing membrane. The ring sequence at the cross passage location was also laid out to center the cuts toward the middle of the segment body and prevent formation of small wedge pieces. The rings were designed as traditionally reinforced, using a 4+2 ring sequence, 5 ft long and 10.5 inches thick. Radial bolts were used with Sofrasar Sof-fix 80 connectors across the circumferential joints.

EXECUTION OF GROUND TREATMENT AT CROSS PASSAGES

Without mitigation, jet grouting is an inherently messy operation. The use of several techniques to control the soil-cement returns was trialed, including

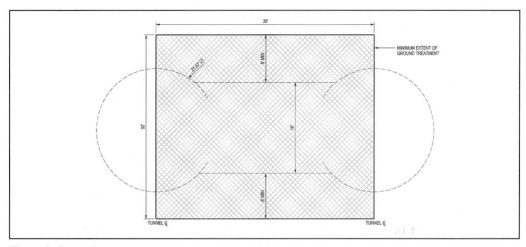

Figure 3. Ground treatment zone at cross passages

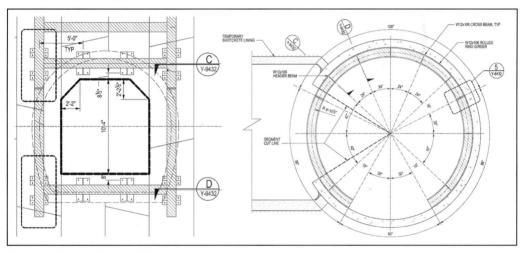

Figure 4. Original layout of the ring support at cross passages

vacuum trucks, inflatable barriers, and concrete curbs to name a few. The best method was a simple solution with trenching around the 32'×40' treatment area to collect the spoils into a pit and pump the slurry into a cement truck, which then transported it to a nearby concrete washout station. Early results from wet grab samples (i.e. the soil+cement slurry ejected to surface during jet grouting) indicated strengths of 700–1000 psi, much higher than needed or expected. While less than 1% of the Type 2 soils were expected to be naturally cemented in place, the influence of the jet grouting operations readily indicated the soils were very aggregate-like in nature. During the course of the jet grouting, it was questioned whether to halt the ground treatment operations and investigate alternative means of grouting the ground. Since

schedule was squarely in the driver's seat at this point, the works continued relatively unchanged. Cores of the treated earth were expected to show the results to be well within the performance parameters, but no one expected the nearly complete and intact cores to show nearly concrete-like qualities, with similar concrete-like compressive strengths obtained with compressive cylinder tests. Photos of samples cores are shown in Figure 5.

As the EPBM passed the cross passage locations, the in-situ ground strength was verified during the cutterhead interventions. Several of the engineers on-site stated the soils looked extremely similar to concrete, absolutely dry, and could not push a knife into the treated ground. Comparatively they were able to readily cut into the untreated side of the face,

Figure 5. Samples of jet grout cores at Cross Passage 1

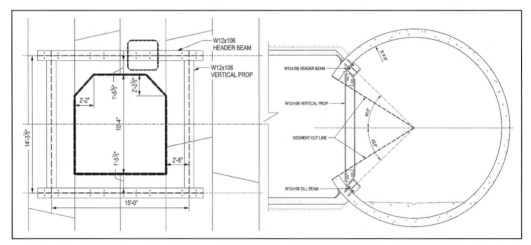

Figure 6. Redesigned steel supports at cross passages

giving further qualitative in-situ information on the expected strength of the ground.

REDESIGN OF THE CROSS PASSAGE, STEEL SUPPORT DESIGN, AND EXCAVATION SEQUENCE

After the high jet grout strengths were gathered, the statistical ACI equivalent strength of the ground was calculated to be 2200 psi. Seeing an opportunity to create a more efficient support, Arup and WSCC brainstormed possible changes to the ring supports and cross passage initial lining. Additional Strand7 runs demonstrated a simple picture frame support, consisting of a header and sill beam and two vertical props would be sufficient to hold the rings open. Mounting plates were eliminated and instead stainless steel bolts were run directly through the flanges at specified locations to make sure rebar was not intercepted. Gusset plates were welded into the beams on the surface to accept the props, allowing quick assembly through plug welds between the prop webs and the gusset plates. This prevented field fit issues with placement tolerances of up to 3". Figure 6 demonstrates the dramatic reduction in the supports in the bored tunnels.

Break-out through the segment rings and into the ground immediately demonstrated that the ground was as expected—a relatively uniform stiff and hard soil+cement medium, requiring a mini-roadheader efficiently excavate. Impact breakers were also used extensively; the roadheader attachment was better for final profiling but was relatively slow for face excavation. Because of its inherent strength, at the direction of the EOR an observational method was adopted with elimination of the SFRS support. Minor sand lenses and discrete windows of untreated soils were encountered, but were not significant nor persistent enough to warrant concern. Water ingress was negligible. At the completion of excavation, a smoothing layer of shotcrete was only needed for placement of waterproofing membrane. Figure 7 is indicative of the quality of ground encountered during excavation.

Figure 7. Typical treated ground at cross passages

CROSS PASSAGE PERMANENT WORKS

Cross passages have been notorious for being sources of leakage, requiring mitigation in prior Metro projects. One of the challenges for Metro is the requirement to use a membrane that prevents entry of water and hydrogen sulfide and methane gas, and is completely resistant against decomposition when exposed to hydrocarbons; currently the only approved material to accomplish this is HDPE sheet. The primary reason for this is its tight crystalline structure. However, this natural barrier also tends to make it stiff, relatively stretch-resistant and therefore subject to tearing if not properly seated or if excessive void space is left behind the lining. Secondly, the creation of complex shapes, such as in cross passages, is also difficult, as are the termination details. Working collaboratively with an HDPE installer, the details at the openings into the bored tunnels were developed to assist in a straightforward assembly, leading to a higher-quality connection to work as intended. One item to note was the designers specified both stainless steel anchors and nuts, which when assembled gall and bind, preventing adjustment of the plates if needed. After recognizing this issue in the field, corrective action to use brass nuts instead was implemented. A termination detail is shown in Figure 8 and a photo showing the

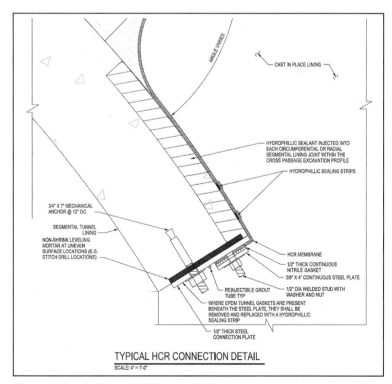

Figure 8. Termination detail at cross passage to bored tunnel

Figure 9. Cross passage with HDPE lining and termination detailing

cross passage waterproofing and detail installed in Figure 9. In addition, the membrane area is compartmentalized using water stops, hydrophilic strips incorporated around all openings, a felt layer added for protection against membrane puncture, and re-groutable hoses to treat any leakage after cast-in-place concrete placement.

Following placement and acceptance of the HDPE waterproofing, final reinforcement was placed and the permanent reinforced concrete frame cast at the interfaces. Final connection with the walkway and the numerous conduits and utilities then followed. The cross passage concreting works were completed in December 2017.

CONCLUSION

Recognizing the potential to reduce cost and schedule due to much higher ground treatment strengths than expected, the design of the cross passage and bored tunnel supports was optimized for the Crenshaw Corridor. Using an observational method in the excavation of the cross passage resulted in the elimination of shotcrete, using only the ground treatment as initial support. When incorporating the permanent water and gas proof HDPE membrane, attention to termination details should be paid in order to result in a leak-free environment.

Walsh-Shea Corridor Constructors, Arup, and HNTB would like to thank the Metro for working in a collaborative environment to deliver the newest extension to the metropolitan rail network in Los Angeles, slated to open in late 2019. In particular, we would like to call attention to the following individuals whose efforts on the underground structures aligned to make this project a success: Matthew Gallagher, Matthew Crow, John Yao, Dana Rogers, Michael Pearce, Carl Christensen, Shemek Oginski, David Girard, Tony Chan, Ryan Henderson, Jon Hurt, Michele Mangione, and Mark Ramsey. Thank you.

REFERENCES

[1] Ong, K. and Crow M., Use of innovative project procurement as a tool in shaping the design of underground megaprojects for the Crenshaw LAX Light Rail Transit Corridor Project in Los Angeles. North American Tunneling Proceedings 2014.

[2] R. Chen, J. Salai, B. Shatz, Station Excavation and TBM Tunnel on Los Angeles Crenshaw Project, RETC Proceedings 2017.

[3] Hatch Mott MacDonald, Geotechnical Baseline Report of Crenshaw/LAX Transit Corridor Project, 2012.

Humboldt Bay Power Plant Decommissioning Shaft—A Case Study in Cutter Soil Mix (CSM) Wall Construction

Zephaniah Varley
WSP

ABSTRACT: The Humboldt Bay Power Plant (HBPP) Decommissioning Project includes the removal of underground facilities previously used for nuclear power generation. Groundwater control and support of excavation are critical elements for controlling demolition and excavation activities during decommissioning. This paper presents the original design concept, as well as the contractor-initiated design modification for shaft construction of a Cutter Soil Mix (CSM) deep shoring system and cutoff wall. The 110-foot diameter, 174-foot deep CSM shoring and cutoff wall allowed for the excavation and decommissioning of the reactor caisson structure beneath the Refuel Building adjacent to Humboldt Bay on the Pacific Ocean.

BACKGROUND

The Pacific Gas & Electric (PG&E), Humboldt Bay Power Plant (HBPP) Unit 3 Nuclear reactor caisson was constructed in the 1960s in Eureka, California next to Humboldt Bay. The plant site included three different power generating units including conventional steam from natural gas or oil combustion and nuclear, utilized adjacent Pacific Ocean bay water as coolant for various power production phases, and had the ability to switch fuel sources depending on power demand and resource availability. The innovative design of the first commercially licensed below-ground nuclear power plant, utilized a 4-foot thick concrete cylinder (caisson) poured in lifts and water jetted into the sand and clay. Internal shield walls and a pressure vessel were placed inside the caisson after the bottom of the cylinder was plugged with tremie concrete for the floor. The caisson design was principally a structural ground support element which also provided secondary containment for the nuclear reactor core. Figure 1 shows an aerial view of the original reactor caisson and adjacent building basement construction on timber piles.

In the 1970s an earthquake in the Eureka area caused minor damage to the nuclear operating unit, forcing a temporary shutdown for evaluation of the caisson's structural stability. The nuclear unit was shut down in 1976 for refueling and seismic upgrades. Repairs subsequently extended the planned shutdown period. In that interval, significant changes were made to nuclear safety standards for reactor operation and design. Ultimately, the decision was made that further modifications were not economical, and the nuclear unit would not be restarted. Since plant operations were still viable with alternate sources, the nuclear cells remained inactive until 1983 when the final decision was made that nuclear operation would no longer ever occur[1]. The nuclear fuel cells were removed and plans were developed for site remediation of the plant facilities and buildings no longer in service. Remediation plans included demolition of various buildings around the site, and the removal of the concrete caisson that had encapsulated the old nuclear fuel cells. A feasibility study was drafted by Kiewit[2], which included draft specifications, plans and concept drawings for site remediation, particularly for caisson removal work.

ORIGINAL DESIGN—THE OBLONG 'COFFIN' CONFIGURATION

The original baseline design approach described in the Kiewit feasibility study included a cement-bentonite backfill slurry wall surrounding the Refuel Building and Turbine Building intended to minimize groundwater infiltration, a sloped soil nail wall for support of the upper caisson excavation, and an 80-foot diameter sheet pile wall and ring beam shoring system for support of the lower caisson excavation. Figure 2 shows the original concept plan view. Figure 3 shows the original concept profile. Figure 4 shows the geotechnical profile along the original slurry wall alignment.

The oblong perimeter slurry wall alignment, around both the Refuel and Turbine Building, was intended to cutoff both shallow groundwater to facilitate the removal of timber piles beneath the buildings, and to cutoff deeper groundwater levels anticipated for the caisson removal shaft. The feasibility study indicated the presence of a clay layer (the Unit F Clay) at an average depth of 170 feet below ground surface (bgs) throughout the footprint of the slurry wall. Additional geotechnical investigation

Figure 1. Caisson construction (circa 1960s)

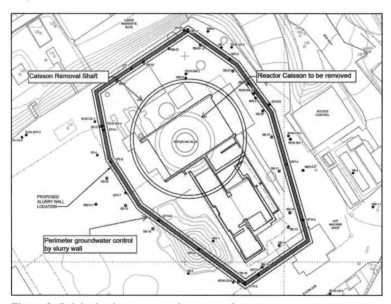

Figure 2. Original caisson removal concept plan

confirmed the presence of the Unit F clay layer to be between 160 and 185 feet bgs along the entire perimeter of the original cutoff wall alignment. The original slurry wall alignment was envisioned to be excavated with a hydromill, given the depth needed to toe into the lower impervious clay layer, with a simple cement-bentonite backfill material to keep groundwater inflow below 1×10^{-6} cm/s. The

original caisson removal shaft, which would provide structural support for excavation and removal of the caisson structure beneath the Refuel Building, was contemplated as a two-part excavation operation. The upper shallow excavation work was contemplated to be excavated with conventional excavators and loaders and supported with a soil nail wall that could be constructed concurrently during adjacent building demolition work. The deeper excavation work was expected to be a sheet pile wall construction operation, with staged excavation allowing for alternating cycles of caisson concrete removal, annular shaft soil excavation and ring beam construction for structural excavation support. All caisson removal work was intended to be done in the "dry" to minimize the potential for any remaining trace nuclear contamination to be picked up by water in the shaft. A deep dewatering well system was part of the original design, as were piezometers and inclinometers for measurement of groundwater levels and shaft wall movements during caisson removal and excavation work.

DESIGN-BUILD INNOVATION—THE THREE-IN-ONE CSM WALL CONCEPT

In 2012 a design-build contract was issued by PGE to Chicago Bridge and Iron (CBI, now known as APTIM), for the entire HBPP site remediation, including building demolition, hazardous material remediation, site grading and caisson removal work. APTIM employed Geosyntec to perform additional geotechnical investigations and to draft slurry wall, dewatering and instrumentation monitoring final design plans and specifications. In 2014, while evaluating potential subcontractors and slurry wall excavation methods, APTIM proposed and developed an alternate for combined groundwater cutoff

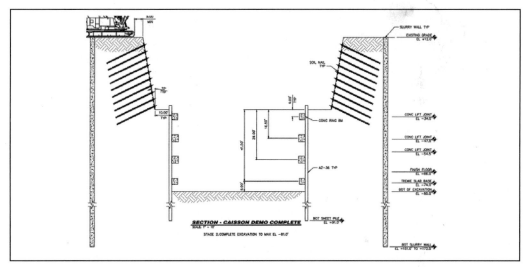

Figure 3. Original caisson removal profile concept

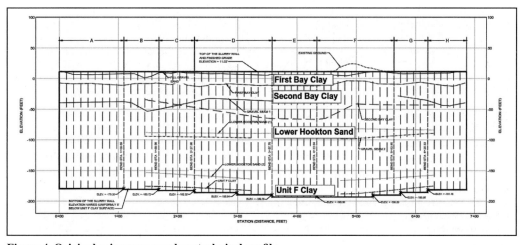

Figure 4. Original caisson removal geotechnical profile

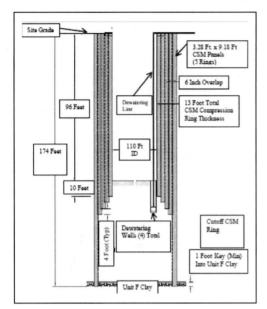

Figure 5. Compression ring section view

and structural excavation support: a Cutter Soil Mix (CSM) wall. ATPIM's drilling and instrumentation/dewatering system subcontractor, Drill Tech Drilling and Shoring, Inc. (Drill Tech) and designer, McMillen Jacobs Associates (MJA), developed a design for a CSM wall constructed with five (5) concentric rings of panels around the caisson, the outermost of which would be toed into the Unit F Clay layer, roughly 174 feet deep. The waler-free shoring system consisting of five rings of panels would build a 13-foot thick "compression ring," with the inner four rings mixed to staggered depths of between 105- and 118-feet to accommodate a 110-foot diameter, 95-foot deep caisson removal excavation. Advantages of the revised configuration included use of a single subcontractor rather than separate slurry wall/soil-nail/sheet piling shoring crews; a smaller footprint compared to the oblong 'coffin'; reuse of soil materials left in place versus removal of ring beam elements upon shaft backfilling; and minimization of time and number of craft in the shaft and their potential exposure/monitoring for trace nuclear contaminants during excavation and backfill operations. Figure 5 shows a profile of the 5-ring shaft concept.

COMPRESSION RING DESIGN

The normal site groundwater water level was at Elevation –5 feet (NGVD) with ground level at Elevation +12, coincident with the operating deck of the Refuel Building/Reactor Building above the caisson. In addition, the site was situated within 100 yards of the Pacific Ocean, therefore subject to tidal

fluctuations. The lateral earth loads with a full height of hydraulic head were used as the basis of design for the compressive loads on the shaft wall ring, and for a check of shaft base stability. Furthermore, the seismic design basis of the CSM shoring was based on a 100-year return period earthquake ground motion (5% probability of exceedance within a 5-year period) and a Peak Ground Acceleration (PGA) of 0.53g. This design basis aligned with industry standards for the design of large dams and was further benchmarked against other CSM deep shoring projects. The shaft design ultimately calculated a table of stress values per depth of shaft and dictated the basic design parameter that CSM wall elements achieve an average compressive strength of 1,000 psi, based on a general Factor of Safety of 3.0.

CSM PANEL LAYOUT AND CONSTRUCTION SEQUENCE

Based on the basic design criteria of 1,000 psi CSM panels DrillTech developed a layout of interconnected panels to create the overall CSM wall shaft. The CSM compression ring layout includes a total of 255 overlapping panels distributed as shown in Table 1.

DrillTech elected to use two Bauer drill rigs, a BG-40 and a BG-50, with fixed Kelly Bar shafts, and BCM10 cutterheads, which produced a nominal 9'-3" long by 3'-4" wide CSM panel. Figure 6 shows one of the cutterheads used on the project.

DrillTech's BG-40 unit was trucked up to the site from the previously completed LA Metro Crenshaw line job that had drilled panels for the TBM launch box and station. DrillTech commissioned the manufacture of a new BG-50 specifically for this project, as the larger rig was needed for the maximum depth of 174 feet required for the exterior Ring E panels for groundwater cutoff.

Construction Sequence

The continuous compression ring was formed with a series of overlapping CSM rings A through E with Ring A being the innermost ring and Ring E being the outermost. Each ring included a series of alternating primary and secondary panels. Primary panels for Rings A, C and E were installed first, then

Table 1. Panel distribution and depth

CSM Ring	Number of Panels	Required Depth (Feet)
A (Inner Ring; shaft face)	47	106
B	49	110
C	51	114
D	53	118
E (Outer Groundwater Cutoff Ring)	55	174

Figure 6. Bauer BCM10 CSM cutterhead

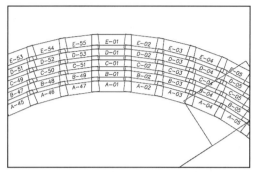

Figure 7. CSM panel numbering (typical)

CSM panels were generally installed from the southwest quadrant clockwise around the compression ring. CSM Panels were installed in a primary/secondary sequence. For example, in any given ring, Panels 1, 3, 5, 7, etc. were installed, followed by 2, 4, 6, 8, etc.

Cutting and Mixing Equipment

During the mixing of slurry used in the backfill mix, real time data acquisition was used to obtain the weight of any dry powders or liquid used to mix the slurry. When the slurries were pumped, the flow rate and total volume of slurry pumped was obtained versus clock time and date using a data acquisition system with real-time monitoring of the pumping process. The pump flow rate and volume was displayed on screen in each CSM drill rig. The downstroke mix design consisted of water and bentonite. The bentonite percentage was varied as required to maintain panel stability during mixing. The upstroke mix design consisted of water and cement with a minimum cement content in the structural zone of twenty pounds per cubic foot of finished panel. The water-cement ratio during the upstroke was between six tenths (0.6) and nine tenths (0.9). Manual checks of all slurry density with a mud balance, and viscosity with a flow cone were performed at each batch plant, once at the start of each shift and once more per shift and whenever the setting on the batch weighing system was changed. Excess spoils were skimmed off the top of adjacent panel trench with excavator buckets, and fines were further segregated out from the slurry mixture using a centrifugal cyclone de-sander set up adjacent to the slurry mixing tanks.

An on-board computer system mounted in the operator's cab recorded the following:

- Time versus depth;
- Inclination of the X (end to end axis) and Y (axis between rings) axis with depth;

secondary panels were installed overlapping the primary panels within the same "Ring." Secondary panels within each Ring were installed when both the adjacent primary panels had cured for more than twelve hours. Panel overlaps shown on the drawings (nominally 9-inches) were used as a guide for construction. Installed adjacent panels were allowed not to overlap so long as no apparent void was apparent between panels, as determined from the panel installation record plots. Once the Rings A, C and E were installed and allowed to cure at least 7 days, drilling began on Rings B and D. Rings B and D followed a similar primary/secondary drilling sequence within respective rings, but had additional "double secondary" panels, where both adjacent B or D ring panels, *and* adjacent A, C and E ring panels had been drilled (i.e. 9-inch overlap in both radial and circumferential directions simultaneously).

In general, the BG-40 installed Ring A panels followed by the Ring C panels then the overlapping Ring B panels. Both the BG-50 and the BG-40 installed the Ring D panels followed by the BG-50 installing the Ring E panels. Ring E panels extended to the Unit F Clay layer which served as the groundwater cutoff wall. Each CSM panel installed was assigned a panel identification number identifying the CSM ring and a sequential number increasing from true north clockwise around the CSM ring.

- Panel deviation with depth of both the X and Y axis;
- Hydraulic pressure in each cutter head plotted versus depth; and
- Grout quantity with depth for both the downstroke and upstroke.

These parameters were viewable by the operator and QC staff during the mixing and, when downloaded, provided the daily shift report for each panel mixed during the shift.

Panel Alignment and Verticality

CSM was a highly effective method for the construction of watertight shafts under adverse soil conditions. The in situ soil-cement mixing process with a Kelly-bar mounted drill rig had the distinct advantage of real-time monitoring of vertical alignment (X- and Y-axes), which is typically not found with other, traditional soil-cement mixing processes. Each Bauer CSM rig was equipped with electronic sensors built into the Kelly bar leads to determine vertical alignment in both fore/aft and left/right directions. The output from the sensors was routed to a console that was visible to the operator during penetration allowing the operator to make real time adjustments as the panel was advanced.

In order to create the 13-ft thick, 110-ft inside diameter shoring cylinder, five overlapping concentric rings of 9'-3" long by 3'-4" wide panels were installed. To create a homogenous shoring wall, each of the 255 individual panels was aligned so that contact with surrounding panels was maintained. This was especially critical in the end-to-end direction for the compression ring to maintain proper load transfer. While it was desirable that continuity be maintained in the side-to-side overlap, this was not as critical. So long as the spacing between the odd numbered panels with depth did not exceed the individual panel length of 9'-3", it was assumed that the end to end continuity was maintained, since the panels on either end acted as a guide. If one panel was harder than the other extra care was taken to steer the cutterhead to ensure that it did not over-cut into an adjacent soft panel. The operators were able to steer the cutterhead in the end-to-end axis by changing the relative speeds and directions of the two cutter wheels. Additionally, if the hardness of the rings acting as guides was significantly different, steering the cutterhead could be accomplished by either tilting the Kelly bar mast or by actuating hydraulic steering flaps on the cutterhead which could push off either side of the panel at depth, thereby pushing the cutterhead in the opposite direction.

All the downloaded panel telemetry data, when plotted, showed all the panels to have overlap in the end-to-end direction for the full depth of the panels.

The specific amount of overlap was considered unimportant; one half inch of overlap has the same effect as 12-inches of overlap. A three-dimensional drawing of the outer cutoff ring was developed to ensure proper tie-in to the Unit F clay layer.

Overall both the BG-40 and BG-50 proved accurate in terms of vertical and horizontal alignment. The less than 1% vertical tolerance (9-inches over 170 feet deep for Ring E) was generally achievable, and where not achieved the effect of adjacent panels overlapping was able to correct for minor deviations within sets of panels and between adjacent Rings.

Quality Control Sampling

Since the CSM wall system was designed for two principle functions—support of excavation and groundwater cutoff, there were two basic parameters for testing the CSM panels for quality and consistency: compressive strength and permeability. For compressive strength, wet "Grab Samples" of the soil mix were taken from each panel with a "bailer type" sampler suspended by a support crane. Grab sample locations in CSM panels were distributed between depths of 40 and 90 feet. Once collected, the wet samples were cast into a set of eight cylinders per sample and stored in a safe location onsite for the initial curing period of 72 hours. No further mixing of the material was permitted. Light tapping of the samples to facilitate consolidation was permitted, and larger clay chunks were strained out with a coarse sieve prior to making each cylinder. After the initial 72-hour curing period was completed, samples were transported to an independent laboratory, stored in a curing room and tested at respectively intervals.

The compressive strength of each set of soil-cement cylinders was tested as follows:

- One cylinder at 7 days of curing;
- One cylinder at 14 days of curing;
- Two cylinders at 28 days of curing;
- Two cylinders after 56 days of curing;
- One cylinder at 112 days of curing; and
- One cylinder at 224 days of curing.

CSM Break Strength Test Result Summary

A definite trend was identified in grab sample compressive strength test results where the secondary Rings B and D, were weaker than the primary Rings A, C and E. It appears that the mixing process, especially with the air required to mix at depths over 100 feet, flushed more of the heavier materials (sands and gravels) out than had been experienced on past projects. The fact that a portion of the Rings B and D panels were mixed twice, or up to three times, (double secondary panels), most likely means

Ring B and D panels consisted of more bentonite and less sand than the Ring A, C, and E panels. They were, therefore, somewhat lighter (wet density) and weaker (compressive strength), on average.

While not specified as an acceptance criterion for panel production, wet density or unit weight of CSM samples was also measured (for consistency of slurry mixing), and there was a correlation between the unit weight and the compressive strength of the samples. The low strength and light-weight samples were predominately in the secondary Rings B and D, and to a lesser extent, in the deep cut-off Ring E. The initial specification compressive strength criteria required the average 56-day strength to be 1,000 psi. The final strength distribution between the various rings and the corresponding depth to maintain a design factor of safety of 3.0 is shown in Table 2.

Taking the results in Table 2 at face value, it appeared the CSM shoring could withstand excavation to a depth of 86 feet (coincident with the planned excavation depth) without compromising the design factor of safety. More than likely, using the excess capacity in Rings A, C and E to compensate for Rings B and D, the potential additional 10' of over-excavation for which the shoring was designed could also be accommodated. However, a detailed look at the distribution of the CSM strengths showed weaker panels are concentrated in the western half of the circular shoring. Ultimately a horizontal core drilling program was developed to obtain additional compressive strength data, and shotcrete was applied in various stages throughout shaft excavation as described in subsequent detail below.

CSM Permeability Test Result Summary

Soil mix below Elevation –94 in the outer cutoff Ring E had a maximum permeability requirement of 1×10^{-6} cm/s. In every fourth panel in the cutoff Ring E three wet-mix samples were taken from between 106-feet and 150-feet below grade specifically for permeability testing. Permeability tests were run on these samples at 28 and 56 days. The average permeability test results for the 14 samples taken was 3.59×10^{-7} cm/s, better than required by

Table 2. 56-Day break average strength required CSM strength versus depth (psi)

Ring	56-Day Strength (psi)	Corresponding Depth to Maintain Design Factor of Safety = 3.0 (Feet)
A	1,256	<96
B	893	87
C	1,111	<96
D	817	86
E	1,041	<96

the specifications. More importantly, drawdown tests of the dewatering well system confirmed the cutoff wall was, for practical purposes, water-tight and would perform as intended.

CSM Deep Shoring and Cutoff Wall Drawdown Test

Once the CSM deep shoring and cutoff wall panels were installed, the dewatering well system was initially turned on to lower the groundwater table inside the CSM compression ring approximately 30 feet. The dewatering system was shut off and the groundwater table allowed to recover within the shaft. Groundwater recovery was measured over time to determine if the CSM panel cutoff wall had cut off groundwater inflow. Recharge to original water levels took about one week. Based on the results of the drawdown test it was estimated that the steady state discharge during excavation would be approximately 20 gpm, and that the CSM wall system was in fact water-tight.

Vertical Core Program

During production CSM panel drilling, DrillTech decided further investigation of the CSM panels was needed by means of vertical core recovery and strength testing of the cores to better understand the wall's in-situ compressive strength. The vertical coring program was intended to supplement wet grab sample data, and was performed by drilling cores through single solid panels, not at the intersection of adjacent panels or rings. When logging cores and performing compressive strength testing, laboratory personnel noticed multiple factors negatively affecting the compressive strength test results. One factor noted was subtle to obvious signs of 'seams' in a few of the core runs which could have indicated overlapping panels. Core sections that showed evidence of visible seams were specifically noted in the lab reports and omitted from the data comparison charts. There were widely varying degrees of abrasion on all core runs due to dislodged aggregate (coarse sand and fine gravel) during coring. Variations in diameter on a single core sample can negatively affect the compressive strength test results. Lab reports and core photos showed this theoretical diameter reduction for all the samples due to abrasion and water injection during drilling. Since the compressive strength of cores was determined by dividing the failure load by the cross-sectional area, it follows that the actual core strengths were all somewhat higher than the values reported above. A direct correlation was found between core sample compressive strength and core sample bulk density. Other comparisons made with the data, specifically compressive strength versus sample depth, and sample depth

Figure 8. Abraded vertical core sample

versus bulk density yielded scattered results with no obvious correlations. The variations in bulk density were likely due to variable material composition and the fact that the heavily-abraded cores simply contained less material, so when divided by the theoretical volume, appear less dense. Unfortunately, there was no apparent correlation between vertical core strength data and wet grab sample core data.

Triaxial Test Program

Triaxial testing was performed on a handful of samples to see if the confining pressure that would be present within the ground increased the compressive strength of CSM samples. Triaxial testing did not show any increase in compressive strengths. It was obvious from looking at plots of the results that the confining pressures were so small as to be inconsequential, as the confining pressures ranged from 1% to 5% of the compressive strength.

Horizontal Core Program

Subsequent to completion of CSM panel drilling, a Horizontal Core Program was implemented to evaluate the strengths of CSM panels between depths of 10 and 50 feet concentrating in the southwest and northwest quadrants of the compression ring. The horizontal cores were intended to provide additional data and direct comparison to wet-grab samples, specifically in the Ring B and Ring D panels, which had shown less than 1,000 psi average strengths. A round of ten six-inch core holes were mapped out for each of the first five ten-foot deep shaft excavation cycles, with slight off-sets each round to obtain

data from multiple panels. Core lengths were drilled 10 feet deep through the Ring A through D panels, as the Ring E panel strengths were not in questions and served as a buffer for keep the shaft watertight at the coring locations. Core data from the first four rounds of drilling was evaluated, and did not show significant deviation from wet-grab sampling. In fact, most horizontal core data was slightly lower than wet grab samples. Coring was discontinued after an evaluation by MJA that directed the use of shotcrete for additional structural stability down to bottom 90-foot shaft excavation level.

SHAFT EXCAVATION, CAISSON DEMOLITION AND SHOTCRETE

Upon completion of CSM drilling activities, the top of the shaft area was graded and setup for excavation and caisson demolition work. A six-inch shotcrete collar was applied to the top 10 feet of the shaft wrapping over the lip of the shaft and covering the top for the Ring A-through Ring E panels. Ventilation equipment was set up and a 275-ton support crane was used to hoist buckets, excavators and demolition equipment into the shaft. Due to radiological testing requirements, shaft excavation lifts were limited to six-foot cycles, and horizontal cores were taken at the first 4 ten-foot lift increments. Detailed review of the seismic analysis led to additional shotcrete thickening (4,000 psi) at the top ten feet to a 12-inch lift, and a full 12-inch lift down to 20 feet, as minor radial cracking of the CSM wall face was predicted to occur from FLAC modelling. In addition, a round of 6×6 W4.0×W4.0 mesh and two rows of 9-foot long rock bolts at 10-foot centers were installed from 20 to 30 feet deep as added protection from potential spalling. Shotcrete was omitted between 30 and 60 feet of excavation, as the shaft model showed adequate structural capacity when compared to measured CSM panel strength data. The bottom 30 feet of shaft excavation was covered with shotcrete for additional structural support and as a quick replacement for rounds of horizontal cores and evaluation thereof. Caisson demolition proceeded concurrently with shaft excavation, and did not encounter any unexpected "treasures" from 1960-eras construction. Figure 9 shows shaft excavation equipment.

CONCLUSION AND LESSONS LEARNED

Demolition and removal of the caisson required a shoring system and cutoff wall installation that met the highest safety standards of PG&E, and was flexible enough to work around ongoing demolition activities and high groundwater at the HBPP site immediately adjacent to Humboldt Bay. The CSM compression ring shoring and cutoff wall proved an effective solution for these conditions. Within the waler-free shoring system, heavy equipment was able

Figure 9. Shaft excavation, caisson demolition and rock bolt drill rig

to safely work around the caisson on solid ground as the demolition progressed. The CSM approach also allowed for concurrent demolition of nearby structures sitting immediately above the caisson. The use of a single subcontractor for panel drilling, instrumentation and dewatering system installation and monitoring, and shaft shotcrete application proved economical and convenient compared to a multi-stage SOE system. Construction of 170+ foot deep panels, with reasonable limits on verticality is achievable, however 1,000 psi panels require diligent cement dosing and are at the high end of achievable range for strength. Panel overlap, particularly at double secondary panels, could cause strength deterioration from multiple mixing cycles. CSM wet grab sample compressive strength gain from 56 to 112 to 224 days was negligible. Low permeability from CSM interlocking panels was achievable with ease. As a verification method for shaft watertightness, a drawdown test proved an effective tool, particularly in conjunction with shaft monitoring instruments (piezometers and inclinometers). These instruments also confirmed that the shaft walls were generally stable during excavation. Overall the CSM shaft concept for this application was a success and the lessons learned herein should serve as an excellent model for future application of this technology.

ACKNOWLEDGMENT

The author would like to acknowledge the following people for their assistance in writing, editing and delivery of engineering documents related to this paper: Carl Zietz, Allen Barkley and Ken Valder (Akana); Bruce Patterson (Kiewit); John Gilbert, Paul Beadle and Alan Brown (APTIM); Mark Lawrence (MacMillan Jacobs Associates); Brett Mainer and Jarred Burrows (DrillTech), and Kerry Rod, Mike Strehlow, Cynthia Cabeceira, Shaina Mason, Bruce Ratcliffe, Bruce Stephens, Bobby Gibson, Kristin Zaitz and all the PG&E engineering, safety, oversight and field staff on site in Eureka.

REFERENCES

[1] A Brief History of Humboldt Bay Power Plant from PG&E website: https://www.pge.com/en_US/about-pge/environment/what-we-are-doing/buildings-and-operations/humboldt-bay-power-plant.page.

[2] Caisson Removal Feasibility Study Final Report, October 29, 2012, prepared by Kiewit for PG&E.

SEM Tunneling of City Trunk Line Across Tujunga Wash

Wolfgang Roth, S. Nesarajah, and Bei Su
AECOM

Philip Lau and Ruwanka Purasinghe
Los Angeles Department of Water & Power

ABSTRACT: Tujunga Wash channel runs diagonally under a street intersection on a bridge. For a 60-in trunk line crossing the channel, a ~10-ft diameter, 250-ft long SEM tunnel was designed "snaking" between driven concrete piles supporting the bridge abutments. The bridge had been constructed in 1948 with a sewer siphon five feet below the channel floor. Ground conditions consist of silty sands and, except during wet season, the groundwater table is generally below tunnel invert. To assess the impact of tunneling-induced ground movements on bridge abutments and sewer, nonlinear soil-structure interaction analysis was performed with FLAC3D. The results indicated negligible settlement of bridge abutments and sewer. Tunnel construction was completed in 2016 with settlements lower than predicted. Tunnel construction was completed in 2016 with settlements lower than predicted.

INTRODUCTION

City Trunk Line South (CTLS) of the Los Angeles Department of Water & Power (LADWP) crosses the Tujunga Wash channel at the intersection of Whitsett Avenue and Riverside Drive in Studio City, California. As shown in Figure 1, the inter-section is located atop a bridge which is oriented about 30 degrees out of alignment with CTLS along Whitsett Avenue. The bridge was constructed in 1948 under the jurisdiction of the US Army Corps of Engineers (USACE), with abutments supported by driven precast concrete piles. Also crossing the channel at this location is a sewer siphon running through an approximately 10-foot wide utility cor-ridor between the piles. The siphon is under the juris-diction of Los Angeles Department of Public Works (LADPW) who is also the agency responsible for maintaining the bridge.

This paper describes the technical approach and results of a soil-structure interaction (SSI) analysis evaluating potential ground movements and effects on adjacent structures induced by exca-vating an approximately 10-foot diameter tunnel with the Sequential Excavation Method (SEM) under the Tujunga Wash channel at the location of the bridge. This analysis was requested by USACE and LADPW as stakeholders, to demonstrate that the tunnel could be constructed without adversely impacting the bridge abutments and sewer siphon, as well as another sewer line crossing the CTLS align-ment north of the channel.

DESIGN CONSIDERATIONS

Four basic options were considered for crossing the channel with the 60-inch diameter steel pipe of the CTLS within the public right of way of Whitsett Avenue:

1. Deep pipe-jacking below the tips of the bridge-abutment piles
2. Shallow pipe-jacking requiring the removal/ underpinning of some piles
3. Shallow tunneling with Sequential Excavation Method (SEM) "snaking" between piles (4) Above-ground pipe cross-ing hung below the bridge structure

After considering the pros and cons of these options, SEM tunneling was selected as the preferred option, with the tunnel to be driven through the narrow util-ity corridor between the piles. A longitudinal section of this option is shown in Figure 2.

The Sequential Excavation Method of Tunneling

The basic principle of SEM tunneling is best described as "tunneling by relying on self-support of the ground" as opposed to "relying on steel or con-crete support." Excessive disturbance of the ground is avoided by sequentially excavating small increments followed by immediate application of shotcrete. Providing a measure of confinement by strength-ening the surface, shotcrete enables the ground to

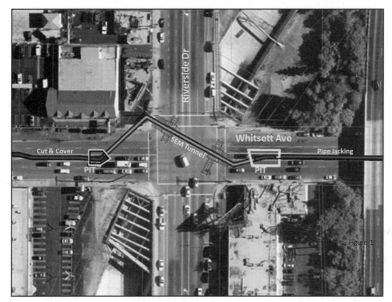

Figure 1. Plan view of CTLS alignment

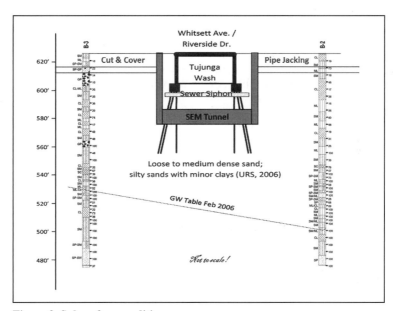

Figure 2. Subsurface conditions

support itself. The length of incremental excavation rounds, shotcrete thickness, and additional support measures (e.g. spiles, etc.) available from a "tool-box," are chosen and continuously adjusted based monitoring of real-time ground response.

With shotcrete having been invented more than a century ago, SEM has gradually grown out of practical field experience and empirical observations.

Because its underlying theoretical framework was first developed and verified by physical model testing in Austria in the 1950s, SEM is also widely known as the New Austrian Tunneling Method (NATM).

Subsurface Conditions

As shown in Figure 2, the site is generally underlain by alluvial sediments consisting of alternating layers of sand and silty sands with minor clays (URS 2006). The sandy soils are generally medium dense to very dense, and the fine-grained materials are very stiff to hard. The groundwater table encountered in the two borings drilled on either side of the channel was well below the proposed tunnel invert. Although historic groundwater levels near the site had been reported as high as 10 feet below ground surface during wet seasons (CDMG 1997), groundwater was not expected to be encountered during tunneling if construction proceeded within the dry season. Even so, the shotcrete liner was designed to withstand hydrostatic loading under the highest historic groundwater table as a stand-alone structure.

Tunnel Cross-Section Constraints

For shallow SEM tunneling to be possible without removing or structurally damaging any piles, the tunnel cross section had to be designed to fit within the narrow utility corridor between the piles. To assess the likelihood of encountering piles driven out of plumb, an informal nationwide opinion survey was conducted. A brief project description was distributed to 22 highly experienced geotechnical and construction engineers, along with the sketch shown in Figure 3, requesting that participants rate the likelihood of encountering various ranges of pile deviation. Based on the data summarized in a table also shown in Figure 3, The Narrowest cross section was designed assuming the piles could be 5% out of plumb in opposite directions. Since occasional deviations of more than 5% could be accommodated without obstructing the 60-inch diameter clearance

needed for installing the carrier pipe, a scenario where any existing pile would have to be removed during tunneling was judged to be highly unlikely. Piles encountered within the excavation envelope of the tunnel were to be wrapped in bond-breaking material to prevent any load transfer between piles and shotcrete liner.

Design Drawings and Specifications

A plan view of the tunnel is shown in Figure 4, and the tunnel-excavation process, which was specified in detail in the contract drawings (DSC 2010), is illustrated in Figure 5. The tunnel was specified to be advanced by three-foot long rounds, each consisting of heading and bench excavations which were immediately supported by a two-inch layer of fast-curing "flashcrete." Following the installation of steel lattice girders, an additional seven-inch thick fiber-reinforced shotcrete liner was applied. Design specifications called 11 rebar spiles to be advanced between 10-o'clock to 2-o'clock positions forming a protective canopy under which the heading for the next round could be safely excavated (Figure 5). Along the segment running parallel and below the existing sewer siphon with only 10 feet of vertical separation, the rebar spiles were to be replaced by self-drilling grouted pipe spiles.

SOIL–STRUCTURE INTERACTION ANALYSIS

The analysis was performed with the computer code FLAC3D, Version 4.0 (Itasca 2009), a three-dimensional explicit finite difference program for

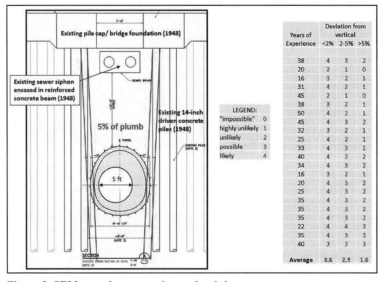

Figure 3. SEM tunnel cross section and opinion survey

Years of Experience	Deviation from vertical		
	<2%	2-5%	>5%
38	4	3	2
20	2	1	0
16	3	2	1
31	4	2	1
45	2	1	0
38	3	2	1
50	4	2	1
45	4	3	2
32	3	2	1
25	4	2	1
33	4	3	1
40	4	3	2
34	4	3	2
16	3	2	1
20	4	3	2
25	4	3	2
35	4	3	2
35	4	3	2
35	4	3	2
22	4	4	3
35	4	3	3
40	3	3	3
Average	3.6	2.5	1.6

LEGEND:
"impossible" 0
highly unlikely 1
unlikely 2
possible 3
likely 4

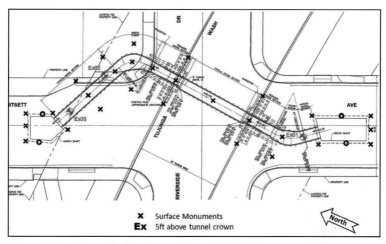

Figure 4. SEM tunnel plan view with geotechnical instrumentation

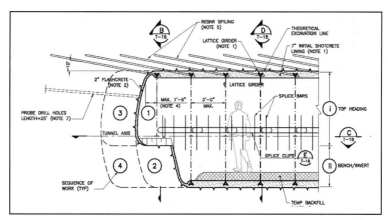

Figure 5. SEM construction sequence

geotechnical engineering and rock mechanics computations. It offers a wide range of capabilities to solve complex problems in geomechanics, including nonlinear static and dynamic stress-strain analysis of soil continua, soil-structure interaction, and groundwater flow.

Basic Model Setup and Material Properties

Figure 6 shows the model mesh of half of the analyzed structure which, for this analysis, was approximated as being symmetric about a vertical plane through the axis of the tunnel. For this analysis soils were represented as continuum elements with an elastic-plastic Mohr-Coulomb constitutive model, with shear strength defined in terms of friction angle and cohesion. The structural elements of the model consist of the following:

- The bridge abutment wall and pile cap are modeled as elastic continua;
- Piles are elastic beams connected with the soil mesh through elastic-plastic shear springs representing pile skin friction;
- The sewer siphon is an elastic beam;
- Grouted pipe spiles are elastic beams; and
- The seven-inch thick shotcrete liner is modeled with elastic shell elements.

Material Properties

Material properties for soil and structural elements used in the analysis are listed in Tables 1 and 2, with soil stiffness increasing linearly with depth. The inherently nonlinear stress-strain behavior of soil was approximated by using small-strain moduli for simulating initial bridge construction, and large-strain moduli for analyzing the effect of tunneling. The former were derived from in-situ shear-wave velocities, Vs, and the latter were estimated from correlations with SPT blow counts.

The time-dependent compressive strength and elastic modulus of shotcrete is shown in the graph of Figure 7 as a function of setup time. With 7% accelerator to be added to the mix, shotcrete reaches a compressive strength of about 600 psi after five hours. This five-hour strength and the associated Young's modulus, E, were used for analyzing ground movements induced by each new round of excavation. For the following next round, the strength of the shotcrete liner placed in previous rounds was then increased to 3,000 psi, and so on.

Modelling Sequence

Pre-tunneling in-situ soil stresses and pile loads where established by analyzing the construction stages shown in Figure 8. The bridge abutment wall and load (30 kips/lft of dead weight) were placed on the pile cap while the top of the wall was horizontally fixed to simulate lateral support provided by the bridge structure. In Stage 4, the bridge abutment was backfilled. Computed pile-cap settlements after applying the abutment wall with bridge load and backfilling the abutment were 0.12 and 0.25 inches, respectively.

The main purpose of the analysis was to evaluate the impact of excavating the tunnel on the piles supporting the bridge abutments. Specifically, the focus was on pile settlements and bending moments due to loss of lateral confinement during tunnel excavation. Since pile configurations were identical for both abutments, this was accomplished by simulating SEM tunnel excavation through only one of the

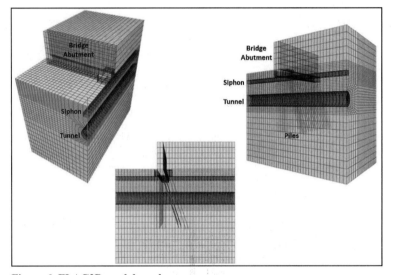

Figure 6. FLAC3D model mesh

Table 1. Soil properties

Unit Weight (pcf)	Friction Angle (deg)	Pile Skin Friction (deg)	Cohesion (psf)	Poisson Ratio	In-situ Vs (ft/s)	Young's Modulus, E (ksf)	Average SPT
125	32	22	200	0.3	540–770	2500–5000 (small) 120–240 (large)	40

Table 2. Structural properties

Structural Element	Unit Weight (pcf)	Young's Modulus, E (ksi)	Moment Inertia (ft⁴)	Yield Moment (kip-ft)	Bridge Load (kips/ft)
Piles	—	3500	0.155	~40	
Siphon	—	2200	7.5	~240	30
Abutment & Pile Cap	150	3500	—	—	
Pipe Spiles	—	30000	0.000016	—	

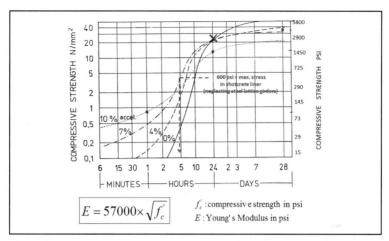

Figure 7. Time-dependent compressive strength of shotcrete

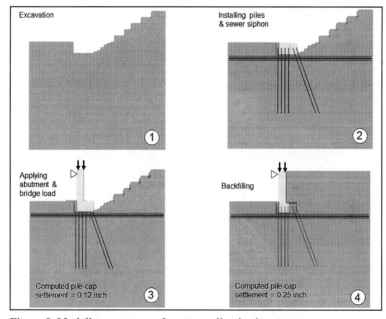

Figure 8. Modeling sequence of pre-tunneling in situ stresses

abutments. So the next step after establishing in-situ stresses, was to simulate SEM tunneling underneath the north abutment. Starting from the face of an "initial" tunnel segment whose shotcrete liner was "wished in place" (Figure 9), a total of nine consecutive excavation rounds were analyzed advancing the tunnel past the north abutment and a 48-in diameter clay-pipe sewer running parallel to, and behind, the north wall of the channel. Though design specifications called for three-ft long rounds to be advanced by heading-and-bench excavation, the rounds in the

model were assumed to be excavated full-face at five-foot lengths.

Analysis Results

The analysis results discussed below address the primary concerns of the stakeholders including: (1) settlement of the bridge abutments due to tunneling adversely affecting/damaging the bridge foundation piles; (2) cracking of the sewer siphon running along the same horizontal alignment as, and merely 10 feet above, the tunnel to be excavated; and (3) cracking of

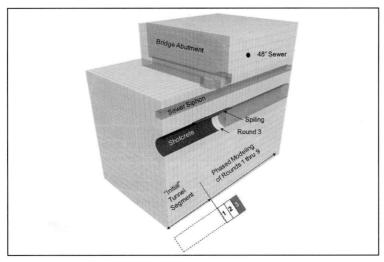

Figure 9. Simulation of SEM tunnel excavation rounds

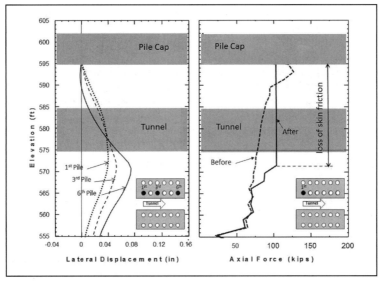

Figure 10. Tunneling-induced pile deflections and axial forces

the shallow 48-inch diameter clay-pipe sewer running parallel to, and behind, the north wall of the channel.

Bridge Foundation Piles

Computed lateral pile displacements (deflections) perpendicular to and closing in on the tunnel are plotted in Figure 10. These deflections are very small ranging from a maximum of merely 0.04 inches for the first pile to a maximum of 0.07 inches for the last (6th) pile affected by the passing tunnel excavation. The latter deflection corresponds to a maximum computed bending moment of 1.25 kip-ft which is

but a small fraction of the piles' moment capacity with a yield moment of 40 kip-ft.

Also shown in Figure 10 is a plot of computed axial forces of the first pile to be affected by the approaching tunnel excavation. To address the concern about loss of axial capacity of piles as the horizontal stresses relax, all piles were modeled without tip resistance, so that the entire load would have to be taken up by skin friction. As shown by this plot, the loss of pile/soil friction was accommodated by transferring a portion of the pile's axial force to deeper soils, while another portion was transferred to other

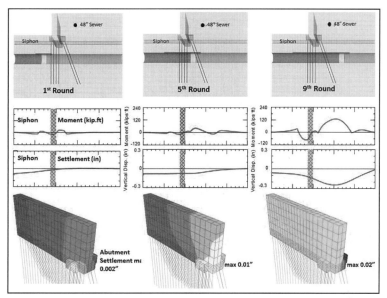

Figure 11. Moments and settlements of sewer siphon and settlement contours of bridge abutment

piles within the pile group supporting the bridge abutment. As discussed below, these load transfers resulted in negligibly small settlements of the bridge abutments.

Bridge Abutment & Sewer Siphon

Figure 11 presents contours of computed settlements of the bridge abutment, along with bending moments and settlements of the sewer siphon induced by tunnel excavation Rounds #1 through #9. Settlements of the bridge abutment are essentially zero. Settlements of the sewer siphon reach a maximum of about ¼ inch after excavating Round 9, with a maximum bending moment of 120 kip-ft which amounts to about 50% of the siphon's estimated yield moment.

Sewer Line Along Tujunga Wash

The maximum tunneling-induced settlement of the 48-inch sewer line running parallel to the Tujunga Wash behind the north abutment of the bridge was computed to be less than 0.1 inches.

CONCLUSIONS

The crossing of LA City Trunk Line South under Tujunga Wash required "snaking" an approximately 10-foot diameter tunnel between deep foundation piles of a bridge carrying a busy street intersection, which was approximately 30 degrees out of alignment with the trunk line. The use of SEM tunneling to solve this challenging problem depended on being able to convincingly demonstrate to LADWP

and other stakeholder agencies involved that it could be done without undue risk of damaging the bridge and nearby structures and utilities. The SSI analysis described in this paper accomplished this task, and the successful completion of this tunnel with SEM set an example for other complex tunneling projects in the future.

ACKNOWLEDGMENT

The authors thank Wolfgang Pichler (Dr. Sauer & Partners) for his contribution to this paper.

REFERENCES

CDMG 1997. California Division of Mines and Geology, DMG Special Publication 117.

DSC, 2010. Design Drawings, Trunk Line South— Unit 4/Phase 2; Dr. G. Sauer Corporation, 5/24/2010.

Itasca, 2009. "Fast Lagrangian Analysis of Continua in 3 Dimensions—FLAC3D, Version. 4.0". ITASCA Consulting Group, Minneapolis, Minnesota, USA.

URS, 2010. *Analysis of Tunneling under Tujunga Wash at Riverside Drive and Whitsett Ave, Studio City, CA.* Los Angeles Department of Water & Power; Report May 21, 2010.

URS, 2006. *Geotechnical Investigation Proposed Trunk Line South-Unit 4, Los Angeles, CA.* Los Angeles Department of Water & Power; Report August 8, 2006.

TRACK 4: CASE STUDIES

Session 5: Sewer/Water 3

Rick Vincent and James Parkes, Chairs

Preparation for Tunneling, Blacklick Creek Sanitary Interceptor Sewer Project in Columbus, Ohio

Ehsan Alavi
Jay Dee Contractors, Inc.

Ed Whitman
Michels Corporation

Valerie Wollet
H.R. Gray

Matthew Anderson
Black & Veatch Corporation

ABSTRACT: The City of Columbus Blacklick Creek Sanitary Interceptor Sewer Project (BCSIS), located in northeast Franklin County, Ohio, includes 6.9km of sewer tunnel. This project also requires the excavation of two shafts at the launch point of the TBM run, 6 intermediate shafts, and two shafts at the receiving point of the TBM. A joint venture of Michels Corporation and JayDee Contractors, Inc. (Blacklick Constructors, LLC) is constructing this project. The tunnel is being excavated using a Herrenknecht EPB TBM. This paper summarizes the preparation work that has been done prior to the launch of the TBM. Additionally, several major issues arose at the beginning of the project which were resolved through collaborative partnership between the contractor and the Construction Management Team. Lessons learned throughout this process are discussed.

INTRODUCTION

The BCSIS will extend the existing 1.7m (66 inch) sewer to the north with a new 3m (120 inch) gravity sewer, from Blacklick Ridge Blvd north to Morse Road. This extension is required to support further development and services to the City of New Albany and Jefferson Water and Sewer District. Furthermore, this extension will allow a future connection from the Rocky Fork Diversion to redirect sewer flow from the Big Walnut sewershed into the Blacklick sewershed.

The tunnel alignment starts at its outlet into the existing Manhole 12, just south of the Blacklick Ridge Boulevard and east of Reynoldsburg-New Albany Road. The alignment extends in a northerly direction near parallel to Reynoldsburg-New Albany Road towards Morse Road as shown in Figure 1. The total length of the alignment is 7,016m (23,020 linear feet), with 6,895m (22,620 linear feet) to be constructed using an Earth Pressure Balance Tunnel Boring Machine (EPB TBM) and the rest by cut-and-cover. There is a new junction chamber at Manhole 12 south of Blacklick Ridge Boulevard. The completed tunnel will collect flows at the intersection of Reynoldsburg-New Albany and Morse Roads through a drop structure conceived to accept future flow from the New Albany, Rocky Fork, and City of Columbus service areas. Along the sewer alignment, there are two intermediate shafts with permanent drop structures for future tie-ins with the Jefferson Township service area, as well as for tunnel maintenance access purposes. These shafts will serve as TBM access and maintenance during the tunnel construction. Other appurtenances include construction of connection pipes, manholes and other ancillary facilities necessary to the operation and maintenance of the new sewer.

Blacklick Constructors LLC (BCL) was the successful bidder out of four candidates on November 18, 2015 with a bid price of $108.9 million. The engineers estimate for this project was $113.7 million including contingency.

The shallow open cut and shaft excavations are expected to encounter fill and alluvial soils. The soil portion of the tunneling excavation is generally in glacial till and outwash deposits, which are significantly overridden and consolidated. The transition between the soils and bedrock typically consist of highly weathered to decomposed rock. Glacially-deposited erratic boulders are also anticipated along the transition zones. There are two major bedrock

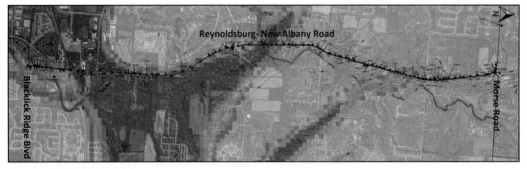

Figure 1. BCSIS tunnel alignment

formations that will be encountered during tunneling and shaft excavations, the Cuyahoga and Sunbury Shale. The Cuyahoga Formation has been subdivided into the Upper and Lower Cuyahoga, with the Upper Cuyahoga containing fine sandstone and siltstone which is absent from the Lower Cuyahoga. Geotechnical Baseline Report, classifies the ground into Fill, Alluvial, Glaciolacustrine, Glaciofluvial, Glacial Till, Transition material, Upper Cuyahoga, Lower Cuyahoga, and Sunbury Shale. 39 percent of the tunnel alignment is in soil, 46 percent in rock, and the rest through Transition material.

GEOLOGY AND WATER INFLOW ALLOWANCE

There are a significant number of residential wells along the tunnel alignment and near the project shaft sites. To limit the impact to residential wells, dewatering has been limited to the main site and the soft ground portion of Shaft 2. The remaining works have an allowable sustained inflow limit of 40LPM (10GPM) with a punctual initial 48hr maximum limit set at 110LPM (30GPM).

The Contract also has provisioned an allowance to respond to, and repair or replace, affected private water wells impacted by the project. This response includes providing temporary water, lowering existing wells, providing new filters, or replacing damaged wells.

SITE WORK

The BCSIS project includes work at four major sites—Shaft 1, Shaft 2, Shaft 3, and Shaft 4—and a significant portion of the work accomplished during the first 12 months of the project involved getting these sites prepared to support the underground work to come.

Shaft 1 and Manhole 12 Sites

The main project site is the Shaft 1 area, a two hectare diamond shape lot (200m × 200m), along the

northbound lanes of Reynoldsburg-New Albany Road. This site is the main operation site for the entire tunnel drive. All of the TBM tunnel spoils are extracted from Shaft 1. The Manhole 12 site, where the BCSIS ties into the existing 1.7m (66-inch) sewer, is located on Blacklick Ridge Blvd on the south side of the Shaft 1. The existing Manhole 12 is to be replaced with a new junction chamber under the BCSIS Project. Figure 2 shows Shaft 1 and Manhole 12 sites.

Since dewatering is allowed at both the Shaft 1 and Manhole 12 sites, to facilitate the cut-and-cover work as well as the installation of the shafts, both Shaft 1 and Manhole 12 were installed using soldier piles and wood lagging. Shaft 1 is 5.5m wide by 20m long rectangular shape, by 15m deep. Similarly, the Manhole 12 shaft was constructed using soldier piles and lagging and was supplemented with several soil nails due to its unusual shape and to preserve the necessary clearance inside the shaft walers to erect the final transition structure. BCL oversized this 7.5m wide, 12m long, by 10m deep shaft to not only expedite the TBM assembly, but also facilitate the first 130m of tunneling, and optimize the tunneling cycle time in this critical startup phase.

The connection between Shaft 1 and Manhole 12 was initially intended to be constructed by cut-and-cover. However road closures on a busy commuter road, doubling as an emergency medical services (EMS) artery, lined by multiple sensitive utilities including high voltage buried transmission lines, sewer and water mains, led BCL and the Construction Management Team (CMT) to seek less disruptive alternatives. The original installation method was altered to incorporate a section of hand excavated tunnel under the Blacklick Ridge Blvd. These efforts are discussed in greater detail in a later section of this paper.

Shortly after clearing and stabilizing the site, BCL began installing the environmental Best Management Practices (BMPs) due to the close proximity to Blacklick Creek and the Notice To Proceed (May 2015) coinciding with the start of the rainy season. BCL elected to supplement the prescribed environmental site development measures and BMPs at the main work site. Improvements included stabilizing the entire site, refining the muck handling and loading methodology, designing and paving a one-way delivery/muck loading circuit to limit tracking and improve site safety, adding an automated wheel wash system, improving the prescribed water treatment system, re-grading the site to promote surface runoff percolation and reduce the volume of the discharges to the creek, and preserving existing peripheral buffer areas not necessary to the operation.

Shaft 2 and Shaft 3 Sites

Three distinct manholes are installed at each of the intermediate shaft locations: one for tunnel maintenance access (2A and 3A) located directly over the crown of the tunnel and connected to the tunnel; one

Figure 2. Shaft 1 and Manhole 12 sites

housing the drop and energy dissipater pool structures connected to the springline of the tunnel; and the third acting as the flow collector. The finished diameter of these manholes ranges from 1.5m to 2.4m, and ranges in depth from 7.5m to 30m. The three manholes in each shaft locations were connected to each other by two horizontal adits. Figure 3 shows profile views of Shaft 2 and Shaft 3.

Due to small sizes of these manholes, BCL chose to drill the individual manholes at the Shaft 2 and Shaft 3 locations. At the Shaft 3 site, a combination of open-cut, for shallow manholes, and drilling, for deeper manhole was used. All drilled holes were then lined using a steel casing grouted in place.

Once the drilling was complete, adits were excavated in rock using hand tunneling methods, and supported by steel horseshoe sets and timber lagging.

Since the allowable water inflow in those locations was extremely low, BCL used a combination of pre-grouting, and in-the-wet techniques to install the shafts and avoid the necessity to dewater during the drilling operation. During the hand-tunneling phases, the water inflow was controlled using chemical grouting whenever necessary. Those measures appear to have been successful in minimizing the impact to neighboring residential wells since less than 3 percent of the contingency allowance had been expended with the TBM 37 percent into the drive.

Shaft 4

The Shaft 4 site receives two manholes. Manhole 4B is a 12m deep hole with a finished diameter of 2m. Manhole 4A is 44m deep and is intended to be used as a TBM retrieval shaft before installing a concrete baffle drop structure with a finished inner diameter of 4.2m as shown in Figure 4.

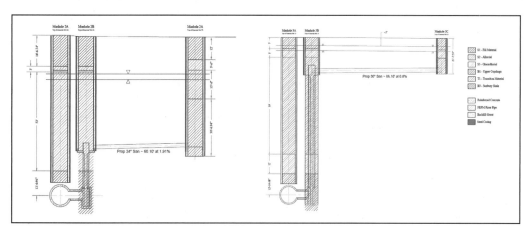

Figure 3. Profile view of Shaft 2 (left) and Shaft 3 (right)

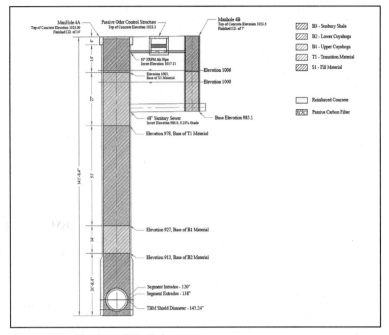

Figure 4. Profile view of Shaft 4

To facilitate the TBM retrieval, BCL chose to maintain an 8m inner diameter while excavating Shaft 4A. Secant piles have been used as the support of excavation means in the soft ground portion of the Shaft 4A. The remainder of the shaft is the only planned drill and blast work on the project. To limit the nuisance to the local residents, BCL has avoided using blasting whenever possible.

PREPARATION FOR TUNNELING

Throughout TBM manufacturing and delivery, BCL installed and commissioned the rest of the project ancillary specialty equipment at the Shaft 1 site. This equipment included a breathable compressed air plant, a medical lock for hyperbaric interventions, water treatment systems, a dry house, shop, and office trailers. Concurrently, utilities were installed, including the main power drop, a substation, and electric switchgear to support the tunnel boring machine, the surface, and subsurface plants. In addition, Sagami Servo Grout plants, a water cooling system for the TBM, muck bin, parking lots, wheel wash and miscellaneous plants were also installed and commissioned.

Breathable air plant. The breathable air plant and the medical lock consist of a Hi-Air compressor farm, medical compressed air chamber able to accommodate 15 divers, an emergency rail mounted compressed air vessel (DART), a control center,

and all the necessary appurtenances for hyperbaric interventions.

Water treatment plant. Satisfactory treatment of site construction waters prior to their release in the designated discharge locations on the project is essential due to the environmental sensitivity of some of the sites. A sediment basin and baffled settlement tanks were utilized to process storm and construction waters prior to being discharged into Blacklick Creek. Individual filter bags were also utilized to further pre-treat the construction waters during shaft construction.

Muck Bin. A bean shaped 1400m³ capacity muck bin was constructed next to Shaft 1. The muck bin is about 2m deep. The muck bin was designed to have enough capacity to store up to two days of tunneling at peak production. A 160T crane hoists and tips the 6m³ muck cars into the muck bin area and a material handler is used for loading the trucks out.

Segment Design. Due to the extremely low water leakage criteria in the contract specifications, BCL was particularly attentive to the tunnel's precast concrete liner design. Despite some setbacks at the beginning of the project with the prescribed hybrid cage/fiber reinforcement design, BCL and the CMT worked in concert towards a solution. Another area of improvements brought to the tunnel liner was the use of embedded gaskets in lieu of the traditional glued on type. These embedded gaskets, along with the use of blind lifting inserts as annular grout ports

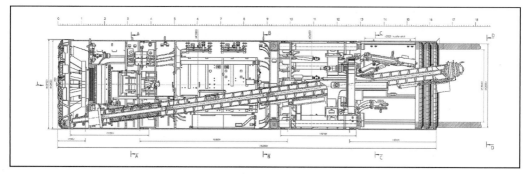

Figure 5. Section view of the Herrenknecht TBM

on the segments, and guide rods along the radial joints of the segments, have all proved to be worthy ameliorations. The tunnel constructed thus far is remarkably dry, and minimal segment damage has been reported.

Segment Storage and Handling. A large portion of the available space in the jobsite is allocated to segment storage and handling. The segments for this project are manufactured in Maryland and must be transported to the Shaft 1 site by truck. Maintaining a constant supply of segments during the winter months could become a challenging endeavor. To mitigate this risk, BCL chose to allocate as much area as possible to segment storage. About 1,000 rings can be stored on the jobsite.

TUNNEL BORING MACHINE PREPARATIONS

A new Herrenknecht (HK) EPB TBM was purchased by BCL for this project. After the receipt of the Notice of Intent from the City of Columbus in January 2015 and prior to receipt of Notice To Proceed in May 2015, BCL requested HK to initiate the design phase and ordered the TBM long lead items to expedite the TBM manufacturing and delivery process. Throughout the design and manufacturing process, BCL and HK collaborated to develop and optimize the TBM for the numerous and disparate challenges it had to overcome. The TBM was manufactured and factory tested in November 2015 and was delivered to job site in Mid-January 2016.

General specifications and layout. The Herrenknecht EPB TBM was designed to tackle a wide range of mixed soft ground conditions including glacial till and running sands, in addition to Shale. A durable mixed ground cutterhead with reliable cutting tools was paramount (see Figure 5 and Table 1).

Due to the small diameter of this TBM (3,705mm) and the requirement of having an airlock for interventions, the TBM body is 16.3m (53.5 feet) long and consists of a tail can, thrust shield, man-lock body, and a front body. Total length of the TBM including TBM body and 11 gantries is 113m (370 feet).

Cutterhead. To handle the changing soft ground conditions, BCL selected tungsten carbide insert pressure compensated mono block disc cutters that perform in the diverse geologies as the main cutting tools. In addition, and based on the BCL's previous experience, additional pre-cutters manufactured by Japanese supplier Staraloy, were welded to the cutterhead, as face pre-cutters and over-cutters, to further protect and armor the cutterhead. "State-of-the-art" engineering and remarkable efforts were put in place to develop a cutterhead for the diverse array of geologic conditions expected throughout this project. In addition, the TBM shields have been fitted with EAG probes to measure the overcut in various locations along the shields. These measurements serve as wear, and overall health, indicators of the cutterhead. Figure 6 shows the cutter layout on the cutterhead.

TBM Guidance. BCL opted for a gyroscopic based TBM guidance system in lieu of the more common Measuring While Driving (MWD) type system (a.k.a. Total Station/laser system). Due to the configuration of the project (30 percent of the drive in curves ranging in radius between 300m and 900m, tunnel finished ID of 3m), using a MWD system would have required a large number of move-ups and likely introduced proportional amounts of errors, and accommodating a usable line of sight on the TBM (laser window) was virtually impossible. On the BCSIS project, the gyroscopic guidance system also presented several advantages over a MWD system:

1. The Gyroscopic system is typically less susceptible to unplanned stoppages:
   a. no instrument or backsight adjustments necessary,
   b. fewer re-calibrations needed,

Table 1. Specifications for the Blacklick HK TBM

Herrenknecht		
Dimension	Overcut (with disc cutters and soft ground tools) diameter	3,795 mm (12.45 ft)
	Front section shield diameter	3,705 mm (12.16 ft)
	Tail skin diameter	3,695 mm (12.12 ft)
	Segment outer diameter	3,505 mm (11.5 ft)
	Segment inner diameter	3,048 mm (10 ft)
	Overall length (assembled)	15,539 mm (51 ft)
Cutterhead	Type	Bidirectional, Mixed Ground
	Opening ratio	< 20 percent
	Cutterhead drive	Hydraulic
	Cutterhead power	400 kW
	Cutterhead speed	0 ~ 6.5 rpm
Torque	Cutterhead working torque	1200 kN-m (Max 1600 KN-m)
Thrust	Trust jack stroke	2,200 mm
	Maximum thrust	19,000 kN
Electrical	Primary voltage	13,200V
	Protection	Class 1, Div 2 (Essential services: lighting/ gas detection/ dewatering/ventilation)
Conveyors	Screw conveyor diameter	500 mm Shafted (1.6 ft)
	maximum grain size	130 mm (5.1 inch)
	Screw conveyor type	Reversible/ telescopic
	Speed	0–33 rpm
	Torque	30 kN-m (Max 37 KN-m)
	Back-up conveyor belt width	500 mm (1.6 ft)
Weights	TBM weight (approx.)	200 tonnes (220 ton)
	Back-up weight (approx.)	100 tonnes (110 ton)

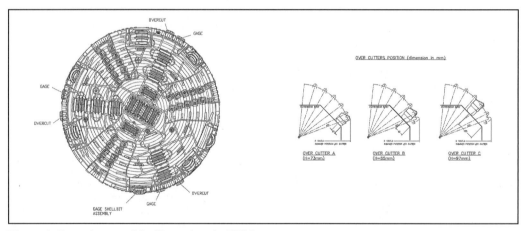

Figure 6. Cutter layout of the Herrenknecht TBM

c. less susceptible to EMI interference from High Voltage Gear due to its location in the TBM shields,

d. the instrument is less likely to be accidentally bumped, since it is stationary and can be installed unobtrusively,

e. eliminates line of sight and radio communication issues between the instrument and the base system in an already extremely constricted space (small I.D. tunnel TBM gantries),

2. Its measurements are immune to refraction, one of the biggest challenges in small diameter tunnel optical survey observations, not only on the horizontal plane, but also vertically; since the system also uses a hydrostatic

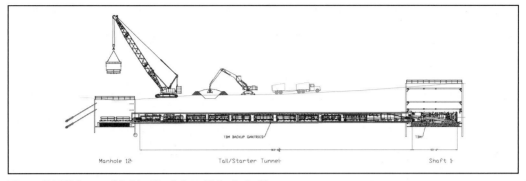

Figure 7. TBM assembled in Shaft 1 and Manhole 12

level monitoring system for the elevation component of the guidance system,

3. This hydrostatic system has a single reference sensor to relocate during the move-up routine, which not only expedites the operation, but also minimizes the possible introduction of errors, thus improving the system accuracy.

The only noteworthy impediment this type of guidance system may have over a MWD is the frequency of the required scheduled maintenance intervals. Indeed, while it is not uncommon to have a MWD system remain effective for several hundred meters in a straight section of tunnel, a gyroscopic based guidance system could require daily adjustments when tunnel production nears the 30m/day mark, due to the instrument characteristic drift. While the error induced by this drift is marginal over distances of less than 30m, this error develops into an exponential function beyond the 30m range.

TBM ASSEMBLY

Further to the efforts to accelerate the design and procurement of the TBM explained above, BCL sequenced the construction of Shaft 1, Manhole 12, and the 130m of tail/starter tunnel to complete simultaneously to expedite the TBM assembly and commissioning. This allowed BCL to assemble the TBM main shield bodies in Shaft 1, while concurrently lowering the TBM gantries through Manhole 12 and sliding them into position through the tail tunnel to connect them to the TBM shields in Shaft 1.

The TBM assembly started on January 6, 2017 and finished on March 27th for total duration of 54 work days. This allowed BCL to test, commission, and successfully launch the TBM on April 12, 2017, over 7 months ahead of the prescribed project milestone. Figure 7 shows the TBM assembled in Shaft 1 and Manhole 12.

TBM LAUNCH AND TUNNELING

Initial 130m tunneling phase. In the initial 130m tunneling phase, and prior to burying the TBM body and ancillaries, the TBM operation was supported from the Manhole 12 shaft. During this phase, benefiting of its modifications to the original shaft dimensions, BCL was able to start tunneling using two muck boxes, two segment cars and a loci to support the operation. As a result, the learning curve was surmounted sooner as cycle times reduced. This operation, albeit not optimum, a shove typically requiring 3.5 to four muck boxes, allowed this typically slow phase of a project to be completed diligently.

Setup change and Production tunneling. Once the TBM gantries cleared Shaft 1, the operation was halted to reconfigure Shaft 1 for the production mining phase. The blind rings across Shaft 1 were removed, a California switch was installed in the tail tunnel, and the utilities were re-arranged in their new and definitive configuration. A flood protection bulkhead was installed at the south end of the tail tunnel in the Manhole 12 Shaft, and the servicing operation of the tunnel moved to its permanent location at Shaft 1.

Intermediate manholes / safe havens. As the TBM approached the first intermediate manhole (Manhole 2A), the hand tunneling activities between the bottom for Shaft 2A and drop-Shaft 2B were completed. The bottom of Shaft 2A was prepared for the TBM hole through. A pre-break pattern was installed and the shaft bottom was filled with sand and flooded to approximately one meter above the crown of the tunnel. Since the TBM was tunneling through rock, it was expected to enter the shaft in open-mode. Once in the shaft, the TBM cutterhead was inspected. While the wear to the tools was nominal (approx. five~ten percent wear), BCL determined it was prudent to redress the face with new cutting tools given the 3km separating the machine from its next safe haven. The TBM then relaunched, and stopped once more when the gantries had sufficiently cleared the Shaft 2 area to allow for the crown connection to be completed. BCL utilized this scheduled downtime to move the tunnel ventilation to Shaft 2, and install its second California switch between Shaft 1 and Shaft 2, precluding the tunnel ventilation duct from interfering with already challenging train traffic crossing conditions over the switch.

COLLABORATION BETWEEN THE CONTRACTOR AND THE CONSTRUCTION MANAGEMENT TEAM

Major issues arose at the beginning of the project that were resolved through collaborative partnership between BCL and the City's CMT. Several lessons were learned throughout this process.

Open cut versus Hand Mining under Blacklick Ridge Blvd. The original design intended for approximately 130m of open-cut (cut and over) Reinforced Concrete Pipe (RCP) to be installed downstream of the proposed working shaft which could be utilized as a tail tunnel allowing a more efficient TBM launch. The design provided for utility by-pass and support, and allowed for closure of Blacklick Ridge Boulevard due to the depth of excavation in the area being approximately 12m (40 feet). Early in the planning phases of the work after bid time, the project team determined that any possible detours resulting from closing the road would cause critical disruption for emergency vehicle response times.

When these issues arose, BCL was open to suggestion and took the initiative to solicit advice and additional quotations from the subcontractors performing the work rather than moving the contractual process into a claim or differing site condition and waiting for direction. BCL and the CMT collaborated to evaluate the restrictions, alternate means of providing emergency services, and the best possible approach to completing the approximately 37m section under the road and utility easements with the least risk and disruption to the public. They met together with stakeholders to understand and address their concerns. Ultimately, to mitigate these issues, it was decided that installation would proceed by excavating a 4.4m (168 inch) hand tunnel with the 3m (120 inch) reinforced concrete pipe grouted in place in the tunnel under the road.

BCL and the CMT worked together to evaluate possible options and ultimately developed a plan for hand tunneling. During this period, BCL brought in a local hand tunneling subcontractor that was familiar with this type of work in similar ground and the team was able to collectively create a plan.

While a plan for the tail tunnel was being developed, BCL was creating the preliminary baseline schedule for the project. These ongoing design changes had the potential to significantly affect several work activity sequencing options. The team's collaborative spirit and quick action allowed the initial change from open-cut to hand-mining to be processed without adding additional time to the contract.

Segment Manufacturing. During the bidding phase of the project, the contractor intended to manufacture the segments according to the contract documents using a combination of wire mesh and fiber reinforcement. Per the American Iron and Steel (AIS) provisions, all projects funded by a Clean Water State Revolving Fund or a Drinking Water State Revolving Fund are required by federal United States law to use iron and steel products that are produced in the United States. After contract execution, it became evident that manufacture of the specified hybrid segments faced two potential issues with utilizing domestically produced steel products. Automated fusion welding equipment capable of producing the required steel cages were no longer available in the United States to the selected segment manufacturer, and production of domestically produced steel fibers meeting the requirements of the specifications were not yet available.

BCL proactively pursued alternate options and maintained open communication with the CMT to allow all parties to participate in evaluation of the alternatives. In consideration for project cost and schedule, a decision was made to allow the use of 100 percent fiber reinforced concrete segments. To remain consistent with the specified products and further minimize schedule slip due to retooling of the manufacturing facility, the revised segment design was based on the use of imported Dramix fibers manufactured by Bekaert Maccaferri. The CMT took the lead to apply for a Public Interest Waiver of American Iron and Steel Requirements for the fiber reinforcement.

Through the combined efforts of the Owner, Contractor and Construction Management Team, a public interest waiver was prepared and presented to the USEPA outlining the rational for use of imported steel fibers. The waiver was granted by the USEPA on November 15, 2016 allowing segment production to commence without project delay.

REMAINING WORK

The final completion of the BCSIS project is expected to occur in July 2020. As of the publishing of this paper, about 40 percent of the tunneling has been completed. The three shafts at the Shaft 2 site and two out of three shafts at the Shaft 3 site have been excavated. Hand tunneling between the intermediate shafts at the Shaft 2 site has been completed. BCL just initiated the excavation of Shaft 4A. Based on the current progress of the work, BCL believes that they can meet and probably improve the final completion date of this contract.

The Columbus OARS CSO Tunnel from Design Through Construction—An Experience in Solution-Filled Karst

Paul Smith and Matthew Anderson
Black & Veatch

Raisa Pesina
City of Columbus Department of Public Utilities

Jeff Coffey
DLZ Corporation

Michael Hall
H.R. Gray

ABSTRACT: The alignment of the 7.01 m (23 ft) diameter bored 7.1 km (23,300 ft) long Columbus OSIS Augmentation Relief Sewer (OARS) CSO Tunnel passes through solution-filled karst limestone and dolomite at depths up to 58.0 m (190 ft) and traverses north from the Jackson Pike Wastewater Treatment Plant (JPWWTP) to the downtown Columbus Arena District. This paper will discuss planning, final design and the use of innovative methods to overcome difficult geologic conditions encountered while excavating and constructing multiple large diameter shafts, the main tunnel, a connecting tunnel and adit connections in the karst limestone. The paper will also discuss project completion, project start-up and delivery of the CSO tunnel system to the City of Columbus.

BACKGROUND

The City of Columbus submitted a Wet Weather Management Plan (WWMP) in July 2005 after receiving and negotiating two Consent Decree orders from the Ohio EPA. The first order, for Sanitary Sewer Overflow (SSO) issued August 2002, mandated the development of a Capacity, Management, Operation, and Maintenance program (CMOM) and a System Evaluation and Capacity Assurance Plan (SECAP). The second order, for Combined Sewer Overflow (CSO) issued September 2004, mandated the development of a Long Term Control Plan (LTCP) per requirements of U.S. EPA Combined Sewer Overflow Control Policy.

The Olentangy Scioto Interceptor Sewer (OSIS) is the City of Columbus' main combined sewer that conveys wastewater and storm water flow north of the City south to the Jackson Pike Wastewater Treatment Plant (JPWWTP). The OARS Tunnel is the main component of the mandated LTCP mitigation and was designed to provide relief to the OSIS by collecting combined sewer overflows from the OSIS and tributary sewers during storm events, provide inline storage during critical periods to reduce/eliminate combined sewer overflows from the critical downtown regulators, and convey the flows to JPWWTP and/or SWWTP for biological treatment. The OARS Tunnel is intended to provide adequate storage capacity to mitigate downtown-area overflows through the year 2047 for all wet weather storms contained within a "typical year" for up to a 10-year wet weather event and is anticipated to remove nearly 2 billion gallons of CSO discharges annually from the Scioto River. The OARS Tunnel is a large undertaking for the City of Columbus that included several years of planning, risk management and mitigation, financial planning, detailed design and partnerships with many stakeholders and consultants throughout the Columbus Metro area.

THE PROJECT

The Columbus OARS Project, valued $370 million, was designed and bid as two contracts or phases with construction beginning in September 2010 and final completion scheduled for March 2018. OARS Phase 1, with a contract value of $264,506,000, was awarded to Kenny/Obayashi, JV (KOJV) in 2010. OARS Phase 2, with a contract value of $76,919,700, was awarded to Trumbull Corporation in 2011. The project includes 7.1 km (23,330 ft) of 7.0 m (23 ft)

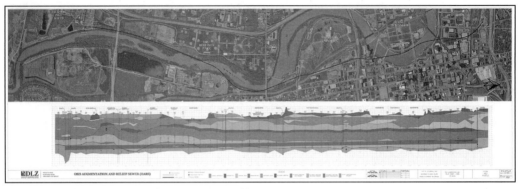

Figure 1. OARS CSO tunnel alignment and profile

diameter excavated tunnel and six large diameter shafts approximately 55 m (180 ft) deep through 30.5 m (100 ft) ± of glaciated overburden requiring watertight support of excavation (SOE) down into bed rock. Pre-excavation grouting, drilling and blasting were required for excavating the shafts, starter tunnel and tail tunnel through bedrock to design grade.

Phase 1

Phase 1 consists of the 7.1 km (23,330 ft) long tunnel that starts at Shafts 1 and 2 by JPWWTP, travels north along and under the Scioto River and ends at Shaft 6 in the Arena District as shown on Figure 1. The tunnel, with 15 curves (427 m (1,400 ft) minimum radius) is excavated through bedding planes of solution filled karstic limestone with a 6.1 m (20 ft) diameter gasketed precast concrete (PCC) segmental liner. Phase 1 also includes one 15.8 m (52 ft) finished diameter OARS Diversion Structure Pump Station (ODS) identified as Shaft 1; one 12.8 m (42 ft) finished diameter OARS Screen Structure (OSS) identified as Shaft 2 with a Screen Electric Building (SEB); the West Gate Chamber (WGC); the Flow Diversion Structure (FDS) extension and one 14.6 m (48 ft) finished diameter access retrieval shaft identified as Shaft 6 with a relief/by-pass structure designed to divert flow from the OSIS to a tangential inlet (TGI) approach channel with vortex drop shaft designed to deliver flow to the tunnel. During the planning and design of Shafts 1 and 2, hydraulics, size of the screening chamber and screens and availability of land were considered. Also considered was the center to center distance between the shafts [68.2 m (224 ft)] along with the 33 m (108 ft) length for the screening structure (starter tunnel) that allowed for TBM erection and launching.

Phase 2

Phase 2 consists of the construction of Shafts 3, 4, & 5; a River Outfall Structure (ROS); two regulator

Figure 2. ODS wet well with pumps and piping installed

overflow connection sewers at Shafts 4 & 5; installation of deep lift submersible pumps and piping in the ODS that included four pumps at 227,125 m³/day [15 mgd each (60 mgd max)] in the tunnel level, two pumps at 151,416 m³/day [20 mgd each (40 mgd max)] in the shaft level and two 200 mm (8 in.) grit pumps rated at 7,500 m³/day [1 mgd each (2 mgd max)], and one 254 mm (10 in.) shaft mixing pump (See Figure 2). Phase 2 also includes final site electric and controls for the ODS pumps, the Scioto Main Relief (SMR) diversion structure; gates; and final adit connections between Shafts 3, 4 and 5 and the OARS Tunnel.

Due to Project delays the three adit connections were contractually moved to Phase 1 for completion. Shaft 3 is a 3 m (10 ft) finished diameter drop shaft with a TGI approach channel for future connection. Shafts 4 and 5 are 9.1 m (30 ft) finished diameter shafts. Flows at both shaft sites are designed to be diverted from the OSIS into the OARS CSO Tunnel via relief/by-pass structures and TGI approach channels with vortex drop shafts. Phase 2 also includes commissioning, start-up and operational demonstration of the OARS CSO Tunnel system.

CONCEPT AND PLANNING

The concept and planning for the OARS Project began after receiving the two consent decree mandates from the EPA. An extensive geotechnical field investigation program, including borings, rock cores, monitoring wells, observation wells, and test wells, was developed and undertaken by the design professional (DP) in order to help determine the optimal horizontal and vertical alignment for the tunnel. Early in the project planning process, a Value Engineering Workshop was held where the decision was made to switch to a deep tunnel system and complete the originally planned two phases of shallow, open excavated sewer in a single tunnel phase. This allowed for the system to be operational eight years earlier than the 2025 deadline required by the CSO consent order. The additional storage volume provided by the larger tunnel also eliminated the need for future above-ground storage structures and a future high-rate treatment system, saving an estimated $175 million.

During the investigation a major concern was the potential for solution features in contact with the overburden and the amount of ground water anticipated due to karstic bedding planes and joints. Cross connections between the bedding planes and joints were noted in borings along the alignment and in down-hole video images of these solution features depicting various sizes in the joints and bedding planes. The connections were verified in fluctuations of the water levels in established monitoring and observation wells in the bedrock and overburden and the nearest river gauge levels. Interestingly, the DP encountered a loss of water during drilling in every borehole drilled into the bedrock. It was assumed that the loss of water was due to bedrock fractures and joints coupled with the known high permeability and transmissivity of the bedrock.

Another example of the connection between bedding planes and joints was from a previous published paper on this project, "...*also noted during air rotary drilling at the Jackson Pike Site, when cross connection was encountered at approximately 230 feet (El.483). Air injected during drilling created a 20 foot column of water to be ejected from an 8-inch well approximately 100 feet away...*"

DESIGN

The design of the OARS Project focused on risk management and the mitigation of groundwater inflows that were anticipated to be encountered during construction. Shaft excavations would be in soils ranging from granular to non-cohesive and cohesive

with depths ranging from 12.2 m (40 ft) at Shaft 3 to 30.5 m (100 ft) at Shaft 1. In addition the static ground water elevations averaged approximately 45.7 m (150 ft) above the tunnel alignment from EL 204.3 m (670 ft) to EL 213.4 m (700 ft), which made groundwater a major consideration during design and construction.

The tunnel was designed to flow by gravity from north to south at a slope of 0.13% for the entire 7.1 km (23,330 ft) alignment. Competent and consistent geology for tunneling was determined to be approximately 54.9 m (180 ft) below the ground surface. The location and depths of the shafts and tunnel required construction mainly in limestone, dolomite and shale. While the local shale is relatively impervious, the limestone and dolomite were characterized as karstic with solution features that transmit high groundwater flows. To mitigate groundwater inflows in the overburden, "water-tight" SOE was required for all drop structure and shaft construction. To mitigate groundwater in the bedrock, a pre-excavation grouting program was required for excavation and construction of the shafts, deaeration chambers, adit connections, TBM starter tunnel and TBM connecting tail tunnel.

Geotechnical Data Report and Geotechnical Baseline Report

A Geotechnical Data Report (GDR) was compiled and developed as part of the construction contract documents that documents the data obtained as the geotechnical field investigations and results were completed. In addition to the GDR, a Project Hydrogeological Report that included slug tests, packer tests, and longer term pump well testing was published. Also a groundwater flow model was developed to estimate potential ground water inflows into the tunnel and shaft excavations. Down-hole video technology was also utilized in selected boreholes to further document the solution features within the bedrock.

A Geotechnical Baseline Report (GBR) was developed as part of the contract documents that included a description of the project, a summary of geologic and geotechnical information for the project, the geologic project setting, previous construction experiences in similar/local geology, ground characterization, subsurface conditions at construction sites, subsurface conditions and foundation recommendations for near surface structures, and construction considerations. The GBR was also developed to help define the grout requirements that would be required to manage groundwater inflow in the shaft and tunnel excavations. The design intent was for the contractor to seal off groundwater inflow in order to gain access into the excavation chamber to inspect the cutterhead and disc cutters.

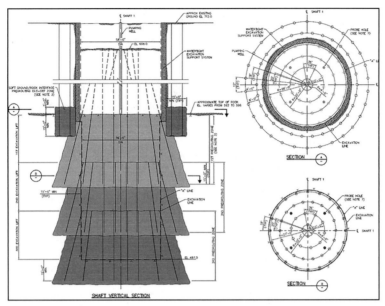

Figure 3. Suggested staged pre-excavation grout approach

CONSTRUCTION

Pre-Excavation Grouting in Bedrock

Pre-excavation grouting of the bedrock for the vertical rock excavation of the shafts was required and specified in the contract documents for Phase 1 and Phase 2. The DP suggested a three-stage grout plug approach in bedrock with inclined grout holes and a soft ground/rock interface pre-grouting zone as shown in Figure 3. Pre-excavation grouting of the bedrock for horizontal excavations with angled grout holes was also suggested in the connecting tunnel, screen chamber, de-aeration chambers, and adits.

The inclined grout hole spacing and minimum grout limits (4.6 m [15 ft] beyond excavation limits) were noted on the contract drawings and grout mix designs including cement types, minimal grout strengths, admixtures, etc. were specified in the contract documents. Grout refusal for both Phase 1 and Phase 2 was specified to be less than 1 cubic foot over two minutes or less than 0.85 m³/hr (3.74 gpm) at maximum injection pressure.

Phase 1—Shafts 1, 2 and 6

KOJV's grouting subcontractor, Nicholson Construction Co. (NCC), submitted an alternative pre-excavation grout plan that incorporated double wall grout curtains installed from the surface outside of the shaft excavations. The outside curtain wall (Line A) was drilled at predetermined radial angles to angle the grout holes around each shaft to intersect anticipated high angled joints, vertical joints

and horizontal bedded planes in the bedrock. Line A holes (primary, secondary and tertiary) were drilled and cased down through the overburden into bedrock and then rock drilled down through the bedrock to 7.6 m (25 ft) below shaft invert. The inner curtain wall (Line B) was drilled vertically through PVC casing installed in the slurry walls down through bedrock to 7.6 m (25 ft) below shaft invert. For both curtain walls at each shaft, NCC utilized downstage grouting at the rock/overburden interface followed by upstage grouting to complete each hole.

After installing all Line A casings through the overburden into bedrock, NCC downstage grouted the first stage zone 7.62 m (25 ft) of each hole (primary, secondary and tertiary) by grouting to 0.003 MPa (0.5 psi) per vertical foot of cover from 1.5 m (5 ft) above the rock/overburden interface. This created a plug in the first stage that could be upstage grouted against. NCC drilled all holes in the slurry wall to 7.6 m (25 ft) below shaft invert and upstage grouted each 7.6 m (25 ft) stage until all holes were completed. NCC completed all primary, secondary, tertiary and quaternary holes as needed using grout pressures of 0.006 MPa (1.0 psi) per vertical foot of cover.

NCC reduced their refusal rate in the field to as low as 0.02 m³/hr (0.1 gpm) in order to meet the specified "watertight" performance test after pre-excavation grouting was completed. If measured flow exceeded 1.13 m³/hr (5 gpm) from a single hole or 2.27 m³/hr (10 gpm) from the four holes combined, then additional grouting was required until the requirement was met. NCC experienced high grout

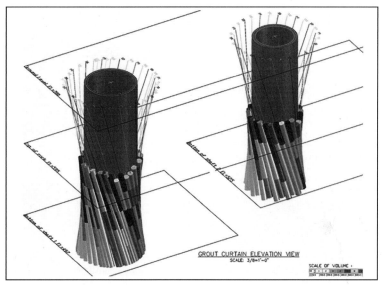

Figure 4. Graphical as-built of pre-excavation grout volume at Shaft 1 and Shaft 2

takes most frequently in the upper portion of the bedrock, but also in every grout zone. NCC drilled and grouted additional full depth quaternary holes in Line A at Shafts 1, 2, and 6 in an effort to meet the performance criteria for the grout curtain. The quaternary holes allowed NCC to reach their refusal rate with a lower volume of grout pumped per zone. NCC utilized an automated data collection system to monitor, measure, record and display graphically as real time data. NCC was able to provide a graphical as-built of the grout stages in each hole including grout takes and volumes placed as shown in Figure 4.

At Shaft 6, NCC changed the sequence of construction by installing the diaphragm slurry walls prior to drilling and grouting the Line A and Line B grout curtains. While drilling and grouting vertical Line B through the slurry wall, NCC was able to drill and grout the questionable construction joints of the slurry wall panels. Two of the nine slurry wall panel end stops with waterstop came loose at the bottom during the tremie concrete placement. The end stops were out-of-plumb and the concrete panels were milled down with the hydramill to allow the rebar cages for the adjacent panels to be placed correctly. The sequence change resulted in a good watertight excavation support system that was verified with the probe hole performance test. Based on geophysical test data in the GBR, GDR and project hydrogeologic report for Shaft 6, values for transmissivity and hydraulic conductivity of the bedrock were anticipated to be high along with an elevated high number of voids. The high grout takes at Shaft 6 verified the anticipated conditions at Shaft 6 as NCC pumped

more pre-excavation grout at Shaft 6 than Shaft 1 and Shaft 2 combined.

Phase 2—Shafts 3, 4 and 5

Trumbull Corporation's grouting subcontractor, Moretrench, also submitted an alternative pre-excavation grout plan that incorporated double wall grout curtains installed form the surface outside of the shaft excavations. The alternative grout plan, approved by the DP, was developed for Shafts 3, 4 and 5. The outside curtain wall (Outside Line) was drilled at predetermined radial angles to angle and batter the grout holes around a shaft to intersect anticipated high angled joints, vertical joints and horizontal bedded planes in the bedrock. Outside Line (OL) holes (primary, secondary and tertiary) were drilled and cased down through the overburden into bedrock and then rock drilled down through the bedrock to 4.6 m (15.1 ft) below shaft invert. The inner curtain wall (Inside Line) was drilled vertically through the secant pile SOE walls down to 4.6 m (15.1 ft) below shaft invert as shown in Figure 5.

Moretrench utilized upstage grouting to complete each double curtain wall hole for Shafts 3, 4 and 5. Moretrench grouted each stage zone to 0.006 MPa (1.0 psi) per vertical foot of cover from above the rock/overburden interface creating a plug in the first stage that could be grouted against. Moretrench drilled the OL primary holes to full depth and upstage grouted each 7.62 m (25.0 ft) stage until all OL primary holes were completed. Moretrench repeated this process for secondary and tertiary holes until

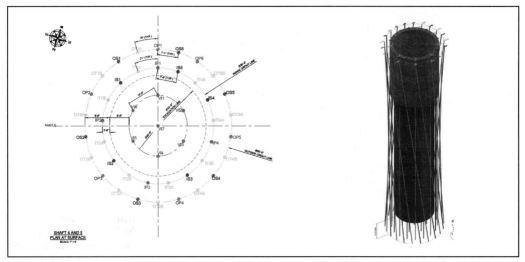

Figure 5. Pre-excavation grouting typical hole layout for Shaft 4 & 5 and rendition of hole pattern

all holes in the OL had been completed, using grout pressures of 0.006 MPa (1.0 psi) per vertical foot of cover. Quaternary holes were drilled and grouted as needed.

For the Inside Line (IL) vertical grout holes, Moretrench drilled the primary holes through the secant piles to 4.6 m (15.1 ft) below shaft invert and upstage grouted each 7.62 m (25.0 ft) stage until all primary holes were completed. Moretrench completed all IL secondary and tertiary holes using the same staging procedure and same grout pressures of 0.006 MPa (1.0 psi) per vertical foot of cover. Quaternary holes were drilled and grouted as needed. Moretrench drilled and grouted a plug at the bottom of each shaft 4.6 m (15.1 ft) below shaft invert. Moretrench also drilled performance test holes in each shaft, three holes in Shaft 3 and seven holes Shafts 4 and 5 for verification. The performance verification test involved probe holes drilled through the entire shaft and groundwater inflow measured. If flow exceeded 1.13 m³/hr (5 gpm) from a single hole or 2.27 m³/hr (10 gpm) from all holes combined, then additional grouting was required until the requirement was met.

SHAFT SUPPORT OF EXCAVATION

Criteria for the design of shaft SOE through the overburden and down into bedrock was provided in the contract documents. The intent was to socket the SOE into the bedrock and grout the overburden bedrock interface zone. KOJV elected to use diaphragm slurry wall construction for the Phase 1 SOE at Shafts 1, 2, and 6 while Trumbull utilized secant

pile wall construction for the Phase 2 SOE at Shafts 3, 4, and 5.

PHASE 1 Slurry Wall SOE at Shafts 1, 2 and 6

NCC designed and constructed the diaphragm slurry wall SOE for KOJV, excavating down through approximately 30.5 m (100 ft) of overburden and into the bedrock for Shafts 1, 2, and 6. The contract documents required the SOE be extended a minimum of 1.5 m (5 ft) into bedrock or into competent rock in order to minimize groundwater infiltration through the rock/overburden transition zone. Based on the geometry of the shafts in the contract drawings, the diaphragm slurry walls were sized at 17.9 m (59.0 ft) for Shaft 1, 14.9 m (49.0 ft) for Shaft 2 and 16.8 m (55.0 ft) for Shaft 6 respectfully.

NCC installed slurry wall panel guide walls and then excavated the diaphragm slurry walls with the following equipment—one hydromill diaphragm wall cutter, a crane equipped with a hydraulic or mechanical clam shell and a chisel to break up hard ground and a support crane (See Figure 6). The trenches were kept open during the excavation by bentonite slurry pumped into the excavation trench and maintained within 0.91 m (3.0 ft) of the top of the guide wall. The construction involved excavating a panel, setting rebar cages in the trench, placing concrete through tremie pipes and pumping the bentonite slurry off as it was displaced by the concrete. The panels were installed sequentially as primary, secondary and closure panels. NCC designed Shaft 1 and Shaft 6 with 9 panels each and Shaft 2 with 8 panels. The panel joints were joined together

Figure 6. Slurry wall panel excavation with hydraulic clam shell and hydromill

with an end stop keyway and a robust waterstop that provided a positive water barrier at each joint.

Phase 1—Cutter Soil Mixing (CSM) SOE at Shaft 1

Additional SOE outside of Shaft 1 was required due to the geometry of the pump station discharge trough and piping near the surface. Weidlinger Associates, Inc. (WAI) designed a 29.6 m (97.0 ft) diameter CSM wall approximately 13.1m (43.0 ft) deep from El. 700 and El. 657 for KOJV and NCC that included wall panels 0.81 m (2.67 ft) thick and 2.8 m (9.2 ft) wide. The CSM wall panels were reinforced by placing W18×76 steel soldier piles in the 2.07 MPa (300 psi) mixture of soil, cement and bentonite and hanging the piles 30.5 cm (1.0 ft) above the bottom of the excavation. KOJV and NCC choose CSM technology based on the geotechnical information provided in the Contract and the cost-effective construction method CSM presents. The CSM wall is constructed like a diaphragm slurry wall with primary, secondary and closure panels.

Phase 1—Additional SOE at Shaft 1—Jet Grouting

NCC used jet grouting at Shaft 1 to a create water tight plug between the outside of the diaphragm wall and the inside of the CSM wall. The jet grout zone was approximately 2.1 m (7.0 ft) thick and extended 10.97 m (36.0 ft) to 13.1 m (43.0 ft) below ground surface at EL 700 from EL 664 to EL 657. The jet grout zone consisted of three concentric circles of overlapping 2.4 m (8.0 ft) diameter columns (total of 135 columns) spaced at 1.98 m (6.5 ft) c/c for Line-A, 1.67 m (5.5 ft) c/c for Line-B and 1.49 m (4.9 ft) c/c for Line-C.

Phase 2—Secant Pile SOE at Shafts 3, 4 and 5

Brierley Associates designed the Shaft 3, 4 and 5 secant pile SOE walls for Trumbull. Case Foundation Company (CFC) was subcontracted by Trumbull to excavate and install the 1.80 m (3.87 ft) diameter overlapping primary (un-reinforced) and secondary (reinforced) secant piles through approximately 18.3 m (60 ft) of overburden and into the bedrock at Shafts 3, 4, and 5. The contract documents required that the SOE to extend a minimum of 1.5 m (5 ft) into the bedrock or into competent rock in order to minimize groundwater infiltration through the rock/overburden transition zone. The bedrock at Shafts 3, 4 and 5 was classified as weathered shale followed by shale and ending in limestone at the shaft bottom.

CFC installed high density styrofoam template guidewall forms at each shaft site for precise location of each secant pile. Primary (un-reinforced) secant piles were excavated and installed by augercast/continuous flight auger method or by sectional casing drill method depending on overburden. Continuous flight auger piles were drilled with a hardened tipped first flight down into the bedrock until refusal. 20.7 MPa (3,000 psi) grout was pumped through the hollow auger stem filling the pile as the auger was withdrawn.

Sectional casing primary secant piles were drilled down into bedrock until refusal. The first section of casing had hardened bits to cut into the bedrock. The overburden spoils inside the casing were removed with an auger down to the bottom and a clean-out bucket was used to completely clean the bottom of the hole. 20.7 MPa (3,000 psi) concrete was tremied from the bottom up filling up the pile while the casing was rotated back out to the surface.

Secondary (reinforced) secant piles were installed with sectional casing rotated and drilled through the overburden down into bedrock until refusal (See Figure 7). The first section of casing was fitted with a cutting shoe designed to core through firm soils, obstructions, and overlapping concrete and into the bedrock. The overburden spoils were removed with an auger down to the bottom and a clean-out bucket was used to completely clean the bottom of the hole. CFC then tremied 20.7 MPa (3,000 psi) concrete from the bottom up, filling up the pile opening while the casing was rotated back out to the surface. CFC finished the pile by placing W27×84 GR50 soldier pile reinforcement in the concrete after the casing was pulled.

SHAFT EXCAVATION

Shaft excavation at all six shafts was difficult and demanding. Excavation of Shaft 1, Shaft 2, the Screen Chamber and the Connecting Tunnel presented unique challenges based on the overburden and bedrock geology present at that site. The innovative process of "Reverse Flow Grouting" was developed to seal off the high groundwater inflow through the high angle clay filled joints at both shafts. The technical details and challenges of that story was discussed in an earlier publication (Gettinger, Hall, et al., 2013).

TUNNEL EXCAVATION

After reviewing the contract bid documents, KOJV procured a 7.0 m (23 ft) diameter "state of the art" Tunnel Boring Machine (TBM) (See Figure 8) from Herrenknecht Tunneling Systems (HTS) capable of mining in multiple modes to complete the tunnel excavation through the difficult complex karst limestone. KOJV anticipated mining 70% of the tunnel in open mode and the remaining 30% in slurry closed mode specifically based on the GBR and the TBM Specification. The mining modes specifically ranged from a non-pressurized or open mode using a screw conveyor discharging onto a belt conveyor to a full pressurized closed slurry mode discharging through a rock crusher and pressure pipe. In between the open and closed modes were four different modes of partially pressurized mining utilizing two discharge gates and one guillotine knife gate.

The HTS TBM was equipped with a double 920 mm (36.2 in) diameter screw conveyor that contained two discharge gates and one guillotine gate located between the two screw conveyors. In the open mode, muck was discharged onto a 762 mm (30 in.) conveyor belt from either of the discharge gates. The guillotine gate could be used to shuttle muck discharge between the two screws or it could be closed completely to eliminate discharging from the rear gate. When switching from open mode to closed mode the forward belt conveyors were designed to be removed from under the screw conveyors and a rotary rock crusher installed onto the bottom side of the rear discharge gate. In the closed mode, the forward discharge gate would be closed and all muck would discharge through the rock crusher and into the slurry discharge pipe.

During the bidding process, the anticipation of karstic limestone and high groundwater flows throughout the tunnel drive resulted in KOJV reaching a strategic agreement with the Shelly Company (Shelly)—essentially exchanging fresh water for excavated limestone. Shelly owns and operates a limestone quarry approximately 1.6 km (1 mile) south of the two shafts (1 & 2) at JPWWTP along the Scioto River. Shelly dewaters the limestone quarry to a known elevation on a daily basis pumping 75,710 m³/day (20 mgd) of groundwater through a separation plant and a permitted outfall into the nearby Scioto River.

Figure 7. Secondary secant pile installation at Shaft 3

Figure 8. Herrenknecht TBM S-674 during factory witness testing in Schwanau, Germany

The land between JPWWTP and Shelly's quarry is owned by the City of Columbus. To move the excavated tunnel rock efficiently and minimize additional truck traffic on heavily traveled SR 104, Shelly reached an agreement with the City of Columbus to move the excavated rock across the property to the quarry. As part of the KOJV/Shelly agreement, KOJV installed two HDPE pipelines to Shelly's quarry—one pipeline for fresh water supply that was connected to the quarry dewatering discharge line and the second pipeline for all tunnel slurry discharge flows and underflows pumped to Shelly's separation plant in the quarry and through the permitted outfall into the Scioto River. In exchange for the inbound and outbound water, Shelly would have access to all excavated rock from the tunnel.

To handle muck removal in the open mode, KOJV utilized a horizontal continuous conveyor belt system in the tunnel as the TBM advanced that discharged onto a vertical conveyor belt in the shaft for discharge onto the surface. The Robbins Company supplied the horizontal conveyor belt system, 762 mm (30 in.) wide, for the tunnel that included an

Figure 9. Radial stacker conveyor, bridge conveyor and horizontal belt storage cassette

advancing tailpiece located on the HTS TBM backup gear, a belt storage cassette with a 149 kW (200 hp) main drive unit on the surface and three 149 kW (200 hp) booster drives in the tunnel. The continuous horizontal conveyor belt, with a 635 t/h (700 tph) capacity, discharged onto a refurbished Lakeshore vertical belt system rated at 725 t/h (800 tph). The 915 mm (36 in.) wide vertical belt discharged onto a bridge conveyor at the surface (See Figure 9) that transferred the muck overland onto a radial stacker conveyor where it was discharged into piles and loaded out by Shelly.

To handle muck removal in the closed mode, KOJV chose a slurry system to handle an anticipated discharge flow rate of 408 m^3/h (1,800 gpm). Twin 200 mm (8 in.) diameter steel pipes were designed to handle both the outbound slurry discharge flow and the inbound fresh water feed in the tunnel and shaft. The slurry discharge system required twelve 8/6 centrifugal slurry pumps, each equipped with a 224 kW (300 hp) VFD controlled motor. The fresh water supply required five 8/6 heavy duty centrifugal clear liquid pumps, each equipped with a VFD controlled 149 kW (200 hp) motor.

The slurry from the tunnel was discharged into a separation plant on the surface with two high capacity shakers rated at 227 m^3/h (1,000 gpm) each, mounted above a 30 m^3 (8,000 gal) holding tank. Rock and solids were separated from the slurry and all underflow water pumped to the quarry. The fresh water feed system consisted of a pump at the quarry that pumped water to a 371 m^3 (98,000 gal) holding tank while a second pump delivered water from the tank into the tunnel.

STARTUP MINING

Delays in the excavation and completion of Shaft 2, Shaft 1, the starter tunnel and the connection tunnel between the shafts due to high water inflows delayed

the delivery and assembly of the HTS TBM. Shaft 2 was used for the assembly TBM and the first three gantries (Nos. 1, 2 & 3). Shaft 1 and the connecting tunnel were used to assemble the remaining five gantries and provide access to the tunnel. The backup system for TBM S-674 consisted of 11 trailing gantries with the Robbins advancing tailpiece and slurry pump number 1 located on Gantry 8. In order to utilize the vertical belt system during the installation and launch of the TBM with the first eight gantries, KOJV excavated an additional 14 m (45 ft) of forward starter tunnel.

KOJV began the tunnel drive on June 4, 2012, starting at Station 3+77. As expected start-up mining was slow as the mining crews learned how the TBM operated and behaved. There was very little groundwater inflow due to the additional 38 m (125 ft) of pre-excavation grouting KOJV elected to complete prior to extending the starter tunnel for assembling the TBM. As the TBM advanced, increasing groundwater inflows and increasing pressure readings on the earth pressure balance (EPB) gauges made it more difficult to use the horizontal continuous conveyor. KOJV made the decision to change the mining operation from open mode to closed mode on August 24, 2012. KOJV installed the remaining three TBM Gantries (Nos. 9, 10 & 11) and the rotary rock crusher under the discharge gate of the upper screw conveyor. Installation of Gantry 11 was critical because it contained an integral component of the slurry system, the telescopic piping system.

On September 5, 2012 KOJV resumed start-up mining in slurry mode at a rate of 7.6 m/day (25 ft/day), stopping at STA 10+77 to complete the final setup for production mining. The scheduled setup included installing rail switches in the shaft, reconfiguring the shaft utility layout and completing the slurry separation system at the surface and at Shelly's quarry. On October 8, 2012 KOJV began production mining.

The challenges and problems encountered during production mining in the karstic limestone including gaining access to the cutterhead for disc cutter maintenance, implementing an innovative cut-off grout program and installing a new revised slurry system while continuing to mine are captured in detail in an earlier publication (Rautenberg, Yamauchi, et al., 2015).

MINING WITH REVISED SLURRY SYSTEM

After advancing the TBM in closed mode for three months, the TBM was encountering groundwater inflows greater than anticipated and it became very apparent that that the slurry system initially selected for closed mode was not sufficient or robust enough to maintain the project schedule. In December 2012, KOJV met with the City and CMT to discuss the schedule and develop a plan to improve production. KOJV presented the idea of completing the mining in closed mode with a new re-designed slurry system.

The OARS project was unique in that it is very rare that a TBM mining in rock would require the use of a pressurized cutting chamber, and even more rare to require the capability to be able to mine as a slurry pressure balance machine in rock. Numerous features were designed into the TBM to give it as much flexibility as possible in hopes of increasing production. However, it quickly became apparent that modifications were going to be required to successfully mine through rock with these near continuous, high-pressure groundwater inflows. The inflows encountered often exceeded 682 to 1,364 m³/h (2,500 to 5,000 gpm) at the face of the machine. Even more problematic was that the TBM did not encounter only isolated areas of high groundwater inflow, but almost continuous groundwater inflow along most of the tunnel. The two critical mining aspects that had to be improved to make it through these extraordinary conditions included using a slurry pumping muck-handling system for the entire project and improving access to the cutter face to permit maintenance and replacement of the rock cutters.

Tunnel muck handling had to be accomplished by pumping the mined rock as a slurry instead of using a conventional conveyor belt system. The improved slurry handling system consisted of 300 mm (12 in.) diameter schedule 80 slurry piping, VFD controlled slurry pumps (in series), dual inline rock crushers, sweeping curves instead of 90-degree bends, and hardened steel components. A telescopic pipe wagon was installed in the TBM that allowed for slurry pipe installation as the work progressed and for slurry piping to be rotated periodically. This greatly extended the life of the piping as significant wear would otherwise occur along the bottom of the pipes. Additionally, feed water piping was also increased to match the slurry piping. This ensured adequate flow through the slurry pipe to maintain the tunnel muck in suspension. It could also be used as back-up slurry piping if needed. These muck handling improvements allowed for peak mining production rates to reach over 27.4 m/day (90 ft/day) with an overall average production rate of approximately 10.7 m/day (35 ft /day).

Typically, a rock tunnel contractor will check the cutters at the face of the machine daily. With constant groundwater and pressures averaging 4.0 bar (about 60 psi) on OARS, access to the cutters was very difficult. Occasionally, it was possible to gain access by opening ports in the cutting chamber to drain the water through the TBM to the tunnel. However, there was frequently more water coming into the excavation chamber than could be drained. The conventional cut-off grout procedure involves

Figure 10. TBM parked in Shaft 3 Safe Haven with all 43 disc cutters removed

drilling and grouting out in front of the machine until the excavation chamber can be dewatered for access. This standard methodology proved very time consuming and inefficient on OARS. Injected grout simply flowed through interconnected karstic features and failed to set up where needed.

KOJV developed a "reverse-flow grout procedure" based off a similar method used when sealing off leaks within a work shaft. Essentially, feed water was directed out of the TBM into the annular space between the shield and the rock until the flow of water through the surrounding rock was going away from the TBM (reversed). Once the water flow was reversed, grout was introduced into the system, where it was carried out into the openings and fissures in the rock where it was needed to seal off inflows. This procedure allowed for the contractor to reduce groundwater inflow by creating an effective water barrier, and was successfully used to routinely access the cutters throughout the tunneling. It is expected that this innovative technique will be used on future tunnel projects within karstic limestone geology.

KOJV stopped mining operations on August 2, 2013 to complete the scheduled three week changeover for the new slurry system. KOJV resumed mining on August 26, 2013, with typical start-up issues. By the third week, production had increased significantly reaching 20 m (65 ft) per day. On January 4, 2014 the TBM had advanced a total of 2,645 m (8,678 ft) to STA 90+55 and reached the safe haven at Shaft 3 (See Figure 10). Safe havens were setup at Shafts 3, 4 and 5 by the Phase 2 contractor, Trumbull Corporation, who was compensated for completing a safe haven for the TBM at each of the three shaft adit connections by pre-excavation grouting and excavating a specified envelope in the tunnel profile.

The scheduled three week shutdown at Shaft 3 resulted in the replacement of all 43 disc cutters, additional grizzly bars welded onto cutterhead buckets, cutterhead cracks welded, slurry discharge pipes in the tunnel rotated 120° and inspection of the two rotary rock crushers. Mining resumed on January 21, 2014.

The TBM pulled into the Shaft 4 safe haven at STA 122+25 on May 9, 2014 after advancing 966 m (3,170 ft). Similar maintenance was performed on the TBM at Shaft 4. KOJV resumed mining on May 30, 2014 and reached the Shaft 5 safe haven at STA 156+29 on September 10, 2014 after the TBM had advanced 1,037 m (3,404 ft). Again, similar maintenance was performed at Shaft 5, including replacing the three wire brush seals. The TBM resumed mining on October 5, 2014 and mined or "holed" through into Shaft 6 at STA 232+71 on September 4, 2015 after advancing the final 2,329 m (7,642 ft) in eleven months through difficult conditions.

TUNNEL CONNECTIONS, SURFACE STRUCTURES AND RESTORATION

After the tunnel excavation was completed KOJV and HTS removed the TBM cutterhead, front shield, middle shield, tail shield and the eleven trailing gantries from Shaft 6. The TBM buy-back arrangement between KOJV and HTS resulted in Herrenknecht salvaging the main sections of the TBM and scrapping the remaining sections.

After the TBM was completely removed from the site KOJV and their site/concrete subcontractor, George Igel Co. (GIC), began the final connection to the OSIS at Shaft 6. This included completing the tangential inlet and vortex drop structure, the 14.6 m (48 ft) diameter final shaft lining, 4.88 m (16 ft) diameter drop shaft (See Figure 11), 6.1 m (20 ft) diameter ventilation shaft inside Shaft 6, precast concrete support beams and shaft cover and final site restoration.

An innovative solution to address surge problems applied on OARS is that the concept of a tall surge tower was integrated into the deep drop shaft design. The drop shaft, vent shaft, and surge storage chambers were all designed and built directly within the work shafts. The upstream shaft was where the TBM was removed, so a large working access shaft had to be excavated. The design utilized the large work shaft area already available such that the drop shaft and vent shaft would be permitted to overflow into a surge storage chamber whenever pressure waves could cause water to surcharge up the shafts. The design of these structures within the drop shafts ensured adequate containment of surge volumes and eliminated impacts to the upstream collection system.

Incorporating these types of structures into a combined sewage relief tunnel shaft had never been done before

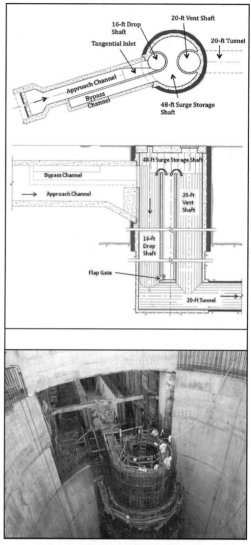

Figure 11. Shaft 6 drop shaft and tangential inlet

GIC had completed construction of the OSIS bypass structure, OSIS relief structure, two large sluice gates, two large weir gates, two large flap valves and a section of the tangential inlet structure prior to the TBM arriving at Shaft 6.

Trumbull completed construction of similar hydraulic structures (bypass, relief, tangential inlet, weir gates, vortex drop shaft, ventilation shaft, shaft liner and shaft cover) at Shafts 3, 4 and 5 on the surface. Trumbull also completed construction of the shafts, deaeration chambers and adit connections short of the tunnel. KOJV completed the final adit connections to the tunnel after the tunnel excavation was completed. Final Restoration was completed at all sites by GIC and their landscaping subcontractor.

COMMISSIONING AND SYSTEM START-UP

The tunnel excavation and completion delay resulted in many schedule sequence changes in both Phase 1 and Phase 2 contracts. Commissioning, start-up and operational demonstration of the CSO tunnel system, pump electrical building, pumps, screen electric building, screen cleaning system, gates and other components were completed in various stages and at various times along the way and were anything but normal. The sequencing of the work required collaboration, clever strategies and coordination.

CONCLUSION

There were an extreme number of valuable lessons learned on this project by everyone involved from beginning the end. The demanding challenges encountered during construction of the OARS CSO Tunnel were solved by collaboration and communication between all of the experienced individuals involved and thinking outside the box became the norm. The City and KOJV were able to maintain an open line of communication throughout that helped keep the project moving forward during the difficult times. Many thanks to everyone involved and their huge effort to complete this challenging project.

REFERENCES

Gettinger, B., Hall, M., Smith, P., Day, D., Fedner, G., Rautenberg, B., Hall, R. 2013. Large Diameter Shaft Construction through Difficult Ground in Columbus, Ohio. *RETC Proceedings,* Washington, DC, pp 250–263.

Rautenberg, B., Yamauchi, K., Smith, P., Hall, M., Fedner, G. 2015. Challenges in Tunneling with a Hard Rock Slurry TBM in Columbus, Ohio. *RETC Proceedings*, New Orleans, LA, pp 922–934.

Day, D., Fedner, G., Rautenberg, R., 2012. OARS, A Tunnel Through Karst Limestone. *North American Tunneling Proceedings*, Indianapolis, IN, pp 571–577.

Urban Hard Rock Tunneling and Blasting in Baltimore City

Todd Brown and Jordan Bradshaw
Bradshaw Construction Corporation

ABSTRACT: Bradshaw Construction recently completed 2,500 feet of tunnel for a 36 inch sanitary sewer in downtown Baltimore City. Given the local geology consisting of very hard rock (up to 37,000 psi) with intermittent veins of highly decomposed rock, the contractor utilized a 72 inch Double Shielded TBM provided by the Robbins Company, upsized from the base 60 inch tunnel design, to provide a greater ability to mine the rock. Access shafts, up to 57 feet deep, were set in congested urban environments. This required extensive utility support during operations and coordination with local residents and businesses, particularly during blasting operations. Through the challenges, the interceptor went online as scheduled.

INTRODUCTION

Like many other major localities across the county, Baltimore, Maryland has been working on a major overhaul of its aging wastewater infrastructure. Since a 2002 consent decree mandate was issued from both the Environmental Protection Agency and the State of Maryland, the City has been tasked with cleaning up local streams and rivers which feed into the Chesapeake Bay. By improving its nearly century old sewer system, Baltimore is working towards the elimination of raw sewage discharges into waterways that are caused by everything from deteriorating pipelines to increased flow from a population larger than the system was originally designed to support.

One component of the wasterwater system improvement by the City of Baltimore Department of Public Works (DPW) was the construction of the Lower Gwynns Run Interceptor. The second phase of the project was completed by Bradshaw Construction Corporation, based in Eldersburg, Maryland, in December 2015. The 36 inch sanitary sewer interceptor was installed using a 72 inch diameter Double Shield Tunnel Boring Machine (TBM) through Baltimore Gneiss, which is composed of very hard rock with intermittent, highly decomposed veins. The new 2,500 foot interceptor transfers sewage from an existing 27 inch sewer below Franklin Street in West Baltimore to a 33 inch sewer with much greater capacity under Baltimore Street, one-half mile to the south. The project traversed a congested urban environment that required access shafts up to 57 feet deep, extensive existing utility support, and a great deal of coordination with local residents and businesses, particularly during blasting operations. The project presented unique challenges and

opportunities for successful collaboration between the owner, engineer, and contractor to improve and complete the project within the contract budget and on schedule.

PRECONSTRUCTION

When first awarded in 2005, the Lower Gwynns Run Interceptor was originally a single project, spanning approximately two miles from Liberty Heights Avenue to Baltimore Street. However, during construction, the final 2,500 feet of 30 inch sanitary sewer pipeline south of Franklin Street was separated into a second project, requiring a redesign by Dewberry Consultants. The Lower Gwynns Run Interceptor—Phase II would be constructed at a lower elevation than originally designed and would thus be constructed entirely within the Gneiss rock formation beneath the congested city streets. As the final carrier pipe was designed to be 30 inches in diameter, the bid allowed for a tunnel excavation as small as 48 inches to accommodate it.

The contractor has been tunneling in the Baltimore area for many decades, with one of their largest recent sewer installations being another Baltimore DPW project called the Upper Jones Falls Interceptor Sewer—Phase II, completed in 2007. Also constructed in hard rock, it consisted of 6,000 feet of 60 inch microtunneling for a 48 inch interceptor sewer, providing immense insight into what methods and techniques do and do not work in similar rock formations. The bidding phase for the Lower Gwynns Run Interceptor—Phase II allowed multiple excavation options including both microtunneling and conventional rock TBM mining. The geotechnical reports for the project noted that the contract should expect very abrasive rock (Cerchar

Figure 1. Tunnel 1 beneath Franklin Street

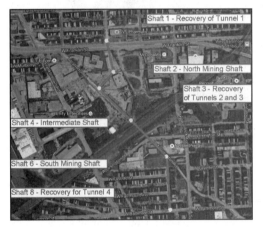

Figure 2. Overall project map

Figure 3. Robbins RH72 double shielded TBM with 14 inch disc cutters

Abrasivity Index values of up to 5.6) with unconfined compressive strength up to 37,000 psi. The contractor's previous local experiences supported these assumptions, having encountered rock up to 45,000 psi in other areas of the city. Such hard rock would require extensive cutter changing and cutter wheel maintenance during any single drive. A minimum 60 inch diameter microtunnel boring machine (MTBM) would be required, regardless of the size of the carrier pipe, because that is the smallest machine available with access to the tunnel face to perform those necessary tasks.

While prior experience on the Upper Jones Falls project showed that a 60 inch diameter machine can successfully mine through such a challenging formation, the experience also lead the contractor to recognize that it may not be the best option for future projects. Accordingly, it was decided that the best approach would be to use a conventional rock TBM, in this case a 72 inch Rockhead Double Shield TBM, manufactured by the Robbins Company. The

larger diameter TBM would provide a greater capacity to excavate the rock at a faster rate, although it would not perform as well as an MTBM in the decomposed zones of the tunnel. The inexact science of utility tunneling in relatively shallow and variable soil matrixes necessitates that the choice of a mining methodology is nearly always a bit of a compromise. For the Lower Gwynns Run project, the anticipated ratio of competent, hard rock tunnel to low quality rock meant that the benefits of this TBM would far exceed the drawbacks.

In addition to the increased excavation capabilities, the TBM offered other benefits. A double shielded TBM propels the cutter wheel forward by stabilizing and gripping the tunnel walls, thus eliminating the need for pipe jacking. By removing jacking force limitations, the contractor was able to propose the elimination of two shafts and reduce the number of tunnels from seven to four, the longest of which being 1,200 feet. The two shafts that were eliminated posed logistical problems. One was located within a residential street and the other behind an active business property. Both shafts would have significant site access problems and would have caused the greatest disturbance to the local community. There was also a cost reduction for the owner as a result of the shaft elimination and, because of the larger tunnel diameter, the changes allowed for an increase to the carrier pipes size from 30 inch to 36 inch without overrunning the project budget.

NORTH TUNNEL CONSTRUCTION

Construction began onsite for the $11.9 million project in August 2014 on the northern half of the job, an area which consisted of the hardest and most competent rock to be expected. The tunnel alignment was the deepest there, averaging approximately 40 feet below surface elevation, and the drive lengths were the shortest due to the right-of-way meandering through a largely industrial area. The first excavation would be for Shaft 2, a bi-directional launch shaft for staging the two northern tunnels, located within the parking lot of a maintenance facility owned by the DPW. The 42 foot deep, 24 foot diameter shaft was excavated by drill-and-blast, with steel ribs and timber lagging supporting the overburden soils and rock bolts, wire mesh and some shotcrete installed in rock zones. These were the typical shaft support methods utilized for each shaft on the project. Blast holes were pre-drilled from the surface elevation to below the shaft invert. However, the contractor found that the inaccuracy of the holes after drilling through approximately 20 feet of soft ground required additional holes be drilled at each blast elevation. Additionally, the time that it took to locate each pre-drilled hole and evacuate the stemming stone for each blast lift was nearly the same amount of time that it would have taken to simply drill out each round of 70 to 80 holes at the rock face. Through the northern shafts were all completed in this manner, the drilling methods on the project's southern shafts were adjusted to compensate for these inefficiencies.

Shaft 2 was less than 50 feet away from the nearest occupied building, a DPW vehicle service garage which maintained everything from police cruisers to snow plows. To limit blast vibrations to acceptable levels, the shaft was blasted in only 4 foot deep lifts. To protect the surrounding areas from potential fly rock, the blast face was covered by a dozen rubber tire blast mats and the surface shaft opening was secured with a timber shaft cover, constructed from a shaft rib to fit the geometry of the opening. Prior to each blast, the offices adjacent to the site within the neighboring buildings were evacuated for the periodic blasts, following almost daily communications with representatives for each building with updates to the blast schedule. Shaft 2 was completed in October 2014 without incident. The TBM recovery

Figure 4. Pre-drilling of blast holes at Shaft 2

Figure 5. Additional drilling at Shaft 3

Figure 6. Header tunnel at Shaft 2

Figure 7. Retraction of TBM from Tunnel 2

shafts, Shaft 1 (32 feet deep) and Shaft 3 (57 feet deep), were then completed using the same methods and communications processes with the neighboring buildings, which were other DPW buildings, personal residences, and a CVS pharmacy. All northern tunnel access shaft construction was completed in January 2015.

Launching the TBM would require more space than the 24 foot diameter shafts provided. Larger launch shafts were not feasible or cost effective due to footprint availability and depth. Therefore, header tunnels were required to install the TBM. In this case, eight feet of additional length was needed and 96 inch horseshoe tunnels were installed by drill-and-blast. These short, small diameter tunnels were drilled by hand, utilizing jack leg drills with four foot deep rounds. Once blasted, the header tunnels were supported with ribs and shotcrete, and an eye was then cast in concrete to match the TBM's grippers to facilitate the launch. Tail tunnels were also utilized on the project's longer drives to increase efficiency of the muck train.

The initial tunnel to be launched was Tunnel 2, which at 250 feet long was the shortest tunnel on the project, running from Shaft 2 to Shaft 3. With the short length and the tunnel being one of the deepest alignments on the project, the contractor viewed this as a great opportunity to utilize the hard rock capabilities of the TBM, which had been successfully used on many other rock tunnels in the Midwestern United States. Both the geotechnical reports and the contractor's experience with rock in Baltimore led to the decision to equip the TBM with a new heavy duty cutter wheel with 14 inch disc cutters, increased from the standard-issue rock cutting wheel equipped with 11.5 inch cutters. However, when Tunnel 2 was ready to be launched in November 2014, the heavy duty wheel was not yet prepared. Given the relatively short length of the tunnel, the decision was made to

launch using the standard cutter wheel. This proved to be the wrong decision. After only 100 feet of mining, which included multiple disc cutter changes, the main drive assembly of the TBM failed. The positive takeaway from the first 100 feet of hard rock mining was that though it proved to be too much for the cutter wheel, the tunnel did not require any supports. This allowed the TBM to be walked back out of the tunnel into the launch shaft utilizing the gripper pads, where it was removed for repairs.

The TBM's main drive assembly was repaired through the collaborative efforts of both the contractor and the TBM manufacturer. It was then equipped with the heavy duty cutter wheel, outfitted with 14 inch disc cutters, and relaunched through the existing Tunnel 2 excavation in January 2015. Given the issues during the first 100 feet of mining with the TBM, great care was taken with the remaining 150 feet of mining, including frequent cutter and maintenance checks. Despite the conservative mining approach, production improved dramatically in the second attempt, improving from an average of 7 feet per shift with the standard cutter wheel to 13 feet per shift with the heavy-duty cutter wheel. In fact, the final mining production days of the tunnel proved to be far and away the best for the drive, averaging approximately 25 feet per shift as the crew became increasingly comfortable with the TBM and it capabilities. With Tunnel 2 successfully completed, the TBM was relaunched for the 400 long Tunnel 1 (Shaft 2 to Shaft 1) in February 2015. The crossing would pass underneath Franklin Street, a heavily traveled commuter route containing a plethora of existing utilities, including the existing 27 inch sanitary sewer the project would intercept.

Tunnel 1 was completed in only three weeks, proving that using 14 inch disc cutters on the cutting wheel was the correct configuration for this Baltimore rock. The bald headed, hard rock tunnel showed that the productions seen at the end of Tunnel 2 were a more typical expectation for this TBM in consistent conditions. However, Tunnel 1 was not a total success. On the last days of mining, the main drive assembly once again started showing signs that it was being heavily taxed. Upon recovery of the TBM and an investigation at the contractor's equipment yard, it was found that the main drive assembly had indeed suffered another similar failure as during Tunnel 2. Had Tunnel 1 been another 50 feet longer, the TBM likely would not have completed the drive. Tunnel 3 was looming on the southern portion of the project, a 1,200 foot reach at an average depth of 50 feet that would cross beneath Calverton Road, Lexington Street, and Railroad right-of-ways for both Amtrak and Norfolk Southern. With this daunting challenge in mind, the contractor and TBM manufacturer again evaluated the structure of the main drive assembly

Figure 8. Tunnel 1 completion

Figure 9. Tunnel 2 completion

and performed a complete overhaul to prepare for the southern tunnels.

SOUTH TUNNEL CONSTRUCTION:

The southern portion of the project began in January 2015 with the construction of Shaft 6. At only 15 feet deep, this shaft would serve as an ideal launch shaft for the two southern tunnels. Tunnel 3 would be mined 1,200 feet to the north, connecting the two sections of the project, and Tunnel 4 would run 650 feet south to the project's tie-in location underneath Baltimore Street. The TBM would be recovered at this location (Shaft 8) and a cast-in-place concrete Bellmouth Structure would be constructed connecting the new interceptor to the existing 33 inch sanitary sewer. The contractor anticipated launching the TBM on Tunnel 4 in April 2015. This would maximize schedule efficiency by allowing construction of the Bellmouth Structure at Shaft 8 while tunneling operations were still progressing on Tunnel 3. It would also allow for a gradual increase in drive length between tunnels, helping mitigate potential risks with the reconstructed TBM. However, upon breaking ground for Shaft 8 on Baltimore Street, plans had to change.

Baltimore Street has historically been one of the primary corridors into the center of the city for both aboveground traffic and underground utilities. While vehicle traffic was able to be managed with detours onto the surrounding streets, the underground utility traffic was quite a different story. The contractor was aware of the existing storm drains, water mains and gas lines that were anticipated within the shaft, but the initial excavation uncovered a massive communications ductbank that would be in direct conflict with the location of the already small window available for recovery of the 65,000 lb. TBM. Given these

Figure 10. Existing utilities within Shaft 8

unanticipated constraints, Tunnel 3 would have to be constructed first while time was taken to coordinate with the owners of the existing utilities.

Given the shallow depths of the southern 1,000 feet of the project, the contractor expected to encounter a higher volume of mixed face/mixed reach conditions composed of both hard and highly decomposed rock strata. Header tunnels were drilled-and-blasted in the same manner as they were done for the northern tunnels, although additional supports, including horseshoe ribs and shotcrete, were required to stabilize the decomposed rock. Additionally, with the southern tunnels being significantly longer than those in the north, tail tunnels were added so that a different spoil transport system could be utilized which consisted of a more powerful electric locomotive and a larger muck car.

Tunnel 3 was launched in April 2015. After the initial 100 feet of Tunnel 3 had been mined, ground conditions deteriorated and periodically required the

Figure 11. Drilling for header and tail tunnels at Shaft 6

Figure 12. Steel rib and timber lagging supports in Tunnel 3

installation of spot rib-and-lagging tunnel support sets to make the tunnel safe from an unstable crown. These variable tunnel conditions were the anticipated drawback of using a TBM rather than an MTBM. Two challenges presented themselves which would have been better addressed by microtunneling. First, an MTBM would have excavated the decomposed rock faster than the TBM. Second, with its propulsion driven by pipe jacking, an MTBM would have been easier to steer as the pipe string would better hold line and grade. Because the TBM propels itself by gripping off of the tunnel walls, lack of competent rock necessitated jacking against the rib-and-lagging supports. As these supports were only located within the decomposed zones, they had to be additionally secured to the tunnel walls in order to achieve the necessary thrust reaction force needed by the TBM to mine the tunnel.

The fight through the weathered zones was worth the endeavor, as after 350 feet the TBM re-entered the competent hard rock found on the northern section of the job. The remaining 850 feet of Tunnel 3 was mined through this rock without the use of supports, with the exception of the installation of steel liner plates through the 225 foot Amtrak Railroad right-of-way zone, per the railroad requirements. Shaft 4, installed near the half-way point of the tunnel on the north side of the railroad tracks for future maintenance access, served as a TBM tune-up site for disc cutter changes and service prior to completing the run just in time for Independence Day, 2015. Again, as with Tunnels 1 and 2, the TBM, the modified drive assembly and the heavy duty cutter wheel performed at their best in the full face of very hard, massive rock.

All throughout the mining of Tunnel 3, the contractor continued to work its way through the web of existing utilities in Shaft 8. The experience gained from mining through 250 feet of decomposed rock on Tunnel 3 made it clear that the TBM would not be able to complete mining on Tunnel 4 without tunnel supports and still be capable of walking itself back to the launch shaft for recovery. Once the rib-and-lagging tunnel supports were installed, traversing the TBM backwards became impossible without jeopardizing the safety of the tunnel. The TBM would need to be removed at Shaft 8, requiring the contractor to find a window in the utilities to do so.

Shaft excavation proceeded at a glacial pace. Vacuum excavation was required to expose the existing lines to prevent possible damage. Existing storm drains had to be taken out of service and temporarily removed and diverted. The water mains and gas lines had to be supported in place to maintain service, while concrete ductbanks had to be removed and their cables suspended across the shaft. When excavation of rock became necessary, it had to be done without blasting in order to protect all of the existing services against fly rock. Excavation began in March 2015 and it took nearly five months to complete the 27 foot deep shaft. The positive development during the shaft excavation was that a window was discovered between the lines which would allow for the TBM to be removed from the tunnel elevation and back to the surface while never compromising the existing utilities or causing a loss of service. The window would involve a complicated, multi-stage pick requiring the TBM be moved around and rotated within the shaft, amongst the utilities. The most important part was the window did exist and the TBM could be recovered without drastic changes to the project.

Tunnel 4 was launched in July 2015 and throughout the 650 feet of excavation, the same decomposed rock conditions from the first 350 feet of Tunnel 3 were encountered. Mining downhill through multiple layers of blocky, weathered ground caused a great deal of cutter wheel wear as well as damage to the TBM's stabilizers and conveyor. Tunnel 4 was a fight from the beginning to the end, until it was ultimately completed in September 2015. Because of the ground conditions, this tunnel would have been the most ideal candidate on the project for microtunneling, but surface access was not favorable for constructing a separation plant or any of the other slurry management systems involved. And despite the relatively large amount of rib-and-lagging supports, arduously installed in the decomposed tunnel zones, the cost of these supports was still far less than a casing that would have been required for a pipe jacking operation. When it was recovered at Shaft 8, the TBM had to be rotated 90-degrees, dragged to the far side of the shaft and removed parallel to the existing sewer line contained in the shaft.

Figure 13. TBM maintenance at Shaft 4

CARRIER PIPE INSTALLATION

Utilizing a TBM with a 74.5 inch cut created numerous advantages for the project, beyond the capabilities provided by the larger disc cutters and greater thrust capacity. The increased diameter allowed workers to simply be able to stand up in the tunnel, a situation which is not as common as those in the utility market would like it to be. It allowed the City of Baltimore to increase the carrier pipe size from 30 inch to 36 inch at a minimal cost. Installing a 42 inch interceptor sewer would have even been possible, though it was determined not to be necessary for this particular sanitary sewer system. The larger tunnel cut also allowed for all of the carrier pipe to be installed within design tolerances despite the TBM sometimes having difficulty steering to those same tolerances in sections where it encountered mixed face conditions that required rib-and-lagging tunnel supports. Even within the 66 inch inside diameter of the supports, the design gap between the ribs and the outside diameter of the 36 inch Hobas Fiberglass Reinforced Polymer (FRP) carrier pipe was greater than a foot, so there was additional tolerance built into the design by extending the cut. The 20 foot lengths of FRP were installed in the tunnels one joint at a time using a pipe dolly on rail and hardwood timber blocking so that the placement of each pipe could to adjusted, when necessary, from the tunnel alignment to create a single, smooth pipeline between manholes. Carrier pipe installation began in March 2015 and was completed the following October.

Tunnels 1, 2 and 3 were backfilled using a fly ash-cement grout, which contained no fine aggregates. This material has always been a preferred grout as it is easily mixed, contains only three ingredients, and can cure to the 200 psi compressive strength required by the project specifications,

Figure 14. Recovery of TBM for Tunnel 4 at Shaft 8

all while maintaining a very low viscosity which allows it to be installed by gravity flow over long distances. Fifteen-hundred cubic yards of the material was successfully placed within 1,850 feet of tunnel up through October 2015. With only the 650 foot long Tunnel 4 left to grout, needing an estimated 550 cubic yards, the local ready-mix supplier who had been producing this material since April informed the contractor that they would no longer be able to supply it for the project. A dwindling amount of Class F fly ash was available on the local market, due to a reduction in coal fired power generation of which fly ash is a byproduct. Fly ash-cement grout typically contains approximately 2,000 pounds of fly ash per cubic yard and the contractor would use an average of 80 cubic yards per day during back fill operations. The supplier simply could not get their

Figure 15. Fabricated FRP curve and manhole at Shaft 3

plant's fly ash silo refilled fast enough to service the project's needs, much less their other clients whose standard concrete mixes would also require fly ash in vastly smaller quantities.

With the project completion deadline looming, the contractor decided to utilize a bentonite-cement grout in lieu of the fly ash based version. Bentonite cement grout would eliminate the fly ash supply issues, as it was becoming increasingly challenging to find a feasible fly ash source independently of the ready mix supplier for onsite batching. Self-batching fly-ash-cement grout would also require enough staging room for the multiple tanker trucks delivering the product 25 tons at a time, which the project site did not readily have. However, due to the contractor's regular pipe jacking operations, wherein bentonite is utilized for casing lubrication, large quantities of bentonite were readily available, and since is makes up only 3 percent of bentonite-cement grout's total weight, only pallets of 50 pound bags would serve the need for batching the grout onsite. Cement was delivered in one ton super-sacks to eliminate the use of tankers and thus tanker storage. As bentonite-cement grout's largest component is water, the hydrant closest to the shaft would provide the majority of "material storage" that was required. The contractor retrofitted a microtunneling slurry separation plant into an onsite grout batching plant, capable of making 20 cubic yards of grout at a time. Tunnel 4 grouting was completed in two weeks in late November 2015.

To limit the size of manholes and still provide long, sweeping bends in the pipeline through changes in the alignment, the DPW specifies the use of brick curve transition structures at those points. The utilization of FRP as the carrier pipe, however, can make this transition much simpler and cost effective, while at the same time providing a better quality sanitary sewer. Rather than build the curves using cast-in-place concrete with brick lined channels, Hobas Pipe USA was able to fabricate a curved section of pipe to match the dimensions of the designed curve. One end of the curve would use a typical coupler to join with the tunnel pipe, while the other end would be inserted into the manhole gasket. Three such curves were successfully used on the project with the longest being 16 feet in length to accommodate a 90 degree turn in the pipeline.

CONCLUSION

The Lower Gwynns Run Interceptor—Phase II project was completed in late December 2015, with the contractor having overcome numerous hurdles while installing the 2,500 foot pipeline, to the result of approximately $400,000 less than the original contract price. The greater Baltimore area contains some incredibly challenging geology. Building such a project in a dense, urban environment only exacerbates those potential issues, particularly when you introduce blasting into the equation. However, blasting is unquestionably the most efficient method of rock excavation for tunnel access shafts and launching the TBM when in these geologic conditions. As long as the proper preparations and notifications are performed in conjunction with local authorities, businesses and residents, blasting will almost always make a project more cost-effective and successful.

When you can encounter some of the hardest, most abrasive rock in the country mere feet away from a completely unstable, decomposed rock zone, the "correct" answer on what the best means and methods for tunneling through it can be hard to determine. By allowing the contractor flexibility in TBM selection, the owner allowed for local, real-world knowledge of the geology to make a calculated decision on the best fit. The Double-Shield TBM was the right fit for this project, despite the difficulties encountered, as no methodology would be without its own problems in such complicated geology. Mining was never without challenges, both those that were expected and those that were not. However, the contractor in coordination with the TBM manufacturer successfully completed all 2,500 feet of tunnel.

ACKNOWLEDGMENTS

Bradshaw Construction Corporation would like to thank all of the parties involved with the project, specifically Dewberry Consultants and the City of Baltimore Department of Public Works, for all of their assistance in the completion of the Lower Gwynns Run Interceptor—Phase II, a project which we were proud to be a part of.

Various Deep Shaft Construction Techniques Used in Atlanta Water Supply Program

Tao Jiang, Wayne Warburton, and Yong Wu
Stantec Consulting Services Inc. (Part of JP2/PRAD, Stantec, and Chester Engineers JV)

Brian Jones
City of Atlanta, Department of Watershed Management

ABSTRACT: The City of Atlanta is converting an over-a-century-old quarry into a 2.4-billion-gallon raw water storage facility through its Water Supply Program, which consists of a deep hard rock tunnel and ten deep shafts with various diameters. The shafts function as pump station shaft, drop shaft, riser shaft, and construction shaft, respectively. Multiple techniques are utilized in the shaft construction, which include conventional drill and blast for the large diameter shafts, blind bore for five 420-foot-deep and 9.5-foot-diameter pump station shafts, and raise bore for a 330-foot-deep and 16.5-foot-diameter riser shaft. The blind bore and raise bore shafts are the largest ever to be built in the southeastern Piedmont geology.

PROJECT BACKGROUND

The current water supply program operated by the City of Atlanta (COA) Department of Watershed Management (DWM) consists of four aged raw water pipelines, one of which dates back to the early 1890s. Based on previous assessments completed by DWM, the entire system is at or will soon reach its recommended useful life. As such, the COA acquired the Bellwood Quarry in 2006 with the long-term goal of converting the quarry to a raw water storage facility with a volume of approximately 2.4 billion gallons. The water storage facility will greatly enhance the reliability of the drinking water supply to the greater Atlanta metropolitan area.

The Chattahoochee River is the source of water supply to the quarry facility. The facility will be operated in an "offline" mode, with raw water being stored in the quarry before being withdrawn for treatment at the Hemphill Water Treatment Plant (HWTP) and/or Chattahoochee Water Treatment Plant (CWTP). The offline operating mode will include routine withdrawal of water from the quarry and replenishment with Chattahoochee River water.

A conveyance system is required to connect the quarry, HWTP, CWTP, and the Chattahoochee River. The conveyance system includes an approximately 4.5-mile-long TBM tunnel with a finished diameter of 10 feet and multiple shafts and connecting adits. The overall project has been divided into two design packages. The first package, termed the Phase 1, connects the quarry to the HWTP, which includes a TBM tunnel of approximately one mile long and four

shafts and associated connecting adits at the quarry site. Quarry highwall stabilization is also part of the Phase 1 package. The stabilization measures include scaling, rock bolting, and installation of a drape system up to 300 feet high. The second package, termed the Phase 1 Extension, connects the HWTP to the CWTP and the Chattahoochee River, which includes a TBM tunnel of approximately 3.5 miles long, five shafts and associated connecting adits at the HWTP site, and one shaft at the end of the TBM tunnel near the CWTP and the Chattahoochee River.

The Construction Manager at Risk (CMAR) model was used as the overall project contracting method. The CMAR is a joint venture of PC Construction and HJ Russell. The tunneling contractor is a joint venture of Atkinson Construction and Technique Construction. The project is currently under construction.

GEOLOGIC SETTINGS

Regional Geologic Setting

The project location is shown in Figure 1, which is generally in the Northwest part of downtown Atlanta, Georgia. It is located in the Piedmont Physiographic Province (McConnell and Abrams, 1984). The geology of the Piedmont in the greater Atlanta area generally consists of medium-grade metamorphic rock with granitic intrusions. These crystalline rocks are some of the oldest rocks in the Southeastern United States. They were generally formed before and during the building of the Appalachian Mountains. Since

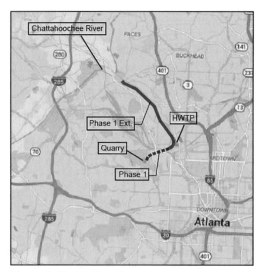

Figure 1. Project location

their origin, the rocks have undergone a complex history of metamorphism, weathering, and deformation.

Various structural features are present in the rocks, including folds, fractures, and lineaments. The high pressures and temperatures at great depths resulted in a full range of deformational styles, ranging from medium-grade metamorphism, through fully-welded ductile shearing and mylonite formation, to brittle fracturing with rocks that commonly contain hydrothermally deposited minerals. At shallower depths, structures like exfoliation fractures were formed in the rocks due to erosion of overburden and unloading. The exfoliation fractures mainly occur along the foliation "planes" and tend to be open and act as conduits for water movement through the rock mass.

Lineaments, which are surface topographic expressions of underlying rock mass or crustal structure, occur throughout the Piedmont. The lineaments are often controlled by weathering associated with discontinuities in the rock. In many cases, the lineaments represent fracture zones in the underlying rock. At depth, the fracture zones are typically cemented with weathered minerals. At shallower depths, erosion of these weathering minerals often results in zones of broken, water-bearing rocks and topographic features such as valleys and draws.

Soil to Rock Transition

The ground conditions at the project site can be divided into a soil zone, an underlying transition zone, and a rock zone. The transition zone and, to some degree, the components of the soil zone, are derived from the underlying rock as a consequence

of weathering. Since weathering is facilitated by fractures, joints, and rock compositions, contacts between these zones are anticipated to be highly irregular.

The soil zone mainly consists of thoroughly degraded residual soils, although layers and lenses of rock and partially weathered rock can occur locally within the soil zone. The transition zone primarily consists of partially weathered rock; however, layers or lenses of soils and/or fresh rock can be present. The rock mass strength of the transition zone is typically much lower as compared to the zone due to the presence of abundant weathered joints and reduced intact strength of partially weathered rock.

The rock zone is dominated by fresh rock. For the southern end at the quarry to the northern end near the Chattahoochee River, the tunnel alignment crosses several geologic units, including the Clairmont Mélange, zoned feldspar gneiss, and the Brevard Zone mylonite (black and white). The Clairmont Mélange has a poorly to well-developed foliation which is typically low angle and undulatory. Lithology within this rock unit is extremely contorted, with quite variable foliations over short distances. In general, random fractures are abundant in the unit, while through-going joint sets are scarce and not well-developed. The zoned feldspar gneiss is composed of epidote, muscovite, biotite, quartz, and feldspar. The rock is fine- to medium-grained with disseminated very coarse zoned feldspar crystals. Deep weathering is characteristic in this rock unit. The Brevard Zone black mylonite is generally composed of biotite, quartz, and feldspar. It is typically extremely fine-grained and weakly foliated. Where the foliation is better developed, the rock is shown to be very contorted. Weathering in this rock unit is generally shallow and uniform in depth. The Brevard Zone white mylonite is interpreted to be sheared granite in the literature. This mylonitized granite is composed of muscovite, quartz, and feldspar. Shearing was pervasive and produced a well-developed shear foliation. The development of a shear foliation weakened the rock and allowed more rapid weathering, resulting in tabular zones of deeper, more intense weathering. Where shear foliation is absent or poorly developed, this rock unit is massive, with few discontinuities and shallow weathering.

Groundwater

In the Atlanta area, the primary groundwater source is infiltration from the ground surface into the overlying soil zone. Precipitation consistently recharges the soil zone. The transition zone typically contains abundant open fractures and can become a major storage source for groundwater. The rock zone has fewer open fractures with depth than the transition

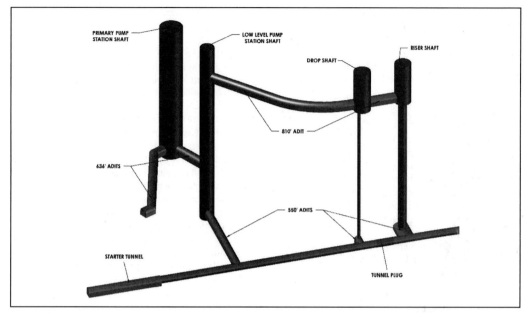

Figure 2. Quarry site layout

zone. However, large fractures with the ability of producing large volumes of water do exist in the rock. Within the rock zone, groundwater is typically conveyed along foliation joints, fractures, veins, and other geologic features that have been enhanced by differential weathering.

DRILL AND BLAST SHAFTS

Overview

The project includes three large-diameter shafts that were constructed using the conventional drill and blast excavation. Two shafts are located at the quarry site (quarry primary pump station shaft and quarry low-level pump station shaft) and the other one is located at the northern end of the tunnel (Peachtree construction/pump station/drop shaft).

The quarry primary pump station shaft(QPPS) is approximately 235 feet deep. The excavated diameter of the shaft is approximately 40 feet with a finished internal shaft diameter of 35 feet. The quarry low-level pump station shaft (QLLPS) is approximately 320 feet deep, with an excavated diameter of about 24 feet and a finished diameter of 20 feet. The QLLPS is connected to the quarry PPS shaft and the TBM tunnel through adits. The Layout of the shafts and adits are shown in Figure 2.

The Peachtree construction/pump station/drop shaft (PPSS) is approximately 250 feet deep. It has an excavated diameter of approximately 34 feet with a finished diameter of 30 feet. The TBM will

be retrieved from this shaft upon completion of the tunnel mining. The shaft will then be converted into a pump station/drop shaft.

Ground Conditions

The LLPS and PPS are located within the Clairmont Mélange unit described above. The ground conditions encountered in the shafts were mapped by a geologist during excavation. The rock was primarily a fresh, granitic gneiss with widely spaced high angle joints with localized areas with more closely spaced jointing. The rock mass quality encountered in the LLPS and PPS, based on the Q-system, ranged from poor to extremely good with typical conditions being very good. The actual Q-values for the two shafts were similar and ranged from approximately 3 to 190 with a typical value of 67. As noted below, these shafts were lined with concrete concurrent with the excavation. The exposed rock in the shafts below the forms, typically about 16 feet in height, was scaled following blasting and no other initial support was required prior to placing concrete. Both shafts encountered minor ground inflows typically less than one gallon per minute and with a total inflow of less than 20 gallons per minute for the entire shaft excavations.

The PPSS is located in the Brevard Mylonite as described above. The Mylonite has more pervasive jointing than the Clairmont Melange. The joints were typically persistent over the exposed rock on the shaft excavation. The jointing formed wedges that dipped into the shaft. In some cases, initial support, consisting of spot positioned dowels and welded wire mesh, were installed prior to placement of the final concrete lining.

Shaft Design and Construction

All three large-diameter shafts are lined with cast-in-place reinforced concrete placed concurrently with excavation using top-down lining methods to a depth approximately 20 feet above bottom of the shaft. The bottom portions of the shafts receive initial ground support consisting of pattern rock dowels and shotcrete. The computer program Unwedge from Rocscience was used to evaluate the initial ground support as well as the maximum allowable unsupported height per excavation round. Locations of the rock dowels were adjusted by the Contractor during excavation to accommodate ground conditions actually encountered. The shaft section with the initial ground support will receive the final concrete lining along with the modified contact grouting in a later stage of the construction contract.

The QLLPS excavation began in March 2016 and excavation and lining of the shaft was completed in December 2016. The QPPS excavation began in March 2016 and excavation and lining of the main shaft was completed on September 2016. The PPSS excavation began in February 2017 and was finished in August 2017.

The shafts were excavated by conventional excavation in the soil and transition zone which comprised approximately seven feet and was supported with steel ring beams and lagging. The rock portion of the shafts were excavated by drilling and blasting. For the QLLPS, the shaft rounds were approximately 24 feet in diameter and round lengths were typically ten to twelve feet. The shaft rounds included approximately 90 holes. Powder factors typically ranged from approximately 3.5 to 4 pounds per cubic yard (lbs/cy). For the QPPS, the shaft rounds were approximately 40 feet in diameter and round lengths were typically 10 to 12 feet. The shaft rounds included approximately 190 holes. Powder factors ranged from approximately 3.5 to 4.5 lbs/cy. For the PPSS, the shaft rounds were approximately 34 feet in diameter and round lengths were typically ten to twelve feet. The shaft rounds included approximately 135 holes. Powder factors typically ranged from approximately 4 to 5 lbs/cy.

Following blasting, the shafts were mucked out using a Caterpillar excavator and a muck bucket. After mucking, the reinforcing steel for the shaft lining was installed followed by lowering of the steel shaft forms. Steel blast proof forms were used in to place the final concrete lining. The concrete was placed in 10-foot to 10.5-foot lifts. The height of lifts was determined by the form size and ground condition.

BLIND BORE SHAFTS

Overview

The project includes five pump station shafts at the HWTP site to provide connections to the TBM tunnel for water access. The center-to-center spacing between adjacent shafts is approximately 20 feet. The shafts are offset approximately 30 feet to the right (east) of the TBM tunnel and are between 100 and 200 feet south of the HWTP Reservoir 2. Each shaft is approximately 420 feet deep with a minimum excavated diameter of 9.5 feet in the rock. Steel liners with an inside diameter of 76 inches are installed in the shafts. Connections between the shafts and the TBM tunnel are through the adits, which have a modified horseshoe-shaped excavation face, approximately 10 feet both in width and in maximum height. The adits are lined with cast-in-place concrete. The finished inside diameter of the adits is eight feet. Blind bore drilling techniques from the ground surface are used to excavate the five shafts at the HWTP site.

Ground Conditions

The subsurface profile at the HWTP shaft site consists of approximately 30 feet of overburden soil underlain by approximately 10 feet of the transition zone followed by rock. Groundwater level is approximately at the top of the transition zone.

Rock conditions were evaluated based on Rock Quality Designation (RQD) and Rock Mass Rating (RMR) measured on the rock cores recovered from the borings drilled at the site. The RQD distribution indicates that approximately 72 percent of the encountered rock are good to excellent, 18 percent are fair, and 10 percent are poor to very poor. The RMR distribution indicates that approximately 30 percent of the rock mass is characterized as very good, 65 percent of the rock mass is characterized as good, and 5 percent of the rock mass is characterized as poor to fair.

A suite of borehole geophysical tests was performed in two of the borings drilled at the site. As shown in Figure 3, analysis of the geophysical data indicates one low-angle foliation joint set and two high-angle non-foliation joint sets. The geophysical data also indicate numerous fractures within the identified joint sets, which contain apertures ranging from 0.25 to 5 inches. These open fractures could act as water communication conduits during the blind bore shaft sinking operations.

Pre-Excavation Grouting

Because of the presence of the open fractures that could act as water communication conduits during the bind bore shaft sinking operations, a pre-excavation

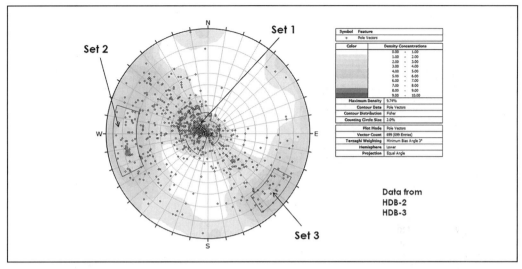

Figure 3. Stereoplot of measured joints at HWTP site

grouting program was developed and implemented at the HWTP site. The program was designed to address several concerns that were raised during the design. First, construction of the shafts could potentially pose significant risks to the HWTP reservoir as the shafts are located in close proximity of the reservoir. The City required that all risks associated with inadvertent dewatering of the HWTP reservoir due to construction of any aspect of the project be kept to an absolute minimum. Second, cuttings of the blind bore drilling are removed from the shafts by reverse circulation. Therefore, the shafts need to be filled with drilling fluid and it is critical to prevent drilling fluid loss during shaft sinking operations. In addition, the shaft excavation is concurrent with the TBM tunnel mining, with some shafts being completed before the TBM arrives at the HWTP site and some of the shafts being drilled after the TBM has mined through the area. A potential for drilling fluid loss and TBM tunnel flooding exists. Considering all the risks and the highly fractured rocks encountered in the borings, a pre-excavation grouting program was deemed necessary. The main purpose of the grouting program is to lower the rock mass permeability such that potential risks for loss of the drilling fluid as well as water communication between the shafts and the HWTP reservoir can be reduced. In addition, properties of the rock within the grouted zone can be improved through consolidation of the rock mass, facilitating shaft sinking and adit excavation operations.

In order to enhance the potential for intersecting the identified joint features as indicated from the geotechnical investigation and analysis, inclined grout holes were selected. The holes were oriented 10° off vertical at a bearing of 260°. Note that due to site restrictions, some of the grout holes were kept vertical. A typical cross section of the pre-excavation grouting is shown in Figure 4.

The grouting program consisted of 42 primary holes and 44 secondary holes (71 inclined and 15 vertical) with a staggered spacing of eight feet between the primary and secondary holes. The area of the target grouted zone is approximately 6,000 square feet. Grouting was generally performed from bottom up in 20-foot stages. The grouting work was completed by Hayward Baker, a subcontractor of the Atkinson/Technique Joint Venue, in October 2016. Analysis of the grout data indicates that favorable results have been achieved and the pre-excavation grouting program has fulfilled the design intent.

Shaft Design and Construction

During normal operations, the shafts at the HWTP site are filled with water and therefore, the steel liners are not subjected to differential water pressure as the internal water pressure would offset the external pressure. However, conditions may occur that the shafts need to be dewatered; therefore, the steel liners are required to be capable of withstanding full external pressure without internal balancing pressure. The critical failure mode in this case is buckling.

Buckling of the steel liner occurs at a critical circumferential stress at which the liner becomes unstable and fails. The groundwater level at the HWTP site is approximately 370 feet above the bottom of the liner; thus a hydrostatic pressure of 240 psi (equivalent to a 370-foot water head multiplied by a factor of safety of 1.5) was used for computing the minimum required steel liner wall thickness.

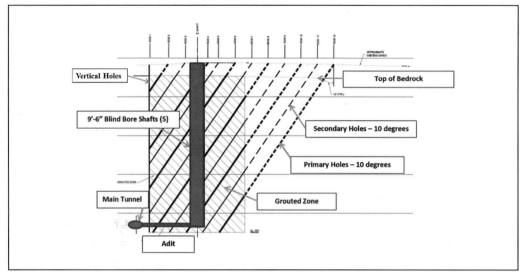

Figure 4. Pre-excavation grouting from ground surface

Analytical methods have been developed for determining the critical buckling pressure for a cylindrical steel liner. It is noted that different analytical methods produce different results; thus, it is prudent to use several methods in a design. For this project, analytical methods developed by Jacobsen (1974) and Vaughan (1956) were used to compute the required steel liner wall thickness in order to withstand an external pressure of 240 psi. Equations for these analytical methods can also be found in USACE (1997) and Hunt et al. (1995). All calculations were implemented in a Mathcad application. Results of the analyses indicate that a minimum wall thickness of ½ inches is required for the 76-inch inside diameter steel liner.

The blind bore shafts were constructed in several steps. The first step involved installing a 144-inch diameter steel casing to support the soil and transition zone for each shaft. This casing was installed by A. H. Beck Foundation Drilling, Inc using an HPM 250 drilling rig mounted on a Caterpillar excavator base. The hole was drilled either dry or using a bentonite slurry to keep the hole open. Subsequently, a 132-inch diameter steel casing were inserted in the holes and grouted in place.

After the surface casings were installed, a pilot hole was drilled in rock for each shaft. These pilot holes were completed by North American Drilling, LLC. The pilot hole consisted of two stages, drilling an 8-inch directionally drilled hole followed by reaming the hole to 17.5 inches. The 8-inch diameter hole was completed using a down-the-hole-hammer (DTTH) with real-time optical guidance system along with a shovel-type bit to provide the directional drilling capability. The holes were surveyed during drilling and the drilled hole was typically within 0.25-inch of the shaft centerline. The 8-inch diameter holes were reamed to 17.5-inches in preparation for the large diameter reaming operation.

The shafts were reamed to the full diameter of 114-inches using rigs manufactured by NAD equipped with hemispherical reaming heads. Two rigs were used to complete the reaming operations. Weights are added to the drill string to provide the required downforce on the drilling head. The cuttings are removed using an air-lift system and the drilling water is run through a settling and re-used to remove the drill cuttings. Penetration rates were typically in the range of about 0.5 feet/hour to 1.0 feet/hour.

The final steel casing was installed after completion of the reaming operations. The casing is a one-inch thick spiral welded steel pipe with an internal coating applied for corrosion protection. The pipe joints are approximately 41 feet long. The pipes were installed by lowering each section into the hole followed by welding to the next pipe joint. The pipe was sealed on the bottom and water was added as the pipe was lowered to offset the buoyancy of the pipe. The annular space around the pipe was grouted in typically 40 to 50 foot stages using tremie pipes. The grout used to grout the annular space had a 28-day compressive strength of 2,500 pounds per square inch.

RAISE BORE SHAFTS

Overview

In addition to the PPS and LLPS, the quarry site also includes a drop shaft and a riser shaft as shown in Figure 2. As shown in Figure 5, the upper portion

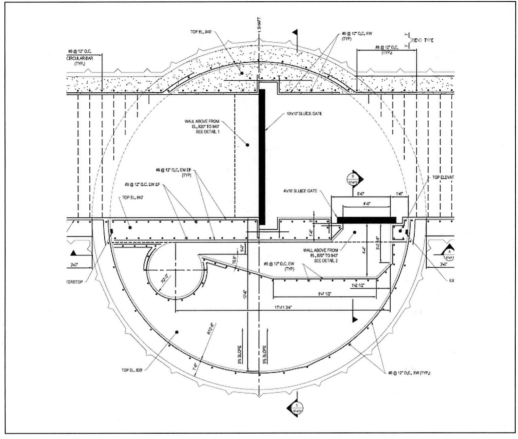

Figure 5. Drop shaft plan view

of the drop shaft (above El. 805) has an excavated diameter of about 29 feet with a finished shaft diameter of 25 feet. The excavated diameter of the lower shaft is approximately eight feet with a finished shaft diameter of 4.5 feet. The total depth of the drop shaft is approximately 335 feet with the upper shaft having a depth of about 80 feet.

Similar to the drop shaft, the riser shaft also consists of an upper portion (above El. 805) and a lower portion. The upper shaft has an excavated diameter of approximately 29 feet with a finished diameter of 25 feet. The lower shaft has an excavated diameter of about 15 feet and a finished diameter of 12 feet. The total depth of the riser shaft is approximately 330 feet with the upper shaft having a depth of about 75 feet. The isometric rendering is illustrated on Figure 6.

The upper portion of the drop shaft and the riser shaft is excavated by the conventional drilling and blasting technique and lined with cast-in-place reinforced concrete. The lower portion of the shafts will be excavated by the raise bore method. For the drop

shaft, a carrier pipe will be installed and the annulus will be backfilled with concrete upon completion of the raise bore. For the riser shaft, the final cast-in-place concrete lining will be installed subsequent to the raise bore excavation.

Ground Conditions and Shaft Construction

The upper portion of the riser shaft (approximately El 880 to El 805) was completed in the soil and transition zone. Initial support consisted of W6×25 steel ring beams and hardwood lagging down to approximately El 805. Water inflows during shaft construction were typically below 20 gallons per minute. The drop shaft is located approximately 50 feet away from riser shaft. In this shaft, the soil and transition zone were encountered down to approximately El 831. This zone was supported with W6×25 steel ring beams and lagging. The remaining port of the shaft encountered fresh to slightly weathered granitic gneiss that was supported with 8-foot long rock

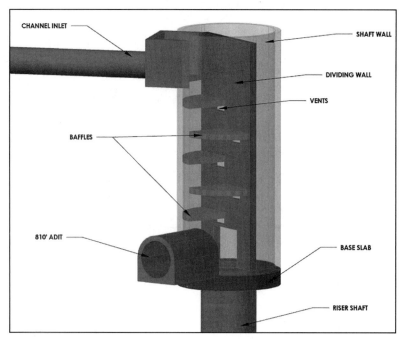

Figure 6. Riser shaft 3D rendering

bolts, and a combination of welded wire fabric and shotcrete.

The remaining portion of the drop shaft and riser shaft will be completed using raise bore. The riser shaft raise bore will be 16.5-feet in diameter while the drop shaft raise bore approximately 8 feet in diameter. During the design, the overall shaft stability was assessed with the McCraken and Stacey empirical shaft stability method (1989) based on Q value of the rock mass, and the localized wedge stability was analyzed using the program Unwedge. 2" thick shotcrete was designed for the rise shaft raise bore to provide the protection during shaft final lining placement, while the drop shaft raise bore was deemed stable with no additional support. According to the construction schedule, the raise bore work is expected to be completed in January 2018.

CONCLUSIONS

This paper presented the deep shaft construction techniques used in the Atlanta Water Supply Program, including blind boring, raise boring and conventional drilling & blasting. The paper also discussed the geotechnical settings, design approaches, and construction considerations in the design of shafts, as well as the findings during shaft construction.

ACKNOWLEDGMENTS

The authors would like to thank the Department of Watershed Management of City of Atlanta for its commitment to make the project possible. The contributions and efforts from other members of Stantec's tunneling team, including Adam Bedell, PG, Konner Horton, EIT and James Jackson, are also greatly appreciated.

REFERENCES

Hunt, S.W., Heuer, R.E., and Safdar, A.G. 1995. Casing Collapse at the CT-8 Dropshaft in Milwaukee. 1995 RETC Proceedings, pp. 197–218.

Jacobsen, S. 1974. Buckling of Circular Rings and Cylindrical Tubes under External Pressure. Water Power, Vol. 26, pp.400–07.

McConnel, K.I. and Abrams, C.E. 1984. Geology of the Greater Atlanta Region. Department of Natural Resources, Environmental Protection Division, Georgia Geologic Survey. Bulletin 96.

U.S. Army Corps of Engineers (USACE), 1997. EM 1110-2-2901 Engineering and Design Tunnels and Shafts in Rock. May 30.

Vaughan, E.W. 1956. Steel Linings for Pressure Shafts in Rock, *J. Power Div., ASCE*, Vol.82, pp. 9491–94940.

South Hartford Conveyance and Storage Tunnel Project—Successful Use of Large-Diameter Secant Piles for Shaft Support

Andrew Perham
The Metropolitan District

Jim Sullivan
AECOM

Mike Brune
Case Foundation

David Belknap and Clay Haynes
Black & Veatch

ABSTRACT: The South Hartford Conveyance and Storage Tunnel (SHCST) is a major component of the Hartford Metropolitan District's Clean Water Project (CWP). The tunnel will capture and store Combined Sewer Overflows (CSO) from the southern portion of Hartford, CT and Sanitary Sewer Overflows (SSO) from West Hartford and Newington, CT. The project includes 21,800 ft of 18 ft final diameter tunnel, several miles of consolidation sewers, eight hydraulic drop structures and a 50 MGD tunnel dewatering pump station. Construction has started on the large diameter shafts which have used large diameter secant piles (59.1-inch diameter) for initial support in the upper sections of the shafts. This paper describes the challenges and methods used to construct the secant pile walls.

PROJECT DESCRIPTION

The purpose of the SHCST project is to eliminate West Hartford and Newington SSO, eliminate Franklin Area CSO discharging to Wethersfield Cove and to reduce CSO discharges to the South Branch Park River.

The SHCST will collect and store CSO and SSO during wet weather events and ultimately convey the overflows to the Hartford Water Pollution Control Facility (HWPCF) for treatment and discharge to the Connecticut River. The eight hydraulic drop shafts are relatively small diameter shafts varying from 2.5 ft to 6.0 ft finished diameter that are being constructed by a Bauer BG-39 drill rig in the soil section and a Wirth drill rig in the rock section. However, the subject of this paper is the large diameter shafts including the Grit/Screening (G/S) Shaft and the Tunnel Pump Station (TPS) Shaft. The retrieval (R) Shaft had not been fully constructed by the time of publication of this paper and is not discussed in detail in this paper.

The G/S Shaft (also known as the Launch Shaft) will contain Bosker traveling screens that will remove large debris from the tunnel water before routing it to the suction header tunnel which will convey the tunnel water to the TPS Shaft. During construction, the tunneling operation for the 18 ft final diameter tunnel will be staged out of the G/S Shaft. The total depth of the G/S Shaft is approximately 203 ft and the final diameter is 39 ft in the upper section of the shaft.

The TPS Shaft will contain wastewater pumps that will lift up to 50 MGD of tunnel water approximately 220 ft to a gravity sewer that connects to the HWPCF headworks. The final diameter of the upper section of the TPS Shaft is 74 ft.

The tunnel boring machine will be retrieved at the R Shaft. The R Shaft will serve as an access point to the tunnel and provides a drop connection to a SSO point on a near surface sewer. The total depth of the R Shaft is approximately 223 feet deep and the final diameter is 32 ft in the upper section of the shaft.

GEOLOGIC SETTING

The project site lies in the Central Lowland physiographic province that extends in a north-south direction in the middle of the state of Connecticut. The central lowland area consists mainly of the sedimentary rock and the associated igneous basalts of

Triassic and Jurassic age. The Hartford Basin of Connecticut and southern Massachusetts is a half graben in structure, 90 miles long, and filled with approximately 13,000 ft of sedimentary rocks, and basaltic lavas and intrusions (Hubert, et al., 1978). The large diameter shafts are being constructed entirely within the Portland Arkose which consists of mainly siltstone with some shale and sandstone.

However, the geologic item of most significance to this paper is the soil overburden. The region has undergone periods of glaciation that has greatly influenced the soil overburden. Glaciers laid down a heterogeneous layer of grounded-up rock (glacial till). This till layer is present over much of lower lying bedrock surfaces. The sediments of the Glacial Lake Hitchcock filled in the deeply-incised Connecticut River Valley. The lake deposits are present in varying forms from Rocky Hill, Connecticut to northern Vermont. Glaciers shaped the topography and left the area with much of the topographic relief present today. More recent alluvial deposits are common along the Connecticut River and Park Rivers and their tributaries.

In the project area, the following soils are present overlying the bedrock, in general order of sequence from the ground surface downwards: artificial fill, alluvium, glaciolacustrine deposits, glaciofluvial deposits and glacial till.

GEOTECHNICAL CHALLENGES

The geotechnical challenges associated with construction of the large diameter shafts were as follows:

- The glaciolacustrine deposits are extremely weak and unconsolidated sediments deposited in a glacial lake bottom. These deposits are up to 30 ft thick in the area of the large diameter shafts. These formations required the following characteristics of the initial support system in the shafts.
  - These extremely weak deposits have no ability to stand after excavation. Thus, the initial support system must be installed before excavation of the soil sections of the shafts.
  - These unconsolidated deposits will compress and settle if dewatered or disturbed. Thus, the initial support system has to be nearly water-tight to prevent soil consolidation.
- The artificial fill deposits have a tendency to contain manmade obstructions such as concrete and wood. These obstructions would be problematic for initial support systems that were installed before excavation.

- The alluvium deposits tend to be coarser grained soils that readily convey groundwater. Although the alluvial deposits could be dewatered with wells, this approach seemed incompatible with pre-excavation support methods.
- The glacial till deposits have cobbles and boulders which makes excavation difficult for drilling machinery.
- The glaciofluvial deposits were only present at the R Shaft location. The glaciofluvial deposits had a tendency to contain cobbles, boulders and wood debris which makes excavation difficult for drilling machinery.

DESIGN OF INITIAL SUPPORT IN THE SOIL SECTION OF THE LARGE-DIAMETER SHAFTS

Noting the numerous challenges presented by the soil strata in the project areas, the design team developed an approved list of soil support systems consisting of secant piles and slurry diaphragm walls. While the design team believed that slurry diaphragm walls might have a slight technical advantage over secant piles, they thought that secant piles with some prescriptive requirements would be a cost-effective alternative to slurry diaphragm walls.

Some prescriptive measures that were added to the secant pile specification were as follows:

- Due to the challenging soil conditions and lack of current and relevant experience in the local marketplace, the designer required minimum relevant project experience for the contractor to construct the secant pile walls.
- The specifications required the secant pile footprint to be pre-excavated to 15 ft below grade to clear manmade obstructions in the artificial fill deposits
- Although it is not common to use steel casings during excavation of secant piles, steel casings were specified for soil excavation to reduce the potential for soil loss.
- Although it is not common to use drilling fluid inside steel casings during excavation of secant piles, drilling fluid was required to reduce the potential for ground heave and the associated ground loss in the base of the secant piles during soil excavation.
- Although it is typical to let the drilling contractor select his embedment depth in bedrock, the designer decided to reduce project risk by specifying a minimum embedment of five ft into bedrock.

PERTINENT DETAILS OF THE CONSTRUCTION METHOD FOR EACH OF THE SHAFTS

As previously mentioned, the secant pile construction method was determined to be a viable option for construction of the large diameter shafts. The contract documents did not specify a minimum secant pile diameter. However, the contract documents did specify a minimum secant pile wall thickness at the intersection of the primary and secondary secant piles. During contractor design and optimization it was determined that 1500 mm (59.1 inch) diameter secant piles with no reinforcement was the most effective and efficient design for the geotechnical conditions and site. Piles were drilled utilizing 1500 mm (59.1 inch) OD segmental casing and were flooded with water as drilling progressed to maintain a sufficient head pressure to prevent blowout of the bottom of the pile. Additionally, a small plug of soil was always maintained within the casing to further prevent ground heave. Table 1 shows the pertinent dimensions of the G/S and TPS Shafts. The secant pile guide wall for the TPS Shaft is shown in Figure 1.

Table 1. Shaft parameters

	G/S Shaft	TPS Shaft
Shaft ID	42'	77'
Soil Depth	70.5'	69'
Rock Embedment	5'	5'
Number of Secants	39	70
Effective Wall Thickness	28.8"	30.4"
Secant Spacing	1152 mm	1120.6 mm

OVERCOMING CONSTRUCTION CHALLENGES

Construction progressed relatively smoothly for the entire duration of the drilling of the G/S and TPS Shafts. Small challenges were encountered throughout construction and included the following:

- Construction of the G/S Shaft was performed within an excavated bowl while waiting for approvals to remove environmentally impacted soil in the area surrounding the shaft. This constraint led to a very wet and muddy site with poor drainage. Due to the excess water, the fill material that was pre-excavated and backfilled became highly saturated and led to some large voids below the guidewall in the top 5–10 ft of the secant piles.
- Schedule constraints required the secant pile construction to begin in January using two 10-hour shifts of drilling. A significant amount of effort was expended to prevent freezing of water lines and damage to the equipment in the cold weather.
- Drilling of 1500 mm (59.1 inch) piles proved to be challenging when drilling secondary piles due to the large overlap between adjacent piles. A larger overlap was designed in these shafts to account for reasonable verticality tolerances and also to provide the required minimum wall thickness. The minimum secant pile wall thickness was critical because no reinforcing steel was provided to account for structural defects in the secant pile wall.

Figure 1. Secant pile guide wall for the TPS Shaft

- Inclinometers were required at multiple locations in both shafts. For piles requiring inclinometers, small rebar cages were fabricated to carry the 6" PVC inclinometer sleeves. Due to the high slump of the concrete and the very light rebar cages, there was a large amount of deflection and movement that occurred in the inclinometer sleeves. Ultimately, several inclinometers were drilled in the secant

pile wall after drilling was completed to provide a straighter hole for inclinometer measurements.

Figure 2 shows the G/S Shaft drilling operations in the winter months. Figure 3 shows soil excavation within the G/S Shaft after completion of the secant pile wall construction.

Figure 2. BG-39 drilling a secant pile

Figure 3. Excavating soil from within the G/S Shaft

THE FINAL RESULTS FOR THE G/S AND TPS SHAFTS

Both the G/S and TPS Shafts were completed on schedule with no major issues. Drilling was completed in less than twelve weeks utilizing two shifts of drilling and one shift of concrete pouring. While several challenges were encountered utilizing 1500 mm secant piles, the large diameter piles proved to be an effective method for construction of secant pile walls.

CONCLUSION

Secant piles, with several prescriptive requirements, were included as an acceptable construction alternative to slurry diaphragm walls for the soil section of the large diameter shafts. After a detailed evaluation of both secant pile walls and slurry diaphragm walls

construction methods, Case Foundation selected secant pile walls as the lowest cost alternative that also provided the lowest risk during construction. The secant pile walls were relatively water-tight and structurally robust initial support systems that were constructed with minimal rework and remediation.

REFERENCES

Hubert, J; Reed, F.; Dowdall, W. and Gilchrist, J. 1978. Guide to the Mesozoic Redbeds of Central Connecticut, Guidebook No. 4, Connecticut Geology and Natural History Survey, Hartford, CT.

Nasri, Verya; Bent, William; Hogan, William 2015. South Hartford CSO Tunnel and Pump Station, Rapid Excavation and Tunneling Conference 2015, New Orleans, LA, June 7 to 10, 2015, pp. 140–151.

An Innovative Approach with Granite Block—Mud Mountain Dam 9-Foot Tunnel Rearmoring Project

Madison Brunk, James Carroll, and Joel James
ILF Consultants, Inc.

Terry Gilliland
Garney Construction

Ellen Engberg
United States Army Corps of Engineers

ABSTRACT: Mud Mountain Dam near the base of Mount Rainier includes two tunnels. One of the tunnels has required ongoing invert maintenance due to the passing of high sediment loads. The United States Army Corps of Engineers (USACE) awarded the design-build contract to rehabilitate the tunnel to Garney Construction teamed with ILF Consultants and Golder Associates. Though originally awarded the project for their mechanically anchored steel armor plating design, the team provided the USACE with a value engineering change proposal (VECP) of natural granite blocks. The granite block design provides an increased design life, lower cost, and shorter construction schedule, with construction beginning in July 2017. This paper discusses details and challenges of the VECP from design through initial construction.

INTRODUCTION

History of the Mud Mountain Dam 9-Foot Tunnel

The Mud Mountain Dam (MMD) was under construction between 1939 and 1942, and it became fully operational in 1948 after being postponed due to World War II. The dam is located on the White River, eight miles southeast of Enumclaw, Washington. It has a 9-foot tunnel and a 23-foot tunnel that carry water 1800 feet through the dam. The 9-foot tunnel specifically is used to move both water and sediment from the upstream side of the dam to the downstream. It passes 450,000 tons of sediment per year on average, ranging from glacial flour to 22 inch diameter boulders. The original invert of the tunnel was concrete reinforced with rails as shown in Figure 1, but the sediment load passing through the tunnel quickly began damaging the invert and eroding the concrete away. From 1968–1993, the rails and grout had to be replaced regularly to restore the integrity of the invert and protect the tunnel from further damage. In 1995, the 9-foot tunnel alignment was adjusted to meet a new intake tower. As part of that work, a new invert with a flat bottom was designed and installed in the tunnel to slow down the wear rate and decrease the frequency of invert repair. The new invert consisted of a 1 inch thick steel liner running the full length of the 9-foot tunnel. In 2006, the United States

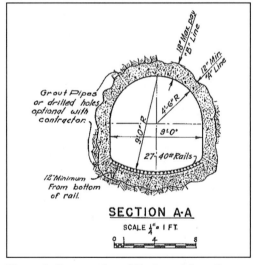

Figure 1. Cross section of the tunnel from the 1939 plan set

Army Corps of Engineers (Owner) noticed the first penetration through the 1 inch steel liner. Damage to the steel liner due to scour is shown in Figure 2. The Owner continued repairing scour holes in the invert. Each time that sediment was run through the tunnel, the damage increased until repairs were being done twice a year and tunnel usage was reduced so that maintenance could keep up with the rate of damage. When the Request for Proposal (RFP) went out for the design-build tunnel rearmoring project in 2016, it was thought that an additional season of use could cause scour to reach the underlying bedrock and the steel invert to be damaged beyond repair.

DESIGN-BUILD BID PROCESS AND TECHNICAL PROPOSAL

In the design-build tender documents, the armor steel design consisted of I-beams welded to the existing invert with two layers of steel plating welded on top. Even with this newly designed armored layer, ongoing maintenance to the top layer of armor plating was anticipated to continue every 15 years. The RFP design included approximately 10,000 feet of welds, and the void space between the existing invert and the plates on top was to be filled with grout. Due to the welded design, repairs would entail hot work within a confined space that included cutting the welds out to remove plates and then welding new plates into place. Repair procedures and estimated repair timelines were to be included as part of the proposal, and the repair process for this steel design would be expensive and labor intensive. A section view of the Owner's design is shown in Figure 3.

Garney Construction (Contractor) teamed with ILF Consultants, Inc (Designer) who created a new design eliminating hot work repairs as shown in Figure 4. The team's design utilized a mechanical fastening system that decreased the hot work required for the initial installation by about 50 percent. The design fastened the armor plates to the I beams with mechanical fasteners so the armor plates could be removed and replaced as needed without cutting any metal or doing hot work in the tunnel. In addition, a section was included in the bid package that proposed an alternative technical concept that used granite stone as an alternate invert material. This was based off similar applications of granite for tunnel inverts in Europe and Japan.

Figure 2. Example of damage to the tunnel liner. Flow to upper right.

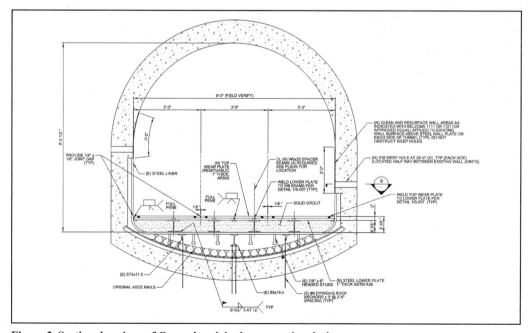

Figure 3. Section drawings of Owner's original re-armoring design

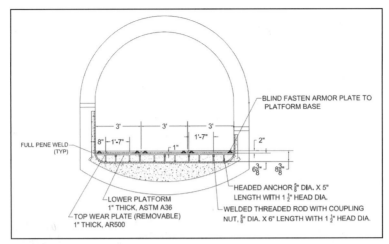

Figure 4. Section drawings of the design team's mechanically fastened design

Figure 5. Granite blocks and pervious concrete subbase in Pfaffensprung

The team was awarded the project for their mechanically fastened design. At the kick off meeting for that design, part of the time was used to present about the granite block alternative. After that discussion, the team requested a 30 day contract extension to put together a VECP.

During the 30 day VECP period, the team produced preliminary design documents and preliminary calculations. Much of the VECP was based on observations of Pfaffensprung tunnel outside of Wassen, Switzerland. Pfaffensprung also passes sediment, and it was commissioned in 1922 with a granite block invert that was replaced approximately 40 years later with different materials. After being disappointed by the performance of the alternative materials, the operators of the Pfaffensprung tunnel decided to return to the original granite invert material because of its superior durability. Therefore,

the current granite invert was installed over several years and completed in 2016. The tunnel's current design consists of Urner granite blocks, a pervious concrete subbase, and drainage pipes as shown in the Figure 5.

The VECP process included abrasion calculations to determine the expected design life of the granite blocks. Granite was to be sourced from the Hardy Island quarry in British Columbia, Canada. Testing of the sourced granite showed it has a higher strength than the Urner Granite used in Pfaffensprung, shown in Table 1. Abrasion data from Pfaffensprung was used to calibrate the MMD abrasion model based on the estimated sediment load and the increased strength of the Hardy Island granite. Three different abrasion models were used to estimate the design life of the granite blocks: Auel, Sklar and Dietrich, and Ishibashi. Abrasion conditions at Pfaffensprung were

Table 1. Strength values of granites used at Pfaffensprung and MMD 9-foot tunnel with percent difference calculated as the percent increase of strength from Urner to Hardy Island Granite

Material Property	Pfaffensprung Urner Granite	MMD Hardy Island Granite	Percent Difference
Compressive Strength, f_c (MPa)	180	223	+24%
Tensile Strength, f_t (MPa)	10.0	11.7	+17%
Young's Modulus, Y_M (GPa)	53.0	64.9	+22%
Material density ρ_m (kg/m^3)	2650	2684	+1%

Table 2. Estimated abrasion depths for MMD granite invert pavers

	Models				
Description	Auel		Sklar and Dietrich		Ishibashi
Abrasion Condition at Pfaffensprung*	Severe	Average	Severe	Average	Average
Average abrasion depths due to normal sediment load					
Sediment supply	Average	Average	Average	Average	Average
Abrasion Rate [in/yr]	*0.019*	*0.013*	*0.011*	*0.006*	*0.013*
Abrasion Depth @ 40 years [in]	*0.74*	*0.54*	*0.43*	*0.24*	*0.52*
Average abrasion depths due to increased sediment load					
Sediment supply (Increased +62%)	Severe	Severe	Severe	Severe	Severe
Abrasion Rate [in/yr]	*0.030*	*0.022*	*0.017*	*0.010*	*0.021*
Abrasion Depth @ 40 years [in]	*1.20*	*0.87*	*0.69*	*0.38*	*0.84*
Local maximum abrasion depths: 3 times higher than Average					
max Abrasion Rate [in/yr]	*0.09*	*0.07*	*0.05*	*0.03*	*0.06*
max Abrasion Depth @ 40 years [in]	*3.60*	*2.60*	*2.07*	*1.15*	*2.53*
Expected lifetime to one-half thickness (5 in) (yr)	> 40	> 40	> 40	> 40	> 40

Source: VAW 2016.

* Calibration made based on Pfaffensprung data using average and severe abrasion depth measurements.

noted as either being Severe or Average and then correlated to the conditions found at MMD. Two types of sediment loads for MMD were considered as being Average or Severe. The Severe condition was analyzed as having an additional 62% bedload from Average, which in turn increased the abrasion rate. A factor of three was used to estimate the local maximum abrasion depths based on the conservative factor found at the Pfaffensprung tunnel. The invert's design life is considered exhausted when the block maximum scour depth reaches 5 inches of its 10 inch thickness. The results of these calculations are found in Table 2 and show that in the worse-case scenario, the Auel method predicts only 3.60 inches of local abrasion in 40 years.

Additionally, block stability was analyzed during the 30 day VECP period. Uplift of the blocks was a concern because of the high velocity of the water flowing over the blocks and the potential for water to jet down between blocks. Preliminary stability calculations were performed during the 30 day period, and later on in the detailed design process, a finite element model was created and refined.

The 4 inch subbase and 10 inch granite blocks are together thicker than the original 8.4 inch steel design, so the Owner expressed concerns that the decreased cross sectional area of the tunnel would reduce its water conveyance. The team asked Golder Associates (hydraulic engineers) to perform a hydraulic study, and the results of the study showed that the amount of water being passed through the dam would not decrease. This was because the proposed design would allow the tunnel to remain operational at higher flood levels when there is increased head.

After the VECP was submitted, there was a question and answer period for the Owner to seek clarification prior to accepting or rejecting the proposal. In December 2016, the Owner gave notice of the acceptance of the VECP with three stipulations: that an additional LiDAR scan be completed as a baseline at the completion of the project, that a monitoring system for block uplift be installed, and that an additional 6 inches of abrasion resistant epoxy coating be installed on the tunnel walls to compensate for the increased granite invert thickness. The team agreed to those stipulations and detailed design began.

VECP DESIGN

Benefits

The VECP has many benefits both for the Owner and for the team. Construction has not been completed,

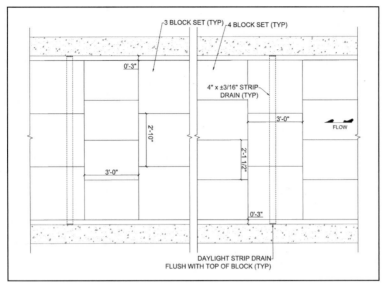

Figure 6. Granite block and strip drain layout

but the design is anticipated to allow for schedule and cost savings that can be quantified at project completion. The tunnel needs to be flushed periodically, requiring a pause in construction and jeopardizing work that has been installed but not completed. The decrease in hot work and the rate at which granite blocks can be installed will potentially result in the job being completed between fewer tunnel flushings, decreasing the need to remobilize or to protect in place work that has already been installed.

Additionally, the estimated design life for the granite block design far exceeds that of the armored steel invert design. The Owner's original concept design for the 1 inch layer of AR500 steel was estimated to need replacement every 15 years. The abrasion models calibrated with Pfaffensprung data estimate that the design life of the granite blocks at MMD will exceed 40 years. It is anticipated that the granite block invert will also be far easier to repair than armored steel plates that would have needed to be cut out and their replacements welded into the tunnel.

Design Considerations

Preliminary stability calculations were performed prior to the VECP acceptance, and these calculations were refined throughout the design process. Finite element models were built to simulate the uplift pressures applied to the granite blocks. When water flows over a material, there is a resultant uplift force. Additionally, if there is a vertical lip between blocks, water can jet down between blocks and contribute to uplift. The block stability models were created to determine safety factors and placement tolerances. The model influenced the design of the invert by leading to the inclusion of strip drains and a joint sealant as described in the Design Components section of this paper.

Abrasion calculations correlated the kinetic energy of sediment moving through the tunnel with its angle of impact with the invert. These calculations affirmed the decision to use Hardy Island granite, as its increased hardness compared to other granite sources would lengthen the design life of the invert. The calculated design life of the granite block remained at 40 years.

Design Components

Granite Blocks

Granite blocks were sourced from British Columbia, Canada and are Hardy Island granite whose compressive strength is approximately 30,000 psi. Blocks are 10 inches thick and approximately 3 feet long. They are placed in alternating rows of 3 and 4 blocks, and each row is 8.5 feet wide. This staggered arrangement as shown in Figure 6 is to eliminate continuous longitudinal joints along the flow path. Observations of other granite lined tunnels show that wear happens along longitudinal joints, so eliminating continuous joints will decrease wear. Joint sealant is installed in all transverse joints to help minimize the water jetting down between joints that contributes to uplift pressures. There is an anticipated 3 inch gap on each side between the edge of the granite block and the side wall of the tunnel. The gaps will be filled with a high strength calcium aluminate concrete (CAC).

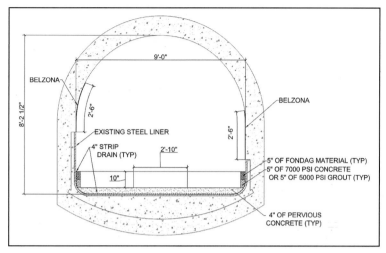

Figure 7. Cross section showing the pervious subbase

This is a pre-blended, high strength concrete that is engineered to withstand high abrasion conditions. The CAC will fill the top 5 inches of the gap and have an additional 5 inches of high strength grout or concrete underneath it as shown in Figure 7. Gaps at the tunnel walls were minimized to decrease the amount of abrasion the CAC would be exposed to. Wear patterns in the previous tunnel floor show abrasion beginning 4 to 5 inches out from the sidewalls. This wear pattern is likely controlled by the geometry of the tunnel and of the boulders that pass through it.

Subbase

A permeable subbase for the blocks is needed to dissipate pressure that could otherwise cause uplift of the blocks. As mentioned, water moving over the blocks naturally creates uplift forces, and water jetting down between blocks can also contribute to uplift. A permeable subbase material allows water that has been jetted between blocks to move through the subbase, thereby decreasing the uplift pressures acting on the blocks. Multiple materials were considered for the subbase material including pervious cellular concrete, gravel, and pervious concrete.

Pervious cellular concrete was initially the preferred material as it can be pumped and screeded and it could have been installed in the entire tunnel prior to installation of the granite blocks. This would have streamlined the granite block installation process while still providing the permeability needed to ensure block stability. However, the two percent grade of the tunnel would have made installation challenging depending on the slump of the mix design. More importantly, no mix design was found for this non-standard application that could provide

the strength needed during construction as well as the high permeability needed during operation.

Gravel was also considered but was eliminated due to concerns about installation and about its stability during operation. Installing a gravel bed in an extended section of the tunnel prior to beginning granite block installation would not be feasible. Choosing gravel would have meant installing bedding material for only one or two rows of blocks at a time, and that would have caused a significant increase in construction time. There were also concerns about the bed's integrity once water began flowing through it. Pressurized water flow could cause liquefaction of the gravel and compromise the tunnel invert.

Because of the potential for liquefaction in a subbase that is not cemented, the team and the Owner both agreed that a cemented solution was necessary. Despite initial concerns regarding constructability, pervious concrete was chosen as the subbase material as in Pfaffensprung. Its strength and permeability both exceed that of pervious cellular concrete, and because it is cemented, liquefaction is not a concern with pervious concrete as it is with gravel. Screed rails welded to the walls of the tunnel ensure that the concrete is installed within the surface tolerance. The pervious subbase is shown in Figure 7.

Strip Drains

As previously articulated based on the mathematical model, strip drains were a necessary addition to the design to overcome uplift pressures. Strip drains installed in the subbase layer increase the overall hydraulic conductivity of the subbase and help to dissipate pressure. They provide a path of decreased resistance for the water within the subbase to daylight above the blocks. Their inclusion increases the factor of safety against uplift throughout the length of the tunnel. Strip drain spacing was determined by the mathematical model, and more frequent strip drains are required at the upstream end of the tunnel where the alignment follows a curve and the cross section of the tunnel decreases. The strip drains are installed directly on the invert prior to the installation of the pervious subbase as shown in Figure 7.

Monitoring System

The Owner determined that the addition of a monitoring system was a requirement for approval of the VECP. Their concern was that if a block were to be lifted out of place, it would expose the already damaged steel liner underneath and create an opportunity for the invert of the tunnel to be further damaged. A monitoring system would notify the government if a block were moved and give them an opportunity to shut down the tunnel and repair the damaged section before the current steel invert could sustain additional damage.

Though initially the team designed a system using fiber optic cables so that the precise location of the failure could be determined, discussion with the Owner made it clear that the additional cost for such a system was not within budget. A far simpler system was designed that would still alert the government of a failure though it is not able to give the precise location of the failure. Direct bury wire loops are fastened to the invert prior to installation of the subbase material and strip drains. If a granite block were to fail and be lifted out of the place, the sediment moving through the tunnel would quickly begin eroding away the exposed subbase material. Without a granite block or subbase material to protect it, the now exposed loop would be damaged and no longer able to pass current. The circuit would be broken and the existing data logger at the toe of the dam would register that there was no longer current moving through the loop. MMD personnel would be alerted and be able to shut down flow through the tunnel before the existing steel invert was damaged further . The total cost for this addition was well below that of the fiber optic system, which was approximately 30 times more expensive. Drawings of the monitoring system are shown in Figure 8 and Figure 9.

LiDAR Scans

The government stipulated that a LiDAR scan of the as-built tunnel was required for acceptance of the VECP in addition to the scan that the team was already planning to perform to use in the design process. Dibit Measuring Technique, Inc performed the scan and used their proprietary software to overlay digital photographs with LiDAR point cloud data to

create three dimensional renderings of the tunnels as shown in Figure 10. This precise rendering was used during the design process to ensure that there would not be interference between the granite blocks and the tunnel walls, as there are cross section deviations throughout the tunnel. An example of this interference check is shown in Figure 11 and Figure 12, and this check served as a project appropriate

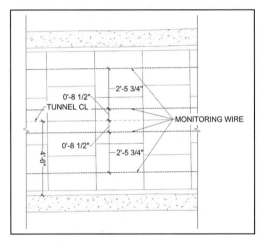

Figure 8. Plan view of monitoring system

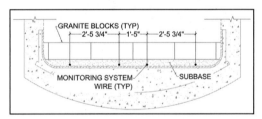

Figure 9. Section view of monitoring system wires

Figure 10. LiDAR scan example image of MMD tunnel

Figure 11. LiDAR scan showing block layout without collision with MMD tunnel walls

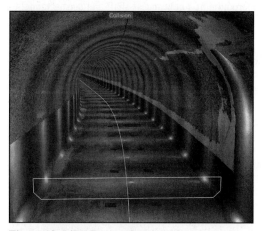

Figure 12. LiDAR scan showing block layout with collision with MMD tunnel walls

replacement for the BIM models normally required on the Owner's projects. In future years, the government will be able to use the as-built scan of the tunnel to quantify abrasion of the invert.

CURRENT STATUS AND OUTLOOK

Construction began in July 2017. The team encountered technical difficultly with the mix design of the cellular concrete that was planned for use as the permeable subbase. Pervious concrete was identified as the preferred solution, and installation of the subbase began in the tunnel. In December 2017, the Owner gave the team notice that the tunnel needed to be used to flush the dam in January 2018, several months earlier than expected. Construction is anticipated to resume in February 2018, beginning with installation of the pervious concrete and followed by placement of the granite blocks. The next flushing is expected to occur in October 2018, and the current schedule projects construction can be completed by that time.

CONCLUSION

The Owner awarded the design-build team the MMD 9-foot Tunnel Rearmoring Project for their innovative mechanically fastened armored steel design. The value engineering change proposal was accepted in December 2016 for the replacement of steel armoring with natural granite stone block. This project was outside the typical realm of engineering and took additional effort by the design team. In-house finite element models, research and development, and calculations were performed to ensure that the invert would be a sustainable solution for the Owner that would decrease the frequency of repairs compared to the armored steel design. The final granite design submitted to the owner included the criteria for scour hole repairs in the existing steel liner, a granite block layout plan, an engineered permeable subbase to prevent block uplift, and a monitoring system to detect invert failure. The design of the granite block invert was finished and submitted in July 2017. The design was approved and construction began that same month once the tunnel ceased flushing operations and was opened for access. The design-build team and Owner are currently working together to implement this unique and innovative design.

REFERENCE

VAW. Mud Mountain Dam 9-foot Tunnel Re-armoring: Abrasion Calculations and Recommendations on the Invert Lining Concept—Expertise report. *VAW Report 4357*, Laboratory of Hydraulics, Hydrology and Glaciology, ETH Zurich, Switzerland, 2016.

Contractor and Engineer Collaborate on Shaft Design, Sewer Tunnel Stabilization Project

David Jurich, Zsolt Horvath, and Evan Friedman
Mott MacDonald

Brett Mainer
Drill Tech Drilling and Shoring

Joseph McDivitt
South Coast Water District

ABSTRACT: After award of the Phase I contract for construction of a new access shaft and access tunnel to support rehabilitation of an aging sewer tunnel in Southern California, the contractor proposed a modification to the shaft design. The owner agreed, and the contractor and engineer collaborated during pre-mobilization to quickly develop a solution that met project performance requirements, better suited the contractor's preferred means and methods, and achieved construction schedule. The re-designed solution integrated initial support with final lining, addressed difficult conditions at the overburden-rock interface, met the challenges of multi-phased construction, and improved sequencing of Phase I and future construction activities.

INTRODUCTION

South Coast Water District provides water and reclaimed water distribution, and wastewater collection and conveyance services to approximately 35,000 residents, 1,000 businesses, and 2 million visitors per year in most of Dana Point, South Laguna Beach, and parts of San Clemente and San Juan Capistrano in Orange County, California (Figure 1). District facilities include 15 reservoirs, 237 km (147 mi) of water lines, 219 km (136 mi) of sewer lines, 7 water pumping stations, 14 sewer pumping stations, and approximately 12,360 water meters.

EXISTING TUNNEL AND PIPELINE

As part of its wastewater collection and conveyance network, the South Coast Water District (District) operates the Beach Interceptor Sewer Tunnel (Tunnel) in South Laguna. A pipeline within the tunnel conveys, on average, 4.7 million L/d (1.25 million gal/d) of wastewater.

Put into service in 1954, the tunnel follows beneath the cliffs fronted by 3.2 km (2 mi) of popular beaches and topped by some of the most desirable and expensive homes along the California coastline. A tunnel and pipeline collapse would have a catastrophic impact to the sensitive coastal environment

Figure 1. Residential properties on coastal bluffs above tunnel with general location map (inset)

Figure 2. Deteriorated timber supported tunnel and pipeline

and the public. Constructed when the area was sparsely populated, the tunnel and sewer have operated in relative obscurity while residential development has completely occupied the surface above.

The 3,192-m-long (10,474-ft-long) tunnel was mined through the San Onofre Breccia bedrock and is supported with nominal 8-inch square timber sets in select intervals. The San Onofre Breccia generally consists of massive to thickly bedded weakly cemented sandstone and breccia and is generally considered one of the more resilient bedrocks within Orange County, however, without engineered support it is susceptible to long term deterioration. Primary access for construction, operation, and maintenance of the tunnel is through a series of 25 portals and adits located at public and private beaches, typically spaced at every few hundred feet. Several of the original access points have collapsed and are abandoned. One portal located at the northern end of the tunnel is accessible by road, the remainder are accessible on foot from public beaches and in many cases through private gated communities.

NEED, PURPOSE, AND PROJECT SCOPE

Shortly after initial project construction, rock slaking and deterioration occurred within the tunnel in several locations and the District sealed and backfilled approximately 226 m (740 ft) of the tunnel in two intervals where there was as much as 3 m (10 ft) of raveling above the tunnel crown. In 1974 several timber supports were replaced, 15 short intervals of deteriorated rock were lined with shotcrete, and the original 600-mm (24-in.) diameter vitrified clay pipe was replaced with a composite fiberglass, polyester resin, and sand pipe known as Techite. From 1974 to 2003, few repairs were made in the tunnel. Beginning in 2003, the District adopted a program to inspect and document conditions, and prioritize

repairs along the tunnel. Since then, tunnel and pipeline repairs have included improving tunnel access, replacing deteriorating portal and adit doors, installing additional timber supports, and updating lateral connections to the pipeline.

The District documented an increase in the frequency of timber repairs and rock fall cleanups and in 2005 hired Mott MacDonald (Engineer) to inspect select portions of the tunnel where there had been a significant number of rock falls and timber support failures (Figure 2). The findings of the inspection led the District to authorize an emergency repair to a 230-m (750-ft) interval of the tunnel.

Planning and design of the emergency repair from 2005 to 2006 by the District, Engineer, and the District's environmental permitting consultant, Denton Mudry, addressed access constraints, protection of the operating pipeline, construction means and methods, and mitigation of potential environmental and public impacts. The emergency repair included:

- Rehabilitation of the portal structure and enlargement of an access adit on a public beach
- Encasement of the composite pipe in reinforced concrete
- Enlargement of the tunnel from a 1.8-m × 1.8-m (6-ft by 6-ft) timber supported trapezoid to a 2.4-m × 2.6-m (7-ft 9-in × 8-ft 8-in) finished horseshoe shape
- Installation of a shotcrete structural lining with steel sets in intervals of poor quality ground

The District contracted with E B Construction and SANCON to make the emergency repairs. Since access to the public beach at the location of the emergency repair is limited to a pedestrian beach access stairway, known as 1,000 steps, from the Coastal Highway, the work was performed using ocean going landing craft dispatched from Long Beach, CA (Figure 3). A total of 63 beach landings were

Figure 3. Boat landing during emergency repair

successfully conducted with a perfect safety record and no adverse impacts to the coastal environment. However, this method of construction access added significant cost and required heightened diligence to proactively manage numerous environmental, safety, and supply chain risks. Access by landing craft limited the size of the construction equipment to hand-held pneumatic spaders and a mini excavator equipped with a hydraulic breaker for excavation, skid steers for muck and materials haulage, and a small mobile batch plant and 100-mm (4-in) slick line for concrete and shotcrete work (Figure 4).

Once the emergency repairs were underway the District directed the Engineer to complete an inspection of the entire tunnel to document conditions and existing ground support, and to identify areas that warranted additional study. The additional inspections and studies included intervals where the tunnel was in close proximity to the bluff or there was evidence of low ground cover, poor quality rock, significant rock falls, and where deep foundations for structures built on top of the tunnel could be affected by an enlargement of the tunnel.

Operational and design criteria for the tunnel stabilization and pipeline replacement included:

- 100-year design life for tunnel and pipeline
- Improved access
- Ability to service lateral connections, make new lateral connections, and maintain the new pipeline
- No adverse impacts due to the project (stability of coastal bluffs, residents, environment)

Since the tunnel was put into service, most of the alignment has been developed and is characterized by narrow roads and residences built with minimal setbacks, condominium complexes, private gated communities, and gated estates. To meet the design criteria of improved access, during planning and design the District and the Engineer considered more than 20 different combinations of improving existing portals and adits, and new access. New access options each included one or more new shaft(s) and new tunnel(s) to connect to the existing tunnel. The District also studied other options such as taking no action, relocating the entire tunnel, and backfilling the tunnel; however, it was the selected tunnel stabilization and pipeline replacement option (Figure 5) that best met the District criteria and needs.

PROJECT DEVELOPMENT

Planning and detailed design of the tunnel stabilization and pipeline replacement took place over a 2-year period during which the District completed several evaluations at critical areas of concern that include detailed surveys, exploratory borings and

Figure 4. Shotcrete application during emergency repair

Figure 5. Photo of stabilized and enlarged tunnel with visualization of new pipeline

test pits from the ground surface and within the tunnel, geophysical surveys from within the tunnel, and laboratory testing of the overburden and rock. The District also conducted numerous public outreach events to solicit input from residents, homeowner associations, governmental agencies, environmental and recreational advocacy groups, and other interested parties. The project followed the California Environmental Quality Act process to obtain permits from the California Coastal Commission, County of Orange, City of South Laguna Beach, and other regulatory agencies. As a result of the public outreach and permitting processes and easement acquisitions, numerous mitigation measures were adopted and implemented in the design and construction documents. Examples of mitigation measures adopted included a ban on explosives and blasting to construct the project, limitations to work hours and time spent on the beaches, and rigorous environmental and geotechnical monitoring programs.

Considerations for location of a new access shaft included easy access to the Coastal Highway, sufficient space for a shaft large enough to accommodate

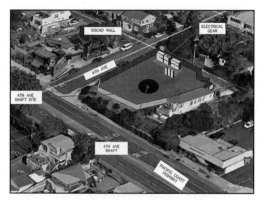

Figure 6. 4th Avenue access shaft site

traditional underground construction equipment, ventilation and hoisting systems, stockpile of materials and spoil, heavy trucks, and temporary medium voltage electrical power equipment. While planning was underway, a private lot located approximately 107 m (350 ft) off alignment, within the middle third of the tunnel, at the corner of 4th Avenue and Coastal Highway became available and was acquired by the District. The new centrally located access shaft site, referred to as the 4th Avenue Access Shaft, is inland of the Coastal Highway and is connected to the existing tunnel with a new access tunnel under the Coastal Highway (Figure 6).

After acquisition of the shaft site, a preliminary layout of improvements was completed and a subsurface exploratory program was performed. Exploratory borings performed by Lawson Geotechnical Group (LGC) at the shaft site included a corehole completed to approximately 6 m (20 ft) below the planned depth of the shaft and two 610-mm (24-in) diameter exploratory bucket auger boreholes completed to refusal at top of rock. Subsurface conditions encountered, and described in the project Geotechnical Data Report (Mott MacDonald, 2016a) and Geotechnical Baseline Report (Mott MacDonald, 2016b), are generally summarized by depth as follows:

- 0–3.6 m (0–12 ft): dry artificial fill composed of medium dense to dense silty sand and clayey sand with little to no gravel; anticipated to exhibit slow raveling behavior.
- 3.6–13.7 m (12–45 ft): dry Upper Terrace alluvial deposits composed of loose to very dense poorly graded sands, sands with silt, clay, silty sands, and clayey sands with little to no gravel; anticipated to exhibit slow raveling and cohesive running behavior.
- 13.7–16.8 m (45–55 ft): saturated Basal Terrace deposit composed of poorly graded sands, poorly graded sand with clap and

clayey gravel with sand at erosional top of rock interface; anticipated to exhibit flowing behavior if saturated, or running and cohesive running behavior when not saturated.
- 16.8–36 m (55–118 ft): San Onofre Breccia, unconfined compressive strength (UCS) range of 1–5.9 MPA (150 to 850 psi) with an average of 4.8 MPA (700 psi), good standup time, and clasts of Catalina Schist with USC from 103–248 MPA (15–36 ksi) with an average of 207 MPA (30 ksi); a limited number of Catalina Schist clasts up to 610-mm (24-in) in diameter anticipated.

Original Design of New Access Shaft

The initial support and final lining of the 4th Ave Access Shaft was designed using numerical modeling and closed form calculations. Numerous load cases were analyzed including construction, normal operation, and seismic events. Seismic loading with bending at the interface between the overburden and rock was the critical load case.

The shaft was sized as a 7.0-m (23-ft) excavated diameter in soil and 6.6-m (21-ft 7-in) excavated diameter in rock, and a 6.4-m (20-ft) inside finished diameter to an excavated depth of 31.7 m (104 ft) measured from top of collar.

Initial ground support design consisted of W6×16 Grade 60 ring beams and grouted 8-gauge corrugated steel liner plate in the artificial fill and Terrance deposits and W6×16 Grade 60 header ring beams and 50 mm (2 in) of fiber-reinforced shotcrete in the San Onofre Breccia.

Based on previous experience with caving of the exploratory borings and nearby foundation caissons, prescriptive pre-excavation grouting was specified for the basal Terrace deposits at the interface between overburden and rock. This grouting was to be performed from the ground surface prior to start of shaft construction.

The base of the shaft was designed with a 1.1-m (42-in) thick reinforced concrete slab. A cast-in-place concrete liner reinforced with #6 and #8 Grade 60 rebars at 200-mm (8-in) centers completed the permanent shaft lining. A removable segmented steel cover with locating pintles and a stainless-steel manway access hatch provided the District the ability to access the shaft for routine inspection and maintenance, as well as future access for large construction equipment.

Value Engineering, Construction Phasing, and Bid

After the original design was completed the District contracted with Parsons to perform an independent Value Engineering (VE) Study of the Project design.

As a result of the VE study, the District directed the Engineer to make two changes to the design:

- Reduce the size of the stabilized tunnel.
- Divide the project into four construction packages, referred to as phases.

Phase I of the project consists of:

- Construction of new access shaft, lump sum payment
- Construction of new access tunnel, payment by linear foot
- Stabilization of 300 feet of the existing tunnel, payment by linear foot

The project delivery method is Design-Bid-Build. The District prequalified potential bidders and advertised the project in September 2016. Drill Tech Drilling and Shoring (Contractor) was awarded Phase I and issued a Notice-To-Proceed in January 2017.

The Issued for Bid drawings and technical specifications for the shaft included a combination of prescriptive ground support measures and performance specifications.

Contractor's Proposed Alternative

During mobilization the Contractor proposed to the District to modify the design of the shaft to better suit their preferred means and methods. The Contractor shared with the District and Engineer design details and photos of recently constructed soft ground shafts in southern California that included elements potentially beneficial to the District. Specific proposed changes included:

- Eliminate the pre-excavation grouting program
- Increase height and thickness of shaft collar from 300 mm (1 ft) to 1.1 m (3.5 ft) to provide more robust fall protection and security in compliance with California OSHA (Cal OSHA) guidelines.
- Add 760 mm (2.5 ft) of removable expanded metal collar extension to bring the overall collar height to the 1.8 m (6 ft) required by Cal OSHA for shafts in excess of 30 m (100 ft) deep.
- Deepen the shaft by 1.2 m (4 ft) to reduce the tunnel grade to allow easier haulage from tunnel
- Replace structural steel brow support with crown spiling at tunnel transition
- Reinforced shotcrete for final liner installed concurrently with initial support, rather than temporary top-down ground support

followed by a permanent cast-in-place concrete liner installed from the bottom up.

The Contractor proposed to prepare the redesign and implement the revised initial support and final lining at no additional cost to the District. As the designer of the revised shaft, the Contractor would seal the drawings and become the Engineer of Record for the structure.

The District and Engineer reviewed the proposed concept and considered the consequences of the revisions, namely:

- Aesthetic impacts of taller concrete collar— During planning and design the District took extensive measures to mitigate the impacts of constructing an access shaft in a mixed residential and commercial neighborhood and the original design included a collar of minimal height. A taller collar would have greater visual impact.
- Shotcrete finish instead of steel form cast-in-place (CIP) concrete finish in shaft—This proved to be insignificant as the concrete finish does not affect performance or impact operations, and aesthetic impacts were minimized by providing a screed finish to the shotcrete surface

The District agreed to the proposed change and authorized the Contractor to begin design.

Revised Shaft Design

To expedite the redesign, the Contractor requested the Engineer provide the shaft design criteria and native files of the original design calculations. The Contractor prepared preliminary calculations for review by the Engineer. During the review process the Engineer proposed to rerun the numerical model for the revised design concept to expedite the redesign. The Contractor accepted the offer and the two companies worked together to adjust the model and support systems, and modify details of shaft wall/base slab construction joint that resulted in a design that met the District's design and operation criteria, and suited the Contractor's preferred means and methods.

The revised numerical model for the analysis of the soil-shaft structure interaction during a seismic event considered a constant 300-mm (12-in) thickness for the proposed single-pass shotcrete lining in both soil and rock. The free-field ground deformations of the soil column were calculated using closed-form solutions assuming a Peak Ground Velocity of 0.45g (LGC Geotechnical, 2010). These deformations were then applied in the numerical model as nodal displacements along the soil-embedded

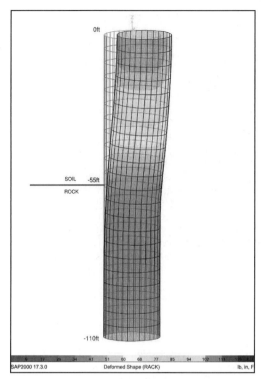

Figure 7. Screen shot of shaft numerical model

portion of the shaft structure and resultant internal forces and reinforcement requirements were determined using finite element analysis (Figure 7).

Once the analyses were completed, the Contractor revised the original AutoCAD design drawings accordingly and submitted the sealed documents to the District for acceptance. The entire redesign process took approximately two weeks to complete.

The revised shaft design consisted of:

- A 1.1-m (3.5-ft) high collar with temporary expanded metal extension above in compliance with Cal OSHA guidelines
- Excavation in 1.5-m (5-ft) lifts
- Installation of lattice girders, welded wire fabric (WWF) and the permanent 300-mm (12-in) thick shotcrete support
- To compensate for the deleted pre-excavation grouting program, when the excavation reached a depth of 1.1 m (3.5 ft) above the anticipated potentially running ground transition zone, drilled a 600-mm (24-in) diameter probe hole
- Delaying placement of the bottom lift of shotcrete until after the base slab was poured,

which naturally keyed the base slab into shaft lining, eliminating the circumferential construction joint and rebar

Shaft Construction

Collar construction began with a 1.8-m (6-ft) sloped excavation, inside of which the steel forms were set. At the bottom of the forms, an initial lattice girder was set and the tie-rods to the next level of lattice girder were driven into the soil below to keep the threads below the concrete pour.

Once the collar was in place, shaft and tunnel utilities were installed. Notable components included a dust collection system and silencer on the ventilation system, and an electric pedestal crane to reduce noise and emissions from the site (Figure 8).

The shaft was advanced through the overburden with a mini excavator, with the general sequence being to excavate a 1.5-m (5-ft) lift one day and place the lattice girder, WWF and 300-mm (12-in) shotcrete layer the next day. The design required additional vertical rebar through the soil and transition zone, but a special order 100 × 100 mm (4×4 in) W4.0×W8.0 WWF was used to provide the increased longitudinal steel without supplemental rebar (Figure 9).

When the shaft was approximately 1.5 m (5 ft) above the potential running transition zone, a small track mounted auger drill was lowered into the shaft and test hole drilled down to the bedrock to verify the conditions. The probe drill encountered about 8 feet of dry, loose sand with embedded cobbles and boulders on top of the bedrock. The excavation proceeded at a slower rate through this zone with reduced lift heights, flash coats, and partial perimeters, but the overall time expended in the transition zone was a small fraction of what the pre-excavation grouting program would have cost.

Once in the bedrock, shaft sinking slowed to a three-day cycle, as two days were required for the mini excavator to excavate a 1.5-m (5-ft) lift in the harder material using a hoe ram. A bucket attachment was used to load the muck box.

Shaft verticality was maintained with plumb bobs and occasionally checked with plummet lasers. The 32.9-m (108-ft) deep shaft was completed with a verticality less than 6 mm (¼ in) off perfectly plumb.

Once the shaft was bottomed out, the base slab was poured, followed by the final lift of shotcrete at the bottom of the shaft wall to restrain the base slab against buoyancy forces. The mini excavator then excavated the first 3 m (10 ft) of tunnel and a 6-m (20-ft) starter tunnel, and the roadheader to be used for the main tunnel drive was inserted into the shaft (Figure 10).

Figure 8. Collar layout with dust collector and silencer in background

Figure 9. Lattice girder and WWF ready for shotcrete

Outcome

In the highly litigious and risk-adverse construction environment of the United States, there is hope for the underground industry. It was a rewarding experience to have multiple parties with different, but overlapping, objectives and priorities work as a team to openly consider a proposed alternative solution, work through the commercial and technical issues, and achieve the project operational objectives for the owner (and District customers) and engineering requirements, while allowing the Contractor to construct the structure using proven methods and equipment. All this collaboration occurred within a short time frame and without incurring additional cost to the District or delay to construction.

One has to ask: Would it have been the same outcome if difficult groundwater conditions and/or very poor-quality ground were encountered? It is the author's collective opinion that the outcome would

Figure 10. Base of shaft and beginning of tunneling

have been the same, as the parties demonstrated a willingness and ability to work together for the shaft design and construction as well as other key elements of the project not discussed herein. Perhaps if a substantial or global differing site condition were the source of the challenges, the commercial and schedule outcome would have been different, but not the approach to completing the work.

Our experience is that in a design-bid-build project delivery method, the design of project elements can be changed after construction contract award to suit contractor preferred means and methods if:

- All vested parties work together
- Project structural and operational performance requirements are met
- Terms of project Permits are met
- All parties can effectively communicate and work quickly and efficiently
- The solution is commercially acceptable to all parties

In many cases, operation/performance requirements and constraints drive the design of underground structures and limit the ability to revise to suit contractor means and methods. However, to bring the best value to each project, we should be alert to potential opportunities to collaborate. The Tunnel Stabilization and Sewer Pipeline Replacement project is an excellent example of an owner, engineer, construction contractor, and construction manager stretching the typical relationships to develop and deliver an alternative solution.

REFERENCES

LGC Geotechnical (2010), *Geotechnical Assessment of the Coastal Bluff Paralleling the South Coast Water District Laguna Beach Sanitary Sewer Interceptor Tunnel*, City of Laguna Beach, California.

Mott MacDonald (2016a), *Geotechnical Data Report, Phase I, Tunnel Stabilization and Sewer Pipeline Replacement Project*, South Coast Water District.

Mott MacDonald (2016b), *Geotechnical Baseline Report, Phase I, Tunnel Stabilization and Sewer Pipeline Replacement Project*, South Coast Water District.

Index

A

Abu Hamour (Musaimeer) Surface & Ground Water Drainage Tunnel, Doha, Qatar, 227–235
Anacostia River Tunnel Project, Washington, D.C., 679–680
Andina PDA Phase I project, Chile, 209–213
Asia
   Rescue Tunnel Twin-Tube, Hong Kong, 70
Atlanta Water Supply Program (WSP), Georgia, 869–874, 995–1002, 1097–1104
atmospheric cutter changing devices (ACCDs), 19–27
Atwater Water Intake and Conveyance Tunnel, Montreal, Quebec, 411–418
Australia
   WestConnex project, Sydney, 401–410
Austria
   Brenner Base Tunnel (BBT), 7–10, 425–433
   Rescue Tunnel, Jenbach, 70–71

B

BART Silicon Valley (BSV) Phase II Extension project, San Jose, California, 596–605, 754–766
Beach Interceptor Sewer Tunnel, Laguna, California, 956–965, 1110–1117
Bergen Point WWTP Outfall Tunnel project, Suffolk County, New York, 587–595
Blacklick Creek Sanitary Interceptor Sewer (BCSIS), Columbus, Ohio, 99–107, 1069–1076

Blue Plains Tunnel Dewatering Pumping Station, Washington, D.C., 621–624
bored tunnels
   BART Silicon Valley (BSV) Phase II Extension project, San Jose, California, 754–766
   Parallel Thimble Shoal Tunnel (CBBT), Virginia, 516–523, 781–792
   Scarborough Subway Extension, Toronto, Ontario, 767–774
   SR99 Bored Tunnel Project, Seattle, Washington, 19–27, 476–481, 532–540, 893–898, 1029–1036
Brenner Base Tunnel (BBT), Austria-Italy, 7–10, 425–433
Building Information Modeling (BIM), 3–10

C

caisson removal, 1049–1057
California
   BART Silicon Valley (BSV) Phase II Extension project, San Jose, 596–605, 754–766
   Beach Interceptor Sewer Tunnel, Laguna, 956–965, 1110–1117
   California High Speed Rail (CHSR) project, 482–489
   California WaterFix tunnels project, 718–726
   Central Bayside System Improvement Project (CBSIP), San Francisco, 503–515
   City Trunk Line South, Los Angeles, 1058–1065

Crenshaw/LAX Transit Corridor Project, Los
  Angeles, 171–184, 437–447, 470–475,
  908–914, 1042–1048
Franciscan Complex, 390–400
Humboldt Bay Power Plant Decommissioning
  Shaft, Eureka, 1049–1057
Interlake Tunnel, Monterey County, 839–848
Los Angeles transit tunnels, 732
Mountain Tunnel, Hetch Hetchy system, San
  Francisco, 236–245, 698–704
North Hollywood Station West Entrance, Los
  Angeles, 922–927
Regional Connector Transit Corridor (RCTC),
  Los Angeles, 1020–1028
San Francisco transit tunnels, 732–733
California High Speed Rail (CHSR) project,
  482–489
California WaterFix tunnels project, 718–726
Canada
  Atwater Water Intake and Conveyance
    Tunnel, Montreal, 411–418
  Canada Line Rapid Transit Project, British
    Columbia, 679
  Coxwell Bypass Tunnel, Toronto, Ontario,
    660–668
  Edmonton Valley Line LRT, Alberta, 934–947
  Elkview Mine, Baldy Ridge Raw Coal
    Conveyor Tunnel, British Columbia,
    250–260
  Louis-Hippolyte-La Fontaine Tunnel, Quebec,
    263–270
  Montreal Transportation Agency Côte-Vertu
    Underground Storage and Maintenance
    Garage, Quebec, 739–745
  Scarborough Subway Extension, Toronto,
    Ontario, 767–774
  Windsor-Detroit Tunnel, 746–753
Canada Line Rapid Transit Project, British
  Columbia, Canada, 679
Central Bayside System Improvement Project
  (CBSIP), San Francisco, California,
  503–515
Central City Stormwater Tunnel System,
  Minneapolis, Minnesota, 271–278
Cheves Hydropower project, Peru, 215–217
Chicago Tunnel and Reservoir Plan (TARP),
  McCook Main Tunnel, Illinois, 875–882,
  883–890
Chile
  Andina PDA Phase I project, 209–213
  El Teniente Ventilation Adits project, 213–215
  subsea tunnels, 625–632

City Trunk Line South, Los Angeles, California,
  1058–1065
closed-form solutions, 490–497
Columbus OSIS Augmentation Relief Sewer
  (OARS) CSO Tunnel, Ohio, 1077–1088
combined sewer overflow (CSO) tunnels. see
  also sewer tunnels
  Chicago Tunnel and Reservoir Plan (TARP),
    McCook Main Tunnel, Illinois, 875–882,
    883–890
  Columbus OSIS Augmentation Relief Sewer
    (OARS) CSO Tunnel, Ohio, 1077–1088
  Coxwell Bypass Tunnel, Toronto, Ontario,
    660–668
  DC Clean Rivers Project (DCCR),
    Washington, D.C., 616–624, 731–732
  DigIndy Deep Rock Tunnel Connector and
    Pump Station, Indianapolis, Indiana,
    951–955
  DigIndy Pleasant Run Deep Tunnel,
    Indianapolis, Indiana, 609–615
  Doan Valley Storage Tunnel, Cleveland, Ohio,
    448–461
  drop shaft design strategies, 448–461
  First Street Tunnel, Washington, D.C.,
    462–469
  Lower and Middle River Des Peres (LMRDP)
    CSO Storage Tunnel, St. Louis, Missouri,
    419–424
  Narragansett Bay Commission (NBC) Phase
    III, Pawtucket Tunnel, Rhode Island,
    643–651
  Ohio Canal Interceptor Tunnel (OCIT), Ohio,
    28–35
  Ohio River Tunnel (ORT), Louisville,
    Kentucky, 633–642
  Ship Canal Water Quality Project (SCWQP),
    Seattle, Washington, 652–659
  South Hartford Conveyance and Storage
    Tunnel (SHCST), Connecticut, 381–389,
    1105–1109
  Three Rivers Protection and Overflow
    Reduction Tunnel (3RPORT), Fort Wayne,
    Indiana, 833–838
communications
  underground innovations, 83–91
  Wi-Fi systems, 72–82
concrete linings. see also precast concrete tunnel
  linings (PCTL)
  fiber reinforced (FRC), 332–340
  for harsh ground and groundwater conditions,
    227–235

steel fiber reinforced (SFRC), 550–555
in tunnels crossing active faults, 11–18,
298–307
Connecticut
South Hartford Conveyance and Storage
Tunnel (SHCST), 381–389, 1105–1109
Construction Impact Assessment Reports
(CIAR), 198–206
Construction Manager at Risk (CMAR) model,
827–828
Atlanta Water Supply Program (WSP),
Georgia, 869–874, 995–1002
construction materials, and sustainability,
714–717
contracts and contracting. *see also* delivery
methods
DB/P3 and risk allocation, 802–811
Three Rivers Protection and Overflow
Reduction Tunnel (3RPORT), Fort Wayne,
Indiana, 833–838
tunnel boring machines (TBMs), 797–798
cost estimating, 559–569, 812–818
Cost Estimation and Validation Process (CEVP),
559–569
Coxwell Bypass Tunnel, Toronto, Ontario,
Canada, 660–668
Crenshaw/LAX Transit Corridor Project, Los
Angeles, California, 171–184, 437–447,
470–475, 908–914, 1042–1048
cross passages
design vs. construction, 1042–1048
ground freezing in excavation of, 570–578,
899–907, 1013–1019
ground-liner interaction, 341–349
mechanized construction of, 66–71
seismic analysis, 490–500
cut-and-cover tunneling
Crenshaw/LAX Transit Corridor Project, Los
Angeles, California, 437–447
Downtown Bellevue Tunnel (DBT),
Washington, 321–331, 350–362
Howard Street Tunnel, Baltimore, Maryland,
291–297
North Hollywood Station West Entrance, Los
Angeles, California, 922–927
Northeast Corridor Superconducting Maglev
(SCMAGLEV) Project, Baltimore-
Washington, 92–96
risk assessment and management strategies,
678–687
SR99 Bored Tunnel Project, Seattle,
Washington, 532–540

in Washington D.C., 729–738
Windsor-Detroit Tunnel, Ontario-Michigan,
746–753
Cutter Soil Mix (CSM) wall systems, 1049–1057

D

data collection and analysis
EPB TBM downtime, 99–107
EPB TBM graphical data, 117–126
geostatistical, 163–170
infrastructure analysis, 671–677
tunnel inspection, 236–245
DC Clean Rivers Project (DCCR), Washington,
D.C., 616–624, 731–732
debris removal, 979–987
delivery methods
Building Information Modeling (BIM), 3–10
Construction Manager at Risk (CMAR),
827–828
current trends in, 825–832
Design-Bid-Build (DBB), 805–806, 819–824,
826–827
Design-Build (DB), 802–811, 819–824,
828–830
Public-Private Partnership (P3), 802–811,
830–831
and TBM procurement, 127–134, 795–801
desalinated water, 625–632
design strategies
active fault zones, 11–18, 298–307
baffle drop structures, 448–461
blast-induced vibration, 482–489
cross passages, 1042–1048
cross passages in seismic zones, 490–500
CSO pumping station structures, 616–624
excavation support systems, 470–475
granite blocks, 1110–1117
groundwater cut-off, 437–447
hard rock tunneling, 1003–1010
high in situ stress, 401–410
reinforcement for PCTL segments, 524–531
slurry walls, 462–469
station caverns, 579–586
tunnel segmental ring geometry, 541–549
tunnel ventilation systems (TVS), 476–481
Design-Bid-Build (DBB) model, 805–806,
819–824, 826–827
Scarborough Subway Extension, Toronto,
Ontario, Canada, 767–774
Ship Canal Water Quality Project (SCWQP),
Seattle, Washington, 652–659

Design-Build (DB) model, 802–811, 819–824, 828–830
  Abu Hamour (Musaimeer) Surface & Ground Water Drainage Tunnel, Doha, Qatar, 227–235
  Crenshaw/LAX Transit Corridor Project, Los Angeles, California, 171–184, 437–447, 470–475
  DC Clean Rivers Project (DCCR), Washington, D.C., 616–624
  First Street Tunnel, Washington, D.C., 462–469
  Interlake Tunnel, Monterey County, California, 839–848
  Norris Cut Force Main Replacement Tunnel, Miami, Florida, 966–978
  North Hollywood Station West Entrance, Los Angeles, California, 922–927
  risk assessment and management strategies, 678–687
  SR99 Bored Tunnel Project, Seattle, Washington, 476–481, 532–540
dewatering, 979–987, 1029–1036
DigIndy Deep Rock Tunnel Connector and Pump Station, Indianapolis, Indiana, 951–955
DigIndy Pleasant Run Deep Tunnel, Indianapolis, Indiana, 609–615
directional drilling, 218–226
discrete element method (DEM) modelling, 49–57
District of Columbia. *see* Washington, D.C.
Doan Valley Storage Tunnel, Cleveland, Ohio, 448–461
Downtown Bellevue Tunnel (DBT), Washington, 321–331, 350–362

E

earth pressure balance (EPB) tunnel boring machines (TBMs)
  chamber air bubbles, 144–153
  cutterhead configurations for glacial geology, 135–143
  cutterhead maintenance, 19–27
  downtime analysis, 99–107
  and dry ground conditions, 908–914
  graphical data analysis, 117–126
  and ground movement, 171–184
  passing under existing tunnels, 1020–1028
  performance prediction, 108–116
  tunnel alignment, 1069–1076
earthquake mitigation. *see also* fault zones
  blast-induced vibration, 482–489

concrete linings, 11–18, 298–307, 503–515
cross passage design, 490–500
East Side Access project, Queens, New York, 163–170
Edmonton Valley Line LRT, Alberta, Canada, 934–947
El Teniente Ventilation Adits project, Chile, 213–215
Elkview Mine, Baldy Ridge Raw Coal Conveyor Tunnel, British Columbia, Canada, 250–260
England
  Thames Tunnel, London, 928–933
environmental considerations
  California WaterFix tunnels project, 718–726
  Lower Meramec Tunnel (LMT), St. Louis, Missouri, 705–713
  sustainable construction materials, 714–717
  tunnel rehabilitation, 698–704
Europe
  Brenner Base Tunnel (BBT), Austria-Italy, 7–10, 425–433
  Gotthard Base Tunnel project, Switzerland, 715–717
  ground freezing in cross passage construction, 570–578
  Kaiser-Friedrich Tunnel, Karlsruhe, Germany, 190–197
  Nazzano Tunnel, Rome, Italy, 1037–1041
  Rescue Tunnel, Hamburg, Germany, 69–70
  Rescue Tunnel, Jenbach, Austria, 70–71
  Slowacki Tunnel, Gdansk, Poland, 575–578
  Soroška tunnel, Slovak Republic, 363–367
  Thames Tunnel, London, England, 928–933
excavation methods. *see also* cut-and-cover tunneling; earth pressure balance (EPB) tunnel boring machines (TBMs); rock tunneling; sequential excavation method (SEM); tunnel boring machines (TBMs)
  dry ground conditions, 908–914
  ground freezing in cross passage construction, 570–578, 899–907, 1013–1019
  single bore, 754–766, 767–774
  small-diameter TBM tunneling, 688–697
  for station caverns, 579–586
excavation support, 470–475, 587–595

F

fault zones
  BART Silicon Valley (BSV) Phase II Extension project, San Jose, California, 754–766

Montreal Transportation Agency Côte-Vertu
Underground Storage and Maintenance
Garage, Quebec, 739–745
SR99 Bored Tunnel Project, Seattle,
Washington, 532–540
fiber reinforced concrete (FRC) linings, 332–340
fire protection strategies, 279–290
fire tests, 263–270
First Street Tunnel, Washington, D.C., 462–469
Florida
Norris Cut Force Main Replacement Tunnel,
Miami, 966–978
Port of Miami Tunnel (POMT), 282–283
formwork, steel, 190–197

G

gates, 875–882
Georgia
Atlanta Water Supply Program (WSP),
869–874, 995–1002, 1097–1104
geostatistical analysis, for ground settlement risk
assessment, 163–170
Germany
Kaiser-Friedrich Tunnel, Karlsruhe, 190–197
Rescue Tunnel, Hamburg, 69–70
Ghomrood water conveyance tunnel, Iran, 59–65
glacial geology, TBM cutterhead configurations
for, 135–143
Gotthard Base Tunnel project, Switzerland,
715–717
granite blocks, 1110–1117
ground freezing, 570–578, 899–907, 1013–1019
ground movement, 171–184, 1029–1036,
1058–1065
grouting, 995–1002, 1042–1048, 1077–1088

H

hard rock tunneling
Atlanta Water Supply Program (WSP),
Georgia, 869–874, 995–1002, 1097–1104
design improvements, 1003–1010
Lower and Middle River Des Peres (LMRDP)
CSO Storage Tunnel, St. Louis, Missouri,
419–424
Lower Gwynns Run Interceptor, Baltimore,
Maryland, 1089–1096
pre-conditioning, 185–189
South Hartford Conveyance and Storage
Tunnel (SHCST), Connecticut, 381–389
highway tunnels. *see* road and highway tunnels
hoists, 858–868

Hong Kong
Rescue Tunnel Twin-Tube, 70
Howard Street Tunnel, Baltimore, Maryland,
291–297
Humboldt Bay Power Plant Decommissioning
Shaft, Eureka, California, 1049–1057
hyperbaric conditions, 190–197

I

Illinois
Chicago Tunnel and Reservoir Plan (TARP),
McCook Main Tunnel, 875–882, 883–890
Indiana
DigIndy Deep Rock Tunnel Connector and
Pump Station, Indianapolis, Indiana,
951–955
DigIndy Pleasant Run Deep Tunnel,
Indianapolis, 609–615
hard rock tunneling in, 1003–1010
Ohio River Bridges East End (ORBEE) cross-
ing tunnel, 283–285
Three Rivers Protection and Overflow
Reduction Tunnel (3RPORT), Fort Wayne,
833–838
industrial control systems (ICS), 36–45
infrastructure analysis, 671–677
inspections
condition assessment, 236–245
National Tunnel Inspection Standards (NTIS),
746–753
instrumentation, 198–206, 934–947, 1029–1036
Interlake Tunnel, Monterey County, California,
839–848
Iran
Ghomrood water conveyance tunnel, 59–65
Karaj-Tehran water conveyance tunnel, 59–65
Italy
Brenner Base Tunnel (BBT), 7–10
Nazzano Tunnel, Rome, 1037–1041

K

Kaiser-Friedrich Tunnel, Karlsruhe, Germany,
190–197
Karaj-Tehran water conveyance tunnel, Iran,
59–65
Kentucky
hard rock tunneling in, 1003–1010
Ohio River Bridges East End (ORBEE) crossing tunnel, 283–285
Ohio River Tunnel (ORT), Louisville,
633–642

L

laser scanning, 49–57
light rail tunnels. *see* transit tunnels
linings
    concrete, 11–18
    concrete mix, 227–235
    designing primary, 363–367
    fiber reinforced concrete (FRC), 332–340
    precast concrete tunnel (PCTL), 298–307,
      503–555
    prefabricated tunnel pipe, 988–994
    shotcrete, 359–360, 425–433
    steel fiber reinforced concrete (SFRC),
      550–555
Louis-Hippolyte-La Fontaine Tunnel, Quebec,
    Canada, 263–270
Lower and Middle River Des Peres (LMRDP)
    CSO Storage Tunnel, St. Louis, Missouri,
    419–424
Lower Gwynns Run Interceptor, Baltimore,
    Maryland, 1089–1096
Lower Meramec Tunnel (LMT), St. Louis,
    Missouri, 705–713

M

marine vessel collision assessment, 516–523
Maryland
    Howard Street Tunnel, Baltimore, 291–297
    Lower Gwynns Run Interceptor, Baltimore,
      1089–1096
    Northeast Corridor Superconducting Maglev
      (SCMAGLEV) Project, 92–96
    Purple Line project, Plymouth Tunnel,
      368–378
Michigan
    Windsor-Detroit Tunnel, 746–753
microtunnel boring machines (MTBM), 851–857
microwave irradiation, 185–189
Middle East
    Abu Hamour (Musaimeer) Surface & Ground
      Water Drainage Tunnel, Doha, Qatar,
      227–235
    Ghomrood water conveyance tunnel, Iran,
      59–65
    Karaj-Tehran water conveyance tunnel, Iran,
      59–65
    Riyadh Metro Line 2 Tunnels project, Saudi
      Arabia, 311–320
    use of steel fiber reinforced concrete (SFRC)
      in, 550–555

mining tunnels
    Andina PDA Phase I project, Chile, 209–213
    El Teniente Ventilation Adits project, Chile,
      213–215
    Elkview Mine, Baldy Ridge Raw Coal
      Conveyor Tunnel, British Columbia,
      Canada, 250–260
Minnesota
    Central City Stormwater Tunnel System,
      Minneapolis, 271–278
Missouri
    Lower and Middle River Des Peres (LMRDP)
      CSO Storage Tunnel, St. Louis, 419–424
    Lower Meramec Tunnel (LMT), St. Louis,
      705–713
modeling
    advance rate, 108–116
    Building Information Modeling (BIM), 3–10
    CEVP-RIAAT process, 559–569
    closed-form solutions, 490–497
    discrete element method (DEM), 49–57
    groundwater drawdown, 371
    numerical, 11–18, 291–297, 321–331,
      363–367, 497–500
    variograms, 163–170
Montreal Transportation Agency Côte-Vertu
    Underground Storage and Maintenance
    Garage, Quebec, Canada, 739–745
Mountain Tunnel, Hetch Hetchy system, San
    Francisco, California, 236–245, 698–704
Mud Mountain Dam Tunnel, Washington,
    1110–1117

N

Narragansett Bay Commission (NBC) Phase III,
    Pawtucket Tunnel, Rhode Island, 643–651
National Tunnel Inspection Standards (NTIS),
    746–753
Nazzano Tunnel, Rome, Italy, 1037–1041
new Austrian tunneling method (NATM). *see*
    sequential excavation method (SEM)
New York
    Bergen Point WWTP Outfall Tunnel project,
      Suffolk County, 587–595
    East Side Access project, Queens, 163–170
    Rondout West Branch Tunnel (RWBT) proj-
      ect, Newburgh, 858–868
    Second Avenue Subway (SAS), New York
      City, 915–921
    transit tunneling in New York City, 732

Norris Cut Force Main Replacement Tunnel, Miami, Florida, 966–978
North America. *see also* United States
    Atwater Water Intake and Conveyance Tunnel, Montreal, Quebec, 411–418
    Canada Line Rapid Transit Project, British Columbia, 679
    Coxwell Bypass Tunnel, Toronto, Ontario, 660–668
    Edmonton Valley Line LRT, Alberta, 934–947
    Elkview Mine, Baldy Ridge Raw Coal Conveyor Tunnel, British Columbia, Canada, 250–260
    ground freezing in cross passage construction, 570–578, 899–907
    Louis-Hippolyte-La Fontaine Tunnel, Quebec, Canada, 263–270
    Montreal Transportation Agency Côte-Vertu Underground Storage and Maintenance Garage, Quebec, 739–745
    Scarborough Subway Extension, Toronto, Ontario, 767–774
    Windsor-Detroit Tunnel, Ontario-Michigan, 746–753
Northeast Corridor Superconducting Maglev (SCMAGLEV) Project, Baltimore-Washington, 92–96
Northgate Link Extension project, Seattle, Washington, 341–349, 573–578, 899–907, 1013–1019
numerical modeling, 11–18, 291–297, 321–331, 363–367, 497–500

O

Ohio
    Blacklick Creek Sanitary Interceptor Sewer (BCSIS), Columbus, 99–107, 1069–1076
    Columbus OSIS Augmentation Relief Sewer (OARS) CSO Tunnel, 1077–1088
    Doan Valley Storage Tunnel, Cleveland, 448–461
    Ohio Canal Interceptor Tunnel (OCIT), 28–35
Ohio Canal Interceptor Tunnel (OCIT), Ohio, 28–35
Ohio River Bridges East End (ORBEE) crossing tunnel, Kentucky-Indiana, 283–285
Ohio River Tunnel (ORT), Louisville, Kentucky, 633–642

P

Parallel Thimble Shoal Tunnel (CBBT), Virginia, 285–287, 516–523, 781–792

Pennsylvania
    West End Trunk Line (WETL), Coatesville, 851–857
Peru
    Cheves Hydropower project, 215–217
pilot holes, 218–226
pipe liners, 988–994
Poland
    Slowacki Tunnel, Gdansk, 575–578
Port of Miami Tunnel (POMT), Florida, 282–283
precast concrete tunnel linings (PCTL)
    Central Bayside System Improvement Project (CBSIP), San Francisco, California, 503–515
    Coxwell Bypass Tunnel, Toronto, Ontario, 660–668
    designing reinforcement for, 524–531
    optimized design of ring geometry, 541–549
    Parallel Thimble Shoal Tunnel (CBBT), Virginia, 516–523
    SR99 Bored Tunnel Project, Washington, 532–540
pre-conditioning hard rock, 185–189
prefabricated tunnel pipe liners, 988–994
procurement. *see* delivery methods
pumping stations, 616–624, 643–651, 951–955
Purple Line project, Plymouth Tunnel, Maryland, 368–378

Q

Qatar
    Abu Hamour (Musaimeer) Surface & Ground Water Drainage Tunnel, Doha, 227–235

R

Regional Connector Transit Corridor (RCTC), Los Angeles, California, 1020–1028
rehabilitation
    Beach Interceptor Sewer Tunnel, Laguna, California, 956–965, 1110–1117
    considerations for, 698–704
    cracking due to sandstone erosion, 271–278
    Mountain Tunnel, Hetch Hetchy system, San Francisco, California, 236–245, 698–704
    Mud Mountain Dam Tunnel, Washington, 1110–1117
    safety upgrades, 263–270
    tunnel support, 250–260
    use of numerical modeling for optimization of, 291–297
    Windsor-Detroit Tunnel, Ontario-Michigan, 746–753

Rescue Tunnel, Austria, 70–71
Rescue Tunnel, Germany, 69–70
Rescue Tunnel Twin-Tube, Hong Kong, 70
Rhode Island
    Narragansett Bay Commission (NBC) Phase
        III, Pawtucket Tunnel, 643–651
ring segmentation and geometry, 541–549
Risk Administration and Analysis Tool (RIAAT),
    559–569, 596–605
risk assessment and management
    contamination, 411–418
    cost estimating, 812–818
    cyber security, 36–45
    and DB/P3 contracts, 802–811
    excavation support design, 587–595
    ground settlement, 163–170
    groundwater control, 1013–1019
    model for cost and schedule integration,
        559–569
    shallow tunnels in urban settings, 678–687
    small-diameter tunneling, 688–697
    subaqueous tunneling, 928–933
    TBM procurement, 795–801
    tunneling methodology comparative analysis,
        596–605
    West End Trunk Line (WETL), Coatesville,
        Pennsylvania case study, 851–857
Riyadh Metro Line 2 Tunnels project, Saudi
    Arabia, 311–320
road and highway tunnels
    Brenner Base Tunnel (BBT), Austria-Italy,
        7–10, 425–433
    Howard Street Tunnel, Baltimore, Maryland,
        291–297
    Louis-Hippolyte-La Fontaine Tunnel, Quebec,
        Canada, 263–270
    Nazzano Tunnel, Rome, Italy, 1037–1041
    Ohio River Bridges East End (ORBEE) cross-
        ing tunnel, Kentucky-Indiana, 283–285
    Parallel Thimble Shoal Tunnel (CBBT),
        Virginia, 285–287, 516–523, 781–792
    Port of Miami Tunnel (POMT), Florida,
        282–283
    Slowacki Tunnel, Gdansk, Poland, 575–578
    Soroška tunnel, Slovak Republic, 363–367
    SR99 Bored Tunnel Project, Seattle,
        Washington, 19–27, 476–481, 532–540,
        893–898, 1029–1036
    WestConnex project, Sydney, Australia,
        401–410
    widening, 1037–1041

Windsor-Detroit Tunnel, Ontario-Michigan,
    746–753
rock mass rating (RMR)
    characterization methods, 392–394
    and TBM downtimes, 58–65
rock tunneling
    Atlanta Water Supply Program (WSP),
        Georgia, 869–874, 995–1002, 1097–1104
    Atwater Water Intake and Conveyance
        Tunnel, Montreal, Quebec, 411–418
    block in matrix (BIM) rock, 390–400
    Brenner Base Tunnel (BBT), Austria-Italy,
        425–433
    Coxwell Bypass Tunnel, Toronto, Ontario,
        660–668
    DigIndy Deep Rock Tunnel Connector and
        Pump Station, Indianapolis, 951–955
    DigIndy Pleasant Run Deep Tunnel,
        Indianapolis, Indiana, 609–615
    Lower and Middle River Des Peres (LMRDP)
        CSO Storage Tunnel, St. Louis, Missouri,
        419–424
    Lower Gwynns Run Interceptor, Baltimore,
        Maryland, 1089–1096
    Narragansett Bay Commission (NBC) Phase
        III, Pawtucket Tunnel, Rhode Island,
        643–651
    South Hartford Conveyance and Storage
        Tunnel (SHCST), Connecticut, 381–389
    WestConnex project, Sydney, Australia,
        401–410
Rondout West Branch Tunnel (RWBT) project,
    Newburgh, New York, 858–868

S

safety and security. see also risk assessment and
        management
    communications technology, 72–82, 83–91
    cross passages, 66–71
    cyber security, 36–45
    fire events, 263–270, 279–290
    marine vessel collision assessment, 516–523
    National Tunnel Inspection Standards (NTIS),
        746–753
    tunnel ventilation systems (TVS), 476–481
Saudi Arabia
    Riyadh Metro Line 2 Tunnels project,
        311–320
SCADA systems, 36–45
Scarborough Subway Extension, Toronto,
    Ontario, Canada, 767–774

SCMAGLEV Project, 92–96

secant piles, 1105–1109

Second Avenue Subway (SAS), New York, New York, 915–921

segmental lining, in tunnels crossing active faults, 298–307

seismic zones. *see also* fault zones

   BART Silicon Valley (BSV) Phase II Extension project, San Jose, California, 754–766

   and cross passage design, 490–500

   design strategies, 11–18, 298–307

   slurry wall structures in, 503–515

sequential excavation method (SEM)

   Brenner Base Tunnel (BBT), Austria-Italy, 425–433

   City Trunk Line South, Los Angeles, California, 1058–1065

   Downtown Bellevue Tunnel (DBT), Washington, 321–331, 350–362

   Edmonton Valley Line LRT, Alberta, 934–947

   Northgate Link Extension project, Seattle, Washington, 341–349

   Purple Line project, Plymouth Tunnel, Maryland, 368–378

   Riyadh Metro Line 2 Tunnels project, Saudi Arabia, 311–320

   shallow cover, 368–378

   Soroška tunnel, Slovak Republic, 363–367

sewer tunnels. *see also* combined sewer overflow (CSO) tunnels

   Anacostia River Tunnel Project, Washington, D.C., 679–680

   Blacklick Creek Sanitary Interceptor Sewer (BCSIS), Columbus, Ohio, 99–107, 1069–1076

   Lower Gwynns Run Interceptor, Baltimore, Maryland, 1089–1096

   Lower Meramec Tunnel (LMT), St. Louis, Missouri, 705–713

   Norris Cut Force Main Replacement Tunnel, Miami, Florida, 966–978

   West End Trunk Line (WETL), Coatesville, Pennsylvania, 851–857

shaft design and construction

   Atlanta Water Supply Program (WSP), Georgia, 1097–1104

   baffle drop structures, 448–461

   Beach Interceptor Sewer Tunnel, Laguna, California, 1110–1117

Blacklick Creek Sanitary Interceptor Sewer (BCSIS), Columbus, Ohio, 1069–1076

Coxwell Bypass Tunnel, Toronto, Ontario, 660–668

DigIndy Pleasant Run Deep Tunnel, Indianapolis, Indiana, 609–615

hoists, 858–868

Humboldt Bay Power Plant Decommissioning Shaft, Eureka, California, 1049–1057

risk-mitigated design for excavation support, 587–595

spider sheeting reinforcement, 246–249

use of secant piles for support, 1105–1109

shallow tunneling. *see* cut-and-cover tunneling

Ship Canal Water Quality Project (SCWQP), Seattle, Washington, 652–659

ship collision assessment, 516–523

shotcrete, 359–360, 425–433

Slovak Republic

   Soroška tunnel, 363–367

Slowacki Tunnel, Gdansk, Poland, 575–578

slurry walls, 462–469, 503–515

soil conditioning, 144–153

Soroška tunnel, Slovak Republic, 363–367

South America

   Andina PDA Phase I project, Chile, 209–213

   Cheves Hydropower project, Peru, 215–217

   El Teniente Ventilation Adits project, Chile, 213–215

   subsea tunnels in Chile, 625–632

South Hartford Conveyance and Storage Tunnel (SHCST), Connecticut, 381–389, 1105–1109

spatial variability, 163–170

spider sheeting, 246–249

SR99 Bored Tunnel Project, Seattle, Washington, 19–27, 476–481, 532–540, 893–898, 1029–1036

station caverns, 579–586

stormwater tunnels

Central City Stormwater Tunnel System, Minneapolis, Minnesota, 271–278

   Waller Creek Tunnel, Austin, Texas, 979–987

structurally controlled instability, 49–57

subaqueous tunneling, 928–933

subsea tunnels, 625–632

subway tunnels. *see* transit tunnels

sustainability, 714–717

Switzerland

   Gotthard Base Tunnel project, 715–717

T

technology
  Building Information Modeling (BIM), 3–10
  communications, 72–82, 83–91
  cyber security, 36–45
  digitization of TBM technology, 157–162
  laser scanning, 49–57
  numerical modeling, 11–18
  and training, 209–217
Texas
  Waller Creek Tunnel, Austin, 979–987
Thames Tunnel, London, England, 928–933
Three Rivers Protection and Overflow Reduction
    Tunnel (3RPORT), Fort Wayne, Indiana,
    833–838
training, 209–217
transit tunnels. *see also* road and highway tunnels
  BART Silicon Valley (BSV) Phase II
    Extension project, San Jose, California,
    596–605, 754–766
  Brenner Base Tunnel, Italy, 7–10
  California High Speed Rail (CHSR) project,
    482–489
  Canada Line Rapid Transit Project, British
    Columbia, 679
  Crenshaw/LAX Transit Corridor Project, Los
    Angeles, California, 171–184, 437–447,
    470–475, 908–914, 1042–1048
  Downtown Bellevue Tunnel (DBT),
    Washington, 321–331, 350–362
  East Side Access project, Queens, New York,
    163–170
  Edmonton Valley Line LRT, Alberta, 934–947
  future of, in Washington, D.C., 729–738
  Gotthard Base Tunnel project, Switzerland,
    715–717
  Kaiser-Friedrich Tunnel, Karlsruhe, Germany,
    190–197
  Los Angeles, 732
  Montreal Transportation Agency Côte-Vertu
    Underground Storage and Maintenance
    Garage, Quebec, 739–745
  New York City, 732
  North Hollywood Station West Entrance, Los
    Angeles, California, 922–927
  Northeast Corridor Superconducting Maglev
    (SCMAGLEV) Project, Baltimore-
    Washington, 92–96
  Northgate Link Extension project, Seattle,
    Washington, 341–349, 573–578, 899–907,
    1013–1019

  Purple Line project, Plymouth Tunnel,
    Maryland, 368–378
  Regional Connector Transit Corridor (RCTC),
    Los Angeles, California, 1020–1028
  Riyadh Metro Line 2 Tunnels project, Saudi
    Arabia, 311–320
  San Francisco, 732–733
  Scarborough Subway Extension, Toronto,
    Ontario, 767–774
  Seattle, 733
  Second Avenue Subway (SAS), New York,
    New York, 915–921
  station caverns, 579–586
  Thames Tunnel, London, England, 928–933
  University Link U230 tunnel project,
    Washington, 108–116
  Washington Metro, Washington, D.C.,
    729–731, 733–738
  West Seattle to Ballard Link Light Rail
    Extension, Washington, 775–780
transportation tunnels. *see* road and highway tun-
    nels; transit tunnels
tunnel boring machines (TBMs). *see also* earth
    pressure balance (EPB) tunnel boring
    machines (TBMs)
  and cross passage construction, 66–71
  cutterhead configurations for glacial geology,
    135–143
  digitization and networking, 157–162
  double shielded, 1089–1096
  downtimes and RMR, 58–65
  hard rock excavation, 185–189
  large-diameter crossover, 28–35
  main beam, 869–874
  microtunnel boring machines (MTBM),
    851–857
  procurement, 127–134, 795–801
  risk assessment and management, 795–801
  single bore tunneling, 754–766, 767–774
  small-diameter tunneling, 688–697
  uncertainty and performance risk, 163–170
tunnel design. *see* design strategies
tunnel ventilation systems (TVS), 476–481

U

underwater tunneling, 625–632, 928–933
United States
  Anacostia River Tunnel Project, Washington,
    D.C., 679–680
  Atlanta Water Supply Program (WSP),
    Georgia, 869–874, 995–1002, 1097–1104

BART Silicon Valley (BSV) Phase II Extension project, San Jose, California, 596–605, 754–766

Beach Interceptor Sewer Tunnel, Laguna, California, 956–965, 1110–1117

Bergen Point WWTP Outfall Tunnel project, Suffolk County, New York, 587–595

Blacklick Creek Sanitary Interceptor Sewer (BCSIS), Columbus, Ohio, 99–107, 1069–1076

Blue Plains Tunnel Dewatering Pumping Station, Washington, D.C., 621–624

California High Speed Rail (CHSR) project, 482–489

California WaterFix tunnels project, 718–726

Central Bayside System Improvement Project (CBSIP), San Francisco, California, 503–515

Central City Stormwater Tunnel System, Minneapolis, Minnesota, 271–278

Chicago Tunnel and Reservoir Plan (TARP), McCook Main Tunnel, Illinois, 875–882, 883–890

City Trunk Line South, Los Angeles, California, 1058–1065

Columbus OSIS Augmentation Relief Sewer (OARS) CSO Tunnel, Ohio, 1077–1088

Crenshaw/LAX Transit Corridor Project, Los Angeles, California, 171–184, 437–447, 470–475, 908–914, 1042–1048

DC Clean Rivers Project (DCCR), Washington, D.C., 616–624, 731–732, 733–738

DigIndy Deep Rock Tunnel Connector and Pump Station, Indianapolis, Indiana, 951–955

DigIndy Pleasant Run Deep Tunnel, Indianapolis, Indiana, 609–615

Doan Valley Storage Tunnel, Cleveland, Ohio, 448–461

Downtown Bellevue Tunnel (DBT), Washington, 321–331, 350–362

East Side Access project, Queens, New York, 163–170

First Street Tunnel, Washington, D.C., 462–469

hard rock tunneling in, 1003–1010

Howard Street Tunnel, Baltimore, Maryland, 291–297

Humboldt Bay Power Plant Decommissioning Shaft, Eureka, California, 1049–1057

Interlake Tunnel, Monterey County, California, 839–848

Lower and Middle River Des Peres (LMRDP) CSO Storage Tunnel, St. Louis, Missouri, 419–424

Lower Gwynns Run Interceptor, Baltimore, Maryland, 1089–1096

Lower Meramec Tunnel (LMT), St. Louis, Missouri, 705–713

Mountain Tunnel, Hetch Hetchy system, San Francisco, California, 236–245, 698–704

Mud Mountain Dam Tunnel, Washington, 1110–1117

Narragansett Bay Commission (NBC) Phase III, Pawtucket Tunnel, Rhode Island, 643–651

Norris Cut Force Main Replacement Tunnel, Miami, Florida, 966–978

North Hollywood Station West Entrance, Los Angeles, California, 922–927

Northeast Corridor Superconducting Maglev (SCMAGLEV) Project, Baltimore-Washington, 92–96

Northgate Link Extension project, Seattle, Washington, 341–349, 573–578, 899–907, 1013–1019

Ohio Canal Interceptor Tunnel (OCIT), Ohio, 28–35

Ohio River Bridges East End (ORBEE) crossing tunnel, Kentucky-Indiana, 283–285

Ohio River Tunnel (ORT), Louisville, Kentucky, 633–642

Parallel Thimble Shoal Tunnel (CBBT), Virginia, 285–287, 516–523, 781–792

Port of Miami Tunnel (POMT), Florida, 282–283

Purple Line project, Plymouth Tunnel, Maryland, 368–378

Regional Connector Transit Corridor (RCTC), Los Angeles, California, 1020–1028

Rondout West Branch Tunnel (RWBT) project, Newburgh, New York, 858–868

Second Avenue Subway (SAS), New York, New York, 915–921

Ship Canal Water Quality Project (SCWQP), Seattle, Washington, 652–659

South Hartford Conveyance and Storage Tunnel (SHCST), Connecticut, 381–389, 1105–1109

SR99 Bored Tunnel Project, Seattle, Washington, 19–27, 476–481, 532–540, 893–898, 1029–1036

Three Rivers Protection and Overflow
Reduction Tunnel (3RPORT), Fort Wayne,
Indiana, 833–838
University Link U230 tunnel project,
Washington, 108–116
Waller Creek Tunnel, Austin, Texas, 979–987
Washington Metro, Washington, D.C.,
729–731
West End Trunk Line (WETL), Coatesville,
Pennsylvania, 851–857
West Seattle to Ballard Link Light Rail
Extension, Washington, 775–780
University Link U230 tunnel project,
Washington, 108–116
urban tunneling
Crenshaw/LAX Transit Corridor Project, Los
Angeles, California, 437–447, 470–475
Downtown Bellevue Tunnel (DBT),
Washington, 321–331, 350–362
Edmonton Valley Line LRT, Alberta, 934–947
geological uncertainties and risk management,
688–697
infrastructure analysis, 671–677
Lower Gwynns Run Interceptor, Baltimore,
Maryland, 1089–1096
North Hollywood Station West Entrance, Los
Angeles, California, 922–927
Purple Line project, Plymouth Tunnel,
Maryland, 368–378
Riyadh Metro Line 2 Tunnels project, Saudi
Arabia, 311–320
Second Avenue Subway (SAS), New York,
New York, 915–921
shallow tunnels, 678–687
transit station caverns, 579–586
West Seattle to Ballard Link Light Rail
Extension, Washington, 775–780

V

ventilation systems, 476–481
vibration studies, 482–489
Virginia
Parallel Thimble Shoal Tunnel (CBBT),
285–287, 516–523, 781–792

W

Waller Creek Tunnel, Austin, Texas, 979–987
Washington (state)
Downtown Bellevue Tunnel (DBT), 321–331,
350–362
Mud Mountain Dam Tunnel, 1110–1117

Northgate Link Extension project, Seattle,
341–349, 573–578, 899–907, 1013–1019
Seattle transit tunnels, 733
Ship Canal Water Quality Project (SCWQP),
Seattle, 652–659
SR99 Bored Tunnel Project, Seattle, 19–27,
476–481, 532–540, 893–898
SR99 Bored Tunnel Project, Seattle,
Washington, 1029–1036
University Link U230 tunnel project, Seattle,
108–116
West Seattle to Ballard Link Light Rail
Extension, 775–780
Washington, D.C.
Anacostia River Tunnel Project, 679–680
Blue Plains Tunnel Dewatering Pumping
Station, 621–624
DC Clean Rivers Project (DCCR), 616–624,
731–732, 733–738
First Street Tunnel, 462–469
Northeast Corridor Superconducting Maglev
(SCMAGLEV) Project, 92–96
transit tunneling in, 729–738
Washington Metro, 729–731
Washington Metro, Washington, D.C., 729–731,
733–738
wastewater tunnels. *see also* sewer tunnels
Beach Interceptor Sewer Tunnel, Laguna,
California, 956–965, 1110–1117
Bergen Point WWTP Outfall Tunnel project,
Suffolk County, New York, 587–595
Central Bayside System Improvement Project
(CBSIP), San Francisco, California,
503–515
water tunnels. *see also* combined sewer overflow
(CSO) tunnels; stormwater tunnels
Abu Hamour (Musaimeer) Surface & Ground
Water Drainage Tunnel, Doha, Qatar,
227–235
Atlanta Water Supply Program (WSP),
Georgia, 869–874, 995–1002, 1097–1104
Atwater Water Intake and Conveyance
Tunnel, Montreal, Quebec, 411–418
California WaterFix tunnels project, 718–726
Cheves Hydropower project, Peru, 215–217
City Trunk Line South, Los Angeles,
California, 1058–1065
Ghomrood water conveyance tunnel, Iran,
59–65
Interlake Tunnel, Monterey County,
California, 839–848

Karaj-Tehran water conveyance tunnel, Iran, 59–65

Mountain Tunnel, Hetch Hetchy system, San Francisco, California, 236–245, 698–704

Mud Mountain Dam Tunnel, Washington, 1110–1117

Rondout West Branch Tunnel (RWBT) project, Newburgh, New York, 858–868

subsea tunnels, 625–632

West End Trunk Line (WETL), Coatesville, Pennsylvania, 851–857

West Seattle to Ballard Link Light Rail Extension, Washington, 775–780

WestConnex project, Sydney, Australia, 401–410

widening tunnels, 1037–1041

Windsor-Detroit Tunnel, Ontario-Michigan, 746–753